A Dynamical Systems Theory of Thermodynamics

PRINCETON SERIES IN APPLIED MATHEMATICS

Edited by

The Princeton Series in Applied Mathematics publishes high quality advanced texts and monographs in all areas of applied mathematics. Books include those of a theoretical and general nature as well as those dealing with the mathematics of specific applications areas and real-world situations.

For a list of books in the series, see Page 721.

A Dynamical Systems Theory of Thermodynamics

Wassim M. Haddad

PRINCETON UNIVERSITY PRESS
PRINCETON AND OXFORD

Published by Princeton University Press
41 William Street, Princeton, New Jersey 08540

In the United Kingdom: Princeton University Press
6 Oxford St, Woodstock, Oxfordshire OX20 1TW

Library of Congress Control Number: 2018954760

ISBN 978-0-691-19014-3

British Library Cataloging-in-Publication Data is available.

This book has been composed in LaTeX.

The publisher would like to acknowledge the author of this volume for providing the camera-ready copy from which this book was printed.

Printed on acid-free paper. ∞

press.princeton.edu

Printed in the United States of America

10 9 8 7 6 5 4 3 2 1

To my wife, Lydia—the unwavering balance and compass guiding us through the path of life. Throughout our life her devotion, kindness, and love are absolute, her insight, awareness, and perception unparalleled, and imagination, creativity, and genius unequaled.

W. M. H.

A new scientific theory or framework does not prevail by persuading its polarized antagonists and making them see its sublime structure, but rather because its calcified oppugners eventually perish, and a new generation of scientists emerge that can appreciate its beauty and stern formalism.

—Wassim M. Haddad

Contents

Preface

Thermodynamics is a physical branch of science that governs the thermal behavior of dynamical systems, from those as simple as refrigerators to those as complex as our expanding universe. The laws of thermodynamics involving conservation of energy and nonconservation of entropy are, without a doubt, two of the most useful and general laws in all of the sciences. The first law of thermodynamics, according to which energy cannot be created or destroyed but merely transformed from one form to another, and the second law of thermodynamics, according to which the *usable* energy in an adiabatically isolated dynamical system is always diminishing in spite of the fact that energy is conserved, have had an impact far beyond science and engineering.

The second law of thermodynamics is intimately connected to the irreversibility of dynamical processes occurring in nature and our observable universe. In particular, the second law asserts that a dynamical system undergoing a transformation from one state to another cannot be restored to its original state and at the same time restore its environment to its original condition. That is, the status quo cannot be restored everywhere. This gives rise to an increasing quantity known as *entropy*.

Entropy permeates the whole of nature, and unlike energy, which describes the state of a dynamical system, entropy is a measure of the quality of that energy and reflects the change in the status quo of a dynamical system. Hence, the law that entropy always increases, the second law of thermodynamics, defines the direction of time flow and shows that a dynamical system state will continually change in the direction of increasing entropy and thus inevitably approach a limiting state corresponding to a state of maximum entropy. It is precisely this irreversibility of all dynamical processes connoting the running down and eventual demise of the universe that has led writers, historians, philosophers, and theologians to ask profound questions such as: How is it possible for life to come into being in a universe governed by a supreme law that impedes the very existence of life?

Even though thermodynamics has provided the foundation for speculation about some of science's most puzzling questions concerning the beginning and the end of the universe, the development of thermodynamics grew out of steam tables and the desire to design and build efficient heat engines, with many scientists and mathematicians expressing concerns about the completeness and clarity of its mathematical foundation over its long and tortuous history. Indeed, many formulations of classical thermodynamics, especially most textbook presentations, poorly amalgamate physics with rigorous mathematics and have had a hard time finding a balance between nineteenth-century steam and heat engine engineering, and twenty-first-century science and mathematics.

In fact, no other discipline in mathematical science is riddled with so many logical and mathematical inconsistencies, differences in definitions, and ill-defined notation as classical thermodynamics. With a few notable exceptions, mathematicians for more than a century have turned away in disquietude from classical thermodynamics, often overlooking its grandiose unsubstantiated claims and allowing it to slip into an abyss of ambiguity.

The development of the theory of thermodynamics follows two conceptually rather different lines of thought. The first (historically), known as *classical thermodynamics*, is based on fundamental laws that are assumed as axioms, which in turn are based on experimental evidence. Conclusions are subsequently drawn from them using the notion of a *thermodynamic state* of a system, which includes temperature, volume, and pressure, among others.

The second, known as *statistical thermodynamics*, has its foundation in classical mechanics. However, since the state of a dynamical system in mechanics is completely specified pointwise in time by each point-mass position and velocity and since thermodynamic systems contain large numbers of particles (atoms or molecules, typically on the order of 10^{23}), an ensemble average of different configurations of molecular motion is considered as the state of the system. In this case, the equivalence between heat and dynamical energy is based on a kinetic theory interpretation reducing all thermal behavior to the statistical motions of atoms and molecules. In addition, the second law of thermodynamics has only statistical certainty wherein entropy is directly related to the relative probability of various states of a collection of molecules.

In this monograph, we utilize the *language* of modern mathematics within a theorem-proof format to develop a general dynamical systems theory for reversible and irreversible equilibrium and nonequilibrium thermodynamics. The monograph is written from a system-theoretic point of view and can be viewed as a contribution to the fields of thermodynamics and mathematical systems theory. In particular, we develop a novel formulation of thermodynamics using a middle-ground theory involving deterministic large-scale dynamical system models that bridges the gap between classical and statistical thermodynamics.

The benefits of such a theory include the advantage of being independent of the simplifying assumptions that are often made in statistical mechanics and at the same time providing a thermodynamic framework with enough detail of how the system really evolves without ever needing to resort to statistical (subjective or informational) probabilities. In particular, we develop a system-theoretic foundation for thermodynamics using a large-scale dynamical systems perspective. Specifically, using compartmental dynamical system energy flow models, we place the universal energy conservation, energy equipartition, temperature equipartition, and entropy nonconservation laws of thermodynamics on a system-theoretic foundation.

Next, we establish the existence of a *new* and dual notion to entropy, namely, *ectropy*, as a measure of the tendency of a dynamical system to do useful work and grow more organized, and we show that conservation of energy in an adiabatically isolated thermodynamic system necessarily leads to nonconservation of ectropy and entropy. In addition, using the system ectropy as a Lyapunov function candidate, we show that our large-scale thermodynamic energy flow model has convergent trajectories to Lyapunov stable equilibria determined by the large-scale system initial subsystem energies. Furthermore, using the system entropy and ectropy functions, we establish a clear connection between irreversibility, the second law of thermodynamics, and the arrow of time.

These results are then generalized to continuum thermodynamics involving infinite-dimensional energy flow conservation models. Since in this case the resulting dynamical system is defined on an infinite-dimensional Banach space that is not locally compact, stability, convergence, and energy equipartition are shown using Sobolev embedding theorems and the notion of generalized (or weak) solutions. In addition, we combine our large-scale thermodynamic system framework with stochastic thermodynamics to develop a stochastic dynamical systems framework of thermodynamics.

Finally, to address the universality of thermodynamics to cosmology we extend our dynamical systems framework of thermodynamics to relativistic thermodynamics. This leads to an entropy dilation principle showing that the rate of change in entropy increase of a moving system decreases as the system's speed increases through space, and hence, motion affects the rate of entropy increase. Furthermore, we elucidate how our proposed dynamical systems framework of thermodynamics can potentially provide deeper insights into some of the most perplexing secrets of the origins and fabric of the universe, including the thermodynamics of living systems, the origins of life, consciousness, the second law and gravity, and illness, aging, and death.

The underlying intention of this monograph is to present a general system-theoretic framework for one of the most useful and general physical branches of science in the language of dynamical systems theory. It is hoped that this monograph will help stimulate increased interaction between physicists and dynamical systems and control theorists. The potential and opportunities for applying and extending this work to further understanding discreteness and continuity, indeterminism and determinism, and quantized curvature and general relativity in the pursuit of the ever elusive unified single field theory are enormous.

The results reported in this monograph were obtained at the School of Aerospace Engineering, Georgia Institute of Technology, Atlanta. The research support provided by the Air Force Office of Scientific Research (AFOSR) over the years has been instrumental in allowing me to explore basic research topics that have led to some of the material in this monograph. I am indebted to AFOSR for its continued support.

I am grateful to many individuals for their valuable discussions and feedback over the years on the fascinating subject of thermodynamics. I especially thank James M. Bailey, Jordan M. Berg, Dennis S. Bernstein, Vijay Chellaboina, Qing Hui, David C. Hyland, Qishuai Liu, Gérard A. Maugin, Sergey G. Nersesov, Tanmay Rajpurohit, Evgenii Rudnyi, and Jan C. Willems.

Atlanta, Georgia, USA, October 2018, *Wassim M. Haddad*

Chapter One

Introduction

1.1 An Overview of Classical Thermodynamics

Energy is a concept that underlies our understanding of all physical phenomena and is a measure of the ability of a dynamical system to produce changes (motion) in its own system state as well as changes in the system states of its surroundings. Thermodynamics is a physical branch of science that deals with laws governing energy flow from one body to another and energy transformations from one form to another. These energy flow laws are captured by the fundamental principles known as the first and second laws of thermodynamics. The first law of thermodynamics gives a precise formulation of the equivalence between *heat* (i.e., the transferring of energy via temperature gradients) and *work* (i.e., the transferring of energy into coherent motion) and states that, among all system transformations, the net system energy is conserved. Hence, energy cannot be created out of nothing and cannot be destroyed; it can merely be transferred from one form to another.

The law of conservation of energy is not a mathematical truth, but rather the consequence of an immeasurable culmination of observations over the chronicle of our civilization and is a fundamental *axiom* of the science of heat. The first law does not tell us whether any particular process can actually occur, that is, it does not restrict the ability to convert work into heat or heat into work, except that energy must be conserved in the process. The second law of thermodynamics asserts that while the system energy is always conserved, it will be degraded to a point where it cannot produce any useful work. More specifically, for any cyclic process that is shielded from heat exchange with its environment, it is impossible to extract work from heat without at the same time discarding some heat, giving rise to an increasing quantity known as *entropy*.

While energy describes the state of a dynamical system, entropy is a measure of the quality of that energy reflecting changes in the status quo of the system and is associated with disorder and the amount of wasted energy in a dynamical (energy) transformation from one state (form) to another. Since the system entropy increases, the entropy of a dynamical system tends to a maximum, and thus time, as determined by system entropy increase [299, 392, 476], flows in one direction only. Even though entropy is a physical property of matter that is not directly observable, it permeates the whole of nature, regulating the *arrow of time*, and is responsible for the enfeeblement and eventual demise of the universe.[1,2] While the laws of thermodynamics form the foundation to basic engineering systems, chemical reaction systems, nuclear reactions, cosmology, and our expanding universe, many mathematicians and scientists have expressed concerns about the completeness and clarity of the different expositions of thermodynamics over its long and tortuous history; see [69, 79, 96, 172, 184, 342, 440, 447, 455].

Since the specific motion of every molecule of a thermodynamic system is impossible to predict, a *macroscopic* model of the system is typically used, with appropriate macroscopic states that include pressure, volume, temperature, internal energy, and entropy, among others. One of the key criticisms of the macroscopic viewpoint of thermodynamics, known as *classical thermodynamics*, is the inability of the model to provide enough detail of how the system really evolves; that is, it is lacking a kinetic mechanism for describing the behavior of heat and work energy.

In developing a kinetic model for heat and dynamical energy, a thermodynamically consistent energy flow model should ensure that the system energy can be modeled by a diffusion equation in the form of a *parabolic* partial differential equation or a divergence structure *first-order hyperbolic* partial differential equation arising in models involving *conservation laws*. Such systems are infinite-dimensional, and hence, finite-dimensional approximations are of very high order, giving rise to large-scale dynamical systems with macroscopic energy transfer dynamics. Since energy is a fundamental concept in the analysis of large-scale dynamical systems, and heat (energy in transition) is a fundamental concept of thermodynamics involving the capacity of hot bodies (more energetic subsystems) to produce work, thermodynamics is a theory of large-scale dynamical systems.

---

[1] Many natural philosophers have associated this ravaging irrecoverability in connection to the second law of thermodynamics with an eschatological terminus of the universe. Namely, the creation of a certain degree of life and order in the universe is inevitably coupled with an even greater degree of death and disorder. A convincing proof of this bold claim has, however, never been given.

[2] The earliest perception of irreversibility of nature and the universe along with time's arrow was postulated by the ancient Greek philosopher Herakleitos ($\sim$ 535–$\sim$ 475 B.C.). Herakleitos' profound statements, *Everything is in a state of flux and nothing is stationary* and *Man cannot step into the same river twice, because neither the man nor the river is the same*, created the foundation for all other speculation on metaphysics and physics. The idea that the universe is in constant change and that there is an underlying order to this change—the *Logos*—postulates the very existence of entropy as a physical property of matter permeating all of nature and the universe.

High-dimensional dynamical systems can arise from both macroscopic and *microscopic* points of view. Microscopic thermodynamic models can have the form of a distributed-parameter model or a large-scale system model comprised of a large number of interconnected Hamiltonian subsystems. For example, in a crystalline solid every molecule in a lattice can be viewed as an undamped vibrational mode comprising a distributed-parameter model in the form of a *second-order hyperbolic* partial differential equation. In contrast to macroscopic models involving the evolution of global quantities (e.g., energy, temperature, entropy), microscopic models are based upon the modeling of local quantities that describe the atoms and molecules that make up the system and their speeds, energies, masses, angular momenta, behavior during collisions, etc. The mathematical formulations based on these quantities form the basis of *statistical mechanics*.

Thermodynamics based on statistical mechanics is known as *statistical thermodynamics* and involves the mechanics of an ensemble of many particles (atoms or molecules) wherein the detailed description of the system state loses importance and only average properties of large numbers of particles are considered. Since microscopic details are obscured on the macroscopic level, it is appropriate to view a macroscopic model as an inherent model of uncertainty. However, for a thermodynamic system the macroscopic and microscopic quantities are related since they are simply different ways of describing the same phenomena. Thus, if the global macroscopic quantities can be expressed in terms of the local microscopic quantities, then the laws of thermodynamics could be described in the language of statistical mechanics.

This interweaving of the microscopic and macroscopic points of view leads to diffusion being a natural consequence of dimensionality and, hence, uncertainty on the microscopic level, despite the fact that there is no uncertainty about the diffusion process per se. Thus, even though as a limiting case a hyperbolic partial differential equation purports to model an infinite number of modes, in reality much of the modal information (e.g., position, velocity, energies) is only poorly known, and hence, such models are largely idealizations. With increased dimensionality comes an increase in uncertainty leading to a greater reliance on macroscopic quantities so that the system model becomes more diffusive in character.

Thermodynamics was spawned from the desire to design and build efficient heat engines, and it quickly spread to speculations about the universe upon the discovery of entropy as a fundamental physical property of matter. The theory of classical thermodynamics was predominantly developed by Carnot, Clausius, Kelvin, Planck, Gibbs, and Carathéodory,[3]

[3]The theory of classical thermodynamics has also been developed over the last one and a half centuries by many other researchers. Notable contributions include the work of Maxwell, Rankine,

and its laws have become one of the most firmly established scientific achievements ever accomplished. The pioneering work of Carnot [80] was the first to establish the impossibility of a *perpetuum mobile* of the second kind[4] by constructing a cyclical process (now known as the Carnot cycle) involving four thermodynamically reversible processes operating between two heat reservoirs at different temperatures, and showing that it is impossible to extract work from heat without at the same time discarding some heat.

Carnot's main assumption (now known as Carnot's principle) was that it is impossible to perform an arbitrarily often repeatable cycle whose only effect is to produce an unlimited amount of positive work. In particular, Carnot showed that the *efficiency* of a reversible cycle[5]—that is, the ratio of the total work produced during the cycle and the amount of heat transferred from a boiler (furnace) to a cooler (refrigerator)—is bounded by a universal maximum, and this maximum is a function only of the temperatures of the boiler and the cooler, and not of the nature of the working substance.

Both heat reservoirs (i.e., furnace and refrigerator) are assumed to have an infinite source of heat so that their state is unchanged by their heat exchange with the engine (i.e., the device that performs the cycle), and hence, the engine is capable of repeating the cycle arbitrarily often. Carnot's result (now known as Carnot's theorem) was remarkably arrived at using the erroneous concept that heat is an indestructible substance, that is, the *caloric theory of heat*.[6] This theory of heat was proposed by Lavoisier and influenced by experiments due to Black involving thermal properties of materials. The theory was based on the incorrect assertion that the temperature of a body was determined by the amount of *caloric* that it contained: an imponderable, indestructible, and highly elastic fluid that surrounded all matter and whose self-repulsive nature was responsible for thermal expansion.

Different notions of the conservation of energy can be traced back to the ancient Greek philosophers Thales ($\sim$ 624–$\sim$ 546 B.C.), Herakleitos ($\sim$ 535–$\sim$ 475 B.C.), and Empedocles ($\sim$ 490–$\sim$ 430 B.C.). Herakleitos postulates

---

Reech, Clapeyron, Bridgman, Kestin, Meixner, and Giles.

[4]A *perpetuum mobile* of the second kind is a cyclic device that would continuously extract heat from the environment and completely convert it into mechanical work. Since such a machine would not create energy, it would not violate the first law of thermodynamics. In contrast, a machine that creates its own energy and thus violates the first law is called a *perpetuum mobile* of the first kind.

[5]Carnot never used the terms *reversible* and *irreversible* cycles, but rather cycles that are performed in an inverse direction and order [319, p. 11]. The term *reversible* was first introduced by Kelvin [437] wherein the cycle can be run backwards.

[6]After Carnot's death, several articles were discovered wherein he had expressed doubt about the caloric theory of heat (i.e., the conservation of heat). However, these articles were not published until the late 1870s, and as such, did not influence Clausius in rejecting the caloric theory of heat and deriving Carnot's results using the energy equivalence principle of Mayer and Joule.

that nothing in nature can be created out of nothing, and nothing that disappears ceases to exist,[7] whereas Empedocles asserts that nothing comes to be or perishes in nature.[8] The mechanical equivalence principle of heat and work energy in its modern form, however, was developed by many scientists in the nineteenth century. Notable contributions include the work of Mayer, Joule, Thomson (Lord Kelvin), Thompson (Count Rumford), Helmholtz, Clausius, Maxwell, and Planck.

Even though many scientists are credited with the law of conservation of energy, it was first discovered independently by Mayer and Joule. Mayer—a surgeon—was the first to state the mechanical equivalence of heat and work energy in its modern form after noticing that his patients' blood in the tropics was a deeper red, leading him to deduce that they were consuming less oxygen, and hence less energy, in order to maintain their body temperature in a hotter climate. This observation in slower human metabolism along with the link between the body's heat release and the chemical energy released by the combustion of oxygen led Mayer to the discovery that heat and mechanical work are interchangeable.

Joule was the first to provide a series of decisive, quantitative studies in the 1840s showing the equivalence between heat and mechanical work. Specifically, he showed that if a thermally isolated system is driven from an initial state to a final state, then the work done is only a function of the initial and final equilibrium states, and is not dependent on the intermediate states or the mechanism doing the work. This path independence property along with the irrelevancy of the method by which the work was done led to the definition of the internal energy function as a new thermodynamic coordinate characterizing the quantity of energy or state of a thermodynamic system. In other words, heat or work do not contribute separately to the internal energy function; only the sum of the two matters.

Using a macroscopic approach and building on the work of Carnot, Clausius [87–90] was the first to introduce the notion of entropy as a physical property of matter and establish the two main laws of thermodynamics involving conservation of energy and nonconservation of entropy.[9] Specifically, using conservation of energy principles, Clausius showed that Carnot's principle is valid. Furthermore, Clausius postulated that it is impossible to

---

[7]Μὲν οὖν φησιν εἶναι τὸ πᾶν διαιρετὸν ἀδιαίρετον, γενητὸν ἀγένητον, θνητὸν ἀθάνατον, λὸγον αἰῶνα, πατέρα υἱὸν, ... ἐστίν ἕν πάντα εἶναι.

[8]Φύσις ουδενός εστίν εόντων αλλά μόνον μίξις τε, διάλλαξίς τε μιγέντων εστί, φύσις δ' επί τοις ονομάζεται ανθρώποισιν—There is no genesis with regard to any of the things in nature but rather a blending and alteration of the mixed elements; man, however, uses the word *nature* to name these events.

[9]Clausius succinctly expressed the first and second laws of thermodynamics as: "Die energie der Welt ist konstant und die entropie der Welt strebt einem maximum zu." Namely, the energy of the Universe is constant and the entropy of the Universe tends to a maximum.

perform a cyclic system transformation whose only effect is to transfer heat from a body at a given temperature to a body at a higher temperature. From this postulate Clausius established the second law of thermodynamics as a statement about entropy increase for *adiabatically isolated systems* (i.e., systems with no heat exchange with the environment).

From this statement Clausius goes on to state what have become known as the most controversial words in the history of thermodynamics and perhaps all of science; namely, the entropy of the universe is tending to a maximum, and the total state of the universe will inevitably approach a limiting state. Clausius' second law decrees that the usable energy in the universe is locked toward a path of degeneration, sliding toward a state of quietus. The fact that the entropy of the universe is a thermodynamically undefined concept led to serious criticism of Clausius' grand universal generalizations by many of his contemporaries as well as numerous scientists, natural philosophers, and theologians who followed.

Clausius' concept of the universe approaching a limiting state was inadvertently based on an analogy between a universe and a finite adiabatically isolated system possessing a finite energy content. His eschatological conclusions are far from obvious for complex dynamical systems with dynamical states far from equilibrium and involving processes beyond a simple exchange of heat and mechanical work. It is not clear where the heat absorbed by the system, if that system is the universe, needed to define the change in entropy between two system states comes from. Nor is it clear whether an infinite and endlessly expanding universe governed by the theory of general relativity has a final equilibrium state.

An additional caveat is the delineation of energy conservation when changes in the curvature of spacetime need to be accounted for. In this case, the energy density tensor in Einstein's field equations is only covariantly conserved (i.e., locally conserved in free-falling coordinates) since it does not account for gravitational energy—an unsolved problem in the general theory of relativity. In particular, conservation of energy and momentum laws, wherein a global time coordinate does not exist, has led to one of the fundamental problems in general relativity. Specifically, in general relativity involving a curved spacetime (i.e., a semi-Riemannian spacetime), the action of the gravitational field is invariant with respect to arbitrary coordinate transformations in semi-Riemannian spacetime with a nonvanishing Jacobian containing a large number of Lie groups.

In this case, it follows from Nöether's theorem [341],[10] which derives

[10]Many conservation laws are a special case of Nöether's theorem, which states that for every one-parameter group of diffeomorphisms defined on an abstract geometrical space (e.g.,

conserved quantities from symmetries and states that every differentiable symmetry of a dynamical action has a corresponding conservation law, that a large number of conservation laws exist, some of which are not physical. In contrast, the classical conservation laws of physics, which follow from time translation invariance, are determined by an invariant property under a particular Lie group with the conserved quantities corresponding to the parameters of the group. And in special relativity, conservation of energy and momentum is a consequence of invariance through the action of infinitesimal translation of the inertial coordinates, wherein the Lorentz transformation relates inertial systems in different inertial coordinates.

In general relativity, the momentum-energy equivalence principle holds only in a local region of spacetime—a flat or Minkowski spacetime. In other words, the energy-momentum conservation laws in gravitation theory involve gauge conservation laws with local time transformations, wherein the covariant transformation generators are canonical horizontal prolongations of vector fields on a world manifold,[11] and hence, in a curved spacetime there does not exist a global energy-momentum conservation law. Nevertheless, the law of conservation of energy is as close to an absolute truth as our incomprehensible universe will allow us to deduce. In his later work [89], Clausius remitted his famous claim that the entropy of the universe is tending to a maximum.

In parallel research Kelvin [240, 438] developed similar, and in some cases identical, results as Clausius, with the main difference being the absence of the concept of entropy. Kelvin's main view of thermodynamics was that of a universal irreversibility of physical phenomena occurring in nature. Kelvin further postulated that it is impossible to perform a cyclic system transformation whose only effect is to transform into work heat from a source that is at the same temperature throughout.[12] Without any supporting mathematical arguments, Kelvin goes on to state that the universe is heading toward a state of eternal rest wherein all life on Earth in

configuration manifolds, Minkowski space, Riemannian space) of a Hamiltonian dynamical system that preserves a Hamiltonian function, there exist first integrals of motion. In other words, the algebra of the group is the set of all Hamiltonian systems whose Hamiltonian functions are the first integrals of motion of the original Hamiltonian system.

[11]A world manifold is a four-dimensional orientable, noncompact, parallelizable manifold that admits a semi-Riemannian metric and a spin structure. Gravitation theories are formulated on tensor bundles that admit canonical horizontal prolongations on a vector field defined on a world manifold. These prolongations are generators of covariant transformations whose vector field components play the role of gauge parameters. Hence, in general relativity the energy-momentum flow collapses to a superpotential of a world vector field defined on a world manifold admitting gauge parameters.

[12]In the case of thermodynamic systems with positive absolute temperatures, Kelvin's postulate can be shown to be equivalent to Clausius' postulate. However, many textbooks erroneously show this equivalence without the assumption of *positive* absolute temperatures. Physical systems possessing a small number of energy levels (i.e., an inverted Boltzmann energy distribution) with negative absolute temperatures are discussed in [121, 260, 264, 307, 373].

the distant future shall perish. This claim by Kelvin involving a universal tendency toward dissipation has come to be known as the *heat death of the universe.*

The universal tendency toward dissipation and the heat death of the universe were expressed long before Kelvin by the ancient Greek philosophers Herakleitos and Leukippos (∼480–∼420 B.C.). In particular, Herakleitos states that this universe, which is the same everywhere, and which no one god or man has made, existed, exists, and will continue to exist as an eternal source of energy set on fire by its own natural laws, and will dissipate under its own laws.[13] Herakleitos' profound statement created the foundation for all metaphysics and physics and marks the beginning of science postulating the big bang theory as the origin of the universe as well as the heat death of the universe. A century after Herakleitos, Leukippos declared that from its genesis, the cosmos has spawned multitudinous worlds that evolve in accordance to a supreme law that is responsible for their expansion, enfeeblement, and eventual demise.[14]

Building on the work of Clausius and Kelvin, Planck [358,362] refined the formulation of classical thermodynamics. From 1897 to 1964, Planck's treatise [358] underwent eleven editions and is considered the definitive exposition on classical thermodynamics. Nevertheless, these editions have several inconsistencies regarding key notions and definitions of reversible and irreversible processes.[15] Planck's main theme of thermodynamics is that entropy increase is a necessary and sufficient condition for irreversibility. Without any proof (mathematical or otherwise), he goes on to conclude that every dynamical system in nature evolves in such a way that the total entropy of all of its parts increases. In the case of reversible processes, he concludes that the total entropy remains constant.

Unlike Clausius' entropy increase conclusion, Planck's increase entropy principle is not restricted to adiabatically isolated dynamical systems. Rather, it applies to all system transformations wherein the initial states of any exogenous system, belonging to the environment and coupled to the transformed dynamical system, return to their initial condition. It is important to note that Planck's entire formulation is restricted to homogeneous systems for which the thermodynamical state is characterized by *two* thermodynamic state variables, that is, a fluid. His formulation of entropy and the second law is not defined for more complex systems that

[13]Κόσμον (τόνδε), τὸν αὐτὸν ἁπάντων, οὔτε τις θεῶν, οὔτε ἀνθρώπων ἐποίησεν, ἀλλ' ἦν ἀεὶ καὶ ἔστιν καὶ ἔσται πῦρ ἀείζωον, ἁπτόμενον μέτρα καὶ ἀποσβεννύμενον μέτρα.

[14]Είναι τε ώσπερ γενέσεις κόσμου, ούτω καί αυξήσεις καί φθίσεις καί φθοράς, κατά τινά ανάγκην.

[15]Truesdell [445, p. 328] characterizes the work as a "gloomy murk," whereas Khinchin [245, p. 142] declares it an "aggregate of logical and mathematical errors superimposed on a general confusion in the definition of the basic quantities."

are not in equilibrium and in an environment that is more complex than one comprising a system of ideal gases.

Unlike the work of Clausius, Kelvin, and Planck involving cyclical system transformations, the work of Gibbs [163] involves system equilibrium states. Specifically, Gibbs assumes a thermodynamic state of a system involving pressure, volume, temperature, energy, and entropy, among others, and proposes that an *isolated system*[16] (i.e., a system with no energy exchange with the environment) is in equilibrium if and only if in all possible variations of the state of the system that do not alter its energy, the variation of the system entropy is negative semidefinite. Thus, the system entropy is maximized at the system equilibrium.

Gibbs also proposed a complementary formulation of his *maximum entropy* principle involving a principle of *minimal energy*. Namely, for an equilibrium of any isolated system, it is necessary and sufficient that in all possible variations of the state of the system that do not alter its entropy, the variation of its energy shall either vanish or be positive. Hence, the system energy is minimized at the system equilibrium.

Gibbs' principles give necessary and sufficient conditions for a thermodynamically stable equilibrium and should be viewed as variational principles defining admissible (i.e., stable) equilibrium states. Thus, they do not provide any information about the dynamical state of the system as a function of time nor any conclusions regarding entropy increase or energy decrease in a dynamical system transformation.

Carathéodory [76, 77] was the first to give a rigorous axiomatic mathematical framework for thermodynamics. In particular, using an *equilibrium* thermodynamic theory, Carathéodory assumes a state space endowed with a Euclidean topology and defines the equilibrium state of the system using thermal and deformation coordinates. Next, he defines an *adiabatic accessibility* relation wherein a reachability condition of an adiabatic process[17] is used such that an empirical statement of the second law characterizes a mathematical structure for an abstract state space. Even though the topology in Carathéodory's thermodynamic framework is induced on $\mathbb{R}^n$ (the space of n-tuples of reals) by taking the metric to be the Euclidean distance function and constructing the corresponding neighborhoods, the metrical properties of the state space do not play a role in his theory as there is no preference for a particular set of system coordinates.

---

[16]Gibbs' principle is weaker than Clausius' principle leading to the second law involving entropy increase since it holds for the more restrictive case of isolated systems.

[17]Carathéodory's definition of an adiabatic process is nonstandard and involves transformations that take place while the system remains in an *adiabatic container*; this allowed him to avoid introducing heat as a primitive variable (i.e., axiomatic element). For details see [76, 77].

Carathéodory's postulate for the second law states that in every open neighborhood of any equilibrium state of a system, there exist equilibrium states such that for some second open neighborhood contained in the first neighborhood, all the equilibrium states in the second neighborhood cannot be reached by adiabatic processes from equilibrium states in the first neighborhood. From this postulate Carathéodory goes on to show that for a special class of systems, which he called *simple systems*, there exists a *locally* defined entropy and an absolute temperature on the state space for every simple system *equilibrium state*. In other words, Carathéodory's postulate establishes the existence of an integrating factor for the heat transfer in an infinitesimal reversible process for a thermodynamic system of an arbitrary number of degrees of freedom that makes entropy an exact (i.e., total) differential.

Unlike the work of Clausius, Kelvin, Planck, and Gibbs, Carathéodory provides a topological formalism for the theory of thermodynamics, which elevates the subject to the level of other theories of modern physics. Specifically, the empirical statement of the second law is replaced by an abstract state space formalism, wherein the second law is converted into a local topological property endowed with a Euclidean metric. This parallels the development of relativity theory, wherein Einstein's original special theory started from empirical principles—e.g., the velocity of light in free space is invariant in all inertial frames—and then was replaced by an abstract geometrical structure: the Minkowski spacetime, wherein the empirical principles are converted into local topological properties of the Minkowski metric. However, one of the key limitations of Carathéodory's work is that his principle is too weak in establishing the existence of a *global* entropy function.

Adopting a microscopic viewpoint, Boltzmann [58] was the first to give a probabilistic interpretation of entropy involving different configurations of molecular motion of the microscopic dynamics. Specifically, Boltzmann reinterpreted thermodynamics in terms of molecules and atoms by relating the *mechanical* behavior of individual atoms with their *thermodynamic* behavior by suitably averaging properties of the individual atoms. In particular, even though individually each molecule and atom obeys Newtonian mechanics, he used the science of statistical mechanics to bridge between the microscopic details and the macroscopic behavior to try to find a mechanical underpinning of the second law.

Even though Boltzmann was the first to give a probabilistic interpretation of entropy as a measure of the disorder of a physical system involving the evolution toward the largest number of possible configurations of the system's states relative to its ordered initial state, Maxwell was the first to use statistical methods to understand the behavior of the kinetic theory of gases. In particular, he postulated that it is not necessary to track the positions and velocities of each individual atom and molecule, but rather it suffices to know their position and velocity distributions; concluding that the second law is merely statistical. His distribution law for the kinetic theory of gases describes an exponential function giving the statistical distribution of the velocities and energies of the gas molecules at thermal equilibrium and provides an agreement with classical (i.e., nonquantum) mechanics.

Although the Maxwell speed distribution law agrees remarkably well with observations for an assembly of weakly interacting particles that are distinguishable, it fails for indistinguishable (i.e., identical) particles at high densities. In these regions, speed distributions predicated on the principles of quantum physics must be used; namely, the Fermi-Dirac and Bose-Einstein distributions. In this case, the Maxwell statistics closely agree with the Bose-Einstein statistics for bosons (photons, α-particles, and all nuclei with an even mass number) and the Fermi-Dirac statistics for fermions (electrons, protons, and neutrons).

Boltzmann, however, further showed that even though individual atoms are assumed to obey the laws of Newtonian mechanics, by suitably averaging over the velocity distributions of these atoms the microscopic (mechanical) behavior of atoms and molecules produced effects visible on a macroscopic (thermodynamic) scale. He goes on to argue that Clausius' thermodynamic entropy (a macroscopic quantity) is proportional to the logarithm of the probability that a system will exist in the state it is in relative to all possible states it could be in. Thus, the entropy of a thermodynamic system state (macrostate) corresponds to the degree of uncertainty about the actual system mechanical state (microstate) when only the thermodynamic system state (macrostate) is known. Hence, the essence of Boltzmann thermodynamics is that thermodynamic systems with a constant energy level will evolve from a less probable state to a more probable state with the equilibrium system state corresponding to a state of maximum entropy (i.e., highest probability).

Interestingly, Boltzmann's original thinking on the subject of entropy increase involved nondecreasing of entropy as an absolute certainty and not just as a statistical certainty. In the 1870s and 1880s, his thoughts on this matter underwent significant refinements and shifted to a probabilistic viewpoint after interactions with Maxwell, Kelvin, Loschmidt, Gibbs, Poincaré, Burbury, and Zermelo; all of whom criticized his original formulation.

In statistical thermodynamics the Boltzmann entropy formula relates the entropy S of an ideal gas to the number of distinct microstates

$\mathcal{W}$ corresponding to a given macrostate as $\mathcal{S} = k \log_e \mathcal{W}$, where k is the Boltzmann constant.[18] Thus, the Boltzmann entropy gives the number of different microscopic configurations of a system's states that leave its macroscopic appearance unchanged, and hence, it connects the Clausius entropy, a macroscopic thermodynamic quantity, to probability, a microscopic statistical quantity.

Even though Boltzmann was the first to link the thermodynamic entropy of a macrostate for some probability distribution of all possible microstates generated by different positions and momenta of various gas molecules [57], it was Planck who first stated (without proof) this entropy formula in his work on blackbody radiation [359]. In addition, Planck was also the first to introduce the precise value of the Boltzmann constant to the formula; Boltzmann merely introduced the proportional logarithmic connection between the entropy $\mathcal{S}$ of an observed macroscopic state, or degree of disorder of a system, to the thermodynamic probability of its occurrence $\mathcal{W}$, never introducing the constant k to the formula.

To further complicate matters, in his original paper [359] Planck stated the formula without derivation or clear justification; a fact that deeply troubled Albert Einstein [130]. Despite the fact that numerous physicists consider $\mathcal{S} = k \log_e \mathcal{W}$ as the second most important formula of physics—second to Einstein's $E = mc^2$—for its unquestionable success in computing the thermodynamic entropy of isolated systems, its theoretical justification remains ambiguous and vague in most statistical thermodynamics textbooks. In this regard, Khinchin [245, p. 142] writes: "All existing attempts to give a general proof of [Boltzmann's entropy formula] must be considered as an aggregate of logical and mathematical errors superimposed on a general confusion in the definition of basic quantities."

In the first half of the twentieth century, the macroscopic (classical) and microscopic (statistical) interpretations of thermodynamics underwent a long and fierce debate. To exacerbate matters, since classical thermodynamics was formulated as a physical theory and not a mathematical theory, many scientists and mathematical physicists expressed concerns about the completeness and clarity of the mathematical foundation of thermodynamics

---

[18] The number of distinct microstates $\mathcal{W}$ can also be regarded as the number of solutions of the Schrödinger equation for the system giving a particular energy distribution. The Schrödinger wave equation describes how a quantum state of a system evolves over time. The solution of the equation characterizes a quantum wave function whose wavelength is related to the system momentum and frequency is related to the system energy. Unlike Planck's discrete quantum transition theory of energy when light interacts with matter, Schrödinger's quantum theory stipulates that quantum transition involves vibrational changes from one form to another; and these vibrational changes are continuous in space and time. Furthermore, if the quantum wave function is known at any given point in time, then Schrödinger's equation uniquely specifies the quantum wave function at any other moment in time, making this constituent part of quantum physics fully deterministic.

[11, 69, 447]. In fact, many fundamental conclusions arrived at by classical thermodynamics can be viewed as paradoxical.

For example, in classical thermodynamics the notion of entropy (and temperature) is only defined for equilibrium states. However, the theory concludes that nonequilibrium states transition toward equilibrium states as a consequence of the law of entropy increase! Furthermore, classical thermodynamics is restricted to systems in equilibrium. The second law infers that for any transformation occurring in an isolated system, the entropy of the final state can never be less than the entropy of the initial state. In this context, the initial and final states of the system are equilibrium states. However, by definition, an equilibrium state is a system state that has the property that whenever the state of the system starts at the equilibrium state it will remain at the equilibrium state for all future time unless an exogenous input acts on the system. Hence, the entropy of the system can only increase if the system is *not* isolated!

Many aspects of classical thermodynamics are riddled with such inconsistencies, and hence it is not surprising that many formulations of thermodynamics, especially most textbook expositions, poorly amalgamate physics with rigorous mathematics. Perhaps this is best eulogized in [447, p. 6], wherein Truesdell describes the present state of the theory of thermodynamics as a "dismal swamp of obscurity." In a desperate attempt to try to make sense of the writings of de Groot, Mazur, Casimir, and Prigogine, he goes on to state that there is "something rotten in the [thermodynamic] state of the Low Countries" [447, p. 134].

Brush [69, p. 581] remarks that "anyone who has taken a course in thermodynamics is well aware, the mathematics used in proving Clausius' theorem ... [has] only the most tenuous relation to that known to mathematicians." And Born [61, p. 119] admits that "I tried hard to understand the classical foundations of the two theorems, as given by Clausius and Kelvin; ... but I could not find the logical and mathematical root of these marvelous results." More recently, Arnold [11, p. 163] writes that "every mathematician knows it is impossible to understand an elementary course in thermodynamics."

As we have outlined, it is clear that there have been many different presentations of classical thermodynamics with varying hypotheses and conclusions. To exacerbate matters, there are also many vaguely defined terms and functions that are central to thermodynamics, such as entropy, enthalpy, free energy, quasi-static, nearly in equilibrium, extensive variables, intensive variables, reversible, irreversible, etc. Furthermore, these functions' domain and codomain are often unspecified and their local and global existence, uniqueness, and regularity properties are unproven.

Moreover, there are no general dynamic equations of motion, no ordinary or partial differential equations, and no general theorems providing mathematical structure and characterizing classes of solutions. Rather, we are asked to believe that a certain differential can be greater than something that is not a differential defying the syllogism of differential calculus, line integrals approximating adiabatic and isothermal paths result in alternative solutions annulling the fundamental theorem of integral calculus, and we are expected to settle for descriptive and unmathematical wordplay in explaining key principles that have far-reaching consequences in engineering, science, and cosmology.

Furthermore, the careless and considerable differences in the definitions of two of the key notions of thermodynamics—namely, the notions of reversibility and irreversibility—have contributed to the widespread confusion and lack of clarity of the exposition of classical thermodynamics over the past one and a half centuries. For example, the concept of reversible processes as defined by Carnot, Clausius, Kelvin, Planck, and Carathéodory have very different meanings. In particular, Carnot never uses the term *reversible*, but rather cycles that can be run backwards. Later he added that these cycles should proceed slowly so that the system remains in equilibrium over the entire cycle. Such system transformations are commonly referred to as *quasi-static* transformations in the thermodynamic literature. Clausius defines reversible (*umkehrbar*) cyclic and noncyclic processes as slowly varying processes wherein successive states of these processes differ by infinitesimals from the equilibrium system states. Alternatively, Kelvin's notions of reversibility involve the ability of a system to completely recover its initial state from the final system state. He does not limit his definition of reversibility to cyclic processes, and hence, a cyclic process can be reversible in the sense of Kelvin but irreversible in the sense of Carnot.

Planck introduced several notions of reversibility. His main notion of reversibility is one of *complete* reversibility and involves recoverability of the original state of the dynamical system while at the same time restoring the environment to its original condition. Unlike Clausius' notion of reversibility, Kelvin's and Planck's notions of reversibility do not require the system to exactly retrace its original trajectory in reverse order. Carathéodory's notion of reversibility involves recoverability of the system state in an adiabatic process resulting in yet another definition of thermodynamic reversibility. These subtle distinctions of (ir)reversibility are often unrecognized in the thermodynamic literature. Notable exceptions to this fact include [65, 448], with [448] providing an excellent exposition of the relation between irreversibility, the second law of thermodynamics, and the arrow of time.

1.2 Thermodynamics and the Arrow of Time

The arrow of time[19] and the second law of thermodynamics is one of the most famous and controversial problems in physics. The controversy between ontological time (i.e., a timeless universe) and the arrow of time (i.e., a constantly changing universe) can be traced back to the famous dialogues between the ancient Greek philosophers Parmenides[20] and Herakleitos on being and becoming. Parmenides, like Einstein, insisted that time is an illusion, that there is nothing new, and that everything is (being) and will forever be. This statement is of course paradoxical since the status quo changed after Parmenides wrote his famous poem *On Nature.*

Parmenides maintained that we all exist *within* spacetime, and time is a one-dimensional continuum in which all events, regardless of when they happen from any given perspective, simply *are.* All events exist endlessly and universally, and occupy ordered points in spacetime, and hence, reality envelops past, present, and future *equally.* More specifically, our picture of the universe at a given moment is identical and contains exactly the same events; we simply have different *conceptions* of what exists at that moment, and hence, different *conceptions* of reality. Conversely, the Heraclitean flux doctrine maintains that nothing ever is, and everything is becoming. In this regard, time gives a *different* ontological status of past, present, and future resulting in an ontological transition, creation, and actualization of events. More specifically, the unfolding of events in the flow of time have counterparts in reality.

Herakleitos' aphorism is predicated on change (becoming); namely, the universe is in a constant state of flux and nothing is stationary—Τα πάντα ρεί καί ούδέν μένει. Furthermore, Herakleitos goes on to state that the universe evolves in accordance with its own laws, which are the only unchangeable things in the universe (e.g., universal conservation and nonconservation laws). His statements that everything is in a state of flux—Τα πάντα ρεί—and that man cannot step into the same river twice, because

---

[19]The phrase *arrow of time* was coined by Eddington in his book *The Nature of the Physical World* [123] and connotes the one-way direction of entropy increase educed from the second law of thermodynamics. Other phrases include the *thermodynamic arrow* and the *entropic arrow of time.* Long before Eddington, however, philosophers and scientists addressed deep questions about time and its direction.

[20]Parmenides (∼515–∼450 B.C.) maintained that there is neither time nor motion. His pupil Zeno of Elea (∼490–∼430 B.C.) constructed four paradoxes—the dichotomy, the Achilles, the flying arrow, and the stadium—to prove that motion is impossible. His logic was "immeasurably subtle and profound" and even though infinitesimal calculus provides a tool that explains Zeno's paradoxes, the paradoxes stand at the intersection of reality and our perception of it; and they remain at the cutting edge of our understanding of space, time, and spacetime [316].

neither the man nor the river is the same—Ποταμείς τοίς αυτοίς εμβαίνομεν τε καί ουκ εμβαίνομεν, είμεν τε καί ουκ είμεν—give the earliest perception of irreversibility of nature and the universe along with time's arrow. The idea that the universe is in constant change and there is an underlying order to this change—the Logos (Λόγος)—postulates the existence of entropy as a physical property of matter permeating the whole of nature and the universe.

Herakleitos' statements are completely consistent with the laws of thermodynamics, which are intimately connected to the irreversibility of dynamical processes in nature. In addition, his aphorisms go beyond the worldview of classical thermodynamics and have deep relativistic ramifications to the spacetime fabric of the cosmos. Specifically, Herakleitos' profound statement—All matter is exchanged for energy, and energy for all matter (Πυρός τε ἀνταμοιβὴ τὰ πάντα καὶ πῦρ ἁπάντων)—is a statement of the law of conservation of mass-energy and is a precursor to the principle of relativity. In describing the nature of the universe Herakleitos postulates that nothing can be created out of nothing, and nothing that disappears ceases to exist. This totality of forms, or mass-energy equivalence, is eternal[21] and unchangeable in a constantly changing universe.

The arrow of time[22] remains one of physics' most perplexing enigmas [122, 178, 220, 254, 297, 362, 377, 463]. Even though time is one of the most familiar concepts mankind has ever encountered, it is the least understood. Puzzling questions of time's mysteries have remained unanswered throughout the centuries.[23] Questions such as, Where does time come from? What would our universe look like without time? Can there be more than one dimension to time? Is time truly a fundamental appurtenance woven into the fabric of the universe, or is it just a useful edifice for organizing our perception of events? Why is the concept of time hardly ever found in the most fundamental physical laws of nature and the universe? Can we go back in time? And if so, can we change past events?

Human experience perceives time flow as unidirectional; the present

[21]It is interesting to note that, despite his steadfast belief in change, Herakleitos embraced the concept of *eternity* as opposed to Parmenides' *endless duration* concept in which all events making up the universe are static and unchanging, eternally occupying fixed points in a frozen immutable future of spacetime.

[22]Perhaps a better expression here is the *geodesic arrow of time*, since, as Einstein's theory of relativity shows, time and space are intricately coupled, and hence one cannot curve space without involving time as well. Thus, time has a shape that goes along with its directionality.

[23]Plato (∼428–∼348 B.C.) writes that time was created as an image of the eternal. While time is everlasting, time is the outcome of change (motion) in the universe. And as night and day and month and the like are all part of time, without the physical universe time ceases to exist. Thus, the creation of the universe has spawned the arrow of time—Χρόνον τε γενέσθαι εἰκόνα τοῦ ἀιδίου. Κἀκεῖνον μὲν ἀεί μένειν, τὴν δὲ τοῦ οὐρανοῦ φορὰν χρόνον εἶναι· καὶ γὰρ νύκτα καὶ ἡμέραν καὶ μῆνα καὶ τὰ τοιαῦτα πάντα χρόνου μέρη εἶναι. Διόπερ ἄνευ τῆς τοῦ κόσμου φύσεως οὐκ εἶναι χρόνον· ἅμα γὰρ ὑπάρχειν αὐτῷ καὶ χρόνον εἶναι.

is forever flowing toward the future and away from a forever fixed past. Many scientists have attributed this *emergence* of the direction of time flow to the second law of thermodynamics due to its intimate connection to the irreversibility of dynamical processes.[24] In this regard, thermodynamics is disjoint from Newtonian and Hamiltonian mechanics (including Einstein's relativistic and Schrödinger's quantum extensions), since these theories are invariant under time reversal, that is, they make no distinction between one direction of time and the other. Such theories possess a *time-reversal symmetry*, wherein, from any given moment of time, the governing laws treat past and future in exactly the same way [258].[25] It is important to stress here that time-reversal symmetry applies to dynamical processes whose reversal is allowed by the physical laws of nature, *not* a reversal of time itself. It is irrelevant whether or not the reversed dynamical process actually occurs in nature; it suffices that the theory allows for the reversed process to occur.

The simplest notion of time-reversal symmetry is the statement wherein the physical theory in question is time-reversal symmetric in the sense that given any solution $x(t)$ to a set of dynamic equations describing the physical laws, then $x(-t)$ is also a solution to the dynamic equations. For example, in Newtonian mechanics this implies that there exists a transformation $\mathcal{R}(q,p)$ such that $\mathcal{R}(q,p) \circ x(t) = x(-t) \circ \mathcal{R}(q,p)$, where $\circ$ denotes the composition operator and $x(-t) = [q(-t), \ -p(-t)]^{\mathrm{T}}$ represents the particles that pass through the same position as $q(t)$, but in reverse order and with reverse velocity $-p(-t)$. It is important to note that if the physical laws describe the dynamics of particles in the presence of a field (e.g., an electromagnetic field), then the reversal of the particle velocities is insufficient for the equations to yield time-reversal symmetry. In this case, it is also necessary to reverse the field, which can be accomplished by modifying the transformation $\mathcal{R}$ accordingly.

As an example of time-reversal symmetry, a film run backwards of a harmonic oscillator over a full period or a planet orbiting the Sun would represent possible events. In contrast, a film run backwards of water in a glass coalescing into a solid ice cube or ashes self-assembling into a log of wood would immediately be identified as an impossible event.

[24] In statistical thermodynamics the arrow of time is viewed as a consequence of high system dimensionality and randomness. However, since in statistical thermodynamics it is not absolutely certain that entropy increases in every dynamical process, the direction of time, as determined by entropy increase, has only statistical certainty and not an absolute certainty. Hence, it cannot be concluded from statistical thermodynamics that time has a unique direction of flow.

[25] There is an exception to this statement involving the laws of physics describing weak nuclear force interactions in Yang-Mills quantum fields [471]. In particular, in certain experimental situations involving high-energy atomic and subatomic collisions, meson particles (K-mesons and B-mesons) exhibit time-reversal asymmetry [85]. However, under a combined transformation involving *charge conjugation* $\mathcal{C}$, which replaces the particles with their antiparticles, *parity* $\mathcal{P}$, which inverts the particles' positions through the origin, and a time-reversal *involution* $\mathcal{R}$, which replaces t with $-t$, the particles' behavior is $\mathcal{CPR}$-invariant. For details see [85].

Over the centuries, many philosophers and scientists shared the views of a Parmenidean frozen river time theory. However, since the advent of the science of thermodynamics in the nineteenth century, philosophy and science took a different point of view with the writings of Hegel, Bergson, Heidegger, Clausius, Kelvin, and Boltzmann; one involving time as our existential dimension. The idea that the second law of thermodynamics provides a physical foundation for the arrow of time has been postulated by many authors [123, 369, 377].[26] However, a convincing mathematical argument of this claim has never been given [178, 254, 448].

The complexities inherent with the afore statement are subtle and are intimately coupled with the universality of thermodynamics, entropy, gravity, and cosmology (see Section 17.4 and Chapter 18). A common misconception of the principle of the entropy increase is surmising that if entropy increases in forward time, then it necessarily decreases in backward time. However, entropy and the second law *do not* alter the (known) laws of physics in any way—the laws have no temporal orientation. In the absence of a unified dynamical systems theory of thermodynamics with Newtonian and Einsteinian mechanics, the second law is *derivative* to the physical laws of motion. Thus, since the (known) laws of nature are autonomous to temporal orientation, the second law implies, with identical certainty, that entropy increases both forward and backward in time from any given moment in time.

This statement, however, is not true in general; it is true only if the primordial state of the universe did *not* begin in a highly ordered, low entropy state. However, quantum fluctuations in Higgs boson particles[27] stretched out by inflation and inflationary cosmology followed by the big bang [183] tells us that the early universe began its trajectory in a highly ordered, low entropy state, which allows us to educe that the entropic arrow of time is *not* a *double-headed* arrow and that the future is indeed in the direction of increasing entropy. This further establishes that the concept of time flow directionality, which almost never enters in any physical theory, is a defining marvel of thermodynamics. Heat (i.e., energy in transition), like gravity, permeates every substance in the universe and its radiation spreads to every part of spacetime. However, unlike gravity, the directional

[26] Conversely, one can also find many authors who maintain that the second law of thermodynamics has nothing to do with irreversibility or the arrow of time [124, 231, 261]; these authors largely maintain that thermodynamic irreversibility and the absence of a temporal orientation of the rest of the laws of physics are disjoint notions. This is due to the fact that classical thermodynamics is riddled with many logical and mathematical inconsistencies with carelessly defined notation and terms. And more importantly, with the notable exception of [195], a dynamical systems foundation of thermodynamics is nonexistent in the literature.

[27] The Higgs boson is an elementary particle (i.e., a particle with an unknown substructure) containing matter (particle mass) and radiation (emission or transmission of energy), and is the finest quantum constituent of the Higgs field. See Chapter 17 for further details.

continuity of entropy and time (i.e., the entropic arrow of time) elevates thermodynamics to a sui generis physical theory of nature.

1.3 Modern Thermodynamics, Information Theory, and Statistical Energy Analysis

In an attempt to generalize classical thermodynamics to nonequilibrium thermodynamics, Onsager [347, 348] developed reciprocity theorems for irreversible processes based on the concept of a local equilibrium that can be described in terms of state variables that are predicated on linear approximations of thermodynamic equilibrium variables. Onsager's theorem pertains to the thermodynamics of linear systems, wherein a symmetric reciprocal relation applies between forces and fluxes. In particular, a flow or flux of matter in thermodiffusion is caused by the force exerted by the thermal gradient. Conversely, a concentration gradient causes a heat flow, an effect that has been experimentally verified for linear transport processes involving thermodiffusion, thermoelectric, and thermomagnetic effects.

Classical irreversible thermodynamics [114, 272, 477], as originally developed by Onsager, characterizes the rate of entropy production of irreversible processes as a sum of the product of fluxes with their associated forces, postulating a linear relationship between the fluxes and forces. The thermodynamic fluxes in the Onsager formulation include the effects of heat conduction, flow of matter (i.e., diffusion), mechanical dissipation (i.e., viscosity), and chemical reactions. Well-known laws of physics confirm Onsager's reciprocity relationships for near equilibrium systems. For example, Fourier's law of heat conduction asserts that heat flow is proportional to a temperature gradient and Fick's law describes a proportional relationship between diffusion and a chemical concentration gradient. Onsager's thermodynamic theory, however, is only correct for near equilibrium processes, wherein a local and linear instantaneous relation between the fluxes and forces holds.

Casimir [82] extended Onsager's principle of macroscopic reversibility to explain the relations between irreversible processes and network theory involving the coupling effects of electrical currents and resistance on entropy production. The Onsager-Casimir reciprocal relations treat only the irreversible aspects of system processes, and thus the theory is an algebraic theory that is primarily restricted to describing (time-independent) system steady states. In addition, the Onsager-Casimir formalism is restricted to linear systems, wherein a linearity restriction is placed on the admissible constitutive relations between the thermodynamic forces and fluxes. Another limitation of the Onsager-Casimir framework is the difficulty in providing a macroscopic description for large-scale complex dynamical systems. In addition, the Onsager-Casimir reciprical relations are not valid on the microscopic thermodynamic level.

Building on Onsager's classical irreversible thermodynamic theory, Prigogine [166, 367, 368] developed a thermodynamic theory of *dissipative nonequilibrium structures*. This theory involves kinetics describing the behavior of systems that are away from equilibrium states. However, Prigogine's thermodynamics lacks functions of the system state, and hence, his concept of entropy for a system away from equilibrium does not have a total differential. Furthermore, Prigogine's characterization of dissipative structures is predicated on a linear expansion of the entropy function about a particular equilibrium, and hence, is limited to the neighborhood of the equilibrium. This is a severe restriction on the applicability of this theory. In addition, his entropy cannot be calculated nor determined [165, 282]. Moreover, the theory requires that locally applied exogenous heat fluxes propagate at infinite velocities across a thermodynamic body, violating both experimental evidence and the principle of causality. To paraphrase Penrose, Prigogine's thermodynamic theory at best should be regarded as a trial or dead end.

In an attempt to extend Onsager's classical irreversible thermodynamic theory beyond a local equilibrium hypothesis, *extended irreversible thermodynamics* was developed in the literature [81, 236] wherein, in addition to the classical thermodynamic variables, dissipating fluxes are introduced as new independent variables providing a link between classical thermodynamics and flux dynamics. These complementary thermodynamic variables involve nonequilibrium quantities and take the form of dissipative fluxes and include heat, viscous pressure, matter, and electric current fluxes, among others. These fluxes are associated with microscopic operators of nonequilibrium statistical mechanics and the kinetic theory of gases, and effectively describe systems with long relaxation times (e.g., low-temperature solids, superfluids, and viscoelastic fluids).

Even though extended irreversible thermodynamics generalizes classical thermodynamics to nonequilibrium systems, the complementary variables are treated on the same level as the classical thermodynamic variables and hence lack any evolution equations. To compensate for this, additional rate equations are introduced for the dissipative fluxes. Specifically, the fluxes are selected as state variables wherein the constitutive equations of Fourier, Fick, Newton, and Ohm are replaced by first-order time evolution equations that include memory and nonlocal effects.

However, unlike the classical thermodynamic variables, which satisfy conservation of mass and energy and are compatible with the second law of thermodynamics, no specific criteria are specified for the evolution equations of the dissipative fluxes. Furthermore, since every dissipative flux is formulated as a thermodynamic variable characterized by a single evolution equation with the system entropy being a function of the fluxes, extended irreversible thermodynamic theories tend to be incompatible with classical thermodynamics. Specifically, the theory yields different definitions for temperature and entropy when specialized to equilibrium thermodynamic systems.

In the last half of the twentieth century, thermodynamics was reformulated as a global nonlinear field theory with the ultimate objective to determine the independent field variables of this theory [92, 333, 393, 446]. This aspect of thermodynamics, which became known as *rational thermodynamics*, was predicated on an entirely new axiomatic approach. As a result of this approach, modern continuum thermodynamics was developed using theories from elastic materials, viscous materials, and materials with memory [91, 106, 107, 182]. The main difference between classical thermodynamics and rational thermodynamics can be traced back to the fact that in rational thermodynamics the second law is not interpreted as a restriction on the transformations a system can undergo, but rather as a restriction on the system's constitutive equations.

Rational thermodynamics is formulated based on nonphysical interpretations of absolute temperature and entropy notions that are not limited to near equilibrium states. Moreover, the thermodynamic system has memory, and hence, the dynamic behavior of the system is determined not only by the present value of the thermodynamic state variables but also by the history of their past values. In addition, the second law of thermodynamics is expressed using the Clausius-Duhem inequality.

Rational thermodynamics is not a thermodynamic theory in the classical sense but rather a theory of thermomechanics of continuous media. This theory, which is also known as *modern continuum thermodynamics*, abandons the concept of a local equilibrium and involves general conservation laws (mass, momentum, energy) for defining a thermodynamic state of a body using a set of postulates and constitutive functionals. These postulates, which include the principles of admissibility (i.e., entropy principle), objectivity or covariance (i.e., reference frame invariance), local action (i.e., influence of a neighborhood), memory (i.e., a dynamic), and symmetry, are applied to the constitutive equations describing the thermodynamic process.

Modern continuum thermodynamics has been extended to account for nonlinear irreversible processes such as the existence of thresholds, plasticity, and hysteresis [118, 241, 309, 310]. These extensions use convex analysis, semigroup theory, and nonlinear programming theory but can lack a clear characterization of the space over which the thermodynamical state variables evolve. The principal weakness of rational thermodynamics is that its range of applicability is limited to closed systems (see Chapter 2) with a single absolute temperature. Thus, it is not applicable to condensed matter physics (e.g., diffusing mixtures or plasma). Furthermore, it does not provide a unique entropy characterization that satisfies the Clausius inequality.

More recently, a contribution to *equilibrium* thermodynamics is given in [280]. This work builds on the work of Carathéodory [76, 77] and Giles [164] by developing a thermodynamic system representation involving a state space on which an adiabatic accessibility relation is defined. The existence and uniqueness of an entropy function is established as a consequence of adiabatic accessibility among *equilibrium states*. As in Carathéodory's work, the authors in [280] also restrict their attention to simple (possibly interconnected) systems in order to arrive at an entropy increase principle. However, it should be noted that the notion of a simple system in [280] is not equivalent to that of Carathéodory's notion of a simple system.

Connections between thermodynamics and systems theory as well as information theory have also been explored in the literature [35,37,66,68,192, 353,463,464,472,478]. Information theory has deep connections to physics in general, and thermodynamics in particular. Many scientists have postulated that information is physical and have suggested that the bit is the irreducible kernel in the universe and it is more fundamental than matter itself, with information forming the very core of existence [167,259]. To produce change (motion) requires energy, whereas to direct this change requires information. In other words, energy takes different forms, but these forms are determined by information. Arguments about the nature of reality is deeply rooted in quantum information, which gives rise to every particle, every force field, and spacetime itself.

In quantum mechanics information can be inaccessible but not annihilated. In other words, information can never be destroyed despite the fact that imperfect system state distinguishability abounds in quantum physics, wherein the Heisenberg uncertainty principle brought the demise of determinism in the microcosm of science. The afore statement concerning the nonannihilation of information is not without its controversy in physics and is at the heart of the black hole information paradox, which resulted from the incomplete unification of quantum mechanics and general relativity.

Specifically, Hawking and Bekenstein [28, 208] argued that general relativity and quantum field theory were inconsistent with the principle that information cannot be lost. In particular, as a consequence of quantum fluctuations near a black hole's event horizon,[28] they showed that black holes radiate particles, and hence, slowly evaporate. And since matter falling into a black hole carries information in its structure, organization, and quantum states, black hole evaporation via radiation obliterates information.

However, using Richard Feynman's sum over histories path integral formulation of quantum theory to the topology of spacetime [146], Hawking later showed that quantum gravity is *unitary* (i.e., the sum of probabilities for all possible outcomes of a given event is unity) and that black holes are never unambiguously black. That is, black holes slowly dissipate before they ever truly form, allowing radiation to contain information, and hence, information is not lost, obviating the information paradox.

In quantum mechanics the Heisenberg uncertainty principle is a consequence of the fact that the outcome of an experiment is affected, or even determined, when observed. The Heisenberg uncertainty principle states that it is impossible to measure both the position and momentum of a particle with absolute precision at a microscopic level, and the product of the uncertainties in these measured values is in the order of the magnitude of the Planck constant. The determination of energy and time is also subject to the same uncertainty principle. The principle is not a statement about our inability to develop accurate measuring instruments, but rather a statement about an intrinsic property of nature; namely, nature has an inherent indeterminacy. And this is a consequence of the fact that any attempt at observing nature will disturb the system under observation, resulting in a lack of precision.

Quantum mechanics provides a probabilistic theory of nature, wherein the equations describe the average behavior of a large collection of identical particles and not the behavior of individual particles. Einstein maintained that the theory was incomplete albeit a good approximation in describing nature. He further asserted that when quantum mechanics had been completed, it would deal with certainties. In a letter to Max Born he states his famous "God does not play dice" dictum, writing: "The theory produces a great deal but hardly brings us closer to the secret of the Old One. I am at all events convinced that He does not play dice" [60, p. 90]. A profound ramification of the Heisenberg uncertainty principle is that the macroscopic principle of causality does not apply at the atomic level.

Information theory addresses the quantification, storage, and communication of information. The study of the effectiveness of communication

[28]In relativistic physics, an *event horizon* is a boundary delineating the set of points in spacetime beyond which events cannot affect an outside observer. In the present context, it refers to the boundary beyond which events cannot escape the black hole's gravitational field.

channels in transmitting information was pioneered by Shannon [406]. Information is encoded, stored (by codes), transmitted through channels of limited capacity, and then decoded. The effectiveness of this process is measured by the *Shannon capacity* of the channel and involves the entropy of a set of events that measure the uncertainty of this set. These channels function as input-output devices that take letters from an input alphabet and transmit letters to an output alphabet with various error probabilities that depend on noise. Hence, entropy in an information-theoretic context is a measure of information uncertainty. Simply put—information is *not* free and is linked to the cost of computing the behavior of matter and energy in our universe [39]. For an excellent exposition of these different facets of thermodynamics see [175].

Thermodynamic principles have also been repeatedly used in coupled mechanical systems to arrive at energy flow models. Specifically, in an attempt to approximate high-dimensional dynamics of large-scale structural (oscillatory) systems with a low-dimensional diffusive (nonoscillatory) dynamical model, structural dynamicists have developed thermodynamic energy flow models using stochastic energy flow techniques. In particular, statistical energy analysis (SEA) predicated on averaging system states over the statistics of the uncertain system parameters has been extensively developed for mechanical and acoustic vibration problems [78, 238, 270, 297, 414, 468]. The aim of SEA is to establish that many concepts of energy flow modeling in high-dimensional mechanical systems have clear connections with statistical mechanics of many particle systems, and hence, the second law of thermodynamics applies to large-scale coupled mechanical systems with modal energies playing the role of temperatures.

Thermodynamic models are derived from large-scale dynamical systems of discrete subsystems involving stored energy flow among subsystems based on the assumption of weak subsystem coupling or identical subsystems. However, the ability of SEA to predict the dynamic behavior of a complex large-scale dynamical system in terms of pairwise subsystem interactions is severely limited by the coupling strength of the remaining subsystems on the subsystem pair. Hence, it is not surprising that SEA energy flow predictions for large-scale systems with strong coupling can be erroneous. From the rigorous perspective of dynamical systems theory, the theoretical foundations of SEA remain inadequate since well-defined mathematical assumptions of the theory are not adequately delineated.

Alternatively, a deterministic thermodynamically motivated energy flow modeling for structural systems is addressed in [247–249]. This approach exploits energy flow models in terms of thermodynamic energy (i.e., the ability to dissipate heat) as opposed to stored energy and is not limited to weak subsystem coupling. A stochastic energy flow *compartmental model* (i.e., a model characterized by energy conservation laws) predicated on averaging system states over the statistics of stochastic system exogenous disturbances is developed in [37]. The basic result demonstrates how linear compartmental models arise from second-moment analysis of state space systems under the assumption of weak coupling. Even though these results can be potentially applicable to linear large-scale dynamical systems with weak coupling, such connections are not explored in [37]. With the notable exception of [78], and more recently [273], none of the aforementioned SEA-related works addresses the second law of thermodynamics involving entropy notions in the energy flow between subsystems.

Motivated by the manifestation of emergent behavior of macroscopic energy transfer in crystalline solids modeled as a lattice of identical molecules involving undamped vibrations, the authors in [44] analyze energy equipartition in linear Hamiltonian systems using average-preserving symmetries. Specifically, the authors consider a Lie group of phase space symmetries of a linear Hamiltonian system and characterize the subgroup of symmetries whose elements are also symmetries of every Hamiltonian system and preserve the time averages of quadratic Hamiltonian functions along system trajectories. In the very specific case of distinct natural frequencies and a two-degree-of-freedom system consisting of an interconnected pair of identical undamped oscillators, the authors show that the time-averaged oscillator energies reach an equipartitioned state. For this limited case, this result shows that time averaging leads to the emergence of damping in lossless Hamiltonian dynamical systems.

1.4 Dynamical Systems

Dynamical systems theory provides a universal mathematical formalism predicated on modern analysis and has become the prevailing language of modern science as it provides the foundation for unlocking many of the mysteries in nature and the universe that involve spatial and temporal evolution. Given that irreversible thermodynamic systems involve a definite direction of evolution, it is natural to merge the two universalisms of thermodynamics and dynamical systems under a single compendium, with the latter providing an ideal language for the former.

A *system* is a combination of components or parts that is perceived as a single entity. The parts making up the system may be clearly or vaguely defined. These parts are related to each other through a particular set of variables, called the *states* of the system, that, together with the knowledge of any system inputs, completely determine the behavior of the system at any given time. A *dynamical system* is a system whose state changes with time. Dynamical systems theory was fathered by Henri Poincaré [363–365], sturdily developed by Birkhoff [50, 51], and has evolved to become one of the most universal mathematical formalisms used to explain system manifestations of nature that involve time.

A dynamical system can be regarded as a mathematical model structure involving an input, state, and output that can capture the dynamical description of a given class of physical systems. Specifically, a *closed* dynamical system consists of three elements—namely, a setting called the *state space*, which is assumed to be Hausdorff[29] and in which the dynamical behavior takes place, such as a torus, topological space, manifold, or locally compact metric space; a mathematical rule or *dynamic*, which specifies the evolution of the system over time; and an initial condition or *state* from which the system starts at some initial time.

An *open* dynamical system interacts with the environment through system *inputs* and system *outputs* and can be viewed as a precise mathematical object that maps exogenous inputs (causes, disturbances) into outputs (effects, responses) via a set of internal variables, the state, which characterizes the influence of past inputs. For dynamical systems described by ordinary differential equations, the independent variable is time, whereas spatially distributed systems described by partial differential equations involve multiple independent variables reflecting, for example, time and space.

The state of a dynamical system can be regarded as an information storage or memory of past system events. The set of (internal) states of a dynamical system must be sufficiently rich to completely determine the behavior of the system for any future time. Hence, the state of a dynamical system at a given time is uniquely determined by the state of the system at the initial time and the present input to the system. In other words, the state of a dynamical system in general depends on both the present input to the system and the past history of the system. Even though it is often assumed that the state of a dynamical system is the *least* set of state variables needed to completely predict the effect of the past upon the future of the system, this is often a convenient simplifying assumption.

Ever since its inception, the basic questions concerning dynamical systems theory have involved qualitative solutions for the properties of a dynamical system; questions such as, For a particular initial system state, does the dynamical system have at least one solution? What are the asymptotic properties of the system solutions? How are the system solutions

[29]A Hausdorff space is a topological space in which there exists a pair of disjoint open neighborhoods for every pair of distinct points in the space.

dependent on the system initial conditions? How are the system solutions dependent on the form of the mathematical description of the dynamic of the system? How do system solutions depend on system parameters? And how do system solutions depend on the properties of the state space on which the system is defined?

Determining the mathematical rule or dynamic that defines the state of physical systems at a given future time from a given present state is one of the central problems of science. Once the flow or dynamic of a dynamical system describing the motion of the system starting from a given initial state is given, dynamical systems theory can be used to describe the behavior of the system states over time for different initial conditions. Throughout the centuries—from the great cosmic theorists of ancient Greece[30] to the present-day quest for a unified field theory—the most important dynamical system is our vicissitudinous universe. By using abstract mathematical models and attaching them to the physical world, astronomers, mathematicians, and physicists have used abstract thought to deduce something that is true about the natural system of the cosmos.

The quest by scientists, such as Brahe, Kepler, Galileo, Newton, Huygens, Euler, Lagrange, Laplace, and Maxwell, to understand the regularities inherent in the distances of the planets from the Sun and their periods and velocities of revolution around the Sun led to the science of dynamical systems as a branch of mathematical physics. Isaac Newton, however, was the first to model the motion of physical systems with differential equations. Newton's greatest achievement was the rediscovery that the motion of the planets and moons of the solar system resulted from a single fundamental source—the gravitational attraction of the heavenly

[30]The Hellenistic period (323–31 B.C.) spawned the scientific revolution leading to today's scientific method and scientific technology, including much of modern science and mathematics in its present formulation. Hellenistic scientists, which included Archimedes, Euclid, Eratosthenes, Eudoxus, Ktesibios, Philo, Apollonios, and many others, were the first to use abstract mathematical models and attach them to the physical world. More importantly, using abstract thought and rigorous mathematics (Euclidean geometry, real numbers, limits, definite integrals), these "modern minds in ancient bodies" were able to deduce complex solutions to practical problems and provide a deep understanding of nature. In his *Forgotten Revolution* [389] Russo convincingly argues that Hellenistic scientists were not just forerunners or anticipators of modern science and mathematics, but rather the true fathers of these disciplines. He goes on to show how science was born in the Hellenistic world and why it had to be reborn.

As in the case of the origins of much of modern science and mathematics, modern engineering can also be traced back to ancient Greece. Technological marvels included Ktesibios' pneumatics, Heron's automata, and arguably the greatest fundamental mechanical invention of all time—the Antikythera mechanism. The Antikythera mechanism, most likely inspired by Archimedes, was built around 76 B.C. and was a device for calculating the motions of the stars and planets, as well as for keeping time and calendar. This first analog computer involving a complex array of meshing gears was a quintessential hybrid dynamical system that unequivocally shows the singular sophistication, capabilities, and imagination of the ancient Greeks, and dispenses with the Western myth that the ancient Greeks developed mathematics but were incapable of creating scientific theories and scientific technology.

bodies. This discovery dates back to Aristarkhos' (310–230 B.C.) heliocentric theory of planetary motion and Hipparkhos' (190–120 B.C.) *dynamical* theory of planetary motion predicated on planetary attractions toward the Sun by a force that is inversely proportional to the square of the distance between the planets and the Sun [389, p. 304].

Many of the concepts of Newtonian mechanics, including relative motion, centrifugal and centripetal force, inertia, projectile motion, resistance, gravity, and the inverse square law, were known to the Hellenistic scientists [389]. For example, Hipparkhos' work *On bodies thrusting down because of gravity* (*Περὶ τῶν διὰ βαρύτητα κάτω φερομένων*) clearly and correctly describes the effects of gravity on projectile motion. And in Ploutarkhos' (46–120 A.D.) work *De facie quae in orbe lunae apparet* (*On the light glowing on the Moon*), he clearly describes the notion of gravitational interaction between heavenly bodies stating that "just as the sun attracts to itself the parts of which it consists, so does the earth ..."[31] [389, p. 304].

Newton himself wrote in his *Classical Scholia* [389, p. 376]: "Pythagoras ... applied to the heavens the proportions found through these experiments [on the pitch of sounds made by weighted strings], and learned from that the harmonies of the spheres. And so, by comparing those weights with the weights of the planets, and the intervals in sound with the intervals of the spheres, and the lengths of string with the distances of the planets [measured] from the center, he understood through the heavenly harmonies that the weights of the planets toward the sun ... are inversely proportional to the squares of their distances." And this admittance of the prior knowledge of the inverse square law predates Hooke's thoughts of explaining Kepler's laws out of the inverse square law communicated in a letter to Newton on January 6, 1680, by over two millennia.

It is important to stress here that what are erroneously called *Newton's laws of motion* in the literature were first discovered by Kepler, Galileo, and Descartes, with the latter first stating the law of inertia in its modern form. Namely, when viewed in an inertial reference frame, a body remains in the same state unless acted upon by a net force; and unconstrained motion follows a rectilinear path. Newton and Leibnitz independently advanced the basic dynamical tool invented two millennia earlier by Archimedes—the calculus;[32] with Euler being the first one to explicitly write down the second law of motion as an equation involving an applied force acting on

---

[31] ὡς γὰρ ὁ ἥλιος εἰς ἑαυτὸν ἐπιστρέφει τὰ μέρη ἐξ ὧν συνέστηκε, καὶ ἡ γῆ (in Ploutarkhos, *De facie quae in orbe lunae apparet*, 924E).

[32] In his treatise on *The Method of Mechanical Theorems* Archimedes (287–212 B.C.) established the foundations of integral calculus using infinitesimals, as well as the foundations of mathematical mechanics. In addition, in one of his problems he constructed the tangent at any given point for a spiral, establishing the origins of differential calculus [29, p. 32].

a body being equal to the time rate of change of its momentum. Newton, however, deduced a *physical* hypothesis—the law of universal gravitation involving an inverse-square law force—in precise mathematical form deriving (at the time) a cosmic dynamic using Euclidian geometry and *not* differential calculus (i.e., differential equations).

In his magnum opus, *Philosophiae Naturalis Principia Mathematica* [337], Newton investigated whether a small perturbation would make a particle moving in a plane around a center of attraction continue to move near the circle, or diverge from it. Newton used his analysis to analyze the motion of the moon orbiting the Earth. Numerous astronomers and mathematicians who followed made significant contributions to dynamical systems theory in an effort to show that the observed deviations of planets and satellites from fixed elliptical orbits were in agreement with Newton's principle of universal gravitation. Notable contributions include the work of Torricelli [443], Euler [137], Lagrange [256], Laplace [271], Dirichlet [116], Liouville [286], Maxwell [311], Routh [386], and Lyapunov [294–296].

Newtonian mechanics developed into the first field of modern science—dynamical systems as a branch of mathematical physics—wherein the circular, elliptical, and parabolic orbits of the heavenly bodies of our solar system were no longer fundamental determinants of motion, but rather approximations of the universal laws of the cosmos specified by governing differential equations of motion. And in the past century, dynamical systems theory has become one of the most fundamental fields of modern science as it provides the foundation for unlocking many of the mysteries in nature and the universe that involve the evolution of time. Dynamical systems theory is used to study ecological systems, geological systems, biological systems, economic systems, neural systems, and physical systems (e.g., mechanics, fluids, magnetic fields, galaxies), to cite but a few examples.

1.5 Dynamical Thermodynamics: A Postmodern Approach

In contrast to mechanics, which is based on a dynamical systems theory, classical thermodynamics (i.e., *thermostatics*) is a physical theory and does not possess equations of motion. Moreover, very little work has been done in obtaining extensions of thermodynamics for systems out of equilibrium. These extensions are commonly known as *thermodynamics of irreversible processes* or *modern irreversible thermodynamics* in the literature [113,367]. Such systems are driven by the continuous flow of matter and energy, are far from equilibrium, and often develop into a multitude of states. Connections between local thermodynamic subsystem interactions of these systems and the globally complex thermo*dynamical* system behavior are often elusive. This statement is true for nature in general and was most eloquently stated first by Herakleitos in his 123rd fragment—Φύσις κρύπτεσθαι φιλεί (Nature loves to hide).

These complex thermodynamic systems involve spatio-temporally evolving structures and can exhibit a hierarchy of emergent system properties. These systems are known as *dissipative systems* [195] and consume energy and matter while maintaining their stable structure by dissipating entropy to the environment. All living systems are dissipative systems; the converse, however, is not necessarily true. Dissipative living systems involve pattern interactions by which life emerges. This nonlinear interaction between the subsystems making up a living system is characterized by *autopoiesis* (self-creation). In the physical universe, billions of stars and galaxies interact to form self-organizing dissipative nonequilibrium structures [252, 369]. The fundamental common phenomenon among nonequilibrium (i.e., dynamical) systems is that they evolve in accordance with the laws of (nonequilibrium) thermodynamics.

Building on the work of nonequilibrium thermodynamic structures [166, 367], Sekimoto [400–404] introduced a stochastic thermodynamic framework predicated on Langevin dynamics in which fluctuation forces are described by Brownian motion. In this framework, the classical thermodynamic notions of heat, work, and entropy production are extended to the level of individual system trajectories of nonequilibrium ensembles. Specifically, system state trajectories are sample continuous and are characterized by a Langevin equation for each individual sample path and a Fokker-Planck equation for the entire ensemble of trajectories.

For such systems, energy conservation holds along fluctuating trajectories of the stochastic Markov process and the second law of thermodynamics is obtained as an ensemble property of the process. In particular, various fluctuation theorems [55,56,102,103,139,153,223,229,230,255,274,401] are derived that constrain the probability distributions for the exchanged heat, mechanical work, and entropy production depending on the nature of the stochastic Langevin system dynamics.

Even though stochastic thermodynamics is applicable to a single realization of the Markov process under consideration with the first and second laws of thermodynamics holding for nonequilibrium systems, the framework only applies to multiple time-scale systems with a few observable slow degrees of freedom. The unobservable degrees of freedom are assumed to be fast, and hence, always constrained to the equilibrium manifold imposed by the instantaneous values of the observed slow degrees of freedom.

Furthermore, if some of the slow variables are not accessible, then the system dynamics are no longer Markovian. In this case, defining a system entropy is virtually impossible. In addition, it is unclear whether fluctuation theorems expressing symmetries of the probability distribution functions for thermodynamic quantities can be grouped into universal classes characterized by asymptotics of these distributions. Moreover, it is also unclear whether there exist system variables that satisfy the transitive equilibration property of the zeroth law of thermodynamics for nonequilibrium stochastic thermodynamic systems.

In an attempt to create a generalized theory of evolution mechanics by unifying classical mechanics with thermodynamics, the authors in [25, 26, 423, 424] developed a framework of system thermodynamics based on the concept of *tribo-fatigue entropy*. This framework, known as *damage mechanics* [25, 26] or *mechanothermodynamics* [423, 424], involves an irreversible entropy function along with its generation rate that captures and quantifies system aging. Specifically, the second law is formulated analytically for organic and inorganic bodies, and the system entropy is determined by a damageability process predicated on mechanical and thermodynamic effects resulting in system state changes.

In [195], the authors develop a postmodern framework for thermodynamics that involves open interconnected dynamical systems that exchange matter and energy with their environment in accordance with the first law (conservation of energy) and the second law (nonconservation of entropy) of thermodynamics. Symmetry can spontaneously occur in such systems by invoking the two fundamental axioms of the science of heat.

Namely, *i*) if the energies in the connected subsystems of an interconnected system are equal, then energy exchange between these subsystems is not possible, and *ii*) energy flows from more energetic subsystems to less energetic subsystems. These axioms establish the existence of a *global* system entropy function as well as *equipartition of energy* [195] in system thermodynamics; an *emergent* behavior in thermodynamic systems. Hence, in complex interconnected thermodynamic systems, higher symmetry (i.e., system decomplexification) is not a property of the system's parts but rather emerges as a result of the nonlinear subsystem interactions.

The goal of the present monograph is directed toward building on the results of [195] to place thermodynamics on a system-theoretic foundation by combining the two universalisms of thermodynamics and dynamical systems theory under a single umbrella so as to harmonize it with classical mechanics. In particular, we develop a novel formulation of thermodynamics that can be viewed as a moderate-sized dynamical systems theory as compared to statistical thermodynamics. This middle-ground theory involves large-scale dynamical system models characterized by ordinary deterministic and stochastic differential equations, as well as infinite-dimensional models characterized by partial differential equations and functional delay differential equations that bridge the gap between classical and statistical thermodynamics.

Specifically, since thermodynamic models are concerned with energy flow among subsystems, we use a state space formulation to develop a nonlinear compartmental dynamical system model that is characterized by energy conservation laws capturing the exchange of energy and matter between coupled macroscopic subsystems. Furthermore, using graph-theoretic notions, we state two thermodynamic axioms consistent with the zeroth and second laws of thermodynamics, which ensure that our large-scale dynamical system model gives rise to a thermodynamically consistent energy flow model. Specifically, using a large-scale dynamical systems theory perspective for thermodynamics, we show that our compartmental dynamical system model leads to a precise formulation of the equivalence between work energy and heat in a large-scale dynamical system.

Since our dynamical thermodynamic formulation is based on a large-scale dynamical systems theory involving the exchange of energy with conservation laws describing transfer, accumulation, and dissipation between subsystems and the environment, our framework goes beyond classical thermodynamics characterized by a purely empirical theory, wherein a physical system is viewed as an input-output *black box* system. Furthermore, unlike classical thermodynamics, which is limited to the description of systems in equilibrium states, our approach addresses nonequilibrium thermodynamic systems. This allows us to connect and unify the behavior of heat as described by the equations of thermal transfer and as described by classical thermodynamics. This exposition further demonstrates that these disciplines of classical physics are derivable from the same principles and are part of the same scientific and mathematical framework.

Our nonequilibrium thermodynamic framework goes beyond classical irreversible thermodynamics developed by Onsager [347, 348] and further extended by Casimir [82] and Prigogine [166, 367, 368], which, as discussed in Section 1.3, fall short of a complete dynamical theory. Specifically, their theories postulate that the local instantaneous thermodynamic variables of the system are the same as that of the system in equilibrium. This implies that the system entropy in a neighborhood of an equilibrium is dependent on the same variables as those at equilibrium, violating Gibbs' maximum entropy principle. In contrast, the proposed system thermodynamic formalism brings classical thermodynamics within the framework of modern nonlinear dynamical systems theory, thus providing information about the dynamical behavior of the thermodynamic state variables between the initial and final equilibrium system states.

Next, we give a deterministic definition of entropy for a large-scale dynamical system that is consistent with the classical thermodynamic definition of entropy, and we show that it satisfies a Clausius-type inequality leading to the law of entropy nonconservation. However, unlike classical thermodynamics, wherein entropy is not defined for arbitrary states out of equilibrium, our definition of entropy holds for nonequilibrium dynamical systems.

Furthermore, we introduce a *new* and dual notion to entropy—namely, *ectropy*[33]—as a measure of the tendency of a large-scale dynamical system to do useful work and grow more organized, and we show that conservation of energy in an adiabatically isolated thermodynamically consistent system necessarily leads to nonconservation of ectropy and entropy. Hence, for every dynamical transformation in an adiabatically isolated thermodynamically consistent system, the entropy of the final system state is greater than or equal to the entropy of the initial system state.

Then, using the system ectropy as a Lyapunov function candidate, we show that in the absence of energy exchange with the environment our thermodynamically consistent large-scale nonlinear dynamical system model possesses a continuum of equilibria and is *semistable*, that is, it has convergent subsystem energies to Lyapunov stable energy equilibria determined by the large-scale system initial subsystem energies. In addition, we show that the steady-state distribution of the large-scale system energies is uniform, leading to system energy equipartitioning corresponding to a minimum ectropy and a maximum entropy equilibrium state.

For our thermodynamically consistent dynamical system model, we further establish the existence of a *unique* continuously differentiable *global* entropy and ectropy function for all equilibrium and nonequilibrium states. Using these global entropy and ectropy functions, we go on to establish a clear connection between thermodynamics and the arrow of time. Specifically, we rigorously show a *state irrecoverability* and hence a *state irreversibility*[34] nature of thermodynamics. In particular, we show that for

---

[33]Ectropy comes from the Greek word $\varepsilon\kappa\tau\rho o\pi\eta$ ($\varepsilon\kappa$ and $\tau\rho o\pi\eta$) for outward transformation connoting evolution or complexification and is the literal antonym of entropy ($\varepsilon\nu\tau\rho o\pi\eta$—$\varepsilon\nu$ and $\tau\rho o\pi\eta$), signifying an inward transformation connoting devolution or decomplexification. The word *entropy* was proposed by Clausius for its phonetic similarity to energy with the additional connotation reflecting change ($\tau\rho o\pi\eta$).

[34]In the terminology of [448], state irreversibility is referred to as *time-reversal non-invariance*. However, since the term *time reversal* is not meant literally (that is, we consider dynamical systems whose trajectory reversal is or is not allowed and *not* a reversal of time itself), state reversibility is a more appropriate expression. And in that regard, a more appropriate expression for the arrow

every nonequilibrium system state and corresponding system trajectory of our thermodynamically consistent large-scale nonlinear dynamical system, there does not exist a state such that the corresponding system trajectory completely recovers the initial system state of the dynamical system and at the same time restores the energy supplied by the environment back to its original condition.

This, along with the existence of a global strictly increasing entropy function on every nontrivial system trajectory, gives a clear *time-reversal asymmetry* characterization of thermodynamics, establishing an emergence of the direction of time flow. In the case where the subsystem energies are proportional to subsystem temperatures, we show that our dynamical system model leads to temperature equipartition, wherein all the system energy is transferred into heat at a uniform temperature. Furthermore, we show that our system-theoretic definition of entropy and the newly proposed notion of ectropy are consistent with Boltzmann's kinetic theory of gases involving an n-body theory of ideal gases divided by diathermal walls. Finally, these results are generalized to continuum thermodynamics, stochastic thermodynamics, and relativistic thermodynamics involving infinite-dimensional, Markovian, and functional energy flow conservation models.

1.6 A Brief Outline of the Monograph

The objective of this monograph is to develop a system-theoretic foundation for thermodynamics using dynamical systems and control notions. The main contents of the monograph are as follows. In Chapter 2, we establish notation and definitions, and we develop several key results on nonnegative and compartmental dynamical systems needed to establish thermodynamically consistent energy flow models. Furthermore, we introduce the notions of (ir)reversible and (ir)recoverable dynamical systems, volume-preserving flows and recurrent dynamical systems, as well as output reversibility in dynamical systems.

In Chapter 3, we use a large-scale dynamical systems perspective to provide a system-theoretic foundation for thermodynamics. Specifically, using a system state space formulation, we develop a nonlinear compartmental dynamical system model characterized by energy conservation laws that is consistent with basic thermodynamic principles. In particular, using the total subsystem energies as a candidate system energy storage function, we show that our thermodynamic system is *lossless*, and hence, can deliver to its surroundings all of its stored subsystem energies and can store all of the work done to all of its subsystems. This leads to the first law of

of time is *system degeneration over time* signifying irrecoverable system changes.

thermodynamics involving conservation of energy and places no limitation on the possibility of transforming heat into work or work into heat.

Next, we show that the classical Clausius equality and inequality for reversible and irreversible thermodynamics are satisfied over cyclic motions for our thermodynamically consistent energy flow model and guarantee the existence of a continuous system entropy function. In addition, we establish the existence of a *unique*, continuously differentiable *global* entropy function for our large-scale dynamical system, which is used to define inverse subsystem temperatures as the derivative of the subsystem entropies with respect to the subsystem energies.

Then we turn our attention to stability and convergence. Specifically, using the system ectropy as a Lyapunov function candidate, we show that in the absence of energy exchange with the environment, the proposed thermodynamic model is semistable with a uniform energy distribution corresponding to a state of minimum ectropy and a state of maximum entropy. Furthermore, using the system entropy and ectropy functions, we develop a clear connection between irreversibility, the second law of thermodynamics, and the entropic arrow of time.

In Chapter 4, we generalize the results of Chapter 3 to the case where the subsystem energies in the large-scale dynamical system model are proportional to subsystem temperatures, and we arrive at temperature equipartition for the proposed thermodynamic model. Furthermore, we provide a kinetic theory interpretation of the steady-state expressions for entropy and ectropy. Moreover, we establish connections between dynamical thermodynamics and classical thermodynamics.

In Chapter 5, we augment our nonlinear compartmental dynamical system model with an additional (deformation) state representing compartmental volumes to arrive at a general statement of the first law of thermodynamics, giving a precise formulation of the equivalence between heat and mechanical work. Furthermore, we define the Gibbs free energy, Helmholtz free energy, and enthalpy functions for our large-scale system thermodynamic model. In addition, we use the proposed augmented nonlinear compartmental dynamical system model in conjunction with a Carnot-like cycle analysis to show the equivalence between the classical Kelvin and Clausius postulates of the second law of thermodynamics.

In Chapter 6, we address the problems of nonnegativity, realizability, reducibility, and semistability of chemical reaction networks. Specifically, we show that mass-action kinetics have nonnegative solutions for initially nonnegative concentrations, we provide a general procedure for reducing the dimensionality of the kinetic equations, and we present stability results based upon Lyapunov methods. Furthermore, we present a state space dynamical system model for chemical thermodynamics. In particular, we use the law of mass action to obtain the dynamics of chemical reaction networks.

In addition, using the notion of the chemical potential, we unify our state space mass-action kinetics model with our dynamical thermodynamic system model involving system energy exchange. Moreover, we show that entropy production during chemical reactions is nonnegative and the dynamical system states of our chemical thermodynamic state space model converge to a state of temperature equipartition and zero affinity (i.e., the difference between the chemical potential of the reactants and the chemical potential of the products in a chemical reaction).

In Chapter 7, we merge the theories of semistability and finite-time stability to develop a rigorous framework for finite-time thermodynamics. Specifically, using a geometric description of homogeneity theory, we develop intercompartmental energy flow laws that guarantee finite-time semistability and energy equipartition for the thermodynamically consistent model developed in Chapter 3.

Next, in Chapter 8, we address the problem of thermodynamic critical phenomena and continuous phase transitions. In particular, to address discontinuities in the derivatives of the thermodynamic state quantities, we consider dynamical systems with Lebesgue measurable and locally essentially bounded vector fields characterized by differential inclusions involving Filippov set-valued maps specifying a set of directions for the system generalized velocities and admitting Filippov solutions with absolutely continuous curves. Moreover, we present Lyapunov-based tests for semistability, finite-time semistability, and energy equipartition for a discontinuous power balance thermodynamic model characterized by differential inclusions.

In Chapter 9, we develop thermodynamic models for discrete-time, large-scale dynamical systems. Specifically, using a framework analogous to Chapter 3, we develop energy flow models possessing discrete energy conservation, energy equipartition, temperature equipartition, and entropy nonconservation principles for discrete-time, large-scale dynamical systems.

To address thermodynamic critical phenomena and discontinuous phase transitions, in Chapter 10 we combine the frameworks of Chapters 3 and 9 to develop hybrid thermodynamic models. Specifically, to capture jump discontinuities in the fundamental thermodynamic state quantities, we develop a hybrid large-scale dynamical system using impulsive compartmental and thermodynamic dynamical system models involving an interacting mixture of continuous and discrete dynamics exhibiting discontinuous flows on appropriate manifolds.

In Chapter 11, we extend the results of Chapter 3 to continuum thermodynamic systems, wherein the subsystems are uniformly distributed over an n-dimensional (not necessarily Euclidean) space. Specifically, we develop a nonlinear distributed-parameter model wherein the system energy is modeled by a conservation equation in the form of a nonlinear partial differential equation. Energy equipartition and semistability are shown using Sobolev embedding theorems and the notion of generalized (or weak) solutions. This exposition shows that the behavior of heat, as described by the equations of thermal transport and as described by classical thermodynamics, is derivable from the same principles and is part of the same scientific discipline, and thus provides a unification between Fourier's theory of heat conduction and classical thermodynamics.

In Chapter 12, we extend the results of Chapter 3 to large-scale dynamical systems driven by Markov diffusion processes to present a unified framework for statistical thermodynamics predicated on a stochastic dynamical systems formalism. Specifically, using a stochastic state space formulation, we develop a nonlinear stochastic compartmental dynamical system model characterized by energy conservation laws that is consistent with statistical thermodynamic principles. In particular, we show that the average stored system energy for our stochastic thermodynamic model is a martingale with respect to the system filtration and is equal to the mean energy that can be extracted from the system and the mean energy that can be delivered to the system in order to transfer it from a zero energy level to an arbitrary nonempty subset in the state space over a finite stopping time.

Next, to effectively address the universality of thermodynamics and the arrow of time to cosmology, we extend our dynamical systems framework of thermodynamics to include relativistic effects. To this end, in Chapter 13 we give a brief exposition of the special and general theories of relativity, and review some basic concepts on relativistic kinematics and relativistic dynamics.

Then, in Chapter 14, we extend our results to thermodynamic systems that are moving relative to a local observer moving with the system and a fixed observer with respect to which the system is in motion. Furthermore, thermodynamic effects in the presence of a strong gravitational field are also discussed. In addition, using the topological isomorphism between entropy and time established in Chapter 3 and Einstein's time dilation assertion that increasing an object's speed through space results in decreasing the object's speed through time, we present an entropy dilation principle, which shows that the change in entropy of a thermodynamic system decreases as the system's speed increases through space.

To account for finite subluminal speed of heat propagation, in Chapter 15 we generalize the results of Chapter 3 to general thermodynamic compartmental systems that account for energy and matter in transit between compartments. Specifically, we develop thermodynamic models that guarantee conservation of energy, semistability, and state equipartitioning with directed and undirected thermal flow as well as flow delays between compartments.

Finally, we draw conclusions in Chapter 16, and in Chapter 17, we present a high-level scientific discussion of several peripheral, albeit key areas of how our dynamical systems framework of thermodynamics can be used to foster the development of new frameworks in explaining the fundamental thermodynamic processes occurring in nature, explore new hypotheses that challenge the use of classical thermodynamics, and develop new assertions that can provide deeper insights into the constitutive mechanisms that describe the acute microcosms and macrocosms of science.

Chapter Two

Dynamical Systems Theory

2.1 Notation, Definitions, and Mathematical Preliminaries

As discussed in Chapter 1, in this monograph we develop thermodynamic system models using large-scale nonlinear compartmental dynamical systems. The mathematical foundation for compartmental modeling is *nonnegative dynamical systems theory* [194, 415], which involves dynamical systems with nonnegative state variables. Since our thermodynamic state equations govern energy, volume, and mass flow between subsystems, it follows from physical arguments that nonnegative system energy, volume, and mass initial conditions give rise to dynamical system trajectories that remain in the nonnegative orthant of the state space. In this chapter, we introduce notation, several definitions, and some key results on nonlinear nonnegative dynamical systems needed for developing the main results of this monograph.

In a definition or when a word is defined in the text, the concept defined is italicized. Italics in the running text are also used for emphasis. The definition of a word, phrase, or symbol is to be understood as an "if and only if" statement. Lower case letters such as x denote vectors, upper case letters such as A denote matrices, upper case script letters such as $\mathcal{S}$ denote sets, and lower case Greek letters such as α denote scalars; however, there are a few exceptions to this convention. The notation $\mathcal{S}_1 \subset \mathcal{S}_2$ means that $\mathcal{S}_1$ is a proper subset of $\mathcal{S}_2$, whereas $\mathcal{S}_1 \subseteq \mathcal{S}_2$ means that either $\mathcal{S}_1$ is a proper subset of $\mathcal{S}_2$ or $\mathcal{S}_1$ is equal to $\mathcal{S}_2$. Throughout the monograph we use two basic types of mathematical statements, namely, *existential* and *universal* statements. An existential statement has the form: there exists $x \in \mathcal{X}$ such that a certain condition C is satisfied; whereas a universal statement has the form: condition C holds for all $x \in \mathcal{X}$. For universal statements we often omit the words "for all" and write: condition C holds, $x \in \mathcal{X}$.

The notation used in this monograph is fairly standard. Specifically, $\mathbb{R}$ (respectively, $\mathbb{C}$) denotes the set of real (respectively, complex) numbers, $\overline{\mathbb{Z}}_+$ denotes the set of nonnegative integers, $\mathbb{Z}_+$ denotes the set of positive integers, $\mathbb{R}^n$ (respectively, $\mathbb{C}^n$) denotes the set of $n \times 1$ real (respectively, complex) column vectors, $\mathbb{R}^{n\times m}$ (respectively, $\mathbb{C}^{n\times m}$) denotes the set of real (respectively, complex) $n \times m$ matrices, $(\cdot)^{\mathrm{T}}$ denotes transpose, $(\cdot)^*$ denotes complex conjugate transpose, $(\cdot)^+$ denotes the Moore-Penrose generalized inverse, $(\cdot)^\#$ denotes the group generalized inverse, $(\cdot)^{\mathrm{D}}$ denotes the Drazin inverse, $(\cdot)^{-1}$ denotes the inverse, $\otimes$ denotes Kronecker product, $\oplus$ denotes Kronecker sum, I_n or I denotes the $n \times n$ identity matrix, $I_{\mathcal{B}}$ denotes the identity operator, and $\mathbf{e}$ denotes the ones vector of order n, that is, $\mathbf{e} = [1, \ldots, 1]^{\mathrm{T}}$. For $x \in \mathbb{R}^q$ we write $x \geq\geq 0$ (respectively, $x >> 0$) to indicate that every component of x is nonnegative (respectively, positive). In this case, we say that x is *nonnegative* or *positive*, respectively.

Likewise, $A \in \mathbb{R}^{p\times q}$ is *nonnegative* or *positive* if every entry of A is nonnegative or positive, respectively, which is written as $A \geq\geq 0$ or $A >> 0$, respectively. The sets $\overline{\mathbb{R}}_+^q$ and $\mathbb{R}_+^q$ denote the nonnegative and positive orthants of $\mathbb{R}^q$, that is, if $x \in \mathbb{R}^q$, then $x \in \overline{\mathbb{R}}_+^q$ and $x \in \mathbb{R}_+^q$ are equivalent, respectively, to $x \geq\geq 0$ and $x >> 0$. Furthermore, C^0 denotes the space of continuous functions, C^r denotes the space of functions with r-continuous derivatives, $\mathcal{L}_2$ denotes the space of square-integrable Lebesgue measurable functions on $[0, \infty)$, and $\mathcal{L}_\infty$ denotes the space of bounded Lebesgue measurable functions on $[0, \infty)$. Similarly, l_2 denotes the space of square-summable sequences on $\overline{\mathbb{Z}}_+$, and l_∞ denotes the space of bounded sequences on $\overline{\mathbb{Z}}_+$. In addition, we denote the boundary, the interior, and the closure of the set $\mathcal{S}$ by $\partial\mathcal{S}$, $\overset{\circ}{\mathcal{S}}$, and $\overline{\mathcal{S}}$, respectively.

We write $\mathrm{Re}\, z$ (respectively, $\mathrm{Im}\, z$) for the real (respectively, imaginary) part of the complex number z, $\|\cdot\|$ for the Euclidean vector norm, $\|\cdot\|_{\mathrm{F}}$ for the Frobenius matrix norm, $\|\cdot\|_{\mathcal{B}}$ for the operator norm of an element in a Banach space $\mathcal{B}$, $\mathcal{R}(A)$ and $\mathcal{N}(A)$ for the range space and the null space of a matrix A, respectively, $\mathrm{spec}(A)$ or $\mathrm{mspec}(A)$ for the spectrum of the square matrix A including multiplicity, $\alpha(A)$ for the spectral abscissa of A (that is, $\alpha(A) = \max\{\mathrm{Re}\,\lambda : \lambda \in \mathrm{spec}(A)\}$), $\rho(A)$ for the spectral radius of A (that is, $\rho(A) = \max\{|\lambda| : \lambda \in \mathrm{spec}(A)\}$), and $\mathrm{ind}(A)$ for the index of A (that is, $\min\{k \in \overline{\mathbb{Z}}_+ : \mathrm{rank}\ A^k = \mathrm{rank}\ A^{k+1}\}$). For vectors $x, y \in \mathbb{R}^p$ and matrices $A, B \in \mathbb{R}^{p\times q}$ we use $x \circ y$ and $A \circ B$ to denote component-by-component and entry-by-entry multiplication, respectively. For a matrix $A \in \mathbb{R}^{p\times q}$, $\mathrm{row}_i(A)$ and $\mathrm{col}_j(A)$ denote the ith row and jth column of A, respectively, and rank A denotes the rank of A.

Furthermore, we write $V'(x)$ for the Fréchet derivative of V at x, $V''(x)$ for the Hessian of V at x, $\mathcal{B}_\varepsilon(x)$, $x \in \mathbb{R}^n$, $\varepsilon > 0$, for the open ball *centered* at x with *radius* ε, $M \geq 0$ (respectively, $M > 0$) to denote the fact that the Hermitian matrix M is nonnegative (respectively, positive) definite, inf to denote infimum (that is, the greatest lower bound), sup to denote supremum (that is, the least upper bound), and $x(t) \to \mathcal{M}$ as $t \to \infty$ to denote that $x(t)$ approaches the set $\mathcal{M}$ (that is, for every $\varepsilon > 0$ there exists $T > 0$ such that $\operatorname{dist}(x(t), \mathcal{M}) < \varepsilon$ for all $t > T$, where $\operatorname{dist}(p, \mathcal{M}) \triangleq \inf_{x \in \mathcal{M}} \|p - x\|$). Finally, the notions of openness, convergence, continuity, and compactness that we use throughout the monograph refer to the topology generated on $\overline{\mathbb{R}}^q_+$ (respectively, $\mathcal{B}$) by the norm $\|\cdot\|$ (respectively, $\|\cdot\|_{\mathcal{B}}$).

In the first part of this chapter, we develop the fundamental results of Lyapunov stability theory for nonnegative dynamical systems. We begin by considering the general nonlinear autonomous dynamical system

$$\dot{x}(t) = f(x(t)), \qquad x(0) = x_0, \qquad t \in \mathcal{I}_{x_0}, \tag{2.1}$$

where $x(t) \in \mathcal{D} \subseteq \mathbb{R}^n$, $t \in \mathcal{I}_{x_0}$, is the system state vector, $\mathcal{D}$ is a relatively open set (see Definition 2.2), $f : \mathcal{D} \to \mathbb{R}^n$ is continuous on $\mathcal{D}$, and $\mathcal{I}_{x_0} = [0, \tau_{x_0})$, $0 \le \tau_{x_0} \le \infty$, is the *maximal interval of existence* for the solution $x(\cdot)$ of (2.1). If $\tau_{x_0} < \infty$, then we say that (2.1) has a *finite-escape* time. A continuously differentiable function $x : \mathcal{I}_{x_0} \to \mathcal{D}$ is said to be a *solution* to (2.1) on the interval $\mathcal{I}_{x_0} \subseteq \mathbb{R}$ with *initial condition* $x(0) = x_0$ if and only if $x(t)$ satisfies (2.1) for all $t \in \mathcal{I}_{x_0}$. We assume that for every initial condition $x(0) \in \mathcal{D}$ and every $\tau_{x_0} > 0$, the dynamical system (2.1) possesses a unique solution $x : [0, \tau_{x_0}) \to \mathcal{D}$ on the interval $[0, \tau_{x_0})$.

We denote the solution to (2.1) with initial condition $x(0) = x_0$ by $s(\cdot, x_0)$, so that the *dynamic* or *flow* of the dynamical system (2.1) given by the map $s : [0, \tau_{x_0}) \times \mathcal{D} \to \mathcal{D}$ is continuous in x and continuously differentiable in t and satisfies the *consistency* property $s(0, x_0) = x_0$ and the *semigroup* property $s(\tau, s(t, x_0)) = s(t + \tau, x_0)$ for all $x_0 \in \mathcal{D}$ and $t, \tau \in [0, \tau_{x_0})$ such that $t + \tau \in [0, \tau_{x_0})$. Given $t \ge 0$ and $x \in \mathcal{D}$, we denote the map $s(t, \cdot) : \mathcal{D} \to \mathcal{D}$ by s_t and the map $s(\cdot, x) : \mathcal{I}_x \to \mathcal{D}$ by s^x. For every $t \in \mathbb{R}$, the map s_t is a homeomorphism and has the inverse flow $s_{-t} \triangleq s_t^{-1} : \mathcal{D}_{-t} \to \mathcal{D}$, where $\mathcal{D}_{-t} \triangleq s_t(\mathcal{D})$ for $t \ge 0$ and $s_t(\mathcal{D})$ denotes the image of $\mathcal{D}$ under the flow s_t. In this case, we say the flow is *reversible* and define the new *time-reversed flow* as $\psi_t = s_{-t}$.

The *orbit* $\mathcal{O}_x$ of a point $x \in \mathcal{D}$ is the set $s^x(\mathcal{I}_x)$. We say that a point $x \in \mathcal{D}$ (or its trajectory s^x) is *attracted* (or *approaches*) $\mathcal{M}$, if $\overline{\mathcal{O}}_x$ is compact and $\omega(x) \in \mathcal{M}$, where $\omega(x)$ is the positive limit set of (2.1) (see Definition 2.5). Unless otherwise stated, we assume $f(\cdot)$ is Lipschitz continuous on $\mathcal{D}$. Furthermore, $x_{\mathrm{e}} \in \mathcal{D}$ is an *equilibrium point* of (2.1) if and only if $f(x_{\mathrm{e}}) = 0$ or, equivalently, $x(t) = x_{\mathrm{e}}$, $t \in \mathcal{I}_{x_0}$. In addition, a subset $\mathcal{D}_{\mathrm{c}} \subseteq \mathcal{D}$ is an *invariant set* relative to (2.1) if $\mathcal{D}_{\mathrm{c}}$ contains the orbits of all its points. Finally, recall that if all solutions to (2.1) are bounded, then it follows from the Peano-Cauchy theorem [191, p. 76] that $\mathcal{I}_{x_0} = \mathbb{R}$, and hence, (2.1) is *forward* and *backward complete*.

The following definition introduces the notion of essentially nonnegative functions and vector fields.

Definition 2.1. Let $f = [f_1, \ldots, f_n]^{\mathrm{T}} : \mathcal{D} \subseteq \overline{\mathbb{R}}_+^n \to \mathbb{R}^n$. Then f is *essentially nonnegative* if $f_i(x) \geq 0$ for all $i = 1, \ldots, n$ and $x \in \overline{\mathbb{R}}_+^n$ such that $x_i = 0$, where x_i denotes the ith component of x.

The following definition and technical lemma are needed.

Definition 2.2. A set $\mathcal{Q} \subseteq \overline{\mathbb{R}}_+^n$ is *open relative to* $\overline{\mathbb{R}}_+^n$ if there exists an open set $\mathcal{R} \subseteq \mathbb{R}^n$ such that $\mathcal{Q} = \mathcal{R} \cap \overline{\mathbb{R}}_+^n$. A set $\mathcal{Q} \subseteq \overline{\mathbb{R}}_+^n$ is *closed relative to* $\overline{\mathbb{R}}_+^n$ if there exists a closed set $\mathcal{R} \subseteq \mathbb{R}^n$ such that $\mathcal{Q} = \mathcal{R} \cap \overline{\mathbb{R}}_+^n$. A set $\mathcal{Q} \subseteq \overline{\mathbb{R}}_+^n$ is *compact relative to* $\overline{\mathbb{R}}_+^n$ if there exists a compact set $\mathcal{R} \subseteq \mathbb{R}^n$ such that $\mathcal{Q} = \mathcal{R} \cap \overline{\mathbb{R}}_+^n$.

Lemma 2.1. Consider the nonlinear dynamical system (2.1) and let $\mathcal{D}_{\mathrm{c}} \subset \mathcal{D}$ be closed relative to $\mathcal{D}$. Then the following statements are equivalent:

i) For all $x \in \mathcal{D}_{\mathrm{c}}$, $\lim_{h \to 0^+} \inf_{y \in \mathcal{D}_{\mathrm{c}}} \|x + hf(x) - y\|/h = 0$.

ii) $\mathcal{D}_{\mathrm{c}}$ is an invariant set with respect to (2.1).

Proof. Assume that *i)* holds. To show *ii)*, let $x_0 \in \mathcal{D}_{\mathrm{c}}$. Since $f(\cdot)$ is Lipschitz continuous it follows that there exist $\varepsilon > 0$ and $L > 0$ such that

$$\|f(x) - f(y)\| \leq L\|x - y\|, \quad x, y \in \mathcal{B}_{2\varepsilon}(x_0). \tag{2.2}$$

Let $T \in [0, \tau_{x_0})$ be such that $s(t, x) \in \mathcal{B}_{2\varepsilon}(x_0)$ and $s(t, y) \in \mathcal{B}_{2\varepsilon}(x_0)$ for all $t \in [0, T)$ and $x, y \in \mathcal{B}_{\varepsilon}(x_0)$. Now, it follows from Gronwall's lemma [191, p. 81] that

$$\|s(t, x) - s(t, y)\| \leq e^{Lt}\|x - y\|, \quad x, y \in \mathcal{B}_{\varepsilon}(x_0), \tag{2.3}$$

for all $t \in [0, T)$.

Next, let $t_1 \in [0, T)$ be such that $\|s(t, x_0) - x_0\| < \varepsilon/3$ for all $t \in (0, t_1)$, and define $\varphi(t) \triangleq \mathrm{dist}(s(t, x_0), \mathcal{D}_{\mathrm{c}}) = \inf_{y \in \mathcal{D}_{\mathrm{c}}} \|s(t, x_0) - y\|$. Note that since $x_0 \in \mathcal{D}_{\mathrm{c}}$, it follows that $\varphi(0) = 0$ and $\varphi(t) \leq \|s(t, x_0) - x_0\| < \varepsilon/3$ for all $t \in (0, t_1)$. Now, let $t \in (0, t_1)$ and let $y_t \in \mathcal{D}_{\mathrm{c}}$ be such that $\|s(t, x_0) - y_t\| - \varphi(t) \leq \varepsilon/3$. Hence,

$$\begin{aligned} \|y_t - x_0\| &= \|y_t - s(t, x_0) + s(t, x_0) - x_0\| \\ &\leq \|s(t, x_0) - x_0\| + \|s(t, x_0) - y_t\| \\ &\leq \|s(t, x_0) - x_0\| + \varphi(t) + \varepsilon/3 \\ &< \varepsilon. \end{aligned}$$

Now, for all $h > 0$ such that $t + h \leq t_1$, since $\|s(t, y_t) - x_0\| < \varepsilon/3 < \varepsilon$ and $\|y_t - x_0\| < \varepsilon$, it follows from (2.3) that

$$
\begin{aligned}
\varphi(t+h) &= \inf_{z \in \mathcal{D}_c} \|s(t+h, x_0) - z\| \\
&\leq \inf_{z \in \mathcal{D}_c} \Big\{ \|s(t+h, x_0) - s(h, y_t)\| + \|s(h, y_t) - y_t - hf(y_t)\| \\
&\quad + \|y_t + hf(y_t) - z\| \Big\} \\
&= \|s(t+h, x_0) - s(h, y_t)\| + \|s(h, y_t) - y_t - hf(y_t)\| \\
&\quad + \mathrm{dist}(y_t + hf(y_t), \mathcal{D}_c) \\
&\leq e^{Lh} \|y_t - s(t, x_0)\| + \|s(h, y_t) - y_t - hf(y_t)\| \\
&\quad + \mathrm{dist}(y_t + hf(y_t), \mathcal{D}_c),
\end{aligned}
\tag{2.4}
$$

which implies that

$$
\begin{aligned}
\frac{\varphi(t+h) - \varphi(t)}{h} &\leq \left(\frac{e^{Lh} - 1}{h} \right) \varphi(t) + \left\| \frac{s(h, y_t) - y_t}{h} - f(y_t) \right\| \\
&\quad + \frac{\mathrm{dist}(y_t + hf(y_t), \mathcal{D}_c)}{h}.
\end{aligned}
$$

Now, letting $h \to 0^+$ yields

$$
\limsup_{h \to 0^+} \frac{\varphi(t+h) - \varphi(t)}{h} \leq L\varphi(t). \tag{2.5}
$$

Next, by Gronwall's lemma [191, p. 81], it follows from (2.5) that $0 \leq \varphi(t) \leq e^{Lt}\varphi(0)$, $t \in (0, t_1)$, and hence, since $\varphi(0) = 0$, it follows that $\varphi(t) = 0$ for all $t \in (0, t_1)$. Now, since $x_0 \in \mathcal{D}_c$ is arbitrary, it follows that, for every $\tau_1 > 0$ such that $\varphi(\tau_1) = 0$, there exists $h > 0$ such that $\varphi(t) = 0$, $t \in [\tau_1, \tau_1 + h)$. Next, let $\tau = \inf\{t > 0 : \varphi(t) > 0\}$ and suppose, *ad absurdum*, that $\tau < \tau_{x_0}$. Since $\varphi(t) = 0$ for all $t \in [0, t_1)$, it follows that $\tau \geq t_1 > 0$ and, by the definition of τ, $\varphi(t) = 0$ for all $t \in [0, \tau)$ or, equivalently, $s(t, x_0) \in \mathcal{D}_c$ for all $t \in [0, \tau)$. Hence, since $s(\tau, x_0) = \lim_{t \to \tau^-} s(t, x)$ and $\mathcal{D}_c$ is relatively closed with respect to $\mathcal{D}$, it follows that $s(\tau, x_0) \in \mathcal{D}_c$. Therefore, $\varphi(\tau) = 0$, which implies that there exists $h > 0$ such that $\varphi(t) = 0$ for all $t \in [\tau, \tau + h)$, contradicting the definition of τ. Thus, $\varphi(t) = 0$, $t \in [0, \tau_{x_0})$, establishing the result.

Conversely, assume $\mathcal{D}_c$ is an invariant set with respect to (2.1) so that, for all $x_0 \in \mathcal{D}_c$ and $h \neq 0$,

$$
\begin{aligned}
\mathrm{dist}(x_0 + hf(x_0), \mathcal{D}_c) &\leq \|s(h, x_0) - x_0 - hf(x_0)\| \\
&= |h| \left\| \frac{s(h, x_0) - x_0}{h} - f(x_0) \right\|.
\end{aligned}
$$

Now, the result follows by letting $h \to 0^+$. □

The flow-invariant set result given by Lemma 2.1 was first proved by Brezis [64] and uses the fact that the vector field f in (2.1) is Lipschitz continuous on $\mathcal{D}$. This result was generalized independently by Crandall [97] and Hartman [207] to the case where f is continuous on $\mathcal{D}$ and (2.1) has a unique right maximally defined solution. The case where $f = f(t,x)$ is a time-varying vector field is subsumed under the case $f = f(x)$ by representing the time-varying dynamical system as an autonomous system, where an additional state is appended to represent time.

Flow-invariant sets and differential inequalities in normed spaces with Lebesgue measurable and locally essentially bounded vector fields, that is, vector fields that are bounded on a bounded neighborhood of every point excluding sets of measure zero, are discussed in [376]. Specifically, extensions to Lemma 2.1 are addressed for unique absolutely continuous solutions to (2.1) with an essentially one-sided Lipschitz condition on f of the Osgood type [375].

Our next result shows that the nonnegative orthant $\overline{\mathbb{R}}_+^n$ is an invariant set with respect to (2.1) if and only if f is essentially nonnegative. In light of the above discussion, this result holds for the case where f is continuous on $\mathcal{D}$ and, more generally, f is measurable and locally essentially bounded on $\mathcal{D}$.

Proposition 2.1. Suppose $\overline{\mathbb{R}}_+^n \subset \mathcal{D}$. Then $\overline{\mathbb{R}}_+^n$ is an invariant set with respect to (2.1) if and only if $f : \mathcal{D} \to \mathbb{R}^n$ is essentially nonnegative.

Proof. Define $\mathrm{dist}(x, \overline{\mathbb{R}}_+^n) \triangleq \inf_{y \in \overline{\mathbb{R}}_+^n} \|x-y\|$, $x \in \mathbb{R}^n$. Now, suppose $f : \mathcal{D} \to \mathbb{R}^n$ is essentially nonnegative and let $x \in \overline{\mathbb{R}}_+^n$. For every $i \in \{1,\ldots,q\}$, if $x_i = 0$, then $x_i + hf_i(x) = hf_i(x) \geq 0$ for all $h \geq 0$, whereas, if $x_i > 0$, then $x_i + hf_i(x) > 0$ for all $|h|$ sufficiently small. Thus, $x + hf(x) \in \overline{\mathbb{R}}_+^n$ for all sufficiently small $h > 0$, and hence, $\lim_{h \to 0^+} \mathrm{dist}(x + hf(x), \overline{\mathbb{R}}_+^n)/h = 0$. It now follows from Lemma 2.1, with $x(0) = x_0$, that $x(t) \in \overline{\mathbb{R}}_+^n$ for all $t \in [0, \tau_{x_0})$.

Conversely, suppose that $\overline{\mathbb{R}}_+^n$ is invariant with respect to (2.1), let $x(0) \in \overline{\mathbb{R}}_+^n$, and suppose, *ad absurdum*, x is such that there exists $i \in \{1,\ldots,q\}$ such that $x_i(0) = 0$ and $f_i(x(0)) < 0$. Then, since f is continuous, there exists sufficiently small $h > 0$ such that $f_i(x(t)) < 0$ for all $t \in [0,h)$, where $x(t)$ is the solution to (2.1). Hence, $x_i(t)$ is strictly decreasing on $[0,h)$, and thus, $x(t) \notin \overline{\mathbb{R}}_+^n$ for all $t \in (0,h)$, which leads to a contradiction. $\square$

It follows from Proposition 2.1 that if $x_0 \geq\geq 0$, then $x(t) \geq\geq 0$, $t \geq 0$, if and only if f is essentially nonnegative. In this case, we say that (2.1) is a *nonnegative dynamical system*. Henceforth, we assume that f is essentially nonnegative so that the nonlinear dynamical system (2.1) is a nonnegative dynamical system.

2.2 Stability Theory for Nonnegative Dynamical Systems

One of the main contributions of this monograph is the connection of ectropy and entropy to the stability properties of thermodynamic processes. System stability is characterized by analyzing the response of a dynamical system to small perturbations in the system states. The most complete stability analysis framework for dynamical systems was developed by Aleksandr Lyapunov [294–296].

Lyapunov's method is based on the construction of a function of the system state coordinates (now known as a Lyapunov function) that serves as a generalized norm of the solution of the dynamical system. Its appeal comes from the fact that conclusions about the behavior of the dynamical system can be drawn without actually computing the system trajectories. As a result, Lyapunov stability theory has become one of the cornerstones of dynamical systems theory.

An equilibrium point of a dynamical system is said to be *stable* or *Lyapunov stable* if, for small values of initial condition disturbances, the perturbed system motion remains in an arbitrarily prescribed small region of the state space. More precisely, stability is equivalent to continuity of solutions as a function of the system initial conditions over a neighborhood of the equilibrium point uniformly in time. If, in addition, all solutions of the dynamical system approach the equilibrium point for large values of time, then the equilibrium point is said to be *asymptotically stable.*

In this monograph, we apply Lyapunov methods [46, 191] to show that our thermodynamically consistent large-scale dynamical system is *semistable*, that is, system trajectories converge to Lyapunov stable equilibrium states that depend upon the system initial conditions. The concept of semistability involves a stability notion that lies perfectly between Lyapunov stability and asymptotic stability, and is vital in addressing the stability of thermodynamic systems having a continuum of equilibria [40, 42].

In this section, we establish key stability results for nonlinear nonnegative dynamical systems. The following definition introduces several types of stability for the equilibrium solution $x(t) \equiv x_{\mathrm{e}} \in \overline{\mathbb{R}}_+^n$ of the nonlinear nonnegative dynamical system (2.1) for $\mathcal{I}_{x_0} = [0, \infty)$.

Definition 2.3. *i*) The equilibrium solution $x(t) \equiv x_{\mathrm{e}} \in \overline{\mathbb{R}}_+^n$ to (2.1) is *Lyapunov stable with respect to* $\overline{\mathbb{R}}_+^n$ if, for all $\varepsilon > 0$, there exists $\delta = \delta(\varepsilon) > 0$

such that if $x_0 \in \mathcal{B}_\delta(x_e) \cap \overline{\mathbb{R}}_+^n$, then $x(t) \in \mathcal{B}_\varepsilon(x_e) \cap \overline{\mathbb{R}}_+^n$, $t \geq 0$.

ii) The equilibrium solution $x(t) \equiv x_e \in \overline{\mathbb{R}}_+^n$ to (2.1) is (*locally*) *asymptotically stable with respect to* $\overline{\mathbb{R}}_+^n$ if it is Lyapunov stable with respect to $\overline{\mathbb{R}}_+^n$ and there exists $\delta > 0$ such that if $x_0 \in \mathcal{B}_\delta(x_e) \cap \overline{\mathbb{R}}_+^n$, then $\lim_{t\to\infty} x(t) = x_e$.

iii) The equilibrium solution $x(t) \equiv x_e \in \overline{\mathbb{R}}_+^n$ to (2.1) is *globally asymptotically stable with respect to* $\overline{\mathbb{R}}_+^n$ if it is Lyapunov stable with respect to $\overline{\mathbb{R}}_+^n$ and, for all $x_0 \in \overline{\mathbb{R}}_+^n$, $\lim_{t\to\infty} x(t) = x_e$.

The following result, known as Lyapunov's direct method, gives sufficient conditions for Lyapunov and asymptotic stability of a nonlinear nonnegative dynamical system. For this result, let $V : \mathcal{D} \to \mathbb{R}$ be a continuously differentiable function with derivative *along the trajectories* of (2.1) given by $\dot{V}(x) \triangleq V'(x)f(x)$. Note that $\dot{V}(x)$ is dependent on the system dynamics (2.1). Since, using the chain rule, $\dot{V}(x) = \frac{\mathrm{d}}{\mathrm{d}t}V(s(t,x))\big|_{t=0} = V'(x)f(x)$, it follows that if $\dot{V}(x)$ is negative, then $V(x)$ decreases along the solution $s(t, x_0)$ of (2.1) through $x_0 \in \mathcal{D}$ at $t = 0$.

Theorem 2.1 (Lyapunov's Theorem). Let $\mathcal{D}$ be an open subset relative to $\overline{\mathbb{R}}_+^n$ that contains x_e. Consider the nonlinear dynamical system (2.1) where f is essentially nonnegative and $f(x_e) = 0$, and assume that there exists a continuously differentiable function $V\colon \mathcal{D} \to \mathbb{R}$ such that

$$V(x_e) = 0, \tag{2.6}$$

$$V(x) > 0, \quad x \in \mathcal{D}, \quad x \neq x_e, \tag{2.7}$$

$$V'(x)f(x) \leq 0, \quad x \in \mathcal{D}. \tag{2.8}$$

Then the equilibrium solution $x(t) \equiv x_e$ to (2.1) is Lyapunov stable with respect to $\overline{\mathbb{R}}_+^n$. If, in addition,

$$V'(x)f(x) < 0, \quad x \in \mathcal{D}, \quad x \neq x_e, \tag{2.9}$$

then the equilibrium solution $x(t) \equiv x_e$ to (2.1) is asymptotically stable with respect to $\overline{\mathbb{R}}_+^n$. Finally, if $V(\cdot)$ is such that

$$V(x) \to \infty \text{ as } \|x\| \to \infty, \tag{2.10}$$

then (2.9) implies that the equilibrium solution $x(t) \equiv x_e$ to (2.1) is globally asymptotically stable with respect to $\overline{\mathbb{R}}_+^n$.

Proof. Let $\varepsilon > 0$ be such that $\mathcal{B}_\varepsilon(x_e) \cap \overline{\mathbb{R}}_+^n \subseteq \mathcal{D}$. Since $\overline{\mathbb{R}}_+^n \cap \partial\mathcal{B}_\varepsilon(x_e)$ is compact and $V(x)$, $x \in \mathcal{D}$, is continuous, it follows that $\alpha \triangleq \min_{x \in \overline{\mathbb{R}}_+^n \cap \partial\mathcal{B}_\varepsilon(x_e)} V(x)$ exists. Note $\alpha > 0$ since $x_e \notin \overline{\mathbb{R}}_+^n \cap \partial\mathcal{B}_\varepsilon(x_e)$ and $V(x) > 0$, $x \in \mathcal{D}$, $x \neq x_e$. Next, let $\beta \in (0, \alpha)$ and define $\mathcal{D}_\beta$ to be the *arcwise connected component* of $\{x \in \mathcal{D} : V(x) \leq \beta\}$ containing $x_{\rm e}$; that is, $\mathcal{D}_\beta$ is the set of all $x \in \mathcal{D}$ such that there exists a continuous function $\psi : [0,1] \to \mathcal{D}$ such that $\psi(0) = x$, $\psi(1) = x_{\rm e}$, and $V(\psi(\mu)) \leq \beta$ for all $\mu \in [0,1]$.[1] Note that $\mathcal{D}_\beta \subset \overline{\mathbb{R}}_+^n \cap \mathcal{B}_\varepsilon(x_{\rm e})$.

To see this, suppose, *ad absurdum*, that $\mathcal{D}_\beta \not\subset \overline{\mathbb{R}}_+^n \cap \mathcal{B}_\varepsilon(x_{\rm e})$. In this case, there exists a point $p \in \mathcal{D}_\beta$ such that $p \in \overline{\mathbb{R}}_+^n \cap \partial\mathcal{B}_\varepsilon(x_{\rm e})$, and hence, $V(p) \geq \alpha > \beta$, which is a contradiction. Now, since $\dot{V}(x) \triangleq V'(x)f(x) \leq 0$, $x \in \mathcal{D}_\beta$, it follows that $V(x(t))$ is a nonincreasing function of time, and hence, $V(x(t)) \leq V(x(0)) \leq \beta$, $t \geq 0$. Hence, $\mathcal{D}_\beta$ is a positively invariant set (see Definition 2.18) with respect to (2.1).

Next, since $V(\cdot)$ is continuous and $V(x_{\rm e}) = 0$, there exists $\delta = \delta(\varepsilon) \in (0, \varepsilon)$ such that $V(x) < \beta$, $x \in \overline{\mathbb{R}}_+^n \cap \mathcal{B}_\delta(x_{\rm e})$. Now, let $x(t)$, $t \geq 0$, satisfy (2.1) with $x(0) \in \overline{\mathbb{R}}_+^n \cap \mathcal{B}_\delta(x_{\rm e})$. Since $\overline{\mathbb{R}}_+^n \cap \mathcal{B}_\delta(x_{\rm e}) \subset \mathcal{D}_\beta \subset \overline{\mathbb{R}}_+^n \cap \mathcal{B}_\varepsilon(x_{\rm e}) \subseteq \mathcal{D}$ and $V'(x)f(x) \leq 0$, $x \in \mathcal{D}$, it follows that

$$V(x(t)) - V(x(0)) = \int_0^t V'(x(s))f(x(s)){\rm d}s \leq 0, \qquad t \geq 0, \tag{2.11}$$

and hence, for all $x(0) \in \overline{\mathbb{R}}_+^n \cap \mathcal{B}_\delta(x_{\rm e})$,

$$V(x(t)) \leq V(x(0)) < \beta, \qquad t \geq 0.$$

Now, since $V(x) \geq \alpha$, $x \in \overline{\mathbb{R}}_+^n \cap \partial\mathcal{B}_\varepsilon(x_{\rm e})$, and $\beta \in (0, \alpha)$, it follows that $x(t) \notin \overline{\mathbb{R}}_+^n \cap \partial\mathcal{B}_\varepsilon(x_{\rm e})$, $t \geq 0$. Hence, since f is essentially nonnegative, for all $\varepsilon > 0$ there exists $\delta = \delta(\varepsilon) > 0$ such that if $x(0) \in \mathcal{B}_\delta(x_{\rm e}) \cap \overline{\mathbb{R}}_+^n$, then $x(t) \in \mathcal{B}_\varepsilon(x_{\rm e}) \cap \overline{\mathbb{R}}_+^n$, $t \geq 0$, which proves Lyapunov stability with respect to $\overline{\mathbb{R}}_+^n$ of the equilibrium solution $x(t) \equiv x_{\rm e}$ to (2.1).

To prove asymptotic stability with respect to $\overline{\mathbb{R}}_+^n$ of the equilibrium solution $x(t) \equiv x_{\rm e}$ to (2.1), suppose that $V'(x)f(x) < 0$, $x \in \mathcal{D}$, $x \neq x_{\rm e}$, and $x(0) \in \overline{\mathbb{R}}_+^n \cap \mathcal{B}_\delta(x_{\rm e})$. Then it follows that $x(t) \in \overline{\mathbb{R}}_+^n \cap \mathcal{B}_\varepsilon(x_{\rm e})$, $t \geq 0$. However, $V(x(t))$, $t \geq 0$, is decreasing and bounded from below by zero. Now, *ad absurdum*, suppose $x(t)$, $t \geq 0$, does not converge to $x_{\rm e}$. This implies that $V(x(t))$, $t \geq 0$, is lower bounded, that is, there exists $L > 0$ such that $V(x(t)) \geq L > 0$, $t \geq 0$. Hence, by continuity of $V(\cdot)$ there exists $\delta' > 0$ such that $V(x) < L$ for $x \in \overline{\mathbb{R}}_+^n \cap \mathcal{B}_{\delta'}(x_{\rm e})$, which further implies that $x(t) \notin \overline{\mathbb{R}}_+^n \cap \mathcal{B}_{\delta'}(x_{\rm e})$ for all $t \geq 0$.

Next, define $L_1 \triangleq \min\{-V'(x)f(x) : \delta' \leq \|x - x_{\rm e}\| \leq \varepsilon, x \in \overline{\mathbb{R}}_+^n\}$. Now,

---

[1] Unless otherwise stated, in the remainder of the monograph we assume that sets of the form $\mathcal{D}_\beta = \{x \in \mathcal{D} : V(x) \leq \beta\}$ correspond to the arcwise connected component of $\{x \in \mathcal{D} : V(x) \leq \beta\}$ containing $x_{\rm e}$. This minor abuse of notation considerably simplifies the presentation.

(2.9) implies $-V'(x)f(x) \geq L_1$, $\delta' \leq \|x - x_{\rm e}\| \leq \varepsilon$, $x \in \overline{\mathbb{R}}_+^n$, or, equivalently,

$$V(x(t)) - V(x(0)) = \int_0^t V'(x(s))f(x(s))\mathrm{d}s \leq -L_1 t,$$

and hence, for all $x(0) \in \overline{\mathbb{R}}_+^n \cap \mathcal{B}_\delta(x_{\rm e})$,

$$V(x(t)) \leq V(x(0)) - L_1 t.$$

Letting $t > \frac{V(x(0))-L}{L_1}$, it follows that $V(x(t)) < L$, which is a contradiction. Hence, $x(t) \to x_{\rm e}$ as $t \to \infty$, establishing asymptotic stability with respect to $\overline{\mathbb{R}}_+^n$.

Finally, to prove global asymptotic stability with respect to $\overline{\mathbb{R}}_+^n$, let $x(0) \in \overline{\mathbb{R}}_+^n$, and let $\beta \triangleq V(x_0)$. Now, the radial unboundedness condition (2.10) implies that there exists $\varepsilon > 0$ such that $V(x) > \beta$ for all $x \in \overline{\mathbb{R}}_+^n$ such that $\|x - x_{\rm e}\| \geq \varepsilon$. Hence, it follows from (2.11) that $V(x(t)) \leq V(x_0) = \beta$, $t \geq 0$, which implies that $x(t) \in \overline{\mathbb{R}}_+^n \cap \mathcal{B}_\varepsilon(x_{\rm e})$, $t \geq 0$. Now, the proof follows as in the proof of the local asymptotic stability result. □

A continuously differentiable function $V(\cdot)$ satisfying (2.6) and (2.7) is called a *Lyapunov function candidate* for the nonnegative dynamical system (2.1). If, additionally, $V(\cdot)$ satisfies (2.8), then $V(\cdot)$ is called a *Lyapunov function* for the nonnegative dynamical system (2.1).

Finally, we note that as in standard nonlinear dynamical systems theory [191], converse Lyapunov theorems for asymptotically stable nonlinear, nonnegative dynamical systems can also be established. Since the statement and proofs of these results are virtually identical to the standard converse Lyapunov stability proofs for nonlinear dynamical systems [191], they are not presented here.

2.3 Invariant Set Stability Theorems

In this section, we introduce the Krasovskii-LaSalle invariance principle for nonnegative dynamical systems to relax one of the conditions on the Lyapunov function $V(\cdot)$ in the theorems given in Section 2.2. In particular, the strict negative-definiteness condition on the Lyapunov derivative can be relaxed while ensuring system asymptotic stability. Specifically, if a continuously differentiable function defined on a compact invariant set with respect to the nonlinear dynamical system (2.1) can be constructed whose derivative along the system's trajectories is negative semidefinite and no system trajectories can stay indefinitely at points where the function's derivative vanishes, then the system's equilibrium point is asymptotically stable. To state and prove the main results of this section, several definitions and a key theorem are needed.

First, we introduce the notion of invariance with respect to the flow of the nonlinear dynamical system (2.1). Given $t \in \mathbb{R}$ we denote the map $s(t, \cdot) : \mathcal{D} \to \mathcal{D}$ by $s_t(x_0)$. Hence, for a fixed $t \in \mathbb{R}$ the set of mappings defined by $s_t(x_0) = s(t, x_0)$ for every $x_0 \in \mathcal{D}$ gives the flow of (2.1). In particular, if $\mathcal{D}_0$ is a collection of initial conditions such that $\mathcal{D}_0 \subset \mathcal{D}$, then the flow $s_t : \mathcal{D} \to \mathcal{D}$ is the motion of all points $x_0 \in \mathcal{D}_0$ or, equivalently, the image of $\mathcal{D}_0 \subset \mathcal{D}$ under the flow s_t, that is, $s_t(\mathcal{D}_0) \subset \mathcal{D}$, where $s_t(\mathcal{D}_0) \triangleq \{y : y = s_t(x_0) \text{ for some } x_0 \in \mathcal{D}_0\}$. Alternatively, if the initial condition $x_0 \in \mathcal{D}$ is fixed and we let $[t_0, t_1] \subset \mathbb{R}$, then the mapping $s(\cdot, x_0) : [t_0, t_1] \to \mathcal{D}$ defines the *solution curve* or *trajectory* of the dynamical system (2.1). Hence, the mapping $s(\cdot, x_0)$ generates a graph in $[t_0, t_1] \times \mathcal{D}$ identifying the trajectory corresponding to the motion along a curve through the point x_0 in a subset $\mathcal{D}$ of the state space. Given $x \in \mathcal{D}$, we denote the map $s(\cdot, x) : \mathbb{R} \to \mathcal{D}$ by $s^x(t)$. Identifying $s(\cdot, x)$ with its graph, the trajectory or *orbit* of a point $x_0 \in \mathcal{D}$ is defined as the motion along the curve

$$\mathcal{O}_{x_0} \triangleq \{x \in \mathcal{D} : \ x = s(t, x_0), \ t \in \mathbb{R}\}. \tag{2.12}$$

For $t \geq 0$, we define the *positive orbit* through the point $x_0 \in \mathcal{D}$ as the motion along the curve

$$\mathcal{O}^+_{x_0} \triangleq \{x \in \mathcal{D} : \ x = s(t, x_0), \ t \geq 0\}. \tag{2.13}$$

Definition 2.4. The trajectory $x(t)$, $t \geq 0$, of (2.1) is *bounded* if there exists $\gamma > 0$ such that $\|x(t)\| < \gamma$, $t \geq 0$.

Definition 2.5. A point $p \in \mathcal{D}$ is a *positive limit point* (or *ω-limit point*) of the trajectory $s(\cdot, x)$ of (2.1) if there exists a monotonic sequence $\{t_n\}_{n=0}^{\infty}$ of positive numbers, with $t_n \to \infty$ as $n \to \infty$, such that $s(t_n, x) \to p$ as $n \to \infty$. The set of all positive limit points of $s(t, x)$, $t \geq 0$, is the *positive limit set* (or *ω-limit set*) $\omega(x)$ of $s(\cdot, x)$ of (2.1). Equivalently, the positive limit set $\omega(x) = \omega(s(t, x))$, $t \geq 0$, is given by $\omega(x) = \cap\{\overline{\mathcal{O}^+_y} : y \in \mathcal{O}^+_x\} = \cap\{\overline{\mathcal{O}^+_{x_n}} : n \in \mathbb{Z}_+\}$.

Definition 2.6. A set $\mathcal{M} \subset \mathcal{D} \subseteq \mathbb{R}^n$ is a *positively invariant set* with respect to the nonlinear dynamical system (2.1) if $s_t(\mathcal{M}) \subseteq \mathcal{M}$ for all $t \geq 0$, where $s_t(\mathcal{M}) \triangleq \{s_t(x) : \ x \in \mathcal{M}\}$. A set $\mathcal{M} \subseteq \mathcal{D}$ is an *invariant set* with respect to the dynamical system (2.1) if $s_t(\mathcal{M}) = \mathcal{M}$ for all $t \in \mathbb{R}$.

Next, we state and prove a key theorem involving positive limit sets. For this result, we use the notation $x(t) \to \mathcal{M} \subseteq \mathcal{D}$ as $t \to \infty$ to denote that $x(t)$ approaches $\mathcal{M}$, that is, for each $\varepsilon > 0$ there exists $T > 0$ such that $\operatorname{dist}(x(t), \mathcal{M}) < \varepsilon$ for all $t > T$, where $\operatorname{dist}(p, \mathcal{M}) \triangleq \inf_{x \in \mathcal{M}} \|p - x\|$.

Theorem 2.2. Consider the nonlinear dynamical system (2.1) where f is essentially nonnegative. Suppose the solution $x(t)$ to (2.1) corresponding to an initial condition $x(0) = x_0 \in \overline{\mathbb{R}}_+^n$ is bounded for all $t \geq 0$. Then the positive limit set $\omega(x_0)$ of $x(t)$, $t \geq 0$, is a nonempty, compact, invariant, and connected subset of $\overline{\mathbb{R}}_+^n$. Furthermore, $x(t) \to \omega(x_0)$ as $t \to \infty$.

Proof. Let $x(t)$, $t \geq 0$, or, equivalently, $s(t, x_0)$, $t \geq 0$, denote the solution to (2.1) corresponding to the initial condition $x(0) = x_0$. Next, since $x(t)$ is bounded for all $t \geq 0$, it follows from the Bolzano-Weierstrass theorem [191, p. 27] that every sequence in the positive orbit $\mathcal{O}_{x_0}^+ \triangleq \{s(t, x) : t \in [0, \infty)\}$ has at least one accumulation point $p \in \mathcal{D}$ as $t \to \infty$, and hence, $\omega(x_0)$ is nonempty. Next, let $p \in \omega(x_0)$ so that there exists an increasing unbounded sequence $\{t_n\}_{n=0}^\infty$, with $t_0 = 0$, such that $\lim_{n\to\infty} x(t_n) = p$. Now, since $x(t_n)$ is uniformly bounded in n, it follows that the limit point p is bounded, which implies that $\omega(x_0)$ is bounded.

To show that $\omega(x_0)$ is closed, let $\{p_i\}_{i=0}^\infty$ be a sequence contained in $\omega(x_0)$ such that $\lim_{i\to\infty} p_i = p$. Now, since $p_i \to p$ as $i \to \infty$ for every $\varepsilon > 0$, there exists an i such that $\|p - p_i\| < \varepsilon/2$. Next, since $p_i \in \omega(x_0)$, there exists $t \geq T$, where T is arbitrary and finite, such that $\|p_i - x(t)\| < \varepsilon/2$. Now, since $\|p - p_i\| < \varepsilon/2$ and $\|p_i - x(t)\| < \varepsilon/2$, it follows that $\|p - x(t)\| \leq \|p_i - x(t)\| + \|p - p_i\| < \varepsilon$, and hence, $p \in \omega(x_0)$. Thus, every accumulation point of $\omega(x_0)$ is an element of $\omega(x_0)$ so that $\omega(x_0)$ is closed. Hence, since $\omega(x_0)$ is closed and bounded, $\omega(x_0)$ is compact.

To show positive invariance of $\omega(x_0)$, let $p \in \omega(x_0)$ so that there exists an increasing unbounded sequence $\{t_n\}_{n=0}^\infty$ such that $x(t_n) \to p$ as $n \to \infty$. Now, let $s(t_n, x_0)$ denote the solution $x(t_n)$ of (2.1) with initial condition $x(0) = x_0$ and note that, since $f : \mathcal{D} \to \mathbb{R}^n$ in (2.1) is Lipschitz continuous on $\mathcal{D}$, $x(t)$, $t \geq 0$, is the unique solution to (2.1) so that by the semigroup property $s(t + t_n, x_0) = s(t, s(t_n, x_0)) = s(t, x(t_n))$. Now, since $x(t)$, $t \geq 0$, is continuous, it follows that, for $t + t_n \geq 0$, $\lim_{n\to\infty} s(t + t_n, x_0) = \lim_{n\to\infty} s(t, x(t_n)) = s(t, p)$, and hence, $s(t, p) \in \omega(x_0)$. Hence, $s_t(\omega(x_0)) \subseteq \omega(x_0)$, $t \geq 0$, establishing positive invariance of $\omega(x_0)$.

To show invariance of $\omega(x_0)$, let $y \in \omega(x_0)$ so that there exists an increasing unbounded sequence $\{t_n\}_{n=0}^\infty$ such that $s(t_n, x_0) \to y$ as $n \to \infty$. Next, let $t \in [0, \infty)$ and note that there exists N such that $t_n > t$, $n \geq N$. Hence, it follows from the semigroup property that $s(t, s(t_n - t, x_0)) = s(t_n, x_0) \to y$ as $n \to \infty$. Now, it follows from the Bolzano-Lebesgue theorem [191, p. 28] that there exists a subsequence $\{z_{n_k}\}_{k=1}^\infty$ of the sequence $z_n = s(t_n - t, x_0)$, $n = N, N+1, \ldots$, such that $z_{n_k} \to z \in \mathcal{D}$ as $k \to \infty$ and, by definition, $z \in \omega(x_0)$. Next, it follows from the continuous dependence property that $\lim_{k\to\infty} s(t, z_{n_k}) = s(t, \lim_{k\to\infty} z_{n_k})$, and hence, $y = s(t, z)$, which implies that $\omega(x_0) \subseteq s_t(\omega(x_0))$, $t \in [0, \infty)$. Now, using positive invariance of $\omega(x_0)$, it follows that $s_t(\omega(x_0)) = \omega(x_0)$, $t \geq 0$, establishing invariance of the positive limit set $\omega(x_0)$.

To show connectedness of $\omega(x_0)$, suppose, *ad absurdum*, that $\omega(x_0)$ is not connected. In this case, there exist two nonempty closed sets $\mathcal{P}_1^+$ and $\mathcal{P}_2^+$ such that $\mathcal{P}_1^+ \cap \mathcal{P}_2^+ = \emptyset$ and $\omega(x_0) = \mathcal{P}_1^+ \cup \mathcal{P}_2^+$. Since $\mathcal{P}_1^+$ and $\mathcal{P}_2^+$ are closed and disjoint, there exist two open sets $\mathcal{S}_1$ and $\mathcal{S}_2$ such that $\mathcal{S}_1 \cap \mathcal{S}_2 = \emptyset$, $\mathcal{P}_1^+ \subset \mathcal{S}_1$, and $\mathcal{P}_2^+ \subset \mathcal{S}_2$. Next, since $f : \mathcal{D} \to \mathbb{R}$ is Lipschitz continuous on $\mathcal{D}$, it follows that the solution $x(t)$, $t \geq 0$, to (2.1) is a continuous function of t. Hence, there exist sequences $\{t_n\}_{n=0}^{\infty}$ and $\{\tau_n\}_{n=0}^{\infty}$ such that $x(t_n) \in \mathcal{S}_1$, $x(\tau_n) \in \mathcal{S}_2$, and $t_n < \tau_n < t_{n+1}$, which implies that there exists a sequence $\{\tau_n\}_{n=0}^{\infty}$, with $t_n < \tau_n < \tau_{n+1}$, such that $x(\tau_n) \notin \mathcal{S}_1 \cup \mathcal{S}_2$. Next, since $x(t)$ is bounded for all $t \geq 0$, it follows that $x(\tau_n) \to \hat{p} \notin \omega(x_0)$ as $n \to \infty$, leading to a contradiction. Hence, $\omega(x_0)$ is connected.

Finally, to show $x(t) \to \omega(x_0)$ as $t \to \infty$, suppose, *ad absurdum*, that $x(t) \not\to \omega(x_0)$ as $t \to \infty$. In this case, there exists a sequence $\{t_n\}_{n=0}^{\infty}$, with $t_n \to \infty$ as $n \to \infty$, such that

$$\inf_{p \in \omega(x_0)} \|x(t_n) - p\| > \varepsilon, \quad n \in \overline{\mathbb{Z}}_+. \tag{2.14}$$

However, since $x(t)$, $t \geq 0$, is bounded, the bounded sequence $\{x(t_n)\}_{n=0}^{\infty}$ contains a convergent subsequence $\{x(t_n^*)\}_{n=0}^{\infty}$ such that $x(t_n^*) \to p^* \in \omega(x_0)$ as $n \to \infty$, which contradicts (2.14). Hence, $x(t) \to \omega(x_0)$ as $t \to \infty$ and, since f is essentially nonnegative, $\omega(x_0) \subseteq \overline{\mathbb{R}}_+^n$. □

Note that Theorem 2.2 implies that if $p \in \mathcal{D}$ is an ω-limit point of a trajectory $s(\cdot, x)$ of (2.1), then all other points of the trajectory $s(\cdot, p)$ of (2.1) through the point p are also ω-limit points of $s(\cdot, x)$, that is, if $p \in \omega(x)$, then $\mathcal{O}_p^+ \subset \omega(x)$. Furthermore, since every equilibrium point $x_e \in \mathcal{D}$ of (2.1) satisfies $s(t, x_e) = x_e$ for all $t \in \mathbb{R}$, all equilibrium points $x_e \in \mathcal{D}$ of (2.1) are their own ω-limit sets. If a trajectory of (2.1) possesses a unique ω-limit point x_e, then it follows from Theorem 2.2 that since $\omega(x_e)$ is invariant with respect to the flow s_t of (2.1), x_e is an equilibrium point of (2.1).

Next, we present the Krasovskii-LaSalle invariance principle for nonnegative dynamical systems.

Theorem 2.3 (Krasovskii-LaSalle Theorem). Consider the nonlinear dynamical system (2.1) where f is essentially nonnegative, assume that $\mathcal{D}_c \subset \mathcal{D} \subseteq \overline{\mathbb{R}}_+^n$ is a compact positively invariant set with respect to (2.1), and assume there exists a continuously differentiable function $V\colon \mathcal{D}_c \to \mathbb{R}$ such that $V'(x)f(x) \leq 0$, $x \in \mathcal{D}_c$. Let $\mathcal{R} \triangleq \{x \in \mathcal{D}_c : V'(x)f(x) = 0\}$ and let $\mathcal{M}$ be the largest invariant set contained in $\mathcal{R}$. If $x(0) \in \mathcal{D}_c$, then $x(t) \to \mathcal{M}$ as

$t \to \infty$.

Proof. Let $x(t)$, $t \geq 0$, be a solution to (2.1) with $x(0) \in \mathcal{D}_c$. Since $V'(x)f(x) \leq 0$, $x \in \mathcal{D}_c$, it follows that

$$V(x(t)) - V(x(\tau)) = \int_\tau^t V'(x(s))f(x(s))\mathrm{d}s \leq 0, \qquad t \geq \tau,$$

and hence $V(x(t)) \leq V(x(\tau))$, $t \geq \tau$, which implies that $V(x(t))$ is a nonincreasing function of t. Next, since $V(\cdot)$ is continuous on the compact set $\mathcal{D}_c$, there exists $\beta \in \mathbb{R}$ such that $V(x) \geq \beta$, $x \in \mathcal{D}_c$. Hence, $\gamma_{x_0} \triangleq \lim_{t\to\infty} V(x(t))$ exists.

Now, for all $p \in \omega(x_0)$ there exists an increasing unbounded sequence $\{t_n\}_{n=0}^\infty$, with $t_0 = 0$, such that $x(t_n) \to p$ as $n \to \infty$. Since $V(x)$, $x \in \mathcal{D}_c$, is continuous, $V(p) = V(\lim_{n\to\infty} x(t_n)) = \lim_{n\to\infty} V(x(t_n)) = \gamma_{x_0}$, and hence, $V(x) = \gamma_{x_0}$ on $\omega(x_0)$. Next, since $\mathcal{D}_c$ is compact and positively invariant, it follows that $x(t)$, $t \geq 0$, is bounded, and hence, it follows from Theorem 2.2 that $\omega(x_0)$ is a nonempty, compact invariant set. Hence, it follows that $V'(x)f(x) = 0$ on $\omega(x_0)$ and thus $\omega(x_0) \subset \mathcal{M} \subset \mathcal{R} \subset \mathcal{D}_c$. Finally, since $x(t) \to \omega(x_0)$ as $t \to \infty$, it follows that $x(t) \to \mathcal{M}$ as $t \to \infty$. □

Next, using Theorem 2.3 we provide a generalization of Theorem 2.1 for local asymptotic stability of a nonlinear dynamical system.

Corollary 2.1. Consider the nonlinear dynamical system (2.1) where f is essentially nonnegative, assume that $\mathcal{D}_c \subset \mathcal{D} \subseteq \overline{\mathbb{R}}_+^n$ is a compact positively invariant set with respect to (2.1) such that $x_e \in \mathcal{D}_c$, and assume that there exists a continuously differentiable function $V\colon\ \mathcal{D}_c \to \mathbb{R}$ such that $V(x_e) = 0$, $V(x) > 0$, $x \neq x_e$, and $V'(x)f(x) \leq 0$, $x \in \mathcal{D}_c$. Furthermore, assume that the set $\mathcal{R} \triangleq \{x \in \mathcal{D}_c\colon\ V'(x)f(x) = 0\}$ contains no invariant set other than the set $\{x_e\}$. Then the equilibrium solution $x(t) \equiv x_e$ to (2.1) is asymptotically stable with respect to $\overline{\mathbb{R}}_+^n$.

Proof. Lyapunov stability with respect to $\overline{\mathbb{R}}_+^n$ of the equilibrium solution $x(t) \equiv x_e$ to (2.1) follows from Theorem 2.1 since $V'(x)f(x) \leq 0$, $x \in \mathcal{D}_c$. Now, it follows from Theorem 2.3 that if $x_0 \in \mathcal{D}_c$, then $\omega(x_0) \subseteq \mathcal{M}$, where $\mathcal{M}$ denotes the largest invariant set contained in $\mathcal{R}$, which implies that $\mathcal{M} = \{x_e\}$. Hence, $x(t) \to \mathcal{M} = \{x_e\}$ as $t \to \infty$, establishing asymptotic stability of the equilibrium solution $x(t) \equiv x_e$ to (2.1) with respect to $\overline{\mathbb{R}}_+^n$. □

In Theorem 2.3 and Corollary 2.1, we explicitly assumed that there exists a compact invariant set $\mathcal{D}_c \subset \mathcal{D} \subseteq \overline{\mathbb{R}}_+^n$ of (2.1). Next, we provide a result that does not require the explicit assumption of the existence of a compact invariant set $\mathcal{D}_c$.

Theorem 2.4. Consider the nonlinear dynamical system (2.1) where f is essentially nonnegative and assume that there exists a continuously differentiable function $V : \overline{\mathbb{R}}_+^n \to \mathbb{R}$ such that

$$\begin{aligned} V(x_e) &= 0, && (2.15)\\ V(x) &> 0, \quad x \in \overline{\mathbb{R}}_+^n, \quad x \neq x_e, && (2.16)\\ V'(x)f(x) &\leq 0, \quad x \in \overline{\mathbb{R}}_+^n. && (2.17) \end{aligned}$$

Let $\mathcal{R} \triangleq \{x \in \overline{\mathbb{R}}_+^n : V'(x)f(x) = 0\}$ and let $\mathcal{M}$ be the largest invariant set contained in $\mathcal{R}$. Then all solutions $x(t)$, $t \geq 0$, of (2.1) that are bounded approach $\mathcal{M}$ as $t \to \infty$.

Proof. Let $x \in \overline{\mathbb{R}}_+^n$ be such that trajectory $s(t, x)$, $t \geq 0$, of (2.1) is bounded. Since f is essentially nonnegative it follows that $s(t, x) \in \overline{\mathbb{R}}_+^n$, $t \geq 0$. Now, with $\mathcal{D}_c = \overline{\mathcal{O}_x^+}$, it follows from Theorem 2.3 that $s(t, x) \to \mathcal{M}$ as $t \to \infty$. □

Next, we present the global invariant set theorem for guaranteeing global asymptotic stability of a nonlinear dynamical system.

Theorem 2.5. Consider the nonlinear dynamical system (2.1) where f is essentially nonnegative and assume there exists a continuously differentiable function $V : \overline{\mathbb{R}}_+^n \to \mathbb{R}$ such that

$$\begin{aligned} V(x_e) &= 0, && (2.18)\\ V(x) &> 0, \quad x \in \overline{\mathbb{R}}_+^n, \quad x \neq x_e, && (2.19)\\ V'(x)f(x) &\leq 0, \quad x \in \overline{\mathbb{R}}_+^n, && (2.20)\\ V(x) &\to \infty \text{ as } \|x\| \to \infty. && (2.21) \end{aligned}$$

Furthermore, assume that the set $\mathcal{R} \triangleq \{x \in \overline{\mathbb{R}}_+^n : V'(x)f(x) = 0\}$ contains no invariant set other than the set $\{x_e\}$. Then the equilibrium solution $x(t) \equiv x_e$ to (2.1) is globally asymptotically stable with respect to $\overline{\mathbb{R}}_+^n$.

Proof. Since (2.18)–(2.20) hold, it follows from Theorem 2.1 that the equilibrium solution $x(t) \equiv x_e$ to (2.1) is Lyapunov stable with respect to $\overline{\mathbb{R}}_+^n$, while the radial unboundedness condition (2.21) implies that all solutions to (2.1) are bounded. Now, Theorem 2.4 implies that $x(t) \to \mathcal{M}$ as $t \to \infty$. However, since $\mathcal{R}$ contains no invariant set other than the set $\{x_e\}$, the set $\mathcal{M}$ is $\{x_e\}$, and hence, global asymptotic stability with respect to $\overline{\mathbb{R}}_+^n$ is immediate. □

2.4 Semistability of Nonnegative Dynamical Systems

As we see in Chapter 3, thermodynamic systems give rise to systems that possess a continuum of equilibria. In this section, we develop a stability analysis framework for systems having a continuum of equilibria. Since every neighborhood of a nonisolated equilibrium contains another equilibrium, a nonisolated equilibrium cannot be asymptotically stable. Hence, asymptotic stability is not the appropriate notion of stability for systems having a continuum of equilibria. Two notions that are of particular relevance to such systems are *convergence* and *semistability*. Convergence is the property whereby every system solution converges to a limit point that may depend on the system initial condition. Semistability is the additional requirement that all solutions converge to limit points that are Lyapunov stable. Semistability for an equilibrium thus implies Lyapunov stability, and is implied by asymptotic stability.

It is important to note that semistability is not merely equivalent to asymptotic stability of the set of equilibria. Indeed, it is possible for a trajectory to converge to the set of equilibria without converging to any one equilibrium point [42]. Conversely, semistability does not imply that the equilibrium set is asymptotically stable in any accepted sense. This is because stability of sets (see [191]) is defined in terms of distance (especially in the case of noncompact sets), and it is possible to construct examples in which the dynamical system is semistable, but the domain of semistability (see Definition 2.8) contains no ε-neighborhood (defined in terms of the distance) of the (noncompact) equilibrium set, thus ruling out asymptotic stability of the equilibrium set. Hence, semistability and set stability of the equilibrium set are independent notions.

The dependence of the limiting state on the initial state is seen in numerous dynamical systems including compartmental systems [227], which arise in chemical kinetics [34], biomedical [226], environmental [343], economic [31], power [412], and thermodynamic systems [195]. For these systems, every trajectory that starts in a neighborhood of a Lyapunov stable equilibrium converges to a (possibly different) Lyapunov stable equilibrium, and hence these systems are semistable. Semistability is especially pertinent to thermodynamic systems that exhibit convergence to an equipartitioned energy state [195]. Semistability was first introduced in [75] for linear systems, whereas references [42] and [45] consider semistability of nonlinear systems and give several stability results for systems having a continuum of equilibria based on nontangency and arc length of trajectories, respectively.

In this section, we develop necessary and sufficient conditions for semistability. We say that the dynamical system (2.1) is *convergent* with respect to $\overline{\mathbb{R}}_+^n$ if $\lim_{t\to\infty} s(t,x)$ exists for every $x \in \overline{\mathbb{R}}_+^n$. The following proposition gives a sufficient condition for a trajectory of (2.1) to converge to a limit.

Proposition 2.2. Consider the nonlinear dynamical system (2.1) where f is essentially nonnegative and let $x \in \overline{\mathbb{R}}_+^n$. If the positive limit set $\omega(x)$ of (2.1) contains a Lyapunov stable (with respect to $\overline{\mathbb{R}}_+^n$) equilibrium point y, then $y = \lim_{t\to\infty} s(t,x)$, that is, $\omega(x) = \{y\}$.

Proof. Suppose $y \in \omega(x)$ is Lyapunov stable with respect to $\overline{\mathbb{R}}_+^n$ and let $\mathcal{N}_\varepsilon \subseteq \overline{\mathbb{R}}_+^n$ be a relatively open neighborhood of y. Since y is Lyapunov stable with respect to $\overline{\mathbb{R}}_+^n$, there exists a relatively open neighborhood $\mathcal{N}_\delta \subset \overline{\mathbb{R}}_+^n$ of y such that $s_t(\mathcal{N}_\delta) \subseteq \mathcal{N}_\varepsilon$ for every $t \geq 0$. Now, since $y \in \omega(x)$, it follows that there exists $\tau \geq 0$ such that $s(\tau,x) \in \mathcal{N}_\delta$. Hence, $s(t+\tau,x) = s_t(s(\tau,x)) \in s_t(\mathcal{N}_\delta) \subseteq \mathcal{N}_\varepsilon$ for every $t > 0$. Since $\mathcal{N}_\varepsilon \subseteq \overline{\mathbb{R}}_+^n$ is arbitrary, it follows that $y = \lim_{t\to\infty} s(t,x)$. Thus, $\lim_{n\to\infty} s(t_n,x) = y$ for every sequence $\{t_n\}_{n=1}^{\infty}$, and hence, $\omega(x) = \{y\}$. □

The following definitions and key proposition are necessary for the main results of this section.

Definition 2.7. An equilibrium solution $x(t) \equiv x_{\mathrm{e}} \in \overline{\mathbb{R}}_+^n$ to (2.1) is *semistable with respect to* $\overline{\mathbb{R}}_+^n$ if it is Lyapunov stable with respect to $\overline{\mathbb{R}}_+^n$ and there exists $\delta > 0$ such that if $x_0 \in \mathcal{B}_\delta(x_{\mathrm{e}}) \cap \overline{\mathbb{R}}_n^+$, then $\lim_{t\to\infty} x(t)$ exists and corresponds to a Lyapunov stable equilibrium point with respect to $\overline{\mathbb{R}}_+^n$. An equilibrium point $x_{\mathrm{e}} \in \overline{\mathbb{R}}^n$ is a *globally semistable equilibrium with respect to* $\overline{\mathbb{R}}_+^n$ if it is Lyapunov stable with respect to $\overline{\mathbb{R}}_+^n$ and, for every $x_0 \in \overline{\mathbb{R}}_n^+$, $\lim_{t\to\infty} x(t)$ exists and corresponds to a Lyapunov stable equilibrium point with respect to $\overline{\mathbb{R}}_+^n$. The system (2.1) is said to be *Lyapunov stable with respect to* $\overline{\mathbb{R}}_+^n$ if every equilibrium point of (2.1) is Lyapunov stable with respect to $\overline{\mathbb{R}}_+^n$. The system (2.1) is said to be *semistable with respect to* $\overline{\mathbb{R}}_+^n$ if every equilibrium point of (2.1) is semistable with respect to $\overline{\mathbb{R}}_+^n$. Finally, (2.1) is said to be *globally semistable with respect to* $\overline{\mathbb{R}}_+^n$ if every equilibrium point of (2.1) is globally semistable with respect to $\overline{\mathbb{R}}_+^n$.

Definition 2.8. The *domain of semistability with respect to* $\overline{\mathbb{R}}_+^n$ is the set of points $x_0 \in \overline{\mathbb{R}}_+^n$ such that if $x(t)$ is a solution to (2.1) with $x(0) = x_0$, $t \geq 0$, then $x(t)$ converges to a Lyapunov stable (with respect to $\overline{\mathbb{R}}_+^n$) equilibrium point in $\overline{\mathbb{R}}_+^n$.

Next, we present alternative equivalent characterizations of semistability of (2.1). For this result, the following definition is needed.

Definition 2.9. A continuous function $\gamma : [0, a) \to [0, \infty)$, where $a \in (0, \infty]$, is of *class* $\mathcal{K}$ if it is strictly increasing and $\gamma(0) = 0$. A continuous function $\gamma : [0, \infty) \to [0, \infty)$ is of *class* $\mathcal{K}_\infty$ if it is strictly increasing, $\gamma(0) = 0$, and $\gamma(s) \to \infty$ as $s \to \infty$. A continuous function $\gamma : [0, \infty) \to [0, \infty)$ is of *class* $\mathcal{L}$ if it is strictly decreasing and $\gamma(s) \to 0$ as $s \to \infty$. Finally, a continuous function $\gamma : [0, a) \times [0, \infty) \to [0, \infty)$ is of *class* $\mathcal{KL}$ if, for each fixed s, $\gamma(r, s)$ is of class $\mathcal{K}$ with respect to r and, for each fixed r, $\gamma(r, s)$ is of class $\mathcal{L}$ with respect to s.

Proposition 2.3. Consider the nonlinear dynamical system $\mathcal{G}$ given by (2.1) where f is essentially nonnegative. Then the following statements are equivalent:

i) $\mathcal{G}$ is semistable with respect to $\overline{\mathbb{R}}_+^n$.

ii) For each $x_e \in f^{-1}(0)$, there exist class $\mathcal{K}$ and $\mathcal{L}$ functions $\alpha(\cdot)$ and $\beta(\cdot)$, respectively, and $\delta = \delta(x_e) > 0$, such that if $\|x_0 - x_e\| < \delta$, then $\|x(t) - x_e\| \leq \alpha(\|x_0 - x_e\|)$, $t \geq 0$, and $\mathrm{dist}(x(t), f^{-1}(0)) \leq \beta(t)$, $t \geq 0$.

iii) For each $x_e \in f^{-1}(0)$, there exist class $\mathcal{K}$ functions $\alpha_1(\cdot)$ and $\alpha_2(\cdot)$, a class $\mathcal{L}$ function $\beta(\cdot)$, and $\delta = \delta(x_e) > 0$, such that if $\|x_0 - x_e\| < \delta$, then $\mathrm{dist}(x(t), f^{-1}(0)) \leq \alpha_1(\|x(t) - x_e\|)\beta(t) \leq \alpha_2(\|x_0 - x_e\|)\beta(t)$, $t \geq 0$.

Proof. To show that *i)* implies *ii)*, suppose (2.1) is semistable and let $x_e \in f^{-1}(0)$. It follows from Lemma 4.5 of [243] that there exist $\delta = \delta(x_e) > 0$ and a class $\mathcal{K}$ function $\alpha(\cdot)$ such that if $\|x_0 - x_e\| \leq \delta$, then $\|x(t) - x_e\| \leq \alpha(\|x_0 - x_e\|)$, $t \geq 0$. Without loss of generality, we may assume that δ is such that $\overline{\mathcal{B}_\delta(x_e)} \cap \overline{\mathbb{R}}_+^n$ is contained in the domain of semistability of (2.1). Hence, for every $x_0 \in \overline{\mathcal{B}_\delta(x_e)} \cap \overline{\mathbb{R}}_+^n$, $\lim_{t\to\infty} x(t) = x^* \in f^{-1}(0)$ and, consequently, $\lim_{t\to\infty} \mathrm{dist}(x(t), f^{-1}(0)) = 0$.

For each $\varepsilon > 0$ and $x_0 \in \overline{\mathcal{B}_\delta(x_e)} \cap \overline{\mathbb{R}}_+^n$, define $T_{x_0}(\varepsilon)$ to be the infimum of T with the property that $\mathrm{dist}(x(t), f^{-1}(0)) < \varepsilon$ for all $t \geq T$, that is, $T_{x_0}(\varepsilon) \triangleq \inf\{T : \mathrm{dist}(x(t), f^{-1}(0)) < \varepsilon, t \geq T\}$. For each $x_0 \in \overline{\mathcal{B}_\delta(x_e)} \cap \overline{\mathbb{R}}_+^n$, the function $T_{x_0}(\varepsilon)$ is nonnegative and nonincreasing in ε, and $T_{x_0}(\varepsilon) = 0$ for sufficiently large ε.

Next, let $T(\varepsilon) \triangleq \sup\{T_{x_0}(\varepsilon) : x_0 \in \overline{\mathcal{B}_\delta(x_e)} \cap \overline{\mathbb{R}}_+^n\}$. We claim that T is well defined. To show this, consider $\varepsilon > 0$ and $x_0 \in \overline{\mathcal{B}_\delta(x_e)} \cap \overline{\mathbb{R}}_+^n$. Since $\mathrm{dist}(s(t, x_0), f^{-1}(0)) < \varepsilon$ for every $t > T_{x_0}(\varepsilon)$, it follows from the continuity of s that, for every $\eta > 0$, there exists an open neighborhood $\mathcal{U}$ of x_0 such that $\mathrm{dist}(s(t, z), f^{-1}(0)) < \varepsilon$ for every $z \in \mathcal{U}$. Hence, $\limsup_{z\to x_0} T_z(\varepsilon) \leq T_{x_0}(\varepsilon)$, implying that the function $x_0 \mapsto T_{x_0}(\varepsilon)$ is upper semicontinuous at the arbitrarily chosen point x_0, and hence on $\overline{\mathcal{B}_\delta(x_e)} \cap \overline{\mathbb{R}}_+^n$. Since an upper semicontinuous function defined on a compact set achieves its supremum, it follows that $T(\varepsilon)$ is well defined. The function $T(\cdot)$ is the pointwise supremum of a collection of nonnegative and nonincreasing functions, and is hence nonnegative and nonincreasing. Moreover, $T(\varepsilon) = 0$ for every $\varepsilon > \max\{\alpha(\|x_0 - x_e\|) : x_0 \in \overline{\mathcal{B}_\delta(x_e)} \cap \overline{\mathbb{R}}_+^n\}$.

Let $\psi(\varepsilon) \triangleq \frac{2}{\varepsilon}\int_{\varepsilon/2}^{\varepsilon} T(\sigma)\mathrm{d}\sigma + \frac{1}{\varepsilon} \geq T(\varepsilon) + \frac{1}{\varepsilon}$. The function $\psi(\varepsilon)$ is positive, continuous, strictly decreasing, and $\psi(\varepsilon) \to 0$ as $\varepsilon \to \infty$. Choose $\beta(\cdot) = \psi^{-1}(\cdot)$. Then $\beta(\cdot)$ is positive, continuous, strictly decreasing, and $\beta(\sigma) \to 0$ as $\sigma \to \infty$. Furthermore, $T(\beta(\sigma)) < \psi(\beta(\sigma)) = \sigma$. Hence, $\mathrm{dist}(x(t), f^{-1}(0)) \leq \beta(t)$, $t \geq 0$.

Next, to show that *ii*) implies *iii*), suppose *ii*) holds and let $x_e \in f^{-1}(0)$. Then it follows from Lemma 4.5 of [243] that x_e is Lyapunov stable with respect to $\overline{\mathbb{R}}_+^n$. Choosing x_0 sufficiently close to x_e, it follows from the inequality $\|x(t) - x_e\| \leq \alpha(\|x_0 - x_e\|)$, $t \geq 0$, that trajectories of (2.1) starting sufficiently close to x_e are bounded, and hence, the positive limit set of (2.1) is nonempty. Since $\lim_{t\to\infty} \mathrm{dist}(x(t), f^{-1}(0)) = 0$, it follows that the positive limit set is contained in $f^{-1}(0)$.

Now, since every point in $f^{-1}(0)$ is Lyapunov stable with respect to $\overline{\mathbb{R}}_+^n$, it follows from Proposition 2.2 that $\lim_{t\to\infty} x(t) = x^*$, where $x^* \in f^{-1}(0)$ is Lyapunov stable with respect to $\overline{\mathbb{R}}_+^n$. If $x^* = x_e$, then it follows using similar arguments as above that there exists a class $\mathcal{L}$ function $\hat{\beta}(\cdot)$ such that $\mathrm{dist}(x(t), f^{-1}(0)) \leq \|x(t) - x_e\| \leq \hat{\beta}(t)$ for every x_0 satisfying $\|x_0 - x_e\| < \delta$ and $t \geq 0$. Hence, $\mathrm{dist}(x(t), f^{-1}(0)) \leq \sqrt{\|x(t) - x_e\|}\sqrt{\hat{\beta}(t)}$, $t \geq 0$.

Next, consider the case where $x^* \neq x_e$ and let $\alpha_1(\cdot)$ be a class $\mathcal{K}$ function. In this case, note that $\lim_{t\to\infty} \mathrm{dist}(x(t), f^{-1}(0))/\alpha_1(\|x(t) - x_e\|) = 0$, and hence, it follows using similar arguments as above that there exists a class $\mathcal{L}$ function $\beta(\cdot)$ such that $\mathrm{dist}(x(t), f^{-1}(0)) \leq \alpha_1(\|x(t) - x_e\|)\beta(t)$, $t \geq 0$. Finally, note that $\alpha_1 \circ \alpha$ is of class $\mathcal{K}$ (by Lemma 4.5 of [243]), and hence, *iii*) follows immediately.

Finally, to show that *iii*) implies *i*), suppose *iii*) holds and let $x_e \in f^{-1}(0)$. Then it follows that $\alpha_1(\|x(t) - x_e\|) \leq \alpha_2(\|x(0) - x_e\|)$, $t \geq 0$, that is, $\|x(t) - x_e\| \leq \alpha(\|x(0) - x_e\|)$, where $t \geq 0$ and $\alpha = \alpha_1^{-1} \circ \alpha_2$ is of class $\mathcal{K}$ (by Lemma 4.2 of [243]). It now follows from Lemma 4.5 of [243] that x_e is Lyapunov stable with respect to $\overline{\mathbb{R}}_+^n$. Since x_e was chosen arbitrarily, it follows that every equilibrium point is Lyapunov stable with respect to $\overline{\mathbb{R}}_+^n$. Furthermore, $\lim_{t\to\infty} \mathrm{dist}(x(t), f^{-1}(0)) = 0$.

Choosing x_0 sufficiently close to x_e, it follows from the inequality $\|x(t) - x_e\| \leq \alpha(\|x_0 - x_e\|)$, $t \geq 0$, that trajectories of (2.1) starting sufficiently close to x_e are bounded, and hence, the positive limit set of (2.1) is nonempty. Since every point in $f^{-1}(0)$ is Lyapunov stable with respect to $\overline{\mathbb{R}}^n_+$, it follows from Proposition 2.2 that $\lim_{t\to\infty} x(t) = x^*$, where $x^* \in f^{-1}(0)$ is Lyapunov stable with respect to $\overline{\mathbb{R}}^n_+$. Hence, by definition, (2.1) is semistable with respect to $\overline{\mathbb{R}}^n_+$. □

Next, we present a sufficient condition for semistability with respect to $\overline{\mathbb{R}}^n_+$.

Theorem 2.6. Consider the nonlinear dynamical system (2.1) where f is essentially nonnegative. Let $\mathcal{Q} \subseteq \overline{\mathbb{R}}^n_+$ be a relatively open neighborhood of $f^{-1}(0) \cap \overline{\mathbb{R}}^n_+$ and assume that there exists a continuously differentiable function $V : \mathcal{Q} \to \mathbb{R}$ such that

$$V'(x)f(x) < 0, \quad x \in \mathcal{Q}\backslash f^{-1}(0). \tag{2.22}$$

If (2.1) is Lyapunov stable with respect to $\overline{\mathbb{R}}^n_+$, then (2.1) is semistable with respect to $\overline{\mathbb{R}}^n_+$. Moreover, if $\mathcal{Q} = \overline{\mathbb{R}}^n_+$ and $V(x) \to \infty$ as $x \to \infty$, then (2.1) is globally semistable with respect to $\overline{\mathbb{R}}^n_+$.

Proof. Since (2.1) is Lyapunov stable with respect to $\overline{\mathbb{R}}^n_+$ by assumption, for every $z \in f^{-1}(0)$, there exists a relatively open neighborhood $\mathcal{V}_z$ of z such that $s([0,\infty) \times \mathcal{V}_z)$ is bounded and contained in $\mathcal{Q}$. The set $\mathcal{V} \triangleq \bigcup_{z\in f^{-1}(0)} \mathcal{V}_z$ is a relatively open neighborhood of $f^{-1}(0)$ contained in $\mathcal{Q}$. Consider $x \in \mathcal{V}$ so that there exists $z \in f^{-1}(0)$ such that $x \in \mathcal{V}_z$ and $s(t,x) \in \mathcal{V}_z$, $t \geq 0$. Since $\mathcal{V}_z$ is bounded, it follows that the positive limit set of x is nonempty and invariant. Furthermore, it follows from (2.22) that $\dot{V}(s(t,x)) \leq 0$, $t \geq 0$, and hence, it follows from Theorem 2.3 that $s(t,x) \to \mathcal{M}$ as $t \to \infty$, where $\mathcal{M}$ is the largest invariant set contained in the set $\mathcal{R} = \{y \in \mathcal{V}_z : V'(y)f(y) = 0\}$. Note that $\mathcal{R} = f^{-1}(0)$ is invariant, and hence, $\mathcal{M} = \mathcal{R}$, which implies that $\lim_{t\to\infty} \mathrm{dist}(s(t,x), f^{-1}(0)) = 0$.

Finally, since every point in $f^{-1}(0)$ is Lyapunov stable with respect to $\overline{\mathbb{R}}^n_+$, it follows from Proposition 2.2 that $\lim_{t\to\infty} s(t,x) = x^*$, where $x^* \in f^{-1}(0)$ is Lyapunov stable with respect to $\overline{\mathbb{R}}^n_+$. Hence, by definition, (2.1) is semistable with respect to $\overline{\mathbb{R}}^n_+$. For $\mathcal{Q} = \overline{\mathbb{R}}^n_+$, global semistablility with respect to $\overline{\mathbb{R}}^n_+$ follows from identical arguments using the radially unbounded condition on $V(\cdot)$. □

Next, we present a slightly more general theorem for semistability wherein we do not assume that all points in the zero-level set of the derivative of V, that is, $\dot{V}^{-1}(0)$, are Lyapunov stable, but rather we assume that all points in the largest invariant subset of $\dot{V}^{-1}(0)$ are Lyapunov stable.

Theorem 2.7. Consider the nonlinear dynamical system (2.1) where f is essentially nonnegative, and let $\mathcal{Q} \subseteq \overline{\mathbb{R}}_+^n$ be a relatively open neighborhood of $f^{-1}(0) \cap \overline{\mathbb{R}}_+^n$. Suppose the orbit $\mathcal{O}_x$ of (2.1) is bounded for all $x \in \mathcal{Q}$ and assume that there exists a continuously differentiable function $V : \mathcal{Q} \to \mathbb{R}$ such that

$$V'(x)f(x) \leq 0, \quad x \in \mathcal{Q}. \tag{2.23}$$

If every point in the largest invariant subset $\mathcal{M}$ of $\{x \in \mathcal{Q} : V'(x)f(x) = 0\}$ is Lyapunov stable with respect to $\overline{\mathbb{R}}_+^n$, then (2.1) is semistable with respect to $\overline{\mathbb{R}}_+^n$. Moreover, if $\mathcal{Q} = \overline{\mathbb{R}}_+^n$ and $V(x) \to \infty$ as $x \to \infty$, then (2.1) is globally semistable with respect to $\overline{\mathbb{R}}_+^n$.

Proof. Since every solution of (2.1) is bounded, it follows from the hypotheses on $V(\cdot)$ that, for every $x \in \mathcal{Q}$, the positive limit set $\omega(x)$ of (2.1) is nonempty and contained in the largest invariant subset $\mathcal{M}$ of $\{x \in \mathcal{Q} : V'(x)f(x) = 0\}$. Since every point in $\mathcal{M}$ is a Lyapunov stable (with respect to $\overline{\mathbb{R}}_+^n$) equilibrium, it follows from Proposition 2.2 that $\omega(x)$ contains a single point for every $x \in \mathcal{Q}$ and $\lim_{t\to\infty} s(t,x)$ exists for every $x \in \mathcal{Q}$. Now, since $\lim_{t\to\infty} s(t,x) \in \mathcal{M}$ is Lyapunov stable with respect to $\overline{\mathbb{R}}_+^n$ for every $x \in \mathcal{Q}$, semistability is immediate. For $\mathcal{Q} = \overline{\mathbb{R}}_+^n$, global semistablility with respect to $\overline{\mathbb{R}}_+^n$ follows from identical arguments using the radially unbounded condition on $V(\cdot)$. □

Example 2.1. Consider the nonlinear dynamical system given by

$$\dot{x}_1(t) = \sigma_{12}(x_2(t)) - \sigma_{21}(x_1(t)), \quad x_1(0) = x_{10}, \quad t \geq 0, \tag{2.24}$$
$$\dot{x}_2(t) = \sigma_{21}(x_1(t)) - \sigma_{12}(x_2(t)), \quad x_2(0) = x_{20}, \tag{2.25}$$

where $x_1, x_2 \in \overline{\mathbb{R}}_+$, $\sigma_{ij}(\cdot)$, $i,j = 1,2$, $i \neq j$, are Lipschitz continuous, $\sigma_{12}(x_2) - \sigma_{21}(x_1) = 0$ if and only if $x_1 = x_2$, and $(x_1 - x_2)(\sigma_{12}(x_2) - \sigma_{21}(x_1)) \leq 0$, $x_1, x_2 \in \overline{\mathbb{R}}_+$. Note that $f^{-1}(0) = \{(x_1,x_2) \in \overline{\mathbb{R}}_+^2 : x_1 = x_2 = \alpha, \alpha \in \overline{\mathbb{R}}_+\}$.

To show that (2.24) and (2.25) is a semistable system with respect to $\overline{\mathbb{R}}_+^2$, consider the Lyapunov function candidate $V(x_1,x_2) = \frac{1}{2}(x_1-\alpha)^2 + \frac{1}{2}(x_2-\alpha)^2$, where $\alpha \in \overline{\mathbb{R}}_+$. Now, it follows that

$$\begin{aligned}
\dot{V}(x_1,x_2) &= (x_1-\alpha)[\sigma_{12}(x_2) - \sigma_{21}(x_1)] + (x_2-\alpha)[\sigma_{21}(x_1) - \sigma_{12}(x_2)] \\
&= x_1[\sigma_{12}(x_2) - \sigma_{21}(x_1)] + x_2[\sigma_{21}(x_1) - \sigma_{12}(x_2)] \\
&= (x_1 - x_2)[\sigma_{12}(x_2) - \sigma_{21}(x_1)] \\
&\leq 0, \quad (x_1,x_2) \in \overline{\mathbb{R}}_+ \times \overline{\mathbb{R}}_+,
\end{aligned} \tag{2.26}$$

which implies that $x_1 = x_2 = \alpha$ is Lyapunov stable with respect to $\overline{\mathbb{R}}_+^2$.

Finally, let $\mathcal{R} \triangleq \{(x_1, x_2) \in \overline{\mathbb{R}}_+^2 : \dot{V}(x_1, x_2) = 0\} = \{(x_1, x_2) \in \overline{\mathbb{R}}_+^2 : x_1 = x_2 = \alpha, \alpha \in \overline{\mathbb{R}}_+\}$. Since $\mathcal{R}$ consists of equilibrium points, it follows that $\mathcal{M} = \mathcal{R}$. Hence, for every $x_1(0), x_2(0) \in \overline{\mathbb{R}}_+$, $(x_1(t), x_2(t)) \to \mathcal{M}$ as $t \to \infty$. Hence, it follows from Theorem 2.7 that $x_1 = x_2 = \alpha$ is semistable with respect to $\overline{\mathbb{R}}_+^2$ for all $\alpha \in \overline{\mathbb{R}}_+$. △

Next, we provide a converse Lyapunov theorem for semistability, which holds with a smooth (i.e., infinitely differentiable) Lyapunov function.

Theorem 2.8. Consider the system (2.1). Suppose (2.1) is semistable with the domain of semistability $\mathcal{D}_0$. Then there exist a smooth nonnegative function $V : \mathcal{D}_0 \to \overline{\mathbb{R}}_+$ and a class $\mathcal{K}_\infty$ function $\alpha(\cdot)$ such that *i)* $V(x) = 0$, $x \in f^{-1}(0)$, *ii)* $V(x) \geq \alpha(\mathrm{dist}(x, f^{-1}(0)))$, $x \in \mathcal{D}_0$, and *iii)* $V'(x)f(x) < 0$, $x \in \mathcal{D}_0 \backslash f^{-1}(0)$.

Proof. For any given solution $x(t)$ of (2.1), the change of time variable from t to $\tau = \int_0^t (1 + \|f(x(s))\|)\mathrm{d}s$ results in the dynamical system

$$\frac{\mathrm{d}\bar{x}}{\mathrm{d}\tau} = \frac{f(\bar{x}(\tau))}{1 + \|f(\bar{x}(\tau))\|}, \quad \bar{x}(0) = x_0, \quad \tau \geq 0, \tag{2.27}$$

where $\bar{x}(\tau) = x(t)$. With a slight abuse of notation, let $\bar{s}(t, x)$, $t \geq 0$, denote the solution of (2.27) starting from $x \in \mathcal{D}_0$. Note that (2.27) implies that $\|\bar{s}(t, x) - \bar{s}(\tau, x)\| \leq |t - \tau|$, $x \in \mathcal{D}_0$, $t, \tau \geq 0$.

Next, define the function $U : \mathcal{D}_0 \to \overline{\mathbb{R}}_+$ by

$$U(x) \triangleq \sup_{t \geq 0} \left\{ \frac{1 + 2t}{1 + t} \mathrm{dist}(\bar{s}(t, x), f^{-1}(0)) \right\}, \quad x \in \mathcal{D}_0. \tag{2.28}$$

Note that $U(\cdot)$ is well defined since (2.27) is semistable. Clearly, *i)* holds with $V(\cdot)$ replaced by $U(\cdot)$. Furthermore, since $U(x) \geq \mathrm{dist}(x, f^{-1}(0))$, $x \in \mathcal{D}_0$, it follows that *ii)* holds with $V(\cdot)$ replaced by $U(\cdot)$.

To show that $U(\cdot)$ is continuous on $\mathcal{D}_0 \backslash f^{-1}(0)$, define $T : \mathcal{D}_0 \backslash f^{-1}(0) \to [0, \infty)$ by $T(z) \triangleq \inf\{h : \mathrm{dist}(\bar{s}(t, z), f^{-1}(0)) < \mathrm{dist}(z, f^{-1}(0))/2$ for all $t \geq h > 0\}$, and denote $\mathcal{W}_\varepsilon \triangleq \{x \in \mathcal{D}_0 : \mathrm{dist}(x, f^{-1}(0)) < \varepsilon\}$. Note that $\mathcal{W}_\varepsilon \supset f^{-1}(0)$ is open. Consider $z \in \mathcal{D}_0 \backslash f^{-1}(0)$ and define $\lambda \triangleq \mathrm{dist}(z, f^{-1}(0)) > 0$ and let $x_\mathrm{e} \triangleq \lim_{t \to \infty} \bar{s}(t, z)$. Since x_e is Lyapunov stable, it follows that there exists a relatively open neighborhood $\mathcal{V}$ of x_e such that all solutions of (2.27) in $\mathcal{V}$ remain in $\mathcal{W}_{\lambda/2}$. Since x_e is semistable, it follows that there exists $h > 0$ such that $\bar{s}(h, z) \in \mathcal{V}$. Consequently, $\bar{s}(h + t, z) \in \mathcal{W}_{\lambda/2}$ for all $t \geq 0$, and hence, it follows that $T(z)$ is well defined.

Next, by continuity of solutions of (2.27) on compact time intervals, it follows that there exists a neighborhood $\mathcal{U}$ of z such that $\mathcal{U}\cap f^{-1}(0)=\emptyset$ and $\bar{s}(T(z),y)\in\mathcal{V}$ for all $y\in\mathcal{U}$. Now, it follows from the choice of $\mathcal{V}$ that $\bar{s}(T(z)+t,y)\in\mathcal{W}_{\lambda/2}$ for all $t\geq 0$ and $y\in\mathcal{U}$. Then, for every $t>T(z)$ and $y\in\mathcal{U}$, $[(1+2t)/(1+t)]\mathrm{dist}(\bar{s}(t,y),f^{-1}(0))\leq 2\mathrm{dist}(\bar{s}(t,y),f^{-1}(0))\leq\lambda$. Therefore, for every $y\in\mathcal{U}$,

$$\begin{aligned} U(z)-U(y) &= \sup_{t\geq 0}\left\{\frac{1+2t}{1+t}\mathrm{dist}(\bar{s}(t,z),f^{-1}(0))\right\} \\ &\quad -\sup_{t\geq 0}\left\{\frac{1+2t}{1+t}\mathrm{dist}(\bar{s}(t,y),f^{-1}(0))\right\} \\ &= \sup_{0\leq t\leq T(z)}\left\{\frac{1+2t}{1+t}\mathrm{dist}(\bar{s}(t,z),f^{-1}(0))\right\} \\ &\quad -\sup_{0\leq t\leq T(z)}\left\{\frac{1+2t}{1+t}\mathrm{dist}(\bar{s}(t,y),f^{-1}(0))\right\}. \end{aligned} \tag{2.29}$$

Hence,

$$\begin{aligned} |U(z)-U(y)| &\leq \sup_{0\leq t\leq T(z)}\left|\frac{1+2t}{1+t}\left(\mathrm{dist}(\bar{s}(t,z),f^{-1}(0))\right.\right. \\ &\quad \left.\left.-\mathrm{dist}(\bar{s}(t,y),f^{-1}(0))\right)\right| \\ &\leq 2\sup_{0\leq t\leq T(z)}\left|\mathrm{dist}(\bar{s}(t,z),f^{-1}(0))-\mathrm{dist}(\bar{s}(t,y),f^{-1}(0))\right| \\ &\leq 2\sup_{0\leq t\leq T(z)}\mathrm{dist}(\bar{s}(t,z),\bar{s}(t,y)),\quad z\in\mathcal{D}_0\backslash f^{-1}(0),\quad y\in\mathcal{U}. \end{aligned} \tag{2.30}$$

Now, it follows from the continuous dependence of solutions $\bar{s}(\cdot,\cdot)$ on system initial conditions (Theorem 3.4 of Chapter I of [202]) and (2.30) that $U(\cdot)$ is continuous at z. Furthermore, it follows from (2.30) that, for every sufficiently small $h>0$,

$$\begin{aligned} |U(\bar{s}(h,z))-U(z)| &\leq 2\sup_{0\leq t\leq T(z)}\|\bar{s}(t,\bar{s}(h,z))-\bar{s}(t,z)\| \\ &= 2\sup_{0\leq t\leq T(z)}\|\bar{s}(t+h,z)-\bar{s}(t,z)\| \\ &\leq 2h, \end{aligned}$$

which implies that $|\dot{U}(z)|\leq 2$. Since $z\in\mathcal{D}_0\backslash f^{-1}(0)$ was chosen arbitrarily, it follows that $U(\cdot)$ is continuous, $|\dot{U}(\cdot)|\leq 2$, and $T(\cdot)$ is well defined on $\mathcal{D}_0\backslash f^{-1}(0)$.

To show that $U(\cdot)$ is continuous on $f^{-1}(0)$, consider $x_{\mathrm{e}}\in f^{-1}(0)$.

Let $\{x_n\}_{n=1}^{\infty}$ be a sequence in $\mathcal{D}_0\backslash f^{-1}(0)$ that converges to x_e. Since x_e is Lyapunov stable, it follows from Lemma 4.5 of [243] that $x(t) \equiv x_e$ is the unique solution to (2.27) with $x_0 = x_e$. By the continuous dependence of solutions $\bar{s}(\cdot,\cdot)$ on system initial conditions (Theorem 3.4 of Chapter I of [202]), $\bar{s}(t, x_n) \to \bar{s}(t, x_e) = x_e$ as $n \to \infty$, $t \geq 0$.

Let $\varepsilon > 0$ and note that it follows from *ii*) of Proposition 3.1 that there exists $\delta = \delta(x_e) > 0$ such that, for every solution of (2.27) in $\mathcal{B}_\delta(x_e)$, there exists $\hat{T} = \hat{T}(x_e, \varepsilon) > 0$ such that $\bar{s}_t(\mathcal{B}_\delta(x_e)) \subset \mathcal{W}_\varepsilon$ for all $t \geq \hat{T}$. Next, note that there exists a positive integer N_1 such that $x_n \in \mathcal{B}_\delta(x_e)$ for all $n \geq N_1$. Now, it follows from (2.28) that for $n \geq N_1$,

$$U(x_n) \leq 2 \sup_{0 \leq t \leq \hat{T}} \operatorname{dist}(\bar{s}(t, x_n), f^{-1}(0)) + 2\varepsilon. \tag{2.31}$$

Next, it follows from Lemma 3.1 of Chapter I of [202] that $\bar{s}(\cdot, x_n)$ converges to $\bar{s}(\cdot, x_e)$ uniformly on $[0, \hat{T}]$. Hence,

$$\begin{aligned} \lim_{n\to\infty} \sup_{0 \leq t \leq \hat{T}} \operatorname{dist}\left(\bar{s}(t, x_n), f^{-1}(0)\right) &= \sup_{0 \leq t \leq \hat{T}} \operatorname{dist}\left(\lim_{n\to\infty} \bar{s}(t, x_n), f^{-1}(0)\right) \\ &= \sup_{0 \leq t \leq \hat{T}} \operatorname{dist}\left(x_e, f^{-1}(0)\right) \\ &= 0, \end{aligned}$$

which implies that there exists a positive integer $N_2 = N_2(x_e, \varepsilon) \geq N_1$ such that $\sup_{0 \leq t \leq \hat{T}} \operatorname{dist}(\bar{s}(t, x_n), f^{-1}(0)) < \varepsilon$ for all $n \geq N_2$. Combining (2.31) with the above result yields $U(x_n) < 4\varepsilon$ for all $n \geq N_2$, which implies that $\lim_{n\to\infty} U(x_n) = 0 = U(x_e)$.

Next, we show that $U(\bar{x}(\tau))$ is strictly decreasing along the solution of (2.27) on $\mathcal{D}\backslash f^{-1}(0)$. Note that for every $x \in \mathcal{D}_0\backslash f^{-1}(0)$ and $0 < h \leq 1/2$ such that $\bar{s}(h, x) \in \mathcal{D}_0\backslash f^{-1}(0)$, it follows from the arguments preceding (2.29) that, for sufficiently small h, the supremum in the definition of $U(\bar{s}(h, x))$ is reached at some time $\hat{t}$ such that $0 \leq \hat{t} \leq T(x)$. Hence,

$$\begin{aligned} U(\bar{s}(h, x)) &= \operatorname{dist}(\bar{s}(\hat{t} + h, x), f^{-1}(0)) \frac{1 + 2\hat{t}}{1 + \hat{t}} \\ &= \operatorname{dist}(\bar{s}(\hat{t} + h, x), f^{-1}(0)) \frac{1 + 2\hat{t} + 2h}{1 + \hat{t} + h} \\ &\quad \cdot \left[1 - \frac{h}{(1 + 2\hat{t} + 2h)(1 + \hat{t})}\right] \\ &\leq U(x) \left[1 - \frac{h}{2(1 + T(x))^2}\right], \end{aligned} \tag{2.32}$$

which implies that $\dot{U}(x) \leq -\frac{1}{2} U(x)(1 + T(x))^{-2} < 0$, $x \in \mathcal{D}_0\backslash f^{-1}(0)$, and hence, *iii*) holds with $V(\cdot)$ replaced by $U(\cdot)$. The function $U(\cdot)$ now satisfies all of the conditions of the theorem except for smoothness.

To obtain smoothness, note that since $|\dot{U}(x)| \leq 2$ for every $x \in \mathcal{D}_0$, it follows that $\dot{U}(x)$ satisfies a boundedness condition in the sense of Wilson [467]. By Theorem 2.5 of [467], there exists a smooth function $W : \mathcal{D}_0 \backslash f^{-1}(0) \to \mathbb{R}$ satisfying $|W(x) - U(x)| < \frac{1}{4}U(x)(1+T(x))^{-2} < \frac{1}{2}U(x)$ and $\dot{W}(x) \leq -\frac{1}{4}U(x)(1+T(x))^{-2} < 0$ for $x \in \mathcal{D}_0 \backslash f^{-1}(0)$. Next, we extend $W(\cdot)$ to all of $\mathcal{D}_0$ by taking $W(z) = 0$ for $z \in f^{-1}(0)$. Now, $W(\cdot)$ is a continuous Lyapunov function that is smooth on $\mathcal{D}_0 \backslash f^{-1}(0)$. Taking $V(x) = W(x)e^{-(W(x))^{-2}}$, and noting that $W(x) > \frac{1}{2}U(x) > \frac{1}{2}\mathrm{dist}(x, f^{-1}(0))$, $x \in \mathcal{D}_0 \backslash f^{-1}(0)$, so that $V(\cdot)$ satisfies *ii*) with $\alpha(r) \triangleq (r/2)e^{-4/r^2}$, we obtain the desired smooth Lyapunov function. □

2.5 Stability Theory for Linear Nonnegative Dynamical Systems

In this section, we develop necessary and sufficient conditions for stability of linear nonnegative dynamical systems. First, however, we introduce several definitions and some key results concerning nonnegative functions and matrices [31, 32, 219, 320] that are necessary for developing the main results of this section.

Definition 2.10. Let $T > 0$. A real function $u : [0, T] \to \mathbb{R}^m$ is a *nonnegative* (respectively, *positive*) *function* if $u(t) \geq\geq 0$ (respectively, $u(t) >> 0$) on the interval $[0, T]$.

Definition 2.11. Let $A \in \mathbb{R}^{n \times n}$. A is a *Z-matrix* if $A_{(i,j)} \leq 0$, $i, j = 1, \ldots, n$, $i \neq j$. A is an *M-matrix* (respectively, a *nonsingular M-matrix*) if A is a Z-matrix and $\mathrm{Re}\,\lambda \geq 0$ (respectively, $\mathrm{Re}\,\lambda > 0$) for all $\lambda \in \mathrm{spec}(A)$. A is *essentially nonnegative* if $-A$ is a Z-matrix, that is, $A_{(i,j)} \geq 0$, $i, j = 1, \ldots, n$, $i \neq j$. A is *compartmental* if A is essentially nonnegative and $\sum_{i=1}^{n} A_{(i,j)} \leq 0$, $j = 1, \ldots, n$, or, equivalently, $A^{\mathrm{T}}\mathbf{e} \leq\leq 0$. Finally, A is *nonnegative*[2] (respectively, *positive*) if $A_{(i,j)} \geq 0$ (respectively, $A_{(i,j)} > 0$), $i, j = 1, 2, \ldots, n$.

The following results are needed for developing several stability results for linear nonnegative dynamical systems.

Theorem 2.9. Let $A \in \mathbb{R}^{n \times n}$ be such that $A \geq\geq 0$. Then $\rho(A) \in \mathrm{spec}(A)$ and there exists $x \geq\geq 0$, $x \neq 0$, such that $Ax = \rho(A)x$.

Proof. See [218, p. 503]. □

[2] In this monograph, it is important to distinguish between a nonnegative (respectively, positive) matrix and a nonnegative-definite (respectively, positive-definite) matrix.

Lemma 2.2. Assume $A \in \mathbb{R}^{n\times n}$ is a Z-matrix. Then the following statements are equivalent:

i) A is an M-matrix.

ii) $\operatorname{Re} \lambda \geq 0$, $\lambda \in \operatorname{spec}(A)$.

iii) There exist a scalar $\alpha > 0$ and an $n \times n$ nonnegative matrix $B \geq\geq 0$ such that $\alpha \geq \rho(B)$ and $A = \alpha I - B$.

iv) If $\lambda \in \operatorname{spec}(A)$, then either $\lambda = 0$ or $\operatorname{Re} \lambda > 0$.

Furthermore, the following statements are equivalent:

v) A is a nonsingular M-matrix.

vi) $\det A \neq 0$ and $A^{-1} \geq\geq 0$.

vii) For every $y \in \mathbb{R}^n$, $y \geq\geq 0$, there exists a unique $x \in \mathbb{R}^n$, $x \geq\geq 0$, such that $Ax = y$.

viii) There exists $x \in \mathbb{R}^n$, $x \geq\geq 0$, such that $Ax >> 0$.

ix) There exists $x \in \mathbb{R}^n$, $x >> 0$, such that $Ax >> 0$.

Proof. The equivalence of *i)* and *ii)* is by definition.

To show that *ii)* implies *iii)*, note that since A is a Z-matrix there exist $\alpha > 0$ and $B \geq\geq 0$ such that $A = \alpha I - B$. Next, it follows from Theorem 2.9 that $\rho(B) \in \operatorname{spec}(B)$ so that $\alpha - \rho(B) \in \operatorname{spec}(A)$. Now, it follows from *ii)* that $\alpha \geq \rho(B)$. Next, to show that *iii)* implies *ii)*, let $\lambda \in \operatorname{spec}(A)$ and since $A = \alpha I - B$ there exists $\beta \in \operatorname{spec}(B)$ such that $\lambda = \alpha - \beta$. Now, since $\alpha \geq \rho(B)$, it follows that $\operatorname{Re}\lambda = \alpha - \operatorname{Re}\beta \geq \alpha - \rho(B) \geq 0$.

To show that *ii)* implies *iv)*, note that if $\lambda \in \operatorname{spec}(A)$, then $\operatorname{Re} \lambda \geq 0$. Suppose there exists $\lambda \in \operatorname{spec}(A)$ such that $\operatorname{Re} \lambda = 0$, that is, $\lambda = \jmath\omega$ for some $\omega \in \mathbb{R}$. Since *ii)* implies *iii)*, it follows that there exist a scalar $\alpha > 0$ and an $n \times n$ nonnegative matrix $B \geq\geq 0$ such that $\alpha \geq \rho(B)$ and $A = \alpha I - B$. Hence, there exists $\beta \in \operatorname{spec}(B)$ such that $\lambda = \alpha - \beta$, which implies that $0 = \operatorname{Re}\lambda = \alpha - \operatorname{Re}\beta \geq \alpha - \rho(B) \geq 0$, which further implies that $\alpha = \beta$ or, equivalently, $\lambda = 0$. Hence, either $\operatorname{Re} \lambda > 0$ or $\lambda = 0$. Statement *ii)* follows trivially from *iv)*, establishing the equivalence of *ii)* and *iv)*.

To show that *v)* implies *vi)*, note that, since A is an M-matrix, it follows from *iii)* that there exist $\alpha > 0$ and $B \geq\geq 0$ such that $A = \alpha I - B$

and $\alpha \geq \rho(B)$. Next, since A is nonsingular and $\alpha - \rho(B) \in \operatorname{spec}(A)$, it follows that $\alpha > \rho(B)$ or, equivalently, $\rho(B/\alpha) < 1$. Hence, $I - \frac{1}{\alpha}B$ is invertible and $(I - \frac{1}{\alpha}B)^{-1} = \sum_{k=0}^{\infty}(\frac{1}{\alpha}B)^k \geq\geq 0$. The result now follows by noting that $A^{-1} = \frac{1}{\alpha}(I - \frac{1}{\alpha}B)^{-1}$.

To show that *vi*) implies *vii*), let $y \geq\geq 0$ and note that, since A is nonsingular, there exists a unique x such that $Ax = y$. Now, since $A^{-1} \geq\geq 0$, it follows that $x = A^{-1}y \geq\geq 0$.

To show that *vii*) implies *viii*), let $y = \mathbf{e}$ and note that it follows from *vii*) that there exists $x \geq\geq 0$ such that $Ax = y = \mathbf{e} >> 0$.

To show that *viii*) implies *ix*), let $\alpha > 0$ and $B \geq\geq 0$ be such that $A = \alpha I - B$. Now, letting $x \geq\geq 0$ be such that $0 << Ax = \alpha x - Bx$, it follows that $\alpha x >> Bx \geq\geq 0$, establishing the result.

To show that *ix*) implies *v*), let $x >> 0$ be such that $Ax >> 0$ and consider the linear dynamical system

$$\dot{y}(t) = -A^{\mathrm{T}}y(t), \quad y(0) = y_0 \geq\geq 0, \quad t \geq 0. \tag{2.33}$$

Note that since A is a Z-matrix it follows that $f(y) = -A^{\mathrm{T}}y$ is essentially nonnegative. Hence, with $V(y) = x^{\mathrm{T}}y$ it follows from Theorem 2.1 that $y(t) = e^{-A^{\mathrm{T}}t}y_0 \to 0$ as $t \to \infty$ for every $y_0 \geq\geq 0$, which implies that $-A^{\mathrm{T}}$ is Hurwitz. Now, the result is immediate by noting that $-A^{\mathrm{T}}$ is Hurwitz if and only if $\operatorname{Re}\lambda > 0$ for all $\lambda \in \operatorname{spec}(A)$ (see Definition 2.12). □

Note that if $f(x) = Ax$, where $A \in \mathbb{R}^{n\times n}$, then f is essentially nonnegative if and only if A is essentially nonnegative. To address linear nonnegative dynamical systems, consider (2.1) with $f(x) = Ax$ so that

$$\dot{x}(t) = Ax(t), \qquad x(0) = x_0, \qquad t \geq 0, \tag{2.34}$$

where $x(t) \in \mathbb{R}^n$, $t \geq 0$, and $A \in \mathbb{R}^{n\times n}$. The solution to (2.34) is standard and is given by $x(t) = e^{At}x(0)$, $t \geq 0$. The following proposition shows that A is essentially nonnegative if and only if the state transition matrix e^{At} is nonnegative on $[0, \infty)$.

Proposition 2.4. Let $A \in \mathbb{R}^{n\times n}$. Then A is essentially nonnegative if and only if e^{At} is nonnegative for all $t \geq 0$. Furthermore, if A is essentially nonnegative and $x_0 \geq\geq 0$, then $x(t) \geq\geq 0$, $t \geq 0$, where $x(t)$, $t \geq 0$, denotes the solution to (2.34).

Proof. The proof is a direct consequence of Proposition 2.1 with $f(x) = Ax$. The proof can also be shown using matrix mathematics [37]. To prove necessity, note that, since A is essentially nonnegative, it follows that

$A_\alpha \triangleq A + \alpha I$ is nonnegative, where $\alpha \triangleq -\min\{A_{(1,1)}, \ldots, A_{(n,n)}\}$. Hence, $e^{A_\alpha t} = e^{(A+\alpha I)t} \geq\geq 0$, $t \geq 0$, and thus $e^{At} = e^{-\alpha t} e^{A_\alpha t} \geq\geq 0$, $t \geq 0$.

Conversely, suppose $e^{At} \geq\geq 0$, $t \geq 0$, and assume, *ad absurdum*, there exist i, j such that $i \neq j$ and $A_{(i,j)} < 0$. Now, since $e^{At} = \sum_{i=1}^{\infty} (k!)^{-1} A^k t^k$, it follows that

$$[e^{At}]_{(i,j)} = I_{(i,j)} + tA_{(i,j)} + \mathcal{O}(t^2),$$

where $\mathcal{O}(t)/t \to 0$ as $t \to 0$. Thus, as $t \to 0$ and $i \neq j$, it follows that $[e^{At}]_{(i,j)} < 0$ for some t sufficiently small, which leads to a contradiction. Hence, A is essentially nonnegative.

Finally, if A is essentially nonnegative and $x_0 \geq\geq 0$, then $x(t) = e^{At} x_0 \geq\geq 0$, $t \geq 0$, is immediate. □

Definition 2.12. Let $A \in \mathbb{R}^{n \times n}$. Then:

i) A is *Lyapunov stable* if $\mathrm{spec}(A) \subset \{s \in \mathbb{C} : \mathrm{Re}\, s \leq 0\}$ and, if $\lambda \in \mathrm{spec}(A)$ and $\mathrm{Re}\,\lambda = 0$, then λ is semisimple.

ii) A is *semistable* if $\mathrm{spec}(A) \subset \{s \in \mathbb{C} : \mathrm{Re}\, s < 0\} \cup \{0\}$ and, if $0 \in \mathrm{spec}(A)$, then 0 is semisimple.

iii) A is *asymptotically stable* or *Hurwitz* if $\mathrm{spec}(A) \subset \{s \in \mathbb{C} : \mathrm{Re}\, s < 0\}$.

The following proposition concerning the Lyapunov stability, semistability, and asymptotic stability of (2.34) is immediate. This result holds whether or not A is an essentially nonnegative matrix.

Proposition 2.5 ([33]). Let $A \in \mathbb{R}^{n \times n}$ and consider the linear dynamical system (2.34). Then the following statements are equivalent:

i) $x_{\mathrm{e}} = 0$ is a Lyapunov stable equilibrium of (2.34).

ii) At least one equilibrium of (2.34) is Lyapunov stable.

iii) Every equilibrium of (2.34) is Lyapunov stable.

iv) A is Lyapunov stable.

v) For every initial condition $x(0) \in \mathbb{R}^n$, $x(t)$ is bounded for all $t \geq 0$.

vi) $\|e^{At}\|$ is bounded for all $t \geq 0$, where $\|\cdot\|$ is a matrix norm on $\mathbb{R}^{n \times n}$.

vii) For every initial condition $x(0) \in \mathbb{R}^n$, $e^{At} x(0)$ is bounded for all $t \geq 0$.

The following statements are equivalent:

viii) A is semistable.

ix) $\lim_{t\to\infty} e^{At}$ exists. In fact, $\lim_{t\to\infty} e^{At} = I - AA^{\#}$.

x) For every initial condition $x(0) \in \mathbb{R}^n$, $\lim_{t\to\infty} x(t)$ exists.

The following statements are equivalent:

xi) $x_{\mathrm{e}} = 0$ is an asymptotically stable equilibrium of (2.34).

xii) A is asymptotically stable.

xiii) $\alpha(A) < 0$.

xiv) For every initial condition $x(0) \in \mathbb{R}^n$, $\lim_{t\to\infty} x(t) = 0$.

xv) For every initial condition $x(0) \in \mathbb{R}^n$, $\lim_{t\to\infty} e^{At}x(0) = 0$.

xvi) $\lim_{t\to\infty} e^{At} = 0$.

The following theorem gives several properties of a nonnegative dynamical system when a Lyapunov-like equation is satisfied for (2.34). Note that it follows from Proposition 2.5 that if A is asymptotically stable, then $\mathcal{N}(A) = \{0\}$.

Theorem 2.10. Let $A \in \mathbb{R}^{n\times n}$ be essentially nonnegative. If there exist vectors $p, r \in \mathbb{R}^n$ such that $p >> 0$ and $r \geq\geq 0$ satisfy

$$0 = A^{\mathrm{T}}p + r, \tag{2.35}$$

then the following statements hold:

i) $-A$ is an M-matrix.

ii) If $\lambda \in \mathrm{spec}(A)$, then either $\mathrm{Re}\,\lambda < 0$ or $\lambda = 0$.

iii) A is semistable and $\lim_{t\to\infty} e^{At} = I - AA^{\#} \geq\geq 0$.

iv) $\mathcal{R}(A) = \mathcal{N}(I - AA^{\#})$ and $\mathcal{N}(A) = \mathcal{R}(I - AA^{\#})$.

v) $\int_0^t e^{A\sigma}\mathrm{d}s = A^{\#}(e^{At} - I) + (I - AA^{\#})t$, $t \geq 0$.

vi) A is nonsingular if and only if $-A$ is a nonsingular M-matrix.

vii) If A is nonsingular, then A is asymptotically stable and $A^{-1} \leq\leq 0$.

Proof. *i*) Consider the linear dynamical system

$$\dot{x}(t) = Ax(t), \quad x(0) = x_0 \geq\geq 0, \quad t \geq 0. \tag{2.36}$$

Note that, since A is an essentially nonnegative matrix, it follows that $f(x) = Ax$ is essentially nonnegative. Hence, with $V(x) = p^{\mathrm{T}}x$ it follows from Theorem 2.1 that the zero solution to (2.36) is Lyapunov stable, and hence, $\operatorname{Re}\lambda \leq 0$ for all $\lambda \in \operatorname{spec}(A)$. Now, *i*) follows by noting A is essentially nonnegative if and only if $-A$ is a Z-matrix.

ii) The proof is a consequence of *i*)–*iv*) of Lemma 2.2.

iii) It follows from *i*) and *ii*) that A is Lyapunov stable and $\operatorname{Re}\lambda < 0$ or $\lambda = 0$ for all $\lambda \in \operatorname{spec}(A)$. Hence, the eigenvalue $\lambda = 0$ (if it exists) is semisimple, and it follows from the Jordan decomposition that there exist invertible matrices $J \in \mathbb{R}^{r\times r}$, where $r = \operatorname{rank} A$, and $S \in \mathbb{R}^{n\times n}$ such that

$$A = S\begin{bmatrix} J & 0 \\ 0 & 0 \end{bmatrix}S^{-1},$$

and J is Hurwitz. Hence, it follows that

$$\begin{aligned} \lim_{t\to\infty} e^{At} &= \lim_{t\to\infty} S\begin{bmatrix} e^{Jt} & 0 \\ 0 & I_{n-r} \end{bmatrix}S^{-1} \\ &= S\begin{bmatrix} 0 & 0 \\ 0 & I_{n-r} \end{bmatrix}S^{-1} \\ &= I_n - S\begin{bmatrix} J & 0 \\ 0 & 0 \end{bmatrix}S^{-1}S\begin{bmatrix} J^{-1} & 0 \\ 0 & 0 \end{bmatrix}S^{-1} \\ &= I_n - AA^{\#}. \end{aligned}$$

Next, since A is essentially nonnegative, it follows from Proposition 2.4 that $e^{At} \geq\geq 0$, $t \geq 0$, which implies that $I - AA^{\#} \geq\geq 0$.

iv) Let $x \in \mathcal{R}(A)$, that is, there exists $y \in \mathbb{R}^n$ such that $x = Ay$. Now, $(I - AA^{\#})x = x - AA^{\#}Ay = x - Ay = 0$, which implies that $\mathcal{R}(A) \subseteq \mathcal{N}(I - AA^{\#})$. Conversely, let $x \in \mathcal{N}(I - AA^{\#})$. Hence, $(I - AA^{\#})x = 0$ or, equivalently, $x = AA^{\#}x$, which implies that $x \in \mathcal{R}(A)$, and hence, proves $\mathcal{R}(A) = \mathcal{N}(I - AA^{\#})$. The equality $\mathcal{N}(A) = \mathcal{R}(I - AA^{\#})$ can be proved in an analogous manner.

v) Note that $A = S\begin{bmatrix} J & 0 \\ 0 & 0 \end{bmatrix}S^{-1}$, and hence,

$$\int_0^t e^{A\sigma}\mathrm{d}\sigma = \int_0^t S\begin{bmatrix} e^{J\sigma} & 0 \\ 0 & I_{n-r} \end{bmatrix}S^{-1}\mathrm{d}\sigma$$

$$
\begin{aligned}
&= S\begin{bmatrix} \int_0^t e^{J\sigma}\mathrm{d}\sigma & 0 \\ 0 & \int_0^t I_{n-r}\mathrm{d}\sigma \end{bmatrix} S^{-1} \\
&= S\begin{bmatrix} J^{-1}(e^{Jt}-I) & 0 \\ 0 & I_{n-r}t \end{bmatrix} S^{-1} \\
&= S\begin{bmatrix} J^{-1} & 0 \\ 0 & 0 \end{bmatrix} S^{-1} S \begin{bmatrix} (e^{Jt}-I_r) & 0 \\ 0 & 0 \end{bmatrix} S^{-1} \\
&\quad + S\begin{bmatrix} 0 & 0 \\ 0 & I_{n-r}t \end{bmatrix} S^{-1} \\
&= A^{\#}(e^{At}-I_n) + (I_n - AA^{\#})t, \quad t \geq 0.
\end{aligned}
$$

vi) The result follows from *i)*.

vii) Asymptotic stability of A is a direct consequence of *ii)*. $A^{-1} \leq\leq 0$ follows from *vi)* of Lemma 2.2. □

It follows from Proposition 2.4 and *iii)* of Theorem 2.10 that

$$\lim_{t\to\infty} x(t) = (I - AA^{\#})x_0 \geq\geq 0.$$

Hence, the set of all equilibria of a semistable linear nonnegative dynamical system lies in $\mathcal{N}(A) = \mathcal{R}(I - AA^{\#})$.

Next, motivated by the fact that for a thermodynamic system the total energy in the system can serve as a Lyapunov function, we give necessary and sufficient conditions for Lyapunov stability, semistability, and asymptotic stability for linear nonnegative dynamical systems using *linear* Lyapunov functions.

Theorem 2.11. Consider the linear dynamical system given by (2.34) where $A \in \mathbb{R}^{n\times n}$ is essentially nonnegative. Then the following statements hold:

i) A is Lyapunov stable if and only if A is semistable.

ii) If there exist vectors $p, r \in \mathbb{R}^n$ such that $p >> 0$ and $r \geq\geq 0$ satisfy

$$0 = A^{\mathrm{T}}p + r, \tag{2.37}$$

then A is semistable (and hence Lyapunov stable).

iii) If A is semistable, then there exist vectors $p, r \in \mathbb{R}^n$ such that $p \geq\geq 0$ and $r \geq\geq 0$ satisfy (2.37).

iv) If there exist vectors $p, r \in \mathbb{R}^n$ such that $p \geq\geq 0$ and $r \geq\geq 0$ satisfy (2.37) and (A, r^{T}) is observable, then $p >> 0$ and (2.34) is asymptotically stable.

Furthermore, the following statements are equivalent:

v) A is asymptotically stable.

vi) There exist vectors $p, r \in \mathbb{R}^n$ such that $p >> 0$ and $r >> 0$ satisfy (2.37).

vii) There exist vectors $p, r \in \mathbb{R}^n$ such that $p \geq\geq 0$ and $r >> 0$ satisfy (2.37).

viii) For every $r \in \mathbb{R}^n$ such that $r >> 0$, there exists $p \in \mathbb{R}^n$ such that $p >> 0$ satisfies (2.37).

Proof. *i*) If A is semistable, then A is Lyapunov stable by definition. Conversely, suppose A is Lyapunov stable and essentially nonnegative. Then it follows from *ii*) of Lemma 2.2 that $-A$ is an M-matrix. Now, it follows from *iv*) of Lemma 2.2 that the real part of each nonzero $\lambda \in \operatorname{spec}(A)$ is negative, and hence, $\operatorname{Re}\lambda < 0$ or $\lambda = 0$. This proves the equivalence between Lyapunov stability and semistability.

ii) The proof is a direct consequence of *iv*) of Theorem 2.10. Alternatively, consider the linear Lyapunov function candidate $V(x) = p^{\mathrm{T}}x$. Note that $V(0) = 0$ and $V(x) > 0$, $x \in \overline{\mathbb{R}}^n_+$, $x \neq 0$. Now, computing the Lyapunov derivative yields

$$\dot{V}(x) \triangleq V'(x)Ax = p^{\mathrm{T}}Ax = -r^{\mathrm{T}}x \leq 0, \qquad x \in \overline{\mathbb{R}}^n_+,$$

establishing Lyapunov stability. Semistability now follows from *i*).

iii) If A is semistable, it follows as in the proof of *i*) that $-A^{\mathrm{T}}$ is an M-matrix. Hence, it follows from *ii*) of Lemma 2.2 that there exist a scalar $\alpha > 0$ and a nonnegative matrix $B \geq\geq 0$ such that $\alpha \geq \rho(B)$ and $A^{\mathrm{T}} = B - \alpha I$. Now, since $B \geq\geq 0$, it follows from Theorem 2.9 that $\rho(B) \in \operatorname{spec}(B)$, and hence, there exists $p \geq\geq 0$ such that $Bp = \rho(B)p$. Thus, $A^{\mathrm{T}}p = Bp - \alpha p = (\rho(B) - \alpha)p \leq\leq 0$, which proves that there exist $p \geq\geq 0$ and $r \geq\geq 0$ such that (2.37) holds.

iv) Assume there exist $p \geq\geq 0$ and $r \geq\geq 0$ such that (2.37) holds and suppose (A, r^{T}) is observable. Now, consider the function $V(x) = p^{\mathrm{T}}x$, $x \in \overline{\mathbb{R}}^n_+$, and note that, since $V(x) \geq 0$, $x \in \overline{\mathbb{R}}^n_+$, and $\dot{V}(x) = p^{\mathrm{T}}Ax = -r^{\mathrm{T}}x \leq 0$, it follows that if $x(0) \in \mathcal{P} \triangleq \{x \in \overline{\mathbb{R}}^n_+ : p^{\mathrm{T}}x = 0\}$, then $V(x(t)) = 0$, $t \geq 0$, which implies that $\frac{\mathrm{d}V(x(t))}{\mathrm{d}t} = 0$. Specifically,

$$\left.\frac{\mathrm{d}V(x(t))}{\mathrm{d}t}\right|_{t=0} = p^{\mathrm{T}}Ax(0) = 0.$$

Hence, if $\hat{x} \in \mathcal{P}$, then $\dot{V}(\hat{x}) = p^{\mathrm{T}}A\hat{x} = -r^{\mathrm{T}}\hat{x} = 0$. Thus, if $\hat{x} \in \mathcal{P}$ then $A\hat{x} \in \mathcal{P}$ and $\hat{x} \in \mathcal{Q} \triangleq \{x \in \overline{\mathbb{R}}_+^n : r^{\mathrm{T}}x = 0\}$. Now, since $A\hat{x} \in \mathcal{P}$, it follows that $A^2\hat{x} \in \mathcal{P}$ and $A\hat{x} \in \mathcal{Q}$. Repeating these arguments yields $A^k\hat{x} \in \mathcal{Q}$, $k = 0, 1, \ldots, n$, or, equivalently, $r^{\mathrm{T}}A^k\hat{x} = 0$, $k = 1, 2, \ldots, n$. Now, since (A, r^{T}) is observable, it follows that $\hat{x} = 0$ and $\mathcal{P} = \{0\}$, which implies that $p >> 0$. Asymptotic stability of (2.34) now follows as a direct consequence of the Krasovskii-LaSalle invariant set theorem with $V(x) = p^{\mathrm{T}}x$ and using the fact that (A, r^{T}) is observable.

To show the equivalence among *v*)–*viii*), first suppose there exist $p \geq\geq 0$ and $r >> 0$ such that (2.37) holds. Now, there exists sufficiently small $\varepsilon > 0$ such that $A^{\mathrm{T}}(p + \varepsilon\mathbf{e}) << 0$ and $p + \varepsilon\mathbf{e} >> 0$, which proves that *vii*) implies *vi*). Since *vi*) implies *vii*), it trivially follows that *vi*) and *vii*) are equivalent. Next, suppose *vi*) holds, that is, there exist $p >> 0$ and $r >> 0$ such that (2.37) holds, and consider the Lyapunov function candidate $V(x) = p^{\mathrm{T}}x$, $x \in \overline{\mathbb{R}}_+^n$. Computing the Lyapunov derivative yields $\dot{V}(x) = p^{\mathrm{T}}Ax = -r^{\mathrm{T}}x < 0$, $x \neq 0$, and hence, it follows that (2.34) is asymptotically stable. Thus, *vi*) implies *v*). Next, suppose (2.34) is asymptotically stable. Hence, $-A^{-\mathrm{T}} \geq\geq 0$ and thus for every $r \in \mathbb{R}_+^n$, $p \triangleq -A^{-\mathrm{T}}r \geq\geq 0$ satisfies (2.37), which proves that *v*) implies *vii*).

Finally, suppose (2.34) is asymptotically stable. Now, as in the proof given above, for every $r \in \mathbb{R}_+^n$, there exists $p \in \overline{\mathbb{R}}_+^n$ such that (2.37) holds. Next, suppose, *ad absurdum*, there exists $x \in \overline{\mathbb{R}}_+^n$, $x \neq 0$, such that $x^{\mathrm{T}}p = 0$, that is, there exists at least one $i \in \{1, 2, \ldots, n\}$ such that $p_i = 0$. Hence, $-x^{\mathrm{T}}A^{-\mathrm{T}}r = 0$. However, since $-A^{\mathrm{T}} \geq\geq 0$, it follows that $-A^{-1}x \geq\geq 0$ and, since $r >> 0$, it follows that $-A^{-1}x = 0$, which implies that $x = 0$, yielding a contradiction. Hence, for every $r \in \mathbb{R}_+^n$, there exists $p \in \mathbb{R}_+^n$ such that (2.37) holds, which proves that *v*) implies *viii*). Since *viii*) implies *vi*) trivially, the equivalence of *v*)–*viii*) is established. □

2.6 Lyapunov Analysis for Continuum Dynamical Systems Defined by Semigroups

In this section, we discuss dynamical systems defined on Banach spaces[3] and present some of the key results on invariant set stability theorems for infinite-dimensional dynamical systems. For infinite-dimensional systems, Lyapunov stability theorems are similar to the Lyapunov theorems just presented with the required conditions verified on the Banach space $\mathcal{B}$ [46]. However, since

[3]A *Banach space* is a vector space of bounded functions defined on a compact set endowed with a Chebyshev norm; that is, the norm placed on bounded functions on a set $\mathcal{D}$ that assigns to each function the supremum of the moduli of the values of the function on $\mathcal{D}$. Equivalently, a Banach space is a complete normed space wherein every Cauchy sequence converges to an element in the space.

norms are not equivalent in infinite-dimensional spaces, Lyapunov stability, semistability, and asymptotic stability are defined with respect to a specific norm $\|\cdot\|_{\mathcal{B}}$ defined on $\mathcal{B}$.

To present the main results of the next several sections, we require some additional notation and definitions. Specifically, we write $\mathrm{ran}(\mathcal{A})$ and $\ker(\mathcal{A})$ for the range and the kernel of the closed, densely defined operator $\mathcal{D}(\mathcal{A}) \subset \mathcal{B} \to \mathcal{B}$. A subset $\mathfrak{D}$ of $\mathbb{R}^n$ is referred to as the spatial domain if it consists of countably many n-tuples $i = (i_1, \ldots, i_n)$. Consider a continuous-time spatially distributed system $\mathcal{G}$ defined on the semi-infinite interval $[0, \infty)$ over a spatial domain $\mathfrak{D}$ given by

$$\frac{\mathrm{d}}{\mathrm{d}t}\psi(t) = (\mathcal{A}\psi)(t), \tag{2.38}$$

where the state $\psi(t)$ is an element of a real separable Banach space $\mathcal{B}$ and the operator $\mathcal{A}$ is *spatially distributed* over $\mathcal{D}(\mathcal{A}) \subset \mathcal{B} \to \mathcal{B}$.

The following definitions are needed.

Definition 2.13 ([104,474]). A one-parameter family of bounded linear operators $\mathcal{T}(t) : \mathcal{B} \to \mathcal{B}$, $t \in [0, \infty)$, is said to be a C^0*-semigroup* (or a *linear semigroup*) if *i)* $\mathcal{T}(0) = I$, *ii)* $\mathcal{T}(t+s) = \mathcal{T}(t)\mathcal{T}(s)$ for every $t, s \in \overline{\mathbb{R}}_+$, and *iii)* $\lim_{t\to 0^+} \mathcal{T}(t)x = x$ for all $x \in \mathcal{B}$.

Definition 2.14 ([98, 323]). Assume that $\mathfrak{D}$ is a subset of a Banach space $\mathcal{B}$. A one-parameter family of (nonlinear) operators $\mathcal{T}(t) : \mathfrak{D} \to \mathfrak{D}$, $t \in [0, \infty)$, is said to be a *nonlinear semigroup* defined on $\mathfrak{D}$ if *i)* $\mathcal{T}(0)(x) = x$ for every $x \in \mathfrak{D}$, *ii)* $\mathcal{T}(t+s)(x) = \mathcal{T}(t)\mathcal{T}(s)(x)$ for every $x \in \mathfrak{D}$ and $t, s \in [0, \infty)$, and *iii)* $\mathcal{T}(t)x$ is continuous in (t, x) on $[0, \infty) \times \mathfrak{D}$.

Recall that a dynamical system $\mathcal{G}$ determined by a C^0-semigroup $\mathcal{T}(t)$ is defined as the set $\mathcal{G} = \{y = s(\cdot, x, t_0) : s(t, x, t_0) \triangleq \mathcal{T}(t - t_0)x, t_0 \in [0, \infty), t \geq t_0, x \in \mathcal{B}\}$ [323]. Likewise, a dynamical system $\mathcal{G}$ determined by a nonlinear semigroup $\mathcal{T}(t)$ is defined as the set $\mathcal{G} = \{y = s(\cdot, x, t_0) : s(t, x, t_0) \triangleq \mathcal{T}(t - t_0)(x), t_0 \in [0, \infty), t \geq t_0, x \in \mathfrak{D}\}$. An *equilibrium point* of $\mathcal{G}$ is a point $x_\mathrm{e} \in \mathcal{D}$ such that $s(t, x_\mathrm{e}) = s(0, x_\mathrm{e})$ for all $t \geq 0$. Let $\mathcal{E}$ denote the set of equilibria of $\mathcal{G}$. Here, we assume that $\mathcal{E} \neq \emptyset$.

The following definition introduces several stability definitions for the equilibrium solution $\psi(t) \equiv x_\mathrm{e} \in \mathcal{B}$ of the nonlinear operator system $\mathcal{G}$ given by (2.38).

Definition 2.15. Let $\mathcal{G}$ be a dynamical system on a Banach space $\mathcal{B}$ with norm $\|\cdot\|_{\mathcal{B}}$ and let $\mathcal{D}$ be a positively invariant set with respect to $\mathcal{G}$. An equilibrium point $x_\mathrm{e} \in \mathcal{D}$ of $\mathcal{G}$ is *Lyapunov stable* if for every relatively open subset $\mathcal{N}_\varepsilon$ of $\mathcal{D}$ containing x_e, there exists a relatively open subset $\mathcal{N}_\delta$ of $\mathcal{D}$ containing x_e such that $s_t(\mathcal{N}_\delta) \subseteq \mathcal{N}_\varepsilon$ for all $t \geq 0$. An equilibrium point $x_e \in \mathcal{D}$ of $\mathcal{G}$ is *semistable* if it is Lyapunov stable and there exists a relatively open subset $\mathcal{U}$ of $\mathcal{D}$ containing x_e such that for all initial conditions in $\mathcal{U}$, the trajectory $s(\cdot,\cdot)$ of $\mathcal{G}$ converges to a Lyapunov stable equilibrium point, that is, for every $z \in \mathcal{U}$, there exists $y \in \mathcal{D}$ such that $\lim_{t\to\infty} s(t,z) = y$, which is a Lyapunov stable equilibrium point of $\mathcal{G}$. An equilibrium point $x_e \in \mathcal{D}$ of $\mathcal{G}$ is *asymptotically stable* if it is Lyapunov stable and there exists a relatively open subset $\mathcal{U}$ of $\mathcal{D}$ containing x_e such that $\lim_{t\to\infty} s(t,z) = x_e$ for all $z \in \mathcal{U}$. Finally, $\mathcal{G}$ is *semistable* if every equilibrium point of $\mathcal{G}$ is semistable.

Definition 2.16. Let $\mathcal{B}$ be a Banach space with norm $\|\cdot\|_{\mathcal{B}}$. A *dynamical system on* $\mathcal{B}$ is the triple $(\mathcal{B}, [t_0,\infty), s)$, where $s : [t_0,\infty) \times \mathcal{B} \to \mathcal{B}$ is such that the following axioms hold:

i) (Continuity): $s(\cdot,\cdot)$ is jointly continuous.

ii) (Consistency): $s(t_0, x_0) = x_0$ for all $t_0 \in \mathbb{R}$ and $x_0 \in \mathcal{B}$.

iii) (Semigroup property): $s(t+\tau, x_0) = s(\tau, s(t, x_0))$ for all $x_0 \in \mathcal{B}$ and $t, \tau \in [t_0, \infty)$.

The above definition of a dynamical system can be generalized to include external system disturbances $u(\cdot) \in \mathcal{U}$, where $\mathcal{U}$ is an input space consisting of bounded continuous $U \subseteq \mathbb{R}^m$-valued functions on $[t_0,\infty)$. In this case, the dynamical system on $\mathcal{B}$ is defined by the pentuple ($\mathcal{B}$, $\mathcal{U}$, $[t_0,\infty)$, s, h), where $s : [t_0,\infty) \times \mathcal{B} \times \mathcal{U} \to \mathcal{B}$ and $h : \mathcal{B} \times U \to Y$ defines a memoryless *read-out map* $y(t) = h(s(t, x_0, u), u(t))$ for all $x_0 \in \mathcal{B}$, $u(\cdot) \in \mathcal{U}$, and $t \in [t_0,\infty)$. Here, $y(\cdot) \in \mathcal{Y}$, where $\mathcal{Y}$ denotes an output space and $y(t)$ belongs to the fixed set $Y \subseteq \mathbb{R}^l$ for all $t \geq t_0$. In this case, to ensure causality of the dynamical system one needs to invoke an additional axiom (determinism axiom), which ensures that the state, and hence the output, of the dynamical system before some time τ are not influenced by the values of the output after time τ. For further details see Definition 2.20.

Henceforth, we denote the dynamical system $(\mathcal{B}, [t_0,\infty), s)$ by $\mathcal{G}$, and we refer to the map $s(\cdot,\cdot)$ as the *flow* or *trajectory* of $\mathcal{G}$ corresponding to $x_0 \in \mathcal{B}$; and for a given $s(t, x_0)$, $t \geq t_0$, we refer to $x_0 \in \mathcal{B}$ as an *initial condition* of $\mathcal{G}$. Given $t \in \mathbb{R}$, we denote the map $s(t,\cdot) : \mathcal{B} \to \mathcal{B}$ by $s_t(x_0)$. Hence, for a fixed $t \in \mathbb{R}$ the set of mappings defined by $s_t(x_0) = s(t, x_0)$ for every $x_0 \in \mathcal{B}$ gives the *flow* of $\mathcal{G}$. In particular, if $\mathcal{B}_0$ is a collection of initial conditions such that $\mathcal{B}_0 \subset \mathcal{B}$, then the flow $s_t : \mathcal{B}_0 \to \mathcal{B}$ is the motion of all points $x_0 \in \mathcal{B}_0$ or, equivalently, the image of $\mathcal{B}_0 \subset \mathcal{B}$ under the flow s_t, that is, $s_t(\mathcal{B}_0) \subset \mathcal{B}$, where $s_t(\mathcal{B}_0) \triangleq \{y : y = s_t(x_0) \text{ for some } x_0 \in \mathcal{B}_0\}$.

Alternatively, if the initial condition $x_0 \in \mathcal{B}$ is fixed and we let $[t_0, t_1] \subset \mathbb{R}$, then the mapping $s(\cdot, x_0) : [t_0, t_1] \to \mathcal{B}$ defines the *solution curve* or *trajectory* of the dynamical system $\mathcal{G}$. Hence, the mapping $s(\cdot, x_0)$ generates a graph in $[t_0, t_1] \times \mathcal{B}$ identifying the trajectory corresponding to the motion along a curve through the point x_0 in a subset $\mathcal{B}$ of the state space. Given $x \in \mathcal{B}$, we denote the map $s(\cdot, x) : \mathbb{R} \to \mathcal{B}$ by $s^x(t)$. Finally, we define a *positive orbit* through the point $x_0 \in \mathcal{B}$ as the motion along the curve

$$\mathcal{O}^+_{x_0} \triangleq \{x \in \mathcal{B} : \ x = s(t, x_0), \ t \geq t_0\}. \tag{2.39}$$

Definition 2.17. Let $\mathcal{G}$ be a dynamical system defined on $\mathcal{B}$. A point $p \in \mathcal{B}$ is a *positive limit point* of the trajectory $s(\cdot, x)$ if there exists a monotonic sequence $\{t_n\}_{n=0}^\infty$ of nonnegative numbers, with $t_n \to \infty$ as $n \to \infty$, such that $s(t_n, x) \to p$ as $n \to \infty$. The set of all positive limit points of $s(t, x)$, $t \geq t_0$, is the *positive limit set* $\omega(x)$ of $\mathcal{G}$.

In the mathematical literature, the positive limit set is often referred to as the ω*-limit set.* Note that if $p \in \mathcal{B}$ is a positive limit point of the trajectory $s(\cdot, x)$, then for all $\varepsilon > 0$ and finite time $T > 0$ there exists $t > T$ such that $\|s(t, x) - p\|_\mathcal{B} < \varepsilon$. This follows from the fact that $\|s(t, x) - p\|_\mathcal{B} < \varepsilon$ for all $\varepsilon > 0$, and some $t > T > 0$ is equivalent to the existence of a sequence $\{t_n\}_{n=0}^\infty$, with $t_n \to \infty$ as $n \to \infty$, such that $s(t_n, x) \to p$ as $n \to \infty$.

Definition 2.18. A set $\mathcal{M} \subset \mathcal{B}$ is a *positively invariant set* with respect to the dynamical system $\mathcal{G}$ if $s_t(\mathcal{M}) \subseteq \mathcal{M}$ for all $t \geq t_0$, where $s_t(\mathcal{M}) \triangleq \{s_t(x) : \ x \in \mathcal{M}\}$ and $s_t(x) \triangleq s(t, x)$, $x \in \mathcal{B}$, $t \geq t_0$. A set $\mathcal{M} \subseteq \mathcal{B}$ is an *invariant set* with respect to the dynamical system $\mathcal{G}$ if $s_t(\mathcal{M}) = \mathcal{M}$ for all $t \in [t_0, \infty)$.

Next, we state a key proposition involving positive limit sets for the infinite-dimensional dynamical system $\mathcal{G}$.

Proposition 2.6 ([201]). Let $\mathcal{G}$ be a dynamical system defined on $\mathcal{B}$ and suppose that the positive orbit $\mathcal{O}^+_x$ through x of $\mathcal{G}$ belongs to a compact[4] subset of $\mathcal{B}$. Then the positive limit set $\omega(x)$ of $\mathcal{O}^+_x$ is a nonempty, compact, connected invariant set.

The following result presents an extension of the Krasovskii-LaSalle invariant set theorem (Theorem 2.3) to infinite-dimensional dynamical systems. This result holds for *undisturbed* dynamical systems (i.e., $u(t) \equiv 0$), as well as for disturbed dynamical systems wherein the input space consists of one constant element only, that is, $u(t) \equiv u^*$. For the statement of this

[4]If $\mathcal{S} \subset \mathcal{B}$, where $\mathcal{B}$ is a Banach space, then $\mathcal{S}$ is *compact* if and only if every open cover of $\mathcal{S}$ contains a finite subcollection of open sets that also covers $\mathcal{S}$.

result, define

$$\dot{V}(x) \triangleq \lim_{h \to 0^+} \frac{1}{h}[V(s(t_0 + h, x)) - V(x)], \quad x \in \mathcal{B}, \tag{2.40}$$

for a given continuous function $V : \mathcal{B} \to \mathbb{R}$ and for every $x \in \mathcal{B}$ such that the limit in (2.40) exists.

Theorem 2.12 ([201]). Consider a dynamical system $\mathcal{G}$ defined on a Banach space $\mathcal{B}$. Let $\mathcal{B}_c \subset \mathcal{B}$ be a closed set, and assume there exists a continuous function $V : \mathcal{B}_c \to \mathbb{R}$ such that $\dot{V}(x) \leq 0$, $x \in \mathcal{B}_c$. Furthermore, let $\mathcal{R} \triangleq \{x \in \mathcal{B}_c : \dot{V}(x) = 0\}$, and let $\mathcal{M}$ denote the largest invariant set (with respect to the dynamical system $\mathcal{G}$) contained in $\mathcal{R}$. Then, for every $x_0 \in \mathcal{B}_c$ such that $\mathcal{O}^+_{x_0} \subset \mathcal{B}_c$ and $\mathcal{O}^+_{x_0}$ is contained in a compact subset of $\mathcal{B}$, $s(t, x_0) \to \mathcal{M}$ as $t \to \infty$.

In order to apply Theorem 2.12, one needs to show that the positive orbit $\mathcal{O}^+_{x_0}$ of $\mathcal{G}$ is contained in a compact subset of $\mathcal{B}$. Even though for finite-dimensional systems this is a direct consequence of boundedness of solutions, for infinite-dimensional systems local boundedness of an orbit of $\mathcal{G}$ does not ensure that the orbit belongs to a compact subset of $\mathcal{B}$. In light of this, we have the following result. For the statement of this result, let $\mathcal{B}$ and $\mathcal{C}$ be Banach spaces and recall that $\mathcal{B}$ is *compactly embedded* in $\mathcal{C}$ if $\mathcal{B} \subset \mathcal{C}$ and a unit ball in $\mathcal{B}$ belongs to a compact subset in $\mathcal{C}$.

Theorem 2.13 ([201]). Let $\mathcal{B}$ and $\mathcal{C}$ be Banach spaces such that $\mathcal{B}$ is compactly embedded in $\mathcal{C}$, and let $\mathcal{G}$ be a dynamical system defined on $\mathcal{B}$ and $\mathcal{C}$. Assume there exist continuous functions $V_{\mathcal{B}} : \mathcal{B} \to \mathbb{R}$ and $V_{\mathcal{C}} : \mathcal{C} \to \mathbb{R}$ such that $\dot{V}_{\mathcal{B}}(x) \leq 0$, $x \in \mathcal{B}_c$, and $\dot{V}_{\mathcal{C}}(x) \leq 0$, $x \in \mathcal{C}_c$, where $\mathcal{B}_c = \{x \in \mathcal{B} : V_{\mathcal{B}}(x) < \eta\}$ and $\mathcal{C}_c = \{x \in \mathcal{C} : V_{\mathcal{C}}(z) < \eta\}$ for some $\eta > 0$ such that $\mathcal{B}_c \subset \mathcal{C}_c$. If $\mathcal{B}_c$ is bounded, then for every $x_0 \in \mathcal{B}_c$, $s(t, x_0) \to \mathcal{M}$ in $\mathcal{C}$ as $t \to \infty$, where $\mathcal{M}$ denotes the largest invariant set contained in $\mathcal{R}$ given by

$$\mathcal{R} = \{x \in \overline{\mathcal{C}_c} : \dot{V}_{\mathcal{C}}(x) = 0\}. \tag{2.41}$$

Theorem 2.13 can also be used to establish existence (in t) of (generalized) solutions of infinite-dimensional dynamical systems $\mathcal{G}$ over the semi-infinite interval $[t_0, \infty)$. In particular, global existence can be obtained by constructing a Lyapunov function $V : \mathcal{B} \to \mathbb{R}$ and invoking the continuation Peano-Cauchy theorem to obtain a dynamical system $\mathcal{G}$ of a subset $\mathcal{B}$ of the space $\mathcal{C}$. For further details see [201].

The next result gives a sufficient condition to guarantee semistability of the equilibria of $\mathcal{G}$.

Theorem 2.14. Consider a dynamical system $\mathcal{G}$ defined on a Banach space $\mathcal{B}$. Let $\mathcal{B}_c \subset \mathcal{B}$ be a closed set and assume that there exists a continuous function $V : \mathcal{B}_c \to \mathbb{R}$ such that $\dot{V}(z) \le 0$, $z \in \mathcal{B}_c$. Furthermore, let $\mathcal{R} \triangleq \{z \in \mathcal{B}_c : \dot{V}(z) = 0\}$, and let $\mathcal{M}$ denote the largest invariant set with respect to the dynamical system $\mathcal{G}$ contained in $\mathcal{R}$. Assume that $\mathcal{O}_z^+ \subseteq \mathcal{B}_c$ for every $z \in \mathcal{B}_c$ and $\mathcal{O}_z^+$ is contained in a compact subset of $\mathcal{B}$. If every point in $\mathcal{M}$ is a Lyapunov stable equilibrium point, then every point in $\mathcal{M}$ is semistable.

Proof. First, it follows from Proposition 2.6 and Theorem 2.12 that, for every $x \in \mathcal{B}_c$, the positive limit set $\omega(x)$ of x is nonempty and contained in the largest invariant subset $\mathcal{M}$ of $\mathcal{R}$. Since every point in $\mathcal{M}$ is a Lyapunov stable equilibrium point, it follows that every point in $\omega(x)$ is a Lyapunov stable equilibrium point.

Next, let $z \in \omega(x)$ and let $\mathcal{U}_\varepsilon$ be an open neighborhood of z. By Lyapunov stability of z, it follows that there exists a relatively open subset $\mathcal{U}_\delta$ containing z such that $s_t(\mathcal{U}_\delta) \subseteq \mathcal{U}_\varepsilon$ for every $t \ge 0$. Since $z \in \omega(x)$, it follows that there exists $h \ge 0$ such that $s(h, x) \in \mathcal{U}_\delta$. Thus, $s(t+h, x) = s_t(s(h, x)) \in s_t(\mathcal{U}_\delta) \subseteq \mathcal{U}_\varepsilon$ for every $t > 0$. Hence, since $\mathcal{U}_\varepsilon$ was chosen arbitrarily, it follows that $z = \lim_{t\to\infty} s(t, x)$.

Now, it follows that $\lim_{i\to\infty} s(t_i, x) \to z$ for every divergent sequence $\{t_i\}_{i=1}^{\infty}$, and hence, $\omega(x) = \{z\}$. Finally, since $\lim_{t\to\infty} s(t, x) \in \mathcal{M}$ is Lyapunov stable for every $x \in \mathcal{B}_c$, it follows from the definition of semistability that every point in $\mathcal{M}$ is semistable. □

In order to apply Theorem 2.14, one needs to show that the positive orbit $\mathcal{O}_x^+$ of $\mathcal{G}$ is contained in a compact subset of $\mathcal{B}$. Using Theorem 2.13, the following result relaxes this assumption.

Proposition 2.7. Let $\mathcal{B}$ and $\mathcal{D}$ be Banach spaces such that $\mathcal{B}$ is compactly embedded in $\mathcal{D}$, and let $\mathcal{G}$ be a dynamical system defined in $\mathcal{B}$ and $\mathcal{D}$. Assume there exist continuous functions $V_\mathcal{B} : \mathcal{B} \to \mathbb{R}$ and $V_\mathcal{D} : \mathcal{D} \to \mathbb{R}$ such that $V_\mathcal{B}(z) \ge 0$, $z \in \mathcal{B}_c$, and $V_\mathcal{D}(z) \ge 0$, $z \in \mathcal{D}_c$, where $\mathcal{B}_c = \{z \in \mathcal{B} : V_\mathcal{B}(z) < \eta\}$ and $\mathcal{D}_c = \{z \in \mathcal{D} : V_\mathcal{D}(z) < \eta\}$ for some $\eta > 0$ such that $\mathcal{B}_c \subset \mathcal{D}_c$. Furthermore, assume that $V_\mathcal{B}(s(t, z_0)) \le V_\mathcal{B}(s(\tau, z_0))$ for all $0 \le \tau \le t$ and $z_0 \in \mathcal{B}_c$, and $V_\mathcal{B}(s(t, z_0)) \le V_\mathcal{B}(s(\tau, z_0))$ for all $0 \le \tau \le t$ and $z_0 \in \mathcal{D}_c$. If $\mathcal{B}_c$ is bounded and every point in the largest invariant subset $\mathcal{M}$ contained in $\mathcal{R}$ given by $\mathcal{R} \triangleq \{z \in \overline{\mathcal{D}_c} : \dot{V}_\mathcal{D}(z) = 0\}$ is a Lyapunov stable equilibrium point of $\mathcal{G}$, then every equilibrium point in $\mathcal{M}$ is semistable.

Proof. First note that the assumptions on $V_\mathcal{B}$ imply that the trajectory $s(t, x)$ of $\mathcal{G}$ remains in $\mathcal{B}_c$ for all $x \in \mathcal{B}_c$ and $t \ge 0$. Furthermore, since $\mathcal{B}$ is compactly embedded in $\mathcal{D}$, $s(t, x)$ is contained in a compact set of $\mathcal{D}_c$

for all $x \in \mathcal{B}_c$ and $t \geq 0$. Now, it follows from Proposition 2.6 and Theorem 2.12 that, for every $x \in \mathcal{B}_c$, the positive limit set $\omega(x)$ of x is nonempty and contained in the largest invariant subset $\mathcal{M}$ of $\mathcal{R}$. The rest of the proof is similar to the proof of Theorem 2.14 and, hence, is omitted. □

Finally, we establish sufficient conditions for semistability of the system $\mathcal{G}$ determined by nonlinear semigroups in the Banach space $\mathcal{B}$. First, the following result is immediate.

Proposition 2.8. Consider the system $\mathcal{G}$. Assume that there exist a continuous function $V : \mathcal{B} \to \mathbb{R}$ and a class $\mathcal{K}$ function $\alpha(\cdot)$ such that

$$\alpha(\|x\|) \leq V(x), \quad x \in \mathcal{B}, \tag{2.42}$$

$$\dot{V}(x) \leq 0, \quad x \in \mathcal{B}. \tag{2.43}$$

If every point in the largest invariant subset $\mathcal{M}$ of $\{x \in \mathcal{B} : \dot{V}(x) = 0\}$ is Lyapunov stable, then the system $\mathcal{G}$ is semistable.

Definition 2.19. The system $\mathcal{G}$ is called a *contraction semigroup* if

$$\|\mathcal{T}(t)(x) - \mathcal{T}(t)(y)\|_{\mathcal{B}} \leq \|x - y\|_{\mathcal{B}} \tag{2.44}$$

holds for all $x, y \in \mathcal{B}$ and $t \geq 0$.

Theorem 2.15. Assume that the system $\mathcal{G}$ is a contraction semigroup. Furthermore, assume that there exists a continuous function $V : \mathcal{B} \to \mathbb{R}$ such that (2.42) holds. If every point in the largest invariant subset $\mathcal{M}$ of $\{x \in \mathcal{B} : \dot{V}(x) = 0\}$ is Lyapunov stable, then the system $\mathcal{G}$ is semistable.

Proof. Let $x_e \in \mathcal{E} \subset \mathcal{B}$ be an equilibrium point of $\mathcal{G}$. By definition, $\mathcal{T}(t)(x_e) = x_e$ for all $t \geq 0$. Now, it follows from (2.44) that

$$\|\mathcal{T}(t)(x) - x_e\|_{\mathcal{B}} \leq \|x - x_e\|_{\mathcal{B}} \tag{2.45}$$

for all $x \in \mathcal{B}$ and $t \geq 0$. Define the open ball $\mathcal{B}_\nu(x_e) \triangleq \{x \in \mathcal{B} : \|x - x_e\|_{\mathcal{B}} < \nu\}$ and the closed ball $\overline{\mathcal{B}}_\varepsilon(x_e) \triangleq \{x \in \mathcal{B} : \|x - x_e\|_{\mathcal{B}} \leq \varepsilon\}$ in the Banach space $\mathcal{B}$, where $\nu > \varepsilon > 0$. Note that $\mathcal{B}_\nu(x_e)$ is an open set and $\overline{\mathcal{B}}_\varepsilon(x_e)$ is a closed set. Clearly, it follows from (2.45) that if $x \in \overline{\mathcal{B}}_\varepsilon(x_e)$, then $s(t, x) = \mathcal{T}(t)(x) \in \overline{\mathcal{B}}_\varepsilon(x_e)$ for all $t \geq 0$. Hence, $\mathcal{O}_x^+ \subseteq \overline{\mathcal{B}}_\varepsilon(x_e)$.

Next, note that since $\overline{\mathcal{B}}_\varepsilon(x_e) \subset \mathcal{B}_\nu(x_e)$, it follows that $\overline{\mathcal{B}}_\varepsilon(x_e)$ has an open subcover $\mathcal{B}_\nu(x_e)$. Now, by definition, $\overline{\mathcal{B}}_\varepsilon(x_e)$ is a compact set and $\mathcal{O}_x^+$ is contained in $\overline{\mathcal{B}}_\varepsilon(x_e)$. With $\mathcal{B}_c = \overline{\mathcal{B}}_\varepsilon(x_e)$, it follows from Theorem 2.14 that every equilibrium point in $\mathcal{M} \cap \mathcal{B}_c$ is semistable. In particular, $x_e \in \mathcal{M} \cap \mathcal{B}_c$ is semistable. Since $x_e \in \mathcal{E}$ is arbitrary, it follows that every equilibrium point in $\mathcal{E}$ is semistable. Thus, by definition, the system $\mathcal{G}$ is semistable. □

2.7 Reversibility, Irreversibility, Recoverability, and Irrecoverability

The notions of reversibility, irreversibility, recoverability, and irrecoverability all play a critical role in thermodynamic processes. In this section, we define the notions of *R-state reversibility*, *state reversibility*, and *state recoverability* of a dynamical system $\mathcal{G}$. R-state reversibility concerns the existence of a system state with the property that a transformed system trajectory through an *involutory operator*[5] R is an image of a given system trajectory of $\mathcal{G}$ on a specified finite time interval. State reversibility concerns the existence of a system state with the property that the resulting system trajectory is the time-reversed image of a given system trajectory of $\mathcal{G}$ on a specified finite time interval. Finally, state recoverability concerns the existence of a system state with the property that the resulting system trajectory completely recovers the initial state of the dynamical system over a finite time interval.

To establish the notions of (ir)reversibility and (ir)recoverability of a dynamical system $\mathcal{G}$ defined on a Banach space $\mathcal{B}$, we require a generalization of Definition 2.16. For this definition, $\mathcal{U}$ is an input space and consists of bounded continuous U-valued functions on $[0, \infty)$. The set $U \subseteq \mathbb{R}^m$ contains the set of input values, that is, at any time $t \geq t_0$, $u(t) \in U$. The space $\mathcal{U}$ is assumed to be closed under the shift operator, that is, if $u \in \mathcal{U}$, then the function u_T defined by $u_T(t) \triangleq u(t+T)$ is contained in $\mathcal{U}$ for all $T \geq 0$. Furthermore, $\mathcal{Y}$ is an output space and consists of continuous Y-valued functions on $[0, \infty)$. The set $Y \subseteq \mathbb{R}^l$ contains the set of output values, that is, each value of $y(t) \in Y$, $t \geq t_0$. The space $\mathcal{Y}$ is assumed to be closed under the shift operator, that is, if $y \in \mathcal{Y}$, then the function y_T defined by $y_T(t) \triangleq y(t+T)$ is contained in $\mathcal{Y}$ for all $T \geq 0$.

Definition 2.20. Let $\mathcal{B}$ be a Banach space with norm $\|\cdot\|_{\mathcal{B}}$. A *dynamical system* on $\mathcal{B}$ is the octuple $(\mathcal{B}, \mathcal{U}, U, \mathcal{Y}, Y, [0, \infty), s, h)$, where $s : [0, \infty) \times \mathcal{B} \times \mathcal{U} \to \mathcal{B}$ and $h : \mathcal{B} \times U \to Y$ are such that the following axioms hold:

i) (Continuity): $s(\cdot, \cdot, u)$ is jointly continuous for all $u \in \mathcal{U}$.

ii) (Consistency): $s(t_0, x_0, u) = x_0$ for all $t_0 \in \mathbb{R}$, $x_0 \in \mathcal{B}$, and $u \in \mathcal{U}$.

iii) (Determinism): $s(t, x_0, u_1) = s(t, x_0, u_2)$ for all $t \in [t_0, \infty)$, $x_0 \in \mathcal{B}$, and u_1, $u_2 \in \mathcal{U}$ satisfying $u_1(\tau) = u_2(\tau)$, $\tau \leq t$.

iv) (Semigroup property): $s(\tau, s(t, x_0, u), u) = s(t+\tau, x_0, u)$ for all $x_0 \in \mathcal{B}$, $u \in \mathcal{U}$, and τ, $t \in [t_0, \infty)$.

---

[5]An involutory operator is an operator $R : \mathcal{B} \to \mathcal{B}$ such that $R^2 = I_{\mathcal{B}}$. That is, if $R(R(x)) = x$ for all $x \in \mathcal{B}$, then R is an involution.

v) (Read-out map): For every $x_0 \in \mathcal{D}$, $u \in \mathcal{U}$, and $t_0 \in \mathbb{R}$, there exists $y \in \mathcal{Y}$ such that $y(t) = h(s(t, x_0, u), u(t))$ for all $t \geq t_0$.

As in Section 2.2, we denote the dynamical system $(\mathcal{B}, \mathcal{U}, U, \mathcal{Y}, Y, [0, \infty), s, h)$ by $\mathcal{G}$. Furthermore, we refer to the map $s(\cdot,\cdot,\cdot)$ as the *flow* or *trajectory* of $\mathcal{G}$ corresponding to $x_0 \in \mathcal{D}$, and for a given $s(t, x_0, u)$, $t \geq t_0$, $u \in \mathcal{U}$, we refer to $x_0 \in \mathcal{D}$ as an *initial condition* of $\mathcal{G}$. Given $t \in \mathbb{R}$, we denote the map $s(t,\cdot,\cdot) : \mathcal{D} \times \mathcal{U} \to \mathcal{D}$ by $s_t(x_0, u)$. Hence, for a fixed $t \in \mathbb{R}$ the set of mappings defined by $s_t(x_0, u) = s(t, x_0, u)$ for every $x_0 \in \mathcal{D}$ and $u \in \mathcal{U}$ gives the *flow* of $\mathcal{G}$. In particular, if $\mathcal{D}_0$ is a collection of initial conditions such that $\mathcal{D}_0 \subset \mathcal{D}$, then the flow $s_t : \mathcal{D}_0 \times \mathcal{U} \to \mathcal{D}$ is the motion of all points $x_0 \in \mathcal{D}_0$ or, equivalently, the image of $\mathcal{D}_0 \subset \mathcal{D}$ under the flow s_t, that is, $s_t(\mathcal{D}_0, \mathcal{U}) \subset \mathcal{D}$, where $s_t(\mathcal{D}_0, \mathcal{U}) \triangleq \{y : y = s_t(x_0, u)$ for some $x_0 \in \mathcal{D}$ and $u \in \mathcal{U}\}$.

Alternatively, if the initial condition $x_0 \in \mathcal{D}$ is fixed and we let $[t_0, t_1] \subset \mathbb{R}$ and $u \in \mathcal{U}$, then the mapping $s(\cdot, x_0, u) : [t_0, t_1] \to \mathcal{D}$ defines the *solution curve* or *trajectory* of the dynamical system $\mathcal{G}$. Hence, the mapping $s(\cdot, x_0, u)$ generates a graph in $[t_0, t_1] \times \mathcal{D}$ identifying the trajectory corresponding to the motion along a curve through the point x_0 with input $u \in \mathcal{U}$ in a subset $\mathcal{D}$ of the state space. Given $x \in \mathcal{D}$ and $u \in \mathcal{U}$, we denote the map $s(\cdot, x, u) : \mathbb{R} \to \mathcal{D}$ by $s^x(t, u)$.

In general, the output of $\mathcal{G}$ depends on both the present input of $\mathcal{G}$ and the past history of $\mathcal{G}$. Hence, the output at some time t_1 depends on the state $s(t_1, x_0, u)$ of $\mathcal{G}$, which effectively serves as an information storage (memory) of past history. Furthermore, the determinism axiom ensures that the state and thus the output before some time t_1 are not influenced by the values of the output after time t_1. Hence, future inputs to $\mathcal{G}$ do not affect past and present outputs of $\mathcal{G}$. This is simply a statement of causality that holds for all physical (nonquantum) systems. Finally, we note that the read-out map is memoryless in the sense that outputs only depend on the instantaneous (present) values of the state and input.

The dynamical system $\mathcal{G}$ is *isolated* if $u(t) \equiv 0$. Furthermore, an *equilibrium point* of the isolated dynamical system $\mathcal{G}$ is a point $x_e \in \mathcal{D}$ satisfying $s(t, x_e, 0) = x_e$, $t \geq t_0$. An equilibrium point $x_e \in \mathcal{D}_c \subseteq \mathcal{D}$ of the isolated dynamical system $\mathcal{G}$ is *Lyapunov stable* with respect to the positively invariant set $\mathcal{D}_c$ if, for every relatively open subset $\mathcal{N}_\varepsilon$ of $\mathcal{D}_c$ containing x_e, there exists a relatively open subset $\mathcal{N}_\delta$ of $\mathcal{D}_c$ containing x_e such that $s_t(\mathcal{N}_\delta, \mathcal{U}) \subset \mathcal{N}_\varepsilon$ for all $t \geq t_0$, where $\mathcal{U} = \{u : \mathbb{R} \to \mathbb{R}^m : u(t) \equiv 0\}$. An equilibrium point $x_e \in \mathcal{D}_c$ of the isolated dynamical system $\mathcal{G}$ is called *semistable* if it is Lyapunov stable and there exists a relatively open subset $\mathcal{N}$ of $\mathcal{D}_c$ containing x_e such that for all initial conditions in $\mathcal{N}$, the trajectory of $\mathcal{G}$ converges to a Lyapunov stable equilibrium point, that is, $\|s(t, x, 0) - y\| \to 0$ as $t \to \infty$, where $y \in \mathcal{D}_c$ is a Lyapunov stable equilibrium point of $\mathcal{G}$ and $x \in \mathcal{N}$. The isolated dynamical system $\mathcal{G}$ is said to be *semistable* if every equilibrium point of $\mathcal{G}$ is semistable.

For the next set of definitions the following notation is needed. For a given interval $[t_0, t_1]$, where $0 \le t_0 < t_1 < \infty$, let $\mathcal{W}_{[t_0,t_1]}$ denote the set of all possible trajectories of $\mathcal{G}$ given by

$$\mathcal{W}_{[t_0,t_1]} \triangleq \{s^x : [t_0, t_1] \times \mathcal{U} \to \mathcal{B} : s^x(\cdot, u(\cdot)) \text{ satisfies Axioms } i) - iv) \text{ of Definition 2.20}, x \in \mathcal{B}, \text{ and } u(\cdot) \in \mathcal{U}\}, \tag{2.46}$$

where $s^x(\cdot, u(\cdot))$ denotes the solution curve or trajectory of $\mathcal{G}$ for a given fixed initial condition $x \in \mathcal{B}$ and input $u(\cdot) \in \mathcal{U}$.

Definition 2.21. Consider the dynamical system $\mathcal{G}$ defined on $\mathcal{B}$. Let $R : \mathcal{B} \to \mathcal{B}$ be an involutive operator (that is, $R^2 = I_{\mathcal{B}}$, where $I_{\mathcal{B}}$ denotes the identity operator on $\mathcal{B}$) and let $s^x(\cdot, u(\cdot)) \in \mathcal{W}_{[t_0,t_1]}$, where $u(\cdot) \in \mathcal{U}$. The function $s^{-x} : [t_0, t_1] \times \mathcal{U} \to \mathcal{B}$ is an *R-reversed trajectory* of $s^x(\cdot, u(\cdot))$ if there exists an input $u^-(\cdot) \in \mathcal{U}$ and a continuous, strictly increasing function $\tau : [t_0, t_1] \to [t_0, t_1]$ such that $\tau(t_0) = t_0$, $\tau(t_1) = t_1$, and

$$s^{-x}(t, u^-(t)) = Rs^x(t_0 + t_1 - \tau(t), u(t_0 + t_1 - \tau(t))), \quad t \in [t_0, t_1]. \tag{2.47}$$

Definition 2.22. Consider the dynamical system $\mathcal{G}$ defined on $\mathcal{B}$. Let $R : \mathcal{B} \to \mathcal{B}$ be an involutive operator, let $r : U \times Y \to \mathbb{R}$, and let $s^x(\cdot, u(\cdot)) \in \mathcal{W}_{[t_0,t_1]}$, where $u(\cdot) \in \mathcal{U}$. $s^x(\cdot, u(\cdot))$ is an *R-reversible trajectory* of $\mathcal{G}$ if there exists an input $u^-(\cdot) \in \mathcal{U}$ such that $s^{-x}(\cdot, u^-(\cdot)) \in \mathcal{W}_{[t_0,t_1]}$ and

$$\int_{t_0}^{t_1} r(u(t), y(t))\mathrm{d}t + \int_{t_0}^{t_1} r(u^-(t), y^-(t))\mathrm{d}t = 0, \tag{2.48}$$

where $y^-(\cdot)$ denotes the read-out map for the R-reversed trajectory of $s^x(\cdot, u(\cdot))$. Furthermore, $\mathcal{G}$ is an *R-state reversible dynamical system* if for every $x \in \mathcal{B}$, $s^x(\cdot, u(\cdot))$, where $u(\cdot) \in \mathcal{U}$, is an R-reversible trajectory of $\mathcal{G}$.

In classical mechanics, R is a transformation that reverses the sign of all system momenta, whereas in classical reversible thermodynamics R can be taken to be the identity operator. Note that if $R = I_{\mathcal{B}}$, then $s^x(\cdot, u(\cdot))$, where $u(\cdot) \in \mathcal{U}$, is an $I_{\mathcal{B}}$-reversible trajectory or, simply, $s^x(\cdot, u(\cdot))$ is a *reversible trajectory*. Furthermore, we say that $\mathcal{G}$ is a *state reversible dynamical system* if and only if for every $x \in \mathcal{B}$, $s^x(\cdot, u(\cdot))$, where $u(\cdot) \in \mathcal{U}$, is a reversible trajectory of $\mathcal{G}$. Note that unlike state reversible systems, R-state reversible dynamical systems need not retrace every stage of the original system trajectory in reverse order, nor is it necessary for the dynamical system to recover the initial system state.

The function $r(u, y)$ in Definition 2.22 is a generalized *power supply* from the environment to the dynamical system through the system's input-output ports (u, y). Hence, (2.48) ensures that the total generalized energy supplied to the dynamical system $\mathcal{G}$ by the environment is returned to the environment over a given R-reversible trajectory starting and ending at any given (not necessarily the same) state $x \in \mathcal{B}$. Furthermore, condition (2.48) ensures that a reversible process completely restores the original dynamic state of a system and at the same time restores the energy supplied by the environment back to its original condition.

The following result provides sufficient conditions for the existence of an R-reversible trajectory of a dynamical system $\mathcal{G}$, and hence, establishes sufficient conditions for R-state reversibility of the dynamical system $\mathcal{G}$.

Theorem 2.16. Consider the dynamical system $\mathcal{G}$ defined on $\mathcal{B}$. Let $R : \mathcal{B} \to \mathcal{B}$ be an involutive operator, and let $s^x(\cdot, u(\cdot)) \in \mathcal{W}_{[t_0,t_1]}$, where $u(\cdot) \in \mathcal{U}$. Assume there exist a continuous function $V : \mathcal{B} \to \mathbb{R}$ and a function $r : U \times Y \to \mathbb{R}$ such that $V(x) = V(Rx)$, $x \in \mathcal{B}$, and for every $x \in \mathcal{B}$ and all $\hat{t}_0$, $\hat{t}_1$, $t_0 \leq \hat{t}_0 < \hat{t}_1 \leq t_1$,

$$V(s^x(\hat{t}_1, u(\hat{t}_1))) \geq V(s^x(\hat{t}_0, u(\hat{t}_0))) + \int_{\hat{t}_0}^{\hat{t}_1} r(u(t), y(t))\mathrm{d}t. \qquad (2.49)$$

Furthermore, assume there exists $\mathcal{M} \subset \mathcal{B}$ such that for all $\hat{t}_0$, $\hat{t}_1$, $t_0 \leq \hat{t}_0 < \hat{t}_1 \leq t_1$, and $s^x(t, u(t)) \notin \mathcal{M}$, $t \in [\hat{t}_0, \hat{t}_1]$, (2.49) holds as a strict inequality. If $s^x(\cdot, u(\cdot))$ is an R-reversible trajectory of $\mathcal{G}$, then $s^x(t, u(t)) \in \mathcal{M}$, $t \in [t_0, t_1]$.

Proof. Let $s^x(\cdot, u(\cdot)) \in \mathcal{W}_{[t_0,t_1]}$, where $u(\cdot) \in \mathcal{U}$, be an R-reversible trajectory of $\mathcal{G}$ so that there exists $u^-(\cdot) \in \mathcal{U}$ such that $s^{-x}(\cdot, u^-(\cdot)) \in \mathcal{W}_{[t_0,t_1]}$. Suppose, *ad absurdum*, there exists $t \in [t_0, t_1]$ such that $s^x(t, u(t)) \notin \mathcal{M}$. Now, it follows that there exists an interval $[\hat{t}_0, \hat{t}_1] \subset [t_0, t_1]$ such that for $t_0 \leq \hat{t}_0 < \hat{t}_1 \leq t_1$,

$$V(s^x(\hat{t}_1, u(\hat{t}_1))) > V(s^x(\hat{t}_0, u(\hat{t}_0))) + \int_{\hat{t}_0}^{\hat{t}_1} r(u(t), y(t))\mathrm{d}t, \qquad (2.50)$$

which further implies that

$$V(s^x(t_1, u(t_1))) > V(s^x(t_0, u(t_0))) + \int_{t_0}^{t_1} r(u(t), y(t))\mathrm{d}t. \qquad (2.51)$$

Next, since $s^{-x}(\cdot, u^-(\cdot)) \in \mathcal{W}_{[t_0,t_1]}$, where $u^-(\cdot) \in \mathcal{U}$, it follows that

$$V(s^{-x}(t_1, u^-(t_1))) \geq V(s^{-x}(t_0, u^-(t_0))) + \int_{t_0}^{t_1} r(u^-(t), y^-(t))\mathrm{d}t. \qquad (2.52)$$

Now, adding (2.51) and (2.52), using the definition of $s^{-x}(\cdot, u^{-}(\cdot))$, using the fact that $V(x) = V(Rx)$, $x \in \mathcal{B}$, and using (2.48) yields

$$V(s^x(t_0, u(t_0))) + V(s^x(t_1, u(t_1))) > V(s^x(t_0, u(t_0))) + V(s^x(t_1, u(t_1))), \tag{2.53}$$

which is a contradiction. Hence, $s^x(t, u(t)) \in \mathcal{M}$, $t \in [t_0, t_1]$. □

It is important to note that since $V : \mathcal{B} \to \mathbb{R}$ in Theorem 2.16 is not sign definite, Theorem 2.16 also holds for the case where the inequality in (2.49) is reversed. The following corollary to Theorem 2.16 is immediate.

Corollary 2.2. Consider the dynamical system $\mathcal{G}$ defined on $\mathcal{B}$. Let $R : \mathcal{B} \to \mathcal{B}$ be an involutive operator, let $\mathcal{M} \subset \mathcal{B}$, and let $s^x(\cdot, u(\cdot)) \in \mathcal{W}_{[t_0,t_1]}$, where $u(\cdot) \in \mathcal{U}$. Assume there exists a continuous function $V : \mathcal{B} \to \mathbb{R}$ such that $V(x) = V(Rx)$, $x \in \mathcal{B}$, and for $s^x(t, u(t)) \notin \mathcal{M}$, $t \in [t_1, t_2]$, $V(s(t, x_0, u(\cdot)))$ is a strictly increasing (respectively, decreasing) function of time. If $s^x(\cdot, u(\cdot))$ is an R-reversible trajectory of $\mathcal{G}$, then $s^x(t, u(t)) \in \mathcal{M}$, $t \in [t_0, t_1]$.

Proof. The proof is a direct consequence of Theorem 2.16 with $r(u, y) \equiv 0$ and the fact that Theorem 2.16 also holds for the case when the inequality in (2.49) is reversed. □

It follows from Corollary 2.2 that if, for a given dynamical system $\mathcal{G}$, there exists an R-reversible trajectory of $\mathcal{G}$, then there does not exist a function of the state of the system that strictly decreases or strictly increases in time on any trajectory of $\mathcal{G}$ lying in $\mathcal{M}$. In this case, the existence of a completely ordered time set having a topological structure involving a closed set homeomorphic to the real line cannot be established. Such systems, which include lossless Newtonian and Hamiltonian systems, are time-reversal symmetric and hence lack an inherent time direction. As we see in Sections 3.5 and 5.1, that is not the case with thermodynamic systems.

Next, we present the notion of state recoverability for a dynamical system $\mathcal{G}$.

Definition 2.23. Consider the dynamical system $\mathcal{G}$ defined on $\mathcal{B}$. Let $r : U \times Y \to \mathbb{R}$ and let $s^x(\cdot, u(\cdot)) \in \mathcal{W}_{[t_0,t_1]}$, where $u(\cdot) \in \mathcal{U}$. Then $s^x(\cdot, u(\cdot))$ is a *recoverable trajectory* of $\mathcal{G}$ if there exists $u^{-}(\cdot) \in \mathcal{U}$ and $t_2 > t_1$ such that $u^{-} : [t_1, t_2] \to U$,

$$s(t_2, s^x(t_1, u(t_1)), u^{-}(t_2)) = s^x(t_0, u(t_0)), \tag{2.54}$$

and

$$\int_{t_0}^{t_1} r(u(t), y(t))\mathrm{d}t + \int_{t_1}^{t_2} r(u^-(t), y^-(t))\mathrm{d}t = 0, \tag{2.55}$$

where $y^-(\cdot)$ denotes the read-out map for the trajectory $s(\cdot, s^x(t_1, u(t_1)), u^-(\cdot))$. Furthermore, $\mathcal{G}$ is a *state recoverable dynamical system* if for every $x \in \mathcal{B}$, $s^x(\cdot, u(\cdot))$ is a recoverable trajectory of $\mathcal{G}$.

It follows from the definition of state recoverability that the way in which the initial dynamical system state is restored may be chosen freely so long as (2.55) is satisfied. Hence, unlike R-state reversibility, it is not necessary for the dynamical system to recover the initial state of the system through an involutive transformation of the system trajectory. Furthermore, unlike state reversibility, it is not necessary for the dynamical system to retrace every stage of the original trajectory in the reverse order. However, condition (2.55) ensures that the recoverable process completely restores the original dynamic state and at the same time restores the energy supplied by the environment back to its original condition. This notion of recoverability is related to Planck's notion of complete reversibility, wherein the initial system state is restored in the "*totality of Nature*" (*die gesamte Natur*).

However, it is important to note here that it is not clear what Planck meant by restoring the initial system state everywhere in the totality of nature. For example, if a system undergoes a thermodynamic process, then would it be necessary to reverse the Earth's rotation, reverse all the photons emitted by the Sun's atoms, reverse the water molecules in every river on the planet, etc. to achieve complete reversibility? In that case, Planck's complete reversibility assumption becomes incongruous.

Rather, one would surmise that the notion of Planck's complete reversibility is restricted to all the bodies that have interacted with the system undergoing the thermodynamic process. Within this realm, Planck's complete reversibility has consequences for all types of interactions (including ones not yet discovered) occurring in nature. This is in stark contrast with Clausius' reversibility (*umkehrbar*) and irreversibility (*nicht umkehrbar*). Furthermore, it is not clear that all conservative mechanical processes are, in fact, reversible using Planck's concept of reversibility (in "der Natur Vorhandenen Reagentien").

The following result provides a sufficient condition for the existence of a recoverable trajectory of a dynamical system $\mathcal{G}$, and hence, establishes sufficient conditions for state recoverability of $\mathcal{G}$.

Theorem 2.17. Consider the dynamical system $\mathcal{G}$ defined on $\mathcal{B}$. Let

$s^x(\cdot, u(\cdot)) \in \mathcal{W}_{[t_0,t_1]}$, where $u(\cdot) \in \mathcal{U}$. Assume there exist a continuous function $V : \mathcal{B} \to \mathbb{R}$ and a function $r : U \times Y \to \mathbb{R}$ such that for every $x \in \mathcal{B}$ and all $\hat{t}_0$, $\hat{t}_1$, $t_0 \le \hat{t}_0 < \hat{t}_1 \le t_1$,

$$V(s^x(\hat{t}_1, u(\hat{t}_1))) \ge V(s^x(\hat{t}_0, u(\hat{t}_0))) + \int_{\hat{t}_0}^{\hat{t}_1} r(u(t), y(t))\mathrm{d}t. \tag{2.56}$$

Furthermore, assume there exists $\mathcal{M} \subset \mathcal{B}$ such that for all $\hat{t}_0$, $\hat{t}_1$, $t_0 \le \hat{t}_0 < \hat{t}_1 \le t_1$, and $s^x(t, u(t)) \notin \mathcal{M}$, $t \in [\hat{t}_0, \hat{t}_1]$, (2.56) holds as a strict inequality. If $s^x(\cdot, u(\cdot))$ is a recoverable trajectory of $\mathcal{G}$, then $s^x(t, u(t)) \in \mathcal{M}$, $t \in [t_0, t_1]$.

Proof. Let $s^x(\cdot, u(\cdot)) \in \mathcal{W}_{[t_0,t_1]}$, where $u(\cdot) \in \mathcal{U}$, be a recoverable trajectory of $\mathcal{G}$ so that there exist $u^-(\cdot) \in \mathcal{U}$ and $t_2 > t_1$ such that $s(t_2, s^x(t_1, u(t_1)), u^-(t_2)) = s^x(t_0, u(t_0))$. Suppose, *ad absurdum*, there exists $t \in [t_0, t_1]$ such that $s^x(t, u(t)) \notin \mathcal{M}$. Now, it follows that there exists an interval $[\hat{t}_0, \hat{t}_1] \subset [t_0, t_1]$ such that for $t_0 \le \hat{t}_0 < \hat{t}_1 \le t_1$,

$$V(s^x(\hat{t}_1, u(\hat{t}_1))) > V(s^x(\hat{t}_0, u(\hat{t}_0))) + \int_{\hat{t}_0}^{\hat{t}_1} r(u(t), y(t))\mathrm{d}t, \tag{2.57}$$

which further implies that

$$V(s^x(t_1, u(t_1))) > V(s^x(t_0, u(t_0))) + \int_{t_0}^{t_1} r(u(t), y(t))\mathrm{d}t. \tag{2.58}$$

Next, it follows from (2.56) with $t_2 > t_1$ that

$$V(s(t_2, s^x(t_1, u(t_1)), u^-(t_2))) \ge V(s(t_1, s^x(t_1, u(t_1)), u^-(t_1))) + \int_{t_1}^{t_2} r(u^-(t), y^-(t))\mathrm{d}t. \tag{2.59}$$

Now, adding (2.58) and (2.59), using the definition of $s(t_2, s^x(t_1, u(t_1), u^-(t_2)))$, and using (2.55) yields

$$V(s^x(t_0, u(t_0))) + V(s^x(t_1, u(t_1))) > V(s^x(t_0, u(t_0))) + V(s^x(t_1, u(t_1))), \tag{2.60}$$

which is a contradiction. Hence, $s^x(t, u(t)) \in \mathcal{M}$, $t \in [t_0, t_1]$. □

The following corollary to Theorem 2.17 is immediate.

Corollary 2.3. Consider the dynamical system $\mathcal{G}$ defined on $\mathcal{B}$. Let $\mathcal{M} \subset \mathcal{B}$ and let $s^x(\cdot, u(\cdot)) \in \mathcal{W}_{[t_0,t_1]}$, where $u(\cdot) \in \mathcal{U}$. Assume there exists a continuous function $V : \mathcal{B} \to \mathbb{R}$ such that for $s^x(t, u(t)) \notin \mathcal{M}$, $t \in [t_0, t_1]$, $V(s(t, x_0, u(\cdot))$ is a strictly increasing (respectively, decreasing) function of time. If $s^x(\cdot, u(\cdot))$ is a recoverable trajectory of $\mathcal{G}$, then $s^x(t, u(t)) \in \mathcal{M}$, $t \in [t_0, t_1]$.

Proof. The proof is a direct consequence of Theorem 2.17 with $r(u, y) \equiv 0$ and the fact that Theorem 2.17 also holds for the case when the inequality in (2.56) is reversed. □

As in the case of R-state reversibility and state reversibility, state recoverability can be used to establish a connection between a dynamical system evolving on a manifold $\mathcal{M} \subset \mathcal{B}$ and the arrow of time. However, in the case of state recoverability, the recoverable dynamical system trajectory need not involve an involutive transformation of the system trajectory, nor is it required to retrace the original system trajectory in recovering the original dynamic state.

It should be noted here that state recoverability is not implied by the concepts of *reachability* and *controllability*,[6] which play a central role in control theory (see Section 3.2). For example, one might envision, albeit with a considerable stretch of the imagination, perfectly controlled inputs that could reassemble a broken egg or even fuse water into solid cubes of ice. In particular, reassembling a broken egg would require a flow of colliding air molecules and microscopic floor vibrations converging onto the impact site from all directions, and forcing every piece of broken shell and drop of yolk to merge toward the collision location. The various molecules that make up the egg would move back in reverse direction, morphing and fusing. The vibrational air and floor waves would have to regulate the molecular motions of every drop of yolk and eggshell to bind the egg back to its original pristine ovoid form.

However, in all such cases, an external source of energy from the environment would be required to operate such an immaculate state recoverable mechanism and would violate condition (2.55). Clearly, state recoverability is a weaker notion than that of state reversibility since state reversibility implies state recoverability; the converse, however, is not generally true. Conversely, state irrecoverability is a logically stronger notion than state irreversibility since state irrecoverability implies state irreversibility. However, as we see in Chapter 3, these notions are equivalent for thermodynamic systems.

2.8 Output Reversibility in Dynamical Systems

The notion of reversibility is one of the fundamental symmetries that arises in natural sciences. Specifically, as discussed in Chapter 1, time-reversal

[6]A dynamical system $\mathcal{G}$ as defined by Definition 2.20 is *reachable* if, for all $x_0 \in \mathcal{B}$, there exist a finite time $t_\mathrm{i} < t_0$ and a square integrable input $u(t)$ defined on $[t_\mathrm{i}, t_0]$ such that the state $x(t)$, $t \geq t_\mathrm{i}$, can be driven from $x(t_\mathrm{i}) = 0$ to $x(t_0) = x_0$. $\mathcal{G}$ is (null) *controllable* if, for all $x_0 \in \mathcal{B}$, there exist a finite time $t_\mathrm{f} > t_0$ and a square integrable input $u(t)$ defined on $[t_0, t_\mathrm{f}]$ such that $x(t)$, $t \geq t_0$, can be driven from $x(t_0) = x_0$ to $x(t_\mathrm{f}) = 0$.

symmetry arises in many physical dynamical systems and, in particular, in classical, relativistic, and quantum mechanics. The governing dynamical system equations for such systems possess reversing symmetries, that is, the concept of time flow does not enter in these physical theories. Such theories possess a *time-reversal symmetry*, wherein, from any given moment of time, the governing dynamical laws treat past and future in exactly the same way [188, 195, 258].

In contrast, thermodynamics describes processes that give rise to *time-reversal non-invariance* [188, 195]. As noted in Chapter 1, the term *time reversal* is not meant literally here; that is, it pertains to whether events that occur *in* time and in a given particular temporal order can also occur in the reverse order. Thermodynamic systems give rise to dynamical systems whose system trajectory reversal is or is not allowed and *not* a reversal of time itself. Nevertheless, many scientists have attributed this emergence of the direction of time flow to the second law of thermodynamics—the law that entropy always increases—due to its intimate connection to the irreversibility of dynamical processes.

In this section, we use system-theoretic notions to give yet another definition of system reversibility for inflow-closed linear dynamical systems. In particular, we consider the free response of a system on a given, finite interval. Specifically, the system is *output reversible* if, for every initial condition and corresponding trajectory, there exists an alternative initial condition such that the corresponding trajectory is the time-reversed image of the original trajectory. Here, for simplicity of exposition, we characterize linear systems that are output reversible.

In [36], output reversibility was addressed for linear dynamical systems with single outputs. As special cases, it was shown that the class of output-reversible systems includes rigid body and Hamiltonian systems. This result suggests that stability and instability play a key role in the arrow of time, independently of dimensionality, nonlinearity, and initial-state sensitivity. In this section, we extend the results of [36] to multioutput systems to obtain a spectral symmetry condition that characterizes output reversible systems. In particular, we show that an inflow-closed linear system is output reversible if and only if its nonimaginary spectrum is symmetric with respect to the imaginary axis.

Time reversibility in closed linear dynamical systems is also studied in [140, 141] using a class of behaviors, which can be described through a set of linear constant coefficient differential equations. Specifically, the authors in [140, 141] consider linear differential equations defined by polynomial matrices. For these systems, the authors define the notion of *J-time-*

reversibility. However, the notion of output reversibility is distinct from the notion of J-time-reversibility defined in [140, 141]. Specifically, for linear systems defined by polynomial matrices that include state space systems as a special case, a system is J-time-reversible if the application of a linear transformation J to the system trajectory yields a trajectory of the *time-reversed system*, that is, the modified system in which t is replaced by $-t$.

For example, consider the unforced single-degree-of-freedom rigid body modeled by

$$\dot{x}(t) = Ax(t), \quad x(0) = x_0, \quad t \geq 0, \tag{2.61}$$

where

$$A = \begin{bmatrix} 0 & 1 \\ 0 & 0 \end{bmatrix}.$$

Then, applying the transformation

$$J = \begin{bmatrix} 1 & 0 \\ 0 & -1 \end{bmatrix}$$

to the trajectory provides a trajectory of the time-reversed system

$$\dot{\hat{x}}(t) = \hat{A}\hat{x}(t), \quad \hat{x}(0) = \hat{x}_0, \quad t \geq t_0, \tag{2.62}$$

where $\hat{x} \triangleq Jx$ and

$$\hat{A} \triangleq JAJ^{-1} = \begin{bmatrix} 0 & -1 \\ 0 & 0 \end{bmatrix}.$$

In contrast, we consider only *forward* trajectories of (2.61) with the goal of determining initial conditions for which the system output $y = Cx$ is the time-reversed image of a given output trajectory. Consequently, output reversibility and J-time-reversibility are distinct notions. In addition, the authors in [140, 141] do not give any connections between the spectral symmetry of the system dynamics and output reversibility.

To present the main results of this section, the following additional notation is needed. We say that the multispectrum of A is *symmetric with respect to the imaginary axis* if $\{\lambda_1, \ldots, \lambda_n\}_{\mathrm{m}} = \{-\lambda_1, \ldots, -\lambda_n\}_{\mathrm{m}}$. Furthermore, we say that, for $A \in \mathbb{R}^{n\times n}$, $\lambda \in \mathrm{mspec}(A)$ is *semisimple* if the algebraic multiplicity of λ is equal to its geometric multiplicity, that is, the complex Jordan form of A is a diagonal matrix.

We begin by considering the nonlinear dynamical system given by

$$\dot{x}(t) = f(x(t)), \quad x(0) = x_0, \quad t \geq 0, \tag{2.63}$$

with system output

$$y(t) = h(x(t)), \tag{2.64}$$

where $x(t) \in \mathbb{R}^n$, $y(t) \in \mathbb{R}^l$, and $f : \mathbb{R}^n \to \mathbb{R}^n$ and $h : \mathbb{R}^n \to \mathbb{R}^l$ are continuous. We assume that solutions of (2.63) exist and are unique on all finite intervals $[0, T)$. For clarity, we write the solution of (2.63) as $x(t, x_0)$ with the output given by $y(t) = y(t, x_0) = h(x(t, x_0))$.

Definition 2.24 ([36]). The dynamical system (2.63) and (2.64) is *output reversible* if, for all $x_0 \in \mathbb{R}^n$ and $t_1 > 0$, there exists an initial condition $\hat{x}_0 \in \mathbb{R}^n$ such that

$$y(t, \hat{x}_0) = y(t_1 - t, x_0), \quad t \in [0, t_1]. \tag{2.65}$$

We wish to determine whether a given system (2.63) and (2.64) is output reversible. Here, we consider the special case of linear systems given by

$$\dot{x}(t) = Ax(t), \quad x(0) = x_0, \quad t \geq 0, \tag{2.66}$$

with system output

$$y(t) = Cx(t), \tag{2.67}$$

where $A \in \mathbb{R}^{n \times n}$ and $C \in \mathbb{R}^{l \times n}$. Furthermore, we assume that (A, C) is observable.[7] It follows from Definition 2.24 that (2.66) and (2.67) is output reversible if and only if, for all $x_0 \in \mathbb{R}^n$ and $t_1 > 0$, there exists $\hat{x}_0 \in \mathbb{R}^n$ such that

$$Ce^{At}\hat{x}_0 = Ce^{A(t_1 - t)}x_0, \quad t \in [0, t_1]. \tag{2.68}$$

Note that output reversibility is a basis-independent property.

The following result shows that if (2.66) and (2.67) is output reversible, then $\hat{x}_0$ satisfying (2.68) is unique.

Proposition 2.9. Let $x_0 \in \mathbb{R}^n$ and $t_1 > 0$, assume that (2.66) and (2.67) is output reversible, and let $\hat{x}_0 \in \mathbb{R}^n$ satisfy (2.68). Then $\hat{x}_0$ satisfies

$$\mathcal{O}\hat{x}_0 = \mathcal{S}\mathcal{O}e^{At_1}x_0 \tag{2.69}$$

and is given uniquely by

$$\hat{x}_0 = \mathcal{O}^\dagger \mathcal{S}\mathcal{O}e^{At_1}x_0, \tag{2.70}$$

[7] The nonlinear system (2.63) and (2.64) is (zero-state) *observable* if $y(t) \equiv 0$ implies $x(t) \equiv 0$. For the linear system (2.66) and (2.67) this definition is equivalent to the pair (A, C) being observable if and only if the set $\{x_0 \in \mathbb{R}^n : y(t) = 0 \text{ for all } t \in [0, t_{\rm f}]\} = \{0\}$.

where $\mathcal{O} \in \mathbb{R}^{nl\times n}$ and $\mathcal{S} \in \mathbb{R}^{nl\times nl}$ are defined by

$$\mathcal{O} \triangleq \begin{bmatrix} C \\ CA \\ \vdots \\ CA^{n-1} \end{bmatrix}, \tag{2.71}$$

$$\mathcal{S} \triangleq \begin{bmatrix} I_l & 0_{l\times l} & \cdots & 0_{l\times l} \\ 0_{l\times l} & -I_l & & \\ & & I_l & \vdots \\ \vdots & & \ddots & 0_{l\times l} \\ 0_{l\times l} & \cdots & 0_{l\times l} & (-1)^{n-1}I_l \end{bmatrix}. \tag{2.72}$$

Proof. Since (2.66) and (2.67) is output reversible, there exists $\hat{x}_0$ satisfying (2.68). Differentiating (2.68) $n-1$ times and setting $t=0$ yields (2.69). Since (A, C) is observable, $\mathcal{O}^\dagger$ is a left inverse of $\mathcal{O}$. Hence, (2.69) implies (2.70). □

Note that since (A, C) is observable, $\operatorname{rank}\mathcal{O} = n$. In addition, if, for some $x_0 \notin \mathcal{N}(A)$ and $t_1 > 0$, there does not exist $\hat{x}_0 \in \mathbb{R}^n$ satisfying (2.69), then (2.68) does not have a solution. In this case, (2.66) and (2.67) is not output reversible.

Corollary 2.4. Let $x_0 \in \mathbb{R}^n$, $t_1 > 0$, and assume that

$$\operatorname{rank}\,[\mathcal{O} \quad \mathcal{S}\mathcal{O}e^{At_1}x_0] > n. \tag{2.73}$$

Then (2.66) and (2.67) is not output reversible.

Proposition 2.10. Assume that $l = n$ and $C \in \mathbb{R}^{n\times n}$ is invertible. Then (2.66) and (2.67) is not output reversible.

Proof. Since $C \in \mathbb{R}^{n\times n}$ is invertible, then (A, C) is observable. Let $x_0 \in \mathbb{R}^n$ and $t_1 > 0$, and note that (2.69) is equivalent to

$$\begin{bmatrix} C \\ CA \\ \vdots \\ CA^{n-1} \end{bmatrix} \hat{x}_0 = \begin{bmatrix} C \\ -CA \\ \vdots \\ (-1)^{n-1}CA^{n-1} \end{bmatrix} e^{At_1}x_0. \tag{2.74}$$

Since $C \in \mathbb{R}^{n\times n}$ is invertible, it follows from (2.74) that

$$\begin{aligned} \hat{x}_0 &= e^{At_1}x_0, \\ A\hat{x}_0 &= -Ae^{At_1}x_0, \\ &\vdots \end{aligned}$$

$$A^{n-1}\hat{x}_0 = (-1)^{n-1}A^{n-1}e^{At_1}x_0,$$

which implies that $A\hat{x}_0 = Ae^{At_1}x_0 = -Ae^{At_1}x_0 = 0$. Thus, $\hat{x}_0 \in \mathcal{N}(A)$ is an equilibrium of (2.66). Furthermore, $x_0 = e^{-At_1}\hat{x}_0$ and, hence, $Ax_0 = Ae^{-At_1}\hat{x}_0 = 0$, which implies that $x_0 \in \mathcal{N}(A)$ is an equilibrium point of (2.66) and, hence, $\hat{x}_0 = x_0$. Thus, (2.69) has only solutions $\hat{x}_0 = x_0$ that are equilibrium points of (2.66). Therefore, (2.66) and (2.67) is not output reversible. □

Proposition 2.10 implies that full state reversibility, that is, $C = I_n$, in linear dynamical systems is impossible. That is, if $C = I_n$, then there does *not* exist an initial condition to generate the solution that will retrace backwards the original solution with time going forward. This conclusion is natural in light of uniqueness of solutions of linear dynamical systems.

Next, we use the fact [33, p. 422] that the state transition operator e^{At} of $A \in \mathbb{R}^{n\times n}$ can be written as a polynomial in A of the form

$$e^{At} = \sum_{i=0}^{n-1} \phi_i(t)A^i. \tag{2.75}$$

The coefficients $\phi_0(t), \ldots, \phi_{n-1}(t)$ are real linear combinations of terms of the form $t^r\mathrm{Re}\ e^{\lambda t}$ and $t^r\mathrm{Im}\ e^{\lambda t}$, where λ is an eigenvalue of A and r is a nonnegative integer. Explicitly, $\phi_i(t)$ is given by the contour integration [33, p. 423]

$$\phi_i(t) = \frac{1}{2\pi\jmath}\oint_{\mathcal{C}} \frac{p^{[i+1]}(z)}{p(z)}e^{tz}\,\mathrm{d}z, \quad i = 0, \ldots, n-1, \tag{2.76}$$

where $\mathcal{C}$ is a clockwise contour enclosing the spectrum of A,

$$p(s) = s^n + \beta_{n-1}s^{n-1} + \cdots + \beta_1 s + \beta_0 \tag{2.77}$$

is the characteristic polynomial of A, that is, $p(s) = \det(sI - A)$, and, for all $i = 0, \ldots, n-1$,

$$p^{[i+1]}(s) = s^{n-i-1} + \beta_{n-1}s^{n-i-2} + \beta_{n-2}s^{n-i-3} + \cdots + \beta_{i+1}. \tag{2.78}$$

Note that $p^{[n]}(s) = 1$. The polynomials $p^{[i+1]}(s)$ satisfy the recursion [33]

$$sp^{[i+1]}(s) = p^{[i]}(s) - \beta_i, \quad i = 0, \ldots, n-1, \tag{2.79}$$

where $p^{[0]}(s) \triangleq p(s)$.

Since (A, C) is observable, it follows from [33, p. 552] that if $l = 1$, then A is cyclic (nonderogatory), and thus, its minimal polynomial coincides with its characteristic polynomial [33, p. 179]. Recall that A is cyclic if and only if A has exactly one Jordan block associated with each distinct eigenvalue.

The next proposition shows that if A is cyclic, then the coefficients satisfying (2.75) are unique.

Proposition 2.11. If $A \in \mathbb{R}^{n\times n}$ is cyclic, then the functions $\phi_0(t), \ldots, \phi_{n-1}(t)$, $t \geq 0$, satisfying (2.75) are unique.

Proof. Let $\hat{\phi}_0(t), \ldots, \hat{\phi}_{n-1}(t)$, $t \geq 0$, satisfy

$$e^{At} = \sum_{i=0}^{n-1} \hat{\phi}_i(t) A^i, \quad t \geq 0. \tag{2.80}$$

Subtracting (2.80) from (2.75) yields

$$\sum_{i=0}^{n-1} [\phi_i(t) - \hat{\phi}_i(t)] A^i = 0, \quad t \geq 0. \tag{2.81}$$

For each $t \geq 0$, the left-hand side of (2.81) represents a polynomial of degree $n-1$ with root A. However, since A is cyclic, its minimal polynomial is equal to its characteristic polynomial; see Proposition 5.5.20 of [33]. Thus, $\phi_i(t) - \hat{\phi}_i(t) = 0$, $t \geq 0$, $i = 0, \ldots, n-1$. □

Define $r \triangleq \operatorname{rank} A$ and note that the dimension of $\mathcal{N}(A)$ is $n - r$ [33, Corollary 2.5.1], that is, $\mathcal{N}(A)$ contains $n - r$ linearly independent vectors.

Lemma 2.3. Let $t^* > 0$ and assume there exist r linearly independent vectors $x_1, \ldots, x_r \in \mathbb{R}^n$ such that $x_i \notin \mathcal{N}(A)$, $i = 1, \ldots, r$, and

$$\operatorname{rank}\,[\mathcal{O} \quad \mathcal{S}\mathcal{O}e^{At^*}x_i] = n, \quad i = 1, \ldots, r. \tag{2.82}$$

Then, for all $x_0 \in \mathbb{R}^n$ and $t_1 > 0$,

$$\operatorname{rank}\,[\mathcal{O} \quad \mathcal{S}\mathcal{O}e^{At_1}x_0] = n. \tag{2.83}$$

Proof. Let $x_{r+1}, \ldots, x_n \in \mathcal{N}(A)$ be linearly independent. Next, let $\hat{x}_0 \in \mathbb{R}^n$ satisfy

$$\mathcal{O}\hat{x}_0 = \mathcal{S}\mathcal{O}e^{At^*}x_i, \quad i = r+1, \ldots, n. \tag{2.84}$$

Since $Ax_i = 0$, $i = r+1, \ldots, n$, it follows that (2.84) is equivalent to

$$\begin{bmatrix} C\hat{x}_0 \\ CA\hat{x}_0 \\ \vdots \\ CA^{n-1}\hat{x}_0 \end{bmatrix} = \begin{bmatrix} Cx_i \\ 0 \\ \vdots \\ 0 \end{bmatrix}, \quad i = r+1, \ldots, n. \tag{2.85}$$

Note that (2.84) holds with $\hat{x}_0 = x_i$ for each $i = r+1, \ldots, n$. It follows from Theorem 2.6.3 of [33] and the fact that (A, C) is observable that, for all $i = r+1, \ldots, n$, $\operatorname{rank}\,[\mathcal{O} \quad \mathcal{S}\mathcal{O}e^{At^*}x_i] = n$, and thus, $\hat{x}_0 = x_i$ is the unique solution to (2.84) for each $i = r+1, \ldots, n$. Thus, it follows from the above arguments and (2.82) that each vector $\mathcal{S}\mathcal{O}e^{At^*}x_i$, $i = 1, \ldots, n$, is a linear combination of the columns of $\mathcal{O} \in \mathbb{R}^{nl \times n}$.

Note that $x_1, \ldots, x_n$ are linearly independent vectors and, thus, form a basis in $\mathbb{R}^n$. To see this, note that every vector $x \in \mathbb{R}^n$ can be represented as $x = y + z$, where $y \in \mathcal{N}(A)$ and $z \in \mathbb{R}^n \backslash \mathcal{N}(A)$. Since $x_1, \ldots, x_r$ form a basis in $\mathbb{R}^n \backslash \mathcal{N}(A)$ and $x_{r+1}, \ldots, x_n$ form a basis in $\mathcal{N}(A)$, it follows that y and z are linear combinations of $x_{r+1}, \ldots, x_n$ and $x_1, \ldots, x_r$, respectively. Hence, $x \in \mathbb{R}^n$ is a linear combination of x_i, $i = 1, \ldots, n$, and hence, $x_1, \ldots, x_n$ form a basis in $\mathbb{R}^n$.

Now, let $x_0 \in \mathbb{R}^n$ and $t_1 > 0$. Since $x_1, \ldots, x_n$ form a basis in $\mathbb{R}^n$, there exist $\alpha_1, \ldots, \alpha_n \in \mathbb{R}$ such that $e^{A(t_1 - t^*)}x_0 = \sum_{i=1}^n \alpha_i x_i$. Furthermore,

$$\mathcal{S}\mathcal{O}e^{At_1}x_0 = \mathcal{S}\mathcal{O}e^{At^*}e^{A(t_1-t^*)}x_0 = \sum_{i=1}^{n} \alpha_i \mathcal{S}\mathcal{O}e^{At^*}x_i. \tag{2.86}$$

Since, for all $i = 1, \ldots, n$, $\mathcal{S}\mathcal{O}e^{At^*}x_i$ is a linear combination of the columns of $\mathcal{O}$, it follows from (2.86) that $\mathcal{S}\mathcal{O}e^{At_1}x_0$ is also a linear combination of the columns of $\mathcal{O}$. Thus, $\operatorname{rank}[\mathcal{O} \quad \mathcal{S}\mathcal{O}e^{At_1}x_0] = n$ for every $x_0 \in \mathbb{R}^n$ and $t_1 > 0$, which proves the result. $\square$

For the following result, define $\Phi(\cdot) \in \mathbb{R}^{nl \times l}$ and $\phi(\cdot) \in \mathbb{R}^n$ given by

$$\Phi(t) \triangleq \begin{bmatrix} \phi_0(t)I_l \\ \vdots \\ \phi_{n-1}(t)I_l \end{bmatrix}, \quad \phi(t) \triangleq \begin{bmatrix} \phi_0(t) \\ \vdots \\ \phi_{n-1}(t) \end{bmatrix}, \tag{2.87}$$

for all $t \geq 0$. Substituting (2.75) into (2.68) yields

$$\Phi^{\mathrm{T}}(t)\mathcal{O}\hat{x}_0 = \Phi^{\mathrm{T}}(-t)\mathcal{O}e^{At_1}x_0, \quad t \geq 0. \tag{2.88}$$

Note that (2.68) and (2.88) are equivalent.

Proposition 2.12. Let $t^* > 0$ and $r = \operatorname{rank}(A)$, and assume there exist r linearly independent vectors $x_1, \ldots, x_r \in \mathbb{R}^n$ such that $x_i \notin \mathcal{N}(A)$, $i = 1, \ldots, r$, and

$$\operatorname{rank}[\mathcal{O} \quad \mathcal{S}\mathcal{O}e^{At^*}x_i] = n, \quad i = 1, \ldots, r. \tag{2.89}$$

The linear system (2.66) and (2.67) is output reversible if and only if

$$\phi(-t) = S\phi(t), \quad t \geq 0, \tag{2.90}$$

where $S \triangleq \operatorname{diag}[1, -1, 1, \cdots, (-1)^{n-1}]$.

Proof. First, note that it follows from Lemma 2.3 that for all $x_0 \in \mathbb{R}^n$

and $t_1 > 0$, $\operatorname{rank}[\mathcal{O} \quad \mathcal{S}\mathcal{O}e^{At_1}x_0] = n$. In addition, note that (2.90) is equivalent to $\Phi(-t) = \mathcal{S}\Phi(t)$, $t \geq 0$, where $\mathcal{S}$ is given by (2.72). To prove necessity, assume that (2.66) and (2.67) is output reversible so that, by Proposition 2.9, $\hat{x}_0$ satisfies (2.69). Substituting (2.69) into (2.88) implies that, for all $x_0 \in \mathbb{R}^n$ and $t_1 > 0$, equality $\Phi^{\mathrm{T}}(t)\mathcal{S}\mathcal{O}e^{At_1}x_0 = \Phi^{\mathrm{T}}(-t)\mathcal{O}e^{At_1}x_0$, $t \geq 0$, holds.

Consequently, it follows that, for all $q \in \mathbb{R}^n$, $\Phi^{\mathrm{T}}(t)\mathcal{S}\mathcal{O}q = \Phi^{\mathrm{T}}(-t)\mathcal{O}q$, $t \geq 0$. Furthermore, since (A, C) is observable, it follows that $\mathcal{O} \in \mathbb{R}^{nl \times n}$ is full rank, and hence, for all $z \in \mathbb{R}^{nl}$, $\Phi^{\mathrm{T}}(t)\mathcal{S}z = \Phi^{\mathrm{T}}(-t)z$, $t \geq 0$, which implies that $\Phi^{\mathrm{T}}(t)\mathcal{S} = \Phi^{\mathrm{T}}(-t)$, $t \geq 0$, which is equivalent to (2.90).

Conversely, it follows from (2.90) that $\Phi^{\mathrm{T}}(-t)\mathcal{O}e^{At_1}x_0 = \Phi^{\mathrm{T}}(t)\mathcal{S}\mathcal{O} e^{At_1}x_0$. Since (A, C) is observable and $\operatorname{rank}[\mathcal{O} \quad \mathcal{S}\mathcal{O}e^{At_1}x_0] = n$, there exists a unique solution $\hat{x}_0 \in \mathbb{R}^n$ satisfying (2.69), and hence,

$$\Phi^{\mathrm{T}}(-t)\mathcal{O}e^{At_1}x_0 = \Phi^{\mathrm{T}}(t)\mathcal{S}\mathcal{O}e^{At_1}x_0 = \Phi^{\mathrm{T}}(t)\mathcal{O}\hat{x}_0, \quad t \geq 0,$$

which implies (2.88). Hence, (2.66) and (2.67) is output reversible. □

In the case of a single output, that is, $C \in \mathbb{R}^{1 \times n}$, with (A, C) observable, condition $\operatorname{rank}[\mathcal{O} \quad \mathcal{S}\mathcal{O}e^{At_1}x_0] = n$ is satisfied since $\mathcal{O} \in \mathbb{R}^{n \times n}$ is invertible. Thus, in the single output case, the output reversibility of (2.66) and (2.67) is independent of C so long as (A, C) is observable.

The following lemma is needed for the main result of this section.

Lemma 2.4. The multispectrum of $A \in \mathbb{R}^{n \times n}$ is symmetric with respect to the imaginary axis if and only if $p(-s) = (-1)^n p(s)$ for all $s \in \mathbb{C}$.

Proof. Sufficiency is immediate. To show necessity, assume that the spectrum of A is symmetric with respect to the imaginary axis. In this case, the multispectrum of A is given by

$$\operatorname{mspec}(A) = \{0, \ldots, 0, \lambda_1, \ldots, \lambda_k, -\lambda_1, \ldots, -\lambda_k\}_{\mathrm{m}},$$

where $\lambda_1, \ldots, \lambda_k \in \mathbb{C}$ are nonzero and the multiplicity of the zero eigenvalue is $r = n - 2k$. Thus,

$$p(s) = s^r \prod_{i=1}^{k} (s - \lambda_i) \prod_{i=1}^{k} (s + \lambda_i).$$

Hence,

$$p(-s) = (-1)^r (-1)^{2k} s^r \prod_{i=1}^{k} (s - \lambda_i) \prod_{i=1}^{k} (s + \lambda_i) = (-1)^{r+2k} p(s) = (-1)^n p(s),$$

which proves the result. □

The following theorem is the main result of the section.

Theorem 2.18. Let $t^* > 0$ and $r = \mathrm{rank}\,(A)$, and assume there exist r linearly independent vectors $x_1, \ldots, x_r \in \mathbb{R}^n$ such that $x_i \notin \mathcal{N}(A)$, $i = 1, \ldots, r$, and

$$\mathrm{rank}\,[\mathcal{O} \quad \mathcal{S}\mathcal{O}e^{At^*}x_i] = n, \quad i = 1, \ldots, r. \tag{2.91}$$

Then (2.66) and (2.67) is output reversible if and only if the multispectrum of A is symmetric with respect to the imaginary axis.

Proof. To prove sufficiency, assume that the spectrum of A is symmetric with respect to the imaginary axis. In this case, it follows from Lemma 2.4 that $p(-s) = (-1)^n p(s)$ for all $s \in \mathbb{C}$. Let $i \in \{0, \ldots, n-1\}$ and $t \geq 0$. Then it follows from (2.76) that

$$\phi_i(-t) = \frac{1}{2\pi\jmath} \oint_{\mathcal{C}} \frac{p^{[i+1]}(z)}{p(z)} e^{-tz}\,\mathrm{d}z = \frac{(-1)^{n-1}}{2\pi\jmath} \oint_{\mathcal{C}} \frac{p^{[i+1]}(-z)}{p(z)} e^{tz}\,\mathrm{d}z,$$

where $p(s)$ and $p^{[i+1]}(s)$ are given by (2.77) and (2.78), respectively.

Next, equating coefficients of equal powers in $p(-s) = (-1)^n p(s)$ yields $\beta_{n-1} = \beta_{n-3} = \cdots = \beta_1 = 0$ if n is even, and $\beta_{n-1} = \beta_{n-3} = \cdots = \beta_0 = 0$ if n is odd. Now, assume that $n - i$ is even. Then

$$p^{[i+1]}(s) = s^{n-i-1} + \beta_{n-2}s^{n-i-3} + \cdots + \beta_{i+2}s. \tag{2.92}$$

Hence,

$$p^{[i+1]}(-s) = -(s^{n-i-1} + \beta_{n-2}s^{n-i-3} + \cdots + \beta_{i+2}s) = -p^{[i+1]}(s).$$

Alternatively, assume that $n - i$ is odd. Then

$$p^{[i+1]}(s) = s^{n-i-1} + \beta_{n-2}s^{n-i-3} + \cdots + \beta_{i+3}s^2 + \beta_{i+1}. \tag{2.93}$$

Thus,

$$p^{[i+1]}(-s) = p^{[i+1]}(s). \tag{2.94}$$

Hence, in both cases,

$$p^{[i+1]}(-s) = (-1)^{n-i-1}p^{[i+1]}(s). \tag{2.95}$$

Thus,

$$\phi_i(-t) = \frac{(-1)^{n-1}(-1)^{n-i-1}}{2\pi\jmath} \oint_{\mathcal{C}} \frac{p^{[i+1]}(z)}{p(z)} e^{tz}\,\mathrm{d}z = (-1)^i \phi_i(t), \quad t \geq 0.$$

Consequently, $\phi(-t) = S\phi(t)$ for all $t \geq 0$. Hence, it follows from Proposition 2.12 that (2.66) and (2.67) is output reversible.

To prove necessity, assume that (2.66) and (2.67) is output reversible. Hence, it follows from Proposition 2.12 that $\phi(-t) = S\phi(t)$, $t \geq 0$, or, equivalently, for every $i = 0, \ldots, n-1$,

$$\begin{aligned}\phi_i(-t) &= \frac{1}{2\pi\jmath}\oint_{\mathcal{C}} \frac{p^{[i+1]}(z)}{p(z)} e^{-tz}\,\mathrm{d}z \\ &= \frac{(-1)^i}{2\pi\jmath}\oint_{\mathcal{C}} \frac{p^{[i+1]}(z)}{p(z)} e^{tz}\,\mathrm{d}z \\ &= \frac{(-1)^{i+1}}{2\pi\jmath}\oint_{\mathcal{C}^-} \frac{p^{[i+1]}(-z)}{p(-z)} e^{-tz}\,\mathrm{d}z, \end{aligned} \tag{2.96}$$

for all $t \geq 0$, where $\mathcal{C}$ and $\mathcal{C}^-$ are contours in $\mathbb{C}$ enclosing the roots of $p(s) = 0$ and $p(-s) = 0$, respectively. Since $p(s) = \det(sI - A)$ is a characteristic polynomial of A, the roots of $p(s) = 0$ are given by $\operatorname{mspec}(A) = \{\lambda_1, \ldots, \lambda_n\}_{\mathrm{m}}$, while the roots of $p(-s) = 0$ are given by $\{-\lambda_1, \ldots, -\lambda_n\}_{\mathrm{m}}$. Therefore, for $i = n-1$, (2.96) is equivalent to

$$\frac{1}{2\pi\jmath}\oint_{\mathcal{C}} \frac{1}{p(z)} e^{-tz}\,\mathrm{d}z = \frac{(-1)^n}{2\pi\jmath}\oint_{\mathcal{C}^-} \frac{1}{p(-z)} e^{-tz}\,\mathrm{d}z, \tag{2.97}$$

where $t \geq 0$, which implies that

$$\sum_{k=1}^{n} \operatorname*{Res}_{z=\lambda_k}\left[\frac{1}{p(z)}\right] e^{-t\lambda_k} = (-1)^n \sum_{k=1}^{n} \operatorname*{Res}_{z=-\lambda_k}\left[\frac{1}{p(-z)}\right] e^{t\lambda_k}, \quad t \geq 0, \tag{2.98}$$

where $\operatorname{Res}[\cdot]$ denotes residue.

If $\operatorname{Re}\lambda_k = 0$, $k = 1, \ldots, n$, then $\operatorname{mspec}(A)$ is symmetric with respect to the imaginary axis. Alternatively, assume there exist $\lambda_l \in \operatorname{mspec}(A)$ such that $\operatorname{Re}\lambda_l \neq 0$ for all $l \in \mathcal{N} \subseteq \{1, \ldots, n\}$. Let λ_m, $m \in \mathcal{N}$, be such that $|\operatorname{Re}\lambda_m| = \max_{l\in\mathcal{N}} |\operatorname{Re}\lambda_l|$. Thus, in order for (2.98) to hold for large $t > 0$, there must exist $\lambda_p \in \operatorname{mspec}(A)$ such that $-\lambda_p = \lambda_m$ and

$$\operatorname*{Res}_{z=\lambda_p}\left[\frac{1}{p(z)}\right] = (-1)^n \operatorname*{Res}_{z=-\lambda_m}\left[\frac{1}{p(-z)}\right], \tag{2.99}$$

which implies that

$$\operatorname*{Res}_{z=\lambda_p}\left[\frac{1}{p(z)}\right] e^{-\lambda_p t} - (-1)^n \operatorname*{Res}_{z=-\lambda_m}\left[\frac{1}{p(-z)}\right] e^{\lambda_m t} = 0, \quad t \geq 0. \tag{2.100}$$

Next, let λ_q, $q \in \mathcal{N}$, be such that the absolute value of its real part is closest to $|\operatorname{Re}\lambda_m|$ to establish the existence of $\lambda_s \in \operatorname{mspec}(A)$ such that $-\lambda_s = \lambda_q$. Recursively repeating this procedure for all λ_l, $l \in \mathcal{N}$, yields

$$\{\lambda_1, \ldots, \lambda_n\} = \{-\lambda_1, \ldots, -\lambda_n\}, \tag{2.101}$$

which implies that the eigenvalues of A are symmetric with respect to the imaginary axis. □

The following corollary specializes Theorem 2.18 to the case where $l = 1$ and recovers Theorem 2.8 of [36].

Corollary 2.5. Assume $C \in \mathbb{R}^{1\times n}$. Then (2.66) and (2.67) is output reversible if and only if the spectrum of A is symmetric with respect to the imaginary axis.

Proof. Since $C \in \mathbb{R}^{1\times n}$ and (A, C) is observable, then $\mathcal{O} \in \mathbb{R}^{n\times n}$ is invertible, and hence, rank $[\mathcal{O} \quad \mathcal{S}\mathcal{O}e^{At_1}x_0] = n$ for all $x_0 \in \mathbb{R}^n$ and $t_1 > 0$. The result now follows from Theorem 2.18. □

2.9 Reversible Dynamical Systems, Volume-Preserving Flows, and Poincaré Recurrence

The notion of R-state reversibility introduced in Section 2.7 is one of the fundamental symmetries that arise in natural science. This notion can also be characterized by the flow of a dynamical system. In particular, consider the dynamical system (2.1) given by

$$\dot{x}(t) = f(x(t)), \quad x(t_0) = x_0, \quad t \in \mathcal{I}_{x_0}, \tag{2.102}$$

where $x(t) \in \mathcal{D} \subseteq \mathbb{R}^q$, $t \in \mathcal{I}_{x_0}$, is the system state vector, $\mathcal{D}$ is an open subset of $\mathbb{R}^q$, and $f : \mathcal{D} \to \mathbb{R}^q$ is locally Lipschitz continuous on $\mathcal{D}$. Since $f(\cdot)$ is locally Lipschitz continuous on $\mathcal{D}$, it follows from Theorem 3.1 of [202, p. 18] that there exists a unique solution to (2.102). In this case, the semigroup property $s(t+\tau, x_0) = s(t, s(\tau, x_0))$, $t+\tau$, $\tau \in \mathcal{I}_{x_0}$ and $t \in \mathcal{I}_{s(\tau,x_0)}$, and the continuity of $s(t,\cdot)$ on $\mathcal{D}$, $t \in \mathcal{I}_{x_0}$, hold.

Now, in terms of the flow $s_t : \mathcal{D} \to \mathcal{D}$ of (2.102), the consistency and semigroup properties of (2.102) can be equivalently written as $s_{t_0}(x_0) = x_0$ and $(s_\tau \circ s_t)(x_0) = s_\tau(s_t(x_0)) = s_{t+\tau}(x_0)$, where "$\circ$" denotes the composition operator. Next, it follows from continuity of solutions and the semigroup property that the map $s_t : \mathcal{D} \to \mathcal{D}$ is a continuous function with a continuous inverse s_{-t}. Thus, s_t, $t \in \mathcal{I}_{x_0}$, generates a one-parameter family of homeomorphisms on $\mathcal{D}$ forming a commutative group under composition.

To show that R-state reversibility can be characterized by the flow of (2.102), let $\mathcal{R} : \mathcal{D} \to \mathcal{D}$ be a continuous map of (2.102) such that

$$\dot{\mathcal{R}}(x(t)) = -f(\mathcal{R}(x(t))), \quad \mathcal{R}(x(t_0)) = \mathcal{R}(x_0), \quad t \in \mathcal{I}_{\mathcal{R}(x_0)}. \tag{2.103}$$

Now, it follows from (2.103) that

$$\mathcal{R} \circ s_t = s_{-t} \circ \mathcal{R}, \quad t \in \mathcal{I}_{x_0}. \tag{2.104}$$

Condition (2.104), with $\mathcal{R}(\cdot)$ satisfying (2.103), defines an R-reversed trajectory of (2.102) in the sense of Definition 2.21 with $\tau(t) = t$.

In the context of classical mechanics involving the *configuration manifold* (space of generalized positions) $\mathcal{Q} = \mathbb{R}^n$ with governing equations given by

$$\dot{q}(t) = \left(\frac{\partial \mathcal{H}(q(t), p(t))}{\partial p(t)}\right)^{\mathrm{T}}, \quad q(t_0) = q_0, \quad t \geq t_0, \tag{2.105}$$

$$\dot{p}(t) = -\left(\frac{\partial \mathcal{H}(q(t), p(t))}{\partial q(t)}\right)^{\mathrm{T}}, \quad p(t_0) = p_0, \tag{2.106}$$

where $q \in \mathbb{R}^n$ denotes generalized system positions, $p \in \mathbb{R}^n$ denotes generalized system momenta, $\mathcal{H} : \mathbb{R}^n \times \mathbb{R}^n \to \mathbb{R}$ is the system Hamiltonian given by $\mathcal{H}(q,p) \triangleq \dot{q}^{\mathrm{T}} p - \mathcal{L}(q, \dot{q})$, $\mathcal{L}(q, \dot{q})$ is the system Lagrangian,[8] and $p(q, \dot{q}) \triangleq \left(\frac{\partial \mathcal{L}(q,\dot{q})}{\partial \dot{q}}\right)^{\mathrm{T}}$, the reversing symmetry $\mathcal{R} : \mathbb{R}^n \times \mathbb{R}^n \to \mathbb{R}^n \times \mathbb{R}^n$ is such that $\mathcal{R}(q,p) = (q, -p)$ and satisfies (2.103). In this case, $\mathcal{R}$ is an involution. This implies that if $(q(t), p(t))$, $t \geq t_0$, is a solution to (2.105) and (2.106), then $(q(-t), -p(-t))$, $t \geq t_0$, is also a solution to (2.105) and (2.106) with initial condition $(q_0, -p_0)$. In the configuration space, this clearly shows the time-reversal nature of lossless mechanical systems.

Reversible dynamical systems tend to exhibit a phenomenon known as *Poincaré recurrence* [10]. Poincaré recurrence states that if a dynamical system has a fixed total energy that restricts its dynamics to bounded subsets of its state space, then the dynamical system will eventually return arbitrarily close to its initial system state infinitely often. More precisely, Poincaré [364] established the fact that if the flow of a dynamical system preserves volume and has only bounded orbits, then for each open set there exist orbits that intersect the set infinitely often. In order to state the Poincaré recurrence theorem, the following definitions are needed.

Definition 2.25. Let $\mathcal{V} \subset \mathbb{R}^q$ be a bounded set. The *volume* $\mathcal{V}_{\mathrm{vol}}$ of $\mathcal{V}$ is defined as

$$\mathcal{V}_{\mathrm{vol}} \triangleq \int_{\mathcal{V}} \mathrm{d}\mathcal{V}. \tag{2.107}$$

Definition 2.26. Let $\mathcal{V} \subset \mathbb{R}^q$ be a bounded set. A map $g : \mathcal{V} \to \mathcal{Q}$, where $\mathcal{Q} \subset \mathbb{R}^q$, is *volume-preserving* if for every $\mathcal{V}_0 \subset \mathcal{V}$, the volume of $g(\mathcal{V}_0)$ is equal to the volume of $\mathcal{V}_0$.

The following theorem, known as Liouville's theorem [10], establishes

[8] Here we assume that the system Lagrangian is *hyperregular* [279] so that the map from the generalized velocities $\dot{q}$ to the generalized momenta p is *bijective* (i.e., one-to-one and onto).

sufficient conditions for volume-preserving flows. For the statement of this theorem, consider the nonlinear dynamical system (2.102) and define the divergence of $f = [f_1, \ldots, f_q]^{\mathrm{T}} : \mathcal{D} \to \mathbb{R}^q$ by

$$\nabla \cdot f(x) \triangleq \sum_{i=1}^{q} \frac{\partial f_i(x)}{\partial x_i}, \tag{2.108}$$

where ∇ denotes the nabla operator, " $\cdot$ " denotes the dot product in $\mathbb{R}^q$, and x_i denotes the ith component of x.

Theorem 2.19. Consider the nonlinear dynamical system (2.102). If $\nabla \cdot f(x) \equiv 0$, then the flow $s_t : \mathcal{D} \to \mathcal{D}$ of (2.102) is volume-preserving.

Proof. Let $\mathcal{V} \subset \mathbb{R}^q$ be a compact set such that its image at time t under the mapping $s_t(\cdot)$ is given by $s_t(\mathcal{V})$. In addition, let $\mathrm{d}\mathcal{S}_{\mathcal{V}}$ denote an infinitesimal surface element of the boundary of the set $\mathcal{V}$ and let $\hat{n}(x)$, $x \in \partial\mathcal{V}$, denote an outward normal vector to the boundary of $\mathcal{V}$. Then the change in volume of $s_t(\mathcal{V})$ at $t = t_0$ is given by

$$\mathrm{d}s_t(\mathcal{V})_{\mathrm{vol}} = \int_{\partial\mathcal{V}} (f(x) \cdot \hat{n}(x)) \mathrm{d}t \mathrm{d}\mathcal{S}_{\mathcal{V}}, \tag{2.109}$$

which, using the divergence theorem, implies that

$$\left. \frac{\mathrm{d}s_t(\mathcal{V})_{\mathrm{vol}}}{\mathrm{d}t} \right|_{t=t_0} = \int_{\partial\mathcal{V}} (f(x) \cdot \hat{n}(x)) \mathrm{d}\mathcal{S}_{\mathcal{V}} = \int_{\mathcal{V}} \nabla \cdot f(x) \mathrm{d}\mathcal{V}. \tag{2.110}$$

Hence, if $\nabla \cdot f(x) \equiv 0$, then $s_t(\cdot)$ is a volume-preserving map. □

Volume preservation is the key conservation law underlying statistical mechanics. The flows of volume-preserving dynamical systems belong to one of the Lie pseudogroups[9] of diffeomorphisms. These systems arise in incompressible fluid dynamics, classical mechanics, and acoustics.

Next, we state the well-known Poincaré recurrence theorem. For the statement of this result let $g^{(n)}(x)$, $n \in \overline{\mathbb{Z}}_+$, denote the n-times composition operator of $g(x)$ with itself and define $g^{(0)}(x) \triangleq x$.

Theorem 2.20. Let $\mathcal{D} \subset \mathbb{R}^q$ be an open bounded set and let $g : \mathcal{D} \to \mathcal{D}$ be a continuous, volume-preserving bijective (one-to-one and onto) map. Then for every open set $\mathcal{N} \subset \mathcal{D}$, there exists $n \in \mathbb{Z}_+$ such that $g^{(n)}(\mathcal{N}) \cap \mathcal{N} \neq \emptyset$. Furthermore, there exists a point $x \in \mathcal{N}$ that returns to $\mathcal{N}$, that is, $g^{(n)}(x) \in \mathcal{N}$ for some $n \in \mathbb{Z}_+$.

[9] A *group* is a set that is closed under an associative binary operation with respect to which there exists a unique identity element within the set and every element in the set has an inverse. A *Lie group* is a topological group that can be given an analytic structure such that the group operation and inversion are analytic. A *Lie pseudogroup* is an infinite-dimensional counterpart of a Lie group.

Proof. The proof of this result is standard; see, for example, [10, p. 72]. For completeness of exposition, however, we provide a proof here. First, note that the images $g^{(p)}(\mathcal{N})$, $p \in \overline{\mathbb{Z}}_+$, under the mapping $g(\cdot)$ of the neighborhood $\mathcal{N} \subset \mathcal{D}$ have the same volume and are all contained in $\mathcal{D}$. Next, define the union of all the images of $\mathcal{N}$ by

$$\mathcal{V} \triangleq \bigcup_{p=0}^{\infty} g^{(p)}(\mathcal{N}) \subset \mathcal{D}. \tag{2.111}$$

Since the volume of a union of disjoint sets is the sum of the individual set volumes, it follows that if $g^{(p)}(\mathcal{N})$, $p \in \overline{\mathbb{Z}}_+$, are disjoint, then $\mathcal{V}_{\mathrm{vol}} = \infty$. However, $\mathcal{V} \subset \mathcal{D}$ and $\mathcal{D}$ is a bounded set by assumption. Hence, there exist $k, l \in \overline{\mathbb{Z}}_+$, with $k > l$, such that $g^{(k)}(\mathcal{N}) \cap g^{(l)}(\mathcal{N}) \neq \emptyset$. Now, applying the inverse $g^{(-1)}$ to this relation l times and using the fact that $g(\cdot)$ is a bijective map, it follows that $g^{(k-l)}(\mathcal{N}) \cap \mathcal{N} \neq \emptyset$. Thus, $g^{(n)}(\mathcal{N}) \cap \mathcal{N} \neq \emptyset$, where $n = k - l$. Hence, there exists a point $x \in \mathcal{N}$ such that $g^{(n)}(x) \in g^{(n)}(\mathcal{N}) \cap \mathcal{N} \subseteq \mathcal{N}$. □

The next result establishes the existence of a point x in $\mathcal{D} \subset \mathbb{R}^q$ such that $\lim_{i\to\infty} g^{(n_i)}(x) = x$ for some sequence $\{n_i\}_{i=1}^{\infty}$, with $n_i \to \infty$ as $i \to \infty$, under a continuous, volume-preserving bijective mapping $g(\cdot)$, which maps a bounded region $\mathcal{D}$ of a Euclidean space onto itself. Hence, x returns infinitely often to every open neighborhood of itself under the mapping $g(\cdot)$.

Theorem 2.21. Let $\mathcal{D} \subset \mathbb{R}^q$ be an open bounded set and let $g : \mathcal{D} \to \mathcal{D}$ be a continuous, volume-preserving bijective map. Then for every open neighborhood $\mathcal{N} \subset \mathcal{D}$, there exists a point $x \in \mathcal{N}$ such that $\lim_{i\to\infty} g^{(n_i)}(x) = x$ for some sequence $\{n_i\}_{i=1}^{\infty}$, with $n_i \to \infty$ as $i \to \infty$. Hence, $x \in \mathcal{N}$ returns to $\mathcal{N}$ infinitely often, that is, there exists a sequence $\{n_i\}_{i=1}^{\infty}$, with $n_i \to \infty$ as $i \to \infty$, such that $g^{(n_i)}(x) \in \mathcal{N}$ for all $i \in \mathbb{Z}_+$.

Proof. Let $\mathcal{N} \subset \mathcal{D}$ be an open set and let $\mathcal{N}_1 \triangleq \mathcal{B}_{\delta_1}(x_1)$ be such that $\overline{\mathcal{N}}_1 \subset \mathcal{N}$ for some $\delta_1 > 0$ and $x_1 \in \mathcal{N}$. Applying Theorem 2.20, with $g(\cdot)$ replaced by $g^{(-1)}(\cdot)$, it follows that there exists $n_1 \in \mathbb{Z}_+$ such that $g^{(-n_1)}(\mathcal{N}_1) \cap \mathcal{N}_1 \neq \emptyset$, which implies that $g^{(-n_1)}(\overline{\mathcal{N}}_1) \cap \overline{\mathcal{N}}_1 \neq \emptyset$. Now, let $\mathcal{N}_2 = \mathcal{B}_{\delta_2}(x_2)$ be such that $\overline{\mathcal{N}}_2 \subset g^{(-n_1)}(\mathcal{N}_1) \cap \mathcal{N}_1$ for some $\delta_2 > 0$ and $x_2 \in g^{(-n_1)}(\mathcal{N}_1) \cap \mathcal{N}_1$. Repeating the above arguments, it follows that there exists $n_2 \in \mathbb{Z}_+$, $n_2 > n_1$, such that $g^{(-n_2)}(\mathcal{N}_2) \cap \mathcal{N}_2 \neq \emptyset$ and $g^{(-n_2)}(\overline{\mathcal{N}}_2) \cap \overline{\mathcal{N}}_2 \neq \emptyset$. Repeating this process recursively, it follows that there exist sequences $\{n_i\}_{i=1}^{\infty}$ and $\{\delta_i\}_{i=1}^{\infty}$, with $n_i \to \infty$ as $i \to \infty$, $\delta_i \to 0$ as $i \to \infty$, and $\delta_i > \delta_{i+1}$, $i = 1, 2, \ldots$, such that $\mathcal{N}_i \supset \mathcal{N}_{i+1}$, $i = 1, 2, \ldots$, and $g^{(-n_i)}(\mathcal{N}_i) \cap \mathcal{N}_i \neq \emptyset$, where $\mathcal{N}_i = \mathcal{B}_{\delta_i}(x_i)$ for some $x_i \in g^{(-n_{i-1})}(\mathcal{N}_{i-1}) \cap \mathcal{N}_{i-1}$ and where $n_0 \triangleq 0$ and $\mathcal{N}_0 \triangleq \mathcal{N}$.

Now, since $\mathcal{N}_i \neq \emptyset$, $i \in \mathbb{Z}_+$, it follows from the Cantor intersection theorem [7, p. 56] that $\mathcal{Z} \triangleq \bigcap_{i=1}^{\infty} \overline{\mathcal{N}}_i \neq \emptyset$. Furthermore, since $\delta_i \to 0$ as $i \to \infty$, it follows that $\mathcal{Z}$ is a singleton. Next, let $x \in \mathcal{Z} = \{z\}$, and since for every $i \in \mathbb{Z}_+$, $\overline{\mathcal{N}}_{i+1} \subset \mathcal{N}_i$, it follows that $x \in \mathcal{N}_i$, $i \in \mathbb{Z}_+$. Now, note that $x \in \mathcal{N}_{i+1} \subset g^{(-n_i)}(\mathcal{N}_i) \cap \mathcal{N}_i$ for all $i \in \mathbb{Z}_+$, which implies that $g^{(n_i)}(x) \in \mathcal{N}_i$, $i \in \mathbb{Z}_+$. Hence, since $\delta_i \to 0$ as $i \to \infty$, it follows that $\lim_{i\to\infty} g^{(n_i)}(x) = x$. □

The next theorem strengthens Poincaré's theorem by showing that for every open neighborhood $\mathcal{N}$ of $\mathcal{D} \subset \mathbb{R}^q$, there exists a subset of $\mathcal{N}$ that is dense[10] in $\mathcal{N}$ so that almost every moving point in $\mathcal{N}$ returns repeatedly to the vicinity of its initial position under a continuous, volume-preserving bijective mapping that maps the bounded region $\mathcal{D}$ onto itself.

Theorem 2.22. Let $\mathcal{D} \subset \mathbb{R}^q$ be an open bounded set and let $g : \mathcal{D} \to \mathcal{D}$ be a continuous, volume-preserving bijective map. Then for every open neighborhood $\mathcal{N} \subset \mathcal{D}$, there exists a dense subset $\mathcal{V} \subset \mathcal{N}$ such that for every point $x \in \mathcal{V}$, $\lim_{i\to\infty} g^{(n_i)}(x) = x$ for some sequence $\{n_i\}_{i=1}^{\infty}$, with $n_i \to \infty$ as $i \to \infty$.

Proof. Let $\mathcal{N} \subset \mathcal{D}$ be an open neighborhood and define $\mathcal{V} \subset \mathcal{N}$ by

$$\mathcal{V} \triangleq \{x \in \mathcal{N} : \text{there exists a sequence } \{n_i\}_{i=1}^{\infty}, \text{ with } n_i \to \infty \text{ as } i \to \infty, \text{ such that } \lim_{i\to\infty} g^{(n_i)}(x) = x\}. \quad (2.112)$$

Now, let $x \in \mathcal{N}$ and let $\{\delta_i\}_{i=1}^{\infty}$ be a strictly decreasing positive sequence with $\delta_i \to 0$ as $i \to \infty$ and $\mathcal{B}_{\delta_1}(x) \subset \mathcal{N}$. It follows from Theorem 2.21 that for every $i \in \mathbb{Z}_+$, there exists $x_i \in \mathcal{B}_{\delta_i}(x)$ such that $\lim_{k\to\infty} g^{(n_k)}(x_i) = x_i$ for some sequence $\{n_k\}_{k=1}^{\infty}$, with $n_k \to \infty$ as $k \to \infty$, which implies that $x_i \in \mathcal{V}$, $i \in \mathbb{Z}_+$. Next, since $\lim_{i\to\infty} x_i = x$, it follows that $x \in \overline{\mathcal{V}}$, which implies that $\mathcal{V} \subseteq \mathcal{N} \subset \overline{\mathcal{V}}$, and hence, $\mathcal{V}$ is a dense subset of $\mathcal{N}$. □

It follows from Theorem 2.22 that almost every point in $\mathcal{D} \subset \mathbb{R}^q$ will return infinitely many times to every open neighborhood of itself under a continuous, volume-preserving bijective mapping that maps a bounded region $\mathcal{D}$ of a Euclidean space onto itself. The following theorem provides several equivalent statements for establishing Poincaré recurrence.

Theorem 2.23. Let $\mathcal{D} \subset \mathbb{R}^q$ be an open bounded set and let $g : \mathcal{D} \to \mathcal{D}$ be a continuous, bijective map. Then the following statements are equivalent:

i) For every open set $\mathcal{N} \subset \mathcal{D}$, there exists a dense subset $\mathcal{V} \subset \mathcal{N}$ such

[10] We say that $\mathcal{V}$ is *dense* in $\mathcal{N}$ if and only if $\mathcal{N}$ is contained in the closure of $\mathcal{V}$.

that, for every point $z \in \mathcal{V}$, $\lim_{i\to\infty} g^{(n_i)}(x) = x$ for some sequence $\{n_i\}_{i=1}^{\infty}$, with $n_i \to \infty$ as $i \to \infty$.

ii) For every open set $\mathcal{N} \subset \mathcal{D}$, there exists a point $x \in \mathcal{N}$ such that $\lim_{i\to\infty} g^{(n_i)}(x) = x$ for some sequence $\{n_i\}_{i=1}^{\infty}$, with $n_i \to \infty$ as $i \to \infty$.

iii) For every open set $\mathcal{N} \subset \mathcal{D}$, there exists a point $x \in \mathcal{N}$ that returns to $\mathcal{N}$ infinitely often, that is, $g^{(n_i)}(x) \in \mathcal{N}$, $i \in \mathbb{Z}_+$, for some sequence $\{n_i\}_{i=1}^{\infty}$, with $n_i \to \infty$ as $i \to \infty$.

iv) For every open set $\mathcal{N} \subset \mathcal{D}$, there exists a point $x \in \mathcal{N}$ that returns to $\mathcal{N}$, that is, $g^{(n)}(x) \in \mathcal{N}$ for some $n \in \mathbb{Z}_+$.

v) For every open set $\mathcal{N} \subset \mathcal{D}$, there exists $n \in \mathbb{Z}_+$ such that $g^{(n)}(\mathcal{N}) \cap \mathcal{N} \neq \emptyset$.

Proof. The implication *i*) implies *ii*) follows trivially and the proof of *ii*) implies *i*) is identical to that of Theorem 2.22. The implications *ii*) implies *iii*), *iii*) implies *iv*), and *iv*) implies *v*) follow trivially. The proof of *v*) implies *ii*) is identical to that of Theorem 2.21. □

Note that it follows from Theorems 2.20, 2.21, and 2.22 that a continuous, bijective map $g : \mathcal{D} \to \mathcal{D}$ exhibits Poincaré recurrence (that is, one of the statements in Theorem 2.23 holds) if $g(\cdot)$ is volume-preserving. For the remainder of this section we consider the nonlinear dynamical system (2.102) and assume that the solutions to (2.102) are defined for all $t \in \mathbb{R}$. Recall that if all solutions to (2.102) are bounded, then it follows from the Peano-Cauchy theorem [202, pp. 16, 17] that $\mathcal{I}_{x_0} = \mathbb{R}$. The following theorem shows that if a dynamical system preserves volume, then almost all trajectories return arbitrarily close to their initial position infinitely often.

Theorem 2.24. Consider the nonlinear dynamical system (2.102). Assume that the flow $s_t : \mathcal{D} \to \mathcal{D}$ of (2.102) is volume-preserving and maps an open bounded set $\mathcal{D}_c \subset \mathbb{R}^q$ onto itself, that is, $\mathcal{D}_c$ is an invariant set with respect to (2.102). Then the nonlinear dynamical system (2.102) exhibits Poincaré recurrence, that is, almost every point $x \in \mathcal{D}_c$ returns to every open neighborhood $\mathcal{N} \subset \mathcal{D}_c$ of x infinitely many times.

Proof. Since $f : \mathcal{D} \to \mathbb{R}^q$ is locally Lipschitz continuous on $\mathcal{D}$ and $s_t(\cdot)$ maps an open bounded set $\mathcal{D}_c \subset \mathbb{R}^n$ onto itself, it follows that the solutions to (2.102) are bounded and unique for all $t \in \mathbb{R}$ and $x_0 \in \mathcal{D}_c$. Thus, the mapping $s_t(\cdot)$ is bijective. Furthermore, since the solutions of (2.102) are continuously dependent on the system's initial conditions, it follows that $s_t(\cdot)$ is continuous. Now, the result follows as a direct consequence of Theorem 2.22 with $g(\cdot) = s_t(\cdot)$ for every $t \geq t_0$. □

It follows from Theorem 2.24 that a nonlinear dynamical system exhibits Poincaré recurrence if one of the statements in Theorem 2.23 holds with $g(\cdot) = s_t(\cdot)$ for every $t \geq t_0$. Note that in this case it follows from *iv*) of Theorem 2.23 that Poincaré recurrence is equivalent to the existence of a point $x \in \mathcal{D}_{\rm c}$ such that x belongs to its positive limit set, that is, $x \in \omega(x)$.

All Hamiltonian dynamical systems of the form (2.105) and (2.106) exhibit Poincaré recurrence since they possess volume-preserving flows and are conservative in the sense that the Hamiltonian function $\mathcal{H}(q,p)$ remains constant along system trajectories. To see this, note that with $z \triangleq [q^{\rm T}, p^{\rm T}]^{\rm T}$, (2.105) and (2.106) can be rewritten as

$$\dot{x}(t) = \mathcal{J}\left(\frac{\partial \mathcal{H}}{\partial x}(x(t))\right)^{\rm T}, \quad x(t_0) = x_0, \quad t \geq t_0, \tag{2.113}$$

where $x_0 \triangleq [q_0^{\rm T}, p_0^{\rm T}]^{\rm T} \in \mathbb{R}^{2n}$ and

$$\mathcal{J} \triangleq \begin{bmatrix} 0_n & I_n \\ -I_n & 0_n \end{bmatrix}. \tag{2.114}$$

Now, since

$$\dot{\mathcal{H}}(x) = \left(\frac{\partial \mathcal{H}}{\partial x}(x)\right)\mathcal{J}\left(\frac{\partial \mathcal{H}}{\partial x}(x)\right)^{\rm T} = 0, \quad x \in \mathbb{R}^{2n}, \tag{2.115}$$

the Hamiltonian function $\mathcal{H}(\cdot)$ is conserved along the flow of (2.113). If $\mathcal{H}(\cdot)$ is bounded from below and is radially unbounded, then every trajectory of the Hamiltonian system (2.113) is bounded. Hence, by choosing the bounded region $\mathcal{D} \triangleq \{x \in \mathbb{R}^{2n} : \mathcal{H}(x) \leq \eta\}$, where $\eta \in \mathbb{R}$ and $\eta > 0$, it follows that the flow $s_t(\cdot)$ of (2.113) maps the bounded region $\mathcal{D}$ onto itself. Since $\eta > 0$ is arbitrary, the region $\mathcal{D}$ can be chosen arbitrarily large.

Furthermore, since (2.113) possesses unique solutions over $\mathbb{R}$, it follows that the mapping $s_t(\cdot)$ is one-to-one and onto. Moreover,

$$\nabla \cdot \mathcal{J}\left(\frac{\partial \mathcal{H}}{\partial x}(x)\right)^{\rm T} = \sum_{i=1}^{n} \frac{\partial^2 \mathcal{H}(q,p)}{\partial q_i \partial p_i} - \sum_{i=1}^{n} \frac{\partial^2 \mathcal{H}(q,p)}{\partial p_i \partial q_i} = 0, \quad x \in \mathbb{R}^{2n}, \tag{2.116}$$

which, by Theorem 2.19, shows that the flow $s_t(\cdot)$ of (2.113) is volume-preserving. Finally, since the flow $s_t(\cdot)$ of (2.113) is volume-preserving, continuous, and bijective, and $s_t(\cdot)$ maps a bounded region of a Euclidean space onto itself, it follows from Theorem 2.24 that the Hamiltonian dynamical system (2.113) exhibits Poincaré recurrence. That is, in any open neighborhood $\mathcal{N}$ of any point $x_0 \in \mathbb{R}^{2n}$ there exists a point $y \in \mathcal{N}$ such that the trajectory $s(t,y)$, $t \geq t_0$, of (2.113) will return to $\mathcal{N}$ infinitely many times.

Poincaré recurrence has been the main source for the long and fierce debate between the microscopic and macroscopic points of view of thermodynamics. In thermodynamic models predicated on statistical mechanics, an isolated dynamical system will return arbitrarily close to its initial state of molecular positions and velocities infinitely often. If the system entropy is determined by the state variables, then it must also return arbitrarily close to its original value, and hence, undergo cyclical changes subverting a monotonic entropy increase. This apparent contradiction between the behavior of a mechanical system of particles and the second law of thermodynamics remains one of the hardest and most controversial problems in statistical physics.

The resolution of this paradox lies in the controversial statement that as system dimensionality increases, the recurrence time increases at an extremely fast rate. Nevertheless, the shortcoming of the mechanistic worldview of thermodynamics is the absence of the emergence of damping in lossless mechanical systems. The emergence of damping is, however, ubiquitous in isolated thermodynamic systems. Hence, the development of a viable dynamical system model for thermodynamics must guarantee the absence of Poincaré recurrence.

It is important to stress here that a key distinction between thermodynamics and mechanics is that thermodynamics is a theory of *open systems*, whereas mechanics is a theory of *closed systems*. The notions, however, of open and closed systems are different in thermodynamics and dynamical systems theory. In particular, thermodynamic systems exchange matter and energy with the environment, and hence, interact with the environment. Such systems are called *open systems* in the thermodynamic literature. Systems that exchange heat (energy in transition) but not matter with the environment are called *closed*, whereas systems that do not exchange energy and matter with the environment are called *isolated*.

Alternatively, in mechanics it is always possible to include interactions with the environment (via feedback interconnecting components) within the system description to obtain an augmented *closed system* in the sense of dynamical systems theory. That is, the system can be described by an evolution law with, possibly, an output equation wherein past trajectories define the future trajectory uniquely and the system output depends on the instantaneous (present) value of the system state.

The next set of results presents sufficient conditions for the absence of Poincaré recurrence for the nonlinear dynamical system (2.102). For these results define the set of equilibria for the nonlinear dynamical system (2.102) in $\mathcal{D}$ by $\mathcal{M}_{\rm e} \triangleq \{x \in \mathcal{D} : f(x) = 0\}$.

Theorem 2.25. Consider the nonlinear dynamical system (2.102) and assume that $\mathcal{D} \setminus \mathcal{M}_{\mathrm{e}} \neq \emptyset$. Assume that there exists a continuous function $V : \mathcal{D} \to \mathbb{R}$ such that for every $x_0 \in \mathcal{D} \setminus \mathcal{M}_{\mathrm{e}}$, $V(s(t, x_0))$, $t \geq t_0$, is an increasing (respectively, decreasing) function of time. Then the nonlinear dynamical system (2.102) does not exhibit Poincaré recurrence in $\mathcal{D} \setminus \mathcal{M}_{\mathrm{e}}$. That is, for some $x \in \mathcal{D}\setminus\mathcal{M}_{\mathrm{e}}$, there exists an open neighborhood $\mathcal{N} \subset \mathcal{D}\setminus\mathcal{M}_{\mathrm{e}}$ of x such that for every $y \in \mathcal{N}$, $y \notin \omega(y)$.

Proof. Suppose, *ad absurdum*, there exists $x \in \mathcal{D} \setminus \mathcal{M}_{\mathrm{e}}$ such that for every open neighborhood $\mathcal{N}$ containing x, there exists a point $y \in \mathcal{N}$ such that $y \in \omega(y)$. Now, let $\{t_i\}_{i=1}^{\infty}$ be such that $t_i \to \infty$ as $i \to \infty$ and $s(t_i, y) \to y$ as $i \to \infty$. Since $V(\cdot)$ is continuous, it follows that $\lim_{i\to\infty} V(s(t_i, y)) = V(y)$. However, since $V(s(\cdot, y))$ is increasing, it follows that $\lim_{i\to\infty} V(s(t_i, y)) > V(y)$, which is a contradiction. The proof for the case where $V(s(t, x_0))$, $t \geq t_0$, is decreasing is identical. $\square$

For the remainder of this section let $\mathcal{D}_{\mathrm{c}} \subseteq \mathcal{D}$ be a closed invariant set with respect to the nonlinear dynamical system (2.102). The following definitions for convergence and stability with respect to a positively invariant set are needed.

Definition 2.27. The nonlinear dynamical system (2.102) is *convergent* with respect to $\mathcal{D}_{\mathrm{c}}$ if $\lim_{t\to\infty} s(t, x)$ exists for every $x \in \mathcal{D}_{\mathrm{c}}$.

Definition 2.28 ([40]). An equilibrium point $x \in \mathcal{D}_{\mathrm{c}} \subseteq \mathcal{D}$ of the nonlinear dynamical system (2.102) is *Lyapunov stable* with respect to the positively invariant set $\mathcal{D}_{\mathrm{c}}$ if, for every relatively open subset $\mathcal{N}_\varepsilon$ of $\mathcal{D}_{\mathrm{c}}$ containing x, there exists a relatively open subset $\mathcal{N}_\delta$ of $\mathcal{D}_{\mathrm{c}}$ containing x such that $s_t(\mathcal{N}_\delta) \subset \mathcal{N}_\varepsilon$ for all $t \geq t_0$. An equilibrium point $x \in \mathcal{D}_{\mathrm{c}}$ of the nonlinear dynamical system (2.102) is *semistable* if it is Lyapunov stable and there exists a relatively open subset $\mathcal{N}$ of $\mathcal{D}_{\mathrm{c}}$ containing x such that for all initial conditions in $\mathcal{N}$, the trajectory of (2.102) converges to a Lyapunov stable equilibrium point, that is, $\|s(t, x) - y\| \to 0$ as $t \to \infty$, where $y \in \mathcal{D}_{\mathrm{c}}$ is a Lyapunov stable equilibrium point of (2.102) and $x \in \mathcal{N}$. The nonlinear dynamical system (2.102) is said to be *semistable* if every equilibrium point of (2.102) is semistable.

If the system (2.102) is convergent with respect to $\mathcal{D}_{\mathrm{c}}$, then the ω-limit set $\omega(x)$ of (2.102) for the trajectory $s^x(t)$ starting at $x \in \mathcal{D}_{\mathrm{c}}$ is a singleton. Furthermore, it follows from continuity of solutions that for every $h \geq 0$, $s_h(\omega(x)) \triangleq \lim_{t\to\infty} s(t + h, x) = \omega(x)$. Thus,

$$\left.\frac{\mathrm{d}s_h(\omega(x))}{\mathrm{d}h}\right|_{h=0} = 0,$$

and hence, $\omega(x)$ is an equilibrium point of (2.102) for all $x \in \mathcal{D}_{\mathrm{c}}$.

The next result relates the continuity of the function $\omega(\cdot)$ at a point x to the stability of the equilibrium point $\omega(x)$.

Proposition 2.13. Suppose the nonlinear dynamical system (2.102) is convergent with respect to $\mathcal{D}_{\mathrm{c}}$. If $\omega(x)$ is a Lyapunov stable equilibrium point for some $x \in \mathcal{D}_{\mathrm{c}}$, then $\omega : \mathcal{D}_{\mathrm{c}} \to \mathcal{D}_{\mathrm{c}}$ is continuous at x.

Proof. The proof of the result appears in [40]. For completeness of exposition, we provide a proof here. Suppose $\omega(x)$ is Lyapunov stable for some $x \in \mathcal{D}_{\mathrm{c}}$, and let $\mathcal{N}_\varepsilon$ be an open neighborhood of $\omega(x)$. Moreover, choose open neighborhoods $\mathcal{N}$ and $\mathcal{N}_\delta$ of $\omega(x)$ such that $\overline{\mathcal{N}} \subset \mathcal{N}_\varepsilon$ and $s_t(\mathcal{N}_\delta) \subseteq \mathcal{N}$ for all $t \geq t_0$, and let $\{x_i\}_{n=1}^{\infty}$ be a sequence in $\mathcal{D}_{\mathrm{c}}$ converging to x. The existence of such neighborhoods follows from the Lyapunov stability of $\omega(x)$.

Next, there exists $h > t_0$ such that $s(h, z) \in \mathcal{N}_\delta$ and, since the solutions to (2.102) are continuously dependent on the system initial conditions, it follows that there exists an open neighborhood $\mathcal{N}_{\hat{\delta}} \triangleq \mathcal{B}_{\hat{\delta}}(x)$, $\hat{\delta} > 0$, of x such that $s(h, y) \in \mathcal{N}_\delta$ for all $y \in \mathcal{N}_{\hat{\delta}}$. Furthermore, it follows from the Lyapunov stability of $\omega(x)$ that $s(t + h, y) \in \mathcal{N}$, $y \in \mathcal{N}_{\hat{\delta}}$, $t \geq 0$, and hence, $\omega(y) \in \overline{\mathcal{N}} \subset \mathcal{N}_\varepsilon$, $y \in \mathcal{N}_{\hat{\delta}}$, which proves that $\omega : \mathcal{D}_{\mathrm{c}} \to \mathcal{D}_{\mathrm{c}}$ is continuous at x. □

The next result gives an alternative sufficient condition for the absence of Poincaré recurrence in a dynamical system.

Theorem 2.26. Consider the nonlinear dynamical system (2.102). Assume that $\mathcal{D}_{\mathrm{c}} \setminus \mathcal{M}_{\mathrm{e}} \neq \emptyset$ and assume (2.102) is convergent and semistable in $\mathcal{D}_{\mathrm{c}}$. Then the nonlinear dynamical system (2.102) does not exhibit Poincaré recurrence in $\mathcal{D}_{\mathrm{c}} \setminus \mathcal{M}_{\mathrm{e}}$. That is, for some $x \in \mathcal{D}_{\mathrm{c}} \setminus \mathcal{M}_{\mathrm{e}}$, there exists an open neighborhood $\mathcal{N} \subset \mathcal{D}_{\mathrm{c}} \setminus \mathcal{M}_{\mathrm{e}}$ of x such that for every $y \in \mathcal{N}$ the trajectory $s(t, y)$, $t \geq t_0$, does not return to $\mathcal{N}$ infinitely many times.

Proof. Let $x \in \mathcal{D}_{\mathrm{c}} \setminus \mathcal{M}_{\mathrm{e}}$ and let $\omega(x) \in \mathcal{M}_{\mathrm{e}}$ be a limiting point for the trajectory $s(t, x)$, $t \geq t_0$, so that $\lim_{t\to\infty} s(t, x) = \omega(x)$. Since (2.102) is convergent and semistable, it follows from Proposition 2.13 that $\omega(x)$, $x \in \mathcal{D}_{\mathrm{c}} \setminus \mathcal{M}_{\mathrm{e}}$, is continuous. Hence, for every $\varepsilon > 0$ there exists $\delta = \delta(\varepsilon) > 0$ such that $\omega(y) \in \mathcal{B}_\varepsilon(\omega(x))$ for all $y \in \mathcal{B}_\delta(x)$. Choose $\varepsilon > 0$ and $\delta > 0$ such that $\overline{\mathcal{B}}_\delta(x) \cap \overline{\mathcal{B}}_\varepsilon(\omega(x)) = \emptyset$. Furthermore, choose $\hat{\varepsilon} > 0$ to be sufficiently small such that

$$\overline{\bigcup_{y \in \mathcal{B}_\delta(x)} \mathcal{B}_{\hat{\varepsilon}}(\omega(y))} \cap \overline{\mathcal{B}}_\delta(x) = \emptyset. \tag{2.117}$$

Since the dynamical system (2.102) is convergent in $\mathcal{D}_c$, it follows that for all $y \in \mathcal{B}_\delta(x)$ and $\hat{\varepsilon} > 0$, there exists $T(\hat{\varepsilon}, y) > t_0$ such that $s(t, y) \in \mathcal{B}_{\hat{\varepsilon}}(\omega(y))$ for all $t > T(\hat{\varepsilon}, y)$. Moreover, it follows from (2.117) that, for all $y \in \mathcal{B}_\delta(x)$, $s(t, y)$, $t \geq t_0$, does not return to $\mathcal{B}_\delta(x)$ infinitely many times, which proves the result with $\mathcal{N} = \mathcal{B}_\delta(x)$. □

Finally, we close this section by noting that the results of this section also apply (with minor modifications) to infinite-dimensional dynamical systems $\mathcal{G}$ with flows $s_t : \mathcal{B} \to \mathcal{B}$ defined on a Banach space $\mathcal{B}$. In particular, we require that the map $g(\cdot)$ in the theorems just presented be a measurable transformation $g : \mathcal{B} \to \mathcal{B}$ that preserves a finite measure $\mu(\cdot)$ on $\mathcal{B}$, that is, $\mu(g^{(-1)}(\mathcal{N})) = \mu(\mathcal{N})$ for every measurable set $\mathcal{N} \subset \mathcal{B}$. In this case, the results in this section also hold for infinite-dimensional dynamical systems with this appropriate minor modification.

2.10 Poincaré Recurrence and Output Reversibility in Linear Dynamical Systems

Reversible dynamical systems, in the sense of Section 2.7, tend to exhibit Poincaré recurrence. Specifically, if the flow of a dynamical system preserves volume and has only bounded orbits, then for every open bounded set there exist orbits that intersect this set infinitely often. It was shown in Section 2.9 that Poincaré recurrence for a nonlinear dynamical system is equivalent to the existence of system orbits whose initial conditions belong to their own positive limit sets. Since the positive limit set is a bounded and invariant set with respect to the dynamical system, it follows that the entire orbit belongs to the positive limit set, and hence, is bounded. Thus, boundedness of the solutions of a dynamical system is an important element of the sufficient conditions establishing whether or not the dynamical system exhibits Poincaré recurrence.

Another important condition for the existence of Poincaré recurrence is volume preservation (Definition 2.26). In particular, if the flow of a dynamical system is volume-preserving, then, assuming that the system orbits are bounded, the image of every open bounded set under the flow of a dynamical system will eventually intersect the original set at some instant of time, which provides the basis for Poincaré recurrence. However, volume preservation is not sufficient for ensuring Poincaré recurrence since an unstable system can have a volume-preserving flow with system trajectories never returning to any neighborhood of their initial condition.

In this section, we establish necessary and sufficient conditions for Poincaré recurrence in linear dynamical systems. Specifically, we show that a linear dynamical system exhibits Poincaré recurrence if and only if the system matrix has purely imaginary, semisimple eigenvalues. Furthermore, for linear dynamical systems, we show that a vanishing trace of the system dynamics is a necessary and sufficient condition for volume preservation. In addition, we show that asymptotically stable and semistable linear systems have volume-decreasing flows, whereas unstable systems can have either volume-increasing, volume-decreasing, or volume-preserving flows.

However, none of the aforementioned systems exhibit Poincaré recurrence, and hence, these systems are irreversible in the sense of Section 2.9. In addition, we show that classical lossless linear Lagrangian and Hamiltonian dynamical systems are volume-preserving and exhibit Poincaré recurrence. Finally, we provide connections between Poincaré recurrence and output reversibility. Specifically, we show that Poincaré recurrence is a sufficient condition for output reversibility in linear dynamical systems.

In this section, we specialize the results of Section 2.9 to linear dynamical systems $\mathcal{G}$ given by

$$\dot{x}(t) = Ax(t), \quad x(0) = x_0, \quad t \geq 0, \tag{2.118}$$

where $x(t) \in \mathbb{R}^n$, $t \geq 0$, and $A \in \mathbb{R}^{n\times n}$. The following theorem presents necessary and sufficient conditions that ensure that the linear dynamical system $\mathcal{G}$ exhibits Poincaré recurrence in $\mathbb{R}^n$.

Theorem 2.27. Consider the linear dynamical system $\mathcal{G}$ given by (2.118). Then $\mathcal{G}$ exhibits Poincaré recurrence in $\mathbb{R}^n$ if and only if $\mathrm{Re}\,\lambda = 0$ and λ is semisimple, where $\lambda \in \mathrm{spec}\,(A)$.

Proof. Consider the Jordan decomposition of A given by $A = SBS^{-1}$, where $S \in \mathbb{C}^{n\times n}$ is a nonsingular matrix and $B = \text{block-diag}[B_1, \ldots, B_m] \in \mathbb{C}^{n\times n}$ is the Jordan form of A [33]. Furthermore, consider the transformed linear dynamical system $\mathcal{G}_{\mathrm{t}}$ given by

$$\dot{y}(t) = By(t), \quad y(0) = y_0, \quad t \geq 0, \tag{2.119}$$

where $y(t) = S^{-1}x(t) \in \mathbb{R}^n$, $t \geq 0$. First, we show that (2.118) exhibits Poincaré recurrence in $\mathcal{D}_{\mathrm{c}} \subseteq \mathbb{R}^n$ if and only if (2.119) exhibits Poincaré recurrence in $\hat{\mathcal{D}}_{\mathrm{c}} \triangleq \{y \in \mathbb{R}^n : y = S^{-1}x,\ x \in \mathcal{D}_{\mathrm{c}}\} \subseteq \mathbb{R}^n$.

To see this, suppose (2.119) exhibits Poincaré recurrence in $\hat{\mathcal{D}}_{\mathrm{c}}$. Let $\mathcal{N} \subset \mathcal{D}_{\mathrm{c}}$ be an open bounded set and consider the set $\hat{\mathcal{N}} \triangleq \{y \in \mathbb{R}^n : y = S^{-1}x,\ x \in \mathcal{N}\}$. Note that since $S \in \mathbb{C}^{n\times n}$ is invertible, the set $\hat{\mathcal{N}}$ is also open and bounded [109]. Since (2.119) exhibits Poincaré recurrence, it follows from Theorem 2.23 that, for the open bounded set $\hat{\mathcal{N}} \subset \hat{\mathcal{D}}_{\mathrm{c}}$, there exists $y_0 \in \hat{\mathcal{N}}$ such that $\hat{s}(t_k, y_0) \in \hat{\mathcal{N}}$ for some sequence $\{t_k\}_{k=1}^{\infty}$, with $t_k \to \infty$ as $k \to \infty$, where $\hat{s}(t, y_0)$, $t \geq 0$, is the solution to (2.119) with the initial condition $y_0 \in \hat{\mathcal{N}}$.

Next, note that since $S \in \mathbb{C}^{n\times n}$ is invertible it follows that $x \in \mathcal{N}$ if and only if $y \in \hat{\mathcal{N}}$, where $y = S^{-1}x$. Thus, it follows that there exists $x_0 = Sy_0 \in \mathcal{N}$ such that $s(t_k, x_0) = S\hat{s}(t_k, y_0) \in \mathcal{N}$ for all t_k, $k = 1, 2, \ldots$, where $s(t, x_0)$, $t \geq 0$, is the solution to (2.118). Since $\mathcal{N} \subset \mathcal{D}_{\rm c}$ is arbitrary it follows that (2.118) exhibits Poincaré recurrence in $\mathcal{D}_{\rm c}$. The converse part of the proof is identical and, hence, is omitted.

To show sufficiency, assume that $\operatorname{Re}\lambda_i = 0$, $i = 1, \ldots, n$, and λ_i, $i = 1, \ldots, n$, are semisimple eigenvalues of A. In this case, the real Jordan form of A is given by $B = \text{block–diag}\,[B_1, \ldots, B_{\frac{n}{2}}] \in \mathbb{R}^{n\times n}$, where

$$B_i = \begin{bmatrix} 0 & |\lambda_i| \\ -|\lambda_i| & 0 \end{bmatrix}, \quad i = 1, \ldots, \frac{n}{2}.$$

Thus, consider the function $V(y) = \frac{1}{2}y^{\rm T}Dy$, where $D = \operatorname{diag}\,[|\lambda_1|, |\lambda_1|, |\lambda_2|, |\lambda_2|, \ldots, |\lambda_{\frac{n}{2}}|, |\lambda_{\frac{n}{2}}|]$, and note that $\dot{V}(y(t)) = 0$, $t \geq 0$, along the trajectories of (2.119).

Next, consider the open bounded set $\hat{\mathcal{D}}_{\rm c} \triangleq \{y \in \mathbb{R}^n : V(y) < \mu\}$, where $\mu \in \mathbb{R}_+$, and note that $\hat{\mathcal{D}}_{\rm c}$ is invariant with respect to (2.119). Hence, $\hat{s}_t : \mathbb{R}^n \to \mathbb{R}^n$ maps a bounded set $\hat{\mathcal{D}}_{\rm c} \subset \mathbb{R}^n$ onto itself. Moreover, it follows from Theorem 2.19 that the flow $\hat{s}_t : \mathbb{R}^n \to \mathbb{R}^n$ of (2.119) is volume-preserving since $\nabla \cdot (By) = \operatorname{tr} B = 0$. Thus, it follows from Theorem 2.24 that (2.119) exhibits Poincaré recurrence in $\hat{\mathcal{D}}_{\rm c}$, and since $\hat{\mathcal{D}}_{\rm c}$ can be chosen arbitrarily large, it further follows that (2.119) exhibits Poincaré recurrence in $\mathbb{R}^n$. Hence, (2.118) exhibits Poincaré recurrence in $\mathbb{R}^n$.

To show necessity, assume that (2.118) exhibits Poincaré recurrence in $\mathbb{R}^n$ or, equivalently, (2.119) exhibits Poincaré recurrence in $\mathbb{R}^n$. Now, suppose, *ad absurdum*, that for some $k \in \{1, \ldots, n\}$, $\operatorname{Re}\lambda_k \neq 0$, where λ_k is an eigenvalue of A. In this case, it follows from the form of the solution to (2.119) that for $y_l(0) \neq 0$, either $y_l(t) \to \infty$ as $t \to \infty$ or $y_l(t) \to 0$ as $t \to \infty$, where $y_l(\cdot)$ denotes the lth component of $y(\cdot)$ associated with the eigenvalue λ_k, which contradicts the assumption of Poincaré recurrence in $\mathbb{R}^n$ for (2.119). Moreover, in this case, (2.119) does not exhibit Poincaré recurrence on the hyperplane $\hat{\mathcal{D}}_{\rm c} \triangleq \{y \in \mathbb{R}^n : y_l = 0\}$ since $\hat{\mathcal{D}}_{\rm c}$ does not contain open subsets of $\mathbb{R}^n$. Thus, $\operatorname{Re}\lambda = 0$, $\lambda \in \operatorname{spec}(A)$.

Next, suppose, *ad absurdum*, that for some $k \in \{1, \ldots, n\}$, λ_k is not semisimple, where λ_k is an eigenvalue of A. Then it follows from the structure of the Jordan matrix $B \in \mathbb{C}^{n\times n}$ that for some $l \in \{1, \ldots, n\}$ such that $y_{l+1}(0) \neq 0$, $y_l(t) \to \infty$ as $t \to \infty$, where $y_l(\cdot)$ is the lth component of $y(\cdot)$ associated with the eigenvalue λ_k, which contradicts the assumption of Poincaré recurrence in $\mathbb{R}^n$ for (2.119). Similarly, (2.119) does not exhibit Poincaré recurrence on the hyperplane $\hat{\mathcal{D}}_{\rm c} \triangleq \{y \in \mathbb{R}^n : y_{l+1} = 0\}$ since $\hat{\mathcal{D}}_{\rm c}$ does not contain open subsets of $\mathbb{R}^n$. Hence, if (2.118) exhibits Poincaré recurrence in $\mathbb{R}^n$, then the eigenvalues of A are purely imaginary and semisimple, which proves the result. □

Note that in the case where $\operatorname{Re}\lambda = 0$ and λ is semisimple, where $\lambda \in \operatorname{spec}(A)$, the solutions to (2.119) are not necessarily periodic since the ratio of eigenvalues of A may be irrational. Nevertheless, in this case, it follows from Theorem 2.27 that (2.119) exhibits Poincaré recurrence in $\mathbb{R}^n$, while in the case of periodic solutions of (2.119), Poincaré recurrence follows trivially.

Theorem 2.27 provides necessary and sufficient conditions for Poincaré recurrence in linear dynamical systems, whereas Theorem 2.24 gives only sufficient conditions for Poincaré recurrence in nonlinear dynamical systems. The next theorem presents a counterpart to Theorem 2.19 for linear systems and shows that a zero divergence of the system dynamics is a necessary *and* sufficient condition for volume preservation. A variation of this result appears in [380] without proof. In particular, [380] establishes sufficient conditions for volume preservation for nonlinear dynamical systems.

Theorem 2.28. Consider the linear dynamical system (2.118). The flow $s_t : \mathcal{D} \to \mathcal{D}$ of (2.118) is volume-preserving if and only if $\operatorname{tr} A = 0$. Alternatively, the flow $s_t : \mathcal{D} \to \mathcal{D}$ of (2.118) is volume-increasing (respectively, volume-decreasing) if and only if $\operatorname{tr} A > 0$ (respectively, $\operatorname{tr} A < 0$).

Proof. Let $\mathcal{V} \subset \mathbb{R}^n$ be a compact set with a nonempty interior such that its image at time t under the mapping $s_t(\cdot)$ is given by $s_t(\mathcal{V})$. In addition, let $\mathrm{d}\mathcal{S}_{\mathcal{V}}$ denote an infinitesimal surface element of the boundary of the set $\mathcal{V}$, and let $\hat{n}(x)$, $x \in \partial\mathcal{V}$, denote an outward normal vector to the boundary of $\mathcal{V}$. Then the change in volume of $s_t(\mathcal{V})$ at $t = t_0$ is given by

$$\mathrm{d}s_t(\mathcal{V})_{\rm vol} = \int_{\partial\mathcal{V}} (Ax \cdot \hat{n}(x))\mathrm{d}t\mathrm{d}\mathcal{S}_{\mathcal{V}}, \tag{2.120}$$

which, using the divergence theorem, implies that

$$\left.\frac{\mathrm{d}s_t(\mathcal{V})_{\rm vol}}{\mathrm{d}t}\right|_{t=t_0} = \int_{\partial\mathcal{V}} (Ax \cdot \hat{n}(x))\mathrm{d}\mathcal{S}_{\mathcal{V}} = \int_{\mathcal{V}} \nabla \cdot Ax\, \mathrm{d}\mathcal{V} = \operatorname{tr} A \operatorname{vol}(\mathcal{V}), \tag{2.121}$$

which proves the result. □

It follows from Theorems 2.27 and 2.28 that if the flow of the linear dynamical system (2.118) is not volume-preserving, then this system does not exhibit Poincaré recurrence in $\mathbb{R}^n$; whereas in the case of the nonlinear dynamical system (2.102), volume preservation along with boundedness of solutions to (2.102) is only a sufficient condition for Poincaré recurrence. In particular, if the system matrix A is either asymptotically stable or semistable, then the flow of (2.118) is volume-decreasing, and hence, (2.118) does not exhibit Poincaré recurrence in $\mathbb{R}^n$. Thus, linear asymptotically stable and semistable systems are irreversible in the sense of Section 2.9. Alternatively, it is possible for unstable linear dynamical systems to have volume-preserving flows; however, Poincaré recurrence does not occur in this case.

Unfortunately, unlike Lyapunov's indirect method [191], which allows us to draw conclusions about the stability of an equilibrium point of a nonlinear system by investigating the stability of the equilibrium point of the linearized system, linearization fails to establish Poincaré recurrence of a nonlinear dynamical system. To see this, consider the nonlinear dynamical system given by

$$\dot{x}_1(t) = -x_2(t) - x_1(t)x_2^2(t), \quad x_1(0) = x_{10}, \quad t \geq 0, \tag{2.122}$$

$$\dot{x}_2(t) = x_1(t) - x_1^2(t)x_2(t), \quad x_2(0) = x_{20}, \tag{2.123}$$

whose linearization is given by (2.118) with

$$A = \begin{bmatrix} 0 & -1 \\ 1 & 0 \end{bmatrix}. \tag{2.124}$$

Theorem 2.27 shows that the linearized system exhibits Poincaré recurrence in $\mathbb{R}^2$. However, a simple Lyapunov analysis involving the Krasovskii-LaSalle invariant set theorem with the Lyapunov function $V(x) = x_1^2 + x_2^2$, $x \in \mathbb{R}^2$, shows that the zero solution $x(t) \equiv 0$ to the nonlinear dynamical system (2.122) and (2.123) is globally asymptotically stable, which precludes Poincaré recurrence in $\mathbb{R}^n$.

Poincaré recurrence for Lagrangian and Hamiltonian dynamical systems has been extensively studied in the classical references [10, 49, 328]. Using Theorem 2.27, we show that all classical lossless linear Lagrangian and Hamiltonian linear systems exhibit Poincaré recurrence, that is, their system trajectories return to every neighborhood of the system initial condition infinitely often. Specifically, consider the n-degree-of-freedom Lagrangian system given by

$$\frac{\mathrm{d}}{\mathrm{d}t}\left[\frac{\partial \mathcal{L}}{\partial \dot{q}}(q(t), \dot{q}(t))\right]^{\mathrm{T}} - \left[\frac{\partial \mathcal{L}}{\partial q}(q(t), \dot{q}(t))\right]^{\mathrm{T}} = 0, \quad q(0) = q_0, \quad \dot{q}(0) = \dot{q}_0, \quad t \geq 0, \tag{2.125}$$

where $q \in \mathbb{R}^n$ represents the generalized system positions, $\dot{q} \in \mathbb{R}^n$ represents the generalized system velocities, $\mathcal{L} : \mathbb{R}^n \times \mathbb{R}^n \to \mathbb{R}$ denotes the system Lagrangian given by $\mathcal{L}(q, \dot{q}) = T(q, \dot{q}) - V(q)$, where $T(q, \dot{q}) = \frac{1}{2}\dot{q}^{\mathrm{T}} M \dot{q}$ is the system kinetic energy, $V(q) = \frac{1}{2}q^{\mathrm{T}} K q$ is the system potential energy, $M \in \mathbb{R}^{n\times n}$ is the system positive definite inertia matrix, and $K \in \mathbb{R}^{n\times n}$ is the system positive definite stiffness matrix.

Now, defining the state variables $x_1 \triangleq q$, $x_2 \triangleq \dot{q}$, and $x \triangleq [x_1^{\mathrm{T}}, x_2^{\mathrm{T}}]^{\mathrm{T}}$, (2.125) can be equivalently written as (2.118) with

$$A \triangleq \begin{bmatrix} 0 & I_n \\ -M^{-1}K & 0 \end{bmatrix}. \tag{2.126}$$

It follows that $A \in \mathbb{R}^{2n\times 2n}$ given by (2.126) can be written as

$$\begin{aligned} A &= \begin{bmatrix} 0 & M^{-\frac{1}{2}} \\ -M^{-1}K^{\frac{1}{2}} & 0 \end{bmatrix} \begin{bmatrix} 0 & K^{\frac{1}{2}}M^{-\frac{1}{2}} \\ -M^{-\frac{1}{2}}K^{\frac{1}{2}} & 0 \end{bmatrix} \\ &\quad \cdot \begin{bmatrix} 0 & -K^{-\frac{1}{2}}M \\ M^{\frac{1}{2}} & 0 \end{bmatrix} \\ &= SBS^{-1}. \end{aligned} \tag{2.127}$$

Since $B = -B^{\mathrm{T}}$, it follows that B is skew-symmetric, and hence, the eigenvalues of B or, equivalently, the eigenvalues of A, are purely imaginary. Furthermore, since B is skew-symmetric, B is normal, and hence, it follows from the Schur decomposition that $\operatorname{rank}(B - \lambda_i I_{2n}) = 2n - \mathrm{am}_B(\lambda_i)$, where λ_i, $i = 1, \ldots, 2n$, is an eigenvalue of B and $\mathrm{am}_B(\lambda_i)$, $i = 1, \ldots, 2n$, is the algebraic multiplicity of λ_i. However, since $\operatorname{rank} Y + \operatorname{def} Y^{\mathrm{T}} = m$ [33, p. 32], where $Y \in \mathbb{R}^{n\times m}$ and $\operatorname{def} Y$ denotes the defect of Y, it follows that $\operatorname{rank}(B - \lambda_i I_{2n}) + \mathrm{gm}_B(\lambda_i) = 2n$, and hence, $\mathrm{gm}_B(\lambda_i) = \mathrm{am}_B(\lambda_i)$, $i = 1, \ldots, 2n$, where $\mathrm{gm}_B(\lambda_i)$ denotes geometric multiplicity of λ_i. Thus, λ_i, $i = 1, \ldots, 2n$, is a semisimple eigenvalue of B, and hence, A and B are semisimple matrices.

Finally, since the eigenvalues of A are purely imaginary and semisimple, it follows from Theorem 2.27 that the Lagrangian system given by (2.118) has a volume-preserving flow and exhibits Poincaré recurrence in $\mathbb{R}^{2n}$. Note that the same result can be arrived at by showing that the flow of (2.118) is volume-preserving and the state trajectories of (2.118) are bounded.

The next result shows that Poincaré recurrence is a sufficient condition for output reversibility in linear dynamical systems.

Theorem 2.29. If the linear dynamical system (2.118) exhibits Poincaré recurrence in $\mathbb{R}^n$, then (2.118) and (2.64) is output reversible.

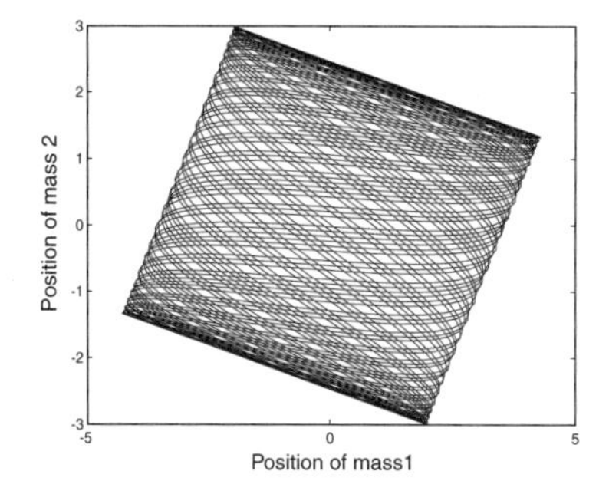

Figure 2.1 Position phase portrait.

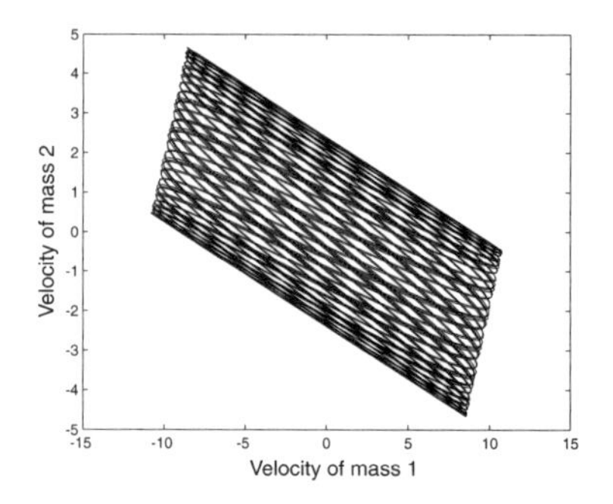

Figure 2.2 Velocity phase portrait.

Proof. The proof is a direct consequence of Theorems 2.27 and 2.18. □

To elucidate Theorem 2.29, consider two coupled oscillators with masses m_1 and m_2, and spring stiffness coefficients k_1 and k_2 so that the inertia, stiffness, and system state matrices are given by

$$M = \begin{bmatrix} m_1 & 0 \\ 0 & m_2 \end{bmatrix}, \qquad K = \begin{bmatrix} k_1 + k_2 & -k_2 \\ -k_2 & k_2 \end{bmatrix}, \tag{2.128}$$

$$A = \begin{bmatrix} 0 & I_2 \\ -M^{-1}K & 0 \end{bmatrix}. \tag{2.129}$$

Note that $\operatorname{spec}(A) = \{\pm j\omega_1, \pm j\omega_2\}$, where $\omega_1 > 0$ and $\omega_2 > 0$. Thus, it follows from Theorem 2.27 that (2.118), with $A \in \mathbb{R}^{4\times4}$ given by (2.129), exhibits Poincaré recurrence in $\mathbb{R}^4$.

Figures 2.1 and 2.2 show the position and velocity phase portraits, respectively, for the initial condition $x_0 = [-2, 3, 0, 0]^{\mathrm{T}}$ and system parameters $m_1 = 2$, $m_2 = 4$, $k_1 = 9$, and $k_2 = 8$. Note that in this case the ratio $\frac{\omega_1}{\omega_2}$ is an irrational number, and hence, the solutions to (2.66) with A given by (2.126) are not periodic. Nevertheless, it follows from Theorem 2.27 that (2.66) exhibits Poincaré recurrence and the state trajectories of (2.66) return to every neighborhood of their initial conditions infinitely often.

Now, consider the reversibility of the relative mass positions, that is, output reversibility with $C = [1, -1, 0, 0]$, and let $m_1 = 2$, $m_2 = 4$, $k_1 = 9$, and $k_2 = 8$. Note that (A, C) is observable since $\operatorname{rank} \mathcal{O} = 4$. Hence, it follows from Theorem 2.29 that (2.118) and (2.64) is output reversible.

For the initial condition $x_0 = [-2, 3, 0, 0]^{\mathrm{T}}$ and $t_1 = 3$, it follows from Proposition 2.9 that $\hat{x}_0 = [1.9951, -2.9333, -1.1973, 0.8646]^{\mathrm{T}}$ will generate the solution to (2.118) that will retrace the original time history of the relative mass positions backwards with time going forward. Figure

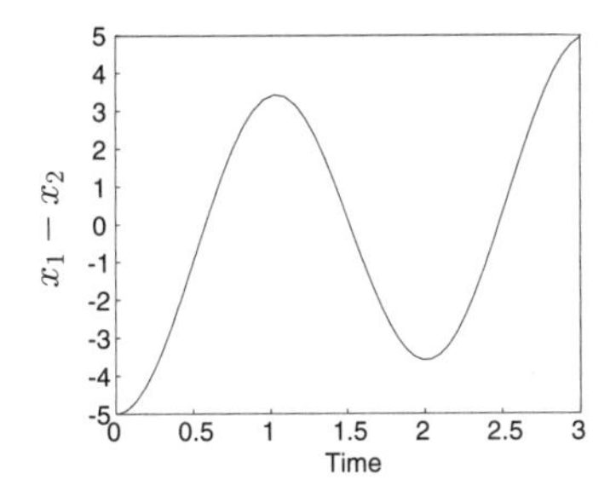

Figure 2.3 Original relative positions.

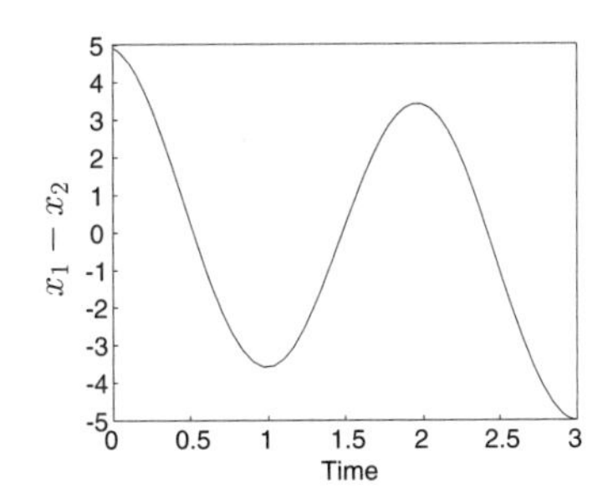

Figure 2.4 Reversed relative positions.

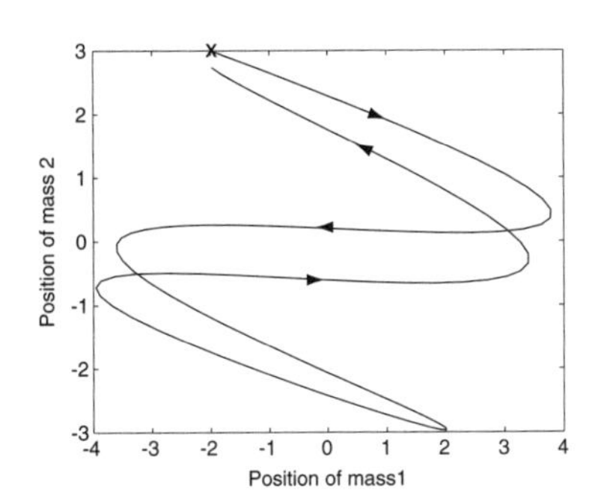

Figure 2.5 Original position phase portrait.

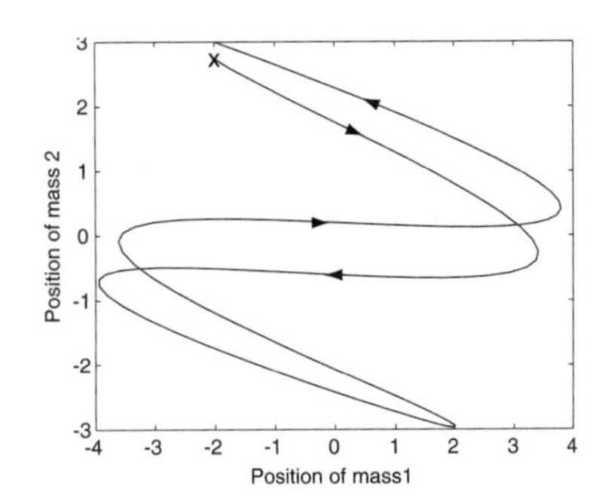

Figure 2.6 Reversed position phase portrait.

2.3 shows the original relative mass positions with the initial condition $x_0 = [-2, 3, 0, 0]^{\mathrm{T}}$ up to $t = 3$, and Figure 2.4 shows relative mass positions with the initial condition $\hat{x}_0 = [1.9951, -2.9333, -1.1973, 0.8646]^{\mathrm{T}}$ up to $t = 3$.

Finally, we consider reversibility of mass positions, that is, output reversibility with

$$C = \begin{bmatrix} 1 & 0 & 0 & 0 \\ 0 & 1 & 0 & 0 \end{bmatrix}. \tag{2.130}$$

Note that for the chosen data $\operatorname{rank} A = 4$. Let $t^* = 1$ and $x_i \in \mathbb{R}^4$, $i = 1, \ldots, 4$, be the ith column of I_4. It can be shown that $\operatorname{rank}\,[\mathcal{O} \quad \mathcal{S}\mathcal{O}e^{At^*}x_i] = 4$, $i = 1, \ldots, 4$. Thus, it follows from Theorem 2.29 that (2.66) and (2.67) with A and C given by (2.126) and (2.130), respectively, is output reversible. For the initial condition $x_0 = [-2, 3, 0, 0]^{\mathrm{T}}$ and $t_1 = 6$, it follows from Proposition 2.9 that $\hat{x}_0 = [-1.9796, 2.7365, 2.3757, -1.6953]^{\mathrm{T}}$ will generate the solution to (2.66) that will retrace the original time history of mass positions backwards with time going forward. Figure 2.5 shows the original phase portrait of mass positions with the initial condition $x_0 = [-2, 3, 0, 0]^{\mathrm{T}}$ up to $t = 6$, and Figure 2.6 shows the phase portrait of mass positions with the initial condition $\hat{x}_0 = [-1.9796, 2.7365, 2.3757, -1.6953]^{\mathrm{T}}$ up to $t = 6$.

Chapter Three

A Dynamical Systems Foundation for Thermodynamics

3.1 Introduction

The fundamental and unifying concept in the analysis of complex large-scale dynamical systems is the concept of energy (i.e., the capacity of a system to do work). As noted in Chapter 1, the energy of a state of a dynamical system is the measure of its ability to produce changes (motion) in its own system state as well as changes in the system states of its surroundings. These changes occur as a direct consequence of the energy flow between different subsystems within the dynamical system. Since heat[1] (energy in transition) is a fundamental concept of thermodynamics involving the capacity of hot bodies (more energetic subsystems) to produce work, thermodynamics is a theory of large-scale dynamical systems.

As in thermodynamic systems, dynamical systems can exhibit energy (due to friction) that becomes unavailable to do useful work. This in turn contributes to an increase in system entropy, a measure of the tendency of a system to lose the ability to do useful work. In this chapter, we use a large-scale dynamical systems perspective to provide a system-theoretic foundation for thermodynamics that bridges the gap between classical and statistical thermodynamics.

To develop a dynamical systems foundation for thermodynamics, we use the state space formalism to construct a mathematical model that is consistent with basic thermodynamic principles. This is in sharp contrast to classical thermodynamics wherein an input-output description of the system is used. However, it is felt that a state space formulation is essential for developing a thermodynamic model with enough detail for describing the thermal behavior of heat and dynamical energy. In addition, such a model is crucial in accounting for internal system properties captured by

[1]Heat is *not* a form of energy but rather a mode of energy transfer due to a temperature gradient. Thus, heat refers to a dynamical process and not to a particular entity.

compartmental system dynamics characterizing conservation laws, wherein subsystem energies can only be transported, stored, or dissipated but not created.

If a physical system possesses these conservation properties externally, then there exists a high possibility that the system also possesses these properties internally. Specifically, if the system possesses conservation properties internally and does not violate any input-output behavior of the physical system established by experimental evidence, then the state space model is theoretically credible. The absence of a state space formalism in classical thermodynamics, and physics in general, is quite disturbing and in our view largely responsible for the monomeric state of classical thermodynamics.

The governing physical laws of nature impose specific relations between the system state variables, with system interconnections captured by shared variables among subsystems. An input-state-output model is ideal in capturing system interconnections with the environment involving the exchange of matter, energy, or information, with inputs serving to capture the influence of the environment on the system, outputs serving to capture the influence of the system on the environment, and internal feedback interconnections—via specific output-to-input assignments—serving to capture interactions between subsystems.

The state space model additionally enforces the fundamental property of *causality*, that is, nonanticipativity of the system, wherein future values of the system inputs do not influence past and present values of the system outputs.[2] More specifically, the system states, and hence, the system outputs before a certain time are not affected by the values of the inputs after that time; a principle holding for all physical systems verified by experiment.

Another important notion of state space modeling is that of *system realization* involving the construction of the state space, the system dynamic, and the system output that yields a dynamical system in state space form and generates the given input-output map—established by experiment—through a suitable choice of the initial system state. This problem has been extensively addressed in dynamical systems theory and leads to the fact that every continuous input-output dynamical system has a continuous realization in state space form; for details see [20, 461, 466].

---

[2]Given an operator $\mathcal{G} : \mathcal{U} \to \mathcal{Y}$ characterizing a dynamical system $\mathcal{G}$, where $\mathcal{U}$ and $\mathcal{Y}$ define input and output spaces, respectively, we say that $\mathcal{G}$ is *causal* if and only if, for every time $T \in [0, \infty)$ and input $u(\cdot) \in \mathcal{U}$, $y_T = (\mathcal{G}(u))_T = (\mathcal{G}(u_T))_T$, where v_T denotes the truncation of the signal $v(\cdot)$ on the interval $[0, T]$. Equivalently, $\mathcal{G}$ is causal if and only if for every pair $u, v \in \mathcal{U}$ and every $T > 0$ such that $u_T = v_T$, $(\mathcal{G}(u))_T = (\mathcal{G}(v))_T$.

In this monograph, we assume that a state space realization for the dynamical thermodynamic system is given. That is, a state space realization is inferred from established experiments by defining a state space, a system dynamic, and a read-out map such that the state of the dynamical system in state space form at time t generates an equivalence class to the input-output pairs for each experiment with system inputs up to time t yielding the same system output after time t regardless of how the system input is applied after time t.

3.2 Conservation of Energy and the First Law of Thermodynamics

To develop a system-theoretic foundation for thermodynamics, we consider a large-scale system model with a combination of subsystems (compartments or parts) that is perceived as a single entity. For each subsystem (compartment) making up the system, we postulate the existence of an *energy* state variable such that the knowledge of these subsystem state variables at any given time $t = t_0$, together with the knowledge of any inputs (heat fluxes) to each of the subsystems for time $t \geq t_0$, completely determines the behavior of the system for any given time $t \geq t_0$. Hence, the (energy) state of our dynamical system at time t is uniquely determined by the state at time t_0 and any external inputs for time $t \geq t_0$ and is independent of the state and inputs before time t_0.

An implicit assumption of our formulation is that our thermodynamic state variables form a set of continuously differentiable coordinates on $\mathcal{D} \subseteq \overline{\mathbb{R}}_+^q$. For systems that possess critical states (e.g., a liquid-vapor critical state) or phase transitions (i.e., transitions between solid, liquid, and gaseous states of matter, and, in the rare cases, plasma), the continuous differentiability assumption on the state variables is limiting. In the first case, the state is *absolutely continuous*, generating a differential equation with a discontinuous right-hand side characterized by differential inclusions involving set-valued (Krasovskii, Filippov, etc.) maps specifying a set of directions for the thermodynamic system state velocities and admitting (Carathéodory, Krasovskii, Filippov, etc.) solutions with absolutely continuous curves [190]. In the second case, the state is *discontinuous*, generating an impulsive differential equation [196]. Using the mathematical frameworks in [190, 196], our proposed thermodynamic formulation is extended in Chapters 8 and 10 to address these generalizations.

In this chapter, we consider a large-scale dynamical system composed of a large number of units with aggregated (or lumped) energy variables representing homogeneous groups of these units. If all the units comprising the system are identical (that is, the system is perfectly homogeneous), then the behavior of the dynamical system can be captured by that of a single plenipotentiary unit. Alternatively, if every interacting system unit is distinct, then the resulting model constitutes a microscopic system.

To develop a middle-ground thermodynamic model placed between complete aggregation (classical thermodynamics) and complete disaggregation (statistical thermodynamics), we subdivide the large-scale dynamical system into a finite number of compartments, each formed by a large number of homogeneous units. Each compartment represents the energy content of the different parts of the dynamical system, and different compartments interact by exchanging heat. Thus, our compartmental thermodynamic model utilizes subsystems or compartments to describe the energy distribution among distinct regions in space with intercompartmental flows representing the heat transfer between these regions. Decreasing the number of compartments results in a more aggregated or homogeneous model, whereas increasing the number of compartments leads to a higher degree of disaggregation resulting in a heterogeneous model.

To formulate our state space thermodynamic model, consider the large-scale dynamical system $\mathcal{G}$ shown in Figure 3.1 involving energy exchange between q interconnected subsystems. Let $E_i : [0, \infty) \to \overline{\mathbb{R}}_+$ denote the energy (and hence a nonnegative quantity[3]) of the ith subsystem, let $S_i : [0, \infty) \to \mathbb{R}$ denote the external power (heat flux) supplied to (or extracted from) the ith subsystem, let $\sigma_{ij} : \overline{\mathbb{R}}_+^q \to \overline{\mathbb{R}}_+$, $i \neq j$, $i, j = 1, \ldots, q$, denote the instantaneous rate of energy (heat) flow from the jth subsystem to the ith subsystem, and let $\sigma_{ii} : \overline{\mathbb{R}}_+^q \to \overline{\mathbb{R}}_+$, $i = 1, \ldots, q$, denote the instantaneous rate of energy (heat) dissipation from the ith subsystem to the environment. Here we assume that $\sigma_{ij} : \overline{\mathbb{R}}_+^q \to \overline{\mathbb{R}}_+$, $i, j = 1, \ldots, q$, are locally Lipschitz continuous on $\overline{\mathbb{R}}_+^q$ and $S_i : [0, \infty) \to \mathbb{R}$, $i = 1, \ldots, q$, are bounded piecewise continuous functions of time.

An *energy balance* for the ith subsystem yields

$$\begin{aligned} E_i(T) &= E_i(t_0) + \sum_{j=1,\, j\neq i}^{q} \int_{t_0}^{T} [\sigma_{ij}(E(t)) - \sigma_{ji}(E(t))]\mathrm{d}t \\ &\quad - \int_{t_0}^{T} \sigma_{ii}(E(t))\mathrm{d}t + \int_{t_0}^{T} S_i(t)\mathrm{d}t, \quad T \geq t_0, \end{aligned} \tag{3.1}$$

[3]Here we assume that subsystem energies are lower bounded so that, without loss of generality, we can shift $E_i(\cdot)$ such that, with a minor abuse of notation, $E_i(t) \geq 0$, $t \geq 0$, $i = 1, \ldots, q$. This assumption corresponds to the existence of a minimal energy state (i.e., the *ground* or *vacuum state*). In quantum physics this assumption is known as *unitarity* and corresponds to a restriction on quantum state evolution that ensures the sum of probabilities for all possible outcomes of a given event is unity—a condition needed for the second law of thermodynamics to hold.

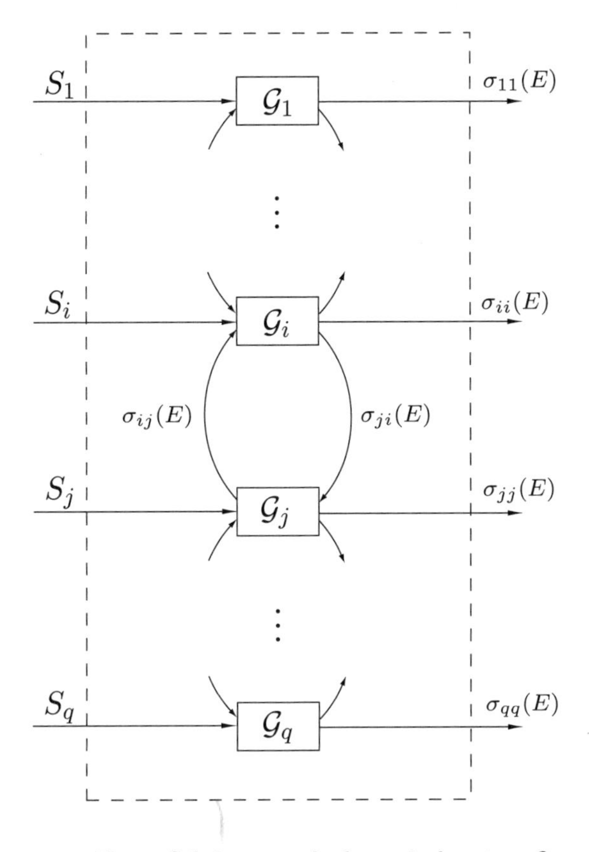

Figure 3.1 Large-scale dynamical system $\mathcal{G}$.

or, equivalently, in vector form,

$$E(T) = E(t_0) + \int_{t_0}^{T} f(E(t))\mathrm{d}t - \int_{t_0}^{T} d(E(t))\mathrm{d}t + \int_{t_0}^{T} S(t)\mathrm{d}t, \ T \geq t_0, \tag{3.2}$$

where $E(t) \triangleq [E_1(t), \ldots, E_q(t)]^{\mathrm{T}}$, $d(E(t)) \triangleq [\sigma_{11}(E(t)), \ldots, \sigma_{qq}(E(t))]^{\mathrm{T}}$, $S(t) \triangleq [S_1(t), \ldots, S_q(t)]^{\mathrm{T}}$, $t \geq t_0$, and $f = [f_1, \ldots, f_q]^{\mathrm{T}} : \overline{\mathbb{R}}_+^q \to \mathbb{R}^q$ is such that

$$f_i(E) = \sum_{j=1,\, j\neq i}^{q} [\sigma_{ij}(E) - \sigma_{ji}(E)], \quad E \in \overline{\mathbb{R}}_+^q. \tag{3.3}$$

It is important to note that the exchange of energy between subsystems in (3.1) is assumed to be a nonlinear function of all the subsystems, that is, $\sigma_{ij} = \sigma_{ij}(E)$, $E \in \overline{\mathbb{R}}_+^q$, $i \neq j$, $i, j = 1, \ldots, q$.

This assumption is made for generality and would depend on the complexity of the diffusion process. For example, thermal processes may include evaporative and radiative heat transfer as well as thermal conduction, giving rise to complex heat transport mechanisms. However, for simple diffusion processes it suffices to assume that $\sigma_{ij}(E) = \sigma_{ij}(E_j)$, wherein the energy flow from the jth subsystem to the ith subsystem is only dependent (possibly nonlinearly) on the energy in the jth subsystem. Similar comments apply to system dissipation.

Note that (3.1) yields a conservation of energy equation and implies that the energy stored in the ith subsystem is equal to the external energy supplied to (or extracted from) the ith subsystem plus the energy gained by the ith subsystem from all other subsystems due to subsystem coupling minus the energy dissipated from the ith subsystem to the environment. Equivalently, (3.1) can be rewritten as

$$\dot{E}_i(t) = \sum_{j=1,\, j\neq i}^{q} [\sigma_{ij}(E(t)) - \sigma_{ji}(E(t))] \; - \; \sigma_{ii}(E(t)) + S_i(t),$$

$$E_i(t_0) = E_{i0}, \quad t \geq t_0, \qquad (3.4)$$

or, in vector form,

$$\dot{E}(t) = f(E(t)) - d(E(t)) + S(t), \quad E(t_0) = E_0, \quad t \geq t_0, \qquad (3.5)$$

where $E_0 \triangleq [E_{10}, \ldots, E_{q0}]^{\mathrm{T}}$, yielding a *power balance* equation that characterizes energy flow between subsystems of the large-scale dynamical system $\mathcal{G}$.

Equation (3.4) shows that the rate of change of energy, or power, in the ith subsystem is equal to the power input (heat flux) to the ith subsystem plus the energy (heat) flow to the ith subsystem from all other subsystems minus the power dissipated from the ith subsystem to the environment. Furthermore, since $f(\cdot) - d(\cdot)$ is locally Lipschitz continuous on $\overline{\mathbb{R}}_+^q$ and $S(\cdot)$ is a bounded piecewise continuous function of time, it follows that (3.5) has a unique solution over the finite time interval $[t_0, \tau_{E_0})$. If, in addition, the power balance equation (3.5) is *input-to-state stable* [242],[4] then $\tau_{E_0} = \infty$.

Equation (3.2) or, equivalently, (3.5) is a statement of the *first law of thermodynamics* as applied to *isochoric transformations* (i.e., constant subsystem volume transformations) for each of the subsystems $\mathcal{G}_i$, $i = 1, \ldots, q$, with $E_i(\cdot)$, $S_i(\cdot)$, $\sigma_{ij}(\cdot)$, $i \neq j$, and $\sigma_{ii}(\cdot)$, $i, j = 1, \ldots, q$, playing the role of the ith subsystem internal energy, rate of heat supplied to (or extracted from) the ith subsystem, heat flow between subsystems due to coupling, and the rate of energy (heat) dissipated to the environment,

[4]The notion of input-to-state stability implies that if an equilibrium solution of a closed nonlinear system is asymptotically stable, then the state of the open system is bounded for every bounded piecewise continuous input.

respectively. To further elucidate that (3.2) is essentially the statement of the principle of the conservation of energy, let the total energy in the large-scale dynamical system $\mathcal{G}$ be given by $U \triangleq \mathbf{e}^{\mathrm{T}}E$, where $\mathbf{e}^{\mathrm{T}} \triangleq [1,\ldots,1]$ and $E \in \overline{\mathbb{R}}_{+}^{q}$, and let the net energy received by the large-scale dynamical system $\mathcal{G}$ over the time interval $[t_1, t_2]$ be given by

$$Q \triangleq \int_{t_1}^{t_2} \mathbf{e}^{\mathrm{T}}[S(t) - d(E(t))]\mathrm{d}t, \tag{3.6}$$

where $E(t)$, $t \geq t_0$, is the solution to (3.5). Then, premultiplying (3.2) by $\mathbf{e}^{\mathrm{T}}$ and using the fact that $\mathbf{e}^{\mathrm{T}}f(E) \equiv 0$, it follows that

$$\Delta U = Q, \tag{3.7}$$

where $\Delta U \triangleq U(t_2) - U(t_1)$ denotes the variation in the total energy of the large-scale dynamical system $\mathcal{G}$ over the time interval $[t_1, t_2]$. This is a statement of the first law of thermodynamics for isochoric transformations of the large-scale dynamical system $\mathcal{G}$ and gives a precise formulation of the equivalence between the variation in system internal energy and heat.

It is important to note that the large-scale dynamical system model (3.5) does not consider work done by the system on the environment nor work done by the environment on the system. Hence, Q can be physically interpreted as the net amount of energy that is received by the system in forms other than work. The extension of addressing work performed by and on the system can be easily addressed by including an additional state equation, coupled to the power balance equation (3.5), involving volume (deformation) states for each subsystem. Since this extension does not alter any of the conceptual results of this chapter, it is not considered in this chapter for simplicity of exposition.

Work performed by the dynamical system on the environment and work done by the environment on the dynamical system as well as work done between subsystems is addressed in Chapter 5. In addition, in its most general form thermodynamics can also involve reacting mixtures and combustion. Specifically, when a chemical reaction occurs, the bonds within molecules of the *reactant* are broken, and atoms and electrons rearrange to form *products*. This extension involving reactive systems is addressed in Chapter 6.

For our large-scale dynamical system model $\mathcal{G}$, we assume that $\sigma_{ij}(E) = 0$, $E \in \overline{\mathbb{R}}_{+}^{q}$, whenever $E_j = 0$, $i, j = 1, \ldots, q$. In this case, $f(E) - d(E)$, $E \in \overline{\mathbb{R}}_{+}^{q}$, is essentially nonnegative. The above constraint implies that if the energy of the jth subsystem of $\mathcal{G}$ is zero, then this subsystem cannot supply any energy to its surroundings nor dissipate energy to the environment. Moreover, we assume that $S_i(t) \geq 0$ whenever $E_i(t) = 0$,

$t \geq t_0$, $i = 1, \ldots, q$, which implies that when the energy of the ith subsystem is zero, then no energy can be extracted from this subsystem.

The following proposition is needed for the main results of this chapter.

Proposition 3.1. Consider the large-scale dynamical system $\mathcal{G}$ with power balance equation given by (3.5). Suppose $\sigma_{ij}(E) = 0$, $E \in \overline{\mathbb{R}}_+^q$, whenever $E_j = 0$, $i, j = 1, \ldots, q$, and $S_i(t) \geq 0$ whenever $E_i(t) = 0$, $t \geq t_0$, $i = 1, \ldots, q$. Then the solution $E(t)$, $t \geq t_0$, to (3.5) is nonnegative for all nonnegative initial conditions $E_0 \in \overline{\mathbb{R}}_+^q$.

Proof. First note that $f(E) - d(E)$, $E \in \overline{\mathbb{R}}_+^q$, is essentially nonnegative. Next, since $S_i(t) \geq 0$ whenever $E_i(t) = 0$, $t \geq t_0$, $i = 1, \ldots, q$, it follows that $\dot{E}_i(t) \geq 0$ for all $t \geq t_0$ and $i = 1, \ldots, q$ whenever $E_i(t) = 0$ and $E_j(t) \geq 0$ for all $j \neq i$ and $t \geq t_0$. This implies that for all nonnegative initial conditions $E_0 \in \overline{\mathbb{R}}_+^q$, the trajectory of $\mathcal{G}$ is directed toward the interior of the nonnegative orthant $\overline{\mathbb{R}}_+^q$ whenever $E_i(t) = 0$, $i = 1, \ldots, q$, and hence remains nonnegative for all $t \geq t_0$. □

Next, premultiplying (3.2) by $\mathbf{e}^{\mathrm{T}}$, using Proposition 3.1, and using the fact that $\mathbf{e}^{\mathrm{T}} f(E) \equiv 0$, it follows that

$$\mathbf{e}^{\mathrm{T}} E(T) = \mathbf{e}^{\mathrm{T}} E(t_0) + \int_{t_0}^{T} \mathbf{e}^{\mathrm{T}} S(t) \mathrm{d}t - \int_{t_0}^{T} \mathbf{e}^{\mathrm{T}} d(E(t)) \mathrm{d}t, \quad T \geq t_0. \tag{3.8}$$

Now, for the large-scale dynamical system $\mathcal{G}$, define the input $u(t) \triangleq S(t)$ and the output $y(t) \triangleq d(E(t))$. Hence, it follows from (3.8) that the large-scale dynamical system $\mathcal{G}$ is *lossless* [191, 464] with respect to the *energy supply rate* $r(u, y) \triangleq \mathbf{e}^{\mathrm{T}} u - \mathbf{e}^{\mathrm{T}} y$ and with the *energy storage function* $U(E) \triangleq \mathbf{e}^{\mathrm{T}} E$, $E \in \overline{\mathbb{R}}_+^q$. The notion of losslessness of a system asserts that the energy $U(E)$ that can be extracted from a system through its input-output ports $(u, y) = (S, d(E))$ is equal to the initial energy stored in the system, and hence, there can be no internal creation of energy; only conservation of energy is possible [191].

The following lemma is required for our next result.

Lemma 3.1. Consider the large-scale dynamical system $\mathcal{G}$ with power balance equation (3.5). Then, for every equilibrium state $E_{\mathrm{e}} \in \overline{\mathbb{R}}_+^q$ and every $\varepsilon > 0$ and $T > 0$, there exist $S_{\mathrm{e}} \in \mathbb{R}^q$, $\alpha > 0$, and $\hat{T} \in [0, T]$ such that for every $\hat{E} \in \overline{\mathbb{R}}_+^q$ with $\|\hat{E} - E_{\mathrm{e}}\| \leq \alpha T$, there exists $S : [0, \hat{T}] \to \mathbb{R}^q$ such that $\|S(t) - S_{\mathrm{e}}\| \leq \varepsilon$, $t \in [0, \hat{T}]$, and $E(t) = E_{\mathrm{e}} + \frac{(\hat{E} - E_{\mathrm{e}})}{\hat{T}} t$, $t \in [0, \hat{T}]$.

Proof. Note that with $S_e \triangleq d(E_e) - f(E_e)$, the state $E_e \in \overline{\mathbb{R}}_+^q$ is an equilibrium state of (3.5). Let $\theta > 0$ and $T > 0$, and define

$$M(\theta, T) \triangleq \sup_{E \in \overline{\mathcal{B}}_1(0),\, t \in [0,T]} \|f(E_e + \theta t E) - d(E_e + \theta t E) + S_e\|. \tag{3.9}$$

Note that for every $T > 0$, $\lim_{\theta \to 0^+} M(\theta, T) = 0$, and for every $\theta > 0$, $\lim_{T \to 0^+} M(\theta, T) = 0$. Next, let $\varepsilon > 0$ and $T > 0$ be given, and let $\alpha > 0$ be such that $M(\alpha, T) + \alpha \le \varepsilon$. (The existence of such an α is guaranteed since $M(\alpha, T) \to 0$ as $\alpha \to 0^+$.) Now, let $\hat{E} \in \overline{\mathbb{R}}_+^q$ be such that $\|\hat{E} - E_e\| \le \alpha T$. With $\hat{T} \triangleq \frac{\|\hat{E} - E_e\|}{\alpha} \le T$ and

$$S(t) = -f(E(t)) + d(E(t)) + \alpha \frac{(\hat{E} - E_e)}{\|\hat{E} - E_e\|}, \quad t \in [0, \hat{T}], \tag{3.10}$$

it follows that

$$E(t) = E_e + \frac{(\hat{E} - E_e)}{\|\hat{E} - E_e\|} \alpha t, \quad t \in [0, \hat{T}], \tag{3.11}$$

is a solution to (3.5).

The result is now immediate by noting that $E(\hat{T}) = \hat{E}$ and

$$\begin{aligned} \|S(t) - S_e\| &\le \left\| f\left(E_e + \tfrac{(\hat{E} - E_e)}{\|\hat{E} - E_e\|} \alpha t\right) - d\left(E_e + \tfrac{(\hat{E} - E_e)}{\|\hat{E} - E_e\|} \alpha t\right) + S_e \right\| + \alpha \\ &\le M(\alpha, T) + \alpha \\ &\le \varepsilon, \quad t \in [0, \hat{T}], \end{aligned} \tag{3.12}$$

which proves the result. □

It follows from Lemma 3.1 that the large-scale dynamical system $\mathcal{G}$ with the power balance equation (3.5) is *reachable* from and *controllable* to the origin in $\overline{\mathbb{R}}_+^q$. Recall that the large-scale dynamical system $\mathcal{G}$ with the power balance equation (3.5) is reachable from the origin in $\overline{\mathbb{R}}_+^q$ if, for all $E_0 = E(t_0) \in \overline{\mathbb{R}}_+^q$, there exists a finite time $t_i \le t_0$ and a square integrable input $S(\cdot)$ defined on $[t_i, t_0]$ such that the state $E(t)$, $t \ge t_i$, can be driven from $E(t_i) = 0$ to $E(t_0) = E_0$. Alternatively, $\mathcal{G}$ is controllable to the origin in $\overline{\mathbb{R}}_+^q$ if, for all $E_0 = E(t_0) \in \overline{\mathbb{R}}_+^q$, there exists a finite time $t_f \ge t_0$ and a square integrable input $S(\cdot)$ defined on $[t_0, t_f]$ such that the state $E(t)$, $t \ge t_0$, can be driven from $E(t_0) = E_0$ to $E(t_f) = 0$.

We let $\mathcal{U}_r$ denote the set of all bounded continuous power inputs (heat fluxes) to the large-scale dynamical system $\mathcal{G}$ such that, for every $T \ge -t_0$, the system energy state can be driven from $E(-T) = 0$ to $E(t_0) = E_0 \in \overline{\mathbb{R}}_+^q$ by $S(\cdot) \in \mathcal{U}_r$, and we let $\mathcal{U}_c$ denote the set of all bounded continuous power inputs (heat fluxes) to the large-scale dynamical system $\mathcal{G}$ such that for every $T \geq t_0$ the system energy state can be driven from $E(t_0) = E_0 \in \overline{\mathbb{R}}_+^q$ to $E(T) = 0$ by $S(\cdot) \in \mathcal{U}_c$. Furthermore, let $\mathcal{U}$ be an input space that is a subset of bounded continuous $\mathbb{R}^q$-valued functions on $\mathbb{R}$. The spaces $\mathcal{U}_r$, $\mathcal{U}_c$, and $\mathcal{U}$ are assumed to be closed under the shift operator, that is, if $S(\cdot) \in \mathcal{U}$ (respectively, $\mathcal{U}_c$ or $\mathcal{U}_r$), then the function S_T defined by $S_T(t) \triangleq S(t+T)$ is contained in $\mathcal{U}$ (respectively, $\mathcal{U}_c$ or $\mathcal{U}_r$) for all $T \geq 0$.

The next result establishes the uniqueness of the internal energy function $U(E)$, $E \in \overline{\mathbb{R}}_+^q$, for our large-scale dynamical system $\mathcal{G}$. For this result define the *available energy* of the large-scale dynamical system $\mathcal{G}$ by

$$U_a(E_0) \triangleq -\inf_{u(\cdot)\in\mathcal{U},\, T\geq t_0} \int_{t_0}^{T} [\mathbf{e}^{\mathrm{T}}u(t) - \mathbf{e}^{\mathrm{T}}y(t)]\mathrm{d}t, \quad E_0 \in \overline{\mathbb{R}}_+^q, \tag{3.13}$$

and the *required energy supply* of the large-scale dynamical system $\mathcal{G}$ by

$$U_r(E_0) \triangleq \inf_{u(\cdot)\in\mathcal{U}_r,\, T\geq -t_0} \int_{-T}^{t_0} [\mathbf{e}^{\mathrm{T}}u(t) - \mathbf{e}^{\mathrm{T}}y(t)]\mathrm{d}t, \quad E_0 \in \overline{\mathbb{R}}_+^q. \tag{3.14}$$

Note that the available energy $U_a(E)$ is the maximum amount of stored energy (net heat) that can be extracted from the large-scale dynamical system $\mathcal{G}$ at any time T, and the required energy supply $U_r(E)$ is the minimum amount of energy (net heat) that can be delivered to the large-scale dynamical system $\mathcal{G}$ to transfer it from a state of minimum potential $E(-T) = 0$ to a given state $E(t_0) = E_0$.

Theorem 3.1. Consider the large-scale dynamical system $\mathcal{G}$ with power balance equation given by (3.5). Then $\mathcal{G}$ is lossless with respect to the energy supply rate $r(u,y) = \mathbf{e}^{\mathrm{T}}u - \mathbf{e}^{\mathrm{T}}y$, where $u(t) \equiv S(t)$ and $y(t) \equiv d(E(t))$, and with the unique energy storage function corresponding to the total energy of the system $\mathcal{G}$ given by

$$\begin{aligned} U(E_0) &= \mathbf{e}^{\mathrm{T}}E_0 \\ &= -\int_{t_0}^{T_+} [\mathbf{e}^{\mathrm{T}}u(t) - \mathbf{e}^{\mathrm{T}}y(t)]\mathrm{d}t \\ &= \int_{-T_-}^{t_0} [\mathbf{e}^{\mathrm{T}}u(t) - \mathbf{e}^{\mathrm{T}}y(t)]\mathrm{d}t, \quad E_0 \in \overline{\mathbb{R}}_+^q, \end{aligned} \tag{3.15}$$

where $E(t)$, $t \geq t_0$, is the solution to (3.5) with admissible input $u(\cdot) \in \mathcal{U}$, $E(-T_-) = 0$, $E(T_+) = 0$, and $E(t_0) = E_0 \in \overline{\mathbb{R}}_+^q$. Furthermore,

$$0 \leq U_a(E_0) = U(E_0) = U_r(E_0) < \infty, \quad E_0 \in \overline{\mathbb{R}}_+^q. \tag{3.16}$$

Proof. Note that it follows from (3.8) that $\mathcal{G}$ is lossless with respect to the energy supply rate $r(u,y) = \mathbf{e}^{\mathrm{T}}u - \mathbf{e}^{\mathrm{T}}y$ and with the energy storage function $U(E) = \mathbf{e}^{\mathrm{T}}E$, $E \in \overline{\mathbb{R}}_+^q$. Since, by Lemma 3.1, $\mathcal{G}$ is reachable from and controllable to the origin in $\overline{\mathbb{R}}_+^q$, it follows from (3.8), with $E(t_0) = E_0 \in \overline{\mathbb{R}}_+^q$ and $E(T_+) = 0$ for some $T_+ \geq t_0$ and $u(\cdot) \in \mathcal{U}$, that

$$\begin{aligned} \mathbf{e}^{\mathrm{T}} E_0 &= -\int_{t_0}^{T_+} [\mathbf{e}^{\mathrm{T}} u(t) - \mathbf{e}^{\mathrm{T}} y(t)]\mathrm{d}t \\ &\leq \sup_{u(\cdot)\in\mathcal{U}, T\geq t_0} \left[-\int_{t_0}^{T} [\mathbf{e}^{\mathrm{T}} u(t) - \mathbf{e}^{\mathrm{T}} y(t)]\mathrm{d}t \right] \\ &= -\inf_{u(\cdot)\in\mathcal{U}, T\geq t_0} \int_{t_0}^{T} [\mathbf{e}^{\mathrm{T}} u(t) - \mathbf{e}^{\mathrm{T}} y(t)]\mathrm{d}t \\ &= U_{\mathrm{a}}(E_0), \quad E_0 \in \overline{\mathbb{R}}_+^q. \end{aligned} \tag{3.17}$$

Alternatively, it follows from (3.8), with $E(-T_-) = 0$ for some $-T_- \leq t_0$ and $u(\cdot) \in \mathcal{U}_{\mathrm{r}}$, that

$$\begin{aligned} \mathbf{e}^{\mathrm{T}} E_0 &= \int_{-T_-}^{t_0} [\mathbf{e}^{\mathrm{T}} u(t) - \mathbf{e}^{\mathrm{T}} y(t)]\mathrm{d}t \\ &\geq \inf_{u(\cdot)\in\mathcal{U}_{\mathrm{r}}, T\geq -t_0} \int_{-T}^{t_0} [\mathbf{e}^{\mathrm{T}} u(t) - \mathbf{e}^{\mathrm{T}} y(t)]\mathrm{d}t \\ &= U_{\mathrm{r}}(E_0), \quad E_0 \in \overline{\mathbb{R}}_+^q. \end{aligned} \tag{3.18}$$

Thus, (3.17) and (3.18) imply that (3.15) is satisfied and

$$U_{\mathrm{r}}(E_0) \leq \mathbf{e}^{\mathrm{T}} E_0 \leq U_{\mathrm{a}}(E_0), \quad E_0 \in \overline{\mathbb{R}}_+^q. \tag{3.19}$$

Conversely, it follows from (3.8) and the fact that $U(E) = \mathbf{e}^{\mathrm{T}} E \geq 0$, $E \in \overline{\mathbb{R}}_+^q$, that, for all $T \geq t_0$ and $u(\cdot) \in \mathcal{U}$,

$$\mathbf{e}^{\mathrm{T}} E(t_0) \geq -\int_{t_0}^{T} [\mathbf{e}^{\mathrm{T}} u(t) - \mathbf{e}^{\mathrm{T}} y(t)]\mathrm{d}t, \quad E(t_0) \in \overline{\mathbb{R}}_+^q, \tag{3.20}$$

which implies that

$$\begin{aligned} \mathbf{e}^{\mathrm{T}} E(t_0) &\geq \sup_{u(\cdot)\in\mathcal{U}, T\geq t_0} \left[-\int_{t_0}^{T} [\mathbf{e}^{\mathrm{T}} u(t) - \mathbf{e}^{\mathrm{T}} y(t)]\mathrm{d}t \right] \\ &= -\inf_{u(\cdot)\in\mathcal{U}, T\geq t_0} \int_{t_0}^{T} [\mathbf{e}^{\mathrm{T}} u(t) - \mathbf{e}^{\mathrm{T}} y(t)]\mathrm{d}t \\ &= U_{\mathrm{a}}(E(t_0)), \quad E(t_0) \in \overline{\mathbb{R}}_+^q. \end{aligned} \tag{3.21}$$

Furthermore, it follows from the definition of $U_{\mathrm{a}}(\cdot)$ that $U_{\mathrm{a}}(E) \geq 0$, $E \in \overline{\mathbb{R}}_+^q$, since the infimum in (3.13) is taken over the set of values containing the zero value ($T = t_0$).

Next, note that it follows from (3.8), with $E(t_0) \in \overline{\mathbb{R}}_+^q$ and $E(-T) = 0$

for all $T \geq -t_0$ and $u(\cdot) \in \mathcal{U}_r$, that

$$\begin{aligned} \mathbf{e}^{\mathrm{T}} E(t_0) &= \int_{-T}^{t_0} [\mathbf{e}^{\mathrm{T}} u(t) - \mathbf{e}^{\mathrm{T}} y(t)] \mathrm{d}t \\ &= \inf_{u(\cdot) \in \mathcal{U}_r, T \geq -t_0} \int_{-T}^{t_0} [\mathbf{e}^{\mathrm{T}} u(t) - \mathbf{e}^{\mathrm{T}} y(t)] \mathrm{d}t \\ &= U_r(E(t_0)), \quad E(t_0) \in \overline{\mathbb{R}}_+^q. \end{aligned} \tag{3.22}$$

Moreover, since the system $\mathcal{G}$ is reachable from the origin, it follows that for every $E(t_0) \in \overline{\mathbb{R}}_+^q$, there exists $T \geq -t_0$ and $u(\cdot) \in \mathcal{U}_r$ such that

$$\int_{-T}^{t_0} [\mathbf{e}^{\mathrm{T}} u(t) - \mathbf{e}^{\mathrm{T}} y(t)] \mathrm{d}t \tag{3.23}$$

is finite, and hence, $U_r(E(t_0)) < \infty$, $E(t_0) \in \overline{\mathbb{R}}_+^q$. Finally, combining (3.19), (3.21), and (3.22), it follows that (3.16) holds. □

It follows from (3.16) and the definitions of available energy $U_a(E_0)$ and the required energy supply $U_r(E_0)$, $E_0 \in \overline{\mathbb{R}}_+^q$, that the large-scale dynamical system $\mathcal{G}$ can deliver to its surroundings all of its stored subsystem energies and can store all of the work done to all of its subsystems. This is in essence a statement of the first law of thermodynamics and places no limitation on the possibility of transforming heat into work or work into heat. In the case where $S(t) \equiv 0$, it follows from (3.8) and the fact that $\sigma_{ii}(E) \geq 0$, $E \in \overline{\mathbb{R}}_+^q$, $i = 1, \ldots, q$, that the zero solution $E(t) \equiv 0$ of the large-scale dynamical system $\mathcal{G}$ with the power balance equation (3.5) is Lyapunov stable with Lyapunov function $U(E)$ corresponding to the total energy in the system.

3.3 Entropy and the Second Law of Thermodynamics

The nonlinear power balance equation (3.5) can exhibit a full range of nonlinear behavior, including bifurcations, limit cycles, and even chaos. However, a thermodynamically consistent energy flow model should ensure that the evolution of the system energy is diffusive (parabolic) in character with convergent subsystem energies. As established in Section 2.9, such a system model would guarantee the absence of the *Poincaré recurrence* phenomenon [10], which states that every finite-dimensional, *isolated* (i.e., $S(t) \equiv 0$ and $d(E) \equiv 0$) dynamical system with *volume-preserving*[5] trajectories (subsystem energies) will return arbitrarily close to its initial system state (energy) infinitely many times. This of course would violate the second law of thermodynamics, since subsystem energies (temperatures) would be allowed to return to their starting state and thereby subvert

[5]A dynamical system is *volume-preserving* if and only if the volume of an arbitrary region of the state space is conserved by the time evolution of the system, even though the shape of the region may change dramatically.

the diffusive character of the dynamical system. Hence, to ensure a thermodynamically consistent energy flow model, we require the following axioms.[6] For the statement of these axioms, we first recall the following graph-theoretic notions.

Definition 3.1 ([31]). A *directed graph* $\mathfrak{G}(\mathcal{C})$ associated with the *connectivity matrix* $\mathcal{C} \in \mathbb{R}^{q\times q}$ has *vertices* $\{1,2,\ldots,q\}$ and an *arc* from vertex i to vertex j, $i \neq j$, if and only if $\mathcal{C}_{(j,i)} \neq 0$. A *graph* $\mathfrak{G}(\mathcal{C})$ associated with the connectivity matrix $\mathcal{C} \in \mathbb{R}^{q\times q}$ is a directed graph for which the *arc set* is symmetric, that is, $\mathcal{C} = \mathcal{C}^{\mathrm{T}}$. We say that $\mathfrak{G}(\mathcal{C})$ is *strongly connected* if for every ordered pair of vertices (i,j), $i \neq j$, there exists a *path* (i.e., a sequence of arcs) leading from i to j.

Recall that the connectivity matrix $\mathcal{C} \in \mathbb{R}^{q\times q}$ is *irreducible*, that is, there does not exist a permutation matrix such that $\mathcal{C}$ is cogredient to a lower-block triangular matrix, if and only if $\mathfrak{G}(\mathcal{C})$ is strongly connected (see Theorem 2.7 of [31]). Let $\phi_{ij}(E) \triangleq \sigma_{ij}(E) - \sigma_{ji}(E)$, $E \in \overline{\mathbb{R}}_+^q$, denote the net energy flow from the jth subsystem $\mathcal{G}_j$ to the ith subsystem $\mathcal{G}_i$ of the large-scale dynamical system $\mathcal{G}$.

Axiom *i*): For the connectivity matrix $\mathcal{C} \in \mathbb{R}^{q\times q}$ associated with the large-scale dynamical system $\mathcal{G}$ defined by

$$\mathcal{C}_{(i,j)} \triangleq \begin{cases} 0, & \text{if } \phi_{ij}(E) \equiv 0, \\ 1, & \text{otherwise,} \end{cases} \quad i \neq j, \quad i,j = 1,\ldots,q, \tag{3.24}$$

and

$$\mathcal{C}_{(i,i)} \triangleq -\sum_{k=1,\, k\neq i}^{q} \mathcal{C}_{(k,i)}, \quad i = j, \quad i = 1,\ldots,q, \tag{3.25}$$

rank $\mathcal{C} = q - 1$, and for $\mathcal{C}_{(i,j)} = 1$, $i \neq j$, $\phi_{ij}(E) = 0$ if and only if $E_i = E_j$.

Axiom *ii*): For $i,j = 1,\ldots,q$, $(E_i - E_j)\phi_{ij}(E) \leq 0$, $E \in \overline{\mathbb{R}}_+^q$.

The fact that $\phi_{ij}(E) = 0$ if and only if $E_i = E_j$, $i \neq j$, implies that subsystems $\mathcal{G}_i$ and $\mathcal{G}_j$ of $\mathcal{G}$ are *connected*; alternatively, $\phi_{ij}(E) \equiv 0$ implies that $\mathcal{G}_i$ and $\mathcal{G}_j$ are *disconnected.* Axiom i) implies that if the energies in the connected subsystems $\mathcal{G}_i$ and $\mathcal{G}_j$ are equal, then energy exchange between these subsystems is not possible. This statement is consistent with the

[6]It can be argued here that a more appropriate terminology is *assumptions* rather than *axioms* since, as will be seen, these are statements taken to be true and used as premises in order to infer certain results, but may not otherwise be accepted. However, as we will see, these statements are equivalent (within our formulation) to the stipulated postulates of the zeroth and second laws of thermodynamics involving transitivity of a thermal equilibrium and heat flowing from *hotter* to *colder* bodies, and as such we refer to them as *axioms.*

zeroth law of thermodynamics, which postulates that temperature equality is a necessary and sufficient condition for thermal equilibrium. Furthermore, it follows from the fact that $\mathcal{C} = \mathcal{C}^{\mathrm{T}}$ and rank $\mathcal{C} = q-1$ that the connectivity matrix $\mathcal{C}$ is irreducible, which implies that for any pair of subsystems $\mathcal{G}_i$ and $\mathcal{G}_j$, $i \neq j$, of $\mathcal{G}$ there exists a sequence of connectors (arcs) of $\mathcal{G}$ that connect $\mathcal{G}_i$ and $\mathcal{G}_j$.

Axiom *ii*) implies that energy flows from more energetic subsystems to less energetic subsystems and is consistent with the *second law of thermodynamics*, which states that heat (energy) must flow in the direction of lower temperatures. Furthermore, note that $\phi_{ij}(E) = -\phi_{ji}(E)$, $E \in \overline{\mathbb{R}}_+^q$, $i \neq j$, $i, j = 1, \ldots, q$, which implies conservation of energy between lossless subsystems. With $S(t) \equiv 0$, Axioms *i*) and *ii*) along with the fact that $\phi_{ij}(E) = -\phi_{ji}(E)$, $E \in \overline{\mathbb{R}}_+^q$, $i \neq j$, $i, j = 1, \ldots, q$, imply that at a given instant of time, energy can only be transported, stored, or dissipated but not created, and the maximum amount of energy that can be transported and/or dissipated from a subsystem cannot exceed the energy in the subsystem.

It is important to note that our formulation of the second law of thermodynamics as given by Axiom *ii*) does not require the mentioning of temperature nor the more primitive subjective notions of hotness or coldness. As we will see later, temperature is defined in terms of the system entropy after we establish the existence of a unique, global, continuously differentiable entropy function for $\mathcal{G}$. However, a thermodynamicist can argue here that the study of thermodynamics is impossible without a clear separation between extensive (energy) and intensive (temperature) properties, and that Axioms *i*) and *ii*) must be expressed in terms of intensive properties.

Recall that an *extensive property* is a function of the size or extent (i.e., quantity of matter) of a system whose magnitude is the sum of the values of its parts. An *intensive property* is not additive and is independent of the size or extent of a system. Thus, intensive properties can be functions of time and space, whereas extensive properties are functions only of time. Mass, volume, energy, and entropy are examples of extensive properties, whereas temperature and pressure are examples of intensive properties. The distinction between extensive and intensive properties in thermodynamics is physically significant but can be ignored from a mathematical perspective.

For the more applied reader Axioms *i*) and *ii*) can be interpreted as $\phi_{ij}(E) = 0$ if and only if $\beta_i E_i = \beta_j E_j$ and $(\beta_i E_i - \beta_j E_j)\phi_{ij}(E) \leq 0$, $E \in \overline{\mathbb{R}}_+^q$, $i, j = 1, \ldots, q$, where $\beta_i > 0$, $i = 1, \ldots, q$, denote the reciprocal of the specific heat (at constant volume) for the ith subsystem, and hence, $T_i \triangleq \beta_i E_i$ is the absolute temperature of the ith subsystem. This is addressed in Chapter 4;

in this chapter, we assume $\beta_i = 1$, $i = 1, \ldots, q$, and hence, every subsystem possesses the same constant heat capacity.

Next, we show that the classical Clausius equality and inequality for reversible and irreversible thermodynamics over cyclic motions are satisfied for our thermodynamically consistent energy flow model. For this result $\oint$ denotes a cyclic integral evaluated along an arbitrary closed path of (3.5) in $\overline{\mathbb{R}}_+^q$; that is, $\oint \triangleq \int_{t_0}^{t_\mathrm{f}}$ with $t_\mathrm{f} \geq t_0$ and $S(\cdot) \in \mathcal{U}$ such that $E(t_\mathrm{f}) = E(t_0) = E_0 \in \overline{\mathbb{R}}_+^q$.

Proposition 3.2. Consider the large-scale dynamical system $\mathcal{G}$ with power balance equation (3.5), and assume that Axioms *i*) and *ii*) hold. Then, for all $E_0 \in \overline{\mathbb{R}}_+^q$, $t_\mathrm{f} \geq t_0$, and $S(\cdot) \in \mathcal{U}$ such that $E(t_\mathrm{f}) = E(t_0) = E_0$,

$$\int_{t_0}^{t_\mathrm{f}} \sum_{i=1}^{q} \frac{S_i(t) - \sigma_{ii}(E(t))}{c + E_i(t)} \mathrm{d}t = \oint \sum_{i=1}^{q} \frac{\mathrm{d}Q_i(t)}{c + E_i(t)} \leq 0, \tag{3.26}$$

where $c > 0$, $\mathrm{d}Q_i(t) \triangleq [S_i(t) - \sigma_{ii}(E(t))]\mathrm{d}t$, $i = 1, \ldots, q$, is the amount of net energy (heat) received by the ith subsystem over the infinitesimal time interval $\mathrm{d}t$, and $E(t)$, $t \geq t_0$, is the solution to (3.5) with initial condition $E(t_0) = E_0$. Furthermore,

$$\oint \sum_{i=1}^{q} \frac{\mathrm{d}Q_i(t)}{c + E_i(t)} = 0 \tag{3.27}$$

if and only if there exists a continuous function $\alpha : [t_0, t_\mathrm{f}] \to \overline{\mathbb{R}}_+$ such that $E(t) = \alpha(t)\mathbf{e}$, $t \in [t_0, t_\mathrm{f}]$.

Proof. Since, by Proposition 3.1, $E(t) \geq\geq 0$, $t \geq t_0$, and $\phi_{ij}(E) = -\phi_{ji}(E)$, $E \in \overline{\mathbb{R}}_+^q$, $i \neq j$, $i, j = 1, \ldots, q$, it follows from (3.5) and Axiom *ii*) that

$$\begin{aligned}
\oint \sum_{i=1}^{q} \frac{\mathrm{d}Q_i(t)}{c + E_i(t)} &= \int_{t_0}^{t_\mathrm{f}} \sum_{i=1}^{q} \frac{\dot{E}_i(t) - \sum_{j=1, j\neq i}^{q} \phi_{ij}(E(t))}{c + E_i(t)} \mathrm{d}t \\
&= \sum_{i=1}^{q} \log_e \left(\frac{c + E_i(t_\mathrm{f})}{c + E_i(t_0)} \right) - \int_{t_0}^{t_\mathrm{f}} \sum_{i=1}^{q} \sum_{j=1, j\neq i}^{q} \frac{\phi_{ij}(E(t))}{c + E_i(t)} \mathrm{d}t \\
&= -\int_{t_0}^{t_\mathrm{f}} \sum_{i=1}^{q} \sum_{j=i+1}^{q} \left(\frac{\phi_{ij}(E(t))}{c + E_i(t)} - \frac{\phi_{ij}(E(t))}{c + E_j(t)} \right) \mathrm{d}t \\
&= -\int_{t_0}^{t_\mathrm{f}} \sum_{i=1}^{q} \sum_{j=i+1}^{q} \frac{\phi_{ij}(E(t))[E_j(t) - E_i(t)]}{(c + E_i(t))(c + E_j(t))} \mathrm{d}t \\
&\leq 0,
\end{aligned} \tag{3.28}$$

which proves (3.26).

To show (3.27), note that it follows from (3.28), Axiom i), and Axiom ii) that (3.27) holds if and only if $E_i(t) = E_j(t)$, $t \in [t_0, t_{\mathrm{f}}]$, $i \neq j$, $i, j = 1, \ldots, q$, or, equivalently, there exists a continuous function $\alpha : [t_0, t_{\mathrm{f}}] \to \overline{\mathbb{R}}_+$ such that $E(t) = \alpha(t)\mathbf{e}$, $t \in [t_0, t_{\mathrm{f}}]$. □

Inequality (3.26) is a generalization of Clausius' inequality for reversible and irreversible thermodynamics as applied to large-scale dynamical systems and restricts the manner in which the system dissipates (scaled) heat over cyclic motions. It follows from Axiom i) and (3.5) that for the *adiabatically isolated* large-scale dynamical system $\mathcal{G}$ (that is, $S(t) \equiv 0$ and $d(E(t)) \equiv 0$), the energy states given by $E_{\mathrm{e}} = \alpha\mathbf{e}$, $\alpha \geq 0$, correspond to the equilibrium energy states of $\mathcal{G}$. Thus, as in classical thermodynamics, we can define an *equilibrium process* as a process in which the trajectory of the large-scale dynamical system $\mathcal{G}$ moves along the equilibrium manifold $\mathcal{M}_{\mathrm{e}} \triangleq \{E \in \overline{\mathbb{R}}_+^q : E = \alpha\mathbf{e}, \alpha \geq 0\}$ corresponding to the set of equilibria of the isolated system $\mathcal{G}$.[7] The power input that can generate such a trajectory can be given by $S(t) = d(E(t)) + u(t)$, $t \geq t_0$, where $u(\cdot) \in \mathcal{U}$ is such that $u_i(t) \equiv u_j(t)$, $i \neq j$, $i, j = 1, \ldots, q$.

Our definition of an equilibrium transformation involves a continuous succession of intermediate states that differ by infinitesimals from equilibrium system states and thus can only connect initial and final states, which are states of equilibrium. This process need not be slowly varying, and hence, equilibrium and quasi-static processes are not synonymous in this monograph. Alternatively, a *nonequilibrium process* is a process that does not lie on the equilibrium manifold $\mathcal{M}_{\mathrm{e}}$. Hence, it follows from Axiom i) that for an equilibrium process $\phi_{ij}(E(t)) = 0$, $t \geq t_0$, $i \neq j$, $i, j = 1, \ldots, q$, and thus, by Proposition 3.2, inequality (3.26) is satisfied as an equality. Alternatively, for a nonequilibrium process it follows from Axioms i) and ii) that (3.26) is satisfied as a strict inequality.

Next, we give a deterministic definition of entropy for the large-scale dynamical system $\mathcal{G}$ that is consistent with the classical thermodynamic definition of entropy.

Definition 3.2. For the large-scale dynamical system $\mathcal{G}$ with power balance equation (3.5), a function $\mathcal{S} : \overline{\mathbb{R}}_+^q \to \mathbb{R}$ satisfying

$$\mathcal{S}(E(t_2)) \geq \mathcal{S}(E(t_1)) + \int_{t_1}^{t_2} \sum_{i=1}^{q} \frac{S_i(t) - \sigma_{ii}(E(t))}{c + E_i(t)} \mathrm{d}t \tag{3.29}$$

[7] Since in this section we are not considering work performed by and on the system, the notions of an *isolated* system and an *adiabatically isolated* system are equivalent.

for every $t_2 \geq t_1 \geq t_0$ and $S(\cdot) \in \mathcal{U}$ is called the *entropy* function of $\mathcal{G}$.

Next, we show that (3.26) guarantees the existence of an entropy function for $\mathcal{G}$. For this result define the *available entropy* of the large-scale dynamical system $\mathcal{G}$ by

$$\mathcal{S}_{\mathrm{a}}(E_0) \triangleq - \sup_{S(\cdot)\in\mathcal{U}_{\mathrm{c}},\, T\geq t_0} \int_{t_0}^{T} \sum_{i=1}^{q} \frac{S_i(t) - \sigma_{ii}(E(t))}{c + E_i(t)} \mathrm{d}t, \tag{3.30}$$

where $E(t_0) = E_0 \in \overline{\mathbb{R}}_+^q$ and $E(T) = 0$, and define the *required entropy supply* of the large-scale dynamical system $\mathcal{G}$ by

$$\mathcal{S}_{\mathrm{r}}(E_0) \triangleq \sup_{S(\cdot)\in\mathcal{U}_{\mathrm{r}},\, T\geq -t_0} \int_{-T}^{t_0} \sum_{i=1}^{q} \frac{S_i(t) - \sigma_{ii}(E(t))}{c + E_i(t)} \mathrm{d}t, \tag{3.31}$$

where $E(-T) = 0$ and $E(t_0) = E_0 \in \overline{\mathbb{R}}_+^q$. Note that the available entropy $\mathcal{S}_{\mathrm{a}}(E_0)$ is the minimum amount of scaled heat (entropy) that can be extracted from the large-scale dynamical system $\mathcal{G}$ in order to transfer it from an initial state $E(t_0) = E_0$ to $E(T) = 0$. Alternatively, the required entropy supply $\mathcal{S}_{\mathrm{r}}(E_0)$ is the maximum amount of scaled heat (entropy) that can be delivered to $\mathcal{G}$ to transfer it from the origin to a given initial state $E(t_0) = E_0$.

Theorem 3.2. Consider the large-scale dynamical system $\mathcal{G}$ with power balance equation (3.5), and assume that Axiom *ii*) holds. Then there exists an entropy function for $\mathcal{G}$. Moreover, $\mathcal{S}_{\mathrm{a}}(E)$, $E \in \overline{\mathbb{R}}_+^q$, and $\mathcal{S}_{\mathrm{r}}(E)$, $E \in \overline{\mathbb{R}}_+^q$, are possible entropy functions for $\mathcal{G}$ with $\mathcal{S}_{\mathrm{a}}(0) = \mathcal{S}_{\mathrm{r}}(0) = 0$. Finally, all entropy functions $\mathcal{S}(E)$, $E \in \overline{\mathbb{R}}_+^q$, for $\mathcal{G}$ satisfy

$$\mathcal{S}_{\mathrm{r}}(E) \leq \mathcal{S}(E) - \mathcal{S}(0) \leq \mathcal{S}_{\mathrm{a}}(E), \quad E \in \overline{\mathbb{R}}_+^q. \tag{3.32}$$

Proof. Since, by Lemma 3.1, $\mathcal{G}$ is controllable to and reachable from the origin in $\overline{\mathbb{R}}_+^q$, it follows from (3.30) and (3.31) that $\mathcal{S}_{\mathrm{a}}(E_0) < \infty$, $E_0 \in \overline{\mathbb{R}}_+^q$, and $\mathcal{S}_{\mathrm{r}}(E_0) > -\infty$, $E_0 \in \overline{\mathbb{R}}_+^q$, respectively. Next, let $E_0 \in \overline{\mathbb{R}}_+^q$, and let $S(\cdot) \in \mathcal{U}$ be such that $E(t_{\mathrm{i}}) = E(t_{\mathrm{f}}) = 0$ and $E(t_0) = E_0$, where $t_{\mathrm{i}} < t_0 < t_{\mathrm{f}}$. In this case, it follows from (3.26) that

$$\int_{t_{\mathrm{i}}}^{t_{\mathrm{f}}} \sum_{i=1}^{q} \frac{S_i(t) - \sigma_{ii}(E(t))}{c + E_i(t)} \mathrm{d}t \leq 0 \tag{3.33}$$

or, equivalently,

$$\int_{t_{\mathrm{i}}}^{t_0} \sum_{i=1}^{q} \frac{S_i(t) - \sigma_{ii}(E(t))}{c + E_i(t)} \mathrm{d}t \leq - \int_{t_0}^{t_{\mathrm{f}}} \sum_{i=1}^{q} \frac{S_i(t) - \sigma_{ii}(E(t))}{c + E_i(t)} \mathrm{d}t. \tag{3.34}$$

Now, taking the supremum on both sides of (3.34) over all $S(\cdot) \in \mathcal{U}_{\rm r}$ and $t_{\rm i} \le t_0$ yields

$$\begin{aligned} \mathcal{S}_{\rm r}(E_0) &= \sup_{S(\cdot)\in\mathcal{U}_{\rm r},\, t_{\rm i}\le t_0} \int_{t_{\rm i}}^{t_0} \sum_{i=1}^{q} \frac{S_i(t) - \sigma_{ii}(E(t))}{c + E_i(t)} {\rm d}t \\ &\le -\int_{t_0}^{t_{\rm f}} \sum_{i=1}^{q} \frac{S_i(t) - \sigma_{ii}(E(t))}{c + E_i(t)} {\rm d}t. \end{aligned} \tag{3.35}$$

Next, taking the infimum on both sides of (3.35) over all $S(\cdot) \in \mathcal{U}_{\rm c}$ and $t_{\rm f} \ge t_0$, we obtain $\mathcal{S}_{\rm r}(E_0) \le \mathcal{S}_{\rm a}(E_0)$, $E_0 \in \overline{\mathbb{R}}_+^q$, which implies that $-\infty < \mathcal{S}_{\rm r}(E_0) \le \mathcal{S}_{\rm a}(E_0) < \infty$, $E_0 \in \overline{\mathbb{R}}_+^q$. Hence, the functions $\mathcal{S}_{\rm a}(\cdot)$ and $\mathcal{S}_{\rm r}(\cdot)$ are well defined.

Next, it follows from the definition of $\mathcal{S}_{\rm a}(\cdot)$ that for every $T \ge t_1$ and $S(\cdot) \in \mathcal{U}_{\rm c}$ such that $E(t_1) \in \overline{\mathbb{R}}_+^q$ and $E(T) = 0$,

$$\begin{aligned} -\mathcal{S}_{\rm a}(E(t_1)) &\ge \int_{t_1}^{t_2} \sum_{i=1}^{q} \frac{S_i(t) - \sigma_{ii}(E(t))}{c + E_i(t)} {\rm d}t \\ &\quad + \int_{t_2}^{T} \sum_{i=1}^{q} \frac{S_i(t) - \sigma_{ii}(E(t))}{c + E_i(t)} {\rm d}t, \quad t_1 \le t_2 \le T, \end{aligned} \tag{3.36}$$

and hence,

$$\begin{aligned} -\mathcal{S}_{\rm a}(E(t_1)) &\ge \int_{t_1}^{t_2} \sum_{i=1}^{q} \frac{S_i(t) - \sigma_{ii}(E(t))}{c + E_i(t)} {\rm d}t \\ &\quad + \sup_{S(\cdot)\in\mathcal{U}_{\rm c},\, T\ge t_2} \int_{t_2}^{T} \sum_{i=1}^{q} \frac{S_i(t) - \sigma_{ii}(E(t))}{c + E_i(t)} {\rm d}t \\ &= \int_{t_1}^{t_2} \sum_{i=1}^{q} \frac{S_i(t) - \sigma_{ii}(E(t))}{c + E_i(t)} {\rm d}t - \mathcal{S}_{\rm a}(E(t_2)), \end{aligned} \tag{3.37}$$

which implies that $\mathcal{S}_{\rm a}(E)$, $E \in \overline{\mathbb{R}}_+^q$, satisfies (3.29). Thus, $\mathcal{S}_{\rm a}(E)$, $E \in \overline{\mathbb{R}}_+^q$, is a possible entropy function for $\mathcal{G}$. Note that with $E(t_0) = E(T) = 0$ it follows from (3.26) that the supremum in (3.30) is taken over the set of negative semidefinite values with one of the values being zero for $S(t) \equiv 0$. Thus, $\mathcal{S}_{\rm a}(0) = 0$.

Similarly, it follows from the definition of $\mathcal{S}_{\rm r}(\cdot)$ that for every $T \ge -t_2$ and $S(\cdot) \in \mathcal{U}_{\rm r}$ such that $E(t_2) \in \overline{\mathbb{R}}_+^q$ and $E(-T) = 0$,

$$\mathcal{S}_{\rm r}(E(t_2)) \ge \int_{-T}^{t_1} \sum_{i=1}^{q} \frac{S_i(t) - \sigma_{ii}(E(t))}{c + E_i(t)} {\rm d}t$$

$$+\int_{t_1}^{t_2}\sum_{i=1}^{q}\frac{S_i(t)-\sigma_{ii}(E(t))}{c+E_i(t)}\mathrm{d}t, \quad -T\le t_1\le t_2, \tag{3.38}$$

and hence,

$$\begin{aligned}\mathcal{S}_{\mathrm{r}}(E(t_2)) \ge& \int_{t_1}^{t_2}\sum_{i=1}^{q}\frac{S_i(t)-\sigma_{ii}(E(t))}{c+E_i(t)}\mathrm{d}t\\ &+\sup_{S(\cdot)\in\mathcal{U}_{\mathrm{r}},T\ge -t_1}\int_{-T}^{t_1}\sum_{i=1}^{q}\frac{S_i(t)-\sigma_{ii}(E(t))}{c+E_i(t)}\mathrm{d}t\\ =& \int_{t_1}^{t_2}\sum_{i=1}^{q}\frac{S_i(t)-\sigma_{ii}(E(t))}{c+E_i(t)}\mathrm{d}t+\mathcal{S}_{\mathrm{r}}(E(t_1)),\end{aligned} \tag{3.39}$$

which implies that $\mathcal{S}_{\mathrm{r}}(E)$, $E\in\overline{\mathbb{R}}_+^q$, satisfies (3.29). Thus, $\mathcal{S}_{\mathrm{r}}(E)$, $E\in\overline{\mathbb{R}}_+^q$, is a possible entropy function for $\mathcal{G}$. Note that with $E(t_0)=E(-T)=0$ it follows from (3.26) that the supremum in (3.31) is taken over the set of negative semidefinite values with one of the values being zero for $S(t)\equiv 0$. Thus, $\mathcal{S}_{\mathrm{r}}(0)=0$.

Next, suppose there exists an entropy function $\mathcal{S}:\overline{\mathbb{R}}_+^q\to\mathbb{R}$ for $\mathcal{G}$, and let $E(t_2)=0$ in (3.29). Then it follows from (3.29) that

$$\mathcal{S}(E(t_1))-\mathcal{S}(0)\le -\int_{t_1}^{t_2}\sum_{i=1}^{q}\frac{S_i(t)-\sigma_{ii}(E(t))}{c+E_i(t)}\mathrm{d}t, \tag{3.40}$$

for all $t_2\ge t_1$ and $S(\cdot)\in\mathcal{U}_{\mathrm{c}}$, which implies that

$$\begin{aligned}\mathcal{S}(E(t_1))-\mathcal{S}(0) \le& \inf_{S(\cdot)\in\mathcal{U}_{\mathrm{c}},\,t_2\ge t_1}\left[-\int_{t_1}^{t_2}\sum_{i=1}^{q}\frac{S_i(t)-\sigma_{ii}(E(t))}{c+E_i(t)}\mathrm{d}t\right]\\ =& -\sup_{S(\cdot)\in\mathcal{U}_{\mathrm{c}},\,t_2\ge t_1}\int_{t_1}^{t_2}\sum_{i=1}^{q}\frac{S_i(t)-\sigma_{ii}(E(t))}{c+E_i(t)}\mathrm{d}t\\ =& \mathcal{S}_{\mathrm{a}}(E(t_1)).\end{aligned} \tag{3.41}$$

Since $E(t_1)$ is arbitrary, it follows that $\mathcal{S}(E)-\mathcal{S}(0)\le\mathcal{S}_{\mathrm{a}}(E)$, $E\in\overline{\mathbb{R}}_+^q$.

Alternatively, let $E(t_1)=0$ in (3.29). Then it follows from (3.29) that

$$\mathcal{S}(E(t_2))-\mathcal{S}(0)\ge\int_{t_1}^{t_2}\sum_{i=1}^{q}\frac{S_i(t)-\sigma_{ii}(E(t))}{c+E_i(t)}\mathrm{d}t \tag{3.42}$$

for all $t_1\le t_2$ and $S(\cdot)\in\mathcal{U}_{\mathrm{r}}$. Hence,

$$\mathcal{S}(E(t_2))-\mathcal{S}(0)\ge\sup_{S(\cdot)\in\mathcal{U}_{\mathrm{r}},\,t_1\le t_2}\int_{t_1}^{t_2}\sum_{i=1}^{q}\frac{S_i(t)-\sigma_{ii}(E(t))}{c+E_i(t)}\mathrm{d}t$$

$$= \mathcal{S}_{\rm r}(E(t_2)), \tag{3.43}$$

which, since $E(t_2)$ is arbitrary, implies that $\mathcal{S}_{\rm r}(E) \leq \mathcal{S}(E) - \mathcal{S}(0)$, $E \in \overline{\mathbb{R}}_+^q$. Thus, all entropy functions for $\mathcal{G}$ satisfy (3.32). □

It is important to note that inequality (3.26) is equivalent to the existence of an entropy function for $\mathcal{G}$. Sufficiency is simply a statement of Theorem 3.2, while necessity follows from (3.29) with $E(t_2) = E(t_1)$. This definition of entropy leads to the second law of thermodynamics being viewed as an axiom in the context of (anti)cyclo-dissipative dynamical systems [214, 463, 465]. A similar remark holds for the definition of ectropy introduced in Section 3.4.

The next result shows that all entropy functions for $\mathcal{G}$ are continuous on $\overline{\mathbb{R}}_+^q$.

Theorem 3.3. Consider the large-scale dynamical system $\mathcal{G}$ with power balance equation (3.5), and let $\mathcal{S} : \overline{\mathbb{R}}_+^q \to \mathbb{R}$ be an entropy function of $\mathcal{G}$. Then $\mathcal{S}(\cdot)$ is continuous on $\overline{\mathbb{R}}_+^q$.

Proof. Let $E_{\rm e} \in \overline{\mathbb{R}}_+^q$ and $S_{\rm e} \in \mathbb{R}^q$ be such that $S_{\rm e} = d(E_{\rm e}) - f(E_{\rm e})$. Note that with $S(t) \equiv S_{\rm e}$, $E_{\rm e}$ is an equilibrium point of the power balance equation (3.5). Next, it follows from Lemma 3.1 that $\mathcal{G}$ is *locally controllable*, that is, for every $T > 0$ and $\varepsilon > 0$, the set of points that can be reached from and to $E_{\rm e}$ in time T using admissible inputs $S : [0, T] \to \mathbb{R}^q$, satisfying $\|S(t) - S_{\rm e}\| < \varepsilon$, contains a neighborhood of $E_{\rm e}$.

Alternatively, this can be shown by considering the linearization of (3.5) at $E = E_{\rm e}$ and $S = S_{\rm e}$ given by

$$\dot{E}(t) = A(E(t) - E_{\rm e}) + B(S(t) - S_{\rm e}), \quad E(t_0) = E_0, \quad t \geq t_0, \tag{3.44}$$

where $A = \left.\frac{\partial f(E)}{\partial E}\right|_{E=E_{\rm e}} - \left.\frac{\partial d(E)}{\partial E}\right|_{E=E_{\rm e}}$ and $B = I_q$. Since $B = I_q$, it follows that

$$\operatorname{rank}\,[B, AB, A^2B, \ldots, A^{q-1}B] = q, \tag{3.45}$$

and hence, the linearized system (3.44) is controllable. Thus, it follows from Proposition 3.3 of [340] that $\mathcal{G}$ is locally controllable.

Next, let $\delta > 0$ and note that it follows from the continuity of $f(\cdot)$ and $d(\cdot)$ that there exist $T > 0$ and $\varepsilon > 0$ such that for every $S : [0, T) \to \mathbb{R}^q$ and $\|S(t) - S_{\rm e}\| < \varepsilon$, $\|E(t) - E_{\rm e}\| < \delta$, $t \in [0, T)$, where $S(\cdot) \in \mathcal{U}$ and $E(t)$, $t \in [0, T)$, denotes the solution to (3.5) with the initial condition $E_{\rm e}$. Furthermore, it follows from the local controllability of $\mathcal{G}$ that for every $\hat{T} \in (0, T]$, there exists a strictly increasing, continuous function $\gamma : \mathbb{R} \to \overline{\mathbb{R}}_+^q$

such that $\gamma(0) = 0$, and for every $E_0 \in \overline{\mathbb{R}}_+^q$ such that $\|E_0 - E_\mathrm{e}\| \le \gamma(\hat{T})$, there exists $\hat{t} \in [0, \hat{T}]$ and an input $S : [0, \hat{T}] \to \mathbb{R}^q$ such that $\|S(t) - S_\mathrm{e}\| < \varepsilon$, $t \in [0, \hat{t})$, and $E(\hat{t}) = E_0$. Hence, there exists $\beta > 0$ such that for every $E_0 \in \overline{\mathbb{R}}_+^q$ such that $\|E_0 - E_\mathrm{e}\| \le \beta$, there exists $\hat{t} \in [0, \gamma^{-1}(\|E_0 - E_\mathrm{e}\|)]$ and an input $S : [t_0, \hat{t}] \to \mathbb{R}^q$ such that $\|S(t) - S_\mathrm{e}\| < \varepsilon$, $t \in [0, \hat{t}]$, and $E(\hat{t}) = E_0$. In addition, it follows from Lemma 3.1 that $S : [0, \hat{t}] \to \mathbb{R}^q$ is such that $E(t) \ge\ge 0$, $t \in [0, \hat{t}]$.

Next, since $\sigma_{ii}(\cdot)$, $i = 1, \ldots, q$, is continuous, it follows that there exists $M \in (0, \infty)$ such that

$$\sup_{\|E - E_\mathrm{e}\| < \delta,\ \|S - S_\mathrm{e}\| < \varepsilon} \left| \sum_{i=1}^{q} \frac{S_i - \sigma_{ii}(E)}{c + E_i} \right| = M. \tag{3.46}$$

Hence, it follows that

$$\begin{aligned} \left| \int_0^{\hat{t}} \sum_{i=1}^{q} \frac{S_i(\sigma) - \sigma_{ii}(E(\sigma))}{c + E_i(\sigma)} \mathrm{d}\sigma \right| &\le \int_0^{\hat{t}} \left| \sum_{i=1}^{q} \frac{S_i(\sigma) - \sigma_{ii}(E(\sigma))}{c + E_i(\sigma)} \right| \mathrm{d}\sigma \\ &\le M\hat{t} \\ &\le M\gamma^{-1}(\|E_0 - E_\mathrm{e}\|). \end{aligned} \tag{3.47}$$

Now, if $\mathcal{S}(\cdot)$ is an entropy function of $\mathcal{G}$, then

$$\mathcal{S}(E(\hat{t})) \ge \mathcal{S}(E_\mathrm{e}) + \int_0^{\hat{t}} \sum_{i=1}^{q} \frac{S_i(\sigma) - \sigma_{ii}(E(\sigma))}{c + E_i(\sigma)} \mathrm{d}\sigma \tag{3.48}$$

or, equivalently,

$$-\int_0^{\hat{t}} \sum_{i=1}^{q} \frac{S_i(\sigma) - \sigma_{ii}(E(\sigma))}{c + E_i(\sigma)} \mathrm{d}\sigma \ge \mathcal{S}(E_\mathrm{e}) - \mathcal{S}(E(\hat{t})). \tag{3.49}$$

If $\mathcal{S}(E_\mathrm{e}) \ge \mathcal{S}(E(\hat{t}))$, then combining (3.47) and (3.49) yields

$$|\mathcal{S}(E_\mathrm{e}) - \mathcal{S}(E(\hat{t}))| \le M\gamma^{-1}(\|E_0 - E_\mathrm{e}\|). \tag{3.50}$$

Alternatively, if $\mathcal{S}(E(\hat{t})) \ge \mathcal{S}(E_\mathrm{e})$, then (3.50) can be derived by reversing the roles of E_e and $E(\hat{t})$. In particular, using the fact that $\mathcal{G}$ is locally controllable from and to E_e or, alternatively, that controllability of (3.44) is equivalent to controllability of the linearization of the time-reversed system

$$\dot{E}(t) = -A(E(t) - E_\mathrm{e}) - B(S(t) - S_\mathrm{e}), \quad E(t_0) = E_0, \quad t \ge t_0, \tag{3.51}$$

similar arguments can be used to show that the set of points that can be steered in small time to E_e contains a neighborhood of $E(\hat{t})$. Hence, since $\gamma(\cdot)$ is continuous and $E(\hat{t})$ is arbitrary, it follows that $\mathcal{S}(\cdot)$ is continuous on $\overline{\mathbb{R}}_+^q$. □

Next, as a direct consequence of Theorem 3.2, we show that all possible entropy functions of $\mathcal{G}$ form a convex set, and hence, there exists a continuum of possible entropy functions for $\mathcal{G}$ ranging from the required entropy supply $\mathcal{S}_{\rm r}(E)$ to the available entropy $\mathcal{S}_{\rm a}(E)$.

Proposition 3.3. Consider the large-scale dynamical system $\mathcal{G}$ with power balance equation (3.5), and assume that Axioms *i*) and *ii*) hold. Then

$$\mathcal{S}(E) \triangleq \alpha \mathcal{S}_{\rm r}(E) + (1-\alpha)\mathcal{S}_{\rm a}(E), \quad \alpha \in [0,1], \tag{3.52}$$

is an entropy function for $\mathcal{G}$.

Proof. The result is a direct consequence of the reachability of $\mathcal{G}$ along with inequality (3.29) by noting that if $\mathcal{S}_{\rm r}(E)$ and $\mathcal{S}_{\rm a}(E)$ satisfy (3.29), then $\mathcal{S}(E)$ satisfies (3.29). □

It follows from Proposition 3.3 that Definition 3.2 does not provide enough information to define the entropy uniquely for nonequilibrium thermodynamic systems with power balance equation (3.5). This difficulty has long been pointed out in the study of thermodynamics [318]. Two particular entropy functions for $\mathcal{G}$ can be computed a priori via the variational problems given by (3.30) and (3.31). For equilibrium thermodynamics, however, uniqueness is not an issue, as shown in the next proposition.

Proposition 3.4. Consider the large-scale dynamical system $\mathcal{G}$ with power balance equation (3.5), and assume that Axioms *i*) and *ii*) hold. Then at every equilibrium state $E = E_{\rm e}$ of the isolated system $\mathcal{G}$, the entropy $\mathcal{S}(E)$, $E \in \overline{\mathbb{R}}_+^q$, of $\mathcal{G}$ is unique (modulo a constant of integration) and is given by

$$\mathcal{S}(E) - \mathcal{S}(0) = \mathcal{S}_{\rm a}(E) = \mathcal{S}_{\rm r}(E) = \mathbf{e}^{\rm T}\mathbf{log}_e(c\mathbf{e} + E) - q\log_e c, \tag{3.53}$$

where $E = E_{\rm e}$ and $\mathbf{log}_e(c\mathbf{e} + E)$ denotes the vector natural logarithm given by $[\log_e(c + E_1), \ldots, \log_e(c + E_q)]^{\rm T}$.

Proof. It follows from Axiom *i*) that for an equilibrium process $\phi_{ij}(E(t)) \equiv 0$, $i \neq j$, $i,j = 1,\ldots,q$. Consider the entropy function $\mathcal{S}_{\rm a}(\cdot)$ given by (3.30), and let $E_0 = E_{\rm e}$ for some equilibrium state $E_{\rm e}$. Then it follows from (3.5) that

$$\begin{aligned}\mathcal{S}_{\rm a}(E_0) &= -\sup_{S(\cdot)\in\mathcal{U}_{\rm c},\, T\geq t_0} \int_{t_0}^{T} \sum_{i=1}^{q} \frac{\dot{E}_i(t) - \sum_{j=1, j\neq i}^{q}\phi_{ij}(E(t))}{c + E_i(t)}{\rm d}t \\ &= -\sup_{S(\cdot)\in\mathcal{U}_{\rm c},\, T\geq t_0} \left[\sum_{i=1}^{q} \log_e\left(\frac{c}{c+E_{i0}}\right)\right.\end{aligned}$$

$$
\begin{aligned}
&\left.-\int_{t_0}^{T}\sum_{i=1}^{q}\sum_{j=1,j\neq i}^{q}\frac{\phi_{ij}(E(t))}{c+E_i(t)}\mathrm{d}t\right] \\
&=\sum_{i=1}^{q}\log_e\left(\frac{c+E_{i0}}{c}\right)+\inf_{S(\cdot)\in\mathcal{U}_{\mathrm{c}},T\geq t_0}\int_{t_0}^{T}\sum_{i=1}^{q}\sum_{j=1,j\neq i}^{q}\frac{\phi_{ij}(E(t))}{c+E_i(t)}\mathrm{d}t \\
&=\sum_{i=1}^{q}\log_e\left(\frac{c+E_{i0}}{c}\right) \\
&\quad+\inf_{S(\cdot)\in\mathcal{U}_{\mathrm{c}},T\geq t_0}\int_{t_0}^{T}\sum_{i=1}^{q}\sum_{j=i+1}^{q}\left[\frac{\phi_{ij}(E(t))}{c+E_i(t)}-\frac{\phi_{ij}(E(t))}{c+E_j(t)}\right]\mathrm{d}t \\
&=\sum_{i=1}^{q}\log_e\left(\frac{c+E_{i0}}{c}\right) \\
&\quad+\inf_{S(\cdot)\in\mathcal{U}_{\mathrm{c}},T\geq t_0}\int_{t_0}^{T}\sum_{i=1}^{q}\sum_{j=i+1}^{q}\frac{\phi_{ij}(E(t))[E_j(t)-E_i(t)]}{(c+E_i(t))(c+E_j(t))}\mathrm{d}t.
\end{aligned}
\tag{3.54}
$$

Since the solution $E(t)$, $t\geq t_0$, to (3.5) is nonnegative for all nonnegative initial conditions, it follows from Axiom *ii*) that the infimum in (3.54) is taken over the set of nonnegative values. However, the zero value of the infimum is achieved on an equilibrium process for which $\phi_{ij}(E(t))\equiv 0$, $i\neq j$, $i,j=1,\ldots,q$. Thus,

$$
\mathcal{S}_{\mathrm{a}}(E_0)=\mathbf{e}^{\mathrm{T}}\mathbf{log}_e(c\mathbf{e}+E_0)-q\log_e c,\quad E_0=E_{\mathrm{e}}. \tag{3.55}
$$

Similarly, consider the entropy function $\mathcal{S}_{\mathrm{r}}(\cdot)$ given by (3.31). Then it follows from (3.5) that, for $E_0=E_{\mathrm{e}}$,

$$
\begin{aligned}
\mathcal{S}_{\mathrm{r}}(E_0) &= \sup_{S(\cdot)\in\mathcal{U}_{\mathrm{r}},T\geq -t_0}\int_{-T}^{t_0}\sum_{i=1}^{q}\frac{\dot{E}_i(t)-\sum_{j=1,j\neq i}^{q}\phi_{ij}(E(t))}{c+E_i(t)}\mathrm{d}t \\
&= \sup_{S(\cdot)\in\mathcal{U}_{\mathrm{r}},T\geq -t_0}\left[\sum_{i=1}^{q}\log_e\left(\frac{c+E_{i0}}{c}\right)\right. \\
&\quad\left.-\int_{-T}^{t_0}\sum_{i=1}^{q}\sum_{j=1,j\neq i}^{q}\frac{\phi_{ij}(E(t))}{c+E_i(t)}\mathrm{d}t\right] \\
&= \sum_{i=1}^{q}\log_e\left(\frac{c+E_{i0}}{c}\right) \\
&\quad-\inf_{S(\cdot)\in\mathcal{U}_{\mathrm{r}},T\geq -t_0}\int_{-T}^{t_0}\sum_{i=1}^{q}\sum_{j=1,j\neq i}^{q}\frac{\phi_{ij}(E(t))}{c+E_i(t)}\mathrm{d}t
\end{aligned}
$$

$$
\begin{aligned}
&= \sum_{i=1}^{q} \log_e \left(\frac{c + E_{i0}}{c} \right) \\
&\quad - \inf_{S(\cdot)\in\mathcal{U}_{\rm r}, T\geq -t_0} \int_{-T}^{t_0} \sum_{i=1}^{q} \sum_{j=i+1}^{q} \left[\frac{\phi_{ij}(E(t))}{c + E_i(t)} - \frac{\phi_{ij}(E(t))}{c + E_j(t)} \right] \mathrm{d}t \\
&= \sum_{i=1}^{q} \log_e \left(\frac{c + E_{i0}}{c} \right) \\
&\quad - \inf_{S(\cdot)\in\mathcal{U}_{\rm r}, T\geq -t_0} \int_{-T}^{t_0} \sum_{i=1}^{q} \sum_{j=i+1}^{q} \frac{\phi_{ij}(E(t))[E_j(t) - E_i(t)]}{(c + E_i(t))(c + E_j(t))} \mathrm{d}t.
\end{aligned}
\tag{3.56}
$$

Now, it follows from Axioms *i*) and *ii*) that the zero value of the infimum in (3.56) is achieved on an equilibrium process and thus

$$
\mathcal{S}_{\rm r}(E_0) = \mathbf{e}^{\rm T}\mathbf{log}_e(c\mathbf{e} + E_0) - q \log_e c, \quad E_0 = E_{\rm e}. \tag{3.57}
$$

Finally, it follows from (3.32) that (3.53) holds. □

The next proposition shows that if (3.29) holds as an equality for some transformation starting and ending at an equilibrium point of the isolated dynamical system $\mathcal{G}$, then this transformation must lie on the equilibrium manifold $\mathcal{M}_{\rm e}$.

Proposition 3.5. Consider the large-scale dynamical system $\mathcal{G}$ with power balance equation (3.5), and assume that Axioms *i*) and *ii*) hold. Let $\mathcal{S}(\cdot)$ denote an entropy of $\mathcal{G}$, and let $E : [t_0, t_1] \to \overline{\mathbb{R}}_+^q$ denote the solution to (3.5) with $E(t_0) = \alpha_0\mathbf{e}$ and $E(t_1) = \alpha_1\mathbf{e}$, where $\alpha_0, \alpha_1 \geq 0$. Then

$$
\mathcal{S}(E(t_1)) = \mathcal{S}(E(t_0)) + \int_{t_0}^{t_1} \sum_{i=1}^{q} \frac{S_i(t) - \sigma_{ii}(E(t))}{c + E_i(t)} \mathrm{d}t \tag{3.58}
$$

if and only if there exists a continuous function $\alpha : [t_0, t_1] \to \overline{\mathbb{R}}_+$ such that $\alpha(t_0) = \alpha_0$, $\alpha(t_1) = \alpha_1$, and $E(t) = \alpha(t)\mathbf{e}$, $t \in [t_0, t_1]$.

Proof. Since $E(t_0)$ and $E(t_1)$ are equilibrium states of the isolated dynamical system $\mathcal{G}$, it follows from Proposition 3.4 that

$$
\mathcal{S}(E(t_1)) - \mathcal{S}(E(t_0)) = q \log_e(c + \alpha_1) - q \log_e(c + \alpha_0). \tag{3.59}
$$

Furthermore, it follows from (3.5) that

$$
\int_{t_0}^{t_1} \sum_{i=1}^{q} \frac{S_i(t) - \sigma_{ii}(E(t))}{c + E_i(t)} \mathrm{d}t
$$

$$= \int_{t_0}^{t_1} \sum_{i=1}^{q} \frac{\dot{E}_i(t) - \sum_{j=1, j\neq i}^{q} \phi_{ij}(E(t))}{c + E_i(t)} \mathrm{d}t$$

$$= q \log_e \left(\frac{c + \alpha_1}{c + \alpha_0} \right)$$

$$- \int_{t_0}^{t_1} \sum_{i=1}^{q} \sum_{j=i+1}^{q} \frac{\phi_{ij}(E(t))[E_j(t) - E_i(t)]}{(c + E_i(t))(c + E_j(t))} \mathrm{d}t. \quad (3.60)$$

Now, it follows from Axioms *i)* and *ii)* that (3.58) holds if and only if $E_i(t) = E_j(t)$, $t \in [t_0, t_1]$, $i \neq j$, $i, j = 1, \ldots, q$, or, equivalently, there exists a continuous function $\alpha : [t_0, t_1] \to \overline{\mathbb{R}}_+$ such that $E(t) = \alpha(t)\mathbf{e}$, $t \in [t_0, t_1]$, $\alpha(t_0) = \alpha_0$, and $\alpha(t_1) = \alpha_1$. □

Even though it follows from Proposition 3.3 that Definition 3.2 does not provide a unique *continuous* entropy function for nonequilibrium systems, the next theorem gives a *unique*, *continuously differentiable* entropy function for $\mathcal{G}$ for equilibrium and nonequilibrium processes. This result answers the long-standing question of how the entropy of a nonequilibrium state of a dynamical process should be defined [272, 318], and establishes its global existence and uniqueness.

Theorem 3.4. Consider the large-scale dynamical system $\mathcal{G}$ with power balance equation (3.5), and assume that Axioms *i)* and *ii)* hold. Then the function $\mathcal{S} : \overline{\mathbb{R}}_+^q \to \overline{\mathbb{R}}_+^q$ given by

$$\mathcal{S}(E) = \mathbf{e}^{\mathrm{T}} \mathbf{log}_e(c\mathbf{e} + E) - q \log_e c, \quad E \in \overline{\mathbb{R}}_+^q, \quad (3.61)$$

where $c > 0$, is a unique (modulo a constant of integration), continuously differentiable entropy function of $\mathcal{G}$. Furthermore, for $E(t) \notin \mathcal{M}_{\mathrm{e}}$, $t \geq t_0$, where $E(t)$, $t \geq t_0$, denotes the solution to (3.5) and $\mathcal{M}_{\mathrm{e}} = \{E \in \overline{\mathbb{R}}_+^q : E = \alpha\mathbf{e}, \ \alpha \geq 0\}$, (3.61) satisfies

$$\mathcal{S}(E(t_2)) > \mathcal{S}(E(t_1)) + \int_{t_1}^{t_2} \sum_{i=1}^{q} \frac{S_i(t) - \sigma_{ii}(E(t))}{c + E_i(t)} \mathrm{d}t \quad (3.62)$$

for every $t_2 \geq t_1 \geq t_0$ and $S(\cdot) \in \mathcal{U}$.

Proof. Since, by Proposition 3.1, $E(t) \geq\geq 0$, $t \geq t_0$, and $\phi_{ij}(E) = -\phi_{ji}(E)$, $E \in \overline{\mathbb{R}}_+^q$, $i \neq j$, $i, j = 1, \ldots, q$, it follows that

$$\dot{\mathcal{S}}(E(t)) = \sum_{i=1}^{q} \frac{\dot{E}_i(t)}{c + E_i(t)}$$

$$= \sum_{i=1}^{q} \left[\frac{S_i(t) - \sigma_{ii}(E(t))}{c + E_i(t)} + \sum_{j=1, j\neq i}^{q} \frac{\phi_{ij}(E(t))}{c + E_i(t)} \right]$$

$$= \sum_{i=1}^{q} \left[\frac{S_i(t) - \sigma_{ii}(E(t))}{c + E_i(t)} + \sum_{j=i+1}^{q} \left(\frac{\phi_{ij}(E(t))}{c + E_i(t)} - \frac{\phi_{ij}(E(t))}{c + E_j(t)} \right) \right]$$

$$= \sum_{i=1}^{q} \frac{S_i(t) - \sigma_{ii}(E(t))}{c + E_i(t)} + \sum_{i=1}^{q} \sum_{j=i+1}^{q} \frac{\phi_{ij}(E(t))[E_j(t) - E_i(t)]}{(c + E_i(t))(c + E_j(t))}$$

$$\geq \sum_{i=1}^{q} \frac{S_i(t) - \sigma_{ii}(E(t))}{c + E_i(t)}, \quad t \geq t_0. \tag{3.63}$$

Now, integrating (3.63) over $[t_1, t_2]$ yields (3.29). Furthermore, in the case where $E(t) \notin \mathcal{M}_\mathrm{e}$, $t \geq t_0$, it follows from Axiom *i*), Axiom *ii*), and (3.63) that (3.62) holds.

To show that (3.61) is a unique, continuously differentiable entropy function of $\mathcal{G}$, let $\mathcal{S}(E)$ be a continuously differentiable entropy function of $\mathcal{G}$ so that $\mathcal{S}(E)$ satisfies (3.29) or, equivalently,

$$\dot{\mathcal{S}}(E(t)) \geq \mu^\mathrm{T}(E(t))[S(t) - d(E(t))], \quad t \geq t_0, \tag{3.64}$$

where $\mu^\mathrm{T}(E) = [\frac{1}{c+E_1}, \ldots, \frac{1}{c+E_q}]$, $E \in \overline{\mathbb{R}}_+^q$, $E(t)$, $t \geq t_0$, denotes the solution to the power balance equation (3.5), and $\dot{\mathcal{S}}(E(t))$ denotes the time derivative of $\mathcal{S}(E)$ along the solution $E(t)$, $t \geq t_0$. Hence, it follows from (3.64) that

$$\mathcal{S}'(E)[f(E) - d(E) + S] \geq \mu^\mathrm{T}(E)[S - d(E)], \quad E \in \overline{\mathbb{R}}_+^q, \quad S \in \mathbb{R}^q, \tag{3.65}$$

which implies that there exist continuous functions $\ell : \overline{\mathbb{R}}_+^q \to \mathbb{R}^p$ and $\mathcal{W} : \overline{\mathbb{R}}_+^q \to \mathbb{R}^{p \times q}$ such that

$$\begin{aligned} 0 = {} & \mathcal{S}'(E)[f(E) - d(E) + S] - \mu^\mathrm{T}(E)[S - d(E)] \\ & -[\ell(E) + \mathcal{W}(E)S]^\mathrm{T}[\ell(E) + \mathcal{W}(E)S], \quad E \in \overline{\mathbb{R}}_+^q, \quad S \in \mathbb{R}^q. \end{aligned} \tag{3.66}$$

Now, equating coefficients of equal powers (of S), it follows that $\mathcal{W}(E) \equiv 0$, $\mathcal{S}'(E) = \mu^\mathrm{T}(E)$, $E \in \overline{\mathbb{R}}_+^q$, and

$$0 = \mathcal{S}'(E)f(E) - \ell^\mathrm{T}(E)\ell(E), \quad E \in \overline{\mathbb{R}}_+^q. \tag{3.67}$$

Hence, $\mathcal{S}(E) = \mathbf{e}^\mathrm{T}\mathbf{log}_e(c\mathbf{e} + E) - q \log_e c$, $E \in \overline{\mathbb{R}}_+^q$, and

$$0 = \mu^\mathrm{T}(E)f(E) - \ell^\mathrm{T}(E)\ell(E), \quad E \in \overline{\mathbb{R}}_+^q. \tag{3.68}$$

Thus, (3.61) is a unique, continuously differentiable entropy function for $\mathcal{G}$. □

Note that it follows from Axiom *i*), Axiom *ii*), and the last equality in (3.63) that the entropy function given by (3.61) satisfies (3.29) as an equality for an equilibrium process and as a strict inequality for a nonequilibrium process. Hence, it follows from Theorem 2.25 that the isolated (i.e., $S(t) \equiv 0$ and $d(E) \equiv 0$) large-scale dynamical system $\mathcal{G}$ does not exhibit Poincaré recurrence in $\overline{\mathbb{R}}_+^q \setminus \mathcal{M}_e$. Furthermore, for any entropy function of $\mathcal{G}$, it follows from Proposition 3.5 that if (3.29) holds as an equality for some transformation starting and ending at equilibrium points of the isolated system $\mathcal{G}$, then this transformation must lie on the equilibrium manifold $\mathcal{M}_e$. However, (3.29) may hold as an equality for nonequilibrium processes starting and ending at nonequilibrium states.

The entropy expression given by (3.61) is identical in form to the Boltzmann entropy for statistical thermodynamics. Due to the fact that the entropy given by (3.61) is indeterminate to the extent of an additive constant, we can place the constant of integration $q \log_e c$ to zero by taking $c = 1$. Since $\mathcal{S}(E)$ given by (3.61) achieves a maximum when all the subsystem energies E_i, $i = 1, \ldots, q$, are equal, the entropy of $\mathcal{G}$ can be thought of as a measure of the tendency of a system to lose the ability to do useful work, lose order, and settle to a more homogeneous state.

Recalling that $\mathrm{d}Q_i(t) = [S_i(t) - \sigma_{ii}(E(t))]\mathrm{d}t$, $i = 1, \ldots, q$, is the infinitesimal amount of the net heat received or dissipated by the ith subsystem of $\mathcal{G}$ over the infinitesimal time interval $\mathrm{d}t$, it follows from (3.29) that

$$\mathrm{d}\mathcal{S}(E(t)) \geq \sum_{i=1}^{q} \frac{\mathrm{d}Q_i(t)}{c + E_i(t)}, \quad t \geq t_0. \tag{3.69}$$

Inequality (3.69) is analogous to the classical thermodynamic inequality for the variation of entropy during an infinitesimal irreversible transformation with the shifted subsystem energies $c + E_i$ playing the role of the ith subsystem thermodynamic (absolute) temperatures.

Specifically, note that since $\frac{\mathrm{d}\mathcal{S}_i}{\mathrm{d}E_i} = \frac{1}{c+E_i}$, where $\mathcal{S}_i = \log_e(c + E_i) - \log_e c$ denotes the unique continuously differentiable ith subsystem entropy, it follows that $\frac{\mathrm{d}\mathcal{S}_i}{\mathrm{d}E_i}$, $i = 1, \ldots, q$, defines the reciprocal of the subsystem thermodynamic temperatures. That is,

$$\frac{1}{T_i} \triangleq \frac{\mathrm{d}\mathcal{S}_i}{\mathrm{d}E_i} \tag{3.70}$$

and $T_i > 0$, $i = 1, \ldots, q$. Hence, in our formulation, temperature is a function derived from entropy and does not involve the primitive subjective notions of hotness and coldness.

It is important to note that in this chapter we view subsystem temperatures to be synonymous with subsystem energies. Even though this does not limit the generality of our theory from a mathematical perspective, it can be physically limiting since it does not allow for the consideration of two subsystems of $\mathcal{G}$ having the same stored energy with one of the subsystems being at a higher temperature (i.e., *hotter*) than the other. This, however, can be easily addressed by assigning different specific heats (i.e., thermal capacities) for each of the compartments of the large-scale system $\mathcal{G}$ as shown in Chapter 4.

Finally, using the system entropy function given by (3.61), we show that our large-scale dynamical system $\mathcal{G}$ with power balance equation (3.5) is state irreversible for every nontrivial (nonequilibrium) trajectory of $\mathcal{G}$. For this result, let $\mathcal{W}_{[t_0,t_1]}$ denote the set of all possible energy trajectories of $\mathcal{G}$ over the time interval $[t_0, t_1]$ given by

$$\mathcal{W}_{[t_0,t_1]} \triangleq \{s^E : [t_0, t_1] \times \mathcal{U} \to \overline{\mathbb{R}}_+^q : s^E(\cdot, S(\cdot)) \text{ satisfies } (3.5)\}, \qquad (3.71)$$

and let $\mathcal{M}_\mathrm{e} \subset \overline{\mathbb{R}}_+^q$ denote the set of equilibria of the isolated system $\mathcal{G}$ given by $\mathcal{M}_\mathrm{e} = \{E \in \overline{\mathbb{R}}_+^q : \alpha\mathbf{e}, \alpha \geq 0\}$.

Theorem 3.5. Consider the large-scale dynamical system $\mathcal{G}$ with power balance equation (3.5), and assume Axioms *i)* and *ii)* hold. Furthermore, let $s^E(\cdot, S(\cdot)) \in \mathcal{W}_{[t_0,t_1]}$, where $S(\cdot) \in \mathcal{U}$. Then $s^E(\cdot, S(\cdot))$ is an I_q-reversible trajectory of $\mathcal{G}$ if and only if $s^E(t, S(t)) \in \mathcal{M}_\mathrm{e}$, $t \in [t_0, t_1]$.

Proof. First, note that it follows from Theorem 3.4 that if $E(t) \notin \mathcal{M}_\mathrm{e}$, $t \geq t_0$, then there exists an entropy function $\mathcal{S}(E)$, $E \in \overline{\mathbb{R}}_+^q$, for $\mathcal{G}$ such that (3.62) holds. Now, sufficiency follows as a direct consequence of Theorem 2.16 with $R = I_q$, $V(z) = \mathcal{S}(E)$, and $r(u, y) = r(S, d(E)) = \sum_{i=1}^q \frac{S_i - \sigma_{ii}(E)}{c+E_i}$.

To show necessity, assume that $s^E(t, S(t)) \in \mathcal{M}_\mathrm{e}$, $t \in [t_0, t_1]$. In this case, it can be shown that $S(t) = d(E(t)) + u(t)$, $t \geq t_0$, where $u(\cdot) \in \mathcal{U}$ is such that $u_i(t) \equiv u_j(t)$, $i \neq j$, $i, j = 1, \ldots, q$. Now, with $S^-(t) = d(E(t)) + u^-(t)$, $t \geq t_0$, where $u^-(t) = -u(t_1 + t_0 - t)$, $t \in [t_0, t_1]$, it follows that $s^E(t, S(t))$ is an I_q-reversible trajectory of $\mathcal{G}$. □

Theorem 3.5 establishes an equivalence between (non)equilibrium and state (ir)reversible thermodynamic systems. Furthermore, Theorem 3.5 shows that for every $E_0 \notin \mathcal{M}_\mathrm{e}$, the large-scale dynamical system $\mathcal{G}$ is state irreversible. In addition, since state irrecoverability implies state irreversibility and, by Theorem 3.5, state irreversibility is equivalent to $E(t) \notin \mathcal{M}_\mathrm{e}$, $t \geq t_0$, it follows from Theorem 2.17 that state (ir)reversibility and state (ir)recoverability are equivalent for our thermodynamically consistent large-scale dynamical system $\mathcal{G}$. Hence, in the remainder of the monograph we use the notions of (non)equilibrium, state (ir)reversible, and state (ir)recoverable dynamical processes interchangeably.

3.4 Ectropy and the Second Law of Thermodynamics

In this section, we introduce a *new* and dual notion to entropy, namely, ectropy, describing the status quo of the large-scale dynamical system $\mathcal{G}$. First, however, we present a dual inequality to inequality (3.26) that holds for our thermodynamically consistent energy flow model over cyclic motions.

Proposition 3.6. Consider the large-scale dynamical system $\mathcal{G}$ with power balance equation (3.5), and assume that Axioms *i*) and *ii*) hold. Then, for all $E_0 \in \overline{\mathbb{R}}_+^q$, $t_\mathrm{f} \geq t_0$, and $S(\cdot) \in \mathcal{U}$ such that $E(t_\mathrm{f}) = E(t_0) = E_0$,

$$\int_{t_0}^{t_\mathrm{f}} \sum_{i=1}^{q} E_i(t)[S_i(t) - \sigma_{ii}(E(t))]\mathrm{d}t = \oint \sum_{i=1}^{q} E_i(t)\mathrm{d}Q_i(t) \geq 0, \tag{3.72}$$

where $E(t)$, $t \geq t_0$, is the solution to (3.5) with initial condition $E(t_0) = E_0$. Furthermore,

$$\oint \sum_{i=1}^{q} E_i(t)\mathrm{d}Q_i(t) = 0 \tag{3.73}$$

if and only if there exists a continuous function $\alpha : [t_0, t_\mathrm{f}] \to \overline{\mathbb{R}}_+$ such that $E(t) = \alpha(t)\mathbf{e}$, $t \in [t_0, t_\mathrm{f}]$.

Proof. Since, by Proposition 3.1, $E(t) \geq\geq 0$, $t \geq t_0$, and $\phi_{ij}(E) = -\phi_{ji}(E)$, $E \in \overline{\mathbb{R}}_+^q$, $i \neq j$, $i, j = 1, \ldots, q$, it follows from (3.5) and Axiom *ii*) that

$$\begin{aligned}
\oint \sum_{i=1}^{q} E_i(t)\mathrm{d}Q_i(t) &= \int_{t_0}^{t_\mathrm{f}} \sum_{i=1}^{q} E_i(t)[\dot{E}_i(t) - \sum_{j=1,\, j\neq i}^{q} \phi_{ij}(E(t))]\mathrm{d}t \\
&= \tfrac{1}{2}E^\mathrm{T}(t_\mathrm{f})E(t_\mathrm{f}) - \tfrac{1}{2}E^\mathrm{T}(t_0)E(t_0) \\
&\quad - \int_{t_0}^{t_\mathrm{f}} \sum_{i=1}^{q} \sum_{j=1,\, j\neq i}^{q} E_i(t)\phi_{ij}(E(t))\mathrm{d}t \\
&= -\int_{t_0}^{t_\mathrm{f}} \sum_{i=1}^{q} \sum_{j=i+1}^{q} \phi_{ij}(E(t))[E_i(t) - E_j(t)]\mathrm{d}t \\
&\geq 0,
\end{aligned} \tag{3.74}$$

which proves (3.72).

To show (3.73), note that it follows from (3.74), Axiom *i*), and Axiom *ii*) that (3.73) holds if and only if $E_i(t) = E_j(t)$, $i \neq j$, $i, j = 1, \ldots, q$, or, equivalently, there exists a continuous function $\alpha : [t_0, t_f] \to \overline{\mathbb{R}}_+$ such that $E(t) = \alpha(t)\mathbf{e}$, $t \in [t_0, t_f]$. □

Inequality (3.72) is an anti–Clausius inequality and restricts the manner in which the system absorbs (scaled) heat over cyclic motions. Note that inequality (3.72) is satisfied as an equality for an equilibrium process and as a strict inequality for a nonequilibrium process.

Next, we present the definition of ectropy for the large-scale dynamical system $\mathcal{G}$.

Definition 3.3. For the large-scale dynamical system $\mathcal{G}$ with power balance equation (3.5), a function $\mathcal{E} : \overline{\mathbb{R}}_+^q \to \mathbb{R}$ satisfying

$$\mathcal{E}(E(t_2)) \leq \mathcal{E}(E(t_1)) + \int_{t_1}^{t_2} \sum_{i=1}^{q} E_i(t)[S_i(t) - \sigma_{ii}(E(t))]\mathrm{d}t \tag{3.75}$$

for every $t_2 \geq t_1 \geq t_0$ and $S(\cdot) \in \mathcal{U}$ is called the *ectropy* function of $\mathcal{G}$.

For the next result, define the *available ectropy* of the large-scale dynamical system $\mathcal{G}$ by

$$\mathcal{E}_\mathrm{a}(E_0) \triangleq - \inf_{S(\cdot)\in\mathcal{U}_\mathrm{c},\, T\geq t_0} \int_{t_0}^{T} \sum_{i=1}^{q} E_i(t)[S_i(t) - \sigma_{ii}(E(t))]\mathrm{d}t, \tag{3.76}$$

where $E(t_0) = E_0 \in \overline{\mathbb{R}}_+^q$ and $E(T) = 0$, and define the *required ectropy supply* of the large-scale dynamical system $\mathcal{G}$ by

$$\mathcal{E}_\mathrm{r}(E_0) \triangleq \inf_{S(\cdot)\in\mathcal{U}_\mathrm{r},\, T\geq -t_0} \int_{-T}^{t_0} \sum_{i=1}^{q} E_i(t)[S_i(t) - \sigma_{ii}(E(t))]\mathrm{d}t, \tag{3.77}$$

where $E(-T) = 0$ and $E(t_0) = E_0 \in \overline{\mathbb{R}}_+^q$. Note that the available ectropy $\mathcal{E}_\mathrm{a}(E_0)$ is the maximum amount of scaled heat (ectropy) that can be extracted from the large-scale dynamical system $\mathcal{G}$ in order to transfer it from an initial state $E(t_0) = E_0$ to $E(T) = 0$. Alternatively, the required ectropy supply $\mathcal{E}_\mathrm{r}(E_0)$ is the minimum amount of scaled heat (ectropy) that can be delivered to $\mathcal{G}$ to transfer it from an initial state $E(-T) = 0$ to a given state $E(t_0) = E_0$.

Theorem 3.6. Consider the large-scale dynamical system $\mathcal{G}$ with power balance equation (3.5), and assume that Axiom *ii*) holds. Then there exists an ectropy function for $\mathcal{G}$. Moreover, $\mathcal{E}_\mathrm{a}(E)$, $E \in \overline{\mathbb{R}}_+^q$, and $\mathcal{E}_\mathrm{r}(E)$, $E \in \overline{\mathbb{R}}_+^q$, are possible ectropy functions for $\mathcal{G}$ with $\mathcal{E}_\mathrm{a}(0) = \mathcal{E}_\mathrm{r}(0) = 0$. Finally, all ectropy functions $\mathcal{E}(E)$, $E \in \overline{\mathbb{R}}_+^q$, for $\mathcal{G}$ satisfy

$$\mathcal{E}_\mathrm{a}(E) \leq \mathcal{E}(E) - \mathcal{E}(0) \leq \mathcal{E}_\mathrm{r}(E), \quad E \in \overline{\mathbb{R}}_+^q. \tag{3.78}$$

Proof. Since, by Lemma 3.1, $\mathcal{G}$ is controllable to and reachable from the origin in $\overline{\mathbb{R}}_+^q$, it follows from (3.76) and (3.77) that $\mathcal{E}_{\rm a}(E_0) > -\infty$, $E_0 \in \overline{\mathbb{R}}_+^q$, and $\mathcal{E}_{\rm r}(E_0) < \infty$, $E_0 \in \overline{\mathbb{R}}_+^q$, respectively. Next, let $E_0 \in \overline{\mathbb{R}}_+^q$, and let $S(\cdot) \in \mathcal{U}$ be such that $E(t_{\rm i}) = E(t_{\rm f}) = 0$ and $E(t_0) = E_0$, where $t_{\rm i} < t_0 < t_{\rm f}$. In this case, it follows from (3.72) that

$$\int_{t_{\rm i}}^{t_{\rm f}} \sum_{i=1}^{q} E_i(t)[S_i(t) - \sigma_{ii}(E(t))]{\rm d}t \geq 0 \tag{3.79}$$

or, equivalently,

$$\int_{t_{\rm i}}^{t_0} \sum_{i=1}^{q} E_i(t)[S_i(t) - \sigma_{ii}(E(t))]{\rm d}t \geq -\int_{t_0}^{t_{\rm f}} \sum_{i=1}^{q} E_i(t)[S_i(t) - \sigma_{ii}(E(t))]{\rm d}t. \tag{3.80}$$

Now, taking the infimum on both sides of (3.80) over all $S(\cdot) \in \mathcal{U}_{\rm r}$ and $t_{\rm i} \leq t_0$ yields

$$\begin{aligned}\mathcal{E}_{\rm r}(E_0) &= \inf_{S(\cdot)\in\mathcal{U}_{\rm r},\, t_{\rm i}\leq t_0} \int_{t_{\rm i}}^{t_0} \sum_{i=1}^{q} E_i(t)[S_i(t) - \sigma_{ii}(E(t))]{\rm d}t \\ &\geq -\int_{t_0}^{t_{\rm f}} \sum_{i=1}^{q} E_i(t)[S_i(t) - \sigma_{ii}(E(t))]{\rm d}t.\end{aligned} \tag{3.81}$$

Next, taking the supremum on both sides of (3.81) over all $S(\cdot) \in \mathcal{U}_{\rm c}$ and $t_{\rm f} \geq t_0$, we obtain $\mathcal{E}_{\rm r}(E_0) \geq \mathcal{E}_{\rm a}(E_0)$, $E_0 \in \overline{\mathbb{R}}_+^q$, which implies that $-\infty < \mathcal{E}_{\rm a}(E_0) \leq \mathcal{E}_{\rm r}(E_0) < \infty$, $E_0 \in \overline{\mathbb{R}}_+^q$. Hence, the functions $\mathcal{E}_{\rm a}(\cdot)$ and $\mathcal{E}_{\rm r}(\cdot)$ are well defined.

Next, it follows from the definition of $\mathcal{E}_{\rm a}(\cdot)$ that, for every $T \geq t_1$ and $S(\cdot) \in \mathcal{U}_{\rm c}$ such that $E(t_1) \in \overline{\mathbb{R}}_+^q$ and $E(T) = 0$,

$$\begin{aligned}-\mathcal{E}_{\rm a}(E(t_1)) \leq{}& \int_{t_1}^{t_2} \sum_{i=1}^{q} E_i(t)[S_i(t) - \sigma_{ii}(E(t))]{\rm d}t \\ &+ \int_{t_2}^{T} \sum_{i=1}^{q} E_i(t)[S_i(t) - \sigma_{ii}(E(t))]{\rm d}t, \quad t_1 \leq t_2 \leq T,\end{aligned} \tag{3.82}$$

and hence,

$$-\mathcal{E}_{\rm a}(E(t_1)) \leq \int_{t_1}^{t_2} \sum_{i=1}^{q} E_i(t)[S_i(t) - \sigma_{ii}(E(t))]{\rm d}t$$

$$+ \inf_{S(\cdot)\in\mathcal{U}_{\mathrm{c}},\, T\geq t_2} \int_{t_2}^{T} \sum_{i=1}^{q} E_i(t)[S_i(t) - \sigma_{ii}(E(t))]\mathrm{d}t$$

$$= \int_{t_1}^{t_2} \sum_{i=1}^{q} E_i(t)[S_i(t) - \sigma_{ii}(E(t))]\mathrm{d}t - \mathcal{E}_{\mathrm{a}}(E(t_2)), \tag{3.83}$$

which implies that $\mathcal{E}_{\mathrm{a}}(E)$, $E \in \overline{\mathbb{R}}_+^q$, satisfies (3.75). Thus, $\mathcal{E}_{\mathrm{a}}(E)$, $E \in \overline{\mathbb{R}}_+^q$, is a possible ectropy function for $\mathcal{G}$. Note that with $E(t_0) = E(T) = 0$ it follows from (3.72) that the infimum in (3.76) is taken over the set of nonnegative values with one of the values being zero for $S(t) \equiv 0$. Thus, $\mathcal{E}_{\mathrm{a}}(0) = 0$.

Similarly, it follows from the definition of $\mathcal{E}_{\mathrm{r}}(\cdot)$ that, for every $T \geq -t_2$ and $S(\cdot) \in \mathcal{U}_{\mathrm{r}}$ such that $E(t_2) \in \overline{\mathbb{R}}_+^q$ and $E(-T) = 0$,

$$\mathcal{E}_{\mathrm{r}}(E(t_2)) \leq \int_{-T}^{t_1} \sum_{i=1}^{q} E_i(t)[S_i(t) - \sigma_{ii}(E(t))]\mathrm{d}t + \int_{t_1}^{t_2} \sum_{i=1}^{q} E_i(t)[S_i(t) - \sigma_{ii}(E(t))]\mathrm{d}t, \quad -T \leq t_1 \leq t_2, \tag{3.84}$$

and hence,

$$\mathcal{E}_{\mathrm{r}}(E(t_2)) \leq \int_{t_1}^{t_2} \sum_{i=1}^{q} E_i(t)[S_i(t) - \sigma_{ii}(E(t))]\mathrm{d}t$$

$$+ \inf_{S(\cdot)\in\mathcal{U}_{\mathrm{r}},\, T\geq -t_1} \int_{-T}^{t_1} \sum_{i=1}^{q} E_i(t)[S_i(t) - \sigma_{ii}(E(t))]\mathrm{d}t$$

$$= \int_{t_1}^{t_2} \sum_{i=1}^{q} E_i(t)[S_i(t) - \sigma_{ii}(E(t))]\mathrm{d}t + \mathcal{E}_{\mathrm{r}}(E(t_1)), \tag{3.85}$$

which implies that $\mathcal{E}_{\mathrm{r}}(E)$, $E \in \overline{\mathbb{R}}_+^q$, satisfies (3.75). Thus, $\mathcal{E}_{\mathrm{r}}(E)$, $E \in \overline{\mathbb{R}}_+^q$, is a possible ectropy function for $\mathcal{G}$. Note that with $E(t_0) = E(-T) = 0$ it follows from (3.72) that the infimum in (3.77) is taken over the set of nonnegative values with one of the values being zero for $S(t) \equiv 0$. Thus, $\mathcal{E}_{\mathrm{r}}(0) = 0$.

Next, suppose there exists an ectropy function $\mathcal{E} : \overline{\mathbb{R}}_+^q \to \mathbb{R}$ for $\mathcal{G}$, and let $E(t_2) = 0$ in (3.75). Then it follows from (3.75) that

$$\mathcal{E}(E(t_1)) - \mathcal{E}(0) \geq - \int_{t_1}^{t_2} \sum_{i=1}^{q} E_i(t)[S_i(t) - \sigma_{ii}(E(t))]\mathrm{d}t \tag{3.86}$$

for all $t_2 \geq t_1$ and $S(\cdot) \in \mathcal{U}_c$, which implies that

$$\begin{aligned}\mathcal{E}(E(t_1)) - \mathcal{E}(0) &\geq \sup_{S(\cdot)\in\mathcal{U}_c,\, t_2\geq t_1} \left[-\int_{t_1}^{t_2} \sum_{i=1}^{q} E_i(t)[S_i(t) - \sigma_{ii}(E(t))]\mathrm{d}t \right] \\ &= -\inf_{S(\cdot)\in\mathcal{U}_c,\, t_2\geq t_1} \int_{t_1}^{t_2} \sum_{i=1}^{q} E_i(t)[S_i(t) - \sigma_{ii}(E(t))]\mathrm{d}t \\ &= \mathcal{E}_a(E(t_1)). \end{aligned} \tag{3.87}$$

Since $E(t_1)$ is arbitrary, it follows that $\mathcal{E}(E) - \mathcal{E}(0) \geq \mathcal{E}_a(E)$, $E \in \overline{\mathbb{R}}_+^q$.

Alternatively, let $E(t_1) = 0$ in (3.75). Then it follows from (3.75) that

$$\mathcal{E}(E(t_2)) - \mathcal{E}(0) \leq \int_{t_1}^{t_2} \sum_{i=1}^{q} E_i(t)[S_i(t) - \sigma_{ii}(E(t))]\mathrm{d}t \tag{3.88}$$

for all $t_1 \leq t_2$ and $S(\cdot) \in \mathcal{U}_r$. Hence,

$$\begin{aligned}\mathcal{E}(E(t_2)) - \mathcal{E}(0) &\leq \inf_{S(\cdot)\in\mathcal{U}_r,\, t_1\leq t_2} \int_{t_1}^{t_2} \sum_{i=1}^{q} E_i(t)[S_i(t) - \sigma_{ii}(E(t))]\mathrm{d}t \\ &= \mathcal{E}_r(E(t_2)), \end{aligned} \tag{3.89}$$

which, since $E(t_2)$ is arbitrary, implies that $\mathcal{E}_r(E) \geq \mathcal{E}(E) - \mathcal{E}(0)$, $E \in \overline{\mathbb{R}}_+^q$. Thus, all ectropy functions for $\mathcal{G}$ satisfy (3.78). □

The next result shows that all ectropy functions for $\mathcal{G}$ are continuous on $\overline{\mathbb{R}}_+^q$.

Theorem 3.7. Consider the large-scale dynamical system $\mathcal{G}$ with power balance equation (3.5), and let $\mathcal{E} : \overline{\mathbb{R}}_+^q \to \mathbb{R}$ be an ectropy function of $\mathcal{G}$. Then $\mathcal{E}(\cdot)$ is continuous on $\overline{\mathbb{R}}_+^q$.

Proof. The proof is identical to the proof of Theorem 3.3. □

The next result is a direct consequence of Theorem 3.6.

Proposition 3.7. Consider the large-scale dynamical system $\mathcal{G}$ with power balance equation (3.5), and assume that Axioms *i*) and *ii*) hold. Then

$$\mathcal{E}(E) \triangleq \alpha\mathcal{E}_a(E) + (1-\alpha)\mathcal{E}_r(E), \quad \alpha \in [0,1], \tag{3.90}$$

is an ectropy function for $\mathcal{G}$.

As in the case of entropy, in the next proposition we show that any ectropy function for $\mathcal{G}$ has a unique form when evaluated on the set of equilibria $\mathcal{M}_e$ for the isolated large-scale dynamical system $\mathcal{G}$.

Proposition 3.8. Consider the large-scale dynamical system $\mathcal{G}$ with power balance equation (3.5), and assume that Axioms i) and ii) hold. Then at every equilibrium state $E = E_{\rm e}$ of the isolated system $\mathcal{G}$, the ectropy $\mathcal{E}(E)$, $E \in \overline{\mathbb{R}}_+^q$, of $\mathcal{G}$ is unique (modulo a constant of integration) and is given by

$$\mathcal{E}(E) - \mathcal{E}(0) = \mathcal{E}_{\rm a}(E) = \mathcal{E}_{\rm r}(E) = \tfrac{1}{2}E^{\rm T}E, \quad E = E_{\rm e}. \tag{3.91}$$

Proof. It follows from Axiom i) that for an equilibrium process $\phi_{ij}(E(t)) \equiv 0$, $i \neq j$, $i,j = 1,\ldots,q$. Consider the ectropy function $\mathcal{E}_{\rm a}(\cdot)$ given by (3.76), and let $E_0 = E_{\rm e}$ for some equilibrium state $E_{\rm e}$. Then it follows from (3.5) that

$$\begin{aligned}
\mathcal{E}_{\rm a}(E_0) &= - \inf_{S(\cdot)\in\mathcal{U}_{\rm c},\, T\geq t_0} \int_{t_0}^{T} \sum_{i=1}^{q} E_i(t)[\dot{E}_i(t) - \sum_{j=1, j\neq i}^{q} \phi_{ij}(E(t))]{\rm d}t \\
&= - \inf_{S(\cdot)\in\mathcal{U}_{\rm c},\, T\geq t_0} \left[-\sum_{i=1}^{q} \tfrac{1}{2}E_{i0}^2 \right. \\
&\qquad \left. - \int_{t_0}^{T} \sum_{i=1}^{q} \sum_{j=1, j\neq i}^{q} E_i(t)\phi_{ij}(E(t)){\rm d}t \right] \\
&= \sum_{i=1}^{q} \tfrac{1}{2}E_{i0}^2 + \sup_{S(\cdot)\in\mathcal{U}_{\rm c},\, T\geq t_0} \int_{t_0}^{T} \sum_{i=1}^{q} \sum_{j=1, j\neq i}^{q} E_i(t)\phi_{ij}(E(t)){\rm d}t \\
&= \tfrac{1}{2}E_0^{\rm T}E_0 \\
&\quad + \sup_{S(\cdot)\in\mathcal{U}_{\rm c},\, T\geq t_0} \int_{t_0}^{T} \sum_{i=1}^{q} \sum_{j=i+1}^{q} [E_i(t) - E_j(t)]\phi_{ij}(E(t)){\rm d}t.
\end{aligned} \tag{3.92}$$

Since the solution $E(t)$, $t \geq t_0$, to (3.5) is nonnegative for all nonnegative initial conditions, it follows from Axiom ii) that the supremum in (3.92) is taken over the set of negative semidefinite values. However, the zero value of the supremum is achieved on an equilibrium process for which $\phi_{ij}(E(t)) \equiv 0$, $i \neq j$, $i,j = 1,\ldots,q$. Thus,

$$\mathcal{E}_{\rm a}(E_0) = \tfrac{1}{2}E_0^{\rm T}E_0, \quad E_0 = E_{\rm e}. \tag{3.93}$$

Similarly, it can be shown that $\mathcal{E}_{\rm r}(E) = \frac{1}{2}E^{\rm T}E$ for $E = E_{\rm e}$. Finally, it follows from (3.78) that (3.91) holds. □

The next proposition shows that if (3.75) holds as an equality for some transformation starting and ending at equilibrium points of the isolated system $\mathcal{G}$, then this transformation lies on the equilibrium manifold $\mathcal{M}_{\rm e}$.

Proposition 3.9. Consider the large-scale dynamical system $\mathcal{G}$ with power balance equation (3.5), and assume that Axioms i) and ii) hold. Let $\mathcal{E}(\cdot)$ denote an ectropy of $\mathcal{G}$, and let $E : [t_0, t_1] \to \overline{\mathbb{R}}_+^q$ denote the solution to (3.5) with $E(t_0) = \alpha_0\mathbf{e}$ and $E(t_1) = \alpha_1\mathbf{e}$, where $\alpha_0, \alpha_1 \geq 0$. Then

$$\mathcal{E}(E(t_1)) = \mathcal{E}(E(t_0)) + \int_{t_0}^{t_1} \sum_{i=1}^{q} E_i(t)[S_i(t) - \sigma_{ii}(E(t))]\mathrm{d}t \tag{3.94}$$

if and only if there exists a continuous function $\alpha : [t_0, t_1] \to \overline{\mathbb{R}}_+$ such that $\alpha(t_0) = \alpha_0$, $\alpha(t_1) = \alpha_1$, and $E(t) = \alpha(t)\mathbf{e}$, $t \in [t_0, t_1]$.

Proof. Since $E(t_0)$ and $E(t_1)$ are equilibrium states of the isolated dynamical system $\mathcal{G}$, it follows from Proposition 3.8 that

$$\mathcal{E}(E(t_1)) - \mathcal{E}(E(t_0)) = \tfrac{1}{2}E^{\mathrm{T}}(t_1)E(t_1) - \tfrac{1}{2}E^{\mathrm{T}}(t_0)E(t_0). \tag{3.95}$$

Furthermore, it follows from (3.5) that

$$\begin{aligned}\int_{t_0}^{t_1} \sum_{i=1}^{q} E_i(t)[S_i(t) - \sigma_{ii}(E(t))]\mathrm{d}t &= \int_{t_0}^{t_1} \sum_{i=1}^{q} E_i(t)[\dot{E}_i(t) - \sum_{j=1, j\neq i}^{q} \phi_{ij}(E(t))]\mathrm{d}t \\ &= \tfrac{1}{2}q\alpha_1^2 - \tfrac{1}{2}q\alpha_0^2 \\ &\quad - \int_{t_0}^{t_1} \sum_{i=1}^{q} \sum_{j=i+1}^{q} \phi_{ij}(E(t))[E_i(t) - E_j(t)]\mathrm{d}t. \end{aligned} \tag{3.96}$$

Now, it follows from Axioms i) and ii) that (3.94) holds if and only if $E_i(t) = E_j(t)$, $t \in [t_0, t_1]$, $i \neq j$, $i, j = 1, \ldots, q$, or, equivalently, there exists a continuous function $\alpha : [t_0, t_1] \to \overline{\mathbb{R}}_+$ such that $E(t) = \alpha(t)\mathbf{e}$, $t \in [t_0, t_1]$, $\alpha(t_0) = \alpha_0$, and $\alpha(t_1) = \alpha_1$. $\square$

The next theorem gives a unique, continuously differentiable ectropy function for $\mathcal{G}$ for equilibrium and nonequilibrium processes.

Theorem 3.8. Consider the large-scale dynamical system $\mathcal{G}$ with power balance equation (3.5), and assume that Axioms i) and ii) hold. Then the function $\mathcal{E} : \overline{\mathbb{R}}_+^q \to \overline{\mathbb{R}}_+$ given by

$$\mathcal{E}(E) = \tfrac{1}{2}E^{\mathrm{T}}E, \quad E \in \overline{\mathbb{R}}_+^q, \tag{3.97}$$

is a unique (modulo a constant of integration), continuously differentiable ectropy function of $\mathcal{G}$. Furthermore, for $E(t) \notin \mathcal{M}_{\mathrm{e}}$, $t \geq t_0$, where $E(t)$, $t \geq t_0$, is the solution to (3.5) and $\mathcal{M}_{\mathrm{e}} = \{E \in \overline{\mathbb{R}}_+^q : E = \alpha\mathbf{e}, \alpha \geq 0\}$, (3.97) satisfies

$$\mathcal{E}(E(t_2)) < \mathcal{E}(E(t_1)) + \int_{t_1}^{t_2} \sum_{i=1}^{q} E_i(t)[S_i(t) - \sigma_{ii}(E(t))]\mathrm{d}t \tag{3.98}$$

for every $t_2 \geq t_1 \geq t_0$ and $S(\cdot) \in \mathcal{U}$.

Proof. Since, by Proposition 3.1, $E(t) \geq\geq 0$, $t \geq t_0$, and $\phi_{ij}(E) = -\phi_{ji}(E)$, $E \in \overline{\mathbb{R}}_+^q$, $i \neq j$, $i,j = 1,\ldots,q$, it follows that

$$\begin{aligned}
\dot{\mathcal{E}}(E(t)) &= \sum_{i=1}^{q} \dot{E}_i(t)E_i(t) \\
&= \sum_{i=1}^{q} E_i(t)[S_i(t) - \sigma_{ii}(E(t))] + \sum_{i=1}^{q} \sum_{j=1,\, j\neq i}^{q} E_i(t)\phi_{ij}(E(t)) \\
&= \sum_{i=1}^{q} E_i(t)[S_i(t) - \sigma_{ii}(E(t))] \\
&\quad + \sum_{i=1}^{q} \sum_{j=i+1}^{q} [E_i(t) - E_j(t)]\phi_{ij}(E(t)) \\
&\leq \sum_{i=1}^{q} E_i(t)[S_i(t) - \sigma_{ii}(E(t))], \quad t \geq t_0.
\end{aligned} \tag{3.99}$$

Now, integrating (3.99) over $[t_1, t_2]$ yields (3.75). Furthermore, in the case where $E(t) \notin \mathcal{M}_\mathrm{e}$, $t \geq t_0$, it follows from Axiom *i*), Axiom *ii*), and (3.99) that (3.98) holds.

To show that (3.97) is a unique, continuously differentiable ectropy function of $\mathcal{G}$, let $\mathcal{E}(E)$ be a continuously differentiable ectropy function of $\mathcal{G}$ so that $\mathcal{E}(E)$ satisfies (3.75) or, equivalently,

$$\dot{\mathcal{E}}(E(t)) \leq E^\mathrm{T}(t)[S(t) - d(E(t))], \quad t \geq t_0, \tag{3.100}$$

where $E(t)$, $t \geq t_0$, denotes the solution to the power balance equation (3.5) and $\dot{\mathcal{E}}(E(t))$ denotes the time derivative of $\mathcal{E}(E)$ along the solution $E(t)$, $t \geq t_0$. Hence, it follows from (3.100) that

$$\mathcal{E}'(E)[f(E) - d(E) + S] \leq E^\mathrm{T}[S - d(E)], \quad E \in \overline{\mathbb{R}}_+^q, \quad S \in \mathbb{R}^q, \tag{3.101}$$

which implies that there exist continuous functions $\ell : \overline{\mathbb{R}}_+^q \to \mathbb{R}^p$ and $\mathcal{W} : \overline{\mathbb{R}}_+^q \to \mathbb{R}^{p\times q}$ such that

$$\begin{aligned}
0 = {} & \mathcal{E}'(E)[f(E) - d(E) + S] - E^\mathrm{T}[S - d(E)] \\
& + [\ell(E) + \mathcal{W}(E)S]^\mathrm{T}[\ell(E) + \mathcal{W}(E)S], \quad E \in \overline{\mathbb{R}}_+^q, \quad S \in \mathbb{R}^q.
\end{aligned}$$

Now, equating coefficients of equal powers (of S), it follows that $\mathcal{W}(E) \equiv 0$, $\mathcal{E}'(E) = E^{\mathrm{T}}$, $E \in \overline{\mathbb{R}}_+^q$, and

$$0 = \mathcal{E}'(E)f(E) + \ell^{\mathrm{T}}(E)\ell(E), \quad E \in \overline{\mathbb{R}}_+^q. \tag{3.102}$$

Hence, $\mathcal{E}(E) = \frac{1}{2}E^{\mathrm{T}}E$, $E \in \overline{\mathbb{R}}_+^q$, and

$$0 = E^{\mathrm{T}}f(E) + \ell^{\mathrm{T}}(E)\ell(E), \quad E \in \overline{\mathbb{R}}_+^q. \tag{3.103}$$

Thus, (3.97) is a unique, continuously differentiable ectropy function for $\mathcal{G}$. □

Inequality (3.99) is known as the *dissipation inequality* and reflects the fact that some of the supplied system ectropy $\mathcal{E}(E)$ to the thermodynamic system $\mathcal{G}$ is stored and some is dissipated. The dissipated ectropy is nonnegative and is given by the difference of what is supplied and what is stored. In addition, the amount of stored ectropy is a function of the energy of the system. Such systems are known as *dissipative systems* [191].

Note that it follows from the last equality in (3.99) that the ectropy function given by (3.97) satisfies (3.75) as an equality for an equilibrium process and as a strict inequality for a nonequilibrium process. Furthermore, it follows from (3.97) that ectropy is a measure of the extent to which the system energy deviates from a homogeneous state. Thus, ectropy is the dual of entropy and is a measure of the tendency of the large-scale dynamical system $\mathcal{G}$ to do useful work and grow more organized. Finally, we note that Theorem 3.5 can also be proved using Theorem 3.8 along with Theorem 2.16, where the inequality in (2.49) is reversed.

3.5 Semistability, Energy Equipartition, Irreversibility, and the Arrow of Time

Inequality (3.29) is a generalization of Clausius' inequality for equilibrium and nonequilibrium thermodynamics as well as reversible and irreversible thermodynamics as applied to large-scale dynamical systems, while inequality (3.75) is an anti–Clausius inequality. Moreover, for the ectropy function defined by (3.97), inequality (3.99) shows that a thermodynamically consistent large-scale dynamical system model is *dissipative* [191, 464] with respect to the *supply rate* $E^{\mathrm{T}}S$ and with *storage function* corresponding to the system ectropy $\mathcal{E}(E)$. For the entropy function given by (3.61), note that $\mathcal{S}(0) = 0$ or, equivalently, $\lim_{E\to 0}\mathcal{S}(E) = 0$, which is consistent with the *third law of thermodynamics* (Nernst's theorem). Nernst's theorem states that the entropy of every system at absolute zero can always be taken to be equal to zero.

The original statement of Nernst's theorem asserted that the change in the *Helmholtz free energy* (see Chapter 5) and the change in the system internal energy of an *isothermal* system (i.e., a system with equal and constant subsystem temperatures) tends to zero as the temperature of the system in thermodynamic equilibrium approaches absolute zero. The modern, and more general, version of the theorem as stated above was formulated by Simon [413] and sequels that it is impossible to reduce the temperature of any system or subsystem to absolute zero in a finite number of cyclic transfer operations.

For the (adiabatically) isolated large-scale dynamical system $\mathcal{G}$, (3.29) yields the fundamental inequality

$$\mathcal{S}(E(t_2)) \geq \mathcal{S}(E(t_1)), \quad t_2 \geq t_1. \tag{3.104}$$

Inequality (3.104) implies that, for any dynamical change in an adiabatically isolated large-scale dynamical system $\mathcal{G}$, the entropy of the final state can never be less than the entropy of the initial state. Inequality (3.104) is often identified with the second law of thermodynamics as a statement about entropy increase. It is important to stress that this result holds for an adiabatically isolated dynamical system. It is, however, possible with power (heat flux) supplied from an external system to reduce the entropy of the dynamical system $\mathcal{G}$. The entropy of both systems taken together, however, cannot decrease.

These observations imply that when the isolated large-scale dynamical system $\mathcal{G}$ with thermodynamically consistent energy flow characteristics (i.e., Axioms *i*) and *ii*) hold) is at a state of maximum entropy consistent with its energy, it cannot be subject to any further dynamical change since any such change would result in a decrease of entropy. This of course implies that the state of *maximum entropy* is the stable state of an isolated system, and this equilibrium state has to be semistable.

Analogously, it follows from (3.75) that the isolated large-scale dynamical system $\mathcal{G}$ satisfies the fundamental inequality

$$\mathcal{E}(E(t_2)) \leq \mathcal{E}(E(t_1)), \quad t_2 \geq t_1, \tag{3.105}$$

which implies that the ectropy of the final state of $\mathcal{G}$ is always less than or equal to the ectropy of the initial state of $\mathcal{G}$. Hence, for the isolated large-scale dynamical system $\mathcal{G}$, the entropy increases if and only if the ectropy decreases. Thus, the state of *minimum ectropy* is the stable state of an isolated system, and this equilibrium state has to be semistable. This result can also be used to show that the isolated large-scale dynamical system $\mathcal{G}$ does not exhibit Poincaré recurrence.

Since our thermodynamic compartmental model involves intercompartmental flows representing energy transfer between compartments, we can use graph-theoretic notions with *undirected graph topologies* (i.e., bidirectional energy flows) to capture the compartmental system interconnections. Graph theory [115, 169] can be useful in the analysis of the connectivity properties of compartmental systems. In particular, a directed graph can be constructed to capture a compartmental model in which the compartments are represented by nodes and the flows are represented by edges or arcs. In this case, the environment must also be considered as an additional node.

Specifically, let $\mathfrak{G} = (\mathcal{V}, \mathcal{E}, \mathcal{A})$ be a *directed graph* (or digraph) denoting the compartmental network with the set of *nodes* (or compartments) $\mathcal{V} = \{1, \ldots, q\}$ involving a finite nonempty set denoting the compartments, the set of *edges* $\mathcal{E} \subseteq \mathcal{V} \times \mathcal{V}$ involving a set of ordered pairs denoting the direction of energy flow, and an *adjacency matrix* $\mathcal{A} \in \mathbb{R}^{q \times q}$ such that $\mathcal{A}_{(i,j)} = 1$, $i, j = 1, \ldots, q$, if $(j, i) \in \mathcal{E}$, while $\mathcal{A}_{(i,j)} = 0$ if $(j, i) \notin \mathcal{E}$. The edge $(j, i) \in \mathcal{E}$ denotes that compartment j can obtain energy from compartment i, but not necessarily vice versa. Moreover, we assume $\mathcal{A}_{(i,i)} = 0$ for all $i \in \mathcal{V}$. A *graph* or *undirected graph* $\mathfrak{G}$ associated with the adjacency matrix $\mathcal{A} \in \mathbb{R}^{q \times q}$ is a directed graph for which the *arc set* is symmetric, that is, $\mathcal{A} = \mathcal{A}^{\mathrm{T}}$. Weighted graphs can also be considered here; however, since this extension does not alter any of the conceptual results, we do not consider this extension here for simplicity of exposition. Finally, we denote the *energy* of the compartment $i \in \{1, \ldots, q\}$ at time t by $E_i(t) \in \overline{\mathbb{R}}_+$.

Proposition 3.10. Consider the large-scale dynamical system $\mathcal{G}$ with power balance equation (3.5) with $d(E) \equiv 0$ and $S(t) \equiv 0$, and assume Axioms *i*) and *ii*) hold. Then $f_i(E) = 0$ for all $i = 1, \ldots, q$ if and only if $E_1 = \cdots = E_q$. Furthermore, $\alpha \mathbf{e}$, $\alpha \geq 0$, is an equilibrium state of (3.5).

Proof. If $E_i = E_j$ for all $(i, j) \in \mathcal{E}$, then $f_i(E) = 0$ for all $i = 1, \ldots, q$ is immediate from Axiom *i*). Next, we show that $f_i(E) = 0$ for all $i = 1, \ldots, q$ implies that $E_1 = \cdots = E_q$. If $f_i(E) = 0$ for all $i = 1, \ldots, q$, then it follows from Axiom *ii*) that

$$\begin{aligned} 0 &= \sum_{i=1}^{q} E_i f_i(E) \\ &= \sum_{i=1}^{q} \sum_{j=1}^{q} E_i \phi_{ij}(E) \\ &= \sum_{i=1}^{q-1} \sum_{j=i+1}^{q} (E_i - E_j) \phi_{ij}(E) \end{aligned}$$

$$\leq 0,$$

where we have used the fact that $\phi_{ij}(E) = -\phi_{ji}(E)$ for all $i, j = 1, \ldots, q$. Hence, $(E_i - E_j)\phi_{ij}(E) = 0$ for all $i, j = 1, \ldots, q$. Now, the result follows from Axiom i).

Alternatively, the proof can also be shown using graph-theoretic concepts. Specifically, if $E_i = E_j$ for all $(i, j) \in \mathcal{E}$, then $f_i(E) = 0$ for all $i = 1, \ldots, q$ is immediate from Axiom i). Next, we show that $f_i(E) = 0$ for all $i = 1, \ldots, q$ implies that $E_1 = \cdots = E_q$. If the values of all nodes are equal, then the result is immediate. Hence, assume there exists a node i^* such that $E_{i^*} \geq E_j$ for all $j \neq i^*$, $j \in \{1, \ldots, q\}$. If $(i, j) \in \mathcal{E}$, then we define a *neighbor* of node i to be node j, and vice versa.

Define the initial node set $\mathcal{J}^{(0)} \triangleq \{i^*\}$ and denote the indices of all the first neighbors of node i^* by $\mathcal{J}^{(1)} = \mathcal{N}_{i^*}$. Then, $f_{i^*}(E) = 0$ implies that $\sum_{j \in \mathcal{N}_{i^*}} \phi_{i^*j}(E_{i^*}, E_j) = 0$. Since $E_j \leq E_{i^*}$ for all $j \in \mathcal{N}_{i^*}$ and, by Axiom ii), $\phi_{ij}(z_i, z_j) \leq 0$ for all $z_i \geq z_j$, it follows that $E_{i^*} = E_j$ for all the first neighbors $j \in \mathcal{J}^{(1)}$. Next, we define the kth neighbor of node i^* and show that the value of node i^* is equal to the values of all kth neighbors of node i^* for $k = 1, \ldots, q-1$. The set of kth neighbors of node i^* is defined by

$$\mathcal{J}^{(k)} \triangleq \mathcal{J}^{(k-1)} \cup \mathcal{N}_{\mathcal{J}^{(k-1)}}, \quad k \geq 1, \quad \mathcal{J}^{(0)} = \{i^*\}, \tag{3.106}$$

where $\mathcal{N}_{\mathcal{J}}$ denotes the set of neighbors of the node set $\mathcal{J} \subseteq \mathcal{V}$. By definition, $\{i^*\} \subset \mathcal{J}^{(k)} \subseteq \mathcal{V}$ for all $k \geq 1$ and $\mathcal{J}^{(k)}$ is a monotonically increasing sequence of node sets in the sense of set inclusions.

Next, we show that $\mathcal{J}^{(q-1)} = \mathcal{V}$. Suppose, *ad absurdum*, $\mathcal{V} \backslash \mathcal{J}^{(q-1)} \neq \emptyset$. Then, by definition, there exists one node $m \in \{1, \ldots, q\}$ disconnected from all the other nodes. Hence, $\mathcal{C}_{(m,i)} = \mathcal{C}_{(i,m)} = 0$, $i = 1, \ldots, q$, which implies that the connectivity matrix $\mathcal{C}$ has a row and a column of zeros. Without loss of generality, assume that $\mathcal{C}$ has the form

$$\mathcal{C} = \begin{bmatrix} \mathcal{C}_{\mathrm{s}} & 0_{(q-1)\times 1} \\ 0_{1\times(q-1)} & 0 \end{bmatrix},$$

where $\mathcal{C}_{\mathrm{s}} \in \mathbb{R}^{(q-1)\times(q-1)}$ denotes the connectivity matrix for the new undirected graph $\mathbb{G}$, which excludes node m from the undirected graph $\mathfrak{G}$. In this case, since rank $\mathcal{C}_{\mathrm{s}} \leq q-2$, it follows that rank $\mathcal{C} < q-1$, which contradicts Axiom i).

Using mathematical induction, we show that the values of all the nodes in $\mathcal{J}^{(k)}$ are equal for $k \geq 1$. This statement holds for $k = 1$. Assuming that the values of all the nodes in $\mathcal{J}^{(k)}$ are equal to the value of node i^*, we show that the values of all the nodes in $\mathcal{J}^{(k+1)}$ are equal to the value of node i^* as well. Note that since $\mathfrak{G}$ is strongly connected, $\mathcal{N}_i \neq \emptyset$

for all $i \in \mathcal{V}$. If $\mathcal{N}_i \cap (\mathcal{J}^{(k+1)} \backslash \mathcal{J}^{(k)}) = \emptyset$ for all i, then it follows that $\mathcal{J}^{(k+1)} = \mathcal{J}^{(k)}$, and hence, the statement holds. Thus, it suffices to show that $E_i = E_{i^*}$ for an arbitrary node $i \in \mathcal{J}^{(k)}$ with $\mathcal{N}_i \cap (\mathcal{J}^{(k+1)} \backslash \mathcal{J}^{(k)}) \neq \emptyset$. For node i, note that $\sum_{j\in\mathcal{N}_i} \phi_{ij}(E_i, E_j) = 0$. Furthermore, note that $\mathcal{N}_i = (\mathcal{N}_i \cap \mathcal{J}^{(k)}) \cup (\mathcal{N}_i \cap (\mathcal{V}\backslash \mathcal{J}^{(k)}))$, $\mathcal{V}\backslash\mathcal{J}^{(k)} = \mathcal{V}\backslash\mathcal{J}^{(k+1)} \cup (\mathcal{J}^{(k+1)} \backslash \mathcal{J}^{(k)})$, $\mathcal{J}^{(k)} \subseteq \mathcal{V}$ for all k, and $\mathcal{J}^{(k+1)}$ contains the set of first neighbors of node i, or $\mathcal{N}_i \subseteq \mathcal{J}^{(k+1)}$. Then it follows that $\mathcal{N}_i \cap (\mathcal{V}\backslash\mathcal{J}^{(k)}) = \mathcal{N}_i \cap (\mathcal{J}^{(k+1)} \backslash \mathcal{J}^{(k)})$ and

$$\sum_{j\in\mathcal{N}_i\cap\mathcal{J}^{(k)}} \phi_{ij}(E_i, E_j) + \sum_{j\in\mathcal{N}_i\cap(\mathcal{J}^{(k+1)}\backslash\mathcal{J}^{(k)})} \phi_{ij}(E_i, E_j) = 0. \tag{3.107}$$

Since $E_j = E_i$ for all nodes $j \in \mathcal{N}_i \cap \mathcal{J}^{(k)} \subseteq \mathcal{J}^{(k)}$, it follows that $\sum_{j\in\mathcal{N}_i\cap\mathcal{J}^{(k)}} \phi_{ij}(E_i, E_j) = 0$, and hence, $\sum_{j\in\mathcal{N}_i\cap(\mathcal{J}^{(k+1)}\backslash\mathcal{J}^{(k)})} \phi_{ij}(E_i, E_j) = 0$. However, since $E_{i^*} = E_i \geq E_j$ for all $i \in \mathcal{J}^{(k)}$ and $j \in \mathcal{V}\backslash\mathcal{J}^{(k)}$, it follows that the values of all nodes in $\mathcal{N}_i \cap (\mathcal{J}^{(k+1)} \backslash \mathcal{J}^{(k)})$ are equal to E_{i^*}. Hence, the values of all nodes i in the node set $\bigcup_{i\in\mathcal{J}^{(k)}} \mathcal{N}_i \cap (\mathcal{J}^{(k+1)} \backslash \mathcal{J}^{(k)}) = \mathcal{J}^{(k+1)} \cap (\mathcal{J}^{(k+1)} \backslash \mathcal{J}^{(k)}) = \mathcal{J}^{(k+1)} \backslash \mathcal{J}^{(k)}$ are equal to E_{i^*}, that is, the values of all the nodes in $\mathcal{J}^{(k+1)}$ are equal. Combining this result with the fact that $\mathcal{J}^{(q-1)} = \mathcal{V}$, it follows that the values of all the nodes in $\mathcal{V}$ are equal.

The second assertion is a direct consequence of the first assertion. □

Theorem 3.9. Consider the large-scale dynamical system $\mathcal{G}$ with power balance equation (3.5) with $S(t) \equiv 0$ and $d(E) \equiv 0$, and assume that Axioms *i*) and *ii*) hold. Then, for every $\alpha \geq 0$, $\alpha\mathbf{e}$ is a semistable equilibrium state of (3.5). Furthermore, $E(t) \to \frac{1}{q}\mathbf{e}\mathbf{e}^{\mathrm{T}}E(t_0)$ as $t \to \infty$ and $\frac{1}{q}\mathbf{e}\mathbf{e}^{\mathrm{T}}E(t_0)$ is a semistable equilibrium state. Finally, if for some $k \in \{1, \ldots, q\}$, $\sigma_{kk}(E) \geq 0$, $E \in \overline{\mathbb{R}}_+^q$, and $\sigma_{kk}(E) = 0$ if and only if $E_k = 0$,[8] then the zero solution $E(t) \equiv 0$ to (3.5) is a globally asymptotically stable equilibrium state of (3.5).

Proof. It follows from Proposition 3.10 that $\alpha\mathbf{e} \in \overline{\mathbb{R}}_+^q$, $\alpha \geq 0$, is an equilibrium state of (3.5). To show Lyapunov stability of the equilibrium state $\alpha\mathbf{e}$, consider the shifted-system ectropy function $\mathcal{E}_{\mathrm{s}}(E) = \frac{1}{2}(E - \alpha\mathbf{e})^{\mathrm{T}}(E - \alpha\mathbf{e})$ as a Lyapunov function candidate. Now, since $\phi_{ij}(E) = -\phi_{ji}(E)$, $E \in \overline{\mathbb{R}}_+^q$, $i \neq j$, $i, j = 1, \ldots, q$, and $\mathbf{e}^{\mathrm{T}}f(E) = 0$, $E \in \overline{\mathbb{R}}_+^q$, it follows from Axiom *ii*) that

$$\begin{aligned} \dot{\mathcal{E}}_{\mathrm{s}}(E) &= (E - \alpha\mathbf{e})^{\mathrm{T}}\dot{E} \\ &= (E - \alpha\mathbf{e})^{\mathrm{T}}f(E) \end{aligned}$$

[8] The assumption $\sigma_{kk}(E) \geq 0$, $E \in \overline{\mathbb{R}}_+^q$, and $\sigma_{kk}(E) = 0$ if and only if $E_k = 0$ for some $k \in \{1, \ldots, q\}$ implies that if the kth subsystem possesses no energy, then this subsystem cannot dissipate energy to the environment. Conversely, if the kth subsystem does not dissipate energy to the environment, then this subsystem has no energy.

$$
\begin{aligned}
&= E^{\mathrm{T}} f(E) \\
&= \sum_{i=1}^{q} E_i \left[\sum_{j=1,\, j\neq i}^{q} \phi_{ij}(E) \right] \\
&= \sum_{i=1}^{q} \sum_{j=i+1}^{q} (E_i - E_j)\phi_{ij}(E) \\
&= \sum_{i=1}^{q} \sum_{j\in\mathcal{K}_i} (E_i - E_j)\phi_{ij}(E) \\
&\le 0, \quad E \in \overline{\mathbb{R}}_+^q, \qquad (3.108)
\end{aligned}
$$

where $\mathcal{K}_i \triangleq \mathcal{N}_i \setminus \cup_{l=1}^{i-1}\{l\}$ and $\mathcal{N}_i \triangleq \{j \in \{1,\ldots,q\} : \phi_{ij}(E) = 0$ if and only if $E_i = E_j\}$, $i = 1,\ldots,q$, which establishes Lyapunov stability of the equilibrium state $\alpha\mathbf{e}$.

To show that $\alpha\mathbf{e}$ is semistable, let $\mathcal{R} \triangleq \{E \in \overline{\mathbb{R}}_+^q : \dot{\mathcal{E}}_{\mathrm{s}}(E) = 0\} = \{E \in \overline{\mathbb{R}}_+^q : (E_i - E_j)\phi_{ij}(E) = 0,\, i = 1,\ldots,q,\, j \in \mathcal{K}_i\}$. Now, by Axiom *i*) the directed graph associated with the connectivity matrix $\mathcal{C}$ for the large-scale dynamical system $\mathcal{G}$ is strongly connected, which implies that $\mathcal{R} = \{E \in \overline{\mathbb{R}}_+^q : E_1 = \cdots = E_q\}$. Since the set $\mathcal{R}$ consists of the equilibrium states of (3.5), it follows that the largest invariant set $\mathcal{M}$ contained in $\mathcal{R}$ is given by $\mathcal{M} = \mathcal{R}$. Hence, it follows from the Krasovskii-LaSalle invariant set theorem that for any initial condition $E(t_0) \in \overline{\mathbb{R}}_+^q$, $E(t) \to \mathcal{M}$ as $t \to \infty$, and hence, $\alpha\mathbf{e}$ is a semistable equilibrium state of (3.5). Next, note that since $\mathbf{e}^{\mathrm{T}}E(t) = \mathbf{e}^{\mathrm{T}}E(t_0)$ and $E(t) \to \mathcal{M}$ as $t \to \infty$, it follows that $E(t) \to \frac{1}{q}\mathbf{e}\mathbf{e}^{\mathrm{T}}E(t_0)$ as $t \to \infty$. Hence, with $\alpha = \frac{1}{q}\mathbf{e}^{\mathrm{T}}E(t_0)$, $\alpha\mathbf{e} = \frac{1}{q}\mathbf{e}\mathbf{e}^{\mathrm{T}}E(t_0)$ is a semistable equilibrium state of (3.5).

Finally, to show that in the case where for some $k \in \{1,\ldots,q\}$, $\sigma_{kk}(E) \ge 0$, $E \in \overline{\mathbb{R}}_+^q$, and $\sigma_{kk}(E) = 0$ if and only if $E_k = 0$, the zero solution $E(t) \equiv 0$ to (3.5) is globally asymptotically stable, consider the system ectropy $\mathcal{E}(E) = \frac{1}{2}E^{\mathrm{T}}E$, $E \in \overline{\mathbb{R}}_+^q$, as a candidate Lyapunov function. Note that $\mathcal{E}(0) = 0$, $\mathcal{E}(E) > 0$, $E \in \overline{\mathbb{R}}_+^q$, $E \neq 0$, and $\mathcal{E}(E)$ is radially unbounded. Now, the Lyapunov derivative along the system energy trajectories of (3.5) is given by

$$
\begin{aligned}
\dot{\mathcal{E}}(E) &= E^{\mathrm{T}}[f(E) - d(E)] \\
&= E^{\mathrm{T}} f(E) - E_k\sigma_{kk}(E) \\
&= \sum_{i=1}^{q} E_i \left[\sum_{j=1, j\neq i}^{q} \phi_{ij}(E) \right] - E_k\sigma_{kk}(E)
\end{aligned}
$$

$$
\begin{aligned}
&= \sum_{i=1}^{q} \sum_{j=i+1}^{q} (E_i - E_j)\phi_{ij}(E) - E_k\sigma_{kk}(E) \\
&= \sum_{i=1}^{q} \sum_{j\in\mathcal{K}_i} (E_i - E_j)\phi_{ij}(E) - E_k\sigma_{kk}(E) \\
&\le 0, \quad E \in \overline{\mathbb{R}}_+^q,
\end{aligned}
\tag{3.109}
$$

which shows that the zero solution $E(t) \equiv 0$ to (3.5) is Lyapunov stable.

To show global asymptotic stability of the zero equilibrium state, let $\mathcal{R} \triangleq \{E \in \overline{\mathbb{R}}_+^q : \dot{\mathcal{E}}(E) = 0\} = \{E \in \overline{\mathbb{R}}_+^q : E_k\sigma_{kk}(E) = 0,\ k \in \{1,\ldots,q\}\} \cap \{E \in \overline{\mathbb{R}}_+^q : (E_i - E_j)\phi_{ij}(E) = 0,\ i = 1,\ldots,q,\ j \in \mathcal{K}_i\}$. Now, since Axiom $i)$ holds and $\sigma_{kk}(E) = 0$ if and only if $E_k = 0$, it follows that $\mathcal{R} = \{E \in \overline{\mathbb{R}}_+^q : E_k = 0,\ k \in \{1,\ldots,q\}\} \cap \{E \in \overline{\mathbb{R}}_+^q : E_1 = E_2 = \cdots = E_q\} = \{0\}$, and hence, the largest invariant set $\mathcal{M}$ contained in $\mathcal{R}$ is given by $\mathcal{M} = \{0\}$. Hence, it follows from the Krasovskii-LaSalle invariant set theorem that for any initial condition $E(t_0) \in \overline{\mathbb{R}}_+^q$, $E(t) \to \mathcal{M} = \{0\}$ as $t \to \infty$, which proves global asymptotic stability of the zero equilibrium state of (3.5). □

Theorem 3.9 shows that the isolated (i.e., $S(t) \equiv 0$ and $d(E) \equiv 0$) large-scale dynamical system $\mathcal{G}$ is semistable. Hence, it follows from Theorem 2.26 that the isolated large-scale dynamical system $\mathcal{G}$ does not exhibit Poincaré recurrence in $\overline{\mathbb{R}}_+^q \setminus \mathcal{M}_{\mathrm{e}}$. In Theorem 3.9 we used the shifted ectropy function to show that for the isolated (i.e., $S(t) \equiv 0$ and $d(E) \equiv 0$) large-scale dynamical system $\mathcal{G}$, $E(t) \to \frac{1}{q}\mathbf{e}\mathbf{e}^{\mathrm{T}}E(t_0)$ as $t \to \infty$ and $\frac{1}{q}\mathbf{e}\mathbf{e}^{\mathrm{T}}E(t_0)$ is a semistable equilibrium state. This result can also be arrived at using the system entropy. Specifically, using the system entropy given by (3.61), we can show attraction of the system trajectories to Lyapunov stable equilibrium points $\alpha\mathbf{e}$, $\alpha \ge 0$, and hence show semistability of these equilibrium states.

To see this, note that since $\mathbf{e}^{\mathrm{T}}f(E) = 0$, $E \in \overline{\mathbb{R}}_+^q$, it follows that $\mathbf{e}^{\mathrm{T}}\dot{E}(t) = 0$, $t \ge t_0$. Hence, $\mathbf{e}^{\mathrm{T}}E(t) = \mathbf{e}^{\mathrm{T}}E(t_0)$, $t \ge t_0$. Furthermore, since $E(t) \ge\ge 0$, $t \ge t_0$, it follows that $0 \le\le E(t) \le\le \mathbf{e}\mathbf{e}^{\mathrm{T}}E(t_0)$, $t \ge t_0$, which implies that all solutions to (3.5) are bounded. Next, since by (3.63) the function $-\mathcal{S}(E(t))$, $t \ge t_0$, is nonincreasing and $E(t)$, $t \ge t_0$, is bounded, it follows from the Krasovskii-LaSalle invariant set theorem that for any initial condition $E(t_0) \in \overline{\mathbb{R}}_+^q$, $E(t) \to \mathcal{M}$ as $t \to \infty$, where $\mathcal{M}$ is the largest invariant set contained in $\mathcal{R} \triangleq \{E \in \overline{\mathbb{R}}_+^q : -\dot{\mathcal{S}}(E) = 0\}$. It now follows from the last inequality of (3.63) that $\mathcal{R} = \{E \in \overline{\mathbb{R}}_+^q : (E_i - E_j)\phi_{ij}(E) = 0,\ i = 1,\ldots,q,\ j \in \mathcal{K}_i\}$, which, since the directed graph associated with the connectivity matrix $\mathcal{C}$ for the large-scale dynamical system $\mathcal{G}$ is strongly connected, implies that $\mathcal{R} = \{E \in \overline{\mathbb{R}}_+^q : E_1 = \cdots = E_q\}$. Since the set $\mathcal{R}$ consists of the equilibrium states of (3.5), it follows that $\mathcal{M} = \mathcal{R}$, which, along with (3.108), establishes semistability of the equilibrium states $\alpha\mathbf{e}$, $\alpha \geq 0$.

Theorem 3.9 implies that the steady-state value of the energy in each subsystem $\mathcal{G}_i$ of the isolated large-scale dynamical system $\mathcal{G}$ is equal, that is, the steady-state energy of the isolated large-scale dynamical system $\mathcal{G}$ given by

$$E_\infty = \frac{1}{q}\mathbf{e}\mathbf{e}^{\mathrm{T}}E(t_0) = \left[\frac{1}{q}\sum_{i=1}^{q} E_i(t_0)\right]\mathbf{e} \tag{3.110}$$

is uniformly distributed over all subsystems of $\mathcal{G}$. This phenomenon is known as *equipartition of energy* [35, 37, 204, 297, 354] and is an emergent behavior in thermodynamic systems.

In classical statistical physics, the equipartition of energy theorem—deduced by Maxwell—states that molecules in a thermal equilibrium state have the same average kinetic energy associated with each degree of freedom of their motion and is given by $\frac{1}{2}kT$, where k is Boltzmann's constant and T is the equilibrium temperature. The phenomenon of equipartition of energy as introduced in this monograph is closely related to the notion of a *monotemperaturic* system discussed in [66].

The next proposition shows that, among all possible energy distributions in the large-scale dynamical system $\mathcal{G}$, energy equipartition corresponds to the minimum value of the system's ectropy and the maximum value of the system's entropy (see Figure 3.2).

Proposition 3.11. Consider the large-scale dynamical system $\mathcal{G}$ with power balance equation (3.5), let $\mathcal{E} : \overline{\mathbb{R}}_+^q \to \overline{\mathbb{R}}_+$ and $\mathcal{S} : \overline{\mathbb{R}}_+^q \to \overline{\mathbb{R}}_+$ denote the ectropy and entropy functions of $\mathcal{G}$ given by (3.97) and (3.61), respectively, and define $\mathcal{D}_{\mathrm{c}} \triangleq \{E \in \overline{\mathbb{R}}_+^q : \mathbf{e}^{\mathrm{T}}E = \beta\}$, where $\beta \geq 0$. Then

$$\arg\min_{E\in\mathcal{D}_{\mathrm{c}}}(\mathcal{E}(E)) = \arg\max_{E\in\mathcal{D}_{\mathrm{c}}}(\mathcal{S}(E)) = E^* = \frac{\beta}{q}\mathbf{e}. \tag{3.111}$$

Furthermore, $\mathcal{E}_{\min} \triangleq \mathcal{E}(E^*) = \frac{1}{2}\frac{\beta^2}{q}$ and $\mathcal{S}_{\max} \triangleq \mathcal{S}(E^*) = q\log_e(c + \frac{\beta}{q}) - q\log_e c$.

Proof. The existence and uniqueness of E^* follows from the fact that $\mathcal{E}(E)$ and $-\mathcal{S}(E)$ are strictly convex continuous functions defined on the compact set $\mathcal{D}_{\mathrm{c}}$. To minimize $\mathcal{E}(E) = \frac{1}{2}E^{\mathrm{T}}E$ subject to $E \in \mathcal{D}_{\mathrm{c}}$, form the Lagrangian $\mathcal{L}(E, \lambda) = \frac{1}{2}E^{\mathrm{T}}E + \lambda(\mathbf{e}^{\mathrm{T}}E - \beta)$, where $\lambda \in \mathbb{R}$ is a Lagrange

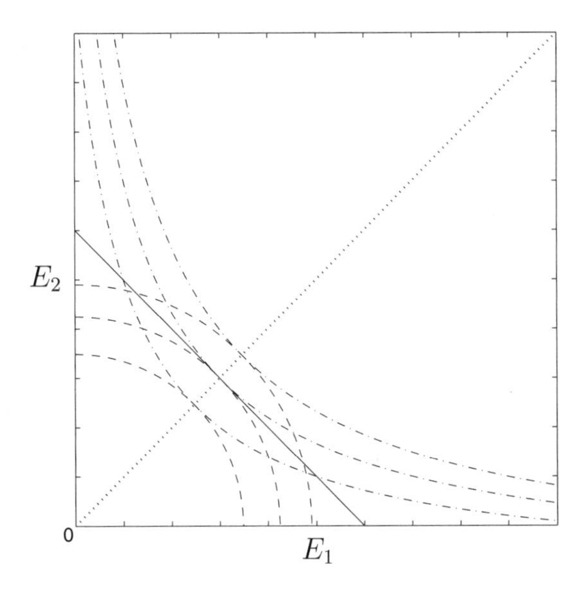

Figure 3.2 Thermodynamic equilibria ($\cdots$), constant energy surfaces (——), constant ectropy surfaces (– –), and constant entropy surfaces (– · –).

multiplier.[9] If E^* solves this minimization problem, then

$$0 = \left.\frac{\partial \mathcal{L}}{\partial E}\right|_{E=E^*} = E^{*\mathrm{T}} + \lambda \mathbf{e}^{\mathrm{T}}, \tag{3.112}$$

and hence, $E^* = -\lambda \mathbf{e}$. Now, it follows from $\mathbf{e}^{\mathrm{T}} E^* = \beta$ that $\lambda = -\frac{\beta}{q}$, which implies that $E^* = \frac{\beta}{q}\mathbf{e} \in \overline{\mathbb{R}}_+^q$. The fact that E^* minimizes the ectropy on the compact set $\mathcal{D}_\mathrm{c}$ can be shown by computing the Hessian of the ectropy for the constrained parameter optimization problem, and showing that the Hessian is positive definite at E^*. $\mathcal{E}_{\min} = \frac{1}{2}\frac{\beta^2}{q}$ is now immediate.

Analogously, to maximize $\mathcal{S}(E) = \mathbf{e}^{\mathrm{T}}\mathbf{log}_e(c\mathbf{e} + E) - q \log_e c$ subject to $E \in \mathcal{D}_\mathrm{c}$, form the Lagrangian $\mathcal{L}(E, \lambda) \triangleq \sum_{i=1}^q \log_e(c + E_i) + \lambda(\mathbf{e}^{\mathrm{T}} E - \beta)$, where $\lambda \in \mathbb{R}$ is a Lagrange multiplier. If E^* solves this maximization problem, then

$$0 = \left.\frac{\partial \mathcal{L}}{\partial E}\right|_{E=E^*} = \left[\frac{1}{c + E_1^*} + \lambda, \ldots, \frac{1}{c + E_q^*} + \lambda\right]. \tag{3.113}$$

Thus, $\lambda = -\frac{1}{c+E_i^*}$, $i = 1, \ldots, q$. If $\lambda = 0$, then the only value of E^* that

[9]Note that we are using the Fritz John Lagrange multiplier theorem [27] to obtain necessary optimality conditions and hence no constraint qualification need be imposed to the optimization problem. Specifically, recall that for a constrained optimization problem one adds an additional nonnegative Lagrange multiplier λ_0 for the objective function and asserts that not all multipliers are simultaneously zero. A constraint qualification ensures that λ_0 is nonzero and, without loss of generality, can be scaled to unity.

satisfies (3.113) is $E^* = \infty$, which does not satisfy the constraint equation $\mathbf{e}^{\mathrm{T}}E = \beta$ for finite $\beta \geq 0$. Hence, $\lambda \neq 0$ and $E_i^* = -(\frac{1}{\lambda} + c)$, $i = 1, \ldots, q$, which implies $E^* = -(\frac{1}{\lambda} + c)\mathbf{e}$. Now, it follows from $\mathbf{e}^{\mathrm{T}}E^* = \beta$ that $-(\frac{1}{\lambda} + c) = \frac{\beta}{q}$ and hence $E^* = \frac{\beta}{q}\mathbf{e} \in \overline{\mathbb{R}}_+^q$. The fact that E^* maximizes the entropy on the compact set $\mathcal{D}_{\mathrm{c}}$ can be shown by computing the Hessian of $\mathcal{S}(E)$ and showing that it is negative definite at E^*. $\mathcal{S}_{\max} = q \log_e(c + \frac{\beta}{q}) - q \log_e c$ is now immediate. □

It follows from (3.104), (3.105), and Proposition 3.11 that conservation of energy in an isolated system necessarily implies nonconservation of ectropy and entropy. Hence, in an isolated large-scale dynamical system $\mathcal{G}$, all the energy, though always conserved, will eventually be degraded (diluted) to the point where it cannot produce any useful work. Hence, all motion would cease and the large-scale dynamical system would be fated to a state of eternal rest (semistability), wherein all subsystems will possess identical energies (energy equipartition). Ectropy would be a minimum and entropy would be a maximum, giving rise to a state of absolute disorder. This is precisely what is known in theoretical physics as the *heat death of the universe*.

As noted in Chapter 1, this *terroristic nimbus* of the second law of thermodynamics was first expressed in the work of Lord Kelvin in 1852 without any supporting mathematical arguments. Kelvin evoked the view that there exists a universal tendency toward energy dissipation—although not annihilation—and that all life on Earth will perish in the distant future.

Next, using the system entropy and ectropy functions given by (3.61) and (3.97), respectively, we show that our large-scale dynamical system $\mathcal{G}$ with power balance equation (3.5) is state irreversible for all nontrivial trajectories of $\mathcal{G}$, establishing a clear connection between our thermodynamic model and the arrow of time.

Theorem 3.10. Consider the large-scale dynamical system $\mathcal{G}$ with power balance equation (3.5) with $S(t) \equiv 0$ and $d(E) \equiv 0$, and assume Axioms *i*) and *ii*) hold. Furthermore, let $s^E(\cdot, 0) \in \mathcal{W}_{[t_0,t_1]}$. Then, for every $E_0 \notin \mathcal{M}_{\mathrm{e}}$, there exists a continuously differentiable function $\mathcal{S} : \overline{\mathbb{R}}_+^q \to \mathbb{R}$ (respectively, $\mathcal{E} : \overline{\mathbb{R}}_+^q \to \mathbb{R}$) such that $\mathcal{S}(s^E(t, 0))$ (respectively, $\mathcal{E}(s^E(t, 0))$) is an increasing (respectively, decreasing) function of time. Furthermore, $s^E(\cdot, 0)$ is an I_q-reversible trajectory of $\mathcal{G}$ if and only if $s^E(t, 0) \in \mathcal{M}_{\mathrm{e}}$, $t \in [t_0, t_1]$.

Proof. The existence of a continuously differentiable function $\mathcal{S} : \overline{\mathbb{R}}_+^q \to \mathbb{R}$ (respectively, $\mathcal{E} : \overline{\mathbb{R}}_+^q \to \mathbb{R}$), which increases (respectively, decreases) on all nontrivial trajectories of $\mathcal{G}$, is a restatement of Theorem

3.4 (respectively, Theorem 3.8) with $S(t) \equiv 0$ and $d(E) \equiv 0$. Now, necessity is immediate, while sufficiency is a direct consequence of Corollary 2.2 with $R = I_q$ and $V(z) = \mathcal{S}(E)$ (respectively, $V(z) = \mathcal{E}(E)$). □

Theorem 3.10 shows that for every $E_0 \notin \mathcal{M}_{\rm e}$, the adiabatically isolated dynamical system $\mathcal{G}$ is state irreversible. This gives a clear connection between our thermodynamic model and the arrow of time. In particular, it follows from Corollary 2.2 and Theorem 3.10 that there exists a function of the system state that strictly increases or strictly decreases in time on any nontrivial trajectory of $\mathcal{G}$ if and only if there does *not* exist a nontrivial reversible trajectory of $\mathcal{G}$. Thus, the existence of the continuously differentiable entropy and ectropy functions given by (3.61) and (3.97) for $\mathcal{G}$ establishes the existence of a completely ordered time set having a topological structure involving a closed set homeomorphic to the real line.

This fact follows from the inverse function theorem of mathematical analysis and the fact that a continuous strictly monotonic function is a topological mapping (i.e., a homeomorphism), and conversely every topological mapping of a strictly monotonic function's domain onto its codomain must be strictly monotonic. More specifically, since the entropy $\mathcal{S}(\cdot)$ of $\mathcal{G}$ is strictly increasing and is continuous on an interval $\mathcal{I} \subset \mathbb{R}$, then $\mathcal{S}^{-1}(\cdot)$ exists, is continuous, and is strictly increasing on the interval $\mathcal{S}(\mathcal{I})$, where $\mathcal{S}(\mathcal{I})$ denotes the image of $\mathcal{I} \subset \mathbb{R}$ under the map $\mathcal{S}$. Thus, $\mathcal{S}(\cdot)$ is a topological mapping. Conversely, every topological mapping of an interval $\mathcal{I}$ onto $\mathcal{S}(\mathcal{I})$ must be a strictly monotonic function [7, p. 95]. This topological property gives a clear time-reversal asymmetry characterization of our thermodynamic model, establishing an emergence of the direction of time flow.

We close this section by showing that our thermodynamically consistent large-scale system $\mathcal{G}$ satisfies *Gibbs' principle* [163, p. 56]. Gibbs' version of the second law of thermodynamics can be stated as follows:

Gibbs' Principle. *For an equilibrium of any isolated system, it is necessary and sufficient that in all possible variations of the state of the system that do not alter its energy, the variation of its entropy shall either vanish or be negative.*

Gibbs' principle is an inference to a maximum entropy principle involving *virtual* variations in an isolated system. Many authors have adopted the principle as a formulation of the second law [71,73,315,449]. As noted in Chapter 1, Gibbs' principle gives necessary and sufficient conditions for a thermal equilibrium and should be viewed as a variational principle defining semistable equilibrium states. Namely, analogous to Hamilton's principle of least action defining the smooth path a dynamical system will execute among all smooth kinematically possible paths, Gibbs' principle defines the physically admissible semistable states and has no connection to dynamic states.

To establish Gibbs' principle for our thermodynamically consistent energy flow model, suppose $E_\mathrm{e} = \alpha \mathbf{e}$, $\alpha \geq 0$, is an equilibrium state of the isolated system $\mathcal{G}$. Now, it follows from Proposition 3.11 that the entropy of $\mathcal{G}$ achieves its maximum at E_e subject to the constant energy level $\mathbf{e}^\mathrm{T} E = \alpha q$, $E \in \overline{\mathbb{R}}_+^q$. Hence, any variation of the state of the system that does not alter its energy leads to a zero or negative variation of the system entropy.

Conversely, suppose that at some point $E^* \in \overline{\mathbb{R}}_+^q$ the variation of the system entropy is either zero or negative for all possible variations in the state of the system that do not alter the system's total energy. Furthermore, *ad absurdum*, let the isolated system $\mathcal{G}$ undergo an irreversible transformation starting at $E^* \notin \mathcal{M}_\mathrm{e}$. Then it follows from Theorem 3.4 that the entropy of $\mathcal{G}$ given by (3.61) increases, which contradicts the above assumption. Hence, the system $\mathcal{G}$ cannot undergo an irreversible transformation starting at $E^* \notin \mathcal{M}_\mathrm{e}$. Alternatively, if the isolated system $\mathcal{G}$ undergoes a reversible transformation starting at $E^* \in \mathcal{M}_\mathrm{e}$, then E^* has to be an equilibrium state of $\mathcal{G}$.

Similarly, using the notion of ectropy, it can be shown that an isolated dynamical system $\mathcal{G}$ is in equilibrium if and only if, in all possible variations of the state of the system that do not alter its energy, the variation of the system ectropy is positive semidefinite. Finally, we note that a dual result to Gibbs' principle can also be established. Specifically, using similar arguments as outlined above, it can be shown that for an equilibrium point of any isolated system it is necessary and sufficient that, in all possible variations of the state of the system that do not alter its entropy (respectively, ectropy), the variation of its energy shall either vanish or be positive (respectively, negative).

3.6 Entropy Increase and the Second Law of Thermodynamics

In the preceding discussion, it was assumed that our large-scale dynamical system model is such that energy flows from more energetic subsystems to less energetic subsystems, that is, heat (energy) flows in the direction of lower temperatures. Although this universal phenomenon can be predicted with virtual certainty, it follows as a manifestation of entropy and ectropy nonconservation for the case of two subsystems.

To see this, consider the isolated large-scale dynamical system $\mathcal{G}$ with power balance equation (3.5) (with $S(t) \equiv 0$ and $d(E) \equiv 0$), and assume that the system entropy given by (3.61) is increasing and hence $\dot{\mathcal{S}}(E(t)) \geq 0$, $t \geq t_0$. Now, since

$$\begin{aligned} 0 &\leq \dot{\mathcal{S}}(E(t)) \\ &= \sum_{i=1}^{q} \frac{\dot{E}_i(t)}{c + E_i(t)} \\ &= \sum_{i=1}^{q} \sum_{j=1,\, j\neq i}^{q} \frac{\phi_{ij}(E(t))}{c + E_i(t)} \\ &= \sum_{i=1}^{q} \sum_{j=i+1}^{q} \left(\frac{\phi_{ij}(E(t))}{c + E_i(t)} - \frac{\phi_{ij}(E(t))}{c + E_j(t)} \right) \\ &= \sum_{i=1}^{q} \sum_{j\in\mathcal{K}_i} \frac{\phi_{ij}(E(t))[E_j(t) - E_i(t)]}{(c + E_i(t))(c + E_j(t))}, \quad t \geq t_0, \end{aligned} \tag{3.114}$$

it follows that for $q = 2$, $(E_1 - E_2)\phi_{12}(E) \leq 0$, $E \in \overline{\mathbb{R}}_+^2$, which implies that energy (heat) flows naturally from a more energetic subsystem (hot object) to a less energetic subsystem (cooler object). The universality of this emergent behavior thus follows from the fact that entropy (respectively, ectropy) transfer, accompanying energy transfer, always increases (respectively, decreases).

In the case where we have multiple subsystems, it is clear from (3.114) that entropy and ectropy nonconservation does not necessarily imply Axiom *ii*). However, if we invoke the additional condition (Axiom *iii*)) that if for any pair of connected subsystems $\mathcal{G}_k$ and $\mathcal{G}_l$, $k \neq l$, with energies $E_k \geq E_l$ (respectively, $E_k \leq E_l$) and for any other pair of connected subsystems $\mathcal{G}_m$ and $\mathcal{G}_n$, $m \neq n$, with energies $E_m \geq E_n$ (respectively, $E_m \leq E_n$), the inequality $\phi_{kl}(E)\phi_{mn}(E) \geq 0$, $E \in \overline{\mathbb{R}}_+^q$, holds, then nonconservation of entropy and ectropy in the isolated large-scale dynamical system $\mathcal{G}$ implies Axiom *ii*).

The inequality $\phi_{kl}(E)\phi_{mn}(E) \geq 0$, $E \in \overline{\mathbb{R}}_+^q$, postulates that the direction of energy flow for any given pair of *energy similar* subsystems is consistent, that is, if for a given pair of connected subsystems at given different energy levels the energy flows in a certain direction, then for any other pair of connected subsystems with an analogous energy level difference, the energy flow direction is consistent with the original pair of subsystems. Note that this assumption does *not* specify the direction of energy flow between subsystems.

To see that $\dot{\mathcal{S}}(E(t)) \geq 0$, $t \geq t_0$, along with Axiom *iii*) implies Axiom *ii*), note that since (3.114) holds for all $t \geq t_0$ and $E(t_0) \in \overline{\mathbb{R}}_+^q$ is arbitrary, (3.114) implies

$$\sum_{i=1}^{q} \sum_{j \in \mathcal{K}_i} \frac{\phi_{ij}(E)(E_j - E_i)}{(c + E_i)(c + E_j)} \geq 0, \quad E \in \overline{\mathbb{R}}_+^q. \tag{3.115}$$

Now, it follows from (3.115) that for any fixed system energy level $E \in \overline{\mathbb{R}}_+^q$ there exists at least one pair of connected subsystems $\mathcal{G}_k$ and $\mathcal{G}_l$, $k \neq l$, such that $\phi_{kl}(E)(E_l - E_k) \geq 0$. Thus, if $E_k \geq E_l$ (respectively, $E_k \leq E_l$), then $\phi_{kl}(E) \leq 0$ (respectively, $\phi_{kl}(E) \geq 0$). Furthermore, it follows from Axiom *iii*) that for any other pair of connected subsystems $\mathcal{G}_m$ and $\mathcal{G}_n$, $m \neq n$, with $E_m \geq E_n$ (respectively, $E_m \leq E_n$) the inequality $\phi_{mn}(E) \leq 0$ (respectively, $\phi_{mn}(E) \geq 0$) holds, which implies that

$$\phi_{mn}(E)(E_n - E_m) \geq 0, \quad m \neq n. \tag{3.116}$$

Thus, it follows from (3.116) that energy (heat) flows naturally from more energetic subsystems (hot objects) to less energetic subsystems (cooler objects).

Of course, since in the isolated large-scale dynamical system $\mathcal{G}$ ectropy decreases if and only if entropy increases, the same result can be arrived at by considering the ectropy of $\mathcal{G}$. Furthermore, since Axiom *ii*) holds, it follows from the conservation of energy and the fact that the large-scale dynamical system $\mathcal{G}$ is strongly connected that nonconservation of entropy and ectropy necessarily implies energy equipartition.

Finally, we close this section by showing that our definition of entropy given by (3.61) satisfies the eight criteria established in [185] for the acceptance of an analytic expression for representing a system entropy function. In particular, note that for a dynamical system $\mathcal{G}$:

i) $\mathcal{S}(E)$ is well defined for every state $E \in \overline{\mathbb{R}}_+^q$ as long as $c > 0$.

ii) If $\mathcal{G}$ is adiabatically isolated, then $\mathcal{S}(E(t))$ is a nondecreasing function of time.

iii) If $\mathcal{S}_i(E_i) = \log_e(c + E_i) - \log_e c$ is the entropy of the ith subsystem of the system $\mathcal{G}$, then $\mathcal{S}(E) = \sum_{i=1}^q \mathcal{S}_i(E_i) = \mathbf{e}^{\mathrm{T}}\mathbf{log}_e(c\mathbf{e} + E) - q\log_e c$, and hence, the system entropy $\mathcal{S}(E)$ is an additive quantity over all subsystems.

iv) For the system $\mathcal{G}$, $\mathcal{S}(E) \geq 0$ for all $E \in \overline{\mathbb{R}}_+^q$.

v) It follows from Proposition 3.11 that for a given value $\beta \geq 0$ of the total energy of the system $\mathcal{G}$, one and only one state, namely, $E^* = \frac{\beta}{q}\mathbf{e}$, corresponds to the largest value of $\mathcal{S}(E)$.

vi) It follows from (3.61) that for the system $\mathcal{G}$, the graph of entropy versus energy is concave and smooth.

vii) For a composite large-scale dynamical system $\mathcal{G}_{\rm C}$ of two dynamical systems $\mathcal{G}_{\rm A}$ and $\mathcal{G}_{\rm B}$, the expression for the composite entropy $\mathcal{S}_{\rm C} = \mathcal{S}_{\rm A} + \mathcal{S}_{\rm B}$, where $\mathcal{S}_{\rm A}$ and $\mathcal{S}_{\rm B}$ are entropies of $\mathcal{G}_{\rm A}$ and $\mathcal{G}_{\rm B}$, respectively, is such that the expression for the equilibrium state where the composite maximum entropy is achieved is identical to those obtained for $\mathcal{G}_{\rm A}$ and $\mathcal{G}_{\rm B}$ individually. Specifically, if $q_{\rm A}$ and $q_{\rm B}$ denote the number of subsystems in $\mathcal{G}_{\rm A}$ and $\mathcal{G}_{\rm B}$, respectively, and $\beta_{\rm A}$ and $\beta_{\rm B}$ denote the total energies of $\mathcal{G}_{\rm A}$ and $\mathcal{G}_{\rm B}$, respectively, then the maximum entropy of $\mathcal{G}_{\rm A}$ and $\mathcal{G}_{\rm B}$ individually is achieved at $E_{\rm A}^* = \frac{\beta_{\rm A}}{q_{\rm A}}\mathbf{e}$ and $E_{\rm B}^* = \frac{\beta_{\rm B}}{q_{\rm B}}\mathbf{e}$, respectively, while the maximum entropy of the composite system $\mathcal{G}_{\rm C}$ is achieved at $E_{\rm C}^* = \frac{\beta_{\rm A}+\beta_{\rm B}}{q_{\rm A}+q_{\rm B}}\mathbf{e}$.

viii) It follows from Theorem 3.9 that for a stable equilibrium state $E = \frac{\beta}{q}\mathbf{e}$, where $\beta \geq 0$ is the total energy of the system $\mathcal{G}$ and q is the number of subsystems of $\mathcal{G}$, the entropy is totally defined by β and q, that is, $\mathcal{S}(E) = q \log_e(c + \frac{\beta}{q}) - q \log_e c$.

Dual criteria to the eight criteria outlined above can also be established for an analytic expression representing system ectropy.

3.7 Interconnections of Thermodynamic Systems

In classical thermodynamics, it is not clear how the environment can be described in thermodynamic terms, and it is often not addressed by the theory. This is the case, for example, in the work of Carnot, Clausius, and Kelvin, wherein they consider the engine performing a cycle as the dynamical system and the reservoir belonging to the environment. This is in contrast to Planck, who views the reservoir as a thermodynamic system with a finite energy content and with the engine belonging to the environment. In this case, the thermodynamic state of the reservoir can change, and hence, the removal of energy from the reservoir via a cyclic engine need not be repeatable. Instead, the definition of a heat reservoir as formulated by Carnot and Kelvin is not at all transparent in the context of dynamical systems theory since it can absorb or emit a finite amount of energy (heat) without changing its thermal (temperature) and deformation (volume) states, and hence, it possesses an infinite heat capacity.

The issue is then whether the thermodynamic state of an infinite heat reservoir changes as it exchanges a finite quantity of heat with the dynamical system. This is not a trivial issue, and its assessment within a given theory has been one of the key demarcation points between physics and mathematics [447, p. 98]. Dynamical systems theory makes no claim about the properties of any system larger than the system under consideration, and any exogenous effects are typically treated as external disturbances. Thus, the effect of the heat reservoir on the thermodynamic process can be viewed as an external disturbance to the system giving rise to an input-output open dynamical system.

In contrast, in classical mechanics it is always possible, at least in principle, to include interactions with the environment via *feedback* interconnecting components, consistent with Newton's third law, to obtain an augmented closed-loop feedback system. A feedback system consists of an interconnection of two systems, a *forward loop* system and a *feedback loop* system. The forward loop system is driven by an input and produces an output that serves as the input to the feedback loop system. The output of the feedback loop system, in turn, serves as the input to the forward loop system. Feedback systems are pervasive in nature and are ideal in capturing the behavior of interconnected dynamical systems.

To harmonize thermodynamics with mechanics, in this section we consider feedback interconnections of two thermodynamically consistent large-scale systems. This interconnection can correspond to a large-scale dynamical system with the environment or an interconnection between two large-scale dynamical systems. In the case where one of the systems corresponds to the environment, this formulation allows us to formally assign a state to the environment. In either case, we show that, under thermodynamically consistent assumptions on the feedback interconnection structure, the closed-loop system is guaranteed to be a thermodynamically consistent system, wherein the energy flow between the large-scale systems is due to the energy inflows and outflows of both systems.

Specifically, consider the large-scale dynamical system $\mathcal{G}$ with power balance equation

$$\dot{E}(t) = f(E(t)) - d(E(t)) + S(t), \quad E(t_0) = E_0, \quad t \geq t_0, \tag{3.117}$$

and outflow

$$y(t) = h(E(t)), \quad t \geq t_0, \tag{3.118}$$

and consider the large-scale dynamical system $\mathcal{G}_{\mathrm{c}}$ with power balance equation

$$\dot{E}_{\mathrm{c}}(t) = f_{\mathrm{c}}(E_{\mathrm{c}}(t)) - d_{\mathrm{c}}(E_{\mathrm{c}}(t)) + S_{\mathrm{c}}(t), \quad E_{\mathrm{c}}(t_0) = E_{\mathrm{c}0}, \quad t \geq t_0, \tag{3.119}$$

and outflow

$$y_{\mathrm{c}}(t) = h_{\mathrm{c}}(E_{\mathrm{c}}(t)), \quad t \geq t_0, \tag{3.120}$$

where $E(t) = [E_1(t), \ldots, E_q(t)]^{\mathrm{T}} \in \overline{\mathbb{R}}_+^q$, $t \geq t_0$, $E_{\mathrm{c}}(t) = [E_{\mathrm{c}1}(t), \ldots, E_{\mathrm{c}q_{\mathrm{c}}}(t)]^{\mathrm{T}} \in \overline{\mathbb{R}}_+^{q_{\mathrm{c}}}$, $t \geq t_0$, $f \triangleq [f_1, \ldots, f_q]^{\mathrm{T}} : \overline{\mathbb{R}}_+^q \to \mathbb{R}^q$, $f_i(E) = \sum_{j=1, j\neq i}^{q} [\sigma_{ij}(E) - \sigma_{ji}(E)]$, $E \in \overline{\mathbb{R}}_+^q$, $i = 1, \ldots, q$, $f_{\mathrm{c}} \triangleq [f_{\mathrm{c}1}, \ldots, f_{\mathrm{c}q_{\mathrm{c}}}]^{\mathrm{T}} : \overline{\mathbb{R}}_+^{q_{\mathrm{c}}} \to \mathbb{R}^{q_{\mathrm{c}}}$, $f_{\mathrm{c}i}(E_{\mathrm{c}}) = \sum_{j=1, j\neq i}^{q_{\mathrm{c}}} [\sigma_{\mathrm{c}ij}(E_{\mathrm{c}}) - \sigma_{\mathrm{c}ji}(E_{\mathrm{c}})]$, $E_{\mathrm{c}} \in \overline{\mathbb{R}}_+^{q_{\mathrm{c}}}$, $i = 1, \ldots, q_{\mathrm{c}}$, $d(E) \triangleq [\sigma_{11}(E), \ldots, \sigma_{qq}(E)]^{\mathrm{T}}$, $E \in \overline{\mathbb{R}}_+^q$, $d_{\mathrm{c}}(E_{\mathrm{c}}) \triangleq [\sigma_{\mathrm{c}11}(E_{\mathrm{c}}), \ldots, \sigma_{\mathrm{c}q_{\mathrm{c}}q_{\mathrm{c}}}(E_{\mathrm{c}})]^{\mathrm{T}}$, $E_{\mathrm{c}} \in \overline{\mathbb{R}}_+^{q_{\mathrm{c}}}$, $S = [S_1, \ldots, S_q]^{\mathrm{T}} : [0, \infty) \to \mathbb{R}^q$, $S_{\mathrm{c}} = [S_{\mathrm{c}1}, \ldots, S_{\mathrm{c}q_{\mathrm{c}}}]^{\mathrm{T}} : [0, \infty) \to \mathbb{R}^{q_{\mathrm{c}}}$, $h = [h_1, \ldots, h_{q_{\mathrm{c}}}]^{\mathrm{T}} : \overline{\mathbb{R}}_+^q \to \overline{\mathbb{R}}_+^{q_{\mathrm{c}}}$, and $h_{\mathrm{c}} = [h_{\mathrm{c}1}, \ldots, h_{\mathrm{c}q}]^{\mathrm{T}} : \overline{\mathbb{R}}_+^{q_{\mathrm{c}}} \to \overline{\mathbb{R}}_+^q$.

To ensure a feedback model consistent with energy conservation laws, the output power functions $h_i(\cdot)$, $i = 1, \ldots, q_{\mathrm{c}}$, and $h_{\mathrm{c}i}(\cdot)$, $i = 1, \ldots, q$, are given by

$$h_i(E) = \sum_{j=1}^{q} \eta_{ij}(E), \quad E \in \overline{\mathbb{R}}_+^q, \quad i = 1, \ldots, q_{\mathrm{c}}, \tag{3.121}$$

and

$$h_{\mathrm{c}i}(E_{\mathrm{c}}) = \sum_{j=1}^{q_{\mathrm{c}}} \eta_{\mathrm{c}ij}(E_{\mathrm{c}}), \quad E_{\mathrm{c}} \in \overline{\mathbb{R}}_+^{q_{\mathrm{c}}}, \quad i = 1, \ldots, q, \tag{3.122}$$

where $\eta_{ij} : \overline{\mathbb{R}}_+^q \to \overline{\mathbb{R}}_+$ denotes the rate of energy flow from the jth subsystem of $\mathcal{G}$ to the ith subsystem of $\mathcal{G}_{\mathrm{c}}$ and $\eta_{\mathrm{c}ij} : \overline{\mathbb{R}}_+^{q_{\mathrm{c}}} \to \overline{\mathbb{R}}_+$ denotes the rate of energy flow from the jth subsystem of $\mathcal{G}_{\mathrm{c}}$ to the ith subsystem of $\mathcal{G}$. The functions $\sigma_{ij} : \overline{\mathbb{R}}_+^q \to \overline{\mathbb{R}}_+$, $i, j = 1, \ldots, q$, have the same meaning and properties as defined in Section 3.2. The function $\sigma_{\mathrm{c}ij} : \overline{\mathbb{R}}_+^{q_{\mathrm{c}}} \to \overline{\mathbb{R}}_+$, $i \neq j$, $i, j = 1, \ldots, q_{\mathrm{c}}$, denotes the rate of energy flow from subsystem $\mathcal{G}_{\mathrm{c}j}$ to subsystem $\mathcal{G}_{\mathrm{c}i}$ of $\mathcal{G}_{\mathrm{c}}$, and $\sigma_{\mathrm{c}ii} : \overline{\mathbb{R}}_+^{q_{\mathrm{c}}} \to \overline{\mathbb{R}}_+$, $i = 1, \ldots, q_{\mathrm{c}}$, denotes the rate of energy dissipation from the subsystem $\mathcal{G}_{\mathrm{c}i}$.

In this case, conservation of energy implies that

$$\sigma_{ii}(E) = \sum_{j=1}^{q_{\mathrm{c}}} \eta_{ji}(E), \quad E \in \overline{\mathbb{R}}_+^q, \quad i = 1, \ldots, q, \tag{3.123}$$

$$\sigma_{\mathrm{c}ii}(E_{\mathrm{c}}) = \sum_{j=1}^{q} \eta_{\mathrm{c}ji}(E_{\mathrm{c}}), \quad E_{\mathrm{c}} \in \overline{\mathbb{R}}_+^{q_{\mathrm{c}}}, \quad i = 1, \ldots, q_{\mathrm{c}}, \tag{3.124}$$

and hence,

$$\sum_{i=1}^{q} \sigma_{ii}(E) = \mathbf{e}^{\mathrm{T}} d(E) = \mathbf{e}_{\mathrm{c}}^{\mathrm{T}} h(E), \quad E \in \overline{\mathbb{R}}_+^q, \tag{3.125}$$

$$\sum_{i=1}^{q_{\mathrm{c}}} \sigma_{\mathrm{c}ii}(E_{\mathrm{c}}) = \mathbf{e}_{\mathrm{c}}^{\mathrm{T}} d_{\mathrm{c}}(E_{\mathrm{c}}) = \mathbf{e}^{\mathrm{T}} h_{\mathrm{c}}(E_{\mathrm{c}}), \quad E_{\mathrm{c}} \in \overline{\mathbb{R}}_+^{q_{\mathrm{c}}}, \tag{3.126}$$

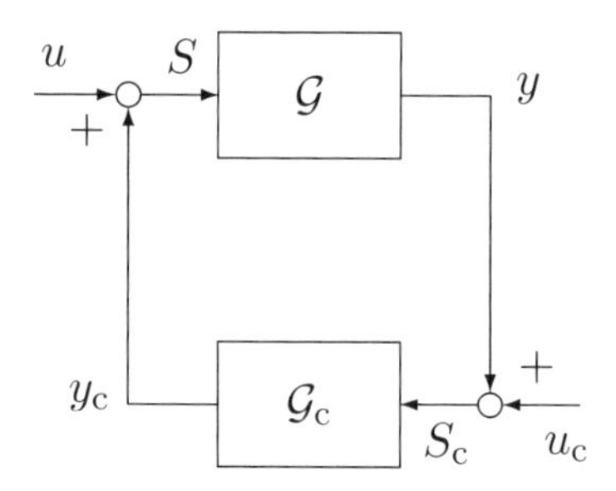

Figure 3.3 Feedback interconnection of large-scale systems $\mathcal{G}$ and $\mathcal{G}_c$.

where $\mathbf{e}$ and $\mathbf{e}_c$ denote ones vectors of compatible dimensions. Here, we assume that $\sigma_{cij} : \overline{\mathbb{R}}_+^{q_c} \to \overline{\mathbb{R}}_+$, $i,j = 1,\ldots,q_c$, and $\eta_{cij} : \overline{\mathbb{R}}_+^{q_c} \to \overline{\mathbb{R}}_+$, $i = 1,\ldots,q$, $j = 1,\ldots,q_c$, are locally Lipschitz continuous on $\overline{\mathbb{R}}_+^{q_c}$, and $\sigma_{ij} : \overline{\mathbb{R}}_+^q \to \overline{\mathbb{R}}_+$, $i,j = 1,\ldots,q$, and $\eta_{ij} : \overline{\mathbb{R}}_+^q \to \overline{\mathbb{R}}_+$, $i = 1,\ldots,q_c$, $j = 1,\ldots,q$, are locally Lipschitz continuous on $\overline{\mathbb{R}}_+^q$.

Next, consider the positive feedback interconnection of the large-scale dynamical systems $\mathcal{G}$ and $\mathcal{G}_c$ shown in Figure 3.3 with $S = u + y_c$ and $S_c = u_c + y$, where $u \in \mathcal{U}$ and $u_c \in \mathcal{U}_{cc}$ are external power inflows to $\mathcal{G}$ and $\mathcal{G}_c$, respectively. Here, $\mathcal{U}$ is a subset of bounded continuous $\mathbb{R}^q$-valued functions on $\mathbb{R}$, and $\mathcal{U}_{cc}$ is a subset of bounded continuous $\mathbb{R}^{q_c}$-valued functions on $\mathbb{R}$.

In this case, the closed-loop system dynamics are given by

$$\dot{\tilde{E}}(t) = \tilde{f}(\tilde{E}(t)) + \tilde{u}(t), \quad \tilde{E}(t_0) = \tilde{E}_0, \quad t \geq t_0, \tag{3.127}$$

where $\tilde{E}(t) \triangleq [E^{\mathrm{T}}(t), E_c^{\mathrm{T}}(t)]^{\mathrm{T}} \in \overline{\mathbb{R}}_+^{\tilde{q}}$, $t \geq t_0$, $\tilde{q} \triangleq q+q_c$, $\tilde{u}(t) \triangleq [u^{\mathrm{T}}(t), u_c^{\mathrm{T}}(t)]^{\mathrm{T}} \in \mathbb{R}^{\tilde{q}}$, $t \geq t_0$, and

$$\tilde{f}(\tilde{E}) \triangleq \begin{bmatrix} f(E) - d(E) + h_c(E_c) \\ f_c(E_c) - d_c(E_c) + h(E) \end{bmatrix}, \quad \tilde{E} \in \overline{\mathbb{R}}_+^{\tilde{q}}. \tag{3.128}$$

Note that the function $\tilde{f} : \overline{\mathbb{R}}_+^{\tilde{q}} \to \mathbb{R}^{\tilde{q}}$ is essentially nonnegative, and hence, it follows from Proposition 2.1 that the solution to (3.127) with $\tilde{u}(t) \equiv 0$ is nonnegative for all nonnegative initial conditions. Furthermore, it follows from (3.125) and (3.126) that $\tilde{\mathbf{e}}^{\mathrm{T}}\tilde{f}(\tilde{E}) = 0$, $\tilde{E} \in \overline{\mathbb{R}}_+^{\tilde{q}}$, where $\tilde{\mathbf{e}} \triangleq [\mathbf{e}^{\mathrm{T}}, \mathbf{e}_c^{\mathrm{T}}]^{\mathrm{T}}$, and hence, energy is conserved in the isolated (i.e., $u(t) \equiv 0$ and $u_c(t) \equiv 0$) closed-loop feedback system consisting of $\mathcal{G}$ and $\mathcal{G}_c$.

Next, let $\tilde{\phi}_{ij}(\tilde{E}) \triangleq \eta_{ij}(E) - \eta_{cji}(E_c)$, $\tilde{E} \in \overline{\mathbb{R}}_+^{\tilde{q}}$, denote the energy flow from the jth subsystem of $\mathcal{G}$ to the ith subsystem of $\mathcal{G}_c$. To ensure that the closed-loop system possesses a thermodynamically consistent energy flow model, we require the following axioms on the interconnection structure for $\mathcal{G}$ and $\mathcal{G}_c$.

Axiom *iv*): $\mathcal{K} \triangleq \{(i,j) \in \{1,\ldots,q_c\} \times \{1,\ldots,q\} : \tilde{\phi}_{ij}(\tilde{E}) = 0$ if and only if $E_{ci} = E_j\} \neq \emptyset$, and if $(i,j) \notin \mathcal{K}$, then $\tilde{\phi}_{ij}(\tilde{E}) \equiv 0$.

Axiom *v*): If $(i,j) \in \mathcal{K}$, then $\tilde{\phi}_{ij}(\tilde{E})(E_{ci} - E_j) \leq 0$, $\tilde{E} \in \overline{\mathbb{R}}_+^{\tilde{q}}$.

Axiom *iv*) is analogous to Axiom *i*) and ensures that the dynamical systems $\mathcal{G}$ and $\mathcal{G}_c$ are strongly connected, that is, for any given ordered pair of subsystems $(\mathcal{G}_j, \mathcal{G}_{ci})$, $i = 1,\ldots,q_c$, $j = 1,\ldots,q$, there exists a path connecting this ordered pair. In other words, the rank of the connectivity matrix $\tilde{\mathcal{C}}$ associated with the feedback interconnection of $\mathcal{G}$ and $\mathcal{G}_c$ is equal to $\tilde{q} - 1$. Axiom *v*) is analogous to Axiom *ii*) and ensures that for any two connected subsystems $\mathcal{G}_j$ and $\mathcal{G}_{ci}$, $i = 1,\ldots,q_c$, $j = 1,\ldots,q$, such that $(i,j) \in \mathcal{K}$, energy flows from higher to lower energy levels. It follows from Axioms *i*) and *iv*) that $\tilde{E} = \alpha\tilde{\mathbf{e}} \in \overline{\mathbb{R}}_+^{\tilde{q}}$, $\alpha \geq 0$, is an equilibrium state of the isolated closed-loop system (3.127) given by the feedback interconnection of $\mathcal{G}$ and $\mathcal{G}_c$.

The next theorem establishes that the equilibrium state $\alpha\tilde{\mathbf{e}} \in \overline{\mathbb{R}}_+^{\tilde{q}}$, $\alpha \geq 0$, of the isolated closed-loop system (3.127) is semistable.

Theorem 3.11. Consider the positive feedback system consisting of the large-scale dynamical systems $\mathcal{G}$ and $\mathcal{G}_c$ given by (3.117)–(3.120) with $u(t) \equiv 0$ and $u_c(t) \equiv 0$, and assume that Axioms *i*), *ii*), *iv*), and *v*) hold. Then, for every $\alpha \geq 0$, $\alpha\tilde{\mathbf{e}} \in \overline{\mathbb{R}}_+^{\tilde{q}}$ is a semistable equilibrium state of (3.127). Furthermore, $\tilde{E}(t) \to \frac{1}{\tilde{q}}\tilde{\mathbf{e}}(\mathbf{e}^{\mathrm{T}}E(t_0) + \mathbf{e}_c^{\mathrm{T}}E_c(t_0))$ as $t \to \infty$ and $\frac{1}{\tilde{q}}\tilde{\mathbf{e}}(\mathbf{e}^{\mathrm{T}}E(t_0) + \mathbf{e}_c^{\mathrm{T}}E_c(t_0))$ is a semistable equilibrium state.

Proof. It follows from Axioms *i*) and *iv*) that $\alpha\tilde{\mathbf{e}} \in \overline{\mathbb{R}}_+^{\tilde{q}}$, $\alpha \geq 0$, is an equilibrium state of the isolated system (3.127). Next, consider the Lyapunov function candidate composed of the sum of the shifted ectropy functions of $\mathcal{G}$ and $\mathcal{G}_c$ given by

$$\tilde{\mathcal{E}}_s(\tilde{E}) = \tfrac{1}{2}(E - \alpha\mathbf{e})^{\mathrm{T}}(E - \alpha\mathbf{e}) + \tfrac{1}{2}(E_c - \alpha\mathbf{e}_c)^{\mathrm{T}}(E_c - \alpha\mathbf{e}_c), \quad \tilde{E} \in \overline{\mathbb{R}}_+^{\tilde{q}}. \tag{3.129}$$

Computing the Lyapunov derivative of (3.129) and using Axioms *i*), *ii*), *iv*), and *v*), it follows that

$$\dot{\tilde{\mathcal{E}}}_s(\tilde{E}) = (E - \alpha\mathbf{e})^{\mathrm{T}}[f(E) - d(E) + h_c(E_c)]$$

$$
\begin{aligned}
&+(E_{\mathrm{c}}-\alpha\mathbf{e}_{\mathrm{c}})^{\mathrm{T}}[f_{\mathrm{c}}(E_{\mathrm{c}})-d_{\mathrm{c}}(E_{\mathrm{c}})+h(E)]\\
&=\sum_{i=1}^{q}\sum_{j=i+1}^{q}\phi_{ij}(E)(E_i-E_j)+\sum_{i=1}^{q_{\mathrm{c}}}\sum_{j=i+1}^{q_{\mathrm{c}}}\phi_{\mathrm{c}ij}(E_{\mathrm{c}})(E_{\mathrm{c}i}-E_{\mathrm{c}j})\\
&\quad+\sum_{i=1}^{q}\sum_{j=1}^{q_{\mathrm{c}}}\tilde{\phi}_{ji}(\tilde{E})(E_{\mathrm{c}j}-E_i)\\
&\leq 0,\quad \tilde{E}\in\overline{\mathbb{R}}_+^{\tilde{q}},
\end{aligned}
\tag{3.130}
$$

which establishes Lyapunov stability of the equilibrium state $\alpha\tilde{\mathbf{e}}\in\overline{\mathbb{R}}_+^{\tilde{q}}$, $\alpha\geq 0$, of the feedback interconnection of $\mathcal{G}$ and $\mathcal{G}_{\mathrm{c}}$.

Next, define the set $\mathcal{R}\triangleq\{\tilde{E}\in\overline{\mathbb{R}}_+^{\tilde{q}}:\dot{\mathcal{E}}_{\mathrm{s}}(\tilde{E})=0\}$, and note that it follows from Axioms i), ii), iv), and v) that $\mathcal{R}=\{\tilde{E}\in\overline{\mathbb{R}}_+^{\tilde{q}}:E_1=\cdots=E_q\}\cap\{\tilde{E}\in\overline{\mathbb{R}}_+^{\tilde{q}}:E_{\mathrm{c}1}=\cdots=E_{\mathrm{c}q_{\mathrm{c}}}\}\cap\{\tilde{E}\in\overline{\mathbb{R}}_+^{\tilde{q}}:E_{\mathrm{c}j}=E_i,\ (j,i)\in\mathcal{K}\}=\{\tilde{E}\in\overline{\mathbb{R}}_+^{\tilde{q}}:E_1=\cdots=E_q=E_{\mathrm{c}1}=\cdots=E_{\mathrm{c}q_{\mathrm{c}}}\}$. Since $\mathcal{R}$ consists of only equilibrium states of the feedback interconnection of $\mathcal{G}$ and $\mathcal{G}_{\mathrm{c}}$, it follows that the largest invariant set $\mathcal{M}$ contained in $\mathcal{R}$ is given by $\mathcal{M}=\mathcal{R}$. Thus, it follows from the Krasovskii-LaSalle invariant set theorem that for every initial condition $\tilde{E}(t_0)\in\overline{\mathbb{R}}_+^{\tilde{q}}$, $\tilde{E}(t)\to\mathcal{M}$ as $t\to\infty$, and hence, $\alpha\tilde{\mathbf{e}}\in\overline{\mathbb{R}}_+^{\tilde{q}}$ is a semistable equilibrium state of the isolated system (3.127).

Finally, note that since $(\mathbf{e}^{\mathrm{T}}E(t)+\mathbf{e}_{\mathrm{c}}^{\mathrm{T}}E_{\mathrm{c}}(t))=(\mathbf{e}^{\mathrm{T}}E(t_0)+\mathbf{e}_{\mathrm{c}}^{\mathrm{T}}E_{\mathrm{c}}(t_0))$, $t\geq t_0$, and $\tilde{E}(t)\to\mathcal{M}$ as $t\to\infty$, it follows that $\tilde{E}(t)\to\frac{1}{\tilde{q}}\tilde{\mathbf{e}}(\mathbf{e}^{\mathrm{T}}E(t_0)+\mathbf{e}_{\mathrm{c}}^{\mathrm{T}}E_{\mathrm{c}}(t_0))$ as $t\to\infty$. Hence, with $\alpha=\frac{1}{\tilde{q}}(\mathbf{e}^{\mathrm{T}}E(t_0)+\mathbf{e}_{\mathrm{c}}^{\mathrm{T}}E_{\mathrm{c}}(t_0))$, $\alpha\tilde{\mathbf{e}}=\frac{1}{\tilde{q}}\tilde{\mathbf{e}}(\mathbf{e}^{\mathrm{T}}E(t_0)+\mathbf{e}_{\mathrm{c}}^{\mathrm{T}}E_{\mathrm{c}}(t_0))$ is a semistable equilibrium state of the isolated closed-loop system (3.127). □

The entropy and ectropy functions for the closed-loop system (3.127) can be constructed by appropriately combining the entropy and ectropy functions of $\mathcal{G}$ and $\mathcal{G}_{\mathrm{c}}$. In particular, the ectropy of the closed-loop system (3.127) is the sum of the individual ectropy functions of $\mathcal{G}$ and $\mathcal{G}_{\mathrm{c}}$ and is given by

$$
\tilde{\mathcal{E}}(\tilde{E})=\mathcal{E}(E)+\mathcal{E}_{\mathrm{c}}(E_{\mathrm{c}})=\tfrac{1}{2}E^{\mathrm{T}}E+\tfrac{1}{2}E_{\mathrm{c}}^{\mathrm{T}}E_{\mathrm{c}}=\tfrac{1}{2}\tilde{E}^{\mathrm{T}}\tilde{E},\quad \tilde{E}\in\overline{\mathbb{R}}_+^{\tilde{q}}.
\tag{3.131}
$$

Hence, it follows that

$$
\dot{\tilde{\mathcal{E}}}(\tilde{E})=\sum_{i=1}^{q}\sum_{j=i+1}^{q}\phi_{ij}(E)(E_i-E_j)+\sum_{i=1}^{q_{\mathrm{c}}}\sum_{j=i+1}^{q_{\mathrm{c}}}\phi_{\mathrm{c}ij}(E_{\mathrm{c}})(E_{\mathrm{c}i}-E_{\mathrm{c}j})
$$

$$
\begin{aligned}
&+\sum_{i=1}^{q}\sum_{j=1}^{q_{\mathrm{c}}}\tilde{\phi}_{ji}(\tilde{E})(E_{\mathrm{c}j}-E_i) \\
&\leq 0, \quad \tilde{E}\in\overline{\mathbb{R}}_+^{\tilde{q}}, \qquad (3.132)
\end{aligned}
$$

which shows that $\tilde{\mathcal{E}}(\tilde{E})$ is a nonincreasing function of time. Note that since the closed-loop system (3.127) satisfies Axioms *i*), *ii*), *iv*), and *v*), it follows from Theorem 3.8 that $\tilde{\mathcal{E}}(\tilde{E})$, $\tilde{E}\in\overline{\mathbb{R}}_+^{q}$, is a unique, continuously differentiable ectropy function of (3.127).

Similarly, the entropy of the isolated closed-loop system (3.127) is the sum of the individual entropy functions of $\mathcal{G}$ and $\mathcal{G}_{\mathrm{c}}$ and is given by

$$
\begin{aligned}
\tilde{\mathcal{S}}(\tilde{E}) &= \mathcal{S}(E)+\mathcal{S}_{\mathrm{c}}(E_{\mathrm{c}}) \\
&= \mathbf{e}^{\mathrm{T}}\mathbf{log}(c\mathbf{e}+E)+\mathbf{e}_{\mathrm{c}}^{\mathrm{T}}\mathbf{log}(c\mathbf{e}_{\mathrm{c}}+E_{\mathrm{c}})-\tilde{q}\log_e c, \quad \tilde{E}\in\overline{\mathbb{R}}_+^{\tilde{q}}. \qquad (3.133)
\end{aligned}
$$

Hence, it follows that

$$
\begin{aligned}
\dot{\tilde{\mathcal{S}}}(\tilde{E}) &= \sum_{i=1}^{q}\sum_{j=i+1}^{q}\frac{\phi_{ij}(E)(E_j-E_i)}{(c+E_i)(c+E_j)}+\sum_{i=1}^{q_{\mathrm{c}}}\sum_{j=i+1}^{q_{\mathrm{c}}}\frac{\phi_{\mathrm{c}ij}(E_{\mathrm{c}})(E_{\mathrm{c}j}-E_{\mathrm{c}i})}{(c+E_{\mathrm{c}i})(c+E_{\mathrm{c}j})} \\
&\quad+\sum_{i=1}^{q}\sum_{j=1}^{q_{\mathrm{c}}}\frac{\tilde{\phi}_{ji}(\tilde{E})(E_i-E_{\mathrm{c}j})}{(c+E_i)(c+E_{\mathrm{c}j})} \\
&\geq 0, \quad \tilde{E}\in\overline{\mathbb{R}}_+^{\tilde{q}}. \qquad (3.134)
\end{aligned}
$$

As in the case of ectropy, it follows from Theorem 3.4 that $\tilde{\mathcal{S}}(\tilde{E})$, $\tilde{E}\in\overline{\mathbb{R}}_+^{\tilde{q}}$, is a unique, continuously differentiable entropy function for the closed-loop system (3.127). Finally, note that it follows from Axioms *i*), *ii*), *iv*), and *v*) that the entropy (respectively, ectropy) of the closed-loop system (3.127) given by (3.133) (respectively, (3.131)) satisfies (3.134) (respectively, (3.132)) as a strict inequality for $\tilde{E}(t)\notin\mathcal{M}_{\mathrm{e}}$, $t\geq t_0$, and as an equality for $\tilde{E}(t)\in\mathcal{M}_{\mathrm{e}}$, $t\geq t_0$, where $\mathcal{M}_{\mathrm{e}}=\{\tilde{E}\in\overline{\mathbb{R}}_+^{\tilde{q}}:\alpha\tilde{\mathbf{e}},\ \alpha\geq 0\}$. Hence, it follows from Theorem 3.10, as applied to the isolated closed-loop system (3.127), that (3.127) is state irreversible for all nontrivial trajectories.

3.8 Monotonicity of System Energies in Thermodynamic Processes

Even though Theorem 3.9 gives sufficient conditions under which the subsystem energies in the large-scale dynamical system $\mathcal{G}$ converge, these subsystem energies may exhibit an oscillatory (hyperbolic) or nonmonotonic behavior prior to convergence. For certain thermodynamical processes, it is desirable to identify system models that guarantee monotonicity of the system energy flows. It is important to note that monotonicity of solutions does not necessarily imply Axiom *ii*), nor does Axiom *ii*) imply monotonicity of solutions. These are two disjoint notions. In this section, we give necessary and sufficient conditions under which the solutions to (3.5) are monotonic.

To develop necessary and sufficient conditions for monotonicity of solutions, note that the power balance equation (3.5) for the large-scale dynamical system $\mathcal{G}$ can be written as

$$\dot{E}(t) = [\mathcal{J}(E(t)) - \mathcal{D}(E(t))]\left(\frac{\partial \mathcal{H}}{\partial E}(E(t))\right)^{\mathrm{T}} + GS(t), \quad E(t_0) = E_0, \quad t \geq t_0, \tag{3.135}$$

where $E(t) \in \overline{\mathbb{R}}_+^q$, $\mathcal{H}(E) = \mathbf{e}^{\mathrm{T}}E$, $S(t) = [S_1(t), \ldots, S_q(t)]^{\mathrm{T}}$, $t \geq t_0$, $\mathcal{J}(E)$ is a skew-symmetric matrix function with $\mathcal{J}_{(i,i)}(E) = 0$ and $\mathcal{J}_{(i,j)}(E) = \sigma_{ij}(E) - \sigma_{ji}(E)$, $i \neq j$, $i,j = 1,\ldots,q$, $\mathcal{D}(E) = \mathrm{diag}[\sigma_{11}(E), \ldots, \sigma_{qq}(E)] \geq 0$, and $G \in \mathbb{R}^{q\times q}$ is a diagonal input matrix that has been included for generality and contains zeros and ones as its entries. Hence, the power balance equation of the large-scale dynamical system $\mathcal{G}$ has a *port-controlled Hamiltonian* structure [291] with a Hamiltonian function $\mathcal{H}(E) = \mathbf{e}^{\mathrm{T}}E = \sum_{i=1}^{q} E_i$ representing the sum of all subsystem energies, $\mathcal{D}(E)$ representing power dissipation in the subsystems, $\mathcal{J}(E) = -\mathcal{J}^{\mathrm{T}}(E)$ representing energy-conserving subsystem coupling, and $S(t)$, $t \geq t_0$, representing supplied system power.

As noted in Section 3.3, the nonlinear power balance equation (3.135) can exhibit a full range of nonlinear behavior, including bifurcations, limit cycles, and even chaos. However, a thermodynamically consistent energy flow model ensures that the evolution of the system energy is diffusive in character with convergent subsystem energies. As further shown in Section 3.3, Axioms *i*) and *ii*) guarantee a thermodynamically consistent energy flow model.

In order to guarantee a thermodynamically consistent energy flow model, we assume Axiom *ii*) holds and seek solutions to (3.135) that exhibit a monotonic behavior of the subsystem energies. This would physically imply that the energy of a subsystem whose initial energy is greater than the average system energy will decrease, while the energy of a subsystem whose initial energy is less than the average system energy will increase. This of course is consistent with the second law of thermodynamics with the additional constraint of monotonic heat flows. The following definition is needed.

Definition 3.4. Consider the large-scale dynamical system $\mathcal{G}$ with power balance equation (3.135). The subsystem energies $E(t)$, $t \geq t_0$, of $\mathcal{G}$ are *monotonic* for all $E_0 \in \mathcal{D}_c \subseteq \overline{\mathbb{R}}_+^q$, where $\mathcal{D}_c$ is a positively invariant set with respect to (3.135), if there exists a weighting matrix $R \in \mathbb{R}^{q\times q}$ such that $R = \operatorname{diag}[r_1, \ldots, r_q]$, $r_i = \pm 1$, $i = 1, \ldots, q$, and, for every $E_0 \in \mathcal{D}_c \subseteq \overline{\mathbb{R}}_+^q$, $RE(t_2) \leq\leq RE(t_1)$, $t_0 \leq t_1 \leq t_2$.

The following result presents necessary and sufficient conditions that guarantee that the subsystem energies of the large-scale dynamical system $\mathcal{G}$ are monotonic. It is important to note that this result holds whether or not Axiom *ii*) holds.

Theorem 3.12. Consider the large-scale dynamical system $\mathcal{G}$ with power balance equation (3.135). Then the following statements hold:

i) If $S(t) \geq\geq 0$, $t \geq t_0$, and there exists a matrix $R \in \mathbb{R}^{q\times q}$ such that $R = \operatorname{diag}[r_1, \ldots, r_q]$, $r_i = \pm 1$, $i = 1, \ldots, q$, $R[\mathcal{J}(E) - \mathcal{D}(E)](\frac{\partial \mathcal{H}}{\partial E}(E))^{\mathrm{T}} \leq\leq 0$, $E \in \overline{\mathbb{R}}_+^q$, and $RG \leq\leq 0$, then the subsystem energies $E(t)$, $t \geq t_0$, of $\mathcal{G}$ are monotonic for all $E_0 \in \overline{\mathbb{R}}_+^q$.

ii) Let $S(t) \equiv 0$ and let $\mathcal{D}_c \subseteq \overline{\mathbb{R}}_+^q$ be a positively invariant set with respect to (3.135). Then the subsystem energies $E(t)$, $t \geq t_0$, of $\mathcal{G}$ are monotonic for all $E_0 \in \mathcal{D}_c \subseteq \overline{\mathbb{R}}_+^q$ if and only if there exists a matrix $R \in \mathbb{R}^{q\times q}$ such that $R = \operatorname{diag}[r_1, \ldots, r_q]$, $r_i = \pm 1$, $i = 1, \ldots, q$, and $R[\mathcal{J}(E) - \mathcal{D}(E)](\frac{\partial \mathcal{H}}{\partial E}(E))^{\mathrm{T}} \leq\leq 0$, $E \in \mathcal{D}_c \subseteq \overline{\mathbb{R}}_+^q$.

Proof. *i*) Let $S(t) \geq\geq 0$, $t \geq t_0$, and assume there exists $R = \operatorname{diag}[r_1, \ldots, r_q]$, $r_i = \pm 1$, $i = 1, \ldots, q$, such that

$$R[\mathcal{J}(E) - \mathcal{D}(E)]\left(\frac{\partial \mathcal{H}}{\partial E}(E)\right)^{\mathrm{T}} \leq\leq 0, \quad E \in \overline{\mathbb{R}}_+^q.$$

Now, it follows from (3.135) that

$$R\dot{E}(t) = R[\mathcal{J}(E(t)) - \mathcal{D}(E(t))]\left(\frac{\partial \mathcal{H}}{\partial E}(E(t))\right)^{\mathrm{T}} + RGS(t),$$
$$E(t_0) = E_0, \quad t \geq t_0, \qquad (3.136)$$

which further implies that

$$RE(t_2) = RE(t_1) + \int_{t_1}^{t_2} R[\mathcal{J}(E(t)) - \mathcal{D}(E(t))]\left(\frac{\partial \mathcal{H}}{\partial E}(E(t))\right)^{\mathrm{T}} \mathrm{d}t$$
$$+ \int_{t_1}^{t_2} RGS(t)\mathrm{d}t. \qquad (3.137)$$

Next, since $[\mathcal{J}(E) - \mathcal{D}(E)](\frac{\partial \mathcal{H}}{\partial E}(E))^{\mathrm{T}}$ is essentially nonnegative and $S(t) \geq\geq 0$, $t \geq t_0$, it follows from Proposition 3.1 that $E(t) \geq\geq 0$, $t \geq t_0$, for all $E_0 \in \overline{\mathbb{R}}_+^q$. Hence, since $R[\mathcal{J}(E) - \mathcal{D}(E)](\frac{\partial \mathcal{H}}{\partial E}(E))^{\mathrm{T}} \leq\leq 0$, $E \in \overline{\mathbb{R}}_+^q$, and $RG \leq\leq 0$, it follows that

$$R[\mathcal{J}(E(t)) - \mathcal{D}(E(t))]\left(\frac{\partial \mathcal{H}}{\partial E}(E(t))\right)^{\mathrm{T}} + RGS(t) \leq\leq 0, \quad t \geq t_0, \tag{3.138}$$

which implies that, for every $E_0 \in \overline{\mathbb{R}}_+^q$, $RE(t_2) \leq\leq RE(t_1)$, $t_0 \leq t_1 \leq t_2$.

ii) To show sufficiency, note that since by assumption $\mathcal{D}_\mathrm{c}$ is positively invariant, then $R[\mathcal{J}(E(t)) - \mathcal{D}(E(t))](\frac{\partial \mathcal{H}}{\partial E}(E(t)))^{\mathrm{T}} \leq\leq 0$, $t \geq t_0$, for all $E_0 \in \mathcal{D}_\mathrm{c} \subseteq \overline{\mathbb{R}}_+^q$. Now, the result follows by using identical arguments as in *i*) with $S(t) \equiv 0$ and $E_0 \in \mathcal{D}_\mathrm{c} \subseteq \overline{\mathbb{R}}_+^q$. To show necessity, assume that (3.135) with $S(t) \equiv 0$ is monotonic for all $E_0 \in \mathcal{D}_\mathrm{c} \subseteq \overline{\mathbb{R}}_+^q$. In this case, (3.136) implies that for every $\tau > t_0$,

$$RE(\tau) = RE_0 + \int_{t_0}^{\tau} R[\mathcal{J}(E(t)) - \mathcal{D}(E(t))]\left(\frac{\partial \mathcal{H}}{\partial E}(E(t))\right)^{\mathrm{T}} \mathrm{d}t. \tag{3.139}$$

Now, suppose, *ad absurdum*, there exist $J \in \{1, \ldots, q\}$ and $E_0 \in \mathcal{D}_\mathrm{c} \subseteq \overline{\mathbb{R}}_+^q$ such that $[R[\mathcal{J}(E_0) - \mathcal{D}(E_0)](\frac{\partial \mathcal{H}}{\partial E}(E_0))^{\mathrm{T}}]_J > 0$. Since the mapping $R[\mathcal{J}(\cdot) - \mathcal{D}(\cdot)](\frac{\partial \mathcal{H}}{\partial E}(\cdot))^{\mathrm{T}}$ and the solution $E(t)$, $t \geq t_0$, to (3.135) are continuous, it follows that there exists $\tau > t_0$ such that

$$\left[R[\mathcal{J}(E(t)) - \mathcal{D}(E(t))]\left(\frac{\partial \mathcal{H}}{\partial E}(E(t))\right)^{\mathrm{T}}\right]_J > 0, \quad t_0 \leq t \leq \tau, \tag{3.140}$$

which implies that $[RE(\tau)]_J > [RE_0]_J$, leading to a contradiction. Hence, $R[\mathcal{J}(E) - \mathcal{D}(E)](\frac{\partial \mathcal{H}}{\partial E}(E))^{\mathrm{T}} \leq\leq 0$, $E \in \mathcal{D}_\mathrm{c} \subseteq \overline{\mathbb{R}}_+^q$. □

It follows from *i*) of Theorem 3.12 that if $G = I_q$ (that is, external power (heat flux) can be injected to all subsystems), then $R = -I_q$, and hence, $[\mathcal{J}(E) - \mathcal{D}(E)](\frac{\partial \mathcal{H}}{\partial E}(E))^{\mathrm{T}} \geq\geq 0$, $E \in \overline{\mathbb{R}}_+^q$. This case would correspond to a power balance equation whose states are all increasing and can only be achieved if $\mathcal{D}(E) = 0$, $E \in \overline{\mathbb{R}}_+^q$. This of course implies that the dynamical system $\mathcal{G}$ cannot dissipate energy, and hence, the transfer of energy (heat) from a lower energy (temperature) level (source) to a higher energy (temperature) level (sink) requires the input of additional heat or energy. This is consistent with Clausius' statement of the second law of thermodynamics.

The following result is a direct consequence of Theorem 3.12 and provides sufficient conditions for convergence of the subsystem energies of the isolated large-scale dynamical system $\mathcal{G}$. Once again, this result holds whether or not Axiom *ii*) holds.

Theorem 3.13. Consider the large-scale dynamical system $\mathcal{G}$ with power balance equation (3.135) and $S(t) \equiv 0$. Let $\mathcal{D}_{\mathrm{c}} \subseteq \overline{\mathbb{R}}_+^q$ be a positively invariant set. If there exists a matrix $R \in \mathbb{R}^{q\times q}$ such that $R = \mathrm{diag}[r_1, \ldots, r_q]$, $r_i = \pm 1$, $i = 1, \ldots, q$, and $R[\mathcal{J}(E) - \mathcal{D}(E)](\frac{\partial \mathcal{H}}{\partial E}(E))^{\mathrm{T}} \leq\leq 0$, $E \in \mathcal{D}_{\mathrm{c}} \subseteq \overline{\mathbb{R}}_+^q$, then, for every $E_0 \in \mathcal{D}_{\mathrm{c}} \subseteq \overline{\mathbb{R}}_+^q$, $\lim_{t\to\infty} E(t)$ exists.

Proof. Since $\mathcal{H}(E) = \mathbf{e}^{\mathrm{T}} E$, $E \in \overline{\mathbb{R}}_+^q$, it follows that

$$\begin{aligned}
\dot{\mathcal{H}}(E) &= \frac{\partial \mathcal{H}}{\partial E}\dot{E} \\
&= \frac{\partial \mathcal{H}}{\partial E}[\mathcal{J}(E) - \mathcal{D}(E)]\left(\frac{\partial \mathcal{H}}{\partial E}\right)^{\mathrm{T}} \\
&= -\frac{\partial \mathcal{H}}{\partial E}\mathcal{D}(E)\left(\frac{\partial \mathcal{H}}{\partial E}\right)^{\mathrm{T}} \\
&\leq 0, \qquad E \in \overline{\mathbb{R}}_+^q,
\end{aligned} \tag{3.141}$$

and hence, $\dot{\mathcal{H}}(E(t)) \leq 0$, $t \geq t_0$, where $E(t)$, $t \geq t_0$, denotes the solution of (3.135). This implies that $\mathcal{H}(E(t)) \leq \mathcal{H}(E_0) = \mathbf{e}^{\mathrm{T}} E_0$, $t \geq t_0$, and hence, for every $E_0 \in \overline{\mathbb{R}}_+^q$, the solution $E(t)$, $t \geq t_0$, of (3.135) is bounded. Hence, for every $i \in \{1, \ldots, q\}$, $E_i(t)$, $t \geq t_0$, is bounded. Furthermore, it follows from Theorem 3.12 that $E_i(t)$, $t \geq t_0$, is monotonic for all $E_0 \in \mathcal{D}_{\mathrm{c}} \subseteq \overline{\mathbb{R}}_+^q$. Now, since $E_i(\cdot)$, $i \in \{1, \ldots, q\}$, is continuous and every bounded nonincreasing or nondecreasing scalar sequence converges to a finite real number, it follows from the monotone convergence theorem that $\lim_{t\to\infty} E_i(t)$, $i \in \{1, \ldots, q\}$, exists. Hence, $\lim_{t\to\infty} E(t)$ exists for all $E_0 \in \mathcal{D}_{\mathrm{c}} \subseteq \overline{\mathbb{R}}_+^q$. □

3.9 The Second Law as a Statement of Entropy Increase

In this section, we reformulate and extend some of the results of Sections 3.3 and 3.5. In particular, unlike the framework presented in the previous sections, wherein we establish the existence and uniqueness of a global entropy function of a specific form for our thermodynamically consistent system model, in this section we assume the existence of a continuously differentiable, strictly concave function that leads to an entropy inequality that can be identified with the second law of thermodynamics as a statement about entropy increase. We then turn our attention to stability and convergence. Specifically, using Lyapunov stability theory and the Krasovskii–LaSalle invariance principle, we show that for an adiabatically isolated system, the proposed interconnected dynamical system model is Lyapunov stable with convergent trajectories to equilibrium states where the temperatures of all subsystems are equal.

For the interconnected dynamical system $\mathcal{G}$ with the power balance equation (3.5), recall the *connectivity matrix* $\mathcal{C} \in \mathbb{R}^{q\times q}$ is such that for $i \neq j$, $i,j = 1,\ldots,q$, $\mathcal{C}_{(i,j)} \triangleq 1$ if $\phi_{ij}(E) \not\equiv 0$ and $\mathcal{C}_{(i,j)} \triangleq 0$ otherwise, and $\mathcal{C}_{(i,i)} \triangleq -\sum_{k=1,\,k\neq i}^{q} \mathcal{C}_{(k,i)}$, $i = 1,\ldots,q$. (The negative of the connectivity matrix, that is, $-\mathcal{C}$, is known as the graph Laplacian in the literature.) Furthermore, recall that if $\operatorname{rank}\mathcal{C} = q-1$, then $\mathcal{G}$ is strongly connected and energy exchange is possible between any two subsystems of $\mathcal{G}$.

The next definition introduces a notion of entropy for the interconnected dynamical system $\mathcal{G}$.

Definition 3.5. Consider the interconnected dynamical system $\mathcal{G}$ with power balance equation (3.5). A continuously differentiable, strictly concave function $\mathcal{S} : \overline{\mathbb{R}}_+^q \to \mathbb{R}$ is called the *entropy* function of $\mathcal{G}$ if

$$\left(\frac{\partial\mathcal{S}(E)}{\partial E_i} - \frac{\partial\mathcal{S}(E)}{\partial E_j}\right)\phi_{ij}(E) \geq 0, \quad E \in \overline{\mathbb{R}}_+^q, \quad i \neq j, \quad i,j = 1,\ldots,q, \tag{3.142}$$

and $\frac{\partial\mathcal{S}(E)}{\partial E_i} = \frac{\partial\mathcal{S}(E)}{\partial E_j}$ if and only if $\phi_{ij}(E) = 0$ with $\mathcal{C}_{(i,j)} = 1$, $i \neq j$, $i,j = 1,\ldots,q$.

It follows from Definition 3.5 that for an *isolated system* $\mathcal{G}$, that is, $S(t) \equiv 0$ and $d(E) \equiv 0$, the entropy function of $\mathcal{G}$ is a nondecreasing function of time. To see this, note that

$$\begin{aligned}
\dot{\mathcal{S}}(E) &= \frac{\partial\mathcal{S}(E)}{\partial E}\dot{E} \\
&= \sum_{i=1}^{q} \frac{\partial\mathcal{S}(E)}{\partial E_i} \sum_{j=1,\,j\neq i}^{q} \phi_{ij}(E) \\
&= \sum_{i=1}^{q} \sum_{j=i+1}^{q} \left(\frac{\partial\mathcal{S}(E)}{\partial E_i} - \frac{\partial\mathcal{S}(E)}{\partial E_j}\right)\phi_{ij}(E) \\
&\geq 0, \quad E \in \overline{\mathbb{R}}_+^q,
\end{aligned} \tag{3.143}$$

where $\frac{\partial\mathcal{S}(E)}{\partial E} \triangleq \left[\frac{\partial\mathcal{S}(E)}{\partial E_1},\ldots,\frac{\partial\mathcal{S}(E)}{\partial E_q}\right]$ and where we used the fact that $\phi_{ij}(E) = -\phi_{ji}(E)$, $E \in \overline{\mathbb{R}}_+^q$, $i \neq j$, $i,j = 1,\ldots,q$.

Proposition 3.12. Consider the isolated (i.e., $S(t) \equiv 0$ and $d(E) \equiv 0$) interconnected dynamical system $\mathcal{G}$ with power balance equation (3.5). Assume that $\operatorname{rank}\mathcal{C} = q-1$ and there exists an entropy function $\mathcal{S} : \overline{\mathbb{R}}_+^q \to \mathbb{R}$ of $\mathcal{G}$. Then $\sum_{j=1}^{q} \phi_{ij}(E) = 0$ for all $i = 1,\ldots,q$ if and only if $\frac{\partial\mathcal{S}(E)}{\partial E_1} = \cdots =$

$\frac{\partial \mathcal{S}(E)}{\partial E_q}$. Furthermore, the set of nonnegative equilibrium states of (3.5) is given by

$$\mathcal{E}_0 \triangleq \left\{ E \in \overline{\mathbb{R}}_+^q : \frac{\partial \mathcal{S}(E)}{\partial E_1} = \cdots = \frac{\partial \mathcal{S}(E)}{\partial E_q} \right\}.$$

Proof. If $\frac{\partial \mathcal{S}(E)}{\partial E_i} = \frac{\partial \mathcal{S}(E)}{\partial E_j}$, then $\phi_{ij}(E) = 0$ for all $i, j = 1, \ldots, q$, which implies that $\sum_{j=1}^q \phi_{ij}(E) = 0$ for all $i = 1, \ldots, q$. Conversely, assume that $\sum_{j=1}^q \phi_{ij}(E) = 0$ for all $i = 1, \ldots, q$, and, since $\mathcal{S}$ is an entropy function of $\mathcal{G}$, it follows that

$$\begin{aligned} 0 &= \sum_{i=1}^q \sum_{j=1}^q \frac{\partial \mathcal{S}(E)}{\partial E_i} \phi_{ij}(E) \\ &= \sum_{i=1}^{q-1} \sum_{j=i+1}^q \left(\frac{\partial \mathcal{S}(E)}{\partial E_i} - \frac{\partial \mathcal{S}(E)}{\partial E_j} \right) \phi_{ij}(E) \\ &\geq 0, \end{aligned}$$

where we have used the fact that $\phi_{ij}(E) = -\phi_{ji}(E)$ for all $i, j = 1, \ldots, q$. Hence,

$$\left(\frac{\partial \mathcal{S}(E)}{\partial E_i} - \frac{\partial \mathcal{S}(E)}{\partial E_j} \right) \phi_{ij}(E) = 0$$

for all $i, j = 1, \ldots, q$. Now, the result follows from the fact that $\operatorname{rank} \mathcal{C} = q - 1$. $\square$

Theorem 3.14. Consider the isolated (i.e., $S(t) \equiv 0$ and $d(E) \equiv 0$) interconnected dynamical system $\mathcal{G}$ with power balance equation (3.5). Assume that $\operatorname{rank} \mathcal{C} = q-1$ and there exists an entropy function $\mathcal{S} : \overline{\mathbb{R}}_+^q \to \mathbb{R}$ of $\mathcal{G}$. Then the isolated system $\mathcal{G}$ is globally semistable with respect to $\overline{\mathbb{R}}_+^q$.

Proof. Since $f(\cdot)$ is essentially nonnegative, it follows from Proposition 2.1 that $E(t) \in \overline{\mathbb{R}}_+^q$, $t \geq t_0$, for all $E_0 \in \overline{\mathbb{R}}_+^q$. Furthermore, note that since $\mathbf{e}^{\mathrm{T}} f(E) = 0$, $E \in \overline{\mathbb{R}}_+^q$, it follows that $\mathbf{e}^{\mathrm{T}} \dot{E}(t) = 0$, $t \geq t_0$. In this case, $\mathbf{e}^{\mathrm{T}} E(t) = \mathbf{e}^{\mathrm{T}} E_0$, $t \geq t_0$, which implies that $E(t)$, $t \geq t_0$, is bounded for all $E_0 \in \overline{\mathbb{R}}_+^q$. Now, it follows from (3.143) that $\mathcal{S}(E(t))$, $t \geq t_0$, is a nondecreasing function of time, and hence, by the Krasovskii-LaSalle theorem, $E(t) \to \mathcal{R} \triangleq \{E \in \overline{\mathbb{R}}_+^q : \dot{\mathcal{S}}(E) = 0\}$ as $t \to \infty$.

Next, it follows from (3.143), Definition 3.5, and the fact that $\operatorname{rank} \mathcal{C} = q - 1$, that

$$\mathcal{R} = \left\{ E \in \overline{\mathbb{R}}_+^q : \frac{\partial \mathcal{S}(E)}{\partial E_1} = \cdots = \frac{\partial \mathcal{S}(E)}{\partial E_q} \right\} = \mathcal{E}_0.$$

Now, let $E_{\mathrm{e}} \in \mathcal{E}_0$ and consider the continuously differentiable function V :

$\mathbb{R}^q \to \mathbb{R}$ defined by

$$V(E) \triangleq \mathcal{S}(E_{\rm e}) - \mathcal{S}(E) - \lambda_{\rm e}(\mathbf{e}^{\rm T}E_{\rm e} - \mathbf{e}^{\rm T}E),$$

where $\lambda_{\rm e} \triangleq \frac{\partial \mathcal{S}}{\partial E_1}(E_{\rm e})$. Next, note that $V(E_{\rm e}) = 0$,

$$\frac{\partial V}{\partial E}(E_{\rm e}) = -\frac{\partial \mathcal{S}}{\partial E}(E_{\rm e}) + \lambda_{\rm e}\mathbf{e}^{\rm T} = 0,$$

and, since $\mathcal{S}(\cdot)$ is a strictly concave function,

$$\frac{\partial^2 V}{\partial E^2}(E_{\rm e}) = -\frac{\partial^2 S}{\partial E^2}(E_{\rm e}) > 0,$$

which implies that $V(\cdot)$ admits a local minimum at $E_{\rm e}$. Thus, $V(E_{\rm e}) = 0$, there exists $\delta > 0$ such that $V(E) > 0$, $E \in \mathcal{B}_\delta(E_{\rm e})\backslash\{E_{\rm e}\}$, and $\dot{V}(E) = -\dot{\mathcal{S}}(E) \leq 0$ for all $E \in \mathcal{B}_\delta(E_{\rm e})\backslash\{E_{\rm e}\}$, which shows that $V(\cdot)$ is a Lyapunov function for $\mathcal{G}$ and $E_{\rm e}$ is a Lyapunov stable equilibrium of $\mathcal{G}$.

Finally, since, for every $E_0 \in \overline{\mathbb{R}}_+^n$, $E(t) \to \mathcal{E}_0$ as $t \to \infty$ and every equilibrium point of $\mathcal{G}$ is Lyapunov stable, it follows from Proposition 2.2 that $\mathcal{G}$ is globally semistable with respect to $\overline{\mathbb{R}}_+^q$. □

In classical thermodynamics, the partial derivative of the system entropy with respect to the system energy defines the reciprocal of the system temperature. Thus, for the interconnected dynamical system $\mathcal{G}$,

$$T_i \triangleq \left(\frac{\partial \mathcal{S}(E)}{\partial E_i}\right)^{-1}, \quad i = 1, \ldots, q, \tag{3.144}$$

represents the temperature of the ith subsystem. Equation (3.142) is a manifestation of the *second law of thermodynamics* and implies that if the temperature of the jth subsystem is greater than the temperature of the ith subsystem, then energy (heat) flows from the jth subsystem to the ith subsystem.

Furthermore, $\frac{\partial \mathcal{S}(E)}{\partial E_i} = \frac{\partial \mathcal{S}(E)}{\partial E_j}$ if and only if $\phi_{ij}(E) = 0$ with $\mathcal{C}_{(i,j)} = 1$, $i \neq j$, $i, j = 1, \ldots, q$, implies that temperature equality is a necessary and sufficient condition for thermal equilibrium. This is a statement of the *zeroth law of thermodynamics*. As a result, Theorem 3.14 shows that, for a strongly connected system $\mathcal{G}$, the subsystem energies converge to the set of equilibrium states where the temperatures of all subsystems are equal; namely, *equipartition of temperature*. In particular, all the system energy is eventually transferred into heat at a uniform temperature, and hence, all dynamical processes in $\mathcal{G}$ (system motions) would cease.

The following result presents a sufficient condition for energy equipartition of the system, that is, the energies of all subsystems are equal. This state of energy equipartition is uniquely determined by the initial energy in the system.

Theorem 3.15. Consider the isolated (i.e., $S(t) \equiv 0$ and $d(E) \equiv 0$) interconnected dynamical system $\mathcal{G}$ with power balance equation (3.5). Assume that $\operatorname{rank} \mathcal{C} = q-1$ and there exists a continuously differentiable, strictly concave function $g : \overline{\mathbb{R}}_+ \to \mathbb{R}$ such that the entropy function $\mathcal{S} : \overline{\mathbb{R}}_+^q \to \mathbb{R}$ of $\mathcal{G}$ is given by $\mathcal{S}(E) = \sum_{i=1}^q g(E_i)$. Then the set of nonnegative equilibrium states of (3.5) is given by $\mathcal{E}_0 = \{\alpha \mathbf{e} : \alpha \geq 0\}$ and $\mathcal{G}$ is semistable with respect to $\overline{\mathbb{R}}_+^q$. Furthermore, $E(t) \to \frac{1}{q}\mathbf{e}\mathbf{e}^{\mathrm{T}} E(t_0)$ as $t \to \infty$ and $\frac{1}{q}\mathbf{e}\mathbf{e}^{\mathrm{T}} E(t_0)$ is a semistable equilibrium state of $\mathcal{G}$.

Proof. First, note that since $g(\cdot)$ is a continuously differentiable, strictly concave function, it follows that

$$\left(\frac{\mathrm{d}g}{\mathrm{d}E_i} - \frac{\mathrm{d}g}{\mathrm{d}E_j}\right)(E_i - E_j) \leq 0, \quad E \in \overline{\mathbb{R}}_+^q, \quad i,j = 1,\ldots,q,$$

which implies that (3.142) is equivalent to

$$(E_i - E_j)\,\phi_{ij}(E) \leq 0, \quad E \in \overline{\mathbb{R}}_+^q, \quad i \neq j, \quad i,j = 1,\ldots,q,$$

and $E_i = E_j$ if and only if $\phi_{ij}(E) = 0$ with $\mathcal{C}_{(i,j)} = 1$, $i \neq j$, $i,j = 1,\ldots,q$. Hence, $-E^{\mathrm{T}}E$ is an entropy function of $\mathcal{G}$.

Next, with $\mathcal{S}(E) = -\frac{1}{2}E^{\mathrm{T}}E$, it follows from Proposition 3.10 that $\mathcal{E}_0 = \{\alpha\mathbf{e} \in \overline{\mathbb{R}}_+^q, \alpha \geq 0\}$. Now, it follows from Theorem 3.14 that $\mathcal{G}$ is globally semistable with respect to $\overline{\mathbb{R}}_+^q$. Finally, since $\mathbf{e}^{\mathrm{T}}E(t) = \mathbf{e}^{\mathrm{T}}E(t_0)$ and $E(t) \to \mathcal{M}$ as $t \to \infty$, it follows that $E(t) \to \frac{1}{q}\mathbf{e}\mathbf{e}^{\mathrm{T}}E(t_0)$ as $t \to \infty$. Hence, with $\alpha = \frac{1}{q}\mathbf{e}^{\mathrm{T}}E(t_0)$, $\alpha\mathbf{e} = \frac{1}{q}\mathbf{e}\mathbf{e}^{\mathrm{T}}E(t_0)$ is a semistable equilibrium state of (3.5). $\square$

If $g(E_i) = \log_e(c+E_i)$, where $c > 0$, so that $\mathcal{S}(E) = \sum_{i=1}^q \log_e(c+E_i)$, then it follows from Theorem 3.15 that $\mathcal{E}_0 = \{\alpha\mathbf{e} : \alpha \geq 0\}$ and the isolated (i.e., $S(t) \equiv 0$ and $d(E) \equiv 0$) interconnected dynamical system $\mathcal{G}$ with the power balance (3.5) is semistable. In this case, the absolute temperature of the ith compartment is given by $c + E_i$. Similarly, if $\mathcal{S}(E) = -\frac{1}{2}E^{\mathrm{T}}E$, then it follows from Theorem 3.15 that $\mathcal{E}_0 = \{\alpha\mathbf{e} : \alpha \geq 0\}$ and the isolated (i.e., $S(t) \equiv 0$ and $d(E) \equiv 0$) interconnected dynamical system $\mathcal{G}$ with the power balance (3.5) is semistable. In both cases, $E(t) \to \frac{1}{q}\mathbf{e}\mathbf{e}^{\mathrm{T}}E(t_0)$ as $t \to \infty$. This shows that the steady-state energy of the isolated interconnected dynamical system $\mathcal{G}$ is given by $\frac{1}{q}\mathbf{e}\mathbf{e}^{\mathrm{T}}E(t_0) = \frac{1}{q}\sum_{i=1}^q E_i(t_0)\mathbf{e}$, and hence, is uniformly distributed over all subsystems of $\mathcal{G}$. This phenomenon is of course *energy equipartition*.

3.10 Thermodynamic Systems with Linear Energy Exchange

In this and the next section, we specialize the key results of this chapter to the case of large-scale dynamical systems with linear energy exchange between subsystems, that is, $f(E) = WE$ and $d(E) = DE$, where $W \in \mathbb{R}^{q\times q}$ and $D \in \mathbb{R}^{q\times q}$. In this case, the vector form of the energy balance equation (3.1), with $t_0 = 0$, is given by

$$E(T) = E(0) + \int_0^T WE(t)\mathrm{d}t - \int_0^T DE(t)\mathrm{d}t + \int_0^T S(t)\mathrm{d}t, \quad T \geq 0, \tag{3.145}$$

or, in power balance form,

$$\dot{E}(t) = WE(t) - DE(t) + S(t), \quad E(0) = E_0, \quad t \geq 0. \tag{3.146}$$

Next, let the net energy flow from the jth subsystem $\mathcal{G}_j$ to the ith subsystem $\mathcal{G}_i$ be parameterized as $\phi_{ij}(E) = \Phi_{ij}^{\mathrm{T}}E$, where $\Phi_{ij} \in \mathbb{R}^q$ and $E \in \overline{\mathbb{R}}_+^q$. In this case, since $f_i(E) = \sum_{i=1,\,j\neq i}^{q} \phi_{ij}(E)$, it follows that

$$W = \left[\sum_{j=2}^{q} \Phi_{1j}, \ldots, \sum_{j=1,\,j\neq i}^{q} \Phi_{ij}, \ldots, \sum_{j=1}^{q-1} \Phi_{qj}\right]^{\mathrm{T}}. \tag{3.147}$$

Since $\phi_{ij}(E) = -\phi_{ji}(E)$, $i,j = 1,\ldots,q$, $i \neq j$, and $E \in \overline{\mathbb{R}}_+^q$, it follows that $\Phi_{ij} = -\Phi_{ji}$, $i \neq j$, $i,j = 1,\ldots,q$. The following proposition gives necessary and sufficient conditions on W so that Axioms i) and ii) hold.

Proposition 3.13. Consider the large-scale dynamical system $\mathcal{G}$ with power balance equation given by (3.146) and with $D = 0$. Then Axioms i) and ii) hold if and only if $W = W^{\mathrm{T}}$, $W\mathbf{e} = 0$, $\operatorname{rank} W = q - 1$, and W is essentially nonnegative.

Proof. Assume Axioms i) and ii) hold. Since by Axiom ii) $(E_i - E_j)\phi_{ij}(E) \leq 0$, $E \in \overline{\mathbb{R}}_+^q$, it follows that $E^{\mathrm{T}}\Phi_{ij}\mathbf{e}_{ij}^{\mathrm{T}}E \leq 0$, $i,j = 1,\ldots,q$, $i \neq j$, where $E \in \overline{\mathbb{R}}_+^q$ and $\mathbf{e}_{ij} \in \mathbb{R}^q$ is a vector whose ith entry is 1, jth entry is -1, and remaining entries are zero. Next, it can be shown that $E^{\mathrm{T}}\Phi_{ij}\mathbf{e}_{ij}^{\mathrm{T}}E \leq 0$, $E \in \overline{\mathbb{R}}_+^q$, $i \neq j$, $i,j = 1,\ldots,q$, if and only if $\Phi_{ij} \in \mathbb{R}^q$ is such that its ith entry is $-\sigma_{ij}$, its jth entry is σ_{ij}, where $\sigma_{ij} \geq 0$, and its remaining entries are zero. Furthermore, since $\Phi_{ij} = -\Phi_{ji}$, $i \neq j$, $i,j = 1,\ldots,q$, it follows that $\sigma_{ij} = \sigma_{ji}$, $i \neq j$, $i,j = 1,\ldots,q$. Hence, W is given by

$$W_{(i,j)} = \begin{cases} -\sum_{k=1,\,k\neq j}^{q} \sigma_{kj}, & i = j, \\ \sigma_{ij}, & i \neq j, \end{cases} \tag{3.148}$$

which implies that W is symmetric (since $\sigma_{ij} = \sigma_{ji}$), essentially nonnegative, and $W\mathbf{e} = 0$.

Now, since by Axiom *i)* $\phi_{ij}(E) = 0$ if and only if $E_i = E_j$ for all $i, j = 1, \ldots, q$, $i \neq j$, such that $\mathcal{C}_{(i,j)} = 1$, it follows that $\sigma_{ij} > 0$ for all $i, j = 1, \ldots, q$, $i \neq j$, such that $\mathcal{C}_{(i,j)} = 1$. Hence, rank $W =$ rank $\mathcal{C} = q - 1$.

The converse is immediate and, hence, is omitted. □

Next, we specialize the energy balance equation (3.146) to the case where $D = \mathrm{diag}[\sigma_{11}, \sigma_{22}, \ldots, \sigma_{qq}]$. In this case, the vector form of the energy balance equation (3.145) is given by

$$E(T) = E(0) + \int_0^T AE(t)\mathrm{d}t + \int_0^T S(t)\mathrm{d}t, \quad T \geq 0, \tag{3.149}$$

or, in power balance form,

$$\dot{E}(t) = AE(t) + S(t), \quad E(0) = E_0, \quad t \geq 0, \tag{3.150}$$

where $A \triangleq W - D$ is such that

$$A_{(i,j)} = \begin{cases} -\sum_{k=1}^{q} \sigma_{kj}, & i = j, \\ \sigma_{ij}, & i \neq j. \end{cases} \tag{3.151}$$

Note that (3.151) implies $\sum_{i=1}^{q} A_{(i,j)} = -\sigma_{jj} \leq 0$, $j = 1, \ldots, q$, and hence, A is a semistable compartmental matrix (see statement *iv)* of Theorem 2.10). If $\sigma_{ii} > 0$, $i = 1, \ldots, q$, then A is an asymptotically stable compartmental matrix.

An important special case of (3.150) is the case where A is symmetric or, equivalently, $\sigma_{ij} = \sigma_{ji}$, $i \neq j$, $i, j = 1, \ldots, q$. In this case, it follows from (3.150) that for each subsystem the power balance equation satisfies

$$\dot{E}_i(t) + \sigma_{ii}E_i(t) + \sum_{j=1,\, j\neq i}^{q} \sigma_{ij}[E_i(t) - E_j(t)] = S_i(t), \quad t \geq 0. \tag{3.152}$$

Note that $\phi_i(E) \triangleq \sum_{j=1,\, j\neq i}^{q} \sigma_{ij}[E_i - E_j]$, $E \in \overline{\mathbb{R}}_+^q$, $i = 1, \ldots, q$, represents the energy flow from the ith subsystem to all other subsystems and is given by the sum of the individual energy flows from the ith subsystem to the jth subsystem. Furthermore, these energy flows are proportional to the energy differences of the subsystems, that is, $E_i - E_j$.

Hence, (3.152) is a power balance equation that governs the energy exchange among coupled subsystems and is completely analogous to the equations of thermal transfer with subsystem energies playing the role of temperatures. Furthermore, note that since $\sigma_{ij} \geq 0$, $i \neq j$, $i, j = 1, \ldots, q$, energy flows from more energetic subsystems to less energetic subsystems, which is consistent with the second law of thermodynamics requiring that heat (energy) *must* flow in the direction of lower temperatures.

The next lemma and proposition are needed for developing expressions for steady-state energy distributions of the large-scale dynamical system $\mathcal{G}$ with the linear power balance equation (3.150).

Lemma 3.2. Let $A \in \mathbb{R}^{q\times q}$ be compartmental and $S \in \mathbb{R}^q$. Then the following properties hold:

i) $-A$ is an M-matrix.

ii) If $\lambda \in \operatorname{spec}(A)$, then either $\operatorname{Re}\lambda < 0$ or $\lambda = 0$.

iii) $\operatorname{ind}(A) \leq 1$.

iv) A is semistable and $\lim_{t\to\infty} e^{At} = I_q - AA^{\#} \geq\geq 0$.

v) $\mathcal{R}(A) = \mathcal{N}(I_q - AA^{\#})$ and $\mathcal{N}(A) = \mathcal{R}(I_q - AA^{\#})$.

vi) $\int_0^t e^{As}\mathrm{d}s = A^{\#}(e^{At} - I_q) + (I_q - AA^{\#})t$, $t \geq 0$.

vii) $\int_0^\infty e^{At}\mathrm{d}t\, S$ exists if and only if $S \in \mathcal{R}(A)$, where $S \in \mathbb{R}^q$.

viii) If $S \in \mathcal{R}(A)$, then $\int_0^\infty e^{At}\mathrm{d}t\, S = -A^{\#}S$.

ix) If $S \in \mathcal{R}(A)$ and $S \geq\geq 0$, then $-A^{\#}S \geq\geq 0$.

x) A is nonsingular if and only if $-A$ is a nonsingular M-matrix.

xi) If A is nonsingular, then A is asymptotically stable and $-A^{-1} \geq\geq 0$.

Proof. The proof of the result appears in [37]. For completeness of exposition, we provide a proof here.

i) Since, by (3.151), $-A^{\mathrm{T}}\mathbf{e} \geq\geq 0$, and $-A$ is a Z-matrix, it follows from Theorem 1 of [32] that $-A^{\mathrm{T}}$, and hence, $-A$ is an M-matrix.

ii) Since $-A$ is an M-matrix, it follows from Theorem 4.6 of [31, p. 150], that $\operatorname{Re}\lambda < 0$ or $\lambda = 0$, where $\lambda \in \operatorname{spec}(A)$.

iii) This follows from the fact that, since $-A^{\mathrm{T}}\mathbf{e} \geq\geq 0$, then $-A$ has "property c" (see [31, pp. 152 and 155]). Hence, since an M-matrix $-A \in \mathbb{R}^{q\times q}$ has "property c" if and only if $\operatorname{ind}(-A) \leq 1$ (see [31, Lemma 4.11, p. 153]), it follows that $\operatorname{ind}(-A) = \operatorname{ind}(A) \leq 1$.

iv) Since $\operatorname{ind}(A) \leq 1$, it follows from the real Jordan decomposition that there exist invertible matrices $J \in \mathbb{R}^{r\times r}$, where $r = \operatorname{rank} A$, and $T \in \mathbb{R}^{q\times q}$ such that

$$A = T\begin{bmatrix} J & 0 \\ 0 & 0 \end{bmatrix}T^{-1} \tag{3.153}$$

and J is asymptotically stable. Hence, it follows that

$$\begin{aligned}\lim_{t\to\infty} e^{At} &= \lim_{t\to\infty} T\begin{bmatrix} e^{Jt} & 0 \\ 0 & I_{q-r} \end{bmatrix}T^{-1} \\ &= T\begin{bmatrix} 0 & 0 \\ 0 & I_{q-r} \end{bmatrix}T^{-1} \\ &= I_q - T\begin{bmatrix} J & 0 \\ 0 & 0 \end{bmatrix}T^{-1}T\begin{bmatrix} J^{-1} & 0 \\ 0 & 0 \end{bmatrix}T^{-1} \\ &= I_q - AA^{\#}. \end{aligned} \tag{3.154}$$

Next, since A is essentially nonnegative, it follows from Proposition 2.4 that $e^{At} \geq\geq 0$, $t \geq 0$, which implies that $I_q - AA^{\#} \geq\geq 0$.

v) Let $x \in \mathcal{R}(A)$, that is, there exists $y \in \mathbb{R}^q$ such that $x = Ay$. Now, $(I - AA^{\#})x = x - AA^{\#}Ay = x - Ay = 0$, which implies that $\mathcal{R}(A) \subseteq \mathcal{N}(I - AA^{\#})$. Conversely, let $x \in \mathcal{N}(I - AA^{\#})$. Hence, $(I - AA^{\#})x = 0$ or, equivalently, $x = AA^{\#}x$, which implies that $x \in \mathcal{R}(A)$, and hence, proves $\mathcal{R}(A) = \mathcal{N}(I - AA^{\#})$. The equality $\mathcal{N}(A) = \mathcal{R}(I - AA^{\#})$ can be proved in an analogous manner.

vi) Note that $A = T\begin{bmatrix} J & 0 \\ 0 & 0 \end{bmatrix}T^{-1}$, and hence,

$$\begin{aligned}\int_0^t e^{A\sigma}\mathrm{d}\sigma &= \int_0^t T\begin{bmatrix} e^{J\sigma} & 0 \\ 0 & I_{q-r} \end{bmatrix}T^{-1}\mathrm{d}\sigma \\ &= T\begin{bmatrix} \int_0^t e^{J\sigma}\mathrm{d}\sigma & 0 \\ 0 & \int_0^t I_{q-r}\mathrm{d}\sigma \end{bmatrix}T^{-1} \\ &= T\begin{bmatrix} J^{-1}(e^{Jt} - I_r) & 0 \\ 0 & I_{q-r}t \end{bmatrix}T^{-1} \\ &= T\begin{bmatrix} J^{-1} & 0 \\ 0 & 0 \end{bmatrix}T^{-1}T\begin{bmatrix} (e^{Jt} - I_r) & 0 \\ 0 & -I_{q-r} \end{bmatrix}T^{-1} \\ &\quad +T\begin{bmatrix} 0 & 0 \\ 0 & I_{q-r}t \end{bmatrix}T^{-1} \\ &= A^{\#}(e^{At} - I_q) + (I_q - AA^{\#})t, \quad t \geq 0. \end{aligned} \tag{3.155}$$

vii) The result is a direct consequence of *iv*), *v*), and *vi*).

viii) The result is a direct consequence of *iv*) and *vi*).

ix) The result follows from *viii*) and the fact that $e^{At} \geq\geq 0$, $t \geq 0$.

x) The result follows from *i*).

xi) Asymptotic stability of A is a direct consequence of *ii*), while $A^{-1} \leq\leq 0$ follows from *ix*) with $S = \mathrm{col}_i(I_q)$, $i = 1, \ldots, q$, where $\mathrm{col}_i(I_q)$ denotes the ith column of I_q. □

Proposition 3.14. Consider the large-scale dynamical system $\mathcal{G}$ with power balance equation given by (3.150). Suppose $E_0 \geq\geq 0$ and $S(t) \geq\geq 0$, $t \geq 0$. Then the solution $E(t)$ to (3.150) is nonnegative for all $t \geq 0$ if and only if A is essentially nonnegative.

Proof. It follows from Lagrange's formula that the solution $E(t)$, $t \geq 0$, to (3.150) is given by

$$E(t) = e^{At}E(0) + \int_0^t e^{A(t-\sigma)}S(\sigma)\mathrm{d}\sigma, \qquad t \geq 0. \tag{3.156}$$

Now, if A is essentially nonnegative, it follows from Proposition 2.4 that $e^{At} \geq\geq 0$, $t \geq 0$, and if $E(0) \in \overline{\mathbb{R}}_+^q$ and $S(t) \geq\geq 0$, $t \geq 0$, then it follows that $E(t) \geq\geq 0$ for all $t \geq 0$.

Conversely, suppose that the solution $E(t)$, $t \geq 0$, to (3.150) is nonnegative for all $E_0 \geq\geq 0$. Then, with $S(t) \equiv 0$, $E(t) = e^{At}E_0$, and hence, e^{At} is nonnegative for all $t \geq 0$. Thus, it follows from Proposition 2.4 that A is essentially nonnegative, which proves the result. □

Next, we develop expressions for the steady-state energy distribution for the large-scale dynamical system $\mathcal{G}$ for the cases where supplied system power $S(t)$ is a periodic function with period $\tau > 0$ (that is, $S(t+\tau) = S(t)$, $t \geq 0$) and $S(t)$ is constant (that is, $S(t) \equiv S$). Define $e(t) \triangleq E(t) - E(t+\tau)$, $t \geq 0$, and note that

$$\dot{e}(t) = Ae(t), \quad e(0) = E(0) - E(\tau), \quad t \geq 0. \tag{3.157}$$

Hence, since

$$e(t) = e^{At}[E(0) - E(\tau)], \quad t \geq 0, \tag{3.158}$$

and A is semistable, it follows from property *iv*) of Lemma 3.2 that

$$\lim_{t\to\infty} e(t) = \lim_{t\to\infty} [E(t) - E(t+\tau)] = (I_q - AA^{\#})[E(0) - E(\tau)], \tag{3.159}$$

which represents a constant offset to the steady-state error energy distribution in the large-scale dynamical system $\mathcal{G}$. For the case where $S(t) \equiv S$,

$\tau \to \infty$, and hence, the following result is immediate. This result first appeared in [37].

Proposition 3.15. Consider the large-scale dynamical system $\mathcal{G}$ with power balance equation given by (3.150). Suppose that $E_0 \geq\geq 0$ and $S(t) \equiv S \geq\geq 0$. Then $E_\infty \triangleq \lim_{t\to\infty} E(t)$ exists if and only if $S \in \mathcal{R}(A)$. In this case,

$$E_\infty = (I_q - AA^{\#})E_0 - A^{\#}S \tag{3.160}$$

and $E_\infty \geq\geq 0$. If, in addition, A is nonsingular, then E_∞ exists for all $S \geq\geq 0$ and is given by

$$E_\infty = -A^{-1}S. \tag{3.161}$$

Proof. Note that it follows from Lagrange's formula that the solution $E(t)$, $t \geq 0$, to (3.150) is given by

$$E(t) = e^{At}E(0) + \int_0^t e^{A(t-\sigma)}S\mathrm{d}\sigma, \quad t \geq 0. \tag{3.162}$$

Now, the result is a direct consequence of Proposition 3.14 and properties *iv*), *vii*), *viii*), and *ix*) of Lemma 3.2. □

3.11 Semistability and Energy Equipartition in Linear Thermodynamic Models

In this section, we show that an isolated large-scale linear dynamical system as well as a nonisolated large-scale linear dynamical system with strong coupling between subsystems and a constant heat flux input has a tendency to uniformly distribute its energy among all of its parts. First, we begin by specializing the result of Proposition 3.15 to the case where there is no energy dissipation from each subsystem $\mathcal{G}_i$ of $\mathcal{G}$, that is, $\sigma_{ii} = 0$, $i = 1, \ldots, q$. Note that in this case $\mathbf{e}^{\mathrm{T}}A = 0$, and hence, $\operatorname{rank} A \leq q - 1$. Furthermore, if $S = 0$, it follows from (3.150) that $\mathbf{e}^{\mathrm{T}}\dot{E}(t) = \mathbf{e}^{\mathrm{T}}AE(t) = 0$, $t \geq 0$, and hence, the total energy of the isolated large-scale dynamical system $\mathcal{G}$ is conserved.

Theorem 3.16. Consider the large-scale dynamical system $\mathcal{G}$ with power balance equation given by (3.150). Assume $\operatorname{rank} A = q - 1$, $\sigma_{ii} = 0$, $i = 1, \ldots, q$, and $A = A^{\mathrm{T}}$. If $E_0 \geq\geq 0$ and $S(t) \equiv 0$, then the equilibrium state $\alpha\mathbf{e}$, $\alpha \geq 0$, of the isolated system $\mathcal{G}$ is semistable and the steady-state energy distribution E_∞ of the isolated large-scale dynamical system $\mathcal{G}$ is given by

$$E_\infty = \left[\frac{1}{q}\sum_{i=1}^{q} E_{i0}\right]\mathbf{e}. \tag{3.163}$$

If, in addition, for some $k \in \{1,\ldots,q\}$, $\sigma_{kk} > 0$, then the zero solution $E(t) \equiv 0$ to (3.150) is globally asymptotically stable.

Proof. Note that since $\mathbf{e}^{\mathrm{T}}A = 0$, it follows from (3.150) with $S(t) \equiv 0$ that $\mathbf{e}^{\mathrm{T}}\dot{E}(t) = 0$, $t \geq 0$, and hence, $\mathbf{e}^{\mathrm{T}}E(t) = \mathbf{e}^{\mathrm{T}}E_0$, $t \geq 0$. Furthermore, since by Proposition 3.14 the solution $E(t)$, $t \geq t_0$, to (3.150) is nonnegative, it follows that $0 \leq E_i(t) \leq \mathbf{e}^{\mathrm{T}}E(t) = \mathbf{e}^{\mathrm{T}}E_0$, $t \geq 0$, $i = 1,\ldots,q$. Hence, the solution $E(t)$, $t \geq 0$, to (3.150) is bounded for all $E_0 \in \overline{\mathbb{R}}_+^q$. Next, note that $\phi_{ij}(E) = \sigma_{ij}(E_j - E_i)$ and $(E_i - E_j)\phi_{ij}(E) = -\sigma_{ij}(E_i - E_j)^2 \leq 0$, $E \in \overline{\mathbb{R}}_+^q$, $i \neq j$, $i,j = 1,\ldots,q$, which implies that Axioms $i)$ and $ii)$ are satisfied, and hence, $\mathcal{G}$ is a thermodynamically consistent linear energy flow model. Thus, $E = \alpha\mathbf{e}$, $\alpha \geq 0$, is the equilibrium state of the isolated large-scale dynamical system $\mathcal{G}$.

To show Lyapunov stability of the equilibrium state $\alpha\mathbf{e}$, consider the shifted-system ectropy function $\mathcal{E}_{\mathrm{s}}(E) = \frac{1}{2}(E - \alpha\mathbf{e})^{\mathrm{T}}(E - \alpha\mathbf{e})$, $E \in \overline{\mathbb{R}}_+^q$, as a Lyapunov function candidate. Then the Lyapunov derivative is given by

$$\begin{aligned}\dot{\mathcal{E}}_{\mathrm{s}}(E) &= (E - \alpha\mathbf{e})^{\mathrm{T}}AE \\ &= E^{\mathrm{T}}AE \\ &= -\sum_{i=1}^{q}\sum_{j=i+1}^{q}\sigma_{ij}(E_i - E_j)^2 \\ &\leq 0, \quad E \in \overline{\mathbb{R}}_+^q,\end{aligned} \tag{3.164}$$

which implies Lyapunov stability of the equilibrium state $\alpha\mathbf{e}$, $\alpha \geq 0$.

Next, consider the set $\mathcal{R} \stackrel{\triangle}{=} \{E \in \overline{\mathbb{R}}_+^q : \dot{\mathcal{E}}_{\mathrm{s}}(E) = 0\} = \{E \in \overline{\mathbb{R}}_+^q : E^{\mathrm{T}}AE = 0\}$. Since A is compartmental and symmetric, it follows from property $ii)$ of Lemma 3.2 that A is a negative semidefinite matrix, and hence, $E^{\mathrm{T}}AE = 0$ if and only if $AE = 0$. Since, by assumption $\operatorname{rank} A = q - 1$, it follows that there exists one and only one linearly independent solution to $AE = 0$ given by $E = \mathbf{e}$. Hence, $\mathcal{R} = \{E \in \overline{\mathbb{R}}_+^q : E = \alpha\mathbf{e}, \alpha \geq 0\}$.

Since $\mathcal{R}$ consists of only equilibrium states of (3.150), it follows that $\mathcal{M} = \mathcal{R}$, where $\mathcal{M}$ is the largest invariant set contained in $\mathcal{R}$. Hence, for every $E_0 \in \overline{\mathbb{R}}_+^q$, it follows from the Krasovskii-LaSalle invariant set theorem that $E(t) \to \alpha\mathbf{e}$ as $t \to \infty$ for some $\alpha \geq 0$, and hence, $\alpha\mathbf{e}$, $\alpha \geq 0$, is a semistable equilibrium state of (3.150). Furthermore, since the energy is conserved in the isolated large-scale dynamical system $\mathcal{G}$, it follows that

$q\alpha = \mathbf{e}^{\mathrm{T}} E_0$. Thus, $\alpha = \frac{1}{q}\sum_{i=1}^{q} E_{i0}$, which implies (3.163).

Finally, to show that in the case where $\sigma_{kk} > 0$ for some $k \in \{1, \ldots, q\}$, the zero solution $E(t) \equiv 0$ to (3.150) is globally asymptotically stable, consider the system ectropy $\mathcal{E}(E) = \frac{1}{2}E^{\mathrm{T}}E$, $E \in \overline{\mathbb{R}}_{+}^{q}$, as a Lyapunov function candidate. Note that Lyapunov stability of the zero equilibrium state follows from the previous analysis with $\alpha = 0$. Next, the Lyapunov derivative is given by

$$\dot{\mathcal{E}}(E) = E^{\mathrm{T}}AE = -\sum_{i=1}^{q}\sum_{j=i+1}^{q} \sigma_{ij}(E_i - E_j)^2 - \sigma_{kk}E_k^2, \quad E \in \overline{\mathbb{R}}_{+}^{q}. \quad (3.165)$$

Consider the set $\mathcal{R} \triangleq \{E \in \overline{\mathbb{R}}_{+}^{q} : \dot{\mathcal{E}}(E) = 0\} = \{E \in \overline{\mathbb{R}}_{+}^{q} : E_1 = \cdots = E_q\} \cap \{E \in \overline{\mathbb{R}}_{+}^{q} : E_k = 0, k \in \{1, \ldots, q\}\} = \{0\}$. Hence, the largest invariant set contained in $\mathcal{R}$ is given by $\mathcal{M} = \mathcal{R} = \{0\}$. Thus, it follows from the Krasovskii-LaSalle invariant set theorem that $E(t) \to \mathcal{M} = \{0\}$ as $t \to \infty$, which proves that the zero solution $E(t) \equiv 0$ to (3.150) is globally asymptotically stable. □

The result of Theorem 3.16 can also be obtained as a direct consequence of Theorem 3.9 with $f(E) = WE$ and $d(E) = DE$. To see this, note that it follows from Proposition 3.13 that the symmetry condition $W = W^{\mathrm{T}}$ along with $W\mathbf{e} = 0$ and $\operatorname{rank} W = q - 1$ ensure that Axioms *i)* and *ii)* are satisfied for the linear energy flow model (3.150). Furthermore, the condition $\operatorname{rank} W = q - 1$ ensures that the directed graph associated with the connectivity matrix $\mathcal{C}$ for $\mathcal{G}$ is strongly connected. The result now follows from Theorem 3.9.

Finally, we examine the steady-state energy distribution for large-scale linear dynamical systems $\mathcal{G}$ in the case of strong coupling between subsystems, that is, $\sigma_{ij} \to \infty$, $i \neq j$, $i, j = 1, \ldots, q$. For this analysis, we assume that A given by (3.151) is symmetric, that is, $\sigma_{ij} = \sigma_{ji}$, $i \neq j$, $i, j = 1, \ldots, q$, and $\sigma_{ii} > 0$, $i = 1, \ldots, q$. Thus, $-A$ is a nonsingular M-matrix for all values of σ_{ij}, $i \neq j$, $i, j = 1, \ldots, q$. Moreover, in this case, it follows that if $\frac{\sigma_{ij}}{\sigma_{kl}} \to 1$ as $\sigma_{ij} \to \infty$, $i \neq j$, and $\sigma_{kl} \to \infty$, $k \neq l$, then

$$\lim_{\sigma_{ij} \to \infty,\, i \neq j} A^{-1} = \lim_{\sigma \to \infty} [-D + \sigma(-qI_q + \mathbf{e}\mathbf{e}^{\mathrm{T}})]^{-1}, \quad (3.166)$$

where $D = \operatorname{diag}[\sigma_{11}, \ldots, \sigma_{qq}] > 0$.

The following lemmas are needed for the next result.

Lemma 3.3. Let $Y \in \mathbb{R}^{q \times q}$ be such that $\operatorname{ind}(Y) \leq 1$. Then $\lim_{\sigma \to \infty}(I_q - \sigma Y)^{-1} = I_q - Y^{\#}Y$.

Proof. Note that

$$\begin{aligned}(I_q - \sigma Y)^{-1} &= I_q + \sigma(I_q - \sigma Y)^{-1} Y \\ &= I_q + \left(\frac{1}{\sigma} I_q - Y\right)^{-1} Y \\ &= I_q - \left(Y - \frac{1}{\sigma} I_q\right)^{-1} Y. \end{aligned} \tag{3.167}$$

Now, using the fact that if $N \in \mathbb{R}^{q\times q}$ and $\operatorname{ind} N \le 1$, then

$$\lim_{\alpha\to 0} (N + \alpha I)^{-1} N = N N^{\#} = N^{\#} N, \tag{3.168}$$

it follows that

$$\lim_{\sigma\to\infty} (I_q - \sigma Y)^{-1} = I_q - \lim_{\frac{1}{\sigma}\to 0} \left(Y - \frac{1}{\sigma} I_q\right)^{-1} Y = I_q - Y^{\#} Y, \tag{3.169}$$

which proves the result. □

Lemma 3.4. Let $D \in \mathbb{R}^{q\times q}$ and $X \in \mathbb{R}^{q\times q}$ be such that $D > 0$ and $X = -qI_q + \mathbf{e}\mathbf{e}^{\mathrm{T}}$. Then

$$I_q - Y^{\#} Y = \frac{D^{\frac{1}{2}} \mathbf{e}\mathbf{e}^{\mathrm{T}} D^{\frac{1}{2}}}{\mathbf{e}^{\mathrm{T}} D \mathbf{e}}, \tag{3.170}$$

where $Y \triangleq D^{-\frac{1}{2}} X D^{-\frac{1}{2}}$.

Proof. Note that

$$Y = D^{-\frac{1}{2}}(-qI_q + \mathbf{e}\mathbf{e}^{\mathrm{T}}) D^{-\frac{1}{2}} = -qD^{-1} + D^{-\frac{1}{2}} \mathbf{e}\mathbf{e}^{\mathrm{T}} D^{-\frac{1}{2}}. \tag{3.171}$$

Now, using the fact that if $N \in \mathbb{R}^{q\times q}$ is nonsingular and symmetric and $b \in \mathbb{R}^q$ is a nonzero vector, then ([324])

$$\begin{aligned}&(N + bb^{\mathrm{T}})^{\#} \\ &= \left(I - \frac{1}{b^{\mathrm{T}} N^{-2} b} N^{-1} b b^{\mathrm{T}} N^{-1}\right) N^{-1} \left(I - \frac{1}{b^{\mathrm{T}} N^{-2} b} N^{-1} b b^{\mathrm{T}} N^{-1}\right),\end{aligned} \tag{3.172}$$

it follows that

$$-Y^{\#} = \frac{1}{q} \left(I_q - \frac{D^{\frac{1}{2}} \mathbf{e}\mathbf{e}^{\mathrm{T}} D^{\frac{1}{2}}}{\mathbf{e}^{\mathrm{T}} D \mathbf{e}}\right) D \left(I_q - \frac{D^{\frac{1}{2}} \mathbf{e}\mathbf{e}^{\mathrm{T}} D^{\frac{1}{2}}}{\mathbf{e}^{\mathrm{T}} D \mathbf{e}}\right). \tag{3.173}$$

Hence,

$$-Y^{\#} Y = -\left(I_q - \frac{D^{\frac{1}{2}} \mathbf{e}\mathbf{e}^{\mathrm{T}} D^{\frac{1}{2}}}{\mathbf{e}^{\mathrm{T}} D \mathbf{e}}\right) D \left(I_q - \frac{D^{\frac{1}{2}} \mathbf{e}\mathbf{e}^{\mathrm{T}} D^{\frac{1}{2}}}{\mathbf{e}^{\mathrm{T}} D \mathbf{e}}\right)$$

$$\cdot\left(D^{-1} - \frac{1}{q}D^{-\frac{1}{2}}\mathbf{e}\mathbf{e}^{\mathrm{T}}D^{-\frac{1}{2}}\right)$$
$$= -\left(I_q - \frac{D^{\frac{1}{2}}\mathbf{e}\mathbf{e}^{\mathrm{T}}D^{\frac{1}{2}}}{\mathbf{e}^{\mathrm{T}}D\mathbf{e}}\right). \tag{3.174}$$

Thus, $I_q - Y^{\#}Y = \frac{D^{\frac{1}{2}}\mathbf{e}\mathbf{e}^{\mathrm{T}}D^{\frac{1}{2}}}{\mathbf{e}^{\mathrm{T}}D\mathbf{e}}$. □

Proposition 3.16. Consider the large-scale dynamical system $\mathcal{G}$ with power balance equation given by (3.150). Let $S(t) \equiv S$, $S \in \mathbb{R}^{q\times q}$, let $A \in \mathbb{R}^{q\times q}$ be compartmental, and assume A is symmetric, $\sigma_{ii} > 0$, $i = 1, \ldots, q$, and $\frac{\sigma_{ij}}{\sigma_{kl}} \to 1$ as $\sigma_{ij} \to \infty$, $i \neq j$, and $\sigma_{kl} \to \infty$, $k \neq l$. Then the steady-state energy distribution E_∞ of the large-scale dynamical system $\mathcal{G}$ is given by

$$E_\infty = \left[\frac{\mathbf{e}^{\mathrm{T}}S}{\sum_{i=1}^{q}\sigma_{ii}}\right]\mathbf{e}. \tag{3.175}$$

Proof. Note that in the case where $\frac{\sigma_{ij}}{\sigma_{kl}} \to 1$ as $\sigma_{ij} \to \infty$, $i \neq j$, and $\sigma_{kl} \to \infty$, $k \neq l$, it follows that $\lim_{\sigma_{ij}\to\infty, i\neq j} A^{-1}$ is given by (3.166). Next, with $D = \mathrm{diag}[\sigma_{11}, \ldots, \sigma_{qq}]$ and $X = -qI_q + \mathbf{e}\mathbf{e}^{\mathrm{T}}$, it follows that $A = -D + \sigma X = -D^{\frac{1}{2}}(I_q - \sigma D^{-\frac{1}{2}}XD^{-\frac{1}{2}})D^{\frac{1}{2}}$. Now, it follows from Lemmas 3.3 and 3.4 that

$$E_\infty = \lim_{\sigma_{ij}\to\infty, i\neq j}(-A^{-1}S) = \frac{\mathbf{e}\mathbf{e}^{\mathrm{T}}}{\mathbf{e}^{\mathrm{T}}D\mathbf{e}}S = \left[\frac{\mathbf{e}^{\mathrm{T}}S}{\sum_{i=1}^{q}\sigma_{ii}}\right]\mathbf{e}, \tag{3.176}$$

which proves the result. □

Proposition 3.16 shows that in the limit of strong coupling, the steady-state energy distribution E_∞ given by (3.161) becomes

$$E_\infty = \lim_{\sigma_{ij}\to\infty, i\neq j}(-A^{-1}S) = \left[\frac{\mathbf{e}^{\mathrm{T}}S}{\sum_{i=1}^{q}\sigma_{ii}}\right]\mathbf{e}, \tag{3.177}$$

which implies energy equipartition. This result first appeared in [37].

3.12 Semistability and Energy Equipartition of Thermodynamic Systems with Directed Energy Flow

In this chapter, we have developed a compartmental dynamical systems framework for thermodynamics wherein each compartment represents the energy content of the different parts of the thermodynamic system and different compartments interact by exchanging heat. A key assumption of our formulation is that intercompartmental energy flows between connected compartments are bidirectional. However, in some applications of thermal sciences the assumption of bidirectional energy flow between compartments can be limiting. In this section, we develop compartmental dynamical system models that guarantee semistability and energy equipartitioning with directed energy flow between compartments.

To address the problem of semistability and energy equipartition of thermodynamic systems with unidirectional energy flow, we use directed graphs to represent intercompartmental connections. Specifically, let $\mathfrak{G} = (\mathcal{V}, \mathcal{E}, \mathcal{A})$ be a weighted directed graph (or digraph) denoting the compartmental thermodynamic network with the set of nodes (or compartments) $\mathcal{V} = \{1, \ldots, q\}$ involving a finite nonempty set denoting the compartments, the set of edges $\mathcal{E} \subseteq \mathcal{V} \times \mathcal{V}$ involving a set of ordered pairs denoting the direction of energy flow, and a *weighted adjacency matrix* $\mathcal{A} \in \mathbb{R}^{q \times q}$ such that $\mathcal{A}_{(i,j)} = \alpha_{ij} > 0$, $i, j = 1, \ldots, q$, if $(j, i) \in \mathcal{E}$, while $\alpha_{ij} = 0$ if $(j, i) \notin \mathcal{E}$.

The edge $(j, i) \in \mathcal{E}$ denotes that compartment $\mathcal{G}_j$ can obtain energy from compartment $\mathcal{G}_i$, but not necessarily vice versa. Moreover, we assume that $\alpha_{ii} = 0$ for all $i \in \mathcal{V}$. Note that if the weights α_{ij}, $i, j = 1, \ldots, q$, are not relevant, then α_{ij} is set to 1 for all $(j, i) \in \mathcal{E}$. In this case, recall that $\mathcal{A}$ is called an adjacency matrix. A graph $\mathfrak{G}$ is *balanced* if $\sum_{j=1}^{q} \alpha_{ij} = \sum_{j=1}^{q} \alpha_{ji}$ for all $i = 1, \ldots, q$.

The thermodynamic energy equipartitioning problem can be characterized as a closed compartmental dynamical system $\mathcal{G}$ given by

$$\dot{E}_i(t) = \sum_{j=1,\, j \neq i}^{q} \phi_{ij}(E_i(t), E_j(t)), \quad E_i(t_0) = E_{i0}, \quad t \geq 0, \quad i = 1, \ldots, q, \tag{3.178}$$

where the energy flow functions $\phi_{ij}(\cdot, \cdot)$, $i, j = 1, \ldots, q$, are locally Lipschitz continuous or, in vector form,

$$\dot{E}(t) = f(E(t)), \quad E(t_0) = E_0, \quad t \geq 0, \tag{3.179}$$

where $E(t) \triangleq [E_1(t), \ldots, E_q(t)]^{\mathrm{T}} \subseteq \overline{\mathbb{R}}_+^n$, $t \geq 0$, and $f = [f_1, \ldots, f_q]^{\mathrm{T}} : \mathcal{D} \to \mathbb{R}^q$ is such that $f_i(E) = \sum_{j=1,\, j \neq i}^{q} \phi_{ij}(E_i, E_j)$, where $\mathcal{D} \subseteq \overline{\mathbb{R}}_+^q$. Here, we assume that $f(\cdot)$ is essentially nonnegative.

Definition 3.6 ([31]). A directed graph $\mathfrak{G}$ is *strongly connected* if for any ordered pair of vertices (i, j), $i \neq j$, there exists a *path* (i.e., sequence of arcs) leading from i to j.

Recall that $\mathcal{A} \in \mathbb{R}^{q \times q}$ is *irreducible*, that is, there does not exist a permutation matrix such that $\mathcal{A}$ is cogredient to a lower-block triangular matrix, if and only if $\mathfrak{G}$ is strongly connected (see Theorem 2.7 of [31]). Furthermore, note that for an undirected graph $\mathcal{A} = \mathcal{A}^{\mathrm{T}}$, and hence, every undirected graph is balanced.

Proposition 3.17. Consider the compartmental dynamical system (3.178) and assume that Axioms *i)* and *ii)* hold. Then $f_i(E) = 0$ for all $i = 1, \ldots, q$ if and only if $E_1 = \cdots = E_q$. Furthermore, $\alpha\mathbf{e}$, $\alpha \in \overline{\mathbb{R}}_+$, is an equilibrium state of (3.178).

Proof. The proof of the first assertion is identical to the proof of Proposition 3.10 with $\mathcal{C}_\mathrm{s}$ in the proof of Proposition 3.10 denoting the connectivity matrix for the new *directed* graph $\mathbb{G}$, which excludes node m from the *directed* graph $\mathfrak{G}$. The second assertion is a direct consequence of the first assertion. $\square$

The following results are needed for the main result of this section. For the statement of these results, $(\cdot)^\mathrm{D}$ denotes the Drazin generalized inverse. Recall that for a diagonal matrix $A \in \mathbb{R}^{q\times q}$ the Drazin inverse $A^\mathrm{D} \in \mathbb{R}^{q\times q}$ is given by $A^\mathrm{D}_{(i,i)} = 0$ if $A_{(i,i)} = 0$ and $A^\mathrm{D}_{(i,i)} = 1/A_{(i,i)}$ if $A_{(i,i)} \neq 0$, $i = 1, \ldots, q$ [33, p. 227].

Proposition 3.18. Let $A \in \mathbb{R}^{q\times q}$ be an essentially nonnegative matrix such that $A = A^\mathrm{T}$. If there exists $p \in \mathbb{R}^q_+$ such that $A^\mathrm{T}p \le\le 0$, then $A \le 0$.

Proof. The proof is a direct consequence of *ii)* of Theorem 2.10 by noting that if A is symmetric, then semistability implies that $A \le 0$. $\square$

Lemma 3.5. Let $X \in \mathbb{R}^{n\times n}$ and $Z \in \mathbb{R}^{m\times m}$ be such that $X = X^\mathrm{T}$ and $Z = Z^\mathrm{T}$, and let $Y \in \mathbb{R}^{n\times m}$ be such that $Y = YZ^\mathrm{D}Z$. Then

$$M \triangleq \begin{bmatrix} X & Y \\ Y^\mathrm{T} & Z \end{bmatrix} \le 0 \tag{3.180}$$

if and only if $Z \le 0$ and $X - YZ^\mathrm{D}Y^\mathrm{T} \le 0$.

Proof. Define

$$T \triangleq \begin{bmatrix} I_n & -YZ^\mathrm{D} \\ 0 & I_m \end{bmatrix}$$

and note that $\det T \neq 0$. Now, noting that $TMT^\mathrm{T} \le 0$ if and only if $M \le 0$, and

$$\begin{aligned} TMT^\mathrm{T} &= \begin{bmatrix} I_n & -YZ^\mathrm{D} \\ 0 & I_m \end{bmatrix} \begin{bmatrix} X & Y \\ Y^\mathrm{T} & Z \end{bmatrix} \begin{bmatrix} I_n & 0 \\ -Z^\mathrm{D}Y^\mathrm{T} & I_m \end{bmatrix} \\ &= \begin{bmatrix} X - YZ^\mathrm{D}Y^\mathrm{T} & 0 \\ 0 & Z \end{bmatrix} \\ &\le 0, \end{aligned}$$

the result follows immediately. $\square$

Lemma 3.6. Let $A \in \mathbb{R}^{q\times q}$ and $A_{\rm d} \in \mathbb{R}^{q\times q}$ be given by either

$$A_{(i,j)} = \begin{cases} -\sum_{k=1,k\neq i}^{q} \sigma_{ik}, & i=j, \\ 0, & i\neq j, \end{cases}$$
$$A_{{\rm d}(i,j)} = \begin{cases} 0, & i=j, \\ \sigma_{ij}, & i\neq j, \end{cases} \quad i,j=1,\ldots,q, \tag{3.181}$$

or

$$A_{(i,j)} = \begin{cases} -\sum_{k=1,k\neq i}^{q} \sigma_{ki}, & i=j, \\ 0, & i\neq j, \end{cases}$$
$$A_{{\rm d}(i,j)} = \begin{cases} 0, & i=j, \\ a_{ij}, & i\neq j, \end{cases} \quad i,j=1,\ldots,q, \tag{3.182}$$

where $\sigma_{ij} \geq 0$, $i,j = 1,\ldots,q$, $i\neq j$. Assume that $\sum_{k=1,k\neq i}^{q} \sigma_{ik} = \sum_{k=1,k\neq i}^{q} \sigma_{ki}$ for each $i=1,\ldots,q$. Then for every $A_{{\rm d}i}$, $i=1,\ldots,q_{\rm d}$, such that $\sum_{i=1}^{q_{\rm d}} A_{{\rm d}i} = A_{\rm d}$, there exist nonnegative definite matrices $Q_i \in \mathbb{R}^{q\times q}$, $i=1,\ldots,q_{\rm d}$, such that

$$2A + \sum_{i=1}^{q_{\rm d}} (Q_i + A_{{\rm d}i}^{\rm T} Q_i^{\rm D} A_{{\rm d}i}) \leq 0. \tag{3.183}$$

Proof. For each $i \in \{1,\ldots,q_{\rm d}\}$, let Q_i be the diagonal matrix defined by

$$Q_{i(l,l)} \triangleq \sum_{m=1,l\neq m}^{q} A_{{\rm d}i(l,m)}, \quad l=1,\ldots,q, \tag{3.184}$$

and note that $A + \sum_{i=1}^{q_{\rm d}} Q_i = 0$, $(A_{{\rm d}i} - Q_i)\mathbf{e} = 0$, and $Q_i Q_i^{\rm D} A_{{\rm d}i} = A_{{\rm d}i}$, $i=1,\ldots,q_{\rm d}$. Hence, $M\mathbf{e} = 0$, where

$$M \triangleq \begin{bmatrix} 2A + \sum_{i=1}^{q_{\rm d}} Q_i & A_{\rm d1}^{\rm T} & A_{\rm d2}^{\rm T} & \cdots & A_{{\rm d}q_{\rm d}}^{\rm T} \\ A_{\rm d1} & -Q_1 & 0 & \cdots & 0 \\ \vdots & \vdots & \vdots & \vdots & \vdots \\ A_{{\rm d}q_{\rm d}} & 0 & 0 & \cdots & -Q_{q_{\rm d}} \end{bmatrix}. \tag{3.185}$$

Now, note that $M = M^{\rm T}$ and $M_{(i,j)} \geq 0$, $i,j=1,\ldots,q$, $i\neq j$. Hence, by statement *ii*) of Theorem 2.10, M is semistable. Thus, by Proposition 3.18, $M \leq 0$. Now, since $Q_i Q_i^{\rm D} A_{{\rm d}i} = A_{{\rm d}i}$, $i=1,\ldots,q_{\rm d}$, it follows from Lemma 3.5 that $M \leq 0$ if and only if (3.183) holds.

Alternatively, if $A \in \mathbb{R}^{q\times q}$ and $A_{\rm d} \in \mathbb{R}^{q\times q}$ are given by (3.182), then let Q_i be the diagonal matrix defined by

$$Q_{i(l,l)} \triangleq \sum_{m=1, l\neq m}^{q} A_{\mathrm{d}i(m,l)}, \quad l = 1, \ldots, q. \tag{3.186}$$

The result now follows using similar arguments as above. □

Next, we consider the case where (3.178) has a nonlinear structure of the form

$$\phi_{ij}(E_i, E_j) = \sigma_{ij}(E_j) - \sigma_{ji}(E_i), \tag{3.187}$$

where $\sigma_{ij} : \mathbb{R} \to \mathbb{R}$, $i, j = 1, \ldots, q$, $i \neq j$, are such that $\sigma_{ij}(0) = 0$ and $\sigma_{ij}(\cdot)$, $i, j = 1, \ldots, q$, $i \neq j$, is strictly increasing. For this result define $f_{\mathrm{c}i}(E_i) \triangleq -\sum_{j=1, j\neq i}^{q} \sigma_{ji}(E_i)$, $f_{\mathrm{d}i}(E) \triangleq \mathbf{e}_i \sum_{j=1}^{q} \sigma_{ij}(E_j)$, $i = 1, \ldots, q$, and $f_{\mathrm{c}}(E) \triangleq [f_{\mathrm{c}1}(E_1), \ldots, f_{\mathrm{c}q}(E_q)]^{\mathrm{T}}$, where $\mathbf{e}_i \in \mathbb{R}^q$ denotes the elementary vector of order q with 1 in the ith component and 0's elsewhere.

Theorem 3.17. Consider the nonlinear dynamical system given by (3.178) or, equivalently, (3.179), where $\phi_{ij}(E_i, E_j)$, $i, j = 1, \ldots, q$, $i \neq j$, is given by (3.187) and $f_{\mathrm{c}i}(\cdot)$, $i = 1, \ldots, q$, is strictly decreasing. Assume that $\mathbf{e}^{\mathrm{T}}[f_{\mathrm{c}}(E) + \sum_{i=1}^{q} f_{\mathrm{d}i}(E)] = 0$, $E \in \overline{\mathbb{R}}_+^q$, and $f_{\mathrm{c}}(E) + \sum_{i=1}^{q} f_{\mathrm{d}i}(E) = 0$ if and only if $E = \alpha\mathbf{e}$ for some $\alpha \in \overline{\mathbb{R}}_+$. Furthermore, assume there exist nonnegative diagonal matrices $P_i \in \mathbb{R}^{q\times q}$, $i = 1, \ldots, q$, such that $P \triangleq \sum_{i=1}^{q} P_i$ is positive definite,

$$P_i^{\mathrm{D}} P_i f_{\mathrm{d}i}(E) = f_{\mathrm{d}i}(E), \quad E \in \overline{\mathbb{R}}_+^q, \quad i = 1, \ldots, q, \tag{3.188}$$

$$\sum_{i=1}^{q} f_{\mathrm{d}i}^{\mathrm{T}}(E) P_i f_{\mathrm{d}i}(E) \leq f_{\mathrm{c}}^{\mathrm{T}}(E) P f_{\mathrm{c}}(E), \quad E \in \overline{\mathbb{R}}_+^q. \tag{3.189}$$

Then, for every $\alpha \in \overline{\mathbb{R}}_+$, $\alpha\mathbf{e}$ is a semistable equilibrium state of (3.179). Furthermore, $E(t) \to \frac{1}{q}\mathbf{e}\mathbf{e}^{\mathrm{T}}E(0)$ as $t \to \infty$ and $\frac{1}{q}\mathbf{e}\mathbf{e}^{\mathrm{T}}E(0)$ is a semistable equilibrium state.

Proof. Consider the nonnegative function given by

$$V(E) = -2\sum_{i=1}^{q} \int_0^{E_i} P_{(i,i)} f_{\mathrm{c}i}(\theta) \mathrm{d}\theta. \tag{3.190}$$

Since $f_{\mathrm{c}i}(\cdot)$, $i = 1, \ldots, q$, is a strictly decreasing function, it follows that

$$V(E) \geq 2\sum_{i=1}^{q} P_{(i,i)}[-f_{\mathrm{c}i}(\delta_i E_i)]E_i > 0$$

for all $E_i \neq 0$, where $0 < \delta_i < 1$, and hence, there exists a class $\mathcal{K}$ function $\alpha(\cdot)$ such that $V(E) \geq \alpha(\|E\|)$.

Now, note that the derivative of $V(E)$ along the trajectories of (3.179) is given by

$$\begin{aligned}\dot{V}(E) &= -2f_{\mathrm{c}}^{\mathrm{T}}(E)Pf_{\mathrm{c}}(E) - 2\sum_{i=1}^{q} f_{\mathrm{c}}^{\mathrm{T}}(E)Pf_{\mathrm{d}i}(E) \\ &\leq -f_{\mathrm{c}}^{\mathrm{T}}(E)Pf_{\mathrm{c}}(E) - 2\sum_{i=1}^{q} f_{\mathrm{c}}^{\mathrm{T}}(E)PP_i^{\mathrm{D}}P_i f_{\mathrm{d}i}(E) \\ &\quad -\sum_{i=1}^{q} f_{\mathrm{d}i}(E)P_iP_i^{\mathrm{D}}P_i f_{\mathrm{d}i}(E) \\ &= -\sum_{i=1}^{q}[Pf_{\mathrm{c}}(E) + P_i f_{\mathrm{d}i}(E)]^{\mathrm{T}}P_i^{\mathrm{D}}[Pf_{\mathrm{c}}(E) + P_i f_{\mathrm{d}i}(E)] \\ &\leq 0, \quad E \in \overline{\mathbb{R}}_+^q, \end{aligned} \tag{3.191}$$

where the first inequality in (3.191) follows from (3.188) and (3.189), and the last equality in (3.191) follows from the fact that

$$f_{\mathrm{c}}^{\mathrm{T}}(E)Pf_{\mathrm{c}}(E) = \sum_{i=1}^{q} f_{\mathrm{c}}^{\mathrm{T}}(E)PP_i^{\mathrm{D}}Pf_{\mathrm{c}}(E), \quad E \in \overline{\mathbb{R}}_+^q.$$

Next, let $\mathcal{R} \triangleq \{E \in \overline{\mathbb{R}}_+^q : Pf_{\mathrm{c}}(E) + P_i f_{\mathrm{d}i}(E) = 0, i = 1,\ldots,q\}$. Then it follows from the Krasovskii-LaSalle theorem that $E(t) \to \mathcal{M}$ as $t \to \infty$, where $\mathcal{M}$ denotes the largest invariant set contained in $\mathcal{R}$. Now, since $\mathbf{e}^{\mathrm{T}}[f_{\mathrm{c}}(E) + \sum_{i=1}^{q} f_{\mathrm{d}i}(E)] = 0$, $E \in \overline{\mathbb{R}}_+^q$, it follows that

$$\begin{aligned}\mathcal{R} \subseteq \hat{\mathcal{R}} &\triangleq \left\{E \in \overline{\mathbb{R}}_+^q : f_{\mathrm{c}}(E) + \sum_{i=1}^{q} f_{\mathrm{d}i}(E) = 0\right\} \\ &= \{E \in \overline{\mathbb{R}}_+^q : E = \alpha\mathbf{e},\ \alpha \in \overline{\mathbb{R}}_+\},\end{aligned} \tag{3.192}$$

which implies that $E(t) \to \hat{\mathcal{R}}$ as $t \to \infty$.

Finally, Lyapunov stability of $\alpha\mathbf{e}$, $\alpha \in \overline{\mathbb{R}}_+$, follows by considering the Lyapunov function candidate

$$V(E) = -2\sum_{i=1}^{q} \int_{\alpha}^{E_i} P_{(i,i)}(f_{\mathrm{c}i}(\theta) - f_{\mathrm{c}i}(\alpha))\mathrm{d}\theta \tag{3.193}$$

and noting that

$$V(E) \geq 2\sum_{i=1}^{q} P_{(i,i)}[f_{\mathrm{c}i}(\alpha) - f_{\mathrm{c}i}(\alpha + \delta_i(E_i - \alpha))](E_i - \alpha) > 0, \quad E \neq \alpha\mathbf{e},$$

where $0 < \delta_i < 1$ and $i = 1,\ldots,q$. In this case, it follows from Theorem 2.7

that, for every $\alpha \in \overline{\mathbb{R}}_+$, $\alpha\mathbf{e}$ is a semistable equilibrium state of (3.179). Furthermore, note that since $\mathbf{e}^{\mathrm{T}}E(t) = \mathbf{e}^{\mathrm{T}}E(0)$, $t \geq 0$, and $E(t) \to \mathcal{M}$ as $t \to \infty$, it follows that $E(t) \to \frac{1}{q}\mathbf{e}\mathbf{e}^{\mathrm{T}}E(0)$ as $t \to \infty$. Hence, with $\alpha = \frac{1}{q}\mathbf{e}^{\mathrm{T}}E(0)$, $\alpha\mathbf{e} = \frac{1}{q}\mathbf{e}\mathbf{e}^{\mathrm{T}}E(0)$ is a semistable equilibrium state of (3.179). □

Theorem 3.18. Consider the nonlinear dynamical system (3.178) or, equivalently, (3.179), and assume that Axioms *i)* and *ii)* hold. Let $\phi_{ij}(E_i, E_j) = \mathcal{C}_{(i,j)}[\sigma(E_j) - \sigma(E_i)]$ for all $i, j = 1, \ldots, q$, $i \neq j$, where $\sigma(0) = 0$ and $\sigma(\cdot)$ is strictly increasing. Assume that $\mathcal{C}^{\mathrm{T}}\mathbf{e} = 0$. Then, for every $\alpha \in \overline{\mathbb{R}}_+$, $\alpha\mathbf{e}$ is a semistable equilibrium state of (3.179). Furthermore, $E(t) \to \frac{1}{q}\mathbf{e}\mathbf{e}^{\mathrm{T}}E(0)$ as $t \to \infty$ and $\frac{1}{q}\mathbf{e}\mathbf{e}^{\mathrm{T}}E(0)$ is a semistable equilibrium state.

Proof. It follows from Lemma 3.6 that there exists Q_i, $i = 1, \ldots, q$, such that (3.183) holds with Q_i given by (3.184), and A and $A_{\mathrm{d}i}$, $i = 1, \ldots, q$, are given by (3.181) with σ_{ij} replaced by $\mathcal{C}_{(i,j)}$. Next, consider the nonnegative function given by

$$V(E) = 2\sum_{i=1}^{q}\int_0^{E_i}\sigma(\theta)\mathrm{d}\theta. \tag{3.194}$$

Since $\sigma(\cdot)$ is a strictly increasing function, it follows that

$$V(E) \geq 2\sum_{i=1}^{q}\sigma(\delta_i E_i)E_i > 0$$

for all $E \neq 0$, where $0 < \delta_i < 1$, and hence, there exists a class $\mathcal{K}$ function $\alpha(\cdot)$ such that $V(E) \geq \alpha(\|E\|)$.

Now, the derivative of $V(E)$ along the trajectories of (3.179) is given by

$$\begin{aligned}
\dot{V}(E) &= 2\hat{\sigma}^{\mathrm{T}}(E)A\hat{\sigma}(E) + 2\sum_{i=1}^{q}\hat{\sigma}^{\mathrm{T}}(E)A_{\mathrm{d}i}\hat{\sigma}(E) \\
&\leq -\sum_{i=1}^{q}[\hat{\sigma}^{\mathrm{T}}(E)Q_i\hat{\sigma}(E) - 2\hat{\sigma}^{\mathrm{T}}(E)A_{\mathrm{d}i}\hat{\sigma}(E) + \hat{\sigma}^{\mathrm{T}}(E)A_{\mathrm{d}i}^{\mathrm{T}}Q_i^{\mathrm{D}}A_{\mathrm{d}i}\hat{\sigma}(E)] \\
&= -\sum_{i=1}^{q}[-Q_i\hat{\sigma}(E) + A_{\mathrm{d}i}\hat{\sigma}(E)]^{\mathrm{T}}Q_i^{\mathrm{D}}[-Q_i\hat{\sigma}(E) + A_{\mathrm{d}i}\hat{\sigma}(E)] \\
&\leq 0, \quad E \in \overline{\mathbb{R}}_+^q,
\end{aligned} \tag{3.195}$$

where $\hat{\sigma} : \overline{\mathbb{R}}_+^q \to \mathbb{R}^q$ is given by $\hat{\sigma}(E) \triangleq [\sigma(E_1), \ldots, \sigma(E_q)]^{\mathrm{T}}$.

Next, let $\mathcal{R} \triangleq \{E \in \overline{\mathbb{R}}_+^q : -Q_i\hat{\sigma}(E) + A_{\mathrm{d}i}\hat{\sigma}(E) = 0, i = 1, \ldots, q\}$. Then it follows from the Krasovskii-LaSalle theorem that $E(t) \to \mathcal{M}$ as

$t \to \infty$, where $\mathcal{M}$ denotes the largest invariant set contained in $\mathcal{R}$. Now, since $A + \sum_{i=1}^{q} Q_i = 0$, it follows that

$$\mathcal{R} \subseteq \hat{\mathcal{R}} \triangleq \left\{ E \in \overline{\mathbb{R}}_+^q : A\hat{\sigma}(E) + \sum_{i=1}^{q} A_{\mathrm{d}i}\hat{\sigma}(E) = 0 \right\}.$$

Hence, since $\mathrm{rank}(A + \sum_{i=1}^{q} A_{\mathrm{d}i}) = q - 1$ and $(A + \sum_{i=1}^{q} A_{\mathrm{d}i})\mathbf{e} = 0$, it follows that the largest invariant set $\hat{\mathcal{M}}$ contained in $\hat{\mathcal{R}}$ is given by $\hat{\mathcal{M}} = \{E \in \overline{\mathbb{R}}_+^q : E = \alpha\mathbf{e}, \alpha \in \mathbb{R}\}$. Furthermore, since $\hat{\mathcal{M}} \subseteq \mathcal{R} \subseteq \hat{\mathcal{R}}$, it follows that $\mathcal{M} = \hat{\mathcal{M}}$.

Finally, Lyapunov stability of $\alpha\mathbf{e}$, $\alpha \in \overline{\mathbb{R}}_+$, follows by considering the Lyapunov function candidate

$$\tilde{V}(E) = 2\sum_{i=1}^{q} \int_{\alpha}^{E_i} [\sigma(\theta) - \sigma(\alpha)]\mathrm{d}\theta \tag{3.196}$$

and noting that

$$\tilde{V}(E) \geq 2\sum_{i=1}^{q} [\sigma(\alpha + \delta_i(E_i - \alpha)) - \sigma(\alpha)](E_i - \alpha) > 0,$$

for all $E_i \neq \alpha$, where $0 < \delta_i < 1$ and $i = 1, \ldots, q$. In this case, it follows from Theorem 2.7 that, for every $\alpha \in \overline{\mathbb{R}}_+$, $\alpha\mathbf{e}$ is a semistable equilibrium state of (3.179). Furthermore, note that since $\mathbf{e}^{\mathrm{T}}E(t) = \mathbf{e}^{\mathrm{T}}E(0)$, $t \geq 0$, and $E(t) \to \mathcal{M}$ as $t \to \infty$, it follows that $E(t) \to \frac{1}{q}\mathbf{e}\mathbf{e}^{\mathrm{T}}E(0)$ as $t \to \infty$. Hence, with $\alpha = \frac{1}{q}\mathbf{e}^{\mathrm{T}}E(0)$, $\alpha\mathbf{e} = \frac{1}{q}\mathbf{e}\mathbf{e}^{\mathrm{T}}E(0)$ is a semistable equilibrium state of (3.179). □

Note that the assumption $\mathcal{C}^{\mathrm{T}}\mathbf{e} = 0$ in Theorem 3.18 implies that the underlying directed graph of $\mathcal{G}$ is balanced. To see this, recall that for a directed graph $\mathfrak{G}$, $\mathcal{A}\mathbf{e} = \mathcal{A}^{\mathrm{T}}\mathbf{e}$ implies that $\mathfrak{G}$ is balanced. Since $\mathcal{C} = \mathcal{A} - \Delta$, where $\mathcal{A}$ denotes the normalized adjacency matrix and

$$\Delta \triangleq \mathrm{diag}\left[\sum_{j=1}^{q} \alpha_{1j}, \ldots, \sum_{j=1}^{q} \alpha_{qj}\right] \in \mathbb{R}^{q \times q},$$

it follows that $\mathcal{A}\mathbf{e} = \mathcal{A}^{\mathrm{T}}\mathbf{e}$ if and only if $\mathcal{C}\mathbf{e} = \mathcal{C}^{\mathrm{T}}\mathbf{e}$. Hence, $\mathcal{C}^{\mathrm{T}}\mathbf{e} = 0$ implies that $\mathfrak{G}$ is balanced.

Theorem 3.18 implies that the steady-state value of the state of each subsystem $\mathcal{G}_i$ of the compartmental dynamical system $\mathcal{G}$ is equal; that is, the steady state of the compartmental dynamical system $\mathcal{G}$ given by

$$E_{\infty} = \frac{1}{q}\mathbf{e}\mathbf{e}^{\mathrm{T}}E(0) = \left[\frac{1}{q}\sum_{i=1}^{q} E_i(0)\right]\mathbf{e}$$

is uniformly distributed over all subsystems of $\mathcal{G}$.

Finally, we specialize Theorem 3.17 to the case where

$$\phi_{ij}(E_i, E_j) = \sigma_{ij}\sigma(E_j) - \sigma_{ji}\sigma(E_i), \tag{3.197}$$

where $\sigma : \overline{\mathbb{R}}_+ \to \mathbb{R}$ is such that $\sigma(u) = 0$ if and only if $u = 0$, $\sigma_{ij} \geq 0$, $i, j = 1, \ldots, q$, $i \neq j$. In this case, (3.179) can be rewritten as

$$\dot{E}(t) = A\hat{\sigma}(E(t)) + \sum_{i=1}^{q} A_{\mathrm{d}i}\hat{\sigma}(E(t)), \quad E(0) = E_0, \quad t \geq 0, \tag{3.198}$$

where $\hat{\sigma} : \overline{\mathbb{R}}_+^q \to \mathbb{R}^q$ is given by $\hat{\sigma}(E) \triangleq [\sigma(E_1), \ldots, \sigma(E_q)]^{\mathrm{T}}$, and A and $A_{\mathrm{d}i}$, $i = 1, \ldots, q$, are given by (3.182).

Theorem 3.19. Consider the compartmental dynamical system given by (3.198), where $\sigma : \overline{\mathbb{R}}_+ \to \mathbb{R}$ is such that $\sigma(0) = 0$ and $\sigma(\cdot)$ is strictly increasing. Assume that $(A + \sum_{i=1}^{q} A_{\mathrm{d}i})^{\mathrm{T}}\mathbf{e} = (A + \sum_{i=1}^{q} A_{\mathrm{d}i})\mathbf{e} = 0$ and $\mathrm{rank}(A + \sum_{i=1}^{q} A_{\mathrm{d}i}) = q - 1$. Then, for every $\alpha \in \overline{\mathbb{R}}_+$, $\alpha\mathbf{e}$ is a semistable equilibrium point of (3.179). Furthermore, $E(t) \to \frac{1}{q}\mathbf{e}\mathbf{e}^{\mathrm{T}}E(0)$ as $t \to \infty$ and $\frac{1}{q}\mathbf{e}\mathbf{e}^{\mathrm{T}}E(0)$ is a semistable equilibrium state.

Proof. It follows from Lemma 3.6 that there exists Q_i, $i = 1, \ldots, q$, such that (3.183) holds with Q_i given by (3.186). Now, since $A = -\sum_{i=1}^{q} Q_i = -\sum_{i=1}^{q} P_i^{\mathrm{D}} = -P^{-1}$, where $P = \sum_{i=1}^{q} P_i$, it follows from (3.183) that, for all $E \in \overline{\mathbb{R}}_+^q$,

$$\begin{aligned} 0 &\geq 2\hat{\sigma}^{\mathrm{T}}(E)A\hat{\sigma}(E) + \hat{\sigma}^{\mathrm{T}}(E)\sum_{i=1}^{q}(Q_i + A_{\mathrm{d}i}^{\mathrm{T}}Q_i^{\mathrm{D}}A_{\mathrm{d}i})\hat{\sigma}(E) \\ &= -f_{\mathrm{c}}^{\mathrm{T}}(E)Pf_{\mathrm{c}}(E) + \sum_{i=1}^{q} f_{\mathrm{d}i}^{\mathrm{T}}(E)P_i f_{\mathrm{d}i}(E), \end{aligned}$$

where $f_{\mathrm{c}}(E) = A\hat{\sigma}(E)$ and $f_{\mathrm{d}i}(E) = A_{\mathrm{d}i}\hat{\sigma}(E)$, $i = 1, \ldots, q$, $E \in \overline{\mathbb{R}}_+^q$. Furthermore, since $P_i^{\mathrm{D}}P_iA_{\mathrm{d}i} = A_{\mathrm{d}i}$, $i = 1, \ldots, q$, it follows that $P_i^{\mathrm{D}}P_if_{\mathrm{d}i}(E) = f_{\mathrm{d}i}(E)$, $i = 1, \ldots, q$, $E \in \overline{\mathbb{R}}_+^q$. Now, the result is an immediate consequence of Theorem 3.17 by noting that $\mathbf{e}^{\mathrm{T}}[f_{\mathrm{c}}(E) + \sum_{i=1}^{q} f_{\mathrm{d}i}(E)] = 0$ and $f_{\mathrm{c}}(E) + \sum_{i=1}^{q} f_{\mathrm{d}i}(E) = 0$ if and only if $E = \alpha\mathbf{e}$ for some $\alpha \in \overline{\mathbb{R}}_+$. □

Chapter Four

Temperature Equipartition and the Kinetic Theory of Gases

4.1 Semistability and Temperature Equipartition

The thermodynamic axioms introduced in Chapter 3 postulate that subsystem energies are synonymous with subsystem temperatures, and hence, every subsystem possesses the same constant heat capacity. In this chapter, we generalize the results of Chapter 3 to the case where the subsystem energies are proportional to the subsystem temperatures with the proportionality variables representing the subsystem *heat* or *thermal capacities* or the subsystem *specific heats.*

The *heat* or *thermal capacity* of a body is an extensive property and is defined as the ratio of the infinitesimal amount of heat absorbed by the body to the infinitesimal increase in temperature produced by this heat. The heat or thermal capacity is a measure of the ability of a substance to absorb the energy supplied to it as heat. In general, the thermal capacity of a body is different if the body is heated at a constant volume or at a constant pressure. For an incompressible substance, however, this distinction is unnecessary.

The *specific heat* is the heat capacity per unit mass of a material and hence is an intensive property of the body. Thus, the specific heat is characteristic of the material of which the body is composed. Neither the heat capacity of a body nor the specific heat of a material is constant; they both depend on the corresponding temperature rise of the body/material with the supplied heat energy. In the limit of temperature increase, that is, as $T_{\rm f} - T_{\rm i} = \Delta T \to 0$, we can define the specific heat $c(T)$ of a body of mass m at a particular temperature T. Specifically, the heat (energy) E that must be added to a body of mass m with specific heat $c(T)$ to increase its temperature from $T_{\rm i}$ to $T_{\rm f}$ is given by

$$E = m \int_{T_{\rm i}}^{T_{\rm f}} c(T) {\rm d}T.$$

In the case where the specific heats of all the subsystems are equal, the results of this section specialize to those of Chapter 3. To include temperature notions in our large-scale dynamical system model, we replace Axioms i) and ii) of Section 3.3 with the following axioms. Let $\beta_i > 0$, $i = 1, \ldots, q$, denote the reciprocal of the specific heat (at constant volume) of the ith subsystem $\mathcal{G}_i$ so that the *absolute temperature* in the ith subsystem is given by $\hat{T}_i = \beta_i E_i$.

Axiom i): For the connectivity matrix $\mathcal{C} \in \mathbb{R}^{q \times q}$ associated with the large-scale dynamical system $\mathcal{G}$ defined by (3.24) and (3.25), rank $\mathcal{C} = q - 1$, and for $\mathcal{C}_{(i,j)} = 1$, $i \neq j$, $\phi_{ij}(E) = 0$ if and only if $\beta_i E_i = \beta_j E_j$.

Axiom ii): For $i, j = 1, \ldots, q$, $(\beta_i E_i - \beta_j E_j)\phi_{ij}(E) \leq 0$, $E \in \overline{\mathbb{R}}_+^q$.

Axiom i) implies that if the temperatures in the connected subsystems $\mathcal{G}_i$ and $\mathcal{G}_j$ are equal, then heat exchange between these subsystems is not possible. This is a statement of the *zeroth law of thermodynamics*, which postulates that temperature equality is a necessary and sufficient condition for *thermal equilibrium*. Axiom ii) implies that heat (energy) must flow in the direction of lower temperatures. This is a statement of the *second law of thermodynamics*, which states that a cyclic system transformation whose only final result is to transfer heat from a body at a given temperature to a body at a higher temperature is impossible.

The following proposition is needed for the statement of the main results of this section.

Proposition 4.1. Consider the large-scale dynamical system $\mathcal{G}$ with power balance equation (3.5), and assume that Axioms i) and ii) hold. Then, for all $E_0 \in \overline{\mathbb{R}}_+^q$, $t_f \geq t_0$, and $S(\cdot) \in \mathcal{U}$ such that $E(t_f) = E(t_0) = E_0$,

$$\int_{t_0}^{t_f} \sum_{i=1}^{q} \frac{S_i(t) - \sigma_{ii}(E(t))}{c + \beta_i E_i(t)} \mathrm{d}t = \oint \sum_{i=1}^{q} \frac{\mathrm{d}Q_i(t)}{c + \beta_i E_i(t)} \leq 0 \tag{4.1}$$

and

$$\int_{t_0}^{t_f} \sum_{i=1}^{q} \beta_i E_i(t)[S_i(t) - \sigma_{ii}(E(t))]\mathrm{d}t = \oint \sum_{i=1}^{q} \beta_i E_i(t)\mathrm{d}Q_i(t) \geq 0, \tag{4.2}$$

where $E(t)$, $t \geq t_0$, is the solution to (3.5) with initial condition $E(t_0) = E_0$. Furthermore,

$$\oint \sum_{i=1}^{q} \frac{\mathrm{d}Q_i(t)}{c + \beta_i E_i(t)} = 0 \tag{4.3}$$

and

$$\oint \sum_{i=1}^{q} \beta_i E_i(t) \mathrm{d}Q_i(t) = 0 \tag{4.4}$$

if and only if there exists a continuous function $\alpha : [t_0, t_{\mathrm{f}}] \to \overline{\mathbb{R}}_+$ such that $E(t) = \alpha(t)\boldsymbol{p}$, $t \in [t_0, t_{\mathrm{f}}]$, where $\boldsymbol{p} \triangleq [1/\beta_1, \ldots, 1/\beta_q]^{\mathrm{T}}$.

Proof. The proof is identical to the proofs of Propositions 3.2 and 3.6. □

Note that with the modified Axiom *i*) the isolated large-scale dynamical system $\mathcal{G}$ has equilibrium energy states given by $E_{\mathrm{e}} = \alpha\boldsymbol{p}$ for $\alpha \geq 0$. As in Section 3.3, we define an equilibrium process as a process in which the trajectory of the system $\mathcal{G}$ moves along the equilibrium manifold $\mathcal{M}_{\mathrm{e}} \triangleq \{E \in \overline{\mathbb{R}}_+^q : E = \alpha\boldsymbol{p}, \alpha \geq 0\}$ corresponding to the set of equilibria for the isolated system $\mathcal{G}$, and we define a nonequilibrium process as a process that does not lie on $\mathcal{M}_{\mathrm{e}}$. Thus, it follows from Axioms *i*) and *ii*) that inequalities (4.1) and (4.2) are satisfied as equalities for an equilibrium process and as strict inequalities for a nonequilibrium process.

Next, in light of our modified axioms, we present a generalized definition for the entropy and ectropy of $\mathcal{G}$.

Definition 4.1. For the large-scale dynamical system $\mathcal{G}$ with power balance equation (3.5), a function $\mathcal{S} : \overline{\mathbb{R}}_+^q \to \mathbb{R}$ satisfying

$$\mathcal{S}(E(t_2)) \geq \mathcal{S}(E(t_1)) + \int_{t_1}^{t_2} \sum_{i=1}^{q} \frac{S_i(t) - \sigma_{ii}(E(t))}{c + \beta_i E_i(t)} \mathrm{d}t \tag{4.5}$$

for every $t_2 \geq t_1 \geq t_0$ and $S(\cdot) \in \mathcal{U}$ is called the *entropy* function of $\mathcal{G}$.

Definition 4.2. For the large-scale dynamical system $\mathcal{G}$ with power balance equation (3.5), a function $\mathcal{E} : \overline{\mathbb{R}}_+^q \to \mathbb{R}$ satisfying

$$\mathcal{E}(E(t_2)) \leq \mathcal{E}(E(t_1)) + \int_{t_1}^{t_2} \sum_{i=1}^{q} \beta_i E_i(t)[S_i(t) - \sigma_{ii}(E(t))] \mathrm{d}t \tag{4.6}$$

for every $t_2 \geq t_1 \geq t_0$ and $S(\cdot) \in \mathcal{U}$ is called the *ectropy* function of $\mathcal{G}$.

For the next result, define the available entropy and available ectropy of the large-scale dynamical system $\mathcal{G}$ by

$$\mathcal{S}_{\mathrm{a}}(E_0) \triangleq - \sup_{S(\cdot) \in \mathcal{U}_{\mathrm{c}}, T \geq t_0} \int_{t_0}^{T} \sum_{i=1}^{q} \frac{S_i(t) - \sigma_{ii}(E(t))}{c + \beta_i E_i(t)} \mathrm{d}t, \tag{4.7}$$

$$\mathcal{E}_{\mathrm{a}}(E_0) \triangleq -\inf_{S(\cdot)\in\mathcal{U}_{\mathrm{c}},\, T\geq t_0} \int_{t_0}^{T} \sum_{i=1}^{q} \beta_i E_i(t)[S_i(t) - \sigma_{ii}(E(t))]\mathrm{d}t, \qquad (4.8)$$

where $E(t_0) = E_0 \in \overline{\mathbb{R}}_+^q$ and $E(T) = 0$, and define the required entropy supply and required ectropy supply of the large-scale dynamical system $\mathcal{G}$ by

$$\mathcal{S}_{\mathrm{r}}(E_0) \triangleq \sup_{S(\cdot)\in\mathcal{U}_{\mathrm{r}},\, T\geq -t_0} \int_{-T}^{t_0} \sum_{i=1}^{q} \frac{S_i(t) - \sigma_{ii}(E(t))}{c + \beta_i E_i(t)}\mathrm{d}t, \qquad (4.9)$$

$$\mathcal{E}_{\mathrm{r}}(E_0) \triangleq \inf_{S(\cdot)\in\mathcal{U}_{\mathrm{r}},\, T\geq -t_0} \int_{-T}^{t_0} \sum_{i=1}^{q} \beta_i E_i(t)[S_i(t) - \sigma_{ii}(E(t))]\mathrm{d}t, \qquad (4.10)$$

where $E(-T) = 0$ and $E(t_0) = E_0 \in \overline{\mathbb{R}}_+^q$.

Theorem 4.1. Consider the large-scale dynamical system $\mathcal{G}$ with power balance equation (3.5), and assume that Axiom *ii*) holds. Then there exist an entropy and an ectropy function for $\mathcal{G}$. Moreover, $\mathcal{S}_{\mathrm{a}}(E)$, $E \in \overline{\mathbb{R}}_+^q$, and $\mathcal{S}_{\mathrm{r}}(E)$, $E \in \overline{\mathbb{R}}_+^q$, are possible entropy functions for $\mathcal{G}$ with $\mathcal{S}_{\mathrm{a}}(0) = \mathcal{S}_{\mathrm{r}}(0) = 0$, and $\mathcal{E}_{\mathrm{a}}(E)$, $E \in \overline{\mathbb{R}}_+^q$, and $\mathcal{E}_{\mathrm{r}}(E)$, $E \in \overline{\mathbb{R}}_+^q$, are possible ectropy functions for $\mathcal{G}$ with $\mathcal{E}_{\mathrm{a}}(0) = \mathcal{E}_{\mathrm{r}}(0) = 0$. Finally, all entropy functions $\mathcal{S}(E)$, $E \in \overline{\mathbb{R}}_+^q$, for $\mathcal{G}$ satisfy

$$\mathcal{S}_{\mathrm{r}}(E) \leq \mathcal{S}(E) - \mathcal{S}(0) \leq \mathcal{S}_{\mathrm{a}}(E), \quad E \in \overline{\mathbb{R}}_+^q, \qquad (4.11)$$

and all ectropy functions $\mathcal{E}(E)$, $E \in \overline{\mathbb{R}}_+^q$, for $\mathcal{G}$ satisfy

$$\mathcal{E}_{\mathrm{a}}(E) \leq \mathcal{E}(E) - \mathcal{E}(0) \leq \mathcal{E}_{\mathrm{r}}(E), \quad E \in \overline{\mathbb{R}}_+^q. \qquad (4.12)$$

Proof. The proof is identical to the proofs of Theorems 3.2 and 3.6. □

The next series of results gives analogous results to the results in Sections 3.3 and 3.4 for the modified definitions of entropy and ectropy given in this chapter.

Theorem 4.2. Consider the large-scale dynamical system $\mathcal{G}$ with power balance equation (3.5), and let $\mathcal{S} : \overline{\mathbb{R}}_+^q \to \mathbb{R}$ and $\mathcal{E} : \overline{\mathbb{R}}_+^q \to \mathbb{R}$ be entropy and ectropy functions of $\mathcal{G}$, respectively. Then $\mathcal{S}(\cdot)$ and $\mathcal{E}(\cdot)$ are continuous on $\overline{\mathbb{R}}_+^q$.

Proof. The proof is identical to the proof of Theorem 3.3. □

For the statement of the next result, recall the definition of $\boldsymbol{p} = [1/\beta_1, \ldots, 1/\beta_q]^{\mathrm{T}}$ given in Proposition 4.1 and define $P \triangleq \mathrm{diag}[\beta_1, \ldots, \beta_q]$.

Proposition 4.2. Consider the large-scale dynamical system $\mathcal{G}$ with power balance equation (3.5), and assume that Axioms i) and ii) hold. Then at every equilibrium state $E_{\rm e} = \alpha \boldsymbol{p}$, $\alpha \geq 0$, of the isolated system $\mathcal{G}$, the entropy $\mathcal{S}(E)$, $E \in \overline{\mathbb{R}}_+^q$, and ectropy $\mathcal{E}(E)$, $E \in \overline{\mathbb{R}}_+^q$, functions of $\mathcal{G}$ are unique (modulo a constant of integration) and are given by

$$\mathcal{S}(E) - \mathcal{S}(0) = \mathcal{S}_{\rm a}(E) = \mathcal{S}_{\rm r}(E) = \boldsymbol{p}^{\rm T}\mathbf{log}_e(c\mathbf{e} + PE) - \mathbf{e}^{\rm T}\boldsymbol{p} \log_e c \tag{4.13}$$

and

$$\mathcal{E}(E) - \mathcal{E}(0) = \mathcal{E}_{\rm a}(E) = \mathcal{E}_{\rm r}(E) = \tfrac{1}{2}E^{\rm T}PE, \tag{4.14}$$

respectively, where $E = E_{\rm e}$ and $\mathbf{log}_e(c\mathbf{e} + PE)$ denotes the vector natural logarithm given by $[\log_e(c + \beta_1 E_1), \ldots, \log_e(c + \beta_q E_q)]^{\rm T}$.

Proof. The proof is identical to the proofs of Propositions 3.4 and 3.8. □

Proposition 4.3. Consider the large-scale dynamical system $\mathcal{G}$ with power balance equation (3.5), and assume that Axioms i) and ii) hold. Let $\mathcal{S}(\cdot)$ and $\mathcal{E}(\cdot)$ denote an entropy and ectropy of $\mathcal{G}$, respectively, and let $E : [t_0, t_1] \to \overline{\mathbb{R}}_+^q$ denote the solution to (3.5) with $E(t_0) = \alpha_0 \boldsymbol{p}$ and $E(t_1) = \alpha_1 \boldsymbol{p}$, where α_0, $\alpha_1 \geq 0$. Then

$$\mathcal{S}(E(t_1)) = \mathcal{S}(E(t_0)) + \int_{t_0}^{t_1} \sum_{i=1}^{q} \frac{S_i(t) - \sigma_{ii}(E(t))}{c + \beta_i E_i(t)} {\rm d}t \tag{4.15}$$

and

$$\mathcal{E}(E(t_1)) = \mathcal{E}(E(t_0)) + \int_{t_0}^{t_1} \sum_{i=1}^{q} \beta_i E_i(t)[S_i(t) - \sigma_{ii}(E(t))] {\rm d}t \tag{4.16}$$

if and only if there exists a continuous function $\alpha : [t_0, t_1] \to \overline{\mathbb{R}}_+$ such that $\alpha(t_0) = \alpha_0$, $\alpha(t_1) = \alpha_1$, and $E(t) = \alpha(t)\boldsymbol{p}$, $t \in [t_0, t_1]$.

Proof. The proof is identical to the proofs of Propositions 3.5 and 3.9. □

Theorem 4.3. Consider the large-scale dynamical system $\mathcal{G}$ with power balance equation (3.5), and assume that Axioms i) and ii) hold. Then the function $\mathcal{S} : \overline{\mathbb{R}}_+^q \to \mathbb{R}$ given by

$$\mathcal{S}(E) = \boldsymbol{p}^{\rm T}\mathbf{log}_e(c\mathbf{e} + PE) - \mathbf{e}^{\rm T}\boldsymbol{p} \log_e c, \quad E \in \overline{\mathbb{R}}_+^q, \tag{4.17}$$

is a unique (modulo a constant of integration), continuously differentiable entropy function of $\mathcal{G}$. Furthermore, the function $\mathcal{E} : \overline{\mathbb{R}}_+^q \to \mathbb{R}$ given by

$$\mathcal{E}(E) = \tfrac{1}{2}E^{\rm T}PE, \quad E \in \overline{\mathbb{R}}_+^q, \tag{4.18}$$

is a unique (modulo a constant of integration), continuously differentiable ectropy function of $\mathcal{G}$. In addition, for $E(t) \notin \mathcal{M}_{\mathrm{e}}$, $t \geq t_0$, where $E(t)$, $t \geq t_0$, denotes the solution to (3.5) and $\mathcal{M}_{\mathrm{e}} = \{E \in \overline{\mathbb{R}}_+^q : E = \alpha \boldsymbol{p}, \alpha \geq 0\}$, (4.17) and (4.18) satisfy

$$\mathcal{S}(E(t_2)) > \mathcal{S}(E(t_1)) + \int_{t_1}^{t_2} \sum_{i=1}^{q} \frac{S_i(t) - \sigma_{ii}(E(t))}{c + \beta_i E_i(t)} \mathrm{d}t \tag{4.19}$$

and

$$\mathcal{E}(E(t_2)) < \mathcal{E}(E(t_1)) + \int_{t_1}^{t_2} \sum_{i=1}^{q} \beta_i E_i(t)[S_i(t) - \sigma_{ii}(E(t))]\mathrm{d}t \tag{4.20}$$

for every $t_2 \geq t_1 \geq t_0$ and $S(\cdot) \in \mathcal{U}$.

Proof. The proof is identical to the proofs of Theorems 3.4 and 3.8. □

It is important to note that Theorem 4.3 establishes the existence of a unique entropy and ectropy function for $\mathcal{G}$ for equilibrium and nonequilibrium processes. Furthermore, it follows from Theorem 4.3 that the entropy and ectropy functions for $\mathcal{G}$ defined by (4.17) and (4.18) satisfy, respectively, (4.5) and (4.6) as equalities for an equilibrium process and as strict inequalities for a nonequilibrium process. Hence, it follows from Theorem 2.25 that the isolated large-scale dynamical system $\mathcal{G}$ does not exhibit Poincaré recurrence in $\overline{\mathbb{R}}_+^q \setminus \mathcal{M}_{\mathrm{e}}$.

Once again, inequality (4.5) is a generalized Clausius inequality for equilibrium and nonequilibrium thermodynamics, while inequality (4.6) is an anti–Clausius inequality. Moreover, for the ectropy function given by (4.18), inequality (4.6) shows that a thermodynamically consistent large-scale dynamical system model is dissipative with respect to the supply rate $E^{\mathrm{T}}PS$ and with storage function corresponding to the system ectropy $\mathcal{E}(E)$.

In addition, if we let $\mathrm{d}Q_i(t) = [S_i(t) - \sigma_{ii}(E(t))]\mathrm{d}t$, $i = 1, \ldots, q$, denote the infinitesimal amount of the net heat received or dissipated by the ith subsystem of $\mathcal{G}$ over the infinitesimal time interval $\mathrm{d}t$ at the (shifted) *absolute* ith *subsystem temperature* $T_i \triangleq c + \beta_i E_i$, then it follows from (4.5) that the system entropy varies by an amount

$$\mathrm{d}\mathcal{S}(E(t)) \geq \sum_{i=1}^{q} \frac{\mathrm{d}Q_i(t)}{c + \beta_i E_i(t)}, \quad t \geq t_0. \tag{4.21}$$

In light of the above definition of temperature, it is important to note that if $\beta_i \neq \beta_j$ for some $i \neq j$, then our thermodynamically consistent large-scale system model allows for the consideration of subsystems that possess the same stored energy, with one subsystem being hotter than the other.

Finally, note that the nonconservation of entropy and ectropy equations (3.104) and (3.105), respectively, for isolated large-scale dynamical systems also hold for the more general definitions of entropy and ectropy given in Definitions 4.1 and 4.2. In addition, using the modified definitions of entropy and ectropy given in Definitions 4.1 and 4.2 and using similar arguments as in Section 3.3, it can be shown that for every $E_0 \notin \mathcal{M}_{\rm e} = \{E \in \overline{\mathbb{R}}_+^q : E = \alpha \boldsymbol{p}, \alpha \geq 0\}$, the nonlinear dynamical system $\mathcal{G}$ with power balance equation (3.5) is state irreversible.

The following theorem is a generalization of Theorem 3.9.

Theorem 4.4. Consider the large-scale dynamical system $\mathcal{G}$ with power balance equation (3.5) with $S(t) \equiv 0$ and $d(E) \equiv 0$, and assume that Axioms *i*) and *ii*) hold. Then, for every $\alpha \geq 0$, $\alpha \boldsymbol{p}$ is a semistable equilibrium state of (3.5). Furthermore, $E(t) \to \frac{1}{\mathbf{e}^{\rm T}\boldsymbol{p}} \boldsymbol{p}\mathbf{e}^{\rm T} E(t_0)$ as $t \to \infty$ and $\frac{1}{\mathbf{e}^{\rm T}\boldsymbol{p}} \boldsymbol{p}\mathbf{e}^{\rm T} E(t_0)$ is a semistable equilibrium state. Finally, if for some $k \in \{1, \ldots, q\}$, $\sigma_{kk}(E) \geq 0$ and $\sigma_{kk}(E) = 0$ if and only if $E_k = 0$, then the zero solution $E(t) \equiv 0$ to (3.5) is a globally asymptotically stable equilibrium state of (3.5).

Proof. It follows from Axiom *i*) that $\alpha \boldsymbol{p} \in \overline{\mathbb{R}}_+^q$, $\alpha \geq 0$, is an equilibrium state of (3.5). To show Lyapunov stability of the equilibrium state $\alpha \boldsymbol{p}$, consider the system-shifted ectropy $\mathcal{E}_{\rm s}(E) = \frac{1}{2}(E - \alpha \boldsymbol{p})^{\rm T} P (E - \alpha \boldsymbol{p})$ as a Lyapunov function candidate. Now, the proof follows as in the proof of Theorem 3.9 by invoking Axiom *ii*) and noting that $\phi_{ij}(E) = -\phi_{ji}(E)$, $E \in \overline{\mathbb{R}}_+^q$, $i \neq j$, $i, j = 1, \ldots, q$, $P\boldsymbol{p} = \mathbf{e}$, and $\mathbf{e}^{\rm T} w(E) = 0$, $E \in \overline{\mathbb{R}}_+^q$. Alternatively, in the case where for some $k \in \{1, \ldots, q\}$, $\sigma_{kk}(E) \geq 0$ and $\sigma_{kk}(E) = 0$ if and only if $E_k = 0$, global asymptotic stability of the zero solution $E(t) \equiv 0$ to (3.5) follows from standard Lyapunov arguments using the system ectropy $\mathcal{E}(E) = \frac{1}{2} E^{\rm T} P E$ as a Lyapunov function candidate. □

It follows from Theorem 4.4 that the steady-state value of the energy in each subsystem $\mathcal{G}_i$ of the isolated large-scale dynamical system $\mathcal{G}$ is given by

$$E_\infty = \frac{1}{\mathbf{e}^{\rm T}\boldsymbol{p}} \boldsymbol{p}\mathbf{e}^{\rm T} E(t_0), \tag{4.22}$$

which implies that

$$E_{i\infty} = \frac{1}{\beta_i \mathbf{e}^{\rm T}\boldsymbol{p}} \mathbf{e}^{\rm T} E(t_0) \tag{4.23}$$

or, equivalently,

$$\hat{T}_{i\infty} = \beta_i E_{i\infty} = \frac{1}{\mathbf{e}^{\rm T}\boldsymbol{p}} \mathbf{e}^{\rm T} E(t_0). \tag{4.24}$$

Hence, the steady-state temperature of the isolated large-scale dynamical system $\mathcal{G}$ given by $\hat{T}_\infty = \frac{1}{\mathbf{e}^{\mathrm{T}}\boldsymbol{p}}\mathbf{e}^{\mathrm{T}}E(t_0)\mathbf{e}$ is uniformly distributed over all the subsystems of $\mathcal{G}$. This phenomenon is known as *temperature equipartition*, wherein all the system energy is eventually transformed into heat at a uniform temperature, and hence, all dynamic processes in $\mathcal{G}$ (system motions) would cease.

Proposition 4.4. Consider the large-scale dynamical system $\mathcal{G}$ with power balance equation (3.5), let $\mathcal{E} : \overline{\mathbb{R}}_+^q \to \overline{\mathbb{R}}_+$ and $\mathcal{S} : \overline{\mathbb{R}}_+^q \to \overline{\mathbb{R}}_+$ denote the ectropy and entropy functions of $\mathcal{G}$ given by (4.18) and (4.17), respectively, and define $\mathcal{D}_{\mathrm{c}} \triangleq \{E \in \overline{\mathbb{R}}_+^q : \mathbf{e}^{\mathrm{T}}E = \beta\}$, where $\beta \geq 0$. Then

$$\arg\min_{E\in\mathcal{D}_{\mathrm{c}}}(\mathcal{E}(E)) = \arg\max_{E\in\mathcal{D}_{\mathrm{c}}}(\mathcal{S}(E)) = E^* = \frac{\beta}{\mathbf{e}^{\mathrm{T}}\boldsymbol{p}}\boldsymbol{p}. \tag{4.25}$$

Furthermore, $\mathcal{E}_{\min} \triangleq \mathcal{E}(E^*) = \frac{1}{2}\frac{\beta^2}{\mathbf{e}^{\mathrm{T}}\boldsymbol{p}}$ and $\mathcal{S}_{\max} \triangleq \mathcal{S}(E^*) = \mathbf{e}^{\mathrm{T}}\boldsymbol{p}\log_e(c + \frac{\beta}{\mathbf{e}^{\mathrm{T}}\boldsymbol{p}}) - \mathbf{e}^{\mathrm{T}}\boldsymbol{p}\log_e c$.

Proof. The proof is identical to the proof of Proposition 3.11 and, hence, is omitted. □

Proposition 4.4 shows that when all the energy of the large-scale dynamical system $\mathcal{G}$ is transformed into heat at a uniform temperature, the system entropy is a maximum and the system ectropy is a minimum.

4.2 Boltzmann Thermodynamics

As noted in Chapter 1, Boltzmann [58] was the first to give a probabilistic interpretation of entropy using a microscopic point of view of molecules in a system. In particular, probability was used in the context of a measure of the variety of ways in which the molecules in a system can be rearranged without changing the macroscopic properties of the system. Specifically, realizing that a system macrostate can be represented by many different microstates involving different configurations of molecular motion, macroscopic phenomena can be derived from the microscopic dynamics. Hence, the entropy of an observed macroscopic state is defined as the logarithmic probability of its occurrence.

Since entropy measures probability and probability, in turn, expresses disorder, entropy is a measure of disorder; or, more specifically, a measure of energy dispersal at a specific temperature. This is perhaps best reflected in Boltzmann's kinetic theory of gases, in which the entropy of a gas, defined in terms of the probability distribution, increases as a more uniformly distributed state is reached when the gas diffuses from a filled container into an empty container. Hence, the entropy of the gas increases until the system reaches the configuration with the largest number of microscopic states, the most probable configuration. Since the final system state can be realized in many more ways than the initial, more organized system state, it has the highest probability and hence the maximal entropy.

In this section, we provide a deterministic kinetic theory interpretation of the steady-state expressions for the entropy and ectropy presented in this chapter. Specifically, we assume that each subsystem $\mathcal{G}_i$ of the large-scale dynamical system $\mathcal{G}$ is a simple system consisting of an ideal gas with rigid walls. Furthermore, we assume that all subsystems $\mathcal{G}_i$ are divided by *diathermal walls* (that is, walls that permit energy flow) and the overall dynamical system is a closed system (that is, the system is separated from the environment by a rigid adiabatic wall). In this case, $\beta_i = k/n_i$, $i = 1, \ldots, q$, where n_i, $i = 1, \ldots, q$, is the number of molecules in the ith subsystem and $k > 0$ is the *Boltzmann constant* (i.e., the gas constant per molecule).

The Boltzmann constant is equal to the ratio of the universal gas constant to Avogadro's number. This constant is physically significant as it provides a measure of the amount of energy transfer (i.e., heat) corresponding to the random thermal motions of the particles making up a substance. However, from a mathematical perspective it can be ignored. Hence, without loss of generality and for simplicity of exposition we set $k = 1$.

In our analysis, we assume that the molecules in the ideal gas are hard elastic spheres, that is, there are no forces between the molecules except during collisions, and the molecules are not deformed by collisions. Thus, there is no internal potential energy, and the system internal energy of the ideal gas is entirely kinetic. Hence, in this case, the temperature of each subsystem $\mathcal{G}_i$ is the average translational kinetic energy per molecule, which is consistent with the kinetic theory of ideal gases.

Definition 4.3. For a given isolated large-scale dynamical system $\mathcal{G}$ in *thermal equilibrium*, define the *equilibrium entropy* of $\mathcal{G}$ by

$$\mathcal{S}_{\mathrm{e}} = n \log_e \Big(c + \frac{\mathbf{e}^{\mathrm{T}} E_\infty}{n}\Big) - n \log_e c$$

and the *equilibrium ectropy* of $\mathcal{G}$ by

$$\mathcal{E}_{\mathrm{e}} = \frac{1}{2} \frac{(\mathbf{e}^{\mathrm{T}} E_\infty)^2}{n},$$

where $\mathbf{e}^{\mathrm{T}} E_\infty$ denotes the total steady-state energy of the large-scale dynamical system $\mathcal{G}$ and n denotes the total number of molecules in $\mathcal{G}$.

Note that the definitions of equilibrium entropy and equilibrium ectropy given in Definition 4.3 are entirely consistent with the equilibrium (maximum) entropy and equilibrium (minimum) ectropy given by Proposition 4.4. Next, assume that each subsystem $\mathcal{G}_i$ is initially in thermal equilibrium. Furthermore, for each subsystem, let E_i and n_i, $i = 1, \ldots, q$, denote the total internal energy and the number of molecules, respectively, in the ith subsystem. Hence, the entropy and ectropy of the ith subsystem are given by $\mathcal{S}_i = n_i \log_e(c + E_i/n_i) - n_i \log_e c$ and $\mathcal{E}_i = \frac{1}{2}\frac{E_i^2}{n_i}$, respectively. Next, note that the entropy and the ectropy of the overall system (after reaching a thermal equilibrium) are given by $\mathcal{S}_\mathrm{e} = n \log_e(c + \frac{\mathbf{e}^\mathrm{T} E_\infty}{n}) - n \log_e c$ and $\mathcal{E}_\mathrm{e} = \frac{1}{2}\frac{(\mathbf{e}^\mathrm{T} E_\infty)^2}{n}$.

Now, it follows from the convexity of $-\log_e(\cdot)$ and conservation of energy that the entropy of $\mathcal{G}$ at thermal equilibrium is given by

$$\begin{aligned} \mathcal{S}_\mathrm{e} &= n \log_e \left(c + \frac{\mathbf{e}^\mathrm{T} E_\infty}{n}\right) - n \log_e c \\ &= n \log_e \left[\sum_{i=1}^q \frac{n_i}{n}\left(c + \frac{E_i}{n_i}\right)\right] - \sum_{i=1}^q n_i \log_e c \\ &\geq n \sum_{i=1}^q \frac{n_i}{n} \log_e \left(c + \frac{E_i}{n_i}\right) - \sum_{i=1}^q n_i \log_e c \\ &= \sum_{i=1}^q \mathcal{S}_i. \end{aligned} \tag{4.26}$$

Furthermore, the ectropy of $\mathcal{G}$ at thermal equilibrium is given by

$$\begin{aligned} \mathcal{E}_\mathrm{e} &= \frac{1}{2}\frac{(\mathbf{e}^\mathrm{T} E_\infty)^2}{n} \\ &= \sum_{i=1}^q \frac{1}{2}\frac{E_i^2}{n_i} - \frac{1}{2n}\sum_{i=1}^q \sum_{j=i+1}^q \frac{(n_j E_i - n_i E_j)^2}{n_i n_j} \\ &\leq \sum_{i=1}^q \frac{1}{2}\frac{E_i^2}{n_i} \\ &= \sum_{i=1}^q \mathcal{E}_i. \end{aligned} \tag{4.27}$$

It follows from (4.26) (respectively, (4.27)) that the equilibrium entropy (respectively, ectropy) of the system (gas) $\mathcal{G}$ is always greater (respectively, less) than or equal to the sum of the entropies (respectively, ectropies) of the individual subsystems $\mathcal{G}_i$. Hence, the entropy (respectively, ectropy) of the gas increases (respectively, decreases) as a more evenly

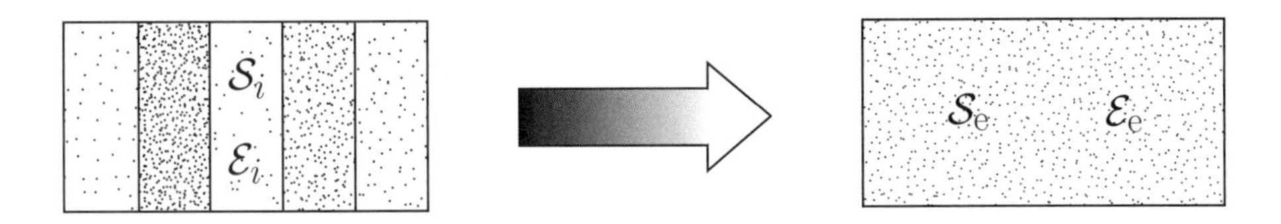

Figure 4.1 Entropy (respectively, ectropy) increases (respectively, decreases) as a more evenly distributed state is reached.

distributed (disordered) state is reached (see Figure 4.1).

Finally, note that it follows from (4.26) and (4.27) that $\mathcal{S}_\mathrm{e} = \sum_{i=1}^{q} \mathcal{S}_i$ and $\mathcal{E}_\mathrm{e} = \sum_{i=1}^{q} \mathcal{E}_i$ if and only if $\frac{E_i}{n_i} = \frac{E_j}{n_j}$, $i \neq j$, $i, j = 1, \ldots, q$, that is, the initial temperatures of all subsystems are equal. Furthermore, it follows from Axioms *i*) and *ii*) that the equality $\frac{E_i(t_0)}{n_i} = \frac{E_j(t_0)}{n_j}$, $i \neq j$, $i, j = 1, \ldots, q$, determines an equilibrium state and hence a state reversible process (i.e., $\frac{E_i(t)}{n_i} = \frac{E_j(t)}{n_j}$, $t \geq t_0$, $i \neq j$, $i, j = 1, \ldots, q$) for the system consisting of q ideal gases.

In light of the above, the following proposition is immediate.

Proposition 4.5. For every state reversible adiabatic process performed on a system consisting of q ideal gases connected by diathermal walls, the total entropy and total ectropy of the system remain constant.

4.3 Connections to Classical Thermodynamic Energy, Entropy, and Thermal Equilibria

In this section, we connect our thermodynamic model to classical thermodynamics for analyzing conductive heat flow in a homogeneous, isotropic, thermally insulated body. In addition, we use the language of classical thermodynamics and, whenever possible, connect the classical thermodynamic concepts to dynamical system-theoretic notions. Specifically, we consider a network of q lumped thermal masses interconnected by links along which heat can flow from one subsystem to another. We refer to an individual subsystem as $\mathcal{G}_i$, or simply subsystem i, and denote the composite system by $\mathcal{G} = \cup_{i=1}^{q} \mathcal{G}_i$.

In classical thermodynamics, *internal energy*, or simply *energy*, may be thought of as the potential of a system to perform work on other systems and on the environment. Energy may be transferred between systems by mass transfer, heat transfer, or work. Subsequently, in this section, energy is transferred only as heat. We denote the internal energy of subsystem i by E_i, and the vector of subsystem energies by $E = [E_1, \ldots, E_q]^{\mathrm{T}}$. The total internal energy of the system is the sum of the subsystem energies and is given by

$$U = \sum_{i=1}^{q} E_i = \mathbf{e}^{\mathrm{T}} E.$$

Recall that an *isolated* system does not exchange energy with any other system or the environment. However, energy can be exchanged between the subsystems comprising an isolated system.

Furthermore, recall that *entropy* is a measure of how well energy is distributed throughout a system. We denote the entropies of subsystem i by $\mathcal{S}_i$, and the vector of subsystem entropies by $\boldsymbol{\mathcal{S}} = [\mathcal{S}_1, \ldots, \mathcal{S}_q]^{\mathrm{T}}$. The system entropy is the sum of the subsystem entropies and is given by

$$\mathcal{S} = \sum_{i=1}^{q} \mathcal{S}_i = \mathbf{e}^{\mathrm{T}} \boldsymbol{\mathcal{S}}.$$

A higher system entropy indicates a more uniform distribution of energy among subsystems; the subsystem entropies themselves have no meaning in isolation. As shown in Chapter 3, for a fixed amount of total energy, the system entropy is maximal when the energy is *equipartitioned*, that is, when every energy storage mode has equal energy.

In classical thermodynamics, *temperature* is well defined only for a system or subsystem in equilibrium. Subsystem i is assumed to be in internal thermal equilibrium, with corresponding temperature $T_i \geq 0$. We denote the vector of subsystem temperatures by $\boldsymbol{T} = [T_1, \ldots, T_q]^{\mathrm{T}}$. Any addition or removal of heat is assumed to be sufficiently slow so that each subsystem remains in internal equilibrium. Since subsystems are generally not in equilibrium with each other, there is generally no meaning of "system temperature." When the subsystems are all at a single common temperature $\bar{T}$, then we say the system is in thermal equilibrium with temperature $\bar{T}$. We write $\bar{\boldsymbol{T}}$ for the temperature vector of a system in thermal equilibrium, that is, $\bar{\boldsymbol{T}} = \mathbf{e}\bar{T}$. In this section, we distinguish the notion of *thermal equilibrium*, meaning the condition of an isolated system at a uniform temperature, from the notion of *dynamic equilibrium*, meaning the condition of a dynamic system with time rate of change equal to zero.

In classical thermodynamics, energy, entropy, and temperature are related by the *fundamental thermodynamic relationship* [63, 252]. In the absence of mechanical work, this relationship can be written as

$$T_i = \left(\frac{\partial \mathcal{S}_i}{\partial E_i}\right)^{-1}. \tag{4.28}$$

We will define entropy as a function of energy, that is, $\mathcal{S}_i(E_i)$ and $\mathcal{S}(E)$. It is also convenient to write the entropies as functions of temperature, using $E_i(T_i)$. Thus, we define $E(\boldsymbol{T}) \triangleq [E_1(T_1), \ldots, E_q(T_q)]^{\mathrm{T}}$, $\tilde{\mathcal{S}}_i(T_i) \triangleq \mathcal{S}_i(E_i(T_i))$, and $\tilde{\mathcal{S}}(\boldsymbol{T}) \triangleq \mathcal{S}(E(\boldsymbol{T})) = \sum_{i=1}^{q} \tilde{\mathcal{S}}_i(T_i)$.

Let $\mathrm{d}Q_i$ be an infinitesimal amount of heat received by subsystem i. Since energy is assumed to be transferred only as heat, $\mathrm{d}E_i = \mathrm{d}Q_i$. The infinitesimal change in entropy that accompanies this heat addition is given by

$$\mathrm{d}\mathcal{S}_i = \left(\frac{\partial \mathcal{S}_i}{\partial E_i}\right)\mathrm{d}E_i = \frac{\mathrm{d}Q_i}{T_i}, \tag{4.29}$$

where subsystem $\mathcal{G}_i$ is in equilibrium at temperature T_i. In terms of heat flow rates, (4.29) can be written as

$$\frac{\mathrm{d}\mathcal{S}_i}{\mathrm{d}t} = \frac{1}{T_i}\frac{\mathrm{d}Q_i}{\mathrm{d}t} = \frac{q_i}{T_i}, \tag{4.30}$$

where $q_i \triangleq \dot{Q}_i$. Here, the time rate of change is assumed to be sufficiently slow so that the system remains in a slowly varying state of equilibrium. This state is referred to as a *quasi-equilibrium state* in classical thermodynamics.

Finally, we use the following classical versions of the zeroth, first, and second laws of thermodynamics.

Zeroth Law of Thermodynamics. If two subsystems are individually in thermal equilibrium with a third subsystem, then the two subsystems are also in thermal equilibrium with each other.

First Law of Thermodynamics. The increase in the internal energy of a subsystem is equal to the heat supplied to the subsystem.[1] The internal energy of an isolated system is constant.

Second Law of Thermodynamics. The entropy of an isolated system does not decrease.

We begin by considering heat transfer between a pair of subsystems. Let $\mathrm{d}Q_{ij}$ denote the heat transferred from subsystem j to subsystem i, and

[1] As noted in Chapter 3, this version of the first law holds only when work performed by the system on the environment and work done by the environment on the system are assumed to be zero.

let $q_{ij} \triangleq \dot{Q}_{ij}$ denote the rate of flow of heat from subsystem j to subsystem i. We denote the corresponding change (respectively, rate of change) in internal energy and entropy as $\mathrm{d}E_{ij}$ and $\mathrm{d}\mathcal{S}_{ij}$ (respectively, $\dot{E}_{ij}$ and $\dot{\mathcal{S}}_{ij}$). Here we note that in Chapter 3, $\dot{E}_{ij}$ is denoted by $\phi_{ij}(E)$ and corresponds to the net energy flow from the jth subsystem $\mathcal{G}_j$ to the ith subsystem $\mathcal{G}_i$. To be consistent with classical thermodynamic notation, however, in this section we use $\dot{E}_{ij}$ to denote this net energy flow.

It is intuitive to think of two subsystems being linked pairwise by a link that conducts heat, but does not store it. This notion is equivalent to pairwise energy conservation, that is, $q_{ji} = -q_{ij}$. If each subsystem is in quasi-equilibrium, then the second law of thermodynamics implies

$$\dot{\mathcal{S}} = \sum_{i=1}^{q} \dot{\mathcal{S}}_i = \sum_{i=1}^{q}\sum_{j=1}^{q} \dot{\mathcal{S}}_{ij} \geq 0.$$

We first consider the case where heat transfer is restricted to be between a single pair of subsystems, namely, subsystem i and subsystem j. This is the case when the system consists of only two subsystems, or when only two subsystems are physically connected. Then the total entropy change is given by

$$\mathrm{d}\mathcal{S} = \mathrm{d}\mathcal{S}_i + \mathrm{d}\mathcal{S}_j = \frac{\mathrm{d}Q_{ij}}{T_i} - \frac{\mathrm{d}Q_{ij}}{T_j} = \left(\frac{T_j - T_i}{T_i T_j}\right)\mathrm{d}Q_{ij},$$

or, equivalently, in terms of the time rate of change

$$\dot{\mathcal{S}} = \left(\frac{T_j - T_i}{T_i T_j}\right) q_{ij}.$$

By the second law of thermodynamics, $\mathrm{d}\mathcal{S} \geq 0$ or, equivalently, $\dot{\mathcal{S}} \geq 0$, implying $\mathrm{sgn}(\mathrm{d}Q_{ij}) = \mathrm{sgn}(T_j - T_i)$ or $\mathrm{sgn}(q_{ij}) = \mathrm{sgn}(T_j - T_i)$, where $\mathrm{sgn}(\sigma) \triangleq \sigma/|\sigma|$, $\sigma \neq 0$, and $\mathrm{sgn}(0) \triangleq 0$. Thus, for heat exchange between subsystem pairs, the second law of thermodynamics implies that if any heat is transferred, then it must move from the subsystem with the higher temperature to the subsystem with the lower temperature.

Definition 4.4. The heat transfer law between a pair of subsystems i and j is of *symmetric Fourier type* if it has the form

$$q_{ij} = \alpha_{ij}(T_j - T_i), \tag{4.31}$$

where $\alpha_{ij} : \overline{\mathbb{R}}_+ \to \overline{\mathbb{R}}_+$ is a function satisfying the following properties: *i*) the *sector bound condition*

$$\delta_1 \leq \frac{\alpha_{ij}(\xi)}{\xi} \leq \delta_2, \quad \xi \neq 0, \tag{4.32}$$

where $\alpha_{ij}(0) = 0$ and $0 < \delta_1 \leq \delta_2$; and *ii*) the *pairwise symmetry condition*

$$\alpha_{ji}(\xi) = -\alpha_{ij}(-\xi). \tag{4.33}$$

The linear Fourier law, in which $q_{ij} = k_{ij}(T_j - T_i)$ and $q_{ji} = k_{ij}(T_i - T_j)$, is a heat transfer law of symmetric Fourier type. For the heat transfer law between two subsystems to be of symmetric Fourier type, the subsystems must be connected; the "heat transfer law" for an unconnected pair of subsystems, $q_{ij} = q_{ji} = 0$, violates the first inequality in the sector bound (4.32).

If the heat transfer law between a pair of subsystems i and j is of symmetric Fourier type, then the energy transfer between those subsystems automatically satisfies the first and second laws of thermodynamics. To see that the first law of thermodynamics is satisfied, note that the time rate of change of the total internal energy of the subsystem pair is given by

$$\begin{aligned} \dot{E}_{ij} + \dot{E}_{ji} &= \alpha_{ij}(T_j - T_i) + \alpha_{ji}(T_i - T_j) \\ &= \alpha_{ij}(T_j - T_i) - \alpha_{ij}(T_j - T_i) \\ &= 0, \end{aligned} \tag{4.34}$$

where the second equality follows from the symmetry condition (4.33).

To see that the second law of thermodynamics is satisfied, note that the time rate of change of the total entropy of the subsystem pair is given by

$$\begin{aligned} \dot{S}_{ij} + \dot{S}_{ji} &= \frac{\alpha_{ij}(T_j - T_i)}{T_i} + \frac{\alpha_{ji}(T_i - T_j)}{T_j} \\ &= \left(\frac{1}{T_i} - \frac{1}{T_j}\right) \alpha_{ij}(T_j - T_i) \\ &= \frac{(T_j - T_i)}{T_i T_j} \alpha_{ij}(T_j - T_i) \\ &\geq 0, \end{aligned} \tag{4.35}$$

where the second equality follows from the symmetry condition (4.33) and the inequality follows from the sector bound condition (4.32). The sector bound condition also implies that equality holds in (4.35) if and only if $T_i = T_j$.

If every pair of subsystems in a system is either disconnected or subject to a heat transfer law of symmetric Fourier type, then the first and second laws of thermodynamics will be satisfied at the system level. To see that the first law of thermodynamics is satisfied, consider the rate of change of the total internal energy of the system given by

$$\begin{aligned}\dot{U} &= \sum_{i=1}^{q}\sum_{j=1}^{q}\dot{E}_{ij} \\ &= \sum_{i=1}^{q}\sum_{j=i+1}^{q}\left(\dot{E}_{ij}+\dot{E}_{ji}\right) \\ &= 0, \end{aligned} \tag{4.36}$$

where the second summand is zero due to (4.34) if the subsystem pair (i, j) is connected, or due to the fact that $\dot{E}_{ij} = -\dot{E}_{ji} = 0$ if the subsystem pair (i, j) is disconnected.

To see that the second law of thermodynamics is satisfied, consider the rate of change of the total entropy of the system given by

$$\begin{aligned}\dot{\mathcal{S}} &= \sum_{i=1}^{q}\sum_{j=1}^{q}\dot{\mathcal{S}}_{ij} \\ &= \sum_{i=1}^{q}\sum_{j=i+1}^{q}\left(\dot{\mathcal{S}}_{ij}+\dot{\mathcal{S}}_{ji}\right) \\ &\geq 0, \end{aligned} \tag{4.37}$$

where the second summand is nonnegative due to (4.35) if the subsystem pair (i, j) is connected, or due to the fact that $\dot{\mathcal{S}}_{ij} = \dot{\mathcal{S}}_{ji} = 0$ if the subsystem pair (i, j) is disconnected. Equality in (4.37) holds if and only if every connected pair of subsystems is at the same temperature.

Equations (4.36) and (4.37) are the first and second laws of thermodynamics as they pertain to interconnected systems composed of pairs of subsystems i and j. The basic principle here is that because each subsystem satisfies the laws of thermodynamics, the interconnected system also satisfies these laws with the internal energy and entropy being the sum of the internal energies and entropies of the subsystems. This corresponds to energy and entropy being extensive quantities; that is, the total system energy and entropy is the sum of the energies and entropies of the individual parts making up the whole system. Furthermore, the interconnection of the thermal subsystems strongly influences the equilibrium properties of the system. For example, if the system consists of two or more disjoint sets of subsystems, then these decoupled components will generally not be in thermal equilibrium.

Next, we use the framework of graph theory to describe the interconnection structure of the subsystems [22, 169]. Every subsystem is a vertex of a directed graph $\mathfrak{G}$, with *vertices* $\mathcal{V}_{\mathfrak{G}}$ and *directed edges*, or *arcs*, $\mathcal{E}_{\mathfrak{G}}$. The total number of vertices in $\mathfrak{G}$ is denoted by $n_{\mathfrak{G}}$. Arcs are written as ordered pairs (j, i). The arc (j, i) is said to *initiate* at j and *terminate* at i. Nodes j and i are called the *tail* and *head*, respectively, of the arc (j, i).

Loops are explicitly forbidden in the graphs considered here, that is, there are no arcs of the form (i, i). The total number of arcs in $\mathfrak{G}$ terminating at node i is the *in-degree* of i, denoted $d_{\mathfrak{G}}^{-}(i)$, and the total number of arcs in $\mathfrak{G}$ initiating at node i is the *out-degree* of i in $\mathfrak{G}$, denoted $d_{\mathfrak{G}}^{+}(i)$. If node j is the tail of an arc that terminates at i we say that j is a *direct predecessor* of i, and if node j is the head of an arc that initiates at i we say that j is a *direct successor* of i in $\mathfrak{G}$. We denote the set of all direct predecessors of i in $\mathfrak{G}$ by $\mathcal{P}_{\mathfrak{G}}(i)$, and we denote the set of all direct successors of i in $\mathfrak{G}$ by $\mathcal{S}_{\mathfrak{G}}(i)$. That is, $\mathcal{P}_{\mathfrak{G}}(i) \triangleq \{j : (j, i) \in \mathcal{E}_{\mathfrak{G}}\}$ and $\mathcal{S}_{\mathfrak{G}}(i) \triangleq \{j : (i, j) \in \mathcal{E}_{\mathfrak{G}}\}$.

A *strong path* in $\mathfrak{G}$ is an ordered sequence of arcs from $\mathcal{E}_{\mathfrak{G}}$ such that the head of any arc is the tail of the next. A strong path in $\mathfrak{G}$ can also be considered as a directed subgraph of $\mathfrak{G}$. A *strong cycle* is a strong path that begins and ends at the same vertex. Every strong cycle $\mathcal{C}$ satisfies $d_{\mathcal{C}}^{-}(i) = d_{\mathcal{C}}^{+}(i)$ for $i \in \mathcal{V}_C$. Vertices may appear in a strong cycle more than once, that is, $d_{\mathcal{C}}^{-}(i) = d_{\mathcal{C}}^{+}(i) \geq 1$ for $i \in \mathcal{V}_C$. If no vertex appears more than once, then the cycle is a *simple strong cycle.* For a simple strong cycle, $d_{\mathcal{C}}^{-}(i) = d_{\mathcal{C}}^{+}(i) = 1$ for $i \in \mathcal{V}_C$. A graph is *strongly connected* if a strong path exists from any vertex to any other vertex.

A property of a strongly connected graph is that it must contain a strong cycle that passes through every vertex in the graph at least once; we call such a cycle a *complete strong cycle.* A complete strong cycle need not include every edge. A strongly connected graph may contain more than one complete strong cycle. A complete strong cycle that contains each vertex exactly once is called a *simple complete strong cycle.* Every simple complete strong cycle satisfies $d_{\mathcal{C}}^{-}(i) = d_{\mathcal{C}}^{+}(i) = 1$ for $i \in \mathcal{V}_G$. The nodes of a simple complete strong cycle $\mathcal{C}$ can be renumbered so that $\mathcal{P}_{\mathcal{C}}(i) = \{i-1\}$ and $\mathcal{S}_{\mathcal{C}}(i-1) = \{i\}$ for $i = 1, \ldots, n_{\mathfrak{G}}$, where, for notational convenience, node 0 is identified with node $n_{\mathfrak{G}}$.

The *adjacency matrix* is $G \triangleq [g_{ij}]$, where $g_{ij} = 1$ if there is an edge initiating at j and terminating at i, that is, if $(j, i) \in \mathcal{E}_{\mathfrak{G}}$. Otherwise, $g_{ij} = 0$. By assumption, $[g_{ii}] = 0$. Though all edges are directed, we say that the system graph is *undirected* if $g_{ji} = g_{ij}$, that is, if G is symmetric. Otherwise, the system graph is said to be *directed.*

Heat transfer on an undirected graph can be considered pairwise on edges (i, j) and (j, i). Thus, the first and second laws of thermodynamics are automatically satisfied for a heat transfer law of symmetric Fourier type on an undirected graph. A linear heat flow law $q_{ij} = k_{ij}(T_j - T_i)$ is associated with the *weighted adjacency matrix* $K = [k_{ij}]$, which we also call the *thermal conductance matrix*. The symmetry condition implies $K = K^{\mathrm{T}}$ for the linear Fourier law.

Next, consider a network of q subsystems, each with thermal mass m_i, temperature T_i, energy E_i, and entropy $\mathcal{S}_i$. Let the interconnection structure be defined by the graph $\mathfrak{G}$. The system is in thermal equilibrium if and only if $\boldsymbol{T} = \bar{\boldsymbol{T}} = \mathbf{e}\bar{T}$ for every $\bar{T} > 0$. In the terminology of dynamical systems theory, every thermal equilibrium is a *nonisolated equilibrium point*,[2] since every thermal equilibrium with a slightly perturbed uniform temperature will also be an equilibrium point in the dynamical systems sense. Thermal systems have the property that, after a small perturbation, the system will return to thermal equilibrium, though typically at a slightly different temperature. In terms of concepts from dynamical systems theory, neither Lyapunov stability nor asymptotic stability capture this behavior. The relevant stability concept is that of semistability, which was introduced in Chapter 2.

We now proceed to analyze the stability of the thermal equilibria of a thermodynamic system. First, we define the entropy of the ith subsystem by

$$\mathcal{S}_i(E_i) \triangleq m_i \log_e(E_i). \tag{4.38}$$

Using (4.28), it follows that

$$T_i = \left(\frac{\partial \mathcal{S}_i}{\partial E_i}\right)^{-1} = \frac{E_i}{m_i}, \tag{4.39}$$

which gives the familiar equation $E_i = m_i T_i$ relating the energy and temperature of a lumped thermal mass. With this relationship, the subsystem and system entropies can be written directly as a function of temperature as

$$\tilde{\mathcal{S}}_i(T_i) \triangleq \mathcal{S}_i(m_i T_i) = m_i \log_e(m_i T_i), \quad i = 1, \ldots, q,$$

and

$$\tilde{\mathcal{S}}(\boldsymbol{T}) \triangleq \mathcal{S}(m_1 T_1, \ldots, m_q T_q) = \sum_{i=1}^{q} \tilde{\mathcal{S}}_i(T_i).$$

We refer to $\mathcal{S}(E)$ or $\tilde{\mathcal{S}}(\boldsymbol{T})$ as the *total entropy function.* Other forms of the entropy function may be chosen, as in Chapter 3; however, the features of those functions will not be required here. One consequence of choosing

[2] An equilibrium point $E_{\mathrm{e}} \in \mathbb{R}^q$ is said to be an *isolated equilibrium point* if and only if there exist $\varepsilon > 0$ such that $\mathcal{B}_\varepsilon(E_{\mathrm{e}})$ contains no equilibrium points other than E_{e}. E_{e} is *nonisolated* if it is not isolated.

(4.38) over the form given in Chapter 3 is that it does not satisfy the third law of thermodynamics (Nernst's theorem), which states that the entropy is zero when the absolute temperature is zero. This is not a significant drawback for the purposes of this classical presentation.

We first consider isolated systems, for which the total system energy is conserved. The set of feasible subsystem temperatures corresponding to a constant system energy U_0 is given by

$$\mathcal{T}(U_0) \triangleq \left\{ \boldsymbol{T} \in \overline{\mathbb{R}}_+^q : \sum_{i=1}^{q} m_i T_i = U_0 \right\}. \tag{4.40}$$

The entropy function has the following key property for isolated systems. For the statement of the following result, define $M \triangleq [m_1, \ldots, m_q]^{\mathrm{T}}$ and $m \triangleq \sum_{i=1}^{n} m_i = \mathbf{e}^{\mathrm{T}} M$.

Theorem 4.5. The total entropy function $\tilde{\mathcal{S}}(\boldsymbol{T})$, with subsystem entropies given by $\tilde{\mathcal{S}}_i(T_i) = m_i \log_e(m_i T_i)$, $i = 1, \ldots, q$, and restricted to $\mathcal{T}(U_0)$, has a unique maximum at $\boldsymbol{T}^* = \mathbf{e} T^*$, where $T^* \triangleq U_0/m$. Equivalently, the total entropy function $\mathcal{S}(E)$ restricted to the set

$$\left\{ E \in \overline{\mathbb{R}}_+^q : \sum_{i=1}^{q} E_i = U_0 \right\}$$

has a unique maximum at $E^* = [E_1^*, \ldots, E_q^*]^{\mathrm{T}}$, where $E_i^* \triangleq m_i T^*$.

Proof. The entropy of the ith subsystem is given by

$$\begin{aligned} \tilde{\mathcal{S}}_i(T_i) &= m_i \log_e(m_i T_i) \\ &= m_i \log_e(m_i T_i) - m_i \log_e(m_i T^*) + m_i \log_e(m_i T^*) \\ &= m_i \log_e(T_i / T^*) + m_i \log_e(m_i T^*). \end{aligned}$$

The corresponding total system entropy is given by

$$\tilde{\mathcal{S}}(\boldsymbol{T}) = \log_e \left(\frac{T_1^{m_1} T_2^{m_2} \cdots T_q^{m_q}}{T^{*(m_1 + m_2 + \cdots + m_q)}} \right) + \mathcal{S}^*,$$

where $\mathcal{S}^* \triangleq \sum_{i=1}^{q} m_i \log_e(m_i T^*)$ is the value of the entropy function at thermal equilibrium.

Now, writing

$$\left(\frac{T_1^{m_1} T_2^{m_2} \cdots T_q^{m_q}}{T^{*(m_1 + m_2 + \cdots + m_q)}} \right) = \left(\frac{T_1^{\mu_1} T_2^{\mu_2} \cdots T_q^{\mu_q}}{T^*} \right)^m,$$

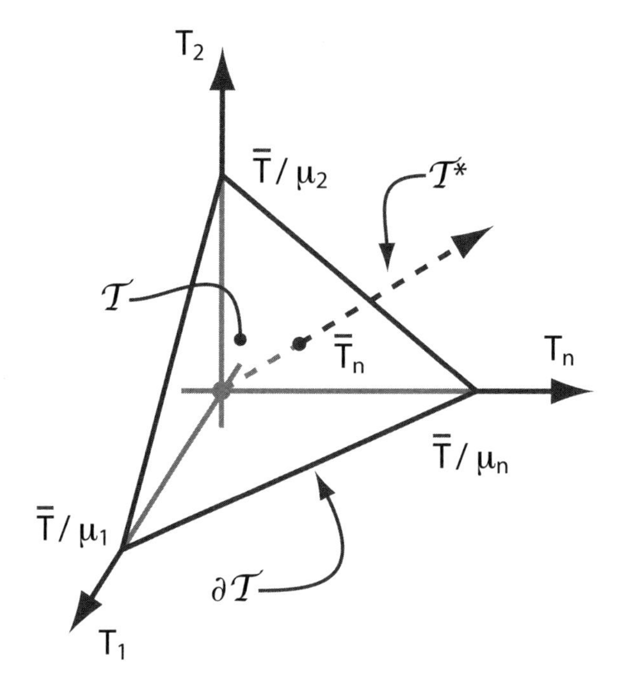

Figure 4.2 The subset of feasible temperatures $\mathcal{T}(U)$ corresponding to total system energy U.

where $\mu_i \triangleq m_i/m$, and noting that

$$T^* = U_0/m = (1/m)\sum_{i=1}^{q} m_i T_i = \sum_{i=1}^{q} \mu_i T_i,$$

it follows that

$$\frac{T_1^{\mu_1} T_2^{\mu_2} \cdots T_q^{\mu_q}}{T^*} = \frac{T_1^{\mu_1} T_2^{\mu_2} \cdots T_q^{\mu_q}}{\mu_1 T_1 + \mu_2 T_2 + \cdots + \mu_q T_q} \le 1,$$

where the inequality follows from the generalized power mean inequality [206], which also implies that equality holds if and only if all the T_i are equal; that is, if and only if $\boldsymbol{T}$ is of the form $\mathbf{e}T$. However, since $T\sum_{i=1}^{q} m_i = Tm = U_0$, the only $\boldsymbol{T}$ of this form that satisfies the energy constraint is $\mathbf{e}T^*$. Hence, $\tilde{\mathcal{S}}(\boldsymbol{T})$ achieves a maximum value of $\mathcal{S}^*$ at $\boldsymbol{T} = \boldsymbol{T}^*$.

The form of the theorem with energy as the independent variable follows from the equality of the two forms of the entropy function $\mathcal{S}(E) = \tilde{\mathcal{S}}(\boldsymbol{T})$, which implies that $\mathcal{S}(E)$ must have a unique maximum at $E^* = [E_1^*, \ldots, E_q^*]^{\mathrm{T}}$, where $E_i^* = m_i T^*$, $i = 1, \ldots, q$. □

One might surmise from Theorem 4.5 that the maximum entropy does not occur when energy is equipartitioned, since the E_i's are not equal, but are instead weighted by the thermal masses m_i. However, energy equipartition refers to uniform distribution of energy over all storage modes, and the subsystems with higher thermal mass have a larger number of storage modes. Maximum entropy corresponds to energy equipartition in this sense.

For a positive system energy U, we denote the temperature vector corresponding to the maximum entropy point by

$$\boldsymbol{T}^*(U) \triangleq (U/m)\mathbf{e}. \tag{4.41}$$

Figure 4.2 depicts the set $\mathcal{T}(U)$, which is the set of feasible subsystem temperatures corresponding to system energy U. We denote the set of feasible temperatures for any positive total system energy, which is the union of $\mathcal{T}(U)$ for all $U \geq 0$, by $\mathcal{T}$. Figure 4.2 also depicts the set $\mathcal{T}^* \triangleq \cup_{U>0}\boldsymbol{T}^*(U)$, that is, the set of all $\boldsymbol{T}^*(U)$ corresponding to a positive total system energy. Finally, the boundary of the feasible set $\mathcal{T}(U)$ is given by

$$\partial\mathcal{T}(U) = \bigcup_{j=1}^{q}\left\{\boldsymbol{T} \in \mathcal{T}(U) \,:\, T_j = 0 \,\text{and}\, \sum_{i=1}^{q}\mu_i T_i = U/m\right\}.$$

Since every point on the boundary $\partial\mathcal{T}(U)$ contains at least one $T_i = 0$, it follows that $\lim_{T\to\partial\mathcal{T}} \tilde{\mathcal{S}}(T) = -\infty$.

The following three properties characterize a thermal equilibrium for isolated thermodynamic systems.

Property 1: The total system energy is conserved, that is, $U(t) = U_0 \triangleq U(0)$ for all $t \geq 0$.

Property 2: $\boldsymbol{T}^*(U_0)$ is the unique equilibrium point of the system in $\mathcal{T}(U_0)$. $\boldsymbol{T}^*(U_0)$ is asymptotically stable in $\mathcal{T}(U_0)$.

Property 3: Every $\boldsymbol{T}^*(U_0)$ is a nonisolated equilibrium point in $\mathcal{T}$. The set $\mathcal{T}^*$ is semistable. Every trajectory starting in $\mathcal{T}$ converges to a point in $\mathcal{T}^*$.

Next, we state our main theorem of this section for thermodynamic systems satisfying a heat transfer law of the symmetric Fourier type.

Theorem 4.6. Consider a network of interconnected thermal masses with a strongly connected and undirected system graph $\mathfrak{G}$. If the energy flow between the subsystems of $\mathcal{G}$ is governed by a heat transfer law of the symmetric Fourier type, then Properties 1–3 hold.

Proof. First, recall that a heat transfer law of the symmetric Fourier type conserves energy if the network is undirected, and hence, Property 1 holds. Next, note that the system of thermal masses with an interconnection law of the symmetric Fourier type has dynamics

$$\dot{T}_i = \frac{1}{m_i} \sum_{j \in \mathcal{P}_{\mathfrak{G}}(i)} \alpha_{ij}(T_j - T_i), \quad i = 1, \ldots, q. \tag{4.42}$$

To show Property 2 holds, first note that every $\boldsymbol{T} = \mathbf{e}T$ is an equilibrium point of (4.42). Furthermore, these points are the *only* equilibrium points. To see this, consider any $\boldsymbol{T}$ not of the form $\mathbf{e}T$, and note that there are a finite number of subsystems. Hence, there must exist at least one subsystem with maximum temperature $T_{\max}$. Since the network is strongly connected, at least one of the subsystems at $T_{\max}$ must have at least one neighbor with a lower temperature T_{lower}. Thus, $\dot{T}_{\text{lower}} < 0$, and hence, this system is not in equilibrium. Finally, since $\boldsymbol{T}^*(U_0) = (U_0/m)\mathbf{e}$ is the only point of the form $\mathbf{e}T$ in $\mathcal{T}(U_0)$, this is the unique equilibrium point.

To show asymptotic stability on $\mathcal{T}(U_0)$, let $\tilde{\mathcal{S}}^* = \tilde{\mathcal{S}}(\boldsymbol{T}^*(U_0))$. By Theorem 4.5, $\tilde{\mathcal{S}}^*$ is a unique maximum for the entropy function, and hence, $\tilde{\mathcal{S}}(\boldsymbol{T}) - \tilde{\mathcal{S}}^* \leq 0$, with equality holding if and only if $\boldsymbol{T} = \boldsymbol{T}^*(U_0)$. The function

$$V_1(\boldsymbol{T}) = \frac{1}{2}\left[\tilde{\mathcal{S}}(\boldsymbol{T}) - \tilde{\mathcal{S}}^*\right]^2$$

is zero at $\boldsymbol{T}^*(U_0)$ and positive at every other point in $\mathcal{T}(U_0)$. Furthermore, $\dot{V}_1(\boldsymbol{T})$ is zero at $\boldsymbol{T}^*(U_0)$ and negative at every other point in $\mathcal{T}(U_0)$. All that remains to show asymptotic stability of $\boldsymbol{T}^*(U_0)$ is to shift the origin of $V_1(\boldsymbol{T})$ to $\boldsymbol{T}^*(U_0)$ and show that the set $\mathcal{T}(U_0)$ is forward invariant.

To see this, define

$$V(\boldsymbol{T}') \triangleq \frac{1}{2}\left[\tilde{\mathcal{S}}(\boldsymbol{T}' + \boldsymbol{T}^*(U_0)) - \tilde{\mathcal{S}}^*\right]^2$$

and note that $V(\boldsymbol{T}')$ is positive definite on $\mathcal{T}(U_0)$ and $\dot{V}(\boldsymbol{T}')$ is negative definite on $\mathcal{T}(U_0)$. To show forward invariance of the set $\mathcal{T}(U_0)$, consider the portion of the boundary of $\mathcal{T}(U_0)$ corresponding to $T_i = 0$ and take the inner product between the vector $\dot{\boldsymbol{T}}$ and the unit vector $\hat{\mathbf{u}}_i$ in the direction of T_i. This inner product is equal to the ith component of $\dot{\boldsymbol{T}}$ evaluated at the boundary point, that is,

$$\dot{\boldsymbol{T}}^{\mathrm{T}}\hat{\mathbf{u}}_i\Big|_{T_i=0} = \frac{1}{m_i} \sum_{j \in \mathcal{N}_{\mathfrak{G}}(i)} \alpha_{ij}(T_j - T_i)\Bigg|_{T_i=0} = \frac{1}{m_i} \sum_{j \in \mathcal{N}_{\mathfrak{G}}(i)} \alpha_{ij}(T_j). \tag{4.43}$$

Now, by the sector bound condition (4.32), $\alpha_{ij}(T_j)/T_j > 0$, and hence, since $T_j \geq 0$ for $j = 1, \ldots, q$, with at least one $T_j > 0$, it follows that

$\dot{\boldsymbol{T}}^{\mathrm{T}}\hat{\mathbf{u}}_i > 0$ at $T_i = 0$. This implies that the trajectories of the system point into $\mathcal{T}(U_0)$ along the boundary $\partial\mathcal{T}(U_0)$, and hence, $\mathcal{T}(U_0)$ is an invariant set of (4.42). Thus, we conclude [191] that $\boldsymbol{T}^*(U_0)$ is asymptotically stable in $\mathcal{T}(U_0)$, with region of attraction equal to all of $\mathcal{T}(U_0)$. This completes the proof of Property 2.

To show the first part of Property 3, consider the equilibrium point $\boldsymbol{T}^*(U_0)$ and note that for every τ that does not result in negative temperatures, every point of the form $\mathbf{e}(T^*(U_0)+\tau)$ is also an equilibrium point of (4.42). Thus, every neighborhood of $\boldsymbol{T}^*(U_0)$ in $\mathcal{T}$ contains another equilibrium point, and hence, $\boldsymbol{T}^*(U_0)$ is nonisolated.

To show that $\boldsymbol{T}^*(U_0)$ is semistable, we first show that $\boldsymbol{T}^*(U_0)$ is Lyapunov stable. To do this, consider the open ball of radius δ_1 in $\mathcal{T}$ centered on $\boldsymbol{T}^*(U_0)$. Furthermore, consider any initial point $\boldsymbol{T}'$ in this ball and denote its energy by U'. Since energy is conserved by a heat transfer law of the symmetric Fourier type, the resulting trajectory will converge to equilibrium point $\boldsymbol{T}^*(U') = \mathbf{e}(U'/m)$. Note that by (4.39), U' satisfies $|U'-U_0| \le \bar{m}\delta_1$, where $\bar{m} \triangleq \max_{i=1,\ldots,q} m_i$. Therefore,

$$\|\boldsymbol{T}^*(U') - \boldsymbol{T}^*(U_0)\| = \|\mathbf{e}(U'-U_0)/m\| = \sqrt{n}|U'-U_0|/m \le \sqrt{n}\delta_1\bar{\mu},$$

where $\bar{\mu} = \bar{m}/m$.

Next, given any $\varepsilon > 0$, choose $\delta_1 = \varepsilon/(2\sqrt{n}\bar{\mu})$. Now, since $\boldsymbol{T}^*(U')$ is asymptotically stable in $\mathcal{T}(U')$, for every $\varepsilon > 0$, choose $\delta_2 > 0$ such that $\|\boldsymbol{T}(t) - \boldsymbol{T}^*(U')\| < (\varepsilon/2)$ for all $t > 0$ and $\|\boldsymbol{T}'\| \le \delta_2$. Then, given any $\varepsilon > 0$, choose δ_1 and δ_2 as above, and require that $\|\boldsymbol{T}' - \boldsymbol{T}(U_0)\| < \min(\delta_1, \delta_2)$, ensuring that

$$\|\boldsymbol{T}(t) - \boldsymbol{T}(U_0)\| \le \|\boldsymbol{T}(t) - \boldsymbol{T}(U')\| + \|\boldsymbol{T}(U') - \boldsymbol{T}(U_0)\| < \varepsilon, \quad t \ge 0.$$

This shows that $\boldsymbol{T}^*(U_0)$ is Lyapunov stable. Semistability of $\boldsymbol{T}^*(U_0)$ now follows immediately from the asymptotic stability of $\boldsymbol{T}^*(U')$ in $\mathcal{T}(U')$ and the Lyapunov stability of $\boldsymbol{T}^*(U')$ in $\mathcal{T}$.

Finally, note that, since the system energy is conserved for every trajectory starting in $\mathcal{T}$, for every initial point in $\mathcal{T}$, the system will converge to a point in $\mathcal{T}^*$, which proves that $\mathcal{T}^*$ is semistable. $\square$

Chapter Five

Work, Heat, and the Carnot Cycle

5.1 On the Equivalence of Work and Heat: The First Law Revisited

In Chapter 3, we showed that the first law of thermodynamics is essentially a statement of the principle of the conservation of energy. Hence, the variation in energy of a dynamical system $\mathcal{G}$ during any transformation is equal to the amount of energy that the system receives from the environment. In Chapter 3, however, the notion of energy that the system receives from the environment and dissipates to the environment was limited to heat and did not include *work*, that is, motion against an opposing force. When external forces act on the dynamical system $\mathcal{G}$, they can produce work on the system, changing the system's internal energy. Thus, addressing work performed by the system on the environment and work done by the environment on the system plays a crucial role in the principle of the conservation of energy for thermodynamic systems.

In this chapter, we augment our nonlinear compartmental dynamical system model with an additional (deformation) state representing compartmental volumes in order to introduce the notion of work into our thermodynamically consistent energy flow model. Specifically, using Figure 5.1, we characterize a power balance equation as well as a volume velocity balance equation such that during a dynamical transformation, the large-scale system $\mathcal{G}$ can perform (positive) work on its surroundings or the surroundings can do (negative) work on $\mathcal{G}$, resulting in subsystem volume changes. Furthermore, we assume that each compartment can perform work on the other compartments.

In this case, the power balance equation (3.5) takes the new form involving energy and deformation states given by

$$\begin{aligned}\dot{E}(t) &= f(E(t), V(t)) + f_{\rm w}(E(t), V(t)) - d_{\rm w}(E(t), V(t)) + S_{\rm w}(t) \\ &\quad -d(E(t), V(t)) + S(t), \quad E(t_0) = E_0, \quad t \geq t_0, \end{aligned} \tag{5.1}$$

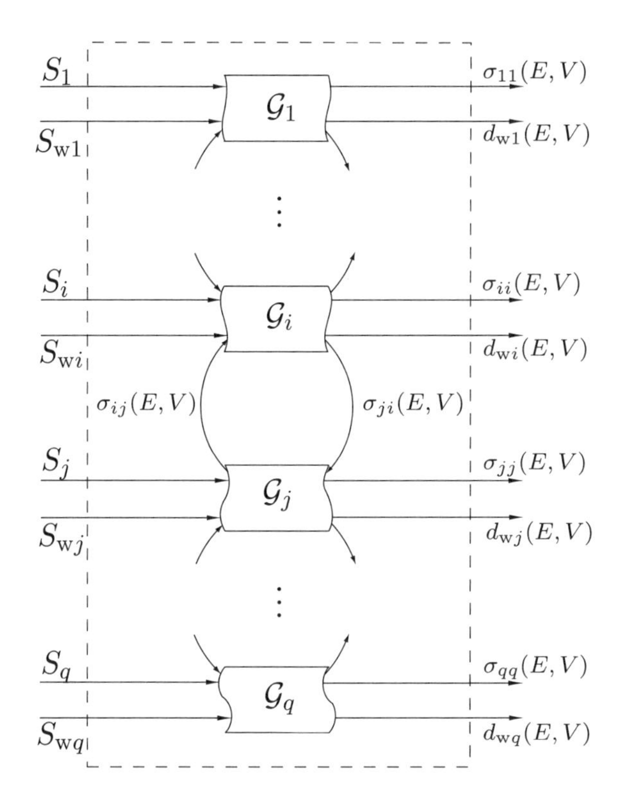

Figure 5.1 Large-scale dynamical system $\mathcal{G}$.

$$\dot{V}_i(t) = \frac{[d_{wi}(E(t),V(t)) - f_{wi}(E(t),V(t)) - S_{wi}(t)]V_i(t)}{(c+E_i(t))},$$
$$V_i(t_0) = V_{i0}, \quad i = 1,\ldots,q, \tag{5.2}$$

where $c > 0$, $V(t) \triangleq [V_1(t),\ldots,V_q(t)]^{\mathrm{T}} \in \mathbb{R}_+^q$, $t \geq t_0$, $V_i : [0,\infty) \to \mathbb{R}_+$, $i = 1,\ldots,q$, denotes the volume of the ith subsystem, $V_{i0} > 0$, $i = 1,\ldots,q$, $d_{\mathrm{w}}(E,V) = [d_{\mathrm{w}1}(E,V),\ldots,d_{\mathrm{w}q}(E,V)]^{\mathrm{T}}$, $d_{\mathrm{w}i} : \overline{\mathbb{R}}_+^q \times \mathbb{R}_+^q \to \overline{\mathbb{R}}_+$, $i = 1,\ldots,q$, denotes the instantaneous rate of work done by the ith subsystem on the environment, $S_{\mathrm{w}}(t) = [S_{\mathrm{w}1}(t),\ldots,S_{\mathrm{w}q}(t)]^{\mathrm{T}}$, $t \geq t_0$, $S_{\mathrm{w}i} : [0,\infty) \to \overline{\mathbb{R}}_+$, $i = 1,\ldots,q$, denotes the instantaneous rate of work done by the environment on the ith subsystem, $d(E,V) = [\sigma_{11}(E,V),\ldots,\sigma_{qq}(E,V)]^{\mathrm{T}}$, $\sigma_{ii} : \overline{\mathbb{R}}_+^q \times \mathbb{R}_+^q \to \overline{\mathbb{R}}_+$, $i = 1,\ldots,q$, denotes the instantaneous rate of energy (heat) dissipation from the ith subsystem to the environment, $S(t) = [S_1(t),\ldots,S_q(t)]^{\mathrm{T}}$, $t \geq t_0$, $S_i : [0,\infty) \to \mathbb{R}$, $i = 1,\ldots,q$, denotes the external power (heat flux) supplied to (or extracted from) the ith subsystem, $f(E,V) \triangleq [f_1(E,V),\ldots,f_q(E,V)]^{\mathrm{T}}$, $f_i(E,V) = \sum_{j=1,\, j\neq i}^{q} [\sigma_{ij}(E,V) - \sigma_{ji}(E,V)]$, $\sigma_{ij} : \overline{\mathbb{R}}_+^q \times \mathbb{R}_+^q \to \overline{\mathbb{R}}_+$, $i \neq j$, $i,j = 1,\ldots,q$, denotes the instantaneous rate of energy (heat) flow from the jth subsystem to the ith subsystem, $f_{\mathrm{w}}(E,V) \triangleq [f_{\mathrm{w}1}(E,V),$

$\ldots, f_{\mathrm{w}q}(E,V)]^{\mathrm{T}}$, $f_{\mathrm{w}i}(E,V) = \sum_{j=1,\, j\neq i}^{q}[p_j(E,V) - p_i(E,V)]v_{ij}(E,V)$, p_i, $i = 1,\ldots,q$, denotes the pressure of ith subsystem, and $v_{ij} : \overline{\mathbb{R}}_+^q \times \mathbb{R}_+^q \to \overline{\mathbb{R}}_+$, $i \neq j$, $i,j = 1,\ldots,q$, denotes the instantaneous time rate of change in volume (i.e., volume velocity) of the ith subsystem due to the pressure exerted by the jth subsystem. Note that $v_{ij}(E,V) = v_{ji}(E,V)$.

As in Chapter 3, we assume that $\sigma_{ij}(E,V) = 0$, $E \in \overline{\mathbb{R}}_+^q$, $V \in \mathbb{R}_+^q$, whenever $E_j = 0$, $i,j = 1,\ldots,q$, and $S_i(t) \geq 0$ whenever $E_i(t) = 0$, $t \geq t_0$, $i = 1,\ldots,q$. Moreover, we assume that $d_{\mathrm{w}i}(E,V) = 0$, $E \in \overline{\mathbb{R}}_+^q$, $V \in \mathbb{R}_+^q$, whenever $E_i = 0$, $i = 1,\ldots,q$, which implies that if the energy of the ith subsystem is zero, then this subsystem cannot perform work on the environment. Finally, we assume that $v_{ij} : \overline{\mathbb{R}}_+^q \times \mathbb{R}_+^q \to \overline{\mathbb{R}}_+$, $i \neq j$, $i,j = 1,\ldots,q$, $\sigma_{ij} : \overline{\mathbb{R}}_+^q \times \mathbb{R}_+^q \to \overline{\mathbb{R}}_+$, $i,j = 1,\ldots,q$, $p_i : \overline{\mathbb{R}}_+^q \times \mathbb{R}_+^q \to \overline{\mathbb{R}}_+$, $i = 1,\ldots,q$, and $d_{\mathrm{w}i} : \overline{\mathbb{R}}_+^q \times \mathbb{R}_+^q \to \overline{\mathbb{R}}_+$, $i = 1,\ldots,q$, are locally Lipschitz continuous on $\overline{\mathbb{R}}_+^q \times \mathbb{R}_+^q$, and $S_i : [0,\infty) \to \mathbb{R}$, $i = 1,\ldots,q$, and $S_{\mathrm{w}i} : [0,\infty) \to \overline{\mathbb{R}}_+$, $i = 1,\ldots,q$, are piecewise continuous and bounded over the semi-infinite interval $[0,\infty)$. The above assumptions guarantee that the solution $[E^{\mathrm{T}}(t), V^{\mathrm{T}}(t)]^{\mathrm{T}}$, $t \geq t_0$, to (5.1) and (5.2) exists and is nonnegative for all nonnegative initial conditions. Finally, note that (5.2) can be written in vector form as

$$\dot{V}(t) = D_p(E(t),V(t))[d_{\mathrm{w}}(E(t),V(t)) - f_{\mathrm{w}}(E(t),V(t)) - S_{\mathrm{w}}(t)], \quad V(t_0) = V_0, \quad t \geq t_0, \quad (5.3)$$

where $V_0 \in \mathbb{R}_+^q$ and $D_p(E,V) \triangleq \mathrm{diag}\left[\frac{V_1}{c+E_1}, \ldots, \frac{V_q}{c+E_q}\right]$, $E \in \overline{\mathbb{R}}_+^q$, $V \in \mathbb{R}_+^q$.

It follows from (5.1) and (5.2) that positive work done by a subsystem on the environment leads to a decrease in the internal energy of the subsystem and an increase in the subsystem volume, which is consistent with the first law of thermodynamics. To see that (5.1) and (5.2) is a statement of the first law of thermodynamics, define the work L done by the large-scale dynamical system $\mathcal{G}$ over the time interval $[t_1, t_2]$ by

$$L \triangleq \int_{t_1}^{t_2} \mathbf{e}^{\mathrm{T}}[d_{\mathrm{w}}(E(t),V(t)) - S_{\mathrm{w}}(t)]\mathrm{d}t, \quad (5.4)$$

where $[E^{\mathrm{T}}(t), V^{\mathrm{T}}(t)]^{\mathrm{T}}$, $t \geq t_0$, is the solution to (5.1) and (5.2). Then, premultiplying (5.1) by $\mathbf{e}^{\mathrm{T}}$ and using the fact that $\mathbf{e}^{\mathrm{T}} f(E,V) \equiv 0$ and $\mathbf{e}^{\mathrm{T}} f_{\mathrm{w}}(E,V) \equiv 0$, it follows that

$$\Delta U = -L + Q, \quad (5.5)$$

where $\Delta U = U(t_2) - U(t_1) \triangleq \mathbf{e}^{\mathrm{T}} E(t_2) - \mathbf{e}^{\mathrm{T}} E(t_1)$ denotes the variation in total energy of the large-scale system $\mathcal{G}$ over the time interval $[t_1, t_2]$ and

$$Q \triangleq \int_{t_1}^{t_2} \mathbf{e}^{\mathrm{T}}[S(t) - d(E(t), V(t))]\mathrm{d}t \tag{5.6}$$

denotes the net energy received ($Q > 0$) or dissipated ($Q < 0$) by $\mathcal{G}$ in forms other than work. This is a statement of the *first law of thermodynamics* for the large-scale dynamical system $\mathcal{G}$ and gives a precise formulation of the equivalence between work and heat. This establishes that heat and mechanical work are two different aspects of energy.

For a cyclic transformation, the initial and final states of $\mathcal{G}$ are the same, and hence, the variation in energy is zero, that is, $\Delta U = 0$. Thus, (5.5) becomes

$$L = Q, \tag{5.7}$$

which shows that the work performed by the system over a cyclic transformation is equal to the net difference of the heat absorbed and the heat surrendered by the system. Finally, note that (5.2) is consistent with the classical thermodynamic equation for the rate of work done by the system on the environment with $\frac{c+E_i}{V_i}$ playing the role of subsystem *pressures*. To see this, note that (5.2) can be equivalently written as

$$\mathrm{d}L_i = \left(\frac{c + E_i}{V_i}\right)\mathrm{d}V_i = p_i\mathrm{d}V_i, \tag{5.8}$$

which, for a single subsystem with volume V and pressure p, has the classical form

$$\mathrm{d}L = p\mathrm{d}V. \tag{5.9}$$

If the total energy of the large-scale dynamical system $\mathcal{G}$ at the initial and final states is fixed, then it follows from (5.5) that the variation[1] δ of the difference between the work done by the large-scale dynamical system $\mathcal{G}$ on the environment and the energy supplied to the large-scale dynamical system $\mathcal{G}$ satisfies

$$\delta(L - Q) = 0. \tag{5.10}$$

Equation (5.10) implies that if during a transformation between two fixed points the large-scale dynamical system $\mathcal{G}$ receives a fixed amount of energy, then the amount of work that the large-scale dynamical system can perform on the environment is also fixed. In other words, for any two paths connecting the initial and final states of the dynamical system

[1]The variation of a function is the supremum of the oscillations of that function over all finite partitions of a given interval.

$\mathcal{G}$ corresponding to identical energy supplies, the work done by the system is the same.

In the case of an adiabatically isolated (i.e., $S(t) \equiv 0$ and $d(E, V) \equiv 0$) dynamical system $\mathcal{G}$, $Q = 0$, and hence, it follows from (5.10) that

$$\delta L = 0. \tag{5.11}$$

This implies that among the set of all possible smooth paths that an adiabatically isolated large-scale dynamical system $\mathcal{G}$ may move between two fixed points over a specified time interval, the only dynamically possible system paths are those that render the work L done by the system on the environment stationary to all variations in the shape of the paths.

This is analogous to *Hamilton's principle of least action* in classical mechanics, which states that among the set of all smooth kinematically possible paths that a dynamical system may move between two fixed points over a specified interval, the only dynamically possible system paths are those that render the integral of the system Lagrangian stationary to all variations in the shape of the paths. It is important to note that the principle of least action does *not* assert that a mechanical system tends to lose action over its course, and hence, it is not a statement about the *time evolution* of mechanical systems.

To guarantee a thermodynamically consistent energy flow model, we assume that Axioms *i)* and *ii)* given in Section 3.3 hold for the large-scale dynamical system $\mathcal{G}$ with $\phi_{ij}(E)$ replaced by $\phi_{ij}(E, V)$, $i \neq j$, $i, j = 1, \ldots, q$. In this case, the results of Section 3.3 pertaining to entropy also hold for the nonlinear thermodynamic model given by (5.1) and (5.2). However, the input spaces $\mathcal{U}$, $\mathcal{U}_{\rm r}$, and $\mathcal{U}_{\rm c}$ consist of subsets of bounded continuous $\mathbb{R}^q \times \mathbb{R}^q$-valued functions on $\mathbb{R}$. Furthermore, all the results regarding equilibrium and nonequilibrium processes also hold. Note that it follows from Axiom *i)* that for the isolated large-scale dynamical system $\mathcal{G}$ given by (5.1) and (5.2) (that is, $S(t) \equiv 0$, $d(E, V) \equiv 0$, $S_{\rm w}(t) \equiv 0$, and $d_{\rm w}(E, V) \equiv 0$), the points $(\alpha \mathbf{e}, \gamma \mathbf{e})$, where $\alpha \geq 0$ and $\gamma > 0$, are the equilibrium states of $\mathcal{G}$. Here, we highlight the fundamental extensions corresponding to this generalization.

For the next result, $\oint$ denotes a cyclic integral evaluated along an arbitrary closed path of (5.1) and (5.2) in $\overline{\mathbb{R}}_+^q \times \mathbb{R}_+^q$; that is, $\oint \triangleq \int_{t_0}^{t_{\rm f}}$ with $t_{\rm f} \geq t_0$ and $(S(\cdot), S_{\rm w}(\cdot)) \in \mathcal{U}$ such that $E(t_{\rm f}) = E(t_0) = E_0 \in \overline{\mathbb{R}}_+^q$ and $V(t_{\rm f}) = V(t_0) = V_0 \in \mathbb{R}_+^q$.

Proposition 5.1. Consider the large-scale dynamical system $\mathcal{G}$ with power and volume velocity balance equations given by (5.1) and (5.2), and assume that Axioms *i)* and *ii)* hold. Then, for all $E_0 \in \overline{\mathbb{R}}_+^q$, $V_0 \in \mathbb{R}_+^q$, $t_{\rm f} \geq t_0$,

$(S(\cdot), S_{\rm w}(\cdot)) \in \mathcal{U}$ such that $E(t_{\rm f}) = E(t_0) = E_0$ and $V(t_{\rm f}) = V(t_0) = V_0$,

$$\int_{t_0}^{t_{\rm f}} \sum_{i=1}^{q} \frac{S_i(t) - \sigma_{ii}(E(t), V(t))}{c + E_i(t)} {\rm d}t = \oint \sum_{i=1}^{q} \frac{{\rm d}Q_i(t)}{c + E_i(t)} \leq 0, \tag{5.12}$$

where $c > 0$, ${\rm d}Q_i(t) \triangleq [S_i(t) - \sigma_{ii}(E(t), V(t))]{\rm d}t$, $i = 1, \ldots, q$, is the amount of the net energy (heat) received by the ith subsystem over the infinitesimal time interval ${\rm d}t$ and $[E^{\rm T}(t), V^{\rm T}(t)]^{\rm T}$, $t \geq t_0$, is the solution to (5.1) and (5.2) with initial condition $[E^{\rm T}(t_0), V^{\rm T}(t_0)]^{\rm T} = [E_0^{\rm T}, V_0^{\rm T}]^{\rm T}$. Furthermore,

$$\oint \sum_{i=1}^{q} \frac{{\rm d}Q_i(t)}{c + E_i(t)} = 0 \tag{5.13}$$

if and only if there exists a continuous function $\alpha : [t_0, t_{\rm f}] \to \overline{\mathbb{R}}_+$ such that $E(t) = \alpha(t)\mathbf{e}$ and $V(t) \in \mathbb{R}_+^q$, $t \in [t_0, t_{\rm f}]$.

Proof. Since the solution to (5.1) and (5.2) is nonnegative and $\phi_{ij}(E, V) = -\phi_{ji}(E, V)$, $E \in \overline{\mathbb{R}}_+^q$, $V \in \mathbb{R}_+^q$, $i \neq j$, $i, j = 1, \ldots, q$, it follows from (5.1), (5.2), and Axiom *ii*) that

$$\begin{aligned}
\oint \sum_{i=1}^{q} \frac{{\rm d}Q_i(t)}{c + E_i(t)} &= \int_{t_0}^{t_{\rm f}} \sum_{i=1}^{q} \frac{\dot{E}_i(t) - \sum_{j=1,\, j\neq i}^{q} \phi_{ij}(E(t), V(t))}{c + E_i(t)} {\rm d}t \\
&\quad + \int_{t_0}^{t_{\rm f}} \sum_{i=1}^{q} \frac{d_{{\rm w}i}(E(t), V(t)) - f_{{\rm w}i}(E(t), V(t)) - S_{{\rm w}i}(t)}{c + E_i(t)} {\rm d}t \\
&= \int_{t_0}^{t_{\rm f}} \sum_{i=1}^{q} \frac{\dot{E}_i(t)}{c + E_i(t)} {\rm d}t + \int_{t_0}^{t_{\rm f}} \sum_{i=1}^{q} \frac{\dot{V}_i(t)}{V_i(t)} {\rm d}t \\
&\quad - \int_{t_0}^{t_{\rm f}} \sum_{i=1}^{q} \sum_{j=1,\, j\neq i}^{q} \frac{\phi_{ij}(E(t), V(t))}{c + E_i(t)} {\rm d}t \\
&= \sum_{i=1}^{q} \log_e \left(\frac{c + E_i(t_{\rm f})}{c + E_i(t_0)} \right) + \sum_{i=1}^{q} \log_e \left(\frac{V_i(t_{\rm f})}{V_i(t_0)} \right) \\
&\quad - \int_{t_0}^{t_{\rm f}} \sum_{i=1}^{q} \sum_{j=i+1}^{q} \left(\frac{\phi_{ij}(E(t), V(t))}{c + E_i(t)} \right. \\
&\quad \left. - \frac{\phi_{ij}(E(t), V(t))}{c + E_j(t)} \right) {\rm d}t \\
&= - \int_{t_0}^{t_{\rm f}} \sum_{i=1}^{q} \sum_{j=i+1}^{q} \frac{\phi_{ij}(E(t), V(t))[E_j(t) - E_i(t)]}{(c + E_i(t))(c + E_j(t))} {\rm d}t \\
&\leq 0,
\end{aligned} \tag{5.14}$$

which proves (5.12).

To show (5.13), note that it follows from (5.14), Axiom i), and Axiom ii) that (5.13) holds if and only if $E_i(t) = E_j(t)$, $t \in [t_0, t_f]$, $i \neq j$, $i, j = 1, \ldots, q$, or, equivalently, there exists a continuous function $\alpha : [t_0, t_f] \to \overline{\mathbb{R}}_+$ such that $E(t) = \alpha(t)\mathbf{e}$ and $V(t) \in \mathbb{R}^q_+$, $t \in [t_0, t_f]$. □

Next, we give a definition of entropy for the large-scale dynamical system $\mathcal{G}$ given by (5.1) and (5.2), which is consistent with the one given in Definition 3.2.

Definition 5.1. For the large-scale dynamical system $\mathcal{G}$ with power and volume velocity balance equations given by (5.1) and (5.2), a function $\mathcal{S} : \overline{\mathbb{R}}^q_+ \times \mathbb{R}^q_+ \to \mathbb{R}$ satisfying

$$\mathcal{S}(E(t_2), V(t_2)) \geq \mathcal{S}(E(t_1), V(t_1)) + \int_{t_1}^{t_2} \sum_{i=1}^{q} \frac{S_i(t) - \sigma_{ii}(E(t), V(t))}{c + E_i(t)} \mathrm{d}t \tag{5.15}$$

for every $t_2 \geq t_1 \geq t_0$ and $(S(\cdot), S_\mathrm{w}(\cdot)) \in \mathcal{U}$ is called the *entropy* function of $\mathcal{G}$.

As in Section 3.3, (5.12) guarantees the existence of an entropy function for $\mathcal{G}$. For the next result, define the available entropy of the large-scale dynamical system $\mathcal{G}$ by

$$\mathcal{S}_\mathrm{a}(E_0, V_0) \triangleq - \sup_{(S(\cdot), S_\mathrm{w}(\cdot)) \in \mathcal{U}_\mathrm{c}, T \geq t_0} \int_{t_0}^{T} \sum_{i=1}^{q} \frac{S_i(t) - \sigma_{ii}(E(t), V(t))}{c + E_i(t)} \mathrm{d}t, \tag{5.16}$$

where $E(t_0) = E_0 \in \overline{\mathbb{R}}^q_+$, $V(t_0) = V_0 \in \mathbb{R}^q_+$, $E(T) = 0$, and $V(T) = V^*$, where $V^* \in \mathbb{R}^q_+$ denotes an arbitrary volume of $\mathcal{G}$ corresponding to the point of minimum system energy, and define the required entropy supply of the large-scale dynamical system $\mathcal{G}$ by

$$\mathcal{S}_\mathrm{r}(E_0, V_0) \triangleq \sup_{(S(\cdot), S_\mathrm{w}(\cdot)) \in \mathcal{U}_\mathrm{r}, T \geq -t_0} \int_{-T}^{t_0} \sum_{i=1}^{q} \frac{S_i(t) - \sigma_{ii}(E(t), V(t))}{c + E_i(t)} \mathrm{d}t, \tag{5.17}$$

where $E(-T) = 0$, $V(-T) = V^*$, $E(t_0) = E_0 \in \overline{\mathbb{R}}^q_+$, and $V(t_0) = V_0 \in \mathbb{R}^q_+$.

Theorem 5.1. Consider the large-scale dynamical system $\mathcal{G}$ with power and volume velocity balance equations given by (5.1) and (5.2), and assume that Axiom ii) holds. Then there exists an entropy function for $\mathcal{G}$. Moreover, $\mathcal{S}_\mathrm{a}(E, V)$, $(E, V) \in \overline{\mathbb{R}}^q_+ \times \mathbb{R}^q_+$, and $\mathcal{S}_\mathrm{r}(E, V)$, $(E, V) \in \overline{\mathbb{R}}^q_+ \times \mathbb{R}^q_+$, are possible entropy functions for $\mathcal{G}$ with $\mathcal{S}_\mathrm{a}(0, V^*) = \mathcal{S}_\mathrm{r}(0, V^*) = 0$. Finally, all entropy functions $\mathcal{S}(E,V)$, $(E,V) \in \overline{\mathbb{R}}_+^q \times \mathbb{R}_+^q$, for $\mathcal{G}$ satisfy

$$\mathcal{S}_{\rm r}(E,V) \leq \mathcal{S}(E,V) - \mathcal{S}(0,V^*) \leq \mathcal{S}_{\rm a}(E,V), \quad (E,V) \in \overline{\mathbb{R}}_+^q \times \mathbb{R}_+^q. \tag{5.18}$$

Proof. The proof is similar to the proof of Theorem 3.2. □

The following theorem shows that all entropy functions for $\mathcal{G}$ are continuous on $\overline{\mathbb{R}}_+^q \times \mathbb{R}_+^q$.

Theorem 5.2. Consider the large-scale dynamical system $\mathcal{G}$ with power and volume velocity balance equations given by (5.1) and (5.2), and let $\mathcal{S} : \overline{\mathbb{R}}_+^q \times \mathbb{R}_+^q \to \mathbb{R}$ be an entropy function of $\mathcal{G}$. Then $\mathcal{S}(\cdot,\cdot)$ is continuous on $\overline{\mathbb{R}}_+^q \times \mathbb{R}_+^q$.

Proof. Let $E_{\rm e} \in \overline{\mathbb{R}}_+^q$, $V_{\rm e} \in \mathbb{R}_+^q$, $S_{\rm we} \in \mathbb{R}^q$, and $S_{\rm e} \in \mathbb{R}^q$ be such that $S_{{\rm we}i} = d_{{\rm w}i}(E_{\rm e}, V_{\rm e}) - f_{{\rm w}i}(E_{\rm e}, V_{\rm e})$ and $S_{{\rm e}i} = \sigma_{ii}(E_{\rm e}, V_{\rm e}) - \sum_{j=1, j\neq i}^q \phi_{ij}(E_{\rm e}, V_{\rm e})$, $i = 1,\ldots,q$. Note that with $S(t) \equiv S_{\rm e}$ and $S_{\rm w}(t) \equiv S_{\rm we}$, $[E_{\rm e}^{\rm T}, V_{\rm e}^{\rm T}]^{\rm T}$ is an equilibrium point of the dynamical system (5.1) and (5.2). Next, define $x \triangleq [E^{\rm T}, V^{\rm T}]^{\rm T}$ and $u \triangleq [S^{\rm T}, S_{\rm w}^{\rm T}]^{\rm T}$, and consider the linearization of (5.1) and (5.2) at $x_{\rm e} = [E_{\rm e}^{\rm T}, V_{\rm e}^{\rm T}]^{\rm T}$ and $u_{\rm e} = [S_{\rm e}^{\rm T}, S_{\rm we}^{\rm T}]^{\rm T}$ given by

$$\dot{x}(t) = A(x(t) - x_{\rm e}) + B(u(t) - u_{\rm e}), \quad x(t_0) = x_0, \quad t \geq t_0, \tag{5.19}$$

where

$$A = \left.\frac{\partial f(x)}{\partial x}\right|_{x=x_{\rm e}}, \tag{5.20}$$

$$f(x) \triangleq [(f(x) + f_{\rm w}(x) - d_{\rm w}(x) - d(x))^{\rm T}, (D_p(x)(d_{\rm w}(x) - f_{\rm w}(x)))^{\rm T}]^{\rm T}, \tag{5.21}$$

$$B = \begin{bmatrix} I_q & I_q \\ 0 & -D_p(x_{\rm e}) \end{bmatrix}. \tag{5.22}$$

Since $\operatorname{rank} B = 2q$ for all $x_{\rm e} \in \overline{\mathbb{R}}_+^q \times \mathbb{R}_+^q$, it follows that

$$\operatorname{rank}[B, AB, A^2B, \ldots, A^{2q-1}B] = 2q, \tag{5.23}$$

and hence, the linearized system (5.19) is controllable. The remainder of the proof now follows identically as in the proof of Theorem 3.3 using a slight generalization of Lemma 3.1. □

The next result is a direct consequence of Theorem 5.1.

Proposition 5.2. Consider the large-scale dynamical system $\mathcal{G}$ with power and volume velocity balance equations given by (5.1) and (5.2), and assume that Axioms *i*) and *ii*) hold. Then

$$\mathcal{S}(E,V) \triangleq \alpha\mathcal{S}_{\rm r}(E,V) + (1-\alpha)\mathcal{S}_{\rm a}(E,V), \quad \alpha \in [0,1], \tag{5.24}$$

is an entropy function for $\mathcal{G}$.

The following propositions address equilibrium processes of $\mathcal{G}$ with the power and volume velocity balance equations (5.1) and (5.2).

Proposition 5.3. Consider the large-scale dynamical system $\mathcal{G}$ with power and volume velocity balance equations given by (5.1) and (5.2), and assume that Axioms *i*) and *ii*) hold. Then at every equilibrium state $(E_{\rm e}, V_{\rm e}) = (\alpha\mathbf{e}, \gamma\mathbf{e})$, where $\alpha \geq 0$ and $\gamma > 0$, of the isolated system $\mathcal{G}$, the entropy $\mathcal{S}(E,V)$, $(E,V) \in \overline{\mathbb{R}}_+^q \times \mathbb{R}_+^q$, of $\mathcal{G}$ is unique (modulo a constant of integration) and is given by

$$\begin{aligned}\mathcal{S}(E,V) - \mathcal{S}(0,V^*) &= \mathcal{S}_{\rm a}(E,V)\\ &= \mathcal{S}_{\rm r}(E,V)\\ &= \mathbf{e}^{\rm T}\mathbf{log}_e(c\mathbf{e}+E) + \mathbf{e}^{\rm T}\mathbf{log}_e V - \mathbf{e}^{\rm T}\mathbf{log}_e V^*\\ &\quad -q\log_e c,\end{aligned} \tag{5.25}$$

where $E = E_{\rm e}$ and $V = V_{\rm e}$.

Proof. The proof is identical to the proof of Proposition 3.4. □

Proposition 5.4. Consider the large-scale dynamical system $\mathcal{G}$ with power and volume velocity balance equations given by (5.1) and (5.2), and assume that Axioms *i*) and *ii*) hold. Let $\mathcal{S}(\cdot,\cdot)$ denote an entropy of $\mathcal{G}$, and let $(E(t), V(t))$, $t \geq t_0$, denote the solution to (5.1) and (5.2) with $E(t_0) = \alpha_0\mathbf{e}$, $E(t_1) = \alpha_1\mathbf{e}$, $V(t_0) = \gamma_0\mathbf{e}$, and $V(t_1) = \gamma_1\mathbf{e}$, where $\alpha_0, \alpha_1 \geq 0$ and $\gamma_0, \gamma_1 > 0$. Then

$$\begin{aligned}\mathcal{S}(E(t_1), V(t_1)) &= \mathcal{S}(E(t_0), V(t_0))\\ &\quad + \int_{t_0}^{t_1} \sum_{i=1}^{q} \frac{S_i(t) - \sigma_{ii}(E(t), V(t))}{c + E_i(t)} {\rm d}t\end{aligned} \tag{5.26}$$

if and only if there exist continuous functions $\alpha : [t_0, t_1] \to \overline{\mathbb{R}}_+$ and $\gamma : [t_0, t_1] \to \mathbb{R}_+$ such that $\alpha(t_0) = \alpha_0$, $\alpha(t_1) = \alpha_1$, $E(t) = \alpha(t)\mathbf{e}$, $\gamma(t_0) = \gamma_0$, $\gamma(t_1) = \gamma_1$, and $V(t) = \gamma(t)\mathbf{e}$, $t \in [t_0, t_1]$.

Proof. The proof is identical to the proof of Proposition 3.5. □

The next result gives a unique, continuously differentiable entropy function for $\mathcal{G}$ for equilibrium and nonequilibrium processes.

Theorem 5.3. Consider the large-scale dynamical system $\mathcal{G}$ with power and volume velocity balance equations given by (5.1) and (5.2), and assume that Axioms i) and ii) hold. Then the function $\mathcal{S} : \overline{\mathbb{R}}^q_+ \times \mathbb{R}^q_+ \to \mathbb{R}$ given by

$$\mathcal{S}(E, V) = \mathbf{e}^{\mathrm{T}}\mathbf{log}_e(c\mathbf{e} + E) + \mathbf{e}^{\mathrm{T}}\mathbf{log}_e V - \mathbf{e}^{\mathrm{T}}\mathbf{log}_e V^* \; - \; q \log_e c, \quad (E, V) \in \overline{\mathbb{R}}^q_+ \times \mathbb{R}^q_+, \tag{5.27}$$

where $c > 0$ and $V^* \in \mathbb{R}^q_+$, is a unique (modulo a constant of integration), continuously differentiable entropy function of $\mathcal{G}$. Furthermore, for $(E(t), V(t)) \notin \mathcal{M}_{\mathrm{e}}$, $t \geq t_0$, where $(E(t), V(t))$, $t \geq t_0$, denotes the solution to (5.1) and (5.2), and $\mathcal{M}_{\mathrm{e}} = \{(E, V) \in \overline{\mathbb{R}}^q_+ \times \mathbb{R}^q_+ : E = \alpha\mathbf{e}, \alpha \geq 0, V = \gamma\mathbf{e}, \gamma > 0\}$, (5.27) satisfies

$$\mathcal{S}(E(t_2), V(t_2)) > \mathcal{S}(E(t_1), V(t_1)) + \int_{t_1}^{t_2} \sum_{i=1}^{q} \frac{S_i(t) - \sigma_{ii}(E(t), V(t))}{c + E_i(t)} \mathrm{d}t \tag{5.28}$$

for every $t_2 \geq t_1 \geq t_0$ and $(S(\cdot), S_{\mathrm{w}}(\cdot)) \in \mathcal{U}$.

Proof. Since the solution $[E^{\mathrm{T}}(t), V^{\mathrm{T}}(t)]^{\mathrm{T}}$, $t \geq t_0$, to (5.1) and (5.2) is nonnegative for all nonnegative initial conditions and $\phi_{ij}(E, V) = -\phi_{ji}(E, V)$, $E \in \overline{\mathbb{R}}^q_+$, $V \in \mathbb{R}^q_+$, $i \neq j$, $i, j = 1, \ldots, q$, it follows that

$$\begin{aligned}
\dot{\mathcal{S}}(E(t), V(t)) &= \sum_{i=1}^{q} \frac{\dot{E}_i(t)}{c + E_i(t)} + \sum_{i=1}^{q} \frac{\dot{V}_i(t)}{V_i(t)} \\
&= \sum_{i=1}^{q} \frac{S_i(t) - \sigma_{ii}(E(t), V(t))}{c + E_i(t)} \\
&\quad + \sum_{i=1}^{q} \frac{S_{\mathrm{w}i}(t) + f_{\mathrm{w}i}(E(t), V(t)) - d_{\mathrm{w}i}(E(t), V(t))}{c + E_i(t)} \\
&\quad + \sum_{i=1}^{q} \sum_{j=1, j \neq i}^{q} \frac{\phi_{ij}(E(t), V(t))}{c + E_i(t)} \\
&\quad + \sum_{i=1}^{q} \frac{d_{\mathrm{w}i}(E(t), V(t)) - f_{\mathrm{w}i}(E(t), V(t)) - S_{\mathrm{w}i}(t)}{c + E_i(t)} \\
&= \sum_{i=1}^{q} \frac{S_i(t) - \sigma_{ii}(E(t), V(t))}{c + E_i(t)} \\
&\quad + \sum_{i=1}^{q} \sum_{j=i+1}^{q} \left(\frac{\phi_{ij}(E(t), V(t))}{c + E_i(t)} - \frac{\phi_{ij}(E(t), V(t))}{c + E_j(t)} \right) \\
&= \sum_{i=1}^{q} \frac{S_i(t) - \sigma_{ii}(E(t), V(t))}{c + E_i(t)}
\end{aligned}$$

$$+\sum_{i=1}^{q}\sum_{j=i+1}^{q}\frac{\phi_{ij}(E(t),V(t))[E_j(t)-E_i(t)]}{(c+E_i(t))(c+E_j(t))}$$

$$\geq \sum_{i=1}^{q}\frac{S_i(t)-\sigma_{ii}(E(t),V(t))}{c+E_i(t)}, \quad t\geq t_0. \tag{5.29}$$

Now, integrating (5.29) over $[t_1, t_2]$ yields (5.15). Furthermore, in the case where $(E(t),V(t))\notin\mathcal{M}_{\rm e}$, $t\geq t_0$, it follows from Axiom *i*), Axiom *ii*), and (5.29) that (5.28) holds.

Uniqueness (modulo a constant of integration) of (5.27) follows using identical arguments as in the proof of Theorem 3.4. □

Note that for an adiabatically isolated large-scale dynamical system $\mathcal{G}$, (5.15) yields the inequality

$$\mathcal{S}(E(t_2),V(t_2))\geq\mathcal{S}(E(t_1),V(t_1)), \quad t_2\geq t_1. \tag{5.30}$$

This inequality is a generalization of Clausius' version of the entropy principle, which states that for every irreversible (*nicht umkehrbar*) process in an adiabatically isolated system beginning and ending at an equilibrium state, the entropy of the final state is greater than or equal to the entropy of the initial state. In addition, for the entropy function given by (5.27), inequality (5.30) is satisfied as a strict inequality for all $(E,V)\in\overline{\mathbb{R}}_+^q\times\mathbb{R}_+^q\setminus\mathcal{M}_{\rm e}$. Hence, it follows from Theorem 2.25 that the adiabatically isolated large-scale dynamical system $\mathcal{G}$ does not exhibit Poincaré recurrence in $\overline{\mathbb{R}}_+^q\times\mathbb{R}_+^q\setminus\mathcal{M}_{\rm e}$.

Furthermore, using a similar analysis as given in the proof of Proposition 3.11, it can be shown that the entropy function given by (5.27) achieves its maximum among all the states of $\mathcal{G}$ with the fixed total energy and the fixed total volume of the system at $E=\alpha\mathbf{e}$ and $V=\gamma\mathbf{e}$, $\alpha\geq 0$, $\gamma>0$. Hence, the maximum system entropy is attained when the energies and volumes of all subsystems of $\mathcal{G}$ are equal. Finally, for the entropy function given by (5.27), note that $\mathcal{S}(0,V^*)=0$ or, equivalently, $\lim_{E\to 0}\mathcal{S}(E,V^*)=\mathcal{S}(0,V^*)=0$, which is consistent with the third law of thermodynamics.

The following result is a slight extension to Theorem 3.5 and shows that the dynamical system $\mathcal{G}$ with power and volume velocity balance equations (5.1) and (5.2) is state irreversible for every nontrivial trajectory of $\mathcal{G}$. For this result, define the augmented state $x\triangleq[E^{\rm T},V^{\rm T}]^{\rm T}$ and the augmented input $u\triangleq[S^{\rm T},S_{\rm w}^{\rm T}]^{\rm T}$, and let $\mathcal{W}_{[t_0,t_1]}$ denote the set of all possible energy and volume trajectories of $\mathcal{G}$ over the time interval $[t_0,t_1]$ given by

$$\mathcal{W}_{[t_0,t_1]}\triangleq\{s^x:[t_0,t_1]\times\mathcal{U}\to\overline{\mathbb{R}}_+^q\times\mathbb{R}_+^q:$$

$$s^x(\cdot,u(\cdot)) \text{ satisfies (5.1) and (5.2)}\}. \quad (5.31)$$

Furthermore, let $\mathcal{M}_{\rm e} \subset \overline{\mathbb{R}}_+^q \times \mathbb{R}_+^q$ denote the set of equilibria of $\mathcal{G}$ given by $\mathcal{M}_{\rm e} = \{(E,V) \in \overline{\mathbb{R}}_+^q \times \mathbb{R}_+^q : E = \alpha\mathbf{e}, \alpha \geq 0, V = \gamma\mathbf{e}, \gamma > 0\}$.

Theorem 5.4. Consider the large-scale dynamical system $\mathcal{G}$ with power and volume velocity balance equations given by (5.1) and (5.2), and assume that Axioms i) and ii) hold. Furthermore, let $s^x(\cdot,u(\cdot)) \in \mathcal{W}_{[t_0,t_1]}$, where $u(\cdot) \in \mathcal{U}$. Then $s^x(\cdot,u(\cdot))$ is an I_{2q}-reversible trajectory of $\mathcal{G}$ if and only if $s^x(t,u(t)) \in \mathcal{M}_{\rm e}$, $t \in [t_0,t_1]$.

Proof. The proof is similar to the proof of Theorem 3.5 and follows from Theorem 5.3 and Theorem 2.16. □

Finally, using the system entropy function given by (5.27), we show a clear connection between our expanded thermodynamic model given by (5.1) and (5.2) and the entropic arrow of time.

Theorem 5.5. Consider the large-scale dynamical system $\mathcal{G}$ with power and volume velocity balance equations (5.1) and (5.2) with $S(t) \equiv 0$ and $d(E,V) \equiv 0$, and assume that Axioms i) and ii) hold. Furthermore, let $s^x(\cdot,u(\cdot)) \in \mathcal{W}_{[t_0,t_1]}$, where $u(\cdot) \in \mathcal{U}$. Then, for every $x_0 \notin \mathcal{M}_{\rm e}$ and $u(\cdot) \in \mathcal{U}$, there exists a continuously differentiable function $\mathcal{S} : \overline{\mathbb{R}}_+^q \times \mathbb{R}_+^q \to \mathbb{R}$ such that $\mathcal{S}(s^x(t,u(t)))$ is a strictly increasing function of time. Furthermore, $s^x(\cdot,u(\cdot))$ is an I_{2q}-reversible trajectory of $\mathcal{G}$ if and only if $s^x(t,u(t)) \in \mathcal{M}_{\rm e}$, $t \in [t_0,t_1]$.

Proof. The existence of a continuously differentiable function $\mathcal{S} : \overline{\mathbb{R}}_+^q \times \mathbb{R}_+^q \to \mathbb{R}$, which strictly increases on all nontrivial trajectories of $\mathcal{G}$, is a restatement of Theorem 5.3 with $S(t) \equiv 0$ and $d(E,V) \equiv 0$. The proof now follows from Theorem 5.4 and Corollary 2.2. □

The existence of a continuously differentiable entropy function on nontrivial irreversible trajectories of the adiabatically isolated large-scale dynamical system $\mathcal{G}$ establishes the existence of a completely ordered time set having a topological structure involving a closed set homeomorphic to the real line.

5.2 Work Energy, Gibbs Free Energy, Helmholtz Free Energy, Enthalpy, and Entropy

In this section, we connect our expanded thermodynamic model developed in Section 5.1 to classical thermodynamics. Specifically, as outlined in Section 5.1, we assume that each subsystem can perform (positive) work on the environment and the environment can perform (negative) work on the subsystems. Furthermore, we assume that each subsystem can perform work on the other subsystems. The rate of work done by the ith subsystem on the environment is denoted by $d_{\mathrm{w}i} : \overline{\mathbb{R}}_+^q \times \mathbb{R}_+^q \to \overline{\mathbb{R}}_+$, $i = 1, \ldots, q$, the rate of work done by the environment on the ith subsystem is denoted by $S_{\mathrm{w}i} : [0, \infty) \to \overline{\mathbb{R}}_+$, $i = 1, \ldots, q$, and the volume of the ith subsystem is denoted by $V_i : [0, \infty) \to \mathbb{R}_+$, $i = 1, \ldots, q$. The net work done by each subsystem on the environment and the other subsystems satisfies

$$p_i(E, V)\mathrm{d}V_i = [d_{\mathrm{w}i}(E, V) - f_{\mathrm{w}i}(E, V) - S_{\mathrm{w}i}(t)]\mathrm{d}t, \tag{5.32}$$

where $p_i(E, V)$, $i = 1, \ldots, q$, denotes the *pressure* in the ith subsystem and $V \triangleq [V_1, \ldots, V_q]^{\mathrm{T}}$.

Furthermore, in the presence of work, the energy balance (3.4) for each subsystem can be rewritten as

$$\mathrm{d}E_i = f_i(E, V)\mathrm{d}t - [d_{\mathrm{w}i}(E, V) - f_{\mathrm{w}i}(E, V) - S_{\mathrm{w}i}(t)]\mathrm{d}t - \sigma_{ii}(E, V)\mathrm{d}t + S_i(t)\mathrm{d}t, \tag{5.33}$$

where $f_i(E, V) \triangleq \sum_{j=1,\, j\neq i}^{q} \phi_{ij}(E, V)$, $\phi_{ij} : \overline{\mathbb{R}}_+^q \times \mathbb{R}_+^q \to \mathbb{R}$, $i \neq j$, $i, j = 1, \ldots, q$, denotes the net instantaneous rate of energy (heat) flow from the jth subsystem to the ith subsystem, $f_{\mathrm{w}i}(E, V) \triangleq \sum_{j=1,\, j\neq i}^{q} (p_j - p_i) v_{ij}(E, V)$, $v_{ij} : \overline{\mathbb{R}}_+^q \times \mathbb{R}_+^q \to \overline{\mathbb{R}}_+$, $i \neq j$, $i, j = 1, \ldots, q$, denotes the net instantaneous time rate of change in volume of the ith subsystem due to the pressure exerted by the jth subsystem, $\sigma_{ii} : \overline{\mathbb{R}}_+^q \times \mathbb{R}_+^q \to \overline{\mathbb{R}}_+$, $i = 1, \ldots, q$, denotes the instantaneous rate of energy dissipation from the ith subsystem to the environment, and, as in Section 3.2, $S_i : [0, \infty) \to \mathbb{R}$, $i = 1, \ldots, q$, denotes the external power supplied to (or extracted from) the ith subsystem.

It follows from (5.32) and (5.33) that positive work done by a subsystem on the environment leads to a decrease in the internal energy of the subsystem and an increase in the subsystem volume, which is consistent with the first law of thermodynamics. The definition of entropy $\mathcal{S}(E, V)$ for $\mathcal{G}$ in the presence of work is given by Definition 5.1.

Next, consider the ith subsystem of $\mathcal{G}$ and assume that E_j and V_j, $j \neq i$, $i = 1, \ldots, q$, are constant. In this case, note that

$$\frac{\mathrm{d}\mathcal{S}}{\mathrm{d}t} = \frac{\partial \mathcal{S}}{\partial E_i}\frac{\mathrm{d}E_i}{\mathrm{d}t} + \frac{\partial \mathcal{S}}{\partial V_i}\frac{\mathrm{d}V_i}{\mathrm{d}t} \tag{5.34}$$

and

$$p_i(E, V) = \left(\frac{\partial \mathcal{S}}{\partial E_i}\right)^{-1} \left(\frac{\partial \mathcal{S}}{\partial V_i}\right), \quad i = 1, \ldots, q. \tag{5.35}$$

It follows from (5.32) and (5.33) that, in the presence of work energy, the energy and deformation states satisfy (5.1) and (5.2) or, in vector form,

$$\dot{E}(t) = f(E(t),V(t)) + f_{\rm w}(E(t),V(t)) - d_{\rm w}(E(t),V(t)) + S_{\rm w}(t) \\ -d(E(t),V(t)) + S(t), \quad E(t_0) = E_0, \quad t \geq t_0, \tag{5.36}$$

$$\dot{V}(t) = D(E(t),V(t))[d_{\rm w}(E(t),V(t)) - f_{\rm w}(E(t),V(t)) - S_{\rm w}(t)], \\ V(t_0) = V_0, \tag{5.37}$$

where

$$\begin{aligned} f(E,V) &\triangleq [f_1(E,V),\ldots,f_q(E,V)]^{\rm T}, \\ f_{\rm w}(E,V) &\triangleq [f_{\rm w1}(E,V),\ldots,f_{{\rm w}q}(E,V)]^{\rm T}, \\ d_{\rm w}(E,V) &\triangleq [d_{\rm w1}(E,V),\ldots,d_{{\rm w}q}(E,V)]^{\rm T}, \\ S_{\rm w}(t) &\triangleq [S_{\rm w1}(t),\ldots,S_{{\rm w}q}(t)]^{\rm T}, \\ d(E,V) &\triangleq [\sigma_{11}(E,V),\ldots,\sigma_{qq}(E,V)]^{\rm T}, \\ S(t) &\triangleq [S_1(t),\ldots,S_q(t)]^{\rm T}, \end{aligned}$$

and

$$D(E,V) \triangleq \operatorname{diag}\left[\left(\frac{\partial \mathcal{S}}{\partial E_1}\right)\left(\frac{\partial \mathcal{S}}{\partial V_1}\right)^{-1},\ldots,\left(\frac{\partial \mathcal{S}}{\partial E_q}\right)\left(\frac{\partial \mathcal{S}}{\partial V_q}\right)^{-1}\right]. \tag{5.38}$$

Note that

$$\left(\frac{\partial \mathcal{S}(E,V)}{\partial V}\right) D(E,V) = \frac{\partial \mathcal{S}(E,V)}{\partial E}. \tag{5.39}$$

As noted in Section 5.1, the power and volume velocity balance equations given by (5.36) and (5.37) represent a statement of the first law of thermodynamics. Specifically, defining the work L done by the interconnected dynamical system $\mathcal{G}$ over the time interval $[t_1, t_2]$ by

$$L \triangleq \int_{t_1}^{t_2} \mathbf{e}^{\rm T}[d_{\rm w}(E(t),V(t)) - S_{\rm w}(t)]{\rm d}t, \tag{5.40}$$

where $[E^{\rm T}(t), V^{\rm T}(t)]^{\rm T}$, $t \geq t_0$, is the solution to (5.36) and (5.37), premultiplying (5.36) by $\mathbf{e}^{\rm T}$, and using the fact that $\mathbf{e}^{\rm T} f(E,V) = 0$ and $\mathbf{e}^{\rm T} f_{\rm w}(E,V) = 0$, it follows that

$$\Delta U = -L + Q, \tag{5.41}$$

where $\Delta U = U(t_2) - U(t_1) \triangleq \mathbf{e}^{\rm T}E(t_2) - \mathbf{e}^{\rm T}E(t_1)$ denotes the variation in the total energy[2] of the interconnected system $\mathcal{G}$ over the time interval $[t_1, t_2]$

[2]The mass-energy equivalence from the theory of relativity gives an additional relationship connecting energy and mass by the equation $\Delta U = \Delta mc^2$, where ΔU is the increase in system energy, Δm is the increase in system mass, and c is the velocity of light. Even though the first law of thermodynamics gives only information of changes in system energy content without providing a unique zero point of the energy content, the generalized relativistic energy relation $E = mc^2$

and

$$Q \triangleq \int_{t_1}^{t_2} \mathbf{e}^{\mathrm{T}}[S(t) - d(E(t), V(t))]\mathrm{d}t \tag{5.42}$$

denotes the net energy received ($Q > 0$) or dissipated ($Q < 0$) by $\mathcal{G}$ in forms other than work.

This of course is a statement of the *first law of thermodynamics* for the interconnected dynamical system $\mathcal{G}$ and gives a precise formulation of the equivalence between work and heat, which establishes that heat and mechanical work are two different aspects of energy. Finally, note that (5.37) is consistent with the classical thermodynamic equation for the rate of work done by the system $\mathcal{G}$ on the environment. To see this, note that (5.37) can be equivalently written as

$$\mathrm{d}L = \mathbf{e}^{\mathrm{T}} D^{-1}(E, V)\mathrm{d}V, \tag{5.43}$$

which, for a single subsystem with volume V and pressure p, has the classical form

$$\mathrm{d}L = p\mathrm{d}V. \tag{5.44}$$

It follows from Definition 5.1 and (5.36)–(5.39) that the time derivative of the entropy function satisfies

$$\begin{aligned}
\dot{\mathcal{S}}(E, V) &= \frac{\partial \mathcal{S}(E, V)}{\partial E}\dot{E} + \frac{\partial \mathcal{S}(E, V)}{\partial V}\dot{V} \\
&= \frac{\partial \mathcal{S}(E, V)}{\partial E} f(E, V) - \frac{\partial \mathcal{S}(E, V)}{\partial E}[d_{\mathrm{w}}(E, V) - f_{\mathrm{w}}(E, V) - S_{\mathrm{w}}(t)] \\
&\quad - \frac{\partial \mathcal{S}(E, V)}{\partial E}(d(E, V) - S(t)) \\
&\quad + \frac{\partial \mathcal{S}(E, V)}{\partial V} D(E, V)[d_{\mathrm{w}}(E, V) - f_{\mathrm{w}}(E, V) - S_{\mathrm{w}}(t)] \\
&= \sum_{i=1}^{q} \frac{\partial \mathcal{S}(E, V)}{\partial E_i} \sum_{j=1,\, j\neq i}^{q} \phi_{ij}(E, V) \\
&\quad + \sum_{i=1}^{q} \frac{\partial \mathcal{S}(E, V)}{\partial E_i}[S_i(t) - d_i(E, V)] \\
&= \sum_{i=1}^{q} \sum_{j=i+1}^{q} \left(\frac{\partial \mathcal{S}(E, V)}{\partial E_i} - \frac{\partial \mathcal{S}(E, V)}{\partial E_j} \right) \phi_{ij}(E, V) \\
&\quad + \sum_{i=1}^{q} \frac{\partial \mathcal{S}(E, V)}{\partial E_i}[S_i(t) - d_i(E, V)]
\end{aligned}$$

infers that a point of zero energy content corresponds to the absence of all system mass.

$$\geq \sum_{i=1}^{q} \frac{\partial \mathcal{S}(E,V)}{\partial E_i}[S_i(t) - d_i(E,V)], \quad (E,V) \in \overline{\mathbb{R}}_+^q \times \mathbb{R}_+^q. \tag{5.45}$$

Noting that $\mathrm{d}Q_i \triangleq [S_i - \sigma_{ii}(E)]\mathrm{d}t$, $i = 1, \ldots, q$, is the infinitesimal amount of the net heat received or dissipated by the ith subsystem of $\mathcal{G}$ over the infinitesimal time interval $\mathrm{d}t$, it follows from (5.45) that

$$\mathrm{d}\mathcal{S}(E) \geq \sum_{i=1}^{q} \frac{\mathrm{d}Q_i}{T_i}. \tag{5.46}$$

Inequality (5.46) is the classical *Clausius inequality* for the variation of entropy during an infinitesimal irreversible transformation.

Next, we define the *Gibbs free energy*, the *Helmholtz free energy*, and the *enthalpy* functions for the interconnected dynamical system $\mathcal{G}$. For this exposition, we assume that the entropy of $\mathcal{G}$ is a sum of individual entropies of subsystems of $\mathcal{G}$, that is,

$$\mathcal{S}(E,V) = \sum_{i=1}^{q} \mathcal{S}_i(E_i, V_i), \quad (E,V) \in \overline{\mathbb{R}}_+^q \times \mathbb{R}_+^q.$$

In this case, the Gibbs free energy of $\mathcal{G}$ is defined by

$$G(E,V) \triangleq \mathbf{e}^{\mathrm{T}}E - \sum_{i=1}^{q} \left(\frac{\partial \mathcal{S}(E,V)}{\partial E_i}\right)^{-1} \mathcal{S}_i(E_i, V_i) + \sum_{i=1}^{q} \left(\frac{\partial \mathcal{S}(E,V)}{\partial E_i}\right)^{-1} \cdot \left(\frac{\partial \mathcal{S}(E,V)}{\partial V_i}\right) V_i, \quad (E,V) \in \overline{\mathbb{R}}_+^q \times \mathbb{R}_+^q, \tag{5.47}$$

the Helmholtz free energy of $\mathcal{G}$ is defined by

$$F(E,V) \triangleq \mathbf{e}^{\mathrm{T}}E - \sum_{i=1}^{q} \left(\frac{\partial \mathcal{S}(E,V)}{\partial E_i}\right)^{-1} \mathcal{S}_i(E_i, V_i), \quad (E,V) \in \overline{\mathbb{R}}_+^q \times \mathbb{R}_+^q, \tag{5.48}$$

and the enthalpy of $\mathcal{G}$ is defined by

$$H(E,V) \triangleq \mathbf{e}^{\mathrm{T}}E + \sum_{i=1}^{q} \left(\frac{\partial \mathcal{S}(E,V)}{\partial E_i}\right)^{-1} \left(\frac{\partial \mathcal{S}(E,V)}{\partial V_i}\right) V_i, \quad (E,V) \in \overline{\mathbb{R}}_+^q \times \mathbb{R}_+^q. \tag{5.49}$$

Note that the above definitions for Gibbs free energy, Helmholtz free energy, and enthalpy are consistent with the classical thermodynamic definitions given by $G(E,V) = U + pV - TS$, $F(E,V) = U - TS$, and $H(E,V) = U + pV$, respectively. Free energy is the energy in the system that can perform work alone. More specifically, the change in the Gibbs free energy is the amount of nonexpansive work that a dynamical process can perform at constant temperature and pressure, whereas the change in the Helmholtz free energy is the maximum amount of work that can be extracted from a dynamical process occurring at constant temperature and volume. Alternatively, enthalpy is the amount of energy that is released as heat when the system is free to expand onto an environment exerting a constant pressure on the system. Furthermore, note that if the interconnected system $\mathcal{G}$ is *isothermal* and *isobaric*, that is, the temperatures of subsystems of $\mathcal{G}$ are equal and remain constant with

$$\left(\frac{\partial \mathcal{S}(E,V)}{\partial E_1}\right)^{-1} = \cdots = \left(\frac{\partial \mathcal{S}(E,V)}{\partial E_q}\right)^{-1} = T > 0, \tag{5.50}$$

and the pressure $p_i(E,V)$ in each subsystem of $\mathcal{G}$ remains constant, respectively, then any transformation in $\mathcal{G}$ is reversible.

The time derivative of $G(E,V)$ along the trajectories of (5.36) and (5.37) is given by

$$\begin{aligned}
\dot{G}(E,V) &= \mathbf{e}^{\mathrm{T}}\dot{E} - \sum_{i=1}^{q}\left(\frac{\partial \mathcal{S}(E,V)}{\partial E_i}\right)^{-1}\left[\frac{\partial \mathcal{S}(E,V)}{\partial E_i}\dot{E}_i + \frac{\partial \mathcal{S}(E,V)}{\partial V_i}\dot{V}_i\right] \\
&\quad + \sum_{i=1}^{q}\left(\frac{\partial \mathcal{S}(E,V)}{\partial E_i}\right)^{-1}\left(\frac{\partial \mathcal{S}(E,V)}{\partial V_i}\right)\dot{V}_i \\
&= 0,
\end{aligned} \tag{5.51}$$

which is consistent with classical thermodynamics in the absence of chemical reactions.

For an isothermal interconnected dynamical system $\mathcal{G}$, the time derivative of $F(E,V)$ along the trajectories of (5.36) and (5.37) is given by

$$\begin{aligned}
\dot{F}(E,V) &= \mathbf{e}^{\mathrm{T}}\dot{E} - \sum_{i=1}^{q}\left(\frac{\partial \mathcal{S}(E,V)}{\partial E_i}\right)^{-1}\left[\frac{\partial \mathcal{S}(E,V)}{\partial E_i}\dot{E}_i + \frac{\partial \mathcal{S}(E,V)}{\partial V_i}\dot{V}_i\right] \\
&= -\sum_{i=1}^{q}\left(\frac{\partial \mathcal{S}(E,V)}{\partial E_i}\right)^{-1}\left(\frac{\partial \mathcal{S}(E,V)}{\partial V_i}\right)\dot{V}_i \\
&= -\sum_{i=1}^{q}[d_{\mathrm{w}i}(E,V) - S_{\mathrm{w}i}(t)] \\
&= -L,
\end{aligned} \tag{5.52}$$

where L is the net amount of work done by the subsystems of $\mathcal{G}$ on the environment. Furthermore, note that if, in addition, the interconnected system $\mathcal{G}$ is *isochoric*, that is, the volumes of each of the subsystems of $\mathcal{G}$

remain constant, then $\dot{F}(E,V)=0$. As we see in Chapter 6, in the presence of chemical reactions the interconnected system $\mathcal{G}$ evolves such that the Helmholtz free energy is minimized.

Finally, for the isolated ($S(t)\equiv 0$ and $d(E,V)\equiv 0$) interconnected dynamical system $\mathcal{G}$, the time derivative of $H(E,V)$ along the trajectories of (5.36) and (5.37) is given by

$$\begin{aligned}\dot{H}(E,V) &= \mathbf{e}^{\mathrm{T}}\dot{E}+\sum_{i=1}^{q}\left(\frac{\partial\mathcal{S}(E,V)}{\partial E_i}\right)^{-1}\left(\frac{\partial\mathcal{S}(E,V)}{\partial V_i}\right)\dot{V}_i\\ &= \mathbf{e}^{\mathrm{T}}\dot{E}+\sum_{i=1}^{q}[d_{\mathrm{w}i}(E,V)-f_{\mathrm{w}i}(E,V)-S_{\mathrm{w}i}(t)]\\ &= \mathbf{e}^{\mathrm{T}}f(E,V)\\ &= 0. \end{aligned} \tag{5.53}$$

We close this section by connecting the Gibbs and Helmholtz free energies with classical thermodynamic principles. Here, we restrict our attention to systems that have the same pressure and temperature in all its subsystems so that the behavior of the dynamical system can be captured by that of a single plenipotentiary unit. In this case, (5.47)–(5.49) become

$$G(E,V) = U+pV-T\mathcal{S}=F+pV=H-T\mathcal{S}, \tag{5.54}$$
$$F(E,V) = U-T\mathcal{S}, \tag{5.55}$$
$$H(E,V) = U+pV. \tag{5.56}$$

Furthermore, it follows from (5.41) and (5.46) that the change in energy and entropy, respectively, over an infinitesimal time interval dt is given by the classical relationships

$$\mathrm{d}U = \mathrm{d}Q-\mathrm{d}L, \tag{5.57}$$
$$\mathrm{d}\mathcal{S} \geq \frac{\mathrm{d}Q}{T}. \tag{5.58}$$

If the system is at *constant pressure*, then it follows from (5.56), (5.44), and (5.57) that

$$\mathrm{d}H=\mathrm{d}U+p\mathrm{d}V=\mathrm{d}U+\mathrm{d}L=\mathrm{d}Q, \tag{5.59}$$

which shows that the heat absorbed by the system is equal to an increase in the system enthalpy. Next, if the system is at a *constant temperature*, then it follows from (5.55) that

$$\mathrm{d}F=\mathrm{d}U-T\mathrm{d}\mathcal{S}, \tag{5.60}$$

which, using (5.57) and (5.58), implies

$$\mathrm{d}L \leq -\mathrm{d}F. \tag{5.61}$$

Thus, it follows from (5.61) that the work that the system can perform on the environment when the system is maintained at constant temperature cannot be greater than the system Helmholtz free energy.

Next, if the system is at *constant temperature* and *constant pressure*, then it follows from (5.54) that

$$\mathrm{d}G = \mathrm{d}U - T\mathrm{d}\mathcal{S} + p\mathrm{d}V, \tag{5.62}$$

which, using (5.57), implies

$$\mathrm{d}G \leq 0. \tag{5.63}$$

This shows that at constant temperature and pressure the Gibbs free energy of a dynamical process spontaneously (i.e., having the ability to naturally perform work) diminishes until it attains a minimum value. In classical thermodynamics, this minimum value corresponds to the thermal equilibrium for an isothermal isobaric process. Alternately, if the system is at *constant volume* and *constant temperature*, then $\mathrm{d}L = 0$ and (5.57) becomes

$$\mathrm{d}U = \mathrm{d}Q. \tag{5.64}$$

Now, using (5.58) and (5.60) it follows that

$$\mathrm{d}F \leq 0, \tag{5.65}$$

which gives a necessary condition for any change of the state of the system at constant volume and temperature. Furthermore, this shows that at constant volume and temperature the Helmholtz free energy of a dynamical process spontaneously diminishes until it attains a minimum value, with the minimum value attained at thermal equilibrium.

Finally, we note that the Helmholtz free energy F is the amount of energy that can be extracted from the system to perform useful work. It follows from (5.55) that not all of the internal energy U is available to be extracted as (Helmholtz) free energy. The scaled entropy $T\mathcal{S}$ is a penalty for extracting the available energy from the system and using it to do useful work. Specifically, the higher the temperature T and the entropy $\mathcal{S}$ of the system, the lower the amount of the usable extractable free energy F.

Hence, if $T\mathcal{S} << U$, then $F \approx U$, which constitutes an efficient thermodynamic process resulting in a minimal entropy production. Furthermore, since the system internal energy U is always greater than F, $T\mathcal{S}$ will always be positive, and hence, $T\mathrm{d}\mathcal{S} + \mathcal{S}\mathrm{d}T > 0$. Moreover, note that if $U = T\mathcal{S}$, then there does not exist any (Helmholtz) free energy that can be extracted from the system. This, however, does not necessarily imply a thermodynamic equilibrium state. Analogous remarks hold for the Gibbs free energy.

5.3 The Carnot Cycle and the Second Law of Thermodynamics

The first law of thermodynamics places no limitation on the possibility of transforming heat into work or work into heat, provided that the total amount of heat is equivalent to the total amount of work, and hence, energy is conserved in the process. The second law of thermodynamics, however, places a definite limitation on the possibility of transforming heat into work. If this were not the case, then one would be able to construct a dynamical system $\mathcal{G}$, which, by extracting heat from the environment, completely transforms this heat into mechanical work. Since the supply of thermal energy contained in the universe is virtually unlimited, such a dynamical system would constitute a *perpetuum mobile* of the second kind.

There have been many statements of the second law of thermodynamics, each emphasizing another facet of the law, but all can be shown to be equivalent to one another. In this section, we use the power and volume velocity balance equations (5.1) and (5.2) in conjunction with a *Carnot-like cycle* analysis for a large-scale dynamical system $\mathcal{G}$ to show the equivalence between the classical Kelvin and Clausius statements (postulates) of the second law of thermodynamics [59, 71, 477].

Kelvin. *A cyclic transformation whose only final result is to transform completely into work heat extracted from an energy source at a uniform temperature throughout is impossible.*

Clausius. *A cyclic transformation whose only final result is to transfer heat from a body at a given temperature to a body at a higher temperature is impossible.*

It is important to note that the Kelvin postulate assumes that the complete transformation of heat into work is the *only* final result of the process. In other words, it is impossible to perform a cyclic transformation for which the only effect is to perform work by absorbing heat from a single heat reservoir (i.e., cooling the reservoir). However, it is not impossible to transform into work all the energy supplied to the system $\mathcal{G}$, provided that the state of the system is changed (see, for example, the analysis of leg $0-1$ in Figure 5.2). Furthermore, note that Axiom *ii*) is equivalent to Clausius' postulate as applied to large-scale cyclic dynamical systems.

Next, we show the equivalence of the Kelvin and Clausius statements. First, however, we consider a Carnot-like cycle for the large-scale dynamical

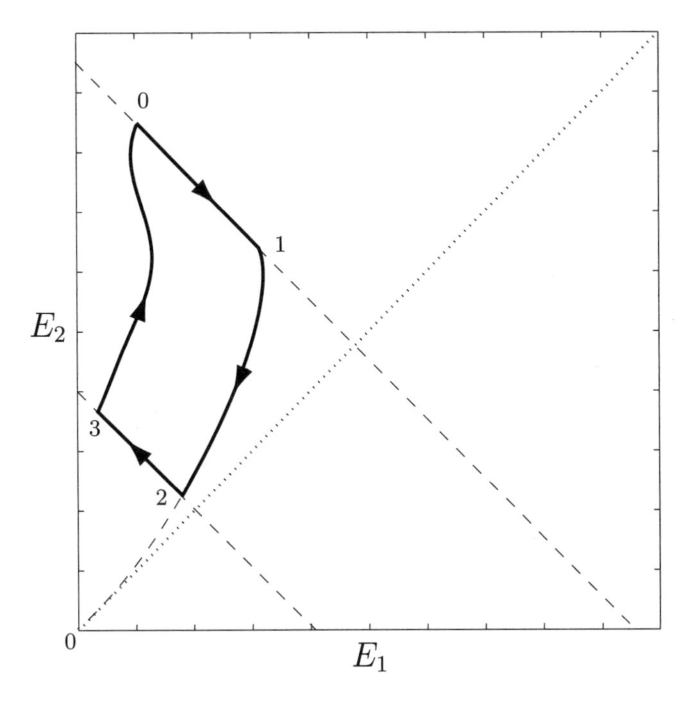

Figure 5.2 Carnot-like cycle: leg $0-1$ (isothermal), leg $1-2$ (adiabatic), leg $2-3$ (isothermal), leg $3-0$ (adiabatic).

system $\mathcal{G}$, which consists of two *isothermal* and two *adiabatic* processes (see Figure 5.2). Recall that a process is isothermal if the total (possibly scaled) energy (temperature) of the dynamical system $\mathcal{G}$ is conserved during this process, while an adiabatic process is a process wherein no external energy in the form of heat is supplied to the system (i.e., $S(t) \equiv 0$ and $d(E, V) \equiv 0$).

First, assume the system undergoes an isothermal transformation while receiving energy (heat) from an external source and performing positive work on the environment so that the dynamics of this leg of the process are given by

$$\begin{aligned}\dot{E}(t) &= f(E(t), V(t)) + f_{\mathrm{w}}(E(t), V(t)) - d_{\mathrm{w}0-1}(E(t), V(t)) + S_{0-1}(t),\\ &\qquad E(t_0) = E_0, \quad t_1 \geq t \geq t_0, \quad (5.66)\end{aligned}$$

or, since $\mathbf{e}^{\mathrm{T}} f(E, V) \equiv 0$, $\mathbf{e}^{\mathrm{T}} f_{\mathrm{w}}(E, V) \equiv 0$, and $\mathbf{e}^{\mathrm{T}} \dot{E}(t) = 0$, $t_1 \geq t \geq t_0$,

$$0 = \mathbf{e}^{\mathrm{T}} \dot{E}(t) = -\mathbf{e}^{\mathrm{T}} d_{\mathrm{w}0-1}(E(t), V(t)) + \mathbf{e}^{\mathrm{T}} S_{0-1}(t), \quad t_1 \geq t \geq t_0. \tag{5.67}$$

Thus,

$$\mathbf{e}^{\mathrm{T}} S_{0-1}(t) = \mathbf{e}^{\mathrm{T}} d_{\mathrm{w}0-1}(E(t), V(t)), \quad t_1 \geq t \geq t_0, \tag{5.68}$$

and hence, the work performed by the system during this leg of the process is, by definition,

$$L_{0-1} = \int_{t_0}^{t_1} \mathbf{e}^{\mathrm{T}} d_{\mathrm{w}0-1}(E(t), V(t))\mathrm{d}t. \tag{5.69}$$

Moreover, it follows from (5.68) that the amount of energy (heat) supplied to the system is $Q_{0-1} = L_{0-1} > 0$.

Next, we thermally isolate the dynamical system $\mathcal{G}$ from the environment and let the system $\mathcal{G}$ perform work on the environment so that the dynamics of the adiabatic process with the initial condition $E(t_1) \in \overline{\mathbb{R}}_+^q$ are given by

$$\dot{E}(t) = f(E(t), V(t)) + f_{\mathrm{w}}(E(t), V(t)) - d_{\mathrm{w}1-2}(E(t), V(t)), \quad t_2 \geq t \geq t_1. \tag{5.70}$$

During this leg of the process, the dynamical system $\mathcal{G}$ performs positive work on the environment by losing its internal energy so that, by (5.70),

$$\begin{aligned} L_{1-2} &= \int_{t_1}^{t_2} \mathbf{e}^{\mathrm{T}} d_{\mathrm{w}1-2}(E(t), V(t))\mathrm{d}t \\ &= -\int_{t_1}^{t_2} \mathbf{e}^{\mathrm{T}} \dot{E}(t)\mathrm{d}t \\ &= -\Delta U \\ &> 0, \end{aligned} \tag{5.71}$$

where $\Delta U \triangleq U(t_2) - U(t_1) = \mathbf{e}^{\mathrm{T}} E(t_2) - \mathbf{e}^{\mathrm{T}} E(t_1)$ is the variation of the total energy in the system $\mathcal{G}$.

Next, we perform an isothermal process during which positive work is being performed on the system $\mathcal{G}$ while the system surrenders energy to the environment. The dynamics of this leg of the process are characterized by

$$\dot{E}(t) = f(E(t), V(t)) + f_{\mathrm{w}}(E(t), V(t)) + S_{\mathrm{w}2-3}(t) - d_{\mathrm{w}2-3}(E(t), V(t)), \quad E(t_2) \in \overline{\mathbb{R}}_+^q, \quad t_3 \geq t \geq t_2, \tag{5.72}$$

where $\mathbf{e}^{\mathrm{T}} \dot{E}(t) \equiv 0$. Thus, the work done by the system $\mathcal{G}$ during this transformation is negative and is given by

$$\begin{aligned} L_{2-3} &= -\int_{t_2}^{t_3} \mathbf{e}^{\mathrm{T}} S_{\mathrm{w}2-3}(t)\mathrm{d}t \\ &= -\int_{t_2}^{t_3} \mathbf{e}^{\mathrm{T}} d_{2-3}(E(t), V(t))\mathrm{d}t \\ &= -Q_{2-3} \\ &< 0, \end{aligned} \tag{5.73}$$

where $Q_{2-3} > 0$ is the amount of energy released to the environment.

Finally, we perform an adiabatic process to drive the system back to its initial state $E_0 \in \overline{\mathbb{R}}_+^q$, $V_0 \in \mathbb{R}_+^q$. During this process, work is performed on the system by the environment so that the dynamics of this leg of the process are given by

$$\dot{E}(t) = f(E(t), V(t)) + f_{\rm w}(E(t), V(t)) + S_{\rm w3-0}(t), \quad E(t_3) \in \overline{\mathbb{R}}_+^q, \quad t_4 \geq t \geq t_3. \tag{5.74}$$

The work done by the dynamical system $\mathcal{G}$ during this transformation is negative and is given by

$$L_{3-0} = -\int_{t_3}^{t_4} \mathbf{e}^{\rm T} S_{\rm w3-0}(t){\rm d}t = -\int_{t_3}^{t_4} \mathbf{e}^{\rm T} \dot{E}(t){\rm d}t = \Delta U < 0. \tag{5.75}$$

Thus, it follows from (5.5) that the total work done by the system during the cycle is given by $L = Q_{0-1} - Q_{2-3}$.

This implies that only part of the heat Q_{0-1} that is absorbed by the large-scale dynamical system from the external source at a higher energy level is transformed into work by the Carnot-like cycle; the rest of the heat Q_{2-3} is surrendered to the external source at a lower energy level. Assuming an infinite energy source and repeating this cycle arbitrarily often establishes the impossibility of a *perpetuum mobile* of the second kind. Finally, since the *efficiency* η of a Carnot-like cycle is given by the ratio of the work done by the system and the energy supplied throughout the cycle, it follows that

$$\eta = \frac{Q_{0-1} - Q_{2-3}}{Q_{0-1}} = 1 - \frac{Q_{2-3}}{Q_{0-1}}. \tag{5.76}$$

To show the equivalence between the Kelvin and the Clausius statements of the second law of thermodynamics, we use a *reductio ad absurdum* argument along with a contrapositive argument, that is, if one of the statements were not valid, then the other statement would also not be valid, and vice versa. Hence, suppose, *ad absurdum*, that in an isolated large-scale system energy can flow from less energetic subsystems to more energetic subsystems. Then, for a Carnot-like cycle we would be able to perform the last isothermal transformation (leg $2-3$ in Figure 5.2) without doing any external work on the system $\mathcal{G}$.

Specifically, in this case the dynamics of the isothermal process would be given by

$$\dot{E}(t) = f(E(t), V(t)) + f_{\rm w}(E(t), V(t)) - d_{2-3}(E(t), V(t)), \quad t_3 \geq t \geq t_2, \tag{5.77}$$

and hence,

$$0 = \mathbf{e}^{\rm T}\dot{E}(t) = -\mathbf{e}^{\rm T} d_{2-3}(E(t), V(t)), \quad t_3 \geq t \geq t_2, \tag{5.78}$$

which implies that

$$Q_{2-3} = \int_{t_2}^{t_3} \mathbf{e}^{\mathrm{T}} d_{2-3}(E(t), V(t))\mathrm{d}t = 0. \tag{5.79}$$

In this case, the efficiency of the Carnot cycle, given by (5.76), is $\eta = 1$. Hence, it would be possible to transform into work all the energy (heat) absorbed by the large-scale system $\mathcal{G}$ without producing any other change in the system. This contradicts the Kelvin postulate.

Conversely, suppose, *ad absurdum*, that it were possible to transform completely into work an amount of energy (heat) supplied to the system $\mathcal{G}$ by an infinite heat source with no other changes in the system. Then it follows from (5.76) that $Q_{2-3} = 0$, and since $d_{2-3}(E(t), V(t)) \geq\geq 0$, $t_3 \geq t \geq t_2$, it follows that $d_{2-3}(E(t), V(t)) \equiv 0$. Moreover, since $S_{\mathrm{w}2-3}(t) \geq\geq 0$, $t_3 \geq t \geq t_2$, it follows from (5.73) that $S_{\mathrm{w}2-3}(t) \equiv 0$.

Here, it suffices to consider the case of two subsystems with $\frac{V_2}{V_1} = \frac{c+E_2}{c+E_1}$ to arrive at a contradiction to Clausius' postulate. In this particular case, $f_{\mathrm{w}}(E(t), V(t)) \equiv 0$ and the dynamics characterized by (5.72) of the second isothermal process become

$$\dot{E}(t) = f(E(t), V(t), \quad t_3 \geq t \geq t_2. \tag{5.80}$$

Premultiplying (5.80) by $E^{\mathrm{T}}(t)$, $t_3 \geq t \geq t_2$, and using the fact that during this isothermal transformation $\dot{E}_1(t) = -\dot{E}_2(t) \leq 0$, $t_3 \geq t \geq t_2$, and $E_1(t) \leq E_2(t)$, $t_3 \geq t \geq t_2$ (see Figure 5.2), it follows that for $t_3 \geq t \geq t_2$,

$$[E_1(t) - E_2(t)]\phi_{12}(E(t), V(t)) = [E_1(t) - E_2(t)]\dot{E}_1(t) \geq 0, \tag{5.81}$$

which implies that during this isothermal transformation energy flows from the less energetic subsystem $\mathcal{G}_1$ (cooler object) to the more energetic subsystem $\mathcal{G}_2$ (hotter object). This contradicts Clausius' postulate. Thus, the Kelvin and Clausius statements of the second law of thermodynamics are equivalent.

Chapter Six

Mass-Action Kinetics and Chemical Thermodynamics

6.1 Introduction

In its most general form, thermodynamics can also involve reacting mixtures and combustion. When a chemical reaction occurs, the bonds within molecules of the *reactant* are broken, and atoms and electrons rearrange to form *products*. The thermodynamic analysis of reactive systems can be addressed as an extension of the compartmental thermodynamic model developed in Chapters 3 and 4. Specifically, in this case the compartments would qualitatively represent different quantities in the same space, and the intercompartmental flows would represent transformation rates in addition to transfer rates.

In particular, the compartments would additionally represent quantities of different chemical substances contained within the compartment, and the compartmental flows would additionally characterize transformation rates of reactants into products. In this case, an additional mass balance equation is included for addressing conservation of energy as well as conservation of mass. This additional mass conservation equation would involve the law of mass action enforcing proportionality between a particular reaction rate and the concentrations of the reactants, and the law of superposition of elementary reactions ensuring that the resultant rates for a particular species is the sum of the elementary reaction rates for the species.

The law of mass action was developed by Guldberg and Waage [181, 293] and states that in a reversible isothermal reaction, the rate of the forward or reverse reaction is proportional to the product of the concentration of the reactants or products, respectively. More specifically, the product of the concentration of the species on one side of the chemical reaction equation divided by the product of the concentration of the species on the other side of the chemical reaction equation is constant, and is independent of the amounts of each chemical substance present at the start of the reaction.

Mass-action kinetics are used in chemistry and chemical engineering to describe the dynamics of systems of chemical reactions, that is, reaction networks [426]. These models are a special form of compartmental systems, which involve mass and energy balance relations [37, 227]. Aside from their role in chemical engineering applications, mass-action kinetics have numerous analytical properties that are of inherent interest from a dynamical systems perspective. For example, mass-action kinetics give rise to systems of differential equations having polynomial nonlinearities. Polynomial systems are notorious for their intricate analytical properties even in low-dimensional cases [228, 250, 253, 397]. Because of physical considerations, however, mass-action kinetics have special properties, such as nonnegative solutions, that are useful for analyzing their behavior [135,157,158,422].

With this motivation in mind, this chapter has several objectives. First, we provide a general construction of the kinetic equations based on the reaction laws. We present this construction in a state space form that is accessible to the dynamical systems community. This presentation is based on the formulation given in [34,84,135]. Next, we consider the nonnegativity of solutions to the kinetic equations. Since the kinetic equations govern the concentrations of the species in the reaction network, it is obvious from physical arguments that nonnegative initial conditions must give rise to trajectories that remain in the nonnegative orthant of the state space.

To demonstrate this fact, we show that the kinetic equations are essentially nonnegative, and we prove that, for all nonnegative initial conditions, the resulting concentrations are nonegative. A related result is mentioned in [135]. In addition, we consider the realizability problem, which is concerned with the inverse problem of constructing a reaction network having specified essentially nonnegative dynamics. In particular, we provide an explicit construction of a reaction network for essentially nonnegative polynomial dynamics involving a scalar state.

Next, we consider the reducibility of the kinetic equations. In certain cases, such as in enzyme kinetics, the kinetic equations can be reduced in dimensionality by using constants involving initial concentrations. We provide a general statement of this procedure. We then consider the stability of the equilibria of the kinetic equations. To do this, we apply Lyapunov methods to the kinetic equations, and we obtain results that guarantee semistability, that is, convergence to a Lyapunov stable equilibrium that depends on the initial concentrations.

Once again, semistability is the appropriate notion of stability for reaction networks, where the limiting concentrations may be nonzero and may depend on the initial concentrations. Then, we revisit the *zero-deficiency* result of [144, 145], which provides rate-independent conditions that guarantee convergence of the species concentrations. In this regard we have two objectives. First, we present the zero-deficiency result for mass-action kinetics in standard matrix terminology, and, second, we prove semistability using the theory developed in Chapter 2.

Finally, we consider an interconnected dynamical system $\mathcal{G}$ where each subsystem represents a substance or species that can exchange energy with other substances as well as undergo chemical reactions with other substances, forming products. Thus, the reactants and products of chemical reactions represent subsystems of $\mathcal{G}$ with the mechanisms of heat exchange between subsystems remaining the same as delineated in Chapter 3. Here, for simplicity of exposition, we do not consider work done by the subsystem on the environment or work done by the environment on the system. This extension can be easily addressed using the formulation in Chapter 5.

6.2 Reaction Networks

We begin by reviewing the general formulation of the kinetic equations that describe chemical reactions with mass-action kinetics. First, consider the familiar reaction

$$2\mathrm{H}_2 + \mathrm{O}_2 \xrightarrow{k} 2\mathrm{H}_2\mathrm{O}. \tag{6.1}$$

The quantities on the left-hand side of the reaction (6.1) are the *reactants*, the quantities on the right-hand side are the *products*, and k denotes the *reaction rate*. The reactants and products are collectively referred to as the *species* of the reaction. Equation (6.1) can be rewritten as

$$\sum_{j=1}^{3} A_j X_j \xrightarrow{k} \sum_{j=1}^{3} B_j X_j, \tag{6.2}$$

where X_1, X_2, and X_3 denote the species H_2, O_2, and $\mathrm{H}_2\mathrm{O}$, respectively, $A_1 = 2$, $A_2 = 1$, $A_3 = 0$, $B_1 = 0$, $B_2 = 0$, and $B_3 = 2$ are the *stoichiometric coefficients*, and k denotes the *reaction rate*. Note that (6.2) can be written in a compact form using the matrix-vector notation

$$AX \xrightarrow{k} BX, \tag{6.3}$$

where $X = [X_1, X_2, X_3]^{\mathrm{T}}$, $A = [A_1, A_2, A_3] = [2, 1, 0]$, and $B = [B_1, B_2, B_3] = [0, 0, 2]$.

Next, consider the *reversible* reaction

$$\mathrm{Na}_2\mathrm{CO}_3 + \mathrm{CaCl}_2 \underset{k_2}{\overset{k_1}{\rightleftharpoons}} \mathrm{CaCO}_3 + 2\mathrm{NaCl}, \tag{6.4}$$

which is a concise notation for the *forward* and *backward* reactions

$$\mathrm{Na_2CO_3} + \mathrm{CaCl_2} \overset{k_1}{\rightarrow} \mathrm{CaCO_3} + 2\mathrm{NaCl}, \tag{6.5}$$

$$\mathrm{CaCO_3} + 2\mathrm{NaCl} \overset{k_2}{\rightarrow} \mathrm{Na_2CO_3} + \mathrm{CaCl_2}, \tag{6.6}$$

where k_1 and k_2 are the reaction rates for the forward and backward reactions, respectively. Now, let X_1, X_2, X_3, and X_4 denote the species Na_2CO_3, $CaCl_2$, $CaCO_3$, and NaCl, respectively, so that (6.4) can be written as

$$X_1 + X_2 \overset{k_1}{\rightarrow} X_3 + 2X_4, \tag{6.7}$$

$$X_3 + 2X_4 \overset{k_2}{\rightarrow} X_1 + X_2, \tag{6.8}$$

or, equivalently, as (6.3), where $X = [X_1, X_2, X_3, X_4]^{\mathrm{T}}$, $k = [k_1, k_2]$, and

$$A = \begin{bmatrix} 1 & 1 & 0 & 0 \\ 0 & 0 & 1 & 2 \end{bmatrix}, \quad B = \begin{bmatrix} 0 & 0 & 1 & 2 \\ 1 & 1 & 0 & 0 \end{bmatrix}.$$

Next, we formulate the kinetic equations for multiple chemical reactions such as (6.7) and (6.8). Specifically, consider s *species* $X_1, \ldots, X_s$, where $s \geq 1$, whose interactions are governed by r *reactions*, where $r \geq 1$, comprising the *reaction network*

$$\sum_{j=1}^{s} A_{ij} X_j \overset{k_i}{\rightarrow} \sum_{j=1}^{s} B_{ij} X_j, \quad i = 1, \ldots, r, \tag{6.9}$$

where, for $i = 1, \ldots, r$, $k_i > 0$ is the *reaction rate* of the ith reaction, $\sum_{j=1}^{s} A_{ij} X_j$ is the *reactant* of the ith reaction, and $\sum_{j=1}^{s} B_{ij} X_j$ is the *product* of the ith reaction. Note that each reaction in the reaction network (6.9) is represented as being irreversible. However, reversible reactions can be modeled by including the reverse reaction as a separate reaction, as in the case of the reaction (6.4). Each *stoichiometric coefficient* A_{ij} and B_{ij} is assumed to be a nonnegative integer. The reaction network (6.9) can be written compactly in matrix-vector form as

$$AX \overset{k}{\rightarrow} BX, \tag{6.10}$$

where $X = [X_1, \ldots, X_s]^{\mathrm{T}}$ is a column vector of species, $k = [k_1, \ldots, k_r]^{\mathrm{T}} \in \overline{\mathbb{R}}_+^r$, and A and B denote the $r \times s$ nonnegative matrices $A = [A_{ij}]$ and $B = [B_{ij}]$.

To avoid vacuous cases, we assume that each species $X_1, \ldots, X_s$ appears in the reaction network (6.10) with at least one nonzero coefficient A_{ij} or B_{ij}. This assumption is equivalent to assuming that none of the columns of

$$\begin{bmatrix} A \\ B \end{bmatrix}$$

is zero. Furthermore, in special cases and only when specifically mentioned, we allow $k_i = 0$, which effectively denotes the fact that the ith reaction is absent. Finally, we assume that, for each $i = 1, \ldots, r$, $\mathrm{row}_i(A) \neq \mathrm{row}_i(B)$ to avoid trivial reactions of the form $X_1 \xrightarrow{k} X_1$ or $X_1 + X_2 \xrightarrow{k} X_1 + X_2$, whose kinetics equations are $\dot{x}_1(t) = 0$ and $\dot{x}_1(t) = 0$, $\dot{x}_2(t) = 0$, respectively.

6.3 The Law of Mass Action and the Kinetic Equations

To derive the dynamics of the reaction network, we invoke the *law of mass action* [426], which states that, for an *elementary reaction*, that is, a reaction in which all of the stoichiometric coefficients of the reactants are 1, the rate of reaction is proportional to the product of the concentrations of the reactants. In particular, consider the reaction

$$X_1 + X_2 \xrightarrow{k} bX_3, \tag{6.11}$$

where X_1, X_2, X_3 are the species and b is a positive integer. Then

$$\begin{aligned} \dot{x}_i(t) &= -kx_1(t)x_2(t), \quad x_i(0) = x_{i0}, \quad t \geq 0, \quad i = 1, 2, && (6.12) \\ \dot{x}_3(t) &= bkx_1(t)x_2(t), \quad x_3(0) = x_{30}, && (6.13) \end{aligned}$$

where $x_i(t)$, $i = 1, 2, 3$, denotes the concentration of the species X_i. Now, writing (6.1) as the elementary reaction

$$\mathrm{H}_2 + \mathrm{H}_2 + \mathrm{O}_2 \xrightarrow{k} 2\mathrm{H}_2\mathrm{O}, \tag{6.14}$$

the law of mass action implies that

$$\begin{aligned} \dot{x}_1(t) &= -2kx_1^2(t)x_2(t), \quad x_1(0) = x_{10}, \quad t \geq 0, && (6.15) \\ \dot{x}_2(t) &= -kx_1^2(t)x_2(t), \quad x_2(0) = x_{20}, && (6.16) \\ \dot{x}_3(t) &= 2kx_1^2(t)x_2(t), \quad x_3(0) = x_{30}, && (6.17) \end{aligned}$$

where $x_1(t)$, $x_2(t)$, and $x_3(t)$ denote the concentrations of H_2, O_2, and $\mathrm{H}_2\mathrm{O}$, respectively, at time t.

Similarly, let $x_i(t)$ denote the concentration of X_i, $i = 1, \ldots, 4$, in (6.7) and (6.8) or, equivalently, the reversible reaction (6.4). In this case, it follows from the law of mass action that

$$\begin{aligned} \dot{x}_1(t) &= -k_1x_1(t)x_2(t) + k_2x_3(t)x_4^2(t), \quad x_1(0) = x_{10}, \quad t \geq 0, && (6.18) \\ \dot{x}_2(t) &= -k_1x_1(t)x_2(t) + k_2x_3(t)x_4^2(t), \quad x_2(0) = x_{20}, && (6.19) \\ \dot{x}_3(t) &= k_1x_1(t)x_2(t) - k_2x_3(t)x_4^2(t), \quad x_3(0) = x_{30}, && (6.20) \\ \dot{x}_4(t) &= 2k_1x_1(t)x_2(t) - 2k_2x_3(t)x_4^2(t), \quad x_4(0) = x_{40}. && (6.21) \end{aligned}$$

To consider general reaction networks, the following notation is needed. For $x = [x_1, \ldots, x_q]^{\mathrm{T}} \in \mathbb{R}^q$ and nonnegative matrix $A = [A_{(i,j)}] \in \mathbb{R}^{p\times q}$, x^A denotes the element of $\mathbb{R}^p$ whose ith component for $i = 1, \ldots, p$ is the product $x_1^{A_{i1}} \cdots x_q^{A_{iq}}$. For example, if

$$A = \begin{bmatrix} 1 & 2 \\ 3 & 4 \end{bmatrix},$$

then

$$x^A = \begin{bmatrix} x_1 x_2^2 \\ x_1^3 x_2^4 \end{bmatrix}.$$

We define $0^0 \triangleq 1$. The matrix exponentiation operation has many convenient properties [33, 278]. For example, if $A, B \in \mathbb{R}^{p\times q}$, then $x^{(A+B)} = x^A x^B$. If $B \in \mathbb{R}^{n\times p}$, then $(x^A)^B = x^{BA}$. Furthermore, $(x \circ y)^A = (x^A) \circ (y^A) = x^A y^A$. Note that $x^{I_p} = x$ and $x^{-A} \circ (x^A) = \mathbf{e}$. Alternatively, if $A \in \mathbb{R}^{p\times p}$, then $x^{-I_p} \circ x^A = x^{A-I_p}$. Furthermore, if $\det A \neq 0$, $x >> 0$, and $y >> 0$, then $x^A = y$ implies that $y = x^{A^{-1}}$. In addition, $\mathbf{log}_e x^A = A\mathbf{log}_e x$ and $e^{A\mathbf{log}_e x} = x^A$, while $x^A = y$ implies $A\mathbf{log}_e x = \mathbf{log}_e y$, where, for $x = [x_1, \ldots, x_s]^{\mathrm{T}} \in \mathbb{R}_+^p$, $\mathbf{log}_e x$ denotes the vector in $\mathbb{R}^p$ whose ith component is $\log_e x_i$. Finally, if $f(x) = x^A$, then $f'(x) = \mathrm{diag}(x^A) A [\mathrm{diag}(x)]^{-1}$, where

$$\mathrm{diag}[x_1, x_2, \ldots, x_n] \triangleq \begin{bmatrix} x_1 & \ldots & 0 \\ \vdots & \ddots & \vdots \\ 0 & \ldots & x_n \end{bmatrix}.$$

For $x = [x_1, \ldots, x_s]^{\mathrm{T}} \in \mathbb{R}^p$, e^x denotes the vector in $\mathbb{R}^p$ whose ith component is e^{x_i}.

Now, more generally, consider the reaction (6.10) and, for $j = 1, \ldots, s$, let $x_j(t)$ denote the concentration of the species X_j at time t. Then, by applying the law of mass action, the dynamics of the reaction network (6.10) are given by the *kinetic equations*

$$\dot{x}(t) = (B - A)^{\mathrm{T}}[k \circ x^A(t)], \quad x(0) = x_0, \quad t \geq 0. \tag{6.22}$$

Defining $K \triangleq \mathrm{diag}[k_1, \ldots, k_r]$, (6.22) can be written as

$$\dot{x}(t) = (B - A)^{\mathrm{T}} K x^A(t), \quad x(0) = x_0, \quad t \geq 0. \tag{6.23}$$

In mass-action kinetics, the *reaction order* $\sum_{j=1}^{s} A_{ij}$ of the ith reaction is the sum of the stoichiometric coefficients of the species appearing in the reactant of the ith reaction. Equation (6.22), which is equivalent to (4.7) of [135], is a matrix-vector formulation of mass-action kinetics. It can be seen that the kinetic equations (6.22) are linear if and only if each row of A contains exactly one 1 with the remaining entries equal to zero, that is, if and only if each reaction is *unimolecular*. In this case, it can be seen that

$x^A = Ax$, and thus (6.22) becomes

$$\dot{x}(t) = Mx(t), \quad x(0) = x_0, \quad t \geq 0, \tag{6.24}$$

where $M \in \mathbb{R}^{s \times s}$ is defined by

$$M \triangleq (B - A)^{\mathrm{T}} K A. \tag{6.25}$$

The reaction network (6.10) is not limited to closed systems for which conservation of mass holds. In fact, (6.10) can also be used to represent open systems in which mass removal and mass addition are allowed. For example, either $A = 0$ or $B = 0$ (but not both) is allowed in the reaction $AX_1 \xrightarrow{k_1} BX_1$. The kinetic equations for the reactions $X_1 \xrightarrow{k_1} 0$ and $0 \xrightarrow{k_1} X_1$, which represent the removal and addition of mass, are $\dot{x}_1(t) = -k_1 x_1(t)$ and $\dot{x}_1(t) = k_1$ with solutions $x_1(t) = x_1(0)e^{-k_1 t}$ and $x_1(t) = k_1 t + x_1(0)$, respectively.

The reactions $X_1 \xrightarrow{k_1} 2X_1$ and $2X_1 \xrightarrow{k_1} 3X_1$, which also represent the addition of mass, have the kinetics $\dot{x}_1(t) = k_1 x_1(t)$ and $\dot{x}_1(t) = k_1 x_1^2(t)$ with solutions $x_1(t) = x_1(0)e^{k_1 t}$ and $x_1(t) = x_1(0)/(1 - k_1 x_1(0)t)$, respectively. Note that the latter solution has finite escape time since it exists only on the interval $[0, 1/(k_1 x_1(0)))$. Finally, the reactions $X \xrightarrow{k} Y$ and $2X \xrightarrow{k} 2Y$, although stoichiometrically equivalent, have different kinetic equations, namely, $\dot{x}(t) = -kx(t)$, $\dot{y}(t) = kx(t)$ and $\dot{x}(t) = -kx^2(t)$, $\dot{y}(t) = kx^2(t)$, respectively. We adopt the convention that the law of mass action applies to the reaction involving the minimum number of molecules necessary for the reaction to occur.

Example 6.1. Consider the reaction network

$$X_1 \xrightarrow{k_1} X_2, \tag{6.26}$$

$$X_2 \xrightarrow{k_2} X_1, \tag{6.27}$$

so that $s = 2$, $r = 2$, and A and B are given by

$$A = \begin{bmatrix} 1 & 0 \\ 0 & 1 \end{bmatrix}, \quad B = \begin{bmatrix} 0 & 1 \\ 1 & 0 \end{bmatrix}. \tag{6.28}$$

The kinetic equations are thus given by

$$\dot{x}_1(t) = -k_1 x_1(t) + k_2 x_2(t), \quad x_1(0) = x_{10}, \quad t \geq 0, \tag{6.29}$$

$$\dot{x}_2(t) = k_1 x_1(t) - k_2 x_2(t), \quad x_2(0) = x_{20}, \tag{6.30}$$

or in linear system form by (6.24), where

$$M = \begin{bmatrix} -k_1 & k_2 \\ k_1 & -k_2 \end{bmatrix} \tag{6.31}$$

and $x = [x_1,\ x_2]^{\mathrm{T}}$. △

Example 6.2. Consider the reaction network

$$X_1 + X_2 \xrightarrow{k_1} 2X_1, \tag{6.32}$$
$$2X_1 \xrightarrow{k_2} X_1 + X_2, \tag{6.33}$$

so that $s = 2$, $r = 2$,

$$A = \begin{bmatrix} 1 & 1 \\ 2 & 0 \end{bmatrix}, \quad B = \begin{bmatrix} 2 & 0 \\ 1 & 1 \end{bmatrix}. \tag{6.34}$$

The kinetic equations are thus given by

$$\dot{x}_1(t) = k_1 x_1(t) x_2(t) - k_2 x_1^2(t), \quad x_1(0) = x_{10}, \quad t \geq 0, \tag{6.35}$$
$$\dot{x}_2(t) = -k_1 x_1(t) x_2(t) + k_2 x_1^2(t), \quad x_2(0) = x_{20}. \tag{6.36}$$

△

Example 6.3. The Lotka-Volterra reaction is given by

$$X_1 \xrightarrow{k_1} 2X_1, \tag{6.37}$$
$$X_1 + X_2 \xrightarrow{k_2} 2X_2, \tag{6.38}$$
$$X_2 \xrightarrow{k_3} 0, \tag{6.39}$$

where x_1 and x_2 denote prey and predator species, respectively, so that $s = 2$ and $r = 3$. Furthermore, A and B are given by

$$A = \begin{bmatrix} 1 & 0 \\ 1 & 1 \\ 0 & 1 \end{bmatrix}, \quad B = \begin{bmatrix} 2 & 0 \\ 0 & 2 \\ 0 & 0 \end{bmatrix}. \tag{6.40}$$

Consequently, the kinetic equations have the form

$$\dot{x}(t) = \begin{bmatrix} 1 & -1 & 0 \\ 0 & 1 & -1 \end{bmatrix} \begin{bmatrix} k_1 x_1(t) \\ k_2 x_1(t) x_2(t) \\ k_3 x_2(t) \end{bmatrix}, \quad x(0) = \begin{bmatrix} x_{10} \\ x_{20} \end{bmatrix}, \quad t \geq 0, \tag{6.41}$$

or, equivalently,

$$\dot{x}_1(t) = k_1 x_1(t) - k_2 x_1(t) x_2(t), \quad x_1(0) = x_{10}, \quad t \geq 0, \tag{6.42}$$
$$\dot{x}_2(t) = -k_3 x_2(t) + k_2 x_1(t) x_2(t), \quad x_2(0) = x_{20}. \tag{6.43}$$

△

Example 6.4. A widely studied reaction network [285] involves the interaction of a substrate S and an enzyme E to produce a product P by means of an intermediate species C. The reactions are given by

$$\mathrm{S}+\mathrm{E} \underset{k_2}{\overset{k_1}{\rightleftharpoons}} \mathrm{C} \overset{k_3}{\rightarrow} \mathrm{P}+\mathrm{E} \tag{6.44}$$

so that $s = 4$ and $r = 3$. Letting $X_1 = \mathrm{S}$, $X_2 = \mathrm{C}$, $X_3 = \mathrm{E}$, and $X_4 = \mathrm{P}$, the corresponding reaction network can be written as

$$X_1 + X_3 \overset{k_1}{\rightarrow} X_2, \tag{6.45}$$

$$X_2 \overset{k_2}{\rightarrow} X_1 + X_3, \tag{6.46}$$

$$X_2 \overset{k_3}{\rightarrow} X_3 + X_4. \tag{6.47}$$

It thus follows that A and B are given by

$$A = \begin{bmatrix} 1 & 0 & 1 & 0 \\ 0 & 1 & 0 & 0 \\ 0 & 1 & 0 & 0 \end{bmatrix}, \quad B = \begin{bmatrix} 0 & 1 & 0 & 0 \\ 1 & 0 & 1 & 0 \\ 0 & 0 & 1 & 1 \end{bmatrix}. \tag{6.48}$$

Consequently, the kinetic equations have the form

$$\dot{x}(t) = \begin{bmatrix} -1 & 1 & 0 \\ 1 & -1 & -1 \\ -1 & 1 & 1 \\ 0 & 0 & 1 \end{bmatrix} \begin{bmatrix} k_1 x_1(t) x_3(t) \\ k_2 x_2(t) \\ k_3 x_2(t) \end{bmatrix}, \quad x(0) = \begin{bmatrix} x_{10} \\ x_{20} \\ x_{30} \\ x_{40} \end{bmatrix}, \quad t \geq 0, \tag{6.49}$$

or, equivalently,

$$\dot{x}_1(t) = k_2 x_2(t) - k_1 x_1(t) x_3(t), \quad x_1(0) = x_{10}, \quad t \geq 0, \tag{6.50}$$

$$\dot{x}_2(t) = -(k_2 + k_3) x_2(t) + k_1 x_1(t) x_3(t), \quad x_2(0) = x_{20}, \tag{6.51}$$

$$\dot{x}_3(t) = (k_2 + k_3) x_2(t) - k_1 x_1(t) x_3(t), \quad x_3(0) = x_{30}, \tag{6.52}$$

$$\dot{x}_4(t) = k_3 x_2(t), \quad x_4(0) = x_{40}. \tag{6.53}$$

△

6.4 Nonnegativity of Solutions

Since the states of the kinetic equations (6.22) represent concentrations, it is natural to expect that, for nonnegative initial concentrations, the concentrations remain nonnegative for as long as the solution exists. In this section, we show that the kinetic equations (6.22) are in fact essentially nonnegative, and hence, nonnegativity of solutions is a direct consequence of Proposition 2.1.

For the following results we consider the system

$$\dot{x}(t) = f(x(t)), \quad x(0) = x_0, \quad t \in [0, \tau_{x_0}), \tag{6.54}$$

where $f : \mathcal{D} \to \mathbb{R}^n$ is locally Lipschitz continuous, $\mathcal{D}$ is an open subset of

$\mathbb{R}^n$, $x_0 \in \mathcal{D}$, and $[0, \tau_{x_0})$, where $0 < \tau_{x_0} \leq \infty$, is the maximal interval of existence for the solution $x(\cdot)$ of (6.54).

Proposition 6.1. Define $f : \mathbb{R}^s \to \mathbb{R}^s$ by $f(x) = (B - A)^{\mathrm{T}}(k \circ x^A)$. Then f is locally Lipschitz continuous and essentially nonnegative.

Proof. Since f is continuously differentiable, it follows that f is locally Lipschitz continuous. Next, let $x \in \overline{\mathbb{R}}_+^n$. For $j \in \{1, \ldots, s\}$ we have

$$f_j(x) = [\mathrm{col}_j(B) - \mathrm{col}_j(A)]^{\mathrm{T}} \begin{bmatrix} k_1 x^{\mathrm{row}_1(A)} \\ \vdots \\ k_r x^{\mathrm{row}_r(A)} \end{bmatrix}$$
$$= \sum_{i=1}^{r} B_{ij} k_i x^{\mathrm{row}_i(A)} - \sum_{i=1}^{r} A_{ij} k_i x^{\mathrm{row}_i(A)}.$$

Note that the first summation is nonnegative since x is nonnegative. Next, note that $A_{ij} k_i x^{\mathrm{row}_i(A)}$ contains the factor $A_{ij} x_j^{A_{ij}}$. Now, to verify essential nonnegativity, let $x_j = 0$. If $A_{ij} > 0$, then $A_{ij} x_j^{A_{ij}} = A_{ij}(0^{A_{ij}}) = 0$, while, if $A_{ij} = 0$, then $A_{ij} x_j^{A_{ij}} = \lim_{x_j \to 0} 0(x_j^0) = \lim_{x_j \to 0} 0(1) = 0$. Consequently, the second summation is zero for all nonnegative $A_{1j}, \ldots, A_{rj}$ whenever $x_j = 0$. Thus, f is essentially nonnegative. □

Theorem 6.1. $\overline{\mathbb{R}}_+^n$ is an invariant set with respect to (6.22).

Proof. The result is an immediate consequence of Propositions 2.1 and 6.1. □

Corollary 6.1. Consider the linear kinetic reaction (6.24), where $M = (B - A)^{\mathrm{T}} K A$ and A has exactly one nonzero entry in each row. Then $f(x) = Mx$ is essentially nonnegative and $\overline{\mathbb{R}}_+^s$ is an invariant set with respect to (6.24).

Proof. Since A is nonnegative, K is nonnegative and diagonal, and A has exactly one nonzero entry in each row, it follows that $A^{\mathrm{T}} K A$ is diagonal and nonnegative. Now, since $B^{\mathrm{T}} K A$ is nonnegative, it follows that M is essentially nonnegative, and hence, $f(x) = Mx$ is essentially nonnegative. The invariance of $\overline{\mathbb{R}}_+^s$ is a direct consequence of Theorem 6.1. □

In the linear case $f(x) = Mx$, where $M \in \mathbb{R}^{n \times n}$ is essentially nonnegative, Theorem 6.1 implies the following result.

Proposition 6.2. Let $M \in \mathbb{R}^{n \times n}$. Then M is essentially nonnegative if and only if $e^{Mt} \geq\geq 0$ for all $t \geq 0$.

Proof. This is a restatement of Proposition 2.4. □

Example 6.5. Consider Example 6.1. For the kinetic equations (6.24) with M given by (6.31) it can be seen that

$$M = \begin{bmatrix} -k_1 & k_2 \\ k_1 & -k_2 \end{bmatrix}$$

is essentially nonnegative. The exponential of M is given by $e^{Mt} = I_2 + \frac{1-e^{-(k_1+k_2)t}}{k_1+k_2}M$, which is nonnegative for all $t \geq 0$. Consequently, if $x(0)$ is nonnegative, then the solution $x(\cdot)$ of (6.24) given by $x(t) = e^{Mt}x(0)$ is nonnegative for all $t \geq 0$. △

Example 6.6. Consider Examples 6.2–6.4. It can be seen that the function f for each of these examples is essentially nonnegative. △

6.5 Realization of Mass-Action Kinetics

In this section, we consider the *realization problem*, which is concerned with the construction of a reaction network whose dynamics are given by specified kinetic equations. In this case, the reaction network is a *realization* of the kinetic equations. Note that the polynomial

$$f(x) = \sum_{i=0}^{\nu} a_i x^i \tag{6.55}$$

in the real scalar x is essentially nonnegative if and only if $a_0 \geq 0$.

Theorem 6.2. Consider the system (6.54), where $n = 1$ and $f : \mathbb{R} \to \mathbb{R}$ is an essentially nonnegative polynomial of degree ν of the form (6.55). Then there exists a reaction network of the form (6.10) with $s = 1$ and $r \leq \nu + 1$, and with stoichiometric coefficient matrices A and B having nonnegative integer entries such that $f(x) = (B - A)^{\mathrm{T}} \left(k \circ x^A\right)$.

Proof. For $i = 1, \ldots, \nu$, define A, B, and $k \in \overline{\mathbb{R}}_+^{\nu+1}$ as

$$A_i \triangleq i, \quad B_i \triangleq (i + \operatorname{sgn} a_i), \quad k_i \triangleq |a_i|, \tag{6.56}$$

where $\operatorname{sgn} 0 \triangleq 0$. Note that $A \geq\geq 0$ and, since $a_0 \geq 0$, it follows that $B \geq\geq 0$. Then the dynamics of the reaction network (6.10) are given by the kinetic equation

$$\begin{aligned} \dot{x}(t) &= (B - A)^{\mathrm{T}} \left[k \circ x^A(t)\right] \\ &= \sum_{i=0}^{\nu} (B_i - A_i) k_i x^{A_i}(t) \end{aligned}$$

$$= \sum_{i=0}^{\nu} (\text{sgn } a_i)|a_i|x^i(t)$$
$$= \sum_{i=0}^{\nu} a_i x^i(t)$$
$$= f(x(t)).$$

Hence, (6.10) is a realization of (6.22), where f is given by (6.55). $\square$

To demonstrate Theorem 6.2, let $\nu = 3$. Then a realization of $\dot{x}_1(t) = a_3 x_1^3(t) + a_2 x_1^2(t) + a_1 x_1(t) + a_0$ is given by the reaction network

$$0 \overset{a_0}{\to} X_1, \tag{6.57}$$
$$X_1 \overset{|a_1|}{\to} (1 + \text{sgn } a_1)X_1, \tag{6.58}$$
$$2X_1 \overset{|a_2|}{\to} (2 + \text{sgn } a_2)X_1, \tag{6.59}$$
$$3X_1 \overset{|a_3|}{\to} (3 + \text{sgn } a_3)X_1, \tag{6.60}$$

where we follow the convention that any reaction with rate constant 0 is removed from the network to avoid trivial reactions of the form $0 \overset{0}{\to} aX$ and $aX \overset{0}{\to} aX$.

If $n \geq 2$ and f is an essentially nonnegative multivariate polynomial in $x_1, \ldots, x_n$, then there does not necessarily exist a reaction network such that $f(x) = (B - A)^{\mathrm{T}} \left(k \circ x^A\right)$. For example, consider the case $n = 2$ and the dynamic equations

$$\dot{x}_1(t) = x_2^2(t) - 2x_2^3(t) + x_2^4(t), \quad x_1(0) = x_{10}, \quad t \geq 0, \tag{6.61}$$
$$\dot{x}_2(t) = 0, \quad x_2(0) = x_{20}. \tag{6.62}$$

Then

$$f(x_1, x_2) = \begin{bmatrix} x_2^2 - 2x_2^3 + x_2^4 \\ 0 \end{bmatrix}$$

is essentially nonnegative. However, (6.61) and (6.62) cannot be realized as a reaction network.

To see this, suppose that (6.61) and (6.62) are the kinetic equations for a reaction network of r reactions involving the species X_1 and X_2. Since $f(\cdot, \cdot)$ is independent of x_1, it follows that the reaction network must have the form

$$a_i X_2 \overset{k_i}{\to} b_i X_1 + c_i X_2, \tag{6.63}$$

where a_i, b_i, and c_i are nonnegative integers and $k_i \geq 0$ for all $i = 1, \ldots, r$. Now, it follows from the law of mass action that the kinetic equations for

(6.63) are given by

$$\dot{x}_1(t) = \sum_{i=1}^{r} b_i k_i x_2^{a_i}(t), \quad x_1(0) = x_{10}, \quad t \geq 0, \tag{6.64}$$

$$\dot{x}_2(t) = \sum_{i=1}^{r} (c_i - a_i) k_i x_2^{a_i}(t), \quad x_2(0) = x_{20}. \tag{6.65}$$

Comparing (6.61) with (6.65), it follows that $a_i \in \{2, 3, 4\}$ for all $i = 1, \ldots, r$. Furthermore, $\sum_{i \in \mathcal{R}} b_i k_i = -2$, where $\mathcal{R} \triangleq \{i \in \{1, \ldots, r\} : a_i = 3\}$, which is a contradiction since $b_i \geq 0$ and $k_i \geq 0$ for all $i = 1, \ldots, r$.

Next, we present a necessary and sufficient condition that guarantees a reaction network realization exists such that $f(x) = (B - A)^{\mathrm{T}} \left(k \circ x^A\right)$.

Theorem 6.3. Consider the system (6.54), where $n > 1$ and $f : \mathbb{R}^n \to \mathbb{R}^n$ is a multivariate polynomial. Then there exists a reaction network of the form (6.10) with $s = n$ such that $f(x) = (B - A)^{\mathrm{T}} \left(k \circ x^A\right)$, where the stoichiometric coefficient matrices A and B have nonnegative integer entries, if and only if for each $j \in \{1, \ldots, n\}$, $f_j(x_1, x_2, \ldots, x_{j-1}, 0, x_{j+1}, \ldots, x_n)$ is a multivariate polynomial with nonnegative integer coefficients.

Proof. To prove sufficiency, let $j \in \{1, \ldots, n\}$. By assumption, $f_j(x)$ is a sum of terms either of the form

$$a_j x_1^{p_1} x_2^{p_2} \cdots x_j^{p_j} \cdots x_n^{p_n}, \tag{6.66}$$

where $p_i \geq 0$ for all $i = 1, \ldots, n$ and $p_j > 0$, or of the form

$$b_j x_1^{q_1} \cdots x_{j-1}^{q_{j-1}} x_{j+1}^{q_{j+1}} \cdots x_n^{q_n}, \tag{6.67}$$

with $b_j > 0$. Next, note that the reaction

$$\sum_{i=1}^{n} p_i X_i \overset{|a_j|}{\to} (p_j + \operatorname{sgn} a_j) X_j + \sum_{i=1, i \neq j}^{n} p_i X_i \tag{6.68}$$

contributes the term (6.66) to $\dot{x}_j$ and no terms to $\dot{x}_i$ for all $i = 1, \ldots, n$ such that $i \neq j$. Similarly, the reaction

$$\sum_{i=1}^{n} q_i X_i \overset{b_j}{\to} X_j + \sum_{i=1, i \neq j}^{n} q_i X_i \tag{6.69}$$

contributes the term (6.67) to the rate of $\dot{x}_j$ and zero terms to $\dot{x}_i$ for all $i = 1, \ldots, n$, $i \neq j$. Hence, for all $j = 1, \ldots, n$, each term of $f_j(x)$ can be realized as a valid reaction, which establishes sufficiency.

To prove necessity, let $x \in \overline{\mathbb{R}}_+^s$ and let $j \in \{1, \ldots, s\}$. Then

$$f_j(x) = \sum_{i=1}^{r} (B_{ij} - A_{ij}) k_i x^{\mathrm{row}_i(A)}.$$

Let $x_j = 0$. If $A_{ij} > 0$, then $x^{\mathrm{row}_i(A)}$ and hence $(B_{ij} - A_{ij}) k_i x^{\mathrm{row}_i(A)} = 0$, whereas, if $A_{ij} = 0$, then

$$(B_{ij} - A_{ij}) k_i x^{\mathrm{row}_i(A)} = \lim_{x_j \to 0^+} B_{ij} k_i x_1^{A_{i1}} \cdots x_{j-1}^{A_{i(j-1)}} x_j^0 x_{j+1}^{A_{i(j+1)}} \cdots x_n^{A_{in}}.$$

Hence,

$$f_j(x) = \sum_{i \in \mathcal{I}_j} B_{ij} k_i x_1^{A_{i1}} \cdots x_{j-1}^{A_{i(j-1)}} x_{j+1}^{A_{i(j+1)}} \cdots x_n^{A_{in}},$$

where $\mathcal{I}_j \triangleq \{i \in \{1, \ldots, r\} : A_{ij} = 0\}$, establishing the result. □

6.6 Reducibility of the Kinetic Equations

In this section, we provide a technique for reducing the number of kinetic equations needed to model the dynamics of the reaction network (6.10). The reduced-order kinetic equations model a subset of the species appearing in the original reaction network. This technique is based on the fact that, while $x(t)$, $t \geq 0$, is confined to the nonnegative orthant for nonnegative initial conditions, the structure of the kinetic equations (6.22) impose an additional constraint on the allowable trajectories.

To state this result we define the *stoichiometric subspace* $\mathcal{S}$ by $\mathcal{S} \triangleq \mathcal{R}((B - A)^{\mathrm{T}})$, which is a subspace of $\mathbb{R}^s$. The dimension of this subspace is given by $q \triangleq \mathrm{rank}((B - A)^{\mathrm{T}}) = \mathrm{rank}(B - A)$, which is the *rank* of the reaction network. Note that $q \leq \min\{r, s\}$. The following result shows that the solution of the kinetic equations (6.22) is confined to an affine subspace that is parallel to the stoichiometric subspace. For convenience, we let $P \in \mathbb{R}^{s \times s}$ denote the unique orthogonal projector whose range is $\mathcal{S}$, and define $P_\perp \triangleq I_s - P$. In terms of the generalized inverse $(\cdot)^+$, P is given by $P = (B - A)^{\mathrm{T}}[(B - A)^{\mathrm{T}}]^+ = (B - A)^+(B - A)$. Note that, if $z \in \mathbb{R}^s$, then $Pz = z$ if and only if $z \in \mathcal{S}$, and therefore $P_\perp z = 0$ if and only if $z \in \mathcal{S}$.

Proposition 6.3. Suppose $x(0) \in \overline{\mathbb{R}}_+^s$. Then, for all $t \in [0, \tau_{x(0)})$, the solution $x(\cdot)$ of (6.22) satisfies

$$x(t) \in (x(0) + \mathcal{S}) \cap \overline{\mathbb{R}}_+^s. \tag{6.70}$$

Proof. It follows from Proposition 6.1 that, for all $t \in [0, \tau_{x(0)})$, $x(t)$ is confined to the nonnegative orthant. To show that $x(t) \in x(0) + \mathcal{S}$ for all $t \in [0, \tau_{x(0)})$, note that $\dot{x}(t) \in \mathcal{S}$ for all $t \in [0, \tau_{x(0)})$, which implies that

$\frac{\mathrm{d}}{\mathrm{d}t}P_\perp[x(t)-x(0)] = P_\perp\dot{x}(t) = 0$ for all $t \in [0, \tau_{x(0)})$. Hence, $P_\perp[x(t)-x(0)]$ is constant for all $t \in [0, \tau_{x(0)})$. Thus, for all $t \in [0, \tau_{x(0)})$, it follows that $P_\perp[x(t)-x(0)] = P_\perp[x(0)-x(0)] = 0$, and hence, $x(t)-x(0) \in \mathcal{S}$, as required. □

Corollary 6.2. Suppose $x(0) \in \overline{\mathbb{R}}_+^s$. Then $(x(0)+\mathcal{S}) \cap \overline{\mathbb{R}}_+^s$ is an invariant set with respect to (6.22).

Proof. Let $\hat{x}(0) \in (x(0)+\mathcal{S}) \cap \overline{\mathbb{R}}_+^s$ so that $\hat{x}(0) = x(0) + w$, where $w \in \mathcal{S}$, and let $\hat{x}(\cdot)$ denote the corresponding solution to (6.22). Then, since $\hat{x}(0) \in \overline{\mathbb{R}}_+^s$, it follows from Proposition 6.3 that, for all $t \in [0, \tau_{\hat{x}(0)})$,

$$\hat{x}(t) \in (\hat{x}(0)+\mathcal{S}) \cap \overline{\mathbb{R}}_+^s = (x(0)+w+\mathcal{S}) \cap \overline{\mathbb{R}}_+^s = (x(0)+\mathcal{S}) \cap \overline{\mathbb{R}}_+^s.$$

This completes the proof. □

Proposition 6.3 shows that the solution $x(\cdot)$ of the kinetic equations (6.22) is confined to the *stoichiometric compatibility class* $(x(0)+\mathcal{S}) \cap \overline{\mathbb{R}}_+^s$, which is a q-dimensional manifold with boundary. (The set $(x(0)+\mathcal{S}) \cap \mathbb{R}_+^s$ is a *positive stoichiometric compatibility class.*) This fact suggests that the dynamics of the reaction network can be represented by a set of q species. In fact, the following result shows that, if $q < s$, then the number of species can be reduced from s to q. Since $q \leq \min\{r, s\}$, this reduction is always possible when $r < s$. For convenience, the following result assumes that the species $x_1, \ldots, x_s$ are labeled such that the first q columns of $B - A$ are linearly independent.

Proposition 6.4. Assume that $q < s$. Furthermore, partition $A = [A_1, A_2]$ and $B = [B_1, B_2]$, where $A_1, B_1 \in \mathbb{R}^{r \times q}$, and assume that $\operatorname{rank}(B_1 - A_1) = q$. In addition, let $F \in \mathbb{R}^{q \times (s-q)}$ satisfy $A_2 - B_2 = (A_1 - B_1)F$. Finally, partition $x = [\hat{x}_1^{\mathrm{T}}, \hat{x}_2^{\mathrm{T}}]^{\mathrm{T}}$, where $\hat{x}_1 \triangleq [x_1, \ldots, x_q]^{\mathrm{T}}$ and $\hat{x}_2 \triangleq [x_{q+1}, \ldots, x_s]^{\mathrm{T}}$. Then

$$\hat{x}_2(t) = F^{\mathrm{T}}\hat{x}_1(t) + \gamma, \quad \hat{x}_2(0) = \hat{x}_{20}, \quad t \geq 0, \tag{6.71}$$

where $\gamma \triangleq \hat{x}_2(0) - F^{\mathrm{T}}\hat{x}_1(0) \in \mathbb{R}^{s-q}$, and $\hat{x}_1(\cdot)$ satisfies

$$\dot{\hat{x}}_1(t) = (B_1 - A_1)^{\mathrm{T}}[k \circ \hat{x}_1^{A_1}(t) \circ (F^{\mathrm{T}}\hat{x}_1(t) + \gamma)^{A_2}], \ \hat{x}_1(0) = \hat{x}_{10}, \ t \geq 0. \tag{6.72}$$

Proof. Left multiplying (6.22) by $[F^{\mathrm{T}}, -I_{s-q}]$ yields $\dot{\hat{x}}_2(t) = F^{\mathrm{T}}\dot{\hat{x}}_1(t)$, which implies (6.71). Next, note that $\dot{\hat{x}}_1(t) = (B_1 - A_1)^{\mathrm{T}}[k \circ x^A(t)] = (B_1 - A_1)^{\mathrm{T}}[k \circ \hat{x}_1^{A_1}(t) \circ \hat{x}_2^{A_2}(t)]$, which, with (6.71), yields (6.72). □

Example 6.7. Consider Example 6.1. Note that $s = 2$, $r = 2$, and $q = 1 < s$, and thus Proposition 6.4 can be applied with $F = -1$. It thus follows that $x_2(t) = -x_1(t) + \gamma$ for all $t \geq 0$, where $\gamma \triangleq x_1(0) + x_2(0)$. Applying Proposition 6.4 with $\hat{x}_1 = x_1$ and $\hat{x}_2 = x_2$, (6.72) yields the scalar kinetic equation

$$\dot{x}_1(t) = -(k_1 + k_2)x_1(t) + k_2\gamma, \quad x_1(0) = x_{10}, \quad t \geq 0, \tag{6.73}$$

which is essentially nonnegative. A reduced reaction network realization for these kinetic equations is given by

$$0 \overset{k_2\gamma}{\to} X_1, \tag{6.74}$$

$$X_1 \overset{k_1+k_2}{\to} 0, \tag{6.75}$$

for which $q = s = 1$ and $r = 2$. △

Example 6.8. Consider Example 6.2. Note that $s = 2$, $r = 2$, and $q = 1 < s$, and thus Proposition 6.4 can be applied with $F = -1$. It thus follows that $x_2(t) = -x_1(t) + \gamma$ for all $t \geq 0$, where $\gamma \triangleq x_1(0) + x_2(0)$. Applying Proposition 6.4 with $\hat{x}_1 = x_1$ and $\hat{x}_2 = x_2$, (6.72) yields the scalar kinetic equation

$$\dot{x}_1(t) = -(k_1 + k_2)x_1^2(t) + k_1\gamma x_1(t), \quad x_1(0) = x_{10}, \quad t \geq 0, \tag{6.76}$$

which is essentially nonnegative. A reaction network realization for this reduced-order kinetic equation is given by

$$X_1 \overset{k_1\gamma}{\to} 2X_1, \tag{6.77}$$

$$2X_1 \overset{k_1+k_2}{\to} X_1, \tag{6.78}$$

for which $q = s = 1$ and $r = 2$. △

Example 6.9. Consider Example 6.3. Note that $s = 2$, $r = 3$, and $q = 2 = s$, and thus reduction is not possible. △

Example 6.10. Consider Example 6.4. Note that $s = 4$, $r = 3$, and $q = 2 < s$, and thus Proposition 6.4 can be applied with

$$F = \begin{bmatrix} 0 & -1 \\ -1 & -1 \end{bmatrix}.$$

It thus follows that $x_3(t) = -x_2(t) + \gamma_1$ and $x_4(t) = -x_1(t) - x_2(t) + \gamma_2$ for all $t \geq 0$, where $\gamma_1 \triangleq x_2(0) + x_3(0)$ and $\gamma_2 \triangleq x_1(0) + x_2(0) + x_4(0)$. Applying Proposition 6.4 with $\hat{x}_1 = [x_1, x_2]^{\mathrm{T}}$ and $\hat{x}_2 = [x_3, x_4]^{\mathrm{T}}$, (6.72) yields the kinetic equations

$$\dot{x}_1(t) = -k_1\gamma_1 x_1(t) + k_2 x_2(t) + k_1 x_1(t)x_2(t), \quad x_1(0) = x_{10}, \quad t \geq 0, \tag{6.79}$$

$$\dot{x}_2(t) = k_1\gamma_1 x_1(t) - (k_2 + k_3)x_2(t) - k_1 x_1(t)x_2(t), \quad x_2(0) = x_{20}, \tag{6.80}$$

which is essentially nonnegative.

The dynamics of the system (6.79) and (6.80) are discussed in [285] and the references given therein. A reaction network realization for these reduced-order kinetic equations is given by

$$X_1 \overset{k_1\gamma_1}{\to} X_2, \tag{6.81}$$

$$X_2 \overset{k_2}{\to} X_1, \tag{6.82}$$

$$X_2 \overset{k_3}{\to} 0, \tag{6.83}$$

$$X_1 + X_2 \overset{k_1}{\to} 2X_1, \tag{6.84}$$

for which $q = s = 2$ and $r = 4$. △

Example 6.11. We now show that not every reduced-order kinetic equation can be realized as a reaction network. For convenience, we relabel the species of Example 6.4 as $X_1 = \mathrm{S}$, $X_2 = \mathrm{P}$, $X_3 = \mathrm{C}$, and $X_4 = \mathrm{E}$. The reaction network (6.45)–(6.47) can now be written as

$$X_1 + X_4 \overset{k_1}{\to} X_3, \tag{6.85}$$

$$X_3 \overset{k_2}{\to} X_1 + X_4, \tag{6.86}$$

$$X_3 \overset{k_3}{\to} X_4 + X_2, \tag{6.87}$$

whose kinetic equations are

$$\dot{x}_1(t) = -k_1x_1(t)x_4(t) + k_2x_3(t), \quad x_1(0) = x_{10}, \quad t \geq 0, \tag{6.88}$$

$$\dot{x}_2(t) = k_3x_3(t), \quad x_2(0) = x_{20}, \tag{6.89}$$

$$\dot{x}_3(t) = k_1x_1(t)x_4(t) - (k_2 + k_2)x_3(t), \quad x_3(0) = x_{30}, \tag{6.90}$$

$$\dot{x}_4(t) = -k_1x_1x_4(t) + (k_2 + k_2)x_3(t), \quad x_4(0) = x_{40}. \tag{6.91}$$

Since $s = 4$, $r = 3$, and $q = 2 < s$, Proposition 6.4 can be applied with $\hat{x}_1 = [x_1, x_2]^{\mathrm{T}}$, $\hat{x}_2 = [x_3, x_4]^{\mathrm{T}}$, and

$$F = \begin{bmatrix} -1 & 1 \\ -1 & 1 \end{bmatrix}.$$

It thus follows that $x_3(t) = -x_1(t) - x_2(t) + \gamma_1$ and $x_4(t) = x_1(t) + x_2(t) + \gamma_2$ for all $t \geq 0$, where $\gamma_1 \triangleq x_1(0) + x_2(0) + x_3(0)$ and $\gamma_2 \triangleq x_4(0) - x_1(0) - x_2(0)$. By applying Proposition 6.4, it follows from (6.72) that

$$\dot{x}_1(t) = -k_1x_1^2(t) - k_1x_1(t)x_2(t) - (k_1\gamma_2 + k_2)x_1(t) - k_2x_2(t) + k_2\gamma_1, \quad x_1(0) = x_{10}, \quad t \geq 0, \tag{6.92}$$

$$\dot{x}_2(t) = -k_3x_1(t) - k_3x_2(t) + k_3\gamma_1, \quad x_2(0) = x_{20}, \tag{6.93}$$

which have nonnegative solutions as long as the initial conditions coincide with the initial conditions of the original kinetic equations (6.88)–(6.91). However, due to the terms $-k_2x_2$ and $-k_3x_1$, (6.92) and (6.93) are not essentially nonnegative, and hence, solutions may become nonnegative. Therefore, (6.92) and (6.93) are not realizable by a reaction network. $\triangle$

The following proposition presents conditions that guarantee nonnegativity of the solutions to reduced-order kinetic equations.

Proposition 6.5. Assume that $q < s$. Furthermore, partition $A = [A_1, A_2]$ and $B = [B_1, B_2]$, where $A_1, B_1 \in \mathbb{R}^{r \times q}$, and assume that $\operatorname{rank}(B_1 - A_1) = q$. In addition, let $F \in \mathbb{R}^{q \times (s-q)}$ satisfy $A_2 - B_2 = (A_1 - B_1)F$. Finally, partition $x = [\hat{x}_1^{\mathrm{T}}, \hat{x}_2^{\mathrm{T}}]^{\mathrm{T}}$, where $\hat{x}_1 \triangleq [x_1, \ldots, x_q]^{\mathrm{T}}$ and $\hat{x}_2 \triangleq [x_{q+1}, \ldots, x_s]^{\mathrm{T}}$. Then, for all $\hat{x}_1(0) \in \overline{\mathbb{R}}_+^q$ and $\gamma \in \mathbb{R}^{s-q}$ such that $\gamma + F^{\mathrm{T}}\hat{x}_1(0) \in \overline{\mathbb{R}}_+^{s-q}$, the solution $\hat{x}_1(t)$ to (6.72) is nonnegative for all $t \geq 0$.

Proof. With $\hat{x}_2(0) = \gamma + F^{\mathrm{T}}\hat{x}_1(0)$, it follows from Proposition 6.4 that the solution to (6.22) is given by $[\hat{x}_1^{\mathrm{T}}(t), \hat{x}_2^{\mathrm{T}}(t)]^{\mathrm{T}}$ for all $t \geq 0$, where $\hat{x}_2(t)$ is given by (6.71). Hence, since $\hat{x}_1(0) \geq\geq 0$ and $\hat{x}_2(0) \geq\geq 0$, it follows that $\hat{x}_1(t) \geq\geq 0$ for all $t \geq 0$. $\square$

6.7 Stability Analysis of Linear and Nonlinear Kinetics

We now consider the stability of equilibria of the kinetic equations (6.22). The following definition defines equilibria for the kinetic equations (6.22).

Definition 6.1. A vector $x_{\mathrm{e}} \in \overline{\mathbb{R}}_+^s$ satisfying

$$(B - A)^{\mathrm{T}}(k \circ x_{\mathrm{e}}^A) = 0 \tag{6.94}$$

is an *equilibrium* of (6.22). If, in addition, $x_{\mathrm{e}} \in \mathbb{R}_+^s$, then x_{e} is a *positive equilibrium* of (6.22).

Let $\mathcal{E}$ denote the set of equilibria of (6.22), and let $\mathcal{E}_+ \subseteq \mathcal{E}$ denote the set of positive equilibria of (6.22). The following result can be used to obtain additional equilibria from known equilibria.

Proposition 6.6. Let $z \in \mathcal{N}(A)$ and let $\lambda \in (0, \infty)$. If $x_{\mathrm{e}} \in \mathcal{E}$, then $\lambda^z \circ x_{\mathrm{e}} \in \mathcal{E}$. Furthermore, if $x_{\mathrm{e}} \in \mathcal{E}_+$, then $\lambda^z \circ x_{\mathrm{e}} \in \mathcal{E}_+$.

Proof. Note that

$$\begin{aligned}
(B - A)^{\mathrm{T}}K(\lambda^z \circ x_{\mathrm{e}})^A &= (B - A)^{\mathrm{T}}K((\lambda^z)^A \circ x_{\mathrm{e}}^A) \\
&= (B - A)^{\mathrm{T}}K(\lambda^{Az} \circ x_{\mathrm{e}}^A) \\
&= (B - A)^{\mathrm{T}}Kx_{\mathrm{e}}^A.
\end{aligned}$$

The proof for the case $x_{\mathrm{e}} \in \mathcal{E}_+$ is identical. $\square$

Note that if x_{e} is an equilibrium but not a positive equilibrium, then at least one of the species has zero concentration for this solution. Furthermore, it can be seen that $x_e = 0$ is an equilibrium of (6.22) if and only if (6.22) has no reaction of the form $0 \xrightarrow{k} C$, where C is a nonzero product and $k > 0$.

Example 6.12. Consider Example 6.1. For this example $\mathcal{E} = \{(x_1, x_2) \in \overline{\mathbb{R}}^2 : x_2 = (k_1/k_2)x_1\}$. △

Example 6.13. Consider Example 6.2. For this example $\mathcal{E} = \{(x_1, x_2) \in \overline{\mathbb{R}}^2 : x_1 = 0 \text{ or } x_2 = (k_2/k_1)x_1\}$. For the reduced system (6.76) $\mathcal{E} = \{0, k_1\gamma/(k_1 + k_2)\}$. △

Example 6.14. Consider Example 6.3. For this example $\mathcal{E} = \{(0, 0), (k_3/k_2, k_1/k_2)\}$. △

Example 6.15. Consider Example 6.4. For this example $\mathcal{E} = \{(x_1, x_2, x_3, x_4) \in \overline{\mathbb{R}}^4 : x_2 = 0 \text{ and } x_1x_3 = 0\}$. For the reduced system (6.79) and (6.80), if $\gamma_1 = x_2(0) + x_3(0) > 0$, then $\mathcal{E} = \{(0, 0)\}$, whereas, if $\gamma_1 = x_2(0) + x_3(0) = 0$, then $\mathcal{E} = \{(x_1, 0) : x_1 \geq 0\}$. △

First, we consider a stability analysis of the linear case, that is, the case in which (6.54) is of the form

$$\dot{x}(t) = Mx(t), \qquad x(0) = x_0, \qquad t \geq 0, \tag{6.95}$$

where $M \in \mathbb{R}^{n\times n}$. In this case, the following results hold. An equilibrium x_e of (6.95) is Lyapunov stable (respectively, semistable) if and only if every equilibrium x_e of (6.95) is Lyapunov stable (respectively, semistable). Furthermore, if an equilibrium of (6.95) is asymptotically stable, then $x_e = 0$. Thus, all three types of stability can be characterized independently of the equilibrium. Specifically, the equilibrium $x_e = 0$ of (6.95) is asymptotically stable if and only if every eigenvalue of M has a negative real part; an equilibrium x_e of (6.95) is semistable if and only if every eigenvalue of M has a negative real part or is zero and, if M is singular, the zero eigenvalue is semisimple; and an equilibrium x_e of (6.95) is Lyapunov stable if and only if every eigenvalue of M has a nonpositive real part and every eigenvalue with zero real part is semisimple.

Now, we consider (6.24) with M given by (6.25). The following result follows from Theorem 2.10. For this proof we construct a linear Lyapunov function that can be interpreted as the mass of the system. To do this, let $\mu_i > 0$, $i = 1, \ldots, s$, denote the molecular mass of the ith species, and define $\mu \triangleq [\mu_1, \ldots, \mu_s]^{\mathrm{T}}$. Then the function $V(x) = \mu^{\mathrm{T}}x$ represents the total mass of the system. Note that arbitrary constants $\mu_i > 0$ can be used, and thus "mass" need not be interpreted literally. Note that V is a positive-definite function with respect to $\overline{\mathbb{R}}_+^s$. We note that the following result makes no use of the structure of M except that it is essentially nonnegative. For the proof of this result, recall that $-M$ is an M-matrix if and only if $-M$ is a Z-matrix and every eigenvalue of M has a nonnegative real part [31].

Proposition 6.7. Consider the following statements:

i) There exists $\mu >> 0$ such that $M^{\mathrm{T}}\mu \leq\leq 0$.

ii) M is Lyapunov stable.

iii) M is semistable.

iv) There exists $\mu \geq\geq 0$, $\mu \neq 0$, such that $M^{\mathrm{T}}\mu \leq\leq 0$.

Then *i*) implies *ii*), *ii*) is equivalent to *iii*), and *iii*) implies *iv*). Furthermore, the following statements are equivalent:

v) M is asymptotically stable.

vi) There exists $\mu >> 0$ such that $M^{\mathrm{T}}\mu << 0$.

vii) There exists $\mu \geq\geq 0$ such that $M^{\mathrm{T}}\mu << 0$.

Proof. The result is a direct consequence of Theorem 2.11. For completeness, however, we provide a proof here. Define $V(x) \triangleq \mu^{\mathrm{T}}x$ so that $V(0) = 0$ and $V(x) > 0$ for all $x \in \overline{\mathbb{R}}_+^s\backslash\{0\}$. Furthermore, $\dot{V}(x) = \mu^{\mathrm{T}}Mx \leq 0$ for all $x \in \overline{\mathbb{R}}_+^s$, which proves that *i*) implies *ii*). The equivalence of *ii*) and *iii*) follows from *i*) of Theorem 2.10. To show that *iii*) implies *iv*), note that since M is semistable, it follows that $-M^{\mathrm{T}}$ is an M-matrix. Hence, it follows from *ii*) of Lemma 2.2 that there exist a scalar $\alpha > 0$ and a nonnegative matrix $Q \geq\geq 0$ such that $\alpha \geq \rho(Q)$ and $M^{\mathrm{T}} = Q - \alpha I_s$. Now, since $Q \geq\geq 0$, it follows from Theorem 2.9 that $\rho(Q) \in \operatorname{spec}(Q)$, and hence, there exists $\mu \geq\geq 0$, $\mu \neq 0$, such that $Q\mu = \rho(Q)\mu$. Thus, $M^{\mathrm{T}}\mu = Q\mu - \alpha\mu = (\rho(Q) - \alpha)\mu \leq\leq 0$, which proves that there exists $\mu \geq\geq 0$, $\mu \neq 0$, such that $M^{\mathrm{T}}\mu \leq\leq 0$.

To show the equivalence of *v*)–*vii*), first suppose there exists $\mu \geq\geq 0$ such that $M^{\mathrm{T}}\mu << 0$. Now, there exists sufficiently small $\varepsilon > 0$ such that $M^{\mathrm{T}}(\mu+\varepsilon\mathbf{e}) << 0$ and $\mu+\varepsilon\mathbf{e} >> 0$, which proves that *vii*) implies *vi*). Since *vi*) implies *vii*), it follows that *vi*) and *vii*) are equivalent. Now, suppose *vi*) holds, that is, there exists $\mu >> 0$ such that $M^{\mathrm{T}}\mu << 0$, and consider the Lyapunov function candidate $V(x) = \mu^{\mathrm{T}}x$, where $x \in \overline{\mathbb{R}}_+^s$. Computing the Lyapunov derivative yields $\dot{V}(x) = \mu^{\mathrm{T}}Mx < 0$ for all $x \in \overline{\mathbb{R}}_+^s\backslash\{0\}$, and hence, it follows that M is asymptotically stable. Thus, *vi*) implies *v*). Next, suppose that M is asymptotically stable. Hence, $-M^{-\mathrm{T}} \geq\geq 0$, and thus, for every $r \in \mathbb{R}_+^s$, it follows that $\mu \triangleq -M^{-\mathrm{T}}r \geq\geq 0$ satisfies $M^{\mathrm{T}}\mu << 0$, which proves that v) implies vii). □

Example 6.16. Consider Example 6.1. Choosing $\mu = [1/k_1, 1/k_2]^{\mathrm{T}} >> 0$, it follows that $M\mu = 0$. Hence, M is semistable. △

The following result uses the Lyapunov function $V(x) = \mu^{\mathrm{T}}x$ to analyze the stability of the zero solution of (6.22). Recall that $x_{\mathrm{e}} = 0$ is an equilibrium of (6.22) if and only if A has no zero rows, that is, if and only if 0 is not a reactant of the reaction network (6.10).

Proposition 6.8. Assume that $x_{\mathrm{e}} = 0$ is an equilibrium of (6.22) and suppose there exists $\mu >> 0$ such that $B\mu \leq\leq A\mu$. Then x_{e} is Lyapunov stable with respect to $\overline{\mathbb{R}}_+^s$. If, in addition, $B\mu << A\mu$, then x_{e} is globally asymptotically stable with respect to $\overline{\mathbb{R}}_+^s$.

Proof. Define $V(x) = \mu^{\mathrm{T}}x$ so that $V(0) = 0$ and $V(x) > 0$ for all $x \in \overline{\mathbb{R}}_+^s \backslash \{0\}$. Since $(B - A)\mu \leq\leq 0$, it follows that

$$\dot{V}(x) = \mu^{\mathrm{T}}(B - A)^{\mathrm{T}}(k \circ x^A) \leq 0, \quad x \in \overline{\mathbb{R}}_+^s. \tag{6.96}$$

Hence, Theorem 2.1 implies that $x_{\mathrm{e}} = 0$ is Lyapunov stable with respect to $\overline{\mathbb{R}}_+^s$. Now suppose that $B\mu << A\mu$. Then $\dot{V}(x) < 0$ for all $x \in \overline{\mathbb{R}}_+^s \backslash \{0\}$. Since V is proper, it follows from Theorem 2.1 that $x_{\mathrm{e}} = 0$ is globally asymptotically stable with respect to $\overline{\mathbb{R}}_+^s$. □

Example 6.17. Consider Example 6.1. Let $\mu = [1/k_1 \ 1/k_2]^{\mathrm{T}}$ so that $(A - B)\mu = 0$. It thus follows from Proposition 6.8 that $x_{\mathrm{e}} = 0$ is Lyapunov stable. Since the kinetic equations are linear, it follows from Proposition 6.7 that M is both Lyapunov stable and semistable. △

Example 6.18. Consider Example 6.2. First note that, because of the structure of the set of equilibria, none of the equilibria are asymptotically stable. Next, we consider an equilibrium x_{e} of the form $(0, \varepsilon)$, where $\varepsilon > 0$. By linearizing the system about this equilibrium, it can be seen that this equilibrium is not Lyapunov stable. Hence, it remains to determine the stability of an equilibrium of the form $(\delta, k_2\delta/k_1)$, where $\delta \geq 0$. To do this, let $\mathcal{Q}$ be the closed set $\mathcal{Q} \triangleq \{(x_1, x_2) \in \overline{\mathbb{R}}_+^2 : x_2 - ax_1 \leq 0\}$, where $a > k_2/k_1$. Note that $\mathcal{Q}$ is invariant since $\frac{\mathrm{d}}{\mathrm{d}t}(x_2 - ax_1)$ is negative on the set $\{(x_1, x_2) : x_2 = ax_1, x_2 \geq 0\}$, while the point $(0, 0)$ is an equilibrium. Note that all of the equilibria contained in $\mathcal{Q}$ are of the form $(\delta, k_2\delta/k_1)$.

Next, define the Lyapunov function candidate $V : \mathcal{Q} \to \mathbb{R}$ by

$$V_\delta(x) = \tfrac{1}{2}(x_1 - \delta + x_2 - k_2\delta/k_1)^2 + \tfrac{1}{2}(k_1x_2 - k_2x_1)^2. \tag{6.97}$$

Then, for all $\delta \geq 0$, we have $V_\delta(\delta, k_2\delta/k_1) = 0$ and $V_\delta(x) > 0$ for all $x \in \mathcal{Q}\backslash\{(\delta, k_2\delta/k_1)\}$. Since $\dot{V}_\delta(x) = -(k_1+k_2)x_1(k_1x_2-k_2x_1)^2 \leq 0$ for all $x \in \mathcal{Q}$, it follows that the equilibrium $(\delta, k_2\delta/k_1)$ is Lyapunov stable with respect to $\mathcal{Q}$ for all $\delta \geq 0$. Finally, to show semistability, define $U(x) = x_1 + x_2$, which satisfies $U(0) = 0$, $U(x) > 0$, $x \in \mathcal{Q}\backslash\{0\}$, and $\dot{U}(x) = 0$, $x \in \mathcal{Q}$. Hence, every trajectory in $\mathcal{Q}$ is bounded. Then $\dot{V}_\delta^{-1}(0) = f^{-1}(0)$, which shows that $\dot{V}_\delta^{-1}(0)$ is an invariant set. Thus, the largest invariant set $\mathcal{M}$ contained in $\dot{V}_\delta^{-1}(0) \cap \mathcal{Q}$ is the set of equilibria $\{(\delta, k_2\delta/k_1) : \delta \geq 0\}$, all of which are Lyapunov stable. Hence, by Theorem 2.7, the kinetic equations are semistable with respect to $\mathcal{Q}$. △

Example 6.19. Consider Example 6.3. By linearizing the kinetic equations about the origin, it can be seen that the origin is not Lyapunov stable. To analyze the stability of the equilibrium $x_e = (k_3/k_2, k_1/k_2)$, consider as in [143, p. 115] the function $U : \mathbb{R}^2_+ \to \mathbb{R}$ defined by $U(x) = k_2(x_1 + x_2) - k_3 \ln x_1 - k_1 \ln x_2$, which satisfies $\dot{U}(x) = 0$ for all $x \in \mathbb{R}^2_+$. It can be seen from the form of the gradient and the Hessian of U that $x = x_e$ is an isolated local minimizer of U. Hence, $V(x) = U(x) - U(x_e)$ satisfies $V(x_e) = 0$ and $V(x) > 0$ for all $x \in \mathcal{D}\backslash\{x_e\}$, where $\mathcal{D}$ is an open neighborhood of x_e. Hence, the equilibrium $x_e = (k_3/k_2, k_1/k_2)$ is Lyapunov stable with respect to $\mathbb{R}^2_+$. Since the solutions consist of closed orbits [143], this equilibrium is not semistable. △

Example 6.20. Consider Example 6.4. For this example let $\mu = [1\ 2\ 1\ 1]^{\mathrm{T}} >> 0$ so that $(A - B)\mu = 0$. It thus follows from Proposition 6.8 that $x_e = 0$ is Lyapunov stable with respect to $\overline{\mathbb{R}}^4_+$. For the reduced kinetic equations (6.79) and (6.80), with $x_2(0) + x_3(0) > 0$, it follows that $x_1 = x_2 = 0$ is the only equilibrium. Now, consider the radially unbounded Lyapunov function $V(x_1, x_2) = \frac{1}{2}k_3x_2^2 + \frac{1}{2}k_1\gamma_1(x_1+x_2)^2$. Since $\dot{V}(x_1, x_2) \leq 0$ for all $x_1, x_2 \geq 0$, global asymptotic stability follows from the Krasovskii-LaSalle invariant set theorem. △

6.8 The Zero-Deficiency Theorem

In this section, we analyze the stability of positive equilibria of the kinetic equations (6.22) using the zero-deficiency theorem [144, 145]. This result provides a sufficient condition for Lyapunov stability and semistability based on the structure of the reaction network and independent of the values of the rate constants. The following definitions are required. A *complex* is either a reactant or a product. For example, in Example 6.3, the complexes include the reactants X_1, $X_1 + X_2$, and X_2 as well as the products $2X_1$, $2X_2$, and 0. Let $m \geq 1$ denote the number of distinct complexes of the reaction network (including the reactant or product 0 if present), and denote the complexes by their corresponding vectors $c_1, \ldots, c_m$ of stoichiometric coefficients. Obviously, $m \leq 2r$. We can identify each complex with a row of A or B so that $c_i \in \mathbb{R}^{1 \times s}$. Thus, m is the number of distinct rows of $\begin{bmatrix} A \\ B \end{bmatrix}$.

In Examples 6.1–6.4, the number of complexes is 2, 2, 6, and 3, respectively. In particular, Example 6.4 involves the three complexes $c_1 = [1, 0, 1, 0]$, $c_2 = [0, 1, 0, 0]$, and $c_3 = [0, 0, 1, 1]$ corresponding to S + E, C, and P + E, respectively. For the following definition, "$c_i \to c_j$" denotes the reaction $c_i X \xrightarrow{k_l} c_j X$, where we assume $k_l > 0$. Recall that reactions of the form $c \to c$ are not allowed.

It is useful to represent the reaction network by a directed graph. Consider a directed graph $\mathfrak{G}$ having m vertices and r edges such that the ith vertex represents the complex c_i, and there exists a directed edge from vertex i to vertex j if and only if the reaction network contains the reaction $c_i \to c_j$. Each edge of c_j is numbered according to the reaction that it represents.

Definition 6.2. Let c_i and c_j be complexes of the reaction network (6.10). Then c_i and c_j are *directly linked* if either $c_i \to c_j$ or $c_j \to c_i$. Furthermore, c_i and c_j are *indirectly linked* if there exist complexes $c_{i_1}, \ldots, c_{i_p}$ such that c_i is directly linked to c_{i_1}, c_{i_1} is directly linked to c_{i_2}, …, c_{i_p} is directly linked to c_j. Finally, c_i and c_j are *linked* if c_i and c_j are either directly or indirectly linked.

The statement that complexes c_i and c_j are linked is an equivalence relation on the set of complexes. This relation induces a partitioning of the set of complexes into disjoint *linkage classes.* These linkage classes are the connected components of the directed graph $\mathfrak{G}$. Let ℓ denote the number of linkage classes of $\mathfrak{G}$, and denote these linkage classes by $\mathcal{C}_1, \ldots, \mathcal{C}_\ell$. Since the reactant and product in each reaction belong to the same linkage class, it follows that $\ell \leq r$. Furthermore, since each linkage class contains at least two complexes, it follows that $\ell \leq m/2$.

As noted in Section 6.6, the rank $q = \operatorname{rank}(B - A)$ of the reaction network (6.22) satisfies $q \leq \min\{r, s\}$. The following result provides a bound for q that is sometimes better. Some additional notation is needed. For $i = 1, \ldots, \ell$, let m_i denote the number of complexes in $\mathcal{C}_i$ so that $\sum_{i=1}^{\ell} m_i = m$. Furthermore, for convenience we order the complexes $c_1, \ldots, c_m$ so that $\mathcal{C}_1 = \{c_1, \ldots, c_{m_1}\}$, $\mathcal{C}_2 = \{c_{m_1+1}, \ldots, c_{m_2}\}$, and so forth.

Next, we reorder the reactions so that the first r_1 rows of $[A, B]$ include the complexes in $\mathcal{C}_1$, rows $r_1 + 1, \ldots, r_1 + r_2$ of $[A, B]$ include the complexes in $\mathcal{C}_2$, and so forth. Hence, $\sum_{i=1}^{\ell} r_i = r$. For $i = 1, \ldots, \ell$, we define the *rank* q_i of the linkage class $\mathcal{C}_i$ to be the number of linearly independent rows in the submatrix of $B - A$ comprised of the rows of $[A, B]$ corresponding to the complexes in $\mathcal{C}_i$. Note that $q \le \sum_{i=1}^{\ell} q_i$. For $i = 1, \ldots, \ell$, it can be seen that $m_i \le r_i + 1$, and thus $m \le r + \ell$. If $q_i = m_i - 1$, then the linkage class $\mathcal{C}_i$ has *full rank.*

Lemma 6.1. Let $i \in \{1, \ldots, \ell\}$. Then $q_i \le m_i - 1$. Furthermore, $q_i = m_i - 1$ if and only if the complexes in $\mathcal{C}_i$ are the vertices of an $(m_i - 1)$-dimensional simplex in $\overline{\mathbb{R}}_+^s$.

Proof. For notational convenience, let $i = 1$ and order the first $m_1 - 1$ reactions so that, for $j = 1, \ldots, m_1 - 1$, the jth reaction is either $c_j \to c_{j+1}$ or $c_{j+1} \to c_j$. The span of the first m_1 rows of $B - A$ is thus equal to the span of $\{c_2 - c_1, \ldots, c_{m_1} - c_{m_1 - 1}\}$. Furthermore, since $\mathcal{C}_1$ is a linkage class, it follows that rows $m_1 + 1, \ldots, r_1$ of $B - A$ are contained in the span of the first m_1 rows of $B - A$. Thus, $q_1 \le m_1 - 1$.

Next, note that the span of $\{c_2 - c_1, \ldots, c_{m_1} - c_{m_1 - 1}\}$ is equal to the span of $\{c_2 - c_1, c_3 - c_1, \ldots, c_{m_1} - c_1\}$, which has dimension $m_1 - 1$ if and only if the complexes in $\mathcal{C}_1$ are the vertices of an $(m_1 - 1)$-dimensional simplex in $\overline{\mathbb{R}}_+^s$ [381, pp. 7, 12]. □

In the terminology of [381], an *affine subspace* is the translate of a subspace. Furthermore, the *affine hull* of a set $\mathcal{S}$ is the smallest affine subspace that contains $\mathcal{S}$. It can be seen that $\mathcal{C}_i$ has full rank if and only if the subspace parallel to the affine hull of $\mathcal{C}_i$ has dimension $m_i - 1$.

Proposition 6.9. $q \le m - \ell$.

Proof. As noted above, $q \le \sum_{i=1}^{\ell} q_i$, while Lemma 6.1 implies that $q_i \le m_i - 1$. Therefore, $q \le \sum_{i=1}^{\ell} q_i \le \sum_{i=1}^{\ell} (m_i - 1) = m - \ell$. □

Definition 6.3. The *deficiency* δ of the reaction network (6.10) is

$$\delta \triangleq m - \ell - q. \tag{6.98}$$

It follows from Proposition 6.9 that the deficiency of a reaction network is a nonnegative integer. If the deficiency of a reaction network is zero, then the reaction network has *zero deficiency.* It can be seen that a reaction network has deficiency zero if and only if *i*) every linkage class has full rank, and *ii*) for every pair $\mathcal{C}_i, \mathcal{C}_j$ of distinct linkage classes, the subspaces parallel to the affine hulls of the linkage classes $\mathcal{C}_i, \mathcal{C}_j$ have trivial intersection.

Example 6.21. Consider Example 6.1. For this reaction network, $m = 2$, $\ell = 1$, $q = 1$, and thus $\delta = 0$. △

Example 6.22. Consider Example 6.2. For this reaction network, $m = 2$, $\ell = 1$, $q = 1$, and thus $\delta = 0$. △

Example 6.23. Consider Example 6.3. For this reaction network, $m = 6$, $\ell = 3$, $q = 2$, and thus $\delta = 1$. △

Example 6.24. Consider Example 6.4. For this reaction network, $m = 3$, $\ell = 1$, $q = 2$, and thus $\delta = 0$. △

Now, define the matrix $C \in \mathbb{R}^{m\times s}$ whose rows are $c_1, \ldots, c_m$. Furthermore, let $\hat{A}, \hat{B} \in \mathbb{R}^{r\times m}$ be the matrices whose rows are unit coordinate vectors in $\mathbb{R}^m$ and that satisfy

$$A = \hat{A}C, \quad B = \hat{B}C. \tag{6.99}$$

It follows that

$$B - A = (\hat{B} - \hat{A})C. \tag{6.100}$$

Note that $\mathcal{N}((\hat{B} - \hat{A})^{\mathrm{T}}) \subseteq \mathcal{N}((B - A)^{\mathrm{T}})$. Next, observe that $\hat{A}_{ij} = 1$ if and only if the complex c_j is the reactant of the ith reaction, that is, if and only if the ith edge of $\mathfrak{G}$ originates from vertex i. Similarly, $\hat{B}_{ij} = 1$ if and only if the ith edge of $\mathfrak{G}$ terminates at vertex j. Consequently, the matrix $(\hat{B} - \hat{A})^{\mathrm{T}}$ is the incidence matrix of the directed graph $\mathfrak{G}$ (see [47, p. 24].)

The following result gives some properties of $\hat{B} - \hat{A}$ and shows that the reverse inclusion $\mathcal{N}((\hat{B} - \hat{A})^{\mathrm{T}}) \supseteq \mathcal{N}((B - A)^{\mathrm{T}})$ holds if $\delta = 0$.

Proposition 6.10. The following statements hold:

i) $\operatorname{rank}(\hat{B} - \hat{A}) = m - \ell$.

ii) $\delta = \dim(\mathcal{R}((\hat{B} - \hat{A})^{\mathrm{T}}) \cap \mathcal{N}(C^{\mathrm{T}}))$.

iii) If $\mu \in \mathbb{R}^s$, then $e^{B\mu} - e^{A\mu} = (\hat{B} - \hat{A})e^{C\mu}$.

iv) $\delta = 0$ if and only if $\mathcal{N}((B - A)^{\mathrm{T}}) = \mathcal{N}((\hat{B} - \hat{A})^{\mathrm{T}})$.

Proof. First, to prove *i*), consider the rows of $\hat{B} - \hat{A}$ corresponding to $\mathcal{C}_1$. As in the proof of Lemma 6.1 we order the first $m_1 - 1$ reactions so that, for $j = 1, \ldots, m_1 - 1$, the jth reaction is either $c_j \to c_{j+1}$ or $c_{j+1} \to c_j$. Therefore, for $j = 1, \ldots, m_1 - 1$, the jth row of $\hat{B} - \hat{A}$ is either $e_j - e_{j+1}$ or $e_{j+1} - e_j$, where e_j is the jth unit coordinate vector in $\mathbb{R}^m$. Thus, the first r_1

rows of $\hat{B} - \hat{A}$ have rank $m_1 - 1$. Using a similar argument for each linkage class and noting that rows of $\hat{B} - \hat{A}$ corresponding to different linkage classes are linearly independent, it follows that $\operatorname{rank}(\hat{B} - \hat{A}) = \sum_{i=1}^{\ell}(m_i - 1) = m - \ell$.

Next, to prove *ii*), it follows from Sylvester's theorem (see Fact 2.10.13 of [33]) that

$$\begin{aligned} q &= \operatorname{rank}(B - A) \\ &= \operatorname{rank}(C^{\mathrm{T}}(\hat{B} - \hat{A})^{\mathrm{T}}) \\ &= \operatorname{rank}((\hat{B} - \hat{A})^{\mathrm{T}}) - \dim(\mathcal{R}((\hat{B} - \hat{A})^{\mathrm{T}}) \cap \mathcal{N}(C^{\mathrm{T}})) \\ &= m - \ell - \dim(\mathcal{R}((\hat{B} - \hat{A})^{\mathrm{T}}) \cap \mathcal{N}(C^{\mathrm{T}})). \end{aligned}$$

To prove *iii*), let $j \in \{1, \ldots, r\}$. Now, since each row of B corresponds to a unique row of C, it follows that $B_j = \operatorname{row}_{k_j}(C)$ for some $k_j \in \{1, \ldots, m\}$. Hence, $B_j = \hat{B}_j C$, where $\hat{B}_{jk} = 1$, $k = k_j$, and $\hat{B}_{jk} = 0$, $k \neq k_j$. Hence,

$$e^{B_j \mu} = e^{\hat{B}_j C \mu} = e^{\operatorname{row}_{k_j}(C)\mu} = \hat{B}_j e^{C\mu}.$$

Similarly, we can show that $\hat{A}_j e^{C\mu} = e^{A_j \mu}$. Hence, $(\hat{B}_j - \hat{A}_j)e^{C\mu} = e^{B_j \mu} - e^{A_j \mu}$.

To prove *iv*), note that

$$\operatorname{rank}((B - A)^{\mathrm{T}}) + \dim(\mathcal{N}((B - A)^{\mathrm{T}}) = r$$

and

$$\operatorname{rank}((\hat{B} - \hat{A})^{\mathrm{T}}) + \dim(\mathcal{N}((\hat{B} - \hat{A})^{\mathrm{T}}) = r.$$

Since $\delta = 0$, it follows from *i*) that $q = \operatorname{rank}((B - A)^{\mathrm{T}}) = m - \ell = \operatorname{rank}(\hat{B} - \hat{A})$, and thus $\dim(\mathcal{N}((\hat{B} - \hat{A})^{\mathrm{T}}) = \dim(\mathcal{N}((B - A)^{\mathrm{T}})$. Since $\mathcal{N}((\hat{B} - \hat{A})^{\mathrm{T}}) \subseteq \mathcal{N}((B - A)^{\mathrm{T}})$, it follows that $\mathcal{N}((B - A)^{\mathrm{T}}) = \mathcal{N}((\hat{B} - \hat{A})^{\mathrm{T}})$. The converse follows by reversing the steps. □

Definition 6.4. Let c_i and c_j be complexes. Then there exists a *direct path* from c_i to c_j if $c_i \to c_j$. Furthermore, there exists an *indirect path* from c_i to c_j if there exist complexes $c_{i_1}, \ldots, c_{i_p}$ such that $c_i \to c_{i_1} \to c_{i_2} \to \cdots \to c_{i_p} \to c_j$. Finally, there exists a *path* from c_i to c_j if there exists either a direct path or an indirect path from c_i to c_j.

Note that the existence of a path from c_i to c_j is stronger than the statement that c_i and c_j are linked since the former condition accounts for the directionality of the reactions.

Definition 6.5. The reaction network (6.10) is *weakly reversible* if, for all pairs of complexes c_i, c_j, the existence of a path from c_i to c_j implies the existence of a path from c_j to c_i.

Note that the existence of a path from c_i to c_j is equivalent to the existence of a directed path from vertex i to vertex j on the graph $\mathfrak{G}$. Consequently, weak reversibility is equivalent to the requirement that every vertex or, equivalently, every edge of $\mathfrak{G}$ must be part of a directed cycle of $\mathfrak{G}$ [47, p. 25]. In the terminology of [218, pp. 357–358], weak reversibility of (6.10) is equivalent to strong connectedness of each connected component of $\mathfrak{G}$.

The following lemmas are needed. Furthermore, for $l = 1, \ldots, \ell$, let $\mathrm{v}_l \in \mathbb{R}^m$ (respectively, $\mathrm{e}_l \in \mathbb{R}^r$) be such that the jth component of v_l (respectively, e_l) is 1 if the jth vertex (respectively, jth edge) of $\mathfrak{G}$ belongs to the lth connected component of $\mathfrak{G}$ and 0 otherwise. It is easy to see that $\hat{A}\mathrm{v}_l = \hat{B}\mathrm{v}_l = \mathrm{e}_l$ for all $l = 1, \ldots, \ell$, which implies that $\mathrm{v}_l \in \mathcal{N}(\hat{B} - \hat{A})$ for all $l = 1, \ldots, \ell$. Next, note that, since each vertex of $\mathfrak{G}$ belongs to exactly one connected component of $\mathfrak{G}$, $\{\mathrm{v}_1, \ldots, \mathrm{v}_\ell\}$ are linearly independent, and hence, since $\operatorname{rank}(\hat{B} - \hat{A}) = m - \ell$, it follows that $\mathcal{N}(\hat{B} - \hat{A})$ is the span of $\{\mathrm{v}_1, \ldots, \mathrm{v}_\ell\}$. Finally, note that

$$e^{\sum_{l=1}^{\ell} \theta_l \mathrm{v}_l} = \sum_{l=1}^{\ell} e^{\theta_l} \mathrm{v}_l, \tag{6.101}$$

where $\theta_1, \ldots, \theta_\ell \in \mathbb{R}$.

Lemma 6.2. Let $\alpha \in \mathbb{R}^r_+$ and define $\Gamma \triangleq (\hat{B} - \hat{A})^{\mathrm{T}}(\alpha \mathbf{e}^{\mathrm{T}} \circ \hat{A}) \in \mathbb{R}^{m \times m}$. Then the following statements hold:

i) The reaction network (6.22) is weakly reversible if and only if there exists $p \in \mathbb{R}^r_+$ such that $(\hat{B} - \hat{A})^{\mathrm{T}} p = 0$.

ii) Assume that the reaction network (6.22) is weakly reversible. Then $\operatorname{rank} \Gamma = m - \ell$ and there exists $p \in \mathbb{R}^m_+$ such that $\Gamma(p \circ \mathrm{v}_l) = 0$ for all $l = 1, \ldots, \ell$.

iii) If the reaction network (6.22) has zero deficiency, then $\operatorname{rank}[C\ \mathrm{v}_1 \cdots \mathrm{v}_\ell] = m$.

Proof. To prove *i*), note that it follows from Theorems 4.5 and 5.2 of [47] that $\mathcal{N}((\hat{B} - \hat{A})^{\mathrm{T}})$ is the span of $\{\eta_1, \ldots, \eta_{n_\mathrm{c}}\}$, where n_c is the number of directed cycles of the graph $\mathfrak{G}$ and η_i is such that the jth component of η_i is 1 if the jth edge is part of the ith directed cycle of $\mathfrak{G}$ and 0 otherwise. Hence, if the reaction network is weakly reversible, then every edge of $\mathfrak{G}$ is part of at least one directed cycle of $\mathfrak{G}$. Now, a positive linear combination of all the cycles of $\mathfrak{G}$ yields $p \in \mathbb{R}^r_+$ such that $(\hat{B} - \hat{A})^{\mathrm{T}} p = 0$. To prove the converse, assume that the reaction network is not weakly reversible or, equivalently, there exists an edge (say, the Jth edge) that does not belong to any cycle of $\mathfrak{G}$. Hence, it follows that the Jth component of all vectors in $\mathcal{N}((\hat{B}-\hat{A})^{\mathrm{T}})$ is zero, which implies that there does not exist $p \in \mathbb{R}_+^r$ such that $(\hat{B}-\hat{A})^{\mathrm{T}} p = 0$.

To prove *ii*), note that $-\Gamma^{\mathrm{T}}$ is the Laplacian of the weighted directed graph [346] obtained by assigning the weight α_i to the ith edge of $\mathfrak{G}$. There exists a permutation matrix $\Pi \in \mathbb{R}^{m\times m}$ such that $\hat{\Gamma} \triangleq \Pi^{\mathrm{T}}\Gamma\Pi$ and $\hat{\Gamma} = \text{block-diag}(\hat{\Gamma}_1, \ldots, \hat{\Gamma}_\ell)$, where $\hat{\Gamma}_l \in \mathbb{R}^{m_l\times m_l}$, $l = 1, \ldots, \ell$, are such that $\sum_{l=1}^{\ell} m_l = m$ and $-\hat{\Gamma}_l^{\mathrm{T}}$ is the Laplacian of $\mathcal{C}_l$. Weak reversibility implies that each connected component of $\mathfrak{G}$ is strongly connected. (Note that $-\hat{\Gamma}^{\mathrm{T}}$ is the Laplacian of $\mathfrak{G}$ in the case where the vertices are reordered such that the lth connected component (linkage class) of $\mathfrak{G}$ contains the vertices (complexes) numbered as $m_{l-1}+1, \ldots, m_l$, $l = 1, \ldots, \ell$, where $m_0 \triangleq 0$.) Hence, it follows from Theorem 1 of [346] that $\operatorname{rank} \hat{\Gamma}_l = m_l - 1$ for all $l = 1, \ldots, \ell$, which implies that $\operatorname{rank} \Gamma = \operatorname{rank} \hat{\Gamma} = m - \ell$.

To prove the second assertion of *ii*), let $l \in \{1, \ldots, \ell\}$, let $c_l \triangleq -\min_{i=1,\ldots,m_l} \gamma_i$, and let $X_l \triangleq \Gamma_l + c_l I_{m_l}$, where γ_i denotes the (i,i)th entry of Γ_l. Now, note that X_l is a nonnegative matrix and, for $i \neq j$, the (i,j)th entry of X_l is positive if and only if there exists an edge from vertex j to vertex i of the linkage class l. Hence, since the reaction network is weakly reversible, it follows from Theorem 6.2.24 of [218] that X_l is an irreducible matrix [218, p. 361], which further implies that there exists $\hat{p}_l \in \mathbb{R}_+^m$ such that $X_l \hat{p}_l = \rho(X_l)\hat{p}_l$ (see Theorem 8.4.4 of [218]). Consequently, $\hat{\Gamma}_l \hat{p}_l = (X_l - c_l I_{m_l})\hat{p}_l = (\rho(X_l) - c_l)\hat{p}_l$ and, since $0 = \mathbf{e}^{\mathrm{T}}\hat{\Gamma}_l\hat{p}_l = (\rho(X_l)-c_l)\mathbf{e}^{\mathrm{T}}\hat{p}_l$ and $\mathbf{e}^{\mathrm{T}}\hat{p}_l > 0$, it follows that $c_l = \rho(X_l)$. Thus, there exists a positive vector $\hat{p}_l \in \mathbb{R}^m$ satisfying $\hat{\Gamma}_l\hat{p}_l = 0$ for all $l = 1, \ldots, \ell$. Now, letting $\hat{p} = [\hat{p}_1^{\mathrm{T}}, \ldots, \hat{p}_\ell^{\mathrm{T}}]^{\mathrm{T}}$ it can be shown that $\hat{p} \circ (\Pi^{\mathrm{T}} \mathrm{v}_l) = [0, \ldots, \hat{p}_l^{\mathrm{T}}, \ldots, 0]^{\mathrm{T}}$ so that $\hat{\Gamma}(\hat{p} \circ (\Pi^{\mathrm{T}}\mathrm{v}_l)) = 0$. Finally, taking $p = \Pi\hat{p}$ implies that $\Gamma(p \circ \mathrm{v}_l) = \Pi\hat{\Gamma}\Pi^{\mathrm{T}}(p \circ \mathrm{v}_l) = \Pi\hat{\Gamma}(\hat{p} \circ (\Pi^{\mathrm{T}}\mathrm{v}_l)) = 0$, establishing the result.

To prove *iii*), let $x \in \mathbb{R}^m$ be such that $x^{\mathrm{T}}[C \ \mathrm{v}_1 \cdots \mathrm{v}_\ell] = 0$ or, equivalently, $x \in \mathcal{N}(C^{\mathrm{T}})$ and $x^{\mathrm{T}}\mathrm{v}_l = 0$ for all $l = 1, \ldots, \ell$. Next, since $\mathcal{N}(\hat{B}-\hat{A})$ is the span of $\{\mathrm{v}_1, \ldots, \mathrm{v}_\ell\}$, it follows that $x \in [\mathcal{N}(\hat{B}-\hat{A})]^{\perp} = \mathcal{R}((\hat{B}-\hat{A})^{\mathrm{T}})$. Hence, $x \in \mathcal{R}((\hat{B}-\hat{A})^{\mathrm{T}}) \cap \mathcal{N}(C^{\mathrm{T}})$ and, since the reaction network has zero deficiency, it follows from statement *ii*) of Proposition 6.10 that $x = 0$, which proves that $\operatorname{rank}[C \ \mathrm{v}_1 \cdots \mathrm{v}_\ell] = m$. □

Lemma 6.3. Assume that the reaction network (6.22) has zero deficiency and assume that there exists $\alpha \in \mathbb{R}_+^r$ such that $(B-A)^{\mathrm{T}}\alpha = 0$. Then $\mu \in \mathbb{R}^s$ satisfies $(B-A)^{\mathrm{T}}(\alpha \circ e^{A\mu}) = 0$ if and only if $\mu \in \mathcal{S}^{\perp}$.

Proof. Since the reaction network (6.22) has zero deficiency, it follows from *iv*) of Proposition 6.10 that $\mathcal{N}((B-A)^{\mathrm{T}}) = \mathcal{N}((\hat{B}-\hat{A})^{\mathrm{T}})$, and hence, $\mathcal{N}((B-A)^{\mathrm{T}})$ is the span of $\{\eta_1,\ldots,\eta_{n_c}\}$ defined in the proof of *i*) of Lemma 6.2. Furthermore, since $\alpha \in \mathcal{N}((B-A)^{\mathrm{T}})$, it follows that $\alpha = \sum_{i=1}^{n_c} \beta_i \eta_i$ for some $\beta_i \in \mathbb{R}$, $i = 1,\ldots,n_c$. Now, note that $\eta_i \circ \mathrm{e}_l = \eta_i$ if the ith cycle of $\mathfrak{G}$ belongs to the lth linkage class of $\mathfrak{G}$ and zero otherwise. In both cases, $(B-A)^{\mathrm{T}}(\eta_i \circ \mathrm{e}_l) = (\hat{B}-\hat{A})^{\mathrm{T}}(\eta_i \circ \mathrm{e}_l) = 0$ for all $i = 1,\ldots,n_c$ and $l = 1,\ldots,\ell$.

To prove necessity, let $\mu \in \mathcal{N}(B-A)$. Hence, $(\hat{B}-\hat{A})C\mu = 0$, which, since $\mathcal{N}(\hat{B}-\hat{A})$ is the span of $\{\mathrm{v}_1,\ldots,\mathrm{v}_\ell\}$, implies that $C\mu = \sum_{l=1}^{\ell} \theta_l \mathrm{v}_l$ for some $\theta_1,\ldots,\theta_\ell \in \mathbb{R}$. Hence, it follows that

$$
\begin{aligned}
(B-A)^{\mathrm{T}}(\alpha \circ e^{A\mu}) &= (B-A)^{\mathrm{T}}(\alpha \circ \hat{A} e^{C\mu}) \\
&= (B-A)^{\mathrm{T}}\Big(\alpha \circ \hat{A} \sum_{l=1}^{\ell} e^{\theta_l} \mathrm{v}_l\Big) \\
&= \sum_{l=1}^{\ell} (B-A)^{\mathrm{T}}(\alpha \circ e^{\theta_l} \mathrm{e}_l) \\
&= \sum_{l=1}^{\ell} \sum_{i=1}^{n_c} (B-A)^{\mathrm{T}}(\beta_i e^{\theta_l} (\eta_i \circ \mathrm{e}_l)) \\
&= 0,
\end{aligned}
$$

where statement *iii*) of Proposition 6.10 is used to obtain the first equality, (6.101) is used to obtain the second equality, and the fact that $\hat{A}\mathrm{v}_l = \mathrm{e}_l$ for all $l = 1,\ldots,\ell$, is used to obtain the third equality.

Conversely, assume that $(B-A)^{\mathrm{T}}(\alpha \circ e^{A\mu}) = 0$, which implies that $(\hat{B}-\hat{A})^{\mathrm{T}}(\alpha \circ e^{A\mu}) = 0$. Hence,

$$
0 = (\hat{B}-\hat{A})^{\mathrm{T}}(\alpha \circ \hat{A} e^{C\mu}) = (\hat{B}-\hat{A})^{\mathrm{T}}(\alpha \mathbf{e}^{\mathrm{T}} \circ \hat{A}) e^{C\mu} = \Gamma e^{C\mu}, \tag{6.102}
$$

where Γ is defined in Lemma 6.2. Next, note that, for all $l = 1,\ldots,\ell$,

$$
\begin{aligned}
\Gamma \mathrm{v}_l &= (\hat{B}-\hat{A})^{\mathrm{T}}(\alpha \mathbf{e}^{\mathrm{T}} \circ \hat{A}) \mathrm{v}_l \\
&= (\hat{B}-\hat{A})^{\mathrm{T}}(\alpha \circ \hat{A} \mathrm{v}_l) \\
&= (\hat{B}-\hat{A})^{\mathrm{T}}(\alpha \circ \mathrm{e}_l) \\
&= \sum_{i=1}^{n_c} \beta_i (\hat{B}-\hat{A})^{\mathrm{T}}(\eta_i \circ \mathrm{e}_l) \\
&= 0.
\end{aligned}
$$

Furthermore, it follows from statement *ii*) of Lemma 6.2 that $\operatorname{rank}\Gamma = m - \ell$, which implies that $\mathcal{N}(\Gamma)$ is the span of $\{\mathrm{v}_1,\ldots,\mathrm{v}_\ell\}$. Hence, it follows from (6.102) that $e^{C\mu} = \sum_{l=1}^{\ell} e^{\theta_l} \mathrm{v}_l$ for some $\theta_1,\ldots,\theta_\ell \in \mathbb{R}$, which implies that

$C\mu = \sum_{l=1}^{\ell} \theta_l \mathrm{v}_l$. Now, the result follows by noting that $(B-A)\mu = (\hat{B} - \hat{A})C\mu = \sum_{l=1}^{\ell} \theta_l (\hat{B} - \hat{A})\mathrm{v}_l = 0$. □

The following result shows that weak reversibility is a necessary and sufficient condition for a reaction network with zero deficiency to have at least one positive equilibrium.

Proposition 6.11. Assume that the reaction network (6.22) has zero deficiency. Then the reaction network (6.22) is weakly reversible if and only if it has a positive equilibrium.

Proof. To prove necessity, let x_{e} be a positive equilibrium of (6.22). Hence, it follows from statement *iv*) of Proposition 6.10 that $(\hat{B} - \hat{A})^{\mathrm{T}} p = 0$, where $p = K x_{\mathrm{e}}^{A} \in \mathbb{R}_{+}^{r}$. Now, it follows from statement *i*) of Lemma 6.2 that the reaction network is weakly reversible.

To prove sufficiency, note that

$$(\hat{B} - \hat{A})^{\mathrm{T}}(k \circ x^{A}) = (\hat{B} - \hat{A})^{\mathrm{T}}(k \circ \hat{A}x^{C}) = (\hat{B} - \hat{A})^{\mathrm{T}}(k\mathbf{e}^{\mathrm{T}} \circ \hat{A})x^{C} = \Gamma x^{C},$$

where $\Gamma \triangleq (\hat{B} - \hat{A})^{\mathrm{T}}(k\mathbf{e}^{\mathrm{T}} \circ \hat{A})$. Now, it follows from *ii*) of Lemma 6.2 that there exists a positive vector $p \in \mathbb{R}^{m}$ such that $\Gamma(p \circ \mathrm{v}_l) = 0$ for all $l = 1, \ldots, \ell$. Next, we show that there exists a positive vector $x \in \mathbb{R}^{s}$ and scalars $\theta_l \in \mathbb{R}$, $l = 1, \ldots, \ell$, such that $x^{C} = p \circ e^{\sum_{l=1}^{\ell} \theta_l \mathrm{v}_l}$. To see this, note that the existence of a positive vector x and scalars θ_l satisfying $x^{C} = p \circ e^{\sum_{l=1}^{\ell} \theta_l \mathrm{v}_l}$ is equivalent to the existence of a solution x to the equation $C \mathbf{log}_e x = \mathbf{log}_e p + \sum_{l=1}^{\ell} \theta_l \mathrm{v}_l$ or, equivalently,

$$[C\ \mathrm{v}_1 \cdots \mathrm{v}_\ell] \begin{bmatrix} \mathbf{log}_e x \\ -\theta_1 \\ \vdots \\ -\theta_\ell \end{bmatrix} = \mathbf{log}_e p. \tag{6.103}$$

Now, since the reaction network has zero deficiency, it follows from *iii*) of Lemma 6.2 that $\operatorname{rank}[C\ \mathrm{v}_1 \cdots \mathrm{v}_\ell] = m$, and hence, (6.103) has a solution, which implies that there exists a positive vector x and scalars θ_l such that $x^{C} = p \circ e^{\sum_{l=1}^{\ell} \theta_l \mathrm{v}_l}$. Next, it follows from (6.101) that

$$(B - A)^{\mathrm{T}}(k \circ x^{A}) = C^{\mathrm{T}} \Gamma x^{C} = C^{\mathrm{T}} \Gamma (p \circ e^{\sum_{l=1}^{\ell} \theta_l \mathrm{v}_l}) = \sum_{l=1}^{\ell} e^{\theta_l} C^{\mathrm{T}} \Gamma (p \circ \mathrm{v}_l) = 0,$$

which implies that x is a positive equilibrium of the reaction network (6.22). □

Next, we show that every positive stoichiometric compatibility class contains exactly one equilibrium for a weakly reversibile reaction network with zero deficiency. The following lemma is needed for this result.

Lemma 6.4. Let $p, \hat{p} \in \mathbb{R}_+^s$, let $\mathcal{X}$ be a subspace of $\mathbb{R}^s$, and define $\mathcal{X}^\perp \triangleq \{x \in \mathbb{R}^s : x^\mathrm{T} y = 0 \text{ for all } y \in \mathcal{X}\}$. Then there exists a unique $\mu \in \mathcal{X}^\perp$ such that $(p \circ e^\mu - \hat{p}) \in \mathcal{X}$.

Proof. Define $\varphi : \mathbb{R}^s \to \mathbb{R}$ by $\varphi(x) \triangleq p^\mathrm{T} e^x - \hat{p}^\mathrm{T} x$. It can be shown that $\lim_{\|x\| \to \infty} \varphi(x) = \infty$. Now, let $r > 0$ and, since $\lim_{\|x\| \to \infty} \varphi(x) = \infty$, it follows that $\mathcal{C}_r \triangleq \{x \in \mathbb{R}^s : \varphi(x) \le r\}$ is a compact set. Hence, $\hat{\mathcal{C}}_r \triangleq \{x \in \mathcal{X}^\perp : \varphi(x) \le r\}$ is also a compact set, which implies that there exists $\mu \in \mathcal{X}^\perp$ such that $\varphi(\mu) \le \varphi(x)$ for all $x \in \hat{\mathcal{C}}_r$. Now, since $\mathcal{X}^\perp = \hat{\mathcal{C}}_r \cup \{x \in \mathcal{X}^\perp : \varphi(x) > r\}$, it follows that $\varphi(\mu) \le \varphi(x)$ for all $x \in \mathcal{X}^\perp$. Specifically, $\varphi(\mu) \le \varphi(\mu + \theta\gamma)$ for all $\theta \in \mathbb{R}$ and $\gamma \in \mathcal{X}^\perp$. Thus, $f(\theta) \triangleq \varphi(\mu + \theta\gamma)$ has a minimum at $\theta = 0$, which implies that

$$0 = \left.\frac{\mathrm{d}f}{\mathrm{d}\theta}\right|_{\theta=0} = \left.\frac{\partial \varphi}{\partial x}\right|_{x=\mu} \gamma.$$

Hence, since $\gamma \in \mathcal{X}^\perp$ is arbitrary, $\left.\frac{\partial \varphi}{\partial x}\right|_{x=\mu} = (p \circ e^\mu - \hat{p}) \in \mathcal{X}$, which establishes existence.

To prove uniqueness, let $\hat{\mu} \in \mathcal{X}^\perp$ be such that $(p \circ e^{\hat{\mu}} - \hat{p}) \in \mathcal{X}$. Since $\mu, \hat{\mu} \in \mathcal{X}^\perp$ and $(p \circ e^\mu - \hat{p}), (p \circ e^{\hat{\mu}} - \hat{p}) \in \mathcal{X}$, it follows that $(\mu - \hat{\mu}) \in \mathcal{X}^\perp$ and $[p \circ (e^\mu - e^{\hat{\mu}})] \in \mathcal{X}$, and hence,

$$0 = (\mu - \hat{\mu})^\mathrm{T} [p \circ (e^\mu - e^{\hat{\mu}})] = \sum_{i=1}^{s} p_i (\mu_i - \hat{\mu}_i)(e^{\mu_i} - e^{\hat{\mu}_i}). \tag{6.104}$$

Next, since the exponential function is an increasing function, it follows that $(\mu_i - \hat{\mu}_i)(e^{\mu_i} - e^{\hat{\mu}_i}) \ge 0$ for all $i = 1, \ldots, s$, and, since $p \in \mathbb{R}_+^s$, it follows from (6.104) that $(\mu_i - \hat{\mu}_i)(e^{\mu_i} - e^{\hat{\mu}_i}) = 0$ for all $i = 1, \ldots, s$, or, equivalently, $\mu = \hat{\mu}$. □

The next result characterizes all positive equilibria of zero-deficiency, weakly reversible reaction networks.

Proposition 6.12. Assume that the reaction network (6.22) has zero deficiency and is weakly reversible, and let x_e be a positive equilibrium. Then

$$\mathcal{E}_+ = \{x \in \mathbb{R}_+^s : \mathbf{log}_e x - \mathbf{log}_e x_\mathrm{e} \in \mathcal{S}^\perp\}. \tag{6.105}$$

Furthermore, every positive stoichiometric compatibility class contains exactly one equilibrium.

Proof. To prove that $\mathcal{E}_+$ has the form (6.105), let x_e be a positive equilibrium, let $x \in \mathbb{R}_+^s$, and define $\mu \triangleq \mathbf{log}_e x - \mathbf{log}_e x_\mathrm{e}$. Then

$$\begin{aligned}(k \circ x^A) &= (k \circ x^A \circ x_\mathrm{e}^{-A} \circ x_\mathrm{e}^A)\\ &= (k \circ e^{A\mathbf{log}_e x} \circ e^{-A\mathbf{log}_e x_\mathrm{e}} \circ x_\mathrm{e}^A)\\ &= (k \circ e^{A\mu} \circ x_\mathrm{e}^A)\\ &= (k \circ x_\mathrm{e}^A \circ e^{A\mu}).\end{aligned}$$

Now, assume that x is also a positive equilibrium so that $(B - A)^\mathrm{T}(k \circ x_\mathrm{e}^A \circ e^{A\mu}) = (B - A)^\mathrm{T}(k \circ x^A) = 0$. Since x_e is an equilibrium, we have $(B-A)^\mathrm{T}(k \circ x_\mathrm{e}^A) = 0$. It thus follows from Lemma 6.3, with $\alpha = k \circ x_\mathrm{e}^A$, that $\mu \in \mathcal{S}^\perp$. Conversely, assume that $\mu \in \mathcal{S}^\perp$. Since $(B - A)^\mathrm{T}(k \circ x_\mathrm{e}^A) = 0$, it follows from Lemma 6.3 that $0 = (B-A)^\mathrm{T}(k \circ x_\mathrm{e}^A \circ e^{A\mu}) = (B-A)^\mathrm{T}(k \circ x^A)$, which shows that x is an equilibrium.

To prove the second assertion, let $\mathcal{S}_p \triangleq \{p + x : x \in \mathcal{S}\}$ denote a stoichiometric compatibility class, where $p \in \mathbb{R}_+^s$. Now, with $\mathcal{X} = \mathcal{S}$, it follows from Lemma 6.4 that there exists a unique $\mu \in \mathcal{S}^\perp$ such that $(x_\mathrm{e} \circ e^\mu - p) \in \mathcal{S}$ or, equivalently, $(x_\mathrm{e} \circ e^\mu) \in \mathcal{S}_p$. Now, the result follows by noting that $\mathcal{E}_+ = \{x_\mathrm{e} \circ e^\mu : \mu \in \mathcal{S}^\perp\} \subset \mathbb{R}_+^s$. □

We now have one of the main results of this section.

Theorem 6.4. If the reaction network (6.22) has zero deficiency, then every positive equilibrium of (6.22) is semistable with respect to $\mathbb{R}_+^s$.

Proof. Let x_e be a positive equilibrium of (6.22) and define the Lyapunov candidate $V : \mathbb{R}_+^s \to \mathbb{R}$ by

$$V(x) \triangleq \sum_{i=1}^{s} [x_i(\log_e x_i - \log_e x_{\mathrm{e}i}) - (x_i - x_{\mathrm{e}i})],$$

where x_i and $x_{\mathrm{e}i}$ are the ith components of x and x_e, respectively. It follows from the inequality $\log_e a \le a - 1$ for all $a > 0$, with $a = x_{\mathrm{e}i}/x_i$, that $V(x) \ge 0$ for all $x \in \mathbb{R}_+^s$. Since $\log_e a = a - 1$ if and only if $a = 1$, it follows that $V(x) = 0$ if and only if $x = x_\mathrm{e}$.

Next, for $x \in \mathbb{R}_+^s$, define $\mu \triangleq \mathbf{log}_e x - \mathbf{log}_e x_\mathrm{e}$, and note that it follows from $\log_e a \le a - 1$, $a > 0$, with $a = e^{\mathrm{row}_i(B\mu)}/e^{\mathrm{row}_i(A\mu)}$, that

$$e^{A\mu} \circ [(B - A)\mu] \le\le e^{B\mu} - e^{A\mu}, \tag{6.106}$$

with equality holding in (6.106) if and only if $(B - A)\mu = 0$. Using (6.106), along with statements *iii*) and *iv*) of Proposition 6.10, yields

$$\dot{V}(x) = \mu^\mathrm{T}(B - A)^\mathrm{T} K x^A$$

$$
\begin{aligned}
&= \mu^{\mathrm{T}}(B-A)^{\mathrm{T}}Ke^{A\mathbf{log}_e x} \\
&= \mu^{\mathrm{T}}(B-A)^{\mathrm{T}}K(e^{A\mathbf{log}_e x_{\mathrm{e}}} \circ e^{A\mu}) \\
&= ([\mu^{\mathrm{T}}(B-A)^{\mathrm{T}}] \circ (e^{A\mu})^{\mathrm{T}})Kx_{\mathrm{e}}^{A} \\
&= (Kx_{\mathrm{e}}^{A})^{\mathrm{T}}(e^{A\mu} \circ [(B-A)\mu]) \\
&\leq (Kx_{\mathrm{e}}^{A})^{\mathrm{T}}(e^{B\mu} - e^{A\mu}) \\
&= (Kx_{\mathrm{e}}^{A})^{\mathrm{T}}(\hat{B} - \hat{A})e^{C\mu} \\
&= [(\hat{B} - \hat{A})^{\mathrm{T}}Kx_{\mathrm{e}}^{A}]^{\mathrm{T}}e^{C\mu} \\
&= 0, \qquad (6.107)
\end{aligned}
$$

which proves that every positive equilibrium of (6.22) is Lyapunov stable.

Next, assume that the reaction network (6.22) has zero deficiency. If $x \in \mathbb{R}_+^s$ satisfies $\dot{V}(x) = 0$, then it follows from (6.107) that $(Kx_{\mathrm{e}}^{A})^{\mathrm{T}}(e^{A\mu} \circ [(B-A)\mu]) = (Kx_{\mathrm{e}}^{A})^{\mathrm{T}}(e^{B\mu} - e^{A\mu})$. Now, since $Kx_{\mathrm{e}}^{A} >> 0$, it follows from (6.106) that $e^{A\mu} \circ [(B-A)\mu] = e^{B\mu} - e^{A\mu}$, which implies that $(B-A)\mu = 0$, and hence, $(\mathbf{log}_e x - \mathbf{log}_e x_{\mathrm{e}}) \in \mathcal{S}^{\perp}$. It now follows from Proposition 6.12 that x is a positive equilibrium of (6.22) and, as shown above, x is Lyapunov stable. Thus, every element of the largest invariant set of $\{x \in \mathbb{R}_+^s : \dot{V}(x) = 0\}$ is a Lyapunov stable equilibrium.

Furthermore, for $\eta > 0$, let $\mathcal{D}_\eta$ denote the closure of the connected component of $\{x \in \mathbb{R}_+^s : V(x) \leq \eta\}$ containing x_{e}. Since $V(\cdot)$ is continuous in $\mathbb{R}_+^s$ and $V(x_{\mathrm{e}}) = 0$, it follows that there exists $\beta > 0$ such that $\mathcal{D}_\beta \subset \mathbb{R}_+^s$ and is compact. Now, with $\mathcal{Q} = \mathcal{D}_\beta$, Theorem 2.7 implies every solution to (6.22) with $x(0) \in \mathcal{D}_\beta$ converges to an equilibrium that is semistable with respect to $\mathcal{D}_\beta$. Finally, the result follows from the definition of semistability with respect to $\mathbb{R}_+^s$ and the fact that $\mathcal{D}_\beta$ has a nonempty interior. □

The following version of Theorem 6.4 is proved in [144, 145].

Theorem 6.5. Assume that the reaction network (6.22) has zero deficiency and is weakly reversible. Then every positive stoichiometric compatibility class contains exactly one equilibrium. This equilibrium is asymptotically stable with respect to the positive stoichiometric compatibility class that it is contained in, and there exist no nontrivial periodic orbits in $\mathbb{R}_+^s$.

Proof. The first assertion is a consequence of Propositions 6.11 and 6.12. The second assertion follows from Theorem 6.4 and uses the facts that every positive stoichiometric compatibility class is invariant, contains exactly one positive equilibrium, and $V(x(t))$ is a (strictly) decreasing function on every nontrivial solution to (6.22) in $\mathbb{R}_+^s$, where $V(\cdot)$ is the Lyapunov function defined in the proof of Theorem 6.4. □

The conclusions of Theorems 6.4 and 6.5 can be strengthened without any additional assumptions. Specifically, it can be shown that, for every initial condition in the nonnegative orthant, the positive limit set of the reaction network (6.22) is a subset of the set of nonnegative equilibria. Furthermore, if every positive stoichiometric compatibility class has no equilibria on its boundary, then every equilibrium is globally asymptotically stable relative to its positive stoichiometric compatibility class.

Example 6.25. Consider Example 6.1. This reaction network has zero deficiency and is weakly reversible. Theorem 6.5 thus implies that every positive stoichiometric compatibility class contains exactly one equilibrium, and this equilibrium is semistable with respect to $\overline{\mathbb{R}}_+^s$. △

Example 6.26. Consider Example 6.2. This reaction network has zero deficiency and is weakly reversible. Theorem 6.5 thus implies that every positive stoichiometric compatibility class contains exactly one equilibrium, and this equilibrium is semistable with respect to $\overline{\mathbb{R}}_+^s$. △

Example 6.27. Consider Example 6.3. This reaction network has deficiency 1 and is not weakly reversible. Hence, Theorem 6.5 does not apply. △

Example 6.28. Consider Example 6.4. Although this reaction network has zero deficiency, it is not weakly reversible. Accordingly, Theorem 6.4 cannot be used to conclude semistability. However, Lyapunov methods, based on nontangency between the vector field and invariant subsets of the level sets of the Lyapunov function $V(x) = \alpha x_1 + x_2$, where $\alpha \in (1, 1+k_3/k_2)$, can be used to conclude semistability of every equilibrium in $\hat{\mathcal{E}} = \{x \in \overline{\mathbb{R}}_+^4 : x_1 = 0,\ x_2 = 0,\ x_3 > 0\}$. For details see [42]. △

The following example is a modification of Example 6.4 to include weak reversibility.

Example 6.29. Consider a modification of Example 6.4 in which all reactions are reversible, that is,

$$\mathrm{S} + \mathrm{E} \underset{k_2}{\overset{k_1}{\rightleftharpoons}} \mathrm{C} \underset{k_4}{\overset{k_3}{\rightleftharpoons}} \mathrm{P} + \mathrm{E}, \tag{6.108}$$

so that $s = 4$ and $r = 4$. It thus follows that A and B are given by

$$A = \begin{bmatrix} 1 & 0 & 1 & 0 \\ 0 & 1 & 0 & 0 \\ 0 & 1 & 0 & 0 \\ 0 & 0 & 1 & 1 \end{bmatrix}, \quad B = \begin{bmatrix} 0 & 1 & 0 & 0 \\ 1 & 0 & 1 & 0 \\ 0 & 0 & 1 & 1 \\ 0 & 1 & 0 & 0 \end{bmatrix}, \tag{6.109}$$

and the kinetic equations have the form

$$\dot{x}_1(t) = k_2x_2(t) - k_1x_1(t)x_3(t), \quad x_1(0) = x_{10}, \quad t \geq 0, \tag{6.110}$$

$$\dot{x}_2(t) = -(k_2 + k_3)x_2(t) + k_1x_1(t)x_3(t) + k_4x_3(t)x_4(t), \quad x_2(0) = x_{20}, \tag{6.111}$$

$$\dot{x}_3(t) = (k_2 + k_3)x_2(t) - k_1x_1(t)x_3(t) - k_4x_3(t)x_4(t), \quad x_3(0) = x_{30}, \tag{6.112}$$

$$\dot{x}_4(t) = k_3x_2(t) - k_4x_3(t)x_4(t), \quad x_4(0) = x_{40}. \tag{6.113}$$

Since this network has zero deficiency and is weakly reversible, Theorem 6.5 implies that every positive stoichiometric compatibility class contains exactly one equilibrium, and this equilibrium is semistable with respect to $\overline{\mathbb{R}}_+^s$. △

6.9 Chemical Equilibria, Chemical Potential, and Chemical Thermodynamics

In this section, we use the law of mass action and chemical reaction networks to develop a state space system model for chemical thermodynamics. Specifically, let $n_j : [0, \infty) \to \overline{\mathbb{R}}_+$, $j = 1, \ldots, q$, denote the *mole number* of the jth species and define $n \triangleq [n_1, \ldots, n_q]^{\mathrm{T}}$. Invoking the law of mass action [426], the species quantities change according to the dynamics

$$\dot{n}(t) = (B - A)^{\mathrm{T}} K n^A(t), \quad n(0) = n_0, \quad t \geq t_0, \tag{6.114}$$

where $K \triangleq \mathrm{diag}[k_1, \ldots, k_r] \in \mathbb{P}^r$ and

$$n^A \triangleq \begin{bmatrix} \prod_{j=1}^q n_j^{A_{1j}} \\ \vdots \\ \prod_{j=1}^q n_j^{A_{rj}} \end{bmatrix} = \begin{bmatrix} n_1^{A_{11}} \cdots n_q^{A_{1q}} \\ \vdots \\ n_1^{A_{r1}} \cdots n_q^{A_{rq}} \end{bmatrix} \in \overline{\mathbb{R}}_+^r. \tag{6.115}$$

Next, let $M_j > 0$, $j = 1, \ldots, q$, denote the *molar mass* (i.e., the mass of one mole of a substance) of the jth species, let $m_j : [0, \infty) \to \overline{\mathbb{R}}_+$, $j = 1, \ldots, q$, denote the mass of the jth species so that $m_j(t) = M_j n_j(t)$, $t \geq t_0$, $j = 1, \ldots, q$, and let $m \triangleq [m_1, \ldots, m_q]^{\mathrm{T}}$. Then, using the transformation $m(t) = Mn(t)$, where $M \triangleq \mathrm{diag}[M_1, \ldots, M_q] \in \mathbb{P}^q$, (6.114) can be rewritten as the *mass flow balance*

$$\dot{m}(t) = M(B - A)^{\mathrm{T}} \tilde{K} m^A(t), \quad m(0) = m_0, \quad t \geq t_0, \tag{6.116}$$

where $\tilde{K} \triangleq \mathrm{diag}\left[\frac{k_1}{\prod_{j=1}^q M_j^{A_{1j}}}, \ldots, \frac{k_r}{\prod_{j=1}^q M_j^{A_{rj}}}\right] \in \mathbb{P}^r$.

In the absence of nuclear reactions, the total mass of the species during each reaction in (6.10) is conserved. Specifically, consider the ith reaction in (6.10) given by (6.9), where the mass of the reactants is $\sum_{j=1}^{q} A_{ij}M_j$ and the mass of the products is $\sum_{j=1}^{q} B_{ij}M_j$. Hence, conservation of mass in the ith reaction is characterized as

$$\sum_{j=1}^{q}(B_{ij} - A_{ij})M_j = 0, \quad i = 1, \ldots, r, \tag{6.117}$$

or, in general for (6.10), as

$$\mathbf{e}^{\mathrm{T}} M(B - A)^{\mathrm{T}} = 0. \tag{6.118}$$

Note that it follows from (6.116) and (6.118) that $\mathbf{e}^{\mathrm{T}}\dot{m}(t) \equiv 0$.

Equation (6.116) characterizes the change in masses of substances in the interconnected dynamical system $\mathcal{G}$ due to chemical reactions. In addition to the change of mass due to chemical reactions, each substance can exchange energy with other substances according to the energy flow mechanism described in Chapter 4; that is, energy flows from substances at a higher temperature to substances at a lower temperature. Furthermore, in the presence of chemical reactions, the exchange of matter affects the change of energy of each substance through the quantity known as the *chemical potential*.

The notion of the chemical potential was introduced by Gibbs in 1875–1878 [161,162] and goes far beyond the scope of chemistry, affecting virtually every process in nature [16, 152, 233]. The chemical potential has a strong connection with the second law of thermodynamics in that *every process in nature evolves from a state of higher chemical potential towards a state of lower chemical potential.* It was postulated by Gibbs [161, 162] that the change in energy of a homogeneous substance is proportional to the change in mass of this substance with the coefficient of proportionality given by the chemical potential of the substance.

To elucidate this, assume the jth substance corresponds to the jth compartment and consider the rate of energy change of the jth substance of $\mathcal{G}$ in the presence of matter exchange. In this case, it follows from (3.4) and Gibbs' postulate that the rate of energy change of the jth substance is given by

$$\dot{E}_j(t) = \left[\sum_{k=1,\, k\neq j}^{q} \phi_{jk}(E(t))\right] - \sigma_{jj}(E(t)) + S_j(t) + \mu_j(E(t), m(t))\dot{m}_j(t),$$

$$E_j(t_0) = E_{j0}, \quad t \geq t_0, \tag{6.119}$$

where $\mu_j : \overline{\mathbb{R}}_+^q \times \overline{\mathbb{R}}_+^q \to \mathbb{R}$, $j = 1, \ldots, q$, is the chemical potential of the jth substance. It follows from (6.119) that $\mu_j(\cdot, \cdot)$ is the chemical potential of a unit mass of the jth substance. We assume that if $E_j = 0$, then $\mu_j(E, m) = 0$, $j = 1, \ldots, q$, which implies that if the energy of the jth substance is zero, then its chemical potential is also zero.

Next, using (6.116) and (6.119), the power and mass flow balances for the interconnected dynamical system $\mathcal{G}$ can be written as

$$\dot{E}(t) = f(E(t)) + P(E(t), m(t))M(B - A)^{\mathrm{T}}\tilde{K}m^{A}(t) - d(E(t)) + S(t),$$
$$E(t_0) = E_0, \quad t \geq t_0, \tag{6.120}$$
$$\dot{m}(t) = M(B - A)^{\mathrm{T}}\tilde{K}m^{A}(t), \quad m(0) = m_0, \tag{6.121}$$

where $P(E, m) \triangleq \operatorname{diag}[\mu_1(E, m), \ldots, \mu_q(E, m)] \in \mathbb{R}^{q \times q}$ and where $f(\cdot)$, $d(\cdot)$, and $S(\cdot)$ are defined as in Section 3.2. It follows from Proposition 2.1 that the dynamics of (6.121) are essentially nonnegative and, since $\mu_j(E, m) = 0$ if $E_j = 0$, $j = 1, \ldots, q$, the dynamics of (6.120) and (6.121) are essentially nonnegative for the isolated dynamical system $\mathcal{G}$ (i.e., $S(t) \equiv 0$ and $d(E) \equiv 0$).

Note that, for the ith reaction in the reaction network given by (6.10), the chemical potentials of the reactants and the products are $\sum_{j=1}^{q} A_{ij}M_j\mu_j(E, m)$ and $\sum_{j=1}^{q} B_{ij}M_j\mu_j(E, m)$, respectively. Thus,

$$\sum_{j=1}^{q} B_{ij}M_j\mu_j(E, m) - \sum_{j=1}^{q} A_{ij}M_j\mu_j(E, m) \leq 0, \quad (E, m) \in \overline{\mathbb{R}}_+^q \times \overline{\mathbb{R}}_+^q, \tag{6.122}$$

is a restatement of the principle that a chemical reaction evolves from a state of a higher chemical potential to that of a lower chemical potential, which is consistent with the second law of thermodynamics. The difference between the chemical potential of the reactants and the chemical potential of the products is called *affinity* [111, 112] and is given by

$$\nu_i(E, m) = \sum_{j=1}^{q} A_{ij}M_j\mu_j(E, m) - \sum_{j=1}^{q} B_{ij}M_j\mu_j(E, m) \geq 0, \quad i = 1, \ldots, r. \tag{6.123}$$

Affinity is a driving force for chemical reactions and is equal to zero at the state of *chemical equilibrium*. A nonzero affinity implies that the system is not in equilibrium and that chemical reactions will continue to occur until the system reaches an equilibrium characterized by zero affinity. The next assumption provides a general form for the inequalities (6.122) and (6.123).

Assumption 6.1. For the chemical reaction network (6.10) with the mass balance (6.121), assume that $\mu(E, m) >> 0$ for all $E \neq 0$ and

$$(B - A)M\mu(E, m) \leq\leq 0, \quad (E, m) \in \overline{\mathbb{R}}_+^q \times \overline{\mathbb{R}}_+^q, \tag{6.124}$$

or, equivalently,

$$\nu(E,m) = (A - B)M\mu(E,m) \geq\geq 0, \quad (E,m) \in \overline{\mathbb{R}}_+^q \times \overline{\mathbb{R}}_+^q, \tag{6.125}$$

where $\mu(E,m) \triangleq [\mu_1(E,m),\ldots,\mu_q(E,m)]^{\rm T}$ is the vector of chemical potentials of the substances of $\mathcal{G}$ and $\nu(E,m) \triangleq [\nu_1(E,m),\ldots,\nu_r(E,m)]^{\rm T}$ is the affinity vector for the reaction network given by (6.10).

Note that equality in (6.124) or, equivalently, in (6.125) characterizes the state of chemical equilibrium when the chemical potentials of the products and reactants are equal or, equivalently, when the affinity of each reaction is equal to zero. In this case, no reaction occurs and $\dot{m}(t) = 0$, $t \geq t_0$.

Next, we characterize the entropy function for the interconnected dynamical system $\mathcal{G}$ with the power and mass flow balances given by (6.120) and (6.121). The definition of entropy for $\mathcal{G}$ in the presence of chemical reactions remains the same as in Definition 3.5 with $\mathcal{S}(E)$ replaced by $\mathcal{S}(E,m)$ and with all other conditions in the definition holding for every $m >> 0$. Consider the jth subsystem of $\mathcal{G}$ and assume that E_k and m_k, $k \neq j$, $k = 1,\ldots,q$, are constant. In this case, note that

$$\frac{{\rm d}\mathcal{S}}{{\rm d}t} = \frac{\partial \mathcal{S}}{\partial E_j}\frac{{\rm d}E_j}{{\rm d}t} + \frac{\partial \mathcal{S}}{\partial m_j}\frac{{\rm d}m_j}{{\rm d}t} \tag{6.126}$$

and recall that

$$\frac{\partial \mathcal{S}}{\partial E}P(E,m) + \frac{\partial \mathcal{S}}{\partial m} = 0. \tag{6.127}$$

Next, it follows from (6.127) that the time derivative of the entropy function $\mathcal{S}(E,m)$ along the trajectories of (6.120) and (6.121) is given by

$$\begin{aligned}
\dot{\mathcal{S}}(E,m) &= \frac{\partial \mathcal{S}(E,m)}{\partial E}\dot{E} + \frac{\partial \mathcal{S}(E,m)}{\partial m}\dot{m} \\
&= \frac{\partial \mathcal{S}(E,m)}{\partial E}f(E) + \left(\frac{\partial \mathcal{S}(E,m)}{\partial E}P(E,m) + \frac{\partial \mathcal{S}(E,m)}{\partial m}\right) \\
&\qquad \cdot M(B-A)^{\rm T}\tilde{K}m^A + \frac{\partial \mathcal{S}(E,m)}{\partial E}S(t) - \frac{\partial \mathcal{S}(E,m)}{\partial E}d(E) \\
&= \frac{\partial \mathcal{S}(E,m)}{\partial E}f(E) + \frac{\partial \mathcal{S}(E,m)}{\partial E}S(t) - \frac{\partial \mathcal{S}(E,m)}{\partial E}d(E) \\
&= \sum_{i=1}^{q}\sum_{j=i+1}^{q}\left(\frac{\partial \mathcal{S}(E,m)}{\partial E_i} - \frac{\partial \mathcal{S}(E,m)}{\partial E_j}\right)\phi_{ij}(E) + \frac{\partial \mathcal{S}(E,m)}{\partial E}S(t) \\
&\qquad - \frac{\partial \mathcal{S}(E,m)}{\partial E}d(E), \quad (E,m) \in \overline{\mathbb{R}}_+^q \times \overline{\mathbb{R}}_+^q.
\end{aligned} \tag{6.128}$$

For the isolated system $\mathcal{G}$ (i.e., $S(t) \equiv 0$ and $d(E) \equiv 0$), the entropy function of $\mathcal{G}$ is a nondecreasing function of time and, using identical arguments as in the proof of Theorem 3.14, it can be shown that $(E(t), m(t)) \to \mathcal{R} \triangleq \left\{(E, m) \in \overline{\mathbb{R}}_+^q \times \overline{\mathbb{R}}_+^q : \frac{\partial \mathcal{S}(E,m)}{\partial E_1} = \cdots = \frac{\partial \mathcal{S}(E,m)}{\partial E_q}\right\}$ as $t \to \infty$ for all $(E_0, m_0) \in \overline{\mathbb{R}}_+^q \times \overline{\mathbb{R}}_+^q$.

The entropy production in the interconnected system $\mathcal{G}$ due to chemical reactions is given by

$$\begin{aligned} \mathrm{d}\mathcal{S}_{\mathrm{i}}(E, m) &= \frac{\partial \mathcal{S}(E,m)}{\partial m} \mathrm{d}m \\ &= -\frac{\partial \mathcal{S}(E,m)}{\partial E} P(E,m) M (B - A)^{\mathrm{T}} \tilde{K} m^A \mathrm{d}t, \\ &\qquad\qquad (E, m) \in \overline{\mathbb{R}}_+^q \times \overline{\mathbb{R}}_+^q. \end{aligned} \tag{6.129}$$

If the interconnected dynamical system $\mathcal{G}$ is isothermal, that is, all subsystems of $\mathcal{G}$ are at the same temperature

$$\left(\frac{\partial \mathcal{S}(E,m)}{\partial E_1}\right)^{-1} = \cdots = \left(\frac{\partial \mathcal{S}(E,m)}{\partial E_q}\right)^{-1} = T, \tag{6.130}$$

where $T > 0$ is the system temperature, then it follows from Assumption 6.1 that

$$\begin{aligned} \mathrm{d}\mathcal{S}_{\mathrm{i}}(E, m) &= -\frac{1}{T} \mathbf{e}^{\mathrm{T}} P(E,m) M (B - A)^{\mathrm{T}} \tilde{K} m^A \mathrm{d}t \\ &= -\frac{1}{T} \mu^{\mathrm{T}}(E,m) M (B - A)^{\mathrm{T}} \tilde{K} m^A \mathrm{d}t \\ &= \frac{1}{T} \nu^{\mathrm{T}}(E,m) \tilde{K} m^A \mathrm{d}t \\ &\geq 0, \quad (E, m) \in \overline{\mathbb{R}}_+^q \times \overline{\mathbb{R}}_+^q. \end{aligned} \tag{6.131}$$

Note that since the affinity of a reaction is equal to zero at the state of a chemical equilibrium, it follows that equality in (6.131) holds if and only if $\nu(E, m) = 0$ for some $E \in \overline{\mathbb{R}}_+^q$ and $m \in \overline{\mathbb{R}}_+^q$.

Theorem 6.6. Consider the isolated (i.e., $S(t) \equiv 0$ and $d(E) \equiv 0$) interconnected dynamical system $\mathcal{G}$ with the power and mass flow balances given by (6.120) and (6.121). Assume that $\operatorname{rank} \mathcal{C} = q - 1$, Assumption 6.1 holds, and there exists an entropy function $\mathcal{S} : \overline{\mathbb{R}}_+^q \times \overline{\mathbb{R}}_+^q \to \mathbb{R}$ of $\mathcal{G}$. Then $(E(t), m(t)) \to \mathcal{R}$ as $t \to \infty$, where $(E(t), m(t))$, $t \geq t_0$, is the solution to (6.120) and (6.121) with the initial condition $(E_0, m_0) \in \overline{\mathbb{R}}_+^q \times \overline{\mathbb{R}}_+^q$ and

$$\begin{aligned} \mathcal{R} = \bigg\{(E, m) \in \overline{\mathbb{R}}_+^q \times \overline{\mathbb{R}}_+^q : \frac{\partial \mathcal{S}(E,m)}{\partial E_1} &= \cdots = \frac{\partial \mathcal{S}(E,m)}{\partial E_q} \\ &\text{and } \nu(E, m) = 0\bigg\}, \end{aligned} \tag{6.132}$$

where $\nu(\cdot,\cdot)$ is the affinity vector of $\mathcal{G}$.

Proof. Since the dynamics of the isolated system $\mathcal{G}$ are essentially nonnegative, it follows from Proposition 2.1 that $(E(t), m(t)) \in \overline{\mathbb{R}}_+^q \times \overline{\mathbb{R}}_+^q$, $t \geq t_0$, for all $(E_0, m_0) \in \overline{\mathbb{R}}_+^q \times \overline{\mathbb{R}}_+^q$. Consider a scalar function $v(E,m) = \mathbf{e}^{\mathrm{T}}E + \mathbf{e}^{\mathrm{T}}m$, $(E,m) \in \overline{\mathbb{R}}_+^q \times \overline{\mathbb{R}}_+^q$, and note that $v(0,0) = 0$ and $v(E,m) > 0$, $(E,m) \in \overline{\mathbb{R}}_+^q \times \overline{\mathbb{R}}_+^q$, $(E,m) \neq (0,0)$. It follows from (6.118), Assumption 6.1, and $\mathbf{e}^{\mathrm{T}}f(E) \equiv 0$ that the time derivative of $v(\cdot,\cdot)$ along the trajectories of (6.120) and (6.121) satisfies

$$\begin{aligned}
\dot{v}(E,m) &= \mathbf{e}^{\mathrm{T}}\dot{E} + \mathbf{e}^{\mathrm{T}}\dot{m} \\
&= \mathbf{e}^{\mathrm{T}}P(E,m)M(B-A)^{\mathrm{T}}\tilde{K}m^A \\
&= \mu^{\mathrm{T}}(E,m)M(B-A)^{\mathrm{T}}\tilde{K}m^A \\
&= -\nu^{\mathrm{T}}(E,m)\tilde{K}m^A \\
&\leq 0, \quad (E,m) \in \overline{\mathbb{R}}_+^q \times \overline{\mathbb{R}}_+^q,
\end{aligned} \tag{6.133}$$

which implies that the solution $(E(t), m(t))$, $t \geq t_0$, to (6.120) and (6.121) is bounded for all initial conditions $(E_0, m_0) \in \overline{\mathbb{R}}_+^q \times \overline{\mathbb{R}}_+^q$.

Next, consider the function $\tilde{v}(E,m) = \mathbf{e}^{\mathrm{T}}E + \mathbf{e}^{\mathrm{T}}m - \mathcal{S}(E,m)$, $(E,m) \in \overline{\mathbb{R}}_+^q \times \overline{\mathbb{R}}_+^q$. Then it follows from (6.128) and (6.133) that the time derivative of $\tilde{v}(\cdot,\cdot)$ along the trajectories of (6.120) and (6.121) satisfies

$$\begin{aligned}
\dot{\tilde{v}}(E,m) &= \mathbf{e}^{\mathrm{T}}\dot{E} + \mathbf{e}^{\mathrm{T}}\dot{m} - \dot{\mathcal{S}}(E,m) \\
&= -\nu^{\mathrm{T}}(E,m)\tilde{K}m^A - \sum_{i=1}^{q}\sum_{j=i+1}^{q}\left(\frac{\partial\mathcal{S}(E,m)}{\partial E_i} - \frac{\partial\mathcal{S}(E,m)}{\partial E_j}\right)\phi_{ij}(E) \\
&\leq 0, \quad (E,m) \in \overline{\mathbb{R}}_+^q \times \overline{\mathbb{R}}_+^q,
\end{aligned} \tag{6.134}$$

which implies that $\tilde{v}(\cdot,\cdot)$ is a nonincreasing function of time, and hence, by the Krasovskii–LaSalle theorem, $(E(t), m(t)) \to \mathcal{R} \triangleq \{(E,m) \in \overline{\mathbb{R}}_+^q \times \overline{\mathbb{R}}_+^q : \dot{\tilde{v}}(E,m) = 0\}$ as $t \to \infty$. Now, it follows from Definition 3.5, Assumption 6.1, and the fact that $\operatorname{rank} \mathcal{C} = q-1$ that

$$\begin{aligned}
\mathcal{R} = &\left\{(E,m) \in \overline{\mathbb{R}}_+^q \times \overline{\mathbb{R}}_+^q : \frac{\partial\mathcal{S}(E,m)}{\partial E_1} = \cdots = \frac{\partial\mathcal{S}(E,m)}{\partial E_q}\right\} \\
&\cap \{(E,m) \in \overline{\mathbb{R}}_+^q \times \overline{\mathbb{R}}_+^q : \nu(E,m) = 0\},
\end{aligned} \tag{6.135}$$

which proves the result. □

Theorem 6.6 implies that the state of the interconnected dynamical system $\mathcal{G}$ converges to the state of thermal and chemical equilibrium when the temperatures of all substances of $\mathcal{G}$ are equal and the masses of all substances reach a state where all reaction affinities are zero, corresponding to a halting of all chemical reactions.

Next, we assume that the entropy of the interconnected dynamical system $\mathcal{G}$ is a sum of individual entropies of subsystems of $\mathcal{G}$, that is, $\mathcal{S}(E,m) = \sum_{j=1}^{q} \mathcal{S}_j(E_j, m_j)$, $(E,m) \in \overline{\mathbb{R}}_+^q \times \overline{\mathbb{R}}_+^q$. In this case, the Helmholtz free energy of $\mathcal{G}$ is given by

$$F(E,m) = \mathbf{e}^{\mathrm{T}}E - \sum_{j=1}^{q} \left(\frac{\partial \mathcal{S}(E,m)}{\partial E_j}\right)^{-1} \mathcal{S}_j(E_j, m_j), \quad (E,m) \in \overline{\mathbb{R}}_+^q \times \overline{\mathbb{R}}_+^q. \tag{6.136}$$

If the interconnected dynamical system $\mathcal{G}$ is isothermal, then the derivative of $F(\cdot,\cdot)$ along the trajectories of (6.120) and (6.121) is given by

$$\begin{aligned} \dot{F}(E,m) &= \mathbf{e}^{\mathrm{T}}\dot{E} - \sum_{j=1}^{q} \left(\frac{\partial \mathcal{S}(E,m)}{\partial E_j}\right)^{-1} \dot{\mathcal{S}}_j(E_j, m_j) \\ &= \mathbf{e}^{\mathrm{T}}\dot{E} - \sum_{j=1}^{q} \left(\frac{\partial \mathcal{S}(E,m)}{\partial E_j}\right)^{-1} \left[\frac{\partial \mathcal{S}_j(E_j,m_j)}{\partial E_j}\dot{E}_j + \frac{\partial \mathcal{S}_j(E_j,m_j)}{\partial m_j}\dot{m}_j\right] \\ &= \mu^{\mathrm{T}}(E,m)M(B-A)^{\mathrm{T}}\tilde{K}m^A \\ &= -\nu^{\mathrm{T}}(E,m)\tilde{K}m^A \\ &\le 0, \quad (E,m) \in \overline{\mathbb{R}}_+^q \times \overline{\mathbb{R}}_+^q, \end{aligned} \tag{6.137}$$

with equality in (6.137) holding if and only if $\nu(E,m) = 0$ for some $E \in \overline{\mathbb{R}}_+^q$ and $m \in \overline{\mathbb{R}}_+^q$, which determines the state of chemical equilibrium.

Hence, the Helmholtz free energy of $\mathcal{G}$ evolves to a minimum when the pressure and temperature of each subsystem of $\mathcal{G}$ are maintained constant, which is consistent with classical thermodynamics. A similar conclusion can be arrived at for the Gibbs free energy if work energy considerations to and by the system are addressed. Thus, the Gibbs and Helmholtz free energies represent a measure of the tendency for a reaction to take place in the interconnected system $\mathcal{G}$, and hence, provide a measure of the work done by the interconnected dynamical system $\mathcal{G}$.

Chapter Seven

Finite-Time Thermodynamics

7.1 Introduction

In the previous chapters, we used a compartmental dynamical systems perspective to provide a system-theoretic approach to thermodynamics. Specifically, using a state space formulation, we developed a nonlinear compartmental dynamical system model characterized by energy conservation laws that is consistent with basic thermodynamic principles. In the case where the compartmental system is isolated, we showed that the dynamical system asymptotically evolves toward a state of energy equipartition. However, in physical systems, energy and temperature equipartition is achieved in finite time rather than merely asymptotically.

In this chapter, we merge the theories of semistability and finite-time stability developed in [41, 42, 45] to develop a rigorous framework for finite-time thermodynamics. First, we present the notions of finite-time convergence and finite-time semistability for nonlinear dynamical systems, and develop several sufficient Lyapunov stability theorems for finite-time semistability. Following [43], we exploit homogeneity as a means for verifying finite-time convergence. Our main result in this direction asserts that a homogeneous system is finite-time semistable if and only if it is semistable and has a negative degree of homogeneity. This main result depends on a converse Lyapunov result for homogeneous semistable systems, which we develop. While our converse result resembles a related result for asymptotically stable systems given in [43, 384], the proof of our result is rendered more difficult by the fact that it does not hold under the notions of homogeneity considered in [43, 384].

More specifically, while previous treatments of homogeneity involved Euler vector fields representing asymptotically stable dynamics, our results involve homogeneity with respect to a semi-Euler vector field representing a semistable system having the same equilibria as the dynamics of interest. Consequently, our theory precludes the use of dilations commonly used in the literature on homogeneous systems (such as [384]), and requires us to adopt a more geometric description of homogeneity (see [43] and references therein). Finally, using this framework we develop intercompartmental energy flow laws that guarantee finite-time semistability and energy equipartition for our thermodynamically consistent dynamical system model developed in Chapter 3.

7.2 Finite-Time Semistability of Nonlinear Nonnegative Dynamical Systems

In this chapter, we consider nonlinear dynamical systems of the form

$$\dot{x}(t) = f(x(t)), \quad x(0) = x_0, \quad t \in \mathcal{I}_{x_0}, \tag{7.1}$$

where $x(t) \in \mathcal{D} \subseteq \overline{\mathbb{R}}_+^n$, $t \in \mathcal{I}_{x_0}$, is the system state vector, $\mathcal{D}$ is a relatively open set with respect to $\overline{\mathbb{R}}_+^n$, $f : \mathcal{D} \to \mathbb{R}$ is continuous and essentially nonnegative on $\mathcal{D}$, $f^{-1}(0) \triangleq \{x \in \mathcal{D} : f(x) = 0\}$ is nonempty, and $\mathcal{I}_{x_0} = [0, \tau_{x_0})$, $0 \leq \tau_{x_0} \leq \infty$, is the maximal interval of existence for the solution $x(\cdot)$ of (7.1). The continuity of f implies that, for every $x_0 \in \mathcal{D}$, there exist $\tau_0 < 0 < \tau_1$ and a solution $x(\cdot)$ of (7.1) defined on (τ_0, τ_1) such that $x(0) = x_0$. A solution x is said to be *right maximally defined* if x cannot be extended on the right (either uniquely or nonuniquely) to a solution of (7.1). Here, we assume that for every initial condition $x_0 \in \mathcal{D}$, (7.1) has a unique right maximally defined solution, and this unique solution is defined on $[0, \infty)$.

Under these assumptions, the solutions of (7.1) define a continuous *global semiflow* on $\mathcal{D}$, that is, $s : [0, \infty) \times \mathcal{D} \to \mathcal{D}$ is a jointly continuous function satisfying the consistency property $s(0, x) = x$ and the semigroup property $s(t, s(\tau, x)) = s(t + \tau, x)$ for every $x \in \mathcal{D}$ and $t, \tau \in [0, \infty)$. Furthermore, we assume that for every initial condition $x_0 \in \mathcal{D} \backslash f^{-1}(0)$, (7.1) has a local unique solution for negative time. Given $t \in [0, \infty)$, we denote the flow $s(t, \cdot) : \mathcal{D} \to \mathcal{D}$ of (7.1) by $s_t(x_0)$ or s_t. Likewise, given $x \in \mathcal{D}$, we denote the solution curve or trajectory $s(\cdot, x) : [0, \infty) \to \mathcal{D}$ of (7.1) by $s^x(t)$ or s^x. The image of $\mathcal{U} \subset \mathcal{D}$ under the flow s_t is defined as $s_t(\mathcal{U}) \triangleq \{y : y = s_t(x_0) \text{ for some } x_0 \in \mathcal{U}\}$. Finally, a set $\mathcal{E} \subseteq \overline{\mathbb{R}}_+^n$ is *connected* if and only if every pair of open sets $\mathcal{U}_i \subseteq \overline{\mathbb{R}}_+^n$, $i = 1, 2$, satisfying $\mathcal{E} \subseteq \mathcal{U}_1 \cup \mathcal{U}_2$ and $\mathcal{U}_i \cap \mathcal{E} \neq \emptyset$, $i = 1, 2$, has a nonempty intersection. A *connected component* of the set $\mathcal{E} \subseteq \overline{\mathbb{R}}_+^n$ is a connected subset of $\mathcal{E}$ that is not properly contained in any connected subset of $\mathcal{E}$.

Next, we establish the notion of finite-time semistability and develop sufficient Lyapunov stability theorems for finite-time semistability.

Definition 7.1. An equilibrium point $x_e \in f^{-1}(0)$ of (7.1) is said to be *finite-time semistable* if there exist a relatively open neighborhood $\mathcal{Q} \subseteq \mathcal{D}$ of x_e and a function $T : \mathcal{Q}\backslash f^{-1}(0) \to (0, \infty)$, called the *settling-time function*, such that the following statements hold:

i) For every $x \in \mathcal{Q}\backslash f^{-1}(0)$, $s(t, x) \in \mathcal{Q}\backslash f^{-1}(0)$ for all $t \in [0, T(x))$, and $\lim_{t \to T(x)} s(t, x)$ exists and is contained in $\mathcal{Q} \cap f^{-1}(0)$.

ii) x_e is semistable.

An equilibrium point $x_e \in f^{-1}(0)$ of (7.1) is said to be *globally finite-time semistable* if it is finite-time semistable with $\mathcal{D} = \mathcal{Q} = \overline{\mathbb{R}}^n_+$. The system (7.1) is said to be *finite-time semistable* if every equilibrium point in $f^{-1}(0)$ is finite-time semistable. Finally, (7.1) is said to be *globally finite-time semistable* if every equilibrium point in $f^{-1}(0)$ is globally finite-time semistable.

It is easy to see from Definition 7.1 that, for all $x \in \mathcal{Q}$,

$$T(x) = \inf\{t \in \overline{\mathbb{R}}_+ : f(s(t, x)) = 0\}, \tag{7.2}$$

where $T(\mathcal{Q} \cap f^{-1}(0)) = \{0\}$.

Lemma 7.1. Suppose (7.1) is finite-time semistable. Let $x_e \in f^{-1}(0)$ be an equilibrium point of (7.1) and let $\mathcal{Q} \subseteq \mathcal{D}$ be as in Definition 7.1. Furthermore, let $T : \mathcal{Q} \to \overline{\mathbb{R}}_+$ be the settling-time function. Then T is continuous on $\mathcal{Q}$ if and only if T is continuous at each $z_e \in \mathcal{Q} \cap f^{-1}(0)$.

Proof. Necessity is immediate. To prove sufficiency, suppose that T is continuous at each $z_e \in \mathcal{Q} \cap f^{-1}(0)$. Let $z \in \mathcal{Q}\backslash f^{-1}(0)$ and consider a sequence $\{z_m\}_{m=1}^{\infty}$ in $\mathcal{Q}$ that converges to z. Let $\tau^- = \liminf_{m\to\infty} T(z_m)$ and $\tau^+ = \limsup_{m\to\infty} T(z_m)$. Note that both τ^- and τ^+ are in $\overline{\mathbb{R}}_+$ and

$$\tau^- \leq \tau^+. \tag{7.3}$$

Next, let $\{z_l^+\}_{l=1}^{\infty}$ be a subsequence of $\{z_m\}_{m=1}^{\infty}$ such that $T(z_l^+) \to \tau^+$ as $l \to \infty$. The sequence $\{(T(z), z_l^+)\}_{l=1}^{\infty}$ converges in $\overline{\mathbb{R}}_+ \times \mathcal{Q}$ to $(T(z), z)$. By continuity and

$$s(T(x) + t, x) = s(T(x), x) \tag{7.4}$$

for all $x \in \mathcal{Q}$ and $t \in \overline{\mathbb{R}}_+$, $s(T(z), z_l^+) \to s(T(z), z) = z_e$ as $l \to \infty$, where $z_e \in \mathcal{Q} \cap f^{-1}(0)$. Since T is assumed to be continuous at each $z_e \in \mathcal{Q} \cap f^{-1}(0)$, $T(s(T(z), z_l^+)) \to T(z_e) = 0$ as $l \to \infty$. Note that

$$T(s(t, x)) = \max\{T(x) - t, 0\} \tag{7.5}$$

for all $x \in \mathcal{Q}$ and $t \in \overline{\mathbb{R}}_+$. Using (7.5) with $t = T(z)$ and $x = z_l^+$, we obtain $\max\{T(z_l^+) - T(z), 0\} \to 0$ as $l \to \infty$. Hence, $\max\{\tau^+ - T(z), 0\} = 0$, that is,

$$\tau^+ \leq T(z). \tag{7.6}$$

Now, let $\{z_l^-\}_{l=1}^\infty$ be a subsequence of $\{z_m\}_{m=1}^\infty$ such that $T(z_l^-) \to \tau^-$ as $l \to \infty$. It follows from (7.3) and (7.6) that $\tau^- \in \mathbb{R}_+$. Therefore, the sequence $\{(T(z_l^-), z_l^-)\}_{l=1}^\infty$ converges in $\mathbb{R}_+ \times \mathcal{Q}$ to (τ^-, z). Since s is continuous, it follows that $s(T(z_l^-), z_l^-) \to s(\tau^-, z)$ as $l \to \infty$. Equation (7.4) implies that $s(T(z_l^-), z_l^-) \in \mathcal{Q} \cap f^{-1}(0)$ for each l. Hence, $s(\tau^-, z) = z_{\rm e}$, $z_{\rm e} \in \mathcal{Q} \cap f^{-1}(0)$ and, by (7.2),

$$T(z) \leq \tau^-. \tag{7.7}$$

It follows from (7.3), (7.6), and (7.7) that $\tau^- = \tau^+ = T(z)$, and hence, $T(z_m) \to T(z)$ as $m \to \infty$. □

Next, we introduce a new definition, which is weaker than finite-time semistability and is needed for the next result.

Definition 7.2. The system (7.1) is said to be *finite-time convergent to* $\mathcal{M} \subseteq f^{-1}(0)$ for $\mathcal{D}_0 \subseteq \mathcal{D}$ if, for every $x_0 \in \mathcal{D}_0$, there exists a finite time $T = T(x_0) > 0$ such that $x(t) \in \mathcal{M}$ for all $t \geq T$.

The next result gives a sufficient condition for characterizing finite-time convergence. For the statement of this result, define

$$\dot{V}(x) \triangleq \lim_{h \to 0^+} \frac{1}{h} \left[V(s(h, x)) - V(x) \right], \quad x \in \mathcal{D}, \tag{7.8}$$

for a given continuous function $V : \mathcal{D} \to \mathbb{R}$ and for every $x \in \mathcal{D}$ such that the limit in (7.8) exists.

Proposition 7.1. Let $\mathcal{D}_0 \subseteq \mathcal{D}$ be positively invariant and $\mathcal{M} \subseteq f^{-1}(0)$. Assume that there exists a continuous function $V : \mathcal{D}_0 \to \mathbb{R}$ such that $\dot{V}(\cdot)$ is defined everywhere on $\mathcal{D}_0$, $V(x) = 0$ if and only if $x \in \mathcal{M} \subset \mathcal{D}_0$, and

$$-c_1 |V(x)|^\alpha \leq \dot{V}(x) \leq -c_2 |V(x)|^\alpha, \quad x \in \mathcal{D}_0 \backslash \mathcal{M}, \tag{7.9}$$

where $c_1 \geq c_2 > 0$ and $0 < \alpha < 1$. Then (7.1) is finite-time convergent to $\mathcal{M}$ for $\{x \in \mathcal{D}_0 : V(x) \geq 0\}$. Alternatively, if V is nonnegative and

$$\dot{V}(x) \leq -c_3 (V(x))^\alpha, \quad x \in \mathcal{D}_0 \backslash \mathcal{M}, \tag{7.10}$$

where $c_3 > 0$, then (7.1) is finite-time convergent to $\mathcal{M}$ for $\mathcal{D}_0$.

Proof. Note that (7.9) is also true for $x \in \mathcal{M}$. Application of the comparison lemma (Theorems 4.1 and 4.2 of [473]) to (7.9) yields $\mu(t, V(x), c_1) \leq V(s(t, x)) \leq \mu(t, V(x), c_2)$, $x \in \{z \in \mathcal{D}_0 : V(z) \geq 0\}$, where μ is given by

$$\mu(t,z,c) \triangleq \begin{cases} (|z|^{1-\alpha} - c(1-\alpha)t)^{\frac{1}{1-\alpha}}, & 0 \le t < \frac{|z|^{1-\alpha}}{c(1-\alpha)}, \quad \alpha < 1, \\ 0, & t \ge \frac{|z|^{1-\alpha}}{c(1-\alpha)}, \quad \alpha < 1. \end{cases} \tag{7.11}$$

Hence, $V(s(t,x)) = 0$ for $t \ge \frac{|V(x)|^{1-\alpha}}{c_2(1-\alpha)}$, which implies that $s(t,x) \in \mathcal{M}$ for $t \ge \frac{|V(x)|^{1-\alpha}}{c_2(1-\alpha)}$. The assertion follows. The second part of the assertion can be proved similarly. □

The next result establishes a relationship between finite-time convergence and finite-time semistability.

Theorem 7.1. Assume that there exists a continuous nonnegative function $V : \mathcal{D} \to \overline{\mathbb{R}}_+$ such that $\dot{V}(\cdot)$ is defined everywhere on $\mathcal{D}$, $V^{-1}(0) = f^{-1}(0)$, and there exists a relatively open neighborhood $\mathcal{Q} \subseteq \mathcal{D}$ such that $\mathcal{Q} \cap f^{-1}(0)$ is nonempty and

$$\dot{V}(x) \le w(V(x)), \quad x \in \mathcal{Q} \backslash f^{-1}(0), \tag{7.12}$$

where $w : \overline{\mathbb{R}}_+ \to \mathbb{R}$ is continuous, $w(0) = 0$, and

$$\dot{z}(t) = w(z(t)), \quad z(0) = z_0 \in \overline{\mathbb{R}}_+, \quad t \ge 0, \tag{7.13}$$

has a unique solution in forward time. If (7.13) is finite-time convergent to the origin for $\overline{\mathbb{R}}_+$ and every point in $\mathcal{Q} \cap f^{-1}(0)$ is a Lyapunov stable equilibrium point of (7.1), then every point in $\mathcal{Q} \cap f^{-1}(0)$ is finite-time semistable. Moreover, the settling-time function of (7.1) is continuous on a relatively open neighborhood of $\mathcal{Q} \cap f^{-1}(0)$. Finally, if $\mathcal{Q} = \mathcal{D}$, then (7.1) is finite-time semistable.

Proof. Consider $x_e \in \mathcal{Q} \cap f^{-1}(0)$. Since $x(t) \equiv x_e$ is Lyapunov stable, it follows that there exists a relatively open positively invariant set $\mathcal{S} \subseteq \mathcal{Q}$ containing x_e. Next, it follows from (7.12) that

$$\dot{V}(s(t,x)) \le w(V(s(t,x))), \quad x \in \mathcal{S}, \quad t \ge 0. \tag{7.14}$$

Now, application of the comparison lemma (Theorem 4.1 of [473]) to the inequality (7.14) with the comparison system (7.13) yields

$$V(s(t,x)) \le \psi(t, V(x)), \quad t \ge 0, \quad x \in \mathcal{S}, \tag{7.15}$$

where $\psi : [0,\infty) \times \mathbb{R} \to \mathbb{R}$ is the global semiflow of (7.13). Since (7.13) is finite-time convergent to the origin for $\overline{\mathbb{R}}_+$, it follows from (7.15) and the nonnegativity of $V(\cdot)$ that

$$V(s(t,x)) = 0, \quad t \ge \hat{T}(V(x)), \quad x \in \mathcal{S}, \tag{7.16}$$

where $\hat{T}(\cdot)$ denotes the settling-time function of (7.13).

Next, since $s(0,x) = x$, $s(\cdot,\cdot)$ is jointly continuous, and $V(s(t,x)) = 0$ is equivalent to $f(s(t,x)) = 0$ on $\mathcal{S}$, it follows that $\inf\{t \in \overline{\mathbb{R}}_+ : f(s(t,x)) = 0\} > 0$ for $x \in \mathcal{S}\backslash f^{-1}(0)$. Furthermore, it follows from (7.16) that $\inf\{t \in \overline{\mathbb{R}}_+ : f(s(t,x)) = 0\} < \infty$ for $x \in \mathcal{S}$. Define $T : \mathcal{S}\backslash f^{-1}(0) \to \overline{\mathbb{R}}_+$ by $T(x) = \inf\{t \in \overline{\mathbb{R}}_+ : f(s(t,x)) = 0\}$. Then it follows that every point in $\mathcal{S} \cap f^{-1}(0)$ is finite-time semistable and T is the settling-time function on $\mathcal{S}$. Furthermore, it follows from (7.16) that $T(x) \leq \hat{T}(V(x))$, $x \in \mathcal{S}$. Since the settling-time function of a one-dimensional finite-time stable system is continuous at the equilibrium, it follows that T is continuous at each point in $\mathcal{S} \cap f^{-1}(0)$. Since $x_{\mathrm{e}} \in \mathcal{Q} \cap f^{-1}(0)$ was chosen arbitrarily, it follows that every point in $\mathcal{Q} \cap f^{-1}(0)$ is finite-time semistable, while Lemma 7.1 implies that T is continuous on a relatively open neighborhood of $\mathcal{Q} \cap f^{-1}(0)$.

The last statement follows by noting that, if $\mathcal{Q} = \mathcal{D}$, then $\mathcal{Q}$ is positively invariant by our assumptions on (7.1), and hence, the preceding arguments hold with $\mathcal{S} = \mathcal{Q}$. □

Theorem 7.2. Assume that there exists a continuous nonnegative function $V : \mathcal{D} \to \overline{\mathbb{R}}_+$ such that $\dot{V}(\cdot)$ is defined everywhere on $\mathcal{D}$, $V^{-1}(0) = f^{-1}(0)$, and there exists a relatively open neighborhood $\mathcal{Q} \subseteq \mathcal{D}$ such that $\mathcal{Q} \cap f^{-1}(0)$ is nonempty and (7.10) holds for all $x \in \mathcal{Q}\backslash f^{-1}(0)$. Furthermore, assume that there exists a continuous nonnegative function $W : \mathcal{Q} \to \overline{\mathbb{R}}_+$ such that $\dot{W}(\cdot)$ is defined everywhere on $\mathcal{Q}$, $W^{-1}(0) = \mathcal{Q} \cap f^{-1}(0)$, and

$$\|f(x)\| \leq -c_0\dot{W}(x), \quad x \in \mathcal{Q}\backslash f^{-1}(0), \tag{7.17}$$

where $c_0 > 0$. Then every point in $\mathcal{Q} \cap f^{-1}(0)$ is finite-time semistable.

Proof. For every $x_{\mathrm{e}} \in \mathcal{Q} \cap f^{-1}(0)$, since $W(x) \geq 0 = W(x_{\mathrm{e}})$ for all $x \in \mathcal{Q}$, it follows from *i*) of Theorem 5.2 of [45] that x_{e} is a Lyapunov stable equilibrium and, hence, every point in $\mathcal{Q} \cap f^{-1}(0)$ is Lyapunov stable. Now, it follows from the second assertion of Proposition 7.1 and Theorem 7.1, with $w(x) = -c_3\mathrm{sgn}(x)|x|^{\alpha}$, that every point in $\mathcal{Q} \cap f^{-1}(0)$ is finite-time semistable. □

7.3 Homogeneity and Finite-Time Semistability

In this section, we develop necessary and sufficient conditions for finite-time semistability of homogeneous dynamical systems. In the sequel, we will need to consider a complete vector field ν on $\overline{\mathbb{R}}^n_+$ such that the solutions of the differential equation $\dot{y}(t) = \nu(y(t))$ define a continuous *global flow* $\psi : \mathbb{R} \times \overline{\mathbb{R}}^n_+ \to \overline{\mathbb{R}}^n_+$ on $\overline{\mathbb{R}}^n_+$, where $\nu^{-1}(0) = f^{-1}(0)$. For each $\tau \in \mathbb{R}$, the map $\psi_\tau(\cdot) = \psi(\tau,\cdot)$ is a homeomorphism and $\psi_\tau^{-1} = \psi_{-\tau}$. We define a function $V : \overline{\mathbb{R}}^n_+ \to \mathbb{R}$ to be *homogeneous of degree* $l \in \mathbb{R}$ *with respect to* ν if and only if $(V \circ \psi_\tau)(x) = e^{l\tau}V(x)$, $\tau \in \mathbb{R}$, $x \in \overline{\mathbb{R}}^n_+$. Our assumptions imply that every connected component of $\overline{\mathbb{R}}_+^n \backslash f^{-1}(0)$ is invariant under ν.

The *Lie derivative* of a continuous function $V : \overline{\mathbb{R}}_+^n \to \mathbb{R}$ with respect to ν is given by $L_\nu V(x) \triangleq \lim_{t \to 0^+} \frac{1}{t}[V(\psi(t,x)) - V(x)]$, whenever the limit on the right-hand side exists. If V is a continuous homogeneous function of degree $l > 0$, then $L_\nu V$ is defined everywhere and satisfies $L_\nu V = lV$. We assume that the vector field ν is a *semi-Euler vector field*, that is, the dynamical system

$$\dot{y}(t) = -\nu(y(t)), \quad y(0) = y_0, \quad t \geq 0, \tag{7.18}$$

is globally semistable with respect to $\overline{\mathbb{R}}_+^n$. Thus, for each $x \in \overline{\mathbb{R}}_+^n$, $\lim_{\tau \to \infty} \psi(-\tau, x) = x^* \in \nu^{-1}(0)$, and for each $x_\mathrm{e} \in \nu^{-1}(0)$, there exists $z \in \overline{\mathbb{R}}_+^n$ such that $x_\mathrm{e} = \lim_{\tau \to \infty} \psi(-\tau, z)$. Finally, we say that the vector field f is *homogeneous of degree* $k \in \mathbb{R}$ *with respect to* ν if and only if $\nu^{-1}(0) = f^{-1}(0)$ and, for every $t \in \overline{\mathbb{R}}_+$ and $\tau \in \mathbb{R}$,

$$s_t \circ \psi_\tau = \psi_\tau \circ s_{e^{k\tau}t}. \tag{7.19}$$

Note that if $V : \overline{\mathbb{R}}_+^n \to \mathbb{R}$ is a homogeneous function of degree l such that $L_f V(x)$ is defined everywhere, then $L_f V(x)$ is a homogeneous function of degree $l + k$. In a geometric, coordinate-free setting, the only link between homogeneity of functions and vector fields is that the Lie derivative of a homogeneous function along a homogeneous vector field is also a homogeneous function. In the special case where the coordinate functions are homogeneous functions, the fact mentioned above can be used to relate the homogeneity of a vector field with that of the components (considered as functions) of its coordinate representation. Such a relation is very familiar in the case of conventional dilations seen in the homogeneity literature [384]. Finally, note that if ν and f are continuously differentiable in a neighborhood of $x \in \overline{\mathbb{R}}_+^n$, then (7.19) holds at x for sufficiently small t and τ if and only if $[\nu, f](x) = kf(x)$ in a neighborhood of $x \in \overline{\mathbb{R}}_+^n$, where the *Lie bracket* $[\nu, f]$ of ν and f can be computed by using $[\nu, f] = \frac{\partial f}{\partial x}\nu - \frac{\partial \nu}{\partial x} f$.

The following lemmas are needed for the main results of this section.

Lemma 7.2. Consider the dynamical system (7.18). Let $\mathcal{D}_\mathrm{c} \subset \overline{\mathbb{R}}_+^n$ be a relatively compact set satisfying $\mathcal{D}_\mathrm{c} \cap \nu^{-1}(0) = \emptyset$. Then for every relatively open set $\mathcal{Q}$ satisfying $\nu^{-1}(0) \subset \mathcal{Q}$, there exist $\tau_1, \tau_2 > 0$ such that $\psi_{-t}(\mathcal{D}_\mathrm{c}) \subset \mathcal{Q}$ for all $t > \tau_1$ and $\psi_\tau(\mathcal{D}_\mathrm{c}) \cap \mathcal{Q} = \emptyset$ for all $\tau > \tau_2$.

Proof. Let $\mathcal{Q}$ be a relatively open neighborhood of $\nu^{-1}(0)$ with respect to $\overline{\mathbb{R}}_+^n$. Since every $z \in \nu^{-1}(0)$ is Lyapunov stable under ν, it follows that there exists a relatively open neighborhood $\mathcal{V}_z$ containing z such that $\psi_{-t}(\mathcal{V}_z) \subseteq \mathcal{Q}$ for all $t \geq 0$. Hence, $\mathcal{V} \triangleq \bigcup_{z \in \nu^{-1}(0)} \mathcal{V}_z$ is relatively open and

$\psi_{-t}(\mathcal{V}) \subseteq \mathcal{Q}$ for all $t \geq 0$.

Next, consider the collection of nested sets $\{\mathcal{D}_t\}_{t>0}$, where $\mathcal{D}_t = \{x \in \mathcal{D}_c : \psi_h(x) \notin \mathcal{V}, h \in [-t, 0]\} = \mathcal{D}_c \cap (\overline{\mathbb{R}}_+^n \backslash (\bigcup_{h\in[-t,0]} \psi_h^{-1}(\mathcal{V})))$, $t > 0$. For each $t > 0$, $\mathcal{D}_t$ is a relatively compact set. Therefore, if $\mathcal{D}_t$ is nonempty for each $t > 0$, then there exists $x \in \bigcap_{t>0} \mathcal{D}_t$, that is, there exists $x \in \mathcal{D}_c$ such that $\psi_{-t}(x) \notin \mathcal{V}$ for all $t > 0$, which contradicts the fact that the domain of semistability of (7.18) is $\overline{\mathbb{R}}_+^n$. Hence, there exists $\tau > 0$ such that $\mathcal{D}_\tau = \emptyset$, that is, $\mathcal{D}_c \subset \bigcup_{h\in[-\tau,0]} \psi_h^{-1}(\mathcal{V})$. Therefore, for every $t > \tau$, $\psi_{-t}(\mathcal{D}_c) \subset \bigcup_{h\in[-\tau,0]} \psi_{-t}(\psi_h^{-1}(\mathcal{V})) = \bigcup_{h\in[-\tau,0]} \psi_{-t-h}(\mathcal{V}) \subseteq \mathcal{Q}$. The second conclusion follows using similar arguments as above. □

Lemma 7.3. Suppose $f : \overline{\mathbb{R}}_+^n \to \mathbb{R}^n$ is homogeneous of degree $k \in \mathbb{R}$ with respect to ν and (7.1) is (locally) semistable. Then the domain of semistability of (7.1) is $\overline{\mathbb{R}}_+^n$.

Proof. Let $\mathcal{A} \subseteq \overline{\mathbb{R}}_+^n$ be the domain of semistability and $x \in \overline{\mathbb{R}}_+^n$. Note that $\mathcal{A}$ is a relatively open neighborhood of $\nu^{-1}(0)$ with respect to $\overline{\mathbb{R}}_+^n$. Since every point in $\nu^{-1}(0)$ is a globally semistable equilibrium under $-\nu$ with respect to $\overline{\mathbb{R}}_+^n$, there exists $\tau > 0$ such that $z = \psi_{-\tau}(x) \in \mathcal{A}$. Then it follows from (7.19) that $s(t, x) = s(t, \psi_\tau(z)) = \psi_\tau(s(e^{k\tau}t, z))$. Since $\lim_{t\to\infty} s(t, z) = x^* \in f^{-1}(0)$, it follows that $\lim_{t\to\infty} s(t, x) = \lim_{t\to\infty} \psi_\tau(s(e^{k\tau}t, z)) = \psi_\tau(\lim_{t\to\infty} s(e^{k\tau}t, z)) = \psi_\tau(x^*) = x^*$, which implies that $x \in \mathcal{A}$. Since $x \in \overline{\mathbb{R}}_+^n$ is arbitrary, $\mathcal{A} = \overline{\mathbb{R}}_+^n$. □

The following theorem presents a converse Lyapunov result for homogeneous semistable systems.

Theorem 7.3. Suppose $f : \overline{\mathbb{R}}_+^n \to \mathbb{R}^n$ is homogeneous of degree $k \in \mathbb{R}$ with respect to ν and (7.1) is semistable. Then for every $l > \max\{-k, 0\}$, there exists a continuous nonnegative function $V : \overline{\mathbb{R}}_+^n \to \overline{\mathbb{R}}_+$ that is homogeneous of degree l with respect to ν, continuously differentiable on $\overline{\mathbb{R}}_+^n \backslash f^{-1}(0)$, and satisfies $V^{-1}(0) = f^{-1}(0)$, $V'(x)f(x) < 0$, $x \in \overline{\mathbb{R}}_+^n \backslash f^{-1}(0)$, and for each $x_e \in f^{-1}(0)$ and each bounded, relatively open neighborhood $\mathcal{D}_0$ containing x_e with respect to $\overline{\mathbb{R}}_+^n$, there exist $c_1 = c_1(\mathcal{D}_0) \geq c_2 = c_2(\mathcal{D}_0) > 0$ such that

$$-c_1[V(x)]^{\frac{l+k}{l}} \leq V'(x)f(x) \leq -c_2[V(x)]^{\frac{l+k}{l}}, \quad x \in \mathcal{D}_0. \tag{7.20}$$

Proof. Choose $l > \max\{-k, 0\}$. First, we prove that there exists a continuous Lyapunov function V on $\overline{\mathbb{R}}_+^n$ that is homogeneous of degree l with respect to ν, continuously differentiable on $\overline{\mathbb{R}}_+^n \backslash f^{-1}(0)$, and $V'(x)f(x) < 0$ for $x \in \overline{\mathbb{R}}_+^n \backslash f^{-1}(0)$. Choose any nondecreasing smooth function $g : \overline{\mathbb{R}}_+ \to$

$[0,1]$ such that $g(s)=0$ for $s\le a$, $g(s)=1$ for $s\ge b$, and $g'(s)>0$ on (a,b), where $0<a<b$ are constants. It follows from Theorem 2.8 and Lemma 7.3 that there exists a continuously differentiable Lyapunov function $U(\cdot)$ on $\overline{\mathbb{R}}_+^n$ satisfying all of the properties in Theorem 2.8.

Next, define

$$V(x)\triangleq\int_{-\infty}^{+\infty}e^{-l\tau}g(U(\psi(\tau,x)))\mathrm{d}\tau,\quad x\in\overline{\mathbb{R}}_+^n. \tag{7.21}$$

Let $\mathcal{Q}$ be a bounded, relatively open set satisfying $\overline{\mathcal{Q}}\cap f^{-1}(0)=\emptyset$. Since every point in $\nu^{-1}(0)$ is a globally semistable equilibrium point under $-\nu$ with respect to $\overline{\mathbb{R}}_+^n$, it follows that for each $x\in\overline{\mathcal{Q}}$, $\lim_{\tau\to+\infty}U(\psi(\tau,x))=+\infty$ and $\lim_{\tau\to+\infty}U(\psi(-\tau,x))=0$. Now, it follows from Lemma 7.2 that there exist time instants $\tau_1<\tau_2$ such that for each $x\in\overline{\mathcal{Q}}$, $U(\psi(\tau,x))\le a$ for all $\tau\le\tau_1$ and $U(\psi(\tau,x))\ge b$ for all $\tau\ge\tau_2$. Hence,

$$V(x)=\int_{\tau_1}^{\tau_2}e^{-l\tau}g(U(\psi(\tau,x)))\mathrm{d}\tau+\frac{e^{-l\tau_2}}{l},\quad x\in\mathcal{Q}, \tag{7.22}$$

which implies that V is well defined, positive, and continuously differentiable on $\mathcal{Q}$.

Next, since $U(\cdot)$ satisfies *i*) and *ii*) of Theorem 2.8, it follows from (7.21) and (7.22) that $V^{-1}(0)=f^{-1}(0)$. Since, for any $\sigma\in\mathbb{R}$ and $x\in\overline{\mathbb{R}}_+^n$,

$$V(\psi(\sigma,x))=\int_{-\infty}^{+\infty}e^{-l\tau}g(U(\psi(\tau+\sigma,x)))\mathrm{d}\tau=e^{l\sigma}V(x), \tag{7.23}$$

by definition, V is homogeneous of degree l. In addition, it follows from (7.19) and (7.22) that

$$\begin{aligned}V'(x)f(x)&=\int_{\tau_1}^{\tau_2}e^{-l\tau}g'(U(\psi(\tau,x)))\frac{\mathrm{d}}{\mathrm{d}t}U(s(e^{-k\tau}t,\psi(\tau,x)))\Big|_{t=0}\mathrm{d}\tau\\&=\int_{\tau_1}^{\tau_2}e^{-(l+k)\tau}g'(U(\psi(\tau,x)))U'(\psi(\tau,x))f(\psi(\tau,x))\mathrm{d}\tau\\&<0,\quad x\in\mathcal{Q},\end{aligned} \tag{7.24}$$

which implies that $V'f$ is negative and continuous on $\mathcal{Q}$. Now, since $\mathcal{Q}$ is arbitrary, it follows that V is well defined and continuously differentiable, and $V'f$ is negative and continuous on $\overline{\mathbb{R}}_+^n\backslash f^{-1}(0)$.

Next, to show continuity at points in $f^{-1}(0)$, we define $T:\overline{\mathbb{R}}_+^n\backslash f^{-1}(0)\to\mathbb{R}$ by $T(x)=\sup\{t\in\mathbb{R}:U(\psi(\tau,x))\le a \text{ for all } \tau\le t\}$, and note that the continuity of U implies that $U(\psi(T(x),x))=a$ for all $x\in\overline{\mathbb{R}}_+^n\backslash f^{-1}(0)$. Let $x_{\mathrm{e}}\in f^{-1}(0)$, and consider a sequence $\{x_k\}_{k=1}^{\infty}$ in $\overline{\mathbb{R}}_+^n\backslash f^{-1}(0)$ converging to x_{e}. We claim that the sequence $\{T(x_k)\}_{k=1}^{\infty}$ has no bounded subsequence so that $\lim_{k\to\infty} T(x_k) = \infty$.

To prove our claim by contradiction, suppose, *ad absurdum*, that $\{T(x_{k_i})\}_{i=1}^{\infty}$ is a bounded subsequence. Without loss of generality, we may assume that the sequence $\{T(x_{k_i})\}_{i=1}^{\infty}$ converges to $h \in \mathbb{R}$. Then, by joint continuity of ψ, $\lim_{i\to\infty} \psi(T(x_{k_i}), x_{k_i}) = \psi(h, x_e) = x_e$, so that $\lim_{i\to\infty} U(\psi(T(x_{k_i}), x_{k_i})) = U(x_e) = 0$. However, this contradicts our observation above that $U(\psi(T(x), x)) = a$ for all $x \in \overline{\mathbb{R}}_+^n \backslash f^{-1}(0)$. The contradiction leads us to conclude that $\lim_{k\to\infty} T(x_k) = \infty$. Now, for each $k = 1, 2, \ldots$, it follows that

$$V(x_k) = \int_{T(x_k)}^{\infty} e^{-l\tau} g(U(\psi(\tau, x_k)))\mathrm{d}\tau \leq \int_{T(x_k)}^{\infty} e^{-l\tau}\mathrm{d}\tau = l^{-1}e^{-lT(x_k)},$$

so that $\lim_{k\to\infty} V(x_k) = 0 = V(x_e)$. Since x_e was chosen arbitrarily, it follows that V is continuous at every $x_e \in f^{-1}(0)$.

To show that V possesses the last property, let $x_e \in f^{-1}(0)$, and choose a bounded, relatively open neighborhood $\mathcal{D}_0$ of x_e with respect to $\overline{\mathbb{R}}_+^n$. Let $\mathcal{W} = \psi(\mathbb{R}_+ \times \mathcal{D}_0)$. For every $\varepsilon > 0$, denote $\mathcal{W}_\varepsilon = \mathcal{W} \cap V^{-1}(\varepsilon)$. For every $\varepsilon > 0$, define the continuous map $\tau_\varepsilon : \overline{\mathbb{R}}_+^n \backslash f^{-1}(0) \to \mathbb{R}$ by $\tau_\varepsilon(x) \triangleq l^{-1}\ln(\varepsilon/V(x))$, and note that, for every $x \in \overline{\mathbb{R}}_+^n \backslash f^{-1}(0)$, $\psi(t, x) \in V^{-1}(\varepsilon)$ if and only if $t = \tau_\varepsilon(x)$. Next, define $\beta_\varepsilon : \overline{\mathbb{R}}_+^n \backslash f^{-1}(0) \to \overline{\mathbb{R}}_+^n$ by $\beta_\varepsilon \triangleq \psi(\tau_\varepsilon(x), x)$. Note that, for every $\varepsilon > 0$, β_ε is continuous, and $\beta_\varepsilon(x) \in V^{-1}(\varepsilon)$ for every $x \in \overline{\mathbb{R}}_+^n \backslash f^{-1}(0)$.

Consider $\varepsilon > 0$. $\mathcal{W}_\varepsilon$ is the union of the images of connected components of $\mathcal{D}_0 \backslash f^{-1}(0)$ under the continuous map β_ε. Since every connected component of $\overline{\mathbb{R}}_+^n \backslash f^{-1}(0)$ is invariant under ν, it follows that the image of each connected component $\mathcal{Q}$ of $\overline{\mathbb{R}}_+^n \backslash f^{-1}(0)$ under β_ε is contained in $\mathcal{Q}$ itself. In particular, the images of connected components of $\mathcal{D}_0 \backslash f^{-1}(0)$ under β_ε are all disjoint. Thus, each connected component of $\mathcal{W}_\varepsilon$ is the image of exactly one connected component of $\mathcal{D}_0 \backslash f^{-1}(0)$ under β_ε. Finally, if ε is small enough so that $V^{-1}(\varepsilon) \cap \mathcal{D}_0$ is nonempty, then $V^{-1}(\varepsilon) \cap \mathcal{D}_0 \subseteq \mathcal{W}_\varepsilon$, and hence, every connected component of $\mathcal{W}_\varepsilon$ has a nonempty intersection with $\mathcal{D}_0 \backslash f^{-1}(0)$.

We claim that $\mathcal{W}_\varepsilon$ is bounded for every $\varepsilon > 0$. It is easy to verify that, for every $\varepsilon_1, \varepsilon_2 \in (0, \infty)$, $\mathcal{W}_{\varepsilon_2} = \psi_h(\mathcal{W}_{\varepsilon_1})$ with $h = l^{-1}\ln(\varepsilon_2/\varepsilon_1)$. Hence, it suffices to prove that there exists $\varepsilon > 0$ such that $\mathcal{W}_\varepsilon$ is bounded. To arrive at a contradiction, suppose, *ad absurdum*, that $\mathcal{W}_\varepsilon$ is unbounded for every $\varepsilon > 0$. Choose a bounded, relatively open neighborhood $\mathcal{V}$ of $\overline{\mathcal{D}}_0$ and a sequence $\{\varepsilon_i\}_{i=1}^{\infty}$ in $(0, \infty)$ converging to 0. By our assumption, for every $i = 1, 2, \ldots$, at least one connected component of $\mathcal{W}_{\varepsilon_i}$ must contain a point in $\overline{\mathbb{R}}_+^n \backslash \mathcal{V}$. On the other hand, for i sufficiently large, every connected component of $\mathcal{W}_{\varepsilon_i}$ has a nonempty intersection with $\mathcal{D}_0 \subset \mathcal{V}$. It follows that $\mathcal{W}_{\varepsilon_i}$ has a nonempty intersection with the boundary of $\mathcal{V}$ for every i sufficiently large. Hence, there exists a sequence $\{x_i\}_{i=1}^{\infty}$ in $\mathcal{D}_0$, and a sequence $\{t_i\}_{i=1}^{\infty}$ in $(0,\infty)$ such that $y_i \triangleq \psi_{t_i}(x_i) \in V^{-1}(\varepsilon_i) \cap \partial\mathcal{V}$ for every $i = 1, 2, \ldots$.

Since $\mathcal{V}$ is bounded, we can assume that the sequence $\{y_i\}_{i=1}^{\infty}$ converges to $y \in \partial\mathcal{V}$. Continuity implies that $V(y) = \lim_{i\to\infty} V(y_i) = \lim_{i\to\infty} \varepsilon_i = 0$. Since $V^{-1}(0) = f^{-1}(0) = \nu^{-1}(0)$, it follows that y is Lyapunov stable under $-\nu$. Since $y \notin \overline{\mathcal{D}}_0$, there exists a relatively open neighborhood $\mathcal{Q}$ of y such that $\mathcal{Q} \cap \mathcal{D}_0 = \emptyset$. The sequence $\{y_i\}_{i=1}^{\infty}$ converges to y while $\psi_{-t_i}(y_i) = x_i \in \mathcal{D}_0 \subset \overline{\mathbb{R}}_+^n \backslash \mathcal{Q}$, which contradicts Lyapunov stability. This contradiction implies that there exists $\varepsilon > 0$ such that $\mathcal{W}_\varepsilon$ is bounded. It now follows that $\mathcal{W}_\varepsilon$ is bounded for every $\varepsilon > 0$.

Finally, consider $x \in \mathcal{D}_0 \backslash f^{-1}(0)$. Choose $\varepsilon > 0$ and note that $\psi_{\tau_\varepsilon(x)}(x) \in \mathcal{W}_\varepsilon$. Furthermore, note that $V'(x)f(x) < 0$ for all $x \in \overline{\mathbb{R}}_+^n \backslash f^{-1}(0)$, $V'(x)f(x)$ is continuous on $\overline{\mathbb{R}}_+^n \backslash f^{-1}(0)$, and $\overline{\mathcal{W}}_\varepsilon \cap f^{-1}(0) = \emptyset$. Then, by homogeneity, $V(\psi_{\tau_\varepsilon(x)}(x)) = \varepsilon$, and hence,

$$\min_{z\in\overline{\mathcal{W}}_\varepsilon} V'(z)f(z) \leq V'(\psi_{\tau_\varepsilon(x)}(x))f(\psi_{\tau_\varepsilon(x)}(x)) \leq \max_{z\in\overline{\mathcal{W}}_\varepsilon} V'(z)f(z). \quad (7.25)$$

Since $V'(\psi_{\tau_\varepsilon(x)}(x))f(\psi_{\tau_\varepsilon(x)}(x))$ is homogeneous of degree $l+k$, it follows that

$$V'(\psi_{\tau_\varepsilon(x)}(x))f(\psi_{\tau_\varepsilon(x)}(x)) = e^{(l+k)\tau_\varepsilon(x)}V'(x)f(x) = \varepsilon^{\frac{l+k}{l}}V(x)^{-\frac{l+k}{l}}V'(x)f(x).$$

Let $c_1 \triangleq -\varepsilon^{-\frac{l+k}{l}} \min_{z\in\overline{\mathcal{W}}_\varepsilon} V'(z)f(z)$ and $c_2 \triangleq -\varepsilon^{-\frac{l+k}{l}} \max_{z\in\overline{\mathcal{W}}_\varepsilon} V'(z)f(z)$. Note that c_1 and c_2 are positive and well defined since $\overline{\mathcal{W}}_\varepsilon$ is compact. Hence, the theorem is proved. □

The following result represents the main application of homogeneity [43] to finite-time semistability.

Theorem 7.4. Suppose f is homogeneous of degree $k \in \mathbb{R}$ with respect to ν. Then (7.1) is finite-time semistable if and only if (7.1) is semistable and $k < 0$. In addition, if (7.1) is finite-time semistable, then the settling-time function $T(\cdot)$ is homogeneous of degree $-k$ with respect to ν and $T(\cdot)$ is continuous on $\overline{\mathbb{R}}_+^n$.

Proof. Since finite-time semistability implies semistability, it suffices to prove that if (7.1) is semistable, then (7.1) is finite-time semistable if and only if $k < 0$. Suppose (7.1) is finite-time semistable and let $l > \max\{-k, 0\}$. Then for each $x_\mathrm{e} \in f^{-1}(0)$, it follows from Theorem 7.3 that there exist a bounded, relatively open, and positively invariant set $\mathcal{S}$ containing x_e, and a continuous nonnegative function $V : \mathcal{S} \to \overline{\mathbb{R}}_+$ that is homogeneous of degree $l+k$ and is such that $V'(x)f(x)$ is continuous, negative on $\mathcal{S}\backslash f^{-1}(0)$, homogeneous of degree $l+k$, and (7.20) holds.

Now, *ad absurdum*, if $k \geq 0$ and $x \in \mathcal{S}\backslash f^{-1}(0)$, then application of the comparison lemma (Theorem 4.2 in [473]) to the first inequality in (7.20) yields $V(s(t,x)) \geq \pi(t, V(x))$, where π is given by

$$\pi(t,x) = \begin{cases} \operatorname{sgn}(x)\left(\frac{1}{|x|^{(\alpha-1)}} + c_1(\alpha-1)t\right)^{-\frac{1}{\alpha-1}}, & \alpha > 1, \\ e^{-c_1 t}x, & \alpha = 1, \end{cases} \tag{7.26}$$

and where $\operatorname{sgn}(x) \triangleq x/|x|$, $x \neq 0$, and $\operatorname{sgn}(0) \triangleq 0$, with $\alpha = 1 + k/l \geq 1$. Since, in this case, $\pi(t, V(x)) > 0$ for all $t \geq 0$, we have $s(t,x) \notin \mathcal{S} \cap f^{-1}(0)$ for every $t \geq 0$; that is, x_e is not a finite-time semistable equilibrium under f, which is a contradiction. Hence, $k < 0$.

Conversely, if $k < 0$, choose $x_\mathrm{e} \in f^{-1}(0)$ and choose a relatively open neighborhood $\mathcal{D}_0$ of x_e such that (7.21) holds. Next, $\mathcal{S}_{x_\mathrm{e}}$ is chosen to be a bounded, positively invariant neighborhood of x_e contained in $\mathcal{D}_0$. Then it follows from Theorem 7.3 that there exists a continuous nonnegative function $V(\cdot)$ such that (7.20) holds on $\mathcal{S}_{x_\mathrm{e}}$. Now, with $c = c_2 > 0$, $0 < \alpha = 1+k/l < 1$, $\mathcal{D}_0 = \mathcal{S}_{x_\mathrm{e}}$, and $w(x) = -c\operatorname{sgn}(x)|x|^\alpha$, it follows from Proposition 7.1 and Theorem 7.1 that x_e is finite-time semistable on $\mathcal{S}_{x_\mathrm{e}}$.

Define $\mathcal{S} \triangleq \bigcup_{x_\mathrm{e}\in f^{-1}(0)} \mathcal{S}_{x_\mathrm{e}}$. Then $\mathcal{S}$ is a relatively open neighborhood of $f^{-1}(0)$ such that every solution in $\mathcal{S}$ converges in finite time to a Lyapunov stable equilibrium. Hence, (7.1) is finite-time semistable. Lemma 7.3 then implies that (7.1) is globally finite-time semistable, and $T(\cdot)$ is defined on $\overline{\mathbb{R}}^n_+$. By Proposition 7.1 with $\mathcal{D}_0 = \mathcal{S}_{x_\mathrm{e}}$, and Theorem 7.1, it follows that $T(\cdot)$ is continuous on $\mathcal{S}_{x_\mathrm{e}}$. Next, since $x_\mathrm{e} \in f^{-1}(0)$ was chosen arbitrarily, it follows from Lemma 7.1 that $T(\cdot)$ is continuous on $\overline{\mathbb{R}}^n_+$.

Finally, let $x \in \overline{\mathbb{R}}^n_+$ and note that, since every point in $\nu^{-1}(0) = f^{-1}(0)$ is a globally semistable equilibrium under $-\nu$ with respect to $\overline{\mathbb{R}}^n_+$, there exists $\tau > 0$ such that $z \triangleq \psi_{-\tau}(x) \in \mathcal{S}$. Then it follows from (7.19) that $s(t,x) = s(t, \psi_\tau(z)) = \psi_\tau(s(e^{k\tau}t, z))$, and hence, $f(s(t,x)) = 0$ if and only if $f(s(e^{k\tau}t, z)) = 0$. Now, it follows that for $x \in \mathcal{S}$, $T(\psi_{-\tau}(x)) = T(z) = e^{k\tau}T(x)$. By definition, it follows that $T(\cdot)$ is homogeneous of degree $-k$ with respect to ν. $\square$

In order to use Theorem 7.4 to prove finite-time semistability of a homogeneous system, *a priori* information of semistability for the system is needed, which is not easy to obtain. To overcome this, we need to develop some sufficient conditions to establish finite-time semistability. Recall that a function $V : \overline{\mathbb{R}}_+^n \to \mathbb{R}$ is said to be *weakly proper* if and only if for every $c \in \mathbb{R}$, every connected component of the set $\{x \in \overline{\mathbb{R}}_+^n : V(x) \le c\} = V^{-1}((-\infty, c])$ is compact [42].

Proposition 7.2. Assume f is homogeneous of degree $k < 0$ with respect to ν. Furthermore, assume that there exists a weakly proper, continuous function $V : \overline{\mathbb{R}}_+^n \to \mathbb{R}$ such that $\dot{V}$ is defined on $\overline{\mathbb{R}}_+^n$ and satisfies $\dot{V}(x) \le 0$ for all $x \in \overline{\mathbb{R}}_+^n$. If every point in the largest invariant subset $\mathcal{N}$ of $\dot{V}^{-1}(0)$ is a Lyapunov stable equilibrium point of (7.1), then (7.1) is finite-time semistable.

Proof. Since $V(\cdot)$ is weakly proper, it follows from Proposition 3.1 of [42] that the positive orbit $s^x([0,\infty))$ of $x \in \overline{\mathbb{R}}_+^n$ is bounded in $\overline{\mathbb{R}}_+^n$. Since every solution is bounded, it follows from the hypotheses on $V(\cdot)$ that for every $x \in \overline{\mathbb{R}}_+^n$, the omega limit set $\omega(x)$ is nonempty and contained in the largest invariant subset $\mathcal{N}$ of $\dot{V}^{-1}(0)$. Since every point in $\mathcal{N}$ is a Lyapunov stable equilibrium point, it follows from Proposition 2.2 that the omega limit set $\omega(x)$ contains a single point for every $x \in \overline{\mathbb{R}}_+^n$. And since $\lim_{t\to\infty} s(t,x) \in \mathcal{N}$ is Lyapunov stable for every $x \in \overline{\mathbb{R}}_+^n$, by definition, the system (7.1) is semistable. Hence, it follows from Theorem 7.4 that (7.1) is finite-time semistable. □

Example 7.1. Consider the nonlinear compartmental dynamical system given by

$$\dot{x}_1(t) = (x_2(t) - x_1(t))^{\frac{1}{3}} + (x_3(t) - x_1(t))^{\frac{1}{3}}, \quad x_1(0) = x_{10}, \quad t \ge 0, \tag{7.27}$$

$$\dot{x}_2(t) = (x_1(t) - x_2(t))^{\frac{1}{3}} + (x_3(t) - x_2(t))^{\frac{1}{3}}, \quad x_2(0) = x_{20}, \tag{7.28}$$

$$\dot{x}_3(t) = (x_1(t) - x_3(t))^{\frac{1}{3}} + (x_2(t) - x_3(t))^{\frac{1}{3}}, \quad x_3(0) = x_{30}, \tag{7.29}$$

where $x_i \in \overline{\mathbb{R}}_+$, $i = 1,2,3$. For each $a \in \overline{\mathbb{R}}_+$, $x_1 = x_2 = x_3 = a$ is the equilibrium point of (7.27)–(7.29). We show that all the equilibrium points of (7.27)–(7.29) are finite-time semistable.

Note that the vector field f of (7.27)–(7.29) is homogeneous of degree -2 with respect to the semi-Euler vector field

$$\nu(x) = (2x_1 - x_2 - x_3)\frac{\partial}{\partial x_1} + (2x_2 - x_1 - x_3)\frac{\partial}{\partial x_2} + (2x_3 - x_1 - x_2)\frac{\partial}{\partial x_3}.$$

Furthermore, $f(\cdot)$ is essentially nonnegative. The differential operator notation in $\nu(x)$ is standard differential geometric notation used to write coordinate expressions for vector fields. This notation is based on the fact that there is a one-to-one correspondence between first-order linear differential operators on real-valued functions and vector fields.

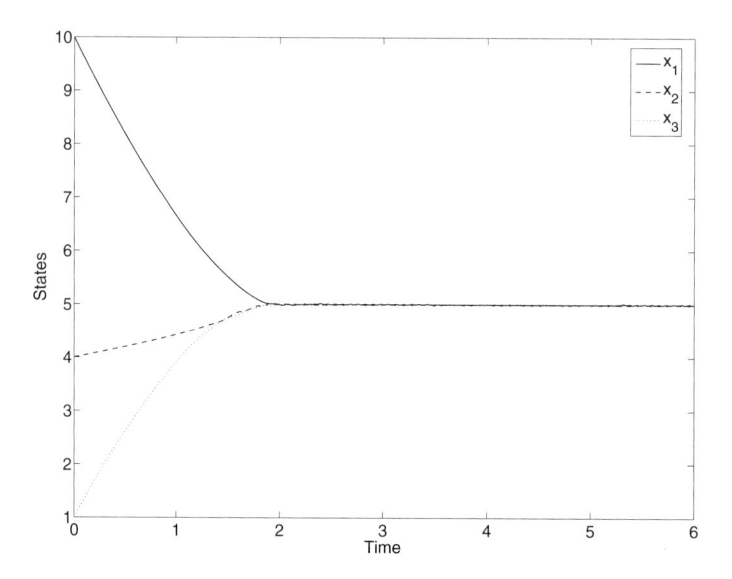

Figure 7.1 State trajectories versus time for Example 7.1.

Next, consider $V(x) = \frac{1}{2}x_1^2 + \frac{1}{2}x_2^2 + \frac{1}{2}x_3^2$. Then $\dot{V}(x(t)) \leq 0$, $t \geq 0$, and $\mathcal{N} = \{x \in \overline{\mathbb{R}}_+^4 : x_1 = x_2 = x_3 = a\}$. Now, it follows from the Lyapunov function candidate $V(x - a\mathbf{e}) = \frac{1}{2}(x_1 - a)^2 + \frac{1}{2}(x_2 - a)^2 + \frac{1}{2}(x_3 - a)^2$ that $\dot{V}(x - a\mathbf{e}) = -(x_1 - x_2)^{\frac{4}{3}} - (x_2 - x_3)^{\frac{4}{3}} - (x_3 - x_1)^{\frac{4}{3}} \leq 0$, which implies that every point in $\mathcal{N}$ is a Lyapunov stable equilibrium point of (7.27)–(7.29). Hence, it follows from Proposition 7.2 that the system (7.27)–(7.29) is finite-time semistable. In fact, $x_1(t) = x_2(t) = x_3(t) = \frac{1}{3}(x_{10} + x_{20} + x_{30})$ for $t \geq T(x_0)$. Figure 7.1 shows the state trajectories versus time. △

7.4 Finite-Time Energy Equipartition in Thermodynamic Systems

In this section, we develop intercompartmental flow laws that guarantee finite-time semistability and energy equipartition for the thermodynamically consistent dynamical system model developed in Chapter 3. Specifically, consider the dynamical system $\mathcal{G}$ given by

$$\dot{E}_i(t) = \sum_{j=1,\, j\neq i}^{q} \phi_{ij}(E_i(t), E_j(t)), \quad E_i(t_0) = E_{i0}, \quad t \geq t_0, \quad i = 1, \ldots, q, \tag{7.30}$$

where $\phi_{ij}(E)$, $E \in \overline{\mathbb{R}}_+^q$, denotes the net energy flow from the jth compartment to the ith compartment defined in Chapter 3. In vector form, (7.30) becomes

$$\dot{E}(t) = f(E(t)), \quad E(t_0) = E_0, \quad t \geq t_0, \tag{7.31}$$

where $E(t) \triangleq [E_1(t), \ldots, E_q(t)]^{\mathrm{T}} \in \overline{\mathbb{R}}_+^q$, $t \geq t_0$, and $f = [f_1, \ldots, f_q]^{\mathrm{T}} : \overline{\mathbb{R}}_+^q \to \mathbb{R}^q$ is such that $f_i(E) = \sum_{j=1,\, j\neq i}^{q} \phi_{ij}(E_i, E_j)$.

Theorem 7.5. Consider the dynamical system (7.31) and assume that Axioms i) and ii) of Chapter 3 hold. Furthermore, assume that $\phi_{ij}(E_i, E_j) = -\phi_{ji}(E_j, E_i)$ for all $i, j = 1, \ldots, q$, $i \neq j$. Then, for every $\alpha \in \overline{\mathbb{R}}_+$, $\alpha\mathbf{e}$ is a semistable equilibrium state of (7.31). Furthermore, $E(t) \to \frac{1}{q}\mathbf{e}\mathbf{e}^{\mathrm{T}}E(t_0)$ as $t \to \infty$ and $\frac{1}{q}\mathbf{e}\mathbf{e}^{\mathrm{T}}E(t_0)$ is a semistable equilibrium state.

Proof. The result is a direct consequence of Proposition 3.10 and Theorem 3.9. □

Theorem 7.5 implies that the steady-state values of the state in each compartment $\mathcal{G}_i$ of the dynamical system $\mathcal{G}$ are equal, that is, the steady-state value of the dynamical system $\mathcal{G}$ given by

$$E_\infty = \frac{1}{q}\mathbf{e}\mathbf{e}^{\mathrm{T}}E(t_0) = \left[\frac{1}{q}\sum_{i=1}^{q} E_i(t_0)\right]\mathbf{e}$$

is uniformly distributed over all compartments of $\mathcal{G}$.

Next, we use the results of Section 7.3 to develop a compartmental model for finite-time thermodynamics. Specifically, consider the dynamical system given by

$$\dot{E}_i(t) = \sum_{j=1, j\neq i}^{q} \phi_{ij}(E_i(t), E_j(t)), \quad E_i(0) = E_{i0}, \quad t \geq 0, \tag{7.32}$$

where, for each $i \in \{1, \ldots, q\}$, $E_i(t) \in \overline{\mathbb{R}}_+$ denotes an energy state for all $t \geq 0$, $\phi_{ij}(\cdot, \cdot)$ satisfies Axioms i) and ii) of Chapter 3, and $\phi_{ij}(E_i, E_j) = -\phi_{ji}(E_j, E_i)$ for all $i, j = 1, \ldots, q$, $i \neq j$.

Theorem 7.6. Consider the dynamical system $\mathcal{G}$ given by (7.32). Assume that Axioms i) and ii) of Chapter 3 hold, and $\phi_{ij}(E_i, E_j) = -\phi_{ji}(E_j, E_i)$ for all $i, j = 1, \ldots, q$, $i \neq j$. Furthermore, assume that the vector field f of the dynamical system (7.32) is homogeneous of degree $k \in \mathbb{R}$ with respect to

$$\nu(E) = -\sum_{i=1}^{q}\left[\sum_{j=1,j\neq i}^{q} \mu_{ij}(E_i, E_j)\right]\frac{\partial}{\partial E_i},$$

where $E \triangleq [E_1, \ldots, E_q]^{\mathrm{T}} \in \overline{\mathbb{R}}_+^q$ and $\mu_{ij}(\cdot, \cdot)$ satisfies Axiom ii), $\mu_{ij}(E_i, E_j) = -\mu_{ji}(E_j, E_i)$, and $\mu_{ij}(E_i, E_j) = 0$ if and only if $E_i = E_j$ for all $i, j = 1, \ldots, q$, $i \neq j$. Then, for every $E_{\mathrm{e}} \in \overline{\mathbb{R}}_+$, $E_{\mathrm{e}}\mathbf{e}$ is a finite-time semistable equilibrium state of $\mathcal{G}$ if and only if $k < 0$. Furthermore, if $k < 0$, then $E(t) = \frac{1}{q}\mathbf{e}\mathbf{e}^{\mathrm{T}}E(0)$

for all $t \geq T(E(0))$ and $\frac{1}{q}\mathbf{e}\mathbf{e}^{\mathrm{T}}E(0)$ is a finite-time semistable equilibrium state, where $T(E(0)) \geq 0$.

Proof. Suppose $k < 0$. It follows from Theorem 7.5 that $E_{\mathrm{e}}\mathbf{e} \in \overline{\mathbb{R}}_+^q$, $E_{\mathrm{e}} \in \overline{\mathbb{R}}_+$, is a semistable equilibrium state of the homogeneous system (7.32). Furthermore, $E(t) \to \frac{1}{q}\mathbf{e}\mathbf{e}^{\mathrm{T}}E(0)$ as $t \to \infty$ and $\frac{1}{q}\mathbf{e}\mathbf{e}^{\mathrm{T}}E(0)$ is a semistable equilibrium state. Next, it can be shown using similar arguments as in the proof of Theorem 7.5 that (7.18) is globally semistable with $\nu(E) = -\sum_{i=1}^{q}\left[\sum_{j=1,j\neq i}^{q}\mu_{ij}(E_i,E_j)\right]\frac{\partial}{\partial E_i}$. Now, it follows from Theorem 7.4 that $E_{\mathrm{e}}\mathbf{e}$ is a finite-time semistable equilibrium state by noting that the vector field $\sum_{j=1,j\neq i}^{q}\phi_{ij}(E_i,E_j)$ is homogeneous of degree $k < 0$ with respect to the semi-Euler vector field $\nu(E) = -\sum_{i=1}^{q}\left[\sum_{j=1,j\neq i}^{q}\mu_{ij}(E_i,E_j)\right]\frac{\partial}{\partial E_i}$. Hence, with $E_{\mathrm{e}} = \frac{1}{q}\mathbf{e}^{\mathrm{T}}E(0)$, $E_{\mathrm{e}}\mathbf{e} = \frac{1}{q}\mathbf{e}\mathbf{e}^{\mathrm{T}}E(0)$ is a finite-time semistable equilibrium state. The converse follows as a direct consequence of Theorem 7.4. □

The following corollary to Theorem 7.6 gives a concrete form for the energy flow function $\phi_{ij}(E_i,E_j)$, $i,j = 1,\ldots,q$, $i \neq j$.

Corollary 7.1. Consider the dynamical system $\mathcal{G}$ given by (7.32) with energy flow function

$$\phi_{ij}(E_i,E_j) = \mathcal{C}_{(i,j)}\,\mathrm{sgn}(E_j - E_i)|E_j - E_i|^{\alpha}, \tag{7.33}$$

where $\alpha > 0$ and $\mathcal{C}_{(i,j)}$ is as in (3.24) with $\mathcal{C} = \mathcal{C}^{\mathrm{T}}$. Assume that Axioms *i)* and *ii)* of Chapter 3 hold. Then, for every $E_{\mathrm{e}} \in \overline{\mathbb{R}}_+$, $E_{\mathrm{e}}\mathbf{e}$ is a finite-time semistable equilibrium state of $\mathcal{G}$ if and only if $\alpha < 1$. Furthermore, if $\alpha < 1$, then $E(t) = \frac{1}{q}\mathbf{e}\mathbf{e}^{\mathrm{T}}E(0)$ for all $t \geq T(E(0))$ and $\frac{1}{q}\mathbf{e}\mathbf{e}^{\mathrm{T}}E(0)$ is a finite-time semistable equilibrium state, where $T(E(0)) \geq 0$.

Proof. First, note that the vector field f of $\mathcal{G}$ is essentially nonnegative. Next, the Lie bracket of $\nu(E) = -\sum_{i=1}^{q}\left[\sum_{j=1,j\neq i}^{q}(E_j - E_i)\right]\frac{\partial}{\partial E_i}$ and the vector field f of the dynamical system (7.32) with (7.33) is given by $[\nu, f] = \left[\sum_{i=1}^{q}\frac{\partial f_1}{\partial E_i}\nu_i - \frac{\partial \nu_1}{\partial E_i}f_i, \ldots, \sum_{i=1}^{q}\frac{\partial f_q}{\partial E_i}\nu_i - \frac{\partial \nu_q}{\partial E_i}f_i\right]^{\mathrm{T}}$. Since for each $i,j = 1,\ldots,q$,

$$\frac{\partial f_j}{\partial E_i}\nu_i - \frac{\partial \nu_j}{\partial E_i}f_i$$
$$= \begin{cases} \mathcal{C}_{(j,i)}\alpha|E_i - E_j|^{\alpha-1}\left[\sum_{s=1,s\neq i}^{q}(E_i - E_s)\right] \\ \quad +\sum_{k=1,k\neq i}^{q}\mathcal{C}_{(i,k)}\mathrm{sgn}(E_k - E_i)|E_k - E_i|^{\alpha}, & i \neq j, \\ \left[\sum_{k=1,k\neq j}^{q}\mathcal{C}_{(j,k)}\alpha|E_k - E_j|^{\alpha-1}\right]\left[\sum_{s=1,s\neq j}^{q}(E_s - E_j)\right] \\ \quad -(q-1)\sum_{k=1,k\neq j}^{q}\mathcal{C}_{(j,k)}\mathrm{sgn}(E_k - E_j)|E_k - E_j|^{\alpha}, & i = j, \end{cases}$$

and noting that $\mathcal{C}_{(i,j)} = \mathcal{C}_{(j,i)}$, $i, j = 1, \ldots, q$, $i \neq j$, it follows that for each $j = 1, \ldots, q$,

$$\begin{aligned}
\sum_{i=1}^{q} \frac{\partial f_j}{\partial E_i}\nu_i - \frac{\partial \nu_j}{\partial E_i} f_i &= \frac{\partial f_j}{\partial E_j}\nu_j - \frac{\partial \nu_j}{\partial E_j} f_j + \sum_{i=1, i\neq j}^{q} \frac{\partial f_j}{\partial E_i}\nu_i - \frac{\partial \nu_j}{\partial E_i} f_i \\
&= \left[\sum_{k=1, k\neq j}^{q} \mathcal{C}_{(j,k)}\alpha|E_k - E_j|^{\alpha-1}\right]\left[\sum_{s=1, s\neq j}^{q} (E_s - E_j)\right] \\
&\quad -(q-1)\sum_{k=1, k\neq j}^{q} \mathcal{C}_{(j,k)}\operatorname{sgn}(E_k - E_j)|E_k - E_j|^{\alpha} \\
&\quad + \sum_{i=1, i\neq j}^{q} \mathcal{C}_{(j,i)}\alpha|E_i - E_j|^{\alpha-1}\left[\sum_{s=1, s\neq i}^{q} (E_i - E_s)\right] \\
&\quad + \sum_{i=1, i\neq j}^{q} \sum_{k=1, k\neq i}^{q} \mathcal{C}_{(i,k)}\operatorname{sgn}(E_k - E_i)|E_k - E_i|^{\alpha} \\
&= \alpha \sum_{k=1, k\neq j}^{q} \mathcal{C}_{(j,k)}\operatorname{sgn}(E_k - E_j)|E_k - E_j|^{\alpha} \\
&\quad + \sum_{k=1, k\neq j}^{q} \sum_{s=1, s\neq j,k}^{q} \mathcal{C}_{(j,k)}\alpha|E_k - E_j|^{\alpha-1}(E_s - E_j) \\
&\quad -(q-1)\sum_{k=1, k\neq j}^{q} \mathcal{C}_{(j,k)}\operatorname{sgn}(E_k - E_j)|E_k - E_j|^{\alpha} \\
&\quad +\alpha \sum_{i=1, i\neq j}^{q} \mathcal{C}_{(j,i)}\operatorname{sgn}(E_i - E_j)|E_i - E_j|^{\alpha} \\
&\quad + \sum_{i=1, i\neq j}^{q} \sum_{s=1, s\neq i,j}^{q} \mathcal{C}_{(j,i)}\alpha|E_i - E_j|^{\alpha-1}(E_i - E_s) \\
&\quad + \sum_{i=1}^{q} \sum_{k=1, k\neq i}^{q} \mathcal{C}_{(i,k)}\operatorname{sgn}(E_k - E_i)|E_k - E_i|^{\alpha} \\
&\quad - \sum_{k=1, k\neq j}^{q} \mathcal{C}_{(j,k)}\operatorname{sgn}(E_k - E_j)|E_k - E_j|^{\alpha} \\
&= 2\alpha \sum_{i=1, i\neq j}^{q} \mathcal{C}_{(j,i)}\operatorname{sgn}(E_i - E_j)|E_i - E_j|^{\alpha}
\end{aligned}$$

$$
\begin{aligned}
&+\alpha \sum_{i=1, i\neq j}^{q} \sum_{s=1, s\neq i,j}^{q} \mathcal{C}_{(j,i)}\mathrm{sgn}(E_i - E_j)|E_i - E_j|^{\alpha} \\
&-q \sum_{k=1, k\neq j}^{q} \mathcal{C}_{(j,k)}\mathrm{sgn}(E_k - E_j)|E_k - E_j|^{\alpha} \\
&= q(\alpha-1) \sum_{i=1, i\neq j}^{q} \mathcal{C}_{(j,i)}\mathrm{sgn}(E_i - E_j)|E_i - E_j|^{\alpha} \\
&= q(\alpha-1) f_j, \qquad (7.34)
\end{aligned}
$$

which implies that the vector field f is homogeneous of degree $k = q(\alpha - 1)$ with respect to the semi-Euler vector field

$$
\nu(E) = -\sum_{i=1}^{q}\left[\sum_{j=1, j\neq i}^{q} (E_j - E_i)\right]\frac{\partial}{\partial E_i}.
$$

Now, the result is a direct consequence of Theorem 7.6. □

Note that Example 7.1 serves as a special case of Corollary 7.1.

Chapter Eight

Critical Phenomena and Continuous Phase Transitions

8.1 Introduction

In the previous chapters, an implicit assumption of our large-scale dynamical system model is that the thermodynamic state variables define a continuously differentiable flow on the nonnegative orthant of the state space. For systems that possess phase transitions and critical states, this assumption is clearly limiting. A *phase transition* is a phenomenon wherein an abrupt change between phases (i.e., a change between one state of matter into another) occurs as the system parameters (e.g., temperature, pressure, chemical composition, electric field, magnetic field) are varied. Phase transitions are not limited to thermodynamic systems, where temperature or pressure are the phase transition driving parameters. They include, for example, quantum, dynamic (i.e., bifurcation), and topological (i.e., structural) phase transitions.

Phase transitions are ubiquitous in nature, with the different phases of water (i.e., vapor, liquid, and ice) perhaps being the most familiar, and with transitions involving change from one phase to the other. Other phase transitions include eutectic transformations, peritectic transformations, spinodal decompositions, mesophase transitions, ferromagnetic-paramagnetic phase transitions of magnetic materials, superconductivity, molecular structure (e.g., polymorph, allotrope, polyamorph) transitions, and quantum condensation.

In the early universe (i.e., $\sim 10^{-35}$ seconds after the burst and through inflationary cosmology and the initial cosmic horizon), symmetry-breaking transitions in the laws of physics due to ravaging pressures and temperatures prevented the permanent formation of elementary particles (e.g., bosons, quarks, leptons, antiquarks, and antileptons). Elementary and composite particles were unable to form stable constituents until the universe cooled beyond the supergravity phase, that is, gravity as predicted by supersymmetric quantum field theory.

If the phase transition driving parameter is temperature, then the higher temperature phase almost always corresponds to a more disordered state, that is, it has a *higher symmetry* than the low temperature phase, and hence, phase transitions involve a change in system entropy. For example, a gas has a higher symmetry than a liquid, which in turn has a higher symmetry than a solid [298, pp. 505–508]. And since the maximum entropy of any dynamical system corresponds to state indistinguishability (i.e., total loss of information storage), highest symmetry, and highest simplicity (i.e., decomplexification), entropy increases during transitions from a solid state to a liquid state to a gas state with increasing temperature, and hence, order corresponds to asymmetry, not symmetry[1] [281]. This change in entropy, as well as other thermodynamic quantities (e.g., energy, enthalpy, volume), can be continuous or discontinuous. A continuous phase transition is manifest by continuous thermodynamic states with discontinuous first-order derivatives of these states, whereas a discontinuous phase transition is manifest by discontinuous thermodynamic states.

The field of *condensed matter physics* is concerned with identifying and classifying these different phases of matter. For example, solid-liquid and liquid-gas transitions at temperatures below the critical state temperature correspond to discontinuous phase transitions, whereas a liquid-gas transition at a temperature above the critical state temperature corresponds to a continuous phase transition. For an excellent exposition of equilibrium and nonequilibrium critical phenomena and phase transitions see [215].

In classical thermodynamics, these phase transition states are known as *critical states* and the theory of *critical phenomena* addresses this anomalous behavior. In nature, however, thermodynamic critical phenomena occurring at thermal equilibria is the exception and not the rule. In the majority of cases, the system initial state is far from a thermal equilibrium, and hence, specific dynamical properties cannot be described using classical thermodynamics. In this chapter, we extend our dynamical systems framework to address thermodynamic critical phenomena with continuous phase transitions; Chapter 10 considers the extension to discontinuous phase transitions.

In the case where the vector field defining the dynamical system is a discontinuous function of the state, system stability can be analyzed using nonsmooth Lyapunov theory involving concepts such as weak and

[1] In other words, every isolated dynamical system spontaneously evolves toward a state of maximum entropy, and hence, maximal symmetry, wherein higher entropy (i.e., lower state information content) is connected to higher symmetry.

strong stability notions, differential inclusions, and generalized gradients of locally Lipschitz continuous functions and proximal subdifferentials of lower semicontinuous functions [93]. The consideration of nonsmooth Lyapunov functions for proving stability of discontinuous systems is an important extension to classical stability theory since, as shown in [408], there exist nonsmooth dynamical systems whose equilibria cannot be proved to be stable using standard continuously differentiable Lyapunov function theory.

To analyze thermodynamic critical phenomena with continuous phase transitions giving rise to discontinuous dynamical systems having a continuum of equilibria, in this chapter we extend the theory of semistability to discontinuous dynamical systems. In particular, we develop sufficient conditions to guarantee weak and strong invariance of Filippov solutions. Moreover, we present Lyapunov-based (i.e., ectropy-based) tests for semistability of thermodynamic systems with discontinuous power balance dynamics using differential inclusions. In addition, we develop sufficient conditions for finite-time semistability for these thermodynamic systems.

8.2 Dynamical Systems with Discontinuous Vector Fields

In this chapter, we consider nonlinear dynamical systems[2] $\mathcal{G}$ of the form

$$\dot{x}(t) = f(x(t)), \quad x(t_0) = x_0, \quad \text{a.e.} \quad t \geq t_0, \tag{8.1}$$

where, for every $t \geq t_0$, $x(t) \in \mathcal{D} \subseteq \mathbb{R}^q$, $f : \mathcal{D} \to \mathbb{R}^q$ is Lebesgue measurable[3] and locally essentially bounded [149] with respect to x, that is, f is bounded on a bounded neighborhood of every point x, excluding sets of measure zero, and admits an equilibrium point at $x_\mathrm{e} \in \mathcal{D}$; that is, $f(x_\mathrm{e}) = 0$.

An *absolutely continuous* function[4] $x : [t_0, \tau] \to \mathbb{R}^q$ is said to be a *Filippov solution* [149] of (8.1) on the interval $[t_0, \tau]$ with initial condition $x(t_0) = x_0$, if $x(t)$ satisfies

$$\dot{x}(t) \in \mathcal{K}[f](x(t)), \quad \text{a.e.} \quad t \in [t_0, \tau], \tag{8.2}$$

[2]In this and the next section, we consider general discontinuous dynamical systems on $\mathcal{D} \subseteq \mathbb{R}^q$. All results hold, mutatis mutandis, for nonnegative discontinuous dynamical systems. For details see [299, 307].

[3]A function $f : \mathcal{D} \to \mathbb{R}^q$ is *Lebesgue measurable* if the inverse image of f on every open (or Borel) set is Lebesgue measurable.

[4]A function $x : [t_0, t_1] \to \mathbb{R}$ is absolutely continuous on $[t_0, t_1]$ if and only if for every $\varepsilon > 0$ there exist $\delta > 0$ such that, for each finite collection $\{(a_1, b_1), \ldots, (a_n, b_n)\}$ of disjoint open intervals with $\sum_{i=1}^n (b_i - a_i) < \delta$, $\sum_{i=1}^n |x(b_i) - x(a_i)| < \varepsilon$. Equivalently, $x(\cdot)$ is absolutely continuous if and only if there exists a Lebesgue integrable function $\kappa : [t_0, t_1] \to \mathbb{R}$ such that $x(t) = x(t_0) + \int_{t_0}^{t_1} \kappa(s)\mathrm{d}s$, $t \in [t_0, t_1]$. Note that every absolutely continuous function is continuous; the converse, however, is not true.

where the *Filippov set-valued map* $\mathcal{K}[f] : \mathbb{R}^n \to 2^{\mathbb{R}^q}$ is defined by

$$\mathcal{K}[f](x) \triangleq \bigcap_{\delta>0} \bigcap_{\mu(\mathcal{Q})=0} \overline{\text{co}}\{f(\mathcal{B}_\delta(x)\backslash \mathcal{Q}\}, \quad x \in \mathbb{R}^n, \tag{8.3}$$

$2^{\mathbb{R}^q}$ denotes the collection of all subsets of $\mathbb{R}^q$, $\mu(\cdot)$ denotes the Lebesgue measure in $\mathbb{R}^n$, "$\overline{\text{co}}$" denotes convex closure, and $\bigcap_{\mu(\mathcal{Q})=0}$ denotes the intersection over all sets $\mathcal{Q}$ of Lebesgue measure zero.[5]

Note that since f is locally essentially bounded, $\mathcal{K}[f](\cdot)$ is upper semicontinuous and has nonempty, compact, and convex values. Thus, Filippov solutions are limits of solutions to $\mathcal{G}$ with f averaged over progressively smaller neighborhoods around the solution point, and hence, allow solutions to be defined at points where f itself is not defined. Hence, the tangent vector to a Filippov solution, when it exists, lies in the convex closure of the limiting values of the system vector field $f(\cdot)$ in progressively smaller neighborhoods around the solution point. Dynamical systems of the form given by (8.1) are called *differential inclusions* in the literature [13], and, for every state $x \in \mathbb{R}^q$, they specify a *set* of possible evolutions of $\mathcal{G}$ rather than a single one.

Since the Filippov set-valued map given by (8.3) is upper semicontinuous with nonempty, convex, and compact values, and $\mathcal{K}[f](\cdot)$ is also locally bounded [149, p. 85], it follows that Filippov solutions to (8.1) exist [149, Thm. 1, p. 77]. Recall that the Filippov solution $t \mapsto x(t)$ to (8.1) is a *right maximal solution* if it cannot be extended (either uniquely or nonuniquely) forward in time. We assume that all right maximal Filippov solutions to (8.1) exist on $[t_0, \infty)$, and hence, we assume that (8.1) is forward complete.

Recall that (8.1) is forward complete if and only if the Filippov solutions to (8.1) are uniformly globally sliding time stable [435, Lem 1, p. 182]. An *equilibrium point* of (8.1) is a point $x_e \in \mathbb{R}^q$ such that $0 \in \mathcal{K}[f](x_e)$. It is easy to see that x_e is an equilibrium point of (8.1) if and only if the constant function $x(\cdot) = x_e$ is a Filippov solution of (8.1). We denote the set of equilibrium points of (8.1) by $\mathcal{E}$. Since the set-valued map $\mathcal{K}[f](\cdot)$ is upper semicontinuous, it follows that $\mathcal{E}$ is closed.

To develop stability properties for discontinuous dynamical systems given by (8.1), we need to introduce the notion of generalized derivatives and gradients. Here we focus on Clarke generalized derivatives and gradients [86].

[5]Alternatively, we can consider Krasovskii solutions of (8.1) wherein the possible misbehavior of the derivative of the state on null measure sets is not ignored; that is, $\mathcal{K}[f](x)$ is replaced with $\mathcal{K}[f](x) = \bigcap_{\delta>0} \overline{\text{co}}\{f(\mathcal{B}_\delta(x))\}$ and where f is assumed to be locally bounded.

Definition 8.1 ([86], [15]). Let $V : \mathbb{R}^q \to \mathbb{R}$ be a locally Lipschitz continuous function. The *Clarke upper generalized derivative* of $V(\cdot)$ at x in the direction of $v \in \mathbb{R}^q$ is defined by

$$V^{o}(x, v) \triangleq \limsup_{y \to x, h \to 0^+} \frac{V(y + hv) - V(y)}{h}. \tag{8.4}$$

The *Clarke generalized gradient* $\partial V : \mathbb{R}^q \to 2^{\mathbb{R}^{1\times n}}$ of $V(\cdot)$ at x is the set

$$\partial V(x) \triangleq \mathrm{co}\left\{ \lim_{i\to\infty} \nabla V(x_i) : x_i \to x,\, x_i \notin \mathcal{N} \cup \mathcal{Q} \right\}, \tag{8.5}$$

where co denotes the convex hull, ∇ denotes the nabla operator, $\mathcal{N}$ is the set of measure zero of points where ∇V does not exist, $\mathcal{Q}$ is any subset of $\mathbb{R}^n$ of measure zero, and the increasing unbounded sequence $\{x_i\}_{i\in\overline{\mathbb{Z}}_+} \subset \mathbb{R}^n$ converges to $x \in \mathbb{R}^n$.

Note that (8.4) always exists. Furthermore, note that it follows from Definition 8.1 that the generalized gradient of V at x consists of all convex combinations of all the possible limits of the gradient at neighboring points where V is differentiable. In addition, note that since $V(\cdot)$ is Lipschitz continuous, it follows from Rademacher's theorem [138, Thm 6, p. 281] that the gradient $\nabla V(\cdot)$ of $V(\cdot)$ exists almost everywhere, and hence, $\nabla V(\cdot)$ is bounded. Specifically, for every $x \in \mathbb{R}^n$, every $\varepsilon > 0$, and every Lipschitz constant L for V on $\overline{\mathcal{B}}_\varepsilon(x)$, $\partial V(x) \subseteq \overline{\mathcal{B}}_L(0)$. Thus, since for every $x \in \mathbb{R}^n$, $\partial V(x)$ is convex, closed, and bounded, it follows that $\partial V(x)$ is compact.

In order to state the main results of this chapter, we need some additional notation and definitions. Given a locally Lipschitz continuous function $V : \mathbb{R}^q \to \mathbb{R}$, the *set-valued Lie derivative* $\mathcal{L}_f V : \mathbb{R}^q \to 2^{\mathbb{R}}$ of V with respect to f at x [15, 94] is defined as

$$\begin{aligned} \mathcal{L}_f V(x) &\triangleq \Big\{ a \in \mathbb{R} : \text{there exists } v \in \mathcal{K}[f](x) \text{ such that} \\ &\qquad\qquad p^{\mathrm{T}} v = a \text{ for all } p^{\mathrm{T}} \in \partial V(x) \Big\} \\ &\subseteq \bigcap_{p^{\mathrm{T}} \in \partial V(x)} p^{\mathrm{T}} \mathcal{K}[f](x). \end{aligned} \tag{8.6}$$

If $\mathcal{K}[f](x)$ is convex with compact values, then $\mathcal{L}_f V(x)$, $x \in \mathbb{R}^q$, is a closed and bounded, possibly empty, interval in $\mathbb{R}$. If $V(\cdot)$ is continuously differentiable at x, then $\mathcal{L}_f V(x) = \{\nabla V(x) \cdot v : v \in \mathcal{K}[f](x)\}$. In the case where $\mathcal{L}_f V(x)$ is nonempty, we use the notion $\max \mathcal{L}_f V(x)$ (respectively, $\min \mathcal{L}_f V(x)$) to denote the largest (respectively, smallest) element of $\mathcal{L}_f V(x)$. Furthermore, we adopt the convention $\max \varnothing = -\infty$. Finally, recall that a function $V : \mathbb{R}^q \to \mathbb{R}$ is *regular* at $x \in \mathbb{R}^q$ [86, Def. 2.3.4] if, for all $v \in \mathbb{R}^q$, the right directional derivative $V'_+(x, v) \triangleq$

$\lim_{h\to 0^+} \frac{1}{h}[V(x+hv) - V(x)]$ exists and $V'_+(x,v) = V^{\rm o}(x,v)$. V is called *regular* on $\mathbb{R}^q$ if it is regular at every $x \in \mathbb{R}^q$.

8.3 Nonsmooth Stability Theory for Discontinuous Dynamical Systems

In this section, we study the stability of discontinuous systems. For stating the main stability theorems we assume that all right maximal Filippov solutions to (8.1) exist on $[0,\infty)$. We say that a set $\mathcal{M}$ is *weakly positively invariant* (respectively, *strongly positively invariant*) with respect to (8.1) if, for every $x_0 \in \mathcal{M}$, $\mathcal{M}$ contains a right maximal solution (respectively, all right maximal solutions) of (8.1) [15, 390]. The set $\mathcal{M} \subseteq \mathbb{R}^q$ is *weakly negatively invariant* if, for every $x \in \mathcal{N}$ and $t \geq 0$, there exist $z \in \mathcal{N}$ and a Filippov solution $\psi(\cdot)$ to (8.1) with $\psi(0) = z$ such that $\psi(t) = x$ and $\psi(\tau) \in \mathcal{N}$ for all $\tau \in [0,t]$. Finally, the set $\mathcal{M} \subseteq \mathbb{R}^q$ is *weakly invariant* if $\mathcal{M}$ is weakly positively invariant as well as weakly negatively invariant.

The next definition introduces the notion of Lyapunov stability, semistability, and asymptotic stability for discontinuous dynamical systems. The adjective "weak" is used in reference to a stability property when the stability property is satisfied by at least one Filippov solution starting from every initial condition in $\mathcal{D}$, whereas "strong" is used when the stability property is satisfied by all Filippov solutions starting from every initial condition in $\mathcal{D}$. In this section, however, we provide strong stability theorems for (8.1), and hence, we omit the adjective "strong" in the statement of our results.

Definition 8.2. Let $\mathcal{D} \subseteq \mathbb{R}^q$ be an open strongly positively invariant set with respect to (8.1). An equilibrium point $x_{\rm e} \in \mathcal{D}$ of (8.1) is *Lyapunov stable* if, for every $\varepsilon > 0$, there exists $\delta = \delta(\varepsilon) > 0$ such that, for every initial condition $x_0 \in \mathcal{B}_\delta(x_{\rm e})$ and every Filippov solution $x(t)$ with the initial condition $x(0) = x_0$, $x(t) \in \mathcal{B}_\varepsilon(x_{\rm e})$ for all $t \geq 0$. An equilibrium point $x_{\rm e} \in \mathcal{D}$ of (8.1) is *semistable* if $x_{\rm e}$ is Lyapunov stable and there exists an open subset $\mathcal{D}_0$ of $\mathcal{D}$ containing $x_{\rm e}$ such that, for all initial conditions in $\mathcal{D}_0$, the Filippov solutions of (8.1) converge to a Lyapunov stable equilibrium point. An equilibrium point $x_{\rm e} \in \mathcal{D}$ of (8.1) is *asymptotically stable* if $x_{\rm e}$ is Lyapunvov stable and there exists $\delta = \delta(\varepsilon) > 0$ such that if $x_0 \in \mathcal{B}_\delta(x_{\rm e})$, then the Filippov solutions of (8.1) converge to $x_{\rm e}$. The system (8.1) is *semistable* (respectively, *asymptotically stable*) *with respect to* $\mathcal{D}$ if every Filippov solution with initial condition in $\mathcal{D}$ converges to a Lyapunov stable equilibrium (respectively, the Lyapunov stable equilibrium $x_{\rm e}$). Finally, (8.1) is said to be *globally semistable* (respectively, *globally asymptotically stable*, *globally exponentially stable*) if (8.1) is semistable (respectively, asymptotically stable) with respect to $\mathbb{R}^q$.

Next, we introduce the definition of finite-time semistability and finite-time stability of (8.1).

Definition 8.3. Let $\mathcal{D} \subseteq \mathbb{R}^q$ be an open strongly positively invariant set with respect to (8.1). An equilibrium point $x_{\rm e} \in \mathcal{E}$ of (8.1) is said to be *finite-time semistable* (respectively, *finite-time stable*) if there exist an open neighborhood $\mathcal{U} \subseteq \mathcal{D}$ of $x_{\rm e}$ and a function $T : \mathcal{U}\backslash\mathcal{E} \to (0, \infty)$, called the *settling-time function*, such that the following statements hold:

i) For every $x \in \mathcal{U}\backslash\mathcal{E}$ and every Filippov solution $\psi(t)$ of (8.1) with $\psi(0) = x$, $\psi(t) \in \mathcal{U}\backslash\mathcal{E}$ for all $t \in [0, T(x))$, and $\lim_{t \to T(x)} \psi(t)$ exists (respectively, $\lim_{t \to T(x)} \psi(t) = x_{\rm e}$) and is contained in $\mathcal{U} \cap \mathcal{E}$.

ii) $x_{\rm e}$ is semistable (respectively, Lyapunov stable and $\mathcal{U} \cap \mathcal{E} = \{x_{\rm e}\}$).

An equilibrium point $x_{\rm e} \in \mathcal{E}$ of (8.1) is said to be *globally finite-time semistable* (respectively, *globally finite-time stable*) if it is finite-time semistable (respectively, finite-time stable) with $\mathcal{D} = \mathcal{U} = \mathbb{R}^q$. The system (8.1) is said to be *finite-time semistable* if every equilibrium point in $\mathcal{E}$ is finite-time semistable. Finally, (8.1) is said to be *globally finite-time semistable* if every equilibrium point in $\mathcal{E}$ is globally finite-time semistable.

Given an absolutely continuous curve $\gamma : [0, \infty) \to \mathbb{R}^q$, *the positive limit set of* γ is the set $\Omega(\gamma)$ of points $y \in \mathbb{R}^q$ for which there exists an increasing divergent sequence $\{t_i\}_{i=1}^{\infty}$ satisfying $\lim_{i \to \infty} \gamma(t_i) = y$. We denote the positive limit set of a Filippov solution $\psi(\cdot)$ of (8.1) by $\Omega(\psi)$. The positive limit set of a bounded Filippov solution of (8.1) is nonempty and weakly invariant with respect to (8.1) [149, Lem. 4, p. 130].

Next, we state sufficient conditions for the stability of discontinuous dynamical systems. Here, we state the stability theorems for only the local case; the global stability theorems are similar except for the additional assumption of properness on the Lyapunov function and nonrestricting the domain of analysis.

Theorem 8.1 ([15,222]). Consider the discontinuous nonlinear dynamical system $\mathcal{G}$ given by (8.1). Let $x_{\rm e}$ be an equilibrium point of $\mathcal{G}$ and let $\mathcal{D} \subseteq \mathbb{R}^q$ be an open and connected set with $x_{\rm e} \in \mathcal{D}$. If $V : \mathcal{D} \to \mathbb{R}$ is a positive-definite, locally Lipschitz continuous, and regular function such that $\max \mathcal{L}_f V(x) \le 0$ (respectively, $\max \mathcal{L}_f V(x) < 0$, $x \neq x_{\rm e}$) for almost all $x \in \mathcal{D}$ such that $\mathcal{L}_f V(x) \neq \varnothing$, then $x_{\rm e}$ is Lyapunov (respectively, asymptotically) stable.

The next result presents an extenson of the Krasovskii-LaSalle invariant set theorem to discontinuous dynamical systems.

Theorem 8.2 ([15,222]). Consider the discontinuous nonlinear dynamical system $\mathcal{G}$ given by (8.1). Let $x_{\rm e}$ be an equilibrium point of $\mathcal{G}$, let $\mathcal{D} \subseteq \mathbb{R}^q$ be an open strongly positively invariant set with respect to (8.1) such that $x_{\rm e} \in \mathcal{D}$, and let $V : \mathcal{D} \to \mathbb{R}$ be locally Lipschitz continuous and regular on $\mathcal{D}$. Assume that, for every $x \in \mathcal{D}$ and every Filippov solution $\psi(\cdot)$ satisfying $\psi(t_0) = x$, there exists a compact subset $\mathcal{D}_{\rm c}$ of $\mathcal{D}$ containing $\psi(t)$ for all $t \geq 0$. Furthermore, assume that $\max \mathcal{L}_f V(x) \leq 0$ for almost all $x \in \mathcal{D}$ such that $\mathcal{L}_f V(x) \neq \varnothing$. Finally, define $\mathcal{R} \triangleq \{x \in \mathcal{D} :\ 0 \in \mathcal{L}_f V(x)\}$ and let $\mathcal{M}$ be the largest weakly positively invariant subset of $\overline{\mathcal{R}} \cap \mathcal{D}$. If $x(t_0) \in \mathcal{D}_{\rm c}$, then $x(t) \to \mathcal{M}$ as $t \to \infty$. If, alternatively, $\mathcal{R}$ contains no invariant set other than $\{x_{\rm e}\}$, then the Filippov solution $x(t) \equiv x_{\rm e}$ of $\mathcal{G}$ is asymptotically stable for all $x_0 \in \mathcal{D}_{\rm c}$.

Next, we develop Lyapunov-based semistability and finite-time semistability theory for discontinuous dynamical systems of the form given by (8.1). The following proposition is needed.

Proposition 8.1. Let $\mathcal{D} \subseteq \mathbb{R}^q$ be an open strongly positively invariant set with respect to (8.1) and let $\psi(\cdot)$ be a Filippov solution of (8.2) with $\psi(0) \in \mathcal{D}$. If $z \in \Omega(\psi) \cap \mathcal{D}$ is a Lyapunov stable equilibrium point, then $z = \lim_{t\to\infty} \psi(t)$ and $\Omega(\psi) = \{z\}$.

Proof. Suppose $z \in \Omega(\psi) \cap \mathcal{D}$ is Lyapunov stable and let $\varepsilon > 0$. Since z is Lyapunov stable, there exists $\delta = \delta(\varepsilon) > 0$ such that, for every $y \in \mathcal{B}_\delta(z)$ and every Filippov solution $\eta(\cdot)$ of (8.2) satisfying $\eta(0) = y$, $\eta(t) \in \mathcal{B}_\varepsilon(z)$ for all $t \geq 0$. Now, since $z \in \Omega(\psi)$, it follows that there exists a divergent sequence $\{t_i\}_{i=1}^{\infty}$ in $[0, \infty)$ such that $\lim_{i\to\infty} \psi(t_i) = z$, and hence, there exists $k \geq 1$ such that $\psi(t_k) \in \mathcal{B}_\delta(z)$. It now follows from our construction of δ that $\psi(t) \in \mathcal{B}_\varepsilon(z)$ for all $t \geq t_k$. Since ε was chosen arbitrarily, it follows that $z = \lim_{t\to\infty} \psi(t)$. Thus, $\lim_{n\to\infty} \psi(t_n) = z$ for every divergent sequence $\{t_n\}_{n=1}^{\infty}$, and hence, $\Omega(\psi) = \{z\}$. □

Next, we present sufficient conditions for the semistability of (8.1).

Theorem 8.3. Let $\mathcal{D} \subseteq \mathbb{R}^q$ be an open strongly positively invariant set with respect to (8.1) and let $V : \mathcal{D} \to \mathbb{R}$ be locally Lipschitz continuous and regular on $\mathcal{D}$. Assume that, for every $x \in \mathcal{D}$ and every Filippov solution $\psi(\cdot)$ satisfying $\psi(0) = x$, there exists a compact subset of $\mathcal{D}$ containing $\psi(t)$ for all $t \geq 0$. Furthermore, assume that $\max \mathcal{L}_f V(x) \leq 0$ for almost all $x \in \mathcal{D}$ such that $\mathcal{L}_f V(x) \neq \varnothing$. Finally, define

$$\mathcal{R} \triangleq \{x \in \mathcal{D} : 0 \in \mathcal{L}_f V(x)\}. \tag{8.7}$$

If every point in the largest weakly positively invariant subset $\mathcal{M}$ of $\overline{\mathcal{R}} \cap \mathcal{D}$ is a Lyapunov stable equilibrium point, then (8.1) is semistable with respect to $\mathcal{D}$.

Proof. Let $x \in \mathcal{D}$, $\psi(\cdot)$ be a Filippov solution to (8.1) with $\psi(0) = x$, and $\Omega(\psi)$ be the positive limit set of ψ. First, we show that $\Omega(\psi) \subseteq \overline{\mathcal{R}}$. Since either $\max \mathcal{L}_f V(x) \leq 0$ or $\mathcal{L}_f V(x) = \varnothing$ for almost all $x \in \mathcal{D}$, it follows from Lemma 1 of [15] that $\frac{\mathrm{d}}{\mathrm{d}t}V(\psi(t))$ exists and is contained in $\mathcal{L}_f V(\psi(t))$ for almost every $t \geq 0$. Now, by assumption, $V(\psi(t)) - V(\psi(\tau)) = \int_\tau^t \frac{\mathrm{d}}{\mathrm{d}t}V(\psi(s))\mathrm{d}s \leq 0$, $t \geq \tau$, and hence, $V(\psi(t)) \leq V(\psi(\tau))$, $t \geq \tau$, which implies that $V(\psi(t))$ is a nonincreasing function of time.

The continuity of V and the boundedness of ψ imply that $V(\psi(\cdot))$ is bounded. Hence, $\gamma_x \triangleq \lim_{t\to\infty} V(\psi(t))$ exists. Next, consider $p \in \Omega(\psi)$. There exists an increasing unbounded sequence $\{t_n\}_{n=1}^\infty$ in $[0,\infty)$ such that $\psi(t_n) \to p$ as $n \to \infty$. Since V is continuous on $\mathcal{D}$, it follows that $V(p) = V(\lim_{n\to\infty} \psi(t_n)) = \lim_{n\to\infty} V(\psi(t_n)) = \gamma_x$, and hence, $V(p) = \gamma_x$ for $p \in \Omega(\psi)$. In other words, $\Omega(\psi)$ is contained in a level set of V.

Let $y \in \Omega(\psi)$. Since $\Omega(\psi)$ is weakly positively invariant, there exists a Filippov solution $\hat{\psi}(\cdot)$ of (8.1) such that $\hat{\psi}(0) = y$ and $\hat{\psi}(t) \in \Omega(\psi)$ for all $t \geq 0$. Since $V(\Omega(\psi)) = \{V(y)\}$, $\frac{\mathrm{d}}{\mathrm{d}t}V(\hat{\psi}(t)) = 0$, and hence, it follows from Lemma 1 of [15] that $0 \in \mathcal{L}_f V(\hat{\psi}(t))$, that is, $\hat{\psi}(t) \in \mathcal{R}$ for almost all $t \in [0, \hat{t}]$. In particular, $y \in \mathcal{R}$. Since $y \in \Omega(\psi)$ was chosen arbitrarily, it follows that $\Omega(\psi) \subseteq \overline{\mathcal{R}}$.

Next, since $\Omega(\psi)$ is weakly positively invariant, it follows that $\Omega(\psi) \subseteq \mathcal{M}$. Moreover, since every point in $\mathcal{M}$ is a Lyapunov stable equilibrium point of (8.1), it follows from Proposition 8.1 that $\Omega(\psi)$ contains a single point and $\lim_{t\to\infty} \psi(t)$ is a Lyapunov stable equilibrium. Now, since $x \in \mathcal{D}$ was chosen arbitrarily, it follows from Definition 8.2 that (8.1) is semistable with respect to $\mathcal{D}$. □

The following corollary to Theorem 8.3 provides sufficient conditions for finite-time semistability of (8.1).

Corollary 8.1. Let $\mathcal{D} \subseteq \mathbb{R}^q$ be an open strongly positively invariant set with respect to (8.1) and let $V : \mathcal{D} \to \mathbb{R}$ be locally Lipschitz continuous and regular on $\mathcal{D}$. Assume that $\max \mathcal{L}_f V(x) < 0$ for almost all $x \in \mathcal{D}\backslash\mathcal{E}$ such that $\mathcal{L}_f V(x) \neq \varnothing$. If every equilibrium in $\mathcal{D}$ is Lyapunov stable, then every equilibrium in $\mathcal{D}$ is semistable. If, in addition, $\max \mathcal{L}_f V(x) \leq -\varepsilon < 0$ for almost every $x \in \mathcal{D}\backslash\mathcal{E}$ such that $\mathcal{L}_f V(x) \neq \varnothing$, then (8.1) is finite-time semistable.

Proof. To prove the first statement, suppose every equilibrium in $\mathcal{D}$, that is, every point in $\mathcal{E} \cap \mathcal{D}$ is Lyapunov stable. By Lyapunov stability, there exists an open set $\mathcal{D}'$ containing $\mathcal{E} \cap \mathcal{D}$ such that $\mathcal{D}'$ is strongly positively invariant with respect to (8.1), and every Filippov solution having initial condition in $\mathcal{D}'$ is bounded. Let $\mathcal{M}$ denote the largest weakly positively invariant subset of the set $\mathcal{R}' \triangleq \{x \in \mathcal{D}' : 0 \in \mathcal{L}_f V(x)\}$. Note that $0 \in \mathcal{L}_f V(x)$ for every $x \in \mathcal{E}$. Since $\mathcal{E} \cap \mathcal{D}$ is weakly positively invariant and contained in $\mathcal{D}'$, it follows that $\mathcal{E} \cap \mathcal{D} \subseteq \mathcal{M}$. Since either $\max \mathcal{L}_f V(x) < 0$ or $\mathcal{L}_f V(x) = \varnothing$ for almost all $x \in \mathcal{D} \backslash \mathcal{E}$, it follows that $\mathcal{R}' \subseteq \mathcal{E}$. Hence, it follows that $\mathcal{M} = \mathcal{E} \cap \mathcal{D}$. Theorem 8.3 now implies that (8.1) is semistable with respect to $\mathcal{D}'$. Since $\mathcal{E} \cap \mathcal{D} = \mathcal{E} \cap \mathcal{D}'$, it follows that every equilibrium in $\mathcal{D}$ is semistable.

If, in addition, $\max \mathcal{L}_f V(x) \le -\varepsilon < 0$ for almost every $x \in \mathcal{D} \backslash \mathcal{E}$ such that $\mathcal{L}_f V(x) \neq \varnothing$, then it follows from Proposition 2.8 of [94] that every Filippov solution originating in $\mathcal{D}'$ reaches $\mathcal{R}'$ in finite time. Thus, it follows from Definition 8.3 that (8.1) is finite-time semistable. □

Example 8.1. Consider the nonlinear switched dynamical system on $\mathcal{D} = \mathbb{R}^2$ given by

$$\dot{x}_1(t) = f_{\sigma(t)}(x_2(t)) - g_{\sigma(t)}(x_1(t)), \quad x_1(0) = x_{10}, \ t \ge 0, \ \sigma(t) \in \Sigma, \tag{8.8}$$

$$\dot{x}_2(t) = g_{\sigma(t)}(x_1(t)) - f_{\sigma(t)}(x_2(t)), \quad x_2(0) = x_{20}, \tag{8.9}$$

where $x_1, x_2 \in \mathbb{R}$, $\sigma : [0, \infty) \to \Sigma$ is a piecewise constant switching signal, Σ is a finite index set, for every $\sigma \in \Sigma$, $f_\sigma(\cdot)$ and $g_\sigma(\cdot)$ are Lipschitz continuous, $f_\sigma(x_2) - g_\sigma(x_1) = 0$ if and only if $x_1 = x_2$, and $(x_1 - x_2)(f_\sigma(x_2) - g_\sigma(x_1)) \le 0$, $x_1, x_2 \in \mathbb{R}$. Note that $f^{-1}(0) = \{(x_1, x_2) \in \mathbb{R}^2 : x_1 = x_2 = \alpha, \alpha \in \mathbb{R}\}$.

To show that (8.8) and (8.9) is semistable, consider the Lyapunov function candidate $V(x_1 - \alpha, x_2 - \alpha) = \frac{1}{2}(x_1 - \alpha)^2 + \frac{1}{2}(x_2 - \alpha)^2$, where $\alpha \in \mathbb{R}$. Now, it follows that

$$\begin{aligned} \dot{V}(x_1 - \alpha, x_2 - \alpha) &= (x_1 - \alpha)[f_\sigma(x_2) - g_\sigma(x_1)] + (x_2 - \alpha)[g_\sigma(x_1) - f_\sigma(x_2)] \\ &= x_1[f_\sigma(x_2) - g_\sigma(x_1)] + x_2[g_\sigma(x_1) - f_\sigma(x_2)] \\ &= (x_1 - x_2)[f_\sigma(x_2) - g_\sigma(x_1)] \\ &\le 0, \quad (x_1, x_2) \in \mathbb{R} \times \mathbb{R}, \end{aligned} \tag{8.10}$$

which, by Theorem 8.1, implies that $x_1 = x_2 = \alpha$ is Lyapunov stable for all $\alpha \in \mathbb{R}$.

Next, we rewrite (8.8) and (8.9) in the form of the differential inclusion (8.2) where $x \triangleq [x_1, x_2]^{\mathrm{T}} \in \mathbb{R}^2$ and $f(x) \triangleq [f_\sigma(x_2) - g_\sigma(x_1), g_\sigma(x_1) - f_\sigma(x_2)]^{\mathrm{T}}$. Let v_x be an arbitrary element of $\mathcal{K}[f](x)$ and note that the Clarke upper generalized derivative of $V(x) = \frac{1}{2}x_1^2 + \frac{1}{2}x_2^2$ along a vector $v_x \in \mathcal{K}[f](x)$ is given by $V^o(x,v_x) = x^{\mathrm{T}}v_x$. Furthermore, note that the set $\mathcal{D}_{\mathrm{c}} \triangleq \{x \in \mathbb{R}^2 : V(x) \le c\}$, where $c > 0$, is a compact set. Next, consider $\max V^o(x,v_x) \triangleq \max_{v_x\in\mathcal{K}[f]}\{x^{\mathrm{T}}v_x\}$. It follows from Theorem 1 of [351] and (8.10) that $x^{\mathrm{T}}\mathcal{K}[f](x) = \mathcal{K}[x^{\mathrm{T}}f](x) = \mathcal{K}\left[(x_1-x_2)(f_\sigma(x_2)-g_\sigma(x_1))\right](x)$, and hence, by definition of $\mathcal{K}[f](x)$, it follows that $\max V^o(x,v_x) = \max\overline{\mathrm{co}}\{(x_1-x_2)(f_\sigma(x_2)-g_\sigma(x_1))\}$. Note that since, by (8.10), $(x_1-x_2)(f_\sigma(x_2)-g_\sigma(x_1)) \le 0$, $x\in\mathbb{R}^2$, it follows that $\max V^o(x,v_x)$ cannot be positive, and hence, the largest value that $\max V^o(x,v_x)$ can achieve is zero.

Finally, let $\mathcal{R} \triangleq \{(x_1,x_2)\in\mathbb{R}^2 : (x_1-x_2)(f_\sigma(x_2)-g_\sigma(x_1)) = 0\} = \{(x_1,x_2)\in\mathbb{R}^2 : x_1 = x_2 = \alpha, \alpha\in\mathbb{R}\}$. Since $\mathcal{R}$ consists of equilibrium points, it follows that $\mathcal{M} = \mathcal{R}$. Note that $\max \mathcal{L}_f V(x) \le \max V^o(x,v_x)$ for every $x\in\mathbb{R}^2$ [15]. Hence, it follows from Theorem 8.3 that $x_1 = x_2 = \alpha$ is semistable for all $\alpha\in\mathbb{R}$. △

Example 8.2. Consider the discontinuous dynamical system on $\mathcal{D} = \mathbb{R}^2$ given by

$$\dot{x}_1(t) = \mathrm{sgn}(x_2(t)-x_1(t)), \quad x_1(0) = x_{10}, \quad t\ge 0, \tag{8.11}$$

$$\dot{x}_2(t) = \mathrm{sgn}(x_1(t)-x_2(t)), \quad x_2(0) = x_{20}, \tag{8.12}$$

where $x_1, x_2 \in\mathbb{R}$, $\mathrm{sgn}(x) \triangleq x/|x|$ for $x\ne 0$, and $\mathrm{sgn}(0)\triangleq 0$. Let $f(x_1,x_2)\triangleq [\mathrm{sgn}(x_2-x_1), \mathrm{sgn}(x_1-x_2)]^{\mathrm{T}}$. Consider $V(x_1,x_2) = \frac{1}{2}(x_1-\alpha)^2+\frac{1}{2}(x_2-\alpha)^2$, where $\alpha\in\mathbb{R}$. Since $V(x_1,x_2)$ is differentiable at $x=(x_1,x_2)$, it follows that $\mathcal{L}_fV(x_1,x_2) = [x_1-\alpha, x_2-\alpha]\mathcal{K}[f](x_1,x_2)$.

Now, it follows from Theorem 1 of [351] that

$$\begin{aligned}
[x_1-\alpha, x_2-\alpha]\mathcal{K}[f](x) &= \mathcal{K}[[x_1-\alpha, x_2-\alpha]f](x)\\
&= \mathcal{K}[-(x_1-x_2)\mathrm{sgn}(x_1-x_2)](x)\\
&= -(x_1-x_2)\mathcal{K}[\mathrm{sgn}(x_1-x_2)](x)\\
&= -(x_1-x_2)\mathrm{SGN}(x_1-x_2)\\
&= -|x_1-x_2|, \quad (x_1,x_2)\in\mathbb{R}^2,
\end{aligned} \tag{8.13}$$

where $\mathrm{SGN}(\cdot)$ is defined by ([351, 408])

$$\mathrm{SGN}(x) \triangleq \begin{cases} -1, & x<0,\\ [-1,1], & x=0,\\ 1, & x>0. \end{cases} \tag{8.14}$$

Hence, $\max\mathcal{L}_fV(x_1,x_2)\le 0$ for almost all $(x_1,x_2)\in\mathbb{R}^2$. Now, it follows from Theorem 8.1 that $(x_1,x_2) = (\alpha,\alpha)$ is Lyapunov stable.

To show semistability, note that $0\in\mathcal{L}_fV(x_1,x_2)$ if and only if $x_1 = x_2$, and hence, $\mathcal{R} = \{(x_1,x_2)\in\mathbb{R}^2 : x_1 = x_2\}$. Since $\mathcal{R}$ is weakly positively invariant and every point in $\mathcal{R}$ is a Lyapunov stable equilibrium, it follows from Theorem 8.3 that (8.11) and (8.12) is semistable.

Finally, we show that (8.11) and (8.12) is finite-time semistable. To see this, consider the nonnegative function $U(x_1, x_2) = |x_1 - x_2|$. Note that

$$\partial U(x_1, x_2) = \begin{cases} \{\mathrm{sgn}(x_1 - x_2)\} \times \{\mathrm{sgn}(x_2 - x_1)\}, & x_1 \neq x_2, \\ [-1, 1] \times [-1, 1], & x_1 = x_2. \end{cases} \tag{8.15}$$

Hence, it follows that

$$\mathcal{L}_f U(x_1, x_2) = \begin{cases} \{-2\}, & x_1 \neq x_2, \\ \{0\}, & x_1 = x_2, \end{cases} \tag{8.16}$$

which implies that $\max \mathcal{L}_f U(x_1, x_2) = -2 < 0$ for almost all $(x_1, x_2) \in \mathbb{R}^2 \backslash \mathcal{R}$. Now, it follows from Corollary 8.1 that (8.11) and (8.12) is globally finite-time semistable. △

Note that Theorem 8.3 and Corollary 8.1 require verifying Lyapunov stability for concluding semistability and finite-time semistability, respectively. However, finding the corresponding Lyapunov function can be a difficult task. To overcome this drawback, we extend the nontangency-based approach of [42] to discontinuous dynamical systems in order to guarantee semistability and finite-time semistability by testing a condition on the vector field f, which avoids proving Lyapunov stability. Before stating our result, we introduce some notation and definitions as well as extended versions of some results from [42].

A set $\mathcal{E} \subseteq \mathbb{R}^q$ is *connected* if and only if every pair of open sets $\mathcal{U}_i \subseteq \mathbb{R}^q$, $i = 1, 2$, satisfying $\mathcal{E} \subseteq \mathcal{U}_1 \cup \mathcal{U}_2$ and $\mathcal{U}_i \cap \mathcal{E} \neq \varnothing$, $i = 1, 2$, has a nonempty intersection. A *connected component* of the set $\mathcal{E} \subseteq \mathbb{R}^q$ is a connected subset of $\mathcal{E}$ that is not properly contained in any connected subset of $\mathcal{E}$. Given a set $\mathcal{E} \subseteq \mathbb{R}^q$, let $\mathrm{coco}\,\mathcal{E}$ denote the convex cone generated by $\mathcal{E}$.

Definition 8.4. Given $x \in \mathbb{R}^q$, the *direction cone* $\mathcal{F}_x$ of f at x is the intersection of closed convex cones of the form $\overline{\bigcap_{\mu(\mathcal{Q})=0} \mathrm{coco}\{f(\mathcal{U} \backslash \mathcal{Q})\}}$, where $\mathcal{U} \subseteq \mathbb{R}^q$ is an open neighborhood of x. Let $\mathcal{E} \subseteq \mathbb{R}^q$. A vector $v \in \mathbb{R}^q$ is *tangent* to $\mathcal{E}$ at $z \in \mathcal{E}$ if there exist a sequence $\{z_i\}_{i=1}^\infty$ in $\mathcal{E}$ converging to z and a sequence $\{h_i\}_{i=1}^\infty$ of positive real numbers converging to zero such that $\lim_{i \to \infty} \frac{1}{h_i}(z_i - z) = v$. The *tangent cone* to $\mathcal{E}$ at z is the closed cone $T_z\mathcal{E}$ of all vectors tangent to $\mathcal{E}$ at z. Finally, the vector field f is *nontangent* to the set $\mathcal{E}$ at the point $z \in \mathcal{E}$ if $T_z\mathcal{E} \cap \mathcal{F}_z \subseteq \{0\}$.

Definition 8.5. Given a point $x \in \mathbb{R}^q$ and a bounded open neighborhood $\mathcal{U} \subset \mathbb{R}^q$ of x, the *restricted prolongation* of x with respect to $\mathcal{U}$ is the set $\mathcal{R}_x^{\mathcal{U}} \subseteq \overline{\mathcal{U}}$ of all subsequential limits of sequences of the form $\{\psi_i(t_i)\}_{i=1}^\infty$, where $\{t_i\}_{i=1}^\infty$ is a sequence in $[0, \infty)$, $\psi_i(\cdot)$ is a Filippov solution to (8.1) with $\psi_i(0) = x_i$, $i = 1, 2, \ldots$, and $\{x_i\}_{i=1}^\infty$ is a sequence in $\mathcal{U}$ converging to x

such that the set $\{z \in \mathbb{R}^q : z = \psi_i(t), t \in [0, t_i]\}$ is contained in $\overline{\mathcal{U}}$ for every $i = 1, 2, \ldots$.

Proposition 8.2. Let $\mathcal{D} \subseteq \mathbb{R}^q$ be an open strongly positively invariant set with respect to (8.1). Furthermore, let $x \in \mathcal{D}$ and let $\mathcal{U} \subseteq \mathcal{D}$ be a bounded open neighborhood of x. Then $\mathcal{R}_x^{\mathcal{U}}$ is connected. Moreover, if x is an equilibrium point of (8.1), then $\mathcal{R}_x^{\mathcal{U}}$ is weakly negatively invariant.

Proof. The proof of connectedness is similar to the proof of the first part of Proposition 6.1 of [42] and, hence, is omitted. To prove weak negative invariance, suppose $x \in \mathcal{D}$ is an equilibrium point of (8.1), and consider $z \in \mathcal{R}_x^{\mathcal{U}}$. Then there exist a sequence $\{t_i\}_{i=1}^{\infty}$ in $[0, \infty)$, a sequence $\{x_i\}_{i=1}^{\infty}$ in $\mathcal{D}$ converging to x, and a sequence $\{\psi_i(\cdot)\}_{i=1}^{\infty}$ of Filippov solutions of (8.1) such that $\lim_{i\to\infty} \psi_i(t_i) = z$ and, for every i, $\psi_i(0) = x_i$ and $\psi_i(h) \in \overline{\mathcal{U}}$ for every $h \in [0, t_i]$.

Now, let $t \geq 0$. First, assume $z = x$. Then $\psi \equiv x$ is a Filippov solution of (8.1) such that $\psi(0) = x$, $\psi(t) = z$, and $\psi(\tau) \in \mathcal{R}_x^{\mathcal{U}}$ for all $\tau \in [0, t]$. Next, consider the case $z \neq x$. First, suppose that the sequence $\{t_i\}_{i=1}^{\infty}$ has a subsequence $\{t_{i_k}\}_{k=1}^{\infty}$ in $[0, t]$. By choosing a subsequence if necessary, we may assume that the subsequence $\{t_{i_k}\}_{k=1}^{\infty}$ converges to T. Necessarily, $T \leq t$. By Lemma 1 in [149, p. 87], a subsequence of the sequence $\{\psi_{i_k}\}_{k=1}^{\infty}$ converges uniformly on compact subsets of $(0, T)$ to a Filippov solution ψ of (8.1). Moreover, the solution ψ satisfies $\psi(0) = x$ and $\psi(T) = z$. For each $s \in [0, T]$, $\psi(s)$ is a subsequential limit of the sequence $\{\psi_{i_k}(s)\}_{k=1}^{\infty}$, and hence, is contained in $\mathcal{R}_x^{\mathcal{U}}$. It is now easy to verify that the function $\beta : [0, t] \to \mathcal{D}$ defined by

$$\beta(s) = \begin{cases} x, & 0 \leq s \leq t - T, \\ \psi(s - t + T), & t - T < s \leq t, \end{cases}$$

is a Filippov solution of (8.1) satisfying $\beta(0) = x$, $\beta(t) = z$, and $\beta(s) \in \mathcal{R}_x^{\mathcal{U}}$ for all $s \in [0, t]$.

Next, suppose that the sequence $\{t_i\}_{i=1}^{\infty}$ has no subsequence in $[0, t]$. Then there exists $N > 0$ such that $t_i > t$ for all $i \geq N$. For each i, define $\beta_i : [0, t] \to \mathcal{D}$ by $\beta(s) = \psi_{i+N}(t_{i+N} - t + s)$. Clearly, each β_i is a Filippov solution of (8.1). Moreover, the sequence $\{\beta_i(t)\}_{i=1}^{\infty}$ converges to z. Let $y \in \mathcal{D}$ be a subsequential limit of the bounded sequence $\{\beta_i(0)\}_{i=1}^{\infty}$. By definition, $y \in \mathcal{R}_x^{\mathcal{U}}$. By Lemma 1 in [149, p. 87], a subsequence of $\{\beta_i\}_{i=1}^{\infty}$ converges uniformly on compact subsets of $(0, t)$ to a Filippov solution β of (8.1). Moreover, we may choose the subsequence such that $\beta(0) = y$ and $\beta(t) = z$.

Finally, for each $s \in [0, t]$, $\beta(s)$ is a subsequential limit of the sequence

$\{\beta_i(s)\}_{i=1}^{\infty}$, and hence, in $\mathcal{R}_x^{\mathcal{U}}$. We have thus shown that there exists a Filippov solution β defined on $[0,t]$ such that $\beta(s) \in \mathcal{R}_x^{\mathcal{U}}$ for all $s \in [0,t]$ and $\beta(t) = z$. Since $t \geq 0$ and $z \in \mathcal{R}_x^{\mathcal{U}}$ were chosen to be arbitrary, it follows that $\mathcal{R}_x^{\mathcal{U}}$ is weakly negatively invariant. □

The following two lemmas and proposition extend related results from [42], and are needed for the main result of this section.

Lemma 8.1. Let $\mathcal{D} \subseteq \mathbb{R}^q$ be an open strongly positively invariant set with respect to (8.1) and let $V : \mathcal{D} \to \mathbb{R}$ be locally Lipschitz continuous and regular on $\mathcal{D}$. Assume that $V(x) \geq 0$, for all $x \in \mathcal{D}$, $V(z) = 0$ for all $z \in \mathcal{E}$, and $\max \mathcal{L}_f V(x) \leq 0$ for almost every $x \in \mathcal{D}$ such that $\mathcal{L}_f V(x) \neq \varnothing$. For every $z \in \mathcal{E}$, let $\mathcal{N}_z$ denote the largest weakly negatively invariant connected subset of $\overline{\mathcal{R}} \cap \mathcal{D}$ containing z, where $\mathcal{R}$ is given by (8.7). Then, for every $x \in \mathcal{E}$ and every bounded open neighborhood $\mathcal{V} \subset \mathcal{D}$ of x, $\mathcal{R}_x^{\mathcal{V}} \subseteq \mathcal{N}_x$.

Proof. Let $x \in \mathcal{E}$ and let $\mathcal{V} \subset \mathcal{D}$ be a bounded open neighborhood of x. Consider $z \in \mathcal{R}_x^{\mathcal{V}}$. Let $\{x_i\}_{i=1}^{\infty}$ be a sequence in $\mathcal{V}$ converging to x and let $\{t_i\}_{i=1}^{\infty}$ be a sequence in $[0,\infty)$ such that the sequence $\{\psi_i(t_i)\}_{i=1}^{\infty}$ converges to z and, for every i, $\psi_i(\tau) \in \overline{\mathcal{V}} \subset \mathcal{D}$ for every $\tau \in [0,t_i]$, where $\psi_i(\cdot)$ is a Filippov solution to (8.1) with $\psi_i(0) = x_i$. Since either $\max \mathcal{L}_f V(y) \leq 0$ or $\mathcal{L}_f V(y) = \varnothing$ for almost every $y \in \mathcal{D}$, it follows from Lemma 1 of [15] that $\frac{\mathrm{d}}{\mathrm{d}t} V(\psi(t))$ exists and is contained in $\mathcal{L}_f V(\psi(t))$ for almost all $t \in [0,\tau]$, where $\psi(\cdot)$ is a Filippov solution to (8.1) with $\psi(0) = y$. Now, by assumption, $V(\psi(\tau)) - V(y) = \int_0^{\tau} \frac{\mathrm{d}}{\mathrm{d}t} V(\psi(s))\mathrm{d}s \leq 0$, $\tau \geq 0$, and hence, $V(\psi(\tau)) \leq V(y)$ for $y \in \mathcal{D}$ and $\tau \geq 0$.

Next, note that $V(z) = \lim_{i\to\infty} V(\psi_i(t_i)) \leq \lim_{i\to\infty} V(x_i) = V(x)$, and hence, $V(z) \leq V(x)$. Since $V(z) \geq 0$ and $V(x) = 0$ by assumption, it follows that $V(z) = V(x) = 0$. Hence, $\mathcal{R}_x^{\mathcal{V}} \subseteq V^{-1}(0) \cap \overline{\mathcal{V}} \subset V^{-1}(0)$. By Proposition 8.2, $\mathcal{R}_x^{\mathcal{V}}$ is weakly negatively invariant and connected, and $x \in \mathcal{R}_x^{\mathcal{V}}$. Hence, $\mathcal{R}_x^{\mathcal{V}} \subseteq \mathcal{M}_x$, where $\mathcal{M}_x$ denotes the largest, weakly negatively invariant connected subset of $V^{-1}(0)$ containing x.

Finally, we show that $\mathcal{M}_x \subseteq \mathcal{N}_x$. Let $z \in \mathcal{M}_x$ and let $t > 0$. By weak negative invariance, there exists $w \in \mathcal{M}_x$ and a Filippov solution $\psi(\cdot)$ to (8.1) satisfying $\psi(0) = w$ such that $\psi(t) = z$ and $\psi(\tau) \in \mathcal{M}_x \subseteq V^{-1}(0)$ for all $\tau \in [0,t]$. Thus, $V(\psi(\tau)) = V(x) = 0$ for every $\tau \in [0,t]$, and hence, by Lemma 1 of [15], $0 \in \mathcal{L}_f V(\psi(\tau))$ for almost every $\tau \in [0,t]$, that is, $\psi(\tau) \in \mathcal{R}$ for almost every $\tau \in [0,t]$. It immediately follows that $z \in \overline{\mathcal{R}}$, and hence, $\mathcal{M}_x \subseteq \overline{\mathcal{R}}$. Since $\mathcal{M}_x$ is weakly negatively invariant, connected, contains x, and is contained in $\mathcal{U}$, it follows that $\mathcal{M}_x \subseteq \mathcal{N}_x$. Hence, $\mathcal{R}_x^{\mathcal{V}} \subseteq \mathcal{M}_x \subseteq \mathcal{N}_x$. □

Lemma 8.2. Let $\mathcal{D} \subseteq \mathbb{R}^q$ be an open strongly positively invariant set with respect to (8.1). Furthermore, let $x \in \mathcal{D}$ and let $\{x_i\}_{i=1}^{\infty}$ be a sequence in $\mathcal{D}$ converging to x. Let $\Sigma_i \subseteq [0,\infty)$, $i = 1, 2, \ldots$, be intervals containing 0, and let $\mathcal{B} \subseteq \mathcal{D}$ be the set of all subsequential limits contained in $\mathcal{D}$ of sequences of the form $\{\psi_i(\tau_i)\}_{i=1}^{\infty}$, where, for each i, $\tau_i \in \Sigma_i$ and $\psi_i : \Sigma_i \to \mathcal{D}$ is a Filippov solution of (8.1) satisfying $\psi_i(0) = x_i$. Then $\mathcal{B} = \{x\}$ if and only if f is nontangent to $\mathcal{B}$ at x.

Proof. First, we note that $x \in \mathcal{B}$ since $x = \lim_{i\to\infty} \psi_i(0)$. Necessity now follows by noting that if $\mathcal{B} = \{x\}$, then $T_x\mathcal{B} = \{0\}$, and hence, $T_x\mathcal{B} \cap \mathcal{F}_x \subseteq \{0\}$.

To prove sufficiency, suppose $z_0 \in \mathcal{B}$, $z_0 \neq x$. Let $\{\mathcal{U}_k\}_{k=1}^{\infty}$ be a nested sequence of bounded open neighborhoods of x in $\mathcal{D}$ such that $\overline{\mathcal{U}_{k+1}} \subset \mathcal{U}_k$ and $x_k \in \mathcal{U}_k$ for every $k = 1, 2, \ldots$, $\bigcap_k \mathcal{U}_k = \{x\}$ and $z_0 \notin \mathcal{U}_1$. Since $z_0 \in \mathcal{B}$, there exists a sequence $\{\tau_i\}_{i=1}^{\infty}$ such that $\tau_i \in \Sigma_i$ for every i and $\lim_{i\to\infty} \psi_i(\tau_i) = z_0 \notin \mathcal{U}_1$. The continuity of Filippov solutions implies that, for every k, there exists a sequence $\{h_j^k\}_{j=k}^{\infty}$ in $[0,\infty)$ such that, for every $j \geq k$, $h_j^k \in \Sigma_j$, $h_j^k \leq \tau_j$, $\psi_j(\tau) \in \mathcal{U}_k$ for every $\tau \in [0, h_j^k)$, and $\psi_j(h_j^k) \in \partial\mathcal{U}_k$. For each k, let $z_k \in \partial\mathcal{U}_k$ be a subsequential limit of the bounded sequence $\{\psi_j(h_j^k)\}_{j=k}^{\infty}$. Then, for every k, it follows that $z_k \in \mathcal{B}$, $z_k \neq x$ and $\lim_{k\to\infty} z_k = x$. Now, consider a subsequential limit v of the bounded sequence $\{\|z_k - x\|^{-1}(z_k - x)\}$. Clearly, $v \in T_x\mathcal{B}$. Also $\|v\| = 1$ so that $v \neq 0$. We claim that $v \in \mathcal{F}_x$.

Let $\mathcal{V} \subseteq \mathcal{D}$ be an open neighborhood of x and consider $\varepsilon > 0$. By construction, there exists k such that $\left\|v - \|z_k - x\|^{-1}(z_k - x)\right\| < \varepsilon/3$. Moreover, since $\bigcap_i \mathcal{U}_i = \{x\}$, we can assume that $\mathcal{U}_k \subseteq \mathcal{V}$. Since z_k belongs to the boundary of an open neighborhood of x, $\delta \triangleq \|z_k - x\| > 0$. Since $z_k = \lim_{i\to\infty} \psi_i(h_i^k)$ and $x = \lim_{i\to\infty} x_i$, there exists i such that $x_i \in \mathcal{V}$, $\|x - x_i\| < \varepsilon\delta/3$, and $\|z_k - \psi_i(h_i^k)\| < \varepsilon\delta/3$. Let $\mathcal{Q} \subset \mathcal{D}$ be a zero-measure set. Then $\mathcal{K}[f](\psi_i(\tau)) \subseteq \mathrm{co}\{f(\mathcal{V}\backslash\mathcal{Q})\}$ for all $\tau \in [0, h_i^k]$, so that $\dot{\psi}_i(\tau) \in \mathrm{co}\{f(\mathcal{V}\backslash\mathcal{Q})\}$ for almost every $\tau \in [0, h_i^k]$. Therefore, it follows from Theorem I.6.13 of [459, p. 145] that $w \triangleq \psi_i(h_i^k) - x_i = \int_0^{h_i^k} \dot{\psi}_i(\tau)d\tau$ is contained in the convex cone generated by $\mathrm{co}\{f(\mathcal{V}\backslash\mathcal{Q})\}$. Since $\mathcal{Q}$ was chosen to be an arbitrary zero-measure set, it follows that $w \in \bigcap_{\mu(\mathcal{Q})=0} \mathrm{coco}\{f(\mathcal{V}\backslash\mathcal{Q})\}$.

Now,

$$\begin{aligned}
\left\|v - \delta^{-1}w\right\| &= \left\|v - \delta^{-1}(z_k - x) - \delta^{-1}(\psi(h_i^k, x_i) - z_k) - \delta^{-1}(x - x_i)\right\| \\
&\leq \left\|v - \|z_k - x\|^{-1}(z_k - x)\right\| + \delta^{-1}\|\psi(h_i^k, x_i) - z_k\| \\
&\quad + \delta^{-1}\|x - x_i\| \\
&< \varepsilon.
\end{aligned}$$

We have thus shown that, for every $\varepsilon > 0$, there exists $w \in \bigcap_{\mu(\mathcal{Q})=0} \mathrm{coco}\{f(\mathcal{V}\backslash\mathcal{Q})\}$ and $\delta > 0$ such that $w \neq 0$ and $\|v - \delta^{-1}w\| < \varepsilon$. It follows that v is contained in the closed convex cone $\overline{\bigcap_{\mu(\mathcal{Q})=0} \mathrm{coco}\{f(\mathcal{V}\backslash\mathcal{Q})\}}$. Since $\mathcal{V}$ was chosen to be an arbitrary open neighborhood of x, it follows that v is contained in $\mathcal{F}_x$. Thus, if $\mathcal{B} \neq \{x\}$, then there exists $v \in \mathbb{R}^q$ such that $v \neq 0$ and $v \in T_x\mathcal{B} \cap \mathcal{F}_x$, that is, f is not nontangent to $\mathcal{B}$ at x. Sufficiency now follows. □

Proposition 8.3. Let $\mathcal{D} \subseteq \mathbb{R}^q$ be an open strongly positively invariant set with respect to (8.1). Furthermore, let $x \in \mathcal{D}$ and let $\mathcal{U} \subseteq \mathcal{D}$ be a bounded open neighborhood of x. If the vector field f of (8.1) is nontangent to $\mathcal{R}_x^{\mathcal{U}}$ at x, then the point x is a Lyapunov stable equilibrium of (8.1).

Proof. Since f is nontangent to $\mathcal{R}_x^{\mathcal{U}}$ at x, by definition, it follows that $T_x\mathcal{R}_x^{\mathcal{U}} \cap \mathcal{F}_x \subseteq \{0\}$. Let $z \in \mathcal{R}_x^{\mathcal{U}}$. Then there exist a sequence $\{x_i\}_{i=1}^{\infty}$ converging to x, a sequence $\{t_i\}_{i=1}^{\infty}$ in $[0, \infty)$, and a sequence $\{\psi_i\}_{i=1}^{\infty}$ of Filippov solutions of (8.2) such that $\psi_i(0) = x_i$ and $\psi([0, t_i]) \subseteq \overline{\mathcal{U}}$ for every $i = 1, 2, \ldots$, and $\lim_{i\to\infty} \psi_i(t_i) = z$.

First, suppose that the sequence $\{t_i\}_{i=1}^{\infty}$ converges to 0. Then it follows from Theorem 11 of [148] that there exists a Filippov solution $\hat{\psi}(\cdot)$ to (8.1) with $\hat{\psi}(0) = x$ such that $\lim_{i\to\infty} \psi_i(t_i) = \hat{\psi}(0) = x$. Next, suppose the sequence $\{t_i\}_{i=1}^{\infty}$ does not converge to 0. Then there exists a subsequence $\{t_{i_k}\}_{k=1}^{\infty}$ of the sequence $\{t_i\}_{i=1}^{\infty}$ such that $\liminf_{k\to\infty} t_{i_k} > 0$. Let $\Sigma_k \triangleq [0, t_{i_k}]$ for each k and let $\mathcal{B} \subseteq \overline{\mathcal{U}}$ denote the set of all subsequential limits of sequences of the form $\{\psi_{i_k}(\tau_k)\}_{k=1}^{\infty}$, where $\tau_k \in \Sigma_k$ for every k. By construction, $z \in \mathcal{B}$ and $\mathcal{B} \subseteq \mathcal{R}_x^{\mathcal{U}}$. Hence, $T_x\mathcal{B} \cap \mathcal{F}_x \subseteq T_x\mathcal{R}_x^{\mathcal{U}} \cap \mathcal{F}_x \subseteq \{0\}$, that is, f is nontangent to $\mathcal{B}$ at x. Now, it follows from Lemma 8.2 that $\mathcal{B} = \{x\}$. Hence, $z = x$. Since $z \in \mathcal{R}_x^{\mathcal{U}}$ is arbitrary, it follows that $\mathcal{R}_x^{\mathcal{U}} = \{x\}$.

Suppose, *ad absurdum*, that x is not a Lyapunov stable equilibrium. Then there exist a bounded open neighborhood $\mathcal{V} \subseteq \mathcal{U}$ of x, a sequence $\{x_i\}_{i=1}^{\infty}$ in $\mathcal{V}$ converging to x, a sequence $\{\psi_i\}_{i=1}^{\infty}$ of Filippov solutions to (8.2), and a sequence $\{t_i\}_{i=1}^{\infty}$ in $[0, \infty)$ such that $\psi_i(x_i) = x_i$ and $\psi_i(t_i) \in \partial\mathcal{V}$ for every i. Without loss of generality, we can assume that the sequence $\{t_i\}_{i=1}^{\infty}$ is chosen such that, for every i, $\psi_i(h) \in \mathcal{V}$ for all $h \in [0, t_i)$. Now, every subsequential limit of the bounded sequence $\{\psi_i(t_i)\}_{i=1}^{\infty}$ is distinct from x by construction and is contained in $\mathcal{R}_x^{\mathcal{U}}$ by definition, which implies that $\mathcal{R}_x^{\mathcal{U}}\backslash\{x\} \neq \varnothing$. This contradicts our earlier conclusion that $\mathcal{R}_x^{\mathcal{U}} = \{x\}$. Hence, x is Lyapunov stable. □

The following theorem gives sufficient conditions for semistability using nontangency of the vector field f.

Theorem 8.4. Let $\mathcal{D} \subseteq \mathbb{R}^q$ be an open strongly positively invariant set with respect to (8.1) and let $V : \mathcal{D} \to \mathbb{R}$ be locally Lipschitz continuous and regular on $\mathcal{D}$. Assume that $V(x) \geq 0$ for all $x \in \mathcal{D}$, $V(z) = 0$ for all $z \in \mathcal{E} \cap \mathcal{D}$, and $\max \mathcal{L}_f V(x) \leq 0$ for almost every $x \in \mathcal{D}$ such that $\mathcal{L}_f V(x) \neq \varnothing$. Furthermore, for every $z \in \mathcal{E}$, let $\mathcal{N}_z$ denote the largest weakly negatively invariant connected subset of $\overline{\mathcal{R}} \cap \mathcal{D}$ containing z, where $\mathcal{R}$ is given by (8.7). If f is nontangent to $\mathcal{N}_z$ at every $z \in \mathcal{E}$, then every equilibrium in $\mathcal{D}$ is semistable.

Proof. Let $\mathcal{V} \subset \mathcal{D}$ be a bounded open neighborhood of $x \in \mathcal{E} \cap \mathcal{D}$. Since f is nontangent to $\mathcal{N}_x$ at the point $x \in \mathcal{E} \cap \mathcal{V}$, it follows that $T_x\mathcal{N}_x \cap \mathcal{F}_x \subseteq \{0\}$. Next, we show that f is nontangent to $\mathcal{R}_x^{\mathcal{V}}$ at the point x. It follows from Lemma 8.1 that $\mathcal{R}_x^{\mathcal{V}} \subseteq \mathcal{N}_x$. Hence, $T_x\mathcal{R}_x^{\mathcal{V}} \cap \mathcal{F}_x \subseteq T_x\mathcal{N}_x \cap \mathcal{F}_x \subseteq \{0\}$, that is, $T_x\mathcal{R}_x^{\mathcal{V}} \cap \mathcal{F}_x \subseteq \{0\}$. By definition, f is nontangent to $\mathcal{R}_x^{\mathcal{V}}$ at the point x. Now, it follows from Proposition 8.3 that x is a Lyapunov stable equilibrium. Since $x \in \mathcal{E} \cap \mathcal{D}$ was chosen arbitrarily, it follows that every equilibrium of (8.1) in $\mathcal{D}$ is Lyapunov stable.

By Lyapunov stability of x, it follows that there exists a strongly positively invariant neighborhood $\mathcal{U} \subset \mathcal{V}$ of x that is open and bounded, and such that $\overline{\mathcal{U}} \subset \mathcal{V}$. Consider $z \in \mathcal{U}$, and let $\psi(\cdot)$ be a Filippov solution of (8.1) with $\psi(0) = z$. Then $\psi(\cdot)$ is bounded in $\mathcal{D}$. Hence, it follows from [149, p. 129] and Theorem 3 of [15] that $\Omega(\psi) \subseteq \overline{\mathcal{U}}$ is nonempty and contained in $\overline{\mathcal{R}}$.

Let $w \in \Omega(\psi)$. The invariance and connectedness of $\Omega(\psi)$ implies that $\Omega(\psi) \subseteq \mathcal{N}_w$. Hence, $T_w\Omega(\psi) \cap \mathcal{F}_w \subseteq T_w\mathcal{N}_w \cap \mathcal{F}_w \subseteq \{0\}$. Now, it follows from Lemma 8.2 (see the proof of Proposition 5.2 of [42]) that $\lim_{t\to\infty} \psi(t)$ exists. Since $z \in \mathcal{U}$ was chosen arbitrarily, it follows that every Filippov solution in $\mathcal{U}$ converges to a limit. The strong invariance of $\mathcal{U}$ implies that the limit of every Filippov solution in $\mathcal{U}$ is contained in $\overline{\mathcal{U}}$. Since every equilibrium in $\overline{\mathcal{U}} \subset \mathcal{V}$ is Lyapunov stable, it follows from Theorem 8.3 that x is semistable. Finally, since $x \in \mathcal{E} \cap \mathcal{D}$ was chosen arbitrarily, it follows that every equilibrium in $\mathcal{D}$ is semistable. □

Example 8.3. Consider the discontinuous dynamical system on $\mathcal{D} = \overline{\mathbb{R}}_+^2$ given by

$$\dot{x}_1(t) = -0.5x_1(t) + f_1(x_1(t), x_2(t)), \quad x_1(0) = x_{10}, \quad t \geq 0, \tag{8.17}$$

$$\dot{x}_2(t) = -x_2(t) + f_2(x_1(t), x_2(t)), \quad x_2(0) = x_{20}, \tag{8.18}$$

where $x_{10} \geq 0$, $x_{20} \geq 0$, and $f_i(\cdot), i = 1, 2$, are given by

$$f_1(x_1, x_2) = \Big[\mathrm{step}\big(x_2 - x_1\big)\big(x_1 - 0.5x_2\big)\Big], \tag{8.19}$$

$$f_2(x_1, x_2) = \left[\text{step}(x_2 - x_1)(2x_1 - x_2)\right], \tag{8.20}$$

where $\text{step}(y) = 1$ for $y \geq 0$ and $\text{step}(y) = 0$, otherwise. For this system, the set of equilibria in $\overline{\mathbb{R}}_+^2$ are given by $\mathcal{E} \triangleq \{(x_1, x_2) \in \overline{\mathbb{R}}_+^2 : x_1 = x_2\}$.

Next, we show that for almost all the initial conditions $(x_{10}, x_{20}) \in \overline{\mathbb{R}}_+^2$, the equilibrium set $\mathcal{E}$ is attractive. To see this, consider the function $V : \mathbb{R}^2 \to \mathbb{R}$ given by $V(x_1, x_2) = \frac{1}{2}\left[(x_1)^2 + (x_2)^2\right]$. Now, it follows that the set-valued Lie derivative $\mathcal{L}_f V(x)$ satisfies

$$\mathcal{L}_f V(x_1, x_2) = \begin{cases} \{ -\frac{1}{2}(x_1)^2 - (x_2)^2 \}, & x_1 > x_2 \geq 0 \,\text{or}\, x_2 > 2x_1 \geq 0, \\ \{ \frac{1}{2}(x_1 - x_2)(x_1 + 4x_2) \}, & 0 \leq x_1 < x_2 < 2x_1, \\ \overline{\text{co}}\{ -\frac{1}{2}(x_1)^2 - (x_2)^2, & \\ \quad \frac{1}{2}(x_1 - x_2)(x_1 + 4x_2) \}, & \text{otherwise, } (x_1, x_2) \in \overline{\mathbb{R}}_+^2. \end{cases}$$

It can be verified that $\max \mathcal{L}_f V(x_1, x_2) \leq 0$ for almost all $(x_1, x_2) \in \overline{\mathbb{R}}_+^2$ and $\mathcal{L}_f V^{-1}(0) = \mathcal{E}$.

Next, we show that the vector field f of the system given by (8.17) and (8.18) is nontangent to $\mathcal{E}$. Let $(x_1, x_2) \in \mathcal{E}$ and note that it follows from the expression of f that the direction cone $\mathcal{F}_{(x_1, x_2)}$ of the vector field f at $(x_1, x_2) \in \mathcal{E}$ is given by

$$\mathcal{F}_{(x_1, x_2)} = \{k[1, 2]^{\mathrm{T}} : k \in \mathbb{R}\}. \tag{8.21}$$

In addition, the tangent cone to $\mathcal{E}$ at $(x_1, x_2) \in \mathcal{E}$ is given by

$$T_{(x_1, x_2)}\mathcal{E} = \{k[1, 1]^{\mathrm{T}} : k \in \mathbb{R}\}. \tag{8.22}$$

Now, it follows from (8.21) and (8.22) that $T_{(x_1, x_2)}\mathcal{E} \cap \mathcal{F}_{(x_1, x_2)} = \{0\}$, and hence, for every $(x_1, x_2) \in \mathcal{E}$, f is nontangent to $\mathcal{E}$ at (x_1, x_2).

Finally, it follows from Theorem 8.4 that almost all solutions of the system converge to Lyapunov stable equilibria in $\mathcal{E}$, and hence, by definition, the system given by (8.17) and (8.18) is multistable. Figure 8.1 shows the state trajectories versus time for the initial condition $[x_{10}, x_{20}] = [1, 2]$. △

8.4 Energy Equipartition for Thermodynamic Systems with Discontinuous Power Balance Dynamics

In this section, we develop intercompartmental flow laws with discontinuous power balance dynamics that guarantee semistability and energy equipartition for a thermodynamically consistent model of the form developed in

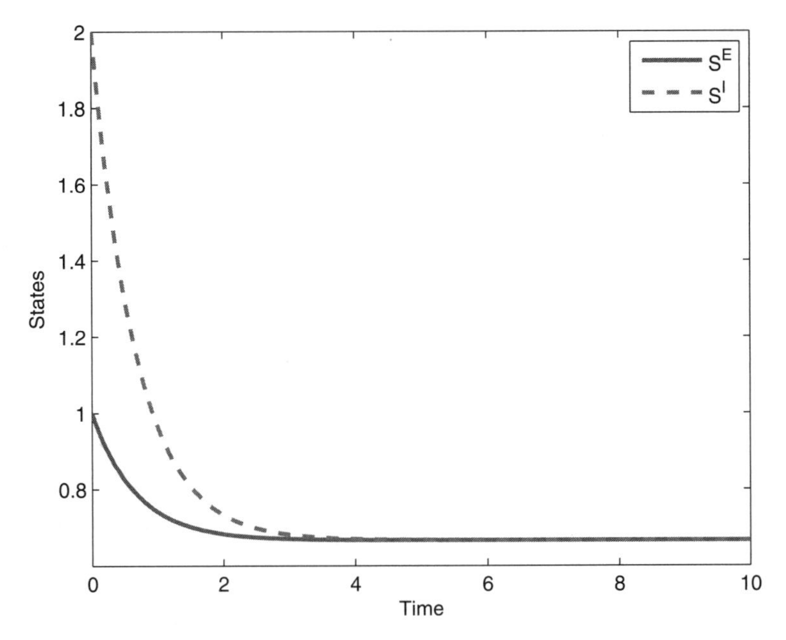

Figure 8.1 State trajectories versus time for Example 8.3 with initial condition $[x_{10}, x_{20}] = [1, 2]$.

Chapter 3. Specifically, consider the dynamical system $\mathcal{G}$ given by

$$\dot{E}_i(t) = \sum_{j=1, j\neq i}^{q} \phi_{ij}(E_i(t), E_j(t)), \quad i = 1, \ldots, q, \tag{8.23}$$

where $\phi_{ij}(\cdot, \cdot)$, $i, j = 1, \ldots, q$, are Lebesgue measurable and locally essentially bounded. Here, $\phi_{ij}(E)$, $E \in \overline{\mathbb{R}}_+^q$, denotes the net energy flow from the jth compartment to the ith compartment as defined in Chapter 3. In vector form, (8.23) becomes

$$\dot{E}(t) = f(E(t)), \quad E(t_0) = E_0, \quad t \geq t_0, \tag{8.24}$$

where $E(t) \triangleq [E_1(t), \ldots, E_q(t)]^{\mathrm{T}} \in \overline{\mathbb{R}}_+^q$, $t \geq t_0$, and $f = [f_1, \ldots, f_q]^{\mathrm{T}} : \overline{\mathbb{R}}_+^q \to \mathbb{R}^q$ is such that $f_i(t) = \sum_{j=1}^{q} \phi_{ij}(E_i, E_j)$.

The following proposition is a restatement of Theorem 7.5.

Proposition 8.4. Consider the dynamical system (8.23) and assume that Axioms *i*) and *ii*) of Chapter 3 hold. Then $f_i(E) = 0$ for all $i = 1, \ldots, q$ if and only if $E_1 = \cdots = E_q$. Furthermore, $\alpha\mathbf{e}$, $\alpha \in \overline{\mathbb{R}}_+$, is an equilibrium state of (8.23).

To address thermodynamic systems with discontinuous power balance dynamics, consider the large-scale dynamical system with the switched power balance equation given by

$$\dot{E}_i(t) = \sum_{j=1, j\neq i}^{q} \phi_{ij}^{\sigma(t)}(E_i(t), E_j(t)), \quad i = 1, \ldots, q, \tag{8.25}$$

where $\sigma : [0, \infty) \to \Sigma$ is a piecewise constant switching signal, Σ is a finite index set, and $\phi_{ij}^{\sigma} : \mathbb{R} \times \mathbb{R} \to \mathbb{R}$ is Lebesgue measurable and locally essentially bounded, and satisfies Axioms i) and ii) of Chapter 3 for every $\sigma \in \Sigma$. We denote by t_i, $i = 1, 2, \ldots$, the consecutive discontinuities of σ. Furthermore, we assume that $\mathcal{C} = \mathcal{C}^{\mathrm{T}}$ in Axiom i), where $\mathcal{C} = \mathcal{C}(t)$, $t \geq 0$. The assumption $\mathcal{C} = \mathcal{C}^{\mathrm{T}}$ implies that the underlying dynamic graph for the dynamical system $\mathcal{G}$ given by (8.23) is undirected.

Theorem 8.5. Consider the switched dynamical system given by (8.25). Assume that Axioms i) and ii) of Chapter 3 hold for every $\sigma \in \Sigma$. Furthermore, assume that $\mathcal{C} = \mathcal{C}^{\mathrm{T}}$, where $\mathcal{C} = \mathcal{C}(t)$, $t \geq 0$, in Axiom i). Then, for every $\alpha \in \overline{\mathbb{R}}_+$, $E_1 = \cdots = E_q = \alpha$ is a semistable state of (8.25). Furthermore, $E_i(t) \to \frac{1}{q}\sum_{i=1}^{q} E_{i0}$ as $t \to \infty$ and $\frac{1}{q}\sum_{i=1}^{q} E_{i0}$ is a semistable equilibrium state.

Proof. Consider the shifted-system ectropy as a Lyapunov function candidate given by

$$V(E) = \frac{1}{2}(E - \alpha\mathbf{e})^{\mathrm{T}}(E - \alpha\mathbf{e}), \tag{8.26}$$

where $E \triangleq [E_1, \ldots, E_q]^{\mathrm{T}} \in \overline{\mathbb{R}}_+^q$ and $\alpha \in \overline{\mathbb{R}}_+$. Next, we rewrite (8.25) as the differential inclusion (8.1). For every $v \in \mathcal{K}[f](E)$, let $V^o(E, v) \triangleq E^{\mathrm{T}}v$ and $\max V^o(E, v) \triangleq \max_{v \in \mathcal{K}[f]}\{E^{\mathrm{T}}v\}$.

Now, it follows from Theorem 1 of [351] that

$$\begin{aligned} E^{\mathrm{T}}\mathcal{K}[f](E) &= \mathcal{K}[E^{\mathrm{T}}f](E) \\ &= \mathcal{K}\left[\sum_{i=1}^{q} E_i \left(\sum_{j=1, j\neq i}^{q} \phi_{ij}^{\sigma}(E_i, E_j)\right)\right](E) \\ &= \mathcal{K}\left[\sum_{i=1}^{q-1}\sum_{j=i+1}^{q} (E_i - E_j)\phi_{ij}^{\sigma}(E_i, E_j)\right](E), \quad E \in \overline{\mathbb{R}}_+^q, \end{aligned} \tag{8.27}$$

and hence, by definition of differential inclusions, it follows that

$$\max V^o(E, v) = \max \overline{\mathrm{co}}\left\{\sum_{i=1}^{q-1}\sum_{j=i+1}^{q} (E_i - E_j)\phi_{ij}^{\sigma}(E_i, E_j)\right\}.$$

Note that since, by Axiom *ii*), $\sum_{i=1}^{q-1}\sum_{j=i+1}^{q}(E_i - E_j)\phi_{ij}^{\sigma}(E_i, E_j) \leq 0$, $E_i \in \overline{\mathbb{R}}_+$, it follows that $\max V^o(E, v)$ cannot be positive, and hence, the largest value $\max V^o(E, v)$ can achieve is zero, which establishes Lyapunov stability of $E \equiv \alpha\mathbf{e}$.

Finally, note that $0 \in \mathcal{L}_f V(E)$ if and only if

$$\sum_{i=1}^{q-1}\sum_{j=i+1}^{q}(E_i - E_j)\phi_{ij}^{\sigma}(E_i, E_j) = 0,$$

and hence,

$$\mathcal{Z} \triangleq \left\{ E \in \overline{\mathbb{R}}_+^q : \sum_{i=1}^{q-1}\sum_{j=i+1}^{q}(E_i - E_j)\phi_{ij}^{\sigma}(E_i, E_j) = 0 \right\}.$$

Now, it follows from Proposition 8.4 that $\mathcal{Z} = \{E \in \overline{\mathbb{R}}_+^q : E_1 = \cdots = E_q\}$. Since $\mathcal{Z}$ consists of equilibrium points, it follows that $\mathcal{M} = \mathcal{Z}$. Hence, it follows from Theorem 8.3 that $E = \alpha\mathbf{e}$ is semistable for all $\alpha \in \overline{\mathbb{R}}_+^q$. □

Note that Example 8.1 serves as a special case of Theorem 8.5.

Next, we give a concrete form for the discontinuous energy flow function $\phi_{ij}(E_i, E_j)$; namely,

$$\phi_{ij}(E_i, E_j) = \mathcal{C}_{(i,j)}\text{sgn}(E_j - E_i), \quad i = 1, \ldots, q. \tag{8.28}$$

It was shown in Chapter 7 that the *continuous* energy flow function $\phi_{ij}(E_i, E_j)$ given by

$$\phi_{ij}(E_i, E_j) = \mathcal{C}_{(i,j)}\text{sgn}(E_j - E_i)|E_j - E_i|^{\alpha}, \quad i = 1, \ldots, q, \tag{8.29}$$

achieves finite-time equipartition of energy for $0 < \alpha < 1$. Here, we show that (8.29) also achieves finite-time energy equipartition for $\alpha = 0$. Note that in this case, (8.29) reduces to (8.28).

Theorem 8.6. Consider the large-scale dynamical system $\mathcal{G}$ given by (8.23) with discontinuous energy flow function (8.28). Assume that Axioms *i*) and *ii*) of Chapter 3 hold. Then, for every $\alpha \in \overline{\mathbb{R}}_+$, $E_1 = \cdots = E_q = \alpha \in \overline{\mathbb{R}}_+$ is a finite-time semistable state of $\mathcal{G}$. Furthermore, $E_i(t) = \frac{1}{q}\sum_{i=1}^{q} E_{i0}$ for $t \geq T(E_{10}, \ldots, E_{q0})$ and $\frac{1}{q}\sum_{i=1}^{q} E_{i0}$ is a semistable equilibrium state.

Proof. Consider the shifted-system ectropy (8.26) as a Lyapunov function candidate. Since $V(E)$ is differentiable at x, it follows that $\mathcal{L}_f V(E) = (E - \alpha\mathbf{e})^{\mathrm{T}}\mathcal{K}[f](E)$. Now, it follows from Theorem 1 of [351] that

$$(E - \alpha\mathbf{e})^{\mathrm{T}}\mathcal{K}[f](E) = \mathcal{K}[(E - \alpha\mathbf{e})^{\mathrm{T}} f](E)$$

$$
\begin{aligned}
&= \mathcal{K}[E^{\mathrm{T}}f](E) \\
&= \mathcal{K}\left[\sum_{i=1}^{q} E_i \sum_{j=1, j\neq i}^{q} \mathcal{C}_{(i,j)}\mathrm{sgn}(E_j - E_i)\right](E) \\
&= \mathcal{K}\left[-\sum_{i=1}^{q}\sum_{j=1, j\neq i}^{q} \mathcal{C}_{(i,j)}(E_i - E_j)\mathrm{sgn}(E_i - E_j)\right](E) \\
&\subseteq -\sum_{i=1}^{q}\sum_{j=1, j\neq i}^{q} \mathcal{C}_{(i,j)}(E_i - E_j)\mathcal{K}[\mathrm{sgn}(E_i - E_j)](E) \\
&= -\sum_{i=1}^{q}\sum_{j=1, j\neq i}^{q} \mathcal{C}_{(i,j)}(E_i - E_j)\mathrm{SGN}(E_i - E_j) \\
&= -\sum_{i=1}^{q}\sum_{j=1, j\neq i}^{q} \mathcal{C}_{(i,j)}|E_i - E_j|, \quad E \in \overline{\mathbb{R}}_+^q,
\end{aligned}
\tag{8.30}
$$

which implies that $\max \mathcal{L}_f V(E) \leq 0$ for almost all $E \in \overline{\mathbb{R}}_+^q$. Hence, it follows from Theorem 8.1 that $E_1 = \cdots = E_q = \alpha$ is Lyapunov stable.

Next, note that since

$$
\begin{aligned}
\mathcal{L}_f V(E) &= \mathcal{K}\left[-\sum_{i=1}^{q}\sum_{j=1, j\neq i}^{q} \mathcal{C}_{(i,j)}(E_i - E_j)\mathrm{sgn}(E_i - E_j)\right](E) \\
&= \mathcal{K}\left[-\sum_{i=1}^{q}\sum_{j=1, j\neq i}^{q} \mathcal{C}_{(i,j)}|E_i - E_j|\right](E),
\end{aligned}
$$

it follows that $0 \in \mathcal{L}_f V(E)$ if and only if $E_1 = \cdots = E_q$, and hence, $\mathcal{Z} = \{E \in \overline{\mathbb{R}}_+^q : E_1 = \cdots = E_q\}$. Since the largest weakly positively invariant subset $\mathcal{M}$ of $\mathcal{Z}$ is given by $\mathcal{M} = \{E \in \overline{\mathbb{R}}_+^q : E_1 = \cdots = E_q = \alpha, \alpha \in \overline{\mathbb{R}}_+\}$, it follows from Theorem 8.3 that $\mathcal{G}$ is semistable.

Finally, we show that $\mathcal{G}$ is finite-time semistable. To see this, consider the nonnegative function $U(E) = \frac{1}{2}\sum_{i=1}^{q}\sum_{j=1, j\neq i}^{q} \mathcal{C}_{(i,j)}|E_i - E_j|$. In this case, using similar arguments as in Example 8.2, it follows that

$$
\mathcal{L}_f U(E) = \begin{cases} \left\{-2\sum_{i=1}^{q}\sum_{j=1, j\neq i}^{q} \mathcal{C}_{(i,j)}\right\}, & E_i \neq E_j,\ i,j = 1,\ldots,q,\ i \neq j, \\ \emptyset, & E_k = E_l \text{ for some } k,l \in \{1,\ldots,q\},\ k \neq l, \\ \{0\}, & E_1 = \cdots = E_q, \end{cases}
\tag{8.31}
$$

which implies that either $\max \mathcal{L}_f U(E) \leq -2\sum_{i=1}^{q}\sum_{j=1,j\neq i}^{q} \mathcal{C}_{(i,j)} < 0$ or $\mathcal{L}_f U(E) = \emptyset$ for almost all $E \in \overline{\mathbb{R}}_+^q \backslash \mathcal{Z}$. Hence, it follows from Corollary 8.1 that $\mathcal{G}$ is globally finite-time semistable. □

Note that Example 8.2 serves as a special case of Theorem 8.6.

Chapter Nine

Thermodynamic Modeling of Discrete Dynamical Systems

9.1 Introduction

In the previous chapters, we developed a continuous-time dynamical systems framework for thermodynamics. In some disciplines, however, a discrete-time framework is more natural in describing the evolving system dynamics. In information theory, for example, discrete information is used as a sequence of symbols, chosen from a finite set, involving quantized signals that are broken into different amplitude levels. Treating these signals as waveform pulses can significantly speed their transmission. In this case, the information flow (i.e., bit transmission) between subsystems is converted into discrete states. In the mathematical theory of communication, the content of information is the same, however, the information density can vary, and Shannon entropy gives a measure of this information density.

In particular, a high Shannon entropy implies that a new character in a transmission sequence contains new information, whereas a low Shannon entropy implies that the next character in the sequence does not provide any new information. Even though discrete and continuous forms of Shannon entropy have been defined in the literature, in its discrete form the value of the Shannon entropy is uniquely determined by the probability measure over the transmitted messages. In contrast, the continuous form of the Shannon entropy depends on the coordinates that are chosen to describe the message, and hence, it does not measure the amount of information content since an information measure is independent of the way in which we describe a message. In addition, the uncertainty reduction in receiving a message of a continuous distribution is infinite, and hence, cannot be measured by the continuous form of the Shannon entropy function.

In this chapter, we give a parallel development to the dynamical system formulation given in Chapter 3 for discrete-time thermodynamic models. Specifically, the present chapter is directed toward developing nonlinear discrete-time compartmental models that are consistent with thermodynamic principles. In particular, since thermodynamic models are concerned with energy flow among subsystems, we develop a nonlinear compartmental dynamical system model that is characterized by energy conservation laws capturing the exchange of energy between coupled macroscopic subsystems. Furthermore, using graph-theoretic notions, we state three thermodynamic axioms consistent with the zeroth and second laws of thermodynamics that ensure that our discrete-time, large-scale dynamical system model gives rise to a thermodynamically consistent energy flow model. Specifically, using a large-scale dynamical systems theory perspective, we show that our compartmental dynamical system model leads to a precise formulation of the equivalence between work energy and heat in a discrete-time, large-scale dynamical system.

Next, we give a deterministic definition of entropy for a discrete-time, large-scale dynamical system that is consistent with the thermodynamic definition of entropy given in Chapter 3 and show that it satisfies a discrete Clausius-type inequality leading to the law of entropy nonconservation. Furthermore, we introduce a discrete-time ectropy function that gives a measure of the tendency of a large-scale dynamical system to do useful work and grow more organized, and show that conservation of energy in an isolated thermodynamically consistent system necessarily leads to nonconservation of ectropy and entropy. Then, using the system ectropy as a Lyapunov function candidate, we show that our thermodynamically consistent large-scale nonlinear dynamical system model possesses a continuum of equilibria and is discrete-time semistable; that is, it has convergent subsystem energies to Lyapunov stable energy equilibria determined by the large-scale system initial subsystem energies.

Finally, we show that the steady-state distribution of the large-scale system energies is uniform, leading to system energy equipartitioning corresponding to a minimum ectropy and a maximum entropy equilibrium state. In the case where the subsystem energies are proportional to subsystem temperatures, we show that our dynamical system model leads to temperature equipartition, wherein all the system energy is transferred into heat at a uniform temperature. Furthermore, we show that our system-theoretic definitions of entropy and ectropy are consistent with Boltzmann's kinetic theory of gases with discrete energy states involving an n-body theory of ideal gases divided by diathermal walls.

9.2 Mathematical Preliminaries

In this section, we introduce some additional notation, several definitions, and some key results needed for developing the main results of this chapter.

We begin by considering the general discrete-time nonlinear dynamical system

$$x(k+1) = f(x(k)), \qquad x(0) = x_0, \qquad k \in \overline{\mathbb{Z}}_+, \tag{9.1}$$

where $x(k) \in \mathcal{D} \subseteq \mathbb{R}^n$, $k \in \overline{\mathbb{Z}}_+$, is the discrete system state vector, $\mathcal{D}$ is a relatively open set, and $f : \mathcal{D} \to \mathbb{R}^n$. Furthermore, we denote the solution to (9.1) with initial condition $x(0) = x_0$ by $s(\cdot, x_0)$, so that the map of the dynamical system given by $s : \overline{\mathbb{Z}}_+ \times \mathcal{D} \to \mathcal{D}$ is continuous on $\mathcal{D}$ and satisfies the consistency property $s(0, x_0) = x_0$ and the semigroup property $s(\kappa, s(k, x_0)) = s(k + \kappa, x_0)$ for all $x_0 \in \mathcal{D}$ and $k, \kappa \in \overline{\mathbb{Z}}_+$. For discrete-time systems we use the notation $s(k, x_0)$, $k \in \overline{\mathbb{Z}}_+$, and $x(k)$, $k \in \overline{\mathbb{Z}}_+$, interchangeably as the solution of the nonlinear discrete-time system (9.1) with initial condition $x(0) = x_0$. Unless otherwise stated, we assume $f(\cdot)$ is continuous on $\mathcal{D}$. Furthermore, $x_e \in \mathcal{D}$ is an *equilibrium point* of (9.1) if and only if $f(x_e) = x_e$.

The following definition introduces the notion of nonnegative vector fields.

Definition 9.1. Let $f = [f_1, \ldots, f_n]^{\mathrm{T}} : \mathcal{D} \subseteq \overline{\mathbb{R}}_+^n \to \mathbb{R}^n$. Then f is *nonnegative* if $f(x) \geq\geq 0$ for all $x \in \overline{\mathbb{R}}_+^n$.

The following proposition shows that $\overline{\mathbb{R}}_+^n$ is an invariant set of (9.1) if and only if f is nonnegative.

Proposition 9.1. Suppose $\overline{\mathbb{R}}_+^n \subset \mathcal{D}$. Then $\overline{\mathbb{R}}_+^n$ is an invariant set with respect to (9.1) if and only if $f : \mathcal{D} \to \mathbb{R}^n$ is nonnegative.

Proof. Suppose $f : \mathcal{D} \to \mathbb{R}^n$ is nonnegative and let $x(0) \in \overline{\mathbb{R}}_+^n$. Then for every $i \in \{1, \ldots, n\}$, it follows that $x_i(k+1) = f_i(x(k)) \geq 0$. Thus, $x(k) \in \overline{\mathbb{R}}_+^n$, $k \in \overline{\mathbb{Z}}_+$. Conversely, suppose $x(k) \in \overline{\mathbb{R}}_+^n$, $k \in \overline{\mathbb{Z}}_+$, for all $x(0) \in \overline{\mathbb{R}}_+^n$ and assume, *ad absurdum*, that there exists $i \in \{1, \ldots, n\}$ and $x_0 \in \overline{\mathbb{R}}_+^n$ such that $f_i(x_0) < 0$. In this case, with $x(0) = x_0$, $x_i(1) = f_i(x(0)) = f_i(x_0) < 0$, which is a contradiction. □

It follows from Proposition 9.1 that if $x_0 \geq\geq 0$, then $x(k) \geq\geq 0$, $k \in \overline{\mathbb{Z}}_+$, if and only if f is nonnegative. In this case, we say that (9.1) is a *discrete-time nonnegative dynamical system*. Henceforth, in this chapter, we assume that f is nonnegative so that the discrete-time nonlinear dynamical system (9.1) is a nonnegative dynamical system.

The following definition introduces several types of stability corresponding to the equilibrium solution $x(k) \equiv x_e$ of the discrete-time system (9.1).

Definition 9.2. *i*) The equilibrium solution $x(k) \equiv x_{\rm e}$ to (9.1) is *Lyapunov stable with respect to* $\overline{\mathbb{R}}_+^n$ if, for all $\varepsilon > 0$, there exists $\delta = \delta(\varepsilon) > 0$ such that if $x_0 \in \mathcal{B}_\delta(x_{\rm e}) \cap \overline{\mathbb{R}}_+^n$, then $x(k) \in \mathcal{B}_\varepsilon(x_{\rm e}) \cap \overline{\mathbb{R}}_+^n$, $k \in \overline{\mathbb{Z}}_+$.

ii) The equilibrium solution $x(k) \equiv x_{\rm e}$ to (9.1) is (*locally*) *asymptotically stable with respect to* $\overline{\mathbb{R}}_+^n$ if it is Lyapunov stable with respect to $\overline{\mathbb{R}}_+^n$ and there exists $\delta > 0$ such that if $x_0 \in \mathcal{B}_\delta(x_{\rm e}) \cap \overline{\mathbb{R}}_+^n$, then $\lim_{k\to\infty} x(k) = x_{\rm e}$.

iii) The equilibrium solution $x(k) \equiv x_{\rm e}$ to (9.1) is *globally asymptotically stable with respect to* $\overline{\mathbb{R}}_+^n$ if it is Lyapunov stable with respect to $\overline{\mathbb{R}}_+^n$ and, for all $x(0) \in \overline{\mathbb{R}}_+^n$, $\lim_{k\to\infty} x(k) = x_{\rm e}$.

The following result gives sufficient conditions for Lyapunov and asymptotic stability of a discrete-time nonlinear dynamical system.

Theorem 9.1. Let $\mathcal{D}$ be a relatively open subset of $\overline{\mathbb{R}}_+^n$ that contains $x_{\rm e}$. Consider the discrete-time nonlinear dynamical system (9.1) where f is nonnegative and $f(x_{\rm e}) = x_{\rm e}$, and assume that there exists a continuous function $V : \mathcal{D} \to \mathbb{R}$ such that

$$V(x_{\rm e}) = 0, \tag{9.2}$$

$$V(x) > 0, \quad x \in \mathcal{D}, \quad x \neq x_{\rm e}, \tag{9.3}$$

$$V(f(x)) - V(x) \le 0, \quad x \in \mathcal{D}. \tag{9.4}$$

Then the equilibrium solution $x(k) \equiv x_{\rm e}$ to (9.1) is Lyapunov stable with respect to $\overline{\mathbb{R}}_+^n$. If, in addition,

$$V(f(x)) - V(x) < 0, \quad x \in \mathcal{D}, \quad x \neq x_{\rm e}, \tag{9.5}$$

then the equilibrium solution $x(k) \equiv x_{\rm e}$ to (9.1) is asymptotically stable with respect to $\overline{\mathbb{R}}_+^n$. Finally, if $V(\cdot)$ is such that

$$V(x) \to \infty \text{ as } \|x\| \to \infty, \tag{9.6}$$

then (9.5) implies that the equilibrium solution $x(k) \equiv x_{\rm e}$ to (9.1) is globally asymptotically stable with respect to $\overline{\mathbb{R}}_+^n$.

Proof. Let $\varepsilon > 0$ be such that $\mathcal{B}_\varepsilon(x_{\rm e}) \cap \overline{\mathbb{R}}_+^n \subseteq \mathcal{D}$. Since $\overline{\mathbb{R}}_+^n \cap \overline{\mathcal{B}}_\varepsilon(x_{\rm e})$ is compact and $f(x)$, $x \in \mathcal{D}$, is continuous, it follows that

$$\eta \triangleq \max\left\{\varepsilon, \max_{x \in \overline{\mathbb{R}}_+^n \cap \overline{\mathcal{B}}_\varepsilon(x_{\rm e})} \|f(x) - x_{\rm e}\|\right\} \tag{9.7}$$

exists. Next, let $\alpha \triangleq \min_{x\in\mathcal{D}:\ \varepsilon \le \|x - x_{\rm e}\| \le \eta} V(x)$. Note that $\alpha > 0$ since $x_{\rm e} \notin \partial\mathcal{B}_\varepsilon(x_{\rm e})$ and $V(x) > 0$, $x \in \mathcal{D}$, $x \neq x_{\rm e}$. Next, let $\beta \in (0, \alpha)$ and define $\mathcal{D}_\beta \triangleq \{x \in \overline{\mathbb{R}}_+^n \cap \mathcal{B}_\varepsilon(x_{\rm e}) :\ V(x) \le \beta\}$. Now, for every $x \in \mathcal{D}_\beta$, it follows from (9.4) that $V(f(x)) \leq V(x) \leq \beta$, and hence, it follows from (9.7) that $\|f(x) - x_{\rm e}\| \leq \eta$, $x \in \mathcal{D}_\beta$.

Next, suppose, *ad absurdum*, that there exists $x \in \mathcal{D}_\beta$ such that $\|f(x) - x_{\rm e}\| \geq \varepsilon$. This implies $V(x) \geq \alpha$, which is a contradiction. Hence, since f is nonnegative, for every $x \in \mathcal{D}_\beta$, it follows that $f(x) \in \overline{\mathbb{R}}_+^n \cap \mathcal{B}_\varepsilon(x_{\rm e}) \subset \mathcal{D}_\beta$, which implies that $\mathcal{D}_\beta$ is a positively invariant set (see Definition 9.4) with respect to (9.1). Next, since $V(\cdot)$ is continuous and $V(x_{\rm e}) = 0$, there exists $\delta = \delta(\varepsilon) \in (0, \varepsilon)$ such that $V(x) < \beta$, $x \in \overline{\mathbb{R}}_+^n \cap \mathcal{B}_\delta(x_{\rm e})$. Now, let $x(k)$, $k \in \overline{\mathbb{Z}}_+$, satisfy (9.1). Since $\overline{\mathbb{R}}_+^n \cap \mathcal{B}_\delta(x_{\rm e}) \subset \mathcal{D}_\beta \subset \overline{\mathbb{R}}_+^n \cap \mathcal{B}_\varepsilon(x_{\rm e}) \subseteq \mathcal{D}$ and $\mathcal{D}_\beta$ is positively invariant with respect to (9.1), it follows that for all $x(0) \in \mathcal{B}_\delta(x_{\rm e}) \cap \overline{\mathbb{R}}_+^n$, $x(k) \in \mathcal{B}_\varepsilon(x_{\rm e}) \cap \overline{\mathbb{R}}_+^n$, $k \in \overline{\mathbb{Z}}_+$, which proves Lyapunov stability of $x_{\rm e}$ with respect to $\overline{\mathbb{R}}_+^n$.

To prove asymptotic stability with respect to $\overline{\mathbb{R}}_+^n$, suppose that $V(f(x)) < V(x)$, $x \in \mathcal{D}$, $x \neq x_{\rm e}$, and $x(0) \in \mathcal{B}_\delta(x_{\rm e}) \cap \overline{\mathbb{R}}_+^n$. Then it follows that $x(k) \in \mathcal{B}_\varepsilon(x_{\rm e}) \cap \overline{\mathbb{R}}_+^n$, $k \in \overline{\mathbb{Z}}_+$. However, $V(x(k))$, $k \in \overline{\mathbb{Z}}_+$, is decreasing and bounded from below by zero. Now, suppose, *ad absurdum*, that $x(k)$, $k \in \overline{\mathbb{Z}}_+$, does not converge to $x_{\rm e}$. This implies that $V(x(k))$, $k \in \overline{\mathbb{Z}}_+$, is lower bounded by a positive number, that is, there exists $L > 0$ such that $V(x(k)) \geq L > 0$, $k \in \overline{\mathbb{Z}}_+$. Hence, by continuity of $V(x)$, $x \in \mathcal{D}$, there exists $\delta' > 0$ such that $V(x) < L$ for $x \in \mathcal{B}_{\delta'}(x_{\rm e}) \cap \overline{\mathbb{R}}_+^n$, which further implies that $x(k) \notin \mathcal{B}_{\delta'}(x_{\rm e}) \cap \overline{\mathbb{R}}_+^n$, $k \in \overline{\mathbb{Z}}_+$.

Next, let $L_1 \triangleq \min\{V(x) - V(f(x)) : \delta' \leq \|x - x_{\rm e}\| \leq \varepsilon, x \in \overline{\mathbb{R}}_+^n\}$. Now, (9.5) implies $V(x) - V(f(x)) \geq L_1$, $\delta' \leq \|x - x_{\rm e}\| \leq \varepsilon$, $x \in \overline{\mathbb{R}}_+^n$, or, equivalently,

$$V(x(k)) - V(x(0)) = \sum_{i=0}^{k-1} [V(f(x(i))) - V(x(i))] \leq -L_1 k,$$

and hence, for all $x(0) \in \overline{\mathbb{R}}_+^n \cap \mathcal{B}_\delta(x_{\rm e})$,

$$V(x(k)) \leq V(x(0)) - L_1 k.$$

Letting $k > \frac{V(x(0)) - L}{L_1}$, it follows that $V(x(k)) < L$, which is a contradiction. Hence, $x(k) \to x_{\rm e}$ as $k \to \infty$, establishing asymptotic stability with respect to $\overline{\mathbb{R}}_+^n$.

Finally, to prove global asymptotic stability with respect to $\overline{\mathbb{R}}_+^n$, let $x_0 \in \overline{\mathbb{R}}_+^n$, and let $\beta \triangleq V(x_0)$. Now, the radial unboundedness condition (9.6) implies that there exists $\varepsilon > 0$ such that $V(x) \geq \beta$ for $\|x - x_{\rm e}\| \geq \varepsilon$, $x \in \mathbb{R}^n$. Hence, since f is nonnegative, it follows from (9.5) that $V(x(k)) \leq V(x(0)) = \beta$, $k \in \overline{\mathbb{Z}}_+$, which implies that $x(k) \in \overline{\mathbb{R}}_+^n \cap \mathcal{B}_\varepsilon(x_{\rm e})$, $k \in \overline{\mathbb{Z}}_+$. Now, the proof follows as in the proof of the local result. □

A continuous function $V(\cdot)$ satisfying (9.2) and (9.3) is called a *Lyapunov function candidate* for the discrete-time nonlinear dynamical system (9.1). If, additionally, $V(\cdot)$ satisfies (9.4), then $V(\cdot)$ is called a *Lyapunov function* for the discrete-time nonlinear dynamical system (9.1).

Next, we use the discrete-time Krasovskii-LaSalle invariance principle to relax one of the conditions on the Lyapunov function $V(\cdot)$ given in Theorem 9.1. In particular, the strict negative-definiteness condition on the Lyapunov difference can be relaxed while ensuring system asymptotic stability. To state the main results of this section, several definitions and a key lemma analogous to the ones given in Section 2.3 are needed.

Definition 9.3. The trajectory $x(k)$, $k \in \overline{\mathbb{Z}}_+$, of (9.1) is *bounded* if there exists $\gamma > 0$ such that $\|x(k)\| < \gamma$, $k \in \overline{\mathbb{Z}}_+$.

Definition 9.4. A set $\mathcal{M} \subset \mathcal{D} \subseteq \mathbb{R}^n$ is a *positively invariant set* for the nonlinear dynamical system (9.1) if $s_k(\mathcal{M}) \subseteq \mathcal{M}$, for all $k \in \overline{\mathbb{Z}}_+$, where $s_k(\mathcal{M}) \triangleq \{s_k(x) : x \in \mathcal{M}\}$. A set $\mathcal{M} \subseteq \mathcal{D} \subseteq \mathbb{R}^n$ is an *invariant set* for the dynamical system (9.1) if $s_k(\mathcal{M}) = \mathcal{M}$ for all $k \in \overline{\mathbb{Z}}_+$.

Definition 9.5. A point $p \in \mathcal{D}$ is a *positive limit point* of the trajectory $x(k)$, $k \in \overline{\mathbb{Z}}_+$, of (9.1) if there exists a monotonic sequence $\{k_n\}_{n=0}^{\infty}$ of nonnegative numbers, with $k_n \to \infty$ as $n \to \infty$, such that $x(k_n) \to p$ as $n \to \infty$. The set of all positive limit points of $x(k)$, $k \in \overline{\mathbb{Z}}_+$, is the *positive limit set* $\omega(x_0)$ of $x(k)$, $k \in \overline{\mathbb{Z}}_+$.

Note that if $p \in \mathcal{D}$ is a positive limit point of the trajectory $x(\cdot)$, then, for all $\varepsilon > 0$ and finite $K \in \mathbb{Z}_+$, there exists $k > K$ such that $\|x(k) - p\| < \varepsilon$. This follows from the fact that $\|x(k) - p\| < \varepsilon$ for all $\varepsilon > 0$ and some $k > K > 0$ is equivalent to the existence of a sequence of integers $\{k_n\}_{n=0}^{\infty}$, with $k_n \to \infty$ as $n \to \infty$, such that $x(k_n) \to p$ as $n \to \infty$.

Next, we state and prove a key lemma involving positive limit sets for discrete-time systems.

Lemma 9.1. Consider the nonlinear dynamical system (9.1) where f is nonnegative. Suppose the solution $x(k)$ to (9.1) corresponding to an initial condition $x(0) = x_0$ is bounded for all $k \in \overline{\mathbb{Z}}_+$. Then the positive limit set $\omega(x_0)$ of $x(k)$, $k \in \overline{\mathbb{Z}}_+$, is a nonempty, compact, invariant subset of $\overline{\mathbb{R}}_+^n$. Furthermore, $x(k) \to \omega(x_0)$ as $k \to \infty$.

Proof. Let $x(k)$, $k \in \overline{\mathbb{Z}}_+$, denote the solution to (9.1) corresponding to the initial condition $x(0) = x_0$. Next, since $x(k)$ is bounded for all

$k \in \overline{\mathbb{Z}}_+$, it follows from the Bolzano-Weierstrass theorem [191, p. 27] that every sequence in the positive orbit $\mathcal{O}^+_{x_0} \triangleq \{s(k, x_0) : k \in \overline{\mathbb{Z}}_+\}$ has at least one accumulation point $p \in \mathcal{D}$ as $k \to \infty$, and hence, $\omega(x_0)$ is nonempty. Next, let $p \in \omega(x_0)$ so that there exists an increasing unbounded sequence $\{k_n\}_{n=0}^{\infty}$, with $k_0 = 0$, such that $\lim_{n\to\infty} x(k_n) = p$. Now, since $x(k_n)$ is uniformly bounded in n, it follows that the limit point p is bounded, which implies that $\omega(x_0)$ is bounded.

To show that $\omega(x_0)$ is closed, let $\{p_i\}_{i=0}^{\infty}$ be a sequence contained in $\omega(x_0)$ such that $\lim_{i\to\infty} p_i = p$. Now, since $p_i \to p$ as $i \to \infty$ for every $\varepsilon > 0$, there exists i such that $\|p - p_i\| < \varepsilon/2$. Next, since $p_i \in \omega(x_0)$, there exists $k \geq K$, where $K \in \overline{\mathbb{Z}}_+$ is arbitrary and finite, such that $\|p_i - x(k)\| < \varepsilon/2$. Now, since $\|p - p_i\| < \varepsilon/2$ and $\|p_i - x(k)\| < \varepsilon/2$, $k \geq K$, it follows that $\|p - x(k)\| \leq \|p_i - x(k)\| + \|p - p_i\| < \varepsilon$. Thus, $p \in \omega(x_0)$. Hence, every accumulation point of $\omega(x_0)$ is an element of $\omega(x_0)$ so that $\omega(x_0)$ is closed. Thus, since $\omega(x_0)$ is closed and bounded, $\omega(x_0)$ is compact.

To show positive invariance of $\omega(x_0)$, let $p \in \omega(x_0)$ so that there exists an increasing sequence $\{k_n\}_{n=0}^{\infty}$ such that $x(k_n) \to p$ as $n \to \infty$. Now, let $s(k_n, x_0)$ denote the solution $x(k_n)$ of (9.1) with initial condition $x(0) = x_0$ and note that, since $f : \mathcal{D} \to \mathcal{D}$ in (9.1) is continuous, $x(k)$, $k \in \overline{\mathbb{Z}}_+$, is the unique solution to (9.1) so that $s(k + k_n, x_0) = s(k, s(k_n, x_0)) = s(k, x(k_n))$. Now, since $s(k, x_0)$, $k \in \overline{\mathbb{Z}}_+$, is continuous with respect to x_0, it follows that, for $k + k_n \geq 0$, $\lim_{n\to\infty} s(k + k_n, x_0) = \lim_{n\to\infty} s(k, x(k_n)) = s(k, p)$, and hence, $s(k, p) \in \omega(x_0)$. Hence, $s_k(\omega(x_0)) \subseteq \omega(x_0)$, $k \in \overline{\mathbb{Z}}_+$, establishing positive invariance of $\omega(x_0)$.

To show invariance of $\omega(x_0)$, let $y \in \omega(x_0)$ so that there exists an increasing unbounded sequence $\{k_n\}_{n=0}^{\infty}$ such that $s(k_n, x_0) \to y$ as $n \to \infty$. Next, let $k \in \overline{\mathbb{Z}}_+$ and note that there exists N such that $k_n > k$, $n \geq N$. Hence, it follows from the semigroup property that $s(k, s(k_n - k, x_0)) = s(k_n, x_0) \to y$ as $n \to \infty$. Now, it follows from the Bolzano-Lebesgue theorem [191, p. 28] that there exists a subsequence z_{n_i} of the sequence $z_n = s(k_n - k, x_0)$, $n = N, N+1, \ldots$, such that $z_{n_i} \to z \in \mathcal{D}$ and, by definition, $z \in \omega(x_0)$. Next, it follows from the continuous dependence property that $\lim_{i\to\infty} s(k, z_{n_i}) = s(k, \lim_{i\to\infty} z_{n_i})$, and hence, $y = s(t, z)$, which implies that $\omega(x_0) \subseteq s_k(\omega(x_0))$, $k \in \overline{\mathbb{Z}}_+$. Now, using positive invariance of $\omega(x_0)$, it follows that $s_k(\omega(x_0)) = \omega(x_0)$, $k \in \overline{\mathbb{Z}}_+$, establishing invariance of the positive limit set $\omega(x_0)$.

Finally, to show $x(k) \to \omega(x_0)$ as $k \to \infty$, suppose, *ad absurdum*, that $x(k) \not\to \omega(x_0)$ as $k \to \infty$. In this case, there exists a sequence $\{k_n\}_{n=0}^{\infty}$, with

$k_n \to \infty$ as $n \to \infty$, such that

$$\inf_{p\in\omega(x_0)} \|x(k_n) - p\| > \varepsilon, \quad n \in \overline{\mathbb{Z}}_+. \tag{9.8}$$

However, since $x(k)$, $k \in \overline{\mathbb{Z}}_+$, is bounded, the bounded sequence $\{x(k_n)\}_{n=0}^{\infty}$ contains a convergent subsequence $\{x(k_n^*)\}_{n=0}^{\infty}$ such that $x(k_n^*) \to p^* \in \omega(x_0)$ as $n \to \infty$, which contradicts (9.8). Hence, $x(k) \to \omega(x_0)$ as $k \to \infty$ and, since f is nonnegative, $\omega(x_0) \subseteq \overline{\mathbb{R}}_+^n$. □

Next, we present a discrete-time version of the Krasovskii-LaSalle invariance principle.

Theorem 9.2. Consider the discrete-time nonlinear dynamical system (9.1) where f is nonnegative, assume $\mathcal{D}_c \subset \mathcal{D} \subseteq \overline{\mathbb{R}}_+^n$ is a compact invariant set with respect to (9.1), and assume that there exists a continuous function $V\colon \mathcal{D}_c \to \mathbb{R}$ such that $V(f(x)) - V(x) \le 0$, $x \in \mathcal{D}_c$. Let $\mathcal{R} \triangleq \{x \in \mathcal{D}_c : V(f(x)) = V(x)\}$ and let $\mathcal{M}$ denote the largest invariant set contained in $\mathcal{R}$. If $x(0) \in \mathcal{D}_c$, then $x(k) \to \mathcal{M}$ as $k \to \infty$.

Proof. Let $x(k)$, $k \in \overline{\mathbb{Z}}_+$, be a solution to (9.1) with $x(0) \in \mathcal{D}_c$. Since $V(f(x)) \le V(x)$, $x \in \mathcal{D}_c$, it follows that

$$V(x(k)) - V(x(\kappa)) = \sum_{i=\kappa}^{k-1}[V(f(x(i))) - V(x(i))] \le 0, \qquad k-1 \ge \kappa,$$

and hence, $V(x(k)) \le V(x(\kappa))$, $k-1 \ge \kappa$, which implies that $V(x(k))$ is a nonincreasing function of k. Next, since $V(x)$ is continuous on the compact set $\mathcal{D}_c$, there exists $L \ge 0$ such that $V(x) \ge L$, $x \in \mathcal{D}_c$. Hence, $\gamma_{x_0} \triangleq \lim_{k\to\infty} V(x(k))$ exists.

Now, for all $p \in \omega(x_0)$ there exists an increasing unbounded sequence $\{k_n\}_{n=0}^{\infty}$, with $k_0 = 0$, such that $x(k_n) \to p$ as $n \to \infty$. Since $V(x)$, $x \in \mathcal{D}$, is continuous, $V(p) = V(\lim_{n\to\infty} x(k_n)) = \lim_{n\to\infty} V(x(k_n)) = \gamma_{x_0}$, and hence, $V(x) = \alpha$ on $\omega(x_0)$. Next, since $\mathcal{D}_c$ is compact and invariant, it follows that $x(k)$, $k \in \overline{\mathbb{Z}}_+$, is bounded, and hence, it follows from Lemma 9.1 that $\omega(x_0)$ is a nonempty, compact invariant set. Hence, it follows that $V(f(x)) = V(x)$ on $\omega(x_0)$, and hence, $\omega(x_0) \subset \mathcal{M} \subset \mathcal{R} \subset \mathcal{D}_c$. Finally, since $x(k) \to \omega(x_0)$ as $k \to \infty$, it follows that $x(k) \to \mathcal{M}$ as $k \to \infty$. □

Now, using Theorem 9.2 we provide a generalization of Theorem 9.1 for local asymptotic stability of a nonlinear dynamical system.

Corollary 9.1. Consider the nonlinear dynamical system (9.1) where f is nonnegative, assume $\mathcal{D}_c \subset \mathcal{D} \subseteq \overline{\mathbb{R}}_+^n$ is a compact invariant set with respect to (9.1) such that $x_e \in \mathcal{D}_c$, and assume that there exists a continuous function $V\colon \mathcal{D}_{\mathrm{c}} \to \mathbb{R}$ such that $V(x_{\mathrm{e}}) = 0$, $V(x) > 0$, $x \neq x_{\mathrm{e}}$, and $V(f(x)) - V(x) \leq 0$, $x \in \mathcal{D}_{\mathrm{c}}$. Furthermore, assume that the set $\mathcal{R} \triangleq \{x \in \mathcal{D}_{\mathrm{c}}\colon\ V(f(x)) = V(x)\}$ contains no invariant set other than the set $\{x_{\mathrm{e}}\}$. Then the equilibrium solution $x(k) \equiv x_{\mathrm{e}}$ to (9.1) is asymptotically stable with respect to $\overline{\mathbb{R}}_+^n$.

Proof. The result is a direct consequence of Theorem 9.2. □

In Theorem 9.2 and Corollary 9.1, we explicitly assumed that there exists a compact invariant set $\mathcal{D}_{\mathrm{c}} \subset \mathcal{D} \subseteq \overline{\mathbb{R}}_+^n$ of (9.1). Next, we provide a result that does not require the existence of a compact invariant set $\mathcal{D}_{\mathrm{c}}$.

Theorem 9.3. Consider the nonlinear dynamical system (9.1) where f is nonnegative and assume that there exists a continuous function $V\colon \overline{\mathbb{R}}_+^n \to \mathbb{R}$ such that

$$V(x_{\mathrm{e}}) = 0, \tag{9.9}$$

$$V(x) > 0, \quad x \in \overline{\mathbb{R}}_+^n, \quad x \neq x_{\mathrm{e}}, \tag{9.10}$$

$$V(f(x)) - V(x) \leq 0, \quad x \in \overline{\mathbb{R}}_+^n. \tag{9.11}$$

Let $\mathcal{R} \triangleq \{x \in \mathbb{R}^n\colon V(f(x)) = V(x)\}$ and let $\mathcal{M}$ be the largest invariant set contained in $\mathcal{R}$. Then all solutions $x(k)$, $k \in \overline{\mathbb{Z}}_+$, to (9.1) that are bounded approach $\mathcal{M}$ as $k \to \infty$.

Proof. Let $x \in \overline{\mathbb{R}}_+^n$ be such that trajectory $s(k, x)$, $k \in \overline{\mathbb{Z}}_+$, of (9.1) is bounded. Since f is nonnegative, it follows that $s(k, x) \in \overline{\mathbb{R}}_+^n$, $k \in \overline{\mathbb{Z}}_+$. Now, with $\mathcal{D}_{\mathrm{c}} = \overline{\mathcal{O}_x^+}$, it follows from Theorem 9.2 that $s(k, x) \to \mathcal{M}$ as $k \to \infty$. □

Next, we present the global invariant set theorem for guaranteeing global asymptotic stability of a discrete-time nonlinear dynamical system.

Theorem 9.4. Consider the nonlinear dynamical system (9.1) where f is nonnegative and assume that there exists a continuous function $V\colon \overline{\mathbb{R}}_+^n \to \mathbb{R}$ such that

$$V(x_{\mathrm{e}}) = 0, \tag{9.12}$$

$$V(x) > 0, \quad x \in \overline{\mathbb{R}}_+^n, \quad x \neq x_{\mathrm{e}}, \tag{9.13}$$

$$V(f(x)) - V(x) \leq 0, \quad x \in \overline{\mathbb{R}}_+^n, \tag{9.14}$$

$$V(x) \to \infty \text{ as } \|x\| \to \infty. \tag{9.15}$$

Furthermore, assume that the set $\mathcal{R} \triangleq \{x \in \mathcal{D}\colon\ V(f(x)) = V(x)\}$ contains no invariant set other than the set $\{x_{\mathrm{e}}\}$. Then the equilibrium solution $x(k) \equiv x_{\mathrm{e}}$ to (9.1) is globally asymptotically stable with respect to $\overline{\mathbb{R}}_+^n$.

Proof. Since (9.12)–(9.14) hold, it follows from Theorem 9.1 that the equilibrium solution $x(k) \equiv x_{\rm e}$ to (9.1) is Lyapunov stable, while the radial unboundedness condition (9.15) implies that all solutions to (9.1) are bounded. Now, Theorem 9.3 implies that $x(k) \to \mathcal{M}$ as $k \to \infty$. However, since $\mathcal{R}$ contains no invariant set other than the set $\{x_{\rm e}\}$, the set $\mathcal{M}$ is $\{x_{\rm e}\}$, and hence, global asymptotic stability is immediate. □

Finally, we introduce the notion of semistability for discrete-time nonnegative dynamical systems.

Definition 9.6. An equilibrium solution $x(k) \equiv x_{\rm e} \in \overline{\mathbb{R}}_+^n$ to (9.1) is *semistable with respect to* $\overline{\mathbb{R}}_+^n$ if it is Lyapunov stable with respect to $\overline{\mathbb{R}}_+^n$ and there exists $\delta > 0$ such that if $x_0 \in \mathcal{B}_\delta(x_{\rm e}) \cap \overline{\mathbb{R}}_+^n$, then $\lim_{k\to\infty} x(k)$ exists and corresponds to a Lyapunov stable equilibrium point with respect to $\overline{\mathbb{R}}_+^n$. An equilibrium point $x_{\rm e} \in \overline{\mathbb{R}}^n$ is a *globally semistable equilibrium with respect to* $\overline{\mathbb{R}}_+^n$ if it is Lyapunov stable with respect to $\overline{\mathbb{R}}_+^n$ and, for every $x_0 \in \overline{\mathbb{R}}_+^n$, $\lim_{k\to\infty} x(k)$ exists and corresponds to a Lyapunov stable equilibrium point with respect to $\overline{\mathbb{R}}_+^n$. The system (9.1) is said to be *Lyapunov stable with respect to* $\overline{\mathbb{R}}_+^n$ if every equilibrium point of (9.1) is Lyapunov stable with respect to $\overline{\mathbb{R}}_+^n$. The system (9.1) is said to be *semistable with respect to* $\overline{\mathbb{R}}_+^n$ if every equilibrium point of (9.1) is semistable with respect to $\overline{\mathbb{R}}_+^n$. Finally, (9.1) is said to be *globally semistable with respect to* $\overline{\mathbb{R}}_+^n$ if every equilibrium point of (9.1) is globally semistable with respect to $\overline{\mathbb{R}}_+^n$.

Theorem 9.5. Let $\mathcal{D}_{\rm c} \subset \overline{\mathbb{R}}_+^n$ be a compact invariant set with respect to (9.1). Suppose there exists a continuous function $V : \mathcal{D}_{\rm c} \to \mathbb{R}$ such that $V(f(x)) - V(x) \leq 0$, $x \in \mathcal{D}_{\rm c}$. Let $\mathcal{R} \triangleq \{x \in \mathcal{D}_{\rm c} : V(f(x)) = V(x)\}$ and let $\mathcal{M}$ denote the largest invariant set contained in $\mathcal{R}$. If every element in $\mathcal{M}$ is a Lyapunov stable equilibrium point with respect to $\mathcal{D}_{\rm c}$, then (9.1) is semistable with respect to $\mathcal{D}_{\rm c}$.

Proof. Since every solution of (9.1) is bounded, it follows from the hypotheses on $V(\cdot)$ that, for every $x \in \mathcal{D}_{\rm c}$, the positive limit set $\omega(x)$ of (9.1) is nonempty and contained in the largest invariant subset $\mathcal{M}$ of $\mathcal{R}$. Since every point in $\mathcal{M}$ is a Lyapunov stable equilibrium point, it follows that every point in $\omega(x)$ is a Lyapunov stable equilibrium point.

Next, let $z \in \omega(x)$ and let $\mathcal{U}_\varepsilon$ be a relatively open neighborhood of z. By Lyapunov stability of z, it follows that there exists a relatively open subset $\mathcal{U}_\delta$ containing z such that $s_k(\mathcal{U}_\delta \cap \overline{\mathbb{R}}_+^n) \subseteq \mathcal{U}_\varepsilon \cap \overline{\mathbb{R}}_+^n$ for every $k \geq k_0$. Since $z \in \omega(x)$, it follows that there exists $h \geq 0$ such that $s(h,x) \in \mathcal{U}_\delta \cap \overline{\mathbb{R}}_+^n$. Thus, $s(k+h,x) = s_k(s(h,x)) \in s_k(\mathcal{U}_\delta \cap \overline{\mathbb{R}}_+^n) \subseteq \mathcal{U}_\varepsilon \cap \overline{\mathbb{R}}_+^n$ for every $k > k_0$. Hence, since $\mathcal{U}_\varepsilon$ was chosen arbitrarily, it follows that $z = \lim_{k\to\infty} s(k,x)$.

Now, it follows that $\lim_{i\to\infty} s(k_i, x) \to z$ for every divergent sequence $\{k_i\}$, and hence, $\omega(x) = \{z\}$. Finally, since $\lim_{k\to\infty} s(k, x) \in \mathcal{M}$ is Lyapunov stable for every $x \in \mathcal{D}_c$, it follows from the definition of semistability that every equilibrium point in $\mathcal{M}$ is semistable. □

Next, we introduce several definitions and some key results concerning nonnegative matrices [31,32,219,320] that are necessary for developing some of the results of this chapter.

Definition 9.7. Let $A \in \mathbb{R}^{n\times n}$. Then A is (*discrete-time*) *compartmental* if A is nonnegative and $\sum_{i=1}^{n} A_{(i,j)} \le 1$, $j = 1, \ldots, n$, or, equivalently, $\mathbf{e}^{\mathrm{T}}(A - I_n) \le\le 0$.

Definition 9.8. A real function $u : \overline{\mathbb{Z}}_+ \to \mathbb{R}^m$ is a *nonnegative* (respectively, *positive*) *function* if $u(k) \ge\ge 0$ (respectively, $u(k) >> 0$), $k \in \overline{\mathbb{Z}}_+$.

The following lemma is needed for developing several of the stability results in Section 9.10.

Lemma 9.2. Let $A \in \mathbb{R}^{n\times n}$ be nonnegative. Then the following statements are equivalent:

i) $I - A$ is an M-matrix.

ii) $\rho(A) \le 1$.

Furthermore, the following statements are equivalent:

iii) $I - A$ is a nonsingular M-matrix.

iv) $\det(I - A) \ne 0$ and $(I - A)^{-1} \ge\ge 0$.

v) For each $y \in \mathbb{R}^n$, $y \ge\ge 0$, there exists a unique $x \in \mathbb{R}^n$, $x \ge\ge 0$, such that $(I - A)x = y$.

vi) There exists $x \in \mathbb{R}^n$, $x \ge\ge 0$, such that $x >> Ax$.

vii) There exists $x \in \mathbb{R}^n$, $x >> 0$, such that $x >> Ax$.

Proof. Since $A \ge\ge 0$, it follows from Theorem 2.9 that $\rho(A) \in \mathrm{spec}(A)$, and hence, $\rho(A) = \max\{\mathrm{Re}\,\lambda : \lambda \in \mathrm{spec}(A)\}$. The equivalence of *i)* and *ii)* follows by noting that $I - A$ is an M-matrix if and only if $\mathrm{Re}\,\lambda \le 1$ for all $\lambda \in \mathrm{spec}(A)$ or, equivalently, $\max\{\mathrm{Re}\,\lambda : \lambda \in \mathrm{spec}(A)\} \le 1$. The equivalence of statements *iii)*–*vii)* follows from *v)*–*ix)* of Lemma 2.2 with A replaced by $I - A$. □

Note that if $f(x) = Ax$, where $A \in \mathbb{R}^{n\times n}$, then f is nonnegative if and only if A is nonnegative. To address discrete-time linear nonnegative dynamical systems, consider (9.1) with $f(x) = Ax$ so that

$$x(k+1) = Ax(k), \quad x(0) = x_0, \quad k \in \overline{\mathbb{Z}}_+, \tag{9.16}$$

where $x(k) \in \mathbb{R}^n$, $k \in \overline{\mathbb{Z}}_+$, and $A \in \mathbb{R}^{n\times n}$. Since the solution to (2.34) is given by $x(k) = A^k x_0$, it follows that $x(k) \geq\geq 0$, $k \in \overline{\mathbb{Z}}_+$, if and only if A is nonnegative. Henceforth, in this chapter we assume that A is nonnegative.

Definition 9.9. Let $A \in \mathbb{R}^{n\times n}$. Then

i) A is (*discrete-time*) *Lyapunov stable* if $\operatorname{spec}(A) \subset \{z \in \mathbb{C} : |z| \leq 1\}$ and, if $\lambda \in \operatorname{spec}(A)$ and $|\lambda| = 1$, then λ is semisimple.

ii) A is (*discrete-time*) *semistable* if $\operatorname{spec}(A) \subset \{z \in \mathbb{C} : |z| < 1\} \cup \{1\}$ and, if $1 \in \operatorname{spec}(A)$, then 1 is semisimple.

iii) A is (*discrete-time*) *asymptotically stable* or *Schur* if $\operatorname{spec}(A) \subset \{z \in \mathbb{C} : |z| < 1\}$.

The following proposition concerning Lyapunov stability, semistability, and asymptotic stability of (9.16) is immediate. This result holds whether or not A is a nonnegative matrix.

Proposition 9.2 ([33]). Let $A \in \mathbb{R}^{n\times n}$ and consider the linear discrete-time dynamical system (9.16). Then the following statements are equivalent:

i) A is Lyapunov stable.

ii) For every initial condition $x(0) \in \mathbb{R}^n$, the sequence $\{\|x(k)\|\}_{k=1}^{\infty}$ is bounded, where $\|\cdot\|$ is a vector norm on $\mathbb{R}^n$.

iii) For every initial condition $x(0) \in \mathbb{R}^n$, the sequence $\{\|A^k x(0)\|\}_{k=1}^{\infty}$ is bounded, where $\|\cdot\|$ is a vector norm on $\mathbb{R}^n$.

iv) The sequence $\{\|A^k\|\}_{k=1}^{\infty}$ is bounded, where $\|\cdot\|$ is a matrix norm on $\mathbb{R}^{n\times n}$.

The following statements are equivalent:

v) A is semistable.

vi) $\lim_{k\to\infty} A^k$ exists. In fact, $\lim_{k\to\infty} A^k = I - (I-A)^{\#}(I-A)$.

vii) For every initial condition $x(0) \in \mathbb{R}^n$, $\lim_{k\to\infty} x(k)$ exists.

The following statements are equivalent:

viii) A is asymptotically stable.

ix) $\rho(A) < 1$.

x) For every initial condition $x(0) \in \mathbb{R}^n$, $\lim_{k\to\infty} x(k) = 0$.

xi) For every initial condition $x(0) \in \mathbb{R}^n$, $\lim_{k\to\infty} A^k x(0) = 0$.

xii) $\lim_{k\to\infty} A^k = 0$.

9.3 Conservation of Discrete Energy and the First Law of Thermodynamics

To develop a discrete-time thermodynamic model, consider a discrete-time, large-scale dynamical system $\mathcal{G}$ analogous to the system shown in Figure 3.1 involving q interconnected subsystems. Let $E_i : \overline{\mathbb{Z}}_+ \to \overline{\mathbb{R}}_+$ denote the energy (and hence a nonnegative quantity) of the ith subsystem, let $S_i : \overline{\mathbb{Z}}_+ \to \mathbb{R}$ denote the external energy supplied to (or extracted from) the ith subsystem, let $\sigma_{ij} : \overline{\mathbb{R}}_+^q \to \overline{\mathbb{R}}_+$, $i \neq j$, $i, j = 1, \ldots, q$, denote the exchange of energy from the jth subsystem to the ith subsystem, and let $\sigma_{ii} : \overline{\mathbb{R}}_+^q \to \overline{\mathbb{R}}_+$, $i = 1, \ldots, q$, denote the energy loss from the ith subsystem. A discrete *energy balance* equation for the ith subsystem yields

$$\Delta E_i(k) = \sum_{j=1,\, j\neq i}^{q} [\sigma_{ij}(E(k)) - \sigma_{ji}(E(k))] - \sigma_{ii}(E(k)) + S_i(k), \quad k \geq k_0, \tag{9.17}$$

where $\Delta E_i(k) \triangleq E_i(k+1) - E_i(k)$, or, equivalently, in vector form,

$$E(k+1) = f(E(k)) - d(E(k)) + S(k), \quad k \geq k_0, \tag{9.18}$$

where $E(k) = [E_1(k), \ldots, E_q(k)]^{\mathrm{T}}$, $S(k) = [S_1(k), \ldots, S_q(k)]^{\mathrm{T}}$, $d(E(k)) = [\sigma_{11}(E(k)), \ldots, \sigma_{qq}(E(k))]^{\mathrm{T}}$, $k \geq k_0$, and $f = [f_1, \ldots, f_q]^{\mathrm{T}} : \overline{\mathbb{R}}_+^q \to \mathbb{R}^q$ is such that

$$f_i(E) = E_i + \sum_{j=1,\, j\neq i}^{q} [\sigma_{ij}(E) - \sigma_{ji}(E)], \quad E \in \overline{\mathbb{R}}_+^q. \tag{9.19}$$

Equation (9.17) yields a discrete-time conservation of energy equation and implies that the change of energy stored in the ith subsystem is equal to the external energy supplied to (or extracted from) the ith subsystem plus the energy gained by the ith subsystem from all other subsystems due to subsystem coupling minus the energy dissipated from the ith subsystem. Note that (9.18) or, equivalently, (9.17) is a statement reminiscent of the *first law of thermodynamics* for each of the subsystems, with $E_i(\cdot)$, $S_i(\cdot)$, $\sigma_{ij}(\cdot)$, $i \neq j$, and $\sigma_{ii}(\cdot)$, $i = 1, \ldots, q$, playing the role of the ith subsystem internal energy, energy supplied to (or extracted from) the ith subsystem, the energy exchange between subsystems due to coupling, and the energy dissipated to the environment, respectively.

To further elucidate that (9.18) is essentially the statement of the principle of the conservation of energy, let the total energy in the discrete-time, large-scale dynamical system $\mathcal{G}$ be given by $U \triangleq \mathbf{e}^{\mathrm{T}}E$, $E \in \overline{\mathbb{R}}_+^q$, and let the energy received by the discrete-time, large-scale dynamical system $\mathcal{G}$ (in forms other than work) over the discrete time interval $\{k_1, \ldots, k_2\}$ be given by $Q \triangleq \sum_{k=k_1}^{k_2} \mathbf{e}^{\mathrm{T}}[S(k) - d(E(k))]$, where $E(k)$, $k \geq k_0$, is the solution to (9.18). Then, premultiplying (9.18) by $\mathbf{e}^{\mathrm{T}}$ and using the fact that $\mathbf{e}^{\mathrm{T}}f(E) = \mathbf{e}^{\mathrm{T}}E$, it follows that

$$\Delta U = Q, \tag{9.20}$$

where $\Delta U \triangleq U(k_2) - U(k_1)$ denotes the variation in the total energy of the discrete-time, large-scale dynamical system $\mathcal{G}$ over the discrete time interval $\{k_1, \ldots, k_2\}$. This is a statement of the first law of thermodynamics for the discrete-time, large-scale dynamical system $\mathcal{G}$ and gives a precise formulation of the equivalence between variation in system internal energy and heat.

It is important to note that our discrete-time, large-scale dynamical system model does not consider work done by the system on the environment nor work done by the environment on the system. Hence, Q can be interpreted physically as the amount of energy that is received by the system in forms other than work. The extension of addressing work performed by and on the system can be easily handled by including an additional state equation, coupled to the energy balance equation (9.18), involving volume states for each subsystem as in Chapter 5. Since this extension does not alter any of the results of this chapter, it is not considered here for simplicity of exposition.

For our large-scale dynamical system model $\mathcal{G}$, we assume that $\sigma_{ij}(E) = 0$, $E \in \overline{\mathbb{R}}_+^q$, whenever $E_j = 0$, $i, j = 1, \ldots, q$. This constraint implies that if the energy of the jth subsystem of $\mathcal{G}$ is zero, then this subsystem cannot supply any energy to its surroundings nor dissipate energy to the environment. Furthermore, for the remainder of this chapter we assume that $E_i \geq \sigma_{ii}(E) - S_i - \sum_{j=1, j\neq i}^{q}[\sigma_{ij}(E) - \sigma_{ji}(E)] = -\Delta E_i$, $E \in \overline{\mathbb{R}}_+^q$, $S \in \mathbb{R}^q$, $i = 1, \ldots, q$. This constraint implies that the energy that can be dissipated, extracted, or exchanged by the ith subsystem cannot exceed the current energy in the subsystem. Note that this assumption implies that $E(k) \geq\geq 0$ for all $k \geq k_0$.

Next, premultiplying (9.18) by $\mathbf{e}^{\mathrm{T}}$ and using the fact that $\mathbf{e}^{\mathrm{T}}f(E) = \mathbf{e}^{\mathrm{T}}E$, it follows that

$$\mathbf{e}^{\mathrm{T}}E(k_1) = \mathbf{e}^{\mathrm{T}}E(k_0) + \sum_{k=k_0}^{k_1-1} \mathbf{e}^{\mathrm{T}}S(k) - \sum_{k=k_0}^{k_1-1} \mathbf{e}^{\mathrm{T}}d(E(k)), \quad k_1 \geq k_0. \quad (9.21)$$

Now, for the discrete-time, large-scale dynamical system $\mathcal{G}$ define the input $u(k) \triangleq S(k)$ and the output $y(k) \triangleq d(E(k))$. Hence, it follows from (9.21) that the discrete-time, large-scale dynamical system $\mathcal{G}$ is *lossless* [191] with respect to the *energy supply rate* $r(u, y) = \mathbf{e}^{\mathrm{T}}u - \mathbf{e}^{\mathrm{T}}y$ and with the *energy storage function* $U(E) \triangleq \mathbf{e}^{\mathrm{T}}E$, $E \in \overline{\mathbb{R}}_+^q$. This implies that (see [191] for details)

$$0 \leq U_{\mathrm{a}}(E_0) = U(E_0) = U_{\mathrm{r}}(E_0) < \infty, \quad E_0 \in \overline{\mathbb{R}}_+^q, \quad (9.22)$$

where

$$U_{\mathrm{a}}(E_0) \triangleq -\inf_{u(\cdot),\, K \geq k_0} \sum_{k=k_0}^{K-1} [\mathbf{e}^{\mathrm{T}}u(k) - \mathbf{e}^{\mathrm{T}}y(k)], \quad (9.23)$$

$$U_{\mathrm{r}}(E_0) \triangleq \inf_{u(\cdot),\, K \geq -k_0+1} \sum_{k=-K}^{k_0-1} [\mathbf{e}^{\mathrm{T}}u(k) - \mathbf{e}^{\mathrm{T}}y(k)], \quad (9.24)$$

and $E_0 = E(k_0) \in \overline{\mathbb{R}}_+^q$.

Since $U_{\mathrm{a}}(E_0)$ is the maximum amount of stored energy that can be extracted from the discrete-time, large-scale dynamical system $\mathcal{G}$ at any discrete-time instant K, and $U_{\mathrm{r}}(E_0)$ is the minimum amount of energy that can be delivered to the discrete-time, large-scale dynamical system $\mathcal{G}$ to transfer it from a state of minimum potential $E(-K) = 0$ to a given state $E(k_0) = E_0$, it follows from (9.22) that the discrete-time, large-scale dynamical system $\mathcal{G}$ can deliver to its surroundings all of its stored subsystem energies and can store all of the work done to all of its subsystems. In the case where $S(k) \equiv 0$, it follows from (9.21) and the fact that $\sigma_{ii}(E) \geq 0$, $E \in \overline{\mathbb{R}}_+^q$, $i = 1, \ldots, q$, that the zero solution $E(k) \equiv 0$ of the discrete-time, large-scale dynamical system $\mathcal{G}$ with the energy balance equation (9.18) is Lyapunov stable with the Lyapunov function $U(E)$ corresponding to the total energy in the system.

The next result shows that the large-scale dynamical system $\mathcal{G}$ is locally controllable.

Proposition 9.3. Consider the discrete-time, large-scale dynamical system $\mathcal{G}$ with energy balance equation (9.18). Then, for every equilibrium state $E_{\mathrm{e}} \in \overline{\mathbb{R}}_+^q$ and every $\varepsilon > 0$ and $T \in \mathbb{Z}_+$, there exist $S_{\mathrm{e}} \in \mathbb{R}^q$, $\alpha > 0$, and $\hat{T} \in \{0, \ldots, T\}$ such that, for every $\hat{E} \in \overline{\mathbb{R}}_+^q$ with $\|\hat{E} - E_{\mathrm{e}}\| \leq \alpha T$, there exists $S:\{0,\ldots,\hat{T}\}\to\mathbb{R}^q$ such that $\|S(k)-S_{\rm e}\|\le\varepsilon$, $k\in\{0,\ldots,\hat{T}\}$, and $E(k)=E_{\rm e}+\frac{(\hat{E}-E_{\rm e})}{\hat{T}}k$, $k\in\{0,\ldots,\hat{T}\}$.

Proof. Note that with $S_{\rm e}=d(E_{\rm e})-f(E_{\rm e})+E_{\rm e}$, the state $E_{\rm e}\in\overline{\mathbb{R}}_+^q$ is an equilibrium state of (9.18). Let $\theta>0$ and $T\in\mathbb{Z}_+$, and define

$$M(\theta,T)\triangleq\sup_{E\in\overline{\mathcal{B}}_1(0),\,k\in\{0,\ldots,T\}}\|f(E_{\rm e}+k\theta E)-f(E_{\rm e})-d(E_{\rm e}+k\theta E)+d(E_{\rm e})-k\theta E\|. \quad (9.25)$$

Note that for every $T\in\mathbb{Z}_+$, $\lim_{\theta\to 0^+}M(\theta,T)=0$. Next, let $\varepsilon>0$ and $T\in\mathbb{Z}_+$ be given, and let $\alpha>0$ be such that $M(\alpha,T)+\alpha\le\varepsilon$. (The existence of such an α is guaranteed since $M(\alpha,T)\to 0$ as $\alpha\to 0^+$). Now, let $\hat{E}\in\overline{\mathbb{R}}_+^q$ be such that $\|\hat{E}-E_{\rm e}\|\le\alpha T$. With $\hat{T}\triangleq\lceil\frac{\|\hat{E}-E_{\rm e}\|}{\alpha}\rceil\le T$, where $\lceil x\rceil$ denotes the smallest integer greater than or equal to x, and

$$S(k)=-f(E(k))+d(E(k))+E(k)+\frac{\hat{E}-E_{\rm e}}{\lceil\frac{\|\hat{E}-E_{\rm e}\|}{\alpha}\rceil},\quad k\in\{0,\ldots,\hat{T}\}, \quad (9.26)$$

it follows that

$$E(k)=E_{\rm e}+\frac{(\hat{E}-E_{\rm e})}{\lceil\frac{\|\hat{E}-E_{\rm e}\|}{\alpha}\rceil}k,\quad k\in\{0,\ldots,\hat{T}\}, \quad (9.27)$$

is a solution to (9.18).

The result is now immediate by noting that $E(\hat{T})=\hat{E}$ and

$$\begin{aligned}\|S(k)-S_{\rm e}\| &\le \left\|w\left(E_{\rm e}+\frac{(\hat{E}-E_{\rm e})}{\lceil\frac{\|\hat{E}-E_{\rm e}\|}{\alpha}\rceil}k\right)-f(E_{\rm e})-d\left(E_{\rm e}+\frac{(\hat{E}-E_{\rm e})}{\lceil\frac{\|\hat{E}-E_{\rm e}\|}{\alpha}\rceil}k\right)\right.\\ &\quad\left.+d(E_{\rm e})-\frac{(\hat{E}-E_{\rm e})}{\lceil\frac{\|\hat{E}-E_{\rm e}\|}{\alpha}\rceil}k\right\|+\alpha\\ &\le M(\alpha,T)+\alpha\\ &\le \varepsilon,\quad k\in\{0,\ldots,\hat{T}\},\end{aligned} \quad (9.28)$$

which proves the result. □

It follows from Proposition 9.3 that the discrete-time, large-scale dynamical system $\mathcal{G}$ with the energy balance equation (9.18) is *reachable* from and *controllable* to the origin in $\overline{\mathbb{R}}_+^q$. Recall that the discrete-time large-scale dynamical system $\mathcal{G}$ with the energy balance equation (9.18) is reachable from the origin in $\overline{\mathbb{R}}_+^q$ if, for all $E_0=E(k_0)\in\overline{\mathbb{R}}_+^q$, there exists a finite time $k_{\rm i}\le k_0$ and an input $S(k)$ defined on $\{k_{\rm i},\ldots,k_0\}$ such that the state $E(k)$, $k\ge k_{\rm i}$, can be driven from $E(k_{\rm i})=0$ to

$E(k_0) = E_0$. Alternatively, $\mathcal{G}$ is controllable to the origin in $\overline{\mathbb{R}}_+^q$ if, for all $E_0 = E(k_0) \in \overline{\mathbb{R}}_+^q$, there exists a finite time $k_f \geq k_0$ and an input $S(k)$ defined on $\{k_0, \ldots, k_f\}$ such that the state $E(k)$, $k \geq k_0$, can be driven from $E(k_0) = E_0$ to $E(k_f) = 0$.

We let $\mathcal{U}_r$ denote the set of all admissible bounded energy inputs to the discrete-time, large-scale dynamical system $\mathcal{G}$ such that for every $K \geq -k_0$, the system energy state can be driven from $E(-K) = 0$ to $E(k_0) = E_0 \in \overline{\mathbb{R}}_+^q$ by $S(\cdot) \in \mathcal{U}_r$, and we let $\mathcal{U}_c$ denote the set of all admissible bounded energy inputs to the discrete-time, large-scale dynamical system $\mathcal{G}$ such that for every $K \geq k_0$, the system energy state can be driven from $E(k_0) = E_0 \in \overline{\mathbb{R}}_+^q$ to $E(K) = 0$ by $S(\cdot) \in \mathcal{U}_c$. Furthermore, let $\mathcal{U}$ be an input space that is a subset of bounded continuous $\mathbb{R}^q$-valued functions on $\mathbb{Z}$. The spaces $\mathcal{U}_r$, $\mathcal{U}_c$, and $\mathcal{U}$ are assumed to be closed under the shift operator; that is, if $S(\cdot) \in \mathcal{U}$ (respectively, $\mathcal{U}_c$ or $\mathcal{U}_r$), then the function S_K defined by $S_K(k) = S(k+K)$ is contained in $\mathcal{U}$ (respectively, $\mathcal{U}_c$ or $\mathcal{U}_r$) for all $K \geq 0$.

9.4 Nonconservation of Discrete Entropy and the Second Law of Thermodynamics

As in the case of the continuous-time power balance equation, the nonlinear discrete-time energy balance equation (9.18) can exhibit a full range of nonlinear behavior, including bifurcations, limit cycles, and even chaos. However, a thermodynamically consistent energy flow model should ensure that the evolution of the system energy is diffusive (parabolic) in character with convergent subsystem energies. Hence, to ensure a thermodynamically consistent energy flow model we require the following axioms. For the statement of these axioms let $\phi_{ij}(E) \triangleq \sigma_{ij}(E) - \sigma_{ji}(E)$, $E \in \overline{\mathbb{R}}_+^q$, denote the net energy exchange between subsystems $\mathcal{G}_i$ and $\mathcal{G}_j$ of the discrete-time, large-scale dynamical system $\mathcal{G}$.

Axiom *i*): For the connectivity matrix $\mathcal{C} \in \mathbb{R}^{q \times q}$ associated with the large-scale dynamical system $\mathcal{G}$ defined by

$$\mathcal{C}_{(i,j)} = \begin{cases} 0, & \text{if } \phi_{ij}(E) \equiv 0, \\ 1, & \text{otherwise,} \end{cases} \quad i \neq j, \quad i,j = 1,\ldots,q, \tag{9.29}$$

and

$$\mathcal{C}_{(i,i)} = -\sum_{k=1,\, k\neq i}^{q} \mathcal{C}_{(k,i)}, \quad i = j, \quad i = 1,\ldots,q, \tag{9.30}$$

rank $\mathcal{C} = q - 1$, and for $\mathcal{C}_{(i,j)} = 1$, $i \neq j$, $\phi_{ij}(E) = 0$ if and only if $E_i = E_j$.

Axiom *ii*): For $i, j = 1, \ldots, q$, $(E_i - E_j)\phi_{ij}(E) \leq 0$, $E \in \overline{\mathbb{R}}_+^q$.

Axiom *iii*): For $i, j = 1, \ldots, q$, $\frac{\Delta E_i - \Delta E_j}{E_i - E_j} \geq -1$, $E_i \neq E_j$.

Axioms *i*) and *ii*) are a restatement of Axioms *i*) and *ii*) of Chapter 3, whereas Axiom *iii*) implies that for any pair of connected subsystems $\mathcal{G}_i$ and $\mathcal{G}_j$, $i \neq j$, the energy difference between consecutive time instants is monotonic; that is, $[E_i(k+1) - E_j(k+1)][E_i(k) - E_j(k)] \geq 0$ for all $E_i \neq E_j$, $k \geq k_0$, $i, j = 1, \ldots, q$.

Next, we establish a version of Clausius' inequality for our thermodynamically consistent energy flow model.

Proposition 9.4. Consider the discrete-time, large-scale dynamical system $\mathcal{G}$ with energy balance equation (9.18) and assume that Axioms *i*), *ii*), and *iii*) hold. Then, for all $E_0 \in \overline{\mathbb{R}}_+^q$, $k_\mathrm{f} \geq k_0$, and $S(\cdot) \in \mathcal{U}$ such that $E(k_\mathrm{f}) = E(k_0) = E_0$,

$$\sum_{k=k_0}^{k_\mathrm{f}-1} \sum_{i=1}^{q} \frac{S_i(k) - \sigma_{ii}(E(k))}{c + E_i(k+1)} = \sum_{k=k_0}^{k_\mathrm{f}-1} \sum_{i=1}^{q} \frac{Q_i(k)}{c + E_i(k+1)} \leq 0, \tag{9.31}$$

where $c > 0$, $Q_i(k) \triangleq S_i(k) - \sigma_{ii}(E(k))$, $i = 1, \ldots, q$, is the amount of net energy (heat) received by the ith subsystem at the kth instant, and $E(k)$, $k \geq k_0$, is the solution to (9.18) with initial condition $E(k_0) = E_0$. Furthermore, equality holds in (9.31) if and only if $\Delta E_i(k) = 0$, $i = 1, \ldots, q$, and $E_i(k) = E_j(k)$, $i, j = 1, \ldots, q$, $i \neq j$, $k \in \{k_0, \ldots, k_\mathrm{f} - 1\}$.

Proof. Since $E(k) \geq\geq 0$, $k \geq k_0$, and $\phi_{ij}(E) = -\phi_{ji}(E)$, $E \in \overline{\mathbb{R}}_+^q$, $i \neq j$, $i, j = 1, \ldots, q$, it follows from (9.18), Axioms *ii*) and *iii*), and the fact that $\frac{x}{x+1} \leq \log_e(1 + x)$, $x > -1$, that

$$\begin{aligned}
&\sum_{k=k_0}^{k_\mathrm{f}-1} \sum_{i=1}^{q} \frac{Q_i(k)}{c + E_i(k+1)} \\
&\quad = \sum_{k=k_0}^{k_\mathrm{f}-1} \sum_{i=1}^{q} \frac{\Delta E_i(k) - \sum_{j=1,\, j\neq i}^{q} \phi_{ij}(E(k))}{c + E_i(k+1)} \\
&\quad = \sum_{k=k_0}^{k_\mathrm{f}-1} \sum_{i=1}^{q} \left[\frac{\Delta E_i(k)}{c + E_i(k)}\right] \left[1 + \frac{\Delta E_i(k)}{c + E_i(k)}\right]^{-1} \\
&\qquad - \sum_{k=k_0}^{k_\mathrm{f}-1} \sum_{i=1}^{q} \sum_{j=1,\, j\neq i}^{q} \frac{\phi_{ij}(E(k))}{c + E_i(k+1)} \\
&\quad \leq \sum_{i=1}^{q} \log_e\left(\frac{c + E_i(k_\mathrm{f})}{c + E_i(k_0)}\right) - \sum_{k=k_0}^{k_\mathrm{f}-1} \sum_{i=1}^{q} \sum_{j=1,\, j\neq i}^{q} \frac{\phi_{ij}(E(k))}{c + E_i(k+1)}
\end{aligned}$$

$$
\begin{aligned}
&= -\sum_{k=k_0}^{k_\mathrm{f}-1}\sum_{i=1}^{q-1}\sum_{j=i+1}^{q}\left(\frac{\phi_{ij}(E(k))}{c+E_i(k+1)} - \frac{\phi_{ij}(E(k))}{c+E_j(k+1)}\right)\\
&= -\sum_{k=k_0}^{k_\mathrm{f}-1}\sum_{i=1}^{q-1}\sum_{j=i+1}^{q}\frac{\phi_{ij}(E(k))[E_j(k+1)-E_i(k+1)]}{(c+E_i(k+1))(c+E_j(k+1))}\\
&\leq 0, \qquad (9.32)
\end{aligned}
$$

which proves (9.31).

Alternatively, equality holds in (9.31) if and only if $\sum_{k=k_0}^{k_\mathrm{f}-1}\frac{\Delta E_i(k)}{c+E_i(k+1)} = 0$, $i = 1,\ldots,q$, and $\phi_{ij}(E(k))(E_j(k+1) - E_i(k+1)) = 0$, $i,j = 1,\ldots,q$, $i \neq j$, $k \geq k_0$. Moreover, $\sum_{k=k_0}^{k_\mathrm{f}-1}\frac{\Delta E_i(k)}{c+E_i(k+1)} = 0$ is equivalent to $\Delta E_i(k) = 0$, $i = 1,\ldots,q$, $k \in \{k_0,\ldots,k_\mathrm{f}-1\}$. Hence, $\phi_{ij}(E(k))(E_j(k+1) - E_i(k+1)) = \phi_{ij}(E(k))(E_j(k) - E_i(k)) = 0$, $i,j = 1,\ldots,q$, $i \neq j$, $k \geq k_0$. Thus, it follows from Axioms $i) - iii)$ that equality holds in (9.31) if and only if $\Delta E_i = 0$, $i = 1,\ldots,q$, and $E_j = E_i$, $i,j = 1,\ldots,q$, $i \neq j$. □

Inequality (9.31) is analogous to Clausius' inequality for reversible and irreversible thermodynamics as applied to discrete-time, large-scale dynamical systems. It follows from Axiom $i)$ and (9.18) that for the *isolated* discrete-time, large-scale dynamical system $\mathcal{G}$, that is, $S(k) \equiv 0$ and $d(E(k)) \equiv 0$, the energy states given by $E_\mathrm{e} = \alpha\mathbf{e}$, $\alpha \geq 0$, correspond to the equilibrium energy states of $\mathcal{G}$. Thus, we can define an *equilibrium process* as a process where the trajectory of the discrete-time, large-scale dynamical system $\mathcal{G}$ stays at the equilibrium point of the isolated system $\mathcal{G}$. The input that can generate such a trajectory can be given by $S(k) = d(E(k))$, $k \geq k_0$.

Alternatively, a *nonequilibrium process* is a process that is not an equilibrium one. Hence, it follows from Axiom $i)$ that for an equilibrium process $\phi_{ij}(E(k)) \equiv 0$, $k \geq k_0$, $i \neq j$, $i,j = 1,\ldots,q$, and thus, by Proposition 9.4 and $\Delta E_i = 0$, $i = 1,\ldots,q$, inequality (9.31) is satisfied as an equality. Thus, for a nonequilibrium process it follows from Axioms $i) - iii)$ that (9.31) is satisfied as a strict inequality.

Next, we give a deterministic definition of entropy for the discrete-time, large-scale dynamical system $\mathcal{G}$ that is consistent with the classical thermodynamic definition of entropy.

Definition 9.10. For the discrete-time, large-scale dynamical system $\mathcal{G}$ with energy balance equation (9.18), a function $\mathcal{S} : \overline{\mathbb{R}}_+^q \to \mathbb{R}$ satisfying

$$
\mathcal{S}(E(k_2)) \geq \mathcal{S}(E(k_1)) + \sum_{k=k_1}^{k_2-1}\sum_{i=1}^{q}\frac{S_i(k) - \sigma_{ii}(E(k))}{c+E_i(k+1)}, \qquad (9.33)
$$

for every $k_2 \geq k_1 \geq k_0$ and $S(\cdot) \in \mathcal{U}$, is called the *entropy* of $\mathcal{G}$.

Next, we show that (9.31) guarantees the existence of an entropy function for $\mathcal{G}$. For this result define the *available entropy* of the large-scale dynamical system $\mathcal{G}$ by

$$\mathcal{S}_{\rm a}(E_0) \triangleq -\sup_{S(\cdot)\in\mathcal{U}_{\rm c},\, K\geq k_0} \sum_{k=k_0}^{K-1}\sum_{i=1}^{q} \frac{S_i(k)-\sigma_{ii}(E(k))}{c+E_i(k+1)}, \tag{9.34}$$

where $E(k_0) = E_0 \in \overline{\mathbb{R}}_+^q$ and $E(K) = 0$, and define the *required entropy supply* of the large-scale dynamical system $\mathcal{G}$ by

$$\mathcal{S}_{\rm r}(E_0) \triangleq \sup_{S(\cdot)\in\mathcal{U}_{\rm r},\, K\geq -k_0+1} \sum_{k=-K}^{k_0-1}\sum_{i=1}^{q} \frac{S_i(k)-\sigma_{ii}(E(k))}{c+E_i(k+1)}, \tag{9.35}$$

where $E(-K) = 0$ and $E(k_0) = E_0 \in \overline{\mathbb{R}}_+^q$. Note that the available entropy $\mathcal{S}_{\rm a}(E_0)$ is the minimum amount of scaled heat (entropy) that can be extracted from the large-scale dynamical system $\mathcal{G}$ in order to transfer it from an initial state $E(k_0) = E_0$ to $E(K) = 0$. Alternatively, the required entropy supply $\mathcal{S}_{\rm r}(E_0)$ is the maximum amount of scaled heat (entropy) that can be delivered to $\mathcal{G}$ to transfer it from the origin to a given initial state $E(k_0) = E_0$.

Theorem 9.6. Consider the discrete-time, large-scale dynamical system $\mathcal{G}$ with energy balance equation (9.18) and assume that Axioms *ii*) and *iii*) hold. Then there exists an entropy function for $\mathcal{G}$. Moreover, $\mathcal{S}_{\rm a}(E)$, $E \in \overline{\mathbb{R}}_+^q$, and $\mathcal{S}_{\rm r}(E)$, $E \in \overline{\mathbb{R}}_+^q$, are possible entropy functions for $\mathcal{G}$ with $\mathcal{S}_{\rm a}(0) = \mathcal{S}_{\rm r}(0) = 0$. Finally, all entropy functions $\mathcal{S}(E)$, $E \in \overline{\mathbb{R}}_+^q$, for $\mathcal{G}$ satisfy

$$\mathcal{S}_{\rm r}(E) \leq \mathcal{S}(E) - \mathcal{S}(0) \leq \mathcal{S}_{\rm a}(E), \quad E \in \overline{\mathbb{R}}_+^q. \tag{9.36}$$

Proof. Since, by Proposition 9.3, $\mathcal{G}$ is controllable to and reachable from the origin in $\overline{\mathbb{R}}_+^q$, it follows from (9.34) and (9.35) that $\mathcal{S}_{\rm a}(E_0) < \infty$, $E_0 \in \overline{\mathbb{R}}_+^q$, and $\mathcal{S}_{\rm r}(E_0) > -\infty$, $E_0 \in \overline{\mathbb{R}}_+^q$, respectively. Next, let $E_0 \in \overline{\mathbb{R}}_+^q$ and let $S(\cdot) \in \mathcal{U}$ be such that $E(k_{\rm i}) = E(k_{\rm f}) = 0$ and $E(k_0) = E_0$, where $k_{\rm i} \leq k_0 \leq k_{\rm f}$. In this case, it follows from (9.31) that

$$\sum_{k=k_{\rm i}}^{k_{\rm f}-1}\sum_{i=1}^{q} \frac{S_i(k)-\sigma_{ii}(E(k))}{c+E_i(k+1)} \leq 0, \tag{9.37}$$

or, equivalently,

$$\sum_{k=k_{\rm i}}^{k_0-1}\sum_{i=1}^{q} \frac{S_i(k)-\sigma_{ii}(E(k))}{c+E_i(k+1)} \leq -\sum_{k=k_0}^{k_{\rm f}-1}\sum_{i=1}^{q} \frac{S_i(k)-\sigma_{ii}(E(k))}{c+E_i(k+1)}. \tag{9.38}$$

Now, taking the supremum on both sides of (9.38) over all $S(\cdot) \in \mathcal{U}_{\mathrm{r}}$ and $k_{\mathrm{i}} + 1 \leq k_0$, we obtain

$$\begin{aligned} \mathcal{S}_{\mathrm{r}}(E_0) &= \sup_{S(\cdot)\in\mathcal{U}_{\mathrm{r}},\, k_{\mathrm{i}}+1\leq k_0} \sum_{k=k_{\mathrm{i}}}^{k_0-1} \sum_{i=1}^{q} \frac{S_i(k) - \sigma_{ii}(E(k))}{c + E_i(k+1)} \\ &\leq -\sum_{k=k_0}^{k_{\mathrm{f}}-1} \sum_{i=1}^{q} \frac{S_i(k) - \sigma_{ii}(E(k))}{c + E_i(k+1)}. \end{aligned} \tag{9.39}$$

Next, taking the infimum on both sides of (9.39) over all $S(\cdot) \in \mathcal{U}_{\mathrm{c}}$ and $k_{\mathrm{f}} \geq k_0$, we obtain $\mathcal{S}_{\mathrm{r}}(E_0) \leq \mathcal{S}_{\mathrm{a}}(E_0)$, $E_0 \in \overline{\mathbb{R}}_+^q$, which implies that $-\infty < \mathcal{S}_{\mathrm{r}}(E_0) \leq \mathcal{S}_{\mathrm{a}}(E_0) < +\infty$, $E_0 \in \overline{\mathbb{R}}_+^q$. Hence, the functions $\mathcal{S}_{\mathrm{a}}(\cdot)$ and $\mathcal{S}_{\mathrm{r}}(\cdot)$ are well defined.

Next, it follows from the definition of $\mathcal{S}_{\mathrm{a}}(\cdot)$ that, for any $K \geq k_1$ and $S(\cdot) \in \mathcal{U}_{\mathrm{c}}$ such that $E(k_1) \in \overline{\mathbb{R}}_+^q$ and $E(K) = 0$,

$$-\mathcal{S}_{\mathrm{a}}(E(k_1)) \geq \sum_{k=k_1}^{k_2-1} \sum_{i=1}^{q} \frac{S_i(k) - \sigma_{ii}(E(k))}{c + E_i(k+1)} + \sum_{k=k_2}^{K-1} \sum_{i=1}^{q} \frac{S_i(k) - \sigma_{ii}(E(k))}{c + E_i(k+1)}, \quad k_1 \leq k_2 \leq K, \tag{9.40}$$

and hence,

$$\begin{aligned} -\mathcal{S}_{\mathrm{a}}(E(k_1)) &\geq \sum_{k=k_1}^{k_2-1} \sum_{i=1}^{q} \frac{S_i(k) - \sigma_{ii}(E(k))}{c + E_i(k+1)} \\ &\quad + \sup_{S(\cdot)\in\mathcal{U}_{\mathrm{c}},\, K\geq k_2} \sum_{k=k_2}^{K-1} \sum_{i=1}^{q} \frac{S_i(k) - \sigma_{ii}(E(k))}{c + E_i(k+1)} \\ &= \sum_{k=k_1}^{k_2-1} \sum_{i=1}^{q} \frac{S_i(k) - \sigma_{ii}(E(k))}{c + E_i(k+1)} - \mathcal{S}_{\mathrm{a}}(E(k_2)), \end{aligned} \tag{9.41}$$

which implies that $\mathcal{S}_{\mathrm{a}}(E)$, $E \in \overline{\mathbb{R}}_+^q$, satisfies (9.33). Thus, $\mathcal{S}_{\mathrm{a}}(E)$, $E \in \overline{\mathbb{R}}_+^q$, is a possible entropy function for $\mathcal{G}$. Note that with $E(k_0) = E(K) = 0$ it follows from (9.31) that the supremum in (9.34) is taken over the set of nonpositive values with one of the values being zero for $S(k) \equiv 0$. Thus, $\mathcal{S}_{\mathrm{a}}(0) = 0$. Similarly, it can be shown that $\mathcal{S}_{\mathrm{r}}(E)$, $E \in \overline{\mathbb{R}}_+^q$, given by (9.35) satisfies (9.33), and hence, is a possible entropy function for the system $\mathcal{G}$ with $\mathcal{S}_{\mathrm{r}}(0) = 0$.

Next, suppose there exists an entropy function $\mathcal{S} : \overline{\mathbb{R}}_+^q \to \mathbb{R}$ for $\mathcal{G}$ and let $E(k_2) = 0$ in (9.33). Then it follows from (9.33) that

$$\mathcal{S}(E(k_1)) - \mathcal{S}(0) \leq -\sum_{k=k_1}^{k_2-1} \sum_{i=1}^{q} \frac{S_i(k) - \sigma_{ii}(E(k))}{c + E_i(k+1)}, \tag{9.42}$$

for all $k_2 \geq k_1$ and $S(\cdot) \in \mathcal{U}_\mathrm{c}$, which implies that

$$\begin{aligned} \mathcal{S}(E(k_1)) - \mathcal{S}(0) &\leq \inf_{S(\cdot)\in\mathcal{U}_\mathrm{c},\, k_2\geq k_1} \left[-\sum_{k=k_1}^{k_2-1} \sum_{i=1}^{q} \frac{S_i(k) - \sigma_{ii}(E(k))}{c + E_i(k+1)} \right] \\ &= -\sup_{S(\cdot)\in\mathcal{U}_\mathrm{c},\, k_2\geq k_1} \sum_{k=k_1}^{k_2-1} \sum_{i=1}^{q} \frac{S_i(k) - \sigma_{ii}(E(k))}{c + E_i(k+1)} \\ &= \mathcal{S}_\mathrm{a}(E(k_1)). \end{aligned} \tag{9.43}$$

Since $E(k_1)$ is arbitrary, it follows that $\mathcal{S}(E) - \mathcal{S}(0) \leq \mathcal{S}_\mathrm{a}(E)$, $E \in \overline{\mathbb{R}}_+^q$.

Alternatively, let $E(k_1) = 0$ in (9.33). Then it follows from (9.33) that

$$\mathcal{S}(E(k_2)) - \mathcal{S}(0) \geq \sum_{k=k_1}^{k_2-1} \sum_{i=1}^{q} \frac{S_i(k) - \sigma_{ii}(E(k))}{c + E_i(k+1)}, \tag{9.44}$$

for all $k_1 + 1 \leq k_2$ and $S(\cdot) \in \mathcal{U}_\mathrm{r}$. Hence,

$$\begin{aligned} \mathcal{S}(E(k_2)) - \mathcal{S}(0) &\geq \sup_{S(\cdot)\in\mathcal{U}_\mathrm{r},\, k_1+1\leq k_2} \sum_{k=k_1}^{k_2-1} \sum_{i=1}^{q} \frac{S_i(k) - \sigma_{ii}(E(k))}{c + E_i(k+1)} \\ &= \mathcal{S}_\mathrm{r}(E(k_2)), \end{aligned} \tag{9.45}$$

which, since $E(k_2)$ is arbitrary, implies that $\mathcal{S}_\mathrm{r}(E) \leq \mathcal{S}(E) - \mathcal{S}(0)$, $E \in \overline{\mathbb{R}}_+^q$. Thus, all entropy functions for $\mathcal{G}$ satisfy (9.36). □

It is important to note that inequality (9.31) is equivalent to the existence of an entropy function for $\mathcal{G}$. Sufficiency is simply a statement of Theorem 9.6, whereas necessity follows from (9.33) with $E(k_2) = E(k_1)$. For a nonequilibrium process with energy balance equation (9.18), Definition 9.10 does not provide enough information to define the entropy uniquely. This difficulty has long been pointed out in [318] for thermodynamic systems. A similar remark holds for the definition of ectropy introduced below.

The next proposition gives a closed-form expression for the entropy of the discrete-time, large-scale dynamical system $\mathcal{G}$.

Proposition 9.5. Consider the discrete-time, large-scale dynamical system $\mathcal{G}$ with energy balance equation (9.18) and assume that Axioms *ii*) and *iii*) hold. Then the function $\mathcal{S} : \overline{\mathbb{R}}_+^q \to \mathbb{R}$ given by

$$\mathcal{S}(E) = \mathbf{e}^{\mathrm{T}}\mathbf{log}_e(c\mathbf{e} + E) - q\log_e c, \quad E \in \overline{\mathbb{R}}_+^q, \tag{9.46}$$

where $c > 0$, is an entropy function of $\mathcal{G}$.

Proof. Since $E(k) \geq\geq 0$, $k \geq k_0$, and $\phi_{ij}(E) = -\phi_{ji}(E)$, $E \in \overline{\mathbb{R}}_+^q$, $i \neq j$, $i,j = 1,\ldots,q$, it follows that

$$\begin{aligned}
\Delta\mathcal{S}(E(k)) &= \sum_{i=1}^{q} \log_e\left[1 + \frac{\Delta E_i(k)}{c + E_i(k)}\right] \\
&\geq \sum_{i=1}^{q} \left[\frac{\Delta E_i(k)}{c + E_i(k)}\right]\left[1 + \frac{\Delta E_i(k)}{c + E_i(k)}\right]^{-1} \\
&= \sum_{i=1}^{q} \frac{\Delta E_i(k)}{c + E_i(k) + \Delta E_i(k)} \\
&= \sum_{i=1}^{q} \frac{\Delta E_i(k)}{c + E_i(k+1)} \\
&= \sum_{i=1}^{q} \left[\frac{S_i(k) - \sigma_{ii}(E(k))}{c + E_i(k+1)} + \sum_{j=1,\, j\neq i}^{q} \frac{\phi_{ij}(E(k))}{c + E_i(k+1)}\right] \\
&= \sum_{i=1}^{q} \frac{S_i(k) - \sigma_{ii}(E(k))}{c + E_i(k+1)} \\
&\quad + \sum_{i=1}^{q-1}\sum_{j=i+1}^{q} \left(\frac{\phi_{ij}(E(k))}{c + E_i(k+1)} - \frac{\phi_{ij}(E(k))}{c + E_j(k+1)}\right) \\
&= \sum_{i=1}^{q} \frac{S_i(k) - \sigma_{ii}(E(k))}{c + E_i(k+1)} \\
&\quad + \sum_{i=1}^{q-1}\sum_{j=i+1}^{q} \frac{\phi_{ij}(E(k))[E_j(k+1) - E_i(k+1)]}{(c + E_i(k+1))(c + E_j(k+1))} \\
&\geq \sum_{i=1}^{q} \frac{S_i(k) - \sigma_{ii}(E(k))}{c + E_i(k+1)}, \quad k \geq k_0,
\end{aligned} \tag{9.47}$$

where in (9.47) we use the fact that $\log_e(1 + x) \geq \frac{x}{x+1}$, $x > -1$. Now, summing (9.47) over $\{k_1,\ldots,k_2 - 1\}$ yields (9.33). □

Note that it follows from the first equality in (9.47) that the entropy function given by (9.46) satisfies (9.33) as an equality for an equilibrium process and as a strict inequality for a nonequilibrium process. The entropy expression given by (9.46) is identical in form to the Boltzmann entropy for statistical thermodynamics and the Shannon entropy characterizing the amount of information. Due to the fact that the entropy is indeterminate to the extent of an additive constant, we can place the constant $q \log_e c$ to zero by taking $c = 1$. Since $\mathcal{S}(E)$ given by (9.46) achieves a maximum when all the subsystem energies E_i, $i = 1, \ldots, q$, are equal, the entropy can be thought of as a measure of the tendency of a system to lose the ability to do useful work, to lose order, and to settle to a more homogeneous state.

9.5 Nonconservation of Discrete Ectropy

In this section, we introduce the dual notion to discrete entropy; namely, discrete ectropy, describing the status quo of the discrete-time, large-scale dynamical system $\mathcal{G}$. First, however, we present a dual inequality to inequality (9.31) that holds for our thermodynamically consistent energy flow model.

Proposition 9.6. Consider the discrete-time, large-scale dynamical system $\mathcal{G}$ with energy balance equation (9.18) and assume that Axioms *i*), *ii*), and *iii*) hold. Then, for all $E_0 \in \overline{\mathbb{R}}_+^q$, $k_\mathrm{f} \geq k_0$, and $S(\cdot) \in \mathcal{U}$ such that $E(k_\mathrm{f}) = E(k_0) = E_0$,

$$\sum_{k=k_0}^{k_\mathrm{f}-1} \sum_{i=1}^{q} E_i(k+1)[S_i(k) - \sigma_{ii}(E(k))] = \sum_{k=k_0}^{k_\mathrm{f}-1} \sum_{i=1}^{q} E_i(k+1)Q_i(k) \geq 0, \quad (9.48)$$

where $E(k)$, $k \geq k_0$, is the solution to (9.18) with initial condition $E(k_0) = E_0$. Furthermore, equality holds in (9.48) if and only if $\Delta E_i = 0$ and $E_i = E_j$, $i, j = 1, \ldots, q$, $i \neq j$.

Proof. Since $E(k) \geq\geq 0$, $k \geq k_0$, and $\phi_{ij}(E) = -\phi_{ji}(E)$, $E \in \overline{\mathbb{R}}_+^q$, $i \neq j$, $i, j = 1, \ldots, q$, it follows from (9.18) and Axioms *ii*) and *iii*) that

$$\begin{aligned}
2 \sum_{k=k_0}^{k_\mathrm{f}-1} \sum_{i=1}^{q} E_i(k+1)Q_i(k) \\
&= \sum_{k=k_0}^{k_\mathrm{f}-1} \sum_{i=1}^{q} E_i^2(k+1) - E_i^2(k) \\
&\quad -2 \sum_{k=k_0}^{k_\mathrm{f}-1} \sum_{i=1}^{q} \sum_{j=1,\, j\neq i}^{q} E_i(k+1)\phi_{ij}(E(k)) \\
&\quad + \sum_{k=k_0}^{k_\mathrm{f}-1} \sum_{i=1}^{q} \left[\sum_{j=1,\, j\neq i}^{q} \phi_{ij}(E(k)) + S_i(k) - \sigma_{ii}(E(k)) \right]^2 \\
&= E^\mathrm{T}(k_\mathrm{f})E(k_\mathrm{f}) - E^\mathrm{T}(k_0)E(k_0)
\end{aligned}$$

$$
\begin{aligned}
& -2\sum_{k=k_0}^{k_\mathrm{f}-1}\sum_{i=1}^{q}\sum_{j=1,\,j\neq i}^{q} E_i(k+1)\phi_{ij}(E(k)) \\
& +\sum_{k=k_0}^{k_\mathrm{f}-1}\sum_{i=1}^{q}\left[\sum_{j=1,\,j\neq i}^{q}\phi_{ij}(E(k))+S_i(k)-\sigma_{ii}(E(k))\right]^2 \\
= & -2\sum_{k=k_0}^{k_\mathrm{f}-1}\sum_{i=1}^{q-1}\sum_{j=i+1}^{q}\phi_{ij}(E(k))[E_i(k+1)-E_j(k+1)] \\
& +\sum_{k=k_0}^{k_\mathrm{f}-1}\sum_{i=1}^{q}\left[\sum_{j=1,\,j\neq i}^{q}\phi_{ij}(E(k))+S_i(k)-\sigma_{ii}(E(k))\right]^2 \\
\geq & \ 0,
\end{aligned}
\tag{9.49}
$$

which proves (9.48).

Alternatively, equality holds in (9.48) if and only if $\phi_{ij}(E(k))(E_i(k+1)-E_j(k+1))=0$ and $\sum_{j=1,\,j\neq i}^{q}\phi_{ij}(E(k))+S_i(k)-\sigma_{ii}(E(k))=0$, $i,j=1,\ldots,q$, $i\neq j$, $k\geq k_0$. Next, $\sum_{j=1,\,j\neq i}^{q}\phi_{ij}(E(k))+S_i(k)-\sigma_{ii}(E(k))=0$ if and only if $\Delta E_i=0$, $i=1,\ldots,q$, $k\geq k_0$. Hence, $\phi_{ij}(E(k))(E_j(k+1)-E_i(k+1))=\phi_{ij}(E(k))(E_j(k)-E_i(k))=0$, $i,j=1,\ldots,q$, $i\neq j$, $k\geq k_0$. Thus, it follows from Axioms *i*) – *iii*) that equality holds in (9.48) if and only if $\Delta E_i=0$, $i=1,\ldots,q$, and $E_j=E_i$, $i,j=1,\ldots,q$, $i\neq j$. □

Note that inequality (9.48) is satisfied as an equality for an equilibrium process and as a strict inequality for a nonequilibrium process. Next, we present the definition of ectropy for the discrete-time, large-scale dynamical system $\mathcal{G}$.

Definition 9.11. For the discrete-time, large-scale dynamical system $\mathcal{G}$ with energy balance equation (9.18), a function $\mathcal{E}:\overline{\mathbb{R}}_+^q\to\mathbb{R}$ satisfying

$$
\mathcal{E}(E(k_2))\leq\mathcal{E}(E(k_1))+\sum_{k=k_1}^{k_2-1}\sum_{i=1}^{q}E_i(k+1)[S_i(k)-\sigma_{ii}(E(k))], \tag{9.50}
$$

for every $k_2\geq k_1\geq k_0$ and $S(\cdot)\in\mathcal{U}$, is called the *ectropy* of $\mathcal{G}$.

For the next result define the *available ectropy* of the discrete-time, large-scale dynamical system $\mathcal{G}$ by

$$
\mathcal{E}_\mathrm{a}(E_0)\triangleq-\inf_{S(\cdot)\in\mathcal{U}_\mathrm{c},\,K\geq k_0}\sum_{k=k_0}^{K-1}\sum_{i=1}^{q}E_i(k+1)[S_i(k)-\sigma_{ii}(E(k))], \tag{9.51}
$$

where $E(k_0)=E_0\in\overline{\mathbb{R}}_+^q$ and $E(K)=0$, and the *required ectropy supply* of the discrete-time, large-scale dynamical system $\mathcal{G}$ by

$$\mathcal{E}_{\rm r}(E_0) \triangleq \inf_{S(\cdot)\in\mathcal{U}_{\rm r},\, K\geq -k_0+1} \sum_{k=-K}^{k_0-1}\sum_{i=1}^{q} E_i(k+1)[S_i(k)-\sigma_{ii}(E(k))], \quad (9.52)$$

where $E(-K) = 0$ and $E(k_0) = E_0 \in \overline{\mathbb{R}}_+^q$. Note that the available ectropy $\mathcal{E}_{\rm a}(E_0)$ is the maximum amount of scaled heat (ectropy) that can be extracted from the large-scale dynamical system $\mathcal{G}$ in order to transfer it from an initial state $E(k_0) = E_0$ to $E(K) = 0$. Alternatively, the required ectropy supply $\mathcal{E}_{\rm r}(E_0)$ is the minimum amount of scaled heat (ectropy) that can be delivered to $\mathcal{G}$ to transfer it from an initial state $E(-K) = 0$ to a given state $E(k_0) = E_0$.

Theorem 9.7. Consider the discrete-time, large-scale dynamical system $\mathcal{G}$ with energy balance equation (9.18) and assume that Axioms *ii*) and *iii*) hold. Then there exists an ectropy function for $\mathcal{G}$. Moreover, $\mathcal{E}_{\rm a}(E)$, $E \in \overline{\mathbb{R}}_+^q$, and $\mathcal{E}_{\rm r}(E)$, $E \in \overline{\mathbb{R}}_+^q$, are possible ectropy functions for $\mathcal{G}$ with $\mathcal{E}_{\rm a}(0) = \mathcal{E}_{\rm r}(0) = 0$. Finally, all ectropy functions $\mathcal{E}(E)$, $E \in \overline{\mathbb{R}}_+^q$, for $\mathcal{G}$ satisfy

$$\mathcal{E}_{\rm a}(E) \leq \mathcal{E}(E) - \mathcal{E}(0) \leq \mathcal{E}_{\rm r}(E), \quad E \in \overline{\mathbb{R}}_+^q. \quad (9.53)$$

Proof. Since, by Proposition 9.3, $\mathcal{G}$ is controllable to and reachable from the origin in $\overline{\mathbb{R}}_+^q$, it follows from (9.51) and (9.52) that $\mathcal{E}_{\rm a}(E_0) > -\infty$, $E_0 \in \overline{\mathbb{R}}_+^q$, and $\mathcal{E}_{\rm r}(E_0) < \infty$, $E_0 \in \overline{\mathbb{R}}_+^q$, respectively. Next, let $E_0 \in \overline{\mathbb{R}}_+^q$ and let $S(\cdot) \in \mathcal{U}$ be such that $E(k_{\rm i}) = E(k_{\rm f}) = 0$ and $E(k_0) = E_0$, where $k_{\rm i} \leq k_0 \leq k_{\rm f}$. In this case, it follows from (9.48) that

$$\sum_{k=k_{\rm i}}^{k_{\rm f}-1}\sum_{i=1}^{q} E_i(k+1)[S_i(k)-\sigma_{ii}(E(k))] \geq 0, \quad (9.54)$$

or, equivalently,

$$\sum_{k=k_{\rm i}}^{k_0-1}\sum_{i=1}^{q} E_i(k+1)[S_i(k)-\sigma_{ii}(E(k))] \geq -\sum_{k=k_0}^{k_{\rm f}-1}\sum_{i=1}^{q} E_i(k+1)[S_i(k)-\sigma_{ii}(E(k))]. \quad (9.55)$$

Now, taking the infimum on both sides of (9.55) over all $S(\cdot) \in \mathcal{U}_{\rm r}$ and $k_{\rm i} + 1 \leq k_0$ yields

$$\mathcal{E}_{\rm r}(E_0) = \inf_{S(\cdot)\in\mathcal{U}_{\rm r},\, k_{\rm i}+1\leq k_0} \sum_{k=k_{\rm i}}^{k_0-1}\sum_{i=1}^{q} E_i(k+1)[S_i(k)-\sigma_{ii}(E(k))]$$

$$\geq -\sum_{k=k_0}^{k_{\mathrm{f}}-1}\sum_{i=1}^{q} E_i(k+1)[S_i(k)-\sigma_{ii}(E(k))]. \tag{9.56}$$

Next, taking the supremum on both sides of (9.56) over all $S(\cdot) \in \mathcal{U}_{\mathrm{c}}$ and $k_{\mathrm{f}} \geq k_0$, we obtain $\mathcal{E}_{\mathrm{r}}(E_0) \geq \mathcal{E}_{\mathrm{a}}(E_0)$, $E_0 \in \overline{\mathbb{R}}_+^q$, which implies that $-\infty < \mathcal{E}_{\mathrm{a}}(E_0) \leq \mathcal{E}_{\mathrm{r}}(E_0) < \infty$, $E_0 \in \overline{\mathbb{R}}_+^q$. Hence, the functions $\mathcal{E}_{\mathrm{a}}(\cdot)$ and $\mathcal{E}_{\mathrm{r}}(\cdot)$ are well defined.

Next, it follows from the definition of $\mathcal{E}_{\mathrm{a}}(\cdot)$ that, for any $K \geq k_1$ and $S(\cdot) \in \mathcal{U}_{\mathrm{c}}$ such that $E(k_1) \in \overline{\mathbb{R}}_+^q$ and $E(K) = 0$,

$$\begin{aligned} -\mathcal{E}_{\mathrm{a}}(E(k_1)) \leq & \sum_{k=k_1}^{k_2-1}\sum_{i=1}^{q} E_i(k+1)[S_i(k)-\sigma_{ii}(E(k))] \\ & + \sum_{k=k_2}^{K-1}\sum_{i=1}^{q} E_i(k+1)[S_i(k)-\sigma_{ii}(E(k))], \quad k_1 \leq k_2 \leq K, \end{aligned} \tag{9.57}$$

and hence,

$$\begin{aligned} -\mathcal{E}_{\mathrm{a}}(E(k_1)) \leq & \sum_{k=k_1}^{k_2-1}\sum_{i=1}^{q} E_i(k+1)[S_i(k)-\sigma_{ii}(E(k))] \\ & + \inf_{S(\cdot)\in\mathcal{U}_{\mathrm{c}},\, K\geq k_2} \sum_{k=k_2}^{K-1}\sum_{i=1}^{q} E_i(k+1)[S_i(k)-\sigma_{ii}(E(k))] \\ = & \sum_{k=k_1}^{k_2-1}\sum_{i=1}^{q} E_i(k+1)[S_i(k)-\sigma_{ii}(E(k))] - \mathcal{E}_{\mathrm{a}}(E(k_2)), \end{aligned} \tag{9.58}$$

which implies that $\mathcal{E}_{\mathrm{a}}(E)$, $E \in \overline{\mathbb{R}}_+^q$, satisfies (9.50). Thus, $\mathcal{E}_{\mathrm{a}}(E)$, $E \in \overline{\mathbb{R}}_+^q$, is a possible ectropy function for the system $\mathcal{G}$. Note that with $E(k_0) = E(K) = 0$ it follows from (9.48) that the infimum in (9.51) is taken over the set of nonnegative values with one of the values being zero for $S(k) \equiv 0$. Thus, $\mathcal{E}_{\mathrm{a}}(0) = 0$. Similarly, it can be shown that $\mathcal{E}_{\mathrm{r}}(E)$, $E \in \overline{\mathbb{R}}_+^q$, given by (9.52) satisfies (9.50), and hence, is a possible ectropy function for the system $\mathcal{G}$ with $\mathcal{E}_{\mathrm{r}}(0) = 0$.

Next, suppose there exists an ectropy function $\mathcal{E} : \overline{\mathbb{R}}_+^q \to \mathbb{R}$ for $\mathcal{G}$ and let $E(k_2) = 0$ in (9.50). Then it follows from (9.50) that

$$\mathcal{E}(E(k_1)) - \mathcal{E}(0) \geq -\sum_{k=k_1}^{k_2-1}\sum_{i=1}^{q} E_i(k+1)[S_i(k)-\sigma_{ii}(E(k))], \tag{9.59}$$

for all $k_2 \geq k_1$ and $S(\cdot) \in \mathcal{U}_c$, which implies that

$$\begin{aligned}
\mathcal{E}(E(k_1)) - \mathcal{E}(0) &\geq \sup_{S(\cdot)\in\mathcal{U}_c,\, k_2\geq k_1} \left[-\sum_{k=k_1}^{k_2-1} \sum_{i=1}^{q} E_i(k+1)[S_i(k) - \sigma_{ii}(E(k))] \right] \\
&= -\inf_{S(\cdot)\in\mathcal{U}_c,\, k_2\geq k_1} \sum_{k=k_1}^{k_2-1} \sum_{i=1}^{q} E_i(k+1)[S_i(k) - \sigma_{ii}(E(k))] \\
&= \mathcal{E}_a(E(k_1)). \qquad (9.60)
\end{aligned}$$

Since $E(k_1)$ is arbitrary, it follows that $\mathcal{E}(E) - \mathcal{E}(0) \geq \mathcal{E}_a(E)$, $E \in \overline{\mathbb{R}}_+^q$.

Alternatively, let $E(k_1) = 0$ in (9.50). Then it follows from (9.50) that

$$\mathcal{E}(E(k_2)) - \mathcal{E}(0) \leq \sum_{k=k_1}^{k_2-1} \sum_{i=1}^{q} E_i(k+1)[S_i(k) - \sigma_{ii}(E(k))], \qquad (9.61)$$

for all $k_1 + 1 \leq k_2$ and $S(\cdot) \in \mathcal{U}_r$. Hence,

$$\begin{aligned}
\mathcal{E}(E(k_2)) - \mathcal{E}(0) &\leq \inf_{S(\cdot)\in\mathcal{U}_r,\, k_1+1\leq k_2} \sum_{k=k_1}^{k_2-1} \sum_{i=1}^{q} E_i(k+1)[S_i(k) - \sigma_{ii}(E(k))] \\
&= \mathcal{E}_r(E(k_2)), \qquad (9.62)
\end{aligned}$$

which, since $E(k_2)$ is arbitrary, implies that $\mathcal{E}_r(E) \geq \mathcal{E}(E) - \mathcal{E}(0)$, $E \in \overline{\mathbb{R}}_+^q$. Thus, all ectropy functions for $\mathcal{G}$ satisfy (9.53). □

The next proposition gives a closed-form expression for the ectropy of the discrete-time, large-scale dynamical system $\mathcal{G}$.

Proposition 9.7. Consider the discrete-time, large-scale dynamical system $\mathcal{G}$ with energy balance equation (9.18) and assume that Axioms *ii)* and *iii)* hold. Then the function $\mathcal{E} : \overline{\mathbb{R}}_+^q \to \mathbb{R}$ given by

$$\mathcal{E}(E) = \tfrac{1}{2} E^{\mathrm{T}} E, \quad E \in \overline{\mathbb{R}}_+^q, \qquad (9.63)$$

is an ectropy function of $\mathcal{G}$.

Proof. Since $E(k) \geq\geq 0$, $k \geq k_0$, and $\phi_{ij}(E) = -\phi_{ji}(E)$, $E \in \overline{\mathbb{R}}_+^q$, $i \neq j$, $i, j = 1, \ldots, q$, it follows that

$$\begin{aligned}
\Delta\mathcal{E}(E(k)) &= \tfrac{1}{2} E^{\mathrm{T}}(k+1)E(k+1) - \tfrac{1}{2} E^{\mathrm{T}}(k)E(k) \\
&= \sum_{i=1}^{q} E_i(k+1)[S_i(k) - \sigma_{ii}(E(k))] \\
&\quad -\tfrac{1}{2} \sum_{i=1}^{q} \left[\sum_{j=1,\, j\neq i}^{q} \phi_{ij}(E(k)) + S_i(k) - \sigma_{ii}(E(k)) \right]^2
\end{aligned}$$

$$
\begin{aligned}
&+\sum_{i=1}^{q}\sum_{j=1,\, j\neq i}^{q} E_i(k+1)\phi_{ij}(E(k)) \\
&= \sum_{i=1}^{q} E_i(k+1)[S_i(k)-\sigma_{ii}(E(k))] \\
&\quad -\tfrac{1}{2}\sum_{i=1}^{q}\left[\sum_{j=1,\, j\neq i}^{q}\phi_{ij}(E(k))+S_i(k)-\sigma_{ii}(E(k))\right]^2 \\
&\quad +\sum_{i=1}^{q-1}\sum_{j=i+1}^{q}[E_i(k+1)-E_j(k+1)]\phi_{ij}(E(k)) \\
&\leq \sum_{i=1}^{q} E_i(k+1)[S_i(k)-\sigma_{ii}(E(k))], \quad k\geq k_0. \qquad (9.64)
\end{aligned}
$$

Now, summing (9.64) over $\{k_1,\ldots,k_2-1\}$ yields (9.50). □

Note that it follows from the last equality in (9.64) that the ectropy function given by (9.63) satisfies (9.50) as an equality for an equilibrium process and as a strict inequality for a nonequilibrium process. It follows from (9.63) that ectropy is a measure of the extent to which the system energy deviates from a homogeneous state. Thus, ectropy is the dual of entropy and is a measure of the tendency of the discrete-time, large-scale dynamical system $\mathcal{G}$ to do useful work and grow more organized.

9.6 Semistability of Discrete-Time Thermodynamic Models

Inequality (9.33) is analogous to Clausius' inequality for equilibrium and nonequilibrium thermodynamics as applied to discrete-time, large-scale dynamical systems, whereas inequality (9.50) is an anti–Clausius inequality. Moreover, for the ectropy function defined by (9.63), inequality (9.64) shows that a thermodynamically consistent discrete-time, large-scale dynamical system is *dissipative* [191] with respect to the supply rate $E^{\mathrm{T}}S$ and with the storage function corresponding to the system ectropy $\mathcal{E}(E)$. For the entropy function given by (9.46) note that $\mathcal{S}(0)=0$ or, equivalently, $\lim_{E\to 0}\mathcal{S}(E)=0$, which is consistent with the *third law of thermodynamics* (Nernst's theorem).

For the isolated discrete-time, large-scale dynamical system $\mathcal{G}$, (9.33) yields the fundamental inequality

$$\mathcal{S}(E(k_2))\geq\mathcal{S}(E(k_1)),\quad k_2\geq k_1. \qquad (9.65)$$

Inequality (9.65) implies that, for any dynamical change in an isolated (i.e., $S(k)\equiv 0$ and $d(E(k))\equiv 0$) discrete-time, large-scale system, the entropy of the final state can never be less than the entropy of the initial state. It is important to stress that this result holds for an isolated dynamical system. It is, however, possible with energy supplied from an external dynamical system (e.g., a controller) to reduce the entropy of the discrete-time, large-scale dynamical system. The entropy of both systems taken together, however, cannot decrease.

The above observations imply that when an isolated discrete-time, large-scale dynamical system with thermodynamically consistent energy flow characteristics (i.e., Axioms $i) - iii)$ hold) is at a state of maximum entropy consistent with its energy, it cannot be subject to any further dynamical change since any such change would result in a decrease of entropy. This of course implies that the state of *maximum entropy* is the stable state of an isolated system and this state has to be semistable.

Analogously, it follows from (9.50) that for an isolated discrete-time, large-scale dynamical system $\mathcal{G}$ the fundamental inequality

$$\mathcal{E}(E(k_2)) \le \mathcal{E}(E(k_1)), \quad k_2 \ge k_1, \tag{9.66}$$

is satisfied, which implies that the ectropy of the final state of $\mathcal{G}$ is always less than or equal to the ectropy of the initial state of $\mathcal{G}$. Hence, for the isolated large-scale dynamical system $\mathcal{G}$ the entropy increases if and only if the ectropy decreases. Thus, the state of *minimum ectropy* is the stable state of an isolated system and this equilibrium state has to be semistable. The next theorem concretizes the above observations.

Theorem 9.8. Consider the discrete-time, large-scale dynamical system $\mathcal{G}$ with energy balance equation (9.18) with $S(k) \equiv 0$ and $d(E) \equiv 0$, and assume that Axioms $i) - iii)$ hold. Then for every $\alpha \ge 0$, $\alpha\mathbf{e}$ is a Lyapunov equilibrium state of (9.18). Furthermore, $E(k) \to \frac{1}{q}\mathbf{e}\mathbf{e}^{\mathrm{T}}E(k_0)$ as $k \to \infty$ and $\frac{1}{q}\mathbf{e}\mathbf{e}^{\mathrm{T}}E(k_0)$ is a semistable equilibrium state. Finally, if for some $m \in \{1,\ldots,q\}$, $\sigma_{mm}(E) \ge 0$, $E \in \overline{\mathbb{R}}_+^q$, and $\sigma_{mm}(E) = 0$ if and only if $E_m = 0$,[1] then the zero solution $E(k) \equiv 0$ to (9.18) is a globally asymptotically stable equilibrium state of (9.18).

Proof. It follows from Axiom $i)$ that $\alpha\mathbf{e} \in \overline{\mathbb{R}}_+^q$, $\alpha \ge 0$, is an equilibrium state for (9.18). To show Lyapunov stability of the equilibrium state $\alpha\mathbf{e}$, consider the system shifted ectropy $\mathcal{E}_{\mathrm{s}}(E) = \frac{1}{2}(E - \alpha\mathbf{e})^{\mathrm{T}}(E - \alpha\mathbf{e})$ as a Lyapunov function candidate. Now, since $\phi_{ij}(E) = -\phi_{ji}(E)$, $E \in \overline{\mathbb{R}}_+^q$, $i \ne j$, $i,j = 1,\ldots,q$, and $\mathbf{e}^{\mathrm{T}}E(k+1) = \mathbf{e}^{\mathrm{T}}E(k)$, $k \ge k_0$, it follows from

[1]The assumption $\sigma_{mm}(E) \ge 0$, $E \in \overline{\mathbb{R}}_+^q$, and $\sigma_{mm}(E) = 0$ if and only if $E_m = 0$ for some $m \in \{1,\ldots,q\}$ implies that if the mth subsystem possesses no energy, then this subsystem cannot dissipate energy to the environment. Conversely, if the mth subsystem does not dissipate energy to the environment, then this subsystem has no energy.

Axioms *ii*) and *iii*) that

$$\begin{aligned}\Delta\mathcal{E}_{\rm s}(E(k)) &= \tfrac{1}{2}(E(k+1)-\alpha\mathbf{e})^{\rm T}(E(k+1)-\alpha\mathbf{e})\\ &\quad -\tfrac{1}{2}(E(k)-\alpha\mathbf{e})^{\rm T}(E(k)-\alpha\mathbf{e})\\ &= \sum_{i=1}^{q}\sum_{j=1,\,j\neq i}^{q} E_i(k+1)\phi_{ij}(E(k)) - \tfrac{1}{2}\sum_{i=1}^{q}\left[\sum_{j=1,\,j\neq i}^{q}\phi_{ij}(E(k))\right]^2\\ &= \sum_{i=1}^{q-1}\sum_{j=i+1}^{q}[E_i(k+1)-E_j(k+1)]\phi_{ij}(E(k))\\ &\quad -\tfrac{1}{2}\sum_{i=1}^{q}\left[\sum_{j=1,\,j\neq i}^{q}\phi_{ij}(E(k))\right]^2\\ &\leq 0,\quad E(k)\in\overline{\mathbb{R}}_+^q,\quad k\geq k_0,\end{aligned}\tag{9.67}$$

which, using Theorem 9.1, establishes Lyapunov stability of the equilibrium state $\alpha\mathbf{e}$.

To show that $\alpha\mathbf{e}$ is semistable, note that

$$\begin{aligned}&\Delta\mathcal{E}_{\rm s}(E(k))\\ &= \sum_{i=1}^{q}\sum_{j=1,\,j\neq i}^{q} E_i(k)\phi_{ij}(E(k)) + \tfrac{1}{2}\sum_{i=1}^{q}\left[\sum_{j=1,\,j\neq i}^{q}\phi_{ij}(E(k))\right]^2\\ &\geq \sum_{i=1}^{q-1}\sum_{j=i+1}^{q}[E_i(k)-E_j(k)]\phi_{ij}(E(k))\\ &= \sum_{i=1}^{q-1}\sum_{j\in\mathcal{K}_i}[E_i(k)-E_j(k)]\phi_{ij}(E(k)),\quad E(k)\in\overline{\mathbb{R}}_+^q,\quad k\geq k_0,\end{aligned}\tag{9.68}$$

where $\mathcal{K}_i \triangleq \mathcal{N}_i\setminus\cup_{l=1}^{i-1}\{l\}$ and $\mathcal{N}_i \triangleq \{j\in\{1,\ldots,q\} : \phi_{ij}(E) = 0 \text{ if and only if } E_i = E_j\}$, $i=1,\ldots,q$.

Next, we show that $\Delta\mathcal{E}_{\rm s}(E)=0$ if and only if $(E_i-E_j)\phi_{ij}(E)=0$, $i=1,\ldots,q$, $j\in\mathcal{K}_i$. First, assume that $(E_i-E_j)\phi_{ij}(E)=0$, $i=1,\ldots,q$, $j\in\mathcal{K}_i$. Then it follows from (9.68) that $\Delta\mathcal{E}_{\rm s}(E)\geq 0$. However, it follows from (9.67) that $\Delta\mathcal{E}_{\rm s}(E)\leq 0$. Hence, $\Delta\mathcal{E}_{\rm s}(E)=0$. Conversely, assume $\Delta\mathcal{E}_{\rm s}(E)=0$. In this case, it follows from (9.67) that $(E_i(k+1)-E_j(k+1))\phi_{ij}(E(k))=0$ and $\sum_{j=1,\,j\neq i}^{q}\phi_{ij}(E(k))=0$, $k\geq k_0$, $i,j=1,\ldots,q$, $i\neq j$. Since

$$\begin{aligned}&[E_i(k+1)-E_j(k+1)]\phi_{ij}(E(k))\\ &\qquad = [E_i(k)-E_j(k)]\phi_{ij}(E(k))\end{aligned}$$

$$+\left[\sum_{h=1,\,h\neq i}^{q}\phi_{ih}(E(k))-\sum_{l=1,\,l\neq j}^{q}\phi_{jl}(E(k))\right]\phi_{ij}(E(k))$$
$$=[E_i(k)-E_j(k)]\phi_{ij}(E(k)),$$
$$k\geq k_0,\quad i,j=1,\ldots,q,\quad i\neq j,\tag{9.69}$$

it follows that $(E_i-E_j)\phi_{ij}(E)=0$, $i=1,\ldots,q$, $j\in\mathcal{K}_i$.

Let $\mathcal{R}\triangleq\{E\in\overline{\mathbb{R}}_+^q:\Delta\mathcal{E}_{\rm s}(E)=0\}=\{E\in\overline{\mathbb{R}}_+^q:(E_i-E_j)\phi_{ij}(E)=0,\ i=1,\ldots,q,\ j\in\mathcal{K}_i\}$. Now, by Axiom $i)$ the directed graph associated with the connectivity matrix $\mathcal{C}$ for the discrete-time, large-scale dynamical system $\mathcal{G}$ is strongly connected, which implies that $\mathcal{R}=\{E\in\overline{\mathbb{R}}_+^q:E_1=\cdots=E_q\}$. Since the set $\mathcal{R}$ consists of the equilibrium states of (9.18), it follows that the largest invariant set $\mathcal{M}$ contained in $\mathcal{R}$ is given by $\mathcal{M}=\mathcal{R}$. Hence, it follows from Corollary 9.1 that for any initial condition $E(k_0)\in\overline{\mathbb{R}}_+^q$, $E(k)\to\mathcal{M}$ as $k\to\infty$, and hence, $\alpha\mathbf{e}$ is a semistable equilibrium state of (9.18). Next, note that since $\mathbf{e}^{\rm T}E(k)=\mathbf{e}^{\rm T}E(k_0)$ and $E(k)\to\mathcal{M}$ as $k\to\infty$, it follows that $E(k)\to\frac{1}{q}\mathbf{e}\mathbf{e}^{\rm T}E(k_0)$ as $k\to\infty$. Hence, with $\alpha=\frac{1}{q}\mathbf{e}^{\rm T}E(k_0)$, $\alpha\mathbf{e}=\frac{1}{q}\mathbf{e}\mathbf{e}^{\rm T}E(k_0)$ is a semistable equilibrium state of (9.18).

Finally, to show that, in the case where for some $m\in\{1,\ldots,q\}$, $\sigma_{mm}(E)\geq 0$, $E\in\overline{\mathbb{R}}_+^q$, and $\sigma_{mm}(E)=0$ if and only if $E_m=0$, the zero solution $E(k)\equiv 0$ to (9.18) is globally asymptotically stable, consider the system ectropy $\mathcal{E}(E)=\frac{1}{2}E^{\rm T}E$ as a candidate Lyapunov function. Note that $\mathcal{E}(0)=0$, $\mathcal{E}(E)>0$, $E\in\overline{\mathbb{R}}_+^q$, $E\neq 0$, and $\mathcal{E}(E)$ is radially unbounded. Now, the Lyapunov difference is given by

$$\Delta\mathcal{E}(E(k))$$
$$=\tfrac{1}{2}E^{\rm T}(k+1)E(k+1)-\tfrac{1}{2}E^{\rm T}(k)E(k)$$
$$=-E_m(k+1)\sigma_{mm}(E(k))-\tfrac{1}{2}\left[\sum_{j=1,\,j\neq m}^{q}\phi_{mj}(E(k))-\sigma_{mm}(E(k))\right]^2$$
$$-\tfrac{1}{2}\sum_{i=1,i\neq m}^{q}\left[\sum_{j=1,\,j\neq i}^{q}\phi_{ij}(E(k))\right]^2+\sum_{i=1}^{q}\sum_{j=1,\,j\neq i}^{q}E_i(k+1)\phi_{ij}(E(k))$$
$$=-E_m(k+1)\sigma_{mm}(E(k))-\tfrac{1}{2}\left[\sum_{j=1,\,j\neq m}^{q}\phi_{mj}(E(k))-\sigma_{mm}(E(k))\right]^2$$
$$-\tfrac{1}{2}\sum_{i=1,i\neq m}^{q}\left[\sum_{j=1,\,j\neq i}^{q}\phi_{ij}(E(k))\right]^2$$

$$+\sum_{i=1}^{q-1}\sum_{j=i+1}^{q}[E_i(k+1)-E_j(k+1)]\phi_{ij}(E(k))$$
$$\leq 0, \quad E(k)\in\overline{\mathbb{R}}_+^q, \quad k\geq k_0, \tag{9.70}$$

which shows that the zero solution $E(k)\equiv 0$ to (9.18) is Lyapunov stable.

To show global asymptotic stability of the zero equilibrium state, note that

$$\begin{aligned}\Delta\mathcal{E}(E(k)) &= \sum_{i=1}^{q-1}\sum_{j=i+1}^{q}[E_i(k)-E_j(k)]\phi_{ij}(E(k))+\tfrac{1}{2}\sum_{i=1,i\neq m}^{q}\left[\sum_{j=1,\,j\neq i}^{q}\phi_{ij}(E(k))\right]^2\\ &\quad -E_m(k)\sigma_{mm}(E(k))+\tfrac{1}{2}\left[\sum_{j=1,\,j\neq m}^{q}\phi_{mj}(E(k))-\sigma_{mm}(E(k))\right]^2\\ &\geq \sum_{i=1}^{q-1}\sum_{j\in\mathcal{K}_i}[E_i(k)-E_j(k)]\phi_{ij}(E(k))-E_m(k)\sigma_{mm}(E(k)),\\ &\qquad E(k)\in\overline{\mathbb{R}}_+^q, \quad k\geq k_0.\end{aligned} \tag{9.71}$$

Next, we show that $\Delta\mathcal{E}(E)=0$ if and only if $(E_i-E_j)\phi_{ij}(E)=0$ and $\sigma_{mm}(E)=0$, $i=1,\ldots,q$, $j\in\mathcal{K}_i$, $m\in\{1,\ldots,q\}$. First, assume that $(E_i-E_j)\phi_{ij}(E)=0$ and $\sigma_{mm}(E)=0$, $i=1,\ldots,q$, $j\in\mathcal{K}_i$, $m\in\{1,\ldots,q\}$. Then it follows from (9.71) that $\Delta\mathcal{E}(E)\geq 0$. However, it follows from (9.70) that $\Delta\mathcal{E}(E)\leq 0$. Thus, $\Delta\mathcal{E}(E)=0$.

Conversely, assume $\Delta\mathcal{E}(E)=0$. Then it follows from (9.70) that $(E_i(k+1)-E_j(k+1))\phi_{ij}(E(k))=0$, $i,j=1,\ldots,q$, $i\neq j$, $\sum_{j=1,\,j\neq i}^{q}\phi_{ij}(E(k))=0$, $i=1,\ldots,q$, $i\neq m$, $k\geq k_0$, and $\sigma_{mm}(E)=0$, $m\in\{1,\ldots,q\}$. Note that in this case it follows that $\sigma_{mm}(E)=\sum_{j=1,j\neq m}^{q}\phi_{mj}(E)=0$, and hence,

$$[E_i(k+1)-E_j(k+1)]\phi_{ij}(E(k))=[E_i(k)-E_j(k)]\phi_{ij}(E(k)),$$
$$k\geq k_0, \quad i,j=1,\ldots,q, \quad i\neq j, \tag{9.72}$$

which implies that $(E_i-E_j)\phi_{ij}(E)=0$, $i=1,\ldots,q$, $j\in\mathcal{K}_i$. Hence, $(E_i-E_j)\phi_{ij}(E)=0$ and $\sigma_{mm}(E)=0$, $i=1,\ldots,q$, $j\in\mathcal{K}_i$, $m\in\{1,\ldots,q\}$ if and only if $\Delta\mathcal{E}(E)=0$.

Let $\mathcal{R}\triangleq\{E\in\overline{\mathbb{R}}_+^q:\Delta\mathcal{E}(E)=0\}=\{E\in\overline{\mathbb{R}}_+^q:\sigma_{mm}(E)=0, m\in\{1,\ldots,q\}\}\cap\{E\in\overline{\mathbb{R}}_+^q:(E_i-E_j)\phi_{ij}(E)=0, i=1,\ldots,q, j\in\mathcal{K}_i\}$. Now, since Axiom i) holds and $\sigma_{mm}(E)=0$ if and only if $E_m=0$ it follows that $\mathcal{R}=\{E\in\overline{\mathbb{R}}_+^q:E_m=0, m\in\{1,\ldots,q\}\}\cap\{E\in\overline{\mathbb{R}}_+^q:E_1=E_2=$

$\cdots = E_q\} = \{0\}$ and the largest invariant set $\mathcal{M}$ contained in $\mathcal{R}$ is given by $\mathcal{M} = \{0\}$. Hence, it follows from Theorem 9.4 that for any initial condition $E(k_0) \in \overline{\mathbb{R}}_+^q$, $E(k) \to \mathcal{M} = \{0\}$ as $k \to \infty$, which proves global asymptotic stability of the zero equilibrium state of (9.18). □

It is important to note that Axiom *iii*) involving monotonicity of solutions is explicitly used to prove semistability for discrete-time compartmental dynamical systems. However, Axiom *iii*) is a sufficient condition and not necessary for guaranteeing semistability. Replacing the monotonicity condition with $\sum_{i=1,j=1,i\neq j}^{q} \alpha_{ij}(E) f_{ij}(E) \geq 0$, where

$$\alpha_{ij}(E) \triangleq \begin{cases} \frac{\phi_{ij}(E)}{E_j - E_i}, & E_i \neq E_j, \\ 0, & E_i = E_j, \end{cases} \tag{9.73}$$

$$f_{ij}(E) \triangleq [E_i(k) - E_j(k)][E_i(k+1) - E_j(k+1)], \tag{9.74}$$

provides a weaker sufficient condition for guaranteeing semistability. However, in this case, to ensure that the entropy of $\mathcal{G}$ is monotonically increasing, we additionally require that $\sum_{i=1,j=1,i\neq j}^{q} \beta_{ij}(E) f_{ij}(E) \geq 0$, where

$$\beta_{ij}(E) \triangleq \begin{cases} \frac{1}{(c+E_i(k+1))(c+E_j(k+1))} \cdot \frac{\phi_{ij}(E(k))}{E_j(k)-E_i(k)}, & E_i \neq E_j, \\ 0, & E_i = E_j. \end{cases} \tag{9.75}$$

Thus, a weaker condition for Axiom *iii*) that combines

$$\sum_{i=1,j=1,i\neq j}^{q} \alpha_{ij}(E) f_{ij}(E) \geq 0$$

and

$$\sum_{i=1,j=1,i\neq j}^{q} \beta_{ij}(E) f_{ij}(E) \geq 0$$

is

$$\sum_{i=1,j=1,i\neq j}^{q} \gamma_{ij}(E) f_{ij}(E) \geq 0,$$

where $\gamma_{ij}(E) \triangleq \alpha_{ij}(E) + \beta_{ij}(E) - \mathrm{sgn}(f_{ij}(E))|\alpha_{ij}(E) - \beta_{ij}(E)|$ and $\mathrm{sgn}(f_{ij}(E)) \triangleq |f_{ij}(E)|/f_{ij}(E)$.

In Theorem 9.8, we used the shifted ectropy function to show that for the isolated (i.e., $S(k) \equiv 0$ and $d(E) \equiv 0$) discrete-time, large-scale dynamical system $\mathcal{G}$ with Axioms *i*) – *iii*), $E(k) \to \frac{1}{q}\mathbf{e}\mathbf{e}^{\mathrm{T}} E(k_0)$ as $k \to \infty$ and $\frac{1}{q}\mathbf{e}\mathbf{e}^{\mathrm{T}} E(k_0)$ is a semistable equilibrium state. This result can also be arrived at using the system entropy for the isolated discrete-time, large-scale dynamical system $\mathcal{G}$ with Axioms *i*) – *iii*).

To see this note that since $\mathbf{e}^{\mathrm{T}} f(E) = \mathbf{e}^{\mathrm{T}} E$, $E \in \overline{\mathbb{R}}_+^q$, it follows that $\mathbf{e}^{\mathrm{T}} \Delta E(k) = 0$, $k \geq k_0$. Hence, $\mathbf{e}^{\mathrm{T}} E(k) = \mathbf{e}^{\mathrm{T}} E(k_0)$, $k \geq k_0$. Furthermore, since $E(k) \geq\geq 0$, $k \geq k_0$, it follows that $0 \leq\leq E(k) \leq\leq \mathbf{e}\mathbf{e}^{\mathrm{T}} E(k_0)$, $k \geq k_0$, which implies that all solutions to (9.18) are bounded. Next, since by (9.65) the entropy $\mathcal{S}(E(k))$, $k \geq k_0$, of $\mathcal{G}$ is monotonically increasing and $E(k)$, $k \geq k_0$, is bounded, the result follows by using similar arguments as in Theorem 9.8 and using the fact that $\frac{x}{1+x} \leq \log_e(1+x) \leq x$ for all $x > -1$.

9.7 Discrete Energy Equipartition

It follows from Theorem 9.8 that the steady-state value of the energy in each subsystem $\mathcal{G}_i$ of the isolated large-scale dynamical system $\mathcal{G}$ is equal; that is, the steady-state energy of the isolated discrete-time, large-scale dynamical system $\mathcal{G}$ given by

$$E_\infty = \frac{1}{q}\mathbf{e}\mathbf{e}^{\mathrm{T}} E(k_0) = \left[\frac{1}{q}\sum_{i=1}^{q} E_i(k_0)\right]\mathbf{e}$$

is uniformly distributed over all subsystems of $\mathcal{G}$. The next proposition shows that among all possible energy distributions in the discrete-time, large-scale dynamical system $\mathcal{G}$, energy equipartition corresponds to the minimum value of the system's ectropy and the maximum value of the system's entropy (see Figure 3.2).

Proposition 9.8. Consider the discrete-time, large-scale dynamical system $\mathcal{G}$ with energy balance equation (9.18), let $\mathcal{E} : \overline{\mathbb{R}}_+^q \to \mathbb{R}$ and $\mathcal{S} : \overline{\mathbb{R}}_+^q \to \mathbb{R}$ denote the ectropy and entropy of $\mathcal{G}$ given by (9.63) and (9.46), respectively, and define $\mathcal{D}_{\mathrm{c}} \triangleq \{E \in \overline{\mathbb{R}}_+^q : \mathbf{e}^{\mathrm{T}} E = \beta\}$, where $\beta \geq 0$. Then

$$\arg\min_{E\in\mathcal{D}_{\mathrm{c}}}(\mathcal{E}(E)) = \arg\max_{E\in\mathcal{D}_{\mathrm{c}}}(\mathcal{S}(E)) = E^* = \frac{\beta}{q}\mathbf{e}. \tag{9.76}$$

Furthermore, $\mathcal{E}_{\min} \triangleq \mathcal{E}(E^*) = \frac{1}{2}\frac{\beta^2}{q}$ and $\mathcal{S}_{\max} \triangleq \mathcal{S}(E^*) = q\log_e(c + \frac{\beta}{q}) - q\log_e c$.

Proof. The proof is identical to the proof of Proposition 3.10 for the continuous-time case. □

9.8 Entropy Increase and the Second Law of Thermodynamics

In the preceding discussion, it was assumed that our discrete-time, large-scale nonlinear dynamical system model is such that energy is exchanged from more energetic subsystems to less energetic subsystems; that is, heat (energy) flows in the direction of lower temperatures. Although this universal phenomenon can be predicted with virtual certainty, it follows as a manifestation of entropy and ectropy nonconservation for the case of two subsystems.

To see this, consider the isolated (i.e., $S(k) \equiv 0$ and $d(E) \equiv 0$) discrete-time, large-scale dynamical system $\mathcal{G}$ with energy balance equation (9.18) and assume that the system entropy is monotonically increasing, and hence, $\Delta\mathcal{S}(E(k)) \geq 0$, $k \geq k_0$. Now, since

$$\begin{aligned}
0 &\leq \Delta\mathcal{S}(E(k)) \\
&= \sum_{i=1}^{q} \log_e \left[1 + \frac{\Delta E_i(k)}{c + E_i(k)}\right] \\
&\leq \sum_{i=1}^{q} \frac{\Delta E_i(k)}{c + E_i(k)} \\
&= \sum_{i=1}^{q} \sum_{j=1, j \neq i}^{q} \frac{\phi_{ij}(E(k))}{c + E_i(k)} \\
&= \sum_{i=1}^{q-1} \sum_{j=i+1}^{q} \left[\frac{\phi_{ij}(E(k))}{c + E_i(k)} - \frac{\phi_{ij}(E(k))}{c + E_j(k)}\right] \\
&= \sum_{i=1}^{q-1} \sum_{j=i+1}^{q} \frac{\phi_{ij}(E(k))[E_j(k) - E_i(k)]}{(c + E_i(k))(c + E_j(k))}, \quad k \geq k_0, \qquad (9.77)
\end{aligned}$$

it follows that for $q = 2$, $(E_1 - E_2)\phi_{12}(E) \leq 0$, $E \in \overline{\mathbb{R}}_+^2$, which implies that energy (heat) flows naturally from a more energetic subsystem (hot object) to a less energetic subsystem (cooler object). The universality of this emergent behavior thus follows from the fact that entropy (respectively, ectropy) transfer, accompanying energy transfer, always increases (respectively, decreases).

In the case where we have multiple subsystems, it is clear from (9.77) that entropy and ectropy nonconservation does not necessarily imply Axiom *ii*). However, if we invoke the additional condition (Axiom *iv*)) that if for any pair of connected subsystems $\mathcal{G}_k$ and $\mathcal{G}_l$, $k \neq l$, with energies $E_k \geq E_l$ (respectively, $E_k \leq E_l$), and for any other pair of connected subsystems $\mathcal{G}_m$ and $\mathcal{G}_n$, $m \neq n$, with energies $E_m \geq E_n$ (respectively, $E_m \leq E_n$), the inequality $\phi_{kl}(E)\phi_{mn}(E) \geq 0$, $E \in \overline{\mathbb{R}}_+^q$, holds, then nonconservation of entropy and ectropy in the isolated discrete-time, large-scale dynamical system $\mathcal{G}$ implies Axiom *ii*).

The above inequality postulates that the direction of energy exchange for any pair of *energy similar* subsystems is consistent; that is, if for a given pair of connected subsystems at given different energy levels the energy flows in a certain direction, then for any other pair of connected subsystems with the same energy level, the energy flow direction is consistent with the original pair of subsystems. Note that this assumption does *not* specify the direction of energy flow between subsystems.

To see that $\Delta\mathcal{S}(E(k)) \geq 0$, $k \geq k_0$, along with Axiom *iv*) implies Axiom *ii*), note that since (9.77) holds for all $k \geq k_0$ and $E(k_0) \in \overline{\mathbb{R}}_+^q$ is arbitrary, (9.77) implies

$$\sum_{i=1}^{q} \sum_{j\in\mathcal{K}_i} \frac{\phi_{ij}(E)(E_j - E_i)}{(c+E_i)(c+E_j)} \geq 0, \quad E \in \overline{\mathbb{R}}_+^q. \tag{9.78}$$

Now, it follows from (9.78) that for any fixed system energy level $E \in \overline{\mathbb{R}}_+^q$ there exists at least one pair of connected subsystems $\mathcal{G}_k$ and $\mathcal{G}_l$, $k \neq l$, such that $\phi_{kl}(E)(E_l - E_k) \geq 0$. Thus, if $E_k \geq E_l$ (respectively, $E_k \leq E_l$), then $\phi_{kl}(E) \leq 0$ (respectively, $\phi_{kl}(E) \geq 0$). Furthermore, it follows from Axiom *iv*) that for any other pair of connected subsystems $\mathcal{G}_m$ and $\mathcal{G}_n$, $m \neq n$, with $E_m \geq E_n$ (respectively, $E_m \leq E_n$), the inequality $\phi_{mn}(E) \leq 0$ (respectively, $\phi_{mn}(E) \geq 0$) holds, which implies that

$$\phi_{mn}(E)(E_n - E_m) \geq 0, \quad m \neq n. \tag{9.79}$$

It follows from (9.79) that energy (heat) flows naturally from more energetic subsystems (hot objects) to less energetic subsystems (cooler objects). Of course, since in the isolated discrete-time, large-scale dynamical system $\mathcal{G}$ ectropy decreases if and only if entropy increases, the same result can be arrived at by considering the ectropy of $\mathcal{G}$. Since Axiom *ii*) holds, it follows from the conservation of energy and the fact that the discrete-time large-scale dynamical system $\mathcal{G}$ is strongly connected that nonconservation of entropy and ectropy necessarily implies energy equipartition.

9.9 Discrete Temperature Equipartition

In this section, we generalize the results of Section 9.3 to the case where the subsystem energies are proportional to the subsystem temperatures with the proportionality constants representing the subsystem *specific heats* or *thermal capacities*. To include temperature notions in our discrete-time, large-scale dynamical system model, we replace Axioms *i*)–*iii*) of Section 9.4 by the following conditions. Let $\beta_i > 0$, $i = 1, \ldots, q$, denote the reciprocal of the specific heat of the ith subsystem $\mathcal{G}_i$ so that the *absolute temperature* in the ith subsystem is given by $\hat{T}_i = \beta_i E_i$.

Axiom *i*): For the connectivity matrix $\mathcal{C} \in \mathbb{R}^{q\times q}$ associated with the discrete-time, large-scale dynamical system $\mathcal{G}$ defined by (9.29) and (9.30), rank $\mathcal{C} = q-1$ and for $\mathcal{C}_{(i,j)} = 1$, $i \neq j$, $\phi_{ij}(E) = 0$ if and only if $\beta_i E_i = \beta_j E_j$.

Axiom *ii*): For $i, j = 1, \ldots, q$, $(\beta_i E_i - \beta_j E_j)\phi_{ij}(E) \leq 0$, $E \in \overline{\mathbb{R}}_+^q$.

Axiom *iii*): For $i, j = 1, \ldots, q$, $\frac{\beta_i \Delta E_i - \beta_j \Delta E_j}{\beta_i E_i - \beta_j E_j} \geq -1$, $\beta_i E_i \neq \beta_j E_j$.

Axioms *i*) and *ii*) are restatements of Axioms *i*) and *ii*) of Chapter 4. Axiom *iii*) implies that for any pair of connected subsystems $\mathcal{G}_i$ and $\mathcal{G}_j$, $i \neq j$, the temperature difference between consecutive time instants is monotonic; that is, $[\beta_i E_i(k+1) - \beta_j E_j(k+1)][\beta_i E_i(k) - \beta_j E_j(k)] \geq 0$ for all $\beta_i E_i \neq \beta_j E_j$, $k \geq k_0$, $i, j = 1, \ldots, q$.

Next, in light of our modified conditions we give a generalized definition for the entropy and ectropy of $\mathcal{G}$. The following proposition is needed for the statement of the main results of this section.

Proposition 9.9. Consider the discrete-time, large-scale dynamical system $\mathcal{G}$ with energy balance equation (9.18) and assume that Axioms *i*), *ii*), and *iii*) hold. Then, for all $E_0 \in \overline{\mathbb{R}}_+^q$, $k_\mathrm{f} \geq k_0$, and $S(\cdot) \in \mathcal{U}$, such that $E(k_\mathrm{f}) = E(k_0) = E_0$,

$$\sum_{k=k_0}^{k_\mathrm{f}-1} \sum_{i=1}^{q} \frac{S_i(k) - \sigma_{ii}(E(k))}{c + \beta_i E_i(k+1)} = \sum_{k=k_0}^{k_\mathrm{f}-1} \sum_{i=1}^{q} \frac{Q_i(k)}{c + \beta_i E_i(k+1)} \leq 0, \tag{9.80}$$

$$\sum_{k=k_0}^{k_\mathrm{f}-1} \sum_{i=1}^{q} \beta_i E_i(k+1)[S_i(k) - \sigma_{ii}(E(k))] = \sum_{k=k_0}^{k_\mathrm{f}-1} \sum_{i=1}^{q} \beta_i E_i(k+1) Q_i(k) \geq 0, \tag{9.81}$$

where $E(k)$, $k \geq k_0$, is the solution to (9.18) with initial condition $E(k_0) = E_0$. Furthermore, equalities hold in (9.80) and (9.81) if and only if $\Delta E_i = 0$ and $\beta_i E_i = \beta_j E_j$, $i, j = 1, \ldots, q$, $i \neq j$.

Proof. The proof is identical to the proofs of Propositions 9.4 and 9.6. □

Note that with the modified Axiom *i*) the isolated discrete-time large-scale dynamical system $\mathcal{G}$ has equilibrium energy states given by $E_\mathrm{e} = \alpha \boldsymbol{p}$, for $\alpha \geq 0$, where $\boldsymbol{p} \triangleq [1/\beta_1, \ldots, 1/\beta_q]^\mathrm{T}$. As in Section 9.3, we define an equilibrium process as a process where the trajectory of the system $\mathcal{G}$ stays at the equilibrium point of the isolated system $\mathcal{G}$ and a nonequilibrium process as a process that is not an equilibrium one. Thus, it follows from Axioms *i*) – *iii*) that inequalities (9.80) and (9.81) are satisfied as equalities for an equilibrium process and as strict inequalities for a nonequilibrium process.

Definition 9.12. For the discrete-time, large-scale dynamical system $\mathcal{G}$ with energy balance equation (9.18), a function $\mathcal{S} : \overline{\mathbb{R}}_+^q \to \mathbb{R}$ satisfying

$$\mathcal{S}(E(k_2)) \geq \mathcal{S}(E(k_1)) + \sum_{k=k_1}^{k_2-1} \sum_{i=1}^{q} \frac{S_i(k) - \sigma_{ii}(E(k))}{c + \beta_i E_i(k+1)}, \tag{9.82}$$

for every $k_2 \geq k_1 \geq k_0$ and $S(\cdot) \in \mathcal{U}$, is called the *entropy* of $\mathcal{G}$.

Definition 9.13. For the discrete-time, large-scale dynamical system $\mathcal{G}$ with energy balance equation (9.18), a function $\mathcal{E} : \overline{\mathbb{R}}_+^q \to \mathbb{R}$ satisfying

$$\mathcal{E}(E(k_2)) \leq \mathcal{E}(E(k_1)) + \sum_{k=k_1}^{k_2-1} \sum_{i=1}^{q} \beta_i E_i(k+1)[S_i(k) - \sigma_{ii}(E(k))], \tag{9.83}$$

for every $k_2 \geq k_1 \geq k_0$ and $S(\cdot) \in \mathcal{U}$, is called the *ectropy* of $\mathcal{G}$.

For the next result define the available entropy and available ectropy of the large-scale dynamical system $\mathcal{G}$ by

$$\mathcal{S}_{\mathrm{a}}(E_0) \triangleq - \sup_{S(\cdot)\in\mathcal{U}_{\mathrm{c}},\, K\geq k_0} \sum_{k=k_0}^{K-1} \sum_{i=1}^{q} \frac{S_i(k) - \sigma_{ii}(E(k))}{c + \beta_i E_i(k+1)}, \tag{9.84}$$

$$\mathcal{E}_{\mathrm{a}}(E_0) \triangleq - \inf_{S(\cdot)\in\mathcal{U}_{\mathrm{c}},\, K\geq k_0} \sum_{k=k_0}^{K-1} \sum_{i=1}^{q} \beta_i E_i(k+1)[S_i(k) - \sigma_{ii}(E(k))], \tag{9.85}$$

where $E(k_0) = E_0 \in \overline{\mathbb{R}}_+^q$ and $E(K) = 0$, and define the required entropy supply and required ectropy supply of the large-scale dynamical system $\mathcal{G}$ by

$$\mathcal{S}_{\mathrm{r}}(E_0) \triangleq \sup_{S(\cdot)\in\mathcal{U}_{\mathrm{r}},\, K\geq -k_0+1} \sum_{k=-K}^{k_0-1} \sum_{i=1}^{q} \frac{S_i(k) - \sigma_{ii}(E(k))}{c + \beta_i E_i(k+1)}, \tag{9.86}$$

$$\mathcal{E}_{\mathrm{r}}(E_0) \triangleq \inf_{S(\cdot)\in\mathcal{U}_{\mathrm{r}},\, K\geq -k_0+1} \sum_{k=-K}^{k_0-1} \sum_{i=1}^{q} \beta_i E_i(k+1)[S_i(k) - \sigma_{ii}(E(k))], \tag{9.87}$$

where $E(-K) = 0$ and $E(k_0) = E_0 \in \overline{\mathbb{R}}_+^q$.

Theorem 9.9. Consider the discrete-time, large-scale dynamical system $\mathcal{G}$ with energy balance equation (9.18) and assume that Axioms *ii*) and *iii*) hold. Then there exists an entropy and an ectropy function for $\mathcal{G}$. Moreover, $\mathcal{S}_{\mathrm{a}}(E)$, $E \in \overline{\mathbb{R}}_+^q$, and $\mathcal{S}_{\mathrm{r}}(E)$, $E \in \overline{\mathbb{R}}_+^q$, are possible entropy functions for $\mathcal{G}$ with $\mathcal{S}_{\mathrm{a}}(0) = \mathcal{S}_{\mathrm{r}}(0) = 0$, and $\mathcal{E}_{\mathrm{a}}(E)$, $E \in \overline{\mathbb{R}}_+^q$, and $\mathcal{E}_{\mathrm{r}}(E)$, $E \in \overline{\mathbb{R}}_+^q$, are possible ectropy functions for $\mathcal{G}$ with $\mathcal{E}_{\mathrm{a}}(0) = \mathcal{E}_{\mathrm{r}}(0) = 0$.

Finally, all entropy functions $\mathcal{S}(E)$, $E \in \overline{\mathbb{R}}_+^q$, for $\mathcal{G}$ satisfy

$$\mathcal{S}_{\rm r}(E) \leq \mathcal{S}(E) - \mathcal{S}(0) \leq \mathcal{S}_{\rm a}(E), \quad E \in \overline{\mathbb{R}}_+^q, \tag{9.88}$$

and all ectropy functions $\mathcal{E}(E)$, $E \in \overline{\mathbb{R}}_+^q$, for $\mathcal{G}$ satisfy

$$\mathcal{E}_{\rm a}(E) \leq \mathcal{E}(E) - \mathcal{E}(0) \leq \mathcal{E}_{\rm r}(E), \quad E \in \overline{\mathbb{R}}_+^q. \tag{9.89}$$

Proof. The proof is identical to the proofs of Theorems 9.6 and 9.7. □

For the statement of the next result recall the definition of $\mathbf{p} = [1/\beta_1, \cdots, 1/\beta_q]^{\rm T}$ and define $P \triangleq \operatorname{diag}[\beta_1, \cdots, \beta_q]$.

Proposition 9.10. Consider the discrete-time, large-scale dynamical system $\mathcal{G}$ with energy balance equation (9.18) and assume that Axioms *i*), *ii*), and *iii*) hold. Then the function $\mathcal{S} : \overline{\mathbb{R}}_+^q \to \mathbb{R}$ given by

$$\mathcal{S}(E) = \mathbf{p}^{\rm T}\mathbf{log}_e(c\mathbf{e} + PE) - \mathbf{e}^{\rm T}\mathbf{p}\log_e c, \quad E \in \overline{\mathbb{R}}_+^q, \tag{9.90}$$

where $\mathbf{log}_e(c\mathbf{e}+PE)$ denotes the vector natural logarithm given by $[\log_e(c+\beta_1 E_1), \ldots, \log_e(c+\beta_q E_q)]^{\rm T}$, is an entropy function of $\mathcal{G}$. Furthermore, the function $\mathcal{E} : \overline{\mathbb{R}}_+^q \to \mathbb{R}$ given by

$$\mathcal{E}(E) = \tfrac{1}{2}E^{\rm T}PE, \quad E \in \overline{\mathbb{R}}_+^q, \tag{9.91}$$

is an ectropy function of $\mathcal{G}$.

Proof. The proof is identical to the proofs of Propositions 9.5 and 9.7. □

As in Section 9.3, it can be shown that the entropy and ectropy functions for $\mathcal{G}$ defined by (9.90) and (9.91) satisfy, respectively, (9.82) and (9.83) as equalities for an equilibrium process and as strict inequalities for a nonequilibrium process.

Once again, inequality (9.82) is analogous to Clausius' inequality for reversible and irreversible thermodynamics, while inequality (9.83) is an anti–Clausius inequality. Moreover, for the ectropy function given by (9.91) inequality (9.83) shows that a thermodynamically consistent large-scale dynamical system model is dissipative with respect to the supply rate $E^{\rm T}PS$ and with the storage function corresponding to the system ectropy $\mathcal{E}(E)$. In addition, if we let $Q_i(k) = S_i(k) - \sigma_{ii}(E(k))$, $i = 1, \ldots, q$, denote the net amount of heat received or dissipated by the ith subsystem of $\mathcal{G}$ at a given time instant at the (shifted) *absolute* ith *subsystem temperature* $T_i \triangleq c + \beta_i E_i$, then it follows from (9.82) that the system entropy varies by an amount

$$\Delta\mathcal{S}(E(k)) \geq \sum_{i=1}^{q} \frac{Q_i(k)}{c+\beta_i E_i(k+1)}, \quad k \geq k_0. \tag{9.92}$$

Finally, note that the nonconservation of entropy and ectropy equations (9.65) and (9.66), respectively, for isolated discrete-time, large-scale dynamical systems also hold for the more general definitions of entropy and ectropy given in Definitions 9.12 and 9.13.

The following theorem is a generalization of Theorem 9.8.

Theorem 9.10. Consider the discrete-time, large-scale dynamical system $\mathcal{G}$ with energy balance equation (9.18) with $S(k) \equiv 0$ and $d(E) \equiv 0$, and assume that Axioms i) – iii) hold. Then, for every $\alpha \geq 0$, $\alpha\boldsymbol{p}$ is a semistable equilibrium state of (9.18). Furthermore, $E(k) \to \frac{1}{\mathbf{e}^{\mathrm{T}}\boldsymbol{p}}\boldsymbol{p}\mathbf{e}^{\mathrm{T}}E(k_0)$ as $k \to \infty$ and $\frac{1}{\mathbf{e}^{\mathrm{T}}\boldsymbol{p}}\boldsymbol{p}\mathbf{e}^{\mathrm{T}}E(k_0)$ is a semistable equilibrium state. Finally, if for some $m \in \{1,\ldots,q\}$, $\sigma_{mm}(E) \geq 0$ and $\sigma_{mm}(E) = 0$ if and only if $E_m = 0$, then the zero solution $E(k) \equiv 0$ to (9.18) is a globally asymptotically stable equilibrium state of (9.18).

Proof. It follows from Axiom i) that $\alpha\boldsymbol{p} \in \overline{\mathbb{R}}_+^q$, $\alpha \geq 0$, is an equilibrium state for (9.18). To show Lyapunov stability of the equilibrium state $\alpha\boldsymbol{p}$, consider the system shifted ectropy $\mathcal{E}_{\mathrm{s}}(E) = \frac{1}{2}(E-\alpha\boldsymbol{p})^{\mathrm{T}}P(E-\alpha\boldsymbol{p})$ as a Lyapunov function candidate. Now, the proof follows as in the proof of Theorem 9.8 by invoking Axioms i) – iii) and noting that $\phi_{ij}(E) = -\phi_{ji}(E)$, $E \in \overline{\mathbb{R}}_+^q$, $i \neq j$, $i,j = 1,\ldots,q$, $P\boldsymbol{p} = \mathbf{e}$, and $\mathbf{e}^{\mathrm{T}}f(E) = \mathbf{e}^{\mathrm{T}}E$, $E \in \overline{\mathbb{R}}_+^q$. Alternatively, in the case where for some $m \in \{1,\ldots,q\}$, $\sigma_{mm}(E) \geq 0$ and $\sigma_{mm}(E) = 0$ if and only if $E_m = 0$, global asymptotic stability of the zero solution $E(k) \equiv 0$ to (9.18) follows from standard Lyapunov arguments using the system ectropy $\mathcal{E}(E) = \frac{1}{2}E^{\mathrm{T}}PE$ as a candidate Lyapunov function. □

It follows from Theorem 9.10 that the steady-state value of the energy in each subsystem $\mathcal{G}_i$ of the isolated discrete-time, large-scale dynamical system $\mathcal{G}$ is given by

$$E_\infty = \frac{1}{\mathbf{e}^{\mathrm{T}}\boldsymbol{p}}\boldsymbol{p}\mathbf{e}^{\mathrm{T}}E(k_0),$$

which implies that

$$E_{i\infty} = \frac{1}{\beta_i\mathbf{e}^{\mathrm{T}}\boldsymbol{p}}\mathbf{e}^{\mathrm{T}}E(k_0)$$

or, equivalently,

$$\hat{T}_{i\infty} = \beta_i E_{i\infty} = \frac{1}{\mathbf{e}^{\mathrm{T}}\boldsymbol{p}}\mathbf{e}^{\mathrm{T}}E(k_0).$$

Hence, the steady-state temperature of the isolated discrete-time, large-scale dynamical system $\mathcal{G}$ given by $\hat{T}_\infty = \frac{1}{\mathbf{e}^{\mathrm{T}}\boldsymbol{p}}\mathbf{e}^{\mathrm{T}}E(k_0)\mathbf{e}$ is uniformly distributed over all the subsystems of $\mathcal{G}$.

Proposition 9.11. Consider the discrete-time, large-scale dynamical system $\mathcal{G}$ with energy balance equation (9.18), let $\mathcal{E} : \overline{\mathbb{R}}_+^q \to \overline{\mathbb{R}}_+$ and $\mathcal{S} : \overline{\mathbb{R}}_+^q \to \mathbb{R}$ denote the ectropy and entropy of $\mathcal{G}$ and be given by (9.91) and (9.90), respectively, and define $\mathcal{D}_\mathrm{c} \triangleq \{E \in \overline{\mathbb{R}}_+^q : \mathbf{e}^{\mathrm{T}}E = \beta\}$, where $\beta \geq 0$. Then

$$\arg\min_{E\in\mathcal{D}_\mathrm{c}}(\mathcal{E}(E)) = \arg\max_{E\in\mathcal{D}_\mathrm{c}}(\mathcal{S}(E)) = E^* = \frac{\beta}{\mathbf{e}^{\mathrm{T}}\boldsymbol{p}}\boldsymbol{p}. \tag{9.93}$$

Furthermore, $\mathcal{E}_{\min} \triangleq \mathcal{E}(E^*) = \frac{1}{2}\frac{\beta^2}{\mathbf{e}^{\mathrm{T}}\boldsymbol{p}}$ and $\mathcal{S}_{\max} \triangleq \mathcal{S}(E^*) = \mathbf{e}^{\mathrm{T}}\boldsymbol{p}\log_e(c + \frac{\beta}{\mathbf{e}^{\mathrm{T}}\boldsymbol{p}}) - \mathbf{e}^{\mathrm{T}}\boldsymbol{p}\log_e c$.

Proof. The proof is identical to the proof of Proposition 3.10. □

Proposition 9.11 shows that when all the energy of a discrete-time, large-scale dynamical system is transformed into heat at a uniform temperature, entropy is a maximum and ectropy is a minimum.

9.10 Discrete Thermodynamic Models with Linear Energy Exchange

In this section, we specialize the results of Section 9.3 to the case of large-scale dynamical systems with linear energy exchange between subsystems; that is, $f(E) = WE$ and $d(E) = DE$, where $W \in \mathbb{R}^{q\times q}$ and $D \in \mathbb{R}^{q\times q}$. In this case, the vector form of the energy balance equation (9.18), with $k_0 = 0$, is given by

$$E(k+1) = WE(k) - DE(k) + S(k), \quad E(0) = E_0, \quad k \geq 0. \tag{9.94}$$

Next, let the net energy exchange from the jth subsystem $\mathcal{G}_j$ to the ith subsystem $\mathcal{G}_i$ be parameterized as $\phi_{ij}(E) = \Phi_{ij}^{\mathrm{T}}E$, where $\Phi_{ij} \in \mathbb{R}^q$ and $E \in \overline{\mathbb{R}}_+^q$. In this case, since $f_i(E) = E_i + \sum_{i=1,j\neq i}^{q}\phi_{ij}(E)$, it follows that

$$W = I_q + \left[\sum_{j=2}^{q}\Phi_{1j}, \ldots, \sum_{j=1,j\neq i}^{q}\Phi_{ij}, \ldots, \sum_{j=1}^{q-1}\Phi_{qj}\right]^{\mathrm{T}}. \tag{9.95}$$

Since $\phi_{ij}(E) = -\phi_{ji}(E)$, $i, j = 1, \ldots, q$, $i \neq j$, $E \in \overline{\mathbb{R}}_+^q$, it follows that $\Phi_{ij} = -\Phi_{ji}$, $i \neq j$, $i, j = 1, \ldots, q$. The following proposition considers the special case where W is symmetric.

Proposition 9.12. Consider the large-scale dynamical system $\mathcal{G}$ with energy balance equation given by (9.94) and with $D = 0$. Then Axioms i)

and ii) hold if and only if $W = W^{\mathrm{T}}$, $(W - I_q)\mathbf{e} = 0$, rank $(W - I_q) = q - 1$, and W is nonnegative. In addition, if $S = 0$ and Axiom iii) holds, then rank $(W + I_q) = q$ and rank $(W^2 - I_q) = q - 1$.

Proof. Assume Axioms i) and ii) hold. Since, by Axiom ii), $(E_i - E_j)\phi_{ij}(E) \leq 0$, $E \in \overline{\mathbb{R}}_+^q$, it follows that $E^{\mathrm{T}}\Phi_{ij}\mathbf{e}_{ij}^{\mathrm{T}}E \leq 0$, $i, j = 1, \ldots, q$, $i \neq j$, where $E \in \overline{\mathbb{R}}_+^q$ and $\mathbf{e}_{ij} \in \mathbb{R}^q$ is a vector whose ith entry is 1, jth entry is -1, and remaining entries are zero. Next, it can be shown that $E^{\mathrm{T}}\Phi_{ij}\mathbf{e}_{ij}^{\mathrm{T}}E \leq 0$, $E \in \overline{\mathbb{R}}_+^q$, $i \neq j$, $i, j = 1, \ldots, q$, if and only if $\Phi_{ij} \in \mathbb{R}^q$ is such that its ith entry is $-\sigma_{ij}$, its jth entry is σ_{ij}, and its remaining entries are zero, where $\sigma_{ij} \geq 0$. Furthermore, since $\Phi_{ij} = -\Phi_{ji}$, $i \neq j$, $i, j = 1, \ldots, q$, it follows that $\sigma_{ij} = \sigma_{ji}$, $i \neq j$, $i, j = 1, \ldots, q$. Hence, W is given by

$$W_{(i,j)} = \begin{cases} 1 - \sum_{k=1, k \neq j}^{q} \sigma_{kj}, & i = j, \\ \sigma_{ij}, & i \neq j, \end{cases} \tag{9.96}$$

which implies that W is symmetric (since $\sigma_{ij} = \sigma_{ji}$) and $(W - I_q)\mathbf{e} = 0$.

Note that since at any given instant of time energy can only be transported or stored but not created and the maximum amount of energy that can be transported cannot exceed the energy in a compartment, it follows that $1 \geq \sum_{k=1, k \neq j}^{q} \sigma_{kj}$. Thus, W is a nonnegative matrix. Now, since by Axiom i), $\phi_{ij}(E) = 0$ if and only if $E_i = E_j$ for all $i, j = 1, \ldots, q$, $i \neq j$, such that $\mathcal{C}_{(i,j)} = 1$, it follows that $\sigma_{ij} > 0$ for all $i, j = 1, \ldots, q$, $i \neq j$, such that $\mathcal{C}_{(i,j)} = 1$. Hence, rank $(W - I_q) =$ rank $\mathcal{C} = q - 1$.

The converse is immediate, and hence, is omitted.

Next, assume Axiom iii) holds. Since, by Axiom iii), $(E_i(k+1) - E_j(k+1))(E_i(k) - E_j(k)) \geq 0$, $i, j = 1, \ldots, q$, $i \neq j$, $k \geq k_0$, it follows that $E^{\mathrm{T}}(k+1)\mathbf{e}_{ij}\mathbf{e}_{ij}^{\mathrm{T}}E(k) \geq 0$ or, equivalently, $E^{\mathrm{T}}(k)W^{\mathrm{T}}\mathbf{e}_{ij}\mathbf{e}_{ij}^{\mathrm{T}}E(k) \geq 0$, $i, j = 1, \ldots, q$, $i \neq j$, $k \geq k_0$, where $E \in \overline{\mathbb{R}}_+^q$. Next, we show that $I_q + W$ is strictly diagonally dominant. Suppose, *ad absurdum*, that $1 + W_{(i,i)} \leq \sum_{l=1, l \neq i}^{q} W_{(i,l)}$ for some i, $1 \leq i \leq q$. Let $E(k_0) = \mathbf{e}_i$, $i = 1, \ldots, q$, where $\mathbf{e}_i \in \overline{\mathbb{R}}_+^q$ is a vector whose ith entry is 1 and remaining entries are zero. Then

$$\begin{aligned} E^{\mathrm{T}}(k_0)W^{\mathrm{T}}\mathbf{e}_{ij}\mathbf{e}_{ij}^{\mathrm{T}}E(k_0) &= \mathbf{e}_i^{\mathrm{T}}W^{\mathrm{T}}\mathbf{e}_{ij}\mathbf{e}_{ij}^{\mathrm{T}}\mathbf{e}_i \\ &= W_{(i,i)} - W_{(i,j)} \\ &= 1 - \sum_{k=1, k \neq j}^{q} \sigma_{kj} - \sigma_{ij} \\ &\geq 0, \quad i, j = 1, \ldots, q, \quad i \neq j. \end{aligned} \tag{9.97}$$

Now, it follows from (9.97) that

$$1 + W_{(i,j)} \leq 1 + W_{(i,i)} \leq \sum_{l=1,l\neq i}^{q} W_{(i,l)}, \quad j = 1,\ldots,q, \quad j \neq i, \quad 1 \leq i \leq q, \tag{9.98}$$

or, equivalently,

$$1 \leq \sum_{l=1,l\neq i,l\neq j}^{q} W_{(i,l)}, \quad j = 1,\ldots,q, \quad j \neq i, \quad 1 \leq i \leq q. \tag{9.99}$$

However, since W is compartmental and symmetric, it follows that

$$\sum_{l=1,l\neq i}^{q} W_{(i,l)} = \sum_{l=1,l\neq i}^{q} W_{(l,i)} = \sum_{l=1,l\neq i}^{q} \sigma_{l,i} \leq 1, \quad i = 1,\ldots,q. \tag{9.100}$$

Now, since $W_{(i,j)} = \sigma_{ij} > 0$ for all $i, j = 1,\ldots,q$, $i \neq j$, it follows that

$$\sum_{l=1,l\neq i,l\neq j}^{q} W_{(i,l)} < \sum_{l=1,l\neq i}^{q} W_{(i,l)} \leq 1, \quad i = 1,\ldots,q, \tag{9.101}$$

which contradicts (9.99).

Next, since $I_q + W$ is strictly diagonally dominant, it follows from Theorem 6.1.10 of [218] that rank $(I_q + W) = q$. Furthermore, since rank $(W^2 - I_q) =$ rank $(W + I_q)(W - I_q)$, it follows from Sylvester's inequality that

$$\begin{aligned} \operatorname{rank}(W + I_q) + \operatorname{rank}(W - I_q) - q & \\ \leq \operatorname{rank}(W^2 - I_q) & \\ \leq \min\{\operatorname{rank}(W + I_q), \operatorname{rank}(W - I_q)\}. & \end{aligned} \tag{9.102}$$

Now, rank $(W^2 - I_q) = q - 1$ follows from (9.102) by noting that rank $(W - I_q) = q - 1$ and rank $(W + I_q) = q$. □

Next, we specialize the energy balance equation (9.94) to the case where $D = \operatorname{diag}[\sigma_{11}, \sigma_{22}, \ldots, \sigma_{qq}]$. In this case, the vector form of the energy balance equation (9.18), with $k_0 = 0$, is given by

$$E(k+1) = AE(k) + S(k), \quad E(0) = E_0, \quad k \in \overline{\mathbb{Z}}_+, \tag{9.103}$$

where $A \triangleq W - D$ is such that

$$A_{(i,j)} = \begin{cases} 1 - \sum_{k=1}^{q} \sigma_{kj}, & i = j, \\ \sigma_{ij}, & i \neq j. \end{cases} \tag{9.104}$$

Note that (9.104) implies $\sum_{i=1}^{q} A_{(i,j)} = 1 - \sigma_{ii} \leq 1$, $j = 1,\ldots,q$, and hence, A is a Lyapunov stable compartmental matrix. If $\sigma_{ii} > 0$, $i = 1,\ldots,q$, then A is an asymptotically stable compartmental matrix.

An important special case of (9.103) is the case where A is symmetric or, equivalently, $\sigma_{ij} = \sigma_{ji}$, $i \neq j$, $i, j = 1, \ldots, q$. In this case, it follows from (9.103) that for each subsystem the energy balance equation satisfies

$$\Delta E_i(k) + \sigma_{ii} E_i(k) + \sum_{j=1,\, j\neq i}^{q} \sigma_{ij}[E_i(k) - E_j(k)] = S_i(k), \quad k \in \overline{\mathbb{Z}}_+. \quad (9.105)$$

Note that $\phi_i(E) \triangleq \sum_{j=1,\, j\neq i}^{q} \sigma_{ij}(E_i - E_j)$, $i = 1, \ldots, q$, represents the energy exchange from the ith subsystem to all other subsystems and is given by the sum of the individual energy exchanges from the ith subsystem to the jth subsystem. Furthermore, these energy exchanges are proportional to the energy differences of the subsystems, that is, $E_i - E_j$.

Hence, (9.105) is an energy balance equation that governs the energy exchange among coupled subsystems and is completely analogous to the equations of thermal transfer with subsystem energies playing the role of temperatures. Furthermore, note that since $\sigma_{ij} \geq 0$, $i, j = 1, \ldots, q$, energy is exchanged from more energetic subsystems to less energetic subsystems, which is consistent with the second law of thermodynamics requiring that heat (energy) *must* flow in the direction of lower temperatures.

The next lemma and proposition are needed for developing expressions for steady-state energy distributions of the discrete-time, large-scale dynamical system $\mathcal{G}$ with linear energy balance equation (9.103).

Lemma 9.3. Let $A \in \mathbb{R}^{q\times q}$ be compartmental and let $S \in \mathbb{R}^q$. Then the following properties hold:

i) $I_q - A$ is an M-matrix.

ii) $|\lambda| \leq 1$, $\lambda \in \operatorname{spec}(A)$.

iii) If A is semistable and $\lambda \in \operatorname{spec}(A)$, then either $|\lambda| < 1$ or $\lambda = 1$ and $\lambda = 1$ is semisimple.

iv) $\operatorname{ind}(I_q - A) \leq 1$ and $\operatorname{ind}(A) \leq 1$.

v) If A is semistable, then $\lim_{k\to\infty} A^k = I_q - (A - I_q)(A - I_q)^{\#} \geq\geq 0$.

vi) $\mathcal{R}(A - I_q) = \mathcal{N}(I_q - (A - I_q)(A - I_q)^{\#})$ and $\mathcal{N}(A - I_q) = \mathcal{R}(I_q - (A - I_q)(A - I_q)^{\#})$.

vii) $\sum_{i=0}^{k} A^i = (A - I_q)^{\#}(A^{k+1} - I_q) + (k+1)[I_q - (A - I_q)(A - I_q)^{\#}]$, $k \in \overline{\mathbb{Z}}_+$.

viii) If A is semistable, then $\sum_{i=0}^{\infty} A^i S$ exists if and only if $S \in \mathcal{R}(A - I_q)$, where $S \in \mathbb{R}^q$.

ix) If A is semistable and $S \in \mathcal{R}(A-I_q)$, then $\sum_{i=0}^{\infty} A^i S = -(A-I_q)^{\#}S$.

x) If A is semistable, $S \in \mathcal{R}(A-I_q)$, and $S \geq\geq 0$, then $-(A-I_q)^{\#}S \geq\geq 0$.

xi) $A-I_q$ is nonsingular if and only if $I_q - A$ is a nonsingular M-matrix.

xii) If A is semistable and $A-I_q$ is nonsingular, then A is asymptotically stable and $(I_q-A)^{-1} \geq\geq 0$.

Proof. *i*) Note that $A^{\mathrm{T}}\mathbf{e} = [-(1-\sum_{i=1}^{q} A_{(i,1)}), -(1-\sum_{i=1}^{q} A_{(i,2)}), \ldots, -(1-\sum_{i=1}^{q} A_{(i,q)})]^{\mathrm{T}} + \mathbf{e}$. Then $(I_q - A)^{\mathrm{T}}\mathbf{e} \geq\geq 0$ and $I_q - A$ is a Z-matrix. It follows from Theorem 1 of [32] that $(I_q-A)^{\mathrm{T}}$, and hence, $I_q - A$ is an M-matrix.

ii) The result follows from *i*) and Lemma 1 of [193].

iii) The result follows from Theorem 2 of [193].

iv) Since $(I_q-A)^{\mathrm{T}}\mathbf{e} \geq\geq 0$, it follows that $I_q - A$ is an M-matrix and has "property c" (see [31]). Hence, it follows from Lemma 4.11 of [31] that $I_q - A$ has "property c" if and only if $\operatorname{ind}(I_q-A) \leq 1$. Next, since $\operatorname{ind}(I_q - A) \leq 1$, it follows from the real Jordan decomposition that there exist invertible matrices $J \in \mathbb{R}^{r\times r}$, where $r = \operatorname{rank}(I_q - A)$, and $U \in \mathbb{R}^{q\times q}$ such that J is diagonal and

$$I_q - A = U \begin{bmatrix} J & 0 \\ 0 & 0 \end{bmatrix} U^{-1}, \tag{9.106}$$

which implies

$$A = U \begin{bmatrix} I_r - J & 0 \\ 0 & I_{q-r} \end{bmatrix} U^{-1}. \tag{9.107}$$

Hence, $\operatorname{ind}(A) \leq 1$.

v) The result follows from Theorem 2 of [193].

vi) Let $x \in \mathcal{R}(A-I_q)$, that is, there exists $y \in \mathbb{R}^q$ such that $x = (A-I_q)y$. Now, $(I_q-(A-I_q)(A-I_q)^{\#})x = x-(A-I_q)(A-I_q)^{\#}(A-I_q)y = x-(A-I_q)y = 0$, which implies that $\mathcal{R}(A-I_q) \subseteq \mathcal{N}(I_q-(A-I_q)(A-I_q)^{\#})$. Conversely, let $x \in \mathcal{N}(I_q-(A-I_q)(A-I_q)^{\#})$. Hence, $(I_q-(A-I_q)(A-I_q)^{\#})x = 0$ or, equivalently, $x = (A-I_q)(A-I_q)^{\#}x$, which implies that $x \in \mathcal{R}(A-I_q)$, which proves $\mathcal{R}(A-I_q) = \mathcal{N}(I_q-(A-I_q)(A-I_q)^{\#})$. The equality $\mathcal{N}(A-I_q) = \mathcal{R}(I_q-(A-I_q)(A-I_q)^{\#})$ can be proved in an analogous manner.

vii) Note that since $A = U \begin{bmatrix} I_r - J & 0 \\ 0 & I_{q-r} \end{bmatrix} U^{-1}$ and J is invertible, it follows that

$$\begin{aligned}
\sum_{i=0}^{k} A^i &= \sum_{i=0}^{k} U \begin{bmatrix} (I_r - J)^i & 0 \\ 0 & I_{q-r} \end{bmatrix} U^{-1} \\
&= U \begin{bmatrix} \sum_{i=0}^{k} (I_r - J)^i & 0 \\ 0 & (k+1) I_{q-r} \end{bmatrix} U^{-1} \\
&= U \begin{bmatrix} -J^{-1}[(I_r - J)^{k+1} - I_r] & 0 \\ 0 & (k+1) I_{q-r} \end{bmatrix} U^{-1} \\
&= U \begin{bmatrix} -J^{-1} & 0 \\ 0 & 0 \end{bmatrix} U^{-1} U \begin{bmatrix} (I_r - J)^{k+1} - I_r & 0 \\ 0 & 0 \end{bmatrix} U^{-1} \\
&\quad + U \begin{bmatrix} 0 & 0 \\ 0 & (k+1) I_{q-r} \end{bmatrix} U^{-1} \\
&= (A - I_q)^{\#} (A^{k+1} - I_q) \\
&\quad + (k+1) \left(I_q - U \begin{bmatrix} J - I_r & 0 \\ 0 & 0 \end{bmatrix} U^{-1} U \begin{bmatrix} (J - I_r)^{-1} & 0 \\ 0 & 0 \end{bmatrix} U^{-1} \right) \\
&= (A - I_q)^{\#} (A^{k+1} - I_q) + (k+1)[I_q - (A - I_q)(A - I_q)^{\#}], \quad k \in \overline{\mathbb{Z}}_+.
\end{aligned} \tag{9.108}$$

viii) The result is a direct consequence of *v*)–*vii*).

ix) The result follows from *v*) and *vii*).

x) The result follows from *ix*).

xi) The result follows from *i*).

xii) Asymptotic stability of A is a direct consequence of *iii*). $(I_q - A)^{-1} \geq\geq 0$ follows from Lemma 1 of [193]. □

Proposition 9.13 ([193]). Consider the discrete-time, large-scale dynamical system $\mathcal{G}$ with energy balance equation given by (9.103). Suppose $E_0 \geq\geq 0$ and $S(k) \geq\geq 0$, $k \in \overline{\mathbb{Z}}_+$. Then the solution $E(k)$, $k \in \overline{\mathbb{Z}}_+$, to (9.103) is nonnegative for all $k \in \overline{\mathbb{Z}}_+$ if and only if A is nonnegative.

Next, we develop expressions for the steady-state energy distribution for a discrete-time, large-scale linear dynamical system $\mathcal{G}$ for the cases where A is semistable, and the supplied system energy $S(k)$ is a periodic function with period $\tau \in \overline{\mathbb{Z}}_+$, $\tau > 0$; that is, $S(k+\tau) = S(k)$, $k \in \overline{\mathbb{Z}}_+$, and $S(k)$ is constant; that is, $S(k) \equiv S$. Define $e(k) \triangleq E(k) - E(k+\tau)$, $k \in \overline{\mathbb{Z}}_+$, and note that

$$e(k+1) = Ae(k), \quad e(0) = E(0) - E(\tau), \quad k \in \overline{\mathbb{Z}}_+. \tag{9.109}$$

Hence, since

$$e(k) = A^k[E(0) - E(\tau)], \quad k \in \overline{\mathbb{Z}}_+, \tag{9.110}$$

and A is semistable, it follows from v) of Lemma 9.3 that

$$\lim_{k\to\infty} e(k) = \lim_{k\to\infty} [E(k) - E(k+\tau)] = [I_q - (A - I_q)(A - I_q)^{\#}][E(0) - E(\tau)], \tag{9.111}$$

which represents a constant offset to the steady-state error energy distribution in the discrete-time, large-scale nonlinear dynamical system $\mathcal{G}$. For the case where $S(k) \equiv S$, $\tau \to \infty$, and hence, the following result is immediate.

Proposition 9.14. Consider the discrete-time, large-scale dynamical system $\mathcal{G}$ with energy balance equation given by (9.103). Suppose that A is semistable, $E_0 \geq\geq 0$, and $S(k) \equiv S \geq\geq 0$. Then $E_\infty \triangleq \lim_{k\to\infty} E(k)$ exists if and only if $S \in \mathcal{R}(A - I_q)$. In this case,

$$E_\infty = [I_q - (A - I_q)(A - I_q)^{\#}]E_0 - (A - I_q)^{\#} S \tag{9.112}$$

and $E_\infty \geq\geq 0$. If, in addition, $A - I_q$ is nonsingular, then E_∞ exists for all $S \geq\geq 0$ and is given by

$$E_\infty = (I_q - A)^{-1} S. \tag{9.113}$$

Proof. Note that it follows from Lagrange's formula that the solution $E(k)$, $k \in \overline{\mathbb{Z}}_+$, to (9.103) is given by

$$E(k) = A^k E_0 + \sum_{i=0}^{k-1} A^{(k-1-i)} S(i), \quad k \in \overline{\mathbb{Z}}_+. \tag{9.114}$$

Now, the result is a direct consequence of Proposition 9.13 and v), $viii$), ix), and x) of Lemma 9.3. □

Next, we specialize the result of Proposition 9.14 to the case where there is no energy dissipation from each subsystem $\mathcal{G}_i$ of $\mathcal{G}$; that is, $\sigma_{ii} = 0$, $i = 1, \ldots, q$. Note that in this case $\mathbf{e}^{\mathrm{T}}(A - I_q) = 0$, and hence, rank $(A - I_q) \leq q - 1$. Furthermore, if $S = 0$, then it follows from (9.103) that $\mathbf{e}^{\mathrm{T}} \Delta E(k) = \mathbf{e}^{\mathrm{T}}(A - I_q)E(k) = 0$, $k \in \overline{\mathbb{Z}}_+$, and hence, the total energy of the isolated discrete-time, large-scale nonlinear dynamical system $\mathcal{G}$ is conserved.

Proposition 9.15. Consider the discrete-time, large-scale dynamical system $\mathcal{G}$ with energy balance equation given by (9.103). Assume rank $(A - I_q) =$ rank $(A^2 - I_q) = q - 1$, $\sigma_{ii} = 0$, $i = 1, \ldots, q$, and $A = A^{\mathrm{T}}$. If $E_0 \geq\geq 0$ and $S = 0$, then the equilibrium state $\alpha\mathbf{e}$, $\alpha \geq 0$, of the isolated system $\mathcal{G}$ is semistable and the steady-state energy distribution E_∞ of the isolated discrete-time, large-scale dynamical system $\mathcal{G}$ is given by

$$E_\infty = \left[\frac{1}{q}\sum_{i=1}^{q} E_{i0}\right]\mathbf{e}. \tag{9.115}$$

If, in addition, for some $m \in \{1,\ldots,q\}$, $\sigma_{mm} > 0$, then the zero solution $E(k) \equiv 0$ to (9.103) is globally asymptotically stable.

Proof. Note that since $\mathbf{e}^{\mathrm{T}}(A - I_q) = 0$, it follows from (9.103) with $S(k) \equiv 0$ that $\mathbf{e}^{\mathrm{T}}\Delta E(k) = 0$, $k \geq 0$, and hence, $\mathbf{e}^{\mathrm{T}}E(k) = \mathbf{e}^{\mathrm{T}}E_0$, $k \geq 0$. Furthermore, since by Proposition 9.13 the solution $E(k)$, $k \geq k_0$, to (9.103) is nonnegative, it follows that $0 \leq E_i(k) \leq \mathbf{e}^{\mathrm{T}}E(k) = \mathbf{e}^{\mathrm{T}}E_0$, $k \geq 0$, $i = 1,...,q$. Hence, the solution $E(k)$, $k \geq 0$, to (9.103) is bounded for all $E_0 \in \overline{\mathbb{R}}_+^q$. Next, note that $\phi_{ij}(E) = \sigma_{ij}(E_j - E_i)$ and $(E_i - E_j)\phi_{ij}(E) = -\sigma_{ij}(E_i - E_j)^2 \leq 0$, $E \in \overline{\mathbb{R}}_+^q$, $i \neq j$, $i,j = 1,...,q$, which implies that Axioms $i)$ and $ii)$ are satisfied. Thus, $E = \alpha\mathbf{e}$, $\alpha \geq 0$, is the equilibrium state of the isolated large-scale dynamical system $\mathcal{G}$.

To show Lyapunov stability of the equilibrium state $\alpha\mathbf{e}$, consider the shifted-system Lyapunov function candidate $\mathcal{E}_{\mathrm{s}}(E) = \frac{1}{2}(E - \alpha\mathbf{e})^{\mathrm{T}}(E - \alpha\mathbf{e})$, $E \in \overline{\mathbb{R}}_+^q$. Since A is compartmental and symmetric, it follows from $ii)$ of Lemma 9.3 that

$$\begin{aligned}\Delta\mathcal{E}_{\mathrm{s}}(E) &= \tfrac{1}{2}(AE - \alpha\mathbf{e})^{\mathrm{T}}(AE - \alpha\mathbf{e}) - \tfrac{1}{2}(E - \alpha\mathbf{e})^{\mathrm{T}}(E - \alpha\mathbf{e})\\ &= \tfrac{1}{2}E^{\mathrm{T}}(A^2 - I_q)E\\ &\leq 0,\end{aligned} \tag{9.116}$$

which implies Lyapunov stability of the equilibrium state $\alpha\mathbf{e}$, $\alpha \geq 0$.

Next, consider the set $\mathcal{R} \triangleq \{E \in \overline{\mathbb{R}}_+^q : \Delta\mathcal{E}_{\mathrm{s}}(E) = 0\} = \{E \in \overline{\mathbb{R}}_+^q : E^{\mathrm{T}}(A^2 - I_q)E = 0\}$. Since A is compartmental and symmetric, it follows from $ii)$ of Lemma 9.3 that $A^2 - I_q$ is a negative-semidefinite matrix, and hence, $E^{\mathrm{T}}(A^2 - I_q)E = 0$ if and only if $(A^2 - I_q)E = 0$. Furthermore, since, by assumption, $\operatorname{rank}(A - I_q) = \operatorname{rank}(A^2 - I_q) = q - 1$, it follows that there exists one and only one linearly independent solution to $(A^2 - I_q)E = 0$ given by $E = \mathbf{e}$. Hence, $\mathcal{R} = \{E \in \overline{\mathbb{R}}_+^q : E = \alpha\mathbf{e}, \alpha \geq 0\}$.

Since $\mathcal{R}$ consists of only equilibrium states of (9.103), it follows that $\mathcal{M} = \mathcal{R}$, where $\mathcal{M}$ is the largest invariant set contained in $\mathcal{R}$. Hence, for every $E_0 \in \overline{\mathbb{R}}_+^q$, it follows from Theorem 9.2 that $E(k) \to \alpha\mathbf{e}$ as $k \to \infty$ for some $\alpha \geq 0$, and hence, $\alpha\mathbf{e}$, $\alpha \geq 0$, is a semistable equilibrium state of (9.103). Furthermore, since the energy is conserved in the isolated large-scale dynamical system $\mathcal{G}$, it follows that $q\alpha = \mathbf{e}^{\mathrm{T}}E_0$. Thus, $\alpha = \frac{1}{q}\sum_{i=1}^{q} E_{i0}$, which implies (9.115).

Finally, to show that in the case where $\sigma_{mm} > 0$ for some $m \in \{1, \ldots, q\}$, the zero solution $E(k) \equiv 0$ to (9.103) is globally asymptotically stable, consider the system ectropy $\mathcal{E}(E) = \frac{1}{2}E^{\mathrm{T}}E$, $E \in \overline{\mathbb{R}}_+^q$, as a candidate Lyapunov function. Note that Lyapunov stability of the zero equilibrium state follows from the previous analysis with $\alpha = 0$. Next, note that

$$\begin{aligned}\Delta\mathcal{E}(E) =& \tfrac{1}{2}E^{\mathrm{T}}(A^2 - I_q)E \\ =& \tfrac{1}{2}E^{\mathrm{T}}[(W - D)^2 - I_q]E \\ =& \tfrac{1}{2}E^{\mathrm{T}}(W^2 - I_q)E - \tfrac{1}{2}E^{\mathrm{T}}(WD + DW - D^2)E \\ =& \tfrac{1}{2}E^{\mathrm{T}}(W^2 - I_q)E - \textstyle\sum_{i=1, i\neq m}^{q} \sigma_{mm}\sigma_{mi}E_m E_i \\ & - \sigma_{mm}(W_{(m,m)} - \sigma_{mm})E_m^2 - \tfrac{1}{2}\sigma_{mm}^2 E_m^2, \quad E \in \overline{\mathbb{R}}_+^q. \end{aligned} \tag{9.117}$$

Consider the set $\mathcal{R} \triangleq \{E \in \overline{\mathbb{R}}_+^q : \Delta\mathcal{E}(E) = 0\} = \{E \in \overline{\mathbb{R}}_+^q : E_1 = \cdots = E_q\} \cap \{E \in \overline{\mathbb{R}}_+^q : E_m = 0, m \in \{1, \ldots, q\}\} = \{0\}$. Hence, the largest invariant set contained in $\mathcal{R}$ is given by $\mathcal{M} = \mathcal{R} = \{0\}$, and thus, it follows from Theorem 9.4 that the zero solution $E(k) \equiv 0$ to (9.103) is globally asymptotically stable. □

Finally, we examine the steady-state energy distribution for large-scale nonlinear dynamical systems $\mathcal{G}$ in the case of strong coupling between subsystems, that is, $\sigma_{ij} \to \infty$, $i \neq j$. For this analysis we assume that A given by (9.103) is symmetric; that is, $\sigma_{ij} = \sigma_{ji}$, $i \neq j$, $i, j = 1, \ldots, q$, and $\sigma_{ii} > 0$, $i = 1, \ldots, q$. Thus, $I_q - A$ is a nonsingular M-matrix for all values of σ_{ij}, $i \neq j$, $i, j = 1, \ldots, q$. Moreover, in this case it follows that if $\frac{\sigma_{ij}}{\sigma_{kl}} \to 1$ as $\sigma_{ij} \to \infty$, $i \neq j$, and $\sigma_{kl} \to \infty$, $k \neq l$, then

$$\lim_{\sigma_{ij} \to \infty, i \neq j} (I_q - A)^{-1} = \lim_{\sigma \to \infty} [D - \sigma(-qI_q + \mathbf{e}\mathbf{e}^{\mathrm{T}})]^{-1}, \tag{9.118}$$

where $D = \mathrm{diag}[\sigma_{11}, \ldots, \sigma_{qq}] > 0$.

Proposition 9.16. Consider the discrete-time, large-scale dynamical system $\mathcal{G}$ with energy balance equation given by (9.103). Let $S(k) \equiv S$, $S \in \mathbb{R}^{q \times q}$, $A \in \mathbb{R}^{q \times q}$ be compartmental and assume A is symmetric, $\sigma_{ii} > 0$, $i = 1, \ldots, q$, $\frac{\sigma_{ij}}{\sigma_{kl}} \to 1$ as $\sigma_{ij} \to \infty$, $i \neq j$, and $\sigma_{kl} \to \infty$, $k \neq l$. Then the steady-state energy distribution E_∞ of the discrete-time, large-scale dynamical system $\mathcal{G}$ is given by

$$E_\infty = \left[\frac{\mathbf{e}^{\mathrm{T}}S}{\sum_{i=1}^{q} \sigma_{ii}}\right]\mathbf{e}. \tag{9.119}$$

Proof. Note that in the case where $\frac{\sigma_{ij}}{\sigma_{kl}} \to 1$ as $\sigma_{ij} \to \infty$, $i \neq j$, and $\sigma_{kl} \to \infty$, $k \neq l$, it follows that the corresponding limit of $(I_q - A)^{-1}$ can be equivalently taken as in (9.118). Next, with $D = \mathrm{diag}[\sigma_{11}, \ldots, \sigma_{qq}]$ and $X = -qI_q + \mathbf{e}\mathbf{e}^{\mathrm{T}}$, it follows that $I_q - A = D - \sigma X = D^{\frac{1}{2}}(I_q - \sigma D^{-\frac{1}{2}}XD^{-\frac{1}{2}})D^{\frac{1}{2}}$.

Now, it follows from Lemmas 3.3 and 3.4 that

$$E_\infty = \lim_{\sigma_{ij}\to\infty,\, i\neq j} (I_q - A)^{-1} S = \frac{\mathbf{e}\mathbf{e}^{\mathrm{T}}}{\mathbf{e}^{\mathrm{T}} D \mathbf{e}} S = \left[\frac{\mathbf{e}^{\mathrm{T}} S}{\sum_{i=1}^{q} \sigma_{ii}} \right] \mathbf{e}, \tag{9.120}$$

which proves the result. □

Proposition 9.16 shows that in the limit of strong coupling the steady-state energy distribution E_∞ given by (9.113) becomes

$$E_\infty = \lim_{\sigma_{ij}\to\infty,\, i\neq j} (I_q - A)^{-1} S = \left[\frac{\mathbf{e}^{\mathrm{T}} S}{\sum_{i=1}^{q} \sigma_{ii}} \right] \mathbf{e}, \tag{9.121}$$

which implies energy equipartition.

Chapter Ten

Critical Phenomena and Discontinuous Phase Transitions

10.1 Introduction

In Chapter 8, we developed a dynamical systems framework for thermodynamics to address thermodynamic critical phenomena and continuous phase transitions. Continuous phase transitions involve a continuous transition across the transition parameter (e.g., temperature) resulting in continuous changes in the thermodynamic state quantities (e.g., internal energy, entropy, enthalpy, volume) with discontinuous first- and higher-order derivatives of these quantities. These phase transitions are known as second- or higher-order transitions in the literature. In this case, the vector field defining the dynamical system is a discontinuous function of the state and can be characterized by differential inclusions involving Filippov set-valued maps specifying a set of directions for the state derivative and admitting Filippov solutions with absolutely continuous curves [190].

Phase transitions can also involve a jump discontinuity in the fundamental thermodynamic state quantities, wherein the transition is accompanied by an instantaneous heat release (i.e., latent heat). These phase transitions are known as first-order transitions in the literature and can be characterized by dynamical systems with an interacting mixture of continuous and discrete dynamics exhibiting discontinuous flows on appropriate manifolds, giving rise to hybrid dynamics.

The mathematical descriptions of many hybrid dynamical systems can be modeled by impulsive differential equations [18, 19, 196, 257, 394]. Impulsive dynamical systems can be viewed as a subclass of hybrid systems and consist of three elements: a continuous-time differential equation, which governs the motion of the dynamical system between impulsive or resetting events; a difference equation, which governs the way the system states are instantaneously changed when a resetting event occurs; and a criterion for determining when the states of the system are to be reset.

Hybrid and impulsive dynamical systems exhibit a very rich dynamical behavior. In particular, the trajectories of hybrid and impulsive dynamical systems can exhibit multiple complex phenomena, such as *Zeno* solutions, *noncontinuability* of solutions or *deadlock*, *beating* or *livelock*, and *confluence* or merging of solutions [196]. A Zeno solution involves a system trajectory with infinitely many resettings in finite time. Deadlock corresponds to a dynamical system state from which no continuation, continuous or discrete, is possible. A hybrid dynamical system experiences beating when the system trajectory encounters the same resetting surface a finite or infinite number of times in zero time. Finally, confluence involves system solutions that coincide after a certain point in time.

These phenomena, along with the breakdown of many of the fundamental properties of classical dynamical systems theory, such as continuity of solutions and continuous dependence of solutions on system initial conditions, make the analysis of hybrid and impulsive dynamical systems extremely challenging. In this chapter, we develop a hybrid thermodynamic model as a special case of nonlinear hybrid compartmental and nonnegative dynamical systems to address thermodynamic critical phenomena with discontinuous phase transitions.

10.2 Stability Theory for Nonlinear Hybrid Nonnegative Dynamical Systems

In this section, we provide sufficient conditions for stability of state-dependent impulsive nonnegative dynamical systems, that is, state-dependent impulsive dynamical systems [196] whose solutions remain in the nonnegative orthant for nonnegative initial conditions. Specifically, we consider nonlinear state-dependent impulsive dynamical systems of the form

$$\dot{x}(t) = f_{\rm c}(x(t)), \qquad x(0) = x_0, \qquad x(t) \notin \mathcal{Z}, \tag{10.1}$$

$$\Delta x(t) = f_{\rm d}(x(t)), \qquad x(t) \in \mathcal{Z}, \tag{10.2}$$

where $t \geq 0$, $x(t) \in \mathcal{D} \subseteq \mathbb{R}^n$, $\mathcal{D}$ is a relatively open subset of $\mathbb{R}^n$ that contains $\overline{\mathbb{R}}_+^n$ with $0 \in \mathcal{D}$, $\Delta x(t) \triangleq x(t^+) - x(t)$, where $x(t^+) \triangleq x(t) + f_{\rm d}(x(t)) = \lim_{\varepsilon \to 0^+} x(t+\varepsilon)$, $f_{\rm c} : \mathcal{D} \to \mathbb{R}^n$ is Lipschitz continuous and satisfies $f_{\rm c}(0) = 0$, $f_{\rm d} : \mathcal{D} \to \mathbb{R}^n$ is continuous, and $\mathcal{Z} \subset \mathcal{D}$ is the *resetting set*. Note that $x_{\rm e} \in \mathcal{D}$ is an equilibrium point of (10.1) and (10.2) if and only if $f_{\rm c}(x_{\rm e}) = 0$ and $f_{\rm d}(x_{\rm e}) = 0$.

We refer to the differential equation (10.1) as the *continuous-time dynamics*, and we refer to the difference equation (10.2) as the *resetting law*. Note that since the resetting set $\mathcal{Z}$ is a subset of the state space $\overline{\mathbb{R}}_+^n$ and is independent of time, state-dependent impulsive dynamical systems are time invariant. For a particular trajectory $x(t)$, we let $\tau_k(x_0)$ denote the kth instant of time at which $x(t)$ intersects $\mathcal{Z}$, and we call the times $\tau_k(x_0)$ the *resetting times*. Thus, the trajectory of the system (10.1) and (10.2) from the initial condition $x(0) = x_0$ is given by $s(t, x_0)$ for $0 < t \leq \tau_1(x_0)$.

If and when the trajectory reaches a state $x_1 \triangleq x(\tau_1(x_0))$ satisfying $x_1 \in \mathcal{Z}$, then the state is instantaneously transferred to $x_1^+ \triangleq x_1 + f_{\rm d}(x_1)$ according to the resetting law (10.2). The trajectory $x(t)$, $\tau_1(x_0) < t \leq \tau_2(x_0)$, is then given by $s(t - \tau_1(x_0), x_1^+)$, and so on. Note that the solution $x(t)$ of (10.1) and (10.2) is left-continuous, that is, it is continuous everywhere except at the resetting times $\tau_k(x_0)$, and

$$x_k \triangleq x(\tau_k(x_0)) = \lim_{\varepsilon \to 0^+} x(\tau_k(x_0) - \varepsilon), \tag{10.3}$$

$$x_k^+ \triangleq x(\tau_k(x_0)) + f_{\rm d}(x(\tau_k(x_0))), \tag{10.4}$$

for $k = 1, 2, \ldots$.

We make the following additional assumptions:

A1. If $x(t) \in \overline{\mathcal{Z}} \backslash \mathcal{Z}$, then there exists $\varepsilon > 0$ such that, for all $0 < \delta < \varepsilon$, $s(\delta, x(t)) \notin \mathcal{Z}$.

A2. If $x \in \mathcal{Z}$, then $x + f_{\rm d}(x) \notin \mathcal{Z}$.

Assumption A1 ensures that if a trajectory reaches the closure of $\mathcal{Z}$ at a point that does not belong to $\mathcal{Z}$, then the trajectory must be directed away from $\mathcal{Z}$; that is, a trajectory cannot enter $\mathcal{Z}$ through a point that belongs to the closure of $\mathcal{Z}$ but not to $\mathcal{Z}$. Furthermore, A2 ensures that when a trajectory intersects the resetting set $\mathcal{Z}$, it instantaneously exits $\mathcal{Z}$. Finally, we note that if $x_0 \in \mathcal{Z}$, then the system initially resets to $x_0^+ = x_0 + f_{\rm d}(x_0) \notin \mathcal{Z}$, which serves as the initial condition for the continuous dynamics (10.1). It follows from A1 and A2 that $\partial\mathcal{Z} \cap \mathcal{Z}$ is closed, and hence, the resetting times $\tau_k(x_0)$ are well defined and distinct. Furthermore, it follows from A2 that if $x^* \in \overline{\mathbb{R}}_+^n$ satisfies $f_{\rm d}(x^*) = 0$, then $x^* \notin \mathcal{Z}$.

To see this, suppose *ad absurdum* that $x^* \in \mathcal{Z}$. Then $x^* + f_{\rm d}(x^*) = x^* \in \mathcal{Z}$, contradicting A2. Thus, if $x = x_{\rm e}$ is an equilibrium point of (10.1) and (10.2), then $x_{\rm e} \notin \mathcal{Z}$, and hence, $x_{\rm e} \in \mathcal{D}$ is an equilibrium point of (10.1) and (10.2) if and only if $f_{\rm c}(x_{\rm e}) = 0$. Finally, since for every $x \in \mathcal{Z}$, $x + f_{\rm d}(x) \in \mathcal{Z}$, it follows that $\tau_2(x) = \tau_1(x) + \tau_1(x + f_{\rm d}(x)) > 0$. For further insights on Assumptions A1 and A2 the interested reader is referred to [196].

Next, we present a result that shows that $\overline{\mathbb{R}}_+^n$ is an invariant set for (10.1) and (10.2) if $f_{\rm c} : \mathcal{D} \to \mathbb{R}^n$ is essentially nonnegative and $f_{\rm d} : \mathcal{D} \to \mathbb{R}^n$

is such that $x + f_{\rm d}(x)$ is nonnegative for all $x \in \overline{\mathbb{R}}_+^n$.

Proposition 10.1. Suppose $\overline{\mathbb{R}}_+^n \subset \mathcal{D}$. If $f_{\rm c} : \mathcal{D} \to \mathbb{R}^n$ is essentially nonnegative and $f_{\rm d} : \mathcal{Z} \to \mathbb{R}^n$ is such that $x + f_{\rm d}(x)$ is nonnegative, then $\overline{\mathbb{R}}_+^n$ is an invariant set with respect to (10.1) and (10.2).

Proof. Consider the continuous-time dynamical system given by

$$\dot{x}_{\rm c}(t) = f_{\rm c}(x_{\rm c}(t)), \qquad x_{\rm c}(0) = x_{\rm c0}, \qquad t \geq 0. \tag{10.5}$$

Now, it follows from Proposition 2.1 that since $f_{\rm c} : \mathcal{D} \to \mathbb{R}^n$ is essentially nonnegative, $\overline{\mathbb{R}}_+^n$ is an invariant set with respect to (10.5), that is, if $x_{\rm c0} \in \overline{\mathbb{R}}_+^n$, then $x_{\rm c}(t) \in \overline{\mathbb{R}}_+^n$, $t \geq 0$. Now, since, with $x_{\rm c0} = x_0$, $x(t) = x_{\rm c}(t)$, $0 \leq t \leq \tau_1(x_0)$, it follows that $x(t) \in \overline{\mathbb{R}}_+^n$, $0 \leq t \leq \tau_1(x_0)$.

Next, since $f_{\rm d} : \mathcal{Z} \to \mathbb{R}^n$ is such that $x + f_{\rm d}(x)$ is nonnegative, it follows that $x_1^+ = x(\tau_1(x_0)) + f_{\rm d}(x(\tau_1(x_0))) \in \overline{\mathbb{R}}_+^n$. Now, since $s(t, x_0) = s(t - \tau_1(x_0), x_1^+)$, $\tau_1(x_0) < t \leq \tau_2(x_0)$, with $x_{\rm c0} = x_1^+$, it follows that $x(t) = x_{\rm c}(t - \tau_1(x_0)) \in \overline{\mathbb{R}}_+^n$, $\tau_1(x_0) < t \leq \tau_2(x_0)$, and hence, $x_2^+ = x(\tau_2(x_0)) + f_{\rm d}(x(\tau_2(x_0))) \in \overline{\mathbb{R}}_+^n$. Repeating this procedure for $\tau_i(x_0)$, $i = 3, 4, \ldots$, it follows that $\overline{\mathbb{R}}_+^n$ is an invariant set with respect to (10.1) and (10.2). $\square$

It is important to note that, unlike continuous-time nonnegative systems and discrete-time nonnegative systems, Proposition 10.1 provides only sufficient conditions ensuring that $\overline{\mathbb{R}}_+^n$ is an invariant set with respect to (10.1) and (10.2). To see this, let $\mathcal{Z} = \partial\overline{\mathbb{R}}_+^n$ and assume $x + f_{\rm d}(x)$, $x \in \mathcal{Z}$, is nonnegative. Then, $\overline{\mathbb{R}}_+^n$ remains invariant with respect to (10.1) and (10.2) irrespective of whether $f_{\rm c}(\cdot)$ is essentially nonnegative or not.

Next, we specialize Proposition 10.1 to linear[1] state-dependent impulsive dynamical systems of the form

$$\dot{x}(t) = A_{\rm c}x(t), \qquad x(0) = x_0, \qquad x(t) \notin \mathcal{Z}, \tag{10.6}$$
$$\Delta x(t) = (A_{\rm d} - I_n)x(t), \qquad x(t) \in \mathcal{Z}, \tag{10.7}$$

where $t \geq 0$, $x(t) \in \overline{\mathbb{R}}_+^n$, $A_{\rm c} \in \mathbb{R}^{n\times n}$ is essentially nonnegative, $A_{\rm d} \in \mathbb{R}^{n\times n}$ is nonnegative, and $\mathcal{Z} \subset \overline{\mathbb{R}}_+^n$. Note that in this case A2 implies that if $x \in \mathcal{Z}$, then $A_{\rm d}x \notin \mathcal{Z}$.

Proposition 10.2. Let $A_{\rm c} \in \mathbb{R}^{n\times n}$ and $A_{\rm d} \in \mathbb{R}^{n\times n}$. If $A_{\rm c}$ is essentially nonnegative and $A_{\rm d}$ is nonnegative, then $\overline{\mathbb{R}}_+^n$ is an invariant set with respect to (10.6) and (10.7).

---

[1]Impulsive dynamical systems with $f_{\rm c}(x) = Ax$ and $f_{\rm d}(x) = (A_{\rm d} - I)x$ are *not* linear. However, this minor abuse in terminology provides a natural way of differentiating between impulsive dynamical systems with nonlinear vector fields versus impulsive dynamical systems with linear vector fields, and considerably simplifies the presentation.

Proof. The proof is a direct consequence of Proposition 10.1 with $f_c(x) = A_c x$ and $f_d(x) = (A_d - I_n)x$. □

Next, we present several key results on the stability of nonlinear hybrid nonnegative dynamical systems. We note that for addressing the stability of the equilibrium solution of a nonnegative impulsive dynamical system the stability definitions introduced in Chapter 2 are valid.

Theorem 10.1. Let $\mathcal{D}$ be an open subset relative to $\overline{\mathbb{R}}_+^n$ that contains x_e. Suppose there exists a continuously differentiable function $V : \overline{\mathbb{R}}_+^n \to [0, \infty)$ satisfying $V(x_e) = 0$, $V(x) > 0$, $x \neq x_e$, and

$$V'(x)f_c(x) \leq 0, \quad x \notin \mathcal{Z}, \tag{10.8}$$

$$V(x + f_d(x)) \leq V(x), \quad x \in \mathcal{Z}. \tag{10.9}$$

Then the equilibrium solution $x(t) \equiv x_e$ of the hybrid nonnegative dynamical system (10.1) and (10.2) is Lyapunov stable. Furthermore, if the inequality (10.8) is strict for all $x \neq x_e$, then the equilibrium solution $x(t) \equiv x_e$ of the hybrid nonnegative dynamical system (10.1) and (10.2) is asymptotically stable. Finally, if $V(x) \to \infty$, as $\|x\| \to \infty$, then asymptotic stability is global.

Proof. Let $\varepsilon > 0$ be such that $\overline{\mathbb{R}}_+^n \cap \mathcal{B}_\varepsilon(x_e) \subseteq \mathcal{D}$. Since $\overline{\mathbb{R}}_+^n \cap \mathcal{B}_\varepsilon(x_e)$ is compact and $f_d(x)$, $x \in \overline{\mathcal{Z}} \subset \overline{\mathbb{R}}_+^n$, is continuous, it follows that

$$\eta \triangleq \max\left\{\varepsilon, \max_{x \in \overline{\mathbb{R}}_+^n \cap \overline{\mathcal{B}}_\varepsilon(0) \cap \overline{\mathcal{Z}}} \|x + f_d(x) - x_e\|\right\} \tag{10.10}$$

exists. Next, let $\alpha \triangleq \min_{x \in \mathcal{D}:\ \varepsilon \leq \|x - x_e\| \leq \eta} V(x)$. Note $\alpha > 0$ since $x_e \notin \overline{\mathbb{R}}_+^n \cap \partial\mathcal{B}_\varepsilon(x_e)$ and $V(x) > 0$, $x \in \mathcal{D}$, $x \neq x_e$. Next, let $\beta \in (0, \alpha)$ and define $\mathcal{D}_\beta \triangleq \{x \in \overline{\mathbb{R}}_+^n \cap \mathcal{B}_\varepsilon(x_e) : V(x) \leq \beta\}$. Now, let $x_0 \in \mathcal{D}_\beta$ and note that it follows from Assumptions A1 and A2 that the resetting times $\tau_k(x_0)$ are well defined and distinct for every trajectory of (10.1) and (10.2).

Prior to the first resetting time, we can determine the value of $V(x(t))$ as

$$V(x(t)) = V(x(0)) + \int_0^t V'(x(\tau))f_c(x(\tau))\mathrm{d}\tau, \quad t \in [0, \tau_1(x_0)]. \tag{10.11}$$

Between consecutive resetting times $\tau_k(x_0)$ and $\tau_{k+1}(x_0)$, we can determine the value of $V(x(t))$ as its initial value plus the integral of its rate of change along the trajectory $x(t)$, that is,

$$V(x(t)) = V(x(\tau_k(x_0)) + f_d(x(\tau_k(x_0)))) + \int_{\tau_k(x_0)}^t V'(x(\tau))f_c(x(\tau))\mathrm{d}\tau,$$

$$t \in (\tau_k(x_0), \tau_{k+1}(x_0)], \quad (10.12)$$

for $k = 1, 2, \ldots$. Adding and subtracting $V(x(\tau_k(x_0)))$ to and from the right-hand side of (10.12) yields

$$\begin{aligned} V(x(t)) &= V(x(\tau_k(x_0))) + [V(x(\tau_k(x_0)) + f_{\rm d}(x(\tau_k(x_0)))) \\ &\quad - V(x(\tau_k(x_0)))] + \int_{\tau_k(x_0)}^{t} V'(x(\tau)) f_{\rm c}(x(\tau)) {\rm d}\tau, \\ &\qquad t \in (\tau_k(x_0), \tau_{k+1}(x_0)], \end{aligned} \quad (10.13)$$

and in particular, at time $\tau_{k+1}(x_0)$,

$$\begin{aligned} V(x(\tau_{k+1}(x_0))) &= V(x(\tau_k(x_0))) + [V(x(\tau_k(x_0)) + f_{\rm d}(x(\tau_k(x_0)))) \\ &\quad - V(x(\tau_k(x_0)))] + \int_{\tau_k(x_0)}^{\tau_{k+1}(x_0)} V'(x(\tau)) f_{\rm c}(x(\tau)) {\rm d}\tau. \end{aligned} \quad (10.14)$$

By recursively substituting (10.14) into (10.13) and ultimately into (10.11), we obtain

$$\begin{aligned} V(x(t)) &= V(x(0)) + \int_0^t V'(x(\tau)) f_{\rm c}(x(\tau)) {\rm d}\tau \\ &\quad + \sum_{i=1}^{k} [V(x(\tau_i(x_0)) + f_{\rm d}(x(\tau_i(x_0)))) - V(x(\tau_i(x_0)))], \\ &\qquad t \in (\tau_k(x_0), \tau_{k+1}(x_0)]. \end{aligned} \quad (10.15)$$

If we allow $t_0 \triangleq 0$ and $\sum_{i=1}^{0} \triangleq 0$, then (10.15) is valid for $k \in \overline{\mathbb{Z}}_+$. From (10.15) and (10.9) we obtain

$$V(x(t)) \leq V(x(0)) + \int_0^t V'(x(\tau)) f_{\rm c}(x(\tau)) {\rm d}\tau, \quad t \geq 0. \quad (10.16)$$

Furthermore, it follows from (10.8) that

$$V(x(t)) \leq V(x(0)) \leq \beta, \quad x(0) \in \mathcal{D}_\beta, \quad t \geq 0. \quad (10.17)$$

Next, suppose, *ad absurdum*, there exists $T > 0$ such that $\|x(T) - x_{\rm e}\| \geq \varepsilon$. Hence, since $\|x_0 - x_{\rm e}\| < \varepsilon$, there exists $t_1 \in (0, T]$ such that either $\|x(t_1) - x_{\rm e}\| = \varepsilon$ or $x(t_1) \in \mathcal{Z}$, $\|x(t_1) - x_{\rm e}\| < \varepsilon$, and $\|x(t_1^+) - x_{\rm e}\| \geq \varepsilon$. If $\|x(t_1) - x_{\rm e}\| = \varepsilon$, then $V(x(t_1)) \geq \alpha > \beta$, which is a contradiction. Alternatively, if $x(t_1) \in \mathcal{Z}$, $\|x(t_1) - x_{\rm e}\| < \varepsilon$, and $\|x(t_1^+) - x_{\rm e}\| \geq \varepsilon$, then $\|x(t_1^+) - x_{\rm e}\| = \|x(t_1) + f_{\rm d}(x(t_1)) - x_{\rm e}\| \leq \eta$, which implies that $V(x(t_1^+)) \geq \alpha > \beta$, contradicting (10.17). Hence, $V(x(t)) \leq \beta$ and $\|x(t) - x_{\rm e}\| < \varepsilon$, $t \geq 0$, which implies that $\mathcal{D}_\beta$ is a positive invariant set (see Definition 2.18) with respect to (10.1) and (10.2). Next, since $V(\cdot)$ is continuous and $V(x_{\rm e}) = 0$, there exists $\delta = \delta(\varepsilon) \in (0, \varepsilon)$ such that $V(x) < \beta$, $x \in \mathcal{B}_\delta(x_e)$. Since $\overline{\mathbb{R}}_+^n \cap \mathcal{B}_\delta(x_e) \subset \mathcal{D}_\beta \subset \overline{\mathbb{R}}_+^n \cap \mathcal{B}_\varepsilon(x_e) \subseteq \mathcal{D}$ and $\mathcal{D}_\beta$ is a positive invariant set with respect to (10.1) and (10.2), it follows that for all $x_0 \in \mathcal{B}_\delta(x_e) \cap \overline{\mathbb{R}}_+^n$, $x(t) \in \overline{\mathbb{R}}_+^n \cap \mathcal{B}_\varepsilon(x_e)$, $t \geq 0$, which establishes Lyapunov stability.

To prove asymptotic stability, suppose (10.9) and (10.15) hold, and let $x_0 \in \overline{\mathbb{R}}_+^n \cap \mathcal{B}_\delta(x_e)$. Then it follows that $x(t) \in \overline{\mathbb{R}}_+^n \cap \mathcal{B}_\varepsilon(x_e)$, $t \geq 0$. However, $V(x(t))$, $t \geq 0$, is monotonically decreasing and bounded from below by zero. Next, it follows from (10.9) and (10.15) that

$$V(x(t)) - V(x(s)) \leq \int_s^t V'(x(\tau)) f_c(x(\tau)) d\tau, \quad t > s, \tag{10.18}$$

and, assuming strict inequality in (10.8), we obtain

$$V(x(t)) < V(x(s)), \quad t > s, \tag{10.19}$$

provided $x(s) \neq x_e$.

Now, suppose, *ad absurdum*, $x(t)$, $t \geq 0$, does not converge to x_e. This implies that $V(x(t))$, $t \geq 0$, is lower bounded by a positive number, that is, there exists $L > 0$ such that $V(x(t)) \geq L > 0$, $t \geq 0$. Hence, by continuity of $V(\cdot)$ there exists $\delta' > 0$ such that $V(x) < L$, $x \in \overline{\mathbb{R}}_+^n \cap \mathcal{B}_{\delta'}(x_e)$, which further implies that $x(t) \notin \overline{\mathbb{R}}_+^n \cap \mathcal{B}_{\delta'}(x_e)$, $t \geq 0$. Next, let $L_1 \triangleq \min_{\delta' \leq \|x - x_e\| \leq \varepsilon} -V'(x) f_c(x)$, which implies that $-V'(x) f_c(x) \geq L_1$, $\delta' \leq \|x - x_e\| \leq \varepsilon$, and hence, it follows from (10.15) that

$$V(x(t)) - V(x_0) \leq \int_0^t V'(x(\tau)) f_c(x(\tau)) d\tau \leq -L_1 t, \tag{10.20}$$

and hence, for all $x_0 \in \overline{\mathbb{R}}_+^n \cap \mathcal{B}_\delta(x_e)$,

$$V(x(t)) \leq V(x_0) - L_1 t.$$

Letting $t > \frac{V(x_0) - L}{L_1}$, it follows that $V(x(t)) < L$, which is a contradiction. Hence, $x(t) \to x_e$ as $t \to \infty$, establishing asymptotic stability.

Finally, global asymptotic stability with respect to $\overline{\mathbb{R}}_+^n$ follows from standard arguments. Specifically, let $x(0) \in \overline{\mathbb{R}}_+^n$, and let $\beta \triangleq V(x(0))$. It follows from the radial unboundedness condition that there exists $\varepsilon > 0$ such that $V(x) > \beta$ for all $x \in \overline{\mathbb{R}}_+^n$ such that $\|x - x_e\| > \varepsilon$. Hence, $\|x(0) - x_e\| \leq \varepsilon$, and, since $V(x(t))$ is strictly decreasing, it follows that $\|x(t) - x_e\| < \varepsilon$, $t > 0$. The remainder of the proof is identical to the proof of asymptotic stability. □

In the proof of Theorem 10.1, we note that assuming strict inequality in (10.8), the inequality (10.19) is obtained *provided* $x(s) \neq x_e$. This proviso is necessary since it may be possible to reset the states to the equilibrium point x_e, in which case $x(s) = x_e$ for a finite value of s. In this case, for $t > s$, we have $V(x(t)) = V(x(s)) = V(x_e) = 0$. This situation does not present a problem, however, since reaching the equilibrium x_e in finite time is a stronger condition than reaching the equilibrium x_e as $t \to \infty$.

Next, we present a generalized Krasovskii-LaSalle invariant set stability theorem for nonlinear hybrid dynamical systems. The following key assumption is needed for the statement of this result. For this assumption, $\mathcal{T}_{x_0} \triangleq [0,\infty)\backslash\{\tau_1(x_0), \tau_2(x_0), \ldots\}$.

Assumption 10.1 ([196]). Consider the impulsive nonnegative dynamical system $\mathcal{G}$ given by (10.1) and (10.2), and let $s(t, x_0), t \geq 0$, denote the solution to (10.1) and (10.2) with initial condition x_0. Then, for every $x_0 \in \mathcal{D}$, there exists a dense subset $\mathcal{T}_{x_0} \subseteq [0,\infty)$ such that $[0,\infty) \setminus \mathcal{T}_{x_0}$ is (finitely or infinitely) countable and, for every $\varepsilon > 0$ and $t \in \mathcal{T}_{x_0}$, there exists $\delta(\varepsilon, x_0, t) > 0$ such that if $\|x_0 - y\| < \delta(\varepsilon, x_0, t)$, $y \in \mathcal{D}$, then $\|s(t, x_0) - s(t, y)\| < \varepsilon$.

Assumption 10.1 is a generalization of the standard continuous dependence property for dynamical systems with continuous flows to dynamical systems with left-continuous flows and ensures continuous dependence over a dense subset of $[0,\infty)$. The following result provides sufficient conditions that guarantee that the nonlinear impulsive dynamical system $\mathcal{G}$ given by (10.1) and (10.2) satisfies Assumption 10.1. For further discussion on Assumption 10.1, see [196].

Proposition 10.3. Consider the impulsive dynamical system $\mathcal{G}$ given by (10.1) and (10.2). Assume that Assumptions A1 and A2 hold, $\tau_1(\cdot)$ is continuous at every $x \notin \overline{\mathcal{Z}}$ such that $0 < \tau_1(x) < \infty$, and if $x \in \mathcal{Z}$, then $x + f_d(x) \in \overline{\mathcal{Z}}\backslash\mathcal{Z}$. Furthermore, for every $x \in \overline{\mathcal{Z}}\backslash\mathcal{Z}$ such that $0 < \tau_1(x) < \infty$, assume that the following statements hold:

i) If a sequence $\{x_i\}_{i=1}^{\infty} \in \mathcal{D}$ is such that $\lim_{i\to\infty} x_i = x$ and $\lim_{i\to\infty} \tau_1(x_i)$ exists, then either $f_d(x) = 0$ and $\lim_{i\to\infty} \tau_1(x_i) = 0$, or $\lim_{i\to\infty} \tau_1(x_i) = \tau_1(x)$.

ii) If a sequence $\{x_i\}_{i=1}^{\infty} \in \overline{\mathcal{Z}}\backslash\mathcal{Z}$ is such that $\lim_{i\to\infty} x_i = x$ and $\lim_{i\to\infty} \tau_1(x_i)$ exists, then $\lim_{i\to\infty} \tau_1(x_i) = \tau_1(x)$.

Then $\mathcal{G}$ satisfies Assumption 10.1.

Proof. Let $x_0 \in \overline{\mathcal{Z}}\backslash\mathcal{Z}$ and let $\{x_i\}_{i=1}^{\infty} \in \mathcal{D}$ be such that $f_d(x_0) = 0$

and $\lim_{i\to\infty}\tau_1(x_i)=0$ hold. Define

$$\begin{aligned} z_i &\triangleq s(\tau_1(x_i),x_i)+f_{\rm d}(s(\tau_1(x_i),x_i))\\ &= \psi(\tau_1(x_i),x_i)+f_{\rm d}(\psi(\tau_1(x_i),x_i)), \quad i=1,2,\ldots, \end{aligned}$$

where $\psi(t,x_0)$ denotes the solution to the continuous-time dynamics (10.1), and note that, since $f_{\rm d}(x_0)=0$ and $\lim_{i\to\infty}\tau_1(x_i)=0$, it follows that $\lim_{i\to\infty}z_i=x_0$. Hence, since by assumption $z_i\in\overline{\mathcal{Z}}\backslash\mathcal{Z}$, $i=1,2,\ldots$, it follows from *ii*) that $\lim_{i\to\infty}\tau_1(z_i)=\tau_1(x_0)$, or, equivalently, $\lim_{i\to\infty}\tau_2(x_i)=\tau_1(x_0)$. Similarly, it can be shown that $\lim_{i\to\infty}\tau_{k+1}(x_i)=\tau_k(x_0)$, $k=2,3,\ldots$. Next, note that

$$\begin{aligned} \lim_{i\to\infty}s(\tau_2(x_i),x_i) &= \lim_{i\to\infty}\psi(\tau_2(x_i)-\tau_1(x_i),s(\tau_1(x_i),x_i)\\ &\quad +f_{\rm d}(s(\tau_1(x_i),x_i)))\\ &= \psi(\tau_1(x_0),x_0)\\ &= s(\tau_1(x_0),x_0). \end{aligned}$$

Now, using mathematical induction it can be shown that $\lim_{i\to\infty}s(\tau_{k+1}(x_i),x_i)=s(\tau_k(x_0),x_0)$, $k=2,3,\ldots$.

Next, let $k\in\{1,2,\ldots\}$ and let $t\in(\tau_k(x_0),\tau_{k+1}(x_0))$. Since

$$\lim_{i\to\infty}\tau_{k+1}(x_i)=\tau_k(x_0),$$

it follows that there exists $I\in\{1,2,\ldots\}$ such that $\tau_{k+1}(x_i)<t$ and $\tau_{k+2}(x_i)>t$ for all $i>I$. Hence, it follows that for every $t\in(\tau_k(x_0),\tau_{k+1}(x_0))$,

$$\begin{aligned} \lim_{i\to\infty}s(t,x_i) &= \lim_{i\to\infty}\psi(t-\tau_{k+1}(x_i),s(\tau_{k+1}(x_i),x_i)\\ &\quad +f_{\rm d}(s(\tau_{k+1}(x_i),x_i)))\\ &= \psi(t-\tau_k(x_0),s(\tau_k(x_0),x_0)+f_{\rm d}(s(\tau_k(x_0),x_0)))\\ &= s(t,x_0). \end{aligned}$$

Alternatively, if $x_0\in\overline{\mathcal{Z}}\backslash\mathcal{Z}$ is such that $\lim_{i\to\infty}\tau_1(x_i)=\tau_1(x_0)$ for $\{x_i\}_{i=1}^\infty\in\overline{\mathcal{Z}}\backslash\mathcal{Z}$, then using identical arguments as above, it can be shown that $\lim_{i\to\infty}s(t,x_i)=s(t,x_0)$ for every $t\in(\tau_k(x_0),\tau_{k+1}(x_0))$, $k=1,2,\ldots$.

Finally, let $x_0\notin\overline{\mathcal{Z}}$, $0<\tau_1(x_0)<\infty$, and assume $\tau_1(\cdot)$ is continuous. In this case, it follows from the definition of $\tau_1(x_0)$ that for every $x_0\notin\overline{\mathcal{Z}}$ and $t\in(\tau_1(x_0),\tau_2(x_0)]$,

$$s(t,x_0)=\psi(t-\tau_1(x_0),s(\tau_1(x_0),x_0)+f_{\rm d}(s(\tau_1(x_0),x_0))). \tag{10.21}$$

Since $\psi(\cdot,\cdot)$ is continuous in both its arguments, $\tau_1(\cdot)$ is continuous at x_0, and $f_{\rm d}(\cdot)$ is continuous, it follows that $s(t,\cdot)$ is continuous at x_0 for every $t\in(\tau_1(x_0),\tau_2(x_0))$. Next, for every sequence $\{x_i\}_{i=1}^\infty\in\mathcal{D}$ such that $\lim_{i\to\infty}x_i=x_0$, it follows that $\lim_{i\to\infty}s(\tau_1(x_i),x_i)=$

$\lim_{i\to\infty}\psi(\tau_1(x_i),x_i) = \psi(\tau_1(x_0),x_0) = s(\tau_1(x_0),x_0)$. Furthermore, note that by assumption $z_i \triangleq s(\tau_1(x_i),x_i) + f_{\rm d}(s(\tau_1(x_i),x_i)) \in \overline{\mathcal{Z}}\backslash\mathcal{Z}$, $i = 0,1,\ldots$. Hence, it follows that for all $t \in (\tau_k(z_0),\tau_{k+1}(z_0))$, $k = 1,2,\ldots$, $\lim_{i\to\infty} s(t,z_i) = s(t,z_0)$, or, equivalently, for all $t \in (\tau_k(x_0),\tau_{k+1}(x_0))$, $k = 2,3,\ldots$, $\lim_{i\to\infty} s(t,x_i) = s(t,x_0)$, which proves the result. □

The following result provides sufficient conditions for establishing continuity of $\tau_1(\cdot)$ at $x_0 \notin \overline{\mathcal{Z}}$ and *sequential continuity* of $\tau_1(\cdot)$ at $x_0 \in \overline{\mathcal{Z}}\backslash\mathcal{Z}$, that is, $\lim_{i\to\infty}\tau_1(x_i) = \tau_1(x_0)$ for $\{x_i\}_{i=1}^{\infty} \notin \mathcal{Z}$ and $\lim_{i\to\infty} x_i = x_0$. For this result, the following definition is needed. First, however, recall that the *Lie derivative* of a smooth function $\mathcal{X} : \mathcal{D} \to \mathbb{R}$ along the vector field of the continuous-time dynamics $f_{\rm c}(x)$ is given by $L_{f_{\rm c}}\mathcal{X}(x) \triangleq \frac{{\rm d}}{{\rm d}t}\mathcal{X}(\psi(t,x))|_{t=0} = \frac{\partial\mathcal{X}(x)}{\partial x} f_{\rm c}(x)$, and the *zeroth* and *higher-order Lie derivatives* are, respectively, defined by $L_{f_{\rm c}}^0\mathcal{X}(x) \triangleq \mathcal{X}(x)$ and $L_{f_{\rm c}}^k\mathcal{X}(x) \triangleq L_{f_{\rm c}}(L_{f_{\rm c}}^{k-1}\mathcal{X}(x))$, where $k \geq 1$.

Definition 10.1. Let $\mathcal{Q} \triangleq \{x \in \mathcal{D} : \mathcal{X}(x) = 0\}$, where $\mathcal{X} : \mathcal{D} \to \mathbb{R}$ is an infinitely differentiable function. A point $x \in \mathcal{Q}$ such that $f_{\rm c}(x) \neq 0$ is *transversal* to (10.1) if there exists $k \in \{1,2,\ldots\}$ such that

$$L_{f_{\rm c}}^r\mathcal{X}(x) = 0, \quad r = 0,\ldots,2k-2, \quad L_{f_{\rm c}}^{2k-1}\mathcal{X}(x) \neq 0. \tag{10.22}$$

Proposition 10.4. Consider the impulsive dynamical system (10.1) and (10.2). Let $\mathcal{X} : \mathcal{D} \to \mathbb{R}$ be an infinitely differentiable function such that $\overline{\mathcal{Z}} = \{x \in \mathcal{D} : \mathcal{X}(x) = 0\}$, and assume that every $x \in \overline{\mathcal{Z}}$ is transversal to (10.1). Then at every $x_0 \notin \overline{\mathcal{Z}}$ such that $0 < \tau_1(x_0) < \infty$, $\tau_1(\cdot)$ is continuous. Furthermore, if $x_0 \in \overline{\mathcal{Z}}\backslash\mathcal{Z}$ is such that $\tau_1(x_0) \in (0,\infty)$ and $\{x_i\}_{i=1}^{\infty} \in \overline{\mathcal{Z}}\backslash\mathcal{Z}$ or $\lim_{i\to\infty}\tau_1(x_i) > 0$, where $\{x_i\}_{i=1}^{\infty} \notin \overline{\mathcal{Z}}$ is such that $\lim_{i\to\infty} x_i = x_0$ and $\lim_{i\to\infty}\tau_1(x_i)$ exists, then $\lim_{i\to\infty}\tau_1(x_i) = \tau_1(x_0)$.

Proof. Let $x_0 \notin \overline{\mathcal{Z}}$ be such that $0 < \tau_1(x_0) < \infty$. It follows from the definition of $\tau_1(\cdot)$ that $s(t,x_0) = \psi(t,x_0)$, $t \in [0,\tau_1(x_0)]$, $\mathcal{X}(s(t,x_0)) \neq 0$, $t \in (0,\tau_1(x_0))$, and $\mathcal{X}(s(\tau_1(x_0),x_0)) = 0$. Without loss of generality, let $\mathcal{X}(s(t,x_0)) > 0$, $t \in (0,\tau_1(x_0))$. Since $\hat{x} \triangleq \psi(\tau_1(x_0),x_0) \in \overline{\mathcal{Z}}$ is transversal to (10.1), it follows that there exists $\theta > 0$ such that $\mathcal{X}(\psi(t,\hat{x})) > 0$, $t \in [-\theta,0)$, and $\mathcal{X}(\psi(t,\hat{x})) < 0$, $t \in (0,\theta]$. (This fact can be easily shown by expanding $\mathcal{X}(\psi(t,x))$ via a Taylor series expansion about $\hat{x}$ and using the fact that $\hat{x}$ is transversal to (10.1).) Hence, $\mathcal{X}(\psi(t,x_0)) > 0$, $t \in [\hat{t}_1,\tau_1(x_0))$, and $\mathcal{X}(\psi(t,x_0)) < 0$, $t \in (\tau_1(x_0),\hat{t}_2]$, where $\hat{t}_1 \triangleq \tau_1(x_0) - \theta$ and $\hat{t}_2 \triangleq \tau_1(x_0) + \theta$.

Next, let $\varepsilon \triangleq \min\{|\mathcal{X}(\psi(\hat{t}_1,x_0))|, |\mathcal{X}(\psi(\hat{t}_2,x_0))|\}$. Now, it follows from the continuity of $\mathcal{X}(\cdot)$ and the continuous dependence of $\psi(\cdot,\cdot)$ on the system initial conditions that there exists $\delta > 0$ such that

$$\sup_{0\leq t\leq \hat{t}_2} |\mathcal{X}(\psi(t,x)) - \mathcal{X}(\psi(t,x_0))| < \varepsilon, \quad x \in \mathcal{B}_\delta(x_0), \tag{10.23}$$

which implies that $\mathcal{X}(\psi(\hat{t}_1, x)) > 0$ and $\mathcal{X}(\psi(\hat{t}_2, x)) < 0$, $x \in \mathcal{B}_\delta(x_0)$. Hence, it follows that $\hat{t}_1 < \tau_1(x) < \hat{t}_2$, $x \in \mathcal{B}_\delta(x_0)$. The continuity of $\tau_1(\cdot)$ at x_0 now follows immediately by noting that θ can be chosen arbitrarily small.

Finally, let $x_0 \in \overline{\mathcal{Z}} \backslash \mathcal{Z}$ be such that $\lim_{i\to\infty} x_i = x_0$ for some sequence $\{x_i\}_{i=1}^{\infty} \in \overline{\mathcal{Z}} \backslash \mathcal{Z}$. Then using similar arguments as above it can be shown that $\lim_{i\to\infty} \tau_1(x_i) = \tau_1(x_0)$. Alternatively, if $x_0 \in \overline{\mathcal{Z}} \backslash \mathcal{Z}$ is such that $\lim_{i\to\infty} x_i = x_0$ and $\lim_{i\to\infty} \tau_1(x_i) > 0$ for some sequence $\{x_i\}_{i=1}^{\infty} \notin \mathcal{Z}$, then it follows that there exists sufficiently small $\hat{t} > 0$ and $I \in \mathbb{Z}_+$ such that $s(\hat{t}, x_i) = \psi(\hat{t}, x_i)$, $i = I, I+1, \ldots$, which implies that $\lim_{i\to\infty} s(\hat{t}, x_i) = s(\hat{t}, x_0)$. Next, define $z_i \triangleq \psi(\hat{t}, x_i)$, $i = 0, 1, \ldots$, so that $\lim_{i\to\infty} z_i = z_0$ and note that it follows from the transversality assumption that $z_0 \notin \overline{\mathcal{Z}}$, which implies that $\tau_1(\cdot)$ is continuous at z_0. Hence, $\lim_{i\to\infty} \tau_1(z_i) = \tau_1(z_0)$. The result now follows by noting that $\tau_1(x_i) = \hat{t} + \tau_1(z_i)$, $i = 1, 2, \ldots$. □

Note that if $x_0 \notin \mathcal{Z}$ is such that $\lim_{i\to\infty} \tau_1(x_i) \neq \tau_1(x_0)$ for some sequence $\{x_i\}_{i=1}^{\infty} \notin \mathcal{Z}$, then it follows from Proposition 10.4 that $\lim_{i\to\infty} \tau_1(x_i) = 0$. The notion of k-transversality introduced in Definition 10.1 differs from the well-known notion of transversality [117, 180] involving an orthogonality condition between a vector field and a differentiable submanifold. In the case where $k = 1$, Definition 10.1 coincides with the standard notion of transversality and guarantees that the solution of the system (10.1) and (10.2) is not tangent to the closure of the resetting set $\mathcal{Z}$ at the intersection with $\overline{\mathcal{Z}}$ [196]. In general, however, k-transversality guarantees that the sign of $\mathcal{X}(x(t))$ changes as the dynamical system trajectory transverses the closure of the resetting set $\mathcal{Z}$ at the intersection with $\overline{\mathcal{Z}}$.

Note that for an impulsive dynamical system a point $p \in \mathcal{D} \subseteq \overline{\mathbb{R}}_+^n$ is a positive limit point of the trajectory $s(t, x_0)$, $t \geq 0$, if and only if there exists a monotonic sequence $\{t_n\}_{n=0}^{\infty} \subset \mathcal{T}_{x_0}$, with $t_n \to \infty$ as $n \to \infty$, such that $s(t_n, x_0) \to p$ as $n \to \infty$. To see this, let $p \in \omega(x_0)$ and recall that $\mathcal{T}_{x_0}$ is a dense subset of the semi-infinite interval $[0, \infty)$. In this case, it follows that there exists an unbounded sequence $\{t_n\}_{n=0}^{\infty}$, with $t_n \to \infty$ as $n \to \infty$, such that $\lim_{n\to\infty} s(t_n, x_0) = p$. Hence, for every $\varepsilon > 0$, there exists $n > 0$ such that $\|s(t_n, x_0) - p\| < \varepsilon/2$. Furthermore, since $s(\cdot, x_0)$ is left-continuous and $\mathcal{T}_{x_0}$ is a dense subset of $[0, \infty)$, there exists $\hat{t}_n \in \mathcal{T}_{x_0}$, $\hat{t}_n \leq t_n$, such that $\|s(\hat{t}_n, x_0) - s(t_n, x_0)\| < \varepsilon/2$, and hence, $\|s(\hat{t}_n, x_0) - p\| \leq \|s(t_n, x_0) - p\| + \|s(\hat{t}_n, x_0) - s(t_n, x_0)\| < \varepsilon$. Using this procedure, with $\varepsilon = 1, 1/2, 1/3, \ldots$, we can construct an unbounded sequence $\{\hat{t}_k\}_{k=1}^{\infty} \subset \mathcal{T}_{x_0}$ such that $\lim_{k\to\infty} s(\hat{t}_k, x_0) = p$. Hence, $p \in \omega(x_0)$ if and only if there exists a monotonic sequence $\{t_n\}_{n=0}^{\infty} \subset \mathcal{T}_{x_0}$, with $t_n \to \infty$ as $n \to \infty$, such that $s(t_n, x_0) \to p$ as $n \to \infty$.

Next, we state and prove a fundamental result on positive limit sets for impulsive dynamical systems. This result generalizes the classical results on positive limit sets to systems with left-continuous flows.

Theorem 10.2. Consider the impulsive dynamical system $\mathcal{G}$ given by (10.1) and (10.2), assume Assumption 10.1 holds, and suppose that for $x_0 \in \mathcal{D} \subseteq \overline{\mathbb{R}}_+^n$ the trajectory $s(t, x_0)$ of $\mathcal{G}$ is bounded for all $t \geq 0$. Then the positive limit set $\omega(x_0)$ of $s(t, x_0)$, $t \geq 0$, is a nonempty, compact invariant subset of $\overline{\mathbb{R}}_+^n$. Furthermore, $s(t, x_0) \to \omega(x_0)$ as $t \to \infty$.

Proof. Let $s(t, x_0)$, $t \geq 0$, denote the solution to $\mathcal{G}$ with initial condition $x_0 \in \mathcal{D}$. Since $s(t, x_0)$ is bounded for all $t \geq 0$, it follows from the Bolzano-Weierstrass theorem [387] that every sequence in the positive orbit $\mathcal{O}_{x_0}^+ \triangleq \{s(t, x_0) : t \in [0, \infty)\}$ has at least one accumulation point $y \in \mathcal{D}$ as $t \to \infty$, and hence, $\omega(x_0)$ is nonempty. Furthermore, since $s(t, x_0)$, $t \geq 0$, is bounded, it follows that $\omega(x_0)$ is bounded.

To show that $\omega(x_0)$ is closed, let $\{y_i\}_{i=0}^{\infty}$ be a sequence contained in $\omega(x_0)$ such that $\lim_{i\to\infty} y_i = y \in \mathcal{D}$. Now, since $y_i \to y$ as $i \to \infty$, it follows that for every $\varepsilon > 0$, there exists i such that $\|y - y_i\| < \varepsilon/2$. Next, since $y_i \in \omega(x_0)$, it follows that for every $T > 0$, there exists $t \geq T$ such that $\|s(t, x_0) - y_i\| < \varepsilon/2$. Hence, it follows that for every $\varepsilon > 0$ and $T > 0$, there exists $t \geq T$ such that $\|s(t, x_0) - y\| \leq \|s(t, x_0) - y_i\| + \|y_i - y\| < \varepsilon$, which implies that $y \in \omega(x_0)$, and hence, $\omega(x_0)$ is closed. Thus, since $\omega(x_0)$ is closed and bounded, $\omega(x_0)$ is compact.

Next, to show positive invariance of $\omega(x_0)$, let $y \in \omega(x_0)$ so that there exists an increasing unbounded sequence $\{t_n\}_{n=0}^{\infty} \subset \mathcal{T}_{x_0}$ such that $s(t_n, x_0) \to y$ as $n \to \infty$. Now, it follows from Assumption 10.1 that for every $\varepsilon > 0$ and $t \in \mathcal{T}_y$, there exists $\delta(\varepsilon, y, t) > 0$ such that $\|y - z\| < \delta(\varepsilon, y, t)$, $z \in \mathcal{D}$, implies $\|s(t, y) - s(t, z)\| < \varepsilon$ or, equivalently, for every sequence $\{y_i\}_{i=1}^{\infty}$ converging to y and $t \in \mathcal{T}_y$, $\lim_{i\to\infty} s(t, y_i) = s(t, y)$. Now, since by assumption there exists a unique solution to $\mathcal{G}$, it follows that the semigroup property $s(\tau, s(t, x_0)) = s(t + \tau, x_0)$ for all $x_0 \in \mathcal{D}$ and $t, \tau \in [0, \infty)$ holds.

Furthermore, since $s(t_n, x_0) \to y$ as $n \to \infty$, it follows from the semigroup property that $s(t, y) = s(t, \lim_{n\to\infty} s(t_n, x_0)) = \lim_{n\to\infty} s(t + t_n, x_0) \in \omega(x_0)$ for all $t \in \mathcal{T}_y$. Hence, $s(t, y) \in \omega(x_0)$ for all $t \in \mathcal{T}_y$. Next, let $t \in [0, \infty) \backslash \mathcal{T}_y$ and note that, since $\mathcal{T}_y$ is dense in $[0, \infty)$, there exists a sequence $\{\tau_n\}_{n=0}^{\infty}$ such that $\tau_n \leq t$, $\tau_n \in \mathcal{T}_y$, and $\lim_{n\to\infty} \tau_n = t$. Now, since $s(\cdot, y)$ is left-continuous, it follows that $\lim_{n\to\infty} s(\tau_n, y) = s(t, y)$. Finally, since $\omega(x_0)$ is closed and $s(\tau_n, y) \in \omega(x_0)$, $n = 1, 2, \ldots$, it follows that $s(t, y) = \lim_{n\to\infty} s(\tau_n, y) \in \omega(x_0)$. Hence, $s_t(\omega(x_0)) \subseteq \omega(x_0)$, $t \geq 0$, establishing positive invariance of $\omega(x_0)$.

Now, to show invariance of $\omega(x_0)$, let $y \in \omega(x_0)$ so that there exists an increasing unbounded sequence $\{t_n\}_{n=0}^{\infty}$ such that $s(t_n, x_0) \to y$ as $n \to \infty$. Next, let $t \in \mathcal{T}_{x_0}$ and note that there exists $N \in \mathbb{Z}_+$ such that $t_n > t$, $n \geq N$. Hence, it follows from the semigroup property that $s(t, s(t_n - t, x_0)) = s(t_n, x_0) \to y$ as $n \to \infty$. Now, it follows from the Bolzano-Weierstrass theorem [387] that there exists a subsequence z_{n_k} of the sequence $z_n = s(t_n - t, x_0)$, $n = N, N+1, \ldots$, such that $z_{n_k} \to z \in \mathcal{D}$ and, by definition, $z \in \omega(x_0)$. Next, it follows from Assumption 10.1 that $\lim_{k\to\infty} s(t, z_{n_k}) = s(t, \lim_{k\to\infty} z_{n_k})$, and hence, $y = s(t, z)$, which implies that $\omega(x_0) \subseteq s_t(\omega(x_0))$, $t \in \mathcal{T}_{x_0}$.

Next, let $t \in [0,\infty)\backslash\mathcal{T}_{x_0}$, let $\hat{t} \in \mathcal{T}_{x_0}$ be such that $\hat{t} > t$, and consider $y \in \omega(x_0)$. Now, there exists $\hat{z} \in \omega(x_0)$ such that $y = s(\hat{t}, \hat{z})$, and it follows from the positive invariance of $\omega(x_0)$ that $z = s(\hat{t} - t, \hat{z}) \in \omega(x_0)$. Furthermore, it follows from the semigroup property of $\mathcal{G}$ (i.e., $s(\tau, s(t, x_0)) = s(t+\tau, x_0)$ for all $x_0 \in \mathcal{D}$ and $t, \tau \in [0,\infty)$) that $s(t, z) = s(t, s(\hat{t} - t, \hat{z})) = s(\hat{t}, \hat{z}) = y$, which implies that for all $t \in [0,\infty)\backslash\mathcal{T}_{x_0}$ and for every $y \in \omega(x_0)$, there exists $z \in \omega(x_0)$ such that $y = s(t, z)$. Hence, $\omega(x_0) \subseteq s_t(\omega(x_0))$, $t \geq 0$. Now, using positive invariance of $\omega(x_0)$, it follows that $s_t(\omega(x_0)) = \omega(x_0)$, $t \geq 0$, establishing invariance of the positive limit set $\omega(x_0)$.

Finally, to show $s(t, x_0) \to \omega(x_0)$ as $t \to \infty$, suppose, *ad absurdum*, $s(t, x_0) \not\to \omega(x_0)$ as $t \to \infty$. In this case, there exists an $\varepsilon > 0$ and a sequence $\{t_n\}_{n=0}^{\infty}$, with $t_n \to \infty$ as $n \to \infty$, such that

$$\inf_{p\in\omega(x_0)} \|s(t_n, x_0) - p\| \geq \varepsilon, \quad n \geq 0.$$

However, since the trajectory $s(t, x_0)$, $t \geq 0$, is bounded, the bounded sequence $\{s(t_n, x_0)\}_{n=0}^{\infty}$ contains a convergent subsequence $\{s(t_n^*, x_0)\}_{n=0}^{\infty}$ such that $s(t_n^*, x_0) \to p^* \in \omega(x_0)$ as $n \to \infty$, which contradicts the original supposition. Hence, $s(t, x_0) \to \omega(x_0)$ as $t \to \infty$, and, since f_c is essentially nonnegative and $x + f_d(x)$ is nonnegative, $\omega(x_0) \subseteq \overline{\mathbb{R}}_+^n$. □

Note that the compactness of the positive limit set $\omega(x_0)$ depends only on the boundedness of the trajectory $s(t, x_0)$, $t \geq 0$, whereas left continuity and Assumption 10.1 are key in proving invariance of the positive limit set $\omega(x_0)$. In classical dynamical systems, where the trajectory $s(\cdot,\cdot)$ is assumed to be continuous in both its arguments, both the left-continuity and the quasi-continuous dependence properties are trivially satisfied. Finally, we note that unlike dynamical systems with continuous flows, the positive limit set of an impulsive dynamical system may not be connected.

Next, we generalize the Krasovskii-LaSalle invariance principle to state-dependent impulsive dynamical systems. This result characterizes impulsive dynamical system limit sets in terms of continuously differentiable functions. In particular, we show that the system trajectories converge to an invariant set contained in a union of level surfaces characterized by the continuous-time dynamics and the resetting system dynamics. Henceforth, we assume that $f_c(\cdot)$, $f_d(\cdot)$, and $\mathcal{Z}$ are such that the dynamical system $\mathcal{G}$ given by (10.1) and (10.2) satisfies Assumption 10.1.

For the next result $V^{-1}(\gamma)$ denotes the γ*-level set* of $V(\cdot)$, that is, $V^{-1}(\gamma) \triangleq \{x \in \mathcal{D}_c : V(x) = \gamma\}$, where $\gamma \in \mathbb{R}$, $\mathcal{D}_c \subseteq \mathcal{D} \subseteq \overline{\mathbb{R}}_+^n$, and $V : \mathcal{D}_c \to \mathbb{R}$ is a continuously differentiable function, and let $\mathcal{M}_\gamma$ denote the largest invariant set (with respect to $\mathcal{G}$) contained in $V^{-1}(\gamma)$.

Theorem 10.3. Consider the hybrid nonnegative dynamical system $\mathcal{G}$ given by (10.1) and (10.2), assume $\mathcal{D}_c \subset \mathcal{D} \subseteq \overline{\mathbb{R}}_+^n$ is a compact positively invariant set with respect to (10.1) and (10.2), and assume that there exists a continuously differentiable function $V : \mathcal{D}_c \to \mathbb{R}$ such that

$$V'(x)f_c(x) \le 0, \quad x \in \mathcal{D}_c, \quad x \notin \mathcal{Z}, \tag{10.24}$$

$$V(x + f_d(x)) \le V(x), \quad x \in \mathcal{D}_c, \quad x \in \mathcal{Z}. \tag{10.25}$$

Let $\mathcal{R} \triangleq \{x \in \mathcal{D}_c : x \notin \mathcal{Z}, V'(x)f_c(x) = 0\} \cup \{x \in \mathcal{D}_c : x \in \mathcal{Z}, V(x + f_d(x)) = V(x)\}$ and let $\mathcal{M}$ denote the largest invariant set contained in $\mathcal{R}$. If $x_0 \in \mathcal{D}_c$, then $x(t) \to \mathcal{M}$ as $t \to \infty$.

Proof. Using identical arguments as in the proof of Theorem 10.1, it follows that for all $t \in (\tau_k(x_0), \tau_{k+1}(x_0)]$,

$$\begin{aligned} V(x(t)) - V(x(0)) &= \int_0^t V'(x(\tau))f_c(x(\tau))\mathrm{d}\tau \\ &\quad + \sum_{i=1}^{k} [V(x(\tau_i(x_0)) + f_d(x(\tau_i(x_0)))) - V(x(\tau_i(x_0)))]. \end{aligned}$$

Hence, it follows from (10.24) and (10.25) that $V(x(t)) \le V(x(0))$, $t \ge 0$. Using a similar argument, it follows that $V(x(t)) \le V(x(\tau))$, $t \ge \tau$, which implies that $V(x(t))$ is a nonincreasing function of time. Since $V(\cdot)$ is continuous on a compact set $\mathcal{D}_c$, there exists $\beta \in \mathbb{R}$ such that $V(x) \ge \beta$, $x \in \mathcal{D}_c$. Furthermore, since $V(x(t))$, $t \ge 0$, is nonincreasing, $\gamma_{x_0} \triangleq \lim_{t\to\infty} V(x(t))$, $x_0 \in \mathcal{D}_c$, exists. Now, for all $y \in \omega(x_0)$ there exists an increasing unbounded sequence $\{t_n\}_{n=0}^\infty$ such that $x(t_n) \to y$ as $n \to \infty$, and, since $V(\cdot)$ is continuous, it follows that $V(y) = V(\lim_{n\to\infty} x(t_n)) = \lim_{n\to\infty} V(x(t_n)) = \gamma_{x_0}$. Hence, $y \in V^{-1}(\gamma_{x_0})$ for all $y \in \omega(x_0)$, or, equivalently, $\omega(x_0) \subseteq V^{-1}(\gamma_{x_0})$.

Next, since $\mathcal{D}_c$ is compact and positively invariant, it follows that $x(t)$, $t \ge 0$, is bounded for all $x_0 \in \mathcal{D}_c$, and hence, it follows from Theorem 10.2 that $\omega(x_0)$ is a nonempty, compact invariant set. Thus, $\omega(x_0)$ is a subset of the largest invariant set contained in $V^{-1}(\gamma_{x_0})$, that is, $\omega(x_0) \subseteq \mathcal{M}_{\gamma_{x_0}}$. Hence, for every $x_0 \in \mathcal{D}_c$, there exists $\gamma_{x_0} \in \mathbb{R}$ such that $\omega(x_0) \subseteq \mathcal{M}_{\gamma_{x_0}}$, where $\mathcal{M}_{\gamma_{x_0}}$ is the largest invariant set contained in $V^{-1}(\gamma_{x_0})$, which implies that $V(x) = \gamma_{x_0}$, $x \in \omega(x_0)$. Now, since $\mathcal{M}_{\gamma_{x_0}}$ is an invariant set, it follows that for all $x(0) \in \mathcal{M}_{\gamma_{x_0}}$, $x(t) \in \mathcal{M}_{\gamma_{x_0}}$, $t \geq 0$, and hence, $\dot{V}(x(t)) \triangleq \frac{\mathrm{d}V(x(t))}{\mathrm{d}t} = V'(x(t))f_c(x(t)) = 0$, for all $x(t) \notin \mathcal{Z}$, and $V(x(t) + f_d(x(t))) = V(x(t))$, for all $x(t) \in \mathcal{Z}$. Thus, $\mathcal{M}_{\gamma_{x_0}}$ is contained in $\mathcal{M}$, which is the largest invariant set contained in $\mathcal{R}$. Hence, $x(t) \to \mathcal{M}$ as $t \to \infty$. □

The following corollaries to Theorem 10.3 present sufficient conditions that guarantee local asymptotic stability of the nonlinear impulsive dynamical system (10.1) and (10.2).

Corollary 10.1. Consider the nonlinear impulsive dynamical system (10.1) and (10.2), assume $\mathcal{D}_c \subset \mathcal{D} \subseteq \overline{\mathbb{R}}_+^n$ is a compact positively invariant set with respect to (10.1) and (10.2) such that $x_e \in \mathcal{D}_c$, and assume there exists a continuously differentiable function $V : \mathcal{D}_c \to \mathbb{R}$ such that $V(x_e) = 0$, $V(x) > 0$, $x \neq x_e$, and (10.24) and (10.25) are satisfied. Furthermore, assume that the set $\mathcal{R} \triangleq \{x \in \mathcal{D}_c : \ x \notin \mathcal{Z},\ V'(x)f_c(x) = 0\} \cup \{x \in \mathcal{D}_c : \ x \in \mathcal{Z},\ V(x + f_d(x)) = V(x)\}$ contains no invariant set other than the set $\{x_e\}$. Then the equilibrium solution $x(t) \equiv x_e$ to (10.1) and (10.2) is asymptotically stable, and $\mathcal{D}_c$ is a subset of the domain of attraction of (10.1) and (10.2).

Proof. Lyapunov stability with respect to $\overline{\mathbb{R}}_+^n$ of the equilibrium solution $x(t) \equiv x_e$ to (10.1) and (10.2) follows from Theorem 10.1. Next, it follows from Theorem 10.3 that if $x_0 \in \mathcal{D}_c$, then $\omega(x_0) \subseteq \mathcal{M}$, where $\mathcal{M}$ denotes the largest invariant set contained in $\mathcal{R}$, which implies that $\mathcal{M} = \{x_e\}$. Hence, $x(t) \to \mathcal{M} = \{x_e\}$ as $t \to \infty$, establishing asymptotic stability of the equilibrium solution $x(t) \equiv x_e$ to (10.1) and (10.2) with respect to $\overline{\mathbb{R}}_+^n$. □

Setting $\mathcal{D} = \overline{\mathbb{R}}_+^n$ and requiring $V(x) \to \infty$ as $\|x\| \to \infty$ in Corollary 10.1, it follows that the equilibrium solution $x(t) \equiv x_e$ to (10.1) and (10.2) is globally asymptotically stable with respect to $\overline{\mathbb{R}}_+^n$. Similar remarks hold for Corollaries 10.2 and 10.3 below.

Corollary 10.2. Consider the nonlinear impulsive dynamical system (10.1) and (10.2), assume $\mathcal{D}_c \subset \mathcal{D} \subseteq \overline{\mathbb{R}}_+^n$ is a compact positively invariant set with respect to (10.1) and (10.2) such that $x_e \in \mathcal{D}_c$, and assume there exists a continuously differentiable function $V : \mathcal{D}_c \to \mathbb{R}$ such that $V(x_e) = 0$, $V(x) > 0$, $x \neq x_e$,

$$V'(x)f_c(x) < 0, \quad x \in \mathcal{D}_c, \quad x \notin \mathcal{Z}, \quad x \neq x_e, \tag{10.26}$$

and (10.25) is satisfied. Then the equilibrium solution $x(t) \equiv x_{\rm e}$ to (10.1) and (10.2) is asymptotically stable with respect to $\overline{\mathbb{R}}_+^n$, and $\mathcal{D}_{\rm c}$ is a subset of the domain of attraction of (10.1) and (10.2).

Proof. It follows from (10.26) that $V'(x)f_{\rm c}(x) = 0$ for all $x \in \mathcal{D}_{\rm c}\backslash\mathcal{Z}$ if and only if $x = x_{\rm e}$. Hence, $\mathcal{R} = \{x_{\rm e}\} \cup \{x \in \mathcal{D}_{\rm c} : x \in \mathcal{Z}, V(x + f_{\rm d}(x)) = V(x)\}$, which contains no invariant set other than $\{x_{\rm e}\}$. Now, the result follows as a direct consequence of Corollary 10.1. □

Corollary 10.3. Consider the nonlinear impulsive dynamical system (10.1) and (10.2), assume $\mathcal{D}_{\rm c} \subset \mathcal{D} \subseteq \overline{\mathbb{R}}_+^n$ is a compact positively invariant set with respect to (10.1) and (10.2) such that $x_{\rm e} \in \mathcal{D}_{\rm c}$, and assume that for all $x_0 \in \mathcal{D}_{\rm c}$, $x_0 \neq x_{\rm e}$, there exists $\tau \geq 0$ such that $x(\tau) \in \mathcal{Z}$, where $x(t)$, $t \geq 0$, denotes the solution to (10.1) and (10.2) with the initial condition x_0. Furthermore, assume there exists a continuously differentiable function $V : \mathcal{D}_{\rm c} \to \mathbb{R}$ such that $V(x_{\rm e}) = 0$, $V(x) > 0$, $x \neq x_{\rm e}$,

$$V(x + f_{\rm d}(x)) - V(x) < 0, \quad x \in \mathcal{D}_{\rm c}, \quad x \in \mathcal{Z}, \tag{10.27}$$

and (10.24) is satisfied. Then the equilibrium solution $x(t) \equiv x_{\rm e}$ to (10.1) and (10.2) is asymptotically stable with respect to $\overline{\mathbb{R}}_+^n$, and $\mathcal{D}_{\rm c}$ is a subset of the domain of attraction of (10.1) and (10.2).

Proof. It follows from (10.27) that $\mathcal{R} = \{x \in \mathcal{D}_{\rm c} : x \notin \mathcal{Z}, V'(x)f_{\rm c}(x) = 0\}$. Since, for all $x_0 \in \mathcal{D}_{\rm c}$, $x_0 \neq x_{\rm e}$, there exists $\tau \geq 0$ such that $x(\tau) \in \mathcal{Z}$, it follows that the largest invariant set contained in $\mathcal{R}$ is $\{x_{\rm e}\}$. Now, the result is a direct consequence of Corollary 10.1. □

10.3 Hybrid Thermodynamic Models

In this section, we combine the results of Chapters 3 and 9 to develop a hybrid thermodynamic system model. To formulate our state space hybrid thermodynamic model, let $E_i(t)$, $i = 1, \ldots, q$, denote the energy (and hence a nonnegative quantity) of the ith subsystem of the hybrid compartmental system shown in Figure 10.1, let $\sigma_{{\rm c}ii}(E) \geq 0$, $E \notin \mathcal{Z}$, denote the rate of flow of energy loss of the ith continuous-time subsystem, let $S_{{\rm c}i}(t) \geq 0$, $t \geq 0$, $i = 1, \ldots, q$, denote the rate of energy inflow supplied to the ith continuous-time subsystem, and let $\phi_{{\rm c}ij}(E(t))$, $t \geq 0$, $i \neq j$, $i, j = 1, \ldots, q$, denote the net energy flow (or power) from the jth continuous-time subsystem to the ith continuous-time subsystem given by $\phi_{{\rm c}ij}(E(t)) = \sigma_{{\rm c}ij}(E(t)) - \sigma_{{\rm c}ji}(E(t))$, where the rates of energy flows are such that $\sigma_{{\rm c}ij}(E) \geq 0$, $E \notin \mathcal{Z}$, $i \neq j$, $i, j = 1, \ldots, q$.

Similarly, for the resetting dynamics, let $\sigma_{{\rm d}ii}(E) \geq 0$, $E \in \mathcal{Z}$, denote the energy loss of the ith discrete-time subsystem, let $S_{{\rm d}i}(t_k) \geq 0$,

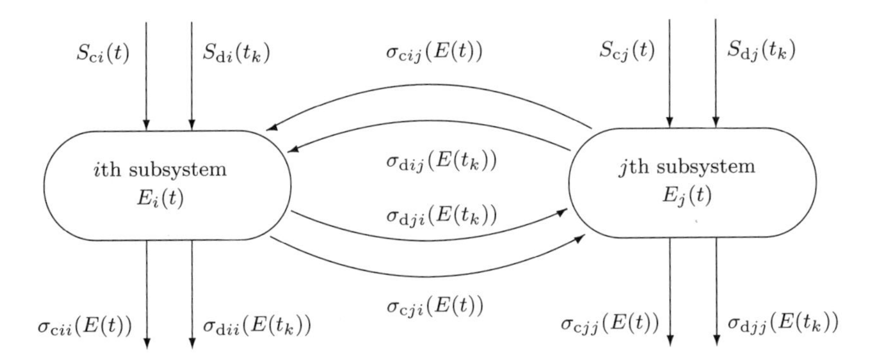

Figure 10.1 Nonlinear hybrid energy flow model.

$i = 1, \ldots, q$, denote the energy inflow supplied to the ith discrete-time subsystem, and let $\phi_{\mathrm{d}ij}(t_k)$, $i \neq j$, $i, j = 1, \ldots, q$, denote the net energy exchange from the jth discrete-time subsystem to the ith discrete-time subsystem given by $\phi_{\mathrm{d}ij}(E(t_k)) = \sigma_{\mathrm{d}ij}(E(t_k)) - \sigma_{\mathrm{d}ji}(E(t_k))$, where $t_k = \tau_k(E_0)$ and the energy flows are such that $\sigma_{\mathrm{d}ij}(E) \geq 0$, $E \in \mathcal{Z}$, $i \neq j$, $i, j = 1, \ldots, q$.

An energy balance for the whole hybrid compartmental thermodynamic system yields

$$\dot{E}_i(t) = -\sigma_{\mathrm{c}ii}(E(t)) + \sum_{j=1, i \neq j}^{q} \phi_{\mathrm{c}ij}(E(t)) + S_{\mathrm{c}i}(t), \quad E(t) \notin \mathcal{Z}, \quad i = 1, \ldots, q, \tag{10.28}$$

$$\Delta E_i(t) = -\sigma_{\mathrm{d}ii}(E(t)) + \sum_{j=1, i \neq j}^{q} \phi_{\mathrm{d}ij}(E(t)) + S_{\mathrm{d}i}(t), \quad E(t) \in \mathcal{Z}, \quad i = 1, \ldots, q, \tag{10.29}$$

or, equivalently,

$$\dot{E}(t) = f_{\mathrm{c}}(E(t)) + S_{\mathrm{c}}(t), \qquad E(0) = E_0, \qquad E(t) \notin \mathcal{Z}, \tag{10.30}$$
$$\Delta E(t) = f_{\mathrm{d}}(E(t)) + S_{\mathrm{d}}(t), \qquad E(t) \in \mathcal{Z}, \tag{10.31}$$

where $E(t) \triangleq [E_1(t), \ldots, E_q(t)]^{\mathrm{T}}$, $S_{\mathrm{c}}(t) \triangleq [S_{\mathrm{c}1}(t), \ldots, S_{\mathrm{c}q}(t)]^{\mathrm{T}}$, $S_{\mathrm{d}}(t) \triangleq [S_{\mathrm{d}1}(t), \ldots, S_{\mathrm{d}q}(t)]^{\mathrm{T}}$, and, for $i, j = 1, \ldots, q$,

$$f_{\mathrm{c}i}(E) = -\sigma_{\mathrm{c}ii}(E) + \sum_{j=1, i \neq j}^{q} [\sigma_{\mathrm{c}ij}(E) - \sigma_{\mathrm{c}ji}(E)], \tag{10.32}$$

$$f_{\mathrm{d}i}(E) = -\sigma_{\mathrm{d}ii}(E) + \sum_{j=1, i\neq j}^{q} [\sigma_{\mathrm{d}ij}(E) - \sigma_{\mathrm{d}ji}(E)]. \tag{10.33}$$

Since all energy flows as well as compartment sizes are nonnegative, it follows that for all $i = 1, \ldots, q$, $f_{\mathrm{c}i}(E) \geq 0$ for all $E \notin \mathcal{Z}$ whenever $E_i = 0$ and whatever the values of E_j, $j \neq i$, and $E_i + f_{\mathrm{d}i}(E) \geq 0$ for all $E \in \mathcal{Z}$. The above physical constraints are implied by $\sigma_{\mathrm{c}ij}(E) \geq 0$, $\sigma_{\mathrm{c}ii}(E) \geq 0$, $E \notin \mathcal{Z}$, $\sigma_{\mathrm{d}ij}(E) \geq 0$, $\sigma_{\mathrm{d}ii}(E) \geq 0$, $E \in \mathcal{Z}$, $S_{\mathrm{c}i} \geq 0$, $S_{\mathrm{d}i} \geq 0$, for all $i, j = 1, \ldots, q$, and if $E_i = 0$, then $\sigma_{\mathrm{c}ii}(E) = 0$ and $\sigma_{\mathrm{c}ji}(E) = 0$ for all $i, j = 1, \ldots, q$, so that $\dot{E}_i \geq 0$. In this case, $f_{\mathrm{c}}(E)$, $E \notin \mathcal{Z}$, is essentially nonnegative and $E + f_{\mathrm{d}}(E) \geq\geq 0$, $E \in \mathcal{Z}$, and hence, the hybrid compartmental model given by (10.28) and (10.29) is a hybrid nonnegative dynamical system.

Taking the total energy of the compartmental system $V(E) = \mathbf{e}^{\mathrm{T}} E = \sum_{i=1}^{q} E_i$ as a Lyapunov function for the inflow-closed (i.e., $S_{\mathrm{c}}(t) \equiv 0$ and $S_{\mathrm{d}}(t_k) \equiv 0$) system (10.28) and (10.29), and assuming $\sigma_{ij}(0) = 0$, $i, j = 1, \ldots, q$, it follows that

$$\begin{aligned}
\dot{V}(E) &= \sum_{i=1}^{q} \dot{E}_i \\
&= -\sum_{i=1}^{q} \sigma_{\mathrm{c}ii}(E) + \sum_{i=1}^{q} \sum_{j=1, i\neq j}^{q} [\sigma_{\mathrm{c}ij}(E) - \sigma_{\mathrm{c}ji}(E)] \\
&= -\sum_{i=1}^{q} \sigma_{\mathrm{c}ii}(E) \\
&\leq 0, \quad E \notin \mathcal{Z},
\end{aligned}$$

and

$$\begin{aligned}
\Delta V(E) &= \sum_{i=1}^{q} \Delta E_i \\
&= -\sum_{i=1}^{q} \sigma_{\mathrm{d}ii}(E) + \sum_{i=1}^{q} \sum_{j=1, i\neq j}^{q} [\sigma_{\mathrm{d}ij}(E) - \sigma_{\mathrm{d}ji}(E)] \\
&= -\sum_{i=1}^{q} \sigma_{\mathrm{d}ii}(E) \\
&\leq 0, \quad E \in \mathcal{Z},
\end{aligned}$$

which, by Theorem 10.1, shows that the zero solution $E(t) \equiv 0$ of the nonlinear hybrid compartmental system given by (10.28) and (10.29) is Lyapunov stable. If (10.28) and (10.29) with $S_{\mathrm{c}}(t) \equiv 0$ and $S_{\mathrm{d}}(t_k) \equiv 0$ has energy losses (outflows) from all compartments over the continuous-time dynamics, then $\sigma_{cii}(E) > 0$, $E \notin \mathcal{Z}$, $E \neq 0$, and hence, by Theorem 10.1, the zero solution $E(t) \equiv 0$ to (10.28) and (10.29) is asymptotically stable.

It is interesting to note that in the linear case $\sigma_{cii}(E) = \sigma_{cii}E_i$, $\phi_{cij}(E) = \sigma_{cij}E_j - \sigma_{cji}E_i$, $\sigma_{dii}(E) = \sigma_{dii}E_i$, and $\phi_{dij}(E) = \sigma_{dij}E_j - \sigma_{dji}E_i$, where $\sigma_{cij} \geq 0$ and $\sigma_{dij} \geq 0$, $i,j = 1,\ldots,q$, (10.30) and (10.31) become

$$\dot{E}(t) = A_{\rm c}E(t) + S_{\rm c}(t), \quad E(0) = E_0, \quad E(t) \notin \mathcal{Z}, \tag{10.34}$$
$$\Delta E(t) = (A_{\rm d} - I_n)E(t) + S_{\rm d}(t), \quad E(t) \in \mathcal{Z}, \tag{10.35}$$

where, for $i,j = 1,\ldots,q$,

$$A_{{\rm c}(i,j)} = \begin{cases} -\sum_{l=1}^{q} \sigma_{{\rm c}li}, & i = j, \\ \sigma_{{\rm c}ij}, & i \neq j, \end{cases} \tag{10.36}$$
$$A_{{\rm d}(i,j)} = \begin{cases} 1 - \sum_{l=1}^{q} \sigma_{{\rm d}li}, & i = j, \\ \sigma_{{\rm d}ij}, & i \neq j. \end{cases} \tag{10.37}$$

Note that, since at any given instant of time compartmental energy can only be transported, stored, or discharged but not created and the maximum amount of energy that can be transported and/or discharged cannot exceed the energy in a compartment, it follows that $1 \geq \sum_{l=1}^{q} \sigma_{{\rm d}li}$. Thus, $A_{\rm c}$ is an essentially nonnegative matrix and $A_{\rm d}$ is a nonnegative matrix, and hence, the hybrid compartmental model given by (10.34) and (10.35) is a hybrid nonnegative dynamical system.

The hybrid compartmental thermodynamic system (10.28) and (10.29) with no energy inflows, that is, $S_{{\rm c}i}(t) \equiv 0$ and $S_{{\rm d}i}(t_k) \equiv 0$, $i = 1,\ldots,q$, is said to be *inflow-closed*. Alternatively, if (10.28) and (10.29) possesses no energy losses (outflows), it is said to be *outflow-closed*. A hybrid compartmental system is said to be *adiabatically isolated* if it is inflow-closed and outflow-closed. Note that for an adiabatically isolated system $\dot{V}(E) = 0$, $E \notin \mathcal{Z}$, and $\Delta V(E) = 0$, $E \in \mathcal{Z}$, which shows that the total energy inside an adiabatically isolated system is conserved.

In the case where $\sigma_{cii}(E) \neq 0$, $E \notin \mathcal{Z}$, $\sigma_{dii}(E) \neq 0$, $E \in \mathcal{Z}$, $S_{{\rm c}i}(t) \neq 0$, and $S_{{\rm d}i}(t_k) \neq 0$, $i = 1,\ldots,q$, it follows that (10.28) and (10.29) can be equivalently written as

$$\dot{E}(t) = [J_{{\rm c}n}(E(t)) - D_{\rm c}(E(t))]\left(\frac{\partial V}{\partial E}(E(t))\right)^{\rm T} + S_{\rm c}(t), \quad E(t) \notin \mathcal{Z}, \tag{10.38}$$

$$\Delta E(t) = [J_{{\rm d}n}(E(t)) - D_{\rm d}(E(t))]\left(\frac{\partial V}{\partial E}(E(t))\right)^{\rm T} + S_{\rm d}(t), \quad E(t) \in \mathcal{Z}, \tag{10.39}$$

where $J_{cn}(E)$ and $J_{dn}(E)$ are skew-symmetric matrix functions with

$$\begin{aligned} J_{cn(i,i)}(E) &= 0, \quad i = 1, \ldots, q, \\ J_{dn(i,i)}(E) &= 0, \quad i = 1, \ldots, q, \\ J_{cn(i,j)}(E) &= \sigma_{cij}(E) - \sigma_{cji}(E), \quad i \neq j, \\ J_{dn(i,j)}(E) &= \sigma_{dij}(E) - \sigma_{dji}(E), \quad i \neq j, \\ D_c(E) &= \operatorname{diag}[\sigma_{c11}(E), \ldots, \sigma_{cnn}(E)] \geq\geq 0, \quad E \in \overline{\mathbb{R}}^n_+, \end{aligned}$$

and

$$D_d(E) = \operatorname{diag}[\sigma_{d11}(E), \ldots, \sigma_{dnn}(E)] \geq\geq 0, \quad E \in \overline{\mathbb{R}}^n_+.$$

Hence, a hybrid compartmental thermodynamic system is a *hybrid port-controlled Hamiltonian system* [200] with a Hamiltonian $\mathcal{H}(E) = V(E) = \mathbf{e}^{\mathrm{T}}E$ representing the total energy in the system, $J_{cn}(E)$, $E \in \overline{\mathbb{R}}^n_+$, representing the energy exchange between subsystems over the continuous-time dynamics, $J_{dn}(E)$, $E \in \overline{\mathbb{R}}^n_+$, representing the energy exchange between subsystems at the resetting instants, $D_c(E)$, $E \in \overline{\mathbb{R}}^n_+$, representing the energy dissipation over the continuous-time dynamics, $D_d(E)$, $E \in \overline{\mathbb{R}}^n_+$, representing the energy dissipation at the resetting instants, $S_c(t)$ representing the supplied power to the system over the continuous-time dynamics, and $S_d(t_k)$ representing the supplied energy to the system at the resetting instants.

Finally, we show that our hybrid compartmental thermodynamic system with measured outputs corresponding to energy and energy rate outflows are lossless with respect to the hybrid energy supply rate $(r_c(u_c, y_c), r_d(u_d, y_d)) = (\mathbf{e}^{\mathrm{T}}u_c - \mathbf{e}^{\mathrm{T}}y_c, \mathbf{e}^{\mathrm{T}}u_d - \mathbf{e}^{\mathrm{T}}y_d)$, where $u_c(t) \triangleq S_c(t)$, $y_c(t) \triangleq D_c(E(t))\mathbf{e}$, $u_d(t_k) \triangleq S_d(t_k)$, and $y_d(t_k) \triangleq D_d(E(t_k))\mathbf{e}$ [196]. Specifically, consider (10.38) and (10.39) with $S_c(t) = u_c(t)$ and $S_d(t_k) = u_d(t_k)$, energy function $V(E) = \mathbf{e}^{\mathrm{T}}E$, and hybrid outputs

$$y_c = D_c(E)\left(\frac{\partial V}{\partial E}\right)^{\mathrm{T}} = [\sigma_{c11}(E), \sigma_{c22}(E), \ldots, \sigma_{cnn}(E)]^{\mathrm{T}}$$

and

$$y_d = D_d(E)\left(\frac{\partial V}{\partial E}\right)^{\mathrm{T}} = [\sigma_{d11}(E), \sigma_{d22}(E), \ldots, \sigma_{dnn}(E)]^{\mathrm{T}}.$$

Now, it follows that

$$\begin{aligned} \dot{V}(E) &= \mathbf{e}^{\mathrm{T}}\left[[J_{cn}(E) - D_c(E)]\left(\frac{\partial V}{\partial E}\right)^{\mathrm{T}} + u_c\right] \\ &= \mathbf{e}^{\mathrm{T}}u_c - \mathbf{e}^{\mathrm{T}}y_c + \mathbf{e}^{\mathrm{T}}J_{cn}(E)\mathbf{e} \end{aligned}$$

$$= \mathbf{e}^{\mathrm{T}} u_{\mathrm{c}} - \mathbf{e}^{\mathrm{T}} y_{\mathrm{c}}, \quad E \notin \mathcal{Z}, \tag{10.40}$$

and

$$\begin{aligned} \Delta V(E) &= \mathbf{e}^{\mathrm{T}} \left[[J_{\mathrm{d}n}(E) - D_{\mathrm{d}}(E)] \left(\frac{\partial V}{\partial E} \right)^{\mathrm{T}} + u_{\mathrm{d}} \right] \\ &= \mathbf{e}^{\mathrm{T}} u_{\mathrm{d}} - \mathbf{e}^{\mathrm{T}} y_{\mathrm{d}} + \mathbf{e}^{\mathrm{T}} J_{\mathrm{d}n}(E) \mathbf{e} \\ &= \mathbf{e}^{\mathrm{T}} u_{\mathrm{d}} - \mathbf{e}^{\mathrm{T}} y_{\mathrm{d}}, \quad E \in \mathcal{Z}, \end{aligned} \tag{10.41}$$

which shows that the hybrid thermodynamic system (10.38) and (10.39) is lossless with respect to the hybrid supply rate $(r_{\mathrm{c}}, r_{\mathrm{d}}) = (\mathbf{e}^{\mathrm{T}} S_{\mathrm{c}} - \mathbf{e}^{\mathrm{T}} y_{\mathrm{c}}, \mathbf{e}^{\mathrm{T}} S_{\mathrm{d}} - \mathbf{e}^{\mathrm{T}} y_{\mathrm{d}})$.

Alternatively, if the hybrid outputs y_{c} and y_{d} correspond to a partial observation of the energy and energy rate outflows, then it can easily be shown that the hybrid compartmental thermodynamic system is dissipative with respect to the hybrid supply rate $(r_{\mathrm{c}}, r_{\mathrm{d}}) = (\mathbf{e}^{\mathrm{T}} u_{\mathrm{c}} - \mathbf{e}^{\mathrm{T}} y_{\mathrm{c}}, \mathbf{e}^{\mathrm{T}} u_{\mathrm{d}} - \mathbf{e}^{\mathrm{T}} y_{\mathrm{d}})$. The notion of hybrid dissipativity provides an interpretation of a generalized hybrid energy balance for a hybrid dynamical system in terms of the stored or accumulated energy, dissipated energy over the continuous-time dynamics, and dissipated energy at the resetting instants [196].

10.4 Conservation of Energy and the Hybrid First Law of Thermodynamics

In this section, we develop a hybrid version of the first law of thermodynamics and establish the uniqueness of the internal energy function $U(E) = V(E) = \mathbf{e}^{\mathrm{T}} E$, $E \in \overline{\mathbb{R}}_+^q$, for our hybrid thermodynamic model

$$\dot{E}(t) = [J_{\mathrm{c}n}(E(t)) - D_{\mathrm{c}}(E(t))] \left(\frac{\partial V}{\partial E}(E(t)) \right)^{\mathrm{T}} + S_{\mathrm{c}}(t), \quad E(t) \notin \mathcal{Z}, \tag{10.42}$$

$$\Delta E(t) = [J_{\mathrm{d}n}(E(t)) - D_{\mathrm{d}}(E(t))] \left(\frac{\partial V}{\partial E}(E(t)) \right)^{\mathrm{T}} + S_{\mathrm{d}}(t), \quad E(t) \in \mathcal{Z}, \tag{10.43}$$

$$y_{\mathrm{c}}(t) = D_{\mathrm{c}}(E(t)) \left(\frac{\partial V}{\partial E}(E(t)) \right)^{\mathrm{T}}, \quad E(t) \notin \mathcal{Z}, \tag{10.44}$$

$$y_{\mathrm{d}}(t) = D_{\mathrm{d}}(E(t)) \left(\frac{\partial V}{\partial E}(E(t)) \right)^{\mathrm{T}}, \quad E(t) \in \mathcal{Z}, \tag{10.45}$$

where, for all $t \geq 0$, $E(t) \in \mathcal{D} \subseteq \overline{\mathbb{R}}_+^q$, $\mathcal{D}$ is a relatively open set with respect to $\overline{\mathbb{R}}_+^q$ and with $0 \in \mathcal{D}$, $\Delta E(t) \triangleq E(t^+) - E(t)$, $S_{\mathrm{c}}(t) \triangleq u_{\mathrm{c}}(t) \in U_{\mathrm{c}} \subseteq \mathbb{R}^q$, $S_{\mathrm{d}}(t_k) \triangleq u_{\mathrm{d}}(t_k) \in U_{\mathrm{d}} \subseteq \mathbb{R}^q$, t_k denotes the kth instant of time at which

$(E(t), u_c(t))$ intersects $\mathcal{Z} \subset \mathcal{D} \times U_c$ for a particular trajectory $E(t)$ and input $u_c(t)$, $y_c(t) \in Y_c \subseteq \mathbb{R}^q$, $y_d(t_k) \in Y_d \subseteq \mathbb{R}^q$, and $\mathcal{Z} \subset \mathcal{D} \times U_c$. Here, we assume that $u_c(\cdot)$ and $u_d(\cdot)$ are restricted to the class of *admissible* inputs consisting of measurable functions such that $(u_c(t), u_d(t_k)) \in U_c \times U_d$ for all $t \geq 0$ and $k \in \mathbb{Z}_{[0,t)} \triangleq \{k : 0 \leq t_k < t\}$, where the constraint set $U_c \times U_d$ is given with $(0,0) \in U_c \times U_d$. Furthermore, we assume that the set $\mathcal{Z} \triangleq \{(E, u_c) : \mathcal{X}(E, u_c) = 0\}$, where $\mathcal{X} : \mathcal{D} \times U_c \to \mathbb{R}$.

More precisely, for the hybrid dynamical system $\mathcal{G}$ given by (10.42)–(10.45) defined on the state space $\mathcal{D} \subseteq \overline{\mathbb{R}}_+^q$, $\mathcal{U} \triangleq \mathcal{U}_c \times \mathcal{U}_d$ and $\mathcal{Y} \triangleq \mathcal{Y}_c \times \mathcal{Y}_d$ define an input and output space, respectively, consisting of left-continuous bounded U-valued and Y-valued functions on the semi-infinite interval $[0, \infty)$. The set $U \triangleq U_c \times U_d$, where $U_c \subseteq \mathbb{R}^q$ and $U_d \subseteq \mathbb{R}^q$, contains the set of input values, that is, for every $u = (u_c, u_d) \in \mathcal{U}$ and $t \in [0, \infty)$, $u(t) \in U$, $u_c(t) \in U_c$, and $u_d(t_k) \in U_d$. The set $Y \triangleq Y_c \times Y_d$, where $Y_c \subseteq \mathbb{R}^{l_c}$ and $Y_d \subseteq \mathbb{R}^{l_d}$, contains the set of output values, that is, for every $y = (y_c, y_d) \in \mathcal{Y}$ and $t \in [0, \infty)$, $y(t) \in Y$, $y_c(t) \in Y_c$, and $y_d(t_k) \in Y_d$.

Furthermore, we let $\mathcal{U}_r$ denote the set of all bounded continuous and discrete inputs (heat fluxes and energy flows) $u = (u_c, u_d)$ to the hybrid large-scale dynamical system $\mathcal{G}$ such that, for every $T \geq -t_0$, the system energy state can be driven from $E(-T) = 0$ to $E(t_0) = E_0 \in \overline{\mathbb{R}}_+^q$ by $u(\cdot) \in \mathcal{U}_r$. The spaces $\mathcal{U}$ and $\mathcal{Y}$ are assumed to be closed under the shift operator, that is, if $u(\cdot) \in \mathcal{U}$ (respectively, $y(\cdot) \in \mathcal{Y}$), then the function u_T (respectively, y_T) defined by $u_T \triangleq u(t + T)$ (respectively, $y_T \triangleq y(t + T)$) is contained in $\mathcal{U}$ (respectively, $\mathcal{Y}$) for all $T \geq 0$.

It follows from Lemma 3.1 and Proposition 9.3 that the hybrid large-scale dynamical system $\mathcal{G}$ given by (10.42)–(10.45) is *reachable* from and *controllable* to the origin in $\overline{\mathbb{R}}_+^q$. In particular, for all $E_0 = E(t_0) \in \overline{\mathbb{R}}_+^q$, there exist a finite time $t_i \leq t_0$, a square integrable input $u_c(t)$ defined on $[t_i, t_0]$, and an input $u_d(t_k)$ defined on $k \in \mathbb{Z}_{[t_i,t_0)}$, such that the state $E(t)$, $t \geq t_i$, can be driven from $E(t_i) = 0$ to $E(t_0) = E_0$. Furthermore, for all $E_0 = E(t_0) \in \overline{\mathbb{R}}_+^q$, there exists a finite time $t_f \geq t_0$, a square integrable input $u_c(t)$ defined on $[t_0, t_f]$, and an input $u_d(t_k)$ defined on $k \in \mathbb{Z}_{[t_i,t_0)}$, such that the state $E(t)$, $t \geq t_0$, can be driven from $E(t_0) = E_0$ to $E(t_f) = 0$.

Next, note that with $U(E) = V(E) = \mathbf{e}^{\mathrm{T}}E$, $E \in \overline{\mathbb{R}}_+^n$, (10.40) and (10.41) are a statement of the *first law of thermodynamics* as applied to *hybrid isochoric transformations*. Specifically, note that (10.40) and (10.41) can be written as the single lossless condition

$$U(E(T)) = U(E(t_0)) + \int_{t_0}^{T} r_c(u_c(t), y_c(t))\mathrm{d}t$$

$$+\sum_{k\in\mathbb{Z}_{[t_0,T)}} r_{\rm d}(u_{\rm d}(t_k),y_{\rm d}(t_k)), \tag{10.46}$$

where $E(t)$, $t \geq t_0$, is a solution to (10.42)–(10.45) with $(u_{\rm c}(t),u_{\rm d}(t_k)) \in U_{\rm c} \times U_{\rm d}$ and $E(t_0) = E_0$. To see this, next we give necessary and sufficient conditions for losslesness over an interval $t \in (t_k, t_{k+1}]$ involving the consecutive resetting times t_k and t_{k+1}.

Theorem 10.4. Consider the hybrid large-scale dynamical system $\mathcal{G}$ given by (10.42)–(10.45). Then $\mathcal{G}$ is lossless with respect to the hybrid supply rate $(r_{\rm c}, r_{\rm d})$ and with energy storage function $U : \overline{\mathbb{R}}_+^n \to \overline{\mathbb{R}}_+$ if and only if, for all $k \in \overline{\mathbb{Z}}_+$,

$$U(E(\hat{t})) - U(E(t)) = \int_t^{\hat{t}} r_{\rm c}(u_{\rm c}(s),y_{\rm c}(s)){\rm d}s, \quad t_k < t \leq \hat{t} \leq t_{k+1}, \tag{10.47}$$

$$U(E(t_k^+)) - U(E(t_k)) = r_{\rm d}(u_{\rm d}(t_k),y_{\rm d}(t_k)). \tag{10.48}$$

Proof. Let $k \in \overline{\mathbb{Z}}_+$ and suppose $\mathcal{G}$ is lossless with respect to the hybrid supply rate $(r_{\rm c}, r_{\rm d})$ and with energy storage function $U : \overline{\mathbb{R}}_+^n \to \overline{\mathbb{R}}_+$. Then (10.46) holds. Now, since for $t_k < t \leq \hat{t} \leq t_{k+1}$, $\mathbb{Z}_{[t,\hat{t})} = \emptyset$, (10.47) is immediate. Next, note that

$$\begin{aligned} U(E(t_k^+)) - U(E(t_k)) \leq & \int_{t_k}^{t_k^+} r_{\rm c}(u_{\rm c}(s),y_{\rm c}(s)){\rm d}s \\ & + r_{\rm d}(u_{\rm d}(t_k),y_{\rm d}(t_k)), \end{aligned} \tag{10.49}$$

which, since $\mathbb{Z}_{[t_k,t_k^+)} = \{k\}$, implies (10.48).

Conversely, suppose (10.47) and (10.48) hold, let $\hat{t} \geq t \geq 0$, and let $\mathbb{Z}_{[t,\hat{t})} = \{i, i+1, \ldots, j\}$. (Note that if $\mathbb{Z}_{[t,\hat{t})} = \emptyset$, then the converse is a direct consequence of (10.46).) In this case, it follows from (10.47) and (10.48) that

$$\begin{aligned} U(E(\hat{t})) - U(E(t)) &= U(E(\hat{t})) - U(E(t_j^+)) + U(E(t_j^+)) - U(E(t_{j-1}^+)) \\ &\quad + U(E(t_{j-1}^+)) - \cdots - U(E(t_i^+)) \\ &\quad + U(E(t_i^+)) - U(E(t)) \\ &= \int_{t_j^+}^{\hat{t}} r_{\rm c}(u_{\rm c}(s),y_{\rm c}(s)){\rm d}s + r_{\rm d}(u_{\rm d}(t_j),y_{\rm d}(t_j)) \\ &\quad + \int_{t_{j-1}^+}^{t_j} r_{\rm c}(u_{\rm c}(s),y_{\rm c}(s)){\rm d}s + \cdots + r_{\rm d}(u_{\rm d}(t_i),y_{\rm d}(t_i)) \end{aligned}$$

$$+\int_{t}^{t_i} r_{\rm c}(u_{\rm c}(s), y_{\rm c}(s)){\rm d}s$$

$$= \int_{t}^{\hat{t}} r_{\rm c}(u_{\rm c}(s), y_{\rm c}(s)){\rm d}s + \sum_{k\in\mathbb{Z}_{[t,\hat{t})}} r_{\rm d}(u_{\rm d}(t_k), y_{\rm d}(t_k)),$$

which implies that $\mathcal{G}$ is lossless with respect to the hybrid supply rate $(r_{\rm c}, r_{\rm d})$.

□

If $U(E(\cdot))$ is continuously differentiable almost everywhere on $[t_0, \infty)$ except on an unbounded closed discrete set $\mathcal{T} = \{t_1, t_2, \ldots\}$, where $\mathcal{T}$ is the set of times when jumps occur for $E(t)$, then it follows from Theorem 10.4 that an equivalent statement for losslessness of the hybrid large-scale dynamical system $\mathcal{G}$ with respect to the hybrid supply rate $(r_{\rm c}, r_{\rm d})$ is

$$\dot{U}(E(t)) = r_{\rm c}(u_{\rm c}(t), y_{\rm c}(t)), \quad t_k < t \le t_{k+1}, \tag{10.50}$$

$$\Delta U(E(t_k)) = r_{\rm d}(u_{\rm d}(t_k), y_{\rm d}(t_k)), \quad k \in \overline{\mathbb{Z}}_+, \tag{10.51}$$

where $\dot{U}(\cdot)$ denotes the total derivative of $U(E(t))$ along the state trajectories $E(t)$, $t \in (t_k, t_{k+1}]$, of the hybrid large-scale dynamical system (10.42)–(10.45) and $\Delta U(E(t_k)) \triangleq U(E(t_k^+)) - U(E(t_k))$, $k \in \overline{\mathbb{Z}}_+$, denotes the difference of the energy function $U(E)$ at the resetting times t_k, $k \in \overline{\mathbb{Z}}_+$, of the hybrid large-scale dynamical system (10.42)–(10.45).

The next result establishes the uniqueness of the internal energy function $U(E) = \mathbf{e}^{\rm T}E$, $E \in \overline{\mathbb{R}}_+^n$, for the hybrid large-scale dynamical system $\mathcal{G}$ given by (10.42)–(10.45). For this result define the *hybrid available energy* $U_{\rm a}(E_0)$ of the hybrid dynamical system $\mathcal{G}$ by

$$U_{\rm a}(E_0) \triangleq - \inf_{(u_{\rm c}(\cdot),u_{\rm d}(\cdot)),\, T\ge t_0} \left[\int_{t_0}^{T} r_{\rm c}(u_{\rm c}(t), y_{\rm c}(t)){\rm d}t + \sum_{k\in\mathbb{Z}_{[t_0,T)}} r_{\rm d}(u_{\rm d}(t_k), y_{\rm d}(t_k)) \right], \tag{10.52}$$

where $E(t)$, $t \ge t_0$, is the solution to (10.42)–(10.45) with admissible inputs $(u_{\rm c}(\cdot), u_{\rm d}(\cdot)) \in \mathcal{U}_{\rm c} \times \mathcal{U}_{\rm d}$ and $E(t_0) = E_0$. Note that $U_{\rm a}(E_0) \ge 0$ for all $E_0 \in \mathcal{D}$ since $U_{\rm a}(E_0)$ is the supremum over a set of numbers containing the zero element $(T = t_0)$.

It follows from (10.52) that the hybrid available energy of a hybrid large-scale dynamical system $\mathcal{G}$ is the maximum amount of stored energy that can be extracted from $\mathcal{G}$ at any time T. Furthermore, note that $U_{\rm a}(E(t))$ is left-continuous on $[t_0, \infty)$ and is continuous everywhere on $[t_0, \infty)$ except on an unbounded closed discrete set $\mathcal{T} = \{t_1, t_2, \ldots\}$, where $\mathcal{T}$ is the set of times when the jumps occur for $E(t)$, $t \ge t_0$.

In addition, define the *hybrid required supply* $U_{\rm r}(E_0)$ of the hybrid large-scale dynamical system $\mathcal{G}$ by

$$U_{\rm r}(E_0) \triangleq \inf_{(u_{\rm c}(\cdot),u_{\rm d}(\cdot)),\,T\leq t_0} \left[\int_T^{t_0} r_{\rm c}(u_{\rm c}(t), y_{\rm c}(t)){\rm d}t + \sum_{k\in\mathbb{Z}_{[T,t_0)}} r_{\rm d}(u_{\rm d}(t_k), y_{\rm d}(t_k)) \right], \qquad (10.53)$$

where $E(t)$, $t \geq T$, is the solution of (10.42)–(10.45) with $E(T) = 0$ and $E(t_0) = E_0$. It follows from (10.53) that the hybrid required supply of the dynamical system $\mathcal{G}$ is the minimum amount of energy that can be delivered to the hybrid dynamical system in order to transfer it from an initial state $E(T) = 0$ to a given state $E(t_0) = E_0$.

Theorem 10.5. Consider the large-scale hybrid dynamical system $\mathcal{G}$ given by (10.42)–(10.45). Then $\mathcal{G}$ is lossless with respect to the hybrid energy supply rate $(r_{\rm c}(u_{\rm c}, y_{\rm c}), r_{\rm d}(u_{\rm d}, y_{\rm d})) = (\mathbf{e}^{\rm T}u_{\rm c} - \mathbf{e}^{\rm T}y_{\rm c}, \mathbf{e}^{\rm T}u_{\rm d} - \mathbf{e}^{\rm T}y_{\rm d})$, where $u_{\rm c}(t) \triangleq S_{\rm c}(t)$, $y_{\rm c}(t) \triangleq D_{\rm c}(E(t))\mathbf{e}$, $u_{\rm d}(t_k) \triangleq S_{\rm d}(t_k)$, and $y_{\rm d}(t_k) \triangleq D_{\rm d}(E(t_k))\mathbf{e}$, and with the unique energy storage function corresponding to the total energy of the system $\mathcal{G}$ given by

$$\begin{aligned} U(E_0) &= \mathbf{e}^{\rm T}E_0 \\ &= -\int_{t_0}^{T_+} r_{\rm c}(u_{\rm c}(t), y_{\rm c}(t)){\rm d}t - \sum_{\mathbb{Z}_{[t_0,T_+)}} r_{\rm d}(u_{\rm d}(t_k), y_{\rm d}(t_k)) \\ &= \int_{-T_-}^{t_0} r_{\rm c}(u_{\rm c}(t), y_{\rm c}(t)){\rm d}t + \sum_{\mathbb{Z}_{[-T_-,t_0)}} r_{\rm d}(u_{\rm d}(t_k), y_{\rm d}(t_k)), \end{aligned} \qquad (10.54)$$

where $E(t)$, $t \geq t_0$, is the solution to (10.42)–(10.45) with admissible hybrid inputs $u(\cdot) = (u_{\rm c}(\cdot), u_{\rm d}(\cdot)) \in \mathcal{U}$, $E(-T_-) = 0$, $E(T_+) = 0$, and $E(t_0) = E_0 \in \overline{\mathbb{R}}_+^q$. Furthermore,

$$0 \leq U_{\rm a}(E_0) = U(E_0) = U_{\rm r}(E_0) < \infty, \quad E_0 \in \overline{\mathbb{R}}_+^q. \qquad (10.55)$$

Proof. Note that it follows from (10.46) that $\mathcal{G}$ is lossless with respect to the hybrid supply rate $(r_{\rm c}, r_{\rm d}) = (\mathbf{e}^{\rm T}u_{\rm c} - \mathbf{e}^{\rm T}y_{\rm c}, \mathbf{e}^{\rm T}u_{\rm d} - \mathbf{e}^{\rm T}y_{\rm d})$ and with the energy storage function $U(E) = \mathbf{e}^{\rm T}E$, $E \in \overline{\mathbb{R}}_+^q$. Since, by Lemma 3.1 and Proposition 9.3, $\mathcal{G}$ is reachable from and controllable to the origin in $\overline{\mathbb{R}}_+^q$, it follows from (10.46), with $E(t_0) = E_0 \in \overline{\mathbb{R}}_+^q$ and $E(T_+) = 0$ for some $T_+ \geq t_0$ and $u(\cdot) \in \mathcal{U}$, that

$$\mathbf{e}^{\rm T}E_0 = -\int_{t_0}^{T_+} r_{\rm c}(u_{\rm c}(t), y_{\rm c}(t)){\rm d}t - \sum_{\mathbb{Z}_{[t_0,T_+)}} r_{\rm d}(u_{\rm d}(t_k), y_{\rm d}(t_k))$$

$$\leq \sup_{u(\cdot)\in\mathcal{U}, T\geq t_0} \left[-\int_{t_0}^{T} r_{\rm c}(u_{\rm c}(t), y_{\rm c}(t)){\rm d}t - \sum_{\mathbb{Z}_{[t_0,T)}} r_{\rm d}(u_{\rm d}(t_k), y_{\rm d}(t_k)) \right]$$
$$= - \inf_{u(\cdot)\in\mathcal{U}, T\geq t_0} \left[\int_{t_0}^{T} r_{\rm c}(u_{\rm c}(t), y_{\rm c}(t)){\rm d}t + \sum_{\mathbb{Z}_{[t_0,T)}} r_{\rm d}(u_{\rm d}(t_k), y_{\rm d}(t_k)) \right]$$
$$= U_{\rm a}(E_0), \quad E_0 \in \overline{\mathbb{R}}_+^q. \tag{10.56}$$

Alternatively, it follows from (10.46), with $E(-T_-) = 0$ for some $-T_- \leq t_0$ and $u(\cdot) \in \mathcal{U}_{\rm r}$, that

$$\mathbf{e}^{\rm T} E_0 = \int_{-T_-}^{t_0} r_{\rm c}(u_{\rm c}(t), y_{\rm c}(t)){\rm d}t + \sum_{\mathbb{Z}_{[-T_-,t_0)}} r_{\rm d}(u_{\rm d}(t_k), y_{\rm d}(t_k))$$
$$\geq \inf_{u(\cdot)\in\mathcal{U}_{\rm r}, T\geq -t_0} \left[\int_{-T_-}^{t_0} r_{\rm c}(u_{\rm c}(t), y_{\rm c}(t)){\rm d}t + \sum_{\mathbb{Z}_{[-T_-,t_0)}} r_{\rm d}(u_{\rm d}(t_k), y_{\rm d}(t_k)) \right]$$
$$= U_{\rm r}(E_0), \quad E_0 \in \overline{\mathbb{R}}_+^q. \tag{10.57}$$

Thus, (10.56) and (10.57) imply that (10.54) is satisfied and

$$U_{\rm r}(E_0) \leq \mathbf{e}^{\rm T} E_0 \leq U_{\rm a}(E_0), \quad E_0 \in \overline{\mathbb{R}}_+^q. \tag{10.58}$$

Conversely, it follows from (10.46) and the fact that $U(E) = \mathbf{e}^{\rm T} E \geq 0$, $E \in \overline{\mathbb{R}}_+^q$, that, for all $T \geq t_0$ and $u(\cdot) \in \mathcal{U}$,

$$\mathbf{e}^{\rm T} E(t_0) \geq -\int_{t_0}^{T} r_{\rm c}(u_{\rm c}(t), y_{\rm c}(t)){\rm d}t - \sum_{\mathbb{Z}_{[t_0,T)}} r_{\rm d}(u_{\rm d}(t_k), y_{\rm d}(t_k)),$$
$$E(t_0) \in \overline{\mathbb{R}}_+^q, \tag{10.59}$$

which implies that

$$\mathbf{e}^{\rm T} E(t_0) \geq \sup_{u(\cdot)\in\mathcal{U}, T\geq t_0} \left[-\int_{t_0}^{T} r_{\rm c}(u_{\rm c}(t), y_{\rm c}(t)){\rm d}t - \sum_{\mathbb{Z}_{[t_0,T)}} r_{\rm d}(u_{\rm d}(t_k), y_{\rm d}(t_k)) \right]$$
$$= - \inf_{u(\cdot)\in\mathcal{U}, T\geq t_0} \left[\int_{t_0}^{T} r_{\rm c}(u_{\rm c}(t), y_{\rm c}(t)){\rm d}t + \sum_{\mathbb{Z}_{[t_0,T)}} r_{\rm d}(u_{\rm d}(t_k), y_{\rm d}(t_k)) \right]$$
$$= U_{\rm a}(E(t_0)), \quad E(t_0) \in \overline{\mathbb{R}}_+^q. \tag{10.60}$$

Furthermore, it follows from the definition of $U_{\rm a}(\cdot)$ that $U_{\rm a}(E) \geq 0$, $E \in \overline{\mathbb{R}}_+^q$, since the infimum in (10.52) is taken over the set of values containing the zero value ($T = t_0$).

Next, note that it follows from (10.46), with $E(t_0) \in \overline{\mathbb{R}}_+^q$ and $E(-T) = 0$ for all $T \geq -t_0$ and $u(\cdot) \in \mathcal{U}_r$, that

$$\begin{aligned} \mathbf{e}^{\mathrm{T}} E(t_0) &= \int_{-T}^{t_0} r_{\mathrm{c}}(u_{\mathrm{c}}(t), y_{\mathrm{c}}(t))\mathrm{d}t + \sum_{\mathbb{Z}_{[-T,t_0)}} r_{\mathrm{d}}(u_{\mathrm{d}}(t_k), y_{\mathrm{d}}(t_k)) \\ &= \inf_{u(\cdot)\in\mathcal{U}_r, T\geq -t_0} \left[\int_{-T}^{t_0} r_{\mathrm{c}}(u_{\mathrm{c}}(t), y_{\mathrm{c}}(t))\mathrm{d}t + \sum_{\mathbb{Z}_{[-T,t_0)}} r_{\mathrm{d}}(u_{\mathrm{d}}(t_k), y_{\mathrm{d}}(t_k)) \right] \\ &= U_{\mathrm{r}}(E(t_0)), \quad E(t_0) \in \overline{\mathbb{R}}_+^q. \end{aligned} \tag{10.61}$$

Moreover, since the system $\mathcal{G}$ is reachable from the origin, it follows that for every $E(t_0) \in \overline{\mathbb{R}}_+^q$, there exists $T \geq -t_0$ and $u(\cdot) \in \mathcal{U}_r$ such that

$$\int_{-T}^{t_0} r_{\mathrm{c}}(u_{\mathrm{c}}(t), y_{\mathrm{c}}(t))\mathrm{d}t + \sum_{\mathbb{Z}_{[-T,t_0)}} r_{\mathrm{d}}(u_{\mathrm{d}}(t_k), y_{\mathrm{d}}(t_k)) \tag{10.62}$$

is finite, and hence, $U_{\mathrm{r}}(E(t_0)) < \infty$, $E(t_0) \in \overline{\mathbb{R}}_+^q$. Finally, combining (10.58), (10.60), and (10.61), it follows that (10.55) holds. □

It follows from (10.55) and the definitions of available energy $U_{\mathrm{a}}(E_0)$ and the required energy supply $U_{\mathrm{r}}(E_0)$, $E_0 \in \overline{\mathbb{R}}_+^q$, that the hybrid large-scale dynamical system $\mathcal{G}$ can deliver to its surroundings all of its stored subsystem energies and can store all of the work done to all of its subsystems. This is in essence a statement of the hybrid version of the first law of thermodynamics and places no limitation on the possibility of transforming heat into work or work into heat.

10.5 Entropy and the Hybrid Second Law of Thermodynamics

In this section, we use the definitions of entropy for the continuous-time and discrete-time thermodynamic systems developed in Chapters 3 and 9 to give a definition of entropy for the hybrid thermodynamic model (10.30) and (10.31) or, equivalently, (10.38) and (10.39). To develop thermodynamically consistent hybrid models, we require some additional notation. Let $\mathcal{O}_i$ denote the set of all compartments with energy (heat) flowing out to the ith compartment, let $\mathcal{I}_i$ denote the set of all compartments receiving energy (heat) from the ith compartment, and let $\mathcal{V} = \{1, \ldots, q\}$ denote the set of vertices representing the intercompartmental connections of the hybrid compartmental thermodynamic system. Furthermore, let $\mathcal{O} = \bigcup_{i\in\mathcal{V}} \mathcal{O}_i$ and

$\mathcal{I} = \bigcup_{i\in\mathcal{V}} \mathcal{I}_i$, and define the resetting set $\mathcal{Z}$ by

$$\mathcal{Z} \triangleq \left\{ E \in \overline{\mathbb{R}}_+^q : \sum_{j\in\mathcal{O}} \mathcal{E}_j(E) - \sum_{i\in\mathcal{I}} \mathcal{E}_i(E) = 0 \,\text{and}\, \phi_{cij}(E) = 0,\ i,j = 1,\ldots,q \right\}. \tag{10.63}$$

The resetting set (10.63) is motivated by thermodynamic principles and guarantees that the energy of the hybrid thermodynamic system is always flowing from regions of higher to lower energies, which is consistent with the second law of thermodynamics. To ensure a thermodynamically consistent energy flow model, we require that Axioms *i*)–*iii*) of Chapter 9 hold for the hybrid large-scale dynamical system $\mathcal{G}$ given by (10.30) and (10.31). Namely, between resetting, the energy flow functions $\phi_{cij}(\cdot)$, $i,j = 1,\ldots,q$, must satisfy the following two axioms:

Axiom *i*): For the connectivity matrix $\mathcal{C} \in \mathbb{R}^{q\times q}$ associated with the hybrid large-scale dynamical system $\mathcal{G}$ defined by

$$\mathcal{C}_{(i,j)} = \begin{cases} 0, & \text{if } \phi_{cij}(E(t)) \equiv 0, \\ 1, & \text{otherwise}, \end{cases} \quad i \neq j,\ i,j = 1,\ldots,q,\ t \geq 0, \tag{10.64}$$

and

$$\mathcal{C}_{(i,i)} = -\sum_{k=1,\, k\neq i}^{q} \mathcal{C}_{(k,i)}, \quad i = j, \quad i = 1,\ldots,q, \tag{10.65}$$

rank $\mathcal{C} = q-1$, and for $\mathcal{C}_{(i,j)} = 1$, $i \neq j$, $\phi_{cij}(E(t)) = 0$ if and only if $E_i(t) = E_j(t)$ for all $E(t) \notin \mathcal{Z}$, $t \geq 0$.

Axiom *ii*): For $i,j = 1,\ldots,q$, $[E_i(t) - E_j(t)]\phi_{cij}(E(t)) \leq 0$, $E(t) \notin \mathcal{Z}$, $t \geq 0$.

Furthermore, across resettings the energy difference must satisfy the following axiom:

Axiom *iii*): For $i,j = 1,\ldots,q$, $[E_i(t_{k+1}) - E_j(t_{k+1})][E_i(t_k) - E_j(t_k)] \geq 0$ for all $E_i(t_k) \neq E_j(t_k)$, $E(t_k) \in \mathcal{Z}$, $k \in \mathbb{Z}_+$.

Axioms *i*)–*iii*) are a restatement of Axioms *i*)–*iii*) of Chapter 9 as applied to the hybrid large-scale dynamical system (10.30) and (10.31).

Next, we give a hybrid definition of entropy for the hybrid large-scale dynamical system $\mathcal{G}$ that generalizes the continuous-time and discrete-time entropy definitions established in Chapters 3 and 9.

Definition 10.2. For the hybrid large-scale dynamical system $\mathcal{G}$ given by (10.30) and (10.31), a function $\mathcal{S} : \overline{\mathbb{R}}_+^q \to \mathbb{R}$ satisfying

$$\mathcal{S}(E(T)) \geq \mathcal{S}(E(t_1)) + \int_{t_1}^{T} \sum_{i=1}^{q} \frac{S_{\mathrm{c}i}(t) - \sigma_{\mathrm{c}ii}(E(t))}{c + E_i(t)} \mathrm{d}t + \sum_{k \in \mathbb{Z}_{[t_1,T)}} \sum_{i=1}^{q} \frac{S_{\mathrm{d}i}(t_k) - \sigma_{\mathrm{d}ii}(E(t_k))}{c + E_i(t_k^+)}, \quad T \geq t_1, \tag{10.66}$$

where $k \in \mathbb{Z}_{[t_1,T)} \triangleq \{k : t_1 \leq t_k < T\}$, $c > 0$, is called an *entropy* function of $\mathcal{G}$.

The next result gives necessary and sufficient conditions for establishing the existence of a hybrid entropy function of $\mathcal{G}$ over an interval $t \in (t_k, t_{k+1}]$ involving the consecutive resetting times t_k and t_{k+1}, $k \in \mathbb{Z}_+$.

Theorem 10.6. Consider the hybrid large-scale dynamical system $\mathcal{G}$ given by (10.30) and (10.31), and assume Axioms *i*)–*iii*) hold. Then a function $\mathcal{S} : \overline{\mathbb{R}}_+^q \to \mathbb{R}$ is an entropy function of $\mathcal{G}$ if and only if

$$\mathcal{S}(E(\hat{t})) \geq \mathcal{S}(E(t)) + \int_{t}^{\hat{t}} \sum_{i=1}^{q} \frac{S_{\mathrm{c}i}(t) - \sigma_{\mathrm{c}ii}(E(t))}{c + E_i(t)} \mathrm{d}t, \quad t_k < t \leq \hat{t} \leq t_{k+1}, \tag{10.67}$$

$$\mathcal{S}(E(t_k^+)) \geq \mathcal{S}(E(t_k)) + \sum_{i=1}^{q} \frac{S_{\mathrm{d}i}(t_k) - \sigma_{\mathrm{d}ii}(E(t_k))}{c + E_i(t_k^+)}, \quad k \in \mathbb{Z}_+. \tag{10.68}$$

Proof. Let $k \in \mathbb{Z}_+$ and suppose $\mathcal{S}(E)$ is an entropy function of $\mathcal{G}$. Then (10.66) holds. Now, since for $t_k < t \leq \hat{t} \leq t_{k+1}$, $\mathbb{Z}_{[t,\hat{t})} = \varnothing$, it follows that $S_{\mathrm{d}i}(t_k) \equiv 0$ and $\sigma_{\mathrm{d}ii}(E(t_k)) \equiv 0$, $k \in \mathbb{Z}_+$. Furthermore, for all $E_0 \in \overline{\mathbb{R}}_+^q$ and $t_\mathrm{f} > t_0$ such that $E(t_\mathrm{f}) = E(t_0) = E_0$, note that

$$\begin{aligned}
\int_{t_0}^{t_\mathrm{f}} \sum_{i=1}^{q} \frac{S_{\mathrm{c}i}(t) - \sigma_{\mathrm{c}ii}(E(t))}{c + E_i(t)} \mathrm{d}t
&= \int_{t_0}^{t_\mathrm{f}} \sum_{i=1}^{q} \frac{\dot{E}_i(t) - \sum_{j=1, j\neq i}^{q} \phi_{\mathrm{c}ij}(E(t))}{c + E_i(t)} \mathrm{d}t \\
&= \sum_{i=1}^{q} \log_e \left(\frac{c + E_i(t_\mathrm{f})}{c + E_i(t_0)} \right) - \int_{t_0}^{t_\mathrm{f}} \sum_{i=1}^{q} \sum_{j=1, i\neq j}^{q} \frac{\phi_{\mathrm{c}ij}(E(t))}{c + E_i(t)} \mathrm{d}t \\
&= - \int_{t_0}^{t_\mathrm{f}} \sum_{i=1}^{q} \sum_{j=i+1}^{q} \left(\frac{\phi_{\mathrm{c}ij}(E(t))}{c + E_i(t)} - \frac{\phi_{\mathrm{c}ij}(E(t))}{c + E_j(t)} \right) \mathrm{d}t
\end{aligned}$$

$$= -\int_{t_0}^{t_{\rm f}} \sum_{i=1}^{q} \sum_{j=i+1}^{q} \frac{\phi_{{\rm c}ij}(E(t))[E_j(t) - E_i(t)]}{(c + E_i(t))(c + E_j(t))} {\rm d}t$$
$$\leq 0, \tag{10.69}$$

which proves (10.67). Next, note that

$$S(E(t_k^+)) \geq S(E(t_k)) + \int_{t_k}^{t_k^+} \sum_{i=1}^{q} \frac{S_{{\rm c}i}(t) - \sigma_{{\rm c}ii}(E(t))}{c + E_i(t)} {\rm d}t$$
$$+ \sum_{i=1}^{q} \frac{S_{{\rm d}i}(t_k) - \sigma_{{\rm d}ii}(E(t_k))}{c + E_i(t_k^+)}, \tag{10.70}$$

which, since $\mathbb{Z}_{[t_k, t_k^+)} = k$, implies (10.68).

Conversely, suppose (10.67) and (10.68) hold, and let $\hat{t} \geq t \geq t_1$ and $\mathbb{Z}_{[t,\hat{t})} = \{i, i+1, \ldots, j\}$. If $\mathbb{Z}_{[t,\hat{t})} = \varnothing$, then it follows from (10.67) that $\mathcal{S}(E)$ is an entropy function $\mathcal{G}$. If $\mathbb{Z}_{[t,\hat{t})} \neq \varnothing$, it follows from (10.67) and (10.68) that

$$\mathcal{S}(E(\hat{t})) - \mathcal{S}(E(t))$$
$$= \mathcal{S}(E(\hat{t})) - \mathcal{S}(E(t_j^+)) + \sum_{m=0}^{j-i-1} [\mathcal{S}(E(t_{j-m}^+)) - \mathcal{S}(E(t_{j-m-1}^+))]$$
$$+ \mathcal{S}(E(t_i^+)) - \mathcal{S}(E(t))$$
$$= \mathcal{S}(E(\hat{t})) - \mathcal{S}(E(t_j^+)) + \sum_{m=0}^{j-i} [\mathcal{S}(E(t_{j-m}^+)) - \mathcal{S}(E(t_{j-m})]$$
$$+ \sum_{m=0}^{j-i-1} [\mathcal{S}(E(t_{j-m})) - \mathcal{S}(E(t_{j-m-1}^+))] + \mathcal{S}(E(t_i)) - \mathcal{S}(E(t))$$
$$\geq \int_{t_j^+}^{\hat{t}} \sum_{i=1}^{q} \frac{S_{{\rm c}i}(t) - \sigma_{{\rm c}ii}(E(t))}{c + E_i(t)} {\rm d}t + \sum_{m=0}^{j-i} \sum_{i=1}^{q} \frac{S_{{\rm d}i}(t_{j-m}) - \sigma_{{\rm d}ii}(E(t_{j-m}))}{c + E_i(t_{j-m}^+)}$$
$$+ \sum_{m=0}^{j-i-1} \int_{t_{j-m-1}^+}^{t_{j-m}} \sum_{i=1}^{q} \frac{S_{{\rm c}i}(t) - \sigma_{{\rm c}ii}(E(t))}{c + E_i(t)} {\rm d}t + \int_{t}^{t_i} \sum_{i=1}^{q} \frac{S_{{\rm c}i}(t) - \sigma_{{\rm c}ii}(E(t))}{c + E_i(t)} {\rm d}t$$
$$= \int_{t}^{\hat{t}} \sum_{i=1}^{q} \frac{S_{{\rm c}i}(t) - \sigma_{{\rm c}ii}(E(t))}{c + E_i(t)} {\rm d}t + \sum_{k \in \mathbb{Z}_{[t,\hat{t})}} \sum_{i=1}^{q} \frac{S_{{\rm d}i}(t_k) - \sigma_{{\rm d}ii}(E(t_k))}{c + E_i(t_k^+)}, \tag{10.71}$$

which implies that $S(E)$ is an entropy function of $\mathcal{G}$. □

The next theorem establishes the existence of a continuously differentiable entropy function for the hybrid large-scale dynamical system $\mathcal{G}$ given by (10.30) and (10.31).

Theorem 10.7. Consider the hybrid large-scale dynamical system $\mathcal{G}$ given by (10.30) and (10.31) with $\mathcal{Z}$ given by (10.63) and assume Axioms *ii*) and *iii*) hold. Then the function $\mathcal{S} : \overline{\mathbb{R}}_+^q \to \mathbb{R}$ given by

$$\mathcal{S}(E) = \mathbf{e}^{\mathrm{T}} \log_e(c\mathbf{e} + E) - q \log_e c, \tag{10.72}$$

where $c > 0$, is a continuously differentiable entropy function of $\mathcal{G}$. In addition,

$$\dot{\mathcal{S}}(E(t)) \geq \sum_{i=1}^{q} \frac{S_{\mathrm{c}i}(t) - \sigma_{\mathrm{c}ii}(E(t))}{c + E_i(t)}, \quad E(t) \notin \mathcal{Z}, \quad t_k < t < t_{k+1}, \tag{10.73}$$

$$\Delta\mathcal{S}(E(t_k)) \geq \sum_{i=1}^{q} \frac{S_{\mathrm{c}i}(t_k) - \sigma_{\mathrm{c}ii}(E(t_k))}{c + E_i(t_k^+)}, \quad E(t_k) \in \mathcal{Z}, \quad k \in \mathbb{Z}_+. \tag{10.74}$$

Proof. Since, by Proposition 10.1, $E(t) \geq\geq 0$, $E(t) \notin \mathcal{Z}, t \in (t_k, t_{k+1}], k \in \mathbb{Z}_+$, and $\phi_{\mathrm{c}ij}(E) = -\phi_{\mathrm{c}ji}(E), i \neq j, i, j = 1, \ldots, q$, it follows that

$$\begin{aligned}
\dot{\mathcal{S}}(E(t)) &= \sum_{i=1}^{q} \frac{\dot{E}_i(t)}{c + E_i(t)} \\
&= \sum_{i=1}^{q} \left[\frac{S_{\mathrm{c}i}(t) - \sigma_{\mathrm{c}ii}(E(t))}{c + E_i(t)} + \sum_{j=1, j\neq i}^{q} \frac{\phi_{\mathrm{c}ij}(E(t))}{c + E_i(t)} \right] \\
&= \sum_{i=1}^{q} \left[\frac{S_{\mathrm{c}i}(t) - \sigma_{\mathrm{c}ii}(E(t))}{c + E_i(t)} + \sum_{j=i+1}^{q} \left(\frac{\phi_{\mathrm{c}ij}(E(t))}{c + E_i(t)} - \frac{\phi_{\mathrm{c}ij}(E(t))}{c + E_j(t)} \right) \right] \\
&= \sum_{i=1}^{q} \frac{S_{\mathrm{c}i}(t) - \sigma_{\mathrm{c}ii}(E(t))}{c + E_i(t)} + \sum_{i=1}^{q-1} \sum_{j=i+1}^{q} \frac{\phi_{\mathrm{c}ij}(E(t))[E_j(t) - E_i(t)]}{(c + E_i(t))(c + E_j(t))} \\
&\geq \sum_{i=1}^{q} \frac{S_{\mathrm{c}i}(t) - \sigma_{\mathrm{c}ii}(E(t))}{c + E_i(t)}, \quad E(t) \notin \mathcal{Z}, \quad t_k < t \leq t_{k+1}.
\end{aligned} \tag{10.75}$$

Furthermore, since $E(t_k) \geq\geq 0, E(t_k) \in \mathcal{Z}, k \in \mathbb{Z}_+$, and $\phi_{\mathrm{d}ij}(E) = -\phi_{\mathrm{d}ji}(E), i \neq j, i, j = 1, \ldots, q$, it follows that

$$\begin{aligned}
&\Delta\mathcal{S}(E(t_k)) \\
&\quad = \sum_{i=1}^{q} \log_e \left[1 + \frac{\Delta E_i(t_k)}{c + E_i(t_k)} \right] \\
&\quad \geq \sum_{i=1}^{q} \left[\frac{\Delta E_i(t_k)}{c + E_i(t_k)} \right] \left[1 + \frac{\Delta E_i(t_k)}{c + E_i(t_k)} \right]^{-1}
\end{aligned}$$

$$
\begin{aligned}
&= \sum_{i=1}^{q} \frac{\Delta E_i(t_k)}{c + E_i(t_k) + \Delta E_i(t_k)} \\
&= \sum_{i=1}^{q} \frac{\Delta E_i(t_k)}{c + E_i(t_k^+)} \\
&= \sum_{i=1}^{q} \left[\frac{S_{\mathrm{d}i}(t_k) - \sigma_{\mathrm{d}ii}(E(t_k))}{c + E_i(t_k^+)} + \sum_{i=1, j\neq i}^{q} \frac{\phi_{\mathrm{d}ij}(E(t_k))}{c + E_i(t_k^+)} \right] \\
&= \sum_{i=1}^{q} \frac{S_{\mathrm{d}i}(t_k) - \sigma_{\mathrm{d}ii}(E(t_k))}{c + E_i(t_k^+)} + \sum_{i=1}^{q-1} \sum_{j=i+1}^{q} \left(\frac{\phi_{\mathrm{d}ij}(E(t_k))}{c + E_i(t_k^+)} - \frac{\phi_{\mathrm{d}ij}(E(t_k))}{c + E_j(t_k^+)} \right) \\
&= \sum_{i=1}^{q} \frac{S_{\mathrm{d}i}(t_k) - \sigma_{\mathrm{d}ii}(E(t_k))}{c + E_i(t_k^+)} + \sum_{i=1}^{q-1} \sum_{j=i+1}^{q} \frac{\phi_{\mathrm{d}ij}(E(t_k))[E_j(t_k^+) - E_i(t_k^+)]}{(c + E_i(t_k^+))(c + E_j(t_k^+))} \\
&\geq \sum_{i=1}^{q} \frac{S_{\mathrm{d}i}(t_k) - \sigma_{\mathrm{d}ii}(E(t_k))}{c + E_i(t_k^+)},
\end{aligned}
\tag{10.76}
$$

where in (10.76) we use the fact that $\frac{x}{1+x} < \log_e(1+x) < x$, $x > -1$, $x \neq 0$. The result is now an immediate consequence of Theorem 10.6. □

10.6 Semistability and Energy Equipartition of Hybrid Thermodynamic Systems

Inequality (10.66) is a generalization of Clausius' inequality for equilibrium and nonequilibrium thermodynamics as well as reversible and irreversible thermodynamics as applied to hybrid large-scale thermodynamic systems. For the (adiabatically) isolated hybrid large-scale dynamical system $\mathcal{G}$, (10.66) yields the hybrid fundamental inequalities

$$
\mathcal{S}(E(\hat{t})) \geq \mathcal{S}(E(t)), \quad E(t) \notin \mathcal{Z}, \quad t_k < t \leq \hat{t} \leq t_{k+1}, \tag{10.77}
$$

$$
\mathcal{S}(E(t_k^+)) \geq \mathcal{S}(E(t_k)), \quad E(t_k) \in \mathcal{Z}, \quad k \in \mathbb{Z}_+. \tag{10.78}
$$

Inequalities (10.77) and (10.78) imply that, for any dynamical change in an isolated hybrid large-scale thermodynamic system, the entropy between resetting events as well as across resetting events is increasing.

The above observations imply that when an isolated hybrid large-scale dynamical system (i.e., $S_{\mathrm{c}}(t) \equiv 0$, $\sigma_{\mathrm{c}ii}(E(t)) \equiv 0$, $S_{\mathrm{d}}(t_k) \equiv 0$, and $\sigma_{\mathrm{d}ii}(E(t_k)) \equiv 0$) with thermodynamically consistent energy flow characteristics (i.e., Axioms *i*)–*iii*) hold) is at a state of maximum entropy consistent with its energy, it cannot be subject to any further dynamical change since any such change would result in a decrease of entropy. This of course implies that the state of *maximum entropy* is the stable state of an isolated hybrid thermodynamic system and this state has to be semistable.

The next theorem concretizes the above observations. First, however, we show that Assumptions A1, A2, and 10.1 hold for the resetting set $\mathcal{Z}$ given by (10.63). Note that for the hybrid large-scale dynamical system $\mathcal{G}$ given by (10.30) and (10.31), Assumption A1 states that if $E \in \overline{\mathcal{Z}}\backslash\mathcal{Z}$, then there exists $\varepsilon > 0$ such that for all $0 < \delta < \varepsilon$, $s(\delta, E) \notin \mathcal{Z}$. To see this, let $\mathcal{X}(E) = \phi_{\mathrm{c}ij}(E)$ and note that

$$\overline{\mathcal{Z}}\backslash\mathcal{Z} = \left\{ E \in \overline{\mathbb{R}}^q : \sum_{j\in\mathcal{O}} \mathcal{E}_j(E) \neq \sum_{i\in\mathcal{I}} \mathcal{E}_i(E) \text{ and } \phi_{\mathrm{c}ij}(E) = 0,\ i,j = 1,\ldots,q \right\}.$$

Now, if $E_0 \in \overline{\mathcal{Z}}\backslash\mathcal{Z}$, it follows from the transversality assumption that there exist $\delta > 0$ such that for all $t \in (0, \delta]$, $L_{f_{\mathrm{c}}}(\mathcal{X}(E(t))) \neq 0$. Hence, since $\mathcal{X}(E(t)) = \mathcal{X}(E(0)) + tL_{f_{\mathrm{c}}}\mathcal{X}(E(\tau))$ for some $\tau \in (0, t]$, it follows that $\mathcal{X}(E(t)) \neq 0$, $t \in (0, \delta]$, which implies that A1 holds.

To show that A2 holds, that is, if $E \in \mathcal{Z}$, then $E + f_{\mathrm{d}}(E) \notin \mathcal{Z}$, note that if $E \in \mathcal{Z}$, then $\mathcal{E}_i(E) = \mathcal{E}_j(E)$ and $\phi_{\mathrm{c}ij}(E) = 0$. Thus, for $m \in \{1, \ldots, q\}$, note that

$$\begin{aligned}
&\Delta\mathcal{E}(E(t_k)) \\
&\quad = \frac{1}{2}E^{\mathrm{T}}(t_k^+)E(t_k^+) - \frac{1}{2}E^{\mathrm{T}}(t_k)E(t_k) \\
&\quad = -E_m(t_k^+)\sigma_{\mathrm{d}mm}(E(t_k)) - \frac{1}{2}\left[\sum_{j=1, j\neq m}^{q} \phi_{\mathrm{d}mj}(E(t_k)) - \sigma_{\mathrm{d}mm}(E(t_k))\right]^2 \\
&\qquad - \frac{1}{2}\sum_{i=1, i\neq m}^{q}\left[\sum_{j=1, j\neq i}^{q} \phi_{\mathrm{d}ij}(E(t_k))\right]^2 + \sum_{i=1}^{q}\sum_{j=1, j\neq i}^{q} E_i(t_k^+)\phi_{\mathrm{d}ij}(E(t_k)) \\
&\quad = -E_m(t_k^+)\sigma_{\mathrm{d}mm}(E(t_k)) - \frac{1}{2}\left[\sum_{j=1, j\neq m}^{q} \phi_{\mathrm{d}mj}(E(t_k)) - \sigma_{\mathrm{d}mm}(E(t_k))\right]^2 \\
&\qquad - \frac{1}{2}\sum_{i=1, i\neq m}^{q}\left[\sum_{j=1, j\neq i}^{q} \phi_{\mathrm{d}ij}(E(t_k))\right]^2 \\
&\qquad + \sum_{i=1}^{q-1}\sum_{j=i+1}^{q} [E_i(t_k^+) - E_j(t_k^+)]\phi_{\mathrm{d}ij}(E(t)) \\
&\quad \leq 0, \quad E(t_k) \in \mathcal{Z}, \quad k \in \mathbb{Z}_+,
\end{aligned} \tag{10.79}$$

which shows that if $\phi_{\mathrm{d}ij}(E) \neq 0$, then $\Delta\mathcal{E}(E) < 0$, $E \in \mathcal{Z}$. Thus, $\sum_{j\in\mathcal{O}} \mathcal{E}_j(E(t)) \neq \sum_{i\in\mathcal{I}} \mathcal{E}_i(E(t))$ at $t = t_k^+$. Furthermore, since $\phi_{\mathrm{c}ij}(E(\cdot))$

is continuous, it follows that $\phi_{cij}(E(t_k^+)) = \phi_{cij}(E(t)) = 0$, which implies that $E(t_k^+) \in \overline{\mathcal{Z}}\backslash\mathcal{Z}$, and hence, Assumption A2 holds.

Finally, to show Assumption 10.1 holds, consider the set

$$\mathcal{M}_\gamma = \left\{ E \in \overline{\mathbb{R}}_+^q : \sum_{j\in\mathcal{O}} \mathcal{E}_j(E) - \sum_{i\in\mathcal{I}} \mathcal{E}_i(E) = \gamma \right\}, \tag{10.80}$$

where $\gamma \geq 0$. It follows from the transversality condition that for every $\gamma \geq 0$, $\mathcal{M}_\gamma$ does not contain any nontrivial trajectory of $\mathcal{G}$. To see this, suppose, *ad absurdum*, that there exists a nontrivial trajectory $E(t) \in \mathcal{M}_\gamma, t \geq 0$, for some $\gamma \geq 0$. In this case, it follows that

$$\begin{aligned} \frac{\mathrm{d}^k}{\mathrm{d}t^k}\left[\sum_{j\in\mathcal{O}} \mathcal{E}_j(E(t)) - \sum_{i\in\mathcal{I}} \mathcal{E}_i(E(t))\right] &= L_{f_c}^k\left[\sum_{j\in\mathcal{O}} \mathcal{E}_j(E(t)) - \sum_{i\in\mathcal{I}} \mathcal{E}_i(E(t))\right] \\ &= 0, \quad k = 1, 2, \ldots, \end{aligned}$$

which contradicts the transversality condition.

Next, we show that for every $E_0 \notin \mathcal{Z}, E_0 \neq 0$, there exists $\tau > 0$ such that $E(\tau) \in \mathcal{Z}$. To see this, suppose, *ad absurdum*, that $E(t) \notin \mathcal{Z}, t \geq 0$, which implies that

$$\sum_{j\in\mathcal{O}} \mathcal{E}_j(E(t)) - \sum_{i\in\mathcal{I}} \mathcal{E}_i(E(t)) \neq 0, \quad t \geq 0, \tag{10.81}$$

or

$$\phi_{cij}(E(t)) \neq 0, \quad t \geq 0. \tag{10.82}$$

If (10.81) holds, then it follows that the system energy will flow from a subsystem with higher energy to a subsystem with lower energy, which implies that $\sum_{j\in\mathcal{O}} \mathcal{E}_j(E(t)) - \sum_{i\in\mathcal{I}} \mathcal{E}_i(E(t)) \to 0$ as $t \to \infty$, and hence, leads to a contradiction. Alternatively, if (10.82) holds, then the system energy will also flow from a subsystem with higher energy to a subsystem with lower energy, which also implies that $\sum_{j\in\mathcal{O}} \mathcal{E}_j(E(t)) - \sum_{i\in\mathcal{I}} \mathcal{E}_i(E(t)) \to 0$ as $t \to \infty$, and hence, leads to a contradiction.

Thus, for every $E_0 \notin \mathcal{Z}$, there exists $\tau > 0$ such that $E(\tau) \in \mathcal{Z}$ and for every $E_0 \notin \mathcal{Z}, 0 < \tau_1(E_0) < \infty$. Now, it follows from Proposition 10.4 that $\tau_1(\cdot)$ is continuous at $E_0 \notin \overline{\mathcal{Z}}$. Furthermore, for all $E_0 \in \overline{\mathcal{Z}}\backslash\mathcal{Z}$ and for every sequence $\{E_{(i)}\}_{i=1}^\infty \in \overline{\mathcal{Z}}\backslash\mathcal{Z}$ converging to $E_0 \in \overline{\mathcal{Z}}\backslash\mathcal{Z}$, it follows from the transversality condition and Proposition 10.4 that $\lim_{i\to\infty} \tau_1(E_{(i)}) = \tau_1(E_0)$. Next, let $E_0 \in \overline{\mathcal{Z}}\backslash\mathcal{Z}$ and let $\{E_{(i)}\}_{i=1}^\infty \in \overline{\mathbb{R}}_+^q$ be such that $\lim_{i\to\infty} E_{(i)} = E_0$ and $\lim_{i\to\infty} \tau_1(E_{(i)})$ exists. In this case, it follows from Proposition 10.4 that either $\lim_{i\to\infty} \tau_1(E_{(i)}) = 0$ or $\lim_{i\to\infty} \tau_1(E_{(i)}) = \tau_1(E_0)$. Furthermore, if $E_0 \in \mathcal{Z}$ and $\phi_{dij}(E) = 0$, then the energy between subsystems does not flow. Thus, $\sum_{j\in\mathcal{O}}\mathcal{E}_j(E)-\sum_{i\in\mathcal{I}}\mathcal{E}_i(E)=0$, and hence, $f_{\rm d}(E_0)=0$. Now, it follows from Proposition 10.3 that Assumption 10.1 holds.

Theorem 10.8. Consider the hybrid large-scale dynamical system $\mathcal{G}$ given by (10.30) and (10.31) with $S_{\rm c}(t)\equiv 0$, $S_{\rm d}(t)\equiv 0$, $\sigma_{{\rm c}ii}(E(t))\equiv 0$, and $\sigma_{{\rm d}ii}(E(t))\equiv 0$, and $\mathcal{Z}$ given by (10.63), and assume that Axioms $i)$ – $iii)$ hold. Then, for every $\alpha\geq 0$, $\alpha\mathbf{e}$ is a semistable equilibrium state of (10.30) and (10.31). Furthermore, $E(t)\to\frac{1}{q}\mathbf{e}\mathbf{e}^{\rm T}E(t_0)$ as $t\to\infty$ and $\frac{1}{q}\mathbf{e}\mathbf{e}^{\rm T}E(t_0)$ is a semistable equilibrium state.

Proof. It follows from Axiom $i)$ that $\alpha\mathbf{e}\in\overline{\mathbb{R}}_+^q,\alpha\geq 0$, is an equilibrium state of (10.30) and (10.31). To show Lyapunov stability of the equilibrium state $\alpha\mathbf{e}$, consider the Lyapunov function candidate $\mathcal{E}(E)=\frac{1}{2}(E-\alpha\mathbf{e})^{\rm T}(E-\alpha\mathbf{e})$. Since $\phi_{{\rm c}ij}(E)=-\phi_{{\rm c}ji}(E), E\in\overline{\mathbb{R}}_+^q, i\neq j, i,j=1,\ldots,q$, and $\mathbf{e}^{\rm T}f_{\rm c}(E)=0$, where $f_{{\rm c}i}(E)=\sum_{j=1,j\neq i}^q[\sigma_{{\rm c}ij}(E)-\sigma_{{\rm c}ji}(E)]$, $i=1,\ldots,q$, it follows from Axiom $ii)$ that

$$\begin{aligned}\dot{\mathcal{E}}(E(t))&=(E(t)-\alpha\mathbf{e})^{\rm T}\dot{E}(t)\\&=(E(t)-\alpha\mathbf{e})^{\rm T}f_{\rm c}(E(t))\\&=E^{\rm T}(t)f_{\rm c}(E(t))\\&=\sum_{i=1}^q E_i(t)\left[\sum_{j=1,j\neq i}^q\phi_{{\rm c}ij}(E(t))\right]\\&=\sum_{i=1}^q\sum_{j=i+1}^q[E_i(t)-E_j(t)]\phi_{{\rm c}ij}(E(t))\\&\leq 0,\quad E(t)\notin\mathcal{Z}.\end{aligned}\tag{10.83}$$

Next, it follows from Axioms $ii)$ and $iii)$ that

$$\begin{aligned}\Delta\mathcal{E}(E(t_k))&=\frac{1}{2}(E(t_k^+)-\alpha\mathbf{e})^{\rm T}(E(t_k^+)-\alpha\mathbf{e})-\frac{1}{2}(E(t_k)-\alpha\mathbf{e})^{\rm T}(E(t_k)-\alpha\mathbf{e})\\&=\sum_{i=1}^q\sum_{j=1,j\neq i}^q E_i(t_k^+)\phi_{{\rm d}ij}(E(t_k))-\frac{1}{2}\sum_{i=1}^q\left[\sum_{j=1,j\neq i}^q\phi_{{\rm d}ij}(E(t_k))\right]^2\\&=\sum_{i=1}^{q-1}\sum_{j=i+1}^q[E_i(t_k^+)-E_j(t_k^+)]\phi_{{\rm d}ij}(E(t_k))\\&\quad-\frac{1}{2}\sum_{i=1}^q\left[\sum_{j=1,j\neq i}^q\phi_{{\rm d}ij}(E(t_k))\right]^2\\&\leq 0,\quad E(t_k)\in\mathcal{Z},\end{aligned}\tag{10.84}$$

which, by Theorem 10.1, establishes Lyapunov stability of the equilibrium state $\alpha\mathbf{e}$.

To show that $\alpha\mathbf{e}$ is semistable, note that

$$\begin{aligned}\dot{\mathcal{E}}(E(t)) &= \sum_{i=1}^{q}\sum_{j=i+1}^{q}[E_i(t)-E_j(t)]\phi_{\mathrm{c}ij}(E(t)) \\ &= \sum_{i=1}^{q}\sum_{j\in\mathcal{K}_i}[E_i(t)-E_j(t)]\phi_{\mathrm{c}ij}(E(t)), \quad E(t)\notin\mathcal{Z},\end{aligned} \tag{10.85}$$

and

$$\begin{aligned}\Delta\mathcal{E}(E(t_k)) &= \sum_{i=1}^{q}\sum_{j=1,j\neq i}^{q}E_i(t_k)\phi_{\mathrm{d}ij}(E(t_k)) + \frac{1}{2}\sum_{i=1}^{q}\left[\sum_{j=1,j\neq i}^{q}\phi_{\mathrm{d}ij}(E(t_k))\right]^2 \\ &\geq \sum_{i=1}^{q-1}\sum_{j=i+1}^{q}[E_i(t_k)-E_j(t_k)]\phi_{\mathrm{d}ij}(E(t_k)) \\ &= \sum_{i=1}^{q-1}\sum_{j\in\mathcal{K}_i}[E_i(t_k)-E_j(t_k)]\phi_{\mathrm{d}ij}(E(t_k)), \quad E(t_k)\in\mathcal{Z},\end{aligned} \tag{10.86}$$

where $\mathcal{K}_i = \mathcal{N}_i\backslash \cup_{l=1}^{i-1}\{l\}$ and $\mathcal{N}_i = \{j\in\{1,\ldots,q\} : \phi_{\mathrm{c}ij}(E) = 0$ and $\phi_{\mathrm{d}ij}(E)=0$ if and only if $E_i=E_j\}$. It follows from (10.85) that $\dot{\mathcal{E}}(E)=0$ if and only if $(E_i-E_j)\phi_{\mathrm{c}ij}(E)=0$, $E\notin\mathcal{Z}$, $i=1,\ldots,q$, $j\in\mathcal{K}_i$.

Next, we show that $\Delta\mathcal{E}(E)=0$ if and only if $(E_i-E_j)\phi_{\mathrm{d}ij}(E)=0$, $E\in\mathcal{Z}$, $i=1,2,\ldots,q$, $j\in\mathcal{K}_i$. Assume that $(E_i-E_j)\phi_{\mathrm{d}ij}(E)=0$, $i=1,\ldots,q$, $j\in\mathcal{K}_i$. Then, it follows from (10.86) that $\Delta\mathcal{E}\geq 0$, and hence, by (10.84), $\Delta\mathcal{E}=0$. In this case, it follows from (10.84) that $[E_i(t_k^+)-E_j(t_k^+)]\phi_{\mathrm{d}ij}(E(t_k))=0$ and $\sum_{j=1,j\neq i}^{q}\phi_{\mathrm{d}ij}(E(t_k))=0$, $E(t_k)\in\mathcal{Z}$, $i,j=1,\ldots,q$, $i\neq j$. Now,

$$\begin{aligned}&\left[E_i(t_k^+)-E_j(t_k^+)\right]\phi_{\mathrm{d}ij}(E(t_k)) \\ &\quad = \left[E_i(t_k)-E_j(t_k)\right]\phi_{\mathrm{d}ij}(E(t_k)) \\ &\qquad + \left[\sum_{h=1,h\neq i}^{q}\phi_{\mathrm{d}ih}(E(t_k)) - \sum_{l=1,l\neq j}^{q}\phi_{\mathrm{d}jl}(E(t_k))\right]\phi_{\mathrm{d}ij}(E(t_k)) \\ &\quad = [E_i(t_k)-E_j(t_k)]\phi_{\mathrm{d}ij}(E(t_k)),\ E(t_k)\in\mathcal{Z},\ \ i,j=1,\ldots,q,\ i\neq j,\end{aligned} \tag{10.87}$$

and hence, $(E_i-E_j)\phi_{\mathrm{d}ij}(E)=0$, $E\in\mathcal{Z}$, $i=1,\ldots,q$, $j\in\mathcal{K}_i$.

Finally, let

$$\begin{aligned}\mathcal{R} &= \{E \in \overline{\mathbb{R}}_+^q : E \notin \mathcal{Z}, \dot{\mathcal{E}}(E) = 0\} \cup \{E \in \overline{\mathbb{R}}_+^q : E \in \mathcal{Z}, \Delta\mathcal{E} = 0\} \\ &= \{E \in \mathbb{R}_+^q : E \notin \mathcal{Z}, (E_i - E_j)\phi_{\mathrm{c}ij}(E) = 0, i = 1, \ldots, q, j \in \mathcal{K}_i\} \\ &\quad \cup \{E \in \mathbb{R}_+^q : E \in \mathcal{Z}, (E_i - E_j)\phi_{\mathrm{d}ij}(E) = 0, i = 1, \ldots, q, j \in \mathcal{K}_i\},\end{aligned}$$

and note that Axiom i) implies that $\mathcal{R} = \{E \in \overline{\mathbb{R}}_+^q : E_1 = \cdots = E_q\}$. Since the set $\mathcal{R}$ consists of the equilibrium state of the system, it follows that the largest invariant set $\mathcal{M}$ contained in $\mathcal{R}$ is given by $\mathcal{M} = \mathcal{R}$. Hence, it follows from Theorem 10.3 that for every initial condition $E(t_0) \in \mathbb{R}_+^q$, $E(t) \to \mathcal{M}$ as $t \to \infty$, and hence, $\alpha\mathbf{e}$ is a semistable equilibrium state of the system. Next, note that since $\mathbf{e}^{\mathrm{T}}E(t) = \mathbf{e}^{\mathrm{T}}E(t_0)$ and $E(t) \to \mathcal{M}$ as $t \to \infty$, it follows that $E(t) \to \frac{1}{q}\mathbf{e}\mathbf{e}^{\mathrm{T}}E(t_0)$ as $t \to \infty$. Hence, with $\alpha = \frac{1}{q}\mathbf{e}^{\mathrm{T}}E(t_0), \alpha\mathbf{e} = \frac{1}{q}\mathbf{e}\mathbf{e}^{\mathrm{T}}E(t_0)$ is a semistable state of the hybrid dynamical system (10.30) and (10.31). □

Chapter Eleven

Continuum Thermodynamics

11.1 Conservation Laws in Continuum Thermodynamics

In this chapter, we extend the results of Chapter 3 to the case of continuum thermodynamic systems, where the subsystems are uniformly distributed over an n-dimensional space. Since these thermodynamic systems involve distributed subsystems, they are described by partial differential equations and hence are infinite-dimensional systems. Our formulation in this chapter involves a unification of the behavior of heat as described by the equations of thermal transfer and classical thermodynamics. With the notable exceptions of [38, 71, 73, 118], the amalgamation of these classical disciplines of physics is virtually nonexistent in the literature.[1]

To extend our dynamical systems framework to continuum thermodynamics, we develop a power balance partial differential equation, with appropriate boundary conditions, describing the motion of an energy density function $u = u(x,t)$ on a function space $\mathcal{X}$. In particular, the system dynamic is expressed by a differential equation involving a nonlinear differential operator on the function space $\mathcal{X}$. The state space $\mathcal{X}$ is assumed to be Hausdorff; for example, the space of continuously differentiable functions defined on a compact Riemannian manifold that vanish on the boundary or, alternatively, having normal derivatives that vanish at the boundary.

More generally, the state space $\mathcal{X}$ can be taken to be an open set in any ordered Banach space whose positive cone has a nonempty interior with a class of systems described by flows that preserve a partial ordering on $\mathcal{X}$. In addition, the function space need *not* be linear with properties such as completeness of a norm, reflexiveness, local convexity, etc. holding. Hence, the state space can be a space of real-valued functions with a natural ordering (i.e., a topological space) with a weak C^r topology. Note that in

[1]Continuum thermodynamics within the context of thermodynamics of thresholds, plasticity, and hysteresis using convex analysis, semigroup theory, and nonlinear programming has been addressed in [241, 309, 310].

this case, the cone of nonnegative functions makes $\mathcal{X}$ an ordered topological space endowed with Banach norms.

For our formulation, we consider infinite-dimensional dynamical systems $\mathcal{G}$ defined over a compact connected set $\mathcal{V} \subset \mathbb{R}^n$ with a smooth (at least C^1) boundary $\partial\mathcal{V}$ and volume $\mathcal{V}_{\mathrm{vol}}$. Furthermore, let $\mathcal{X}$ denote a space of two-times continuously differentiable scalar functions defined on $\mathcal{V}$, let $u(x,t)$, where $u : \mathcal{V} \times [0,\infty) \to \overline{\mathbb{R}}_+$, denote the energy density of the dynamical system $\mathcal{G}$ at the point $x \triangleq [x_1, \ldots, x_n]^{\mathrm{T}} \in \mathcal{V}$ and time instant $t \geq t_0$, let $\phi : \mathcal{V} \times \overline{\mathbb{R}}_+ \times \mathbb{R}^n \to \mathbb{R}^n$ denote the system energy flow within the continuum $\mathcal{V}$, that is, $\phi(x, u(x,t), \nabla u(x,t)) = [\phi_1(x, u(x,t), \nabla u(x,t)), \ldots, \phi_n(x, u(x,t), \nabla u(x,t))]^{\mathrm{T}}$, where $\phi_i(\cdot,\cdot,\cdot)$ denotes the energy flow through a unit area per unit time in the x_i direction for all $i = 1, \ldots, n$, and $\nabla u(x,t) \triangleq [D_1 u(x,t), \ldots, D_n u(x,t)]$, $x \in \mathcal{V}$, $t \geq t_0$, denotes the gradient of $u(\cdot,t)$ with respect to the spatial variable x, and let $s : \mathcal{V} \times [0,\infty) \to \overline{\mathbb{R}}_+$ denote the energy (heat) flow into a unit volume per unit time from sources uniformly distributed over $\mathcal{V}$.

To obtain the power balance equation for a uniformly distributed thermodynamic system, note that for any smooth, bounded region $\mathcal{V} \subset \mathbb{R}^n$, the integral $\int_{\mathcal{V}} u(x,t)\mathrm{d}\mathcal{V}$ denotes the total amount of energy within $\mathcal{V}$ at time t. Hence, the rate of change of energy within $\mathcal{V}$ is governed by the *flux* operator $\phi : \mathcal{V} \times \overline{\mathbb{R}}_+ \times \mathbb{R}^n \to \mathbb{R}^n$ and the external supplied power $s : \mathcal{V} \times [0,\infty) \to \overline{\mathbb{R}}_+$, which control the rate of loss and increase of the total energy through the boundary $\partial\mathcal{V}$ and the interior $\overset{\circ}{\mathcal{V}}$ of $\mathcal{V}$, respectively. Hence, for each time t,

$$\frac{\mathrm{d}}{\mathrm{d}t}\int_{\mathcal{V}} u(x,t)\mathrm{d}\mathcal{V} = -\int_{\partial\mathcal{V}} \phi(x, u(x,t), \nabla u(x,t)) \cdot \hat{n}(x)\mathrm{d}\mathcal{S}_{\mathcal{V}} + \int_{\mathcal{V}} s(x,t)\mathrm{d}\mathcal{V}, \tag{11.1}$$

where $\hat{n}(x)$ denotes the outward normal vector to the boundary $\partial\mathcal{V}$ (at x) of the set $\mathcal{V}$, $\mathrm{d}\mathcal{S}_{\mathcal{V}}$ denotes an infinitesimal surface element of the boundary of the set $\mathcal{V}$, and "$\cdot$" denotes the dot product in $\mathbb{R}^n$.

Using the divergence theorem, it follows from (11.1) that

$$\begin{aligned}\frac{\mathrm{d}}{\mathrm{d}t}\int_{\mathcal{V}} u(x,t)\mathrm{d}\mathcal{V} &= -\int_{\partial\mathcal{V}} \phi(x, u(x,t), \nabla u(x,t)) \cdot \hat{n}(x)\mathrm{d}\mathcal{S}_{\mathcal{V}} \\ &\quad + \int_{\mathcal{V}} s(x,t)\mathrm{d}\mathcal{V} \\ &= -\int_{\mathcal{V}} \nabla \cdot \phi(x, u(x,t), \nabla u(x,t))\mathrm{d}\mathcal{V} + \int_{\mathcal{V}} s(x,t)\mathrm{d}\mathcal{V},\end{aligned} \tag{11.2}$$

where ∇ denotes the nabla operator. Since the region $\mathcal{V} \subset \mathbb{R}^n$ is arbitrary, it follows that the power balance equation over a unit volume within the continuum $\mathcal{V}$ involving the rate of energy density change, the external supplied power (heat flux), and the energy (heat) flow within the continuum is given by

$$\frac{\partial u(x,t)}{\partial t} = -\nabla \cdot \phi(x, u(x,t), \nabla u(x,t)) + s(x,t), \quad x \in \mathcal{V}, \quad t \geq t_0, \tag{11.3}$$

$$u(x,t_0) = u_{t_0}(x), \quad x \in \mathcal{V}, \quad \phi(x, u(x,t), \nabla u(x,t)) \cdot \hat{n}(x) \geq 0, \quad x \in \partial\mathcal{V}, \quad t \geq t_0, \tag{11.4}$$

where $u_{t_0} \in \mathcal{X}$ is a given initial energy density distribution.

The power balance diffusion[2] or conservation equation (11.3) (depending on the structure chosen for the flux operator $\phi(x, u(x,t), \nabla u(x,t)))$ describes the time evolution of the energy density $u(x,t)$ over the region $\mathcal{V}$, while the boundary condition in (11.4) involving the dot product implies that the energy of the system $\mathcal{G}$ can be either stored or dissipated but not supplied through the boundary of $\mathcal{V}$. Here, for simplicity of exposition, we assume that there is no work done by the system on the environment nor is there work done by the environment on the system.

This extension can be easily handled by modifying the natural boundary condition for (11.3). In particular, this case would require that the system (11.3) and (11.4) is such that at every instant of time the domain $\mathcal{V}$ and its boundary $\partial\mathcal{V}$ are defined as $\mathcal{V} = \{x \in \mathbb{R}^n : g(x,t) \leq 0, t \geq t_0\}$ and $\partial\mathcal{V} = \{x \in \mathbb{R}^n : g(x,t) = 0, t \geq t_0\}$, where $g : \mathbb{R}^n \times [0,\infty) \to \mathbb{R}$ is a given continuously differentiable function, and consequently, the outward normal vector to the boundary $\partial\mathcal{V}$ at $x \in \partial\mathcal{V}$ and time $t \geq t_0$ is given by $\hat{n}^{\mathrm{T}}(x,t) = \nabla g(x,t)$.

We denote the energy density distribution over the set $\mathcal{V}$ at time $t \geq t_0$ by $u_t \in \mathcal{X}$ so that for each $t \geq t_0$ the set of mappings generated by $u_t(x) \equiv u(x,t)$ for every $x \in \mathcal{V}$ gives the *flow* of $\mathcal{G}$. We assume that the function $\phi(\cdot,\cdot,\cdot)$ is continuously differentiable so that (11.3) and (11.4) admits a unique solution $u(x,t)$, $x \in \mathcal{V}$, $t \geq t_0$, and $u(\cdot,t) \in \mathcal{X}$, $t \geq t_0$, is continuously dependent on the initial energy density distribution $u_{t_0}(x)$, $x \in$

[2]It is important to note here that, as we see in Chapter 14, adopting a diffusion-type power balance equation leads to a contradiction between thermodynamics and relativity. Specifically, for a Fourier heat diffusion model there does not exist a time delay between heat extraction and heat supply as well as heat flow and temperature difference in an isotropic continuum. However, given that in relativity theory the speed of light is the ultimate speed for all possible velocities between material systems, an instantaneous speed of heat propagation in thermodynamic heat conduction between bodies in direct thermal contact renders parabolic heat diffusion equations untenable for relativistic thermodynamic models. Alternatively, hyperbolic conservation law models employing Cattaneo-Vernotte type modifications of the heat flux function can lead to a finite subluminal speed of thermal propagation in an isotropic medium. For further details see Chapter 15.

$\mathcal{V}$. Specifically, using standard maximum principles and under certain hypotheses [326], parabolic equations of the form (11.3) with *Dirichlet* or *Neumann* boundary conditions can generate *strongly monotone* flows in function spaces $C^1(\mathcal{V})$ and $C^0(\mathcal{V})$, where $C^r(\mathcal{V})$ denotes a function space defined on $\mathcal{V}$ with r-continuous derivatives.

Recall that given a compact n-dimensional submanifold $\mathcal{V} \subset \mathbb{R}^n$ with smooth boundary $\partial\mathcal{V}$, a Dirichlet boundary condition has the form $u_t(x) = 0$ on $\partial\mathcal{V}$, whereas a Neumann boundary condition has the form $\nabla u_t(x) \cdot v = 0$, where ∇u_t denotes the gradient of $u(x,t)$ in x and $v : \partial\mathcal{V} \to \mathbb{R}^n$ is a smooth vector field transverse to $\partial\mathcal{V}$. Furthermore, a map $u_t : \mathcal{V} \to \mathbb{R}$ between ordered spaces is called *monotone* if $x \leq\leq y$ implies $u_t(x) \leq u_t(y)$. The map u_t is *strongly monotone* if $x << y$ implies $u_t(x) < u_t(y)$. Moreover, recall that an *ordered space* is a topological space $\mathcal{X}$ together with a *partial order relation* $\mathcal{R} \subset \mathcal{X} \times \mathcal{X}$ (i.e., $\mathcal{R}$ is reflexive, antisymmetric, and transitive), which is a closed subspace. The ordered space $\mathcal{X}$ is *strongly ordered* if $\mathcal{X}$ is asymmetric instead of reflexive and antisymmetric, and has the property that every nonempty and compact subset of $\mathcal{X}$ has both a greatest lower bound and a least upper bound in $\mathcal{X}$. In this case, the interior of $\mathcal{R}$ is dense in $\mathcal{R}$.

The function spaces $C^0(\mathcal{V})$ and $C^1(\mathcal{V})$ are topological spaces when given the weak $C^r(\mathcal{V})$ topology and possess Banach norms. In addition, the cone of nonnegative functions makes them ordered topological vector spaces. Finally, recall that $C^r(\mathcal{V})$ is strongly ordered, whereas $C_0^0(\mathcal{V}) \triangleq \{u \in C^0(\mathcal{V}) : u_t(x) = 0 \text{ on } \partial\mathcal{V}\}$ is not strongly ordered if $\partial\mathcal{V}$ is nonempty and $C_0^r(\mathcal{V}) \triangleq \{u \in C^r(\mathcal{V}) : u_t(x) = 0 \text{ on } \partial\mathcal{V}\}$, where $r > 0$, is strongly ordered.

In [216, 326], existence, uniqueness, and regularity of solutions for a semilinear parabolic partial differential equation are established over the semi-infinite time interval $[0, \infty)$ on $C^r(\mathcal{V})$ function spaces, and the well-posedness of the problem is addressed. Furthermore, a semidynamical systems framework is constructed over a positively invariant complete normed space—the phase space for the evolution of solutions—and is shown to possess bounded absorbing sets in forward time. Asymptotic compactness of the semigroup of solution operators is then ensured to guarantee the existence of an attractor (i.e., a compact nonempty set that attracts some neighborhood of itself and contains the orbits of all its points). This follows from the well-known smoothing properties of the solution flow in the function space $C^r(\mathcal{V})$, which implies that every orbit that is bounded in the $\mathcal{L}_2$ norm has a compact closure in $C^r(\mathcal{V})$.

The fact that a continuum thermodynamic system described by

a parabolic partial differential equation guarantees the existence of an attractor further implies that the infinite-dimensional power balance equation (11.3) defined in the function space $C^r(\mathcal{V})$ contains a smooth *finite-dimensional* subsystem that contains the orbits of all its points; that is, it contains a finite-dimensional inertial manifold that is invariant under the flow. Moreover, every attractor for the strongly monotone flow determined by the solutions of a semilinear parabolic problem is shown in [216] to contain a Lyapunov stable equilibrium, and hence, contains all the information of the asymptotic behavior of the semilinear parabolic dynamical system. Since an attractor containing a Lyapunov stable equilibrium cannot contain a dense orbit (unless the orbit is a single equilibrium), and any cycles in the attractor cannot be dense, it follows that monotone flows do not possess chaotic dynamics.

Alternatively, it is well known that hyperbolic nonlinear partial differential equations arising in models of conservation laws with quasi-linear first-order flux operators in divergence form need not have smooth differentiable solutions (*classical solutions*), and one has to use the notion of generalized functions (i.e., Schwartz distributions) that provides a framework in which the energy density function $u(x,t)$ may be differentiated in a generalized sense infinitely often [138]. In this case, one has a well-defined notion of solutions that have jump discontinuities, which propagate as shock waves. Thus, one has to deal with *generalized* or *weak* solutions wherein uniqueness is lost. In this case, the *Clausius-Duhem inequality* is invoked for identifying the physically relevant (i.e., thermodynamically admissible) solution.

More specifically, if the flux operator $\phi(x,u(x,t),\nabla u(x,t))$ is nonlinear, leading to nonconstant characteristic speeds (i.e., eigenvalues) of the Jacobian matrix of the flux operator, and u_{t_0} is a two-times continuously differentiable function with compact support and its derivative is sufficiently small on $[t_0,\infty)$, then the classical solution to (11.3) and (11.4) breaks down at a finite time. As a consequence of this, one may only hope to find generalized (or weak) solutions to (11.3) and (11.4) over the semi-infinite interval $[t_0,\infty)$, that is, $\mathcal{L}_\infty$ functions $u(\cdot,\cdot)$ that satisfy (11.3) in the sense of distributions, which provides a framework in which $u(\cdot,\cdot)$ may be differentiated in a general sense infinitely often.

Here, $\mathcal{L}_\infty$ denotes the space of bounded Lebesgue measurable functions on $\mathcal{V}$ and provides the broadest framework for weak solutions. Alternatively, a natural function class for weak solutions is the space $\mathcal{BV}$ consisting of functions of bounded variation. Recall that a bounded measurable function $u(x,t)$ has locally bounded variation if its partial distributional (i.e., Schwartzian) derivatives are locally finite Radon measures.[3]

In this case, the domain of $u(x,t)$ can be partitioned into three pairwise subsets involving a regular set $\mathcal{C}$ supporting an approximate Lebesgue limit of $u(x,t)$ for every point $(x,t) \in \mathcal{C}$, a jump set $\mathcal{J}$ comprising a countable union of $\mathrm{C}^1(\mathcal{V})$ arcs, and a residual set $\mathcal{R}$ with a vanishing one-dimensional Hausdorff measure. Moreover, when $u(x,t)$ is a $\mathcal{BV}$ solution of (11.3) and (11.4), the arcs of $\mathcal{J}$ are interpreted as propagating shock waves, and $u(x,t)$ satisfies (11.3) in the sense of Schwartz distributions with the *Rankine-Hugoniot condition* [138] holding at every point on the jump set $\mathcal{J}$. For further details see [105, 138].

Next, we establish the uniqueness of the internal energy functional $U(u_t)$, $u_t \in \mathcal{X}$, for the dynamical system $\mathcal{G}$ defined by

$$U(u_{t_0}) \triangleq \int_{\mathcal{V}} u(x,t_0)\mathrm{d}\mathcal{V}, \quad u_{t_0} \in \mathcal{X}. \tag{11.5}$$

First, however, the following result on local controllability of our continuum thermodynamic model is required. For this result, let $\mathcal{L}_p = \mathcal{L}_p(\mathcal{V})$ denote a Lebesgue space, that is,

$$\mathcal{L}_p = \{u_t : \mathcal{V} \to \mathbb{R} : u_t \text{ is Lebesgue measurable}^4 \text{ and } \|u_t\|_{\mathcal{L}_p} < \infty\},$$

where

$$\|u_t\|_{\mathcal{L}_p} \triangleq \left[\int_{\mathcal{V}} |u_t|^p \mathrm{d}\mathcal{V}\right]^{1/p}, \quad 1 \le p < \infty, \tag{11.6}$$

and if $p = \infty$,

$$\|u_t\|_{\mathcal{L}_\infty} \triangleq \operatorname*{ess\,sup}_{\mathcal{V}} |u_t|, \tag{11.7}$$

where "ess" denotes essential.

Lemma 11.1. Consider the dynamical system $\mathcal{G}$ with power balance equation (11.3) and (11.4). Then, for every equilibrium state $u_{\mathrm{e}}(\cdot) \in \mathcal{X}$ and every $\varepsilon > 0$ and $T > 0$, there exist $s_{\mathrm{e}} : \mathcal{V} \to \mathbb{R}$, $\alpha > 0$, and $\hat{T} \in [0,T]$ such

[3] A *Radon measure* is a measure defined on a σ-algebra of Borel sets of a Hausdorff topological space such that, for every point in the space, there exist neighborhoods of finite measures endowed with the property that the measure of every set equals the supremum of the inner measure of all its compact subsets.

[4] A function $u_t : \mathcal{V} \to \mathcal{Q}$, where $\mathcal{V}$ and $\mathcal{Q}$ are sets equipped with respective σ-algebras $\mathcal{K}$ and $\mathcal{L}$, is *measurable* if the preimage of $\mathcal{I}$ under u_t is in $\mathcal{K}$ for every $\mathcal{I} \in \mathcal{L}$; that is, $u_t^{-1}(\mathcal{I}) = \{x \in \mathcal{V} : f(x) \in \mathcal{I}\} \in \mathcal{K}$ for every $\mathcal{I} \in \mathcal{L}$. Equivalently, measurable functions are structure-preserving functions between the measurable spaces $(\mathcal{V}, \mathcal{K})$ and $(\mathcal{Q}, \mathcal{L})$. In the case where the domain and codomain represent different σ-algebras on the same underlying set (i.e., $\mathcal{V} = \mathcal{Q}$), then u_t is *Lebesgue measurable*. In this case, the function $u_t : \mathcal{V} \to \mathbb{R}$ is the pointwise limit, except for a set of Lebesgue measure zero, of a sequence of piecewise constant functions on $\mathcal{V}$. Recall that measurable functions are closed under addition and multiplication, but not composition.

that for every $\hat{u}(\cdot) \in \mathcal{X}$ with $\|\hat{u} - u_{\rm e}\|_{\mathcal{L}_p} \le \alpha T$, there exists $s : \mathcal{V} \times [0, \hat{T}] \to \mathbb{R}$ such that $\|s(\cdot, t) - s_{\rm e}(\cdot)\|_{\mathcal{L}_p} \le \varepsilon$, $t \in [0, \hat{T}]$, and $u(x,t) = u_{\rm e}(x) + \frac{\hat{u}(x) - u_{\rm e}(x)}{\hat{T}} t$, $x \in \mathcal{V}$, $t \in [0, \hat{T}]$.

Proof. Note that with $s_{\rm e}(x) = \nabla \cdot \phi(x, u_{\rm e}(x), \nabla u_{\rm e}(x))$, $x \in \mathcal{V}$, the state $u_{\rm e}(\cdot) \in \mathcal{X}$ is an equilibrium state of (11.3) and (11.4). Let $\theta > 0$ and $T > 0$, and define

$$M(\theta, T) \triangleq \sup_{u(\cdot) \in \overline{\mathcal{B}}_1(0),\, t \in [0,T]} \| -\nabla \cdot \phi(\cdot, u_{\rm e}(\cdot) + \theta t u(\cdot), \nabla u_{\rm e}(\cdot) + \theta t \nabla u(\cdot)) + s_{\rm e}(\cdot)\|_{\mathcal{L}_p}, \tag{11.8}$$

where $\overline{\mathcal{B}}_1(0)$ denotes the closed unit ball in $\mathcal{L}_p$. Note that for every $T > 0$, $\lim_{\theta \to 0^+} M(\theta, T) = 0$, and for every $\theta > 0$, $\lim_{T \to 0^+} M(\theta, T) = 0$.

Next, let $\varepsilon > 0$ and $T > 0$ be given, and let $\alpha > 0$ be such that $M(\alpha, T) + \alpha \le \varepsilon$. (The existence of such an α is guaranteed since $M(\alpha, T) \to 0$ as $\alpha \to 0^+$.) Now, let $\hat{u}(\cdot) \in \mathcal{X}$ be such that $\|\hat{u} - u_{\rm e}\|_{\mathcal{L}_p} \le \alpha T$. With $\hat{T} \triangleq \frac{\|\hat{u} - u_{\rm e}\|_{\mathcal{L}_p}}{\alpha} \le T$ and

$$s(x,t) = \nabla \cdot \phi(x, u(x,t), \nabla u(x,t)) + \alpha \frac{\hat{u}(x) - u_{\rm e}(x)}{\|\hat{u} - u_{\rm e}\|_{\mathcal{L}_p}}, \quad x \in \mathcal{V}, \quad t \in [0, \hat{T}], \tag{11.9}$$

it follows that

$$u(x,t) = u_{\rm e}(x) + \frac{\hat{u}(x) - u_{\rm e}(x)}{\|\hat{u} - u_{\rm e}\|_{\mathcal{L}_p}} \alpha t, \quad x \in \mathcal{V}, \quad t \in [0, \hat{T}], \tag{11.10}$$

is a solution to (11.3) and (11.4).

The result is now immediate by noting that $u(x, \hat{T}) = \hat{u}(x)$, $x \in \mathcal{V}$, and

$$\begin{aligned} \|s(\cdot, t) - s_{\rm e}(\cdot)\|_{\mathcal{L}_p} &\le \| - \nabla \cdot \phi(\cdot, u(\cdot, t), \nabla u(\cdot, t)) + s_{\rm e}(\cdot)\|_{\mathcal{L}_p} + \alpha \\ &\le M(\alpha, T) + \alpha \\ &\le \varepsilon, \qquad t \in [0, \hat{T}], \end{aligned} \tag{11.11}$$

which proves the result. □

It follows from Lemma 11.1 that the dynamical system $\mathcal{G}$ given by (11.3) and (11.4) is controllable to the zero energy density distribution, that is, for every $u_{t_0} \in \mathcal{X}$ there exists a finite time $t_{\rm f} \ge t_0$ and $s(\cdot, \cdot) \in \mathcal{U}$ defined on $x \in \mathcal{V}$ and $t \in [t_0, t_{\rm f}]$ such that the energy density distribution $u(x,t)$ can be driven from $u(x, t_0) = u_{t_0}(x)$ to $u(x, t_{\rm f}) = 0$, $x \in \mathcal{V}$. Here, $\mathcal{U}$ denotes the set of all energy inputs $s(\cdot, \cdot)$ to the system $\mathcal{G}$ that are uniformly bounded in x and continuous in x and t. In addition, it follows from Lemma 11.1 that

(11.3) and (11.4) is reachable from the zero energy density distribution, that is, for every $u_{t_0} \in \mathcal{X}$ there exists a finite time $t_{\rm i} \le t_0$ and $s(\cdot,\cdot) \in \mathcal{U}$ defined on $x \in \mathcal{V}$ and $t \in [t_{\rm i}, t_0]$ such that the energy density distribution $u(x,t)$ can be driven from $u(x,t_{\rm i}) = 0$ to $u(x,t_0) = u_{t_0}(x)$, $x \in \mathcal{V}$.

Next, let $\mathcal{U}_{\rm c} \subset \mathcal{U}$ denote the set of all energy inputs to the system $\mathcal{G}$ such that for every $T \ge t_0$ the system energy distribution can be driven from $u(x,t_0) = u_{t_0}(x)$, $u_{t_0} \in \mathcal{X}$, to $u(x,T) = 0$, $x \in \mathcal{V}$, by $s(\cdot,\cdot) \in \mathcal{U}_{\rm c}$, and we let $\mathcal{U}_{\rm r} \subset \mathcal{U}$ denote the set of all energy inputs to the system $\mathcal{G}$ such that for every $T \ge -t_0$ the system energy distribution can be driven from $u(x,-T) = 0$ to $u(x,t_0) = u_{t_0}(x)$, $x \in \mathcal{V}$, $u_{t_0} \in \mathcal{X}$, by $s(\cdot,\cdot) \in \mathcal{U}_{\rm r}$.

For the statement of the next result, define the available energy of the dynamical system $\mathcal{G}$ by

$$\begin{aligned} U_{\rm a}(u_{t_0}) \triangleq {} & - \inf_{s(\cdot,\cdot)\in\mathcal{U},\, T\ge t_0} \int_{t_0}^{T} \left[\int_{\mathcal{V}} s(x,t)\mathrm{d}\mathcal{V} \right. \\ & \left. - \int_{\partial\mathcal{V}} \phi(x,u(x,t),\nabla u(x,t)) \cdot \hat{n}(x)\mathrm{d}\mathcal{S}_{\mathcal{V}} \right] \mathrm{d}t, \quad u_{t_0} \in \mathcal{X}, \end{aligned} \tag{11.12}$$

and the required energy supply of the dynamical system $\mathcal{G}$ by

$$\begin{aligned} U_{\rm r}(u_{t_0}) \triangleq {} & \inf_{s(\cdot,\cdot)\in\mathcal{U}_{\rm r},\, T\ge -t_0} \int_{-T}^{t_0} \left[\int_{\mathcal{V}} s(x,t)\mathrm{d}\mathcal{V} \right. \\ & \left. - \int_{\partial\mathcal{V}} \phi(x,u(x,t),\nabla u(x,t)) \cdot \hat{n}(x)\mathrm{d}\mathcal{S}_{\mathcal{V}} \right] \mathrm{d}t, \quad u_{t_0} \in \mathcal{X}. \end{aligned} \tag{11.13}$$

Note that the available energy $U_{\rm a}(u_{t_0})$ is the maximum amount of stored energy (net heat) that can be extracted from the dynamical system $\mathcal{G}$ at any time T, and the required energy supply $U_{\rm r}(u_{t_0})$ is the minimum amount of energy (net heat) that can be delivered to the dynamical system $\mathcal{G}$ to transfer it from a state of minimum energy density distribution $u_{-T} = u(x,-T) = 0$, $x \in \mathcal{V}$, to a state of given energy density distribution $u_{t_0} \in \mathcal{X}$.

Theorem 11.1. Consider the dynamical system $\mathcal{G}$ with power balance equation (11.3) and (11.4). Then $\mathcal{G}$ is lossless with respect to the energy supply rate $\int_{\mathcal{V}} s(x,t)\mathrm{d}\mathcal{V} - \int_{\mathcal{V}} y(x,t)\mathrm{d}\mathcal{V}$, where $y(x,t) \equiv \nabla \cdot \phi(x,u(x,t), \nabla u(x,t))$, and with the unique energy storage functional corresponding to the total energy of the dynamical system $\mathcal{G}$ given by

$$U(u_{t_0}) = \int_{\mathcal{V}} u(x,t_0)\mathrm{d}\mathcal{V}$$

$$= -\int_{t_0}^{T_+} \left[\int_{\mathcal{V}} s(x,t) \mathrm{d}\mathcal{V} - \int_{\mathcal{V}} y(x,t) \mathrm{d}\mathcal{V} \right] \mathrm{d}t$$

$$= \int_{-T_-}^{t_0} \left[\int_{\mathcal{V}} s(x,t) \mathrm{d}\mathcal{V} - \int_{\mathcal{V}} y(x,t) \mathrm{d}\mathcal{V} \right] \mathrm{d}t, \quad u_{t_0} \in \mathcal{X}, \tag{11.14}$$

where $u(x,t)$, $x \in \mathcal{V}$, $t \geq t_0$, is the solution to (11.3) and (11.4) with admissible input $s(\cdot,\cdot) \in \mathcal{U}$, $u_{-T_-} = 0$, $u_{T_+} = 0$, and $u_{t_0} \in \mathcal{X}$. Furthermore,

$$0 \leq U_{\mathrm{a}}(u_{t_0}) = U(u_{t_0}) = U_{\mathrm{r}}(u_{t_0}) < \infty, \quad u_{t_0} \in \mathcal{X}. \tag{11.15}$$

Proof. First, it follows from Lemma 11.1 that $\mathcal{G}$ is reachable from and controllable to the origin in $\mathcal{X}$. Next, it follows from (11.2) that

$$U(u_t) - U(u_{t_0}) = \int_{t_0}^{t} \left[\int_{\mathcal{V}} s(x,t) \mathrm{d}\mathcal{V} - \int_{\mathcal{V}} \nabla \cdot \phi(x, u(x,t), \nabla u(x,t)) \mathrm{d}\mathcal{V} \right] \mathrm{d}t, \tag{11.16}$$

which shows that the dynamical system $\mathcal{G}$ is lossless with respect to the energy supply rate $\int_{\mathcal{V}} s(x,t)\mathrm{d}\mathcal{V} - \int_{\mathcal{V}} y(x,t)\mathrm{d}\mathcal{V}$ and with the energy storage functional $U(u_{t_0}) = \int_{\mathcal{V}} u(x,t_0)\mathrm{d}\mathcal{V}$, $u_{t_0} \in \mathcal{X}$. The remainder of the proof now follows identically as in the proof of Theorem 3.1 by noting that

$$\int_{\mathcal{V}} \nabla \cdot \phi(x, u(x,t), \nabla u(x,t)) \mathrm{d}\mathcal{V} = \int_{\partial\mathcal{V}} \phi(x, u(x,t), \nabla u(x,t)) \cdot \hat{n}(x) \mathrm{d}\mathcal{S}_{\mathcal{V}}, \tag{11.17}$$

which in turn follows from the divergence theorem. □

It follows from (11.16) that the dynamical system $\mathcal{G}$ is lossless with respect to the net heat supply rate $\int_{\mathcal{V}} s(x,t)\mathrm{d}\mathcal{V} - \int_{\mathcal{V}} y(x,t)\mathrm{d}\mathcal{V}$ and with the unique energy storage functional $U(u_{t_0})$ given by (11.5). This is in essence a statement of the first law of thermodynamics for isochoric transformations of infinite-dimensional systems. As in Chapter 3, to ensure a thermodynamically consistent energy flow infinite-dimensional model, we require the following axioms, which are analogous to Axioms i) and ii).

Axiom *i*)′: For every $x \in \mathcal{V}$ and unit vector $\mathbf{u} \in \mathbb{R}^n$, $\phi(x, u_t(x), \nabla u_t(x)) \cdot \mathbf{u} = 0$ if and only if $\nabla u_t(x)\mathbf{u} = 0$.

Axiom *ii*)′: For every $x \in \mathcal{V}$ and unit vector $\mathbf{u} \in \mathbb{R}^n$, $\phi(x, u_t(x), \nabla u_t(x)) \cdot \mathbf{u} > 0$ if and only if $\nabla u_t(x)\mathbf{u} < 0$, and $\phi(x, u_t(x), \nabla u_t(x)) \cdot \mathbf{u} < 0$ if and only if $\nabla u_t(x)\mathbf{u} > 0$.

Note that Axiom i)′ implies that $\phi_i(x, u_t(x), \nabla u_t(x)) = 0$ if and only if $D_i u_t(x) = 0$, $x \in \mathcal{V}$, $i = 1, \ldots, n$, while Axiom $ii)'$ implies that $\phi_i(x, u_t(x), \nabla u_t(x)) D_i u_t(x) \le 0$, $x \in \mathcal{V}$, $i = 1, \ldots, n$, which further implies that $\nabla u_t(x)\ \phi(x, u_t(x), \nabla u_t(x)) \le 0$, $x \in \mathcal{V}$, that is, energy (heat) flows from regions of higher to lower energy densities. If $s(x,t) \equiv 0$, then Axioms $i)'$ and $ii)'$ along with the fact that $\phi(x, u(x,t), \nabla u(x,t)) \cdot \hat{n}(x) \ge 0$, $x \in \partial\mathcal{V}$, $t \ge t_0$, imply that at a given instant of time the energy of the dynamical system $\mathcal{G}$ can only be transported, stored, or dissipated but not created. We assume that if $u(\hat{x}, \hat{t}) = 0$ for some $\hat{x} \in \partial\mathcal{V}$ and $\hat{t} \ge t_0$, then $\phi(\hat{x}, u(\hat{x}, \hat{t}), \nabla u(\hat{x}, \hat{t})) = 0$, which, along with the boundary condition (11.4), implies that energy dissipation is not possible on the boundary of $\mathcal{V}$ through points with zero energy density.

With this assumption and Axiom $ii)'$ it follows that the solution $u(x,t)$, $x \in \mathcal{V}$, $t \ge t_0$, to (11.3) and (11.4) is nonnegative for all nonnegative initial energy density distributions $u_{t_0}(x) \ge 0$, $x \in \mathcal{V}$. To see this, note that if $u(\hat{x}, \hat{t}) = 0$ for some $\hat{x} \in \overset{\circ}{\mathcal{V}}$ and $\hat{t} \ge t_0$, then it follows from Axiom $ii)'$ that the energy flow $\phi(y, u(y, \hat{t}), \nabla u(y, \hat{t}))$ is directed toward the point $\hat{x}$ for all points y in a sufficiently small neighborhood of $\hat{x}$. This property, along with the fact that $s : \mathcal{V} \times [0, \infty) \to \overline{\mathbb{R}}_+$ is a nonnegative function, implies that $\frac{\partial u(\hat{x}, \hat{t})}{\partial t} \ge 0$.

Alternatively, if $u(\hat{x}, \hat{t}) = 0$ for some $\hat{x} \in \partial\mathcal{V}$ and $\hat{t} \ge t_0$, then it follows from the above assumption that dissipation of energy through the point $\hat{x} \in \mathcal{V}$ is not possible, which, along with Axiom $ii)'$ and the fact that $s : \mathcal{V} \times [0, \infty) \to \overline{\mathbb{R}}_+$ is a nonnegative function, imply that $\frac{\partial u(\hat{x}, \hat{t})}{\partial t} \ge 0$ for $\hat{x} \in \partial\mathcal{V}$ and $\hat{t} \ge t_0$. Thus, the solution to (11.3) and (11.4) is nonnegative for all nonnegative initial energy density distributions.

The following proposition shows that the solution $u(x,t), x \in \mathcal{V}, t \ge t_0$, to (11.3) and (11.4) is nonnegative for all nonnegative initial information density distributions $u_{t_0}(x) \ge 0$, $x \in \mathcal{V}$.

Proposition 11.1. Consider the dynamical system $\mathcal{G}$ given by (11.3) and (11.4). Assume that Axioms $i)'$ and $ii)'$ hold. Furthermore, assume that if $u(\hat{x}, \hat{t}) = 0$ for some $\hat{x} \in \partial\mathcal{V}$ and $\hat{t} \ge t_0$, then $\phi(\hat{x}, u(\hat{x}, \hat{t}), \nabla u(\hat{x}, \hat{t})) = 0$. Then the solution $u(x,t)$, $x \in \mathcal{V}$, $t \ge t_0$, to (11.3) and (11.4) is nonnegative for all nonnegative initial density distributions $u_{t_0}(x) \ge 0$, $x \in \mathcal{V}$.

Proof. Note that if $u(\hat{x}, \hat{t}) = 0$ for some $\hat{x}$ in the interior of $\mathcal{V}$ and $\hat{t} \ge t_0$, then it follows from Axiom $ii)'$ that $\phi(y, u(y, \hat{t}), \nabla u(y, \hat{t}))$ is directed toward the point $\hat{x}$ for all points y in a sufficiently small neighborhood of $\hat{x}$. This property along with (11.3) implies that $\frac{\partial u(\hat{x}, \hat{t})}{\partial t} \ge 0$. Alternatively, if $u(\hat{x}, \hat{t}) = 0$ for some $\hat{x} \in \partial\mathcal{V}$ and $\hat{t} \ge t_0$, then it follows from (11.3) and

Axioms $i)'$ and $ii)'$ that $\frac{\partial u(\hat{x},\hat{t})}{\partial t} \geq 0$. Thus, the solution to (11.3) and (11.4) is nonnegative for all nonnegative initial density distributions. □

For the remainder of this chapter, $\mathrm{d}\mathcal{V}$ represents an infinitesimal volume element of $\mathcal{V}$, $\mathcal{S}_{\mathcal{V}}$ denotes the surface enclosing $\mathcal{V}$, and $\mathrm{d}\mathcal{S}_{\mathcal{V}}$ denotes an infinitesimal boundary element.

11.2 Entropy and Ectropy for Continuum Thermodynamics

In this section, we establish the classical Clausius inequality for our thermodynamically consistent infinite-dimensional energy flow model given by (11.3) and (11.4). For this result, note that it follows from Axiom $i)'$ that for the isolated dynamical system $\mathcal{G}$, that is, $s(x,t) \equiv 0$ and $\phi(x,u(x,t),\nabla u(x,t))\cdot\hat{n}(x) \equiv 0$, the function $u(x,t) = \alpha$, $x \in \mathcal{V}$, $t \geq t_0$, $\alpha \geq 0$, is the solution to (11.3) and (11.4) with $u_{t_0}(x) = \alpha$, $x \in \mathcal{V}$. Thus, as in Chapter 3, we define an equilibrium process for the system $\mathcal{G}$ as a process where the trajectory of $\mathcal{G}$ moves along the equilibrium manifold $\mathcal{M}_{\mathrm{e}} \triangleq \{u_t \in \mathcal{X} : u_t(x) = \alpha,\ x \in \mathcal{V},\ \alpha \geq 0\}$, that is, $u(x,t) = \alpha(t), x \in \mathcal{V}$, $t \geq t_0$, for some $\mathcal{L}_\infty$ function $\alpha : [0,\infty) \to \overline{\mathbb{R}}_+$. A nonequilibrium process is a process that does not lie on $\mathcal{M}_{\mathrm{e}}$.

The next result establishes a Clausius-type inequality for equilibrium and nonequilibrium transformations of the infinite-dimensional dynamical system $\mathcal{G}$.

Proposition 11.2. Consider the dynamical system $\mathcal{G}$ with power balance equation (11.3) and (11.4), and assume that Axioms $i)'$ and $ii)'$ hold. Then, for every initial energy density distribution $u_{t_0} \in \mathcal{X}$, $t_{\mathrm{f}} \geq t_0$, and $s(\cdot,\cdot) \in \mathcal{U}$ such that $u_{t_{\mathrm{f}}}(x) = u_{t_0}(x)$, $x \in \mathcal{V}$,

$$\int_{t_0}^{t_{\mathrm{f}}} \left[\int_{\mathcal{V}} \frac{s(x,t)}{c+u(x,t)} \mathrm{d}\mathcal{V} - \int_{\partial\mathcal{V}} \frac{\phi(x,u(x,t),\nabla u(x,t))\cdot\hat{n}(x)}{c+u(x,t)} \mathrm{d}\mathcal{S}_{\mathcal{V}} \right] \mathrm{d}t \leq 0, \tag{11.18}$$

where $c > 0$ and $u(x,t)$, $x \in \mathcal{V}$, $t \geq t_0$, is the solution to (11.3) and (11.4). Furthermore,

$$\int_{t_0}^{t_{\mathrm{f}}} \left[\int_{\mathcal{V}} \frac{s(x,t)}{c+u(x,t)} \mathrm{d}\mathcal{V} - \int_{\partial\mathcal{V}} \frac{\phi(x,u(x,t),\nabla u(x,t))\cdot\hat{n}(x)}{c+u(x,t)} \mathrm{d}\mathcal{S}_{\mathcal{V}} \right] \mathrm{d}t = 0 \tag{11.19}$$

if and only if there exists an $\mathcal{L}_\infty$ function $\alpha : [t_0, t_{\mathrm{f}}] \to \overline{\mathbb{R}}_+$ such that $u(x,t) = \alpha(t)$, $x \in \mathcal{V}$, $t \in [t_0, t_{\mathrm{f}}]$.

Proof. It follows from (11.3), the Green-Gauss theorem, and Axiom

$ii)'$ that

$$\begin{aligned}\int_{t_0}^{t_\mathrm{f}} &\left[\int_{\mathcal{V}} \frac{s(x,t)}{c+u(x,t)}\mathrm{d}\mathcal{V} - \int_{\partial\mathcal{V}} \frac{\phi(x,u(x,t),\nabla u(x,t))\cdot\hat{n}(x)}{c+u(x,t)}\mathrm{d}\mathcal{S}_{\mathcal{V}}\right]\mathrm{d}t \\
&= \int_{t_0}^{t_\mathrm{f}}\int_{\mathcal{V}} \frac{\frac{\partial u(x,t)}{\partial t} + \nabla\cdot\phi(x,u(x,t),\nabla u(x,t))}{c+u(x,t)}\mathrm{d}\mathcal{V}\mathrm{d}t \\
&\quad - \int_{t_0}^{t_\mathrm{f}}\int_{\partial\mathcal{V}} \frac{\phi(x,u(x,t),\nabla u(x,t))\cdot\hat{n}(x)}{c+u(x,t)}\mathrm{d}\mathcal{S}_{\mathcal{V}}\mathrm{d}t \\
&= \int_{\mathcal{V}} \log_e\left(\frac{c+u(x,t_\mathrm{f})}{c+u(x,t_0)}\right)\mathrm{d}\mathcal{V} \\
&\quad + \int_{t_0}^{t_\mathrm{f}}\int_{\partial\mathcal{V}} \frac{\phi(x,u(x,t),\nabla u(x,t))\cdot\hat{n}(x)}{c+u(x,t)}\mathrm{d}\mathcal{S}_{\mathcal{V}}\mathrm{d}t \\
&\quad + \int_{t_0}^{t_\mathrm{f}}\int_{\mathcal{V}} \frac{\nabla u(x,t)\phi(x,u(x,t),\nabla u(x,t))}{(c+u(x,t))^2}\mathrm{d}\mathcal{V}\mathrm{d}t \\
&\quad - \int_{t_0}^{t_\mathrm{f}}\int_{\partial\mathcal{V}} \frac{\phi(x,u(x,t),\nabla u(x,t))\cdot\hat{n}(x)}{c+u(x,t)}\mathrm{d}\mathcal{S}_{\mathcal{V}}\mathrm{d}t \\
&= \int_{t_0}^{t_\mathrm{f}}\int_{\mathcal{V}} \frac{\nabla u(x,t)\phi(x,u(x,t),\nabla u(x,t))}{(c+u(x,t))^2}\mathrm{d}\mathcal{V}\mathrm{d}t \\
&\leq 0, \end{aligned} \tag{11.20}$$

which proves (11.18).

To show (11.19), note that it follows from (11.20), Axiom $i)'$, and Axiom $ii)'$ that (11.19) holds if and only if $\nabla u(x,t) = 0$ for all $x \in \mathcal{V}$ and $t \in [t_0, t_\mathrm{f}]$ or, equivalently, there exists an $\mathcal{L}_\infty$ function $\alpha : [t_0, t_\mathrm{f}] \to \overline{\mathbb{R}}_+$ such that $u(x,t) = \alpha(t)$, $x \in \mathcal{V}$, $t \in [t_0, t_\mathrm{f}]$. □

Next, we define an entropy functional for the continuum dynamical system $\mathcal{G}$.

Definition 11.1. For the dynamical system $\mathcal{G}$ with power balance equation (11.3) and (11.4), the functional $\mathcal{S} : \mathcal{X} \to \mathbb{R}$ satisfying

$$\mathcal{S}(u_{t_2}) \geq \mathcal{S}(u_{t_1}) + \int_{t_1}^{t_2} q(t)\mathrm{d}t \tag{11.21}$$

for all $s(\cdot,\cdot) \in \mathcal{U}$ and $t_2 \geq t_1 \geq t_0$, where

$$q(t) \triangleq \int_{\mathcal{V}} \frac{s(x,t)}{c+u(x,t)}\mathrm{d}\mathcal{V} - \int_{\partial\mathcal{V}} \frac{\phi(x,u(x,t),\nabla u(x,t))\cdot\hat{n}(x)}{c+u(x,t)}\mathrm{d}\mathcal{S}_{\mathcal{V}} \tag{11.22}$$

and $c > 0$, is called the *entropy* functional of $\mathcal{G}$.

In the next theorem, we show that (11.18) guarantees the existence of an entropy functional for the dynamical system $\mathcal{G}$ given by (11.3) and (11.4). For this result, define the available entropy of the dynamical system $\mathcal{G}$ by

$$\mathcal{S}_{\mathrm{a}}(u_{t_0}) \triangleq - \sup_{s(\cdot,\cdot)\in\mathcal{U}_{\mathrm{c}},\, T\geq t_0} \int_{t_0}^{T} q(t)\mathrm{d}t, \tag{11.23}$$

where $q(t)$ is given by (11.22), $u(x,t_0) = u_{t_0}(x)$, $x \in \mathcal{V}$, $u_{t_0} \in \mathcal{X}$, and $u(x,T) = 0$, $x \in \mathcal{V}$, and define the required entropy supply of the dynamical system $\mathcal{G}$ by

$$\mathcal{S}_{\mathrm{r}}(u_{t_0}) \triangleq \sup_{s(\cdot,\cdot)\in\mathcal{U}_{\mathrm{r}},\, T\geq -t_0} \int_{-T}^{t_0} q(t)\mathrm{d}t, \tag{11.24}$$

where $u(x,-T) = 0$, $x \in \mathcal{V}$, $u(x,t_0) = u_{t_0}(x)$, $x \in \mathcal{V}$, and $u_{t_0} \in \mathcal{X}$.

Theorem 11.2. Consider the dynamical system $\mathcal{G}$ with power balance equation (11.3) and (11.4), and assume that Axiom $ii)'$ holds. Then there exists an entropy functional for $\mathcal{G}$. Moreover, $\mathcal{S}_{\mathrm{a}}(u_{t_0})$, $u_{t_0} \in \mathcal{X}$, and $\mathcal{S}_{\mathrm{r}}(u_{t_0})$, $u_{t_0} \in \mathcal{X}$, are possible entropy functionals for $\mathcal{G}$ with $\mathcal{S}_{\mathrm{a}}(0) = \mathcal{S}_{\mathrm{r}}(0) = 0$. Finally, all entropy functionals $\mathcal{S}(u_{t_0})$, $u_{t_0} \in \mathcal{X}$, for $\mathcal{G}$ satisfy

$$\mathcal{S}_{\mathrm{r}}(u_{t_0}) \leq \mathcal{S}(u_{t_0}) - \mathcal{S}(0) \leq \mathcal{S}_{\mathrm{a}}(u_{t_0}), \quad u_{t_0} \in \mathcal{X}. \tag{11.25}$$

Proof. The proof is identical to the proof of Theorem 3.2. □

The next result shows that all entropy functionals for $\mathcal{G}$ are continuous on $\mathcal{X}$ with norm $\|\cdot\|_{\mathcal{L}_1}$.

Theorem 11.3. Consider the dynamical system $\mathcal{G}$ with power balance equation (11.3) and (11.4), and let $\mathcal{S} : \mathcal{X} \to \mathbb{R}$ be an entropy functional of $\mathcal{G}$. Then $\mathcal{S}(\cdot)$ is continuous on $\mathcal{X}$ with respect to the $\mathcal{L}_1$ norm.

Proof. Let $u_{\mathrm{e}}(\cdot) \in \mathcal{X}$ and $s_{\mathrm{e}} : \mathcal{V} \to \mathbb{R}$ be such that $s_{\mathrm{e}}(x) = \nabla \cdot \phi(x, u_{\mathrm{e}}(x), \nabla u_{\mathrm{e}}(x))$, $x \in \mathcal{V}$. Note that with $s(x,t) \equiv s_{\mathrm{e}}(x)$, $x \in \mathcal{V}$, $u_{\mathrm{e}}(\cdot)$ is an equilibrium state of the power balance equation (11.3) and (11.4). Next, it follows from Lemma 11.1 that for every $\varepsilon > 0$ and $T > 0$, there exist $s_{\mathrm{e}} : \mathcal{V} \to \mathbb{R}$ and $\alpha > 0$ such that for every $\hat{u}(\cdot) \in \mathcal{X}$ with $\|\hat{u} - u_{\mathrm{e}}\|_{\mathcal{L}_1} \leq \alpha T$, there exists $s : \mathcal{V} \times [0, \hat{T}] \to \mathbb{R}$ such that $\|s(\cdot,t) - s_{\mathrm{e}}(\cdot)\|_{\mathcal{L}_1} \leq \varepsilon$, $t \in [0,\hat{T}]$, and $u(x,t) = u_{\mathrm{e}}(x) + \frac{\hat{u}(x) - u_{\mathrm{e}}(x)}{\hat{T}} t$, $x \in \mathcal{V}$, $t \in [0,\hat{T}]$, where $\hat{T} = \frac{\|\hat{u} - u_{\mathrm{e}}\|_{\mathcal{L}_1}}{\alpha}$. Hence, for every $\delta > 0$ and $\varepsilon > 0$, there exist $s_{\mathrm{e}} : \mathcal{V} \to \mathbb{R}$ and $\alpha > 0$ such that for every $\hat{u}(\cdot) \in \mathcal{X}$ with $\|\hat{u} - u_{\mathrm{e}}\|_{\mathcal{L}_1} \leq \delta$, there exists $s : \mathcal{V} \times [0,\hat{T}] \to \mathbb{R}$ such that $\|s(\cdot,t) - s_{\mathrm{e}}(\cdot)\|_{\mathcal{L}_1} \leq \varepsilon$, $t \in [0,\hat{T}]$, and $u(x,t) = u_{\mathrm{e}}(x) + \frac{\hat{u}(x) - u_{\mathrm{e}}(x)}{\hat{T}} t$, $x \in \mathcal{V}$, $t \in [0,\hat{T}]$, where $\hat{T} = \frac{\|\hat{u} - u_{\mathrm{e}}\|_{\mathcal{L}_1}}{\alpha}$.

Next, since $\phi(\cdot,\cdot,\cdot)$ is continuous, it follows that there exists $M \in (0,\infty)$ such that

$$\sup_{\|u-u_{\rm e}\|_{\mathcal{L}_1}<\delta,\ \|s-s_{\rm e}\|_{\mathcal{L}_1}<\varepsilon} \left\| \frac{s(x) - \nabla\cdot\phi(x,u(x),\nabla u(x))}{c+u(x)} + \frac{\nabla u(x)\phi(x,u(x),\nabla u(x))}{(c+u(x))^2} \right\|_{\mathcal{L}_1} = M. \tag{11.26}$$

Hence, it follows that

$$\begin{aligned}
&\left| \int_0^{\hat{T}} \left[\int_{\mathcal{V}} \frac{s(x,t) - \nabla\cdot\phi(x,u(x,t),\nabla u(x,t))}{c+u(x,t)} {\rm d}\mathcal{V} \right.\right. \\
&\qquad \left.\left. + \int_{\mathcal{V}} \frac{\nabla u(x,t)\phi(x,u(x,t),\nabla u(x,t))}{(c+u(x,t))^2} {\rm d}\mathcal{V} \right] {\rm d}t \right| \\
&\quad \leq \int_0^{\hat{T}} \left\| \frac{s(x,t) - \nabla\cdot\phi(x,u(x,t),\nabla u(x,t))}{c+u(x,t)} \right. \\
&\qquad \left. + \frac{\nabla u(x,t)\phi(x,u(x,t),\nabla u(x,t))}{(c+u(x,t))^2} \right\|_{\mathcal{L}_1} {\rm d}t \\
&\quad \leq M\hat{T} \\
&\quad = \frac{M}{\alpha}\|\hat{u} - u_{\rm e}\|_{\mathcal{L}_1}.
\end{aligned} \tag{11.27}$$

Next, if $\mathcal{S}(\cdot)$ is an entropy functional of $\mathcal{G}$, then, since $u(x,\hat{T}) = \hat{u}(x)$, $x \in \mathcal{V}$,

$$\begin{aligned}
\mathcal{S}(\hat{u}) \geq \mathcal{S}(u_{\rm e}) &+ \int_0^{\hat{T}} \int_{\mathcal{V}} \frac{s(x,t)}{c+u(x,t)} {\rm d}\mathcal{V}{\rm d}t \\
&- \int_0^{\hat{T}} \int_{\partial\mathcal{V}} \frac{\phi(x,u(x,t),\nabla u(x,t))\cdot\hat{n}(x)}{c+u(x,t)} {\rm d}\mathcal{S}_{\mathcal{V}}{\rm d}t.
\end{aligned} \tag{11.28}$$

Hence, it follows from the Green-Gauss theorem that

$$\begin{aligned}
\mathcal{S}(u_{\rm e}) - \mathcal{S}(\hat{u}) &\leq -\int_0^{\hat{T}} \int_{\mathcal{V}} \frac{s(x,t)}{c+u(x,t)} {\rm d}\mathcal{V}{\rm d}t \\
&\quad + \int_0^{\hat{T}} \int_{\partial\mathcal{V}} \frac{\phi(x,u(x,t),\nabla u(x,t))\cdot\hat{n}(x)}{c+u(x,t)} {\rm d}\mathcal{S}_{\mathcal{V}}{\rm d}t \\
&= -\int_0^{\hat{T}} \int_{\mathcal{V}} \frac{s(x,t) - \nabla\cdot\phi(x,u(x,t),\nabla u(x,t))}{c+u(x,t)} {\rm d}\mathcal{V}{\rm d}t \\
&\quad - \int_0^{\hat{T}} \int_{\mathcal{V}} \frac{\nabla u(x,t)\phi(x,u(x,t),\nabla u(x,t))}{(c+u(x,t))^2} {\rm d}\mathcal{V}{\rm d}t.
\end{aligned} \tag{11.29}$$

Now, if $\mathcal{S}(u_{\rm e}) \geq \mathcal{S}(\hat{u})$, then combining (11.27) and (11.29) yields

$$|\mathcal{S}(u_{\rm e}) - \mathcal{S}(\hat{u})| \leq \frac{M}{\alpha}\|\hat{u} - u_{\rm e}\|_{\mathcal{L}_1}. \tag{11.30}$$

Alternatively, if $\mathcal{S}(\hat{u}) \geq \mathcal{S}(u_{\rm e})$, then (11.30) can be derived by reversing the roles of $u_{\rm e}(\cdot)$ and $\hat{u}(\cdot)$. Hence, it follows that $\mathcal{S}(\cdot)$ is continuous on $\mathcal{X}$ with norm $\|\cdot\|_{\mathcal{L}_1}$. □

As for the finite-dimensional case, Definition 11.1 does not provide enough information to define the entropy uniquely for nonequilibrium continuum thermodynamics. Specifically, using a similar result to Proposition 3.3, it can be shown that all possible entropy functionals form a convex set, and hence, there exists a continuum of entropy functionals ranging from the required entropy supply $\mathcal{S}_{\rm r}(u_{t_0})$ to the available entropy $\mathcal{S}_{\rm a}(u_{t_0})$.

The following two propositions address processes for equilibrium continuum thermodynamics wherein uniqueness is not an issue.

Proposition 11.3. Consider the dynamical system $\mathcal{G}$ with power balance equation (11.3) and (11.4), and assume that Axioms *i*)′ and *ii*)′ hold. Then, for every equilibrium energy density distribution $u_{t{\rm e}}(x) = \alpha$, $x \in \mathcal{V}$, $\alpha \geq 0$, the entropy $\mathcal{S}(u_t)$, $u_t \in \mathcal{X}$, of $\mathcal{G}$ is unique (modulo a constant of integration) and is given by

$$\mathcal{S}(u_t) - \mathcal{S}(0) = \mathcal{S}_{\rm a}(u_t) = \mathcal{S}_{\rm r}(u_t) = \mathcal{V}_{\rm vol}\log_e(c+\alpha) - \mathcal{V}_{\rm vol}\log_e c, \tag{11.31}$$

where $u_t(x) = u_{t{\rm e}}(x) = \alpha$, $x \in \mathcal{V}$.

Proof. It follows from (11.3) and the Green-Gauss theorem that

$$\begin{aligned} q(t) &= \int_{\mathcal{V}} \frac{\frac{\partial u(x,t)}{\partial t} + \nabla\cdot\phi(x,u(x,t),\nabla u(x,t))}{c+u(x,t)}{\rm d}\mathcal{V} \\ &\quad - \int_{\partial\mathcal{V}} \frac{\phi(x,u(x,t),\nabla u(x,t))\cdot\hat{n}(x)}{c+u(x,t)}{\rm d}\mathcal{S}_{\mathcal{V}} \\ &= \int_{\mathcal{V}} \frac{1}{c+u(x,t)}\frac{\partial u(x,t)}{\partial t}{\rm d}\mathcal{V} \\ &\quad + \int_{\mathcal{V}} \frac{\nabla u(x,t)\phi(x,u(x,t),\nabla u(x,t))}{(c+u(x,t))^2}{\rm d}\mathcal{V}. \end{aligned} \tag{11.32}$$

Next, consider the entropy functional $\mathcal{S}_{\rm a}(u_{t_0})$ given by (11.23), and let $u_{t_0}(x) = u_{t{\rm e}}(x) = \alpha$, $x \in \mathcal{V}$, $\alpha \geq 0$. Then it follows from (11.32) that

$$\mathcal{S}_{\rm a}(u_{t{\rm e}}) = - \sup_{s(\cdot,\cdot)\in\mathcal{U}_{\rm c},\,T\geq t_0} \left[\int_{t_0}^{T}\int_{\mathcal{V}} \frac{1}{c+u(x,t)}\frac{\partial u(x,t)}{\partial t}{\rm d}\mathcal{V}{\rm d}t\right.$$

$$
\begin{aligned}
&+\int_{t_0}^{T}\int_{\mathcal{V}}\frac{\nabla u(x,t)\phi(x,u(x,t),\nabla u(x,t))}{(c+u(x,t))^2}\mathrm{d}\mathcal{V}\mathrm{d}t\Bigg] \\
=&-\sup_{s(\cdot,\cdot)\in\mathcal{U}_{\mathrm{c}},\,T\geq t_0}\Bigg[\int_{\mathcal{V}}\log_e\left(\frac{c}{c+\alpha}\right)\mathrm{d}\mathcal{V} \\
&+\int_{t_0}^{T}\int_{\mathcal{V}}\frac{\nabla u(x,t)\phi(x,u(x,t),\nabla u(x,t))}{(c+u(x,t))^2}\mathrm{d}\mathcal{V}\mathrm{d}t\Bigg] \\
=&\int_{\mathcal{V}}\log_e\left(\frac{c+\alpha}{c}\right)\mathrm{d}\mathcal{V} \\
&-\sup_{s(\cdot,\cdot)\in\mathcal{U}_{\mathrm{c}},\,T\geq t_0}\int_{t_0}^{T}\int_{\mathcal{V}}\frac{\nabla u(x,t)\phi(x,u(x,t),\nabla u(x,t))}{(c+u(x,t))^2}\mathrm{d}\mathcal{V}\mathrm{d}t.
\end{aligned}
\tag{11.33}
$$

It follows from Axiom $ii)'$ that the supremum in (11.33) is taken over the set of negative semidefinite values. However, the zero value of the supremum is achieved on an equilibrium transformation for which $\phi(x,u(x,t),\nabla u(x,t))\equiv 0$, and thus

$$
\mathcal{S}_{\mathrm{a}}(u_{te})=\mathcal{V}_{\mathrm{vol}}\log_e(c+\alpha)-\mathcal{V}_{\mathrm{vol}}\log_e c. \tag{11.34}
$$

Similarly, it can be shown that $\mathcal{S}_{\mathrm{r}}(u_{te})=\mathcal{V}_{\mathrm{vol}}\log_e(c+\alpha)-\mathcal{V}_{\mathrm{vol}}\log_e c$. Finally, it follows from (11.25) that (11.31) holds. □

Proposition 11.4. Consider the dynamical system $\mathcal{G}$ with power balance equation (11.3) and (11.4), and assume that Axioms $i)'$ and $ii)'$ hold. Let $\mathcal{S}(\cdot)$ denote an entropy of $\mathcal{G}$, and let $u(x,t)$, $x\in\mathcal{V}$, $t\geq t_0$, be the solution to (11.3) and (11.4) with $u(x,t_0)=\alpha_0$, $x\in\mathcal{V}$, and $u(x,t_1)=\alpha_1$, $x\in\mathcal{V}$, where α_0, $\alpha_1\geq 0$. Then

$$
\mathcal{S}(u_{t_1})=\mathcal{S}(u_{t_0})+\int_{t_0}^{t_1}q(t)\mathrm{d}t \tag{11.35}
$$

if and only if there exists an $\mathcal{L}_\infty$ function $\alpha:[t_0,t_1]\to\overline{\mathbb{R}}_+$ such that $\alpha(t_0)=\alpha_0$, $\alpha(t_1)=\alpha_1$, and $u(x,t)=\alpha(t)$, $x\in\mathcal{V}$, $t\in[t_0,t_1]$.

Proof. It follows from Proposition 11.3 that

$$
\mathcal{S}(u_{t_1})-\mathcal{S}(u_{t_0})=\mathcal{V}_{\mathrm{vol}}\log_e(c+\alpha_1)-\mathcal{V}_{\mathrm{vol}}\log_e(c+\alpha_0). \tag{11.36}
$$

Furthermore, it follows from (11.32) that

$$
\begin{aligned}
\int_{t_0}^{t_1}q(t)\mathrm{d}t =& \mathcal{V}_{\mathrm{vol}}\log_e\left(\frac{c+\alpha_1}{c+\alpha_0}\right) \\
&+\int_{t_0}^{t_1}\int_{\mathcal{V}}\frac{\nabla u(x,t)\phi(x,u(x,t),\nabla u(x,t))}{(c+u(x,t))^2}\mathrm{d}\mathcal{V}\mathrm{d}t.
\end{aligned}
\tag{11.37}
$$

Now, it follows from Axioms $i)'$ and $ii)'$ that (11.35) holds if and only if

$\nabla u(x,t) = 0$, $x \in \mathcal{V}$, $t \in [t_0, t_1]$, or, equivalently, there exists an $\mathcal{L}_\infty$ function $\alpha : [t_0, t_1] \to \overline{\mathbb{R}}_+$ such that $u(x,t) = \alpha(t)$, $x \in \mathcal{V}$, $t \in [t_0, t_1]$, $\alpha(t_0) = \alpha_0$, and $\alpha(t_1) = \alpha_1$. □

In the next theorem, we present a unique, continuously differentiable entropy functional for the dynamical system $\mathcal{G}$. This result holds for equilibrium and nonequilibrium processes.

Theorem 11.4. Consider the dynamical system $\mathcal{G}$ with power balance equation (11.3) and (11.4), and assume that Axioms $i)'$ and $ii)'$ hold. Then the functional $\mathcal{S} : \mathcal{X} \to \mathbb{R}$ given by

$$\mathcal{S}(u_t) = \int_{\mathcal{V}} \log_e(c + u_t(x))\mathrm{d}\mathcal{V} - \mathcal{V}_{\mathrm{vol}} \log_e c \tag{11.38}$$

is a unique (modulo a constant of integration), continuously differentiable entropy functional of $\mathcal{G}$. Furthermore, if $u_t \notin \mathcal{M}_\mathrm{e}$, $t \geq t_0$, where $u_t = u(x,t)$ denotes the solution to (11.3) and (11.4) and $\mathcal{M}_\mathrm{e} = \{u_t \in \mathcal{X} : u_t = \alpha, \alpha \geq 0\}$, then (11.38) satisfies

$$\mathcal{S}(u_{t_2}) > \mathcal{S}(u_{t_1}) + \int_{t_1}^{t_2} q(t)\mathrm{d}t. \tag{11.39}$$

Proof. It follows from the Green-Gauss theorem, Axiom $ii)'$, and (11.38) that

$$\begin{aligned}
\dot{\mathcal{S}}(u_t) &= \int_{\mathcal{V}} \frac{1}{c + u(x,t)} \frac{\partial u(x,t)}{\partial t} \mathrm{d}\mathcal{V} \\
&= \int_{\mathcal{V}} \frac{1}{c + u(x,t)} \left(-\nabla \cdot \phi(x, u(x,t), \nabla u(x,t)) + s(x,t)\right) \mathrm{d}\mathcal{V} \\
&= -\int_{\mathcal{V}} \frac{\nabla u(x,t)\phi(x, u(x,t), \nabla u(x,t))}{(c + u(x,t))^2} \mathrm{d}\mathcal{V} \\
&\quad - \int_{\partial\mathcal{V}} \frac{\phi(x, u(x,t), \nabla u(x,t)) \cdot \hat{n}(x)}{c + u(x,t)} \mathrm{d}\mathcal{S}_{\mathcal{V}} \\
&\quad + \int_{\mathcal{V}} \frac{s(x,t)}{c + u(x,t)} \mathrm{d}\mathcal{V} \\
&\geq q(t).
\end{aligned} \tag{11.40}$$

Now, integrating (11.40) over $[t_1, t_2]$ yields (11.21). Furthermore, if $u_t \notin \mathcal{M}_\mathrm{e}$, $t \geq t_0$, then it follows from Axiom $i)'$, Axiom $ii)'$, and (11.40) that (11.39) holds.

To show that (11.38) is a unique, continuously differentiable entropy function of $\mathcal{G}$, let $\mathcal{S}(u_t)$ be a continuously differentiable entropy functional of $\mathcal{G}$ so that $\mathcal{S}(u_t)$ satisfies (11.21) or, equivalently,

$$\begin{aligned}\dot{\mathcal{S}}(u_t) &\geq -\int_{\partial\mathcal{V}} \frac{\phi(x,u_t,\nabla u_t)\cdot\hat{n}(x)}{c+u_t}\mathrm{d}\mathcal{S}_{\mathcal{V}} \\ &= -\int_{\mathcal{V}} \nabla\cdot(\mu(u_t)S(x,t))\mathrm{d}\mathcal{V} \\ &= -\mu(u_t)S(x,t), \quad t\geq t_0, \end{aligned} \tag{11.41}$$

where $\mu(u_t) \triangleq \frac{1}{c+u_t}$, $S(x,t) \triangleq \phi(x,u_t,\nabla u_t)$, u_t, $t\geq t_0$, denotes the solution to (11.3) and (11.4), and $\dot{\mathcal{S}}(u_t)$ denotes the time derivative of $\mathcal{S}(u_t)$ along the solution u_t, $t \geq t_0$. Hence, it follows from (11.41) that

$$\mathcal{S}'(u_t)[-\nabla\cdot S(x,t)] \geq -\mu(u_t)S(x,t), \quad u_t\in\overline{\mathbb{R}}_+, \quad x\in\mathcal{V}, \quad t\geq t_0, \tag{11.42}$$

that is,

$$\mathcal{S}'(u_t)\left[-S(x,t)-\int_{\mathcal{V}}\nabla^2 S(x,t)\mathrm{d}\mathcal{V}\right] \geq -\mu(u_t)S(x,t),\ u_t\in\overline{\mathbb{R}}_+,\ x\in\mathcal{V},\ t\geq t_0, \tag{11.43}$$

which implies that there exist continuous functions $\ell:\overline{\mathbb{R}}_+\to\mathbb{R}^p$ and $\mathcal{W}:\overline{\mathbb{R}}_+\to\mathbb{R}^{p\times q}$ such that

$$\begin{aligned}0 = {}& \mathcal{S}'(u_t)\left[-S(x,t)-\int_{\mathcal{V}}\nabla^2 S(x,t)\mathrm{d}\mathcal{V}\right]+\mu(u_t)S(x,t) \\ &-[\ell(u_t)+\mathcal{W}(u_t)S(x,t)]^{\mathrm{T}}[\ell(u_t)+\mathcal{W}(u_t)S(x,t)],\ u_t\in\overline{\mathbb{R}}_+,\\ & x\in\mathcal{V},\ t\geq t_0.\end{aligned} \tag{11.44}$$

Now, equating coefficients of equal powers (of S), it follows that $\mathcal{W}(u_t) \equiv 0$, $\mathcal{S}'(u_t)=\mu(u_t)$, $u_t\in\overline{\mathbb{R}}_+$, and

$$0=\mathcal{S}'(u_t)\int_{\mathcal{V}}\nabla^2 S(x,t)\mathrm{d}\mathcal{V}+\ell^{\mathrm{T}}(u_t)\ell(u_t), \quad u_t\in\overline{\mathbb{R}}_+. \tag{11.45}$$

Hence, $\mathcal{S}(u_t)=\int_{\mathcal{V}}\log_e(c+u_t(x))\mathrm{d}\mathcal{V}-\mathcal{V}_{\mathrm{vol}}\log_e c$, $u_t\in\overline{\mathbb{R}}_+$. Thus, (11.38) is a unique, continuously differentiable entropy functional for $\mathcal{G}$. □

It follows from Theorem 11.4 that if no information flow is allowed into or out of $\mathcal{V}$ (i.e., the system is isolated), then $\mathcal{S}(u_{t_2})\geq\mathcal{S}(u_{t_1})$, $t_2\geq t_1$. This shows that for an adiabatically isolated system, the entropy of the final state is greater than or equal to the entropy of the initial state.

The next result shows that for every nontrivial trajectory of $\mathcal{G}$, the dynamical system $\mathcal{G}$ is state irreversible. For this result, let $\mathcal{W}_{[t_0,t_1]}$ denote the set of all possible energy density distributions of $\mathcal{G}$ over the time interval $[t_0,t_1]$ given by

$$\mathcal{W}_{[t_0,t_1]} \triangleq \{s^u:[t_0,t_1]\times\mathcal{U}\to\mathcal{X}: s^u(\cdot,s(\cdot,\cdot)) \text{ satisfies}$$

(11.3) and (11.4)}. (11.46)

Theorem 11.5. Consider the dynamical system $\mathcal{G}$ with power balance equation (11.3) and (11.4), and assume that Axioms $i)'$ and $ii)'$ hold. Furthermore, let $s^u(\cdot, s(\cdot,\cdot)) \in \mathcal{W}_{[t_0,t_1]}$, where $s(\cdot,\cdot) \in \mathcal{U}$. Then $s^u(\cdot, s(\cdot,\cdot))$ is an $I_{\mathcal{X}}$-reversible trajectory of $\mathcal{G}$ if and only if $s^u(t, s(x,t)) \in \mathcal{M}_{\rm e}$, $t \in [t_0, t_1]$.

Proof. The proof is similar to the proof of Theorem 3.5. □

Next, we establish a dual inequality to inequality (11.18) that is satisfied for our thermodynamically consistent energy flow model.

Proposition 11.5. Consider the dynamical system $\mathcal{G}$ with power balance equation (11.3) and (11.4), and assume that Axioms $i)'$ and $ii)'$ hold. Then, for every initial energy density distribution $u_{t_0} \in \mathcal{X}$, $t_{\rm f} \geq t_0$, and $s(\cdot,\cdot) \in \mathcal{U}$ such that $u_{t_{\rm f}}(x) = u_{t_0}(x)$, $x \in \mathcal{V}$,

$$\int_{t_0}^{t_{\rm f}} \int_{\mathcal{V}} u(x,t)s(x,t){\rm d}\mathcal{V}{\rm d}t - \int_{t_0}^{t_{\rm f}} \int_{\partial\mathcal{V}} u(x,t)\phi(x,u(x,t),\nabla u(x,t)) \cdot \hat{n}(x){\rm d}\mathcal{S}_{\mathcal{V}}{\rm d}t \geq 0, \quad (11.47)$$

where $u(x,t)$, $x \in \mathcal{V}$, $t \geq t_0$, is the solution to (11.3) and (11.4). Furthermore,

$$\int_{t_0}^{t_{\rm f}} \int_{\mathcal{V}} u(x,t)s(x,t){\rm d}\mathcal{V}{\rm d}t - \int_{t_0}^{t_{\rm f}} \int_{\partial\mathcal{V}} u(x,t)\phi(x,u(x,t),\nabla u(x,t)) \cdot \hat{n}(x){\rm d}\mathcal{S}_{\mathcal{V}}{\rm d}t = 0 \quad (11.48)$$

if and only if there exists an $\mathcal{L}_\infty$ function $\alpha : [t_0, t_{\rm f}] \to \overline{\mathbb{R}}_+$ such that $u(x,t) = \alpha(t)$, $x \in \mathcal{V}$, $t \in [t_0, t_{\rm f}]$.

Proof. It follows from (11.3), the Green-Gauss theorem, and Axiom $ii)'$ that

$$\begin{aligned}
&\int_{t_0}^{t_{\rm f}} \int_{\mathcal{V}} u(x,t)s(x,t){\rm d}\mathcal{V}{\rm d}t - \int_{t_0}^{t_{\rm f}} \int_{\partial\mathcal{V}} u(x,t)\phi(x,u(x,t),\nabla u(x,t)) \cdot \hat{n}(x){\rm d}\mathcal{S}_{\mathcal{V}}{\rm d}t \\
&= \int_{t_0}^{t_{\rm f}} \int_{\mathcal{V}} u(x,t)\left(\frac{\partial u(x,t)}{\partial t} + \nabla \cdot \phi(x,u(x,t),\nabla u(x,t))\right){\rm d}\mathcal{V}{\rm d}t \\
&\quad - \int_{t_0}^{t_{\rm f}} \int_{\partial\mathcal{V}} u(x,t)\phi(x,u(x,t),\nabla u(x,t)) \cdot \hat{n}(x){\rm d}\mathcal{S}_{\mathcal{V}}{\rm d}t
\end{aligned}$$

$$
\begin{aligned}
&= \int_{\mathcal{V}} \left[\tfrac{1}{2}u^2(x,t_{\mathrm{f}}) - \tfrac{1}{2}u^2(x,t_0)\right] \mathrm{d}\mathcal{V} \\
&\quad + \int_{t_0}^{t_{\mathrm{f}}} \int_{\partial\mathcal{V}} u(x,t)\phi(x,u(x,t),\nabla u(x,t)) \cdot \hat{n}(x)\mathrm{d}\mathcal{S}_{\mathcal{V}}\mathrm{d}t \\
&\quad - \int_{t_0}^{t_{\mathrm{f}}} \int_{\mathcal{V}} \nabla u(x,t)\phi(x,u(x,t),\nabla u(x,t))\mathrm{d}\mathcal{V}\mathrm{d}t \\
&\quad - \int_{t_0}^{t_{\mathrm{f}}} \int_{\partial\mathcal{V}} u(x,t)\phi(x,u(x,t),\nabla u(x,t)) \cdot \hat{n}(x)\mathrm{d}\mathcal{S}_{\mathcal{V}}\mathrm{d}t \\
&= - \int_{t_0}^{t_{\mathrm{f}}} \int_{\mathcal{V}} \nabla u(x,t)\phi(x,u(x,t),\nabla u(x,t))\mathrm{d}\mathcal{V}\mathrm{d}t \\
&\geq 0,
\end{aligned} \tag{11.49}
$$

which proves (11.47).

To show (11.48), note that it follows from (11.49), Axiom $i)'$, and Axiom $ii)'$ that (11.48) holds if and only if $\nabla u(x,t) = 0$ for all $x \in \mathcal{V}$ and $t \in [t_0, t_{\mathrm{f}}]$ or, equivalently, there exists an $\mathcal{L}_\infty$ function $\alpha : [t_0, t_{\mathrm{f}}] \to \overline{\mathbb{R}}_+$ such that $u(x,t) = \alpha(t)$, $x \in \mathcal{V}$, $t \in [t_0, t_{\mathrm{f}}]$. □

Definition 11.2. For the dynamical system $\mathcal{G}$ with power balance equation (11.3) and (11.4), the functional $\mathcal{E} : \mathcal{X} \to \mathbb{R}$ satisfying

$$
\mathcal{E}(u_{t_2}) \leq \mathcal{E}(u_{t_1}) + \mathcal{V}_{\mathrm{vol}} \int_{t_1}^{t_2} \hat{q}(t)\mathrm{d}t \tag{11.50}
$$

for all $s(\cdot,\cdot) \in \mathcal{U}$ and $t_2 \geq t_1 \geq t_0$, where

$$
\begin{aligned}
\hat{q}(t) \triangleq& \int_{\mathcal{V}} u(x,t)s(x,t)\mathrm{d}\mathcal{V} \\
&- \int_{\partial\mathcal{V}} u(x,t)\phi(x,u(x,t),\nabla u(x,t)) \cdot \hat{n}(x)\mathrm{d}\mathcal{S}_{\mathcal{V}},
\end{aligned} \tag{11.51}
$$

is called the *ectropy* functional of $\mathcal{G}$.

The next theorem shows that (11.47) guarantees the existence of an ectropy functional for the dynamical system $\mathcal{G}$ given by (11.3) and (11.4). For this result, define the available ectropy of the dynamical system $\mathcal{G}$ by

$$
\mathcal{E}_{\mathrm{a}}(u_{t_0}) \triangleq -\mathcal{V}_{\mathrm{vol}} \inf_{s(\cdot,\cdot)\in\mathcal{U}_{\mathrm{c}}, T\geq t_0} \int_{t_0}^{T} \hat{q}(t)\mathrm{d}t, \tag{11.52}
$$

where $\hat{q}(t)$ is given by (11.51), $u(x,t_0) = u_{t_0}(x)$, $x \in \mathcal{V}$, $u_{t_0} \in \mathcal{X}$, and $u(x,T) = 0$, $x \in \mathcal{V}$, and define the required ectropy supply of the dynamical system $\mathcal{G}$ by

$$\mathcal{E}_{\rm r}(u_{t_0}) \stackrel{\triangle}{=} \mathcal{V}_{\rm vol} \inf_{s(\cdot,\cdot)\in\mathcal{U}_{\rm r},\,T\geq -t_0} \int_{-T}^{t_0} \hat{q}(t){\rm d}t, \tag{11.53}$$

where $u(x,-T)=0$, $x\in\mathcal{V}$, $u(x,t_0)=u_{t_0}(x)$, $x\in\mathcal{V}$, and $u_{t_0}\in\mathcal{X}$.

Theorem 11.6. Consider the dynamical system $\mathcal{G}$ with power balance equation (11.3) and (11.4), and assume that Axiom $ii)'$ holds. Then there exists an ectropy functional for $\mathcal{G}$. Moreover, $\mathcal{E}_{\rm a}(u_{t_0})$, $u_{t_0}\in\mathcal{X}$, and $\mathcal{E}_{\rm r}(u_{t_0})$, $u_{t_0}\in\mathcal{X}$, are possible ectropy functionals for $\mathcal{G}$ with $\mathcal{E}_{\rm a}(0)=\mathcal{E}_{\rm r}(0)=0$. Finally, all ectropy functionals $\mathcal{E}(u_{t_0})$, $u_{t_0}\in\mathcal{X}$, for $\mathcal{G}$ satisfy

$$\mathcal{E}_{\rm a}(u_{t_0}) \leq \mathcal{E}(u_{t_0}) - \mathcal{E}(0) \leq \mathcal{E}_{\rm r}(u_{t_0}), \quad u_{t_0}\in\mathcal{X}. \tag{11.54}$$

Proof. The proof is identical to the proof of Theorem 3.6. □

The next theorem shows that all ectropy functionals for $\mathcal{G}$ are continuous on $\mathcal{X}$ with norm $\|\cdot\|_{\mathcal{L}_1}$.

Theorem 11.7. Consider the dynamical system $\mathcal{G}$ with power balance equation (11.3) and (11.4), and let $\mathcal{E}:\mathcal{X}\to\mathbb{R}$ be an ectropy functional of $\mathcal{G}$. Then $\mathcal{E}(\cdot)$ is continuous on $\mathcal{X}$ with respect to the $\mathcal{L}_1$ norm.

Proof. The proof is identical to the proof of Theorem 11.3. □

The following two propositions are dual to Propositions 11.3 and 11.4 and address equilibrium processes for continuum thermodynamics using ectropy notions.

Proposition 11.6. Consider the dynamical system $\mathcal{G}$ with power balance equation (11.3) and (11.4), and assume that Axioms $i)'$ and $ii)'$ hold. Then, for every energy density distribution $u_{\rm te}(x)=\alpha$, $x\in\mathcal{V}$, $\alpha\geq 0$, the ectropy $\mathcal{E}(u_t)$, $u_t\in\mathcal{X}$, of $\mathcal{G}$ is unique (modulo a constant of integration) and is given by

$$\mathcal{E}(u_t) - \mathcal{E}(0) = \mathcal{E}_{\rm a}(u_t) = \mathcal{E}_{\rm r}(u_t) = \frac{(\alpha\mathcal{V}_{\rm vol})^2}{2}, \tag{11.55}$$

where $u_t(x)=u_{\rm te}(x)=\alpha$, $x\in\mathcal{V}$.

Proof. The proof is identical to the proof of Proposition 11.3. □

Proposition 11.7. Consider the dynamical system $\mathcal{G}$ with power balance equation (11.3) and (11.4), and assume that Axioms $i)'$ and $ii)'$ hold. Let $\mathcal{E}(\cdot)$ denote an ectropy of $\mathcal{G}$, and let $u(x,t)$, $x\in\mathcal{V}$, $t\geq t_0$, be the solution to (11.3) and (11.4) with $u(x,t_0)=\alpha_0$, $x\in\mathcal{V}$, and $u(x,t_1)=\alpha_1$,

$x \in \mathcal{V}$, where α_0, $\alpha_1 \geq 0$. Then

$$\mathcal{E}(u_{t_1}) = \mathcal{E}(u_{t_0}) + \mathcal{V}_{\rm vol}\int_{t_0}^{t_1} \hat{q}(t){\rm d}t \tag{11.56}$$

if and only if there exists an $\mathcal{L}_\infty$ function $\alpha : [t_0, t_1] :\to \overline{\mathbb{R}}_+$ such that $\alpha(t_0) = \alpha_0$, $\alpha(t_1) = \alpha_1$, and $u(x,t) = \alpha(t)$, $x \in \mathcal{V}$, $t \in [t_0, t_1]$.

Proof. The proof is identical to the proof of Proposition 11.4. □

In the next theorem, we present a unique, continuously differentiable ectropy functional for the dynamical system $\mathcal{G}$. This result holds for equilibrium and nonequilibrium processes.

Theorem 11.8. Consider the dynamical system $\mathcal{G}$ with power balance equation (11.3) and (11.4), and assume that Axioms *i*)′ and *ii*)′ hold. Then the functional $\mathcal{E} : \mathcal{X} \to \mathbb{R}$ given by

$$\mathcal{E}(u_t) = \frac{\mathcal{V}_{\rm vol}}{2}\int_{\mathcal{V}} u_t^2(x){\rm d}\mathcal{V} \tag{11.57}$$

is a unique (modulo a constant of integration), continuously differentiable ectropy functional of $\mathcal{G}$. Furthermore, if $u_t \notin \mathcal{M}_{\rm e}$, $t \geq t_0$, where $u_t = u(x,t)$ denotes the solution to (11.3) and (11.4) and $\mathcal{M}_{\rm e} = \{u_t \in \mathcal{X} : u_t = \alpha, \alpha \geq 0\}$, then (11.57) satisfies

$$\mathcal{E}(u_{t_2}) < \mathcal{E}(u_{t_1}) + \mathcal{V}_{\rm vol}\int_{t_1}^{t_2} \hat{q}(t){\rm d}t. \tag{11.58}$$

Proof. It follows from the Green-Gauss theorem, Axiom *ii*)′, (11.3), and (11.57) that

$$\begin{aligned}
\dot{\mathcal{E}}(u_t) &= \mathcal{V}_{\rm vol}\int_{\mathcal{V}} u(x,t)\frac{\partial u(x,t)}{\partial t}{\rm d}\mathcal{V} \\
&= \mathcal{V}_{\rm vol}\int_{\mathcal{V}} u(x,t)\left(-\nabla\cdot\phi(x,u(x,t),\nabla u(x,t)) + s(x,t)\right){\rm d}\mathcal{V} \\
&= -\mathcal{V}_{\rm vol}\int_{\partial\mathcal{V}} u(x,t)\phi(x,u(x,t),\nabla u(x,t))\cdot\hat{n}(x){\rm d}\mathcal{S}_{\mathcal{V}} \\
&\quad +\mathcal{V}_{\rm vol}\int_{\mathcal{V}} \nabla u(x,t)\phi(x,u(x,t),\nabla u(x,t)){\rm d}\mathcal{V} \\
&\quad +\mathcal{V}_{\rm vol}\int_{\mathcal{V}} u(x,t)s(x,t){\rm d}\mathcal{V} \\
&\leq \mathcal{V}_{\rm vol}\hat{q}(t).
\end{aligned} \tag{11.59}$$

Now, integrating (11.59) over $[t_1, t_2]$ yields (11.50). Furthermore, if $u_t \notin \mathcal{M}_{\rm e}$, $t \geq t_0$, then it follows from Axiom *i*)′, Axiom *ii*)′, and (11.59) that

(11.58) holds.

The uniqueness of the ectropy functional (11.57) follows as in the proof of Theorem 3.8. □

Inequality (11.21) is a generalization of Clausius' inequality for equilibrium and nonequilibrium thermodynamics as applied to infinite-dimensional systems, while inequality (11.50) is an anti–Clausius inequality that shows that a thermodynamically consistent infinite-dimensional dynamical system is dissipative with respect to the supply rate $\mathcal{V}_{\rm vol}\hat{q}(t)$ and with the storage functional corresponding to the system ectropy. In addition, note that it follows from (11.21) that the infinitesimal increment in the entropy of $\mathcal{G}$ over the infinitesimal time interval dt satisfies

$$\mathrm{d}\mathcal{S}(u_t) \geq \left[\int_{\mathcal{V}} \frac{s(x,t)}{c+u(x,t)}\mathrm{d}\mathcal{V} - \int_{\partial\mathcal{V}} \frac{\phi(x,u(x,t),\nabla u(x,t))\cdot\hat{n}(x)}{c+u(x,t)}\mathrm{d}\mathcal{S}_{\mathcal{V}}\right]\mathrm{d}t, \quad (11.60)$$

where the shifted energy density $c + u(x,t)$ plays the role of absolute temperature at the spatial coordinate x and time t.

For an isolated dynamical system $\mathcal{G}$ (that is, $s(x,t) \equiv 0$ and $\phi(x,u(x,t),\nabla u(x,t)) \cdot \hat{n}(x) \equiv 0$, $x \in \partial\mathcal{V}$), (11.21) and (11.50) yield the fundamental inequalities

$$\mathcal{S}(u_{t_2}) \geq \mathcal{S}(u_{t_1}), \quad t_2 \geq t_1, \quad (11.61)$$

and

$$\mathcal{E}(u_{t_2}) \leq \mathcal{E}(u_{t_1}), \quad t_2 \geq t_1. \quad (11.62)$$

Hence, for an isolated infinite-dimensional system $\mathcal{G}$, the entropy increases if and only if the ectropy decreases. It is important to note that (11.62) also holds in the case where $\phi(x,u(x,t),\nabla u(x,t)) \cdot \hat{n}(x) \not\equiv 0$, $x \in \partial\mathcal{V}$, whereas (11.61) does not necessarily hold in that case.

11.3 Semistability and Energy Equipartition in Continuum Thermodynamics

In this section, we show that the infinite-dimensional thermodynamic energy flow model has convergent flows to Lyapunov stable uniform equilibrium energy density distributions determined by the system initial energy density distribution. However, since our continuous dynamical system $\mathcal{G}$ is defined on the infinite-dimensional space $\mathcal{X}$, bounded orbits of $\mathcal{G}$ may not lie in a compact subset of $\mathcal{X}$, which is crucial to being able to invoke the invariance principle for infinite-dimensional dynamical systems [201]. This is in contrast to the dynamical system $\mathcal{G}$ considered in the previous chapters arising from a power balance (ordinary differential) equation defined on a finite-dimensional space $\overline{\mathbb{R}}^q_+$, wherein local boundedness of an orbit of $\mathcal{G}$ ensures that the orbit belongs to a compact subset of $\overline{\mathbb{R}}^q_+$.

Hence, to ensure that bounded orbits of $\mathcal{G}$ lie in compact sets, we construct a larger space $\mathcal{H}$ as a Sobolev space so that $\mathcal{X} \subset \mathcal{H}$, and by the Sobolev embedding theorem [418, 454], there exists a Banach space $\mathcal{B} \supset \mathcal{H}$ such that the unit ball in $\mathcal{H}$ belongs to a compact set in $\mathcal{B}$, that is, $\mathcal{H}$ is *compactly embedded* in $\mathcal{B}$. In this case, it follows from Proposition 2.6 that a bounded orbit of the dynamical system $\mathcal{G}$ defined on $\mathcal{H}$ has a nonempty, compact, connected invariant omega limit set in $\mathcal{B}$.

For the next result, $\mathcal{L}_2$ denotes the space of square-integrable Lebesgue measurable functions on $\mathcal{V}$, and the $\mathcal{L}_2$ operator norm $\|\cdot\|_{\mathcal{L}_2}$ on $\mathcal{X}$ is used for the definitions of Lyapunov, semi-, and asymptotic stability. Furthermore, we introduce the Sobolev spaces[5]

$$\mathcal{W}_2^1(\mathcal{V}) \triangleq \{u_t : \mathcal{V} \to \mathbb{R} : u_t \in \mathrm{C}^1(\mathcal{V}) \cap \mathcal{L}_2(\mathcal{V}),\ (\nabla u_t)^{\mathrm{T}} \in \mathcal{L}_2(\mathcal{V})\}_{\mathrm{co}} \tag{11.63}$$

and

$$\mathcal{W}_2^0(\mathcal{V}) \triangleq \{u_t : \mathcal{V} \to \mathbb{R} : u_t \in \mathrm{C}^0(\mathcal{V}) \cap \mathcal{L}_2(\mathcal{V})\}_{\mathrm{co}} \subset \mathcal{L}_2(\mathcal{V}), \tag{11.64}$$

where $\mathrm{C}^r(\mathcal{V})$ denotes a function space defined on $\mathcal{V}$ with r-continuous derivatives and $\{\cdot\}_{\mathrm{co}}$ denotes completion[6] of $\{\cdot\}$ in $\mathcal{L}_2$ in the sense of [454], with norms

$$\|u_t\|_{\mathcal{W}_2^1} \triangleq \left[\int_{\mathcal{V}} \left(u_t^2(x) + \nabla u_t(x)\,(\nabla u_t(x))^{\mathrm{T}}\right) \mathrm{d}\mathcal{V}\right]^{\frac{1}{2}}, \tag{11.65}$$

$$\|u_t\|_{\mathcal{W}_2^0} \triangleq \|u_t\|_{\mathcal{L}_2} = \left[\int_{\mathcal{V}} u_t^2(x)\mathrm{d}\mathcal{V}\right]^{\frac{1}{2}}, \tag{11.66}$$

defined on $\mathcal{W}_2^1(\mathcal{V})$ and $\mathcal{W}_2^0(\mathcal{V})$, respectively, where the gradient $\nabla u_t(x)$ in (11.65) is interpreted in the sense of a generalized gradient [454]. Note that since the solutions to (11.3) and (11.4) are assumed to be two-times continuously differentiable functions on a compact set $\mathcal{V}$, it follows that $u_t(x)$, $t \geq t_0$, belongs to both $\mathcal{W}_2^1(\mathcal{V})$ and $\mathcal{W}_2^0(\mathcal{V})$.

Theorem 11.9. Consider the dynamical system given by (11.3) and

---

[5]A *Sobolev space* is a vector space of locally summable functions having weak derivatives of various orders and endowed with a norm formed from a combination of $\mathcal{L}_p$-norms.

[6]The space $\{\cdot\}$ defined as part of (11.63) is not complete with respect to the norm generated by the inner product (11.65). This space can be completed by adding the limit points of all Cauchy sequences in $\{\cdot\}$. In this way, $\{\cdot\}$ is embedded in the larger normed space $\{\cdot\}_{\mathrm{co}}$, which is complete. Of course, by the Riesz-Fischer theorem [387, p. 125], $\mathcal{L}_2$ is complete with respect to the norm generated by the inner product (11.66).

(11.4), and assume that Axioms $i)'$ and $ii)'$ hold. If

$$\phi(x, u(x,t), \nabla u(x,t)) \cdot \hat{n}(x) = 0, \quad x \in \partial\mathcal{V}, \quad t \geq t_0, \tag{11.67}$$

then $u(x,t) \equiv \alpha$, $\alpha \geq 0$, is Lyapunov stable.

Proof. It follows from Axiom $i)'$ that $u(x,t) \equiv \alpha$, $\alpha \geq 0$, is an equilibrium state for (11.3) and (11.4). To show Lyapunov stability of the equilibrium state $u(x,t) \equiv \alpha$, consider the shifted-system scaled ectropy as a Lyapunov functional candidate given by

$$\mathcal{E}(u_t - \alpha) = \frac{1}{2}\int_{\mathcal{V}} (u_t(x) - \alpha)^2 \mathrm{d}\mathcal{V} = \frac{1}{2}\|u_t - \alpha\|^2_{\mathcal{L}_2}. \tag{11.68}$$

Now, it follows from the Green-Gauss theorem and Axioms $i)'$ and $ii)'$ that

$$\begin{aligned}
\dot{\mathcal{E}}(u_t - \alpha) &= \int_{\mathcal{V}} (u(x,t) - \alpha)\frac{\partial u(x,t)}{\partial t}\mathrm{d}\mathcal{V} \\
&= -\int_{\mathcal{V}} u(x,t)\nabla \cdot \phi(x, u(x,t), \nabla u(x,t))\mathrm{d}\mathcal{V} \\
&\quad +\alpha\int_{\mathcal{V}} \nabla \cdot \phi(x, u(x,t), \nabla u(x,t))\mathrm{d}\mathcal{V} \\
&= \int_{\mathcal{V}} \nabla u(x,t)\phi(x, u(x,t), \nabla u(x,t))\mathrm{d}\mathcal{V} \\
&\quad -\int_{\partial\mathcal{V}} u(x,t)\phi(x, u(x,t), \nabla u(x,t)) \cdot \hat{n}(x)\mathrm{d}\mathcal{S}_{\mathcal{V}} \\
&\quad +\alpha\int_{\partial\mathcal{V}} \phi(x, u(x,t), \nabla u(x,t)) \cdot \hat{n}(x)\mathrm{d}\mathcal{S}_{\mathcal{V}} \\
&= \int_{\mathcal{V}} \nabla u(x,t)\phi(x, u(x,t), \nabla u(x,t))\mathrm{d}\mathcal{V} \\
&\leq 0, \quad u_t \in \mathcal{W}_2^0(\mathcal{V}),
\end{aligned} \tag{11.69}$$

which establishes Lyapunov stability of the equilibrium state $u(x,t) \equiv \alpha$. □

Next, we show that the total $\mathcal{L}_2$ norm of the energy of (11.3) and (11.4) is nonincreasing.

Proposition 11.8. Consider the dynamical system given by (11.3) and (11.4), and assume that Axioms $i)'$ and $ii)'$ hold. If either $u(x,t) = 0$ for all $x \in \partial\mathcal{V}$ and $t \geq t_0$ or (11.67) holds, then $\|u_t\|_{\mathcal{W}_2^0} \leq \|u_\tau\|_{\mathcal{W}_2^0}$ for all $t_0 \leq \tau \leq t$.

Proof. Assume $u(x,t) = 0$ for all $x \in \partial\mathcal{V}$ and $t \geq t_0$, and consider the functional

$$\mathcal{E}(u_t) = \|u_t\|^2_{\mathcal{W}_2^0}. \tag{11.70}$$

Now, it follows from the Green-Gauss theorem and Axioms $i)'$ and $ii)'$ that

$$\begin{aligned}\frac{1}{2}\dot{\mathcal{E}}(u_t) &= \int_{\mathcal{V}} u(x,t)\frac{\partial u(x,t)}{\partial t}\mathrm{d}\mathcal{V} \\ &= -\int_{\mathcal{V}} u(x,t)\nabla\cdot\phi(x,u(x,t),\nabla u(x,t))\mathrm{d}\mathcal{V} \\ &= \int_{\mathcal{V}} \nabla u(x,t)\phi(x,u(x,t),\nabla u(x,t))\mathrm{d}\mathcal{V} \\ &\quad -\int_{\partial\mathcal{V}} u(x,t)\phi(x,u(x,t),\nabla u(x,t))\cdot\hat{n}(x)\mathrm{d}\mathcal{S}_{\mathcal{V}} \\ &\leq 0, \quad u_t\in\mathcal{W}_2^0(\mathcal{V}),\end{aligned} \tag{11.71}$$

which implies that $\|u_t\|_{\mathcal{W}_2^0} \leq \|u_\tau\|_{\mathcal{W}_2^0}$ for all $t_0 \leq \tau \leq t$.

Alternatively, if (11.67) holds, then

$$\frac{1}{2}\dot{\mathcal{E}}(u_t) = \int_{\mathcal{V}} \nabla u(x,t)\phi(x,u(x,t),\nabla u(x,t))\mathrm{d}\mathcal{V} \leq 0, \quad u_t\in\mathcal{W}_2^0(\mathcal{V}), \tag{11.72}$$

which implies that $\|u_t\|_{\mathcal{W}_2^0} \leq \|u_\tau\|_{\mathcal{W}_2^0}$ for all $t_0 \leq \tau \leq t$. □

Next, we present necessary and sufficient conditions for semistability of our continuum thermodynamic model (11.3) and (11.4).

Theorem 11.10. Consider the dynamical system given by (11.3) and (11.4). Assume that Axioms $i)'$ and $ii)'$ hold and $D(u_t,u_t) \leq D(u_\tau,u_\tau)$ for all $t_0 \leq \tau \leq t$. Then, for every $\alpha \geq 0$, $u(x,t) \equiv \alpha$ is a semistable equilibrium state of (11.3) and (11.4) if and only if (11.67) holds. In this case, $u(x,t) \to \frac{1}{\mathcal{V}_{\mathrm{vol}}}\int_{\mathcal{V}} u_{t_0}(x)\mathrm{d}\mathcal{V}$ as $t\to\infty$ for every initial condition $u_{t_0}\in\mathcal{W}_2^1(\mathcal{V})$ and every $x\in\mathcal{V}$; moreover, $\frac{1}{\mathcal{V}_{\mathrm{vol}}}\int_{\mathcal{V}} u_{t_0}(x)\mathrm{d}\mathcal{V}$ is a semistable equilibrium state of (11.3) and (11.4).

Proof. Assume that (11.67) holds. Then it follows from Theorem 11.9 that $u(x,t)\equiv\alpha$, $\alpha\geq 0$, is Lyapunov stable. Next, to show semistability of this equilibrium state, consider the Lyapunov functionals (11.70) and

$$V(u_t) = \|u_t\|^2_{\mathcal{W}_2^1}, \quad u_t\in\mathcal{W}_2^1(\mathcal{V}). \tag{11.73}$$

It follows from Proposition 11.8 that $\mathcal{E}(u_t)$ is a nonincreasing functional of time for all $u_{t_0}\in\mathcal{W}_2^0(\mathcal{V})$. Furthermore, note that $V(u_t) = \mathcal{E}(u_t) + D(u_t,u_t)$. Hence, by assumption, $\mathcal{E}(u_t)$ is a nonincreasing functional of time for all $u_{t_0}\in\mathcal{W}_2^1(\mathcal{V})$.

Next, since the functionals $V(u_t)$ and $\mathcal{E}(u_t)$ are nonincreasing and bounded from below by zero, it follows that $V(u_t)$ and $\mathcal{E}(u_t)$ are bounded functionals for every $u_{t_0}\in\mathcal{W}_2^1(\mathcal{V})$. This implies that the positive orbit

$\mathcal{O}_{u_{t_0}}^+ \triangleq \{u_t \in \mathcal{W}_2^1(\mathcal{V}) : u_t(x) = u(x,t), x \in \mathcal{V}, t \in [t_0, \infty)\}$ of (11.3) and (11.4) is bounded in $\mathcal{W}_2^1(\mathcal{V})$ for all $u_{t_0} \in \mathcal{W}_2^1(\mathcal{V})$. Furthermore, it follows from Sobolev's embedding theorem [418,454] that $\mathcal{W}_2^1(\mathcal{V})$ is compactly embedded in $\mathcal{W}_2^0(\mathcal{V})$, and hence, $\mathcal{O}_{u_{t_0}}^+$ is contained in a compact subset of $\mathcal{W}_2^0(\mathcal{V})$.

Now, define the sets $\mathcal{D}_{\mathcal{W}_2^1} = \{u_t \in \mathcal{W}_2^1(\mathcal{V}) : V(u_t) < \eta\}$ and $\mathcal{D}_{\mathcal{W}_2^0} = \{u_t \in \mathcal{W}_2^0(\mathcal{V}) : \mathcal{E}(u_t) < \eta\}$ for some arbitrary $\eta > 0$. Note that $\mathcal{D}_{\mathcal{W}_2^1}$ and $\mathcal{D}_{\mathcal{W}_2^0}$ are invariant sets with respect to (11.3) and (11.4). Moreover, it follows from the definition of $V(u_t)$ and $\mathcal{E}(u_t)$ that $\mathcal{D}_{\mathcal{W}_2^1}$ and $\mathcal{D}_{\mathcal{W}_2^0}$ are bounded sets in $\mathcal{W}_2^1(\mathcal{V})$ and $\mathcal{W}_2^0(\mathcal{V})$, respectively, and $\mathcal{D}_{\mathcal{W}_2^1} \subset \mathcal{D}_{\mathcal{W}_2^0}$. Next, let $\mathcal{R} \triangleq \{u_t \in \overline{\mathcal{D}}_{\mathcal{W}_2^0} : \dot{V}(u_t) = 0\} = \{u_t \in \overline{\mathcal{D}}_{\mathcal{W}_2^0} : \nabla u_t(x)\phi(x, u_t(x), \nabla u_t(x)) = 0, x \in \mathcal{V}\}$. Now, it follows from Axiom $i)'$ that $\mathcal{R} = \{u_t \in \overline{\mathcal{D}}_{\mathcal{W}_2^0} : \nabla u_t(x) = 0, x \in \mathcal{V}\}$ or $\mathcal{R} = \{u_t \in \mathcal{W}_2^0(\mathcal{V}) : u_t(x) \equiv \sigma, 0 \le \sigma \le \sqrt{\frac{\eta}{\mathcal{V}_{\mathrm{vol}}}}\}$, that is, $\mathcal{R}$ is the set of uniform density distributions, which are the equilibrium states of (11.3) and (11.4).

Since the set $\mathcal{R}$ consists of only the equilibrium states of (11.3) and (11.4), it follows that the largest invariant set $\mathcal{M}$ contained in $\mathcal{R}$ is given by $\mathcal{M} = \mathcal{R}$. Hence, noting that $\mathcal{M}$ belongs to the set of generalized (weak) solutions of (11.3) and (11.4) defined on $\mathcal{R}$, it follows from Theorem 2.4 that $u(x,t) \equiv \alpha$ is a semistable equilibrium state of (11.3) and (11.4). Moreover, since $\eta > 0$ can be arbitrarily large but finite and $V(u_t)$ is radially unbounded, the previous statement holds for all $u_{t_0} \in \mathcal{W}_2^1(\mathcal{V})$. Next, note that since, by the divergence theorem,

$$\begin{aligned} \int_{\mathcal{V}} \frac{\partial u(x,t)}{\partial t} \mathrm{d}\mathcal{V} &= -\int_{\mathcal{V}} \nabla \cdot \phi(x, u(x,t), \nabla u(x,t)) \mathrm{d}\mathcal{V} \\ &= -\int_{\partial\mathcal{V}} \phi(x, u(x,t), \nabla u(x,t)) \cdot \hat{n}(x) \mathrm{d}\mathcal{S}_{\mathcal{V}} \\ &= 0, \end{aligned} \tag{11.74}$$

it follows that $\int_{\mathcal{V}} u(x,t)\mathrm{d}\mathcal{V} = \int_{\mathcal{V}} u_{t_0}(x)\mathrm{d}\mathcal{V}$, $t \ge t_0$, which implies that $u(x,t) \to \frac{1}{\mathcal{V}_{\mathrm{vol}}} \int_{\mathcal{V}} u_{t_0}(x)\mathrm{d}\mathcal{V}$ as $t \to \infty$.

Conversely, assume that, for every $\alpha \ge 0$, $u(x,t) \equiv \alpha$ is a semistable equilibrium state of (11.3) and (11.4). Suppose, *ad absurdum*, there exists at least one point $x_{\mathrm{p}} \in \partial\mathcal{V}$ such that $\phi(x_{\mathrm{p}}, u_t(x_{\mathrm{p}}, \nabla u_t(x_{\mathrm{p}}))) \cdot \mathbf{n}(x_{\mathrm{p}}) > 0$. Consider the Lyapunov functional (11.70) and note that the Lyapunov derivative of $\mathcal{E}(u_t)$ is given by (11.71). Let

$$\begin{aligned} \mathcal{R} &\triangleq \{u_t \in \overline{\mathcal{D}}_{\mathcal{W}_2^0} : \dot{\mathcal{E}}(u_t) = 0\} \\ &= \{u_t \in \overline{\mathcal{D}}_{\mathcal{W}_2^0} : \nabla u_t(x)\phi(x, u_t(x), \nabla u_t(x)) = 0, x \in \mathcal{V}\} \\ &\quad \cap \{u_t \in \overline{\mathcal{D}}_{\mathcal{W}_2^0} : u(x,t)\phi(x, u_t(x), \nabla u_t(x)) \cdot \hat{n}(x) = 0, x \in \partial\mathcal{V}\}. \end{aligned}$$

Now, since Axiom $i)'$ holds, it follows that

$$\begin{aligned}\mathcal{R} &= \{u_t \in \overline{\mathcal{D}}_{\mathcal{W}_2^0} : \nabla u_t(x) = 0, x \in \mathcal{V}\} \cap \{u_t \in \overline{\mathcal{D}}_{\mathcal{W}_2^0} : u_t(x_\mathrm{p}) = 0, x_\mathrm{p} \in \partial\mathcal{V}\} \\ &= \{0\},\end{aligned}$$

and the largest invariant set $\mathcal{M}$ contained in $\mathcal{R}$ is given by $\mathcal{M} = \{0\}$. By assumption, $V(u_t)$ is a nonincreasing functional of time for all $u_{t_0} \in \mathcal{W}_2^1(\mathcal{V})$, and since $V(u_t)$ is bounded from below by zero, the positive orbit $\mathcal{O}_{u_{t_0}}^+$ of (11.3) and (11.4) is bounded in $\mathcal{W}_2^1(\mathcal{V})$.

Hence, since $\mathcal{W}_2^1(\mathcal{V})$ is compactly embedded in $\mathcal{W}_2^0(\mathcal{V})$, it follows from Sobolev's embedding theorem [418,454] that $\mathcal{O}_{u_{t_0}}^+$ is contained in a compact subset of $\mathcal{W}_2^0(\mathcal{V})$. Thus, it follows from Theorem 2.13 that for every initial density distribution $u_{t_0} \in \mathcal{D}_{\mathcal{W}_2^0}$, $u(x,t) \to \mathcal{M} = \{0\}$ as $t \to \infty$ with respect to the norm $\|\cdot\|_{\mathcal{W}_2^0}$, which shows asymptotic stability of the zero equilibrium state of (11.3) and (11.4). However, since asymptotic stability of (11.3) and (11.4) is equivalent to semistability of (11.3) and (11.4) if and only if the equilibrium state of (11.3) and (11.4) is zero, this contradicts the assumption that for every $\alpha \geq 0$, $u(x,t) \equiv \alpha$ is an equilibrium state of (11.3) and (11.4). Hence, (11.67) holds. □

The following corollary to Theorem 11.10 is immediate.

Corollary 11.1. Consider the dynamical system $\mathcal{G}$ given by (11.3) and (11.4). Assume that Axioms $i)'$ and $ii)'$ hold and

$$\nabla^2 u_t(x) \nabla \cdot \phi(x, u_t(x), \nabla u_t(x)) \leq 0, \quad x \in \mathcal{V}, \quad u_t \in \mathcal{W}_2^1(\mathcal{V}), \tag{11.75}$$

where $\nabla^2 \triangleq \nabla \cdot \nabla$ denotes the Laplace operator. Then, for every $\alpha \geq 0$, $u(x,t) \equiv \alpha$ is a semistable equilibrium state of (11.3) and (11.4) if and only if (11.67) holds. In this case, $u(x,t) \to \frac{1}{\mathcal{V}_\mathrm{vol}} \int_\mathcal{V} u_{t_0}(x) \mathrm{d}\mathcal{V}$ as $t \to \infty$ for every initial condition $u_{t_0} \in \mathcal{W}_2^1(\mathcal{V})$ and every $x \in \mathcal{V}$; moreover, $\frac{1}{\mathcal{V}_\mathrm{vol}} \int_\mathcal{V} u_{t_0}(x) \mathrm{d}\mathcal{V}$ is a semistable equilibrium state of (11.3) and (11.4).

Proof. The result is a direct consequence of Theorem 11.10 by showing that the Dirichlet integral $D(u_t, u_t)$ of u_t is nonincreasing. To see this, note that it follows from the Green-Gauss theorem and (11.67) that

$$\begin{aligned}\frac{1}{2}\dot{D}(u_t, u_t) &= \int_\mathcal{V} \nabla u(x,t) \frac{\partial}{\partial t} (\nabla u(x,t))^\mathrm{T} \mathrm{d}\mathcal{V} \\ &= \int_{\partial\mathcal{V}} \frac{\partial u(x,t)}{\partial t} D_{\hat{n}(x)} u(x,t) \mathrm{d}\mathcal{S}_\mathcal{V} \\ &\quad + \int_\mathcal{V} \nabla^2 u(x,t) \nabla \cdot \phi(x, u(x,t), \nabla u(x,t)) \mathrm{d}\mathcal{V},\end{aligned} \tag{11.76}$$

where $D_{\hat{n}(x)} u(x,t) \triangleq \nabla u(x,t) \hat{n}(x)$ denotes the directional derivative of

$u(x,t)$ along $\hat{n}(x)$ at $x \in \partial\mathcal{V}$. Next, it follows from (11.67) and Axiom $i)'$, with $\mathbf{u} = \hat{n}(x)$, that $D_{\hat{n}(x)}u(x,t) = 0$, $x \in \partial\mathcal{V}$. Hence, it follows from (11.75) and (11.76) that $\dot{D}(u_t, u_t) \le 0$, $t \ge t_0$, for every $u_{t_0} \in \mathcal{W}_2^1(\mathcal{V})$. □

The following theorem combines Theorem 11.10 and Corollary 11.1. For completeness we provide a self-contained proof of this result.

Theorem 11.11. Consider the dynamical system $\mathcal{G}$ with power balance equation (11.3) and (11.4) with $s(x,t) \equiv 0$ and $\phi(x, u(x,t), \nabla u(x,t)) \cdot \hat{n}(x) \equiv 0$, $x \in \partial\mathcal{V}$. Assume that Axioms $i)'$ and $ii)'$ hold and

$$\nabla^2 u_t(x) \nabla \cdot \phi(x, u_t(x), \nabla u_t(x)) \le 0, \quad x \in \mathcal{V}, \quad u_t \in \mathcal{W}_2^1(\mathcal{V}), \tag{11.77}$$

where $\nabla^2 \triangleq \nabla \cdot \nabla$ denotes the Laplacian operator. Then, for every $\alpha \ge 0$, $u(x,t) \equiv \alpha$ is a semistable equilibrium state of (11.3) and (11.4). Furthermore, $u(x,t) \to \frac{1}{\mathcal{V}_{\rm vol}} \int_{\mathcal{V}} u_{t_0}(x) \mathrm{d}\mathcal{V}$ as $t \to \infty$ for every initial energy density distribution $u_{t_0} \in \mathcal{W}_2^1(\mathcal{V})$ and every $x \in \mathcal{V}$; moreover, $\frac{1}{\mathcal{V}_{\rm vol}} \int_{\mathcal{V}} u_{t_0}(x) \mathrm{d}\mathcal{V}$ is a semistable equilibrium distribution state of (11.3) and (11.4). Finally, if $s(x,t) \equiv 0$ and there exists at least one point $x_{\rm p} \in \partial\mathcal{V}$ such that $\phi(x_{\rm p}, u_t(x_{\rm p}), \nabla u_t(x_{\rm p})) \cdot \hat{n}(x_{\rm p}) > 0$ and $\phi(x_{\rm p}, u_t(x_{\rm p}), \nabla u_t(x_{\rm p})) \cdot \hat{n}(x_{\rm p}) = 0$ if and only if $u_t(x_{\rm p}) = 0$, then the zero solution $u(x,t) \equiv 0$ to (11.3) and (11.4) is a globally asymptotically stable equilibrium state of (11.3) and (11.4).

Proof. It follows from Axiom $i)'$ that $u(x,t) \equiv \alpha$, $\alpha \ge 0$, is an equilibrium state for (11.3) and (11.4) with $s(x,t) \equiv 0$ and $\phi(x, u(x,t), \nabla u(x,t)) \cdot \hat{n}(x) \equiv 0$. To show Lyapunov stability of the equilibrium state $u(x,t) \equiv \alpha$, consider the shifted-system scaled ectropy $\mathcal{E}_{\rm s}(u_t) = \frac{1}{2} \int_{\mathcal{V}} (u_t(x) - \alpha)^2 \mathrm{d}\mathcal{V} = \frac{1}{2} \|u_t - \alpha\|_{\mathcal{L}_2}^2$ as a Lyapunov functional candidate. Now, it follows from the Green-Gauss theorem and Axiom $ii)'$ that

$$\begin{aligned}
\dot{\mathcal{E}}_{\rm s}(u_t) &= \int_{\mathcal{V}} (u(x,t) - \alpha) \frac{\partial u(x,t)}{\partial t} \mathrm{d}\mathcal{V} \\
&= -\int_{\mathcal{V}} u(x,t) \nabla \cdot \phi(x, u(x,t), \nabla u(x,t)) \mathrm{d}\mathcal{V} \\
&\quad + \alpha \int_{\mathcal{V}} \nabla \cdot \phi(x, u(x,t), \nabla u(x,t)) \mathrm{d}\mathcal{V} \\
&= \int_{\mathcal{V}} \nabla u(x,t) \phi(x, u(x,t), \nabla u(x,t)) \mathrm{d}\mathcal{V} \\
&\quad - \int_{\partial\mathcal{V}} u(x,t) \phi(x, u(x,t), \nabla u(x,t)) \cdot \hat{n}(x) \, \mathrm{d}\mathcal{S}_{\mathcal{V}} \\
&\quad + \alpha \int_{\partial\mathcal{V}} \phi(x, u(x,t), \nabla u(x,t)) \cdot \hat{n}(x) \, \mathrm{d}\mathcal{S}_{\mathcal{V}}
\end{aligned}$$

$$
\begin{aligned}
&= \int_{\mathcal{V}} \nabla u(x,t)\phi(x,u(x,t),\nabla u(x,t))\mathrm{d}\mathcal{V} \\
&\leq 0, \quad u_t \in \mathcal{W}_2^0(\mathcal{V}),
\end{aligned} \tag{11.78}
$$

which establishes Lyapunov stability of the equilibrium state $u(x,t) \equiv \alpha$.

Next, to show semistability of this equilibrium state, consider the following (scaled) ectropy and ectropy-like Lyapunov functionals

$$
\mathcal{E}_0(u_t) = \|u_t\|^2_{\mathcal{W}_2^0}, \quad u_t \in \mathcal{W}_2^0(\mathcal{V}), \tag{11.79}
$$

$$
\mathcal{E}_1(u_t) = \|u_t\|^2_{\mathcal{W}_2^1}, \quad u_t \in \mathcal{W}_2^1(\mathcal{V}). \tag{11.80}
$$

It follows from (11.50) with $s(x,t) \equiv 0$ that $\mathcal{E}_0(u_t)$ is a nonincreasing functional of time for all $u_{t_0} \in \mathcal{W}_2^0(\mathcal{V})$. Furthermore, it follows from the Green-Gauss theorem and the boundary condition $\phi(x,u(x,t),\nabla u(x,t)) \cdot \hat{n}(x) \equiv 0$, $x \in \partial\mathcal{V}$, that

$$
\begin{aligned}
\tfrac{1}{2}\dot{\mathcal{E}}_1(u_t) &= \int_{\mathcal{V}} \left(u(x,t)\frac{\partial u(x,t)}{\partial t} + \nabla u(x,t)\frac{\partial}{\partial t}\left(\nabla u(x,t)\right)^{\mathrm{T}} \right) \mathrm{d}\mathcal{V} \\
&= \int_{\mathcal{V}} \nabla u(x,t)\phi(x,u(x,t),\nabla u(x,t))\mathrm{d}\mathcal{V} \\
&\quad - \int_{\partial\mathcal{V}} u(x,t)\phi(x,u(x,t),\nabla u(x,t)) \cdot \hat{n}(x)\,\mathrm{d}\mathcal{S}_{\mathcal{V}} \\
&\quad + \int_{\partial\mathcal{V}} \frac{\partial u(x,t)}{\partial t} D_{\hat{n}(x)}u(x,t)\,\mathrm{d}\mathcal{S}_{\mathcal{V}} \\
&\quad + \int_{\mathcal{V}} \nabla^2 u(x,t)\nabla \cdot \phi(x,u(x,t),\nabla u(x,t))\mathrm{d}\mathcal{V} \\
&= \int_{\mathcal{V}} \nabla u(x,t)\phi(x,u(x,t),\nabla u(x,t))\mathrm{d}\mathcal{V} \\
&\quad + \int_{\partial\mathcal{V}} \frac{\partial u(x,t)}{\partial t} D_{\hat{n}(x)}u(x,t)\,\mathrm{d}\mathcal{S}_{\mathcal{V}} \\
&\quad + \int_{\mathcal{V}} \nabla^2 u(x,t)\nabla \cdot \phi(x,u(x,t),\nabla u(x,t))\mathrm{d}\mathcal{V},
\end{aligned} \tag{11.81}
$$

where $D_{\hat{n}(x)}u(x,t) \triangleq \nabla u(x,t)\hat{n}(x)$ denotes the directional derivative of $u(x,t)$ along $\hat{n}(x)$ at $x \in \partial\mathcal{V}$.

Next, note that for the isolated dynamical system $\mathcal{G}$ with the boundary condition $\phi(x,u(x,t),\nabla u(x,t)) \cdot \hat{n}(x) \equiv 0$, $x \in \partial\mathcal{V}$, it follows from Axiom $i)'$, with $\mathbf{u} = \hat{n}(x)$, that $D_{\hat{n}(x)}u(x,t) \equiv 0$, $x \in \partial\mathcal{V}$. Hence, it follows from Axiom $ii)'$, (11.77), and (11.81) that $\dot{\mathcal{E}}_1(u_t) \leq 0$, $t \geq t_0$, for any $u_{t_0} \in \mathcal{W}_2^1(\mathcal{V})$. Furthermore, since the functionals $\mathcal{E}_1(u_t)$ and $\mathcal{E}_0(u_t)$ are nonincreasing and bounded from below by zero, it follows that $\mathcal{E}_1(u_t)$ and $\mathcal{E}_0(u_t)$ are bounded functionals for every $u_{t_0} \in \mathcal{W}_2^1(\mathcal{V})$. This implies that the positive orbit

$\mathcal{O}_{u_{t_0}}^+ \triangleq \{u_t \in \mathcal{W}_2^1(\mathcal{V}) : u_t(x) = u(x,t),\, x \in \mathcal{V},\, t \in [t_0, \infty)\}$ of $\mathcal{G}$ is bounded in $\mathcal{W}_2^1(\mathcal{V})$ for all $u_{t_0} \in \mathcal{W}_2^1(\mathcal{V})$. Furthermore, it follows from Sobolev's embedding theorem [418,454] that $\mathcal{W}_2^1(\mathcal{V})$ is compactly embedded in $\mathcal{W}_2^0(\mathcal{V})$, and hence, $\mathcal{O}_{u_{t_0}}^+$ is contained in a compact subset of $\mathcal{W}_2^0(\mathcal{V})$.

Next, define the sets $\mathcal{D}_{\mathcal{W}_2^1} = \{u_t \in \mathcal{W}_2^1(\mathcal{V}) : \mathcal{E}_1(u_t) < \eta\}$ and $\mathcal{D}_{\mathcal{W}_2^0} = \{u_t \in \mathcal{W}_2^0(\mathcal{V}) : \mathcal{E}_0(u_t) < \eta\}$ for some arbitrary $\eta > 0$. Note that $\mathcal{D}_{\mathcal{W}_2^1}$ and $\mathcal{D}_{\mathcal{W}_2^0}$ are invariant sets with respect to the dynamical system $\mathcal{G}$. Moreover, it follows from the definition of $\mathcal{E}_1(u_t)$ and $\mathcal{E}_0(u_t)$ that $\mathcal{D}_{\mathcal{W}_2^1}$ and $\mathcal{D}_{\mathcal{W}_2^0}$ are bounded sets in $\mathcal{W}_2^1(\mathcal{V})$ and $\mathcal{W}_2^0(\mathcal{V})$, respectively, and $\mathcal{D}_{\mathcal{W}_2^1} \subset \mathcal{D}_{\mathcal{W}_2^0}$. Next, let

$$\begin{aligned}\mathcal{R} &\triangleq \{u_t \in \overline{\mathcal{D}}_{\mathcal{W}_2^0} : \dot{\mathcal{E}}_0(u_t) = 0\}\\ &= \{u_t \in \overline{\mathcal{D}}_{\mathcal{W}_2^0} : \nabla u_t(x)\phi(x, u_t(x), \nabla u_t(x)) = 0,\ x \in \mathcal{V}\}.\end{aligned}$$

Now, it follows from Axioms $i)'$ and $ii)'$ that $\mathcal{R} = \{u_t \in \overline{\mathcal{D}}_{\mathcal{W}_2^0} : \nabla u_t(x) = 0,\, x \in \mathcal{V}\}$ or $\mathcal{R} = \{u_t \in \mathcal{W}_2^0(\mathcal{V}) : u_t(x) \equiv \sigma,\, 0 \le \sigma \le \sqrt{\frac{\eta}{\mathcal{V}_{\rm vol}}}\}$, that is, $\mathcal{R}$ is the set of uniform energy density distributions, which are the equilibrium states of (11.3) and (11.4).

Since the set $\mathcal{R}$ consists of only the equilibrium states of (11.3) and (11.4), it follows that the largest invariant set $\mathcal{M}$ contained in $\mathcal{R}$ is given by $\mathcal{M} = \mathcal{R}$. Hence, noting that $\mathcal{M}$ belongs to the set of generalized (or weak) solutions to (11.3) and (11.4) defined on $\mathcal{R}$, it follows from Theorem 2.13 that for any initial energy density distribution $u_{t_0} \in \mathcal{D}_{\mathcal{W}_2^1}$, $u(x,t) \to \mathcal{M}$ as $t \to \infty$ with respect to the norm $\|\cdot\|_{\mathcal{W}_2^0}$, and hence, $u(x,t) \equiv \alpha$ is a semistable equilibrium state of (11.3) and (11.4). Moreover, since $\eta > 0$ can be arbitrarily large but finite and $\mathcal{E}_1(u_t)$ is radially unbounded, the previous statement holds for all $u_{t_0} \in \mathcal{W}_2^1(\mathcal{V})$. Next, note that since, by the divergence theorem,

$$\begin{aligned}\int_{\mathcal{V}} \frac{\partial u(x,t)}{\partial t} \mathrm{d}\mathcal{V} &= -\int_{\mathcal{V}} \nabla \cdot \phi(x, u(x,t), \nabla u(x,t)) \mathrm{d}\mathcal{V}\\ &= -\int_{\partial\mathcal{V}} \phi(x, u(x,t), \nabla u(x,t)) \cdot \hat{n}(x)\, \mathrm{d}\mathcal{S}_{\mathcal{V}}\\ &= 0,\end{aligned}\tag{11.82}$$

it follows that $\int_{\mathcal{V}} u(x,t)\mathrm{d}\mathcal{V} = \int_{\mathcal{V}} u_{t_0}(x)\mathrm{d}\mathcal{V}$, $t \ge t_0$, which implies that $u(x,t) \to \frac{1}{\mathcal{V}_{\rm vol}} \int_{\mathcal{V}} u_{t_0}(x)\mathrm{d}\mathcal{V}$ as $t \to \infty$.

Finally, we show that if $s(x,t) \equiv 0$ and there exists at least one point $x_{\rm p} \in \partial\mathcal{V}$ such that $\phi(x_{\rm p}, u_t(x_{\rm p}), \nabla u_t(x_{\rm p})) \cdot \hat{n}(x_{\rm p}) > 0$ and $\phi(x_{\rm p}, u_t(x_{\rm p}), \nabla u_t(x_{\rm p})) \cdot \hat{n}(x_{\rm p}) = 0$ if and only if $u_t(x_{\rm p}) = 0$, then the zero solution $u(x,t) \equiv 0$ to (11.3) and (11.4) is a globally asymptotically stable equilibrium state. Note that it follows from the above analysis with $\alpha = 0$ that the zero solution $u(x,t) \equiv 0$ is semistable, and hence, a Lyapunov stable equilibrium state of (11.3) and (11.4). Furthermore, it follows from Axiom $ii)'$ with $\mathbf{u} = \hat{n}(x_{\rm p})$ that $D_{\hat{n}(x_{\rm p})}u(x_{\rm p},t) = \nabla u(x_{\rm p},t)\hat{n}(x_{\rm p}) < 0$ and $D_{\hat{n}(x_{\rm p})}u(x_{\rm p},t) = 0$ if and only if $u(x_{\rm p},t) = 0$.

In this case, using Axiom $ii)'$, it follows that the energy flow is directed towards the point $x_{\rm p} \in \partial\mathcal{V}$, and hence, $\frac{\partial u(x_{\rm p},t)}{\partial t} > 0$ and $D_{\hat{n}(x_{\rm p})}u(x_{\rm p},t)\frac{\partial u(x_{\rm p},t)}{\partial t} < 0$. Thus, it follows from Axiom $ii)'$, (11.77), and (11.81) that $\mathcal{E}_1(u_t)$ is a nonincreasing functional of time for all $u_{t_0} \in \mathcal{W}_2^1(\mathcal{V})$, and since $\mathcal{E}_1(u_t)$ is bounded from below by zero, the positive orbit $\mathcal{O}^+_{u_{t_0}}$ of $\mathcal{G}$ is bounded in $\mathcal{W}_2^1(\mathcal{V})$. Hence, since $\mathcal{W}_2^1(\mathcal{V})$ is compactly embedded in $\mathcal{W}_2^0(\mathcal{V})$, it follows from Sobolev's embedding theorem [418, 454] that $\mathcal{O}^+_{u_{t_0}}$ is contained in a compact subset of $\mathcal{W}_2^0(\mathcal{V})$.

Next, consider the (scaled) ectropy Lyapunov functional $\mathcal{E}_0(u_t)$ and note that the Lyapunov derivative is given by

$$\begin{aligned}
\tfrac{1}{2}\dot{\mathcal{E}}_0(u_t) &= \int_{\mathcal{V}} u(x,t)\frac{\partial u(x,t)}{\partial t}\mathrm{d}\mathcal{V} \\
&= -\int_{\mathcal{V}} u(x,t)\nabla\cdot\phi(x,u(x,t),\nabla u(x,t))\mathrm{d}\mathcal{V} \\
&= \int_{\mathcal{V}} \nabla u(x,t)\phi(x,u(x,t),\nabla u(x,t))\mathrm{d}\mathcal{V} \\
&\quad - \int_{\partial\mathcal{V}} u(x,t)\phi(x,u(x,t),\nabla u(x,t))\cdot\hat{n}(x)\,\mathrm{d}\mathcal{S}_{\mathcal{V}} \\
&\le 0, \quad u_t \in \mathcal{W}_2^0(\mathcal{V}).
\end{aligned} \tag{11.83}$$

Furthermore, let

$$\begin{aligned}
\mathcal{R} &\triangleq \{u_t \in \overline{\mathcal{D}}_{\mathcal{W}_2^0} : \dot{\mathcal{E}}_0(u_t) = 0\} \\
&= \{u_t \in \overline{\mathcal{D}}_{\mathcal{W}_2^0} : \nabla u_t(x)\phi(x,u_t(x),\nabla u_t(x)) \equiv 0,\ x \in \mathcal{V}\} \\
&\quad \cap \{u_t \in \overline{\mathcal{D}}_{\mathcal{W}_2^0} : \phi(x,u_t(x),\nabla u_t(x))\cdot\hat{n}(x) = 0,\ x \in \partial\mathcal{V}\}.
\end{aligned}$$

Now, since Axioms $i)'$ and $ii)'$ hold,

$$\begin{aligned}
\mathcal{R} &= \{u_t \in \overline{\mathcal{D}}_{\mathcal{W}_2^0} : \nabla u_t(x) = 0,\ x \in \mathcal{V}\} \\
&\quad \cap \{u_t \in \overline{\mathcal{D}}_{\mathcal{W}_2^0} : u_t(x_{\rm p}) = 0 \text{ for some } x_{\rm p} \in \partial\mathcal{V}\} = \{0\},
\end{aligned}$$

and the largest invariant set $\mathcal{M}$ contained in $\mathcal{R}$ is given by $\mathcal{M} = \{0\}$. Hence, it follows from Theorem 2.13 that for any initial energy density distribution $u_{t_0} \in \mathcal{D}_{\mathcal{W}_2^1}$, $u(x,t) \to \mathcal{M} = \{0\}$ as $t \to \infty$ with respect to the norm $\|\cdot\|_{\mathcal{W}_2^0}$, which, since $\eta > 0$ is arbitrary and $\mathcal{E}_1(u_t)$ is radially unbounded, proves global asymptotic stability of the zero equilibrium state of (11.3) and (11.4). □

Condition (11.77) physically implies that for an energy density distribution $u_t(x)$, $x \in \mathcal{V}$, the energy flow $\phi(x, u_t(x), \nabla u_t(x))$ at $x \in \mathcal{V}$ is proportional to the energy density at this point. Note that for the linear energy flow model corresponding to the heat equation, that is, $\phi(x, u_t(x), \nabla u_t(x)) = -k\,[\nabla u_t(x)]^{\mathrm{T}}$, where $k > 0$ is a conductivity constant, condition (11.77) is automatically satisfied since

$$\nabla^2 u_t(x) \nabla \cdot \phi(x, u_t(x), \nabla u_t(x)) = -k[\nabla^2 u_t(x)]^2 \leq 0, \quad x \in \mathcal{V}.$$

Theorem 11.11 shows that the isolated dynamical system $\mathcal{G}$ is semistable. Hence, it follows from the infinite-dimensional version of Theorem 2.26 that the isolated dynamical system $\mathcal{G}$ does not exhibit Poincaré recurrence in $\mathcal{X} \setminus \mathcal{M}_{\mathrm{e}}$. This result can also be arrived at using (11.39) or (11.58) along with the infinite-dimensional version of Theorem 2.25.

Equation (11.67) plays a critical role in the behavior of (11.3) and (11.4). In particular, (11.67) along with Axioms $i)'$ and $ii)'$ give a criterion for guaranteeing semistability of (11.3) and (11.4). Next, we analyze the behavior of (11.3) and (11.4) using (11.67). First, we consider a Dirichlet boundary condition. The Dirichlet boundary condition for (11.3) and (11.4) involves the form

$$u(x,t) = U_{\mathrm{d}}(x,t), \quad x \in \partial\mathcal{V}, \quad t \geq t_0. \tag{11.84}$$

It follows from (11.67) and Axiom $i)'$ that for the Dirichlet boundary condition, the input $U_{\mathrm{d}}(x,t)$ should be chosen to satisfy

$$\nabla g(x) \nabla^{\mathrm{T}} U_{\mathrm{d}}(x,t) = 0, \quad x \in \partial\mathcal{V}, \quad t \geq t_0. \tag{11.85}$$

Next, we consider a Neumann boundary condition for (11.3) and (11.4). The Neumann boundary condition for (11.3) and (11.4) involves the form

$$\frac{\partial u(x,t)}{\partial \hat{n}(x)} = U_{\mathrm{n}}(x,t), \quad x \in \partial\mathcal{V}, \quad t \geq t_0. \tag{11.86}$$

However, since $\frac{\partial u(x,t)}{\partial \hat{n}(x)} = \nabla u_t(x) \cdot \hat{n}(x)$, it follows from (11.67) and Axiom $i)'$ that $U_{\mathrm{n}}(x,t) = 0$, $x \in \partial\mathcal{V}$, $t \geq t_0$, resulting in a trivial Neumann boundary condition.

Finally, we consider a linear form of (11.3) and (11.4). Specifically, consider the linear (heat) equation given by

$$\frac{\partial u(x,t)}{\partial t} = \nabla^2 u(x,t), \quad x \in \mathcal{V}, \quad t \geq t_0, \quad u(x,t_0) = u_{t_0}(x), \quad x \in \mathcal{V}, \tag{11.87}$$

where $u : \mathbb{R} \times [0, \infty) \to \overline{\mathbb{R}}_+$. It can be easily shown that Axioms $i)'$ and $ii)'$ hold, and (11.75) holds for (11.87). Now, using the Neumann boundary condition

$$\nabla u(x,t) \cdot \hat{n}(x) = 0, \quad x \in \partial\mathcal{V}, \quad t \geq t_0, \tag{11.88}$$

it follows that all the equilibrium points of (11.87) are given by $u(x,t) \equiv \alpha \in \mathbb{R}$ [138, p. 346]. Hence, it follows from Corollary 11.1 that the linear equation (11.87) achieves uniform temperature distributions over $\mathcal{V}$. The boundary condition (11.88) implies that there is no heat flow into or out of $\mathcal{V}$, that is, the boundary $\partial\mathcal{V}$ is insulated.

Finally, we consider the Neumann boundary condition given by

$$U_{\mathrm{n}}(x,t) = -c(u(x,t) - u_{\mathrm{e}}), \quad x \in \partial\mathcal{V}, \quad t \geq t_0, \tag{11.89}$$

where $c > 0$ and $u_{\mathrm{e}} \geq 0$. This form is known as *Newton's law of cooling* in the literature [150, p. 155] and guarantees that, outside $\mathcal{V}$, the temperature $u(x,t)$ is maintained at u_{e} and the rate of heat flow across the boundary is proportional to $u - u_{\mathrm{e}}$. More precisely, Newton's law of cooling states that when a body is losing heat under forced convection, the time rate of change of heat loss $\dot{Q}$ is proportional to the surface area A of the body and the temperature difference of the body T and the environment T_{e}. That is,

$$\dot{Q} = hA(T - T_{\mathrm{e}}),$$

where h is the convective heat transfer coefficient.

Proposition 11.9. Consider the linear equation (11.87) with the boundary condition (11.89). Then $u(x,t) \equiv u_{\mathrm{e}}$ is an asymptotically stable equilibrium state of (11.87) and (11.89).

Proof. Consider the Lyapunov functional candidate $\mathcal{E}(u_t - u_{\mathrm{e}}) = \frac{1}{2}\int_{\mathcal{V}}(u_t(x) - u_{\mathrm{e}})^2 \mathrm{d}\mathcal{V} = \frac{1}{2}\|u_t - u_{\mathrm{e}}\|^2_{\mathcal{L}_2}$. Now, it follows from the Green-Gauss theorem that

$$\begin{aligned}
\dot{\mathcal{E}}(u_t - u_{\mathrm{e}}) &= \int_{\mathcal{V}} (u(x,t) - u_{\mathrm{e}}) \frac{\partial u(x,t)}{\partial t} \mathrm{d}\mathcal{V} \\
&= \int_{\mathcal{V}} u(x,t) \nabla^2 u(x,t) \mathrm{d}\mathcal{V} - u_{\mathrm{e}} \int_{\mathcal{V}} \nabla^2 u(x,t) \mathrm{d}\mathcal{V} \\
&= -\int_{\mathcal{V}} \nabla u(x,t) \nabla^{\mathrm{T}} u(x,t) \mathrm{d}\mathcal{V} + \int_{\partial\mathcal{V}} u(x,t) \nabla u(x,t) \cdot \hat{n}(x) \mathrm{d}\mathcal{S}_{\mathcal{V}} \\
&\quad -u_{\mathrm{e}} \int_{\partial\mathcal{V}} \nabla u(x,t) \cdot \hat{n}(x) \mathrm{d}\mathcal{S}_{\mathcal{V}} \\
&= -D(u_t, u_t) + \int_{\partial\mathcal{V}} (u(x,t) - u_{\mathrm{e}}) \nabla u(x,t) \cdot \hat{n}(x) \mathrm{d}\mathcal{S}_{\mathcal{V}}
\end{aligned}$$

$$= -D(u_t, u_t) - c\int_{\partial\mathcal{V}} (u(x,t) - u_{\rm e})^2 {\rm d}\mathcal{S}_{\mathcal{V}}$$
$$< 0, \quad u_t \in \mathcal{W}_2^0(\mathcal{V}), \quad u_t \neq u_{\rm e}, \tag{11.90}$$

which establishes asymptotic stability of the equilibrium state $u(x,t) \equiv u_{\rm e}$. □

Example 11.1. Consider the diffusion equation model for heat transport in a uniform bar. Using Fourier's law of heat conduction, which states that the time rate of change of heat transfer through a body is proportional to the temperature gradient and the area through which heat flows, it follows that the relation between the rate of heat exchanged with the environment $q(x,t)$ (with $q(x,t) > 0$ when heat is absorbed by the bar) and the temperature $T(x,t)$ is given by the partial differential equation

$$\rho\frac{\partial}{\partial t}T(x,t) = \gamma\frac{\partial^2}{\partial x^2}T(x,t) + q(x,t), \tag{11.91}$$

where ρ is (proportional to) the specific heat coefficient of the material and γ is the heat diffusion coefficient. More specifically, the factor $\alpha \triangleq \gamma/\rho$ is a property of the material known as *thermal diffusivity* and is a measure of the quantity of heat per unit time that passes through the body.

Assuming that the length of the bar is L, the temperature at the ends of the bar is fixed to T_0, and there is no heat transport at the ends of the bar, the boundary conditions are given by

$$T(0,\cdot) = T(L,\cdot) = T_0, \quad \frac{\partial}{\partial x}T(0,\cdot) = \frac{\partial}{\partial x}T(L,\cdot) = 0. \tag{11.92}$$

For simplicity of exposition, we assume that the units have been chosen such that $\rho = 1$, $\gamma = 1$, and $L = 1$.

Next, note that for every $(T,q) : \mathbb{R} \times [0,1] \to \mathbb{R}_+ \times \mathbb{R}$ satisfying (11.91) with boundary conditions (11.92),

$$\frac{\rm d}{{\rm d}t}\int_0^1 T(x,t){\rm d}x = \int_0^1 q(x,t){\rm d}x \tag{11.93}$$

holds, where $\int_0^1 q(x,t){\rm d}x$ is the power delivered to the bar at time t, and hence, $\int_0^1 T(x,t){\rm d}x$ is the system stored energy. This follows from the fact that, up to an additive constant, $\int_0^1 T(x,t){\rm d}x$ is the unique time function whose derivative along solutions of (11.91) is equal to $\int_0^1 q(x,t){\rm d}x$.

Next, note that $(T,q) : \mathbb{R} \times [0,1] \to \mathbb{R}_+ \times \mathbb{R}$ also satisfies

$$\frac{\rm d}{{\rm d}t}\int_0^1 \log_e T(x,t){\rm d}x = \int_0^1 \left(\frac{1}{T(x,t)}\frac{\partial}{\partial x}T(x,t)\right)^2 {\rm d}x + \int_0^1 \frac{q(x,t)}{T(x,t)}{\rm d}x,$$

and hence,

$$\frac{\mathrm{d}}{\mathrm{d}t}\int_0^1 \log_e T(x,t)\mathrm{d}x \geq \int_0^1 \frac{q(x,t)}{T(x,t)}\mathrm{d}x, \tag{11.94}$$

which implies that $\int_0^1 \log_e T(x,t)\mathrm{d}x$ is the system entropy. This follows from the fact that, up to an additive constant, $\int_0^1 \log_e T(x,t)\mathrm{d}x$ is the unique function whose time derivative is greater than or equal to $\int_0^1 \frac{q(x,t)}{T(x,t)}\mathrm{d}x$.

Now, assume that we apply a cyclic heat input, wherein the bar is heated at initial time t_i in a temperature distribution $T(\cdot,t_\mathrm{i})$ and ending at final time $t_\mathrm{f} > t_\mathrm{i}$ in the same temperature distribution $T(\cdot,t_\mathrm{f}) = T(\cdot,t_\mathrm{i})$. Note that over the temporal interval $[t_\mathrm{i},t_\mathrm{f}]$ and spatial interval $[0,1]$, $q(x,t)$ can be positive, negative, or zero. However, in all such cases it follows from (11.93) and (11.94) that

$$\int_{t_\mathrm{i}}^{t_\mathrm{f}}\left(\int_0^1 q(x,t)\mathrm{d}x\right)\mathrm{d}t = 0 \tag{11.95}$$

and

$$\int_{t_\mathrm{i}}^{t_\mathrm{f}}\left(\int_0^1 \frac{q(x,t)}{T(x,t)}\mathrm{d}x\right)\mathrm{d}t \leq 0. \tag{11.96}$$

Now, it follows from (11.95) and (11.96) that

$$\max_{x\in[0,1],t\in[t_\mathrm{i},t_\mathrm{f}]}\{T(x,t) : q(x,t) > 0\} \geq \min_{x\in[0,1],t\in[t_\mathrm{i},t_\mathrm{f}]}\{T(x,t) : q(x,t) < 0\}, \tag{11.97}$$

which gives Clausius' version of the second law. Note that (11.97) does not involve the system entropy.

The equality

$$\int_{t_\mathrm{i}}^{t_\mathrm{f}}\left(\int_0^1 q(x,t)\mathrm{d}x\right)\mathrm{d}t = 0$$

implies that the net effect of the (T,q) history is to transport exactly the same amount of heat from places and times where it is delivered by the environment to the bar, to places and times where it is delivered by the bar to the environment. Energy (heat) is merely redistributed. This is a statement of the conservation of energy. However, the inequality

$$\max_{x\in[0,1],t\in[t_\mathrm{i},t_\mathrm{f}]}\{T(x,t) : q(x,t) > 0\} \geq \min_{x\in[0,1],t\in[t_\mathrm{i},t_\mathrm{f}]}\{T(x,t) : q(x,t) < 0\}$$

implies that the coldest point where heat flows into the bar cannot have a higher temperature than the hottest point where heat flows out of the bar. This is the statement of the second law of thermodynamics, which implies that the bar cannot be used to transport heat from cold to hot.

Next, let $\mathcal{B} = \mathcal{L}_2[0,1]$ and $q(x,t) \equiv 0$, and note that (11.91) can be rewritten as (2.38) with $\mathcal{A} = \frac{\mathrm{d}^2}{\mathrm{d}x^2}$. Furthermore, it can be shown (see [104]) that $\mathcal{A}$ is the infinitesimal generator of the C^0-semigroup

$$\mathcal{T}(t)z = \sum_{n=0}^{\infty} e^{\lambda_n t}\langle z, \phi_n\rangle\phi_n,$$

where $\lambda_n = -n^2\pi^2$, $n \geq 0$, $\phi_n(x) = \sqrt{2}\cos(n\pi x)$, $n \geq 1$, and $\phi_0(x) = 1$.

Next, we show that (11.91) with $q(x,t) \equiv 0$ is semistable. To see this, note that

$$\begin{aligned}\|T(t)z - T(t)y\|_{\mathcal{B}} &= \|T(t)(z-y)\|_{\mathcal{B}} \\ &= \left\|\sum_{n=0}^{\infty} e^{\lambda_n t}(z-y)^{\mathrm{T}}\phi_n\phi_n\right\|_{\mathcal{B}} \\ &\leq \sum_{n=0}^{\infty} |(z-y)^{\mathrm{T}}\phi_n|\|\phi_n\|_{\mathcal{B}}, \quad z, y \in \mathcal{B}.\end{aligned}$$

Now, using the Cauchy-Schwarz inequality, it follows that

$$\begin{aligned}\sum_{n=0}^{\infty} |(z-y)^{\mathrm{T}}\phi_n|\|\phi_n\|_{\mathcal{B}} &\leq \sum_{n=0}^{\infty} \|z-y\|_{\mathcal{B}}\|\phi_n\|_{\mathcal{B}}^2 \\ &= \|z-y\|_{\mathcal{B}}\sum_{n=0}^{\infty}\|\phi_n\|_{\mathcal{B}}^2 \\ &= \|z-y\|_{\mathcal{B}},\end{aligned}$$

and hence, $\mathcal{T}$ is a contraction semigroup.

Next, consider the ectropy function $\mathcal{E}(T) = \|T\|_{\mathcal{B}}^2$, $T \in \mathcal{B}$, as a Lyapunov function candidate and note that $\mathcal{E}(\cdot)$ is continuous. With this Lyapunov function candidate it follows that $\dot{\mathcal{E}}(T) = -2\|\nabla T\|_{\mathcal{B}}^2 \leq 0$. Define $\mathcal{R} = \{T \in \mathcal{B} : \nabla T = 0\}$ and note that $\mathcal{R}$ consists of only the equilibrium states of (11.87), which implies that the largest invariant set $\mathcal{M}$ contained in $\mathcal{R}$ is given by $\mathcal{M} = \mathcal{R} = \{T \in \mathcal{B} : T = T_{\mathrm{e}} \in \mathbb{R}\}$. Finally, let $\mathcal{E}(T - T_{\mathrm{e}}) = \|T - T_{\mathrm{e}}\|_{\mathcal{B}}^2$ and note that $\dot{\mathcal{E}}(T - T_{\mathrm{e}}) = -2\|\nabla T\|_{\mathcal{B}}^2 \leq 0$. Hence, every point in $\mathcal{M}$ is Lyapunov stable. Now, it follows from Theorem 2.15 that (11.91) with $q(x,t) \equiv 0$ is semistable. △

Next, we give an analogous proposition to Proposition 3.11 for infinite-dimensional systems.

Proposition 11.10. Consider the dynamical system $\mathcal{G}$ with power balance equation (11.3) and (11.4), let $\mathcal{E} : \mathcal{X} \to \overline{\mathbb{R}}_+$ and $\mathcal{S} : \mathcal{X} \to \overline{\mathbb{R}}_+$ denote the ectropy and entropy functionals of $\mathcal{G}$ given by (11.57) and (11.38), respectively, and define $\mathcal{D}_{\mathrm{c}} \triangleq \{u_t \in \mathcal{X} : \int_{\mathcal{V}} u_t(x)\mathrm{d}\mathcal{V} = \beta\}$, where $\beta \geq 0$.

Then

$$\arg\min_{u_t \in \mathcal{D}_c}(\mathcal{E}(u_t)) = \arg\max_{u_t \in \mathcal{D}_c}(\mathcal{S}(u_t)) = u_t^* = \frac{\beta}{\mathcal{V}_{\mathrm{vol}}}. \quad (11.98)$$

Furthermore, $\mathcal{E}_{\min} \triangleq \mathcal{E}(u_t^*) = \frac{\beta^2}{2}$ and $\mathcal{S}_{\max} \triangleq \mathcal{S}(u_t^*) = \mathcal{V}_{\mathrm{vol}}[\log_e(c + \frac{\beta}{\mathcal{V}_{\mathrm{vol}}}) - \log_e c]$.

Proof. The proof is similar to the proof of Proposition 3.11, and hence, is omitted. The only difference here is that $\mathcal{E}(u_t)$ and $-\mathcal{S}(u_t)$ are real-valued convex *functionals* defined on the function space $\mathcal{X}$, and $\int_{\mathcal{V}} u_t(x)\mathrm{d}\mathcal{V}$ is a convex mapping from a convex subset of $\mathcal{X}$ into a normed space. The result thus follows as a direct consequence of global theory for constrained optimization of functionals [292]. □

Next, we use the entropy functional (respectively, ectropy functional) given by (11.38) (respectively, (11.57)) to show a clear connection between our continuum thermodynamic model given by (11.3) and (11.4), and the arrow of time.

Theorem 11.12. Consider the dynamical system $\mathcal{G}$ with power balance equation (11.3) and (11.4) with $s(x,t) \equiv 0$ and $\phi(x, u(x,t), \nabla u(x,t)) \cdot \hat{n}(x) \equiv 0$, $x \in \partial\mathcal{V}$, and assume Axioms $i)'$ and $ii)'$ hold. Furthermore, let $s^u(\cdot, 0) \in \mathcal{W}_{[t_0,t_1]}$. Then, for every $u_{t_0} \notin \mathcal{M}_{\mathrm{e}}$, there exists a continuously differentiable functional $\mathcal{S} : \mathcal{X} \to \mathbb{R}$ (respectively, $\mathcal{E} : \mathcal{X} \to \mathbb{R}$) such that $\mathcal{S}(s^u(t,0))$ (respectively, $\mathcal{E}(s^u(t,0)))$ is an increasing (respectively, decreasing) function of time. Furthermore, $s^u(\cdot, 0)$ is an $I_{\mathcal{X}}$-reversible trajectory of $\mathcal{G}$ if and only if $s^u(t,0) \in \mathcal{M}_{\mathrm{e}}$, $t \in [t_0, t_1]$.

Proof. The proof is similar to the proof of Theorem 3.10 and follows from Corollary 2.2. □

11.4 Advection-Diffusion Dynamics

The nonlinear partial differential equation (11.3) describes a general conservation equation. An important special case of (11.3) is an *advection-diffusion* model describing a *diffusion* process: a universal process leading to the elimination of concentration gradients in solids, liquids, and gases. The macroscopic laws that describe diffusion are known as Fick's laws and relate the net diffusive flux of matter from regions of high concentrations to regions of low concentrations with changes in concentration over time.

In this section, we turn our attention to a specific form of (11.3) involving an advection-diffusion model [179, 345] defined over a compact connected set $\mathcal{V} \subset \mathbb{R}^n$ with a smooth boundary $\partial\mathcal{V}$ and volume $\mathcal{V}_{\mathrm{vol}}$ given by

$$\frac{\partial\rho(x,t)}{\partial t} = -\nabla\cdot(\rho(x,t)v(x,t)) + \nabla\cdot\left(B(x,t)\nabla^{\mathrm{T}}\rho(x,t)\right), \quad (11.99)$$

$$\rho(x,t_0) = \rho_{t_0}(x), \quad x\in\mathcal{V}, \quad t\geq t_0, \quad (11.100)$$

where $\rho : \mathcal{V}\times[0,\infty)\to\overline{\mathbb{R}}_+$ denotes the density distribution of a species concentration (e.g., mass, fluid, energy) at the point $x=[x_1,\ldots,x_n]^{\mathrm{T}}\in\mathcal{V}$ and time instant $t\geq t_0$, $v:\mathcal{V}\times[0,\infty)\to\mathbb{R}^n$ is a density-dependent advection velocity, and $B:\mathcal{V}\times[0,\infty)\to\mathbb{R}^{n\times n}$ is a diffusion operator. Here, we consider the case where $v(x,t)$ is given by

$$v(x,t) = -k\nabla^{\mathrm{T}}\rho(x,t), \quad x\in\mathcal{V}, \quad t\geq t_0, \quad (11.101)$$

where $k\in\mathbb{R}$ and $B(x,t)=\lambda I_n\in\mathbb{R}^{n\times n}$ for all $x\in\mathcal{V}$ and $t\geq t_0$, where $\lambda\in\mathbb{R}$.

Theorem 11.13. Consider the dynamical system given by (11.99) and (11.100) with $B(x,t)\equiv\lambda I_n$. Assume that $v(x,t)$ satisfies (11.101). If $k,\lambda\geq 0$ are such that $k^2+\lambda^2\neq 0$, then for every $\alpha\in\overline{\mathbb{R}}_+$, $\rho(x,t)\equiv\alpha$ is a semistable equilibrium state of (11.99) and (11.100) if and only if $\nabla\rho(x,t)\cdot\hat{n}(x)=0$, where $x\in\partial\mathcal{V}$ and $t\geq t_0$. In this case, $\rho(x,t)\to\frac{1}{\mathcal{V}_{\mathrm{vol}}}\int_{\mathcal{V}}\rho_{t_0}(x)\mathrm{d}\mathcal{V}$ as $t\to\infty$ for every initial condition $\rho_{t_0}\in\mathcal{W}_2^1(\mathcal{V})$ and every $x\in\mathcal{V}$; moreover, $\frac{1}{\mathcal{V}_{\mathrm{vol}}}\int_{\mathcal{V}}\rho_{t_0}(x)\mathrm{d}\mathcal{V}$ is a semistable equilibrium state of (11.99) and (11.100).

Proof. First, let $k\geq 0$ and $\lambda>0$. In this case, $\phi(x,\rho(x,t),\nabla\rho(x,t)) = -(k\rho(x,t)+\lambda)\nabla^{\mathrm{T}}\rho(x,t)$, and hence, Axioms $i)'$ and $ii)'$ hold. Furthermore,

$$\begin{aligned}\nabla^2\rho_t(x)\nabla\cdot\phi(x,\rho_t(x),\nabla\rho_t(x)) &\\ = \nabla^2\rho_t(x)\left[-k\nabla\rho_t(x)\nabla^{\mathrm{T}}\rho_t(x)-(k\rho_t(x)+\lambda)\nabla^2\rho_t(x)\right] &\\ = -k[\nabla^2\rho_t(x)]^2-(k\rho_t(x)+\lambda)[\nabla^2\rho_t(x)]^2 &\\ \leq 0, \quad x\in\mathcal{V}, & \quad (11.102)\end{aligned}$$

and hence, (11.75) holds. Now, the result is a direct consequence of Corollary 11.1.

Alternatively, let $k>0$ and $\lambda=0$, and assume that $\rho(x,t)\nabla\rho(x,t)\cdot\hat{n}(x)=0$ for $x\in\partial\mathcal{V}$ and $t\geq t_0$. To show Lyapunov stability of $\rho(x,t)\equiv\alpha$, consider the Lyapunov functional (11.68) with $u(x,t)$ replaced by $\rho(x,t)$. Now, it follows from the Green-Gauss theorem that

$$\begin{aligned}\dot{\mathcal{E}}(\rho_t-\alpha) &= \int_{\mathcal{V}}(\rho(x,t)-\alpha)\frac{\partial\rho(x,t)}{\partial t}\mathrm{d}\mathcal{V}\\ &= -\int_{\mathcal{V}}\rho(x,t)\nabla\cdot(\rho(x,t)v(x,t))\mathrm{d}\mathcal{V}+\alpha\int_{\mathcal{V}}\nabla\cdot(\rho(x,t)v(x,t))\mathrm{d}\mathcal{V}\\ &= \int_{\mathcal{V}}\nabla\rho(x,t)\rho(x,t)v(x,t)\mathrm{d}\mathcal{V}-\int_{\partial\mathcal{V}}\rho(x,t)\rho(x,t)v(x,t)\cdot\hat{n}(x)\mathrm{d}\mathcal{S}_{\mathcal{V}}\end{aligned}$$

$$
\begin{aligned}
&+\alpha \int_{\partial \mathcal{V}} \rho(x,t)v(x,t)\cdot \hat{n}(x)\mathrm{d}\mathcal{S}_{\mathcal{V}} \\
&= -\int_{\mathcal{V}} \rho(x,t)\nabla\rho(x,t)\nabla^{\mathrm{T}}\rho(x,t)\mathrm{d}\mathcal{V} \\
&\le 0, \quad \rho_t \in \mathcal{W}_2^0(\mathcal{V}),
\end{aligned} \tag{11.103}
$$

which proves Lyapunov stability of $\rho(x,t)\equiv\alpha$.

To show semistability of $\rho(x,t)\equiv\alpha$, consider the Lyapunov functionals (11.70) and (11.73). Now, it follows from (11.103), with $\alpha=0$, that $\mathcal{E}(\rho_t)$ is a nonincreasing functional of time for all $\rho_{t_0}\in\mathcal{W}_2^0(\mathcal{V})$. Furthermore, it follows from the Green-Gauss theorem that

$$
\begin{aligned}
\frac{1}{2}\dot{D}(\rho_t,\rho_t) &= \int_{\mathcal{V}} \nabla\rho(x,t)\frac{\partial}{\partial t}(\nabla\rho(x,t))^{\mathrm{T}}\mathrm{d}\mathcal{V} \\
&= \int_{\partial\mathcal{V}} \frac{\partial\rho(x,t)}{\partial t}D_{\hat{n}(x)}\rho(x,t)\mathrm{d}\mathcal{S}_{\mathcal{V}} - k\int_{\mathcal{V}}[\nabla^2\rho(x,t)]^2\mathrm{d}\mathcal{V}.
\end{aligned} \tag{11.104}
$$

Next, using similar arguments as in the proof of Corollary 11.1, it can be shown that $D(\rho_t,\rho_t)$ is a nonincreasing functional of time for all $\rho_{t_0}\in\mathcal{W}_2^1(\mathcal{V})$. Furthermore, note that $V(\rho_t)=\mathcal{E}(\rho_t)+D(\rho_t,\rho_t)$. Hence, $\mathcal{E}(\rho_t)$ is a nonincreasing functional of time for all $\rho_{t_0}\in\mathcal{W}_2^1(\mathcal{V})$. The rest of the sufficiency part of the proof now follows as in the proof of Theorem 11.10.

The converse follows as in the proof of Theorem 11.10. □

We close this section by noting that the results of this chapter can be easily generalized to the case where the energy density at a point $x\in\mathcal{V}$ is proportional to the temperature, that is, $\hat{T}(x,t)=\beta(x)u(x,t)$, where $\hat{T}(x,t)$ is the temperature distribution over the continuum and $\beta(x)$ is the reciprocal of the specific heat at the spatial coordinate x. In this case, results analogous to those of Section 4.1 can be easily derived for the infinite-dimensional thermodynamic model.

Finally, it is important to note that the results of this section apply to an arbitrary (not necessarily Cartesian) n-dimensional space. In particular, we could consider a coordinate transformation $y=Y(x)$, where $Y(0)=0$ and $Y:\mathcal{V}\to\mathbb{R}^n$ is a diffeomorphism in the neighborhood of the origin, so that y is defined on the image of $\mathcal{V}\subset\mathbb{R}^n$ under the mapping Y. In this case, however, the nabla and gradient operators need to be redefined appropriately [6, pp. 350–351].

Chapter Twelve

Stochastic Thermodynamics: A Dynamical Systems Approach

12.1 Introduction

In an attempt to generalize classical thermodynamics to irreversible nonequilibrium thermodynamics, a relatively new framework has been developed that combines stochasticity and nonequilibrium dynamics. This framework is known as *stochastic thermodynamics* [400–404] and goes beyond linear irreversible thermodynamics, addressing transport properties and entropy production in terms of forces and fluxes via linear system response theory [113, 347, 348, 367]. Stochastic thermodynamics is applicable to nonequilibrium systems, extending the validity of the laws of thermodynamics beyond the linear response regime by providing a system thermodynamic paradigm formulated on the level of individual system state realizations that are arbitrarily far from equilibrium. The thermodynamic variables of heat, work, and entropy, along with the concomitant first and second laws of thermodynamics, are formulated on the level of individual dynamical system trajectories using stochastic differential equations.

The nonequilibrium conditions in stochastic thermodynamics are imposed by an exogenous stochastic disturbance or an initial system state that is far from the system equilibrium, resulting in an open (i.e., driven) or relaxation dynamical process. More specifically, the exogenous disturbance is modeled as an independent standard Wiener process (i.e., Brownian motion) defined on a complete filtered probability space, wherein the current state is only dependent on the most recent event. The stochastic system dynamics are described by an overdamped Langevin equation [403, 404] in which fluctuation and dissipation forces obey the Einstein relation expressing that diffusion is a result of both thermal fluctuations and frictional dissipation [125].

Brownian motion refers to the irregular movement of microscopic particles suspended in a liquid and was discovered by the botanist Robert

Brown [67].[1] This random motion is explained as the result of collisions between the suspended particles (i.e., Brownian particles) and the molecules of the liquid. Einstein was the first to formulate the theory of Brownian motion by assuming that the particles suspended in the liquid contribute to the thermal fluctuations of the medium and, in accordance with the principle of equipartition of energy [237], the average translational kinetic energy of each particle [125]. Thus, Brownian motion results from collisions by molecules of the fluid, wherein the suspended particles acquire the same average kinetic energy as the molecules of the fluid. This theory suggested that all matter consists of atoms (or molecules) and heat is the energy of motion (i.e., kinetic energy) of the atoms.

The use of statistical methods in developing a general molecular theory of heat predicated on random motions of Newtonian atoms led to the connection between the dynamics of heat flow and the behavior of electromagnetic radiation. A year after Einstein published his theory on Brownian motion, Smoluchovski [417] confirmed the relation between friction and diffusion. In an attempt to simplify Einstein's theory of Brownian motion, Langevin [269] was the first to model the effect of Brownian motion using a stochastic differential equation (now known as a Langevin equation) wherein spherical particles are suspended in a medium and acted upon by external forces.

In stochastic thermodynamics, the Langevin equation captures the coupling between the system particle damping and the energy input to the particles via thermal effects. Namely, the frictional forces extract the particle kinetic energy, which in turn is injected back to the particles in the form of thermal fluctuations. This captures the phenomenological behavior of a Brownian particle suspended in a fluid medium, which can be modeled as a continuous Markov process.[2] Specifically, since collisions between the fluid molecules and a Brownian particle are more inelastic at higher viscosities, and temperature decreases with increasing viscosity in a fluid, additional heat is transferred to the fluid to maintain its temperature in accordance

[1]Even though Robert Brown [67] is credited with the discovery of Brownian motion, the chaotic motion of atoms was first conjectured by Leukippos—the ancient Greek philosopher to first develop the theory of atomism. In Book II of his poem *De Rerum Natura* (*On the Nature of the Universe*), Lucretius (99–55 B.C.) attributes the observed disordered motion of dust lit by a sun ray to Leukippos. He goes on to state that Leukippos asserted that the irregular and extremely fast motion of atoms is the cause for the slower motion of the larger dust particles [389, p. 23].

[2]It is important to note here that Brownian motion is not a Markov process on arbitrary time scales since the Markov assumption (i.e., the conditional probability distribution of the future states depends only on the current state) does not hold for the detailed dynamics of the Brownian particle collisions. However, the outcome of each Brownian particle collision depends only on the initial condition of the collision, and hence, on the most recent collision, which is precisely the Markov assumption. Thus, on a time scale of the same order of magnitude as the mean time between particle collisions (i.e., the Markov-Einstein time scale), Brownian motion can be assumed to be Markovian.

with the equipartition theorem. This heat is transferred to the Brownian particle through an increased disturbance intensity by the fluid molecules. These collisions between the Brownian particle and fluid molecules result in the observed persistent irregular and random motion of the particles.

The balance between damping (i.e., deceleration) of the particles due to frictional effects resulting in local heating of the fluid, and consequently entropy *production*, and the energy injection of the particles due to thermal fluctuations resulting in local cooling of the fluid, and consequently entropy *consumption*, is quantified by fluctuation theorems [55,56,102,103,139,153, 223,229,230,255,274]. Thus, even though, on average, the entropy is positive (i.e., entropy production), there exist sample paths wherein the entropy decreases, albeit with an exponentially lower probability than that of entropy production. In other words, a stochastic thermodynamic system exhibits a symmetry in the probability distribution of the entropy production in the asymptotic nonequilibrium process.

Fluctuation theorems give a precise prediction for the cases in which entropy decreases in stochastic thermodynamic systems and provide a key relation between entropy production and irreversibility. Specifically, the entropy production of individual sample path trajectories of a stochastic thermodynamic system described by a Markov process is not restricted by the second law, but rather the average entropy production is determined to be positive. Furthermore, the notions of heat and work in stochastic thermodynamic systems allow for a formulation of the first law of thermodynamics on the level of individual sample path trajectories with microscopic states (i.e., positions and velocities) governed by a stochastic Langevin equation and macroscopic states governed by a Fokker-Planck equation [452] (or a Kolmogorov forward equation, depending on context) describing the evolution of the probability density function of the microscopic (stochastic) states.

In this chapter, we combine our large-scale thermodynamic system model developed in Chapter 3 with stochastic thermodynamics to develop a stochastic dynamical systems framework of thermodynamics. Specifically, we develop a large-scale dynamical system model driven by Markov diffusion processes to present a unified framework for statistical thermodynamics predicated on a stochastic dynamical systems formalism. In particular, using a stochastic state space formulation, we develop a nonlinear stochastic compartmental dynamical system model characterized by energy conservation laws that is consistent with statistical thermodynamic principles. Moreover, we show that the difference between the supplied system energy and the stored system energy for our stochastic thermodynamic model is a martingale with respect to the system filtration. In addition, we show that the average stored system energy is equal to the mean energy that can be extracted from the system and the mean energy that can be delivered to the system in order to transfer it from a zero energy level to an arbitrary nonempty subset in the state space over a finite stopping time.

Next, using the system ectropy as a Lyapunov function candidate, we show that in the absence of energy exchange with the environment the proposed stochastic thermodynamic model is stochastically semistable in the sense that all sample path trajectories converge almost surely to a set of equilibrium solutions, wherein every equilibrium solution in the set is almost surely Lyapunov stable. Finally, we show that the steady-state distribution of the large-scale sample path system energies is uniform, leading to system energy equipartitioning corresponding to a maximum entropy equilibrium state.

12.2 Stochastic Dynamical Systems

To extend our dynamical thermodynamic formulation to stochastic thermodynamics, we require additional notation, definitions, and mathematical machinery. A review of some basic results on nonlinear stochastic dynamical systems is given in [9,244,302,304,344]. Recall that given a sample space Ω, a *σ-algebra* $\mathcal{F}$ on Ω is a collection of subsets of Ω such that $\varnothing \in \mathcal{F}$, if $F \in \mathcal{F}$, then $\Omega \backslash F \in \mathcal{F}$, and if $F_1, F_2, \ldots \in \mathcal{F}$, then $\bigcup_{i=1}^{\infty} F_i \in \mathcal{F}$ and $\bigcap_{i=1}^{\infty} F_i \in \mathcal{F}$. The pair $(\Omega, \mathcal{F})$ is called a *measurable space*, and the *probability measure* $\mathbb{P}$ defined on $(\Omega, \mathcal{F})$ is a function $\mathbb{P} : \mathcal{F} \to [0,1]$ such that $\mathbb{P}(\varnothing) = 0$, $\mathbb{P}(\Omega) = 1$, and if $F_1, F_2, \ldots \in \mathcal{F}$ and $F_i \cap F_j = \varnothing$, $i \neq j$, then $\mathbb{P}(\bigcup_{i=1}^{\infty} F_i) = \sum_{i=1}^{\infty} \mathbb{P}(F_i)$. The triple $(\Omega, \mathcal{F}, \mathbb{P})$ is called a *probability space* if $\mathcal{F}$ contains all subsets of Ω with $\mathbb{P}$-outer measure[3] zero [344].

The subsets $\mathcal{F}$ of Ω belonging to $\mathcal{F}$ are called *$\mathcal{F}$-measurable sets.* If $\Omega = \mathbb{R}^n$ and $\mathfrak{B}^n$ is the family of all open sets in $\mathbb{R}^n$, then $\mathfrak{B}^n$ is called the *Borel σ-algebra* and the elements $\mathcal{B}$ of $\mathfrak{B}^n$ are called *Borel sets.* If $(\Omega, \mathcal{F}, \mathbb{P})$ is a given probability space, then the real-valued function (random variable) $x : \Omega \to \mathbb{R}$ is *$\mathcal{F}$-measurable* if $\{\omega \in \Omega : x(\omega) \in \mathcal{B}\} \in \mathcal{F}$ for all Borel sets $\mathcal{B} \subset \mathbb{R}^n$. Given the probability space $(\Omega, \mathcal{F}, \mathbb{P})$, a *filtration* is a family $\{\mathcal{F}_t\}_{t \geq 0}$ of σ-algebras $\mathcal{F}_t \subset \mathcal{F}$ such that $\mathcal{F}_t \subset \mathcal{F}_s$ for all $0 \leq t < s < \infty$.

In this chapter, we define a complete probability space as $(\Omega, \mathcal{F}, \mathbb{P})$, where Ω denotes the sample space, $\mathcal{F}$ denotes a σ-algebra, and $\mathbb{P}$ defines a probability measure on the σ-algebra $\mathcal{F}$; that is, $\mathbb{P}$ is a nonnegative countably additive set function on $\mathcal{F}$ such that $\mathbb{P}(\Omega) = 1$ [9]. Furthermore, we assume that $w(\cdot)$ is a standard d-dimensional Wiener process defined by

[3]The $\mathbb{P}$-outer measure on a set Ω is an isotone (i.e., order-preserving), countably additive, extended real-valued set function defined for all subsets of Ω with $\mathbb{P}(\varnothing) = 0$.

$(w(\cdot),\Omega,\mathcal{F},\mathbb{P}^{w_0})$, where $\mathbb{P}^{w_0}$ is the classical Wiener measure [344, p. 10], with a continuous-time filtration $\{\mathcal{F}_t\}_{t\geq 0}$ generated by the Wiener process $w(t)$ up to time t.

We denote by $\mathcal{G}$ a stochastic dynamical system generating a filtration $\{\mathcal{F}_t\}_{t\geq 0}$ adapted to the stochastic process $x:\overline{\mathbb{R}}_+\times\Omega\to\mathcal{D}$ on $(\Omega,\mathcal{F},\mathbb{P}^{x_0})$ satisfying $\mathcal{F}_\tau\subset\mathcal{F}_t$, $0\leq\tau<t$, such that $\{\omega\in\Omega: x(t,\omega)\in\mathcal{B}\}\in\mathcal{F}_t$, $t\geq 0$, for all Borel sets $\mathcal{B}\subset\mathbb{R}^n$ contained in the Borel σ-algebra $\mathfrak{B}^n$. We say that the stochastic process $x:\overline{\mathbb{R}}_+\times\Omega\to\mathcal{D}$ is *$\mathcal{F}_t$-adapted* if $x(t)$ is $\mathcal{F}_t$-measurable for every $t\geq 0$. Furthermore, we say that $\mathcal{G}$ satisfies the *Markov property* if the conditional probability distribution of the future states of the stochastic process generated by $\mathcal{G}$ depends only on the present state. In this case, $\mathcal{G}$ generates a *Markov process*, which results in a decoupling of the past from the future in the sense that the present value of the state of $\mathcal{G}$ contains sufficient information so as to encapsulate the past system inputs. Here we use the notation $x(t)$ to represent the stochastic process $x(t,\omega)$, omitting its dependence on ω. Furthermore, $\mathfrak{B}^n$ denotes the σ-algebra of Borel sets in $\mathcal{D}\subseteq\mathbb{R}^n$ and $\mathfrak{S}$ denotes a σ-algebra generated on a set $\mathcal{S}\subseteq\mathbb{R}^n$.

We denote the set of equivalence classes of measurable, integrable, and square-integrable $\mathbb{R}^n$ or $\mathbb{R}^{n\times m}$ (depending on context) valued random processes on $(\Omega,\mathcal{F},\mathbb{P})$ over the semi-infinite parameter space $[0,\infty)$ by $\mathcal{L}^0(\Omega,\mathcal{F},\mathbb{P})$, $\mathcal{L}^1(\Omega,\mathcal{F},\mathbb{P})$, and $\mathcal{L}^2(\Omega,\mathcal{F},\mathbb{P})$, respectively, where the equivalence relation is the one induced by $\mathbb{P}$-almost-sure equality. In particular, elements of $\mathcal{L}^0(\Omega,\mathcal{F},\mathbb{P})$ take finite values $\mathbb{P}$-almost surely (a.s.) or with probability one. Hence, depending on the context, $\mathbb{R}^n$ will denote either the set of $n\times 1$ real variables or the subspace of $\mathcal{L}^0(\Omega,\mathcal{F},\mathbb{P})$ comprising $\mathbb{R}^n$ random processes that are constant almost surely. All inequalities and equalities involving random processes on $(\Omega,\mathcal{F},\mathbb{P})$ are to be understood to hold $\mathbb{P}$-almost surely. Furthermore, $\mathbb{E}[\,\cdot\,]$ and $\mathbb{E}^{x_0}[\,\cdot\,]$ denote, respectively, the expectation with respect to the probability measure $\mathbb{P}$ and with respect to the classical Wiener measure $\mathbb{P}^{x_0}$.

Given $x\in\mathcal{L}^0(\Omega,\mathcal{F},\mathbb{P})$, $\{x=0\}$ denotes the set $\{\omega\in\Omega: x(t,\omega)=0\}$, and so on. Given $x\in\mathcal{L}^0(\Omega,\mathcal{F},\mathbb{P})$ and $\mathcal{E}\in\mathcal{F}$, we say x is nonzero on $\mathcal{E}$ if $\mathbb{P}(\{x=0\}\cap\mathcal{E})=0$. Furthermore, given $x\in\mathcal{L}^1(\Omega,\mathcal{F},\mathbb{P})$ and a σ-algebra $\mathcal{E}\subseteq\mathcal{F}$, $\mathbb{E}^{\mathbb{P}}[x]$ and $\mathbb{E}^{\mathbb{P}}[x|\mathcal{E}]$ denote, respectively, the expectation of the random variable x and the conditional expectation of x given $\mathcal{E}$, with all moments taken under the measure $\mathbb{P}$. In formulations wherein it is clear from the context which measure is used, we omit the symbol $\mathbb{P}$ in denoting expectation, and similarly for conditional expectation. Specifically, in such cases we denote the expectation with respect to the probability space $(\Omega,\mathcal{F},\mathbb{P})$ by $\mathbb{E}[\,\cdot\,]$, and similarly for conditional expectation.

A stochastic process $x : \overline{\mathbb{R}}_+ \times \Omega \to \mathcal{D}$ on $(\Omega, \mathcal{F}, \mathbb{P}^{x_0})$ is called a *martingale* with respect to the filtration $\{\mathcal{F}_t\}_{t\geq 0}$ if and only if $x(t)$ is an $\mathcal{F}_t$-measurable random vector for all $t \geq 0$, $\mathbb{E}[x(t)] < \infty$, and $x(\tau) = \mathbb{E}[x(t)|\mathcal{F}_\tau]$ for all $t \geq \tau \geq 0$. Thus, a martingale has the property that the expectation of the next value of the martingale is equal to its current value given all previous values of the dynamical process. If we replace the equality in $x(\tau) = \mathbb{E}\left[x(t)|\mathcal{F}_\tau\right]$ with "$\leq$" (respectively, "$\geq$"), then $x(\cdot)$ is a *supermartingale* (respectively, *submartingale*). Note that every martingale is both a submartingale and supermartingale.

A random variable $\tau : \Omega \to [0, \infty]$ is called a *stopping time* with respect to $\mathcal{F}_t$ if and only if $\{\omega \in \Omega : \tau(\omega) \leq t\} \in \mathcal{F}_t$, $t \geq 0$. Thus, the set of all $\omega \in \Omega$ such that $\tau(\omega) \leq t$ is an $\mathcal{F}_t$-measurable set. Note that $\tau(\omega)$ can take on finite as well as infinite values and characterizes whether at each time t an event at time $\tau(\omega) < t$ has occurred using only the information in $\mathcal{F}_t$.

Finally, we write $\mathrm{tr}(\cdot)$ for the trace operator and $\mathcal{H}_n$ for the Hilbert space of random vectors $x \in \mathbb{R}^n$ with finite average power, that is, $\mathcal{H}_n \triangleq \{x : \Omega \to \mathbb{R}^n : \mathbb{E}[x^\mathrm{T}x] < \infty\}$. For an open set $\mathcal{D} \subseteq \mathbb{R}^n$, $\mathcal{H}_n^\mathcal{D} \triangleq \{x \in \mathcal{H}_n : x : \Omega \to \mathcal{D}\}$ denotes the set of all the random vectors in $\mathcal{H}_n$ induced by $\mathcal{D}$. Similarly, for every $x_0 \in \mathbb{R}^n$, $\mathcal{H}_n^{x_0} \triangleq \{x \in \mathcal{H}_n : x \overset{\text{a.s.}}{=} x_0\}$. Furthermore, recall that C^2 denotes the space of real-valued functions $V : \mathcal{D} \to \mathbb{R}$ that are two-times continuously differentiable with respect to $x \in \mathcal{D} \subseteq \mathbb{R}^n$.

Definition 12.1. Let $(S, \mathfrak{S})$ and $(T, \mathfrak{T})$ be measurable spaces, and let $\mu : S \times \mathfrak{T} \to \overline{\mathbb{R}}_+$. If the function $\mu(s, \mathcal{B})$ is $\mathfrak{S}$-measurable in $s \in S$ for a fixed $\mathcal{B} \in \mathfrak{T}$ and $\mu(s, \mathcal{B})$ is a probability measure in $\mathcal{B} \in \mathfrak{T}$ for a fixed $s \in S$, then μ is called a (*probability*) *kernel* from S to T. Furthermore, for $s \leq t$, the function $\mu_{s,t} : S \times \mathfrak{S} \to \mathbb{R}$ is called a *regular conditional probability measure* if $\mu_{s,t}(\cdot, \mathfrak{S})$ is measurable, $\mu_{s,t}(S, \cdot)$ is a probability measure, and $\mu_{s,t}(\cdot,\cdot)$ satisfies

$$\mu_{s,t}(x(s), \mathcal{B}) = \mathbb{P}(x(t) \in \mathcal{B}|x(s)) = \mathbb{P}(x(t) \in \mathcal{B}|\mathcal{F}_s), \quad x(\cdot) \in \mathcal{H}_n, \qquad (12.1)$$

where $\mathbb{P}(x(t) \in \mathcal{B}|x(s)) = \mathbb{P}(0, x, t, \mathcal{B})$, $x \in \mathbb{R}^n$, and $\mathbb{P}(s, x, t, \mathcal{B})$, $t \geq s$, is the *transition probability* of the point $x \in \mathbb{R}^n$ at time instant s into all Borel subsets $\mathcal{B} \subset \mathbb{R}^n$ at time instant t.

Any family of regular conditional probability measures $\{\mu_{s,t}\}_{s\leq t}$ satisfying the Chapman-Kolmogorov equation ([9])

$$\mathbb{P}(s, x, t, \mathcal{B}) = \int_{\mathbb{R}^n} \mathbb{P}(s, x, \sigma, \mathrm{d}z)\mathbb{P}(s, z, t, \mathcal{B}), \qquad (12.2)$$

or, equivalently,

$$\mu_{s,t}(x,\mathcal{B}) = \int_{\mathbb{R}^n} \mu_{s,\sigma}(x,\mathrm{d}z)\mu_{\sigma,t}(z,\mathcal{B}), \tag{12.3}$$

where $0 \le s \le \sigma \le t < \infty$, $x, z \in \mathbb{R}^n$, and $\mathcal{B} \in \mathfrak{B}^n$, is called a *semigroup of Markov kernels.* The Markov kernels are called *time homogeneous* if and only if $\mu_{s,t} = \mu_{0,t-s}$ holds for all $s \le t$.

Consider the nonlinear stochastic dynamical system $\mathcal{G}$ given by

$$\mathrm{d}x(t) = f(x(t))\mathrm{d}t + D(x(t))\mathrm{d}w(t), \quad x(0) \overset{\text{a.s.}}{=} x_0, \qquad t \in \mathcal{I}_{x(0)}, \tag{12.4}$$

where, for every $t \in \mathcal{I}_{x_0}$, $x(t) \in \mathcal{H}_n^{\mathcal{D}}$ is an $\mathcal{F}_t$-measurable random state vector, $x(0) \in \mathcal{H}_n^{x_0}$, $\mathcal{D} \subseteq \mathbb{R}^n$ is a relatively open set with $0 \in \mathcal{D}$, $w(\cdot)$ is a d-dimensional independent standard Wiener process (i.e., Brownian motion) defined on a complete filtered probability space $(\Omega, \mathcal{F}, \{\mathcal{F}_t\}_{t\ge 0}, \mathbb{P})$, $x(0)$ is independent of $(w(t) - w(0)), t \ge 0$, $f : \mathcal{D} \to \mathbb{R}^n$ and $D : \mathcal{D} \to \mathbb{R}^{n\times d}$ are continuous, $\mathcal{E} \triangleq f^{-1}(0) \cap D^{-1}(0) \triangleq \{x \in \mathcal{D} : f(x) = 0 \text{ and } D(x) = 0\}$ is nonempty, and $\mathcal{I}_{x(0)} = [0, \tau_{x(0)})$, $0 \le \tau_{x(0)} \le \infty$, is the maximal interval of existence for the solution $x(\cdot)$ of (12.4).

An *equilibrium point* of (12.4) is a point $x_\mathrm{e} \in \mathbb{R}^n$ such that $f(x_\mathrm{e}) = 0$ and $D(x_\mathrm{e}) = 0$. It is easy to see that x_e is an equilibrium point of (12.4) if and only if the constant stochastic process $x(\cdot) \overset{\text{a.s.}}{=} x_\mathrm{e}$ is a solution of (12.4). We denote the set of equilibrium points of (12.4) by $\mathcal{E} \triangleq \{\omega \in \Omega : x(t,\omega) = x_\mathrm{e}\} = \{x_\mathrm{e} \in \mathcal{D} : f(x_\mathrm{e}) = 0 \text{ and } D(x_\mathrm{e}) = 0\}$.

The filtered probability space $(\Omega, \mathcal{F}, \{\mathcal{F}_t\}_{t\ge 0}, \mathbb{P})$ is clearly a real vector space with addition and scalar multiplication defined componentwise and pointwise. An $\mathbb{R}^n$-valued stochastic process $x : [0,\tau] \times \Omega \to \mathcal{D}$ is said to be a *solution* of (12.4) on the time interval $[0,\tau]$ with initial condition $x(0) \overset{\text{a.s.}}{=} x_0$ if $x(\cdot)$ is *progressively measurable* (i.e., $x(\cdot)$ is nonanticipating and measurable in t and ω) with respect to the filtration $\{\mathcal{F}_t\}_{t\ge 0}$, $f \in \mathcal{L}^1(\Omega, \mathcal{F}, \mathbb{P})$, $D \in \mathcal{L}^2(\Omega, \mathcal{F}, \mathbb{P})$, and

$$x(t) = x_0 + \int_0^t f(x(\sigma))\mathrm{d}\sigma + \int_0^t D(x(\sigma))\mathrm{d}w(\sigma) \ \text{ a.s.}, \quad t \in [0,\tau], \tag{12.5}$$

where the integrals in (12.5) are Itô integrals [156]. If the map $t \to w(t,\omega)$, $\omega \in \Omega$, had a bounded variation, then the natural definition for the integrals in (12.5) would be the Lebesgue-Stieltjes integral where ω is viewed as a parameter. However, since sample Wiener paths are nowhere differentiable and not of bounded variation for almost all $\omega \in \Omega$, the integrals in (12.5) need to be defined as Itô integrals [224, 225].

Note that for each fixed $t \ge 0$, the random variable $\omega \mapsto x(t,\omega)$

assigns a vector $x(\omega)$ to every outcome $\omega \in \Omega$ of an experiment, and for each fixed $\omega \in \Omega$, the mapping $t \mapsto x(t,\omega)$ is the *sample path* of the stochastic process $x(t)$, $t \geq 0$. A pathwise solution $t \mapsto x(t)$ of (12.4) in $(\Omega, \{\mathcal{F}_t\}_{t\geq 0}, \mathbb{P}^{x_0})$ is said to be *right maximally* defined if x cannot be extended (either uniquely or nonuniquely) forward in time. We assume that all right maximal pathwise solutions to (12.4) in $(\Omega, \{\mathcal{F}_t\}_{t\geq 0}, \mathbb{P}^{x_0})$ exist on $[0,\infty)$, and hence, we assume that (12.4) is *forward complete*. Sufficient conditions for forward completeness or *global solutions* of (12.4) are given in [156,304].

Furthermore, we assume that $f : \mathcal{D} \to \mathbb{R}^n$ and $D : \mathcal{D} \to \mathbb{R}^{n\times d}$ satisfy the uniform Lipschitz continuity condition

$$\|f(x) - f(y)\| + \|D(x) - D(y)\|_{\mathrm{F}} \leq L\|x - y\|, \quad x, y \in \mathcal{D}\backslash\{0\}, \tag{12.6}$$

and the growth restriction condition

$$\|f(x)\|^2 + \|D(x)\|_{\mathrm{F}}^2 \leq L^2(1 + \|x\|^2), \quad x \in \mathcal{D}\backslash\{0\}, \tag{12.7}$$

for some Lipschitz constant $L > 0$, and hence, since $x(0) \in \mathcal{H}_n^{\mathcal{D}}$ and $x(0)$ is independent of $(w(t) - w(0)), t \geq 0$, it follows that there exists a unique solution $x \in \mathcal{L}^2(\Omega, \mathcal{F}, \mathbb{P})$ of (12.4) forward in time for all initial conditions in the following sense. For every $x \in \mathcal{H}_n^{\mathcal{D}}\backslash\{0\}$ there exists $\tau_x > 0$ such that if $x_1 : [0, \tau_1] \times \Omega \to \mathcal{D}$ and $x_2 : [0, \tau_2] \times \Omega \to \mathcal{D}$ are two solutions of (12.4); that is, if $x_1, x_2 \in \mathcal{L}^2(\Omega, \mathcal{F}, \mathbb{P})$ with continuous sample paths almost surely solve (12.4), then $\tau_x \leq \min\{\tau_1, \tau_2\}$ and $\mathbb{P}\big(x_1(t) = x_2(t),\ 0 \leq t \leq \tau_x\big) = 1$.

The uniform Lipschitz continuity condition (12.6) guarantees uniqueness of solutions, whereas the linear growth condition (12.7) rules out finite escape times. A weaker sufficient condition for the existence of a unique solution to (12.4) using a notion of (finite or infinite) escape time under the local Lipschitz continuity condition (12.6) without the growth condition (12.7) is given in [470]. Alternatively, existence and uniqueness of solutions even when the uniform Lipschitz continuity condition (12.6) does not hold are given in [156, p. 152].

The unique solution to (12.4) determines an $\mathbb{R}^n$-valued, time homogeneous Feller continuous Markov process $x(\cdot)$, and hence, its stationary Feller transition probability function is given by $\big($[244, Thm. 3.4], [9, Thm. 9.2.8]$\big)$

$$\mathbb{P}(x(t) \in \mathcal{B}|x(t_0) \overset{\text{a.s.}}{=} x_0) = \mathbb{P}(0, x_0, t - t_0, \mathcal{B}), \quad x_0 \in \mathbb{R}^n, \tag{12.8}$$

for all $t \geq t_0$ and all Borel subsets $\mathcal{B}$ of $\mathbb{R}^n$, where $\mathbb{P}(\sigma, x, t, \mathcal{B}), t \geq \sigma$, denotes the probability of transition of the point $x \in \mathbb{R}^n$ at time instant s into the set $\mathcal{B} \subset \mathbb{R}^n$ at time instant t. Recall that every continuous process with a Feller transition probability function is also a strong Markov process [244, p. 101]. Finally, we say that the dynamical system (12.4) is *convergent in probability* with respect to the closed set $\mathcal{H}_n^{\mathcal{D}_c} \subseteq \mathcal{H}_n^{\mathcal{D}}$ if and only if the pointwise $\lim_{t\to\infty} s(t,x,\omega)$ exists for every $x \in \mathcal{D}_c \subseteq \mathbb{R}^n$ and $\omega \in \Omega$.

Here, the measurable map $s : [0,\tau_x) \times \mathcal{D} \times \Omega \to \mathcal{D}$ is the *dynamic* or *flow* of the stochastic dynamical system (12.2) and, for all $t, \tau \in [0,\tau_x)$, satisfies the *cocycle* property $s(\tau, s(t,x),\omega) = s(t+\tau, x, \omega)$ and the *identity* (on $\mathcal{D}$) property $s(0,x,\omega) = x$ for all $x \in \mathcal{D}$ and $\omega \in \Omega$. The measurable map $s_t \triangleq s(t,\cdot,\omega) : \mathcal{D} \to \mathcal{D}$ is continuously differentiable for all $t \in [0,\tau_x)$ outside a $\mathbb{P}$-nullset and the sample path trajectory $s^x \triangleq s(\cdot,x,\omega) : [0,\tau_x) \to \mathcal{D}$ is continuous in $\mathcal{D}$ for all $t \in [0,\tau_x)$. Thus, for every $x \in \mathcal{D}$, there exists a trajectory of measures defined for all $t \in [0,\tau_x)$ satisfying the dynamical processes (12.4) with initial condition $x(0) \overset{\text{a.s.}}{=} x_0$. For simplicity of exposition we write $s(t,x)$ for $s(t,x,\omega)$, omitting its dependence on ω.

Definition 12.2. A point $p \in \mathcal{D}$ is a *limit point* of the trajectory $s(\cdot,x)$ of (12.4) if there exists a monotonic sequence $\{t_n\}_{n=0}^{\infty}$ of positive numbers, with $t_n \to \infty$ as $n \to \infty$, such that $s(t_n,x) \overset{\text{a.s.}}{\to} p$ as $n \to \infty$. The set of all limit points of $s(t,x), t \geq 0$, is the *limit set* $\omega(x)$ of $s(\cdot,x)$ of (12.4).

It is important to note that the ω-limit set of a stochastic dynamical system is a ω-limit set of a trajectory of measures, that is, $p \in \omega(x)$ is a weak limit of a sequence of measures taken along every sample continuous bounded trajectory of (12.4). It can be shown that the ω-limit set of a stationary stochastic dynamical system attracts bounded sets and is measurable with respect to the σ-algebra of invariant sets. Thus, the measures of the stochastic process $x(\cdot)$ tend to an invariant set of measures, and $x(t)$ asymptotically tends to the closure of the support set (i.e., kernel) of this set of measures almost surely.

However, unlike deterministic dynamical systems, wherein ω-limit sets serve as global attractors, in stochastic dynamical systems stochastic invariance (see Definition 12.4) leads to ω-limit sets being defined for each fixed sample $\omega \in \Omega$ of the underlying probability space $(\Omega, \mathcal{F}, \mathbb{P})$, and hence, are pathwise attractors. This is due to the fact that a cocycle property rather than a semigroup property holds for stochastic dynamical systems. For details see [70, 99, 100].

Definition 12.3 ([344, Def. 7.7]). Let $x(\cdot)$ be a time-homogeneous Markov process in $\mathcal{H}_n^{\mathcal{D}}$ and let $V : \mathcal{D} \to \mathbb{R}$. Then the *infinitesimal generator* $\mathcal{L}$ of $x(t)$, $t \geq 0$, with $x(0) \overset{\text{a.s.}}{=} x_0$, is defined by

$$\mathcal{L}V(x_0) \triangleq \lim_{t\to 0^+} \frac{\mathbb{E}^{x_0}[V(x(t))] - V(x_0)}{t}, \quad x_0 \in \mathcal{D}, \tag{12.9}$$

where $\mathbb{E}^{x_0}$ denotes the expectation with respect to the transition probability measure $\mathbb{P}^{x_0}(x(t) \in \mathcal{B}) \triangleq \mathbb{P}(0, x_0, t, \mathcal{B})$.

If $V \in \mathrm{C}^2$ and has a compact support,[4] and $x(t)$, $t \geq 0$, satisfies (12.4), then the limit in (12.9) exists for all $x \in \mathcal{D}$ and the infinitesimal generator $\mathcal{L}$ of $x(t)$, $t \geq 0$, can be characterized by the system *drift* and *diffusion* functions $f(x)$ and $D(x)$ defining the stochastic dynamical system (12.4) and is given by ([344, Thm. 7.9])

$$\mathcal{L}V(x) \triangleq \frac{\partial V(x)}{\partial x} f(x) + \frac{1}{2} \mathrm{tr}\, D^{\mathrm{T}}(x) \frac{\partial^2 V(x)}{\partial x^2} D(x), \quad x \in \mathcal{D}. \tag{12.10}$$

Next, we extend Proposition 2.1 to stochastic dynamical systems. First, however, the following definition on stochastic invariance is needed.

Definition 12.4. A relatively open set $\mathcal{D} \subset \mathbb{R}^n$ is *invariant with respect to* (12.4) if $\mathcal{D}$ is Borel and, for all $x_0 \in \mathcal{D}$, $\mathbb{P}^{x_0}(x(t) \in \mathcal{D}) = 1$, $t \geq 0$.

Proposition 12.1. Suppose $\overline{\mathbb{R}}_+^n \subset \mathcal{D}$. Then $\overline{\mathbb{R}}_+^n$ is an invariant set with respect to (12.4) if and only if $f : \mathcal{D} \to \mathbb{R}^n$ is essentially nonnegative and $D_{(i,j)}(x) = 0$, $j = 1, \ldots, d$, whenever $x_i = 0$, $i = 1, \ldots, n$.

Proof. Define $\mathrm{dist}(x, \overline{\mathbb{R}}_+^n) \triangleq \inf_{y \in \overline{\mathbb{R}}_+^n} \|x - y\|$, $x \in \mathbb{R}^n$. Now, suppose $f : \mathcal{D} \to \mathbb{R}^n$ is essentially nonnegative and let $x \in \overline{\mathbb{R}}_+^n$. For every $i \in \{1, \ldots, q\}$, if $x_i = 0$, then $x_i + hf_i(x) + \mathrm{row}_i(D(x))[w(h, \omega) - w(0, \omega)] = hf_i(x) \geq 0$ for all $h \geq 0$ and all $\omega \in \Omega$, whereas, if $x_i > 0$, then it follows from the continuity of $D(\cdot)$ and the sample continuity of $w(\cdot)$ that $x_i + hf_i(x) + \mathrm{row}_i(D(x))[w(h, \omega) - w(0, \omega)] \geq 0$ for all $|h|$ sufficiently small and all $\omega \in \Omega$. Thus, $x + hf(x) + \mathrm{row}_i(D(x))[w(h, \omega) - w(0, \omega)] \in \overline{\mathbb{R}}_+^n$ for all sufficiently small $h > 0$ and all $\omega \in \Omega$, and hence, $\lim_{h \to 0^+} \mathrm{dist}(x + hf(x) + \mathrm{row}_i(D(x))[w(h, \omega) - w(0, \omega)], \overline{\mathbb{R}}_+^n)/h = 0$. It now follows from Lemma 2.1, with $x(0) \stackrel{\text{a.s.}}{=} x_0$, that $\mathbb{P}^{x_0}\left(x(t) \in \overline{\mathbb{R}}_+^n\right) = 1$ for all $t \in [0, \tau_{x_0})$.

Conversely, suppose that $\overline{\mathbb{R}}_+^n$ is invariant with respect to (12.4), let $\mathbb{P}^{x_0}(x(0) \in \overline{\mathbb{R}}_+^n) = 1$, and suppose, *ad absurdum*, x is such that there exists $i \in \{1, \ldots, q\}$ such that $x_i(0) \stackrel{\text{a.s.}}{=} 0$ and $f_i(x(0))h + \mathrm{row}_i(D(x))[w(h, \omega) - w(0, \omega)] < 0$ for all $\omega \in \Omega$. Then, since f and D are continuous and a Wiener process $w(\cdot)$ can be positive or negative with equal probability, there exists sufficiently small $h > 0$ such that $\mathbb{P}^{x_0}(f_i(x(t))\mathrm{d}t + \mathrm{row}_i(D(x(t)))\mathrm{d}w(t) < 0) \neq 0$ for all $t \in [0, h)$, where $x(t)$ is the solution to (12.4). Hence, $x_i(t)$ is strictly decreasing on $[0, h)$ with nonzero probability, and thus, $\mathbb{P}^{x_0}\left(x(t) \in \overline{\mathbb{R}}_+^n\right) \neq 1$ for all $t \in (0, h)$, which leads to a contradiction. □

[4]The *support* of a function is the unique smallest closed set for which the complement of the set has measure zero. A function has a *compact support* if its support is a compact set.

It follows from Proposition 12.1 that if $x_0 \geq\geq 0$, then $x(t) \overset{\text{a.s.}}{\geq\geq} 0$, $t \geq 0$, if and only if f is essentially nonnegative and $D_{(i,j)}(x) = 0$, $j = 1, \ldots, d$, whenever $x_i = 0$, $i = 1, \ldots, n$. In this case, we say that (12.4) is a *stochastic nonnegative dynamical system*. Henceforth, we assume that f and D are such that the nonlinear stochastic dynamical system (12.4) is a stochastic nonnegative dynamical system.

12.3 Stability Theory for Stochastic Nonnegative Dynamical Systems

In this section, we establish key stability results in probability for stochastic nonnegative dynamical systems. The mathematical machinery used is supermartingale theory and ergodic theory of Markov processes [244]. Specifically, deterministic stability theory is extended to stochastic dynamical systems by establishing supermartingale properties of Lyapunov functions. The following definition introduces several notions of stability in probability for the equilibrium solution $x(t) \overset{\text{a.s.}}{\equiv} x_\text{e} \in \overline{\mathbb{R}}_+^n$ of the stochastic nonnegative dynamical system (12.4) for $\mathcal{I}_{x(0)} = [0, \infty)$.

Definition 12.5. *i*) The equilibrium solution $x(t) \overset{\text{a.s.}}{\equiv} x_\text{e} \in \overline{\mathbb{R}}_+^n$ to (12.4) is *Lyapunov stable in probability with respect to* $\overline{\mathbb{R}}_+^n$ if, for every $\varepsilon > 0$,

$$\lim_{x_0 \to x_\text{e}} \mathbb{P}^{x_0} \left(\sup_{t \geq 0} \|x(t) - x_\text{e}\| > \varepsilon \right) = 0. \tag{12.11}$$

Equivalently, the equilibrium solution $x(t) \overset{\text{a.s.}}{\equiv} x_\text{e} \in \overline{\mathbb{R}}_+^n$ to (12.4) is Lyapunov stable in probability with respect to $\overline{\mathbb{R}}_+^n$ if, for every $\varepsilon > 0$ and $\rho \in (0, 1)$, there exists $\delta = \delta(\rho, \varepsilon) > 0$ such that, for all $x_0 \in \mathcal{B}_\delta(x_\text{e}) \cap \overline{\mathbb{R}}_+^n$,

$$\mathbb{P}^{x_0} \left(\sup_{t \geq 0} \|x(t) - x_\text{e}\| > \varepsilon \right) \leq \rho. \tag{12.12}$$

ii) The equilibrium solution $x(t) \overset{\text{a.s.}}{\equiv} x_\text{e} \in \overline{\mathbb{R}}_+^n$ to (12.4) is *asymptotically stable in probability with respect to* $\overline{\mathbb{R}}_+^n$ if it is Lyapunov stable in probability with respect to $\overline{\mathbb{R}}_+^n$ and

$$\lim_{x_0 \to x_\text{e}} \mathbb{P}^{x_0} \left(\lim_{t \to \infty} \|x(t) - x_\text{e}\| = 0 \right) = 1. \tag{12.13}$$

Equivalently, the equilibrium solution $x(t) \overset{\text{a.s.}}{\equiv} x_\text{e} \in \overline{\mathbb{R}}_+^n$ to (12.4) is asymptotically stable in probability with respect to $\overline{\mathbb{R}}_+^n$ if it is Lyapunov stable in probability with respect to $\overline{\mathbb{R}}_+^n$ and, for every $\rho \in (0, 1)$, there exists $\delta = \delta(\rho) > 0$ such that if $x_0 \in \mathcal{B}_\delta(x_e) \cap \overline{\mathbb{R}}_+^n$, then

$$\mathbb{P}^{x_0}\left(\lim_{t\to\infty} \|x(t) - x_e\| = 0\right) \geq 1 - \rho. \tag{12.14}$$

iii) The equilibrium solution $x(t) \overset{\text{a.s.}}{\equiv} x_e \in \overline{\mathbb{R}}_+^n$ to (12.4) is *globally asymptotically stable in probability with respect to* $\overline{\mathbb{R}}_+^n$ if it is Lyapunov stable in probability with respect to $\overline{\mathbb{R}}_+^n$ and, for all $x_0 \in \overline{\mathbb{R}}_+^n$,

$$\mathbb{P}^{x_0}\left(\lim_{t\to\infty} \|x(t) - x_e\| = 0\right) = 1. \tag{12.15}$$

As in deterministic stability theory, for a given $\varepsilon > 0$ the subset $\mathcal{B}_\varepsilon(x_e) \cap \overline{\mathbb{R}}_+^n$ defines a cylindrical region in the (t, x)-space wherein the trajectory $x(t)$, $t \geq 0$, belongs. However, in stochastic stability theory, for every $x_0 \in \mathcal{B}_\delta(x_e) \cap \overline{\mathbb{R}}_+^n$, there exists a probability of less than or equal to ρ that the system solution $s(t, x_0)$ leaves the subset $\mathcal{B}_\varepsilon(x_e) \cap \overline{\mathbb{R}}_+^n$; and for $x_0 = x_e$ this probability is zero. In other words, the probability of escape is continuous at $x_0 = x_e$ with small deviations from the equilibrium implying a small probability of escape.

The following lemma gives an equivalent characterization of Lyapunov and asymptotic stability in probability with respect to $\overline{\mathbb{R}}_+^n$ in terms of class $\mathcal{K}$, $\mathcal{K}_\infty$, and $\mathcal{KL}$ functions.

Lemma 12.1. *i*) The equilibrium solution $x(t) \overset{\text{a.s.}}{\equiv} x_e$ to (12.4) is Lyapunov stable in probability with respect to $\overline{\mathbb{R}}_+^n$ if and only if for every $\rho > 0$ there exist a class $\mathcal{K}$ function $\alpha_\rho(\cdot)$ and a constant $c = c(\rho) > 0$ such that, for all $x_0 \in \mathcal{B}_c(x_e) \cap \overline{\mathbb{R}}_+^n$,

$$\mathbb{P}^{x_0}\left(\|x(t) - x_e\| > \alpha_\rho(\|x_0 - x_e\|)\right) \leq \rho, \quad t \geq 0. \tag{12.16}$$

ii) The equilibrium solution $x(t) \overset{\text{a.s.}}{\equiv} x_e$ to (12.4) is asymptotically stable in probability with respect to $\overline{\mathbb{R}}_+^n$ if and only if for every $\rho > 0$ there exist a class $\mathcal{KL}$ function $\beta_\rho(\cdot, \cdot)$ and a constant $c = c(\rho) > 0$ such that, for all $x_0 \in \mathcal{B}_c(x_e) \cap \overline{\mathbb{R}}_+^n$,

$$\mathbb{P}^{x_0}\left(\|x(t) - x_e\| > \beta_\rho(\|x_0 - x_e\|, t)\right) \leq \rho, \quad t \geq 0. \tag{12.17}$$

Proof. *i*) Suppose that there exist a class $\mathcal{K}$ function $\alpha_\rho(\cdot)$ and a constant $c = c(\rho) > 0$ such that, for every $\rho > 0$ and $x_0 \in \mathcal{B}_c(x_e) \cap \overline{\mathbb{R}}_+^n$,

$$\mathbb{P}^{x_0}\left(\|x(t) - x_e\| > \alpha_\rho(\|x_0 - x_e\|)\right) \leq \rho, \quad t \geq 0. \tag{12.18}$$

Now, given $\varepsilon > 0$, let $\delta(\rho, \varepsilon) = \min\{c(\rho), \alpha_\rho^{-1}(\varepsilon)\}$. Then, for $x_0 \in \mathcal{B}_\delta(x_e) \cap$

$\overline{\mathbb{R}}_+^n$ and $t \geq 0$,

$$\begin{aligned}\mathbb{P}^{x_0}\left(\|x(t)-x_{\rm e}\| > \alpha_\rho(\|x_0-x_{\rm e}\|)\right) &\geq \mathbb{P}^{x_0}\left(\|x(t)-x_{\rm e}\| > \alpha_\rho(\delta)\right)\\ &\geq \mathbb{P}^{x_0}\left(\|x(t)-x_{\rm e}\| > \alpha_\rho(\alpha_\rho^{-1}(\varepsilon))\right)\\ &\geq \mathbb{P}^{x_0}\left(\|x(t)-x_{\rm e}\| > \varepsilon\right).\end{aligned}$$

Therefore, for every given $\varepsilon > 0$ and $\rho > 0$, there exists $\delta > 0$ such that, for all $x_0 \in \mathcal{B}_\delta(x_{\rm e}) \cap \overline{\mathbb{R}}_+^n$,

$$\mathbb{P}^{x_0}\left(\sup_{t\geq 0}\|x(t)-x_{\rm e}\| > \varepsilon\right) \leq \rho,$$

which proves that the equilibrium solution $x(t) \overset{\rm a.s.}{\equiv} x_{\rm e}$ is Lyapunov stable in probability with respect to $\overline{\mathbb{R}}_+^n$.

Conversely, for every given ε and ρ, let $\bar{\delta}(\varepsilon,\rho)$ be the supremum of all admissible $\delta(\varepsilon,\rho)$. Note that the function $\delta(\cdot,\cdot)$ is positive and nondecreasing in its first argument, but not necessarily continuous. For every $\rho > 0$ choose a class $\mathcal{K}$ function $\gamma_\rho(r)$ such that $\gamma_\rho(r) \leq k\bar{\delta}(r,\rho)$, $0 < k < 1$. Let $c(\rho) = \lim_{r\to\infty}\gamma_\rho(r)$ and $\alpha_\rho(r) = \gamma_\rho^{-1}(r)$, and note that $\alpha_\rho(\cdot)$ is class $\mathcal{K}$ [243, Lemma 4.2]. Next, for every $\rho > 0$ and $x_0 \in \mathcal{B}_{c(\rho)}(x_{\rm e}) \cap \overline{\mathbb{R}}_+^n$, let $\varepsilon = \alpha_\rho(\|x_0 - x_{\rm e}\|)$. Then $\|x_0 - x_{\rm e}\| < \bar{\delta}(\varepsilon,\rho)$ and

$$\mathbb{P}^{x_0}\left(\sup_{t\geq 0}\|x(t)-x_{\rm e}\| > \varepsilon\right) \leq \rho \tag{12.19}$$

imply

$$\mathbb{P}^{x_0}\left(\|x(t)-x_{\rm e}\| > \alpha_\rho(\|x_0-x_{\rm e}\|)\right) \leq \rho, \quad t \geq 0. \tag{12.20}$$

ii) Suppose that there exists a class $\mathcal{KL}$ function $\beta(r,s)$ such that (12.17) is satisfied. Then

$$\mathbb{P}^{x_0}\left(\|x(t)-x_{\rm e}\| > \beta_\rho(\|x_0-x_{\rm e}\|,0)\right) \leq \rho, \quad t \geq 0,$$

which implies that the equilibrium solution $x(t) \overset{\rm a.s.}{\equiv} x_{\rm e}$ is Lyapunov stable in probability with respect to $\overline{\mathbb{R}}_+^n$. Moreover, for $x_0 \in \mathcal{B}_{c(\rho)}(x_{\rm e}) \cap \overline{\mathbb{R}}_+^n$, the solution to (12.4) satisfies

$$\mathbb{P}^{x_0}\left(\|x(t)-x_{\rm e}\| > \beta_\rho(\|c(\rho)\|,t)\right) \leq \rho, \quad t \geq 0.$$

Now, letting $t \to \infty$ yields $\mathbb{P}^{x_0}\left(\lim_{t\to\infty}\|x(t)-x_{\rm e}\| > 0\right) \leq \rho$ for every $\rho > 0$, and hence, $\mathbb{P}^{x_0}\left(\lim_{t\to\infty}\|x(t)-x_{\rm e}\| = 0\right) \geq 1-\rho$, which implies that the equilibrium solution $x(t) \overset{\rm a.s.}{\equiv} x_{\rm e}$ is asymptotically stable in probability with respect to $\overline{\mathbb{R}}_+^n$.

Conversely, suppose that the equilibrium solution $x(t) \overset{\rm a.s.}{\equiv} x_{\rm e}$ is asymptotically stable in probability with respect to $\overline{\mathbb{R}}_+^n$. In this case, for every $\rho > 0$ there exist a constant $c(\rho) > 0$ and a class $\mathcal{K}$ function $\alpha_\rho(\cdot)$ such that, for every $r \in (0, c(\rho)]$, the solution $x(t)$, $t \geq 0$, to (12.4) satisfies

$$\mathbb{P}^{x_0}\left(\sup_{t\geq 0}\|x(t) - x_{\mathrm{e}}\| > \alpha_\rho(r)\right) \leq \mathbb{P}^{x_0}\left(\sup_{t\geq 0}\|x(t) - x_{\mathrm{e}}\| > \alpha_\rho(\|x_0 - x_{\mathrm{e}}\|)\right) \leq \rho \tag{12.21}$$

for all $\|x_0 - x_{\mathrm{e}}\| < r$. Moreover, given $\eta > 0$ there exists $T = T_\rho(\eta, r) \geq 0$ such that

$$\mathbb{P}^{x_0}\left(\sup_{t\geq T_\rho(\eta,r)}\|x(t) - x_{\mathrm{e}}\| > \eta\right) \leq \rho.$$

Let $\overline{T}_\rho(\eta, r)$ be the infimum of all admissible $T_\rho(\eta, r)$ and note that $\overline{T}_\rho(\eta, r)$ is nonnegative and nonincreasing in η, nondecreasing in r, and $\overline{T}_\rho(\eta, r) = 0$ for all $\eta \geq \alpha(r)$. Now, let

$$W_{r,\rho}(\eta) = \frac{2}{\eta}\int_{\frac{\eta}{2}}^{\eta}\overline{T}_\rho(s, r)\mathrm{d}s + \frac{r}{\eta} \geq \overline{T}_\rho(\eta, r) + \frac{r}{\eta}$$

and note that $W_{r,\rho}(\eta)$ is positive and has the following properties: *i*) for every fixed r and ρ, $W_{r,\rho}(\eta)$ is continuous, strictly decreasing, and $\lim_{\eta\to\infty} W_{r,\rho}(\eta) = 0$; and *ii*) for every fixed η and ρ, $W_{r,\rho}(\eta)$ is strictly increasing in r.

Next, let $U_{r,\rho} = W_{r,\rho}^{-1}$ and note that $U_{r,\rho}$ satisfies properties *i*) and *ii*) of $W_{r,\rho}$, and $\overline{T}_\rho(U_{r,\rho}(\sigma), r) < W_{r,\rho}(U_{r,\rho}(\sigma)) = \sigma$. Therefore,

$$\mathbb{P}^{x_0}\left(\|x(t) - x_{\mathrm{e}}\| > U_{r,\rho}(t)\right) \leq \rho, \quad t \geq 0, \tag{12.22}$$

for all $\|x_0 - x_{\mathrm{e}}\| < r$. Now, using (12.21) and (12.22) it follows that

$$\mathbb{P}^{x_0}\left(\|x(t) - x_{\mathrm{e}}\| > \sqrt{\alpha_\rho(\|x_0 - x_{\mathrm{e}}\|)U_{c(\rho),\rho}(t)}\right) \leq \rho, \ \|x_0 - x_{\mathrm{e}}\| < c(\rho), \ t \geq 0.$$

Thus, inequality (12.17) is satisfied with $\beta_\rho(\|x_0 - x_{\mathrm{e}}\|, t) = \sqrt{\alpha_\rho(\|x_0 - x_{\mathrm{e}}\|)} \cdot \sqrt{U_{c(\rho),\rho}(t)}$. □

Next, we present sufficient conditions for Lyapunov and asymptotic stability in probability for nonlinear stochastic nonnegative dynamical systems. First, however, the following definition of a recurrent process relative to a domain $\mathcal{D}_{\mathrm{r}}$ is needed.

Definition 12.6. A Markov process $x(\cdot)$ in $\mathcal{H}_n^{\mathcal{D}}$ is *recurrent relative to the domain* $\mathcal{D}_{\mathrm{r}}$ or, equivalently, $\mathcal{D}_{\mathrm{r}} \subset \mathcal{D}$ is *recurrent in* $\mathcal{D}$, if there exists a finite time $t > 0$ such that

$$\mathbb{P}^{x}\left(x(t) \in \mathcal{D}_{\mathrm{r}}\right) = 1. \tag{12.23}$$

In addition, $\mathcal{D}_\mathrm{r}$ is *positive recurrent* if

$$\sup_{x\in\mathcal{D}_\mathrm{c}} \mathbb{E}^x \inf\{t \geq 0 : x(t) \in \mathcal{D}_\mathrm{r}\} < \infty \tag{12.24}$$

for every compact set $\mathcal{D}_\mathrm{c} \subset \mathcal{D}$.

Theorem 12.1. Let $\mathcal{D}$ be an open subset relative to $\overline{\mathbb{R}}_+^n$ that contains x_e. Consider the nonlinear stochastic dynamical system (12.4) where f is essentially nonnegative and $f(x_\mathrm{e}) = 0$, $D_{(i,j)}(x) = 0$, $j = 1, \ldots, d$, whenever $x_i = 0$, $i = 1, \ldots, n$, and $D(x_\mathrm{e}) = 0$. Assume that there exists a two-times continuously differentiable function $V : \mathcal{D} \to \mathbb{R}$ such that

$$V(x_\mathrm{e}) = 0, \tag{12.25}$$

$$V(x) > 0, \qquad x \in \mathcal{D}, \qquad x \neq x_\mathrm{e}, \tag{12.26}$$

$$\frac{\partial V(x)}{\partial x} f(x) + \frac{1}{2}\mathrm{tr}\, D^\mathrm{T}(x)\frac{\partial^2 V(x)}{\partial x^2} D(x) \leq 0, \qquad x \in \mathcal{D}. \tag{12.27}$$

Then the equilibrium solution $x(t) \equiv x_\mathrm{e}$ to (12.4) is Lyapunov stable in probability with respect to $\overline{\mathbb{R}}_+^n$. If, in addition,

$$\frac{\partial V(x)}{\partial x} f(x) + \frac{1}{2}\mathrm{tr}\, D^\mathrm{T}(x)\frac{\partial^2 V(x)}{\partial x^2} D(x) < 0, \qquad x \in \mathcal{D}, \qquad x \neq x_\mathrm{e}, \tag{12.28}$$

then the equilibrium solution $x(t) \equiv x_\mathrm{e}$ to (12.4) is asymptotically stable in probability with respect to $\overline{\mathbb{R}}_+^n$. Finally, if $\mathcal{D} = \overline{\mathbb{R}}_+^n$ and $V(\cdot)$ is radially unbounded, then the equilibrium solution $x(t) \equiv x_\mathrm{e}$ to (12.4) is globally asymptotically stable in probability with respect to $\overline{\mathbb{R}}_+^n$.

Proof. Let $\delta > 0$ be such that $\mathcal{B}_\delta(x_\mathrm{e}) \cap \overline{\mathbb{R}}_+^n \subseteq \mathcal{D}$, define $V_\delta \triangleq \inf_{x\in\mathcal{D}\backslash\mathcal{B}_\delta(x_\mathrm{e})\cap\overline{\mathbb{R}}_+^n} V(x) > 0$, and let τ_δ be the stopping time wherein the trajectory $x(t)$, $t \geq 0$, of (12.4) exits the bounded domain $\mathcal{B}_\delta(x_\mathrm{e}) \cap \overline{\mathbb{R}}_+^n \subseteq \mathcal{D}$ with $\tau_\delta(t) \triangleq \min\{t, \tau_\delta\}$. Since $V(\cdot)$ is two-times continuously differentiable and (12.27) holds, it follows from Lemma 5.4 of [244] that

$$\mathbb{E}^x\left[V(x(\tau_\delta(t)))\right] \leq V(x) \tag{12.29}$$

for all $x \in \mathcal{B}_\delta(x_\mathrm{e}) \cap \overline{\mathbb{R}}_+^n$ and $t \geq 0$. Now, using Chebyshev's inequality [244, p. 14] yields

$$\mathbb{P}^x\left(\sup_{0\leq s\leq t} \|x(s) - x_\mathrm{e}\| > \delta\right) \leq \frac{\mathbb{E}^x\left[V(x(\tau_\delta(t)))\right]}{V_\delta} \leq \frac{V(x)}{V_\delta}. \tag{12.30}$$

Next, taking the limit as $t \to \infty$, (12.30) yields

$$\mathbb{P}^x\left(\sup_{s\geq 0} \|x(s) - x_\mathrm{e}\| > \delta\right) \leq \frac{V(x)}{V_\delta}, \tag{12.31}$$

and hence, Lyapunov stability in probability with respect to $\overline{\mathbb{R}}_+^n$ follows from the continuity of $V(\cdot)$ and (12.25).

To prove asymptotic stability in probability with respect to $\overline{\mathbb{R}}_+^n$, note that the stochastic process $V(x(\tau_\delta(t)))$ is a supermartingale [244, Lemma 5.4], and hence, it follows from Theorem 5.1 of [244] that

$$\lim_{t\to\infty} V(x(\tau_\delta(t))) \overset{\text{a.s.}}{=} \nu. \tag{12.32}$$

Let $\mathfrak{B}^x$ denote the set of all sample trajectories of (12.4) starting from $x \in \overline{\mathbb{R}}_+^n$ for which $\tau_\delta = \infty$. Since the equilibrium solution $x(t) \equiv x_\text{e}$ to (12.4) is Lyapunov stable in probability with respect to $\overline{\mathbb{R}}_+^n$, it follows that

$$\lim_{x\to x_\text{e}} \mathbb{P}^x(\mathfrak{B}^x) = 1. \tag{12.33}$$

Next, it follows from Theorem 3.9 of [244] and (12.28) that all sample trajectories contained in $\mathfrak{B}^x$, except for a set of trajectories with measure zero, satisfy $\inf_{t>0} \|x(t) - x_\text{e}\| = 0$. Moreover, it follows from Lemma 5.3 of [244] that

$$\liminf_{t\to\infty} \|x(t) - x_\text{e}\| = 0, \tag{12.34}$$

and hence, using (12.25), $\liminf_{t\to\infty} V(x(t)) = 0$. Now, (12.32) implies

$$\lim_{t\to\infty} V(x(\tau_\delta(t))) = \lim_{t\to\infty} V(x(t)) \tag{12.35}$$

for almost all sample trajectories in $\mathfrak{B}^x$, and hence,

$$\lim_{t\to\infty} V(x(t)) = \liminf_{t\to\infty} V(x(t)) = 0, \tag{12.36}$$

which, using (12.25) and (12.26), further implies that

$$\lim_{t\to\infty} \|x(t) - x_\text{e}\| = 0. \tag{12.37}$$

Now, asymptotic stability in probability with respect to $\overline{\mathbb{R}}_+^n$ is a direct consequence of (12.33) and (12.37).

Finally, to prove global asymptotic stability in probability with respect to $\overline{\mathbb{R}}_+^n$, note that it follows from Lyapunov stability in probability with respect to $\overline{\mathbb{R}}_+^n$ that, for every $\varepsilon > 0$ and $\rho = \varepsilon$, there exists $\delta > 0$ such that, for all $x \in \mathcal{B}_\delta(x_\text{e}) \cap \overline{\mathbb{R}}_+^n$,

$$\mathbb{P}^x\left(\sup_{t>0} \|x(t) - x_\text{e}\| > \varepsilon\right) < \varepsilon. \tag{12.38}$$

Moreover, it follows from Lemma 3.9, Theorem 3.9 of [244], and the radial unboundedness of $V(\cdot)$ that the solution $x(t)$, $t \geq 0$, of (12.4) is recurrent relative to the domain $\mathcal{B}_\varepsilon(x_\text{e}) \cap \overline{\mathbb{R}}_+^n$ for every $\varepsilon > 0$. Thus, $\tilde{\tau}_\delta \overset{\text{a.s.}}{<} \infty$, where $\tilde{\tau}_\delta$ is the first hitting time of the trajectories starting from the set $\overline{\mathbb{R}}_+^n \backslash \mathcal{B}_\delta(x_\text{e})$ and transitioning into the set $\mathcal{B}_\delta(x_\text{e}) \cap \overline{\mathbb{R}}_+^n$.

Now, using the strong Markov property of solutions and choosing $\delta > 0$ such that $x \in \overline{\mathbb{R}}_+^n \setminus \mathcal{B}_\delta(x_{\rm e})$ yields

$$\begin{aligned}
&\mathbb{P}^x\left(\limsup_{t\to\infty} \|x(t) - x_{\rm e}\| > \varepsilon\right) \\
&= \int_{\sigma=0}^{\infty}\int_{y\in\partial\mathcal{B}_\delta(x_{\rm e})\cap\overline{\mathbb{R}}_+^n} \mathbb{P}\left(\tilde{\tau}_\delta \in {\rm d}\sigma,\, x(\tilde{\tau}_\delta)\in {\rm d}y\right)\mathbb{P}^y\left(\limsup_{t\to\infty}\|x(t) - x_{\rm e}\| > \varepsilon\right) \\
&= \int_{\sigma=0}^{\infty}\int_{y\in\partial\mathcal{B}_\delta(x_{\rm e})\cap\overline{\mathbb{R}}_+^n} \mathbb{P}\left(\tilde{\tau}_\delta \in {\rm d}\sigma,\, x(\tilde{\tau}_\delta)\in {\rm d}y\right)\mathbb{P}^y\left(\sup_{t>0}\|x(t) - x_{\rm e}\| > \varepsilon\right) \\
&\le \varepsilon,
\end{aligned}\tag{12.39}$$

which proves global asymptotic stability in probability with respect to $\overline{\mathbb{R}}_+^n$. □

A more general stochastic stability notion can also be introduced here involving stochastic stability and convergence to an invariant (stationary) distribution. In this case, state convergence is not to an equilibrium point but rather to a stationary distribution. This framework can relax the vanishing perturbation assumption $D(x_{\rm e}) = 0$ at the equilibrium point $x_{\rm e}$ and requires a more involved analysis framework showing stability of the underlying Markov semigroup [321].

As in nonlinear stochastic dynamical systems theory [244], converse Lyapunov theorems for Lyapunov and asymptotic stability in probability for stochastic nonnegative dynamical systems can also be established. However, in this case, a nondegeneracy condition on $D(x)$, $x \in \mathcal{D}$, is required [244].

Finally, we establish a stochastic version of the Krasovskii-LaSalle stability theorem for nonnegative dynamical systems. For nonlinear stochastic dynamical systems this result is due to Mao [303].

Theorem 12.2. Consider the nonlinear stochastic nonnegative dynamical system (12.4). Let $\mathcal{D} \subseteq \overline{\mathbb{R}}_+^n$ be an invariant set with respect to (12.4) and assume that there exists a two-times continuously differentiable function $V : \mathcal{D} \to \overline{\mathbb{R}}_+$ and a continuous function $\eta : \overline{\mathbb{R}}_+ \to \overline{\mathbb{R}}_+$ such that

$$\frac{\partial V(x)}{\partial x} f(x) + \frac{1}{2}{\rm tr}\, D^{\rm T}(x)\frac{\partial^2 V(x)}{\partial x^2}D(x) \le -\eta(V(x)), \qquad x\in\mathcal{D}. \tag{12.40}$$

Then, for every $x_0 \in \mathcal{D}$, $\lim_{t\to\infty} V(x(t))$ exists and is finite almost surely, and

$$\lim_{t\to\infty}\eta(V(x(t))) \overset{\rm a.s.}{=} 0. \tag{12.41}$$

Proof. Since $\mathcal{D} \subseteq \overline{\mathbb{R}}_+^n$ is invariant with respect to (12.4), it follows that, for all $x_0 \in \mathcal{D}$, $\mathbb{P}^{x_0}(x(t) \in \mathcal{D}) = 1$, $t \geq 0$. Furthermore, using Itô's chain rule formula and (12.40) we have

$$\begin{aligned} V(x(t)) &= V(x_0) + \int_0^t \mathcal{L}V(x(\sigma))\mathrm{d}\sigma + \int_0^t \frac{\partial V(x(\sigma))}{\partial x} D(x(\sigma))\mathrm{d}w(\sigma) \\ &\leq V(x_0) - \int_0^t \eta(V(x(\sigma)))\mathrm{d}\sigma + \int_0^t \frac{\partial V(x(\sigma))}{\partial x} D(x(\sigma))\mathrm{d}w(\sigma). \end{aligned} \tag{12.42}$$

Now, it follows from Theorem 7 of [287, p. 139] that $\lim_{t\to\infty} V(x(t))$ exists and is finite almost surely, and

$$\lim_{t\to\infty} \int_0^t \eta(V(x(\sigma)))\mathrm{d}\sigma \overset{\text{a.s.}}{<} \infty. \tag{12.43}$$

To show that $\lim_{t\to\infty} \eta(V(x(t))) \overset{\text{a.s.}}{=} 0$ suppose, *ad absurdum*, that there exists a sample space $\bar{\Omega} \subset \Omega$ such that $\mathbb{P}(\bar{\Omega}) > 0$ and

$$\limsup_{t\to\infty} \eta(V(x(t,\omega))) > 0, \quad \omega \in \bar{\Omega}. \tag{12.44}$$

Let $\{t_n\}_{n=0}^{\infty}$, $n \in \mathbb{Z}_+$, be a monotonic sequence with $t_n + 1 < t_{n+1}$ and note that there exist $\varepsilon > 0$ and $N \in \mathbb{Z}_+$ such that

$$\eta(V(x(t_n,\omega))) > \varepsilon, \quad n \geq N. \tag{12.45}$$

Now, it follows from the continuity of $\eta(\cdot)$ and $V(\cdot)$ and the sample continuity of $x(\cdot)$ that there exist $\delta > 0$, $\delta_1 > 0$, and $\delta_2 > 0$, such that if $|V(x(t_n,\omega)) - V(x(t,\omega))| \leq \delta_2$, then

$$|\eta(V(x(t_n,\omega))) - \eta(V(x(t,\omega)))| \leq \frac{\varepsilon}{2}, \tag{12.46}$$

if $\|x(t_n,\omega) - x(t,\omega)\| \leq \delta_1$, then

$$|V(x(t_n,\omega)) - V(x(t,\omega))| \leq \delta_2, \tag{12.47}$$

and if $|t_n - t| \leq \delta$, then

$$\|x(t_n,\omega) - x(t,\omega)\| \leq \delta_1. \tag{12.48}$$

Thus, using (12.45) and (12.46) it follows that, for all $|t_n - t| \leq \delta$ and $n \geq N$,

$$\eta(V(x(t,\omega))) \geq \eta(V(x(t_n,\omega))) - |\eta(V(x(t_n,\omega))) - \eta(V(x(t,\omega)))| > \frac{\varepsilon}{2}, \tag{12.49}$$

and hence,

$$\lim_{t\to\infty} \int_0^t \eta(V(x(\sigma)))\mathrm{d}\sigma \geq \sum_{n=N}^{\infty} \int_{t_n}^{t_n+\delta} \eta(V(x(\sigma)))\mathrm{d}\sigma \geq \sum_{n=N}^{\infty} \frac{\varepsilon\delta}{2} = \infty,$$

(12.50)

which contradicts (12.43). Thus, $\lim_{t\to\infty}\eta(V(x(t)))\stackrel{\text{a.s.}}{=}0$. □

Note that if we define $\mathcal{D}_\eta \triangleq \{v\geq 0:\eta(v)=0\}$, then it can be shown that $\mathcal{D}_\eta\neq\varnothing$ and (12.41) implies

$$\lim_{t\to\infty}\text{dist}(V(x(t)),\mathcal{D}_\eta)\stackrel{\text{a.s.}}{=}0, \tag{12.51}$$

that is, $V(x(t))$ will asymptotically approach the set $\mathcal{D}_\eta$ with probability one. Thus, if $\eta(V(x))=x_{\text{e}}$ if and only if $x=x_{\text{e}}$ and $V(\cdot)$ is positive definite with respect to x_{e}, then it follows from Theorems 12.1 and 12.2 that the equilibrium solution $x(t)\stackrel{\text{a.s.}}{\equiv}x_{\text{e}}$ to (12.4) is asymptotically stable in probability with respect to $\overline{\mathbb{R}}_+^n$.

12.4 Semistability of Stochastic Nonnegative Dynamical Systems

In this section, we present necessary and sufficient conditions for stochastic semistability. First, we present several key propositions. The following proposition gives a sufficient condition for a trajectory of (12.4) to converge to a limit point. For this result, $\mathcal{D}_{\text{c}}\subseteq\mathcal{D}\subseteq\overline{\mathbb{R}}_+^n$ denotes a positively invariant set with respect to (12.4), and $s_t(\mathcal{H}_n^{\mathcal{D}_{\text{c}}})$ denotes the image of $\mathcal{H}_n^{\mathcal{D}_{\text{c}}}\subset\mathcal{H}_n^{\mathcal{D}}$ under the flow $s_t:\mathcal{H}_n^{\mathcal{D}_{\text{c}}}\to\mathcal{H}_n^{\mathcal{D}}$, that is, $s_t(\mathcal{H}_n^{\mathcal{D}_{\text{c}}})\triangleq\{y:y=s_t(x_0)\text{ for some }x(0)\stackrel{\text{a.s.}}{=}x_0\in\mathcal{H}_n^{\mathcal{D}_{\text{c}}}\}$.

Proposition 12.2. Consider the nonlinear stochastic nonnegative dynamical system (12.4) and let $x\in\mathcal{D}_{\text{c}}$. If the limit set $\omega(x)$ of (12.4) contains an equilibrium point y that is Lyapunov stable in probability (with respect to $\overline{\mathbb{R}}_+^n$), then $\lim_{x\to y}\mathbb{P}^x\big(\|\lim_{t\to\infty}s(t,x)-y\|=0\big)=1$, that is, $\omega(x)\stackrel{\text{a.s.}}{=}\{y\}$ as $x\to y$.

Proof. Suppose $y\in\omega(x)$ is Lyapunov stable in probability with respect to $\overline{\mathbb{R}}_+^n$ and let $\mathcal{N}_\varepsilon\subseteq\mathcal{D}_{\text{c}}$ be a relatively open neighborhood of y. Since y is Lyapunov stable in probability with respect to $\overline{\mathbb{R}}_+^n$, there exists a relatively open neighborhood $\mathcal{N}_\delta\subset\mathcal{D}_{\text{c}}$ of y such that $s_t(\mathcal{H}_n^{\mathcal{N}_\delta})\subseteq\mathcal{H}_n^{\mathcal{N}_\varepsilon}$ as $x\to y$ for every $t\geq 0$. Now, since $y\in\omega(x)$, it follows that there exists $\tau\geq 0$ such that $s(\tau,x)\in\mathcal{H}_n^{\mathcal{N}_\delta}$. Hence, $s(t+\tau,x)=s_t(s(\tau,x))\in s_t(\mathcal{H}_n^{\mathcal{N}_\delta})\subseteq\mathcal{H}_n^{\mathcal{N}_\varepsilon}$ for every $t>0$. Since $\mathcal{N}_\varepsilon\subseteq\mathcal{D}_{\text{c}}$ is arbitrary, it follows that $y\stackrel{\text{a.s.}}{=}\lim_{t\to\infty}s(t,x)$. Thus, $\lim_{n\to\infty}s(t_n,x)\stackrel{\text{a.s.}}{=}y$ as $x\to y$ for every sequence $\{t_n\}_{n=1}^\infty$, and hence, $\omega(x)\stackrel{\text{a.s.}}{=}\{y\}$ as $x\to y$. □

The following definition introduces the notion of stochastic semistability.

Definition 12.7. An equilibrium solution $x(t) \stackrel{\text{a.s.}}{\equiv} x_{\mathrm{e}} \in \mathcal{E}$ of (12.4) is *stochastically semistable with respect to* $\overline{\mathbb{R}}_+^n$ if the following statements hold.

i) For every $\varepsilon > 0$, $\lim_{x_0 \to x_{\mathrm{e}}} \mathbb{P}^{x_0}\left(\sup_{0 \leq t < \infty} \|x(t) - x_{\mathrm{e}}\| > \varepsilon\right) = 0$. Equivalently, for every $\varepsilon > 0$ and $\rho \in (0,1)$, there exist $\delta = \delta(\varepsilon, \rho) > 0$ such that, for all $x_0 \in \mathcal{B}_\delta(x_{\mathrm{e}}) \cap \overline{\mathbb{R}}_+^n$,

$$\mathbb{P}^{x_0}\left(\sup_{0 \leq t < \infty} \|x(t) - x_{\mathrm{e}}\| > \varepsilon\right) \leq \rho.$$

ii) $\lim_{\mathrm{dist}(x_0, \mathcal{E}) \to 0} \mathbb{P}^{x_0}\left(\lim_{t \to \infty} \mathrm{dist}(x(t), \mathcal{E}) = 0\right) = 1$. Equivalently, for every $\rho \in (0,1)$, there exist $\delta = \delta(\rho) > 0$ such that if $\mathrm{dist}(x_0, \mathcal{E}) \leq \delta$, then $\mathbb{P}^{x_0}\left(\lim_{t \to \infty} \mathrm{dist}(x(t), \mathcal{E}) = 0\right) \geq 1 - \rho$.

The dynamical system (12.4) is *stochastically semistable with respect to* $\overline{\mathbb{R}}_+^n$ if every equilibrium solution of (12.4) is stochastically semistable with respect to $\overline{\mathbb{R}}_+^n$. Finally, the dynamical system (12.4) is *globally stochastically semistable with respect to* $\overline{\mathbb{R}}_+^n$ if *i)* holds and $\mathbb{P}^{x_0}\left(\lim_{t \to \infty} \mathrm{dist}(x(t), \mathcal{E}) = 0\right) = 1$ for all $x_0 \in \overline{\mathbb{R}}^n$.

Note that if $x(t) \stackrel{\text{a.s.}}{\equiv} x_{\mathrm{e}} \in \mathcal{E}$ only satisfies *i)* in Definition 12.7, then the equilibrium solution $x(t) \stackrel{\text{a.s.}}{\equiv} x_{\mathrm{e}} \in \mathcal{E}$ of (12.4) is Lyapunov stable in probability with respect to $\overline{\mathbb{R}}_+^n$.

Definition 12.8. For a given $\rho \in (0,1)$, the *ρ-domain of semistability with respect to* $\overline{\mathbb{R}}_+^n$ is the set of points $x_0 \in \mathcal{D} \subseteq \overline{\mathbb{R}}^n$ such that if $x(t)$, $t \geq 0$, is a solution to (12.4) with $x(0) \stackrel{\text{a.s.}}{=} x_0$, then $x(t)$ converges to an equilibrium point in $\mathcal{D}$ that is Lyapunov stable in probability (with respect to $\overline{\mathbb{R}}_+^n$) with probability greater than or equal to $1 - \rho$.

Note that if (12.4) is stochastically semistable, then its ρ-domain of semistability contains the set of equilibria in its interior.

Next, we present alternative equivalent characterizations for stochastic semistability of (12.4).

Proposition 12.3. Consider the nonlinear stochastic nonnegative dynamical system $\mathcal{G}$ given by (12.4). Then the following statements are equivalent:

i) $\mathcal{G}$ is stochastically semistable with respect to $\overline{\mathbb{R}}_+^n$.

ii) For every $x_{\mathrm{e}} \in \mathcal{E}$ and $\rho > 0$, there exist class $\mathcal{K}$ and $\mathcal{L}$ functions $\alpha_\rho(\cdot)$ and $\beta_\rho(\cdot)$, respectively, and $\delta = \delta(x_{\mathrm{e}}, \rho) > 0$ such that, if $x_0 \in \mathcal{B}_\delta(x_{\mathrm{e}}) \cap \overline{\mathbb{R}}_+^n$, then

$$\mathbb{P}^{x_0}\left(\|x(t) - x_{\rm e}\| > \alpha_\rho(\|x_0 - x_{\rm e}\|)\right) \le \rho, \quad t \ge 0,$$

and $\mathbb{P}^{x_0}\left(\mathrm{dist}(x(t), \mathcal{E}) > \beta_\rho(t)\right) \le \rho$, $t \ge 0$.

iii) For every $x_{\rm e} \in \mathcal{E}$ and $\rho > 0$, there exist class $\mathcal{K}$ functions $\alpha_{1\rho}(\cdot)$ and $\alpha_{2\rho}(\cdot)$, a class $\mathcal{L}$ function $\beta_\rho(\cdot)$, and $\delta = \delta(x_{\rm e}, \rho) > 0$ such that, if $x_0 \in \mathcal{B}_\delta(x_{\rm e}) \cap \overline{\mathbb{R}}_+^n$, then

$$\begin{aligned}\mathbb{P}^{x_0}&\left(\mathrm{dist}(x(t), \mathcal{E}) > \alpha_{2\rho}(\|x_0 - x_{\rm e}\|)\beta_\rho(t)\right)\\ &\le \mathbb{P}^{x_0}\left(\alpha_{1\rho}(\|x(t) - x_{\rm e}\|) > \alpha_{2\rho}(\|x_0 - x_{\rm e}\|)\right) \le \rho, \quad t \ge 0.\end{aligned}$$

Proof. To show that *i)* implies *ii)*, suppose (12.4) is stochastically semistable with respect to $\overline{\mathbb{R}}_+^n$ and let $x_{\rm e} \in \mathcal{E}$. It follows from Lemma 12.1 that for every $\rho > 0$ there exists $\delta = \delta(x_{\rm e}, \rho) > 0$ and a class $\mathcal{K}$ function $\alpha_\rho(\cdot)$ such that if $\|x_0 - x_{\rm e}\| \le \delta$, then $\mathbb{P}^{x_0}\left(\|x(t) - x_{\rm e}\| > \alpha_\rho(\|x_0 - x_{\rm e}\|)\right) \le \rho$, $t \ge 0$. Without loss of generality, we can assume that δ is such that $\overline{\mathcal{B}_\delta(x_{\rm e})} \cap \overline{\mathbb{R}}_+^n$ is contained in the ρ-domain of semistability of (12.4). Hence, for every $x_0 \in \overline{\mathcal{B}_\delta(x_{\rm e})} \cap \overline{\mathbb{R}}_+^n$, $\lim_{t\to\infty} x(t) \overset{\rm a.s.}{=} x^* \in \mathcal{E}$ and, consequently, $\mathbb{P}^{x_0}\left(\lim_{t\to\infty} \mathrm{dist}(x(t), \mathcal{E}) = 0\right) = 1$.

For every $\varepsilon > 0$, $\rho > 0$, and $x_0 \in \overline{\mathcal{B}_\delta(x_{\rm e})} \cap \overline{\mathbb{R}}_+^n$, define $T_{x_0}(\varepsilon, \rho)$ to be the infimum of T with the property that $\mathbb{P}^{x_0}\left(\sup_{t\ge T} \mathrm{dist}(x(t), \mathcal{E}) > \varepsilon\right) \le \rho$, that is,

$$T_{x_0}(\varepsilon, \rho) \triangleq \inf\left\{T : \mathbb{P}^{x_0}\left(\sup_{t\ge T} \mathrm{dist}(x(t), \mathcal{E}) > \varepsilon\right) \le \rho\right\}.$$

For each $x_0 \in \overline{\mathcal{B}_\delta(x_{\rm e})} \cap \overline{\mathbb{R}}_+^n$ and ρ, the function $T_{x_0}(\varepsilon, \rho)$ is nonnegative and nonincreasing in ε, and $T_{x_0}(\varepsilon, \rho) = 0$ for sufficiently large ε.

Next, let $T(\varepsilon, \rho) \triangleq \sup\{T_{x_0}(\varepsilon, \rho) : x_0 \in \overline{\mathcal{B}_\delta(x_{\rm e})} \cap \overline{\mathbb{R}}_+^n\}$. We claim that T is well defined. To show this, consider $\varepsilon > 0$, $\rho > 0$, and $x_0 \in \overline{\mathcal{B}_\delta(x_{\rm e})} \cap \overline{\mathbb{R}}_+^n$. Since $\mathbb{P}^{x_0}\left(\sup_{t\ge T_{x_0}(\varepsilon,\rho)} \mathrm{dist}(x(t), \mathcal{E}) > \varepsilon\right) \le \rho$, it follows from the sample continuity of s that, for every $\varepsilon > 0$ and $\rho > 0$, there exists an open neighborhood $\mathcal{U}$ of x_0 such that $\mathbb{P}^{x_0}\left(\sup_{t\ge T_z(\varepsilon,\rho)} \mathrm{dist}(s(t,z), \mathcal{E}) > \varepsilon\right) \le \rho$ for every $z \in \mathcal{U}$. Hence, $\limsup_{z\to x_0} T_z(\varepsilon, \rho) \le T_{x_0}(\varepsilon, \rho)$, implying that the function $x_0 \mapsto T_{x_0}(\varepsilon, \rho)$ is upper semicontinuous at the arbitrarily chosen point x_0, and hence on $\overline{\mathcal{B}_\delta(x_{\rm e})} \cap \overline{\mathbb{R}}_+^n$. Since an upper semicontinuous function defined on a compact set achieves its supremum, it follows that $T(\varepsilon, \rho)$ is well defined. The function $T(\cdot)$ is the pointwise supremum of a collection of nonnegative and nonincreasing functions, and hence is nonnegative and nonincreasing. Moreover, $T(\varepsilon, \rho) = 0$ for every $\varepsilon > \max\{\alpha_\rho(\|x_0 - x_{\rm e}\|) : x_0 \in \overline{\mathcal{B}_\delta(x_{\rm e})} \cap \overline{\mathbb{R}}_+^n\}$.

Let $\psi_\rho(\varepsilon) \triangleq \frac{2}{\varepsilon}\int_{\varepsilon/2}^{\varepsilon} T(\sigma,\rho)\mathrm{d}\sigma + \frac{1}{\varepsilon} \geq T(\varepsilon,\rho) + \frac{1}{\varepsilon}$. The function $\psi_\rho(\varepsilon)$ is positive, continuous, strictly decreasing, and $\psi_\rho(\varepsilon) \to 0$ as $\varepsilon \to \infty$. Choose $\beta_\rho(\cdot) = \psi^{-1}(\cdot)$. Then $\beta_\rho(\cdot)$ is positive, continuous, strictly decreasing, and $\lim_{\sigma\to\infty}\beta_\rho(\sigma) = 0$. Furthermore, $T(\beta_\rho(\sigma),\rho) < \psi_\rho(\beta_\rho(\sigma)) = \sigma$. Hence, $\mathbb{P}^{x_0}\left(\mathrm{dist}(x(t),\mathcal{E}) > \beta_\rho(t)\right) \leq \rho$, $t \geq 0$.

Next, to show that *ii*) implies *iii*), suppose *ii*) holds and let $x_\mathrm{e} \in \mathcal{E}$. Then it follows from *i*) of Lemma 12.1 that x_e is Lyapunov stable in probability with respect to $\overline{\mathbb{R}}_+^n$. For every $\rho > 0$, choosing x_0 sufficiently close to x_e, it follows from the inequality $\mathbb{P}^{x_0}\left(\|x(t) - x_\mathrm{e}\| > \alpha_\rho(\|x_0 - x_\mathrm{e}\|)\right) \leq \rho$, $t \geq 0$, that trajectories of (12.4) starting sufficiently close to x_e are bounded, and hence, the positive limit set of (12.4) is nonempty. Since $\mathbb{P}^{x_0}\left(\lim_{t\to\infty}\mathrm{dist}(x(t),\mathcal{E}) = 0\right) = 1$ as $\mathrm{dist}(x_0,\mathcal{E}) \to 0$, it follows that the positive limit set is contained in $\mathcal{E}$ as $\mathrm{dist}(x_0,\mathcal{E}) \to 0$.

Now, since every point in $\mathcal{E}$ is Lyapunov stable in probability with respect to $\overline{\mathbb{R}}_+^n$, it follows from Proposition 12.2 that $\lim_{t\to\infty} x(t) \overset{\text{a.s.}}{=} x^*$ as $x_0 \to x^*$, where $x^* \in \mathcal{E}$ is Lyapunov stable in probability with respect to $\overline{\mathbb{R}}_+^n$. If $x^* = x_\mathrm{e}$, then it follows using similar arguments as above that there exists a class $\mathcal{L}$ function $\hat{\beta}_\rho(\cdot)$ such that

$$\mathbb{P}^{x_0}\left(\mathrm{dist}(x(t),\mathcal{E}) > \hat{\beta}_\rho(t)\right) \leq \mathbb{P}^{x_0}\left(\|x(t) - x_\mathrm{e}\| > \hat{\beta}_\rho(t)\right) \leq \rho$$

for every x_0 satisfying $\|x_0 - x_\mathrm{e}\| < \delta$ and $t \geq 0$. Hence,

$$\mathbb{P}^{x_0}\left(\mathrm{dist}(x(t),\mathcal{E}) > \sqrt{\|x(t) - x_\mathrm{e}\|}\sqrt{\hat{\beta}_\rho(t)}\right) \leq \rho, \quad t \geq 0.$$

Next, consider the case where $x^* \neq x_\mathrm{e}$ and let $\alpha_{1\rho}(\cdot)$ be a class $\mathcal{K}$ function. In this case, note that

$$\mathbb{P}^{x_0}\left(\lim_{t\to\infty}\mathrm{dist}(x(t),\mathcal{E})/\alpha_{1\rho}(\|x(t) - x_\mathrm{e}\|) = 0\right) \geq 1 - \rho,$$

and hence, it follows using similar arguments as above that there exists a class $\mathcal{L}$ function $\beta_\rho(\cdot)$ such that

$$\mathbb{P}^{x_0}\left(\mathrm{dist}(x(t),\mathcal{E}) > \alpha_{1\rho}(\|x(t) - x_\mathrm{e}\|)\beta_\rho(t)\right) \leq \rho, \quad t \geq 0.$$

Now, note that $\alpha_{1\rho} \circ \alpha_\rho$ is of class $\mathcal{K}$ (by [243, Lemma 4.2]), and hence, *iii*) follows immediately.

Finally, to show that *iii*) implies *i*), suppose *iii*) holds and let $x_\mathrm{e} \in \mathcal{E}$. Then it follows that for every $\rho > 0$,

$$\mathbb{P}^{x_0}\left(\alpha_{1\rho}(\|x(t) - x_\mathrm{e}\|) > \alpha_{2\rho}(\|x(0) - x_\mathrm{e}\|)\right) \leq \rho, \quad t \geq 0,$$

that is, $\mathbb{P}^{x_0}[\|x(t) - x_\mathrm{e}\| > \alpha_\rho(\|x(0) - x_\mathrm{e}\|)] \leq \rho$, where $t \geq 0$ and

$\alpha_\rho = \alpha_{1\rho}{}^{-1} \circ \alpha_{2\rho}$ is of class $\mathcal{K}$ (by [243, Lemma 4.2]). It now follows from *i*) of Lemma 12.1 that $x_{\rm e}$ is Lyapunov stable in probability with respect to $\overline{\mathbb{R}}_+^n$. Since $x_{\rm e}$ was chosen arbitrarily, it follows that every equilibrium point is Lyapunov stable in probability with respect to $\overline{\mathbb{R}}_+^n$. Furthermore, $\mathbb{P}^{x_0}\left(\lim_{t\to\infty} \operatorname{dist}(x(t), \mathcal{E}) = 0\right) \geq 1 - \rho$.

Choosing x_0 sufficiently close to $x_{\rm e}$, it follows from the inequality $\mathbb{P}^{x_0}\left(\|x(t) - x_{\rm e}\| > \alpha_\rho(\|x_0 - x_{\rm e}\|)\right) \leq \rho$, $t \geq 0$, that trajectories of (12.4) are almost surely bounded as $x_0 \to x_{\rm e}$, and hence, the positive limit set of (12.4) is nonempty as $x_0 \to x_{\rm e}$. Since every point in $\mathcal{E}$ is Lyapunov stable in probability with respect to $\overline{\mathbb{R}}_+^n$, it follows from Proposition 12.2 that $\lim_{t\to\infty} x(t) \overset{\rm a.s.}{=} x^*$ as $x_0 \to x^*$, where $x^* \in \mathcal{E}$ is Lyapunov stable in probability with respect to $\overline{\mathbb{R}}_+^n$. Hence, by Definition 12.7, (12.4) is stochastically semistable with respect to $\overline{\mathbb{R}}_+^n$. □

Next, we develop necessary and sufficient conditions for stochastic semistability. First, we present sufficient conditions for stochastic semistability.

Theorem 12.3. Consider the nonlinear stochastic nonnegative dynamical system (12.4). Let $\mathcal{Q} \subseteq \overline{\mathbb{R}}_+^n$ be a relatively open neighborhood of $\mathcal{E}$ and assume that there exists a two-times continuously differentiable function $V : \mathcal{Q} \to \overline{\mathbb{R}}_+$ such that

$$V'(x)f(x) + \frac{1}{2}\operatorname{tr} D^{\rm T}(x)V''(x)D(x) < 0, \quad x \in \mathcal{Q}\backslash\mathcal{E}. \tag{12.52}$$

If every equilibrium point of (12.4) is Lyapunov stable in probability with respect to $\overline{\mathbb{R}}_+^n$, then (12.4) is stochastically semistable with respect to $\overline{\mathbb{R}}_+^n$. Moreover, if $\mathcal{Q} = \overline{\mathbb{R}}_+^n$ and $V(x) \to \infty$ as $\|x\| \to \infty$, then (12.4) is globally stochastically semistable with respect to $\overline{\mathbb{R}}_+^n$.

Proof. Since every equilibrium point of (12.4) is Lyapunov stable in probability with respect to $\overline{\mathbb{R}}_+^n$ by assumption, for every $z \in \mathcal{E}$, there exists a relatively open neighborhood $\mathcal{V}_z$ of z such that $s([0,\infty) \times \mathcal{V}_z \cap \mathcal{B}_\varepsilon(z))$, $\varepsilon > 0$, is bounded and contained in $\mathcal{Q}$ as $\varepsilon \to 0$. The set $\mathcal{V}_\varepsilon \triangleq \bigcup_{z\in\mathcal{E}} \mathcal{V}_z \cap \mathcal{B}_\varepsilon(z)$, $\varepsilon > 0$, is a relatively open neighborhood of $\mathcal{E}$ contained in $\mathcal{Q}$. Consider $x \in \mathcal{V}_\varepsilon$ so that there exists $z \in \mathcal{E}$ such that $x \in \mathcal{V}_z \cap \mathcal{B}_\varepsilon(z)$ and $s(t,x) \in \mathcal{H}_n^{\mathcal{V}_z \cap \mathcal{B}_\varepsilon(z)}$, $t \geq 0$, as $\varepsilon \to 0$. Since $\mathcal{V}_z \cap \mathcal{B}_\varepsilon(z)$ is bounded and invariant with respect to the solution of (12.4) as $\varepsilon \to 0$, it follows that $\mathcal{V}_\varepsilon$ is invariant with respect to the solution of (12.4) as $\varepsilon \to 0$. Furthermore, it follows from (12.52) that $\mathcal{L}V(s(t,x)) < 0$, $t \geq 0$, and hence, since $\mathcal{V}_\varepsilon$ is bounded, it follows from Theorem 12.2 that $\lim_{t\to\infty} \mathcal{L}V(s(t,x)) \overset{\rm a.s.}{=} 0$ as $\varepsilon \to 0$.

It is easy to see that $\mathcal{L}V(x) \neq 0$ by assumption and $\mathcal{L}V(x_{\rm e}) = 0$,

$x_{\rm e} \in \mathcal{E}$. Therefore, $s(t,x) \stackrel{\rm a.s.}{\to} \mathcal{E}$ as $t \to \infty$ and $\varepsilon \to 0$, which implies that $\lim_{{\rm dist}(x,\mathcal{E})\to 0} \mathbb{P}^x(\lim_{t\to\infty} {\rm dist}(s(t,x),\mathcal{E}) = 0) = 1$. Finally, since every point in $\mathcal{E}$ is Lyapunov stable in probability with respect to $\overline{\mathbb{R}}_+^n$, it follows from Proposition 12.2 that $\lim_{t\to\infty} s(t,x) \stackrel{\rm a.s.}{=} x^*$ as $x \to x^*$, where $x^* \in \mathcal{E}$ is Lyapunov stable in probability with respect to $\overline{\mathbb{R}}_+^n$. Hence, by Definition 12.7, (12.4) is semistable. For $\mathcal{Q} = \overline{\mathbb{R}}_+^n$, global stochastic semistability with respect to $\overline{\mathbb{R}}_+^n$ follows from identical arguments using the radially unbounded condition on $V(\cdot)$. □

Next, we present a slightly more general theorem for stochastic semistability wherein we do not assume that all points in $\mathcal{L}V^{-1}(0)$ are Lyapunov stable in probability with respect to $\overline{\mathbb{R}}_+^n$ but rather we assume that all points in $(\eta\circ V)^{-1}(0)$ are Lyapunov stable in probability with respect to $\overline{\mathbb{R}}_+^n$ for some continuous function $\eta : \overline{\mathbb{R}}_+ \to \overline{\mathbb{R}}_+$.

Theorem 12.4. Consider the nonlinear stochastic nonnegative dynamical system (12.4) and let $\mathcal{Q} \subseteq \overline{\mathbb{R}}_+^n$ be a relatively open neighborhood of $\mathcal{E}$. Assume that there exist a two-times continuously differentiable function $V : \mathcal{Q} \to \overline{\mathbb{R}}_+$ and a continuous function $\eta : \overline{\mathbb{R}}_+ \to \overline{\mathbb{R}}_+$ such that

$$V'(x)f(x) + \frac{1}{2}{\rm tr}\ D^{\rm T}(x)V''(x)D(x) \le -\eta(V(x)), \quad x \in \mathcal{Q}. \tag{12.53}$$

If every point in the set $\mathcal{M} \triangleq \{x \in \mathcal{Q} : \eta(V(x)) = 0\}$ is Lyapunov stable in probability with respect to $\overline{\mathbb{R}}_+^n$, then (12.4) is stochastically semistable with respect to $\overline{\mathbb{R}}_+^n$. Moreover, if $\mathcal{Q} = \overline{\mathbb{R}}_+^n$ and $V(x) \to \infty$ as $\|x\| \to \infty$, then (12.4) is globally stochastically semistable with respect to $\overline{\mathbb{R}}_+^n$.

Proof. Since, by assumption, (12.4) is Lyapunov stable in probability with respect to $\overline{\mathbb{R}}_+^n$ for all $z \in \mathcal{M}$, there exists a relatively open neighborhood $\mathcal{V}_z$ of z such that $s([0,\infty) \times \mathcal{V}_z \cap \mathcal{B}_\varepsilon(z))$, $\varepsilon > 0$, is bounded and contained in $\mathcal{Q}$ as $\varepsilon \to 0$. The set $\mathcal{V}_\varepsilon \triangleq \bigcup_{z\in\mathcal{M}} \mathcal{V}_z \cap \mathcal{B}_\varepsilon(z)$ is a relatively open neighborhood of $\mathcal{M}$ contained in $\mathcal{Q}$. Consider $x \in \mathcal{V}_\varepsilon$ so that there exists $z \in \mathcal{M}$ such that $x \in \mathcal{V}_z \cap \mathcal{B}_\varepsilon(z)$ and $s(t,x) \in \mathcal{H}_n^{\mathcal{V}_z\cap\mathcal{B}_\varepsilon(z)}$, $t \ge 0$, as $\varepsilon \to 0$. Since $\mathcal{V}_z$ is bounded it follows that $\mathcal{V}_\varepsilon$ is invariant with respect to the solution of (12.4) as $\varepsilon \to 0$. Furthermore, it follows from (12.53) that $\mathcal{L}V(s(t,x)) \le -\eta(V(s(t,x)))$, $t \ge 0$, and hence, since $\mathcal{V}_\varepsilon$ is bounded and invariant with respect to the solution of (12.4) as $\varepsilon \to 0$, it follows from Theorem 12.2 that $\lim_{t\to\infty} \eta(V(s(t,x))) \stackrel{\rm a.s.}{=} 0$ as $\varepsilon \to 0$. Therefore, $s(t,x) \stackrel{\rm a.s.}{\to} \mathcal{M}$ as $t \to \infty$ and $\varepsilon \to 0$, which implies that $\lim_{{\rm dist}(x,\mathcal{M})\to 0} \mathbb{P}^x\, (\lim_{t\to\infty} {\rm dist}(s(t,x),\mathcal{M}) = 0) = 1$.

Finally, since every point in $\mathcal{M}$ is Lyapunov stable in probability with respect to $\overline{\mathbb{R}}_+^n$, it follows from Proposition 12.2 that $\lim_{t\to\infty} s(t,x) \stackrel{\rm a.s.}{=} x^*$ as $x \to x^*$, where $x^* \in \mathcal{M}$ is Lyapunov stable in probability with respect to $\overline{\mathbb{R}}_+^n$.

Hence, by definition, (12.4) is semistable. For $\mathcal{Q} = \overline{\mathbb{R}}_+^n$, global stochastic semistability with respect to $\overline{\mathbb{R}}_+^n$ follows from identical arguments using the radially unbounded condition on $V(\cdot)$. □

Example 12.1. Consider the nonlinear stochastic nonnegative dynamical system on $\mathcal{H}_2$ given by

$$\mathrm{d}x_1(t) = [\sigma_{12}(x_2(t)) - \sigma_{21}(x_1(t))]\mathrm{d}t + \gamma(x_2(t) - x_1(t))\mathrm{d}w(t), \qquad x_1(0) \overset{\text{a.s.}}{=} x_{10}, \quad t \geq 0, \tag{12.54}$$

$$\mathrm{d}x_2(t) = [\sigma_{21}(x_1(t)) - \sigma_{12}(x_2(t))]\mathrm{d}t + \gamma(x_1(t) - x_2(t))\mathrm{d}w(t), \qquad x_2(0) \overset{\text{a.s.}}{=} x_{20}, \tag{12.55}$$

where $\sigma_{ij}(\cdot)$, $i, j = 1, 2$, $i \neq j$, are Lipschitz continuous and $\gamma > 0$. Equations (12.54) and (12.55) represent the collective dynamics of two subsystems that interact by exchanging energy. The energy states of the subsystems are described by the scalar random variables x_1 and x_2. The unity coefficients scaling $\sigma_{ij}(\cdot)$, $i, j \in \{1, 2\}$, $i \neq j$, appearing in (12.54) and (12.55) represent the topology of the energy exchange between the subsystems. More specifically, given $i, j \in \{1, 2\}$, $i \neq j$, a coefficient of 1 denotes that subsystem j receives energy from subsystem i, and a coefficient of zero denotes that subsystem i and j are disconnected, and hence, cannot exchange energies.

The connectivity between the subsystems can be represented by a graph $\mathfrak{G}$ having two nodes such that $\mathfrak{G}$ has a directed edge from node i to node j if and only if subsystem j can receive energy from subsystem i. Since the coefficients scaling $\sigma_{ij}(\cdot)$, $i, j \in \{1, 2\}$, $i \neq j$, are constants, the graph topology is fixed. Furthermore, note that the directed graph $\mathfrak{G}$ is *weakly connected* since the underlying undirected graph is connected; that is, every subsystem receives energy from, or delivers energy to, at least one other subsystem.

Note that (12.54) and (12.55) can be cast in the form of (12.4) with

$$f(x) = \begin{bmatrix} \sigma_{12}(x_2) - \sigma_{21}(x_1) \\ \sigma_{21}(x_1) - \sigma_{12}(x_2) \end{bmatrix}, \quad D(x) = \begin{bmatrix} \gamma(x_2 - x_1) \\ \gamma(x_1 - x_2) \end{bmatrix},$$

where the stochastic term $D(x)\mathrm{d}w$ represents probabilistic variations in the energy transfer between the two subsystems. Furthermore, note that since

$$\mathbf{e}_2^{\mathrm{T}}\mathrm{d}x(t) = \mathbf{e}_2^{\mathrm{T}} f(x(t))\mathrm{d}t + \mathbf{e}_2^{\mathrm{T}} D(x(t))\mathrm{d}w(t) = 0, \quad x(0) \overset{\text{a.s.}}{=} x_0, \quad t \geq 0,$$

where $\mathbf{e}_2 \triangleq [1\ 1]^{\mathrm{T}}$, it follows that $\mathrm{d}x_1(t) + \mathrm{d}x_2(t) = 0$, which implies that the total system energy is conserved.

In this example, we use Theorem 12.3 to analyze the collective behavior of (12.54) and (12.55). Specifically, we are interested in the energy equipartitioning behavior of the subsystems. For this purpose, we make the assumptions $\sigma_{ij}(x_j) - \sigma_{ji}(x_i) = 0$ if and only if $x_i = x_j$, $i \neq j$, and $(x_i - x_j)[\sigma_{ij}(x_j) - \sigma_{ji}(x_i)] \leq -\gamma^2(x_1 - x_2)$ for $i, j \in \{1, 2\}$.

The first assumption implies that if the energies in the connected subsystems i and j are equal, then energy exchange between the subsystems is not possible. This statement is reminiscent of the *zeroth law of thermodynamics*, which postulates that temperature equality is a necessary and sufficient condition for thermal equilibrium. The second assumption implies that energy flows from more energetic subsystems to less energetic subsystems and is reminiscent of the *second law of thermodynamics*, which states that heat (energy) must flow in the direction of lower temperatures. It is important to note here that due to the stochastic term $D(x)\mathrm{d}w$ capturing probabilistic variations in the energy transfer between the subsystems, the second assumption requires that the scaled net energy flow $(x_i - x_j)[\sigma_{ij}(x_j) - \sigma_{ji}(x_i)]$ is bounded by the negative intensity of the diffusion coefficient given by $\frac{1}{2}\mathrm{tr}\, D(x)D^{\mathrm{T}}(x)$.

To show that (12.54) and (12.55) is stochastically semistable with respect to $\overline{\mathbb{R}}_+^2$, note that $\mathcal{E} \triangleq f^{-1}(0) \cap D^{-1}(0) = \{(x_1, x_2) \in \overline{\mathbb{R}}_+^2 : x_1 = x_2 = \alpha, \alpha \in \overline{\mathbb{R}}_+\}$ and consider the Lyapunov function candidate $V(x_1, x_2) = \frac{1}{2}(x_1 - \alpha)^2 + \frac{1}{2}(x_2 - \alpha)^2$, where $\alpha \in \overline{\mathbb{R}}_+$. Now, it follows that

$$
\begin{aligned}
\mathcal{L}V(x_1, x_2) &= (x_1 - \alpha)[\sigma_{12}(x_2) - \sigma_{21}(x_1)] + (x_2 - \alpha)[\sigma_{21}(x_1) - \sigma_{12}(x_2)] \\
&\quad + \tfrac{1}{2}[(\gamma(x_2 - x_1))^2 + (\gamma(x_1 - x_2))^2] \\
&= x_1[\sigma_{12}(x_2) - \sigma_{21}(x_1)] + x_2[\sigma_{21}(x_1) - \sigma_{12}(x_2)] + (\gamma(x_1 - x_2))^2 \\
&= (x_1 - x_2)[\sigma_{12}(x_2) - \sigma_{21}(x_1) + \gamma^2(x_1 - x_2)] \\
&\leq 0, \quad (x_1, x_2) \in \overline{\mathbb{R}}_+ \times \overline{\mathbb{R}}_+,
\end{aligned} \tag{12.56}
$$

which implies that $x_1 = x_2 = \alpha$ is Lyapunov stable in probability with respect to $\overline{\mathbb{R}}_+^2$.

Next, it is easy to see that $\mathcal{L}V(x_1, x_2) \neq 0$ when $x_1 \neq x_2$, and hence, $\mathcal{L}V(x_1, x_2) < 0$, $(x_1, x_2) \in \overline{\mathbb{R}}_+^2 \backslash \mathcal{E}$. Therefore, it follows from Theorem 12.3 that $x_1 = x_2 = \alpha$ is stochastically semistable with respect to $\overline{\mathbb{R}}_+^2$ for all $\alpha \in \overline{\mathbb{R}}_+$. Furthermore, note that $\mathbf{e}_2^{\mathrm{T}}\mathrm{d}x(t) \overset{\text{a.s.}}{=} 0$, $t \geq 0$, implies

$$
x(t) \overset{\text{a.s.}}{\to} \frac{1}{2}\mathbf{e}_2\mathbf{e}_2^{\mathrm{T}}x(0) \overset{\text{a.s.}}{=} \frac{1}{2}[x_1(0) + x_2(0)]\mathbf{e}_2 \quad \text{as } t \to \infty.
$$

Note that an identical assertion holds for the collective dynamics of n subsystems with a connected undirected energy graph topology. △

Finally, we provide a converse Lyapunov theorem for stochastic semistability. For this result, recall that $\mathcal{L}V(x_{\rm e}) = 0$ for every $x_{\rm e} \in \mathcal{E}$. Also note that it follows from (12.9) that $\mathcal{L}V(x) = \mathcal{L}V(s(0,x))$.

Theorem 12.5. Consider the nonlinear stochastic nonnegative dynamical system (12.4). Suppose (12.4) is stochastically semistable with a ρ-domain of semistability $\mathcal{D}_0$. Then there exist a continuous nonnegative function $V : \mathcal{D}_0 \to \overline{\mathbb{R}}_+$ and a class $\mathcal{K}_\infty$ function $\alpha(\cdot)$ such that *i)* $V(x) = 0$, $x \in \mathcal{E}$, *ii)* $V(x) \geq \alpha(\mathrm{dist}(x,\mathcal{E}))$, $x \in \mathcal{D}_0$, and *iii)* $\mathcal{L}V(x) < 0$, $x \in \mathcal{D}_0\backslash\mathcal{E}$.

Proof. Let $\mathfrak{B}^{x_0}$ denote the set of all sample trajectories of (12.4) for which $\lim_{t\to\infty}\mathrm{dist}(x(t,\omega),\mathcal{E}) = 0$ and $x(\{t \geq 0\},\omega) \in \mathfrak{B}^{x_0}$, $\omega \in \Omega$, and let $\mathbb{1}_{\mathfrak{B}^{x_0}}(\omega)$, $\omega \in \Omega$, denote the indicator function defined on the set $\mathfrak{B}^{x_0}$, that is,

$$\mathbb{1}_{\mathfrak{B}^{x_0}}(\omega) \triangleq \begin{cases} 1, & \text{if } x(\{t \geq 0\},\omega) \in \mathfrak{B}^{x_0}, \\ 0, & \text{otherwise.} \end{cases}$$

Note that by definition $\mathbb{P}^{x_0}\left(\mathfrak{B}^{x_0}\right) \geq 1-\rho$ for all $x_0 \in \mathcal{D}_0$. Define the function $V : \mathcal{D}_0 \to \overline{\mathbb{R}}_+$ by

$$V(x) \triangleq \sup_{t\geq 0}\left\{\frac{1+2t}{1+t}\mathbb{E}\left[\mathrm{dist}(s(t,x),\mathcal{E})\mathbb{1}_{\mathfrak{B}^{x}}(\omega)\right]\right\}, \quad x \in \mathcal{D}_0, \quad (12.57)$$

and note that $V(\cdot)$ is well defined since (12.4) is stochastically semistable with respect to $\overline{\mathbb{R}}_+^n$. Clearly, *i)* holds. Furthermore, since $V(x) \geq \mathrm{dist}(x,\mathcal{E})$, $x \in \mathcal{D}_0$, it follows that *ii)* holds with $\alpha(r) = r$.

To show that $V(\cdot)$ is continuous on $\mathcal{D}_0\backslash\mathcal{E}$, define $T : \mathcal{D}_0\backslash\mathcal{E} \to [0,\infty)$ by $T(z) \triangleq \inf\{h : \mathbb{E}\left[\mathrm{dist}(s(h,z),\mathcal{E})\mathbb{1}_{\mathfrak{B}^{z}}(\omega)\right] < \mathrm{dist}(z,\mathcal{E})/2 \text{ for all } t \geq h > 0\}$, and denote

$$\mathcal{W}_\varepsilon \triangleq \left\{x \in \mathcal{D}_0 : \mathbb{P}^x\left(\sup_{t\geq 0}\mathrm{dist}(s(t,x),\mathcal{E}) \leq \varepsilon\right) \geq 1-\rho\right\}. \quad (12.58)$$

Note that $\mathcal{W}_\varepsilon \supset \mathcal{E}$ is open and contains an open neighborhood of $\mathcal{E}$. Consider $z \in \mathcal{D}_0\backslash\mathcal{E}$ and define $\lambda \triangleq \mathrm{dist}(z,\mathcal{E}) > 0$. Then it follows from stochastic semistability of (12.4) that there exists $h > 0$ such that $\mathbb{P}^z\left(s(h,z) \in \mathcal{W}_{\lambda/2}\right) \geq 1-\rho$. Consequently, $\mathbb{P}^z\left(s(h+t,z) \in \mathcal{W}_{\lambda/2}\right) \geq 1-\rho$ for all $t \geq 0$, and hence, it follows that $T(z)$ is well defined. Since $\mathcal{W}_{\lambda/2}$ is open, there exists a neighborhood $\mathcal{B}_\sigma(s(T(z),z)$ such that $\mathbb{P}^z\left(\mathcal{B}_\sigma(s(T(z),z)) \subset \mathcal{W}_{\lambda/2}\right) \geq 1-\rho$. Hence, $\mathcal{N} \subset \mathcal{D}_0$ is a neighborhood of z such that $s_{T(z)}(\mathcal{H}_n^{\mathcal{N}}) \triangleq \mathcal{B}_\sigma(s(T(z),z))$.

Next, choose $\eta > 0$ such that $\eta < \lambda/2$ and $\mathcal{B}_\eta(z) \subset \mathcal{N}$. Then, for every $t > T(z)$ and $y \in \mathcal{B}_\eta(z)$,

$$[(1+2t)/(1+t)]\mathbb{E}\left[\mathrm{dist}(s(t,y),\mathcal{E})\mathbb{1}_{\mathfrak{B}^{y}}(\omega)\right] \leq 2\mathbb{E}\left[\mathrm{dist}(s(t,y),\mathcal{E})\mathbb{1}_{\mathfrak{B}^{y}}(\omega)\right] \leq \lambda.$$

Therefore, for every $y \in \mathcal{B}_\eta(z)$,

$$\begin{aligned} V(z) - V(y) &= \sup_{t\geq 0}\left\{\frac{1+2t}{1+t}\mathbb{E}\left[\operatorname{dist}(s(t,z),\mathcal{E})\mathbb{1}_{\mathfrak{B}^z}(\omega)\right]\right\} \\ &\quad - \sup_{t\geq 0}\left\{\frac{1+2t}{1+t}\mathbb{E}\left[\operatorname{dist}(s(t,y),\mathcal{E})\mathbb{1}_{\mathfrak{B}^y}(\omega)\right]\right\} \\ &= \sup_{0\leq t\leq T(z)}\left\{\frac{1+2t}{1+t}\mathbb{E}\left[\operatorname{dist}(s(t,z),\mathcal{E})\mathbb{1}_{\mathfrak{B}^z}(\omega)\right]\right\} \\ &\quad - \sup_{0\leq t\leq T(z)}\left\{\frac{1+2t}{1+t}\mathbb{E}\left[\operatorname{dist}(s(t,y),\mathcal{E})\mathbb{1}_{\mathfrak{B}^y}(\omega)\right]\right\}. \quad (12.59) \end{aligned}$$

Hence,

$$\begin{aligned} &|V(z) - V(y)| \\ &\quad\leq \sup_{0\leq t\leq T(z)}\left|\frac{1+2t}{1+t}\left(\mathbb{E}\left[\operatorname{dist}(s(t,z),\mathcal{E})\mathbb{1}_{\mathfrak{B}^z}(\omega)\right]\right.\right. \quad (12.60) \\ &\qquad\qquad \left.\left. -\mathbb{E}\left[\operatorname{dist}(s(t,y),\mathcal{E})\mathbb{1}_{\mathfrak{B}^y}(\omega)\right]\right)\right| \\ &\quad\leq 2\sup_{0\leq t\leq T(z)}\left|\mathbb{E}\left[\operatorname{dist}(s(t,z),\mathcal{E})\mathbb{1}_{\mathfrak{B}^z}(\omega)\right] - \mathbb{E}\left[\operatorname{dist}(s(t,y),\mathcal{E})\mathbb{1}_{\mathfrak{B}^y}(\omega)\right]\right| \\ &\quad\leq 2\sup_{0\leq t\leq T(z)}\mathbb{E}\left[\operatorname{dist}(s(t,z),s(t,y))\right],\ z\in\mathcal{D}_0\backslash\mathcal{E},\ y\in\mathcal{B}_\eta(z). \quad (12.61) \end{aligned}$$

Now, since $f(\cdot)$ and $D(\cdot)$ satisfy (12.6) and (12.7), it follows from continuous dependence of solutions $s(\cdot,\cdot)$ on system initial conditions [9, Thm. 7.3.1] and (12.61) that $V(\cdot)$ is continuous on $\mathcal{D}_0\backslash\mathcal{E}$.

To show that $V(\cdot)$ is continuous on $\mathcal{E}$, consider $x_\mathrm{e} \in \mathcal{E}$. Let $\{x_n\}_{n=1}^\infty$ be a sequence in $\mathcal{D}_0\backslash\mathcal{E}$ that converges to x_e. Since x_e is Lyapunov stable in probability with respect to $\overline{\mathbb{R}}_+^n$, it follows that $x(t) \overset{\text{a.s.}}{\equiv} x_\mathrm{e}$ is the unique solution to (12.4) with $x(0) \overset{\text{a.s.}}{=} x_\mathrm{e}$. By continuous dependence of solutions $s(\cdot,\cdot)$ on system initial conditions [9, Thm. 7.3.1], $s(t,x_n) \overset{\text{a.s.}}{\to} s(t,x_\mathrm{e}) \overset{\text{a.s.}}{=} x_\mathrm{e}$ as $n\to\infty$, $t\geq 0$.

Let $\varepsilon > 0$ and note that it follows from *ii*) of Proposition 12.3 that there exists $\delta = \delta(x_\mathrm{e}) > 0$ such that for every solution of (12.4) in $\mathcal{B}_\delta(x_\mathrm{e})$ there exists $\hat{T} = \hat{T}(x_\mathrm{e},\varepsilon) > 0$ such that $\mathbb{P}\left(s_t(\mathcal{H}_n^{\mathcal{B}_\delta(x_\mathrm{e})}) \subset \mathcal{W}_\varepsilon\right) \geq 1-\rho$ for all $t\geq\hat{T}$. Next, note that there exists a positive integer N_1 such that $x_n \in \mathcal{B}_\delta(x_\mathrm{e})$ for all $n \geq N_1$. Now, it follows from (12.57) that

$$V(x_n) \leq 2\sup_{0\leq t\leq\hat{T}}\mathbb{E}[\operatorname{dist}(s(t,x_n),\mathcal{E})\mathbb{1}_{\mathfrak{B}^{x_n}}(\omega)] + 2\varepsilon, \quad n\geq N_1. \quad (12.62)$$

Next, it follows from [9, Thm. 7.3.1] that $\mathbb{E}[|s(\cdot, x_n)|]$ converges to $\mathbb{E}[|s(\cdot, x_{\rm e})|]$ uniformly on $[0, \hat{T}]$. Hence,

$$\begin{aligned}\lim_{n\to\infty} \sup_{0\le t\le \hat{T}} \mathbb{E}\left[{\rm dist}(s(t,x_n),\mathcal{E})\mathbb{1}_{\mathfrak{B}^{x_n}}(\omega)\right] &= \sup_{0\le t\le \hat{T}} \mathbb{E}\left[\lim_{n\to\infty} {\rm dist}(s(t,x_n),\mathcal{E})\mathbb{1}_{\mathfrak{B}^{x_n}}(\omega)\right]\\ &\le \sup_{0\le t\le \hat{T}} {\rm dist}(x_{\rm e},\mathcal{E})\\ &= 0, \end{aligned}\tag{12.63}$$

which implies that there exists a positive integer $N_2 = N_2(x_{\rm e}, \varepsilon) \ge N_1$ such that

$$\sup_{0\le t\le \hat{T}} \mathbb{E}\left[{\rm dist}(s(t,x_n),\mathcal{E})\mathbb{1}_{\mathfrak{B}^{x_n}}(\omega)\right] < \varepsilon$$

for all $n \ge N_2$. Combining (12.62) with the above result yields $V(x_n) < 4\varepsilon$ for all $n \ge N_2$, which implies that $\lim_{n\to\infty} V(x_n) = 0 = V(x_{\rm e})$.

Finally, we show that $\mathcal{L}V(x(t))$ is negative along the solution of (12.4) on $\mathcal{D}_0\backslash\mathcal{E}$. Note that for every $x \in \mathcal{D}_0\backslash\mathcal{E}$ and $0 < h \le 1/2$ such that $\mathbb{P}\left(s(h,x) \in \mathcal{D}_0\backslash\mathcal{E}\right) \ge 1 - \rho$, it follows from the definition of $T(\cdot)$ that $\mathbb{E}\left[V(s(h,x))\right]$ is reached at some time $\hat{t}$ such that $0 \le \hat{t} \le T(x)$. Hence, it follows from the law of iterated expectation that

$$\begin{aligned}&\mathbb{E}\left[V(s(h,x))\right]\\ &\quad= \mathbb{E}\left[\mathbb{E}\left[{\rm dist}(s(\hat{t}+h,x),\mathcal{E})\mathbb{1}_{\mathfrak{B}^{s(h,x)}}(\omega)\right]\frac{1+2\hat{t}}{1+\hat{t}}\right]\\ &\quad= \mathbb{E}\left[{\rm dist}(s(\hat{t}+h,x),\mathcal{E})\mathbb{1}_{\mathfrak{B}^{x}}(\omega)\right]\frac{1+2\hat{t}+2h}{1+\hat{t}+h}\left[1-\frac{h}{(1+2\hat{t}+2h)(1+\hat{t})}\right]\\ &\quad\le V(x)\left[1-\frac{h}{2(1+T(x))^2}\right], \end{aligned}\tag{12.64}$$

which implies that

$$\mathcal{L}V(x) = \lim_{h\to 0^+} \frac{\mathbb{E}\left[V(s(h,x))\right] - V(x)}{h} \le -\frac{1}{2}V(x)(1+T(x))^{-2} < 0, \ x \in \mathcal{D}_0\backslash\mathcal{E},$$

and hence, *iii*) holds. □

12.5 Conservation of Energy and the First Law of Thermodynamics: A Stochastic Perspective

In this section, we extend the thermodynamic model proposed in Chapter 3 to include probabilistic variations in the instantaneous rate of energy dissipation as well as probabilistic variations in the energy transfer between

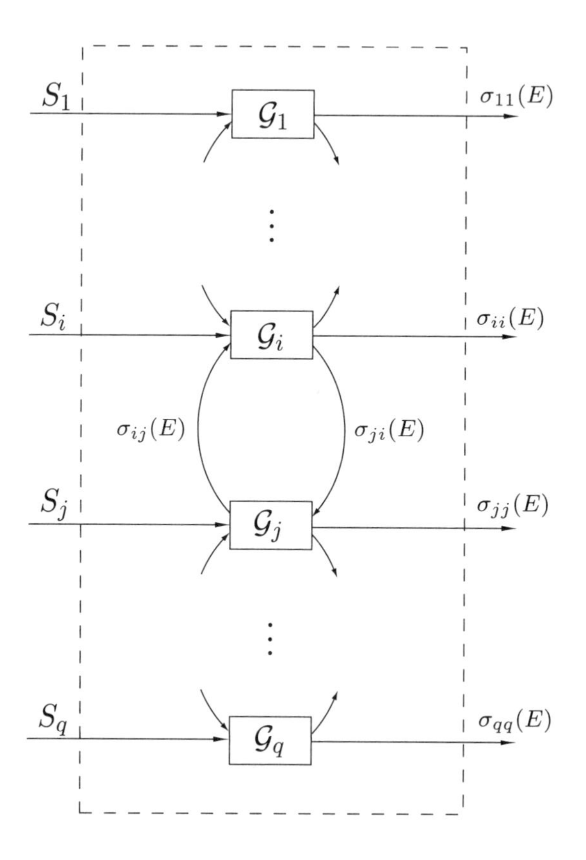

Figure 12.1 Large-scale dynamical system $\mathcal{G}$ with $D(E) = 0$ and $J(E) = 0$.

the subsystems. To formulate our state space stochastic thermodynamic model, we consider the large-scale stochastic dynamical system $\mathcal{G}$ shown in Figure 12.1 involving energy exchange between q interconnected subsystems and use the notation developed in Chapter 3.

Specifically, $E_i : [0, \infty) \to \overline{\mathbb{R}}_+$ denotes the energy (and hence a nonnegative quantity) of the ith subsystem, $S_i : [0, \infty) \to \mathbb{R}$ denotes the external power (heat flux) supplied to (or extracted from) the ith subsystem, $\sigma_{ij} : \overline{\mathbb{R}}_+^q \to \overline{\mathbb{R}}_+$, $i \neq j$, $i, j = 1, \ldots, q$, denotes the instantaneous rate of energy (heat) flow from the jth subsystem to the ith subsystem, $J_{(i,k)} : \overline{\mathbb{R}}_+^q \to \overline{\mathbb{R}}_+$, $i = 1, \ldots, q$, $k = 1, \ldots, d_1$, denotes the instantaneous rate of energy (heat) received or delivered to the ith subsystem from all other subsystems due to the stochastic disturbance $w_{1k}(\cdot)$, $\sigma_{ii} : \overline{\mathbb{R}}_+^q \to \overline{\mathbb{R}}_+$, $i = 1, \ldots, q$, denotes the instantaneous rate of energy (heat) dissipation from the ith subsystem to the environment, and $D_{(i,l)} : \overline{\mathbb{R}}_+^q \to \overline{\mathbb{R}}_+$, $i = 1, \ldots, q$, $l = 1, \ldots, d_2$, denotes the instantaneous rate of energy (heat) dissipation from the ith subsystem to the environment due to the stochastic disturbance $w_{2l}(\cdot)$. Here we assume that $\sigma_{ij} : \overline{\mathbb{R}}_+^q \to \overline{\mathbb{R}}_+$, $i, j = 1, \ldots, q$, $J_{(i,k)} : \overline{\mathbb{R}}_+^q \to \overline{\mathbb{R}}_+$, $i = 1, \ldots, q$, $k = 1, \ldots, d_1$, and $D_{(i,l)} : \overline{\mathbb{R}}_+^q \to \overline{\mathbb{R}}_+$, $i = 1, \ldots, q$, $l = 1, \ldots, d_2$, are locally Lipschitz continuous on $\overline{\mathbb{R}}_+^q$ and satisfy a linear growth condition, and $S_i : [0, \infty) \to \mathbb{R}$, $i = 1, \ldots, q$, are bounded piecewise continuous functions of time.

An *energy balance* for the ith subsystem yields

$$\begin{aligned} E_i(T) = {} & E_i(t_0) + \sum_{j=1,\, j\neq i}^{q} \int_{t_0}^{T} [\sigma_{ij}(E(t)) - \sigma_{ji}(E(t))]\mathrm{d}t \\ & + \int_{t_0}^{T} \mathrm{row}_i(J(E(t)))\mathrm{d}w_1(t) - \int_{t_0}^{T} \sigma_{ii}(E(t))\mathrm{d}t \\ & - \int_{t_0}^{T} \mathrm{row}_i(D(E(t)))\mathrm{d}w_2(t) + \int_{t_0}^{T} S_i(t)\mathrm{d}t, \quad T \geq t_0, \end{aligned} \tag{12.65}$$

or, equivalently, in vector form,

$$\begin{aligned} E(T) = {} & E(t_0) + \int_{t_0}^{T} f(E(t))\mathrm{d}t + \int_{t_0}^{T} J(E(t))\mathrm{d}w_1(t) - \int_{t_0}^{T} d(E(t))\mathrm{d}t \\ & - \int_{t_0}^{T} D(E(t))\mathrm{d}w_2(t) + \int_{t_0}^{T} S(t)\mathrm{d}t, \quad T \geq t_0, \end{aligned} \tag{12.66}$$

where $E(t) \triangleq [E_1(t), \ldots, E_q(t)]^{\mathrm{T}}$, $w_1(\cdot)$ and $w_2(\cdot)$ are, respectively, a d_1-dimensional and d_2-dimensional independent standard Wiener process (i.e., Brownian motion) defined on a complete filtered probability space $(\Omega, \mathcal{F}, \{\mathcal{F}_t\}_{t\geq t_0}, \mathbb{P})$, $E(t_0)$ is independent of $(w_1(t) - w_1(t_0))$, $t \geq t_0$, and $(w_2(t) - w_2(t_0))$, $t \geq t_0$, and

$$\begin{aligned} d(E(t)) & \triangleq [\sigma_{11}(E(t)), \ldots, \sigma_{qq}(E(t))]^{\mathrm{T}}, \\ S(t) & \triangleq [S_1(t), \ldots, S_q(t)]^{\mathrm{T}}, \\ f(E) & = [f_1(E), \ldots, f_q(E)]^{\mathrm{T}} : \overline{\mathbb{R}}_+^q \to \mathbb{R}^q, \\ J(E) & = [\mathrm{row}_1(J(E)), \ldots, \mathrm{row}_q(J(E))]^{\mathrm{T}} : \overline{\mathbb{R}}_+^q \to \mathbb{R}^q \times \mathbb{R}^{d_1}, \\ D(E) & = [\mathrm{row}_1(D(E)), \ldots, \mathrm{row}_q(D(E))]^{\mathrm{T}} : \overline{\mathbb{R}}_+^q \to \mathbb{R}^q \times \mathbb{R}^{d_2}. \end{aligned}$$

Here, the stochastic disturbance $J(E)\mathrm{d}w_1$ in (12.66) captures probabilistic variations in the energy transfer rates between compartments, and the stochastic disturbance $D(E)\mathrm{d}w_2$ captures probabilistic variations in the instantaneous rate of energy dissipation.

Equivalently, (12.65) can be rewritten as

$$\mathrm{d}E_i(t) = \sum_{j=1,\, j\neq i}^{q} [\sigma_{ij}(E(t)) - \sigma_{ji}(E(t))]\mathrm{d}t + \mathrm{row}_i(J(E(t)))\mathrm{d}w_1(t)$$

$$-\sigma_{ii}(E(t))\mathrm{d}t - \mathrm{row}_i(D(E(t)))\mathrm{d}w_2(t) + S_i(t)\mathrm{d}t,$$
$$E_i(t_0) \overset{\mathrm{a.s.}}{=} E_{i0}, \quad t \geq t_0, \quad (12.67)$$

or, in vector form,

$$\begin{aligned} \mathrm{d}E(t) &= f(E(t))\mathrm{d}t + J(E(t))\mathrm{d}w_1(t) - d(E(t))\mathrm{d}t \\ &\quad -D(E(t))\mathrm{d}w_2(t) + S(t)\mathrm{d}t, \quad E(t_0) \overset{\mathrm{a.s.}}{=} E_0, \quad t \geq t_0, \end{aligned} \quad (12.68)$$

where $E_0 \triangleq [E_{10}, \ldots, E_{q0}]^{\mathrm{T}}$, yielding a *differential energy balance* equation that characterizes energy flow between subsystems of the large-scale stochastic dynamical system $\mathcal{G}$. Here we assume that $S(\cdot)$ satisfies sufficient regularity conditions such that (12.68) has a unique solution forward in time. Specifically, we assume that the external power (heat flux) $S(\cdot)$ supplied to the large-scale stochastic dynamical system $\mathcal{G}$ consists of measurable functions $S(\cdot)$ adapted to the filtration $\{\mathcal{F}_t\}_{t \geq t_0}$ such that $S(t) \in \mathcal{H}_q$, $t \geq t_0$, for all $t \geq s$, $w(t) - w(s)$ is independent of $S(\tau)$, $w(\tau)$, $\tau \leq s$, and $E(t_0)$, where $w(t) \triangleq [w_1^{\mathrm{T}}(t), w_2^{\mathrm{T}}(t)]^{\mathrm{T}}$, and hence, $S(\cdot)$ is nonanticipative. Furthermore, we assume that $S(\cdot)$ takes values in a compact metrizable set. In this case, it follows from Theorem 2.2.4 of [8] that there exists a pathwise unique solution to (12.68) in $(\Omega, \{\mathcal{F}_t\}_{t \geq t_0}, \mathbb{P}^{E_0})$.

Equation (12.66) or, equivalently, (12.68) is a statement of the *first law for stochastic thermodynamics* as applied to *isochoric transformations* (i.e., constant subsystem volume transformations) for each of the subsystems $\mathcal{G}_i$, $i = 1, \ldots, q$. To see this, let the total energy in the large-scale stochastic dynamical system $\mathcal{G}$ be given by $U \triangleq \mathbf{e}^{\mathrm{T}} E$, where $\mathbf{e}^{\mathrm{T}} \triangleq [1, \ldots, 1]$ and $E \in \overline{\mathbb{R}}_+^q$, and let the net energy received by the large-scale dynamical system $\mathcal{G}$ over the time interval $[t_1, t_2]$ be given by

$$Q \triangleq \int_{t_1}^{t_2} \mathbf{e}^{\mathrm{T}}[S(t) - d(E(t))]\mathrm{d}t - \int_{t_1}^{t_2} \mathbf{e}^{\mathrm{T}} D(E(t))\mathrm{d}w_2(t), \quad (12.69)$$

where $E(t)$, $t \geq t_0$, is the solution to (12.68). Then, premultiplying (12.66) by $\mathbf{e}^{\mathrm{T}}$ and using the fact that $\mathbf{e}^{\mathrm{T}} f(E) \equiv 0$ and $\mathbf{e}^{\mathrm{T}} J(E) \equiv 0$, it follows that

$$\Delta U = Q, \quad (12.70)$$

where $\Delta U \triangleq U(t_2) - U(t_1)$ denotes the variation in the total energy of the large-scale stochastic dynamical system $\mathcal{G}$ over the time interval $[t_1, t_2]$.

For our large-scale stochastic dynamical system model $\mathcal{G}$, we assume that $\sigma_{ij}(E) = 0$, $E \in \overline{\mathbb{R}}_+^q$, $\sigma_{jj}(E) = 0$, $E \in \overline{\mathbb{R}}_+^q$, $J_{(j,k)}(E) = 0$, $E \in \overline{\mathbb{R}}_+^q$, $k = 1, \ldots, d_1$, and $D_{(j,l)}(E) = 0$, $E \in \overline{\mathbb{R}}_+^q$, $l = 1, \ldots, d_2$, whenever $E_j = 0$, $j = 1, \ldots, q$. In this case, $f(E) - d(E)$, $E \in \overline{\mathbb{R}}_+^q$, is essentially nonnegative. The above constraint implies that if the energy of the jth subsystem of $\mathcal{G}$ is zero, then this subsystem cannot supply any energy to its surroundings nor dissipate energy to the environment. Moreover, we assume that $S_i(t) \geq 0$ whenever $E_i(t) = 0$, $t \geq t_0$, $i = 1, \ldots, q$, which implies that when the energy of the ith subsystem is zero, then no energy can be extracted from this subsystem.

The following proposition is needed for the main results of this chapter.

Proposition 12.4. Consider the large-scale stochastic dynamical system $\mathcal{G}$ with differential energy balance equation given by (12.68). Suppose $\sigma_{ij}(E) = 0$, $E \in \overline{\mathbb{R}}^q_+$, $J_{(j,k)}(E) = 0$, $E \in \overline{\mathbb{R}}^q_+$, $k = 1, \ldots, d_1$, and $D_{(j,l)}(E) = 0$, $E \in \overline{\mathbb{R}}^q_+$, $l = 1, \ldots, d_2$, whenever $E_j = 0$, $j = 1, \ldots, q$, and $S_i(t) \geq 0$ whenever $E_i(t) = 0$, $t \geq t_0$, $i = 1, \ldots, q$. Then the solution $E(t)$, $t \geq t_0$, to (12.68) is nonnegative for all nonnegative initial conditions $E_0 \in \overline{\mathbb{R}}^q_+$.

Proof. First note that $f(E) - d(E)$, $E \in \overline{\mathbb{R}}^q_+$, is essentially nonnegative, $J_{(j,k)}(E) = 0$, $E \in \overline{\mathbb{R}}^q_+$, $k = 1, \ldots, d_1$, and $D_{(j,l)}(E) = 0$, $E \in \overline{\mathbb{R}}^q_+$, $l = 1, \ldots, d_2$, whenever $E_j = 0$, $j = 1, \ldots, q$. Next, since $S_i(t) \geq 0$ whenever $E_i(t) = 0$, $t \geq t_0$, $i = 1, \ldots, q$, it follows that $\mathrm{d}E_i(t) \geq 0$ for all $t \geq t_0$ and $i = 1, \ldots, q$ whenever $E_i(t) = 0$ and $E_j(t) \geq 0$ for all $j \neq i$ and $t \geq t_0$. This implies that for all nonnegative initial conditions $E_0 \in \overline{\mathbb{R}}^q_+$, every sample trajectory of $\mathcal{G}$ is directed towards the interior of the nonnegative orthant $\overline{\mathbb{R}}^q_+$ whenever $E_i(t) = 0$, $i = 1, \ldots, q$, and hence, remains almost surely nonnegative for all $t \geq t_0$. □

Next, premultiplying (12.66) by $\mathbf{e}^{\mathrm{T}}$, using Proposition 12.4, and using the fact that $\mathbf{e}^{\mathrm{T}} f(E) \equiv 0$ and $\mathbf{e}^{\mathrm{T}} J(E) \equiv 0$, it follows that

$$\begin{aligned} \mathbf{e}^{\mathrm{T}} E(T) &= \mathbf{e}^{\mathrm{T}} E(t_0) + \int_{t_0}^{T} \mathbf{e}^{\mathrm{T}} S(t)\mathrm{d}t - \int_{t_0}^{T} \mathbf{e}^{\mathrm{T}} d(E(t))\mathrm{d}t \\ &\quad - \int_{t_0}^{T} \mathbf{e}^{\mathrm{T}} D(E(t))\mathrm{d}w_2(t), \quad T \geq t_0. \end{aligned} \tag{12.71}$$

Now, for the large-scale stochastic dynamical system $\mathcal{G}$, define the input $u(t) \triangleq S(t)$ and the output $y(t) \triangleq d(E(t))$. Hence, it follows from (12.71) that for any two $\mathcal{F}_t$-stopping times τ_1 and τ_2 such that $\tau_1 \geq \tau_2$ almost surely,

$$\begin{aligned} &\mathbb{E}\left[\mathbf{e}^{\mathrm{T}} E(\tau_2) | \mathcal{F}_{\tau_1}\right] \\ &\quad = \mathbf{e}^{\mathrm{T}} E(\tau_1) + \mathbb{E}\left[\int_{\tau_1}^{\tau_2} \mathbf{e}^{\mathrm{T}} S(t)\mathrm{d}t | \mathcal{F}_{\tau_1}\right] - \mathbb{E}\left[\int_{\tau_1}^{\tau_2} \mathbf{e}^{\mathrm{T}} d(E(t))\mathrm{d}t | \mathcal{F}_{\tau_1}\right] \\ &\qquad - \mathbb{E}\left[\int_{\tau_1}^{\tau_2} \mathbf{e}^{\mathrm{T}} D(E(t))\mathrm{d}w_2(t) | \mathcal{F}_{\tau_1}\right] \\ &\quad = \mathbf{e}^{\mathrm{T}} E(\tau_1) + \mathbb{E}\left[\int_{\tau_1}^{\tau_2} \left[\mathbf{e}^{\mathrm{T}} S(t) - \mathbf{e}^{\mathrm{T}} d(E(t))\right] \mathrm{d}t | \mathcal{F}_{\tau_1}\right]. \end{aligned} \tag{12.72}$$

Thus, the large-scale stochastic dynamical system $\mathcal{G}$ is *stochastically lossless* [372] with respect to the *energy supply rate* $r(u, y) \triangleq \mathbf{e}^{\mathrm{T}}u - \mathbf{e}^{\mathrm{T}}y$ and with the *energy storage function* $U(E) \triangleq \mathbf{e}^{\mathrm{T}}E$, $E \in \overline{\mathbb{R}}_+^q$. In other words, the difference in the supplied system energy and the stored system energy is a martingale with respect to the differential energy balance system filtration.

The following lemma is required for our next result.

Lemma 12.2. Consider the large-scale stochastic dynamical system $\mathcal{G}$ with differential energy balance equation (12.68). Then, for every equilibrium state $E_{\mathrm{e}} \in \mathcal{H}_q^+$ and every $\varepsilon > 0$ and $\tau \overset{\text{a.s.}}{>} 0$, there exist $S_{\mathrm{e}} \in \mathcal{H}_q$, $\alpha > 0$, and $\tau \overset{\text{a.s.}}{>} \hat{\tau} \overset{\text{a.s.}}{>} 0$ such that, for every $\hat{E} \in \mathcal{H}_q^+$ with $\|\hat{E} - E_{\mathrm{e}}\| \overset{\text{a.s.}}{\leq} \alpha\tau$, there exists $S : \overline{\mathbb{R}}_+ \to \mathcal{H}_q$ such that $\|S(t) - S_{\mathrm{e}}\| \overset{\text{a.s.}}{\leq} \varepsilon$, $t \in [0, \hat{\tau}]$, and $E(t) = E_{\mathrm{e}} + \frac{(\hat{E} - E_{\mathrm{e}})}{\hat{\tau}}t$, $t \in [0, \hat{\tau}]$.

Proof. Note that with $S_{\mathrm{e}} \triangleq d(E_{\mathrm{e}}) - f(E_{\mathrm{e}})$, the state $E_{\mathrm{e}} \in \overline{\mathbb{R}}_+^q$ is an equilibrium state of (12.68). Let $\theta > 0$ and $\tau \overset{\text{a.s.}}{>} 0$, and define

$$M(\theta, \tau) \triangleq \sup_{E \in \mathcal{H}_q^{\overline{\mathcal{B}}_1(0)}, t \in [0, \tau]} \|f(E_{\mathrm{e}} + \theta t E) - d(E_{\mathrm{e}} + \theta t E) + S_{\mathrm{e}}\|, \quad (12.73)$$

$$M_J(\theta, \tau) \triangleq \sup_{E \in \mathcal{H}_q^{\overline{\mathcal{B}}_1(0)}, t \in [0, \tau]} \left\| J\left(E_{\mathrm{e}} + \frac{(\hat{E} - E_{\mathrm{e}})}{\|\hat{E} - E_{\mathrm{e}}\|}\alpha t\right)\right\|, \quad (12.74)$$

$$M_D(\theta, \tau) \triangleq \sup_{E \in \mathcal{H}_q^{\overline{\mathcal{B}}_1(0)}, t \in [0, \tau]} \left\| D\left(E_{\mathrm{e}} + \frac{(\hat{E} - E_{\mathrm{e}})}{\|\hat{E} - E_{\mathrm{e}}\|}\alpha t\right)\right\|. \quad (12.75)$$

Note that for every $\tau \overset{\text{a.s.}}{>} 0$, $\lim_{\theta \to 0^+} M(\theta, \tau) \overset{\text{a.s.}}{=} 0$, $\lim_{\theta \to 0^+} M_J(\theta, \tau) \overset{\text{a.s.}}{=} 0$, and $\lim_{\theta \to 0^+} M_D(\theta, \tau) \overset{\text{a.s.}}{=} 0$, and for every $\theta > 0$, $\lim_{\tau \overset{\text{a.s.}}{\to} 0^+} M(\theta, \tau) \overset{\text{a.s.}}{=} 0$, $\lim_{\tau \overset{\text{a.s.}}{\to} 0^+} M_J(\theta, \tau) \overset{\text{a.s.}}{=} 0$, and $\lim_{\tau \overset{\text{a.s.}}{\to} 0^+} M_D(\theta, \tau) \overset{\text{a.s.}}{=} 0$. Moreover, it follows from Lévy's modulus of continuity theorem [277] that for sufficiently small $\mathrm{d}t > 0$, $\|\mathrm{d}w_1(t)\| \overset{\text{a.s.}}{\leq} M_W(\mathrm{d}t)\mathrm{d}t$ and $\|\mathrm{d}w_2(t)\| \overset{\text{a.s.}}{\leq} M_W(\mathrm{d}t)\mathrm{d}t$, where $M_W(\mathrm{d}t) \triangleq \sqrt{2\mathrm{d}t \log_e\left(\frac{1}{\mathrm{d}t}\right)}$.

Next, let $\varepsilon > 0$ and $\tau \overset{\text{a.s.}}{>} 0$ be given and, for sufficiently small $\mathrm{d}t > 0$, let $\alpha > 0$ be such that

$$M(\alpha, \tau) + \alpha + M_J(\alpha, \tau)M_W(\mathrm{d}t) + M_D(\alpha, \tau)M_W(\mathrm{d}t) \overset{\text{a.s.}}{\leq} \varepsilon.$$

(The existence of such an α is guaranteed since $M(\alpha, \tau) \overset{\text{a.s.}}{\to} 0$, $M_J(\alpha, \tau) \overset{\text{a.s.}}{\to} 0$, and $M_D(\alpha, \tau) \overset{\text{a.s.}}{\to} 0$ as $\alpha \to 0^+$.) Now, let $\hat{E} \in \mathcal{H}_q^+$ be such that $\|\hat{E} - E_{\mathrm{e}}\| \overset{\text{a.s.}}{\leq}$

$\alpha\tau$. With $\hat{\tau} \triangleq \frac{\|\hat{E}-E_{\rm e}\|}{\alpha} \stackrel{\rm a.s.}{\leq} \tau$ and

$$\begin{aligned} S(t){\rm d}t &= \left[-f(E(t)) + d(E(t)) + \alpha\frac{(\hat{E}-E_{\rm e})}{\|\hat{E}-E_{\rm e}\|}\right]{\rm d}t \\ &\quad -J(E(t)){\rm d}w_1(t) + D(E(t)){\rm d}w_2(t), \quad t \in [0,\hat{\tau}], \end{aligned} \tag{12.76}$$

it follows that

$$E(t) = E_{\rm e} + \frac{(\hat{E}-E_{\rm e})}{\|\hat{E}-E_{\rm e}\|}\alpha t, \quad t \in [0,\hat{\tau}], \tag{12.77}$$

is a solution to (12.68).

The result is now immediate by noting that $E(\hat{\tau}) \stackrel{\rm a.s.}{=} \hat{E}$ and

$$\begin{aligned} \|S(t)-S_{\rm e}\|{\rm d}t &\stackrel{\rm a.s.}{\leq} \left\|f\left(E_{\rm e} + \tfrac{(\hat{E}-E_{\rm e})}{\|\hat{E}-E_{\rm e}\|}\alpha t\right) - d\left(E_{\rm e} + \tfrac{(\hat{E}-E_{\rm e})}{\|\hat{E}-E_{\rm e}\|}\alpha t\right) + S_{\rm e}\right\|{\rm d}t \\ &\quad +\alpha{\rm d}t + \left\|J\left(E_{\rm e} + \tfrac{(\hat{E}-E_{\rm e})}{\|\hat{E}-E_{\rm e}\|}\alpha t\right)\right\| |{\rm d}w_1(t)| \\ &\quad + \left\|D\left(E_{\rm e} + \tfrac{(\hat{E}-E_{\rm e})}{\|\hat{E}-E_{\rm e}\|}\alpha t\right)\right\| |{\rm d}w_2(t)| \\ &\stackrel{\rm a.s.}{\leq} [M(\alpha,\tau) + \alpha + M_J(\alpha,\tau)M_W({\rm d}t) \\ &\quad +M_D(\alpha,\tau)M_W({\rm d}t)]{\rm d}t, \quad t \in [0,\hat{\tau}], \end{aligned} \tag{12.78}$$

and hence,

$$\begin{aligned} \|S(t)-S_{\rm e}\| &\stackrel{\rm a.s.}{\leq} M(\alpha,\tau) + \alpha + M_J(\alpha,\tau)M_W({\rm d}t) + M_D(\alpha,\tau)M_W({\rm d}t) \\ &\stackrel{\rm a.s.}{\leq} \varepsilon, \end{aligned}$$

which proves the result. □

It follows from Lemma 12.2 that the large-scale stochastic dynamical system $\mathcal{G}$ with the differential energy balance equation (12.68) is *stochastically reachable* from and *stochastically controllable* to the origin in $\overline{\mathbb{R}}_+^q$ [372]. Recall that the large-scale stochastic dynamical system $\mathcal{G}$ with the differential energy balance equation (12.68) is stochastically reachable from the origin in $\overline{\mathbb{R}}_+^q$ if, for all $E_0 \in \overline{\mathbb{R}}_+^q$ and $\varepsilon > 0$, there exist a finite random variable $\tau_{\mathcal{B}_\varepsilon(E_0)} \stackrel{\rm a.s.}{\geq} t_0$, called the *first hitting time*, defined by

$$\tau_{\mathcal{B}_\varepsilon(E_0)}(\omega) \triangleq \inf\{t \geq t_0 : E(t,\omega) \in \mathcal{B}_\varepsilon(E_0)\},$$

and an $\mathcal{F}_t$-adapted square integrable input $S(\cdot)$ defined on $[t_0, \tau_{\mathcal{B}_\varepsilon(E_0)}]$ such that the state $E(t)$, $t \geq t_0$, can be driven from $E(t_0) \stackrel{\rm a.s.}{=} 0$ to $E(\tau_{\mathcal{B}_\varepsilon(E_0)})$ and $\mathbb{E}\left[\tau_{E_0}\right] < \infty$, where $\tau_{E_0} \triangleq \sup_{\varepsilon>0} \tau_{\mathcal{B}_\varepsilon(E_0)}$ and the supremum is taken pointwise.

Alternatively, $\mathcal{G}$ is stochastically controllable to the origin in $\overline{\mathbb{R}}_+^q$ if, for all $E(t_0) \overset{\text{a.s.}}{=} E_0$, $E_0 \in \overline{\mathbb{R}}_+^q$, there exist a finite random variable $\tilde{\tau}_{\mathcal{B}_\varepsilon(E_0)} \overset{\text{a.s.}}{\geq} t_0$ defined by

$$\tilde{\tau}_{\mathcal{B}_\varepsilon(E_0)}(\omega) \triangleq \inf\{t \geq t_0 : E(t,\omega) \in \mathcal{B}_\varepsilon(0)\},$$

and an $\mathcal{F}_t$-adapted square integrable input $S(\cdot)$ defined on $[t_0, \tilde{\tau}_{\mathcal{B}_\varepsilon(E_0)}]$ such that the state $E(t)$, $t \geq t_0$, can be driven from $E(t_0) \overset{\text{a.s.}}{=} E_0$ to $E(\tilde{\tau}_{\mathcal{B}_\varepsilon(E_0)}) \in \mathcal{B}_\varepsilon(0)$ and $\tilde{\tau}_{E_0} \triangleq \sup_{\varepsilon>0} \tilde{\tau}_{\mathcal{B}_\varepsilon(E_0)}$ with a pointwise supremum.

We let $\mathcal{U}_\mathrm{r}$ denote the set of measurable bounded $\mathcal{H}_q^+$-valued stochastic processes on the semi-infinite interval $[t_0, \infty)$ consisting of power inputs (heat fluxes) to the large-scale stochastic dynamical system $\mathcal{G}$ such that for every $\tau_{E_0} \overset{\text{a.s.}}{\geq} t_0$ the system energy state can be driven from $E(t_0) \overset{\text{a.s.}}{=} 0$ to $E(\tau_{E_0})$ by $S(\cdot) \in \mathcal{U}_\mathrm{r}$. Furthermore, we let $\mathcal{U}_\mathrm{c}$ denote the set of measurable bounded $\mathcal{H}_q^+$-valued stochastic processes on the semi-infinite interval $[t_0, \infty)$ consisting of power inputs (heat fluxes) to the large-scale stochastic dynamical system $\mathcal{G}$ such that the system energy state can be driven from $E(t_0) \overset{\text{a.s.}}{=} E_0$, $E_0 \in \overline{\mathbb{R}}_+^q$, to $E(\tilde{\tau}_{E_0})$ by $S(\cdot) \in \mathcal{U}_\mathrm{c}$. Finally, let $\mathcal{U}$ be an input space that is a subset of measurable bounded $\mathcal{H}_q^+$-valued stochastic processes on $\mathbb{R}$. The spaces $\mathcal{U}_\mathrm{r}$, $\mathcal{U}_\mathrm{c}$, and $\mathcal{U}$ are assumed to be closed under the shift operator, that is, if $S(\cdot) \in \mathcal{U}$ (respectively, $\mathcal{U}_\mathrm{c}$ or $\mathcal{U}_\mathrm{r}$), then the function S_T defined by $S_T(t) \triangleq S(t+T)$ is contained in $\mathcal{U}$ (respectively, $\mathcal{U}_\mathrm{c}$ or $\mathcal{U}_\mathrm{r}$) for all $T \geq 0$.

The next result establishes the uniqueness of the internal energy function $U(E)$, $E \in \overline{\mathbb{R}}_+^q$, for our large-scale stochastic dynamical system $\mathcal{G}$. For this result define the *available energy* of the large-scale stochastic dynamical system $\mathcal{G}$ by

$$U_\mathrm{a}(E_0) \triangleq -\inf_{u(\cdot)\in\mathcal{U},\, \tau \overset{\text{a.s.}}{\geq} t_0} \mathbb{E}\left[\mathbb{E}\left[\int_{t_0}^{\tau} [\mathbf{e}^\mathrm{T} u(t) - \mathbf{e}^\mathrm{T} y(t)]\mathrm{d}t | E(t_0) \overset{\text{a.s.}}{=} E_0\right]\right], \quad E_0 \in \overline{\mathbb{R}}_+^q, \tag{12.79}$$

where $E(t)$, $t \geq t_0$, is the solution to (12.68) with $E(t) \overset{\text{a.s.}}{=} E_0$ and admissible inputs $S(\cdot) \in \mathcal{U}$. The infimum in (12.79) is taken over all $\mathcal{F}_t$-measurable inputs $S(\cdot)$, all finite $\mathcal{F}_t$-stopping times $\tau \overset{\text{a.s.}}{\geq} 0$, and all system sample paths with initial value $E(t_0) \overset{\text{a.s.}}{=} E_0$ and terminal value left free. Furthermore, define the *required energy supply* of the large-scale stochastic dynamical system $\mathcal{G}$ by

$$U_\mathrm{r}(E_0) \triangleq \inf_{u(\cdot)\in\mathcal{U}_\mathrm{r},\, \tau_{E_0} \overset{\text{a.s.}}{\geq} 0} \mathbb{E}\left[\mathbb{E}\left[\int_0^{\tau_{E_0}} [\mathbf{e}^\mathrm{T} u(t) - \mathbf{e}^\mathrm{T} y(t)]\mathrm{d}t | E(0) \overset{\text{a.s.}}{=} 0\right]\right], \quad E_0 \in \overline{\mathbb{R}}_+^q. \tag{12.80}$$

The infimum in (12.80) is taken over all system sample paths starting from

$E(t_0) \overset{\text{a.s.}}{=} 0$ and ending at $E(\tau_{E_0}) \overset{\text{a.s.}}{=} E_0$ at time $t = \tau_{E_0}$, and all times $t \geq t_0$.

Note that the available energy $U_{\mathrm{a}}(E)$ is the maximum amount of stored energy (net heat) that can be extracted from the large-scale stochastic dynamical system $\mathcal{G}$ at any finite stopping time τ, and the required energy supply $U_{\mathrm{r}}(E)$ is the minimum amount of energy (net heat) that can be delivered to the large-scale stochastic dynamical system $\mathcal{G}$ such that, for all $\varepsilon > 0$, $\mathbb{P}^0\left(\lim_{t\to\tau_{E_0}} E(t) \in \mathcal{B}_\varepsilon(E_0)\right) = 1$.

Theorem 12.6. Consider the large-scale stochastic dynamical system $\mathcal{G}$ with differential energy balance equation given by (12.68). Then $\mathcal{G}$ is stochastically lossless with respect to the energy supply rate $r(u, y) = \mathbf{e}^{\mathrm{T}}u - \mathbf{e}^{\mathrm{T}}y$, where $u(t) \equiv S(t)$ and $y(t) \equiv d(E(t))$, and with the unique energy storage function corresponding to the total energy of the system $\mathcal{G}$ given by

$$\begin{aligned} U(E_0) &= \mathbf{e}^{\mathrm{T}}E_0 \\ &= -\mathbb{E}\left[\mathbb{E}\left[\int_0^{\tau_0}[\mathbf{e}^{\mathrm{T}}u(t) - \mathbf{e}^{\mathrm{T}}y(t)]\mathrm{d}t | E(0) \overset{\text{a.s.}}{=} E_0\right]\right] \\ &= \mathbb{E}\left[\mathbb{E}\left[\int_0^{\tau_{E_0}}[\mathbf{e}^{\mathrm{T}}u(t) - \mathbf{e}^{\mathrm{T}}y(t)]\mathrm{d}t | E(0) \overset{\text{a.s.}}{=} 0\right]\right], \quad E_0 \in \overline{\mathbb{R}}_+^q, \end{aligned} \tag{12.81}$$

where $E(t)$, $t \geq t_0$, is the solution to (12.68) with admissible input $u(\cdot) \in \mathcal{U}$, $E(\tau_0) \overset{\text{a.s.}}{=} 0$, and $E(\tau_{E_0}) \overset{\text{a.s.}}{=} E_0 \in \overline{\mathbb{R}}_+^q$. Furthermore,

$$0 \leq U_{\mathrm{a}}(E_0) = U(E_0) = U_{\mathrm{r}}(E_0) < \infty, \quad E_0 \in \overline{\mathbb{R}}_+^q. \tag{12.82}$$

Proof. Note that it follows from (12.71) that $\mathcal{G}$ is stochastically lossless with respect to the energy supply rate $r(u, y) = \mathbf{e}^{\mathrm{T}}u - \mathbf{e}^{\mathrm{T}}y$ and with the energy storage function $U(E) = \mathbf{e}^{\mathrm{T}}E$, $E \in \overline{\mathbb{R}}_+^q$. Since, by Lemma 12.2, $\mathcal{G}$ is reachable from and controllable to the origin in $\overline{\mathbb{R}}_+^q$, it follows from (12.71), with $E(t_0) \overset{\text{a.s.}}{=} E_0 \in \overline{\mathbb{R}}_+^q$ and $E(\tau_+) \overset{\text{a.s.}}{=} 0$ for some $\tau_+ \overset{\text{a.s.}}{\geq} t_0$ and $u(\cdot) \in \mathcal{U}$, that

$$\begin{aligned} \mathbf{e}^{\mathrm{T}}E_0 &= -\mathbb{E}\left[\mathbb{E}\left[\int_{t_0}^{\tau_+}[\mathbf{e}^{\mathrm{T}}u(t) - \mathbf{e}^{\mathrm{T}}y(t)]\mathrm{d}t | E(t_0) \overset{\text{a.s.}}{=} E_0\right]\right] \\ &\leq \sup_{u(\cdot)\in\mathcal{U},\, \tau_+ \overset{\text{a.s.}}{\geq} t_0} -\mathbb{E}\left[\mathbb{E}\left[\int_{t_0}^{\tau_+}[\mathbf{e}^{\mathrm{T}}u(t) - \mathbf{e}^{\mathrm{T}}y(t)]\mathrm{d}t | E(t_0) \overset{\text{a.s.}}{=} E_0\right]\right] \\ &= -\inf_{u(\cdot)\in\mathcal{U},\, \tau_+ \overset{\text{a.s.}}{\geq} t_0} \mathbb{E}\left[\mathbb{E}\left[\int_{t_0}^{\tau_+}[\mathbf{e}^{\mathrm{T}}u(t) - \mathbf{e}^{\mathrm{T}}y(t)]\mathrm{d}t | E(t_0) \overset{\text{a.s.}}{=} E_0\right]\right] \\ &= U_{\mathrm{a}}(E_0), \quad E_0 \in \overline{\mathbb{R}}_+^q. \end{aligned} \tag{12.83}$$

Alternatively, it follows from (12.71), with $E(0) \overset{\text{a.s.}}{=} 0$ for some $\tau_- \overset{\text{a.s.}}{\geq} 0$

and $u(\cdot) \in \mathcal{U}_{\mathrm{r}}$, that

$$\begin{aligned} \mathbf{e}^{\mathrm{T}} E_0 &= \mathbb{E}\left[\mathbb{E}\left[\int_0^{\tau_-} [\mathbf{e}^{\mathrm{T}} u(t) - \mathbf{e}^{\mathrm{T}} y(t)]\mathrm{d}t | E(0) \overset{\text{a.s.}}{=} 0\right]\right] \\ &\geq \inf_{u(\cdot)\in\mathcal{U}_{\mathrm{r}},\, \tau_- \overset{\text{a.s.}}{\geq} 0} \mathbb{E}\left[\mathbb{E}\left[\int_0^{\tau_-} [\mathbf{e}^{\mathrm{T}} u(t) - \mathbf{e}^{\mathrm{T}} y(t)]\mathrm{d}t | E(0) \overset{\text{a.s.}}{=} 0\right]\right] \\ &= U_{\mathrm{r}}(E_0), \quad E_0 \in \overline{\mathbb{R}}_+^q. \end{aligned} \tag{12.84}$$

Thus, (12.83) and (12.84) imply that (12.81) is satisfied and

$$U_{\mathrm{r}}(E_0) \leq \mathbf{e}^{\mathrm{T}} E_0 \leq U_{\mathrm{a}}(E_0), \quad E_0 \in \overline{\mathbb{R}}_+^q. \tag{12.85}$$

Conversely, it follows from (12.71) and the fact that $U(E) = \mathbf{e}^{\mathrm{T}} E \geq 0$, $E \in \overline{\mathbb{R}}_+^q$, that, for all $\tau \overset{\text{a.s.}}{\geq} t_0$ and $u(\cdot) \in \mathcal{U}$,

$$\mathbf{e}^{\mathrm{T}} E_0 \geq -\mathbb{E}\left[\mathbb{E}\left[\int_{t_0}^{\tau} [\mathbf{e}^{\mathrm{T}} u(t) - \mathbf{e}^{\mathrm{T}} y(t)]\mathrm{d}t | E(t_0) \overset{\text{a.s.}}{=} E_0\right]\right], \quad E_0 \in \overline{\mathbb{R}}_+^q, \tag{12.86}$$

which implies that

$$\begin{aligned} \mathbf{e}^{\mathrm{T}} E(t_0) &\geq \sup_{u(\cdot)\in\mathcal{U},\, \tau \overset{\text{a.s.}}{\geq} t_0} -\mathbb{E}\left[\mathbb{E}\left[\int_{t_0}^{\tau} [\mathbf{e}^{\mathrm{T}} u(t) - \mathbf{e}^{\mathrm{T}} y(t)]\mathrm{d}t | E(t_0) \overset{\text{a.s.}}{=} E_0\right]\right] \\ &= -\inf_{u(\cdot)\in\mathcal{U},\, \tau \overset{\text{a.s.}}{\geq} t_0} \mathbb{E}\left[\mathbb{E}\left[\int_{t_0}^{\tau} [\mathbf{e}^{\mathrm{T}} u(t) - \mathbf{e}^{\mathrm{T}} y(t)]\mathrm{d}t | E(t_0) \overset{\text{a.s.}}{=} E_0\right]\right] \\ &= U_{\mathrm{a}}(E_0), \quad E_0 \in \overline{\mathbb{R}}_+^q. \end{aligned} \tag{12.87}$$

Furthermore, it follows from the definition of $U_{\mathrm{a}}(\cdot)$ that $U_{\mathrm{a}}(E) \geq 0$, $E \in \overline{\mathbb{R}}_+^q$, since the infimum in (12.79) is taken over the set of values containing the zero value ($\tau \overset{\text{a.s.}}{=} t_0$).

Next, note that it follows from (12.71), with $E(0) \overset{\text{a.s.}}{=} 0$ and $E(\tau) \overset{\text{a.s.}}{=} E_0$, $E_0 \in \overline{\mathbb{R}}_+^q$, for all $\tau \overset{\text{a.s.}}{\geq} 0$ and $u(\cdot) \in \mathcal{U}_{\mathrm{r}}$, that

$$\begin{aligned} \mathbf{e}^{\mathrm{T}} E_0 &= \mathbb{E}\left[\mathbb{E}\left[\int_0^{\tau} [\mathbf{e}^{\mathrm{T}} u(t) - \mathbf{e}^{\mathrm{T}} y(t)]\mathrm{d}t | E(0) \overset{\text{a.s.}}{=} 0\right]\right] \\ &= \inf_{u(\cdot)\in\mathcal{U}_{\mathrm{r}},\, \tau \overset{\text{a.s.}}{\geq} 0} \mathbb{E}\left[\mathbb{E}\left[\int_0^{\tau} [\mathbf{e}^{\mathrm{T}} u(t) - \mathbf{e}^{\mathrm{T}} y(t)]\mathrm{d}t | E(0) \overset{\text{a.s.}}{=} 0\right]\right] \\ &= U_{\mathrm{r}}(E_0), \quad E_0 \in \overline{\mathbb{R}}_+^q. \end{aligned} \tag{12.88}$$

Moreover, since the system $\mathcal{G}$ is reachable from the origin, it follows that for every $E_0 \in \overline{\mathbb{R}}_+^q$, there exists $\tau \overset{\text{a.s.}}{\geq} 0$ and $u(\cdot) \in \mathcal{U}_{\rm r}$ such that

$$\mathbb{E}\left[\mathbb{E}\left[\int_0^{\tau}[\mathbf{e}^{\rm T}u(t) - \mathbf{e}^{\rm T}y(t)]\mathrm{d}t|E(0) \overset{\text{a.s.}}{=} 0\right]\right] \tag{12.89}$$

is finite, and hence, $U_{\rm r}(E_0) < \infty$, $E_0 \in \overline{\mathbb{R}}_+^q$. Finally, combining (12.85), (12.87), and (12.88), it follows that (12.82) holds. □

It follows from (12.82) and the definitions of available energy $U_{\rm a}(E_0)$ and the required energy supply $U_{\rm r}(E_0)$, $E_0 \in \overline{\mathbb{R}}_+^q$, that the large-scale stochastic dynamical system $\mathcal{G}$ can deliver to its surroundings all of its stored subsystem energies and can store all of the work done to all of its subsystems. This is in essence a statement of the first law of stochastic thermodynamics and places no limitation on the possibility of transforming heat into work or work into heat. In the case where $S(t) \equiv 0$, it follows from (12.71) and the fact that $\sigma_{ii}(E) \geq 0$, $E \in \overline{\mathbb{R}}_+^q$, $i = 1, \ldots, q$, that the zero solution $E(t) \equiv 0$ of the large-scale stochastic dynamical system $\mathcal{G}$ with the differential energy balance equation (12.68) is Lyapunov stable in probability with respect to $\overline{\mathbb{R}}_+^n$ with Lyapunov function $U(E)$ corresponding to the total energy in the system.

12.6 Entropy and the Second Law of Thermodynamics

As for the deterministic power balance equation (3.5), the nonlinear differential energy balance equation (12.68) can exhibit a full range of nonlinear behavior, including bifurcations, limit cycles, and even chaos. To ensure a thermodynamically consistent energy flow model, we require the following axioms.

Axiom *i*): For the connectivity matrix $\mathcal{C} \in \mathbb{R}^{q\times q}$ associated with the large-scale stochastic dynamical system $\mathcal{G}$ defined by

$$\mathcal{C}_{(i,j)} \triangleq \begin{cases} 0, & \text{if } \phi_{ij}(E) \equiv 0, \\ 1, & \text{otherwise,} \end{cases} \quad i \neq j, \quad i, j = 1, \ldots, q, \tag{12.90}$$

and

$$\mathcal{C}_{(i,i)} \triangleq -\sum_{k=1,\, k\neq i}^{q} \mathcal{C}_{(k,i)}, \quad i = j, \quad i = 1, \ldots, q, \tag{12.91}$$

rank $\mathcal{C} = q - 1$, and for $\mathcal{C}_{(i,j)} = 1$, $i \neq j$, $\phi_{ij}(E) = 0$ if and only if $E_i = E_j$.

Axiom *ii*): For $i, j = 1, \ldots, q$, $(E_i - E_j)\phi_{ij}(E) \overset{\text{a.s.}}{\leq} 0$, $E \in \overline{\mathbb{R}}_+^q$, and, for all $c > 0$,

$$\sum_{j=1,\, j\neq i}^{q} (E_i - E_j)\phi_{ij}(E)\frac{c+E_i}{c+E_j} \leq -\mathrm{row}_i(J(E))\mathrm{row}_i^{\mathrm{T}}(J(E)), \quad i = 1,\ldots,q.$$

Axioms *i)* and *ii)* are the stochastic analog of Axioms *i)* and *ii)* of Chapter 3. It is important to note here that due to the stochastic term $J(E)\mathrm{d}w_1$ capturing probabilistic variations in the heat transfer between the subsystems, Axiom *ii)* requires that the scaled net energy flow is bounded by the negative intensity of the system diffusion.

Next, we show that the classical Clausius equality and inequality for reversible and irreversible thermodynamics over cyclic motions are satisfied for our stochastic thermodynamically consistent energy flow model. For this result $\oint$ denotes a cyclic integral evaluated along an arbitrary closed path of (12.68) in $\overline{\mathbb{R}}_+^q$; that is, $\oint \triangleq \int_{t_0}^{\tau_{\mathrm{f}}}$ with $\tau_{\mathrm{f}} \overset{\text{a.s.}}{\geq} t_0$ and $S(\cdot) \in \mathcal{U}$ such that $E(\tau_{\mathrm{f}}) \overset{\text{a.s.}}{=} E(t_0) \overset{\text{a.s.}}{=} E_0 \in \overline{\mathbb{R}}_+^q$.

Proposition 12.5. Consider the large-scale stochastic dynamical system $\mathcal{G}$ with differential energy balance equation (12.68), and assume that Axioms *i)* and *ii)* hold. Then, for all $E_0 \in \overline{\mathbb{R}}_+^q$, $\tau_{\mathrm{f}} \geq t_0$, and $S(\cdot) \in \mathcal{U}$ such that $E(\tau_{\mathrm{f}}) \overset{\text{a.s.}}{=} E(t_0) \overset{\text{a.s.}}{=} E_0$,

$$\begin{aligned}\mathbb{E}^{E_0}&\left[\int_{t_0}^{\tau_{\mathrm{f}}}\sum_{i=1}^{q}\left[\frac{S_i(t)-\sigma_{ii}(E(t))}{c+E_i(t)}\mathrm{d}t\right]\right]\\ &= \mathbb{E}^{E_0}\left[\oint\sum_{i=1}^{q}\frac{\mathrm{d}Q_i(t)}{c+E_i(t)}\right]\\ &\leq \mathbb{E}^{E_0}\left[\int_{t_0}^{\tau_{\mathrm{f}}}\sum_{i=1}^{q}\frac{1}{2}\frac{\mathrm{row}_i(D(E(t)))\mathrm{row}_i^{\mathrm{T}}(D(E(t)))}{(c+E_i(t))^2}\mathrm{d}t\right], \quad (12.92)\end{aligned}$$

where $c > 0$, $\mathrm{d}Q_i(t) \triangleq [S_i(t) - \sigma_{ii}(E(t))]\mathrm{d}t$, $i = 1,\ldots,q$, is the amount of net energy (heat) received by the ith subsystem over the infinitesimal time interval $\mathrm{d}t$, and $E(t)$, $t \geq t_0$, is the solution to (12.68) with initial condition $E(t_0) \overset{\text{a.s.}}{=} E_0$. Furthermore,

$$\mathbb{E}^{E_0}\left[\oint\sum_{i=1}^{q}\frac{\mathrm{d}Q_i(t)}{c+E_i(t)}\right] = \mathbb{E}^{E_0}\left[\int_{t_0}^{\tau_{\mathrm{f}}}\sum_{i=1}^{q}\frac{1}{2}\frac{\mathrm{row}_i(D(E(t)))\mathrm{row}_i^{\mathrm{T}}(D(E(t)))}{(c+E_i(t))^2}\mathrm{d}t\right] \quad (12.93)$$

if and only if there exists a continuous function $\alpha : [t_0, t_{\mathrm{f}}] \to \overline{\mathbb{R}}_+$ such that

$E(t) = \alpha(t)\mathbf{e}$, $t \in [t_0, t_\mathrm{f}]$.

Proof. Since, by Proposition 12.4, $E(t) \geq\geq 0$, $t \geq t_0$, and $\phi_{ij}(E) = -\phi_{ji}(E)$, $E \in \overline{\mathbb{R}}_+^q$, $i \neq j$, $i, j = 1, \ldots, q$, it follows from (12.68), Ito's lemma, and Axiom *ii*) that, for all $\tau_\mathrm{f} \overset{\text{a.s.}}{\geq} t_0$,

$$
\begin{aligned}
\mathbb{E}^{E_0}\left[\oint \sum_{i=1}^{q} \frac{\mathrm{d}Q_i(t)}{c + E_i(t)}\right]
&= \mathbb{E}^{E_0}\left[\int_{t_0}^{\tau_\mathrm{f}} \sum_{i=1}^{q} \frac{\mathrm{d}E_i(t) - \sum_{j=1,\, j\neq i}^{q} \phi_{ij}(E(t))\mathrm{d}t}{c + E_i(t)}\right] \\
&= \mathbb{E}^{E_0}\left[\int_{t_0}^{\tau_\mathrm{f}} \sum_{i=1}^{q} \frac{\mathrm{d}E_i(t)}{c + E_i(t)}\right] - \mathbb{E}^{E_0}\left[\int_{t_0}^{\tau_\mathrm{f}} \sum_{i=1}^{q} \frac{\sum_{j=1,\, j\neq i}^{q} \phi_{ij}(E(t))\mathrm{d}t}{c + E_i(t)}\right] \\
&= \mathbb{E}^{E_0}\Bigg[\int_{t_0}^{\tau_\mathrm{f}} \sum_{i=1}^{q} \Bigg[\mathrm{d}\log_e\left(c + E_i(t)\right) \\
&\quad + \frac{1}{2}\frac{\mathrm{row}_i(J(E(t)))\mathrm{row}_i^\mathrm{T}(J(E(t)))}{(c + E_i(t))^2}\mathrm{d}t \\
&\quad + \frac{1}{2}\frac{\mathrm{row}_i(D(E(t)))\mathrm{row}_i^\mathrm{T}(D(E(t)))}{(c + E_i(t))^2}\mathrm{d}t\Bigg]\Bigg] \\
&\quad - \mathbb{E}^{E_0}\left[\int_{t_0}^{\tau_\mathrm{f}} \sum_{i=1}^{q} \sum_{j=1,\, j\neq i}^{q} \frac{1}{2}\left(\frac{\phi_{ij}(E(t))}{c + E_i(t)} - \frac{\phi_{ij}(E(t))}{c + E_j(t)}\right)\mathrm{d}t\right] \\
&= \mathbb{E}^{E_0}\left[\sum_{i=1}^{q} \log_e\left(\frac{c + E_i(\tau_\mathrm{f})}{c + E_i(t_0)}\right)\right] \\
&\quad - \mathbb{E}^{E_0}\left[\int_{t_0}^{\tau_\mathrm{f}} \sum_{i=1}^{q} \sum_{j=1,\, j\neq i}^{q} \frac{1}{2}\frac{\phi_{ij}(E(t))[E_j(t) - E_i(t)]}{(c + E_i(t))(c + E_j(t))}\mathrm{d}t\right] \\
&\quad + \mathbb{E}^{E_0}\left[\int_{t_0}^{\tau_\mathrm{f}} \sum_{i=1}^{q} \frac{1}{2}\frac{\mathrm{row}_i(J(E(t)))\mathrm{row}_i^\mathrm{T}(J(E(t)))}{(c + E_i(t))^2}\mathrm{d}t\right] \\
&\quad + \mathbb{E}^{E_0}\left[\int_{t_0}^{\tau_\mathrm{f}} \sum_{i=1}^{q} \frac{1}{2}\frac{\mathrm{row}_i(D(E(t)))\mathrm{row}_i^\mathrm{T}(D(E(t)))}{(c + E_i(t))^2}\mathrm{d}t\right] \\
&= \mathbb{E}^{E_0}\Bigg[\int_{t_0}^{\tau_\mathrm{f}} \sum_{i=1}^{q} \frac{1}{2}\frac{1}{(c + E_i(t))^2}\Bigg[\sum_{j=1,\, j\neq i}^{q} \phi_{ij}(E(t))[E_i(t) - E_j(t)] \\
&\quad \cdot\frac{c + E_i(t)}{c + E_j(t)} + \mathrm{row}_i(J(E(t)))\mathrm{row}_i^\mathrm{T}(J(E(t)))\Bigg]\mathrm{d}t\Bigg]
\end{aligned}
$$

$$+\mathbb{E}^{E_0}\left[\int_{t_0}^{\tau_{\rm f}}\sum_{i=1}^{q}\frac{1}{2}\frac{\mathrm{row}_i(D(E(t)))\mathrm{row}_i^{\rm T}(D(E(t)))}{(c+E_i(t))^2}\mathrm{d}t\right]$$

$$\leq \mathbb{E}^{E_0}\left[\int_{t_0}^{\tau_{\rm f}}\sum_{i=1}^{q}\frac{1}{2}\frac{\mathrm{row}_i(D(E(t)))\mathrm{row}_i^{\rm T}(D(E(t)))}{(c+E_i(t))^2}\mathrm{d}t\right], \qquad (12.94)$$

which proves (12.92).

To show (12.93), note that it follows from (12.94), Axiom i), and Axiom ii) that (12.93) holds if and only if $E_i(t) \overset{\text{a.s.}}{=} E_j(t)$, $t \in [t_0, \tau_{\rm f}]$, $i \neq j$, $i, j = 1, \ldots, q$, or, equivalently, there exists a continuous function $\alpha : [t_0, \tau_{\rm f}] \to \overline{\mathbb{R}}_+$ such that $E(t) \overset{\text{a.s.}}{=} \alpha(t)\mathbf{e}$, $t \in [t_0, \tau_{\rm f}]$. □

Inequality (12.92) is a generalization of Clausius' inequality for reversible and irreversible thermodynamics as applied to large-scale stochastic dynamical systems and restricts the manner in which the system dissipates (scaled) heat over cyclic motions. Note that the Clausius inequality for the stochastic thermodynamic model is stronger than the Clausius inequality for the deterministic thermodynamic model derived in Chapter 3.

It follows from Axiom i) and (12.68) that for the *adiabatically isolated* large-scale stochastic dynamical system $\mathcal{G}$ (that is, $S(t) \equiv 0$ and $D(E(t)) \equiv 0$), the energy states given by $E_{\rm e} = \alpha\mathbf{e}$, $\alpha \geq 0$, correspond to the equilibrium energy states of $\mathcal{G}$. Thus, as in classical thermodynamics, we can define an *equilibrium process* as a process in which the trajectory of the large-scale stochastic dynamical system $\mathcal{G}$ moves along the equilibrium manifold $\mathcal{M}_{\rm e} \triangleq \{E \in \overline{\mathbb{R}}_+^q : E = \alpha\mathbf{e}, \alpha \geq 0\}$ corresponding to the set of equilibria of the isolated system $\mathcal{G}$. Alternatively, a *nonequilibrium process* is a process that does not lie on the equilibrium manifold $\mathcal{M}_{\rm e}$. Hence, it follows from Axiom i) that for an equilibrium process $\phi_{ij}(E(t)) = 0$, $t \geq t_0$, $i \neq j$, $i, j = 1, \ldots, q$, and thus, by Proposition 12.5, inequality (12.92) is satisfied as an equality. Alternatively, for a nonequilibrium process it follows from Axioms i) and ii) that (12.92) is satisfied as a strict inequality.

Next, we give a stochastic definition of entropy for the large-scale stochastic dynamical system $\mathcal{G}$ that is consistent with the classical thermodynamic definition of entropy.

Definition 12.9. For the large-scale stochastic dynamical system $\mathcal{G}$ with differential energy balance equation (12.68), a function $\mathcal{S} : \overline{\mathbb{R}}_+^q \to \mathbb{R}$ satisfying

$$\mathbb{E}\left[\mathcal{S}(E(\tau_2))|\mathcal{F}_{\tau_1}\right] \geq \mathcal{S}(E(\tau_1)) + \mathbb{E}\left[\int_{\tau_1}^{\tau_2}\sum_{i=1}^{q}\left[\frac{S_i(t) - \sigma_{ii}(E(t))}{c+E_i(t)}\right.\right.$$

$$-\frac{1}{2}\frac{\mathrm{row}_i(D(E(t)))\mathrm{row}_i^{\mathrm{T}}(D(E(t)))}{(c+E_i(t))^2}\Bigg]\mathrm{d}t|\mathcal{F}_{\tau_1}\Bigg] \quad (12.95)$$

for every $\mathcal{F}_t$-stopping times $\tau_2 \overset{\text{a.s.}}{\geq} \tau_1 \overset{\text{a.s.}}{\geq} t_0$ and $S(\cdot) \in \mathcal{U}$ is called the *entropy* function of $\mathcal{G}$.

Note that it follows from Definition 12.9 that the difference between the system entropy production and the stored system entropy is a submartingale with respect to the differential energy balance system filtration.

Next, we show that (12.92) guarantees the existence of an entropy function for $\mathcal{G}$. For this result define the *available entropy* of the large-scale stochastic dynamical system $\mathcal{G}$ by

$$\begin{aligned}\mathcal{S}_{\mathrm{a}}(E_0) \triangleq -\sup_{S(\cdot)\in\mathcal{U}_{\mathrm{c}},\,\tau_0 \overset{\text{a.s.}}{\geq} t_0} \mathbb{E}\Bigg[\mathbb{E}\Bigg[\int_{t_0}^{\tau_0}\sum_{i=1}^{q}\Bigg[\frac{S_i(t)-\sigma_{ii}(E(t))}{c+E_i(t)} \\ -\frac{1}{2}\frac{\mathrm{row}_i(D(E(t)))\mathrm{row}_i^{\mathrm{T}}(D(E(t)))}{(c+E_i(t))^2}\Bigg]\mathrm{d}t|E(t_0) \overset{\text{a.s.}}{=} E_0\Bigg]\Bigg], \quad (12.96)\end{aligned}$$

where $E_0 \in \overline{\mathbb{R}}_+^q$ and $E(\tau_0) \overset{\text{a.s.}}{=} 0$, and define the *required entropy supply* of the large-scale stochastic dynamical system $\mathcal{G}$ by

$$\begin{aligned}\mathcal{S}_{\mathrm{r}}(E_0) \triangleq \sup_{S(\cdot)\in\mathcal{U}_{\mathrm{r}},\,\tau_{E_0} \overset{\text{a.s.}}{\geq} t_0} \mathbb{E}\Bigg[\mathbb{E}\Bigg[\int_{t_0}^{\tau_{E_0}}\sum_{i=1}^{q}\Bigg[\frac{S_i(t)-\sigma_{ii}(E(t))}{c+E_i(t)} \\ -\frac{1}{2}\frac{\mathrm{row}_i(D(E(t)))\mathrm{row}_i^{\mathrm{T}}(D(E(t)))}{(c+E_i(t))^2}\Bigg]\mathrm{d}t|E(t_0) \overset{\text{a.s.}}{=} 0\Bigg]\Bigg], \quad (12.97)\end{aligned}$$

where $E(\tau_{E_0}) \overset{\text{a.s.}}{=} E_0 \in \overline{\mathbb{R}}_+^q$. Note that the available entropy $\mathcal{S}_{\mathrm{a}}(E_0)$ is the minimum amount of scaled heat (entropy) that can be extracted from the large-scale stochastic dynamical system $\mathcal{G}$ in order to transfer it from an initial state $E(t_0) = E_0$ to $E(T) = 0$. Alternatively, the required entropy supply $\mathcal{S}_{\mathrm{r}}(E_0)$ is the maximum amount of scaled heat (entropy) that can be delivered to $\mathcal{G}$ to transfer it from the origin to a given subset in the state space containing the initial state $E(t_0) = E_0$ over a finite stopping time.

Theorem 12.7. Consider the large-scale stochastic dynamical system $\mathcal{G}$ with differential energy balance equation (12.68), and assume that Axiom *ii*) holds. Then there exists an entropy function for $\mathcal{G}$. Moreover, $\mathcal{S}_{\mathrm{a}}(E)$, $E \in \overline{\mathbb{R}}_+^q$, and $\mathcal{S}_{\mathrm{r}}(E)$, $E \in \overline{\mathbb{R}}_+^q$, are possible entropy functions for $\mathcal{G}$ with $\mathcal{S}_{\mathrm{a}}(0) = \mathcal{S}_{\mathrm{r}}(0) = 0$. Finally, all entropy functions $\mathcal{S}(E)$, $E \in \overline{\mathbb{R}}_+^q$, for $\mathcal{G}$ satisfy

$$\mathcal{S}_{\mathrm{r}}(E) \leq \mathcal{S}(E) - \mathcal{S}(0) \leq \mathcal{S}_{\mathrm{a}}(E), \quad E \in \overline{\mathbb{R}}_+^q. \quad (12.98)$$

Proof. Since, by Lemma 12.2, $\mathcal{G}$ is stochastically controllable to and stochastically reachable from the origin in $\overline{\mathbb{R}}_+^q$, it follows from (12.96) and (12.97) that $\mathcal{S}_{\rm a}(E_0) < \infty$, $E_0 \in \overline{\mathbb{R}}_+^q$, and $\mathcal{S}_{\rm r}(E_0) > -\infty$, $E_0 \in \overline{\mathbb{R}}_+^q$, respectively. Next, let $E_0 \in \overline{\mathbb{R}}_+^q$, and let $S(\cdot) \in \mathcal{U}$ be such that $E(\tau_{\rm i}) \stackrel{\rm a.s.}{=} E(\tau_{\rm f}) \stackrel{\rm a.s.}{=} 0$ and $E(\tau_0) \stackrel{\rm a.s.}{=} E_0$, where $\tau_{\rm i} \stackrel{\rm a.s.}{<} \tau_0 \stackrel{\rm a.s.}{<} \tau_{\rm f}$. In this case, it follows from (12.92) that, for all $\tau_{\rm i} \stackrel{\rm a.s.}{<} \tau_0 \stackrel{\rm a.s.}{<} \tau_{\rm f}$,

$$\mathbb{E}\left[\mathbb{E}\left[\int_{\tau_{\rm i}}^{\tau_{\rm f}} \sum_{i=1}^q \left[\frac{S_i(t) - \sigma_{ii}(E(t))}{c + E_i(t)} - \frac{1}{2}\frac{\mathrm{row}_i(D(E(t)))\mathrm{row}_i^{\rm T}(D(E(t)))}{(c+E_i(t))^2}\right] \mathrm{d}t \,\middle|\, E(\tau_{\rm i}) \stackrel{\rm a.s.}{=} 0\right]\right] \leq 0. \quad (12.99)$$

Next, using the strong Markov property we have

$$\begin{aligned}
&\mathbb{E}\left[\mathbb{E}\left[\int_{\tau_{\rm i}}^{\tau_{\rm f}} \sum_{i=1}^q \left[\frac{S_i(t) - \sigma_{ii}(E(t))}{c + E_i(t)} - \frac{1}{2}\frac{\mathrm{row}_i(D(E(t)))\mathrm{row}_i^{\rm T}(D(E(t)))}{(c+E_i(t))^2}\right] \mathrm{d}t \,\middle|\, E(\tau_{\rm i})\right]\right] \\
&\quad = \mathbb{E}\left[\mathbb{E}\left[\int_{\tau_{\rm i}}^{\tau_0} \sum_{i=1}^q \left[\frac{S_i(t) - \sigma_{ii}(E(t))}{c + E_i(t)} - \frac{1}{2}\frac{\mathrm{row}_i(D(E(t)))\mathrm{row}_i^{\rm T}(D(E(t)))}{(c+E_i(t))^2}\right] \mathrm{d}t\right.\right. \\
&\qquad \left.\left. + \int_{\tau_0}^{\tau_{\rm f}} \sum_{i=1}^q \left[\frac{S_i(t) - \sigma_{ii}(E(t))}{c + E_i(t)} - \frac{1}{2}\frac{\mathrm{row}_i(D(E(t)))\mathrm{row}_i^{\rm T}(D(E(t)))}{(c+E_i(t))^2}\right] \mathrm{d}t \,\middle|\, E(\tau_{\rm i})\right]\right] \\
&\quad = \mathbb{E}\left[\mathbb{E}\left[\int_{\tau_{\rm i}}^{\tau_0} \sum_{i=1}^q \left[\frac{S_i(t) - \sigma_{ii}(E(t))}{c + E_i(t)} - \frac{1}{2}\frac{\mathrm{row}_i(D(E(t)))\mathrm{row}_i^{\rm T}(D(E(t)))}{(c+E_i(t))^2}\right] \mathrm{d}t \,\middle|\, E(\tau_{\rm i})\right]\right] \\
&\qquad + \mathbb{E}\left[\mathbb{E}\left[\int_{\tau_0}^{\tau_{\rm f}} \sum_{i=1}^q \left[\frac{S_i(t) - \sigma_{ii}(E(t))}{c + E_i(t)} - \frac{1}{2}\frac{\mathrm{row}_i(D(E(t)))\mathrm{row}_i^{\rm T}(D(E(t)))}{(c+E_i(t))^2}\right] \mathrm{d}t \,\middle|\, \mathcal{F}_{\tau_0}\right]\right] \\
&\quad = \mathbb{E}\left[\mathbb{E}\left[\int_{\tau_{\rm i}}^{\tau_0} \sum_{i=1}^q \left[\frac{S_i(t) - \sigma_{ii}(E(t))}{c + E_i(t)}\right.\right.\right.
\end{aligned}$$

$$
-\frac{1}{2}\frac{\mathrm{row}_i(D(E(t)))\mathrm{row}_i^{\mathrm{T}}(D(E(t)))}{(c+E_i(t))^2}\Bigg]\mathrm{d}t|E(\tau_{\mathrm{i}})\Bigg]\Bigg]
$$

$$
+\mathbb{E}\Bigg[\mathbb{E}\Bigg[\int_{\tau_0}^{\tau_{\mathrm{f}}}\sum_{i=1}^{q}\Bigg[\frac{S_i(t)-\sigma_{ii}(E(t))}{c+E_i(t)}
-\frac{1}{2}\frac{\mathrm{row}_i(D(E(t)))\mathrm{row}_i^{\mathrm{T}}(D(E(t)))}{(c+E_i(t))^2}\Bigg]\mathrm{d}t|E(\tau_0)\Bigg]\Bigg], \quad (12.100)
$$

and hence, (12.99) implies

$$
\mathbb{E}\Bigg[\mathbb{E}\Bigg[\int_{\tau_{\mathrm{i}}}^{\tau_0}\sum_{i=1}^{q}\Bigg[\frac{S_i(t)-\sigma_{ii}(E(t))}{c+E_i(t)}
-\frac{1}{2}\frac{\mathrm{row}_i(D(E(t)))\mathrm{row}_i^{\mathrm{T}}(D(E(t)))}{(c+E_i(t))^2}\Bigg]\mathrm{d}t|E(\tau_{\mathrm{i}})\Bigg]\Bigg]
$$

$$
\leq -\mathbb{E}\Bigg[\mathbb{E}\Bigg[\int_{\tau_0}^{\tau_{\mathrm{f}}}\sum_{i=1}^{q}\Bigg[\frac{S_i(t)-\sigma_{ii}(E(t))}{c+E_i(t)}
-\frac{1}{2}\frac{\mathrm{row}_i(D(E(t)))\mathrm{row}_i^{\mathrm{T}}(D(E(t)))}{(c+E_i(t))^2}\Bigg]\mathrm{d}t|E(\tau_0)\Bigg]\Bigg]. \quad (12.101)
$$

Now, taking the supremum on both sides of (12.101) over all $S(\cdot)\in\mathcal{U}_{\mathrm{r}}$ and $\tau_{\mathrm{i}}\overset{\mathrm{a.s.}}{\leq}\tau_0$ yields

$$
\mathcal{S}_{\mathrm{r}}(E_0) = \sup_{S(\cdot)\in\mathcal{U}_{\mathrm{r}},\,\tau_{\mathrm{i}}\overset{\mathrm{a.s.}}{\leq}\tau_0}\mathbb{E}\Bigg[\mathbb{E}\Bigg[\int_{\tau_{\mathrm{i}}}^{\tau_0}\sum_{i=1}^{q}\Bigg[\frac{S_i(t)-\sigma_{ii}(E(t))}{c+E_i(t)}
-\frac{1}{2}\frac{\mathrm{row}_i(D(E(t)))\mathrm{row}_i^{\mathrm{T}}(D(E(t)))}{(c+E_i(t))^2}\mathrm{d}t|E(\tau_{\mathrm{i}})\Bigg]\Bigg]
$$

$$
\leq -\mathbb{E}\Bigg[\mathbb{E}\Bigg[\int_{\tau_0}^{\tau_{\mathrm{f}}}\sum_{i=1}^{q}\Bigg[\frac{S_i(t)-\sigma_{ii}(E(t))}{c+E_i(t)}
-\frac{1}{2}\frac{\mathrm{row}_i(D(E(t)))\mathrm{row}_i^{\mathrm{T}}(D(E(t)))}{(c+E_i(t))^2}\Bigg]\mathrm{d}t|E(\tau_0)\Bigg]\Bigg]. \quad (12.102)
$$

Next, taking the infimum on both sides of (12.102) over all $S(\cdot)\in\mathcal{U}_{\mathrm{c}}$ and $\tau_{\mathrm{f}}\overset{\mathrm{a.s.}}{\geq}\tau_0$, we obtain $\mathcal{S}_{\mathrm{r}}(E_0)\leq\mathcal{S}_{\mathrm{a}}(E_0)$, $E_0\in\overline{\mathbb{R}}_+^q$, which implies that $-\infty<\mathcal{S}_{\mathrm{r}}(E_0)\leq\mathcal{S}_{\mathrm{a}}(E_0)<\infty$, $E_0\in\overline{\mathbb{R}}_+^q$. Hence, the functions $\mathcal{S}_{\mathrm{a}}(\cdot)$ and $\mathcal{S}_{\mathrm{r}}(\cdot)$ are well defined.

Next, it follows from the definition of $\mathcal{S}_{\mathrm{a}}(\cdot)$, the law of iterated expectation, and the strong Markov property that, for every stopping time

$\mathcal{T} \overset{\text{a.s.}}{\geq} \tau_1$ and $S(\cdot) \in \mathcal{U}_{\text{c}}$ such that $E(\tau_1) \in \overline{\mathcal{H}}_q^+$ and $E(\mathcal{T}) \overset{\text{a.s.}}{=} 0$,

$$
\begin{aligned}
-\mathcal{S}_{\text{a}}(E(\tau_1)) = & \sup_{S(\cdot)\in\mathcal{U}_{\text{c}},\, \mathcal{T}\overset{\text{a.s.}}{\geq}\tau_1} \mathbb{E}\bigg[\int_{\tau_1}^{\mathcal{T}} \sum_{i=1}^{q} \bigg[\frac{S_i(t) - \sigma_{ii}(E(t))}{c + E_i(t)} \\
& -\frac{1}{2}\frac{\text{row}_i(D(E(t)))\text{row}_i^{\text{T}}(D(E(t)))}{(c + E_i(t))^2}\bigg] \text{d}t | E(\tau_1)\bigg], \\
\geq & \ \mathbb{E}\bigg[\int_{\tau_1}^{\tau_2} \sum_{i=1}^{q} \bigg[\frac{S_i(t) - \sigma_{ii}(E(t))}{c + E_i(t)} \\
& -\frac{1}{2}\frac{\text{row}_i(D(E(t)))\text{row}_i^{\text{T}}(D(E(t)))}{(c + E_i(t))^2}\bigg] \text{d}t | \mathcal{F}_{\tau_1}\bigg] \\
& + \sup_{S(\cdot)\in\mathcal{U}_{\text{c}},\, \mathcal{T}\overset{\text{a.s.}}{\geq}\tau_2} \mathbb{E}\bigg[\int_{\tau_2}^{\mathcal{T}} \sum_{i=1}^{q} \bigg[\frac{S_i(t) - \sigma_{ii}(E(t))}{c + E_i(t)} \\
& -\frac{1}{2}\frac{\text{row}_i(D(E(t)))\text{row}_i^{\text{T}}(D(E(t)))}{(c + E_i(t))^2}\bigg] \text{d}t | \mathcal{F}_{\tau_1}\bigg], \\
= & \ \mathbb{E}\bigg[\int_{\tau_1}^{\tau_2} \sum_{i=1}^{q} \bigg[\frac{S_i(t) - \sigma_{ii}(E(t))}{c + E_i(t)} \\
& -\frac{1}{2}\frac{\text{row}_i(D(E(t)))\text{row}_i^{\text{T}}(D(E(t)))}{(c + E_i(t))^2}\bigg] \text{d}t | \mathcal{F}_{\tau_1}\bigg] \\
& + \mathbb{E}\bigg[\sup_{S(\cdot)\in\mathcal{U}_{\text{c}},\, \mathcal{T}\overset{\text{a.s.}}{\geq}\tau_2} \mathbb{E}\bigg[\int_{\tau_2}^{\mathcal{T}} \sum_{i=1}^{q} \bigg[\frac{S_i(t) - \sigma_{ii}(E(t))}{c + E_i(t)} \\
& -\frac{1}{2}\frac{\text{row}_i(D(E(t)))\text{row}_i^{\text{T}}(D(E(t)))}{(c + E_i(t))^2}\bigg] \text{d}t | \mathcal{F}_{\tau_2}\bigg] | \mathcal{F}_{\tau_1}\bigg], \\
= & \ \mathbb{E}\bigg[\int_{\tau_1}^{\tau_2} \sum_{i=1}^{q} \bigg[\frac{S_i(t) - \sigma_{ii}(E(t))}{c + E_i(t)} \\
& -\frac{1}{2}\frac{\text{row}_i(D(E(t)))\text{row}_i^{\text{T}}(D(E(t)))}{(c + E_i(t))^2}\bigg] \text{d}t | \mathcal{F}_{\tau_1}\bigg] \\
& -\mathbb{E}\left[\mathcal{S}_{\text{a}}(E(\tau_2)) | \mathcal{F}_{\tau_1}\right], \quad \tau_1 \overset{\text{a.s.}}{\leq} \tau_2 \overset{\text{a.s.}}{\leq} \mathcal{T},
\end{aligned}
\tag{12.103}
$$

which implies that $\mathcal{S}_{\text{a}}(E)$, $E \in \overline{\mathbb{R}}_+^q$, satisfies (12.95). Thus, $\mathcal{S}_{\text{a}}(E)$, $E \in \overline{\mathbb{R}}_+^q$, is a possible entropy function for $\mathcal{G}$. Note that with $E(\tau_0) \overset{\text{a.s.}}{=} E(\mathcal{T}) \overset{\text{a.s.}}{=} 0$ it follows from (12.92) that the supremum in (12.96) is taken over the set of negative semidefinite values with one of the values being zero for $S(t) \overset{\text{a.s.}}{\equiv} 0$. Thus, $\mathcal{S}_{\text{a}}(0) = 0$.

Similarly, it follows from the definition of $\mathcal{S}_{\text{r}}(\cdot)$ that for every stopping time $\mathcal{T} \le \tau_2$ and $S(\cdot) \in \mathcal{U}_{\rm r}$ such that $E(\tau_2) \in \overline{\mathcal{H}}_q^+$ and $E(\mathcal{T}) \overset{\rm a.s.}{=} 0$,

$$\begin{aligned}
\mathcal{S}_{\rm r}(E(\tau_2)) &= \sup_{S(\cdot)\in\mathcal{U}_{\rm r},\, \mathcal{T} \overset{\rm a.s.}{\le} \tau_2} \int_{\mathcal{T}}^{\tau_2} \sum_{i=1}^{q} \left[\frac{S_i(t) - \sigma_{ii}(E(t))}{c + E_i(t)} \right. \\
&\qquad \left. - \frac{1}{2} \frac{\mathrm{row}_i(D(E(t)))\mathrm{row}_i^{\rm T}(D(E(t)))}{(c + E_i(t))^2} \right] \mathrm{d}t, \\
&\ge \sup_{S(\cdot)\in\mathcal{U}_{\rm r},\, \mathcal{T} \overset{\rm a.s.}{\le} \tau_1} \int_{\mathcal{T}}^{\tau_1} \left[\sum_{i=1}^{q} \frac{S_i(t) - \sigma_{ii}(E(t))}{c + E_i(t)} \right. \\
&\qquad \left. - \frac{1}{2} \frac{\mathrm{row}_i(D(E(t)))\mathrm{row}_i^{\rm T}(D(E(t)))}{(c + E_i(t))^2} \right] \mathrm{d}t \\
&\quad + \int_{\tau_1}^{\tau_2} \sum_{i=1}^{q} \left[\frac{S_i(t) - \sigma_{ii}(E(t))}{c + E_i(t)} \right. \\
&\qquad \left. - \frac{1}{2} \frac{\mathrm{row}_i(D(E(t)))\mathrm{row}_i^{\rm T}(D(E(t)))}{(c + E_i(t))^2} \right] \mathrm{d}t, \\
&= \mathcal{S}_{\rm r}(E(\tau_1)) + \int_{\tau_1}^{\tau_2} \sum_{i=1}^{q} \left[\frac{S_i(t) - \sigma_{ii}(E(t))}{c + E_i(t)} \right. \\
&\qquad \left. - \frac{1}{2} \frac{\mathrm{row}_i(D(E(t)))\mathrm{row}_i^{\rm T}(D(E(t)))}{(c + E_i(t))^2} \right] \mathrm{d}t, \\
&\qquad\qquad \mathcal{T} \overset{\rm a.s.}{\le} \tau_1 \overset{\rm a.s.}{\le} \tau_2, \qquad (12.104)
\end{aligned}$$

which implies that $\mathcal{S}_{\rm r}(E)$, $E \in \overline{\mathbb{R}}_+^q$, satisfies (12.95). Thus, $\mathcal{S}_{\rm r}(E)$, $E \in \overline{\mathbb{R}}_+^q$, is a possible entropy function for $\mathcal{G}$. Note that with $E(t_0) \overset{\rm a.s.}{=} E(\mathcal{T}) \overset{\rm a.s.}{=} 0$ it follows from (12.92) that the supremum in (12.97) is taken over the set of negative semidefinite values with one of the values being zero for $S(t) \overset{\rm a.s.}{\equiv} 0$. Thus, $\mathcal{S}_{\rm r}(0) = 0$.

Next, suppose there exists an entropy function $\mathcal{S} : \overline{\mathbb{R}}_+^q \to \mathbb{R}$ for $\mathcal{G}$, and let $E(\tau_2) \overset{\rm a.s.}{=} 0$ in (12.95). Then it follows from (12.95) that

$$\begin{aligned}
\mathcal{S}(E(\tau_1)) - \mathcal{S}(0) \le -\mathbb{E}\Bigg[\int_{\tau_1}^{\tau_2} \sum_{i=1}^{q} & \left[\frac{S_i(t) - \sigma_{ii}(E(t))}{c + E_i(t)} \right. \\
& \left. - \frac{1}{2} \frac{\mathrm{row}_i(D(E(t)))\mathrm{row}_i^{\rm T}(D(E(t)))}{(c + E_i(t))^2} \right] \mathrm{d}t \Big| \mathcal{F}_{\tau_1} \Bigg]
\end{aligned} \qquad (12.105)$$

for all $\tau_2 \overset{\text{a.s.}}{\geq} \tau_1$ and $S(\cdot) \in \mathcal{U}_{\text{c}}$, which implies that

$$\begin{aligned}
\mathcal{S}(E(\tau_1)) - \mathcal{S}(0) &\leq \inf_{S(\cdot)\in\mathcal{U}_{\text{c}},\, \tau_2 \overset{\text{a.s.}}{\geq} \tau_1} \Bigg[-\mathbb{E}\Bigg[\int_{\tau_1}^{\tau_2} \sum_{i=1}^{q} \Bigg[\frac{S_i(t) - \sigma_{ii}(E(t))}{c + E_i(t)} \\
&\quad - \frac{1}{2} \frac{\text{row}_i(D(E(t)))\text{row}_i^{\text{T}}(D(E(t)))}{(c + E_i(t))^2} \Bigg] \text{d}t \big| \mathcal{F}_{\tau_1} \Bigg]\Bigg] \\
&= - \sup_{S(\cdot)\in\mathcal{U}_{\text{c}},\, \tau_2 \overset{\text{a.s.}}{\geq} \tau_1} \mathbb{E}\Bigg[\int_{\tau_1}^{\tau_2} \sum_{i=1}^{q} \Bigg[\frac{S_i(t) - \sigma_{ii}(E(t))}{c + E_i(t)} \\
&\quad - \frac{1}{2} \frac{\text{row}_i(D(E(t)))\text{row}_i^{\text{T}}(D(E(t)))}{(c + E_i(t))^2} \Bigg] \text{d}t \big| \mathcal{F}_{\tau_1} \Bigg] \\
&= \mathcal{S}_{\text{a}}(E(\tau_1)). \qquad (12.106)
\end{aligned}$$

Since $E(\tau_1)$ is arbitrary, it follows that $\mathcal{S}(E) - \mathcal{S}(0) \leq \mathcal{S}_{\text{a}}(E)$, $E \in \overline{\mathbb{R}}_+^q$.

Alternatively, let $E(\tau_1) \overset{\text{a.s.}}{=} 0$ in (12.95). Then it follows from (12.95) that

$$\begin{aligned}
\mathcal{S}(E(\tau_2)) - \mathcal{S}(0) &\geq \mathbb{E}\Bigg[\int_{\tau_1}^{\tau_2} \sum_{i=1}^{q} \Bigg[\frac{S_i(t) - \sigma_{ii}(E(t))}{c + E_i(t)} \\
&\quad - \frac{1}{2} \frac{\text{row}_i(D(E(t)))\text{row}_i^{\text{T}}(D(E(t)))}{(c + E_i(t))^2} \Bigg] \text{d}t \big| \mathcal{F}_{\tau_1} \Bigg] \qquad (12.107)
\end{aligned}$$

for all $\tau_1 \overset{\text{a.s.}}{\leq} \tau_2$ and $S(\cdot) \in \mathcal{U}_{\text{r}}$. Hence,

$$\begin{aligned}
\mathcal{S}(E(\tau_2)) - \mathcal{S}(0) &\geq \sup_{S(\cdot)\in\mathcal{U}_{\text{r}},\, \tau_1 \overset{\text{a.s.}}{\leq} \tau_2} \mathbb{E}\Bigg[\int_{\tau_1}^{\tau_2} \sum_{i=1}^{q} \Bigg[\frac{S_i(t) - \sigma_{ii}(E(t))}{c + E_i(t)} \\
&\quad - \frac{1}{2} \frac{\text{row}_i(D(E(t)))\text{row}_i^{\text{T}}(D(E(t)))}{(c + E_i(t))^2} \Bigg] \text{d}t \big| \mathcal{F}_{\tau_1} \Bigg] \\
&= \mathcal{S}_{\text{r}}(E(\tau_2)), \qquad (12.108)
\end{aligned}$$

which, since $E(\tau_2)$ is arbitrary, implies that $\mathcal{S}_{\text{r}}(E) \leq \mathcal{S}(E) - \mathcal{S}(0)$, $E \in \overline{\mathbb{R}}_+^q$. Thus, all entropy functions for $\mathcal{G}$ satisfy (12.98). □

It is important to note that inequality (12.92) is equivalent to the existence of an entropy function for $\mathcal{G}$. Sufficiency is simply a statement of Theorem 12.7, while necessity follows from (12.95) with $E(t_2) \overset{\text{a.s.}}{=} E(t_1)$. This definition of entropy leads to the second law for stochastic thermodynamics being viewed as an axiom in the context of stochastic (anti)cyclo-dissipative dynamical systems [372].

The next result shows that all entropy functions for $\mathcal{G}$ are continuous on $\overline{\mathbb{R}}_+^q$.

Theorem 12.8. Consider the large-scale stochastic dynamical system $\mathcal{G}$ with differential energy balance equation (12.68), and let $\mathcal{S} : \overline{\mathbb{R}}_+^q \to \mathbb{R}$ be an entropy function of $\mathcal{G}$. Then $\mathcal{S}(\cdot)$ is continuous on $\overline{\mathbb{R}}_+^q$.

Proof. Let $E_{\rm e} \in \overline{\mathbb{R}}_+^q$ and $S_{\rm e} \in \mathbb{R}^q$ be such that $S_{\rm e} = d(E_{\rm e}) - f(E_{\rm e})$. Note that with $S(t) \overset{\rm a.s.}{\equiv} S_{\rm e}$, $E_{\rm e}$ is an equilibrium point of the differential energy balance equation (12.68). Next, it follows from Lemma 12.2 that $\mathcal{G}$ is *locally stochastically controllable*, that is, for every $\tau \overset{\rm a.s.}{>} 0$ and $\varepsilon > 0$, the set of points that can be reached from and to $E_{\rm e}$ in time T using admissible inputs $S : [0, \tau] \to \mathcal{H}_q$, satisfying $\|S(t) - S_{\rm e}\| \overset{\rm a.s.}{<} \varepsilon$, contains a neighborhood of $E_{\rm e}$.

Next, let $\delta > 0$ and note that it follows from the continuity of $f(\cdot)$, $d(\cdot)$, $J(\cdot)$, and $D(\cdot)$ that there exist $\tau > 0$ and $\varepsilon > 0$ such that for every $S : [0, \tau) \to \mathbb{R}^q$ and $\|S(t) - S_{\rm e}\| \overset{\rm a.s.}{<} \varepsilon$, $\|E(t) - E_{\rm e}\| \overset{\rm a.s.}{<} \delta$, $t \in [0, \tau)$, where $S(\cdot) \in \mathcal{U}$ and $E(t)$, $t \in [0, \tau)$, denotes the solution to (12.68) with the initial condition $E_{\rm e}$. Furthermore, it follows from the local controllability of $\mathcal{G}$ that for every $\hat{\tau} \in (0, \tau]$, there exists a strictly increasing, continuous function $\gamma : \overline{\mathbb{R}}_+^q \to \overline{\mathbb{R}}_+^q$ such that $\gamma(0) = 0$, and for every $E_0 \in \mathcal{H}_q^+$ such that $\|E_0 - E_{\rm e}\| \overset{\rm a.s.}{\leq} \gamma(\hat{\tau})$, there exist $0 \overset{\rm a.s.}{\leq} \tilde{\tau} \overset{\rm a.s.}{\leq} \hat{\tau}$ and an input $S : [0, \hat{\tau}] \to \mathcal{H}_q$ such that $\|S(t) - S_{\rm e}\| < \varepsilon$, $t \in [0, \tilde{\tau})$, and $E(\hat{t}) \overset{\rm a.s.}{=} E_0$. Hence, there exists $\beta > 0$ such that for every $E_0 \in \mathcal{H}_q^+$ such that $\|E_0 - E_{\rm e}\| \overset{\rm a.s.}{\leq} \beta$, there exist $0 \overset{\rm a.s.}{\leq} \hat{\tau} \overset{\rm a.s.}{\leq} \gamma^{-1}(\|E_0 - E_{\rm e}\|)]$ and an input $S : [t_0, \hat{\tau}] \to \mathcal{H}_q$ such that $\|S(t) - S_{\rm e}\| \overset{\rm a.s.}{<} \varepsilon$, $t \in [0, \hat{t}]$, and $E(\hat{t}) \overset{\rm a.s.}{=} E_0$. In addition, it follows from Lemma 12.2 that $S : [0, \hat{\tau}] \to \mathcal{H}_q$ is such that $E(t) \overset{\rm a.s.}{\geq\geq} 0$, $t \in [0, \hat{\tau}]$.

Next, since $\sigma_{ii}(\cdot)$, $i = 1, \ldots, q$, is continuous, it follows that there exists $M \in \mathcal{H}_1^+$ such that

$$\sup_{\|E - E_{\rm e}\| \overset{\rm a.s.}{<} \delta,\ \|S - S_{\rm e}\| \overset{\rm a.s.}{<} \varepsilon} \left| \sum_{i=1}^q \left[\frac{S_i - \sigma_{ii}(E)}{c + E_i} - \frac{1}{2} \frac{\mathrm{row}_i(D(E))\mathrm{row}_i^{\rm T}(D(E))}{(c + E_i)^2} \right] \right| = M. \tag{12.109}$$

Hence, it follows that

$$\left| \int_0^{\hat{\tau}} \sum_{i=1}^q \left[\frac{S_i(\sigma) - \sigma_{ii}(E(\sigma))}{c + E_i(\sigma)} - \frac{1}{2} \frac{\mathrm{row}_i(D(E(\sigma)))\mathrm{row}_i^{\rm T}(D(E(\sigma)))}{(c + E_i(\sigma))^2} \right] \mathrm{d}\sigma \right|$$

$$
\begin{aligned}
&\overset{\text{a.s.}}{\leq} \int_0^{\hat{\tau}} \left| \sum_{i=1}^{q} \left[\frac{S_i(\sigma) - \sigma_{ii}(E(\sigma))}{c + E_i(\sigma)} - \frac{1}{2} \frac{\mathrm{row}_i(D(E(\sigma)))\mathrm{row}_i^{\mathrm{T}}(D(E(\sigma)))}{(c + E_i(\sigma))^2} \right] \right| \mathrm{d}\sigma \\
&\overset{\text{a.s.}}{\leq} M\hat{\tau} \\
&\overset{\text{a.s.}}{\leq} M\gamma^{-1}(\|E_0 - E_{\mathrm{e}}\|). \qquad (12.110)
\end{aligned}
$$

Now, if $\mathcal{S}(\cdot)$ is an entropy function of $\mathcal{G}$, then

$$
\begin{aligned}
\mathbb{E}\left[\mathcal{S}(E(\hat{\tau}))|\mathcal{F}_0\right] \overset{\text{a.s.}}{\geq} \mathcal{S}(E_{\mathrm{e}}) + \mathbb{E}\Bigg[\int_0^{\hat{\tau}} \sum_{i=1}^{q} \Bigg[& \frac{S_i(\sigma) - \sigma_{ii}(E(\sigma))}{c + E_i(\sigma)} \\
& - \frac{1}{2} \frac{\mathrm{row}_i(D(E(\sigma)))\mathrm{row}_i^{\mathrm{T}}(D(E(\sigma)))}{(c + E_i(\sigma))^2} \Bigg] \mathrm{d}\sigma | \mathcal{F}_0 \Bigg]
\end{aligned}
\qquad (12.111)
$$

or, equivalently,

$$
\begin{aligned}
-\mathbb{E}\left[\int_0^{\hat{\tau}} \sum_{i=1}^{q} \left[\frac{S_i(\sigma) - \sigma_{ii}(E(\sigma))}{c + E_i(\sigma)} - \frac{1}{2} \frac{\mathrm{row}_i(D(E(\sigma)))\mathrm{row}_i^{\mathrm{T}}(D(E(\sigma)))}{(c + E_i(\sigma))^2} \right] \mathrm{d}\sigma | \mathcal{F}_0 \right] \\
\overset{\text{a.s.}}{\geq} \mathcal{S}(E_{\mathrm{e}}) - \mathbb{E}\left[\mathcal{S}(E(\hat{\tau}))|\mathcal{F}_0\right]. \qquad (12.112)
\end{aligned}
$$

If $\mathcal{S}(E_{\mathrm{e}}) \overset{\text{a.s.}}{\geq} \mathcal{S}(E(\hat{\tau}))$, then combining (12.110) and (12.112) yields

$$
|\mathcal{S}(E_{\mathrm{e}}) - \mathbb{E}\left[\mathcal{S}(E(\hat{\tau}))|\mathcal{F}_0\right]| \overset{\text{a.s.}}{\leq} \mathbb{E}\left[M\gamma^{-1}(\|E_0 - E_{\mathrm{e}}\|)|\mathcal{F}_0\right]. \qquad (12.113)
$$

Alternatively, if $\mathcal{S}(E(\hat{\tau})) \overset{\text{a.s.}}{\geq} \mathcal{S}(E_{\mathrm{e}})$, then (12.113) can be derived by reversing the roles of E_{e} and $E(\hat{\tau})$. Specifically, for $E_0 \in \overline{\mathbb{R}}_+^q$ and $E(\hat{\tau}) \overset{\text{a.s.}}{=} E_0$, (12.113) becomes

$$
|\mathcal{S}(E_0) - \mathcal{S}(E_{\mathrm{e}})| \leq \mathbb{E}[M]\gamma^{-1}(\|E_0 - E_{\mathrm{e}}\|).
$$

Hence, since $\gamma(\cdot)$ is continuous and $E(\hat{\tau})$ is arbitrary, it follows that $\mathcal{S}(\cdot)$ is continuous on $\overline{\mathbb{R}}_+^q$. □

Next, as a direct consequence of Theorem 12.7, we show that all possible entropy functions of $\mathcal{G}$ form a convex set, and hence, there exists a continuum of possible entropy functions for $\mathcal{G}$ ranging from the required entropy supply $\mathcal{S}_{\mathrm{r}}(E)$ to the available entropy $\mathcal{S}_{\mathrm{a}}(E)$.

Proposition 12.6. Consider the large-scale stochastic dynamical system $\mathcal{G}$ with differential energy balance equation (12.68), and assume that Axioms *i*) and *ii*) hold. Then

$$
\mathcal{S}(E) \triangleq \alpha\mathcal{S}_{\mathrm{r}}(E) + (1 - \alpha)\mathcal{S}_{\mathrm{a}}(E), \quad \alpha \in [0, 1], \qquad (12.114)
$$

is an entropy function for $\mathcal{G}$.

Proof. The result is a direct consequence of the reachability of $\mathcal{G}$ along with inequality (12.95) by noting that if $\mathcal{S}_{\rm r}(E)$ and $\mathcal{S}_{\rm a}(E)$ satisfy (12.95), then $\mathcal{S}(E)$ satisfies (12.95). □

It follows from Proposition 12.6 that Definition 12.9 does not provide enough information to define the entropy uniquely for nonequilibrium thermodynamic systems with differential energy balance equation (12.68). As noted in Chapter 3, this difficulty has long been pointed out in [318]. Two particular entropy functions for $\mathcal{G}$ can be computed a priori via the variational problems given by (12.96) and (12.97). For equilibrium thermodynamics, however, uniqueness is not an issue, as shown in the next proposition.

Proposition 12.7. Consider the large-scale stochastic dynamical system $\mathcal{G}$ with differential energy balance equation (12.68), and assume that Axioms *i)* and *ii)* hold. Then at every equilibrium state $E = E_{\rm e}$ of the isolated system $\mathcal{G}$, the entropy $\mathcal{S}(E)$, $E \in \overline{\mathbb{R}}_+^q$, of $\mathcal{G}$ is unique (modulo a constant of integration) and is given by

$$\mathcal{S}(E) - \mathcal{S}(0) = \mathcal{S}_{\rm a}(E) = \mathcal{S}_{\rm r}(E) = \mathbf{e}^{\rm T}\mathbf{log}_e(c\mathbf{e} + E) - q\log_e c, \quad (12.115)$$

where $E = E_{\rm e}$ and $\mathbf{log}_e(c\mathbf{e} + E)$ denotes the vector natural logarithm given by $[\log_e(c + E_1), \ldots, \log_e(c + E_q)]^{\rm T}$.

Proof. It follows from Axiom *i)* and Axiom *ii)* that for an equilibrium process $\phi_{ij}(E(t)) \stackrel{\rm a.s.}{\equiv} 0$, $i \neq j$, $i, j = 1, \ldots, q$, $D(E(t)) \stackrel{\rm a.s.}{\equiv} 0$, and $J(E(t)) \stackrel{\rm a.s.}{\equiv} 0$. Consider the entropy function $\mathcal{S}_{\rm a}(\cdot)$ given by (12.96), and let $E_0 = E_{\rm e}$ for some equilibrium state $E_{\rm e}$. Then it follows from (12.68) that

$$\begin{aligned}
\mathcal{S}_{\rm a}(E_0) &= -\sup_{S(\cdot)\in\mathcal{U}_{\rm c},\,\mathcal{T}\stackrel{\rm a.s.}{\geq} t_0} \mathbb{E}\left[\mathbb{E}\left[\int_{t_0}^{\mathcal{T}} \sum_{i=1}^{q} \left[\frac{S_i(t) - \sigma_{ii}(E(t))}{c + E_i(t)}\right.\right.\right. \\
&\qquad \left.\left.\left. - \frac{1}{2}\frac{\mathrm{row}_i(D(E(t)))\mathrm{row}_i^{\rm T}(D(E(t)))}{(c + E_i(\sigma))^2}\right]\mathrm{d}t \,\Big|\, E(t_0) \stackrel{\rm a.s.}{=} E_0\right]\right] \\
&= -\sup_{S(\cdot)\in\mathcal{U}_{\rm c},\,\mathcal{T}\stackrel{\rm a.s.}{\geq} t_0} \mathbb{E}\left[\mathbb{E}\left[\int_{t_0}^{\mathcal{T}} \sum_{i=1}^{q} \left[\frac{\mathrm{d}E_i(t) - \sum_{j=1, j\neq i}^{q}\phi_{ij}(E(t))\mathrm{d}t}{c + E_i(t)}\right.\right.\right. \\
&\qquad \left.\left.\left. - \frac{1}{2}\frac{\mathrm{row}_i(D(E(t)))\mathrm{row}_i^{\rm T}(D(E(t)))}{(c + E_i(\sigma))^2}\mathrm{d}t\right] \,\Big|\, E(t_0) \stackrel{\rm a.s.}{=} E_0\right]\right] \\
&= -\sup_{S(\cdot)\in\mathcal{U}_{\rm c},\,\mathcal{T}\stackrel{\rm a.s.}{\geq} t_0} \mathbb{E}\left[\mathbb{E}\left[\sum_{i=1}^{q} \log_e\left(\frac{c}{c + E_{i0}}\right)\right.\right.
\end{aligned}$$

$$+\int_{t_0}^{\mathcal{T}}\sum_{i=1}^{q}\frac{1}{2}\frac{\mathrm{row}_i(J(E(t)))\mathrm{row}_i^{\mathrm{T}}(J(E(t)))}{(c+E_i(t))^2}\mathrm{d}t$$

$$-\int_{t_0}^{T}\sum_{i=1}^{q}\sum_{j=1,j\neq i}^{q}\frac{\phi_{ij}(E(t))}{c+E_i(t)}\mathrm{d}t|E(t_0)\overset{\mathrm{a.s.}}{=}E_0\Bigg]\Bigg]$$

$$=-\sup_{S(\cdot)\in\mathcal{U}_{\mathrm{c}},\mathcal{T}\overset{\mathrm{a.s.}}{\geq}t_0}\mathbb{E}\Bigg[\mathbb{E}\Bigg[\sum_{i=1}^{q}\log_e\left(\frac{c}{c+E_{i0}}\right)$$

$$+\int_{t_0}^{\mathcal{T}}\sum_{i=1}^{q}\frac{1}{2}\frac{\mathrm{row}_i(J(E(t)))\mathrm{row}_i^{\mathrm{T}}(J(E(t)))}{(c+E_i(t))^2}\mathrm{d}t$$

$$-\int_{t_0}^{T}\sum_{i=1}^{q}\sum_{j=1,j\neq i}^{q}\frac{1}{2}\left(\frac{\phi_{ij}(E(t))}{c+E_i(t)}-\frac{\phi_{ij}(E(t))}{c+E_j(t)}\right)\mathrm{d}t|E(t_0)\overset{\mathrm{a.s.}}{=}E_0\Bigg]\Bigg]$$

$$=\sum_{i=1}^{q}\log_e\left(\frac{c+E_{i0}}{c}\right)$$

$$+\inf_{S(\cdot)\in\mathcal{U}_{\mathrm{c}},\mathcal{T}\overset{\mathrm{a.s.}}{\geq}t_0}\mathbb{E}\Bigg[\mathbb{E}\Bigg[\int_{t_0}^{\mathcal{T}}\sum_{i=1}^{q}-\frac{1}{2}\frac{1}{(c+E_i(t))^2}\Bigg[\sum_{j=1,\,j\neq i}^{q}\phi_{ij}(E(t))$$

$$\cdot[E_i(t)-E_j(t)]\frac{c+E_i(t)}{c+E_j(t)}+\mathrm{row}_i(J(E(t)))$$

$$\cdot\mathrm{row}_i^{\mathrm{T}}(J(E(t)))\Bigg]\mathrm{d}t|E(t_0)\overset{\mathrm{a.s.}}{=}E_0\Bigg]\Bigg]. \tag{12.116}$$

Since the solution $E(t)$, $t\geq t_0$, to (12.68) is nonnegative for all nonnegative initial conditions, it follows from Axiom *ii)* that the infimum in (12.116) is taken over the set of nonnegative values. However, the zero value of the infimum is achieved on an equilibrium process for which $\phi_{ij}(E(t))\overset{\mathrm{a.s.}}{\equiv}0$, $i\neq j$, $i,j=1,\ldots,q$. Thus,

$$\mathcal{S}_{\mathrm{a}}(E_0)=\mathbf{e}^{\mathrm{T}}\mathbf{log}_e(c\mathbf{e}+E_0)-q\log_e c,\quad E_0=E_{\mathrm{e}}. \tag{12.117}$$

Similarly, consider the entropy function $\mathcal{S}_{\mathrm{r}}(\cdot)$ given by (12.97). Then, it follows from (12.68) that, for $E_0=E_{\mathrm{e}}$,

$$\mathcal{S}_{\mathrm{r}}(E_0)=\sup_{S(\cdot)\in\mathcal{U}_{\mathrm{r}},\mathcal{T}\overset{\mathrm{a.s.}}{\geq}t_0}\mathbb{E}\Bigg[\mathbb{E}\Bigg[\int_{t_0}^{\mathcal{T}}\sum_{i=1}^{q}\Bigg[\frac{S_i(t)-\sigma_{ii}(E(t))}{c+E_i(t)}$$

$$-\frac{1}{2}\frac{\mathrm{row}_i(D(E(t)))\mathrm{row}_i^{\mathrm{T}}(D(E(t)))}{(c+E_i(t))^2}\Bigg]\mathrm{d}t|E(t_0)\overset{\mathrm{a.s.}}{=}0\Bigg]\Bigg]$$

$$
\begin{aligned}
&= \sup_{S(\cdot)\in\mathcal{U}_{\mathrm{r}},\,\mathcal{T}\overset{\text{a.s.}}{\geq} t_0} \mathbb{E}\left[\mathbb{E}\left[\int_{t_0}^{\mathcal{T}}\sum_{i=1}^{q}\left[\frac{\mathrm{d}E_i(t)-\sum_{j=1,j\neq i}^{q}\phi_{ij}(E(t))\mathrm{d}t}{c+E_i(t)}\right.\right.\right.\\
&\qquad \left.\left.\left.-\frac{1}{2}\frac{\mathrm{row}_i(D(E(t)))\mathrm{row}_i^{\mathrm{T}}(D(E(t)))}{(c+E_i(t))^2}\mathrm{d}t\right]\,|E(t_0)\overset{\text{a.s.}}{=}0\right]\right]\\
&= \sup_{S(\cdot)\in\mathcal{U}_{\mathrm{r}},\,\mathcal{T}\overset{\text{a.s.}}{\geq} t_0} \mathbb{E}\left[\mathbb{E}\left[\sum_{i=1}^{q}\log_e\left(\frac{c+E_{i0}}{c}\right)\right.\right.\\
&\qquad +\int_{t_0}^{\mathcal{T}}\sum_{i=1}^{q}\frac{1}{2}\frac{\mathrm{row}_i(J(E(t)))\mathrm{row}_i^{\mathrm{T}}(J(E(t)))}{(c+E_i(t))^2}\mathrm{d}t\\
&\qquad \left.\left.-\int_{t_0}^{\mathcal{T}}\sum_{i=1}^{q}\sum_{j=1,j\neq i}^{q}\frac{\phi_{ij}(E(t))}{c+E_i(t)}\mathrm{d}t|E(t_0)\overset{\text{a.s.}}{=}0\right]\right]\\
&= \sum_{i=1}^{q}\log_e\left(\frac{c+E_{i0}}{c}\right)\\
&\qquad + \sup_{S(\cdot)\in\mathcal{U}_{\mathrm{r}},\,\mathcal{T}\overset{\text{a.s.}}{\geq} t_0} \mathbb{E}\left[\mathbb{E}\left[\int_{t_0}^{\mathcal{T}}\sum_{i=1}^{q}\frac{1}{2}\frac{1}{(c+E_i(t))^2}\right.\right.\\
&\qquad \cdot\left[\sum_{j=1,\,j\neq i}^{q}\phi_{ij}(E(t))[E_i(t)-E_j(t)]\frac{c+E_i(t)}{c+E_j(t)}\right.\\
&\qquad \left.\left.\left.+\mathrm{row}_i(J(E(t)))\mathrm{row}_i^{\mathrm{T}}(J(E(t)))\right]\mathrm{d}t|E(t_0)\overset{\text{a.s.}}{=}0\right]\right].
\end{aligned}
\tag{12.118}
$$

Now, it follows from Axioms *i*) and *ii*) that the zero value of the supremum in (12.118) is achieved on an equilibrium process and thus

$$
\mathcal{S}_{\mathrm{r}}(E_0)=\mathbf{e}^{\mathrm{T}}\mathbf{log}_e(c\mathbf{e}+E_0)-q\log_e c,\quad E_0=E_{\mathrm{e}}. \tag{12.119}
$$

Finally, it follows from (12.98) that (12.115) holds. □

The next proposition shows that if (12.95) holds as an equality for some transformation starting and ending at an equilibrium point of the isolated dynamical system $\mathcal{G}$, then this transformation must lie on the equilibrium manifold $\mathcal{M}_{\mathrm{e}}$.

Proposition 12.8. Consider the large-scale stochastic dynamical system $\mathcal{G}$ with differential energy balance equation (12.68), and assume that Axioms *i*) and *ii*) hold. Let $\mathcal{S}(\cdot)$ denote an entropy of $\mathcal{G}$, and let $E:[t_0,t_1]\to\overline{\mathbb{R}}_+^q$ denote the solution to (12.68) with $E(t_0)\overset{\text{a.s.}}{=}\alpha_0\mathbf{e}$ and $E(t_1)\overset{\text{a.s.}}{=}\alpha_1\mathbf{e}$, where α_0, $\alpha_1 \geq 0$. Then

$$\mathbb{E}\left[\mathcal{S}(E(t_1))|\mathcal{F}_{t_0}\right] = \mathcal{S}(E(t_0)) + \mathbb{E}\Bigg[\int_{t_0}^{t_1}\sum_{i=1}^{q}\Bigg[\frac{S_i(t)-\sigma_{ii}(E(t))}{c+E_i(t)} - \frac{1}{2}\frac{\mathrm{row}_i(D(E(t)))\mathrm{row}_i^{\mathrm{T}}(D(E(t)))}{(c+E_i(t))^2}\Bigg]\mathrm{d}t|\mathcal{F}_{t_0}\Bigg] \quad (12.120)$$

if and only if there exists a continuous function $\alpha : [t_0, t_1] \to \overline{\mathbb{R}}_+$ such that $\alpha(t_0) = \alpha_0$, $\alpha(t_1) = \alpha_1$, and $E(t) \overset{\mathrm{a.s.}}{=} \alpha(t)\mathbf{e}$, $t \in [t_0, t_1]$.

Proof. Since $E(t_0)$ and $E(t_1)$ are equilibrium states of the isolated dynamical system $\mathcal{G}$, it follows from Proposition 12.7 that

$$\mathbb{E}\left[\mathcal{S}(E(t_1))|\mathcal{F}_{t_0}\right] - \mathcal{S}(E(t_0)) \overset{\mathrm{a.s.}}{=} q\log_e(c+\alpha_1) - q\log_e(c+\alpha_0). \quad (12.121)$$

Furthermore, it follows from (12.68) that

$$\begin{aligned}
\mathbb{E}&\Bigg[\int_{t_0}^{t_1}\Bigg[\sum_{i=1}^{q}\frac{S_i(t)-\sigma_{ii}(E(t))}{c+E_i(t)} - \frac{1}{2}\frac{\mathrm{row}_i(D(E(t)))\mathrm{row}_i^{\mathrm{T}}(D(E(t)))}{(c+E_i(t))^2}\Bigg]\mathrm{d}t|\mathcal{F}_{t_0}\Bigg] \\
&= \mathbb{E}\Bigg[\int_{t_0}^{t_1}\sum_{i=1}^{q}\frac{\mathrm{d}E_i(t) - \sum_{j=1,j\neq i}^{q}\phi_{ij}(E(t))\mathrm{d}t}{c+E_i(t)}|\mathcal{F}_{t_0}\Bigg] \\
&\quad -\mathbb{E}\Bigg[\int_{t_0}^{t_1}\sum_{i=1}^{q}\frac{1}{2}\frac{\mathrm{row}_i(D(E(t)))\mathrm{row}_i^{\mathrm{T}}(D(E(t)))}{(c+E_i(t))^2}\mathrm{d}t|\mathcal{F}_{t_0}\Bigg] \\
&= q\log_e\left(\frac{c+\alpha_1}{c+\alpha_0}\right) \\
&\quad +\mathbb{E}\Bigg[\int_{t_0}^{t_1}\sum_{i=1}^{q}\frac{1}{2}\frac{1}{(c+E_i(t))^2}\Bigg[\sum_{j=1,\,j\neq i}^{q}\phi_{ij}(E(t))[E_i(t)-E_j(t)] \\
&\quad \cdot\frac{c+E_i(t)}{c+E_j(t)} + \mathrm{row}_i(J(E(t)))\mathrm{row}_i^{\mathrm{T}}(J(E(t)))\Bigg]\mathrm{d}t|\mathcal{F}_{t_0}\Bigg]. \quad (12.122)
\end{aligned}$$

Now, it follows from Axioms *i*) and *ii*) that (12.120) holds if and only if $E_i(t) = E_j(t)$, $t \in [t_0, t_1]$, $i \neq j$, $i, j = 1, \ldots, q$, or, equivalently, there exists a continuous function $\alpha : [t_0, t_1] \to \overline{\mathbb{R}}_+$ such that $E(t) \overset{\mathrm{a.s.}}{=} \alpha(t)\mathbf{e}$, $t \in [t_0, t_1]$, $\alpha(t_0) = \alpha_0$, and $\alpha(t_1) = \alpha_1$. □

Even though it follows from Proposition 12.6 that Definition 12.9 does not provide a unique *continuous* entropy function for nonequilibrium systems, the next theorem gives a *unique, two-times continuously differentiable* entropy function for $\mathcal{G}$ for equilibrium and nonequilibrium processes. As for deterministic dynamical processes, this result answers the long-standing question of how the entropy of a nonequilibrium state of a stochastic dynamical process should be defined [195,272,318], and establishes its global existence and uniqueness.

Theorem 12.9. Consider the large-scale stochastic dynamical system $\mathcal{G}$ with differential energy balance equation (12.68), and assume that Axioms i) and ii) hold. Then the function $\mathcal{S} : \overline{\mathbb{R}}_+^q \to \overline{\mathbb{R}}_+^q$ given by

$$\mathcal{S}(E) = \mathbf{e}^{\mathrm{T}}\mathbf{log}_e(c\mathbf{e} + E) - q\log_e c, \quad E \in \overline{\mathbb{R}}_+^q, \tag{12.123}$$

where $c > 0$, is a unique (modulo a constant of integration), two-times continuously differentiable entropy function of $\mathcal{G}$. Furthermore, for $E(t) \notin \mathcal{H}_q^{\mathcal{M}_e}$, $t \geq t_0$, where $E(t)$, $t \geq t_0$, denotes the solution to (12.68) and $\mathcal{M}_e = \{E \in \overline{\mathbb{R}}_+^q : E = \alpha\mathbf{e}, \alpha \geq 0\}$, (12.123) satisfies

$$\mathbb{E}\left[\mathcal{S}(E(t_2))|\mathcal{F}_{t_1}\right] > \mathcal{S}(E(t_1)) + \mathbb{E}\left[\int_{t_1}^{t_2}\sum_{i=1}^q\left[\frac{S_i(t) - \sigma_{ii}(E(t))}{c + E_i(t)} - \frac{1}{2}\frac{\mathrm{row}_i(D(E(t)))\mathrm{row}_i^{\mathrm{T}}(D(E(t)))}{(c + E_i(t))^2}\right]\mathrm{d}t\Big|\mathcal{F}_{t_1}\right] \tag{12.124}$$

for every $t_2 \geq t_1 \geq t_0$ and $S(\cdot) \in \mathcal{U}$.

Proof. Since, by Proposition 3.1, $E(t) \geq\geq 0$, $t \geq t_0$, and $\phi_{ij}(E) = -\phi_{ji}(E)$, $E \in \overline{\mathbb{R}}_+^q$, $i \neq j$, $i, j = 1, \ldots, q$, it follows that

$$\begin{aligned}
&\mathbb{E}\left[\mathcal{S}(E(t_2))|\mathcal{F}_{t_1}\right] - \mathcal{S}(E(t_1)) \\
&= \mathbb{E}\left[\int_{t_1}^{t_2}\mathrm{d}\mathcal{S}(E(t))|\mathcal{F}_{t_1}\right] \\
&= \mathbb{E}\left[\int_{t_1}^{t_2}\sum_{i=1}^q\frac{\mathrm{d}E_i(t)}{c + E_i(t)} - \frac{1}{2}\left[\frac{\mathrm{row}_i(J(E(t)))\mathrm{row}_i^{\mathrm{T}}(J(E(t)))}{(c + E_i(t))^2}\right.\right. \\
&\qquad \left.\left. + \frac{\mathrm{row}_i(D(E(t)))\mathrm{row}_i^{\mathrm{T}}(D(E(t)))}{(c + E_i(t))^2}\right]\mathrm{d}t\Big|\mathcal{F}_{t_1}\right] \\
&= \mathbb{E}\left[\int_{t_1}^{t_2}\sum_{i=1}^q\left[\frac{S_i(t) - \sigma_{ii}(E(t))}{c + E_i(t)}\right.\right. \\
&\qquad \left.\left. - \frac{1}{2}\frac{\mathrm{row}_i(D(E(t)))\mathrm{row}_i^{\mathrm{T}}(D(E(t)))}{(c + E_i(t))^2}\right]\mathrm{d}t\Big|\mathcal{F}_{t_1}\right] \\
&\qquad + \mathbb{E}\left[\int_{t_1}^{t_2}\sum_{i=1}^q\left[\sum_{j=1,\, j\neq i}^q\frac{\phi_{ij}(E(t))}{c + E_i(t)}\right.\right. \\
&\qquad \left.\left. - \frac{1}{2}\frac{\mathrm{row}_i(J(E(t)))\mathrm{row}_i^{\mathrm{T}}(J(E(t)))}{(c + E_i(t))^2}\right]\mathrm{d}t\Big|\mathcal{F}_{t_1}\right]
\end{aligned}$$

$$
\begin{aligned}
&= \mathbb{E}\left[\int_{t_1}^{t_2} \sum_{i=1}^{q} \left[\frac{S_i(t) - \sigma_{ii}(E(t))}{c + E_i(t)}\right.\right. \\
&\quad \left.\left. -\frac{1}{2}\frac{\mathrm{row}_i(D(E(t)))\mathrm{row}_i^{\mathrm{T}}(D(E(t)))}{(c + E_i(t))^2}\right] \mathrm{d}t|\mathcal{F}_{t_1}\right] \\
&\quad -\mathbb{E}\left[\int_{t_1}^{t_2} \sum_{i=1}^{q} \frac{1}{2}\frac{1}{(c + E_i(t))^2}\left[\sum_{j=1,\, j\neq i}^{q} \phi_{ij}(E(t))[E_i(t) - E_j(t)]\frac{c + E_i(t)}{c + E_j(t)}\right.\right. \\
&\quad \left.\left. +\mathrm{row}_i(J(E(t)))\mathrm{row}_i^{\mathrm{T}}(J(E(t)))\right] \mathrm{d}t|\mathcal{F}_{t_1}\right] \\
&\geq \mathbb{E}\left[\int_{t_1}^{t_2} \sum_{i=1}^{q} \left[\frac{S_i(t) - \sigma_{ii}(E(t))}{c + E_i(t)}\right.\right. \\
&\quad \left.\left. -\frac{1}{2}\frac{\mathrm{row}_i(D(E(t)))\mathrm{row}_i^{\mathrm{T}}(D(E(t)))}{(c + E_i(t))^2}\right] \mathrm{d}t|\mathcal{F}_{t_1}\right], \quad t \geq t_0. \qquad (12.125)
\end{aligned}
$$

Furthermore, in the case where $E(t) \notin \mathcal{H}_q^{\mathcal{M}_{\mathrm{e}}}$, $t \geq t_0$, it follows from Axiom *i*), Axiom *ii*), and (12.125) that (12.124) holds.

To show that (12.123) is a unique, two-times continuously differentiable entropy function of $\mathcal{G}$, let $\mathcal{S}(E)$ be a continuously differentiable entropy function of $\mathcal{G}$ so that $\mathcal{S}(E)$ satisfies (12.95) or, equivalently,

$$
\mathfrak{L}\mathcal{S}(E) \geq \mu_1^{\mathrm{T}}(E)[S - d(E)] - \frac{1}{2}\mathrm{tr}\, \mu_2(E)D(E)D^{\mathrm{T}}(E), \quad E \in \overline{\mathbb{R}}_+^q, \quad S \in \mathbb{R}^q, \qquad (12.126)
$$

where $\mu_1^{\mathrm{T}}(E) = [\frac{1}{c+E_1}, \ldots, \frac{1}{c+E_q}]$ and $\mu_2(E) = \mathrm{diag}[\frac{1}{(c+E_1)^2}, \ldots, \frac{1}{(c+E_q)^2}]$, $E \in \overline{\mathbb{R}}_+^q$, $E(t)$, $t \geq t_0$, denotes the solution to the differential energy balance equation (12.68), and $\mathfrak{L}\mathcal{S}(E(t))$ denotes the infinitesimal generator of $\mathcal{S}(E)$ along the solution $E(t)$, $t \geq t_0$. Hence, it follows from (12.126) that

$$
\begin{aligned}
&\mathcal{S}'(E)[f(E) - d(E) + S] + \frac{1}{2}\mathrm{tr}\, \mathcal{S}''(E)[J(E)J^{\mathrm{T}}(E) + D(E)D^{\mathrm{T}}(E)] \\
&\quad \geq \mu_1^{\mathrm{T}}(E)[S - d(E)] - \frac{1}{2}\mathrm{tr}\, \mu_2(E)D(E)D^{\mathrm{T}}(E), \quad E \in \overline{\mathbb{R}}_+^q, \quad S \in \mathbb{R}^q, \qquad (12.127)
\end{aligned}
$$

which implies that there exist continuous functions $\ell : \overline{\mathbb{R}}_+^q \to \mathbb{R}^p$ and $\mathcal{W} : \overline{\mathbb{R}}_+^q \to \mathbb{R}^{p\times q}$ such that

$$
\begin{aligned}
0 &= \mathcal{S}'(E)[f(E) - d(E) + S] + \frac{1}{2}\mathrm{tr}\, \mathcal{S}''(E)[J(E)J^{\mathrm{T}}(E) + D(E)D^{\mathrm{T}}(E)] \\
&\quad -\mu_1^{\mathrm{T}}(E)[S - d(E)] + \frac{1}{2}\mathrm{tr}\, \mu_2(E)D(E)D^{\mathrm{T}}(E)
\end{aligned}
$$

$$-[\ell(E)+\mathcal{W}(E)S]^{\mathrm{T}}[\ell(E)+\mathcal{W}(E)S], \quad E\in\overline{\mathbb{R}}_{+}^{q}, \quad S\in\mathbb{R}^{q}. \quad (12.128)$$

Now, equating coefficients of equal powers (of S and D), it follows that $\mathcal{W}(E)\equiv 0$, $\mathcal{S}'(E)=\mu^{\mathrm{T}}(E)$, $\mathcal{S}''(E)=-\mu_2(E)$, $E\in\overline{\mathbb{R}}_{+}^{q}$, and

$$0=\mathcal{S}'(E)f(E)+\frac{1}{2}\mathrm{tr}\,\mathcal{S}''(E)J(E)J^{\mathrm{T}}(E)-\ell^{\mathrm{T}}(E)\ell(E), \quad E\in\overline{\mathbb{R}}_{+}^{q}. \quad (12.129)$$

Hence, $\mathcal{S}(E)=\mathbf{e}^{\mathrm{T}}\mathbf{log}_e(c\mathbf{e}+E)-q\log_e c$, $E\in\overline{\mathbb{R}}_{+}^{q}$, and

$$0=\mu_1^{\mathrm{T}}(E)f(E)-\frac{1}{2}\mathrm{tr}\,\mu_2(E)J(E)J^{\mathrm{T}}(E)-\ell(E)\ell^{\mathrm{T}}(E), \quad E\in\overline{\mathbb{R}}_{+}^{q}. \quad (12.130)$$

Thus, (12.123) is a unique, two-times continuously differentiable entropy function for $\mathcal{G}$. □

Note that it follows from Axiom *i*), Axiom *ii*), and the last equality in (12.125) that the entropy function given by (12.123) satisfies (12.95) as an equality for an equilibrium process and as a strict inequality for a nonequilibrium process. Furthermore, for any entropy function of $\mathcal{G}$, it follows from Proposition 12.8 that if (12.95) holds as an equality for some transformation starting and ending at equilibrium points of the isolated system $\mathcal{G}$, then this transformation must lie on the equilibrium manifold $\mathcal{M}_{\mathrm{e}}$. However, (12.95) may hold as an equality for nonequilibrium processes starting and ending at nonequilibrium states.

The entropy expression given by (12.123) is identical in form to the Boltzmann entropy for statistical thermodynamics. Due to the fact that the entropy given by (12.123) is indeterminate to the extent of an additive constant, we can place the constant of integration $q\log_e c$ to zero by taking $c=1$. Since $\mathcal{S}(E)$ given by (12.123) achieves a maximum when all the subsystem energies E_i, $i=1,\ldots,q$, are equal, the entropy of $\mathcal{G}$ can be thought of as a measure of the tendency of a system to lose the ability to do useful work, lose order, and settle to a more homogeneous state.

Recalling that $\mathbb{E}\,[\mathrm{d}Q_i(t)|\mathcal{F}_t]=[S_i(t)-\sigma_{ii}(E(t))]\mathrm{d}t$, $i=1,\ldots,q$, is the infinitesimal amount of the net heat received or dissipated by the ith subsystem of $\mathcal{G}$ over the infinitesimal time interval $\mathrm{d}t$, it follows from (12.95) that

$$\mathbb{E}\,[\mathrm{d}\mathcal{S}(E(t))|\mathcal{F}_t]\geq\sum_{i=1}^{q}\left[\frac{\mathrm{d}Q_i(t)}{c+E_i(t)}-\frac{1}{2}\frac{\mathrm{row}_i(D(E(t)))\mathrm{row}_i^{\mathrm{T}}(D(E(t)))}{(c+E_i(t))^2}\right], \quad t\geq t_0. \quad (12.131)$$

Inequality (12.131) is analogous to the classical thermodynamic inequality for the variation of entropy during an infinitesimal irreversible transformation with the shifted subsystem energies $c+E_i$ playing the role of the ith subsystem thermodynamic (absolute) temperatures.

It is important to note that as in Chapter 3, in this chapter we view subsystem temperatures to be synonymous with subsystem energies. Even though this does not limit the generality of our theory from a mathematical perspective, it can be physically limiting since it does not allow for the consideration of two subsystems of $\mathcal{G}$ having the same stored energy with one of the subsystems being at a higher temperature (i.e., *hotter*) than the other. This, however, can be easily addressed by assigning different specific heats (i.e., thermal capacities) for each of the compartments of the large-scale system $\mathcal{G}$ as shown in Chapter 4.

12.7 Stochastic Semistability and Energy Equipartition

For the (adiabatically) isolated large-scale stochastic dynamical system $\mathcal{G}$, (12.95) yields the fundamental inequality

$$\mathbb{E}\left[\mathcal{S}(E(\tau_2))|\mathcal{F}_{\tau_1}\right] \geq \mathcal{S}(E(\tau_1)), \quad \tau_2 \overset{\text{a.s.}}{\geq} \tau_1. \tag{12.132}$$

Inequality (12.132) implies that, for any dynamical change in an adiabatically isolated large-scale stochastic dynamical system $\mathcal{G}$, the entropy of the final state can never be less than the entropy of the initial state. Inequality (12.132) is the second law of thermodynamics for stochastic systems and gives a statement about entropy increase. It is important to stress that this result holds for an adiabatically isolated stochastic dynamical system. It is, however, possible with power (heat flux) supplied from an external system to reduce the entropy of the dynamical system $\mathcal{G}$. The entropy of both systems taken together, however, cannot decrease.

This observation implies that when the isolated large-scale dynamical system $\mathcal{G}$ with thermodynamically consistent energy flow characteristics (i.e., Axioms *i*) and *ii*) hold) is at a state of maximum entropy consistent with its energy, it cannot be subject to any further dynamical change since any such change would result in a decrease of entropy. This of course implies that the state of *maximum entropy* is the stable state of an isolated system, and this equilibrium state has to be stochastically semistable.

Theorem 12.10. Consider the large-scale stochastic dynamical system $\mathcal{G}$ with differential energy balance equation (12.68) with $S(t) \overset{\text{a.s.}}{\equiv} 0$, $D(E(t)) \overset{\text{a.s.}}{\equiv} 0$, and $d(E(t)) \overset{\text{a.s.}}{\equiv} 0$, and assume that Axioms *i*) and *ii*) hold. Then, for every $\alpha \geq 0$, $\alpha\mathbf{e}$ is a stochastic semistable equilibrium state of (12.68). Furthermore, $E(t) \overset{\text{a.s.}}{\to} \frac{1}{q}\mathbf{e}\mathbf{e}^{\mathrm{T}}E(t_0)$ as $t \to \infty$ and $\frac{1}{q}\mathbf{e}\mathbf{e}^{\mathrm{T}}E(t_0)$ is a semistable equilibrium state. Finally, if for some $k \in \{1, \ldots, q\}$, $\sigma_{kk}(E) \geq 0$,

$E \in \overline{\mathbb{R}}_+^q$, and $\sigma_{kk}(E) = 0$ if and only if $E_k = 0$,[5] then the zero solution $E(t) \equiv 0$ to (12.68) is an equilibrium state of (12.68) that is globally asymptotically stable in probability.

Proof. It follows from Axioms i) and ii) that $\alpha\mathbf{e} \in \overline{\mathbb{R}}_+^q$, $\alpha \geq 0$, is an equilibrium state of (12.68). To show Lyapunov stability of the equilibrium state $\alpha\mathbf{e}$, consider $V(E) = \frac{1}{2}(E - \alpha\mathbf{e})^{\mathrm{T}}(E - \alpha\mathbf{e})$ as a Lyapunov function candidate. Note that for $c >> \max\{E_i, E_j\}$, $i \neq j$, $i, j = 1, \ldots, q$,

$$\frac{c + E_i}{c + E_j} = \frac{1 + E_i/c}{1 + E_j/c} \approx 1. \tag{12.133}$$

Since Axiom ii) holds for all $c > 0$, we have

$$\sum_{j=1,\, j\neq i}^{q} (E_i - E_j)\phi_{ij}(E) \leq -\mathrm{row}_i(J(E))\mathrm{row}_i^{\mathrm{T}}(J(E)), \quad i = 1, \ldots, q. \tag{12.134}$$

Now, since $\phi_{ij}(E) = -\phi_{ji}(E)$, $E \in \overline{\mathbb{R}}_+^q$, $i \neq j$, $i, j = 1, \ldots, q$, $\mathbf{e}^{\mathrm{T}} f(E) = 0$, $E \in \overline{\mathbb{R}}_+^q$, and $\mathbf{e}^{\mathrm{T}} J(E) = 0$, $E \in \overline{\mathbb{R}}_+^q$, it follows from (12.134) that

$$\begin{aligned}
\mathfrak{L}V(E) &= (E - \alpha\mathbf{e})^{\mathrm{T}} f(E) + \frac{1}{2}\mathrm{tr} J(E) J^{\mathrm{T}}(E) \\
&= E^{\mathrm{T}} f(E) + \frac{1}{2} J(E) J^{\mathrm{T}}(E) \\
&= \sum_{i=1}^{q} E_i \left[\sum_{j=1,\, j\neq i}^{q} \phi_{ij}(E) \right] + \frac{1}{2} \sum_{i=1}^{q} \mathrm{row}_i(J(E))\mathrm{row}_i^{\mathrm{T}}(J(E)) \\
&= \frac{1}{2} \sum_{i=1}^{q} \left[\sum_{j=1,\, j\neq i}^{q} (E_i - E_j)\phi_{ij}(E) + \mathrm{row}_i(J(E))\mathrm{row}_i^{\mathrm{T}}(J(E)) \right] \\
&\leq 0, \quad E \in \overline{\mathbb{R}}_+^q,
\end{aligned} \tag{12.135}$$

which establishes Lyapunov stability in probability of the equilibrium state $\alpha\mathbf{e}$.

To show that $\alpha\mathbf{e}$ is stochastically semistable, let $\mathcal{R} \triangleq \{E \in \overline{\mathbb{R}}_+^q : \mathfrak{L}V(E) = 0\}$. Now, by Axioms i) and ii) the directed graph associated with the connectivity matrix $\mathcal{C}$ for the large-scale dynamical system $\mathcal{G}$ is strongly connected, which implies that $\mathcal{R} = \{E \in \overline{\mathbb{R}}_+^q : E_1 = \cdots = E_q\}$. Since $\overline{\mathbb{R}}_+^q$ is invariant and $V(E)$ is radially unbounded, it follows from Theorem 12.2 that, for every initial condition $E(t_0) \in \overline{\mathbb{R}}_+^q$, $E(t) \overset{\text{a.s.}}{\to} \mathcal{R}$ as $t \to \infty$,

[5] As noted in Chapter 3, the assumption $\sigma_{kk}(E) \geq 0$, $E \in \overline{\mathbb{R}}_+^q$, and $\sigma_{kk}(E) = 0$ if and only if $E_k = 0$ for some $k \in \{1, ..., q\}$ implies that if the kth subsystem possesses no energy, then this subsystem cannot dissipate energy to the environment. Conversely, if the kth subsystem does not dissipate energy to the environment, then this subsystem has no energy.

and hence, $\alpha\mathbf{e}$ is a stochastic semistable equilibrium state of (12.68). Next, note that since $\mathbf{e}^{\mathrm{T}}E(t) = \mathbf{e}^{\mathrm{T}}E(t_0)$ and $E(t) \stackrel{\text{a.s.}}{\to} \mathcal{R}$ as $t \to \infty$, it follows that $E(t) \stackrel{\text{a.s.}}{\to} \frac{1}{q}\mathbf{e}\mathbf{e}^{\mathrm{T}}E(t_0)$ as $t \to \infty$. Hence, with $\alpha = \frac{1}{q}\mathbf{e}^{\mathrm{T}}E(t_0)$, $\alpha\mathbf{e} = \frac{1}{q}\mathbf{e}\mathbf{e}^{\mathrm{T}}E(t_0)$ is a semistable equilibrium state of (12.68).

To show that in the case where for some $k \in \{1,\ldots,q\}$, $\sigma_{kk}(E) \geq 0$, $E \in \overline{\mathbb{R}}_+^q$, and $\sigma_{kk}(E) = 0$ if and only if $E_k = 0$, the zero solution $E(t) \equiv 0$ to (12.68) is globally asymptotically stable in probability, consider $V(E) = \frac{1}{2}E^{\mathrm{T}}E$, $E \in \overline{\mathbb{R}}_+^q$, as a candidate Lyapunov function. Note that $V(0) = 0$, $V(E) > 0$, $E \in \overline{\mathbb{R}}_+^q$, $E \neq 0$, and $V(E)$ is radially unbounded. Now, the infinitesimal generator of the Lyapunov function along the system energy trajectories of (12.68) is given by

$$
\begin{aligned}
&\mathfrak{L}V(E)\\
&\quad = E^{\mathrm{T}}[f(E) - d(E)] + \frac{1}{2}\mathrm{tr}J(E)J^{\mathrm{T}}(E)\\
&\quad = E^{\mathrm{T}}f(E) + \frac{1}{2}\mathrm{tr}J(E)J^{\mathrm{T}}(E) - E_k\sigma_{kk}(E)\\
&\quad = \sum_{i=1}^{q} E_i\left[\sum_{j=1, j\neq i}^{q}\phi_{ij}(E)\right] + \frac{1}{2}\sum_{i=1}^{q}\mathrm{row}_i(J(E))\mathrm{row}_i^{\mathrm{T}}(J(E)) - E_k\sigma_{kk}(E)\\
&\quad = \frac{1}{2}\sum_{i=1}^{q}\left[\sum_{j=1, j\neq i}^{q}(E_i - E_j)\phi_{ij}(E) + \mathrm{row}_i(J(E))\mathrm{row}_i^{\mathrm{T}}(J(E))\right] - E_k\sigma_{kk}(E)\\
&\quad \leq 0, \quad E \in \overline{\mathbb{R}}_+^q,
\end{aligned}
\tag{12.136}
$$

which shows that the zero solution $E(t) \stackrel{\text{a.s.}}{\equiv} 0$ to (12.68) is Lyapunov stable in probability.

Finally, to show global asymptotic stability in probability of the zero equilibrium state, let $\mathcal{R} \triangleq \{E \in \overline{\mathbb{R}}_+^q : \mathfrak{L}V(E) = 0\}$. Now, since Axiom i) holds and $\sigma_{kk}(E) = 0$ if and only if $E_k = 0$, it follows that $\mathcal{R} = \{E \in \overline{\mathbb{R}}_+^q : E_k = 0, k \in \{1,\ldots,q\}\} \cap \{E \in \overline{\mathbb{R}}_+^q : E_1 = E_2 = \cdots = E_q\} = \{0\}$. Hence, it follows from Theorem 12.2 that for every initial condition $E(t_0) \in \overline{\mathbb{R}}_+^q$, $E(t) \stackrel{\text{a.s.}}{\to} \mathcal{R} = \{0\}$ as $t \to \infty$, which proves global asymptotic stability in probability of the zero equilibrium state of (12.68). □

In Theorem 3.9 we used the energy Lyapunov function to show that for the isolated (i.e., $S(t) \stackrel{\text{a.s.}}{\equiv} 0$, $d(E) \stackrel{\text{a.s.}}{\equiv} 0$, and $D(E) \stackrel{\text{a.s.}}{\equiv} 0$) large-scale stochastic dynamical system $\mathcal{G}$, $E(t) \stackrel{\text{a.s.}}{\to} \frac{1}{q}\mathbf{e}\mathbf{e}^{\mathrm{T}}E(t_0)$ as $t \to \infty$ and $\frac{1}{q}\mathbf{e}\mathbf{e}^{\mathrm{T}}E(t_0)$ is a stochastic semistable equilibrium state. This result can also be arrived at using the system entropy. Specifically, using the system entropy given by (12.123), we can show attraction of the system trajectories to Lyapunov in probability stable equilibrium points $\alpha\mathbf{e}$, $\alpha \geq 0$, and hence, show stochastic semistability of these equilibrium states.

To see this, note that since $\mathbf{e}^{\mathrm{T}} f(E) = 0$, $E \in \overline{\mathbb{R}}_+^q$, and $\mathbf{e}^{\mathrm{T}} J(E) = 0$, $E \in \overline{\mathbb{R}}_+^q$, it follows that $\mathbf{e}^{\mathrm{T}} \mathrm{d}E(t) = 0$, $t \geq t_0$. Hence, $\mathbf{e}^{\mathrm{T}} E(t) \overset{\text{a.s.}}{=} \mathbf{e}^{\mathrm{T}} E(t_0)$, $t \geq t_0$. Furthermore, since $E(t) \geq\geq 0$, $t \geq t_0$, it follows that $0 \leq\leq E(t) \leq\leq \mathbf{e}\mathbf{e}^{\mathrm{T}} E(t_0)$ a.s., $t \geq t_0$, which implies that all solutions to (12.68) are almost surely bounded. Next, since by (12.125) the function $-\mathcal{S}(E(t))$, $t \geq t_0$, is a supermartingale and $E(t)$, $t \geq t_0$, is bounded, it follows from Theorem 12.2 that for every initial condition $E(t_0) \in \overline{\mathbb{R}}_+^q$, $E(t) \overset{\text{a.s.}}{\to} \mathcal{R}$ as $t \to \infty$, where $\mathcal{R} \triangleq \{E \in \overline{\mathbb{R}}_+^q : \mathfrak{L}\mathcal{S}(E) = 0\}$.

It now follows from the last inequality of (12.125) that $\mathcal{R} = \{E \in \overline{\mathbb{R}}_+^q : (E_i - E_j)\phi_{ij}(E) = 0, i = 1, \ldots, q, j \in \mathcal{K}_i\}$, which, since the directed graph associated with the connectivity matrix $\mathcal{C}$ for the large-scale dynamical system $\mathcal{G}$ is strongly connected, implies that $\mathcal{R} = \{E \in \overline{\mathbb{R}}_+^q : E_1 = \cdots = E_q\}$. Since the set $\mathcal{R}$ consists of the equilibrium states of (12.68), it follows that $\mathcal{M} = \mathcal{R}$, which, along with (12.135), establishes stochastic semistability of the equilibrium states $\alpha\mathbf{e}$, $\alpha \geq 0$.

Theorem 3.9 implies that the steady-state value of the energy in each subsystem $\mathcal{G}_i$ of the isolated stochastic large-scale dynamical system $\mathcal{G}$ is equal, that is, the steady-state energy of the isolated large-scale stochastic dynamical system $\mathcal{G}$ given by

$$E_\infty \overset{\text{a.s.}}{=} \frac{1}{q}\mathbf{e}\mathbf{e}^{\mathrm{T}} E(t_0) \overset{\text{a.s.}}{=} \left[\frac{1}{q}\sum_{i=1}^{q} E_i(t_0)\right]\mathbf{e} \tag{12.137}$$

is uniformly distributed over all subsystems of $\mathcal{G}$. This is of course *equipartition of energy* [35,37,204,297,354] and, as for deterministic thermodynamic systems, is an emergent behavior in stochastic thermodynamic systems.

Finally, it is important to note that using the realizations of each sample path of the stochastic energy variables characterized by the stochastic differential energy balance dynamical model, we can describe the probability density function of our large-scale stochastic thermodynamic model by a continuous-time and continuous-space Fokker-Planck evolution equation to give a thermodynamic interpretation between the stationary solution of the Fokker-Planck equation and the canonical thermodynamic equilibrium distribution. Since our stochastic thermodynamic model does not restrict the second law on each individual sample path trajectory, this formulation can give the second law as a fluctuation theorem to give a precise prediction of the cases in which the system entropy decreases over a given time interval for our model.

Chapter Thirteen

Relativistic Mechanics

13.1 Introduction

Our universe is the quintessential large-scale dynamical system, and classical physics has provided a rigorous underpinning of its governing laws that are consistent with human experience and intuition. Newtonian mechanics and Lagrangian and Hamiltonian dynamics provide a rigorous mathematical framework for classical mechanics—the oldest of the physical sciences. The theory of Newtonian mechanics embodies the science of kinematics and dynamics of moving bodies and embedded within its laws is the Galilean principle of relativity, which states that the laws of mechanics are equally valid for all observers moving uniformly with respect to each other.

This invariance principle of mechanics, which was known from the time of Galileo, postulated that the laws of mechanics are equivalent in all inertial (i.e., nonaccelerating) frames. In other words, all inertial observers undergoing uniform relative motion would observe the same laws of mechanics involving the conservation of energy, conservation of linear momentum, conservation of angular momentum, etc., while possibly measuring different values of velocities, energies, momenta, etc.

In the late seventeenth century, Newton went on to make his greatest discovery—the law of universal gravitation. He postulated that gravity exists in all bodies universally, and showed that bodies attract each other with a force that is directly proportional to the product of their masses and inversely proportional to the square of their separation. He maintained that this universal principle pervades the entire universe affecting everything in it, and its influence is direct, *immediate*, and definitive. With this law of universal gravitation, Newton demonstrated that terrestrial physics and celestial physics can be derived from the same basic principles and are part of the same scientific discipline.

Even though the laws of classical (i.e., Newtonian) mechanics provide an isomorphism between an idealized, exact description of the Keplerian motion of heavenly bodies in our solar system deduced from direct astronomical observations, wherein the circular, elliptical, and parabolic orbits of these bodies are no longer fundamental determinants of motion, the integral concepts of space and time necessary in forming the underpinning of classical mechanics are flawed resulting in an approximation of the governing universal laws of mechanics. In particular, in his *Principia Mathematica* [337] Newton asserted that space and time are absolute and unchanging entities in the fabric of the cosmos.

Specifically, he writes: "I do not define time, space, and motion, as being well known to all .... Absolute, true and mathematical time, of itself and from its own nature, flows equably without relation to anything external, and by another name is called duration." He goes on to state: "Absolute space, in its own nature, without reference to anything external, remains always similar and unmovable." The absolute structure of time and space of Newtonian mechanics asserts that a body being at rest is at rest with respect to absolute space; and a body is accelerating when it is accelerating with respect to absolute space.

In classical mechanics, Kepler, Galileo, Newton, and Huygens sought to completely identify the fundamental laws that would define the kinematics and dynamics of moving bodies. These laws have become known as Newton's laws of motion and in the seventeenth through the nineteenth centuries the physical approach of classical mechanics was replaced by mathematical theories involving abstract theoretical structures by giants such as Euler, Lagrange, and Hamilton.

However, what has become known as the most significant scientific event of the nineteenth century is Maxwell's discovery of the laws of electrodynamics, which sought to do the same thing for electromagnetism as what Newton's laws of motion did for classical mechanics. Maxwell's equations completely characterize the electric and magnetic fields arising from distributions of electric charges and currents, and describe how these fields change in time [313]. His equations describe how changing magnetic fields produce electric fields, assert the nonexistence of magnetic monopoles, and describe how electric and magnetic fields interact and propagate.[1]

Maxwell's mathematical grounding of electromagnetism included Gauss' flux theorem [159], relating the distribution of electric charge to the resulting electric field, and cemented Faraday's profound intuitions based on decades of experimental observations [142]. Faraday was the first to

[1]As in classical mechanics, which serves as an approximation theory to the theory of relativity, classical electromagnetism predicated on Maxwell's field equations is an approximation to the relativistic field theory of electrodynamics describing how light and matter interact through the exchange of photons.

recognize that changes in a magnetic force produce electricity, and the amount of electricity that is produced is proportional to the rate of change of the magnetic force. In Maxwell's equations the laws of Gauss and Faraday remain unchanged. However, Ampère's theorem—asserting that magnetic circulation along concentric paths around a straight wire carrying a current is unchanged and is proportional to the current through the permittivity of free space—is modified to account for varying electric fields through the introduction of a displacement current [312]. As in classical mechanics, wherein Euler provided the mathematical formulation of Newton's $F = ma$,[2] Maxwell provided the mathematical foundation of Faraday's research, which lacked mathematical sophistication.

The theory of electromagnetism explained the way a changing magnetic field generates an electric current, and conversely the way an electric field generates a magnetic field. This coupling between electricity and magnetism demonstrated that these physical entities are simply different aspects of a single physical phenomenon—electromagnetism—and not two distinct phenomena as was previously thought. This unification between electricity and magnetism further led to the realization that an electric current through a wire produces a magnetic field around the wire, and conversely a magnetic field induced around a closed loop is proportional to the electric current and the displacement current it encloses. Hence, electricity in motion results in magnetism, and magnetism in motion produces electricity. This profound insight led to experiments showing that the displacement of electricity is relative; absolute motion cannot be determined by electrical and magnetic experiments.

In particular, in the mid-nineteenth century scientists performed an electrical and magnetic experiment, wherein the current in a conductor moving at a constant speed with respect to the magnet was calculated with respect to the reference frame of the magnet and the reference frame of the conductor. In either case, the current was the same, indicating that only relative motion matters and there is no absolute standard of rest. Specifically, the magnet displaced through a loop of wire at a fixed speed relative to the wire produced the same current as when the loop of wire was

---

[2]Even though Newton stated that "A change in motion is proportional to the motive force impressed and takes place along the straight line in which the force is impressed," it was Euler, almost a century later, who expressed this statement as a mathematical equation involving force and change in momentum. Contrary to mainstream perception, the mathematical formalism and rigor used by Newton was elementary as compared to the mathematics of Euclid, Apollonios, and Archimedes. One needs only to compare how Newton presents his formulation on the limit of the ratio of two infinitesimals, which he calls the "ultimate proportion of evanescent quantities," in [337, Book I, Section I], with Archimedes' Proposition 5 *On Spirals* [330, pp. 17–18], where he uses infinitesimals of different orders to determine the tangential direction of an arbitrary point of a spiral. This comparison clearly shows that Newton lacked the mathematical sophistication developed two thousand years earlier by Hellenistic mathematicians. This is further substantiated by comparing the mathematical formalism in [337] with [329–332].

moved around the magnet in the reverse direction and at the same speed.

In other words, if the relative speed of the magnet and loop of wire is the same, then the current is the same irrespective of whether the magnet or the wire is moving. This implies that there does not exist a state of absolute rest since it is impossible for an observer to deduce whether the magnet or the conductor (i.e., the wire) is moving by observing the current. If the same experiment is repeated in an inertial frame, then there would be no way to determine the speed of the moving observer from the experiment.

However, it follows from Maxwell's equations that the charge in the conductor will be subject to a magnetic force in the reference frame of the conductor and an electric force in the reference frame of the observer, leading to two different descriptions of the same phenomenon depending on the observer reference frame. This inconsistency is due to the fact that Newtonian mechanics is predicated on a Galilean invariance transformation for the *forces* driving the charge that produces the current, whereas Maxwellian electrodynamics predicts that the *fields* generating these forces transform according to a Lorentzian invariance.

Maxwell's equations additionally predicted that perturbations to electromagnetic fields propagate as electromagnetic waves of electrical and magnetic forces traveling through space at the speed of light. From this result, Maxwell proposed that light is a traveling wave of electromagnetic energy.[3] This led to numerous scientists searching for the material medium needed to transmit light waves from a source to a receiver. In conforming with the translation invariance principle of Newtonian mechanics requiring the existence of an absolute time and an absolute space, physicists reasoned, by analogy to classical wave propagation theory requiring a material medium for wave transmission, that light travels in a *luminiferous aether*, which defines the state of absolute rest for electromagnetic phenomena.

Numerous experiments[4] that followed in attempting to determine

[3]Later Planck and Einstein modified this view of the nature of light to one involving a wave-particle duality. Einstein's photon (particle) theory of light [126] asserted that energy flow is not continuous but rather evolves in indivisible packets or *quanta*, and light behaves at times as a wave and at other times as a particle. And this behavior depends on what an observer chooses to measure. This wave-particle duality of the nature of light led to the foundations of quantum physics, the Heisenberg uncertainty principle, and the demise of determinism in the microcosm of science.

[4]The most famous of these experiments was the Michelson-Morley experiment performed in the spring and summer of 1887 [322]. In an attempt to detect the relative motion of matter through the stationary luminiferous aether and find a state of absolute rest for electromagnetic phenomena, Michelson and Morley compared the speed of light in perpendicular directions. They could not find any difference in the speed of electromagnetic waves in any direction in the presumed aether. Over the years, many Michelson-Morley–type experiments have been performed with increased sensitivity and all resulting in negative results, ruling out the existence of a stationary aether.

a state of absolute rest for electromagnetic phenomena showed that the velocity of light is isotropic. In other words, the speed of light in free space is invariant regardless of the motion of the observer or that of the light's source. This result highlighted the inconsistency between the principle of relative motion and the principle of the consistency of the speed of light, leading to one of science's most profound repudiations of Occam's razor.

With the advent of the special theory of relativity [127, 132], Einstein dismantled Newton's notions of absolute time and absolute space. In particular, Einstein developed the special theory of relativity from two postulates. He conjectured that the laws of physics (and not just mechanics) are the same in all inertial (i.e., nonaccelerating) frames; no preferred inertial frame exists, and hence, observers in constant motion are completely equivalent. In other words, all observers moving at uniform speed will arrive at identical laws of physics. In addition, he conjectured that the speed of light[5] c is the same in all inertial frames, and hence, the speed of light in free space is the same as measured by all observers, independent of their own speed of motion. Einstein's second postulate asserts that light does not adhere to a particle-motion model, but rather light is an electromagnetic wave described by Maxwell's field equations, with the caveat that no medium against which the speed of light is fixed is necessary.

The derivation of special relativity makes use of several additional assumptions, including spatial homogeneity (i.e., translation invariance of space), spatial isotropy (i.e., rotational invariance of free space), and memorylessness (i.e., past actions cannot be inferred from invariants). These assumptions are assertions about the structure of spacetime. Namely, space and time form a four-dimensional continuum and there exists a *global* inertial frame that covers all space and time, and hence, the geometry of spacetime is flat. (In general relativity, this assumption is replaced with a weaker postulate asserting the existence of a *local* spacetime.) Furthermore, Einstein asserted that Newtonian mechanics hold for slow velocities so that the energy-momentum four-vector in special relativity relating temporal and spatial components collapses to a one energy scalar and one momentum three-vector.

Even though in special relativity velocities, distances across space, and time intervals are relative, *spacetime* is an absolute concept, and hence, relativity theory is an invariance theory. Specifically, space and time are intricately coupled and cannot be defined separately from each other leading to a consistency of the speed of light regardless of the observer's velocity. Hence, a body moves through space as well as through time, wherein a stationary (i.e., not moving through space) body's motion is entirely motion

[5] The currently accepted value of the speed of light is 2.99792458×10^8 m/s.

through time. And as the body speeds away from a stationary observer, its motion through time is transformed into motion through space, slowing down its motion through time for an observer moving with the body relative to the stationary observer.

In other words, time slows down for the moving observer relative to the stationary observer leading to what is known as the *time dilation* principle. Thus, special relativity asserts that the combined speed of a body's motion through spacetime, that is, its motion through space and its motion through time, is always equal to the speed of light. Hence, the maximum speed of the body through space is attained when all the light speed motion through time is transformed into light speed motion through space.

The bifurcation between Newtonian mechanics (absolute space and time) and Einsteinian mechanics (absolute spacetime) is observed when a body's speed through space is a significant fraction of the speed of light, with time stopping when the body is traveling at light speed through space. Special relativity leads to several counterintuitive observable effects, including time dilation, length contraction, relativistic mass, mass-energy equivalence, speed limit universality, and relativity of simultaneity [378,433].

Time dilation involves the non-invariance between time intervals of two events from one observer to another; length contraction corresponds to motion affecting measurements; and relativity of simultaneity corresponds to nonsynchronization of simultaneous events in inertial frames that are moving relative to one another. Furthermore, in contrast to classical mechanics predicated on the Galilean transformation of velocities, wherein relative velocities add, relativistic velocity addition predicated on the Lorentz transformations results in the speed of light as the universal speed limit, which further implies that no information signal or matter can travel faster than light in a vacuum. This results in the fact that, to all observers, cause will precede effect, upholding the principle of causality.

The universal principle of the special theory of relativity is embedded within the postulate that the laws of physics are invariant with respect to the Lorentz transformation characterizing coordinate transformations between two reference frames that are moving at constant velocity relative to each other. In each reference frame, an observer measures distance and time intervals with respect to a local coordinate system and a local clock, and the transformations connect the space and time coordinates of an event as measured by the observer in each frame. At relativistic speeds the Lorentz transformation displaces the Galilean transformation of Newtonian physics based on absolute space and time, and forms a one-parameter group of linear mappings.[6] This group is described by the Lorentz transformation wherein a spacetime event is fixed at the origin and can be considered as a hyperbolic rotation of a Minkowski (i.e., flat) spacetime with the spacetime geometry measured by a Minkowski metric.

The special theory of relativity applies to physics with weak gravitational fields wherein the geometric curvature of spacetime due to gravity is negligible. In the presence of a strong gravitational field, special relativity is inadequate in capturing cosmological physics. In this case, the general theory of relativity is used to capture the fundamental laws of physics, wherein a non-Euclidean Riemannian geometry is invoked to address gravitational effects on the curvature of spacetime [131, 325].

In Riemannian spacetime a four-dimensional, semi-Riemannian manifold[7] is used to represent spacetime and a semi-Riemannian metric tensor is used to define the measurements of lengths and angles in spacetime admitting a Levi-Civita connection,[8] which locally approximates a Minkowski spacetime. In other words, over small inertial coordinates the metric is Minkowskian with vanishing first partial derivatives and Levi-Civita connection coefficients. This specialization leads to special relativity approximating general relativity in the case of weak gravitational fields, that is, for sufficiently small scales and in coordinates of free fall.

Unlike the inverse-square law of Newton's universal gravitation, wherein gravity is a force whose influence is instantaneous and its source is mass, in the general theory of relativity absolute time and space do not exist, and gravitation is not a force but rather a property of spacetime, with spacetime regulating the motion of matter and matter regulating the curvature of spacetime. Hence, the effect of gravity can only travel through space at the speed of light and is *not* instantaneous.

The mathematical formulation of the general theory of relativity is characterized by Einstein's field equations [131], which involve a system of ten coupled nonlinear partial differential equations relating the presence of matter and energy to the curvature of spacetime. The solution of

---

[6]The Lorentz transformations only describe transformations wherein the spacetime event at the origin of the coordinate system is fixed, and hence, they are a special case of the Poincaré group of symmetry transformations, which include translation of the origin.

[7]A semi-Riemannian manifold is a generalization of a Riemannian manifold (i.e., a real smooth manifold endowed with an inner product on a tangent space) wherein the metric tensor is degenerate (i.e., not necessarily positive definite). Recall that every tangent space on a semi-Riemannian manifold is a semi-Euclidean space characterized by a (possibly isotropic) quadratic form.

[8]An affine connection is a geometric object (e.g., points, vectors, arcs, functions, curves) defined on a smooth manifold connecting tangent spaces that allows differentiability of tangent vector fields. The Levi-Civita (affine) connection is a torsion-free connection on the tangent bundle of a manifold that preserves a semi-Riemannian metric.

Einstein's field equations provides the components of the semi-Riemannian metric tensor of spacetime describing the geometry of the spacetime and the geodesic trajectories of spacetime.

Einstein's field equations lead to the equivalence between gravity and acceleration. In particular, the force an observer experiences due to gravity and the force an observer experiences due to acceleration are the same, and hence, if an observer experiences gravity's influence, then the observer is accelerating. This *equivalence principle* leads to a definition of a stationary observer as one who experiences no forces.

In Newtonian mechanics as well as in special relativity, space and spacetime provide a reference for defining accelerated motion. Since space and spacetime are absolute in Newtonian mechanics and the special theory of relativity, acceleration is also absolute. In contrast, in general relativity space and time are coupled to mass and energy, and hence, they are dynamic and *not* unchangeable. And this dynamic warping and curving of spacetime is a function of the gravitational field leading to a relational notion of acceleration. In other words, acceleration is relative to the gravitational field.[9]

As in the case of special relativity, general relativity also predicts several subtle cosmological effects. These include gravitational time dilation and frequency shifting, light deflection and gravitational time delay, gravitational waves, precession of apsides, orbital decay, and geodetic precession and frame dragging [325, 434]. Gravitational time dilation refers to the fact that dynamic processes closer to massive bodies evolve slower as compared to processes in free space, whereas frequency shifting refers to the fact that light frequencies are increased (blueshifting) when light travels towards an increased gravitational potential (moving towards a gravity well) and decreased (redshifting) when light travels away from a decreased gravitational potential (moving away from a gravity well).

Light deflection refers to the bending of light trajectories in a gravitational field towards a massive body, whereas gravitational time delay refers to the fact that light signals take longer to move through a gravitational field due to the influence of gravity on the geometry of space. General relativity additionally predicts the existence of gravitational waves; ripples in the metric of the spacetime continuum propagating at the speed of light. Finally, general relativity also predicts precession of planetary orbits, orbital decay due to gravitational wave emission, and frame dragging or

[9]Since a zero gravitational field is a well-defined field which can be measured and changed, it provides a gravitational field (i.e., the zero gravitational field) to which acceleration can be relative to. Thus, special relativity can be viewed as a special case of general relativity for which the gravitational field is zero.

gravitomagnetic effects (i.e., relativity of direction) in a neighborhood of a rotating mass.

13.2 Relativistic Kinematics

In this section, we briefly outline the underlying principles for relativistic kinematics of Einstein's special theory of relativity. Specifically, we present the two postulates of relativity and derive the Lorentz transformation equations connecting the physical variables of two uniformly moving (inertial) systems in relative motion. To present the main results of this chapter, we require some additional notation and definitions. We define a *point* as having zero size and zero mass, whereas a *particle* has zero size and nonnegative mass. A *reference point* is a point relative to which the position of other points are defined. The *location* and *motion* of a point or particle is *relative* to other points or particles and can be directly measured by a local observer (or measurer). A *discrete* (respectively, *continuum*) *body* is composed of a finite (respectively, infinite) number of particles; a body is *rigid* if its shape is constant.

Here, we need to clarify what is meant by a rigid body in the context of relativistic mechanics. Since any physical body has elastic properties, it will incur some distortion during accelerated motion, and hence, ideal rigid bodies do not exist in the physical world and simply provide a useful abstraction in classical mechanics. These elastic deformations are ignored in classical physics by assuming that the body is rigid enough so that transmission of external forces exerted on the body are instantaneous, leading to simultaneous motion from one end of the body to another. Hence, the body does not change its shape (i.e., dimensions) as observed from any reference frame.

However, in relativistic mechanics this abstraction needs further clarification since a rigid body requires the instantaneous transmission of signals through the body; a proviso that we will see is untenable in relativity. Even though we can ignore the relativistic time delay in electromagnetic signals and forces in much the same way we ignore the transmission speed of external forces governed by the speed of sound in classical rigid body mechanics, the relativistic length of a moving body changes with motion and hence needs to be addressed. Thus, a *relativistic rigid body* is a body that maintains its rest frame dimensions in translational motion. In other words, relativistic rigid body motion leads to changes in relativistic length in *every* frame relative to which the body is moving. For further details of rigid body motion in special relativity see [352].

A *frame* $\mathcal{F}$ consists of a linearly independent and (usually) mutually perpendicular axes defining a hyperspace of a four-dimensional Euclidean (flat) spacetime continuum. Since the physical quantities x, y, z, and t describing spatial and temporal occurrences in hyperspace have no physical locations or true time, a frame has no location and no absolute time. We assign the *origin* of the frame as a reference point, which can be used to define positions and time relative to other spatial and temporal occurrences.

Velocities and accelerations of the physical quantities x, y, and z depend on the frame with respect to which spatial and temporal changes are measured by a local observer. Hence, the derivatives (with respect to relative time) of the physical quantities x, y, and z are defined with respect to frames. Finally, an *inertial frame* is a frame in which the law of inertia (i.e., Newton's first law) holds; that is, the relative acceleration with respect to the frame of every pair of unforced particles is zero. Inertial frames describe unaccelerated systems since a body that is acted on by a zero net external force will move with a constant velocity.

The special theory of relativity is based on two key postulates developed by Albert Einstein. Namely:

Postulate 1): The laws of physics are the same in all inertial systems; no preferred inertial system exists.

Postulate 2): The speed of light in free space (i.e., in a vacuum) is the same in all inertial systems.

The first postulate of special relativity is known as the *principle of relativity* and states that it is impossible to measure unaccelerated translatory motion of a system through free space or through any medium pervading space. Hence, it is only meaningful to speak of the relative velocity of two systems by comparing measurements *between* inertial frames; it is meaningless to speak of the absolute velocity of a single system through free space. In other words, it is impossible for any experiment to ascertain whether an inertial system is intrinsically stationary or in motion with respect to any other inertial system from observations confined to a *single* reference frame; we can only assert the relative motion of the two systems by comparing observations *between* reference frames.

The second postulate of special relativity is known as the *principle of the constancy of the speed of light*[10] and states that the velocity of light in

[10]The principle of the constancy of the speed of light contradicts the Galilean transformation of velocities and is consistent with the Michelson-Morley experiment as well as numerous subsequent experiments. This principle leads to the fact that there does not exist a universal time which is the same for all observers, and hence, *spatial and temporal* events measured by observers in relative motion are coupled. In other words, location measurements and time interval measurements are

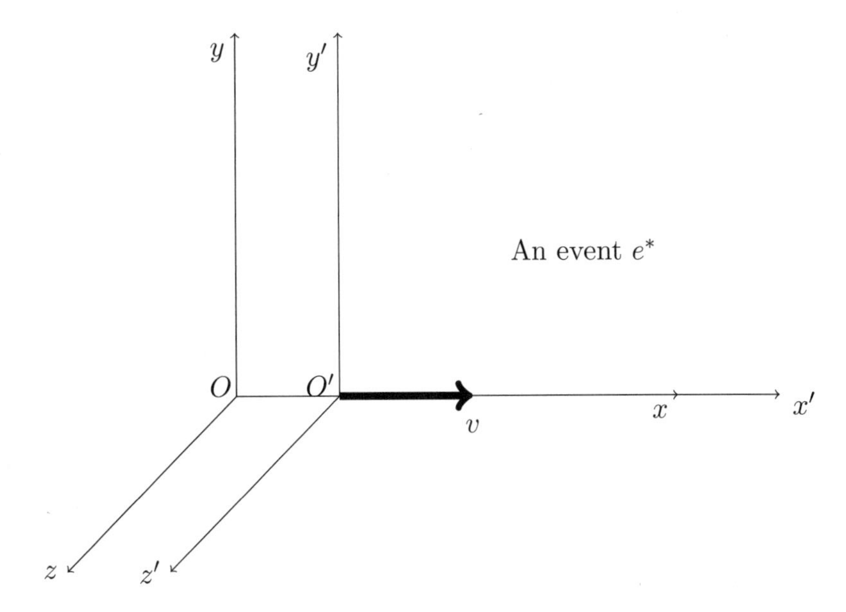

Figure 13.1 The primed reference frame $\mathcal{F}'$ is moving with a constant velocity v relative to the unprimed reference frame $\mathcal{F}$ as measured by a stationary observer in $\mathcal{F}$. Alternatively, by the principle of relativity, a stationary observer in $\mathcal{F}'$ measures a velocity $-v$ relative to $\mathcal{F}$.

free space is the same for all observers, and is independent of the relative velocity of the light source and the observer. In other words, the velocity of light is independent of the velocity of its source. Coupled with the first postulate, this leads to the fact that it is meaningless to refer to the absolute velocity of a light source; it is only meaningful to refer to the relative velocity of the source and the observer.

Special relativity is based on the fundamental idea that all inertial observers are equivalent and is restricted to the description of events in inertial reference frames. However, it is important to stress that the motion of the bodies we analyze can be accelerating with respect to the inertial frames. The special theory of relativity is derived from the two aforementioned postulates along with the assumptions that space and time are *homogeneous* and space is *isotropic*. The homogeneity assumption asserts that all points in space and time are equivalent, and hence, a measurement of a length interval or time interval between any two events in the spacetime continuum is not dependent on the location or the time instant the event occurs in the reference frame. The isotropy assumption asserts that space has the same properties in all directions.

Relativity is a theory of the measurement of space and time, and how

---

relative and depend on the reference frame of the observer.

motion affects their measurements. To develop the relations connecting spatial and temporal measurements between two observers in relative motion, consider the two systems of spacetime coordinates $\mathcal{F}$ and $\mathcal{F}'$ shown in Figure 13.1. We assume that the inertial frame $\mathcal{F}'$ is moving at a constant velocity $\boldsymbol{v}$ with respect to $\mathcal{F}$. For simplicity of exposition, we choose three linearly independent and mutually perpendicular axes, and consider relative motion between the origins of $\mathcal{F}$ and $\mathcal{F}'$ to be along the common x-x' axis so that $\boldsymbol{v} = [v\ 0\ 0]^{\mathrm{T}}$.

This orientational simplification along with the chosen relative velocity of the frames does not affect the physical principles of the results we will derive. The two observers use location measurement devices, which have been compared and calibrated against one another, and time measurement devices (clocks), which have been synchronized and calibrated against one another.[11] Furthermore, at $t = 0$ and $t' = 0$, the two origins O and O' of the two reference frames $\mathcal{F}$ and $\mathcal{F}'$ coincide.

Let e^* denote an event where spatial and temporal coordinates are measured by different observers in each inertial frame. Specifically, an observer attached to the origin of the reference frame $\mathcal{F}$ measures the space and time coordinates (x, y, z, t) for this event, whereas an observer attached to the origin of the reference frame $\mathcal{F}'$ measures the space and time coordinates (x', y', z', t') for the same event. Now, using the homogeneity assumption of space and time along with the assumption that the two origins O and O' are in coincidence at $t = 0$ and $t' = 0$, it follows that the most general form for describing the kinematical occurrence of any given spatiotemporal event from one reference system to the other is given by the linear equations

$$x' = a_{11}x + a_{12}y + a_{13}z + a_{14}t, \tag{13.1}$$

$$y' = a_{21}x + a_{22}y + a_{23}z + a_{24}t, \tag{13.2}$$

$$z' = a_{31}x + a_{32}y + a_{33}z + a_{34}t, \tag{13.3}$$

$$t' = a_{41}x + a_{42}y + a_{43}z + a_{44}t. \tag{13.4}$$

It is important to note that the most general transformation describing a *given* kinematical occurrence by an observer moving with the reference system $\mathcal{F}'$ relative to an observer describing the *same* kinematical occurrence moving with reference frame $\mathcal{F}$ is *affine* in the spacetime coordinates. However, since t and t' are equal to zero when the two origins O and

[11]Since in relativity theory two separate events that are simultaneous with respect to one reference frame are not necessarily simultaneous with respect to another reference frame, length measurements as well as time interval measurements depend on the reference frame of the observer, and hence, are relative. Thus, location and time measurement devices (clocks) need to be compared, calibrated, and synchronized against one another so that observers moving relative to each other measure the same speed of light. For further details see [378].

O' are in coincidence, the constant terms in each of the terms in (13.1)–(13.4) are zero. Furthermore, homogeneity of spacetime asserts that a linear transformation is necessary to guarantee a nonphysical preference of the spacetime coordinates. In addition, the transformation equations (13.1)–(13.4) suggest that there does not exist a single universal time uniformly suitable for all observers, but rather kinematical changes involving spatial measurements from one set of Cartesian axes to another should be connected to changes in temporal measurements if the laws of physics are to be invariant in the two systems of coordinates.

Under the assumptions of the principle of relativity and the principle of the constancy of the speed of light, in a four-dimensional manifold, corresponding to the four spacetime coordinates (x, y, z, t), it can be shown that (13.1)–(13.4) transform in accordance to a tensor of rank two defined by a collection of sixteen quantities for each given point in the spacetime continuum [420]. It follows from the theory of tensor analysis [420] that tensor equations are invariant with respect to a given category of coordinate transformations. Moreover, the condition that a scalar quantity (or magnitude) is invariant to a transformation of coordinates is key in developing the transformation laws of special relativity.

Recall that the $\mathcal{F}$ and $\mathcal{F}'$ frames are coincident at $t = t' = 0$, and subsequently the $\mathcal{F}'$ frame moves parallel to the x-axis at a uniform velocity v. Now, if a light pulse is sent out from the origin O of the $\mathcal{F}$ frame at $t = 0$, then the arrival time t of the pulse at a detector located at the point (x, y, z) satisfies

$$x^2 + y^2 + z^2 - (\mathrm{c}t)^2 = x^2 + y^2 + z^2 + (\jmath \mathrm{c}t)^2 = 0. \tag{13.5}$$

Similarly, in the $\mathcal{F}'$ frame, the light wave front is described by

$$x'^2 + y'^2 + z'^2 - (\mathrm{c}t')^2 = x'^2 + y'^2 + z'^2 + (\jmath \mathrm{c}t')^2 = 0, \tag{13.6}$$

since, by the second postulate of special relativity, the velocity of light in free space must measure the same for different observers in inertial reference frames.

Alternatively, (13.5) and (13.6) are a consequence of the fact that if all the components of a tensor vanish in one coordinate system, then they necessarily must vanish in all other admissible[12] coordinate systems [420, Theorem, p. 66]. In (13.5) and (13.6) the left (and middle) terms of the equations represent the square of a length in a four-dimensional spacetime continuum; the square of a length is a scalar (invariant), and hence, a tensor of rank zero.

[12] Admissible coordinate systems are formed using a group of (spacetime) transformations whose new set of coordinates are introduced as functions of the original coordinates and whose values are transformed in accordance with certain definite rules.

Next, if we regard the left (and middle) terms in (13.5) and (13.6) as the square of the interval in the four-dimensional spacetime continuum of the $\mathcal{F}$ and $\mathcal{F}'$ frames, respectively, then we postulate that these square intervals must be equal since they are true scalars. This is a statement of the fact that the square of the length is invariant; that is, the square of the length is the same in all admissible coordinate systems whatever value it might have, including zero. Hence,

$$x^2 + y^2 + z^2 - (\mathrm{c}t)^2 = x'^2 + y'^2 + z'^2 - (\mathrm{c}t')^2, \tag{13.7}$$

or, in differential form,

$$\mathrm{d}x^2 + \mathrm{d}y^2 + \mathrm{d}z^2 - \mathrm{c}^2\mathrm{d}t^2 = \mathrm{d}x'^2 + \mathrm{d}y'^2 + \mathrm{d}z'^2 - \mathrm{c}^2\mathrm{d}t'^2, \tag{13.8}$$

where the differentials $(\mathrm{d}x, \mathrm{d}y, \mathrm{d}z, \mathrm{d}t)$ and $(\mathrm{d}x', \mathrm{d}y', \mathrm{d}z', \mathrm{d}t')$ give the infinitesimal spatiotemporal interval in the two reference systems corresponding to differences in position and time of *given* pairs of neighboring events.

Now, taking the differential arc element in the four-dimensional hyperspace as

$$\mathrm{d}s^2 = -\mathrm{d}x^2 - \mathrm{d}y^2 - \mathrm{d}z^2 + \mathrm{c}^2\mathrm{d}t^2, \tag{13.9}$$

it follows that $\mathrm{d}s$ completely characterizes the geometry corresponding to the spacetime continuum modulo additional assumptions on connectivity and identification of points. Furthermore, note that it follows from (13.8) and (13.9) that the differential arc element in space $\mathrm{d}s$ exists independent of any particular choice of axes and is invariant for all transformation of coordinates.

It follows from (13.9) that

$$\mathrm{d}s = 0 \tag{13.10}$$

characterizes every differential component of the four-dimensional trajectory of a light pulse, traveling with the velocity c in the spacetime continuum, since setting $\mathrm{d}s = 0$ in (13.9) yields

$$\left(\frac{\mathrm{d}x}{\mathrm{d}t}\right)^2 + \left(\frac{\mathrm{d}y}{\mathrm{d}t}\right)^2 + \left(\frac{\mathrm{d}z}{\mathrm{d}t}\right)^2 = \mathrm{c}^2. \tag{13.11}$$

Transforming to any new set of interval coordinates (x', y', z', t') relative to (x, y, z, t) and using (13.8) and (13.10) yields

$$\left(\frac{\mathrm{d}x'}{\mathrm{d}t'}\right)^2 + \left(\frac{\mathrm{d}y'}{\mathrm{d}t'}\right)^2 + \left(\frac{\mathrm{d}z'}{\mathrm{d}t'}\right)^2 = \mathrm{c}^2, \tag{13.12}$$

as is required by the principle of the constancy of the speed of light.

Returning to Figure 13.1, we can readily deduce that the y'-axis in the $\mathcal{F}'$ frame always remains parallel to the y-axis in the $\mathcal{F}$ frame; and, similarly, the z'-axis in the $\mathcal{F}'$ frame always remains parallel to the z-axis in the $\mathcal{F}$ frame. Hence,

$$y^2 = y'^2 \tag{13.13}$$

and

$$z^2 = z'^2, \tag{13.14}$$

which implies

$$y = y', \tag{13.15}$$

$$z = z'. \tag{13.16}$$

Next, using (13.7), (13.13), and (13.14) yields

$$x^2 - (\mathrm{c}t)^2 = x'^2 - (\mathrm{c}t')^2. \tag{13.17}$$

Now, using (13.1) and (13.4), the most general linear connection between (x', t') and (x, t) such that (13.17) holds is given by

$$x' = a_{11}x + a_{14}t, \tag{13.18}$$

$$t' = a_{41}x + a_{44}t. \tag{13.19}$$

Since at $v = 0$, x' is only dependent on x and not on t; and t' is only dependent on t and not on x, it follows that

$$\lim_{v\to 0} \begin{bmatrix} a_{11} \\ a_{14} \\ a_{41} \\ a_{44} \end{bmatrix} = \begin{bmatrix} 1 \\ 0 \\ 0 \\ 1 \end{bmatrix}. \tag{13.20}$$

Noting that the origin of the $\mathcal{F}'$ frame moves with a velocity v relative to the origin of the frame $\mathcal{F}$, its position as measured by an observer in $\mathcal{F}$ is given by $x = vt$ so that when $x' = 0$ it follows from (13.18) that

$$0 = x' = a_{11}vt + a_{14}t = (a_{11}v + a_{14})t, \tag{13.21}$$

and hence,

$$a_{14} = -a_{11}v. \tag{13.22}$$

Now, substituting (13.18), (13.19), and (13.22) into (13.17) yields

$$x^2 - (\mathrm{c}t)^2 = (a_{11}x - a_{11}vt)^2 - \mathrm{c}^2(a_{44}t + a_{41}x)^2, \tag{13.23}$$

or, equivalently,

$$x^2-(\mathrm{c}t)^2 = (a_{11}^2-a_{41}^2\mathrm{c}^2)x^2-2(a_{11}^2v+a_{41}a_{44}\mathrm{c}^2)xt+(a_{11}^2v^2-a_{44}^2\mathrm{c}^2)t^2. \tag{13.24}$$

Equating the corresponding coefficients in (13.24) yields

$$a_{11}^2 - a_{41}^2\mathrm{c}^2 = 1, \tag{13.25}$$

$$a_{11}^2v^2 - a_{44}^2\mathrm{c}^2 = -\mathrm{c}^2, \tag{13.26}$$

$$a_{11}^2v + a_{41}a_{44}\mathrm{c}^2 = 0. \tag{13.27}$$

Next, solving for a_{11}, a_{41}, and a_{44} we obtain

$$a_{11} = a_{44} = \frac{1}{\sqrt{1-(v/c)^2}}, \tag{13.28}$$

$$a_{41} = -\frac{v}{c^2}a_{11} = -\frac{v}{c^2}\frac{1}{\sqrt{1-(v/c)^2}}. \tag{13.29}$$

Finally, substituting (13.22), (13.28), and (13.29) into (13.18) and (13.19), and using (13.15) and (13.16) we obtain the transformation equations

$$x' = \frac{x - vt}{\sqrt{1-(v/c)^2}}, \tag{13.30}$$

$$y' = y, \tag{13.31}$$

$$z' = z, \tag{13.32}$$

$$t' = \frac{t - (v/c^2)x}{\sqrt{1-(v/c)^2}}. \tag{13.33}$$

These equations are known as the *Lorentz transformation equations*[13] and form the core of special relativity theory.[14] Note that for relative velocities v between the reference systems $\mathcal{F}$ and $\mathcal{F}'$ greater than the velocity of light c, the Lorentz transformation equations became imaginary. As we see in Section 13.4, this is consistent with the fact that c is an upper limit for all possible relative velocities between material systems.

If we exchange our frames of reference or, equivalently, consider the spacetime event being observed in $\mathcal{F}'$ rather than in $\mathcal{F}$, the *inverse transformations* are obtained by interchanging the primed and unprimed quantities, and by replacing v by $-v$. In this case, the $\mathcal{F}$ frame moves to the left of the $\mathcal{F}'$ frame as opposed to the $\mathcal{F}'$ frame moving to the right of the $\mathcal{F}$ frame. This process yields

$$x = \frac{x' + vt'}{\sqrt{1-(v/c)^2}}, \tag{13.34}$$

[13]The Lorentz transformation equations are also known as Lorentz-Fitzgerald and the Lorentz-Einstein transformation equations in the literature. Poincaré originally gave Lorentz's name to the equations who had proposed the equations before Einstein in his classical theory of electrons. Furthermore, Fitzgerald and Lorentz used the equations to obtain an ad hoc explanation of the negative result of the Michelson-Morley experiment in an attempt to preserve the erroneous concept of an absolute aether frame relative to which the velocity v is defined.

[14]The Lorentz transformation equations can also be derived directly from (13.1)–(13.4) by determining the sixteen coefficients using the first and second postulates of relativity, and assuming the homogeneity of space and time as well as the isotropy of space. For details see [378].

$$y = y', \tag{13.35}$$

$$z = z', \tag{13.36}$$

$$t = \frac{t' + (v/\mathrm{c}^2)x'}{\sqrt{1-(v/\mathrm{c})^2}}. \tag{13.37}$$

Note that the Lorentz transformations (13.30)–(13.33) and the inverse Lorentz transformations (13.34)–(13.37) have exactly the same form modulo the sign of the relative velocity v. This is completely consistent with the first postulate of relativity since absolute velocity is meaningless as the two reference systems $\mathcal{F}$ and $\mathcal{F}'$ must be equivalent in the description of physical occurrences. Furthermore, using the differential form of (13.34)–(13.37), it follows that

$$\begin{aligned} &\mathrm{d}x^2 + \mathrm{d}y^2 + \mathrm{d}z^2 - \mathrm{c}^2\mathrm{d}t^2 \\ &\quad = \left(\frac{\mathrm{d}x' + v\mathrm{d}t'}{\sqrt{1-(v/\mathrm{c})^2}}\right)^2 + \mathrm{d}y'^2 + \mathrm{d}z'^2 - \mathrm{c}^2\left(\frac{\mathrm{d}t' + (v/\mathrm{c}^2)\mathrm{d}x'}{\sqrt{1-(v/\mathrm{c})^2}}\right)^2 \\ &\quad = \mathrm{d}x'^2 + \mathrm{d}y'^2 + \mathrm{d}z'^2 - \mathrm{c}^2\mathrm{d}t'^2, \end{aligned} \tag{13.38}$$

establishing the invariance of the left-hand side of (13.38) to the Lorentz transformations and showing that the velocity of light in free space is the same for all observers as required by the second postulate of relativity.

Finally, we close this section by showing that when the relative velocity v between the two reference systems $\mathcal{F}$ and $\mathcal{F}'$ is small as compared to that of the speed of light c, that is, $v/\mathrm{c} \ll 1$, the Lorentz transformation equations collapse to the classical *Galilean transformation equations*. Specifically, for $v/\mathrm{c} \ll 1$, (13.30)–(13.33) specialize to

$$x' = x - vt, \tag{13.39}$$

$$y' = y, \tag{13.40}$$

$$z' = z, \tag{13.41}$$

$$t' = t. \tag{13.42}$$

Equations (13.40) and (13.41) are a restatement of (13.31) and (13.32). Equation (13.39) follows immediately from (13.30) as $v/c \to 0$. To show (13.42), let the motion of the origin O' of the frame $\mathcal{F}'$ relative to the frame $\mathcal{F}$ be given by $x = vt$. Then it follows from (13.33) that

$$t' = \frac{t - v^2t/\mathrm{c}^2}{\sqrt{1-(v/\mathrm{c})^2}} = t\sqrt{1-(v/\mathrm{c})^2}, \tag{13.43}$$

which implies that $t = t'$ as $v/c \to 0$.

13.3 Length Contraction and Time Dilation

In this section, we present key consequences on length and time measurements of the Lorentz transformation equations. Computing the differential form of (13.30)–(13.33) yields

$$\mathrm{d}x' = \frac{\mathrm{d}x - v\mathrm{d}t}{\sqrt{1-(v/\mathrm{c})^2}}, \tag{13.44}$$

$$\mathrm{d}y' = \mathrm{d}y, \tag{13.45}$$

$$\mathrm{d}z' = \mathrm{d}z, \tag{13.46}$$

$$\mathrm{d}t' = \frac{\mathrm{d}t - (v/\mathrm{c}^2)\mathrm{d}x}{\sqrt{1-(v/\mathrm{c})^2}}. \tag{13.47}$$

Next, consider a rod, parallel to the x-x' axis, at rest in the $\mathcal{F}'$ frame. An observer in the $\mathcal{F}'$ frame measures a length $\mathrm{d}x'$ for the rod. However, an observer in the $\mathcal{F}$ frame, measuring the same rod, which is moving with respect to the $\mathcal{F}$ frame, measures a length $\mathrm{d}x$ for the same rod. Since the moving rod is measured instantaneously in the $\mathcal{F}$ frame by the observer, $\mathrm{d}t = 0$.

Now, using (13.44) and the fact that $\mathrm{d}t = 0$, it follows that

$$\mathrm{d}x' = \frac{\mathrm{d}x}{\sqrt{1-(v/\mathrm{c})^2}}, \tag{13.48}$$

or, equivalently,

$$\mathrm{d}x = \mathrm{d}x'\sqrt{1-(v/\mathrm{c})^2}. \tag{13.49}$$

The fact that $\mathrm{d}x < \mathrm{d}x'$ is known as *length contraction*[15] and asserts that a body's measured length, moving with a velocity v relative to an observer, is contracted in the direction of its motion by a factor of $\sqrt{1-(v/\mathrm{c})^2}$. It follows from (13.45) and (13.46) that the body's dimensions perpendicular to the direction of motion are not affected, and hence, there is no disagreement in length measurements between the two systems of coordinates at right angles to the line of motion.

Next, we show that length contraction is reciprocal in the sense that a body's length will measure longer when the body is at rest relative to the

[15]The magnitude of the contraction $\sqrt{1-(v/\mathrm{c})^2}$ is the same as that postulated by Lorentz and Fitzgerald in connection with their study of electric fields in a moving spherical charge. However, Lorentz and Fitzgerald considered their contraction as a *real contraction* produced by the passage of bodies through the luminiferous aether. In contrast, Einstein's interpretation of the contraction is a property of the spacetime manifold subjected to the Lorentz transformation equations (13.30)–(13.33).

observer. Specifically, consider the same rod, parallel to the x-x' axis, but now at rest in the $\mathcal{F}$ frame. In this frame, the body measures a length of $\mathrm{d}x$. However, an observer in the $\mathcal{F}'$ frame measures a length $\mathrm{d}x'$ for the same rod with $\mathrm{d}t' = 0$. That is, the measurement is instantaneous in the $\mathcal{F}'$ frame. In this case, it follows from (13.47) that

$$\mathrm{d}t' = \frac{\mathrm{d}t - (v/\mathrm{c}^2)\mathrm{d}x}{\sqrt{1 - (v/\mathrm{c})^2}} = 0, \tag{13.50}$$

and hence, (13.44) yields

$$\mathrm{d}x' = \mathrm{d}x\sqrt{1 - (v/\mathrm{c})^2}. \tag{13.51}$$

Hence, interchanging the frames results in $\mathrm{d}x' < \mathrm{d}x$ asserting that motion affects measurement. Furthermore, it is important to note that the two experiments are symmetrical and in agreement.

Next, we turn our attention on how motion affects time. Let an observer in the $\mathcal{F}'$ frame measure two events that occur at the same place (i.e., $\mathrm{d}x' = 0$) in $\mathcal{F}'$, and let $\mathrm{d}t'$ be the time interval between the two events. For an observer in the frame $\mathcal{F}$, the measuring device (i.e., clock) in $\mathcal{F}'$ is moving. Since the two events occur at the same point in the frame $\mathcal{F}'$, it follows from (13.44) that

$$\mathrm{d}x' = \frac{\mathrm{d}x - v\mathrm{d}t}{\sqrt{1 - (v/\mathrm{c})^2}} = 0, \tag{13.52}$$

and hence, (13.47) yields

$$\mathrm{d}t' = \mathrm{d}t\sqrt{1 - (v/\mathrm{c})^2}, \tag{13.53}$$

or, equivalently,

$$\mathrm{d}t = \frac{\mathrm{d}t'}{\sqrt{1 - (v/\mathrm{c})^2}}. \tag{13.54}$$

The fact that $\mathrm{d}t > \mathrm{d}t'$ is known as *time dilation* and asserts that a body's measured time rate, moving with a velocity v relative to an observer, slows down by a factor of $\sqrt{1 - (v/\mathrm{c})^2}$. In other words, a clock moving at a constant velocity relative to an inertial frame that contains synchronized clocks will measure a slower rate when compared to these synchronized clocks.

As in length contraction, time dilation is reciprocal in the sense that a body's measured time rate will measure faster when it is at rest relative to the observer. To see this, simply interchange the frames so that an observer in the $\mathcal{F}$ frame measures two events that occur at the same place (i.e., $\mathrm{d}x = 0$) in $\mathcal{F}$, and let $\mathrm{d}t$ be the time interval between the two events. For an observer in the frame $\mathcal{F}'$, the clock in $\mathcal{F}$ is moving. In this case, setting $\mathrm{d}x = 0$ in (13.47) yields

$$\mathrm{d}t' = \frac{\mathrm{d}t}{\sqrt{1-(v/c)^2}}, \tag{13.55}$$

which asserts that $\mathrm{d}t < \mathrm{d}t'$. In both cases, the time duration is longer when measured by clocks relative to which the first clock is moving.

The results of this section show that whichever frame we assign as a *proper frame*,[16] the observer in the nonproper (i.e., the other) frame will measure a contracted length and a dilated time interval. Hence, in relativity theory separated events that are simultaneous with respect to one reference frame need not be simultaneous with respect to another reference frame.

13.4 Relativistic Velocity and Acceleration Transformations

It follows from classical physics that if a particle is moving at a velocity $\boldsymbol{v}$ with respect to an inertial reference frame $\mathcal{F}$ and an observer moves with a velocity $\boldsymbol{u}'$ with respect to an inertial reference frame $\mathcal{F}'$, then the observer's velocity $\boldsymbol{u}$ relative to $\mathcal{F}$ is given by

$$\boldsymbol{u} = \boldsymbol{u}' + \boldsymbol{v}. \tag{13.56}$$

This result is a direct consequence of the Galilean transformation equations (13.39)–(13.42) with

$$\boldsymbol{u} = \left[\frac{\mathrm{d}x}{\mathrm{d}t}\ \frac{\mathrm{d}y}{\mathrm{d}t}\ \frac{\mathrm{d}z}{\mathrm{d}t}\right]^{\mathrm{T}} = \left[u_x\ u_y\ u_z\right]^{\mathrm{T}},$$

$$\boldsymbol{u}' = \left[\frac{\mathrm{d}x'}{\mathrm{d}t}\ \frac{\mathrm{d}y'}{\mathrm{d}t}\ \frac{\mathrm{d}z'}{\mathrm{d}t}\right]^{\mathrm{T}} = \left[u_x'\ u_y'\ u_z'\right]^{\mathrm{T}},$$

and $\boldsymbol{v} = [v\ 0\ 0]^{\mathrm{T}}$, and is known as the *Galilean velocity addition theorem*. The fact that the observer's velocity $\boldsymbol{u}$ relative to $\mathcal{F}$ is the vector sum of the two velocities $\boldsymbol{u}'$ and $\boldsymbol{v}$ follows from the fact that in classical physics time is absolute, that is, $t = t'$. In relativistic physics, however, $t \neq t'$, resulting in a different formula for relativistic addition of velocities.

To develop the transformation equations for relativistic velocities first

[16] In relativity theory, a *proper frame* is a frame in which a variable (e.g., length, time interval, mass, temperature) of a body is measured in the inertial frame in which the body is at rest. Such variables are called *proper variables* (e.g., proper length, proper time interval, proper mass, proper temperature).

note that it follows from (13.47) that

$$\frac{\mathrm{d}t'}{\mathrm{d}t} = \frac{1-(v/\mathrm{c}^2)\frac{\mathrm{d}x}{\mathrm{d}t}}{\sqrt{1-(v/\mathrm{c})^2}} = \frac{1-(v/\mathrm{c}^2)u_x}{\sqrt{1-(v/\mathrm{c})^2}}, \tag{13.57}$$

where $u_x \triangleq \frac{\mathrm{d}x}{\mathrm{d}t}$. Equation (13.57) connects differential time intervals in the two reference systems $\mathcal{F}'$ and $\mathcal{F}$ between neighboring events occurring at neighboring points in space.

Now, differentiating (13.30)–(13.32) with respect to time in the respective system of coordinates and using (13.57), it follows that

$$u'_x = \frac{\mathrm{d}x'}{\mathrm{d}t'} = \frac{\mathrm{d}x'}{\mathrm{d}t}\frac{\mathrm{d}t}{\mathrm{d}t'} = \frac{\frac{\mathrm{d}x}{\mathrm{d}t} - v}{1-(v/\mathrm{c}^2)\frac{\mathrm{d}x}{\mathrm{d}t}} = \frac{u_x - v}{1-(v/\mathrm{c}^2)u_x}, \tag{13.58}$$

$$u'_y = \frac{\mathrm{d}y'}{\mathrm{d}t'} = \frac{\mathrm{d}y'}{\mathrm{d}t}\frac{\mathrm{d}t}{\mathrm{d}t'} = \frac{\frac{\mathrm{d}y}{\mathrm{d}t}\sqrt{1-(v/c)^2}}{1-(v/\mathrm{c}^2)\frac{\mathrm{d}x}{\mathrm{d}t}} = \frac{u_y\sqrt{1-(v/c)^2}}{1-(v/\mathrm{c}^2)u_x}, \tag{13.59}$$

$$u'_z = \frac{\mathrm{d}z'}{\mathrm{d}t'} = \frac{\mathrm{d}z'}{\mathrm{d}t}\frac{\mathrm{d}t}{\mathrm{d}t'} = \frac{\frac{\mathrm{d}z}{\mathrm{d}t}\sqrt{1-(v/c)^2}}{1-(v/\mathrm{c}^2)\frac{\mathrm{d}x}{\mathrm{d}t}} = \frac{u_z\sqrt{1-(v/c)^2}}{1-(v/\mathrm{c}^2)u_x}. \tag{13.60}$$

Hence, if a particle in the reference system $\mathcal{F}$ is moving with uniform velocity $\boldsymbol{u} = [u_x \; u_y \; u_z]^{\mathrm{T}}$, then the velocity $\boldsymbol{u}' = [u'_x \; u'_y \; u'_z]^{\mathrm{T}}$ of the particle as measured by an observer in the reference system $\mathcal{F}'$ is given by (13.58)–(13.60). These equations are known as the *Einstein velocity addition theorem*.

The inverse transformations are obtained by interchanging the primed and unprimed velocities, and by replacing v by $-v$. In particular,

$$u_x = \frac{u'_x + v}{1+(v/\mathrm{c}^2)u'_x}, \tag{13.61}$$

$$u_y = \frac{u'_y\sqrt{1-(v/c)^2}}{1+(v/\mathrm{c}^2)u'_x}, \tag{13.62}$$

$$u_z = \frac{u'_z\sqrt{1-(v/c)^2}}{1+(v/\mathrm{c}^2)u'_x}. \tag{13.63}$$

Note that the perpendicular (i.e., transverse) components u_y and u_z of the velocity $\boldsymbol{u}$ of a particle as measured in the $\mathcal{F}$ frame are functions of both the transverse (i.e., u'_y and u'_z) and parallel (i.e., u'_x) components of the velocity $\boldsymbol{u}'$ of the particle as measured in the $\mathcal{F}'$ frame.

Furthermore, if the components of the velocity $\boldsymbol{u}'$ and the velocity v are small as compared to c, then (13.61)–(13.63) collapse to (13.56). In addition, it follows from (13.61) that the velocity of light c gives an upper limit to all possible velocities. To see this, assume that a body in frame $\mathcal{F}'$

is moving with a velocity $u'_x = \mathrm{c}$ in the same direction with the frame, which is also moving at a velocity c with respect to $\mathcal{F}$. The measured velocity of the body with respect to the system $\mathcal{F}$, using (13.61), is given by

$$u_x = \frac{\mathrm{c} + \mathrm{c}}{1 + \mathrm{c}^2/\mathrm{c}^2} = \mathrm{c}, \tag{13.64}$$

asserting that the velocity of light is an upper limit for all possible velocities between material systems.[17] It follows from (13.64) that all observers measure the same speed c for the speed of light, which implies that the velocity of light is independent of the velocity of the light source.

If the succession of cause and effect is to be preserved in all reference frames, then it is impossible for signals to be transmitted faster than the speed of light. This follows from the fact that if a time-ordered cause-and-effect event occurs in a reference frame $\mathcal{F}$, where the effect is propagated faster than the speed of light, then it follows from Einstein's velocity addition theorem that there exists an inertial frame $\mathcal{F}'$, moving relative to $\mathcal{F}$ with a velocity less than the speed of light, in which the effect will precede the cause. Thus, relativity is completely consistent with the principle of causality in the sense that observed ordered events of physical processes are not affected by transformations between inertial frames. In other words, unlike the notion of simultaneity, which is observer dependent, the time ordering of events are not observer dependent. This is further discussed in Section 13.5.

Finally, we give the transformation equations for relativistic accelerations. In particular, differentiating (13.58)–(13.60) with respect to time in the respective system of coordinates and using $a'_x \triangleq \frac{\mathrm{d}u'_x}{\mathrm{d}t'} = \frac{\mathrm{d}u'_x}{\mathrm{d}t}\frac{\mathrm{d}t}{\mathrm{d}t'}$ (and similarly for a'_y and a'_z) we obtain

$$a'_x = \left[1 - \left(\frac{v}{\mathrm{c}^2}\right) u_x\right]^{-3} \left[1 - \left(\frac{v}{\mathrm{c}}\right)^2\right]^{3/2} a_x, \tag{13.65}$$

$$\begin{aligned} a'_y = {} & \left[1 - \left(\frac{v}{\mathrm{c}^2}\right) u_x\right]^{-2} \left[1 - \left(\frac{v}{\mathrm{c}}\right)^2\right] a_y \\ & + u_y \left(\frac{v}{\mathrm{c}^2}\right) \left[1 - \left(\frac{v}{\mathrm{c}^2}\right) u_x\right]^{-3} \left[1 - \left(\frac{v}{\mathrm{c}}\right)^2\right] a_x, \end{aligned} \tag{13.66}$$

$$\begin{aligned} a'_z = {} & \left[1 - \left(\frac{v}{\mathrm{c}^2}\right) u_x\right]^{-2} \left[1 - \left(\frac{v}{\mathrm{c}}\right)^2\right] a_z \\ & + u_z \left(\frac{v}{\mathrm{c}^2}\right) \left[1 - \left(\frac{v}{\mathrm{c}^2}\right) u_x\right]^{-3} \left[1 - \left(\frac{v}{\mathrm{c}}\right)^2\right] a_x, \end{aligned} \tag{13.67}$$

where $a_x \triangleq \frac{\mathrm{d}u_x}{\mathrm{d}t}$, $a_y \triangleq \frac{\mathrm{d}u_y}{\mathrm{d}t}$, and $a_z \triangleq \frac{\mathrm{d}u_z}{\mathrm{d}t}$.

[17] Even though information, matter, or energy cannot have speeds that are greater than the speed of light, there are certain kinematical processes that *can* have superluminal speeds. For details see [385].

Similar expressions can be obtained for the inverse transformations by interchanging the primed and unprimed velocities and accelerations, and by replacing v by $-v$. Note that unlike relativistic velocities, wherein a constant velocity in system $\mathcal{F}$ results in a constant velocity in system $\mathcal{F}'$, it follows from (13.65)–(13.67) that a constant acceleration with respect to the reference system $\mathcal{F}$ does not necessarily imply a constant acceleration with respect to the reference system $\mathcal{F}'$. This is due to the fact that the component accelerations a'_x, a'_y, and a'_z in $\mathcal{F}'$ are functions of both the component accelerations a_x, a_y, and a_z as well as the possibly changing in time component velocities u_x, u_y, and u_z of the reference system $\mathcal{F}$. This clearly shows that even though the reference frames in special relativity are inertial (i.e., nonaccelerating), the motion of the bodies we analyze can be accelerating with respect to these frames.

13.5 Special Relativity, Minkowski Space, and the Spacetime Continuum

The Minkowski space, also known as the Minkowski spacetime, combines the three-dimensional Euclidean space with the temporal coordinate of time into a hyperspace of four dimensions (i.e., a four-dimensional manifold), wherein a spacetime interval between any two events is invariant with respect to the inertial frame that the events are measured. In other words, all inertial reference frames will record an identical total metric in spacetime between events, and hence, spacetime is an absolute concept providing a mathematical structure for which special relativity can be formulated as an invariance theory.

It is important to stress that all directions in the Minkowski spacetime (i.e., the hyperspace of four dimensions) are *not* equivalent as the temporal dimension of time is treated differently from the three spatial dimensions, and hence, the Minkowski spacetime is distinct from the four-dimensional Euclidean space. This distinction is often emphasized in the literature by referring to the hyperspace as the spacetime continuum of three plus one dimensions (i.e., $\mathbb{R}^{3+1}$) rather than a four-dimensional (i.e., $\mathbb{R}^4$) space.

To further elucidate this distinction, recall that in a three-dimensional Euclidean space the isometry group (i.e., the homeomorphic group that preserves Euclidean distance) is the Euclidean group and consists of rotations, reflections, and transformations. When a temporal component is appended as a fourth dimension to form a four-dimensional Euclidean space, the isometry group includes all Galilean transformations and *separately* preserves the three-dimensional Euclidean (purely spatial) distance (metric) and the one-dimensional temporal (time) differences. In contrast, in the Minkowski space the spacetime (i.e., the four-dimensional hyperspace) is equipped with a nonpositive and nondegenerate metric (i.e., the Minkowski metric) given by (13.9), which couples space and time. In this case, the isometry group preserving the spacetime differential arc interval $\mathrm{d}s^2$ is the Poincaré group and as shown in (13.38) includes the Lorentz transformations.

The introduction of an imaginary time coordinate $\jmath ct$, but *not* an imaginary time (see the middle equation in (13.5) and (13.6)), can be used to morph the time coordinate so as to mimic the space coordinates. This is known as a *Wick rotation*, wherein the Minkowski metric becomes a Euclidean distance (i.e., an inner product). Since in this case the Euclidean group preserves the Euclidean distance extended from $\mathbb{R}^3$ to $\mathbb{R}^3 \times \mathbb{C}$, an *explicit* formulation of a metric tensor is not required. However, it is important to note that even though solutions in an $\mathbb{R}^{3+1}$ Minkowski space can be related to solutions in an $\mathbb{R}^3 \times \mathbb{C}$ Euclidean space for the flat spacetime of special relativity, this relation breaks down in the curved spacetime of general relativity. Furthermore, this formulation conceals the indefinite nature of the Minkowski metric (i.e., a pseudo inner product) as well as the nonrotational nature of the Lorentz transformations (i.e., the *Lorentz boost*). For details see [325].

Since the Minkowski metric (13.9) is a scalar (a tensor of rank zero), it is independent of any particular choice of axes, and hence, it is invariant under *all* possible coordinate changes from one inertial frame to another. For example, the Galilean transformations (13.39)–(13.42) as well as the Lorentz transformations (13.30)–(13.33) leave the form of (13.9) unchanged. Furthermore, all transformations involving spatial rotation of the coordinate axes leave (13.9) invariant. For example, the Minkowski metric remains unchanged to the orthogonal rotation transformation (through the angle θ) given by

$$x' = x\cos\theta + y\sin\theta, \tag{13.68}$$
$$y' = y\cos\theta - x\sin\theta, \tag{13.69}$$
$$z' = z, \tag{13.70}$$
$$t' = t. \tag{13.71}$$

Note that since (13.30)–(13.33) can be equivalently expressed as

$$x' = x\cosh\phi - ct\sinh\phi, \tag{13.72}$$
$$y' = y, \tag{13.73}$$
$$z' = z, \tag{13.74}$$
$$ct' = ct\cosh\phi - x\sinh\phi, \tag{13.75}$$

where

$$\phi = \cosh^{-1}\left(\frac{1}{\sqrt{1-(v/c)^2}}\right), \tag{13.76}$$

the Lorentz transformation equations can be regarded as a rotation of coordinates in a four-dimensional space with three coordinates representing space and one coordinate representing time. This assertion follows by comparing (13.68)–(13.71) and (13.72)–(13.75), and noting that the Lorentz transformation equations can be viewed as an imaginary rotation in the xt-plane. Furthermore, note that the Minkowski *metric signature* $(-1,-1,-1,+1)$ involving the three negative signs and the one positive sign in (13.9) preserves the distinction between spatial and temporal coordinates for *every* real transformation of coordinates.

The Minkowski space $\mathcal{M} \subseteq \mathbb{R}^{3+1}$ is thus a four-dimensional real vector space equipped with a nondegenerate, symmetric bilinear metric (13.9) on the tangent space $T_p\mathcal{M}$ at each point $p \in \mathcal{M}$. Since the geometry corresponding to spacetime in special relativity is flat, rectilinear coordinates preserve the invariance of (13.9). Furthermore, the components $p = (x, y, z, t)$ of the Minkowski space $\mathcal{M}$ correspond to *events* in spacetime with $\mathcal{M}$ being the simplest special case of a semi-Riemannian manifold equipped with the Lorentz metric (13.9) characterizing a four-dimensional flat spacetime with a chosen signature; namely, three negative (respectively, positive) signs for the spatial components and one positive (respectively, negative) sign for the temporal component. The symmetry group (i.e., the group of all transformations preserving the differential arc interval (13.9)) with a given metric signature choice is isomorphic with the symmetry group preserving the other signature choice. These groups of transformations include the Poincaré group, which allow for origin translation, as well as the Lorentz group of transformations, which leave the origin fixed.

Events in the spacetime continuum are classified according to $c^2dt^2 - dx^2 - dy^2 - dz^2$. Specifically, we say that an event $p = (x, y, z, t)$ is *spacelike* if $c^2dt^2 < dx^2 + dy^2 + dz^2$, *timelike* if $c^2dt^2 > dx^2 + dy^2 + dz^2$, and *lightlike* or *singular* if $c^2dt^2 = dx^2 + dy^2 + dz^2$. These classifications can be used to further characterize space separation as well as time order of events in the spacetime continuum. Specifically, using the Lorentz transformation equations, we can transform our system to a set of *proper coordinates*, wherein the differential arc interval ds is characterized by the aforementioned classifications. In particular, if the interval ds is spacelike, then we can always find a set of proper coordinates in which the time component is zero, whereas if the interval ds is timelike, then we can always find a set of proper coordinates in which the spatial components are zero.

To see this, without loss of generality consider the x-axis and ignore the y- and z-axes. This can be achieved by a spatial rotation of the axes to eliminate the y and z components. In this case, the differential arc interval is given by

$$\mathrm{d}s^2 = -\mathrm{d}x^2 + \mathrm{c}^2\mathrm{d}t^2. \tag{13.77}$$

Now, using the transformations (13.44)–(13.47) and keeping the dimensions of the coordinates the same we obtain

$$\mathrm{d}x' = \frac{\mathrm{d}x - (v/\mathrm{c})\mathrm{c}\mathrm{d}t}{\sqrt{1-(v/\mathrm{c})^2}}, \tag{13.78}$$

$$\mathrm{c}\mathrm{d}t' = \frac{\mathrm{c}\mathrm{d}t - (v/\mathrm{c})\mathrm{d}x}{\sqrt{1-(v/\mathrm{c})^2}}. \tag{13.79}$$

If (13.77) is spacelike, then $\mathrm{d}x^2 > \mathrm{c}^2\mathrm{d}t^2$ and, hence, choosing $|v/\mathrm{c}| \leq 1$ such that $\mathrm{c}\mathrm{d}t' = 0$ gives

$$\mathrm{d}s^2 = -\mathrm{d}x'^2 \tag{13.80}$$

in the primed coordinates. Conversely, if (13.77) is timelike, then $\mathrm{c}^2\mathrm{d}t^2 > \mathrm{d}x^2$ and, hence, choosing $|v/\mathrm{c}| \leq 1$ such that $\mathrm{d}x' = 0$ gives

$$\mathrm{d}s^2 = \mathrm{d}t'^2 \tag{13.81}$$

in the primed coordinates.

The above analysis shows that if a region in spacetime is spacelike, then there exists a reference frame $\mathcal{F}'$ such that two events appear simultaneous in $\mathcal{F}'$. In this case, the differential arc interval between events in $\mathcal{F}'$ is given by

$$\mathrm{d}s^2 = -\mathrm{d}x^2 + \mathrm{c}^2\mathrm{d}t^2 = -\mathrm{d}x'^2 + \mathrm{c}^2\mathrm{d}t'^2 = -\mathrm{d}x'^2, \tag{13.82}$$

and hence, $\mathrm{c}^2\mathrm{d}t'^2 = 0$ in $\mathcal{F}'$ so that $\mathrm{d}s = \mathrm{d}x'$. Alternatively, if a region in spacetime is timelike, then there exists a reference frame $\mathcal{F}'$ such that two events occur at the same place in $\mathcal{F}'$. In this case, the differential arc interval between events in $\mathcal{F}'$ is given by

$$\mathrm{d}s^2 = -\mathrm{d}x^2 + \mathrm{c}^2\mathrm{d}t^2 = -\mathrm{d}x'^2 + \mathrm{c}^2\mathrm{d}t'^2 = \mathrm{c}^2\mathrm{d}t'^2, \tag{13.83}$$

and hence, $\mathrm{d}x'^2 = 0$ in $\mathcal{F}'$ so that $\mathrm{d}s = \mathrm{c}\mathrm{d}t'$.

Next, we use the above analysis to give a geometric representation of the spacetime continuum in order to provide further insights into simultaneity and time order of events. First, note that for a given reference frame $\mathcal{F}$ with differential arc interval (13.77) and orthogonal axes x and $\mathrm{c}t$, the motion of a particle in $\mathcal{F}$ traces a *trajectory* or *curve* that gives the loci of all the spacetime points or events corresponding to the motion of the particle. Minkowski referred to the spacetime continuum as *the world* with *world points* giving events in spacetime and *world lines* giving the set of all events characterizing the history of motion of the particle.

Note that the tangent to a world line at any given point is given by

$$\frac{\mathrm{d}x}{\mathrm{d}(\mathrm{c}t)} = \frac{1}{\mathrm{c}}\frac{\mathrm{d}x}{\mathrm{d}t}. \tag{13.84}$$

Since $\frac{\mathrm{d}x}{\mathrm{d}(\mathrm{c}t)} = \frac{u}{\mathrm{c}}$ and $u < \mathrm{c}$ for every particle with nonzero mass, it follows that the slope $\frac{\mathrm{d}x}{\mathrm{d}(\mathrm{c}t)}$ is always less than 45°. Furthermore, note that for $u = \mathrm{c}$ the world line of a light wave corresponds to a straight line in the x-ct axis with a slope of 45°.

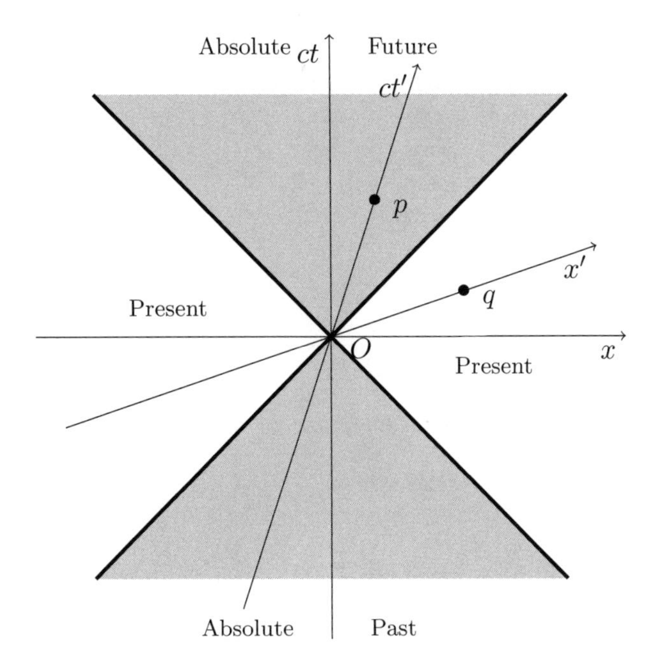

Figure 13.2 Light cone in a two-dimensional spacetime.

Consider the *light cone* shown in Figure 13.2 wherein a pulse of light emanating at O travels in all directions through the spacetime continuum. This event o is localized to a single point in space and a single instant of time. It is clear from Figure 13.2 that in the shaded region, circumscribed by world lines of light waves, there exists a reference frame $\mathcal{F}'$ for which the events o and p occur at the same place (i.e., $\mathrm{d}x' = 0$) and are separated by time. Clearly, event p in the shaded upper half of the light cone occurs after event o in time as measured in $\mathcal{F}'$. This holds for every event p in the shaded upper half of the light cone shown in Figure 13.2. This region is known as the *absolute future* since every event in this region occurs in the future relative to event o.

Alternatively, if event p is in the shaded lower half of the light cone, then event p precedes event o in time as measured in $\mathcal{F}'$. This holds for every event p in the shaded lower half of the light cone shown in Figure 13.2. This region is known as the *absolute past* since every event in this region occurs in the past relative to event o. Hence, for both regions there exist reference frames for which events o and p occur at the same place in space with an absolute time order determining the events. Hence, both spacetime regions are timelike.

Next, consider the unshaded regions in Figure 13.2. It is clear from the figure that in the unshaded region, circumscribed by world lines of light waves, there exists an inertial reference frame $\mathcal{F}'$ for which events o and q occur at the same time (i.e., $\mathrm{d}t' = 0$) and are separated only in space. This holds for every event q in the unshaded left and right regions of the light cone shown in Figure 13.2. This region is known as the *present* since there always exists an inertial frame $\mathcal{F}'$ for which the events o and q occur simultaneously in time. Hence, present regions in spacetime are spacelike. It is important to stress here that there also exist other inertial frames for which the events o and q are not simultaneous and the time ordering of these events is not absolute but rather relative. Furthermore, note that events occurring in the present are absolutely separated from event o, whereas events occurring in the absolute past or absolute future are not ordered spatially relative to the event o.

Events occurring on the boundary of the light cone correspond to a lightlike or singular region of spacetime. In this case, the null geodesics lie along a dual cone (see Figure 13.2) characterized by

$$\mathrm{d}s^2 = -\mathrm{d}x^2 + \mathrm{c}^2\mathrm{d}t^2 = -\mathrm{d}x'^2 + \mathrm{c}^2\mathrm{d}t'^2 = 0, \tag{13.85}$$

and hence, $\mathrm{d}x' = \mathrm{c}\mathrm{d}t'$ in $\mathcal{F}'$ so that $\mathrm{d}s = 0$. Hence, in this case the differential arc element between events in $\mathcal{F}'$ vanishes and the boundary lines represent the world lines of light rays and characterize the limiting velocity $v = \mathrm{c}$ in relativity theory. Furthermore, note that in the shaded region of spacetime (i.e., the absolute future and absolute past) the differential arc element is real resulting in a proper time interval that is real, whereas in the unshaded region of spacetime (i.e., the present) the differential arc element is imaginary resulting in a proper time interval that is imaginary. This latter case corresponds to velocities that are greater in magnitude than the speed of light.

The above observations lead to an intersting assertion regarding the principle of causality in relativity theory. Specifically, it follows from the analysis given above that in the *present* region of the spacetime continuum there does not exist an absolute time order of events. In particular, event o

can precede event q in one inertial frame, whereas the same event o can follow event q in another inertial frame. To deduce cause and effect the events need to be examined at the same spatial point in the spacetime continuum. However, the spacelike region of the spacetime continuum (i.e., the present) requires velocities faster than the speed of light for signals traveling from one event to another. Thus, there does not exist an inertial reference frame with respect to which the two events can occur at the same place, and hence, causality cannot be tested.

This assertion also follows from the fact that all causal events should interact physically. However, since no signal in the universe can travel faster than the velocity of light, events in the spacelike region cannot interact physically. Alternatively, in the absolute past and absolute future (i.e., the timelike region of the spacetime continuum) relativity theory yields a univocal time ordering of events occurring at the same place with these events interacting physically at given world line points. Thus, relativity is completely consistent with the principle of causality.

Finally, we close this section by noting that the four-dimensional spacetime geometry provides a rigorous and compact mathematical structure using tensor calculus for describing the flat spacetime continuum of special relativity as well as the curved spacetime continuum of general relativity [325]. Tensor equations consist of multilinear differential forms that are invariant under a given category of coordinate transformations in n-dimensional space, and hence, they provide an ideal mathematical structure for studying special and general relativity. For example, tensor index notation immediately manifests the covariance of the Minkowski spacetime under the Poincaré group without the need of lengthy calculations for checking this fact.

More specifically, tensor calculus provides the necessary mathematical structure for addressing transformation of invariance as well as transformation by covariances and contravariances. Covariant and contravariant vectors and tensors[18] provide the foundation for expressing the position, velocity, acceleration, force, and momentum four-vectors in special relativity as well as the curvature and energy-momentum tensors in general relativity. However, even though tensor analysis provides a considerable simplification to the calculus of special and general relativity, it also requires a considerable overhead in new notation and mathematical machinery [420].

---

[18] *Covariant tensors* are tensors whose components transform as their basis vectors, whereas *contravariant tensors* are tensors whose components transform inversely to their basis vectors. More specifically, a covariant tensor of order ν is a component of a tensor product of the dual of a vector space with itself ν times, whereas a contravariant tensor of order μ is a component of a tensor product of a vector space with itself μ times.

13.6 Relativistic Dynamics

In the previous sections, we considered the effects of the special theory of relativity to kinematical phenomena. In this and the next two sections, we consider dynamic phenomena and provide the necessary modifications to the laws of classical mechanics so that they are consistent with relativity theory. Since the laws of classical mechanics predicated on Newtonian dynamics are invariant under a Galilean transformation and *not* under a Lorentz transformation, they are inconsistent with relativity theory.

In addition, in classical mechanics an external force acting on a body can accelerate the body to an infinite speed, whereas the Einstein velocity addition theorem asserts that the velocity of the body cannot exceed the speed of light. Another inconsistency between Newtonian mechanics and Einsteinian mechanics is that the former allows for action-at-a distance force with action and reaction forces being equal (Newton's third law). With the exception of contact forces, this is meaningless in Einsteinian mechanics due to the relative nature of simultaneity being intimately connected with the flow of time. Specifically, even though we can assert that ordered time between two events occurring at the *same place* can be absolutely determined, if those events are separated in space, then the simultaneity of these events is a relative concept.

To develop the concepts of relativistic dynamics we confine our attention to a discrete body of finite particles and examine the principles of conservation of mass, momentum, and energy. Conservation of mass asserts that the total mass of a system of particles remains constant as the particles interact, whereas conservation of momentum asserts that the total momentum of the system of particles in the x, y, and z directional coordinates also remains constant. In accordance with the principles of relativity, these conservation laws must hold in all sets of coordinates in uniform relative motion. In other words, we require that the conservation laws are valid relativistically, and hence, are invariant under the Lorentz transformation equations.

Since the principles of conservation of mass, momentum, and energy, as well as the relativity of motion, also hold under the Galilean transformation equations in classical mechanics in all sets of coordinates in uniform motion (i.e., these laws are invariant under a Galilean transformation), it is clear that the definition of mass, momentum, and/or energy need to be redefined in such a way—reducing to their classical forms as $v/\mathrm{c} \to 0$—so that these quantities remain invariant under a Lorentz transformation. As we see next, this is achieved by assuming that the mass of the particle is a function of its velocity or, equivalently, by redefining the notion of momentum in such a way so that the law of conservation of momentum is invariant under a Lorentz transformation.

To develop the form of the mass of a moving particle first note that, using (13.61)–(13.63), the *Lorentz contraction factor* for a body moving with a velocity $u = \sqrt{u_x^2 + u_y^2 + u_z^2}$ with respect to a given system of coordinates (x, y, z, t) in $\mathcal{F}$ relative to a velocity u' in $\mathcal{F}'$ with coordinates (x', y', z', t') is given by

$$\sqrt{1 - (u/c)^2} = \frac{\sqrt{1 - (u'/c)^2}\sqrt{1 - (v/c)^2}}{1 + (v/c^2)u'_x}. \tag{13.86}$$

Next, we consider the principles of conservation of mass and momentum in the reference frames $\mathcal{F}$ and $\mathcal{F}'$. For simplicity of exposition, we consider two identical elastic particles with mass m_0 in the reference frame $\mathcal{F}'$ moving parallel to the x'-axis with velocity u' and $-u'$, respectively.

Since the particles are elastic and identical, and are moving along the same path with identical speeds but opposite direction, it follows that at the instant of collision the particles will come to rest and then rebound under the action of the elastic forces, retracing their original trajectory with velocities $-u'$ and u', respectively. Obviously, it follows from the principles of conservation of mass and momentum that the total mass and total momentum of the two particles are the same before and at the instant of collision in the reference frame $\mathcal{F}'$.

Next, we consider the same event being observed in $\mathcal{F}$ moving relative to $\mathcal{F}'$ in the x-axis with a velocity $-v$. With respect to this new coordinate system (x, y, z, t), the relativistic velocities of the two particles are u_1 and u_2, respectively, before the collision and are given by

$$u_1 = \frac{u' + v}{1 + (v/c^2)u'} \tag{13.87}$$

and

$$u_2 = \frac{-u' + v}{1 - (v/c^2)u'}. \tag{13.88}$$

Now, noting that the two velocities are not identical in either magnitude or direction and allowing for the possibility of the particle mass to be a function of the velocity, let m_1 and m_2 be the masses of the particles before the collision. Furthermore, let $M = m_1 + m_2$ denote the sum of the masses of the two particles with respect to $\mathcal{F}$. Note that at the instant of collision the two particles will come to *relative* rest in $\mathcal{F}'$, have a mass M, and have a velocity v with respect to the reference frame $\mathcal{F}$.

Requiring that the total mass and total momentum of the two particles be the same before the collision and at the instant of *relative* rest in $\mathcal{F}'$, it follows that for the reference frame $\mathcal{F}$,

$$m_1 + m_2 = M \tag{13.89}$$

and

$$m_1 u_1 + m_2 u_2 = M v. \tag{13.90}$$

Next, using (13.87)–(13.90), we obtain

$$\frac{m_1}{m_2} = \frac{1 + (v/c^2)u'}{1 - (v/c^2)u'}, \tag{13.91}$$

or, using the Lorentz contraction factor (13.86),

$$\frac{m_1}{m_2} = \frac{\sqrt{1 - (u_2/c)^2}}{\sqrt{1 - (u_1/c)^2}}. \tag{13.92}$$

Since at rest (as measured in the reference frame $\mathcal{F}'$) the *proper mass* or *rest mass* of the two particles is the same and is equal to m_0, it follows from (13.92) that when particle two is at rest relative to the reference frame $\mathcal{F}$, $m_2 = m_0$ and $u_2 = 0$, and hence, the relativistic mass $m_1 \triangleq m$ with speed $u_1 = u$ is given by[19]

$$m = \frac{m_0}{\sqrt{1 - (u/c)^2}}. \tag{13.93}$$

Here, u is the speed of the moving particle relative to the reference frame $\mathcal{F}$ and does not necessarily have any connection with changing reference frames. In other words, (13.93) is applicable in one reference frame in which all measurements are made; in physics this is called the *laboratory frame*. Finally, note that (13.93) preserves the form of the classical momentum conservation law with the momentum vector given by

$$\boldsymbol{p} = m\boldsymbol{u} = \frac{m_0 \boldsymbol{u}}{\sqrt{1 - (u/c)^2}} \tag{13.94}$$

and with the new definition of mass m given by (13.93) being necessary for guaranteeing relativistic invariance.

[19] Although the derivation for the relativistic mass was restricted to the special case of a head-on collision of two moving particles, it can be shown that an identical expression to (13.93), with u representing the total velocity of the particle (i.e., $u^2 = u_x^2 + u_y^2 + u_z^2$), can be obtained to guarantee conservation of mass and momentum for all system coordinates and for any kind of collision between particles.

13.7 Force, Work, and Kinetic Energy

Since the mass of a moving particle is a function of its velocity, in relativistic mechanics Newton's second law is generalized to

$$\boldsymbol{F} = \frac{\mathrm{d}}{\mathrm{d}t}\boldsymbol{p} = \frac{\mathrm{d}}{\mathrm{d}t}\left[\frac{m_0\boldsymbol{u}}{\sqrt{1-(u/\mathrm{c})^2}}\right], \tag{13.95}$$

or, in scalar form,

$$F_x = \frac{\mathrm{d}}{\mathrm{d}t}(mu_x) = \frac{\mathrm{d}}{\mathrm{d}t}\left[\frac{m_0u_x}{\sqrt{1-(u/\mathrm{c})^2}}\right], \tag{13.96}$$

$$F_y = \frac{\mathrm{d}}{\mathrm{d}t}(mu_y) = \frac{\mathrm{d}}{\mathrm{d}t}\left[\frac{m_0u_y}{\sqrt{1-(u/\mathrm{c})^2}}\right], \tag{13.97}$$

$$F_z = \frac{\mathrm{d}}{\mathrm{d}t}(mu_z) = \frac{\mathrm{d}}{\mathrm{d}t}\left[\frac{m_0u_z}{\sqrt{1-(u/\mathrm{c})^2}}\right], \tag{13.98}$$

where the magnitude u of the total velocity appears in the denominator of each component of vector equation (13.95). Note that it follows from (13.96)–(13.98) that unlike Newtonian mechanics, wherein force and acceleration act in the same direction, in Einsteinian mechanics relativistic force does not necessarily act in the same direction as acceleration.

Next, using the relativistic force law (13.95) and the fact that the work done on a particle is equal to the applied force on the particle multiplied by the distance through which the particle is displaced in the direction of the applied force, it follows that

$$\mathrm{d}W = \boldsymbol{F}^{\mathrm{T}}\mathrm{d}\boldsymbol{r}, \tag{13.99}$$

where $\boldsymbol{r}$ is the position vector determining the position of the particle in the laboratory frame. Now, since the work done by a resultant force acting on a particle is equal to the change in the particle's kinetic energy E_{KE}, it follows from (13.95) and (13.99) that

$$\begin{aligned}\mathrm{d}E_{\mathrm{KE}} &= m\left(\frac{\mathrm{d}\boldsymbol{u}}{\mathrm{d}t}\right)^{\mathrm{T}}\mathrm{d}\boldsymbol{r} + \frac{\mathrm{d}m}{\mathrm{d}t}\boldsymbol{u}^{\mathrm{T}}\mathrm{d}\boldsymbol{r}\\ &= m\boldsymbol{u}^{\mathrm{T}}\mathrm{d}\boldsymbol{u} + \|\boldsymbol{u}\|_2^2\mathrm{d}m\\ &= mu\mathrm{d}u + u^2\mathrm{d}m.\end{aligned} \tag{13.100}$$

Using (13.93) and noting that

$$\frac{\mathrm{d}m}{\mathrm{d}u} = \frac{m_0\left(u/\mathrm{c}^2\right)}{\left[1-(u/\mathrm{c})^2\right]^{3/2}}, \tag{13.101}$$

it follows from (13.100) that

$$\begin{aligned} \mathrm{d}E_{\mathrm{KE}} &= \frac{m_0 u}{\sqrt{1-(u/\mathrm{c})^2}}\mathrm{d}u + \frac{m_0\left(u^3/\mathrm{c}^2\right)}{\left[1-(u/\mathrm{c})^2\right]^{3/2}}\mathrm{d}u \\ &= \frac{m_0 u}{\left[1-(u/\mathrm{c})^2\right]^{3/2}}\mathrm{d}u, \end{aligned} \tag{13.102}$$

which shows that the change in the relativistic kinetic energy of a particle is a function of the change in the particle's velocity and this change is path independent.

Next, using (13.93), it follows that

$$m^2\mathrm{c}^2 - m^2u^2 = m_0^2\mathrm{c}^2, \tag{13.103}$$

which, in differential form, gives

$$mu\mathrm{d}u + u^2\mathrm{d}m = \mathrm{c}^2\mathrm{d}m. \tag{13.104}$$

Now, using (13.104) and integrating (13.100) from an initial zero velocity to an arbitrary terminal velocity u yields

$$E_{\mathrm{KE}} = \int_0^u \mathrm{c}^2\mathrm{d}m = \mathrm{c}^2\int_{m_0}^m \mathrm{d}m = m\mathrm{c}^2 - m_0\mathrm{c}^2, \tag{13.105}$$

or, equivalently, using (13.93),

$$E_{\mathrm{KE}} = \frac{m_0\mathrm{c}^2}{\sqrt{1-(u/\mathrm{c})^2}} - m_0\mathrm{c}^2. \tag{13.106}$$

Equation (13.106) gives an expression for the relativistic kinetic energy of a particle with rest mass m_0 moving with a velocity u. Furthermore, it follows from (13.105) that

$$\mathrm{d}E_{\mathrm{KE}} = \mathrm{c}^2\mathrm{d}m, \tag{13.107}$$

which shows that the change in the relativistic kinetic energy of a particle is proportional to a change in its inertial mass.

Defining $m\mathrm{c}^2 \triangleq E$, where E denotes the *total energy* associated with the relativistic mass m, (13.106) can be rewritten as

$$E = m\mathrm{c}^2 = E_{\mathrm{KE}} + m_0\mathrm{c}^2, \tag{13.108}$$

where $m_0\mathrm{c}^2$ is the *rest energy* of the particle. Note that in contrast to classical physics, wherein the energy of a particle is unique modulo a constant of integration, relativity theory fixes the arbitrary constant to $E_0 = m_0\mathrm{c}^2$ corresponding to the energy of the particle at rest.

As for the other transformation equations we developed, the expression for the relativistic kinetic energy of a particle given by (13.106) reduces to the classical form $E_{\mathrm{KE}} = \frac{1}{2}m_0u^2$ when $u/\mathrm{c} \ll 1$. To see this, note that using the binomial expansion in u/c, (13.106) can be written as

$$\begin{aligned} E_{\mathrm{KE}} &= m_0\mathrm{c}^2\left[\left(1-\frac{u^2}{\mathrm{c}^2}\right)^{-1/2}-1\right] \\ &= m_0\mathrm{c}^2\left[1+\frac{1}{2}\left(\frac{u}{\mathrm{c}}\right)^2+\frac{3}{8}\left(\frac{u}{\mathrm{c}}\right)^4+\cdots-1\right] \\ &= \frac{1}{2}m_0u^2+\frac{3}{8}\frac{u^4}{\mathrm{c}^2}+\ldots, \end{aligned} \tag{13.109}$$

which, taking the first term as the only significant term for low velocities, yields

$$E_{\mathrm{KE}} = \frac{1}{2}m_0u^2. \tag{13.110}$$

We close this section by examining the ramifications of (13.107) to the equivalence of mass and energy. In particular, (13.105) and (13.107) assert the existence of a unified concept of *mass-energy*, wherein relativistic mass can be converted into total energy and total energy into relativistic mass. This mass-energy invariance is analogous to the equivalence between work and heat asserted by the first law of thermodynamics. This equivalence between mass and energy couples the principle of conservation of energy with the principle of conservation of mass. This unified concept of mass-energy is critical in understanding the conversion between rest mass-energy and thermal mass-energy as well as nuclear reactions involving fission and fusion, wherein neither mass nor energy, as conceived classically, is conserved but where mass-energy is conserved.

Finally, we obtain an important relation between the total mass-energy E of a particle and the magnitude of its momentum $p = \sqrt{p_x^2+p_y^2+p_z^2}$. In particular, multiplying (13.103) by c^2 and using $E = m\mathrm{c}^2$ and $p = mu$ yields

$$E^2 = p^2\mathrm{c}^2 + m_0^2\mathrm{c}^4. \tag{13.111}$$

Furthermore, an important consequence of the mass-energy equivalence is that a transfer of energy will invariably involve the presence of momentum. Specifically, it follows from (13.94) and $E = m\mathrm{c}^2$ that

$$\boldsymbol{p} = m\boldsymbol{u} = \frac{E}{\mathrm{c}^2}\boldsymbol{u}. \tag{13.112}$$

More generally, an analogous expression holds for momentum density in spacetime describing the density of energy flow for all forms of energy transfer and not just transfer of energy by particle motion.

13.8 Relativistic Momentum, Energy, Mass, and Force Transformations

In the previous two sections, we introduced the concepts of relativistic mass and momentum as well as total energy and rest mass-energy. The formulas that were developed for these concepts were applicable for a single reference frame (i.e., the laboratory frame) and do not have any connection with changing reference frames. In this section, we present the transformation equations for relativistic momentum, energy, mass, and force connecting the values of these variables in a reference frame $\mathcal{F}$ to their corresponding values in a reference frame $\mathcal{F}'$ moving uniformly, with velocity v, with respect to $\mathcal{F}$ along the common x-x' axis.

It follows from (13.94) that the components of momentum vector $\boldsymbol{p}$ and the total energy $E = mc^2$ in the $\mathcal{F}$ frame are given by

$$p_x = \frac{m_0 u_x}{\sqrt{1-(u/c)^2}}, \tag{13.113}$$

$$p_y = \frac{m_0 u_y}{\sqrt{1-(u/c)^2}}, \tag{13.114}$$

$$p_z = \frac{m_0 u_z}{\sqrt{1-(u/c)^2}}, \tag{13.115}$$

$$E = \frac{m_0 c^2}{\sqrt{1-(u/c)^2}}, \tag{13.116}$$

whereas the corresponding expressions in the $\mathcal{F}'$ frame are given by

$$p'_x = \frac{m_0 u'_x}{\sqrt{1-(u'/c)^2}}, \tag{13.117}$$

$$p'_y = \frac{m_0 u'_y}{\sqrt{1-(u'/c)^2}}, \tag{13.118}$$

$$p'_z = \frac{m_0 u'_z}{\sqrt{1-(u'/c)^2}}, \tag{13.119}$$

$$E' = \frac{m_0 c^2}{\sqrt{1-(u'/c)^2}}. \tag{13.120}$$

Now, using (13.86) and (13.61)–(13.63) it can be shown, after some algebra, that the interrelations between (13.113)–(13.116) and (13.117)–(13.120) are given by

$$p_x = \frac{1}{\sqrt{1-(v/c)^2}}\left(p'_x + \frac{E'v}{c^2}\right), \tag{13.121}$$

$$p_y = p'_y, \tag{13.122}$$

$$p_z = p'_z, \tag{13.123}$$

$$E = \frac{1}{\sqrt{1-(v/c)^2}}\left(E' + vp'_x\right). \tag{13.124}$$

The inverse transformations are obtained by interchanging the primed and unprimed momenta and energies, and by replacing v by $-v$. Specifically,

$$p'_x = \frac{1}{\sqrt{1-(v/c)^2}}\left(p_x - \frac{Ev}{c^2}\right), \tag{13.125}$$

$$p'_y = p_y, \tag{13.126}$$

$$p'_z = p_z, \tag{13.127}$$

$$E' = \frac{1}{\sqrt{1-(v/c)^2}}\left(E - vp_x\right). \tag{13.128}$$

A salient feature in the momentum and energy transformation equations is the interdependence of energy and momentum of a particle in relativity theory. These equations assert that if energy and momentum are conserved in one inertial frame, then they are also conserved in all other inertial frames. In addition, if momentum (respectively, energy) is conserved, then so is energy (respectively, momentum).

It is interesting to note that if (13.125)–(13.128) are compared to the Lorentz transformation equations (13.30)–(13.33) transforming the coordinates (x, y, z, t) into the new coordinates (x', y', z', t'), then it can be seen that the components of the momentum vector (p_x, p_y, p_z) and the scaled total energy E/c^2 transform exactly as the spacetime coordinates (x, y, z, t). Specifically, rewriting (13.125) and (13.128) as

$$p'_x = \frac{p_x - v\left(E/c^2\right)}{\sqrt{1-(v/c)^2}} \tag{13.129}$$

and

$$\frac{E'}{c^2} = \frac{\left(E/c^2\right) - \left(v/c^2\right)p_x}{\sqrt{1-(v/c)^2}}, \tag{13.130}$$

it follows that (13.125)–(13.128) have exactly the same form as (13.30)–(13.33). Thus, when relativity is viewed as a four-dimensional spacetime, energy and momentum, which are separate concepts in Newtonian mechanics, form a four-vector relating the temporal component (the energy) to the spatial component (the momentum) through the Lorentz transformation equations. As we see below, a similar observation holds for power and force.

Next, using (13.116) and (13.120) along with (13.124) and (13.125), it can be shown that

$$m = \frac{m' \left[1 + \left(v/c^2\right) u'_x\right]}{\sqrt{1 - (v/c)^2}}, \tag{13.131}$$

with its inverse form given by

$$m' = \frac{m \left[1 - \left(v/c^2\right) u_x\right]}{\sqrt{1 - (v/c)^2}}. \tag{13.132}$$

Equations (13.131) and (13.132) give the transformation and inverse transformation equations for mass from one inertial frame to another. Note that differentiating (13.131) with respect to time and using $\frac{dm'}{dt} = \frac{dm'}{dt'}\frac{dt'}{dt}$ along with (13.55) yields

$$\frac{dm}{dt} = \frac{dm'}{dt'} + \frac{m'v}{c^2} \left[1 + \left(v/c^2\right) u'_x\right]^{-1} \frac{du'_x}{dt'}, \tag{13.133}$$

which gives a transformation equation for the time rate of change of mass of a particle as a function of the rate at which the particle velocity changes.

Finally, using (13.96)–(13.98), (13.131), and (13.133) it can be shown that the transformation equations for the relativistic forces are given by

$$F_x = F'_x + \frac{vu'_y}{c^2 + vu'_x} F'_y + \frac{vu'_z}{c^2 + vu'_x} F'_z, \tag{13.134}$$

$$F_y = \frac{c^2 \sqrt{1 - (v/c)^2}}{c^2 + vu'_x} F'_y, \tag{13.135}$$

$$F_z = \frac{c^2 \sqrt{1 - (v/c)^2}}{c^2 + vu'_x} F'_z. \tag{13.136}$$

Similar expressions can be obtained for the inverse transformations by interchanging the primed and unprimed velocities and forces, and by replacing v by $-v$. A salient feature in the force transformation equations is the interdependence of forces and velocities. This suggests that a force in one reference frame is related to the power generated by a force in another reference frame.

13.9 The Principle of Equivalence and General Relativity

The theory of relativity is based on the principle that absolute motion is meaningless and that all motion is relative; that is, we can only detect and measure the motion of a given body relative to another body. The special theory of relativity confines this principle to uniform translatory motion in a region of free space where strong gravitational fields are neglected. However, to effectively address the universality of relativity to cosmology, gravitational effects need to also be considered. And in this case, as noted in Section 13.1, Newton's law of gravitation needs to be modified.

This assertion follows from special relativity and the fact that Newton's law of universal gravitation is predicated on an action-at-a-distance theory. In particular, Newton's law of gravitation asserts that gravitational force interactions between bodies are transmitted instantaneously. This, however, requires that gravitational effects propagate with infinite speed in violation of the relativistic requirement that nothing in the universe can travel faster than the velocity of light.

After completing his special theory of relativity, Albert Einstein published his general theory of relativity to include reference frames in nonuniform (i.e., accelerated) motion [131]. The general theory of relativity reformulated the laws of physics so that they are invariant with respect to noninertial reference frames and constrained the propagation speed of gravitational effects to the speed of light. At the heart of general relativity is the *principle of equivalence*.

Principle of Equivalence: The laws of physics cannot distinguish between motion in a uniform gravitational field and uniform acceleration.

This principle implies that inertial forces and gravitational forces are indistinguishable, and hence, gravity and inertia are equivalent. In other words, given an inertial (i.e., nonaccelerating) reference frame $\mathcal{F}$ in a uniform gravitational field and a noninertial (i.e., accelerating) reference frame $\mathcal{F}'$ that is uniformly accelerating with respect to $\mathcal{F}$ and not subjected to a gravitational field, then the two reference frames are physically equivalent. Hence, observations carried out in the two reference frames $\mathcal{F}$ and $\mathcal{F}'$ will yield identical results.

This equivalence also holds when analogous changes are made in the states of motion in each reference frame. Specifically, if a body in $\mathcal{F}$ is accelerating in the direction of the uniform gravitational field with the magnitude of acceleration equal to the strength of the field, then all the particles of the body will behave identically as if they are in an inertial reference frame with no gravitational field. Thus, it is impossible for an observer to deduce whether the forces on the body are due to inertial effects of the accelerating body or due to the gravitational effects. Therefore, it is meaningless to speak of absolute acceleration of a system through free space; it is only meaningful to speak of relative accelerations of two systems between reference frames.

The fact that the principle of equivalence asserts that in a uniform gravitational field it is always possible to transform the spacetime coordinates such that the effects of gravity disappear, it follows that inertial mass and gravitational mass are equal. This assertion follows immediately from the fact that since all bodies that are acted on by a zero net external force will move with a uniform velocity relative to an inertial frame, the motion of these same bodies in a noninertial (i.e., accelerating) frame will fall with an identical acceleration in a uniform gravitational field. Thus, depending on the observer's viewpoint, these effects are gravitational or inertial, which leads to the equivalence between inertial and gravitational mass.

In the statement of the principle of equivalence, we included the proviso qualifier "uniform" to both the gravitational field and acceleration. In the general case of nonuniform gravitational fields, the statement of the principle of equivalence needs to be modified since the strength of the gravitational field is nonuniform, and hence, the acceleration due to gravity will be different in different parts of the field. In this case, barring any gravitational singularities, the principle of equivalence takes on a *local* nature, wherein in a sufficiently small region of the spacetime continuum the effects of gravitation can be eliminated by using local free-falling coordinates.

In other words, the statement of the principle of equivalence is reformulated so that any spacetime point can be transformed to coordinates wherein the gravitational effects will disappear over a differential region in a neighborhood of that point. The fact, however, that the principle of equivalence creates the possibility of gravitational fields that cannot be *globally* eliminated by a suitable coordinate system transformation leads to the intricate relation between geometry and gravity (i.e., metric and gravitation). In particular, space and time can warp into each other and are inextricably linked by gravitation through massive bodies shaping the curvature of spacetime.

Using the fundamental idea that relativity applies to all motion along with the principle of equivalence, and assuming that the laws of physics can be expressed in a form that is independent of a particular spacetime coordinate system, Einstein developed the field equations of general relativity relating the curvature of spacetime to the distribution of energy and momentum as

$$R_{\mu\nu} - \frac{1}{2}Rg_{\mu\nu} = \frac{8\pi G}{c^4}T_{\mu\nu}, \tag{13.137}$$

where $R_{\mu\nu}$ is the contracted Riemann-Christoffel tensor (also known as the Riemann curvature tensor), R is the invariant of $R_{\mu\nu}$ obtained through further contraction, $g_{\mu\nu}$ is the fundamental metrical tensor (gravitational potential functions), $T_{\mu\nu}$ is the energy-momentum tensor defined in any set of natural coordinates, G is Newton's gravitational constant, c is the speed of light, and the indices μ, ν take on values of $1, 2, 3,$ and 4.

This equation is derived using the *principle of covariance* from tensor calculus, wherein the laws of physics are expressed in a form that is independent of any particular choice of spacetime coordinate system with covariant expressions for the geometrical and physical relations. A complete derivation of Einstein's field equations (13.137) using the theory of tensor calculus, as developed by Ricci and Levi-Civita, appears in [325].

A year after Einstein published his general theory of relativity, he reformulated his field equations (13.137) to include the additional tensor term $\Lambda g_{\mu\nu}$ in the right-hand side of (13.137); namely,

$$R_{\mu\nu} - \frac{1}{2} R g_{\mu\nu} + \Lambda g_{\mu\nu} = \frac{8\pi G}{c^4} T_{\mu\nu}, \tag{13.138}$$

where the scalar constant Λ is known as the *cosmological constant* [128] and corresponds to the value of the energy density of the vacuum of space. The extemporaneous inclusion of the term $\Lambda g_{\mu\nu}$ in (13.137) served to, what Einstein had thought, suppress gravity from expanding the universe and resulting in a field equation model for a static universe.[20] A few years later, Einstein removed the term $\Lambda g_{\mu\nu}$ from his theory after Hubble's 1929 discovery that all galaxies outside the group containing the Milky Way are moving away from each other, leading to an expanding universe.

It is reputed that Einstein called the addition of the ad hoc term $\Lambda g_{\mu\nu}$ to his field equations (13.137) the "biggest blunder" of his life [154, p. 44]. Ironically, however, seventy years later cosmologists restored this factor to Einstein's field equations to explain the observational data of an accelerating universe from distant supernovae [379], showing that approximately 68% of the mass-energy density of the universe corresponds to dark energy. Equation (13.138) is the current standard model of cosmology and is known as the lambda cold dark matter (Lambda-CDM) model. This model explains data from measurements of many cosmological observations and leads to several interesting and surprising consequences, including the *holographic principle* [428] (i.e., a finite maximum entropy for the observable universe) and *quintessence* [374] (i.e., the postulation that dark energy is responsible for the accelerating rate of expansion of the universe).

The Riemann curvature tensor field $R_{\mu\nu}$ appearing in (13.137) and (13.138) captures the curvature of spacetime in the coordinates $p = (x^1, x^2, x^3, x^4)$, and is characterized as a semi-Riemannian manifold $\mathcal{M}$

---

[20]Contrary to what Einstein surmised, (13.138) possesses an unstable equilibrium, and hence, does not result in a static universe model at the given equilibrium. A perturbation analysis of (13.138) shows that a small expansion in the model will result in the release of vacuum energy leading to additional expansion [391, p. 59].

equipped with a nondegenerate (i.e., isomorphic) and nonpositive-definite symmetric metric tensor $g_{\mu\nu} = g_{\mu\nu}(p)$ on the tangent space $T_p\mathcal{M}$, where $p \in \mathcal{M}$. In particular, the Riemann curvature tensor field connects the metric tensor $g_{\mu\nu}$ to each point on the spacetime manifold $\mathcal{M}$, providing a measure of the degree to which $g_{\mu\nu}$ is not locally isometric (i.e., nonholonomic) to an $\mathbb{R}^4$ Euclidean flat (i.e., $\mathbb{R}^{3+1}$ Minkowski) spacetime. More specifically, $R_{\mu\nu}$ is given in terms of the Levi-Civita connection[21] (i.e., an endomorphism) and captures the noncommutativity of the covariant derivatives of the tensor field (i.e., the derivatives along the tangent vectors of $\mathcal{M}$), giving an integrability obstruction measure for the existence of an isometry (i.e., a distance-preserving transformation) to a Euclidean (flat) spacetime.

More precisely, (13.137) delineates how a given amount of mass and energy warps spacetime. The tensor $R_{\mu\nu} - \frac{1}{2}Rg_{\mu\nu}$ is called the *Einstein tensor*, which is a specific divergence-free combination of the Ricci tensor and its metric, and describes the curvature of spacetime whose effect we *perceive* as a gravitational force. The term $T_{\mu\nu}$ is called the *energy-momentum tensor* and determines the behavior of matter and energy distributed through specetime.

Even though (13.137) appears deceptively simple, it involves ten coupled nonlinear partial differential equations that are quite formidable to solve.[22] In particular, since $\mu, \nu = 1, 2, 3, 4$, (13.137) corresponds to sixteen partial differential equations. However, since for $\mu = i$ and $\nu = j$, $i, j \in \{1, 2, 3, 4\}$, (13.137) gives the same equations with $\mu = j$ and $\nu = i$, (13.137) comprises a system of ten coupled nonlinear partial differential equations.

The index terms μ, ν with values equal to 1,2, and 3 correspond to the spatial coordinates (x, y, z), whereas the index terms μ, ν with values equal to 4 correspond to the temporal coordinate t. Hence, (13.137) with $\mu \in \{1, 2, 3\}$ and $\nu = 4$ or $\mu = 4$ and $\nu = \{1, 2, 3\}$ relates space and time.

---

[21] A *Levi-Civita connection* is a torsion-free metric connection (i.e., a geometric object) defined on the tangent bundle $T\mathcal{M}$ (i.e., the set of disjoint unions of the tangent spaces $T_p\mathcal{M}$ of the semi-Riemannian manifold $\mathcal{M}$) that preserves a given Riemannian metric $g_p : T_p\mathcal{M} \times T_p\mathcal{M} \to \mathbb{R}$, $p \in \mathcal{M}$. Relative to the coordinates $p = (x^1, x^2, x^3, x^4)$, the components of the metric tensor at each point p are given by $g_{\mu\nu} \triangleq g_p\left(\frac{\partial}{\partial x^\mu}\Big|_p, \frac{\partial}{\partial x^\nu}\Big|_p\right)$ or, equivalently, in terms of the dual basis $(\mathrm{d}x^1, \mathrm{d}x^2, \mathrm{d}x^3, \mathrm{d}x^4)$ of the cotangent bundle as the multilinear function of tensors $\sum_{\mu,\nu=1}^{4,4} g_{\mu\nu}\mathrm{d}x^\mu \mathrm{d}x^\nu \triangleq g_{\mu\nu}\mathrm{d}x^\mu \otimes \mathrm{d}x^\nu$, where $\otimes$ denotes the tensor product [325].

[22] Recall that in a four-dimensional continuum corresponding to the four generalized coordinates (x^1, x^2, x^3, x^4), a covariant tensor $T_{\alpha\beta}$ of rank two is defined as a collection of sixteen quantities associated with a given point in the continuum whose values transform as $T'_{\mu\nu} = \frac{\partial x^\alpha}{\partial x'^\mu}\frac{\partial x^\beta}{\partial x'^\nu}$, where (x'^1, x'^2, x'^3, x'^4) corresponds to the new set of coordinates, which are functions of the original coordinates (x^1, x^2, x^3, x^4). Note that since the transformation factors $\frac{\partial x^i}{\partial x'^j}, i, j \in \{\alpha, \beta, \mu, \nu\}$, are in general different at different points in the continuum, higher-rank tensors (i.e., nonscalar tensors) are associated with a given point in the continuum. For further details see [420].

In this case, the energy-momentum tensor characterizes the momentum of matter moving through space, with the Einstein tensor describing how this motion induces a warping and curving of spacetime.

Alternatively, (13.137) with $\mu, \nu \in \{1, 2, 3\}$ relates only space. In this case, the energy-momentum tensor characterizes the stress (i.e., pressure and shear) that moving matter engenders in space, with the Einstein tensor describing how that matter dilates space. Finally, (13.137) with $\mu = \nu = 4$ relates only time. In this case, the energy-momentum tensor captures the effects of energy accelerating or decelerating time, with the Einstein tensor describing the change in this flow of time.

In the special case of the absence of a large gravitational mass, it can be shown that the metrical properties of the interval element[23] $\mathrm{d}s^2$ are constant, and hence, the Riemann curvature tensor $R_{\mu\nu}$ in (13.137) vanishes on the spacetime manifold. In this case, the geodesics in the spacetime curvature are determined by rectilinear (straight) lines [420]. Thus, the distinction between the spatial coordinates (x, y, z) and the temporal coordinate t is maintained, resulting in a Euclidean metric space for spacetime. Moreover, Newton's laws of motion, with mass as a function of velocity, are invariant with respect to the group of Lorentz transformations. In contrast, if the spacetime manifold with the metrical properties of the interval element $\mathrm{d}s^2$ need to account for curvilinear (i.e., nonrectilinear) geodesics corresponding to the trajectories of particles in the presence of a large gravitational mass, then the Riemann curvature tensor $R_{\mu\nu}$ does not vanish.

The solution to the ten nonlinear partial differential equations (13.137) for the ten unknown potential functions $g_{\mu\nu}$ is extremely challenging. In fact, no general solution for the system of equations given by (13.137) in known. However, under several simplifying assumptions, including the isotropy of space and homogeneity of spacetime, several solutions have been developed in the literature. These include the Schwarzschild solution [396], the Kerr solution [239], and the Friedmann-Lemaître-Robertson-Walker solution [30].

The simplest solution, however, corresponds to the case where *every* point in an inhomogeneous gravitational field without any gravitational singularities is transformed or (equivalently) characterized by a *different*

[23]The metric or potential functions of a four-dimensional manifold $\mathcal{M}$, with coordinates $p = (x^1, x^2, x^3, x^4)$, characterized by the metrical tensor $g_{\mu\nu} = g_{\mu\nu}(p)$ define an interval element $\mathrm{d}s$ corresponding to the infinitesimal vector $\mathrm{d}x^\mu$ in $\mathcal{M}$ given by the quadratic form $\mathrm{d}s^2 = g_{\mu\nu}(p)\mathrm{d}x^\mu \otimes \mathrm{d}x^\nu$, where $g_{\mu\nu} = g_{\nu\mu}$ and $\mu, \nu = 1, 2, 3, 4$. The potential functions $g_{\mu\nu}$ of the curved spacetime manifold are chosen so that particle trajectories in $\mathcal{M}$ satisfy geodesic equations [420, p. 275]. In the special case where the spacetime coordinates are orthogonal and Cartesian, the interval element $\mathrm{d}s^2$ in $\mathcal{M} = \mathbb{R}^4$ reduces to the canonical form $\mathrm{d}s^2 = \mathrm{c}^2\mathrm{d}t^2 - \mathrm{d}x^\mu \otimes \mathrm{d}x^\mu$. In this case, the Riemann curvature tensor $R_{\mu\nu}$ vanishes on $\mathcal{M}$, and hence, the geodesics in $\mathcal{M} = \mathbb{R}^{3+1}$ are determined by straight lines.

noninertial frame replacing infinitesimal points in the field. In this case, the solution to (13.137) gives the uncurved Minkowski spacetime solution of special relativity, wherein the invariance of the laws of physics with respect to the group of Lorentz transformations applies to infinitesimal regions of spacetime. And in the case where one seeks nonlocal solutions, the solutions show that matter causes spacetime to warp with spacetime curvature replacing the gravitational field in classical Newtonian theory.

The more general solutions given in [30, 239, 396] describe particular geometries of spacetime. Specifically, the solutions given in [396] describe the geometry around a nonrotating spherical mass and compute the curvature of the Schwarzschild metric. Using this metric the geodesics are computed, leading to the observed predictions of the advance precession of the perihelion of Mercury and the bending of light by matter. The solutions given in [239] describe the geometry of a rotating black hole, whereas the solution in [30] gives the trajectories for an expanding universe model.

Even though solutions to (13.137) additionally predict several other cosmological effects, such as gravitational waves, gravitational time dilation, and gravitational redshifting, perhaps the most important aspect of (13.137) is fact that it is derived by postulating that the laws of physics are invariant with respect to a group of transformations between *all* reference frames; irrespective of whether they are inertial or noninertial. In other words, *all* observers are equivalent. In addition, (13.137) yields a theory of gravitation in which non-Euclidean geometry is linked to gravitation. More specifically, spacetime curvature is determined by the presence of matter, and the warping of spacetime replaces the gravitational field of classical physics with gravitational effects propagating at the speed of light. This leads to an expanding universe wherein mass-energy creates the space it expands into and engenders time flow generating entropy as it evolves.

Chapter Fourteen

Relativistic Thermodynamics

14.1 Introduction

Classical thermodynamics as well as the dynamical systems framework formulation of thermodynamics presented in this monograph are developed for systems that are assumed to be at rest with respect to a local observer and in the absence of strong gravitational fields. To effectively address the universality of thermodynamics and the arrow of time to cosmology, the dynamical systems framework of thermodynamics developed in Chapters 2–12 needs to be extended to thermodynamic systems that are moving relative to a local observer moving with the system and a stationary observer with respect to which the system is in motion. In addition, the gravitational effects on the thermodynamics of moving systems need to be considered.

The earliest work on relativistic thermodynamics can be traced back to Einstein [129] and Planck [360], wherein they deduced that entropy is a relativistic invariant and the Lorentz transformation law gives the relativistic temperature change in a moving body as[1]

$$T' = T\sqrt{1-(v/c)^2}, \tag{14.1}$$

with the heat flux transforming as

$$\mathrm{d}Q' = \mathrm{d}Q\sqrt{1-(v/c)^2}, \tag{14.2}$$

where here we are using the coordinate reference frame notation established in Chapter 13. Specifically, T and $\mathrm{d}Q$ are the Kelvin temperature and heat flux as measured in the rest system of coordinates (i.e., proper coordinates) and T' and $\mathrm{d}Q'$ are the corresponding temperature and heat flux detected in the moving inertial system.

[1]Even though Einstein and Planck are credited with developing the first relativistic thermodynamic theory, it was von Mosengeil [457] who was the first to arrive at the relativistic temperature transformation expression of (14.1) from strictly electromagnetic considerations while determining the dynamical properties of a moving enclosure suffused with blackbody radiation.

Thus, a moving body in the Einstein-Planck theory would appear *colder* to a moving observer and the heat flux would correspondingly *diminish.* Ever since Einstein and Planck attempted to unify special relativity theory with thermodynamics, the Einstein-Planck relativistic thermodynamic theory has been subjected to a heated controversy and considerable debate over the past one hundred years; see for example [3, 12, 14, 21, 54, 74, 95, 119, 120, 155, 187, 251, 262, 263, 265–268, 283, 306, 335, 349, 436, 442, 450–452, 456, 458, 475].

This is primarily due to the fact that thermodynamic variables such as entropy, temperature, heat, and work are global quantities, and hence, they need to be defined with respect to a specific class of hyperplanes in the Minkowski spacetime. Moreover, unlike the thermodynamics of stationary systems, wherein the system energy can be uniquely decomposed into the system internal energy, the energy due to heat transfer, and the energy due to mechanical work, in relativistic thermodynamics this decomposition is not covariant since heat (a mode of energy) exchange is accompanied by momentum flow, and hence, there exist nonunique ways in defining heat and work leading to an ambiguity in the Lorentz transformation of thermal energy and temperature.

If the first and second laws of thermodynamics are required to satisfy the principle of relativity, then it is necessary to separate heat transfer from mechanical work for the thermodynamics of moving systems. In other words, a specific definition for heat transfer needs to be decided upon that further reduces to the classical definition of heat transfer for the thermodynamics of stationary systems. However, in inertially moving thermodynamic systems, the system energy transfer cannot be uniquely separated into heat and mechanical work; this arbitrary division in defining heat and mechanical work for relativistic systems has led to different formulations of relativistic thermodynamics.

The source for the nonuniqueness of how temperature transforms under the Lorentz transformation law stems from the fact that energy and momentum are interdependent in relativity theory. Furthermore, and quite surprisingly, there does not exist a mathematically precise definition of temperature in the literature. Thus, attempting to define temperature (defined only for equilibrium processes in classical thermodynamics) from the spatiotemporal (dynamic) energy-momentum tensor results in a nonunique decomposition for heat and work in relativistic systems. This observation went unnoticed by Einstein and Planck as well as many of the early architects of relativity theory, including Tolman [442], Pauli [352], and von Laue [456].

In the later stages of his life, however, Einstein expressed doubt about the thermodynamic transformation laws leading to (14.1). In a series of letters to Max von Laue in the early 1950s, Einstein changed his view from moving bodies becoming cooler, to moving bodies becoming hotter by the Lorentz factor, to finally concluding that "the temperature should be treated in every case as an invariant" [288, 289].

A little over five decades after Einstein and Planck formulated their relativistic thermodynamic theory, Ott [349] challenged the Einstein-Planck theory by alleging that the momentum due to heat transfer had been neglected in their formulation, and hence, since conservation of momentum is an integral part of conservation of energy in relativity theory, Ott claimed that their temperature transformation for relativistic thermodynamics was incorrect. Specifically, his formulation postulated the invariance of entropy and asserted that the relativistic temperature transforms as

$$T' = \frac{T}{\sqrt{1-(v/\mathrm{c})^2}}, \tag{14.3}$$

with the heat flux transforming as

$$\mathrm{d}Q' = \frac{\mathrm{d}Q}{\sqrt{1-(v/\mathrm{c})^2}}. \tag{14.4}$$

In the Ott formulation, a moving body would appear *hotter* to a moving observer and the heat flux would correspondingly be *larger*. This formulation was actually arrived at fifteen years earlier by Blanusa [54] but had gone unnoticed in the mainstream physics literature. A few years after Ott's result, similar conclusions were also independently arrived at by Arzelies [12]. Ever since the Blanusa-Ott formulation of relativistic thermodynamics appeared in the literature, numerous papers followed (see, for example, [3, 14, 74, 95, 119, 120, 155, 187, 251, 262, 263, 268, 283, 335, 436, 450, 451, 475] and the references therein) adopting the Blanusa-Ott theory as well as asserting new temperature transformation laws.

In an attempt to reconcile the two competing relativistic thermodynamic theories, Landsberg [262, 263] maintained that temperature has a real physical meaning only in a laboratory frame (i.e., the frame of reference in which the laboratory is at rest) and asserted that temperature is invariant with respect to the speed of an inertial frame. Furthermore, he asserted that it is impossible to test for thermal equilibrium between two relatively moving inertial systems. Thus, to circumvent the apparent contradiction in the temperature transformation between the Einstein-Planck and Blanusa-Ott theories, Landsberg postulated that

$$T' = T, \tag{14.5}$$

with the heat flux transforming as in (14.2).

Van Kampen [450,451] went one step further by arguing that it is impossible to establish a good thermal contact of a moving body at relativistic speeds, and hence, it is impossible to measure its temperature. This is a blatantly vacuous statement and in essence declares that thermodynamics does not exist in relativity theory. It is just as difficult to measure distance in a fast moving body with respect to a stationary reference frame. This, however, does not preclude the use of radar methods and the simultaneity concept to compare electromagnetic waves and consequently define length. Alternatively, an optical pyrometer can be used to measure the temperature of a moving body from a distance without the need for establishing thermal contact.

In the Van Kampen formulation of relativistic thermodynamics, a covariant thermal energy-momentum transfer four-vector is introduced that defines an inverse temperature four-vector resulting in Lorentz invariance of entropy, heat, and temperature; that is, $\mathcal{S}' = \mathcal{S}$, $\mathrm{d}Q' = \mathrm{d}Q$, and $T' = T$. In this case, a moving body would appear neither hotter or colder to a moving observer.

In addition to the aforementioned relativistic temperature transformations, Landsberg and Johns [268] characterize a plethora of possible temperature transformations, all of which appear plausible in unifying thermodynamics with special relativity. They conclude that satisfactory temperature measurements in moving thermodynamic systems must lead to $T' = T$. However, their result leads to a violation of the principle of relativity as applied to the relativistic second law of thermodynamics. Balescu [21] gives a summary of the controversy between relativity and temperature, with a more recent discussion of the relativistic temperature transformation and its ramifications to statistical thermodynamics given in [265–267].

With the notable exception of [14, 306], all of the aforementioned relativistic thermodynamic formalisms are based on a Lorentz invariance of the system entropy, that is, $\mathcal{S}' = \mathcal{S}$. This erroneous deduction is based on the Boltzmann interpretation of entropy and can be traced back to the original work of Planck [360]. Namely, Planck argued that since entropy corresponds to the number of possible thermodynamic system states, this quantity cannot be a function of velocity with which an observer is moving past the thermodynamic system, and hence, entropy must be "naturally" Lorentz invariant.

The definition of entropy involving the natural logarithm of a discrete number of system states, however, is confined to thermostatics and does not account for nonequilibrium thermodynamic systems. Furthermore, all of the proposed theories are confined to a unification of thermodynamics with special relativity, and with Tolman [442] providing the most elaborate extension of thermodynamics to general relativity. Moreover, all of the relativistic thermodynamic formalisms are predicated on equilibrium systems characterized by time-invariant tensor densities in different inertial (for special relativity) and noninertial (for general relativity) reference frames.

Because of the relativistic equivalence of mass and energy, the mass associated with heat transfer in the Einstein-Planck relativistic thermodynamic theory is perceived as momentum, and hence, mechanical work by an observer in a moving frame. Alternatively, in the Ott-Arzelies relativistic thermodynamic theory this same work is viewed as part of the heat transfer. In other words, in the Einstein-Planck formulation the exchange of heat takes place at constant velocity whereas in the Ott-Arzelies formulation heat transfer takes place at constant momentum. And since both theories postulate a Lorentz invariance for entropy, it makes no physical difference which relativistic *thermostatic* formalism is adopted. In other words, there does not exist an experiment that can uniquely define heat and work in relativistic thermostatic systems.

However, as noted above, to effectively address the universality of thermodynamics to cosmogony and cosmology involving the genesis and *dynamic* structure of the cosmos, a relativistic thermodynamic formalism needs to account for nonequilibrium thermodynamic systems. And since, as shown in Chapter 3, in a dynamical systems formulation of thermodynamics the existence of a global strictly increasing entropy function on every nontrivial trajectory establishes the existence of a completely ordered time set that has a topological structure involving a closed set homeomorphic to the real line, entropy invariance is untenable.

In other words, not only are time and space intricately coupled, but given the topological isomorphism between entropy and time, and Einstein's time dilation assertion that increasing a body's speed through space results in decreasing the body's speed through time, relativistic thermodynamics leads to an *entropy dilation principle*, wherein the rate of change in entropy increase of a moving system would decrease as the system's speed increases through space.

More succinctly, motion affects the rate of entropy increase. In particular, an inertial observer moving with a system at a constant velocity relative to a stationary observer with respect to whom the thermodynamic system is in motion will experience a slower change in entropy increase when compared to the entropy of the system as deduced by the experimental findings obtained by the stationary observer. This deduction removes some of the ambiguity of relativistic thermodynamic systems predicated on thermostatics, asserts that the Boltzmann constant is *not* invariant with the speed of the inertial frame, and leads to several new and profound ramifications of relativistic thermodynamics.

14.2 Special Relativity and Thermodynamics

In this section, we use the results of Chapter 13 to develop a thermodynamic theory of moving systems. Specifically, we present the relativistic transformations for the thermodynamic variables of volume, pressure, energy, work, heat, temperature, and entropy of thermodynamic systems that are in motion relative to an inertial reference frame. Specifically, we relate the thermodynamic variables of interest in a given inertial frame $\mathcal{F}'$ with the thermodynamic variables as measured in a proper frame $\mathcal{F}$ by a local observer moving with the thermodynamic system in question.

Here we use the thermodynamic relation notions as established in Chapter 5 and the coordinate reference frame notation as established in Chapter 13. Furthermore, for simplicity of exposition, we consider a single subsystem (i.e., compartment) in our analysis characterized by two thermodynamic state variables, that is, the working substance exerting pressure on its surroundings is a fluid. The multicompartment case as delineated in Chapter 5 can be treated in an identical manner as was done in Section 5.2, with the analogous scalar-vector definitions for compartmental volume, pressure, energy, work, heat, temperature, and entropy.

Using the principle of relativity, the first and second laws of thermodynamics retain their usual forms in every inertial frame. Specifically, the first law gives

$$\Delta U = -L + Q, \tag{14.6}$$

which states that the change in energy ΔU of a thermodynamic system is a function of the system state and involves an energy balance between the heat Q received by the system and the work L done by the system. Furthermore, the second law leads to

$$\mathrm{d}S \geq \frac{\mathrm{d}Q}{T}, \tag{14.7}$$

which states that the system entropy is a function of the system states and its change $\mathrm{d}S$ is bounded from below by the infinitesimal amount of the net heat absorbed $\mathrm{d}Q$ or dissipated $-\mathrm{d}Q$ by the system at a specific temperature T. Note that both the first and second laws of thermodynamics involve statements of changes in energy and entropy, and do not provide a unique zero point for the system energy and entropy.

The mass-energy equivalence from the theory of relativity, however, gives an additional relationship connecting the change in energy and mass via the relativistic relation

$$\Delta U = \Delta m c^2, \tag{14.8}$$

where Δm denotes the increase in system mass over the system rest mass. Thus, even though the first law of thermodynamics for stationary systems gives only information of changes in system energy content without providing a unique zero point of the energy content, the relativistic energy relation

$$U = mc^2 \tag{14.9}$$

infers that a point of zero energy content corresponds to the absence of all system mass. As discussed in Chapter 3, the zero point of entropy content is provided by the third law of thermodynamics, which associates a zero entropy value to all pure perfectly crystalline substances at an absolute zero temperature.

To develop the relativistic Lorentz transformations for the thermodynamic variables, we assume that the thermodynamic system is at rest relative to the proper inertial reference frame $\mathcal{F}$. Furthermore, the thermodynamic quantities of internal energy U', heat Q', work W', volume V', pressure p', temperature T', and entropy $\mathcal{S}'$ are measured or deduced with respect to the inertial reference frame $\mathcal{F}'$ moving at constant velocity $\boldsymbol{u}$, with magnitude u, relative to the reference frame $\mathcal{F}$. Here, once again, for simplicity of exposition we choose three linearly independent and mutually perpendicular axes, and consider relative motion between the origins of $\mathcal{F}$ and $\mathcal{F}'$ to be along the common x-x' axis as shown in Figure 13.1.

The Lorentz transformation equations for the thermodynamic quantities in $\mathcal{F}'$ moving with uniform velocity relative to $\mathcal{F}$, as measured by a stationary observer in $\mathcal{F}$, should require that the laws of thermodynamics are valid relativistically, and hence, are invariant under the Lorentz transformation equations. Thus, the first and second laws of thermodynamics in the frame $\mathcal{F}'$ should be equivalent in form to a proper coordinate frame $\mathcal{F}$, wherein the thermodynamic system is at rest.

Here, we develop the transformation equations in a form relating the thermodynamic quantities in the reference frame $\mathcal{F}'$ with their respective quantities as measured or deduced in the proper frame $\mathcal{F}$ by a local observer moving with the thermodynamic system. In other words, we ask the question, If a body is moving with a constant velocity u relative to a rest (i.e., proper) system of coordinates $\mathcal{F}$, then how are its thermodynamic quantities detected in $\mathcal{F}'$ related to the thermodynamic quantities in $\mathcal{F}$ as measured or deduced in the rest system of coordinates $\mathcal{F}$?

First, we relate the compartmental volume V' of a thermodynamic system moving with uniform velocity u, as measured by a stationary observer in $\mathcal{F}$, to the compartment volume V as measured by the observer in $\mathcal{F}$. Here we assume that the thermodynamic system state is defined by energy and volume or temperature and pressure variables with the material substance in the compartment exerting equal pressure in all directions and not supporting shear. Since there is no disagreement in the length measurement between the two systems of coordinates at right angles to the line of motion and, by Lorentz contraction, a body's length will measure shorter in the direction of motion, it follows from the length contraction principle that

$$V' = V\sqrt{1-(u/c)^2}, \tag{14.10}$$

where V is the compartmental volume as measured in the proper coordinate frame $\mathcal{F}$.

Next, to relate the pressure p' in the compartment of the moving thermodynamic system relative to the compartmental pressure p as measured in the proper coordinates, recall that pressure is defined as the applied force per unit area. Now, since the relative motion between the origins of $\mathcal{F}$ and $\mathcal{F}'$ is along the common x-x' axis and the forces F_x, F_y, and F_z acting on the surfaces of the compartmental system are perpendicular to the (x, y, z) and (x', y', z') axes, it follows from the force transformation equations (13.134)–(13.136) that

$$F'_x = F_x, \tag{14.11}$$

$$F'_y = F_y\sqrt{1-(u/c)^2}, \tag{14.12}$$

$$F'_z = F_z\sqrt{1-(u/c)^2}, \tag{14.13}$$

where F_x, F_y, and F_z are the forces acting on the surfaces of the compartmental system as measured in the reference frame $\mathcal{F}$. Now, since the compartment area in the direction of motion will not be affected by length contraction, whereas the areas perpendicular to the y'- and z'-axes will undergo a length contraction of a ratio of $\sqrt{1-(u/c)^2}$ to 1, it follows from (14.11)–(14.13) and the definition of pressure as force per unit area that

$$p' = p, \tag{14.14}$$

where p is the compartmental pressure as measured in the proper coordinate frame $\mathcal{F}$.

Next, we use (13.112) to obtain an expression for the energy of the moving thermodynamic compartment. We assume that the mass density and momentum density of the compartment as measured by an inertial observer moving with the compartment are ρ' and $\boldsymbol{p}'_{\mathrm{d}}$, respectively. Here, we start the system at a state of rest and calculate the work necessary to drive the system to a given velocity $\boldsymbol{u}$. In our formulation, we define the work done by the thermodynamic system on the environment as the energy transfer due to (macroscopic) external forces, and all other modes of energy transfer are characterized as heat flow.

First, we assume that the acceleration to the given velocity $\boldsymbol{u}$ takes place adiabatically, wherein changes in the velocity take place without the flow of heat or any changes in the internal system conditions as measured by a local observer. Then, in analogy to Planck's formulation [360], we use the relation between the mass, energy, and momentum to develop an expression for the momentum of the moving compartment. Specifically, because of the mass-energy equivalence of special relativity, the system momentum will in general change as a function of energy even if the system velocity remains constant.

In particular, the total momentum density of the compartment is given by the sum of the momentum density of the moving mass in the compartment and the momentum density associated with the energy flow (i.e., power) density resulting from the work done on the moving material by the stress forces (i.e., pressure) acting on the moving compartmental walls, giving

$$\boldsymbol{p}'_{\mathrm{d}} = \rho' \boldsymbol{u} + \frac{p'}{\mathrm{c}^2} \boldsymbol{u}. \tag{14.15}$$

Now, using $\rho' = \frac{m'}{V'} = \frac{U'}{\mathrm{c}^2 V'}$, it follows from (14.15) that

$$\boldsymbol{p}'_{\mathrm{d}} = \left(\frac{U' + p'V'}{\mathrm{c}^2 V'} \right) \boldsymbol{u}, \tag{14.16}$$

and hence, the momentum of the moving compartment with volume V' is given by

$$\boldsymbol{p}' = \left(\frac{U' + p'V'}{\mathrm{c}^2} \right) \boldsymbol{u}. \tag{14.17}$$

Note that $U'\boldsymbol{u}/\mathrm{c}^2$ gives the momentum due to energy transport and $p'V'\boldsymbol{u}/\mathrm{c}^2$ gives the momentum associated with the energy flow resulting from work done on the moving compartment through the action of external pressure.

Now, the force $\boldsymbol{F}'$ generated by accelerating the compartment from a state of rest (i.e., zero velocity) to the final velocity $\boldsymbol{u}$ is given by the rate of change of momentum, and hence,

$$\boldsymbol{F}' = \frac{\mathrm{d}}{\mathrm{d}t'} \boldsymbol{p}' = \frac{\mathrm{d}}{\mathrm{d}t'} \left[\left(\frac{U' + p'V'}{\mathrm{c}^2} \right) \boldsymbol{u} \right]. \tag{14.18}$$

Next, to compute the work done along with the energy increase when the compartment moves from a state of rest to a velocity $\boldsymbol{u}$, note that the rate of change in energy is given by

$$\frac{\mathrm{d}U'}{\mathrm{d}t'} = \boldsymbol{F}'^{\mathrm{T}}\boldsymbol{u} - p'\frac{\mathrm{d}V'}{\mathrm{d}t'}, \tag{14.19}$$

where $\boldsymbol{F}'^{\mathrm{T}}\boldsymbol{u}$ denotes the power generated by the applied force $\boldsymbol{F}'$ producing the acceleration and $-p'\frac{\mathrm{d}V'}{\mathrm{d}t'}$ denotes the power dissipated by stress forces (i.e., pressure) acting on the compartment volume, which, due to the Lorentz contraction, is *decreasing* in its length in the direction of motion, that is, in the x-x' axis.

Now, using (14.10) and (14.14), noting that $p = p'$ and V are constants, and substituting (14.18) for $\boldsymbol{F}'$ in (14.19) yields

$$\begin{aligned}
\frac{\mathrm{d}U'}{\mathrm{d}t'} &= \frac{\mathrm{d}}{\mathrm{d}t'}\left[\left(\frac{U'+p'V'}{\mathrm{c}^2}\right)\boldsymbol{u}^{\mathrm{T}}\right]\boldsymbol{u} - p'\frac{\mathrm{d}V'}{\mathrm{d}t'} \\
&= \left[\frac{1}{\mathrm{c}^2}\frac{\mathrm{d}U'}{\mathrm{d}t'}\boldsymbol{u}^{\mathrm{T}} + \frac{p'}{\mathrm{c}^2}\frac{\mathrm{d}V'}{\mathrm{d}t'}\boldsymbol{u}^{\mathrm{T}} + \left(\frac{U'+p'V'}{\mathrm{c}^2}\right)\frac{\mathrm{d}\boldsymbol{u}^{\mathrm{T}}}{\mathrm{d}t'}\right]\boldsymbol{u} - p'\frac{\mathrm{d}V'}{\mathrm{d}t'} \\
&= \frac{1}{\mathrm{c}^2}\frac{\mathrm{d}U'}{\mathrm{d}t'}\boldsymbol{u}^{\mathrm{T}}\boldsymbol{u} + \frac{p'}{\mathrm{c}^2}\frac{\mathrm{d}V'}{\mathrm{d}t'}\boldsymbol{u}^{\mathrm{T}}\boldsymbol{u} + \left(\frac{U'+p'V'}{\mathrm{c}^2}\right)\frac{\mathrm{d}\boldsymbol{u}^{\mathrm{T}}}{\mathrm{d}t'}\boldsymbol{u} - p'\frac{\mathrm{d}V'}{\mathrm{d}t'} \\
&= \frac{u^2}{\mathrm{c}^2}\frac{\mathrm{d}U'}{\mathrm{d}t'} + \frac{pu^2}{\mathrm{c}^2}\frac{\mathrm{d}V'}{\mathrm{d}t'} + \left(\frac{U'+p'V'}{\mathrm{c}^2}\right)u\frac{\mathrm{d}u}{\mathrm{d}t'} - p'\frac{\mathrm{d}V'}{\mathrm{d}t'},
\end{aligned} \tag{14.20}$$

or, equivalently,

$$\left(1-\frac{u^2}{\mathrm{c}^2}\right)\frac{\mathrm{d}}{\mathrm{d}t'}\left(U'+p'V'\right) = \frac{U'+p'V'}{\mathrm{c}^2}u\frac{\mathrm{d}u}{\mathrm{d}t'}. \tag{14.21}$$

Next, rearranging (14.21) and integrating from a zero initial velocity at time $t = t' = 0$ to a final terminal velocity u at time t' yields

$$\int_0^{t'} \frac{\mathrm{d}\left(U'+p'V'\right)}{U'+p'V'} = \int_0^{u} \frac{\sigma}{\mathrm{c}^2-\sigma^2}\mathrm{d}\sigma, \tag{14.22}$$

which implies

$$\log_e\left(U'+p'V'\right) - \log_e\left(U+pV\right) = \frac{1}{2}\log_e \mathrm{c}^2 - \frac{1}{2}\left(\mathrm{c}^2-u^2\right), \tag{14.23}$$

or, equivalently,

$$\log_e\left(\frac{U'+p'V'}{U+pV}\right) = \log_e\left(\frac{\mathrm{c}}{\sqrt{\mathrm{c}^2-u^2}}\right). \tag{14.24}$$

Hence,

$$U'+p'V' = \frac{U+pV}{\sqrt{1-(u/\mathrm{c})^2}}. \tag{14.25}$$

Finally, using (14.10) and (14.14), (14.25) yields

$$U' = \frac{U + pV\left(u^2/c^2\right)}{\sqrt{1-(u/c)^2}}. \tag{14.26}$$

We now turn our attention to the differential work dW' done by the compartmental system through the external applied force necessary to maintain a constant velocity $\boldsymbol{u}$. Here, recall that it follows from relativistic dynamics that the equivalence between mass and energy couples the conservation of energy with the conservation of momentum. Hence, the momentum of the compartmental system can change even if the velocity of the system is constant when the compartmental energy changes.

Thus, to maintain a constant velocity, it is necessary to include an external force that balances the differential work done by the stress forces due to pressure. Specifically, the differential work done by the compartmental system at *constant* velocity is given by

$$dW' = p'dV' - \boldsymbol{u}^{\mathrm{T}} d\boldsymbol{p}', \tag{14.27}$$

where $p'dV'$ denotes the differential work done by the compartmental system due to stress forces (i.e., pressure) and $\boldsymbol{u}^{\mathrm{T}} d\boldsymbol{p}'$ denotes the differential work done to the compartmental system by an external force necessary to maintain the constant velocity $\boldsymbol{u}$.

Next, using (14.17) it follows from (14.27) that

$$\begin{aligned} dW' &= p'dV' - \boldsymbol{u}^{\mathrm{T}} d\left(\frac{U' + p'V'}{c^2}\right)\boldsymbol{u} \\ &= p'dV' - \frac{u^2}{c^2} d\left(U' + p'V'\right). \end{aligned} \tag{14.28}$$

Now, using (14.10), (14.14), and (14.26) yields

$$dW' = pdV\sqrt{1-(u/c)^2} - \frac{u^2/c^2}{\sqrt{1-(u/c)^2}} d\left(U + pV\right), \tag{14.29}$$

or, equivalently,

$$dW' = dW\sqrt{1-(u/c)^2} - \frac{u^2/c^2}{\sqrt{1-(u/c)^2}} d\left(U + pV\right), \tag{14.30}$$

which gives an expression for the differential work done by the moving compartment in terms of the differential work, internal energy, pressure, and volume as measured in the proper coordinate frame $\mathcal{F}$.

Using the principle of relativity, the first law of thermodynamics holds for all inertial frames, and hence, it follows from (14.6) that

$$\mathrm{d}Q' = \mathrm{d}U' + \mathrm{d}W' \tag{14.31}$$

and

$$\mathrm{d}Q = \mathrm{d}U + \mathrm{d}W. \tag{14.32}$$

Next, using (14.26) it follows that

$$\mathrm{d}U' = \frac{\mathrm{d}U + \mathrm{d}\,(pV)\,(u^2/\mathrm{c}^2)}{\sqrt{1-(u/\mathrm{c})^2}}, \tag{14.33}$$

and hence, (14.33), (14.29), and (14.31) imply

$$\begin{aligned} \mathrm{d}Q' &= \frac{\mathrm{d}U + \mathrm{d}\,(pV)\,(u^2/\mathrm{c}^2)}{\sqrt{1-(u/\mathrm{c})^2}} + \mathrm{d}W\sqrt{1-(u/\mathrm{c})^2} \\ &\quad - \frac{u^2/\mathrm{c}^2}{\sqrt{1-(u/\mathrm{c})^2}}\left[\mathrm{d}U + \mathrm{d}\,(pV)\right] \\ &= (\mathrm{d}U + \mathrm{d}W)\sqrt{1-(u/\mathrm{c})^2}. \end{aligned} \tag{14.34}$$

Now, using (14.32) it follows that

$$\mathrm{d}Q' = \mathrm{d}Q\sqrt{1-(u/\mathrm{c})^2}. \tag{14.35}$$

Finally, we turn our attention to relativistic entropy and temperature. Given the topological isomorphism between entropy and time, and Einstein's time dilation principle asserting that a body's measured time rate will measure faster when it is at rest relative to the time rate of a moving body, it follows that, in analogy to (13.54) and assuming a linear isomorphism between entropy and time,

$$\mathrm{d}\mathcal{S}' = \mathrm{d}\mathcal{S}\sqrt{1-(u/\mathrm{c})^2}. \tag{14.36}$$

Now, since by the principle of relativity the second law of thermodynamics must hold for all inertial frames, it follows from (14.7) that

$$\mathrm{d}\mathcal{S}' \geq \frac{\mathrm{d}Q'}{T'} \tag{14.37}$$

and

$$\mathrm{d}\mathcal{S} \geq \frac{\mathrm{d}Q}{T}. \tag{14.38}$$

Thus, using (14.35) and (14.36), and (14.37) and (14.38), it immediately follows that

$$T' = T, \tag{14.39}$$

which shows that temperature is invariant with the speed of the inertial reference frame.

Our formulation here presents a key departure from virtually all of the existing relativistic thermodynamic formulations in the literature, which assume that entropy is Lorentz invariant.[2] In particular, almost all relativistic thermodynamic formulations in the literature assume that the system velocity or system momentum changes through a *quasi-static* reversible adiabatic acceleration, which leaves the system internal state as well as the proper entropy $\mathcal{S}$ and the entropy $\mathcal{S}'$ with respect to the moving coordinate system unchanged. In our formulation of dynamical thermodynamics, we are not subservient to equilibrium and quasi-static processes; our concept of dynamical entropy is defined for a nonequilibrium state of a dynamical process.

It is important to note that even though we assumed that the relational change between the proper and nonproper entropy transforms exactly as the relational change between the proper and nonproper time interval, entropy dilation and time dilation with respect to a stationary observer differ by an invariant factor. In other words, motion affects the rate at which entropy increases in the same way as motion affects the rate at which time increases. Specifically, the change in system entropy slows down with motion analogously to moving clocks running slow.

Thus, since time dilation with respect to a stationary observer in relativity theory applies to every natural phenomenon involving the flow of time, biological functions such as heart rate, pulse rate, cell apoptosis, etc. would run at a slower rate with motion, resulting in a slowdown of aging. And if motion affects the rate of a biological clock in the same way it affects a physical clock, then one would expect motion to also affect entropy, which is responsible for the breakdown of every living organism and nonliving organized system in nature, in the same way. In other words, motion affects the rates of both physical and biological clocks, leading to an entropy dilation with respect to a stationary observer relative to a moving observer.

This observation is completely consistent with the correct interpretation of the twin paradox thought experiment [378]. Recall that the perceived logical contradiction in Einstein's twin paradox thought experiment follows from the notion that in relativity theory either twin can be regarded as the traveler, and hence, depending on which twin is viewed as the stationary observer, time dilation leads to each twin finding the other younger. This seemingly paradoxical conclusion is predicated on the erroneous assumption

[2] A notable exception is [14] and [306], which will be discussed in Section 14.3.

that both reference frames are symmetrical and interchangeable. However, the fact that one of the twins would have to execute a round-trip trajectory implies that the reference frame attached to the traveling twin is *not* inertial since the traveling twin would have to accelerate and decelerate over their journey. This leads to a nonsymmetrical situation wherein one twin is always in an inertial frame whereas the other twin is not.

One can also argue that since general relativity stipulates that all observers (inertial and noninertial) are equivalent, then the twin paradox is not resolved. However, because of the principle of equivalence one can easily show that this reformulation of the twin paradox is also unsound. Specifically, if we assume that the Earth moves away from the traveling twin, then the entire universe must move away with the Earth; and in executing a round-trip trajectory, the whole universe would have to accelerate, decelerate, and then accelerate again. This accelerating universe results in a gravitational field in the reference frame attached to the nontraveling twin slowing down time, whereas the traveling twin making the round-trip journey feels no accelerations. When computing the frequency shifts of light in this gravitational field, we arrive at the same conclusion as we would have obtained if we were to assume that the traveling twin executed the round-trip journey.

In arriving at (14.39) we assumed that the homeomorphic relationship between the entropy dilation and time dilation is linear, and hence, using (13.54) we deduced (14.36). The fact, however, that there exists a topological isomorphism between entropy and time does not necessarily imply that the relativistic transformation of the change in entropy scales as the relativistic transformation for time dilation. In general, (14.36) can have the more general form

$$\mathrm{d}\mathcal{S}' = f(\beta)\mathrm{d}\mathcal{S}, \tag{14.40}$$

where $f : [0, 1] \to \mathbb{R}_+$ is a continuous function of $\beta \triangleq v/c$. In this case, since (14.35) holds, the temperature will have to scale as

$$T' = g(\beta)T, \tag{14.41}$$

where $g : [0, 1] \to \mathbb{R}_+$ is continuous.

Thus, it follows from (14.35), (14.37), (14.38), (14.40), and (14.41) that

$$\mathrm{d}\mathcal{S}' = f(\beta)\mathrm{d}\mathcal{S} \geq \frac{\mathrm{d}Q'}{T'} = \frac{\mathrm{d}Q\sqrt{1-\beta^2}}{g(\beta)T}, \tag{14.42}$$

and hence,

$$\mathrm{d}\mathcal{S} \geq \frac{\mathrm{d}Q}{T}\frac{\sqrt{1-\beta^2}}{f(\beta)g(\beta)}. \tag{14.43}$$

Now, since (14.38) holds, it follows from (14.43) that

$$f(\beta)g(\beta) = \sqrt{1-\beta^2}. \tag{14.44}$$

The indeterminacy manifested through the extra degree of freedom in the inability to uniquely specify the functions $f(\beta)$ and $g(\beta)$ in (14.44) is discussed in Section 14.3.

From the above formulation, the thermodynamics of moving systems is characterized by the thermodynamic equations involving the thermodynamic transformation variables given by

$$V' = V\sqrt{1-(u/c)^2}, \tag{14.45}$$

$$p' = p, \tag{14.46}$$

$$U' = \frac{U + pV\left(u^2/c^2\right)}{\sqrt{1-(u/c)^2}}, \tag{14.47}$$

$$dW' = dW\sqrt{1-(u/c)^2} - \frac{u^2/c^2}{\sqrt{1-(u/c)^2}} d\left(U + pV\right), \tag{14.48}$$

$$dQ' = dQ\sqrt{1-(u/c)^2}, \tag{14.49}$$

$$dS' = f(\beta)dS, \tag{14.50}$$

$$T' = g(\beta)T. \tag{14.51}$$

Since these equations constitute a departure from the thermodynamic equations for stationary systems and involve second-order or higher factors of u/c, it is clear that it is impossible to experimentally verify relativistic thermodynamic theory given the present state of our scientific and technological capabilities.

14.3 Relativity, Temperature Invariance, and the Entropy Dilation Principle

In Section 14.2, we developed the relativistic transformations for the thermodynamic variables of moving systems. These equations were given by (14.45)–(14.51) and provide the relationship between the thermodynamic quantities as measured in the rest system of coordinates (i.e., proper coordinates) to the corresponding quantities detected or deduced in an inertially moving system.

However, as discussed in Section 14.1, there exist several antinomies between the relativistic transformations of temperature in the literature that have not been satisfactorily addressed and that have profound ramifications in relativistic thermodynamics and cosmology. Furthermore, with the notable exception of our formulation along with the suppositions in [14,306], all of the relativistic transformations of temperature in the literature are predicated on the erroneous conjecture that entropy is a Lorentz invariant. The topological isomorphism between entropy and time shows that this conjecture is invalid.

Even though the authors in [14, 306] arrive at similar conclusions involving a Lorentz invariance of temperature and a relativistic transformation of entropy, their hypothesis is *not* based on the mathematical formalism established in Chapter 3 yielding a topological equivalence between entropy and time, and leading to an emergence between the change in entropy and the direction of time flow. Specifically, the author in [14] conjectures without any proof (mathematical or otherwise) that temperature is invariant with speed.

This assertion is arrived at through a series of physical arguments based on phenomenological considerations that seem to lead to the invariance of temperature under a Lorentz transformation. The main argument provided alleges that if the universe is expanding with distant galaxies moving away from each other at relatively high speeds and (14.1) holds, then these galaxies should be cold and invisible. Alternatively, if (14.3) holds, then these galaxies should be infinitely hot and bright. And since neither of these are observable phenomena in the universe, [14] claims that temperature must be invariant with the speed of the inertial reference frame.

The caveat in the aforementioned argument lies in the fact that if the energy density in Einstein's field equation is positive capturing the effect of an expanding universe with distant galaxies moving at very high speeds, then the associated negative pressure captured by the Riemann curvature tensor results in an *accelerated* expansion of the universe. Thus, even though the temperature invariance assumption will hold for a noninertial observer *outside* the system, that does *not* imply that temperature is invariant if the system observed is accelerating and/or the system is subjected to a strong gravitational field.

Alternatively, if we assume that the expansion speed is constant and use Planck's formula [361] for characterizing the intensity of light as a function of frequency and temperature, then there is a key difference between the frequency shift of light due to the relativistic Doppler effect and the shift of light due to the change in temperature. Specifically, the relativistic Doppler effect would account for changes in the frequency of light due to relativistic motion between a moving light source and an observer, where time dilation results in a larger number of oscillations in time when compared to the number of oscillations in time of the moving inertial frame.

Even if we separate the light intensity change due to the relativistic Doppler effect from that of the temperature, then the shift attributed to the temperature is a product of both the temperature T *and* the Boltzmann constant k. Thus, since entropy is no longer invariant in this formulation, the product kT, and not just the Boltzmann constant k, can relativistically scale leading to a similar indeterminacy as in (14.44). This can produce the observed effect we see as distant galaxies move away in our expanding universe without relativistic temperature invariance necessarily holding.

The authors in [306] also arrive at similar conclusions as in [14] using a set-theoretic definition of temperature involving a bijective (i.e., one-to-one and onto) continuous mapping of a manifold $\mathcal{H}$ defined on a simply connected subset $\mathcal{I}$ of the reals $\mathbb{R}$. Specifically, it is shown that $\mathcal{H}$ is topologically equivalent to $\mathbb{R}$ and contains a countably dense and unbounded subset $\mathcal{F}_{\rm th}$ of all the *thermometric* fixed points contained in $\mathcal{H}$.

Thermometric fixed points characterize observable qualitative properties of a physical system state such as the boiling and freezing points of water at atmospheric pressure and provide a countable dense subset in $\mathcal{H}$ having a topological structure involving a closed set homeomorphic to the real line. The properties of the set $\mathcal{F}_{\rm th}$ and its relation to the manifold $\mathcal{H}$ are characterized by a set of postulates satisfying empirical observations. Furthermore, the mapping of the manifold $\mathcal{H}$ induces a homeomorphism on the set of positive real numbers characterizing the Kelvin temperature scale.

Using the assertion that thermometric fixed points have the same behavior in all inertial frames—a statement first proffered in [14]—the authors in [306] claim that temperature must be a Lorentz invariant. Specifically, the authors in [306] argue that water boiling in a system at rest cannot "simultaneously" *not* be boiling if observed from an inertially moving system. Hence, they conclude that every thermometric fixed point has to "correspond to the same hotness level" regardless of which inertial frame is used to observe the event.

What has been overlooked here, however, is that relativity is a theory of quantitative measurements, and how motion affects these measurements, and not what we perceive through observations. In other words, what is perceived through observation might not necessarily correspond to what is measured. This statement is equally valid for stationary systems.

For example, in Rayleigh flow the stagnation temperature is a variable, and hence, in regions wherein the flow accelerates faster than heat is added, the addition of heat to the system causes the system temperature to decrease. This is known as the Rayleigh effect and is a clear example that what is deduced through observation does not necessarily correspond to what is measured. Furthermore, since simultaneity is *not* independent of the frame of reference used to describe the water boiling event, the Lorentz transformation of time plays a key role in the observed process. In addition, even though the pressure controlling the boiling point of water is Lorentz invariant, the volume is not.

To further elucidate the afore discussion, we turn our attention to a perfect (i.e., ideal) monatomic gas. The relation between the pressure, volume, and temperature of such a gas follows from the perfect (ideal) gas law and is given by

$$pV = NkT, \tag{14.52}$$

where p is the pressure of the gas, V is the volume of the gas, N is the number of molecules present in the given volume V, k is the Boltzmann constant relating the temperature and kinetic energy in the gas, and T is the absolute temperature of the gas. Recall that the Boltzmann constant k is given by the ratio of the gas law constant R for one mole of the gas to Avogadro's number $N_{\rm A}$ quantifying the number of molecules in a mole, that is,

$$k = \frac{R}{N_{\rm A}}. \tag{14.53}$$

Now, relating the number of moles (i.e., the number density) to the number of molecules N in a given volume and the number of molecules in a mole $N_{\rm A}$ by $n = N/N_{\rm A}$, (14.52) can be equivalently written as

$$pV = nRT. \tag{14.54}$$

It is important to stress here that a perfect monatomic gas is an idealization and provides a useful abstraction in physics that has been used to develop the ideal gas temperature scale through (14.52). This temperature scale is totally arbitrary and has been concomitantly associated with the definition and measurement of temperature for historical and practical reasons. Furthermore, the molecular equation (14.52) is derived from statistical mechanics and does *not* provide a phenomenological definition of temperature as used in mainstream physics, chemistry, and engineering predicated on macroscopic sensors (e.g., thermometers). Invoking probability theory and the badly understood concepts of statistical mechanics in an attempt to define the already enigmatic physical concept of temperature has led to a lack of a satisfactory definition of temperature in the literature [305].

Since by the principle of relativity the ideal gas law must hold for all reference frames, it follows from (14.52) that

$$p'V' = Nk'T' \tag{14.55}$$

and

$$pV = NkT, \tag{14.56}$$

where the number of molecules N present in a given volume is relativistically invariant for every inertial frame. Now, using (14.45) and (14.46) it follows from (14.55) and (14.56) that

$$\begin{aligned} p'V' &= pV\sqrt{1-(u/c)^2} \\ &= NkT\sqrt{1-(u/c)^2} \\ &= Nk'T', \end{aligned} \tag{14.57}$$

which yields

$$k'T' = kT\sqrt{1-(u/c)^2}. \tag{14.58}$$

Note that (14.58) has the same form as (14.35) confirming that the relativistic transformation of the change in the product of entropy and temperature $\mathrm{d}\mathcal{S}'T'$, which can be characterized by (14.58), is equivalent to the relativistic change in heat $\mathrm{d}Q'$. Next, using (14.44) and (14.51) it follows from (14.58) that

$$k' = f(\beta)k, \tag{14.59}$$

which shows that the Boltzmann constant is *not* invariant with speed. Furthermore, since the Boltzmann constant k is related to the gas law constant R through the number of molecules N_{A} in a mole and N_{A} is a relativistic invariant, it follows from (14.53) and (14.59) that

$$R' = f(\beta)R. \tag{14.60}$$

If we can ascertain that temperature is a relativistic invariant, then $T' = T$, and hence, $g(\beta) = 1$ and $f(\beta) = \sqrt{1-(u/c)^2}$. In this case, it follows from (14.59) and (14.60) that

$$k' = k\sqrt{1-(u/c)^2} \tag{14.61}$$

and

$$R' = R\sqrt{1-(u/c)^2}. \tag{14.62}$$

This would imply that the relativistic change in entropy transforms exactly as the relativistic change in time. However, in the absence of experimental verification of relativistic temperature invariance given the present state of our scientific and technological capabilities, as well as the indeterminacy condition manifested through the mathematical analysis of the relativistic second law leading to (14.44), the assertion of relativistic temperature invariance is unverifiable.[3]

Both (14.59) and (14.61) have far-reaching consequences in physics and cosmology. The Boltzmann constant relates the translational kinetic energy of the molecules of an ideal gas at temperature T by[4] ([237])

$$E_{\mathrm{KE}} = \frac{3}{2} NkT. \tag{14.63}$$

In classical physics, k is a universal constant that is independent of the chemical composition, mass of the molecules, substance phase, etc. Relativistic thermodynamics, however, asserts that even though the Boltzmann constant can be assumed to be independent of such physical variables, it is *not* Lorentz invariant with respect to the relative velocities of inertially moving bodies.

Finally, we close this section by showing that when the relative velocity u between the two reference frames $\mathcal{F}$ and $\mathcal{F}'$ approaches the speed of light c, the rate of change in entropy increase of the moving system decreases as the system's speed increases through space. Specifically, note that it follows from (14.50) that

$$\frac{\mathrm{d}\mathcal{S}'}{\mathrm{d}t'} = \frac{\mathrm{d}\mathcal{S}}{\mathrm{d}t}\frac{\mathrm{d}t}{\mathrm{d}t'} f(\beta), \tag{14.64}$$

which, using (13.57), yields

$$\frac{\mathrm{d}\mathcal{S}'}{\mathrm{d}t'} = \left[\frac{\sqrt{1-(u/\mathrm{c})^2}}{1-(u/\mathrm{c}^2)u_x}\right] f(\beta)\frac{\mathrm{d}\mathcal{S}}{\mathrm{d}t}. \tag{14.65}$$

Thus,

$$\lim_{u\to\mathrm{c}} \frac{\mathrm{d}\mathcal{S}'}{\mathrm{d}t'} = 0, \tag{14.66}$$

which shows that motion affects the rate of entropy increase. And in particular, the change in system entropy is observer dependent and is fastest when the system is at rest relative to an observer. Hence, the change in entropy of an inertially moving system relative to a stationary observer decreases leading to an entropy contraction. Conversely, an observer in a nonproper reference frame detects a dilated change in entropy as measured in a proper coordinate frame by a local observer moving with the thermodynamic system.

---

[3]If one were to adopt Einstein's famous maxim in that a theory should be "as simple as possible, but no simpler," then one would surmise that temperature is a relativistic invariant.

[4]For sufficiently light molecules at high temperatures (14.63) becomes $E_{\mathrm{KE}} = 3NkT$ [441].

14.4 General Relativity and Thermodynamics

The extension of thermodynamics to general relativity is far more complex and subtle than the unification of special relativity with thermodynamics presented in Section 14.2. The most complete formulation attempt of general relativity and thermodynamics was developed by Tolman [442]. However, Tolman's general relativistic thermodynamics, along with its incremental extensions in the literature, are restricted to equilibrium thermodynamic systems and are almost all exclusively predicated on the archaistic assertion that entropy is a Lorentz invariant. Furthermore, the theory is based on homogeneous models for the distribution of material leading to time-invariant tensor densities in different reference frames. These assumptions, along with the unsound assertion of the invariance of entropy and the demise of the gas law and Boltzmann constants as universal constants, place these formulations of general relativistic thermodynamics in serious peril.

To viably extend thermodynamics to general relativity and address the universality of thermodynamics to cosmology, the dynamical systems framework formulation of thermodynamics developed in this monograph needs to be merged with general relativity theory, wherein the tensor currents in the Einstein field equation are explicitly dependent on time for every noninertial reference frame. This will lead to time-varying light cone observables further leading to spatial quantum fluctuations. Furthermore, since the thermodynamic variables of entropy, temperature, heat, and work are global quantities, and matter (i.e., energy and momentum) distorts (i.e., curves) spacetime in the presence of a gravitational field, spacetime curvature (i.e., confinement) can violate global conservation laws, leading to singularities when integrating tensor densities to obtain these global thermodynamic variables.

The extension of the dynamical systems framework of thermodynamics to general relativity is beyond the scope of this monograph as it requires several extensions of the developed thermodynamic system framework as well as additional machinery using tensor calculus. The aforementioned complexities notwithstanding, the notions of a thermal equilibrium in the presence of a gravitational field, the direction of heat flow in the presence of momentum transfer, and the continued progression of an isolated thermodynamic process that does not (asymptotically or in finite time) converge to an equipartitioned energy state corresponding to a state of maximum entropy all need to be reviewed.

Even the method of measurement for assessing the existence of a thermal equilibrium between moving *inertial* systems can be brought into question. If heat is exchanged between relatively moving inertial systems, then by the energy-momentum conservation principle there must also exist a momentum exchange between the systems. In this case, does heat necessarily flow in the direction of lower temperatures?

A strong gravitational field brings the notion of a uniform temperature distribution (i.e., Axiom *i*) of Chapter 4) as a necessary condition for thermal equilibrium into question as well. Specifically, in the presence of a strong gravitational field, a temperature gradient is necessary for preventing the flow of heat from a higher to a lower gravitational potential because of the equivalence between mass and energy. This is due to the fact that all forms of energy, including heat, have inertia, and hence by the equivalence principle also have weight. Thus, heat flow from a higher to a lower gravitational potential must be accompanied by a decrease in potential energy. In this case, a uniform temperature distribution would no longer be a necessary condition for the establishment of a thermal equilibrium in a gravitational field. In other words, in general relativity a temperature gradient is necessary at thermal equilibrium for preventing heat flow from a region of a higher gravitational potential to a region of a lower gravitational potential.

Since general relativity theory predicts the formation of super ultra-dense regions in the universe wherein gravitational singularities are developed during gravitational collapse of massive stars, *gravitational entropy*, that is, matter entropy generated by strong gravitational fields, needs to be accounted for in a general relativistic thermodynamic theory. Sufficient matter clustering due to a nonuniform, nonhomogeneous strong gravitational field can create black holes containing unassailable high entropy wells in the universe. Hence, black hole thermodynamics [28, 208, 209], which postulates that the measure of a black hole's entropy is proportional to the surface area of it's event horizon, needs to be merged with relativistic dynamical thermodynamics.[5] More specifically, in the presence of a strong gravitational field the rate of change of entropy through a null hypersurface[6] in a semi-Riemannian spacetime is proportional to the area of a spacelike cross section of the null hypersurface (i.e., its light sheet) [62]. More simply stated, for

[5]It is shown in [439] that for a globally semi-Riemannian spacetime topology containing compact Cauchy surfaces and satisfying certain generic restrictions on the energy-momentum tensor [210, pp. 254–255] and timelike convergence conditions on the global causal structure of spacetime [210, p. 95], the spacetime does not admit Poincaré recurrence. Since Poincaré recurrence does not hold for gravitational fields due to spacetime singularities forcing time in general relativity to be linear rather than cyclic, it is possible to define a dynamic entropy function which increases monotonically in time, and hence, we can define global thermodynamic variables and state space variables simultaneously for a closed universe governed by general relativity.

[6]A *null hypersurface* in a semi-Riemannian space $\mathcal{M}$ is a hypersurface (i.e., a submanifold of $\mathcal{M}$) whose normal vector at every point on the hypersurface has zero length with respect to the local semi-Riemannian metric. Hence, a null hypersurface in a semi-Riemannain spacetime $\mathcal{M}$ is a codimension one submanifold of $\mathcal{M}$ with a degenerate pullback (i.e., precomposition) local metric tensor to the submanifold.

any given region in space under the influence of a strong gravitational field the entropy is proportional to the region's surface area and not its volume as is the case in the absence of a strong gravitational field. This would require a change in the form of the entropy function given in Chapter 5.

Another important consideration is establishing a physically accurate kinetic equation for the description of heat transfer with finite thermal propagation speeds in the spacetime continuum. It is well known that a parabolic wave diffusion equation for characterizing thermal transport allows for the physically flawed prediction of an infinite wave speed of heat conduction between bodies in direct thermal contact [17, 52, 350, 388, 410]. Since parabolic heat transfer equations in the temperature field can be formulated as a consequence of the first and second laws of thermodynamics [409, 410], they confirm a thermodynamical prediction of infinite thermal propagation in heat conduction.

More specifically, Fourier's heat diffusion equation describes the energy flow distance from the origin of a fixed inertial frame increasing proportionally with the square root of the proper time, resulting in a proper energy transport velocity increasing inversely proportionally with the square root of the proper time [52, p. 146]. Over short time scales, this leads to a contradiction between thermodynamics and relativity theory. In particular, the fact that relativity theory asserts that the speed of light is an upper limit for all possible velocities between material systems, superluminal heat propagation speeds contradict the establishment of a virtually instantaneous thermal equilibrium between bodies in short proximity as predicted by classical thermodynamics [410].

In an attempt to develop a consistent description of heat conduction with finite speed of heat propagation, several heat transport models have been introduced in the literature [17, 176, 327, 350, 388]. The most common hyperbolic heat conduction model yielding a finite speed of thermal propagation was developed by Cattaneo [83] and Vernotte [453]. The Cattaneo-Vernotte transport law involves a modification of the classical Fourier law that transforms the parabolic heat equation into a damped wave equation by including a regularization term in the equation with a heat flux relaxation constant.

Even though several hyperbolic heat conduction models have been developed in the literature to capture finite thermal wave speeds, many of these models display the unphysical behavior that, in an adiabatically isolated system, heat can flow from regions of lower to higher temperatures over finite time intervals, contradicting the second law of thermodynamics. In addition, some models give temperature predictions below absolute zero.

This motivated authors to develop alternative hyperbolic heat conduction equations consistent with rational extended thermodynamics based on nonphysical interpretations of absolute temperature and entropy notions [236, 334]. A notable exception to this is the model developed in [410], which gives a classical thermodynamically consistent description of heat conduction with finite speed of heat propagation.

Finally, it is also not clear whether an infinite and endlessly expanding universe governed by the theory of general relativity has a final equilibrium state corresponding to a state of maximum entropy. The energy density tensor in Einstein's field equation is only covariantly conserved when changes in the curvature of the spacetime continuum need to be addressed since it does not account for gravitational energy. Moreover, it follows from relativistic mechanics that the total *proper* energy (i.e., rest energy) in an isolated system need *not* be constant. Thus, cosmological models can undergo cyclical expansions and contractions as long as there exists a sufficient amount of external energy to continue this cyclical morphing while maintaining an increase in both energy and entropy, with the latter never reaching a maximum value. In other words, cosmological models can be constructed that do not possess a limitation on the total proper energy, and hence, would also not possess an upper bound on the system entropy.

In particular, Tolman [442] shows that when coupled with expansion, matter and radiation in the universe will change their entropy contents. For example, energy will flow from high temperature radiation to lower temperature matter leading to an entropy increase in the universe. And as long as the relaxation time of such processes is greater than the expansion time scale, then these processes will generate entropy leading to an unbounded cosmological entropy generation. Alternatively, given the inexorable expansion of the universe and postulating that entropy covaries with the size of the universe, other cosmological models assert an unbounded entropy-increasing behavior of the universe; associating gravitational homogeneity with low entropy and inhomogeneity with high entropy, and with inflation serving as a low entropy initial condition state of the cosmos giving it the enormous impetus to evolve.

Chapter Fifteen

Thermodynamic Models with Subluminal Heat Propagation Speeds

15.1 Introduction

As seen in Chapter 3, nonnegative and compartmental system models play a key role in modeling thermodynamic systems. Such models comprise homogeneous interconnected subsystems (or compartments) that exchange variable quantities of energy with conservation laws describing transfer, accumulation, and outflows between compartments and the environment. The power balance equation that governs the energy exchange among coupled compartmental subsystems was shown to be completely analogous to the equations of thermal transport. A key physical limitation of such systems is that energy transfer between compartments is not instantaneous, and realistic models for capturing the dynamics of such systems should account for energy in transit between compartments [186, 226, 301].

As discussed in Section 14.4, over short time scales power balance equations predicated on Fourier heat diffusion–type equations lead to a flawed formulation of relativistic thermodynamics [52]. Specifically, Fourier heat diffusion models lead to a thermodynamical prediction of instantaneous heat extraction and heat supply as well as heat flow and temperature difference between bodies in direct thermal contact [17, 350, 388, 410]. Given that in relativity theory the propagation speed of any physical signal is constrained by the speed of light, the thermal state of any system cannot change instantaneously or superluminally.

To further elucidate the phenomenon of instantaneous speed of heat propagation in heat conduction predicated on parabolic diffusion models, consider for simplicity of exposition and without loss of generality two isolated subsystems in direct thermal contact divided by a diathermal wall and with energy transferred only as heat. Here we use the notation developed in Section 4.3 where, for $i, j \in \{1, 2\}$, $i \neq j$, E_i denotes the internal energy of subsystem i, $\mathcal{S}_i$ denotes the entropy of subsystem i, T_i denotes the temperature of subsystem i, and $\mathrm{d}Q_{ij}$ denotes the heat transferred from subsystem j to subsystem i. Furthermore, we denote by $q_{ij} \triangleq \dot{Q}_{ij}$ the rate of flow of heat from subsystem j to subsystem i with corresponding change (respectively, rate of change) in internal energy and entropy as $\mathrm{d}Q_{ij}$ and $\mathrm{d}\mathcal{S}_{ij}$ (respectively, q_{ij} and $\dot{\mathcal{S}}_{ij}$).

Now, following the analysis in [410], it follows from the first law of thermodynamics that the total internal energy U of the system is conserved, and hence,

$$\dot{U}(t) = \dot{E}_1(t) + \dot{E}_2(t) = q_{12}(t) + q_{21}(t) = 0, \quad t \geq 0, \tag{15.1}$$

or, equivalently,

$$q_{12}(t) = -q_{21}(t), \quad t \geq 0. \tag{15.2}$$

Note that (15.2) also holds in the case where the indices are interchanged. Next, it follows from the second law of thermodynamics that the total change in entropy $\mathrm{d}\mathcal{S} = \mathrm{d}\mathcal{S}_{12} + \mathrm{d}\mathcal{S}_{21}$ of the system must increase, and hence,

$$\dot{\mathcal{S}}(t) = \dot{\mathcal{S}}_1(t) + \dot{\mathcal{S}}_2(t) = \frac{q_{12}(t)}{T_1(t)} + \frac{q_{21}(t)}{T_2(t)} \geq 0, \quad t \geq 0, \tag{15.3}$$

which, using (15.2), yields

$$q_{21}(t)[T_1(t) - T_2(t)] \geq 0, \quad t \geq 0. \tag{15.4}$$

Note that (15.2) implies that heat is instantaneously propagated from subsystem 1 to subsystem 2, and hence, there does not exist a time delay between heat received and heat dissipated. Thus, (15.2) leads to an infinite speed of heat propagation in any heat transfer process involving subsystems in direct thermal contact.

To further analyze this phenomenon, consider a heat transfer scenario involving a time delay between heat supply $q_{ij}(t) > 0$, $t \geq 0$, and heat dissipation $q_{ij}(t) < 0$, $t \geq 0$, $i, j \in \{1, 2\}$, $i \neq j$. In particular, consider the three stages of heat transfer corresponding to $q_{12}(t) < 0$ and $q_{21}(t) = 0$ for $t \in [0, \tau_1)$, $q_{12}(t) < 0$ and $q_{21}(t) > 0$ for $t \in [\tau_1, \tau_2)$, and $q_{12}(t) = 0$ and $q_{21}(t) > 0$ for $t \in [\tau_2, \tau_3)$. In this case, it follows that

$$\dot{U}(t) = \dot{E}_1(t) + \dot{E}_2(t) = q_{12}(t) + q_{21}(t) < 0, \quad t \in [0, \tau_1), \tag{15.5}$$

which violates the first law of thermodynamics. Furthermore,

$$\dot{\mathcal{S}}(t) = \dot{\mathcal{S}}_1(t) + \dot{\mathcal{S}}_2(t) = \frac{q_{12}(t)}{T_1(t)} + \frac{q_{21}(t)}{T_2(t)} < 0, \quad t \in [0, \tau_1), \tag{15.6}$$

which violates the second law of thermodynamics. Equations (15.5) and (15.6) imply that there cannot exist a time delay between heat received and heat dissipated between bodies in direct thermal contact, and hence, $\tau_1 = 0$, which further implies an infinite heat propagation speed.

Next, note that a necessary and sufficient condition for (15.4) to hold is

$$q_{21}(t) = \alpha[T_1(t) - T_2(t)] \geq 0, \quad t \geq 0, \tag{15.7}$$

where α is a positive (not necessarily constant) coefficient. Equation (15.7) is Newton's law for heat transfer and shows that the instantaneous heat transfer rate between two subsystems is proportional to the instantaneous subsystem temperature difference. This shows that there does not exist a time delay between heat flow rates and temperature.

To show that Fourier's law of heat conduction follows directly from the first and second laws of thermodynamics, consider (15.4) with $q_{12} \triangleq q_i$ denoting the rate at which heat is transferred by conduction over a compact connected set $\mathcal{V} \subset \mathbb{R}^n$ with a smooth (at least C^1) boundary $\partial\mathcal{V}$ and volume $\mathcal{V}_{\mathrm{vol}}$. Now, using the notation developed in Chapter 11 for a thermodynamic continuum, it follows from the limiting form of (15.4) that the heat flows $q_i(x,t)$, $i = 1, \ldots, n$, over the divided volume of $\mathcal{V}$ parallel to the coordinate axes x_i, $i = 1, \ldots, n$, satisfy

$$q_i(x,t)\frac{\partial T}{\partial x_i}(x,t) \leq 0, \quad x \in \mathcal{V}, \quad t \geq 0, \tag{15.8}$$

and hence,

$$\nabla T(x,t)q(x,t) = \|\nabla T(x,t)\|\|q(x,t)\|\cos\theta \leq 0, \quad x \in \mathcal{V}, \quad t \geq 0, \tag{15.9}$$

where $\theta \in [0, \pi]$ is the angle between the vectors $\nabla T(x,t) \triangleq [D_1 T(x,t), \ldots, D_n T(x,t)]$ and $q(x,t) \triangleq [q_1(x,t), \ldots, q_n(x,t)]^{\mathrm{T}}$. It is important to note that (15.8) applies to an arbitrary (not necessarily Cartesian) n-dimensional space.

Note that (15.8) and (15.9) hold if and only if $\theta = \pi$. To see this, it need only be noted that necessity follows by assuming $\theta = \pi$ and showing that (15.8) holds with $q(x,t) = -\kappa[\nabla T(x,t)]^{\mathrm{T}}$, where κ is a positive scalar, whereas sufficiency follows by assuming, *ad absurdum*, $\theta \neq \pi$ and showing that in this case there exists $i \in \{1, \ldots, n\}$ such that

$$q_i(x,t)\frac{\partial T}{\partial x_i}(x,t) > 0, \quad x \in \mathcal{V}, \quad t \geq 0.$$

Hence, with $\theta = \pi$, it follows from (15.8) and (15.9) that

$$q(x,t) = -\kappa[\nabla T(x,t)]^{\mathrm{T}}, \quad x \in \mathcal{V}, \quad t \geq 0, \tag{15.10}$$

which gives Fourier's law of heat conduction and where κ is a local material property of heat conduction in the continuum.

Next, invoking conservation of energy over the divided volume of $\mathcal{V}$ by noting that the rate of energy conducted into $\mathcal{V}$ plus the rate of energy generated inside $\mathcal{V}$ must balance the rate of energy conducted out of $\mathcal{V}$ plus the rate of energy stored in $\mathcal{V}$ yields the first law of thermodynamics for a solid body as

$$\rho c \frac{\partial T(x,t)}{\partial t} = -\nabla \cdot q(x,t) + s(x,t), \quad x \in \mathcal{V}, \quad t \geq 0, \tag{15.11}$$

where ρ is the material density of the body, c is the heat capacity of the body, and $s(x,t)$ is the rate of energy generated inside $\mathcal{V}$ per unit volume and unit time. Now, using (15.10), it follows from (15.11) that

$$\frac{\partial T(x,t)}{\partial t} = \gamma \nabla^2 T(x,t) + s(x,t), \quad x \in \mathcal{V}, \quad t \geq 0, \tag{15.12}$$

where γ is the thermal diffusivity of the body.

Equation (15.12) is the classical heat conduction equation and confirms the thermodynamic prediction of an infinite thermal wave speed between bodies in direct thermal contact for thermal parabolic diffusion models. And as discussed in Section 14.4, the Fourier heat diffusion equation describes the propagation of energy flow increasing proportionally with the square root of time leading to a thermal transport speed increasing inversely proportionally with the square root of time, which can exceed the speed of light over short time scales [52, p. 146].

To accurately describe the evolution of thermodynamic systems while accounting for finite heat propagation, it is necessary to include in any mathematical model of the system dynamics some information about the past system states. This of course leads to (infinite-dimensional) delay dynamical systems [203, 339]. In this chapter, we extend the thermodynamic framework developed in Chapter 3 to account for heat flow in transit between compartments. Specifically, we develop thermodynamic models that guarantee conservation of energy, semistability, and energy equipartioning in the face of energy flow delays between compartments.

15.2 Lyapunov Stability Theory for Time Delay Nonnegative Dynamical Systems

To extend our dynamical thermodynamic formulation to account for finite speed of heat propagation, we require some additional notation, definitions, and mathematical machinery. We begin by considering the general nonlinear autonomous time delay dynamical system

$$\dot{x}(t) = f(x(t)) + f_{\mathrm{d}}(x(t-\tau_{\mathrm{d}})), \quad x(\theta) = \eta(\theta), \quad -\tau_{\mathrm{d}} \leq \theta \leq 0, \tag{15.13}$$

where $x(t) \in \mathbb{R}^n$, $f : \mathbb{R}^n \to \mathbb{R}^n$, $f_{\mathrm{d}} : \mathbb{R}^n \to \mathbb{R}^n$, $f(0) = f_{\mathrm{d}}(0) = 0$, and $\eta \in \mathcal{C} = \mathcal{C}([-\tau_{\mathrm{d}}, 0], \mathbb{R}^n)$, where $\mathcal{C}$ is a Banach space of continuous functions mapping the interval $[-\tau_d, 0]$ into $\mathbb{R}^n$ with topology of uniform convergence and designated norm given by $|||\eta||| = \sup_{-\tau_d \le \theta \le 0} \|\eta(\theta)\|$. Here, $\eta : [-\tau_d, 0] \to \mathbb{R}^n$ is a continuous vector-valued function specifying the initial state of the system.

Furthermore, let $x_t \in \mathcal{C}$ defined by $x_t(\theta) \triangleq x(t+\theta)$, $\theta \in [-\tau_d, 0]$, denote the (infinite-dimensional) state of (15.13) at time t corresponding to the *piece of trajectories* x between $t-\tau_d$ and t or, equivalently, the *element* x_t in the space of continuous functions defined on the interval $[-\tau_d, 0]$ and taking values in $\mathbb{R}^n$. Unless otherwise stated, we assume $f(\cdot)$ and $f_d(\cdot)$ are locally Lipschitz continuous on $\mathbb{R}^n$. Finally, $x_e \in \mathbb{R}^n$ is an *equilibrium point* of (15.13) if and only if $f(x_e) = 0$ and $f_d(x_e) = 0$.

The following proposition shows that $\overline{\mathbb{R}}_+^n$ is an invariant set of (15.13) if and only if f is essentially nonnegative and f_d is nonnegative.

Proposition 15.1. $\overline{\mathbb{R}}_+^n$ is an invariant set with respect to (15.13) if and only if $f : \mathbb{R}^n \to \mathbb{R}^n$ is essentially nonnegative and $f_d : \mathbb{R}^n \to \mathbb{R}^n$ is nonnegative.

Proof. Assume $\overline{\mathbb{R}}_+^n$ is invariant with respect to (15.13), $\eta(0) = x \in \overline{\mathbb{R}}_+^n$, $\eta(\theta) = 0$ for all $\theta \in [-\tau_d, 0)$, and suppose, *ad absurdum*, that x is such that there exists $i \in \{1, \ldots, q\}$ such that $x_i = 0$ and $f_i(x) < 0$. Then, since f and f_d are continuous and $f_d(0) = 0$, there exists sufficiently small $h > 0$ such that $\dot{x}_i(t) < 0$ for all $t \in [0, h)$. Hence, $x_i(t)$ is strictly decreasing on $[0, h)$, and thus $x(t) \notin \overline{\mathbb{R}}_+^n$ for all $t \in (0, h)$, which leads to a contradiction.

Next, assume that $\eta(-\tau_d) = x \in \overline{\mathbb{R}}_+^n$, $\eta(\theta) = 0$ for all $\theta \in (-\tau_d, 0]$, and suppose, *ad absurdum*, x is such that $f_{di}(x) < 0$. Hence, there exists sufficiently small $h > 0$ such that $\dot{x}_i(t) < 0$ for all $t \in [0, h)$ and $x_i(t)$ is strictly decreasing on $[0, h)$, and thus, $x(t) \notin \overline{\mathbb{R}}_+^n$ for all $t \in (0, h)$, which leads to a contradiction.

Conversely, assume that f is essentially nonnegative and f_d is nonnegative. Now, consider the nonlinear dynamical system

$$\dot{y}(t) = f(y(t)) + w(t), \quad y(0) = \eta(0), \quad t \ge 0, \tag{15.14}$$

where

$$w(t) = \begin{cases} f_d(\eta(t-\tau_d)), & 0 \le t \le \tau_d, \\ 0, & t > \tau_d. \end{cases}$$

Note that the solution $x(t)$, $t \ge 0$, to (15.13) satisfies (15.14) for all $t \in [0, \tau_d]$. Furthermore, since $f(\cdot)$ is Lipschitz continuous it follows that $y(t) = x(t)$, $t \in [0, \tau_d]$.

Next, since $f(\cdot)$ is essentially nonnegative and $w(t) \geq\geq 0$ for all $t \geq 0$, it follows from Proposition 4.3 of [194] that $y(t) \geq\geq 0$, $t \geq 0$. Hence, $x(t) \geq\geq 0$, $t \in [0, \tau_{\rm d}]$. Now, repeating the above procedure by replacing t with $t - \tau_{\rm d}$ in (15.13), we can show that $x(t) \geq\geq 0$, $t \in [\tau_{\rm d}, 2\tau_{\rm d}]$. Repeating this procedure iteratively, it follows that $x(t) \geq\geq 0$, $t \geq 0$. □

It follows from Proposition 15.1 that if $\eta(\theta) \geq\geq 0$, $\theta \in [-\tau_{\rm d}, 0]$, then $x(t) \geq\geq 0$, $t \geq 0$, if and only if f is essentially nonnegative and $f_{\rm d}$ is nonnegative. In this case, we say that (15.13) is a *nonnegative time delay dynamical system.* In this chapter, we assume that f is essentially nonnegative and $f_{\rm d}$ is nonnegative so that the nonlinear time-delay dynamical system (15.13) is a nonnegative time delay dynamical system.

The following definition introduces several types of stability for the equilibrium solution $x_t \equiv x_{\rm e} \in \overline{\mathbb{R}}_+^n$ of the nonlinear nonnegative time delay dynamical system (15.13). Define $\mathcal{C}_+ \triangleq \{\psi : [-\tau_{\rm d}, 0] \to \overline{\mathbb{R}}_+^n : \psi \in \mathcal{C}\}$ and, for $x_{\rm e} \in \mathbb{R}^n$, define $\hat{\mathcal{B}}_\delta(x_{\rm e}) \triangleq \{\psi \in \mathcal{C} : \|\psi(\theta) - x_{\rm e}\| \leq \delta, \ -\tau_{\rm d} \leq \theta \leq 0\}$.

Definition 15.1. *i*) The equilibrium solution $x_t \equiv x_{\rm e} \in \overline{\mathbb{R}}_+^n$ to (15.13) is *Lyapunov stable with respect to* $\overline{\mathbb{R}}_+^n$ if, for every $\varepsilon > 0$, there exists $\delta = \delta(\varepsilon) > 0$ such that if $\eta \in \hat{\mathcal{B}}_\delta(x_{\rm e}) \cap \mathcal{C}_+$, then $x(t, \eta(\theta)) \in \mathcal{B}_\varepsilon(x_{\rm e}) \cap \mathcal{C}_+$, $t \geq 0$.

ii) The equilibrium solution $x_t \equiv x_{\rm e} \in \overline{\mathbb{R}}_+^n$ to (15.13) is *(locally) asymptotically stable with respect to* $\overline{\mathbb{R}}_+^n$ if it is Lyapunov stable with respect to $\overline{\mathbb{R}}_+^n$ and there exists $\delta > 0$ such that if $\eta \in \hat{\mathcal{B}}_\delta(x_{\rm e}) \cap \mathcal{C}_+$, then $\lim_{t\to\infty} x(t, \eta(\theta)) = x_{\rm e}$.

iii) The equilibrium solution $x_t \equiv x_{\rm e} \in \overline{\mathbb{R}}_+^n$ to (15.13) is *globally asymptotically stable with respect to* $\overline{\mathbb{R}}_+^n$ if it is Lyapunov stable with respect to $\overline{\mathbb{R}}_+^n$ and, for every $\eta \in \mathcal{C}_+$, $\lim_{t\to\infty} x(t, \eta(\theta)) = x_{\rm e}$.

The following result gives sufficient conditions for Lyapunov and asymptotic stability of a nonlinear nonnegative dynamical system with time delay. For this result, let $V : \mathcal{C} \to \mathbb{R}$ be a continuously differentiable functional with derivative *along the trajectories* of (15.13) given by

$$\dot{V}(\psi) \triangleq \lim_{h\to 0} \frac{1}{h}[V(x_h(\psi)) - V(\psi)],$$

where $x_h(\psi)$ denotes the state of (15.13) at $t = h$ with initial condition ψ. Note that $\dot{V}(\psi)$ is dependent on the system dynamics (15.13). First, the following definition is needed.

Definition 15.2. A set $\mathcal{Q} \subseteq \mathcal{C}_+$ is *open relative to* $\mathcal{C}_+$ if there exists an open set $\mathcal{R} \subseteq \mathcal{C}$ such that $\mathcal{Q} = \mathcal{R} \cap \mathcal{C}_+$. A set $\mathcal{Q} \subseteq \mathcal{C}_+$ is *closed relative to* $\mathcal{C}_+$ if there exists a closed set $\mathcal{R} \subseteq \mathcal{C}$ such that $\mathcal{Q} = \mathcal{R} \cap \mathcal{C}_+$. A set $\mathcal{Q} \subseteq \mathcal{C}_+$ is *compact relative to* $\mathcal{C}_+$ if there exists a compact set $\mathcal{R} \subseteq \mathcal{C}$ such that $\mathcal{Q} = \mathcal{R} \cap \mathcal{C}_+$.

Theorem 15.1. Let $\mathcal{D}$ be an open subset relative to $\mathcal{C}_+$ that contains $x_{\rm e}$. Consider the nonlinear dynamical system (15.13) where f is essentially nonnegative, $f_{\rm d}$ is nonnegative, $f(x_{\rm e}) = 0$, and $f_{\rm d}(x_{\rm e}) = 0$, and assume that there exist a continuously differentiable functional $V\colon\ \mathcal{D} \to \mathbb{R}$ and a class $\mathcal{K}$ function $\alpha(\cdot)$ such that

$$V(x_{\rm e}) = 0, \tag{15.15}$$

$$\alpha(\|\psi(0) - x_{\rm e}\|) \le V(\psi), \quad \psi \in \mathcal{D}, \tag{15.16}$$

$$\dot{V}(\psi) \le 0, \quad \psi \in \mathcal{D}. \tag{15.17}$$

Then the equilibrium solution $x_t \equiv x_{\rm e}$ to (15.13) is Lyapunov stable with respect to $\overline{\mathbb{R}}^n_+$. If, in addition, there exists a class $\mathcal{K}$ function $\gamma(\cdot)$ such that

$$\dot{V}(\psi) \le -\gamma(\|\psi(0) - x_{\rm e}\|), \quad \psi \in \mathcal{D}, \tag{15.18}$$

then the equilibrium solution $x_t \equiv x_{\rm e}$ to (15.13) is asymptotically stable with respect to $\overline{\mathbb{R}}^n_+$. Finally, if, in addition, $\mathcal{D} = \mathcal{C}_+$ and $\alpha(\cdot)$ is a class $\mathcal{K}_\infty$ function, then the equilibrium solution $x_t \equiv x_{\rm e}$ to (15.13) is globally asymptotically stable with respect to $\overline{\mathbb{R}}^n_+$.

Proof. Let $\varepsilon > 0$ be such that $\hat{\mathcal{B}}_\varepsilon(x_{\rm e}) \cap \mathcal{C}_+ \subset \mathcal{D}$, define $\eta \triangleq \alpha(\varepsilon)$, and define $\mathcal{D}_\eta \triangleq \{\psi \in \hat{\mathcal{B}}_\varepsilon(x_{\rm e}) \cap \mathcal{C}_+ : V(\psi) < \eta\}$. Since $V(\cdot)$ is continuous and $V(x_{\rm e}) = 0$, it follows that $\mathcal{D}_\eta$ is nonempty and there exists $\delta = \delta(\varepsilon) > 0$ such that $V(\psi) < \eta$, $\psi \in \hat{\mathcal{B}}_\delta(x_{\rm e}) \cap \mathcal{C}_+$. Hence, $\hat{\mathcal{B}}_\delta(x_{\rm e}) \cap \mathcal{C}_+ \subseteq \mathcal{D}_\eta$.

Next, since $\dot{V}(\psi) \le 0$, it follows that $V(x_t)$ is a nonincreasing function of time, and hence, for every $\eta \in \hat{\mathcal{B}}_\delta(x_{\rm e}) \cap \mathcal{C}_+ \subseteq \mathcal{D}_\eta$, it follows that

$$\alpha(\|x(t) - x_{\rm e}\|) \le V(x_t) \le V(\eta) < \eta = \alpha(\varepsilon).$$

Thus, since f is essentially nonnegative and $f_{\rm d}$ is nonnegative it follows from Proposition 15.1 that $\mathcal{C}_+$ is invariant with respect to (15.13). Hence, for every $\eta \in \hat{\mathcal{B}}_\delta(x_{\rm e}) \cap \mathcal{C}_+$, $x(t, \eta(\theta)) \in \mathcal{B}_\varepsilon(x_{\rm e})$, $t \ge 0$, establishing Lyapunov stability with respect to $\overline{\mathbb{R}}^n_+$.

Next, assume that (15.18) holds and note that Lyapunov stability follows from the first assertion. Now, to prove asymptotic stability with respect to $\overline{\mathbb{R}}^n_+$, let $\varepsilon > 0$ and $\delta = \delta(\varepsilon)$ be such that for every $\eta \in \hat{\mathcal{B}}_\delta(x_{\rm e}) \cap \mathcal{C}_+$, $x(t, \eta(\theta)) \in \hat{\mathcal{B}}_\varepsilon(x_{\rm e}) \cap \mathcal{C}_+$, $t \ge 0$ (the existence of such a (δ, ε) pair follows from Lyapunov stability with respect to $\overline{\mathbb{R}}^n_+$). Note that $V(x_t)$ is a nonincreasing function of time and, since $V(\cdot)$ is bounded from below, it follows from the monotone convergence theorem [191, p. 37] that there exists $L \geq 0$ such that $\lim_{t\to\infty} V(x_t) = L$.

Now, suppose that for some $\psi \in \hat{\mathcal{B}}_\delta(x_e) \cap \mathcal{C}_+$, *ad absurdum*, $L > 0$ so that $\mathcal{D}_L \triangleq \{\psi \in \hat{\mathcal{B}}_\varepsilon(x_e) \cap \mathcal{C}_+ : V(\psi) \leq L\}$ is nonempty and $x(t, \eta(\theta)) \notin \mathcal{D}_L$, $t \geq 0$. Thus, as in the proof of Lyapunov stability, there exists $\hat{\delta} > 0$ such that $\hat{\mathcal{B}}_{\hat{\delta}}(x_e) \cap \mathcal{C}_+ \subset \mathcal{D}_L$. Hence, it follows from (15.18) that, for a given $\eta \in \hat{\mathcal{B}}_\delta(x_e) \cap \mathcal{C}_+ \backslash \mathcal{D}_L$ and $t \geq 0$,

$$\begin{aligned} V(x_t) &= V(\eta) + \int_0^t \dot{V}(x_s)\mathrm{d}s \\ &\leq V(\eta) - \int_0^t \gamma(\|x(s) - x_e\|)\mathrm{d}s \\ &\leq V(\eta) - \gamma(\hat{\delta})t. \end{aligned}$$

Letting $t > \frac{V(\eta)-L}{\gamma(\hat{\delta})}$, it follows that $V(x_t) < L$, which is a contradiction.

Thus, $L = 0$, and, since $\eta \in \hat{\mathcal{B}}_\delta(x_e) \cap \mathcal{C}_+$ was chosen arbitrarily, it follows that $V(x_t) \to 0$ as $t \to \infty$ for all $\eta \in \hat{\mathcal{B}}_\delta(x_e) \cap \mathcal{C}_+$. Now, since $V(x_t) \geq \alpha(\|x(t) - x_e\|) \geq 0$, it follows that $\alpha(\|x(t) - x_e\|) \to 0$ or, equivalently, $x(t, \eta(\theta)) \to x_e$ as $t \to \infty$, establishing asymptotic stability with respect to $\overline{\mathbb{R}}_+^n$.

Finally, to prove global asymptotic stability with respect to $\overline{\mathbb{R}}_+^n$, let $\delta > 0$ be such that $\eta \in \hat{\mathcal{B}}_\delta(x_e) \cap \mathcal{C}_+$ and assume $\alpha(\cdot)$ is a class $\mathcal{K}_\infty$ function. Since $\alpha(\cdot)$ is a class $\mathcal{K}_\infty$ function it follows that there exists $\varepsilon > 0$ such that $V(\eta) < \alpha(\varepsilon)$. Now, (15.18) implies that $V(x_t)$ is a nonincreasing function of time, and hence, it follows from (15.16) that

$$\alpha(\|x(t) - x_e\|) \leq V(x_t) \leq V(\eta) < \alpha(\varepsilon), \quad t \geq 0.$$

Hence, $x_t \in \mathcal{B}_\varepsilon(x_e) \cap \mathcal{C}_+$, $t \geq 0$. Now, the proof follows as in the proof of the local asymptotic stability result. □

15.3 Invariant Set Stability Theorems

In this section, we introduce the Krasovskii-LaSalle invariance principle for time delay dynamical systems. To state the main results of this section several definitions and a key theorem are needed.

First, we introduce the notion of invariance with respect to the flow $s_t(\psi)$ of a nonlinear dynamical system with time delay. Consider the nonlinear dynamical system (15.13) and for $\psi \in \mathcal{D}$ let the map $s(\cdot, \psi) : \mathbb{R} \to \mathcal{D}$ denote the solution curve or trajectory of (15.13) through the point

ψ in $\mathcal{D}$. Identifying $s(\cdot,\psi)$ with its graph, the trajectory or *orbit* of a point $\eta \in \mathcal{D}$ is defined as the motion along the curve

$$\mathcal{O}_\eta \triangleq \{\psi \in \mathcal{D} : \ \psi = s(t,\eta), \ t \in \mathbb{R}\}. \tag{15.19}$$

For $t \geq 0$, we define the *positive orbit* through the point $\eta \in \mathcal{D}$ as the motion along the curve

$$\mathcal{O}_\eta^+ \triangleq \{\psi \in \mathcal{D} : \ \psi = s(t,\eta), \ t \geq 0\}. \tag{15.20}$$

Finally, recall that if the positive orbit $\mathcal{O}_\eta^+$ of (15.13) is bounded, then $\mathcal{O}_\eta^+$ is *precompact* [201], that is, $\mathcal{O}_\eta^+$ can be enclosed in the union of a finite number of ε-balls around elements of $\mathcal{O}_\eta^+$.

Definition 15.3. A point $p \in \mathcal{D}$ is a *positive limit point* of the trajectory $s(\cdot,\psi)$ of (15.13) if there exists a monotonic sequence $\{t_n\}_{n=0}^\infty$ of positive numbers, with $t_n \to \infty$ as $n \to \infty$, such that $s(t_n,\psi) \to p$ as $n \to \infty$. The set of all positive limit points of $s(t,\psi)$, $t \geq 0$, is the *positive limit set* $\omega(\psi)$ of $s(\cdot,\psi)$ of (15.13).

Definition 15.4. A set $\mathcal{M} \subset \mathcal{D} \subseteq \mathcal{C}$ is a *positively invariant set* with respect to the nonlinear dynamical system (15.13) if $s_t(\mathcal{M}) \subseteq \mathcal{M}$ for all $t \geq 0$, where $s_t(\mathcal{M}) \triangleq \{s_t(\psi) : \ \psi \in \mathcal{M}\}$. A set $\mathcal{M} \subseteq \mathcal{D}$ is an *invariant set* with respect to the dynamical system (15.13) if $s_t(\mathcal{M}) = \mathcal{M}$ for all $t \in \mathbb{R}$.

Next, we state a key theorem involving positive limit sets. Furthermore, we use the notation $x_t \to \mathcal{M} \subseteq \mathcal{D}$ as $t \to \infty$ to denote that x_t approaches $\mathcal{M}$, that is, for each $\varepsilon > 0$ there exists $T > 0$ such that $\mathrm{dist}(x_t,\mathcal{M}) < \varepsilon$ for all $t > T$, where $\mathrm{dist}(p,\mathcal{M}) \triangleq \inf_{\psi\in\mathcal{M}} |\!|\!|p-\psi|\!|\!|$.

Theorem 15.2. Consider the nonlinear dynamical system (15.13) where f is essentially nonnegative and f_d is nonnegative. Suppose the solution x_t to (15.13) corresponding to an initial condition $\eta \in \mathcal{C}_+$ is bounded for all $t \geq 0$. Then the positive limit set $\omega(\eta)$ of x_t, $t \geq 0$, is a nonempty, compact, and invariant subset of $\mathcal{C}_+$. Furthermore, $x_t \to \omega(\eta)$ as $t \to \infty$.

Proof. It follows from Lemma 1.4 of [203, p. 103] that for every $\eta \in \mathcal{C}_+$, $\overline{\mathcal{O}}_\eta^+$ is compact. Furthermore, note that it is easy to show that

$$\omega(\eta) = \bigcap \{\overline{\mathcal{O}}_\psi^+ : \psi = s(t,\eta), \ t \geq 0\}. \tag{15.21}$$

Hence, it follows from (15.21) that $\omega(\eta)$ is nonempty and compact.

To show positive invariance of $\omega(\eta)$, let $p \in \omega(\eta)$ so that there exists an increasing unbounded sequence $\{t_n\}_{n=0}^\infty$ such that $s(t_n,\eta) \to p$ as $n \to \infty$. Now, it follows from the semigroup property that $s(t+t_n,\eta) = s(t, s(t_n,\eta))$ for all $t \geq 0$. Next, since $s(\cdot,\eta)$ is continuous, it follows that, for $t+t_n \geq 0$,

$\lim_{n\to\infty} s(t+t_n,\eta) = \lim_{n\to\infty} s(t, s(t_n,\eta)) = s(t,p)$, and hence, $s(t,p) \in \omega(\eta)$, establishing positive invariance of $\omega(\eta)$.

To show invariance of $\omega(\eta)$ let $y \in \omega(\eta)$ so that there exists an increasing unbounded sequence $\{t_n\}_{n=0}^{\infty}$ such that $s(t_n,\eta) \to y$ as $n \to \infty$. Next, let $t \in [0,\infty)$ and note that there exists N such that $t_n > t$, $n \geq N$. Hence, it follows from the semigroup property that $s(t, s(t_n - t, \eta)) = s(t_n,\eta) \to y$ as $n \to \infty$. Now, since $\overline{\mathcal{O}}_\eta^+$ is compact there exists a subsequence $\{z_{n_k}\}_{k=1}^{\infty}$ of the sequence $z_n = s(t_n - t, \eta)$, $n = N, N+1, \ldots$, such that $z_{n_k} \to z \in \mathcal{D}$ as $k \to \infty$ and, by definition, $z \in \omega(\eta)$.

Next, it follows from the continuous dependence property that

$$\lim_{k\to\infty} s(t, z_{n_k}) = s\Big(t, \lim_{k\to\infty} z_{n_k}\Big),$$

and hence, $y = s(t,z)$, which implies that $\omega(\eta) \subseteq s_t(\omega(\eta))$, $t \in [0,\infty)$. Now, using positive invariance of $\omega(\eta)$ it follows that $s_t(\omega(\eta)) = \omega(\eta)$, $t \geq 0$, establishing invariance of the positive limit set $\omega(\eta)$.

Finally, to show $s(t,\eta) \to \omega(\eta)$ as $t \to \infty$, suppose, *ad absurdum*, $s(t,\eta) \not\to \omega(\eta)$ as $t \to \infty$. In this case, there exists a sequence $\{t_n\}_{n=0}^{\infty}$, with $t_n \to \infty$ as $n \to \infty$, and $\varepsilon > 0$ such that

$$\operatorname{dist}(x(t_n), \omega(\eta)) > \varepsilon, \quad n \in \overline{\mathbb{Z}}_+. \tag{15.22}$$

However, since $\overline{\mathcal{O}}_\eta^+$ is compact there exists a convergent subsequence $\{s(t_n^*,\eta)\}_{n=0}^{\infty}$ such that $s(t_n^*,\eta) \to p^* \in \omega(\eta)$ as $n \to \infty$, which contradicts (15.22). Hence, $s(t,\eta) \to \omega(\eta)$ as $t \to \infty$. □

Next, we present the Krasovskii-LaSalle theorem for nonnegative dynamical systems with time delay.

Theorem 15.3. Consider the nonlinear dynamical system (15.13) where f is essentially nonnegative and $f_{\rm d}$ is nonnegative, assume that $\mathcal{D}_{\rm c} \subset \mathcal{D} \subseteq \mathcal{C}_+$ is a closed positively invariant set with respect to (15.13), and assume there exists a continuously differentiable functional $V\colon \mathcal{D}_{\rm c} \to \mathbb{R}$ such that $V(\cdot)$ is bounded from below and $\dot{V}(\psi) \leq 0$, $\psi \in \mathcal{D}_{\rm c}$. Let $\mathcal{R} \triangleq \{\psi \in \mathcal{D}_{\rm c} : \dot{V}(\psi) = 0\}$ and let $\mathcal{M}$ be the largest invariant set contained in $\mathcal{R}$. If $\eta \in \mathcal{D}_{\rm c}$ and x_t, $t \geq 0$, is bounded, then $x_t \to \mathcal{M}$ as $t \to \infty$.

Proof. Let x_t, $t \geq 0$, be a solution to (15.13) with $\eta \in \mathcal{D}_{\rm c}$. Since $\dot{V}(\psi) \leq 0$, $\psi \in \mathcal{D}_{\rm c}$, it follows that

$$V(x_t) - V(x_\tau) = \int_\tau^t \dot{V}(x_s){\rm d}s \leq 0, \quad t \geq \tau,$$

and hence, $V(x_t) \leq V(x_\tau)$, $t \geq \tau$, which implies that $V(x_t)$ is a nonincreasing function of t. Next, since $V(\cdot)$ is continuous on the closed set $\mathcal{D}_c$ and bounded from below, there exists $\beta \in \mathbb{R}$ such that $V(\psi) \geq \beta$, $\psi \in \mathcal{D}_c$. Hence, $\gamma_\eta \triangleq \lim_{t\to\infty} V(x_t)$ exists.

Now, for all $p \in \omega(\eta)$, there exists an increasing unbounded sequence $\{t_n\}_{n=0}^{\infty}$, with $t_0 = 0$, such that $s(t_n, \eta) \to p$ as $n \to \infty$. Since $V(\psi)$, $\psi \in \mathcal{D}_c$, is continuous, $V(p) = V(\lim_{n\to\infty} s(t_n, \eta)) = \lim_{n\to\infty} V(s(t_n, \eta)) = \gamma_\eta$, and hence, $V(\psi) = \gamma_\eta$ on $\omega(\eta)$. Now, since $x(t)$, $t \geq 0$, is bounded it follows from Theorem 15.2 that $\omega(\eta)$ is a nonempty invariant set. Hence, it follows that $\dot{V}(\psi) = 0$ on $\omega(\eta)$ and thus $\omega(\eta) \subset \mathcal{M} \subset \mathcal{R} \subset \mathcal{D}_c$. Finally, since $x_t \to \omega(\eta)$ as $t \to \infty$, it follows that $x_t \to \mathcal{M}$ as $t \to \infty$. □

Next, using Theorem 15.3 we provide a generalization of Theorem 15.1 for local asymptotic stability of a nonlinear dynamical system with time delay.

Corollary 15.1. Consider the nonlinear dynamical system (15.13) where f is essentially nonnegative and f_d is nonnegative, assume that $\mathcal{D}_c \subset \mathcal{D} \subseteq \mathcal{C}_+$ is a positively invariant set with respect to (15.13) such that $x_e \in \mathcal{D}_c$, and assume that there exist a continuously differentiable functional $V\colon \mathcal{D}_c \to \mathbb{R}$ and a class $\mathcal{K}$ function $\alpha(\cdot)$ such that $V(x_e) = 0$, $V(\psi) \geq \alpha(\|\psi(0) - x_e\|)$, and $\dot{V}(\psi) \leq 0$, $\psi \in \mathcal{D}_c$. Furthermore, assume that the set $\mathcal{R} \triangleq \{\psi \in \mathcal{D}_c\colon\ \dot{V}(\psi) = 0\}$ contains no invariant set other than the set $\{x_e\}$. Then the equilibrium solution $x_t \equiv x_e$ to (15.13) is asymptotically stable with respect to $\overline{\mathbb{R}}_+^n$.

Proof. Since $\dot{V}(\psi) \leq 0$, $\psi \in \mathcal{D}_c$, it follows from Theorem 15.3 that if $\eta \in \mathcal{D}_c$, then $\omega(\eta) \subseteq \mathcal{M}$, where $\mathcal{M}$ denotes the largest invariant set contained in $\mathcal{R}$, which implies that $\mathcal{M} = \{x_e\}$. Hence, $x_t \to \mathcal{M} = \{x_e\}$ as $t \to \infty$, establishing asymptotic stability of the equilibrium solution $x_t \equiv x_e$ to (15.13) with respect to $\overline{\mathbb{R}}_+^n$. □

Finally, we present a global invariant set theorem for guaranteeing global asymptotic stability of a nonlinear nonnegative dynamical system with time delay.

Theorem 15.4. Consider the nonlinear dynamical system (15.13) where f is essentially nonnegative and f_d is nonnegative, and assume there exist a continuously differentiable functional $V : \mathcal{C}_+ \to \mathbb{R}$ and class $\mathcal{K}_\infty$ functions $\alpha(\cdot)$ and $\gamma(\cdot)$ such that

$$V(x_e) = 0, \tag{15.23}$$

$$\alpha(\|\psi(0) - x_e\|) \leq V(\psi), \quad \psi \in \mathcal{C}_+, \tag{15.24}$$

$$\dot{V}(\psi) \leq -\gamma(\|\psi(0) - x_e\|), \quad \psi \in \mathcal{C}_+. \tag{15.25}$$

Furthermore, assume that the set $\mathcal{R} \triangleq \{\psi \in \mathcal{C}_+ : \dot{V}(\psi) = 0\}$ contains no invariant set other than the set $\{x_e\}$. Then the equilibrium solution $x_t \equiv x_e$ to (15.13) is globally asymptotically stable with respect to $\overline{\mathbb{R}}_+^n$.

Proof. Since (15.23)–(15.25) hold and $\alpha(\cdot)$ is a class $\mathcal{K}_\infty$ function, it follows from Theorem 15.1 that the equilibrium solution $x_t \equiv x_e$ to (15.13) is Lyapunov stable with respect to $\overline{\mathbb{R}}_+^n$ and bounded. Now, Theorem 15.3 implies that $x_t \to \mathcal{M}$ as $t \to \infty$. However, since $\mathcal{R}$ contains no invariant set other than the set $\{x_e\}$, the set $\mathcal{M}$ is $\{x_e\}$, and hence, global asymptotic stability with respect to $\overline{\mathbb{R}}_+^n$ is immediate. □

15.4 Linear and Nonlinear Nonnegative Dynamical Systems with Time Delay

In this section, we consider linear and nonlinear nonnegative dynamical systems with time delay along with their specialization to compartmental thermodynamic systems accounting for energy in transit between compartments. First, we consider a linear time delay dynamical system $\mathcal{G}$ of the form

$$\dot{x}(t) = Ax(t) + A_{\rm d}x(t-\tau), \quad x(\theta) = \eta(\theta), \quad -\tau \leq \theta \leq 0, \quad t \geq 0, \quad (15.26)$$

where $x(t) \in \mathbb{R}^n$, $t \geq 0$, $A \in \mathbb{R}^{n\times n}$, $A_{\rm d} \in \mathbb{R}^{n\times n}$, $\tau \geq 0$, and $\eta(\cdot) \in \mathcal{C} = \mathcal{C}([-\tau, 0], \mathbb{R}^n)$ is a continuous vector-valued function specifying the initial state of the system. Note that since $\eta(\cdot)$ is continuous, it follows from Theorem 2.1 of [203, p. 14] that there exists a unique solution $x(\eta)$ defined on $[-\tau, \infty)$ that coincides with η on $[-\tau, 0]$ and satisfies (15.26) for $t \geq 0$.

The following definition is needed for the main results of this section.

Definition 15.5. The linear time delay dynamical system $\mathcal{G}$ given by (15.26) is *nonnegative* if for every $\eta(\cdot) \in \mathcal{C}_+$, where $\mathcal{C}_+ \triangleq \{\psi(\cdot) \in \mathcal{C} : \psi(\theta) \geq\geq 0,\ \theta \in [-\tau, 0]\}$, the solution $x(t)$, $t \geq 0$, to (15.26) is nonnegative.

Proposition 15.2. The linear time delay dynamical system $\mathcal{G}$ given by (15.26) is nonnegative if and only if $A \in \mathbb{R}^{n\times n}$ is essentially nonnegative and $A_{\rm d} \in \mathbb{R}^{n\times n}$ is nonnegative.

Proof. The proof is a direct consequence of Proposition 15.1. The proof can also be shown using matrix mathematics. Specifically, the solution to (15.26) is given by

$$x(t) = e^{At}x(0) + \int_0^t e^{A(t-\theta)} A_{\rm d} x(\theta - \tau) {\rm d}\theta$$

$$= e^{At}\eta(0) + \int_{-\tau}^{t-\tau} e^{A(t-\tau-\theta)} A_{\mathrm{d}} x(\theta)\mathrm{d}\theta. \qquad (15.27)$$

Now, if A is essentially nonnegative, then it follows from Proposition 2.4 that $e^{At} \geq\geq 0$, $t \geq 0$; and if $\eta(\cdot) \in \mathcal{C}_+$ and A_{d} is nonnegative, then it follows that

$$x(t) = e^{At}\eta(0) + \int_{-\tau}^{t-\tau} e^{A(t-\tau-\theta)} A_{\mathrm{d}} \eta(\theta)\mathrm{d}\theta \geq\geq 0, \qquad t \in [0, \tau).$$

Alternatively, for all $\tau < t$,

$$x(t) = e^{A\tau} x(t-\tau) + \int_0^{\tau} e^{A(\tau-\theta)} A_{\mathrm{d}} x(t+\theta-2\tau)\mathrm{d}\theta,$$

and hence, since $x(t) \geq\geq 0$, $t \in [-\tau, \tau)$, it follows that $x(t) \geq\geq 0$, $\tau \leq t < 2\tau$. Repeating this procedure iteratively, it follows that $x(t) \geq\geq 0$, $t \geq 0$.

Conversely, assume $\mathcal{G}$ is nonnegative and suppose, *ad absurdum*, that A is not essentially nonnegative. That is, suppose there exist $I, J \in \{1, 2, \ldots, n\}$, $I \neq J$, such that $A_{(I,J)} < 0$. Now, let $\eta(\cdot) \in \mathcal{C}_+$ be such that $\eta(\theta) = 0$, $-\tau \leq \theta \leq \tau - \eta$, and $\eta(0) = e_J$, where $\tau > \eta > 0$ and $e_J \in \mathbb{R}^n$ is a vector of zeros with one in the Jth component. Next, it follows from (15.27) that

$$x(t) = e^{At} e_J, \qquad 0 \leq t < \eta.$$

Hence, for sufficiently small $T > 0$, $M_{(I,J)} < 0$, where $M \triangleq e^{AT}$, which implies that $x_I(T) < 0$, leading to a contradiction.

Now, suppose, *ad absurdum*, A_{d} is not nonnegative, that is, there exist $I, J \in \{1, 2, \ldots, n\}$ such that $A_{\mathrm{d}(I,J)} < 0$. Next, let $\{v_n\}_{n=1}^{\infty} \subset \mathcal{C}_+$ denote a sequence of functions such that $\lim_{n\to\infty} v_n(\theta) = e_J \delta(\theta + \eta - \tau)$, where $0 < \eta < \tau$ and $\delta(\cdot)$ denotes the Dirac delta function. In this case, it follows from (15.27) that

$$x_n(\eta) = e^{A\eta} v_n(0) + \int_0^{\eta} e^{A(\eta-\theta)} A_{\mathrm{d}} x(\theta - \tau)\mathrm{d}\theta,$$

where $x_n(\cdot)$ denotes the solution to (15.26) with $\phi(\theta) = v_n(\theta)$, which further implies that $x(\eta) = \lim_{n\to\infty} x_n(\eta) = e^{A\eta} A_{\mathrm{d}} e_J$. Now, by choosing η sufficiently small it follows that $x_I(\eta) < 0$, which is a contradiction. $\square$

Note that it follows from Proposition 15.2 that, since A_{d} in (15.26) is required to be nonnegative, the linear time delay dynamical system given by

$$\dot{x}(t) = Ax(t-\tau), \quad x(\theta) = \eta(\theta), \quad -\tau \leq \theta \leq 0, \quad t \geq 0, \qquad (15.28)$$

where $x(t) \in \mathbb{R}^n$, $t \geq 0$, $A \in \mathbb{R}^{n\times n}$ is essentially nonnegative, $\tau \geq 0$, and $\eta(\cdot) \in \mathcal{C}_+$, fails to yield a nonnegative trajectory under arbitrarily small delays $\tau > 0$. As an example, consider the scalar ($n = 1$) linear time delay dynamical system $\mathcal{G}$ given by

$$\dot{x}(t) = -x(t-\tau), \quad x(\theta) = \eta(\theta), \quad -\tau \leq \theta \leq 0, \quad t \geq 0.$$

Now, note that $\mathcal{G}$ is nonnegative if $\tau = 0$, whereas $\mathcal{G}$ is not nonnegative if $\tau > 0$ with $\eta(\theta) = -\theta$. To see this, it need only be noted that

$$x(\tau) = x(0) + \int_{-\tau}^{0} -x(\theta)\mathrm{d}\theta = -\tau^2/2.$$

For the remainder of this section, we assume that A is essentially nonnegative and A_d is nonnegative so that for every $\eta(\cdot) \in \mathcal{C}_+$, the linear time delay dynamical system $\mathcal{G}$ given by (15.26) is nonnegative.

Next, we present necessary and sufficient conditions for asymptotic stability for the linear time delay nonnegative dynamical system (15.26).

Theorem 15.5. Consider the linear nonnegative time delay dynamical system $\mathcal{G}$ given by (15.26) where $A \in \mathbb{R}^{n\times n}$ is essentially nonnegative and $A_\mathrm{d} \in \mathbb{R}^{n\times n}$ is nonnegative, and let $\bar{\tau} > 0$. Then $\mathcal{G}$ is asymptotically stable for all $\tau \in [0, \bar{\tau}]$ if and only if there exist $p, r \in \mathbb{R}^n$ such that $p >> 0$ and $r >> 0$ satisfy

$$0 = (A + A_\mathrm{d})^\mathrm{T} p + r. \tag{15.29}$$

Proof. To prove necessity, assume that the linear time delay dynamical system $\mathcal{G}$ given by (15.26) is asymptotically stable for all $\tau \in [0, \bar{\tau}]$. In this case, it follows that the linear nonnegative dynamical system

$$\dot{x}(t) = (A + A_\mathrm{d})x(t), \qquad x(0) = x_0 \in \overline{\mathbb{R}}_+^n, \qquad t \geq 0, \tag{15.30}$$

or, equivalently, (15.26), with $\tau = 0$, is asymptotically stable. Now, it follows from Theorem 2.11 that there exist $p >> 0$ and $r >> 0$ such that (15.29) is satisfied.

Conversely, to prove sufficiency, assume that (15.29) holds and consider the candidate Lyapunov-Krasovskii functional $V : \mathcal{C}_+ \to \mathbb{R}$ given by

$$V(\psi) = p^\mathrm{T}\psi(0) + \int_{-\tau}^{0} p^\mathrm{T} A_\mathrm{d}\psi(\theta)\mathrm{d}\theta, \qquad \psi(\cdot) \in \mathcal{C}_+.$$

Now, note that $V(\psi) \geq p^\mathrm{T}\psi(0) \geq \alpha\|\psi(0)\|$, where $\alpha \triangleq \min_{i\in\{1,2,\ldots,n\}} p_i > 0$. Next, using (15.29), it follows that the Lyapunov-Krasovskii directional derivative along the trajectories of (15.26) is given by

$$\begin{aligned}\dot{V}(x_t) &= p^\mathrm{T}\dot{x}(t) + p^\mathrm{T}A_\mathrm{d}[x(t) - x(t-\tau)]\\ &= p^\mathrm{T}(A + A_\mathrm{d})x(t)\end{aligned}$$

$$= -r^{\mathrm{T}}x(t)$$
$$\leq -\beta\|x(t)\|,$$

where $\beta \triangleq \min_{i\in\{1,2,\ldots,n\}} r_i > 0$ and $x_t(\theta) = x(t+\theta)$, $\theta \in [-\tau, 0]$, denotes the (infinite-dimensional) state of the time delay dynamical system $\mathcal{G}$. Now, it follows from Theorem 15.1 that the linear nonnegative time delay dynamical system $\mathcal{G}$ is asymptotically stable for all $\tau \in [0, \bar{\tau}]$. □

The results presented in Proposition 15.2 and Theorem 15.5 can be easily extended to systems with multiple delays of the form

$$\dot{x}(t) = Ax(t) + \sum_{i=1}^{n_{\mathrm{d}}} A_{\mathrm{d}i}x(t-\tau_i), \quad x(\theta) = \eta(\theta), \quad -\bar{\tau} \leq \theta \leq 0, \quad t \geq 0, \tag{15.31}$$

where $x(t) \in \mathbb{R}^n$, $t \geq 0$, $A \in \mathbb{R}^{n\times n}$ is essentially nonnegative, $A_{\mathrm{d}i} \in \mathbb{R}^{n\times n}$, $i = 1, \ldots, n_{\mathrm{d}}$, is nonnegative, $\bar{\tau} = \max_{i\in\{1,\ldots,n_{\mathrm{d}}\}} \tau_i$, and $\eta(\cdot) \in \{\psi(\cdot) \in \mathcal{C}([-\bar{\tau}, 0], \mathbb{R}^n) : \psi(\theta) \geq\geq 0, \ \theta \in [-\bar{\tau}, 0]\}$. In this case, (15.29) becomes

$$0 = \left(A + \sum_{i=1}^{n_{\mathrm{d}}} A_{\mathrm{d}i}\right)^{\mathrm{T}} p + r, \tag{15.32}$$

with asymptotic stability of (15.31) proved by the Lyapunov-Krasovskii functional

$$V(\psi) = p^{\mathrm{T}}\psi(0) + \sum_{i=1}^{n_{\mathrm{d}}} \int_{-\tau_i}^{0} p^{\mathrm{T}} A_{\mathrm{d}i}\psi(\theta)\mathrm{d}\theta. \tag{15.33}$$

Next, we show that inflow-closed, linear thermodynamic systems as developed in Section 3.10 with the additional constraint of time delays are a special case of the linear nonnegative time delay system given by (15.26). To see this, for $i = 1, \ldots, n$, let $E_i(t)$, $t \geq 0$, denote the energy (and hence a nonnegative quantity) of the ith subsystem of the compartmental system shown in Figure 15.1, let $\sigma_{ii} \geq 0$ denote the energy loss rate of the ith subsystem, and let $\phi_{ij}(t-\tau)$, $i \neq j$, denote the net energy flow (or heat flux) from the jth subsystem to the ith subsystem given by $\phi_{ij}(t-\tau) = \sigma_{ij}E_j(t-\tau) - \sigma_{ji}E_i(t)$, where the energy transfer rate $\sigma_{ij} \geq 0$, $i \neq j$, and τ is the fixed time it takes for the energy to flow from the jth subsystem to the ith subsystem. For simplicity of exposition we have assumed that all energy transfer times between compartments are given by τ. The more general multiple delay case can be addressed as shown in (15.31).

Now, an energy balance for the whole compartmental system yields

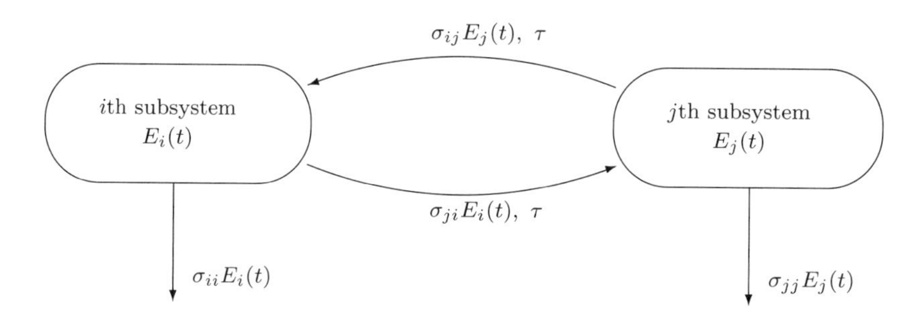

Figure 15.1 Linear inflow-closed thermodynamic model with time delay.

the delay power balance equation

$$\dot{E}_i(t) = -\left[\sigma_{ii} + \sum_{j=1, i\neq j}^{n} \sigma_{ji}\right] E_j(t) + \sum_{j=1, i\neq j}^{n} \sigma_{ij}E_j(t-\tau), \quad t \geq 0,$$
$$i = 1, \ldots, n, \quad (15.34)$$

or, equivalently, in vector form

$$\dot{E}(t) = AE(t) + A_{\mathrm{d}}E(t-\tau), \quad E(\theta) = \eta(\theta), \quad -\tau \leq \theta \leq 0, \quad t \geq 0, \quad (15.35)$$

where $E(t) = [E_1(t), \ldots, E_n(t)]^{\mathrm{T}}$, $\eta(\cdot) \in \mathcal{C}_+$, and, for $i, j = 1, \ldots, n$,

$$A_{(i,j)} = \begin{cases} -\sum_{k=1}^{n} \sigma_{ki}, & i = j, \\ 0, & i \neq j, \end{cases} \quad A_{\mathrm{d}(i,j)} = \begin{cases} 0, & i = j, \\ \sigma_{ij}, & i \neq j. \end{cases} \quad (15.36)$$

Note that A is essentially nonnegative and A_{d} is nonnegative. Furthermore, $A + A_{\mathrm{d}}$ is a compartmental matrix, and hence, it follows from Theorem 2.10 that Re $\lambda < 0$ or $\lambda = 0$, where λ is an eigenvalue of $A + A_{\mathrm{d}}$. Now, it follows from Theorems 2.11 and 15.5 that the zero solution $E_t \equiv 0$ to (15.35) is asymptotically stable for all $\tau \in [0, \bar{\tau}]$ if and only if $A + A_{\mathrm{d}}$ is Hurwitz.

Alternatively, asymptotic stability of (15.35) for all $\tau \in [0, \bar{\tau}]$ can be deduced using the Lyapunov-Krasovskii energy functional

$$V(\psi) = \mathbf{e}^{\mathrm{T}}\psi(0) + \int_{-\tau}^{0} \mathbf{e}^{\mathrm{T}}A_{\mathrm{d}}\psi(\theta)\mathrm{d}\theta, \quad \psi(\cdot) \in \mathcal{C}_+, \quad (15.37)$$

which captures the total energy of the system at $t = 0$ plus the integral of the energy flow in transit between compartments over the time intervals it takes for the energy to flow through the intercompartmental connections. In this case, it follows that $\dot{V}(E_t) \leq -\beta\|E(t)\|$, where $\beta \triangleq \min_{i\in\{1,\ldots,n\}} \sigma_{ii}$ and $E_t(\theta) = E(t+\theta)$, $\theta \in [-\tau, 0]$. This result is not surprising, since for an inflow-closed compartmental system the law of conservation of energy eliminates the possibility of unbounded solutions.

Next, we present a nonlinear extension of Proposition 15.2 and Theorem 15.5. Specifically, we consider nonlinear time delay dynamical systems $\mathcal{G}$ of the form

$$\dot{x}(t) = Ax(t) + f_{\rm d}(x(t-\tau)), \quad x(\theta) = \eta(\theta), \quad -\tau \leq \theta \leq 0, \quad t \geq 0, \tag{15.38}$$

where $x(t) \in \mathbb{R}^n$, $t \geq 0$, $A \in \mathbb{R}^{n \times n}$, $f_{\rm d} : \mathbb{R}^n \to \mathbb{R}^n$ is locally Lipschitz and $f_{\rm d}(0) = 0$, $\tau \geq 0$, and $\eta(\cdot) \in \mathcal{C}$. Once again, since $\eta(\cdot)$ is continuous, the existence and uniqueness of solutions to (15.38) follow from Theorem 2.3 of [203, p. 44]. For the nonlinear time delay dynamical system (15.38), the definition of nonnegativity holds with (15.26) replaced by (15.38).

Proposition 15.3. Consider the nonlinear time delay dynamical system $\mathcal{G}$ given by (15.38). If $A \in \mathbb{R}^{n \times n}$ is essentially nonnegative and $f_{\rm d} : \mathbb{R}^n \to \mathbb{R}^n$ is nonnegative, then $\mathcal{G}$ is nonnegative.

Proof. The solution to (15.38) is given by

$$\begin{aligned} x(t) &= e^{At}x(0) + \int_0^t e^{A(t-\theta)} f_{\rm d}(x(\theta - \tau)){\rm d}\theta \\ &= e^{At}\eta(0) + \int_{-\tau}^{t-\tau} e^{A(t-\tau-\theta)} f_{\rm d}(x(\theta)){\rm d}\theta. \end{aligned} \tag{15.39}$$

Now, if A is essentially nonnegative, then it follows from Proposition 2.4 that $e^{At} \geq\geq 0$, $t \geq 0$; and if $\eta(\cdot) \in \mathcal{C}_+$ and $f_{\rm d}$ is nonnegative, then it follows that

$$x(t) = e^{At}\eta(0) + \int_{-\tau}^{t-\tau} e^{A(t-\tau-\theta)} f_{\rm d}(\eta(\theta)){\rm d}\theta \geq\geq 0, \quad t \in [0, \tau).$$

Alternatively, for all $\tau < t$,

$$x(t) = e^{At}x(t-\tau) + \int_0^{\tau} e^{A(\tau-\theta)} f_{\rm d}(x(t+\theta-2\tau)){\rm d}\theta,$$

and hence, since $x(t) \geq\geq 0$, $t \in [-\tau, \tau)$, it follows that $x(t) \geq\geq 0$, $\tau \leq t < 2\tau$. Repeating this procedure iteratively, it follows that $x(t) \geq\geq 0$, $t \geq 0$. □

Next, we present sufficient conditions for asymptotic stability for nonlinear nonnegative dynamical systems given by (15.38).

Theorem 15.6. Consider the nonlinear nonnegative time delay dynamical system $\mathcal{G}$ given by (15.38) where $A \in \mathbb{R}^{n \times n}$ is essentially nonnegative, $f_{\rm d} : \mathbb{R}^n \to \mathbb{R}^n$ is nonnegative, and $f_{\rm d}(x) \leq\leq \gamma x$, $x \in \overline{\mathbb{R}}_+^n$, where $\gamma > 0$, and let $\bar{\tau} > 0$. If there exist $p, r \in \mathbb{R}^n$ such that $p >> 0$ and $r >> 0$ satisfy

$$0 = (A + \gamma I_n)^{\mathrm{T}} p + r, \qquad (15.40)$$

then $\mathcal{G}$ is asymptotically stable for all $\tau \in [0, \bar{\tau}]$.

Proof. Consider the candidate Lyapunov-Krasovskii functional $V : \mathcal{C}_+ \to \mathbb{R}$ given by

$$V(\psi) = p^{\mathrm{T}} \psi(0) + \int_{-\tau}^{0} p^{\mathrm{T}} f_{\mathrm{d}}(\psi(\theta)) \mathrm{d}\theta, \qquad \psi(\cdot) \in \mathcal{C}_+.$$

Now, note that $V(\psi) \geq p^{\mathrm{T}} \psi(0) \geq \alpha \|\psi(0)\|$, where $\alpha \triangleq \min_{i \in \{1,2,\ldots,n\}} p_i > 0$. Next, using (15.40), it follows that the Lyapunov-Krasovskii directional derivative along the trajectories of (15.38) is given by

$$\begin{aligned} \dot{V}(x_t) &= p^{\mathrm{T}} \dot{x}(t) + p^{\mathrm{T}} [f_{\mathrm{d}}(x(t)) - f_{\mathrm{d}}(x(t-\tau))] \\ &= p^{\mathrm{T}} (Ax(t) + f_{\mathrm{d}}(x(t))) \\ &\leq p^{\mathrm{T}} Ax(t) + \gamma p^{\mathrm{T}} x(t) \\ &= -r^{\mathrm{T}} x(t) \\ &\leq -\beta \|x(t)\|, \end{aligned}$$

where $\beta \triangleq \min_{i \in \{1,2,\ldots,n\}} r_i > 0$. Now, it follows from Theorem 15.1 that the nonlinear nonnegative time delay dynamical system $\mathcal{G}$ is asymptotically stable for all $\tau \in [0, \bar{\tau}]$. □

The structural constraint $f_{\mathrm{d}}(x) \leq\leq \gamma x$, $x \in \overline{\mathbb{R}}_+^n$, where $\gamma > 0$, in the statement of Theorem 15.6 is naturally satisfied for many compartmental thermodynamic systems. For example, in nonlinear thermodynamic systems the thermal transport across a conducting surface may be captured by a heat flux described by the saturable form

$$f_{\mathrm{d}i}(x_i, x_j) = \eta_{\max}[(x_i^\alpha/(x_i^\alpha + \beta) - (x_j^\alpha/(x_j^\alpha + \beta)],$$

where x_i and x_j are the energies of the ith and jth compartments, respectively, and $\eta_{\max}$, α, and β are model parameters. This nonlinear intercompartmental energy flow model satisfies the structural constraint of Theorem 15.6.

15.5 Conservation of Energy for Thermodynamic Systems with Time Delay

In this section, we develop the first law of thermodynamics for a compartmental energy flow model with time delay. For simplicity of exposition, we restrict our attention to a linear thermodynamic model. Nonlinear retarded thermodynamic models are developed in Section 15.7. For $i = 1, \ldots, n$, let $E_i(t)$, $t \geq 0$, denote the energy (and hence a nonnegative quantity) of

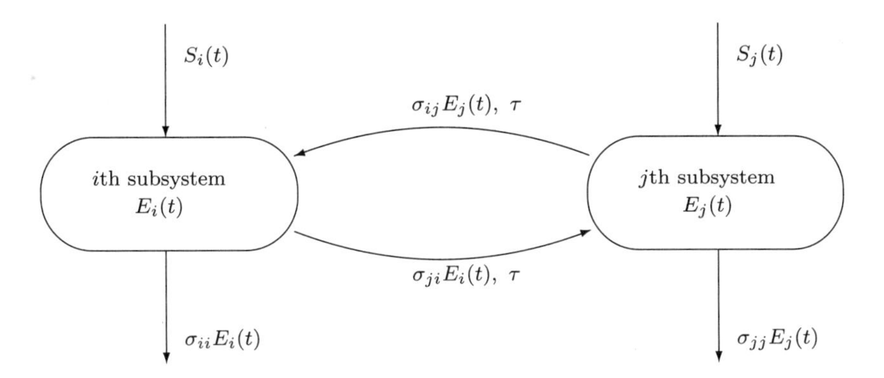

Figure 15.2 Linear thermodynamic model with time delay.

the ith subsystem of the compartmental system shown in Figure 15.2, let $\sigma_{ii} \geq 0$ denote the energy loss rate of the ith subsystem, let $\phi_{ij}(t-\tau)$, $i \neq j$, denote the net energy flow (or heat flux) from the jth subsystem to the ith subsystem given by $\phi_{ij}(t-\tau) = \sigma_{ij}E_j(t-\tau) - \sigma_{ji}E_i(t)$, where the energy transfer rate $\sigma_{ij} \geq 0$, $i \neq j$, and τ is the fixed time it takes for the energy to flow from the jth subsystem to the ith subsystem, and let $S_i(t) \geq 0$, $t \geq 0$, denote the input energy flux to the ith compartment. For simplicity of exposition we have assumed that all transfer times between compartments are given by τ. The more general multiple delay case can be addressed as shown in Section 15.7.

Now, an energy balance for the whole compartmental system yields the delay power balance equation

$$\dot{E}_i(t) = -\left[\sigma_{ii} + \sum_{j=1, i\neq j}^{n} \sigma_{ji}\right] E_i(t) + \sum_{j=1, i\neq j}^{n} \sigma_{ij}E_j(t-\tau) + S_i(t), \quad t \geq 0, \quad i = 1, \ldots, n, \tag{15.41}$$

or, equivalently, in vector form

$$\dot{E}(t) = AE(t) + A_{\mathrm{d}}E(t-\tau) + S(t), \quad E(\theta) = \eta(\theta), \quad -\tau \leq \theta \leq 0, \quad t \geq 0, \tag{15.42}$$

where $E(t) = [E_1(t), \ldots, E_n(t)]^{\mathrm{T}}$, $S(t) = [S_1(t), \ldots, S_n(t)]^{\mathrm{T}}$, $\eta(\cdot) \in \mathcal{C}_+$, and for $i, j = 1, \ldots, n$,

$$A_{(i,j)} = \begin{cases} -\sum_{k=1}^{n} \sigma_{ki}, & i = j, \\ 0, & i \neq j, \end{cases} \quad A_{\mathrm{d}(i,j)} = \begin{cases} 0, & i = j, \\ \sigma_{ij}, & i \neq j. \end{cases} \tag{15.43}$$

Note that A is essentially nonnegative and A_{d} is nonnegative. Furthermore, note that $A + A_{\rm d}$ is a compartmental matrix and

$$(A + A_{\rm d})E = [J_n(E) - D(E)]\mathbf{e}, \tag{15.44}$$

where $J_n(E)$ is a skew-symmetric matrix function with $J_{n(i,i)}(E) = 0$ and $J_{n(i,j)}(E) = \sigma_{ij}E_j - \sigma_{ji}E_i$, $i \neq j$, and

$$D(E) = \mathrm{diag}[\sigma_{11}E_1, \ldots, \sigma_{nn}E_n] \geq\geq 0, \quad E \in \overline{\mathbb{R}}_+^n.$$

To show that all compartmental thermodynamic systems of the form (15.42) with outputs corresponding to

$$y = D(E)\mathbf{e} = [\sigma_{11}E_1, \ldots, \sigma_{nn}E_n]^{\rm T}$$

are lossless, consider the energy storage functional

$$V_{\rm s}(\psi) = \mathbf{e}^{\rm T}\psi(0) + \mathbf{e}^{\rm T}\int_{-\tau}^{0} A_{\rm d}\psi(\theta){\rm d}\theta. \tag{15.45}$$

Note that the energy storage functional $V_{\rm s}(\psi)$ captures the total energy of the system at $t = 0$ plus the integral of the energy flow in transit between the compartments over the time intervals it takes for the mass to flow through the intercompartmental connections.

Now, it follows that

$$\begin{aligned} \dot{V}_{\rm s}(E_t) &= \mathbf{e}^{\rm T}\dot{E}(t) + \mathbf{e}^{\rm T}A_{\rm d}[E(t) - E(t-\tau)] + \mathbf{e}^{\rm T}S(t) \\ &= \mathbf{e}^{\rm T}[(A + A_{\rm d})E(t) + S(t)] \\ &= \mathbf{e}^{\rm T}S(t) - \mathbf{e}^{\rm T}y(t) + \mathbf{e}^{\rm T}J_n(E(t))\mathbf{e} \\ &= \mathbf{e}^{\rm T}S(t) - \mathbf{e}^{\rm T}y(t), \quad E_t \in \mathcal{C}_+, \end{aligned} \tag{15.46}$$

which shows that all compartmental thermodynamic systems of the form given by (15.42) with inputs $u = S$ and outputs $y = D(E)\mathbf{e}$ are lossless with respect to the supply rate $r(u, y) = \mathbf{e}^{\rm T}u - \mathbf{e}^{\rm T}y$ [197]. Note that in the case where the system is isolated, that is, $u(t) \equiv 0$ and $y(t) \equiv 0$, $\dot{V}_{\rm s}(E_t) = 0$, $E_t \in \mathcal{C}_+$, which corresponds to conservation of energy in the system. This is a statement of the first law of thermodynamics for systems with finite heat propagation speeds.

15.6 Semistability and Equipartition of Energy for Linear Thermodynamic Systems with Time Delay

In this and the next section, we extend some key results from Chapter 3 to thermodynamic systems with time delay. We first consider linear, time delay dynamical systems $\mathcal{G}$ of the form

$$\dot{E}(t) = AE(t) + \sum_{i=1}^{q_{\mathrm{d}}} A_{\mathrm{d}i} E(t - \tau_i), \quad E(\theta) = \eta(\theta), \quad -\bar{\tau} \le \theta \le 0, \quad t \ge 0, \tag{15.47}$$

where $E(t) \in \overline{\mathbb{R}}_+^q$, $t \ge 0$, $A \in \mathbb{R}^{q \times q}$, $A_{\mathrm{d}i} \in \mathbb{R}^{q \times q}$, $\tau_i \in \mathbb{R}$, $i = 1, \ldots, q_{\mathrm{d}}$, $\bar{\tau} = \max_{i \in \{1,\ldots,q_{\mathrm{d}}\}} \tau_i$, $\eta(\cdot) \in \mathcal{C}_+ \triangleq \{\psi(\cdot) \in \mathcal{C}([-\bar{\tau}, 0], \mathbb{R}^q) : \psi(\theta) \ge\ge 0, \theta \in [-\bar{\tau}, 0]\}$ is a continuous vector-valued function specifying the initial state of the system, and $\mathcal{C}([-\bar{\tau}, 0], \mathbb{R}^q)$ denotes a Banach space of continuous functions mapping the interval $[-\bar{\tau}, 0]$ into $\mathbb{R}^q$ with the topology of uniform convergence. In addition, note that since $\eta(\cdot)$ is continuous it follows from Theorem 2.1 of [203, p. 14] that there exists a unique solution $E(\eta)$ defined on $[-\bar{\tau}, \infty)$ that coincides with η on $[-\bar{\tau}, 0]$ and satisfies (15.47) for all $t \ge 0$. As we see at the end of this section, (15.47) can characterize a linear thermodynamic model with delayed intercompartmental energy flow.

Definition 15.6. The linear time delay dynamical system (15.47) is called a *compartmental dynamical system* if $A \in \mathbb{R}^{q \times q}$ is essentially nonnegative, $A_{\mathrm{d}i} \in \mathbb{R}^{q \times q}$, $i = 1, \ldots, q_{\mathrm{d}}$, is nonnegative, and $A + \sum_{i=1}^{q_{\mathrm{d}}} A_{\mathrm{d}i}$ is a compartmental matrix.

Note that the linear time delay dynamical system (15.47) is compartmental if A and $A_{\mathrm{d}} \triangleq \sum_{i=1}^{q_{\mathrm{d}}} A_{\mathrm{d}i}$ are given by

$$A_{(i,j)} = \begin{cases} -\sum_{k=1}^{n} \sigma_{ki}, & i = j, \\ 0, & i \ne j, \end{cases} \qquad A_{\mathrm{d}(i,j)} = \begin{cases} 0, & i = j, \\ \sigma_{ij}, & i \ne j, \end{cases} \tag{15.48}$$

where $\sigma_{ii} \ge 0$, $i \in \{1, \ldots, q\}$, denotes the loss coefficients of the ith compartment and $\sigma_{ij} \ge 0$, $i \ne j$, $i, j \in \{1, \ldots, q\}$, denotes the transfer coefficients from the jth compartment to the ith compartment.

Next, we present sufficient conditions for semistability and system state equipartition for linear compartmental dynamical systems with time delay. The following lemma is needed for the main theorem of this section.

Lemma 15.1. Let $A \in \mathbb{R}^{q \times q}$ and $A_{\mathrm{d}} = \sum_{i=1}^{q_{\mathrm{d}}} A_{\mathrm{d}i}$ be given by (15.48). Assume that $(A + \sum_{i=1}^{q_{\mathrm{d}}} A_{\mathrm{d}i})\mathbf{e} = 0$. Then there exist nonnegative definite matrices $Q_i \in \mathbb{R}^{q \times q}$, $i = 1, \ldots, q_{\mathrm{d}}$, such that

$$A + A^{\mathrm{T}} + \sum_{i=1}^{q_{\mathrm{d}}} (Q_i + A_{\mathrm{d}i}^{\mathrm{T}} Q_i^{\mathrm{D}} A_{\mathrm{d}i}) \le 0. \tag{15.49}$$

Proof. For each $i \in \{1, \ldots, q_{\mathrm{d}}\}$, let Q_i be the diagonal matrix defined by

$$Q_{i(l,l)} \triangleq \sum_{m=1,l\neq m}^{q_{\rm d}} A_{{\rm d}i(l,m)}, \tag{15.50}$$

and note that it follows from (15.50) and the definition of the Drazin inverse that $(A_{{\rm d}i} - Q_i)\mathbf{e} = 0$ and $Q_i Q_i^{\rm D} A_{{\rm d}i} = A_{{\rm d}i}$, $i = 1, \ldots, q_{\rm d}$. Since A and Q_i, $i = 1, \ldots, q_{\rm d}$, are diagonal and $(A + \sum_{i=1}^{q_{\rm d}} A_{{\rm d}i})\mathbf{e} = 0$ it follows that $A + \sum_{i=1}^{q_{\rm d}} Q_i = 0$. Hence, $M\mathbf{e} = 0$, where

$$M \triangleq \begin{bmatrix} A + A^{\rm T} + \sum_{i=1}^{q_{\rm d}} Q_i & A_{\rm d1}^{\rm T} & A_{\rm d2}^{\rm T} & \cdots & A_{{\rm d}q_{\rm d}}^{\rm T} \\ A_{\rm d1} & -Q_1 & 0 & \cdots & 0 \\ \vdots & \vdots & \vdots & \vdots & \vdots \\ A_{{\rm d}q_{\rm d}} & 0 & 0 & \cdots & -Q_{q_{\rm d}} \end{bmatrix}. \tag{15.51}$$

Next, note that $M = M^{\rm T}$ and $M_{(i,j)} \geq 0$, $i, j = 1, \ldots, q_{\rm d}$, $i \neq j$. Hence, by *ii*) of Theorem 2.10, M is semistable. Now, it follows from Proposition 3.18 that $M \leq 0$ and since $Q_i Q_i^{\rm D} A_{{\rm d}i} = A_{{\rm d}i}$, $i = 1, \ldots, q_{\rm d}$, it follows from Lemma 3.5 that $M \leq 0$ if and only if (15.49) holds. □

Theorem 15.7. Consider the linear time delay dynamical system given by (15.47) where A and $A_{\rm d}$ are given by (15.48). Assume that $(A + \sum_{i=1}^{q_{\rm d}} A_{{\rm d}i})^{\rm T}\mathbf{e} = (A + \sum_{i=1}^{q_{\rm d}} A_{{\rm d}i})\mathbf{e} = 0$ and $\operatorname{rank}(A + \sum_{i=1}^{q_{\rm d}} A_{{\rm d}i}) = q - 1$. Then, for every $\alpha \geq 0$, $\alpha\mathbf{e}$ is a semistable equilibrium point of (15.47). Furthermore, $E(t) \to \alpha^*\mathbf{e}$ as $t \to \infty$, where

$$\alpha^* = \frac{\mathbf{e}^{\rm T}\eta(0) + \sum_{i=1}^{q_{\rm d}} \int_{-\tau_i}^{0} \mathbf{e}^{\rm T} A_{{\rm d}i}\eta(\theta){\rm d}\theta}{q + \sum_{i=1}^{q_{\rm d}} \tau_i \mathbf{e}^{\rm T} A_{{\rm d}i}\mathbf{e}}. \tag{15.52}$$

Proof. It follows from Lemma 15.1 that there exist nonnegative matrices Q_i, $i = 1, \ldots, q_{\rm d}$, such that (15.49) holds. Now, consider the Lyapunov-Krasovskii functional $V : \mathcal{C}_+ \to \mathbb{R}$ given by

$$V(\psi(\cdot)) = \psi^{\rm T}(0)\psi(0) + \sum_{i=1}^{q_{\rm d}} \int_{-\tau_i}^{0} \psi^{\rm T}(\theta) A_{{\rm d}i}^{\rm T} Q_i^{\rm D} A_{{\rm d}i}\psi(\theta){\rm d}\theta, \tag{15.53}$$

and note that the directional derivative of $V(E_t)$ along the trajectories of (15.47) is given by

$$\begin{aligned} \dot{V}(E_t) &= 2E^{\rm T}(t)\dot{E}(t) + \sum_{i=1}^{q_{\rm d}} E^{\rm T}(t) A_{{\rm d}i}^{\rm T} Q_i^{\rm D} A_{{\rm d}i} E(t) \\ &\quad - \sum_{i=1}^{q_{\rm d}} E^{\rm T}(t-\tau_i) A_{{\rm d}i}^{\rm T} Q_i^{\rm D} A_{{\rm d}i} E(t-\tau_i) \end{aligned}$$

$$
\begin{aligned}
&= 2E^{\mathrm{T}}(t)AE(t) + 2E^{\mathrm{T}}(t)\sum_{i=1}^{q_{\mathrm{d}}} A_{\mathrm{d}i}E(t-\tau_i) \\
&\quad + \sum_{i=1}^{q_{\mathrm{d}}} E^{\mathrm{T}}(t)A_{\mathrm{d}i}^{\mathrm{T}}Q_i^{\mathrm{D}}A_{\mathrm{d}i}E(t) \\
&\quad - \sum_{i=1}^{q_{\mathrm{d}}} E^{\mathrm{T}}(t-\tau_i)A_{\mathrm{d}i}^{\mathrm{T}}Q_i^{\mathrm{D}}A_{\mathrm{d}i}E(t-\tau_i) \\
&\leq -\sum_{i=1}^{q_{\mathrm{d}}} [x^{\mathrm{T}}(t)Q_iE(t) - 2x^{\mathrm{T}}(t)A_{\mathrm{d}i}E(t-\tau_i) \\
&\quad + E^{\mathrm{T}}(t-\tau_i)A_{\mathrm{d}i}^{\mathrm{T}}Q_i^{\mathrm{D}}A_{\mathrm{d}i}E(t-\tau_i)] \\
&= -\sum_{i=1}^{q_{\mathrm{d}}} [-Q_iE(t) + A_{\mathrm{d}i}E(t-\tau_i)]^{\mathrm{T}}Q_i^{\mathrm{D}}[-Q_iE(t) + A_{\mathrm{d}i}E(t-\tau_i)] \\
&\leq 0, \quad t \geq 0.
\end{aligned} \tag{15.54}
$$

Next, let $\mathcal{R} \triangleq \{\psi(\cdot) \in \mathcal{C}_+ : -Q_i\psi(0) + A_{\mathrm{d}i}\psi(-\tau_i) = 0, i = 1, \ldots, q_{\mathrm{d}}\}$ and note that, since the positive orbit $\mathcal{O}_\eta^+$ of (15.47) is bounded, $\mathcal{O}_\eta^+$ belongs to a compact subset of $\mathcal{C}_+$, and hence, it follows from Theorem 15.3 that $E_t \to \mathcal{M}$, where $\mathcal{M}$ denotes the largest invariant set contained in $\mathcal{R}$. Now, since $A + \sum_{i=1}^{q_{\mathrm{d}}} Q_i = 0$, it follows that $\mathcal{R} \subset \hat{\mathcal{R}} \triangleq \{\psi(\cdot) \in \mathcal{C}_+ : A\psi(0) + \sum_{i=1}^{q_{\mathrm{d}}} A_{\mathrm{d}i}\psi(-\tau_i) = 0\}$. Hence, since $\operatorname{rank}(A + \sum_{i=1}^{q_{\mathrm{d}}} A_{\mathrm{d}i}) = q - 1$ and $(A + \sum_{i=1}^{q_{\mathrm{d}}} A_{\mathrm{d}i})\mathbf{e} = 0$, it follows that the largest invariant set $\hat{\mathcal{M}}$ contained in $\hat{\mathcal{R}}$ is given by $\hat{\mathcal{M}} = \{\psi \in \mathcal{C}_+ : \psi(\theta) = \alpha\mathbf{e}, \theta \in [-\bar{\tau}, 0], \alpha \geq 0\}$. Furthermore, since $\hat{\mathcal{M}} \subset \mathcal{R} \subset \hat{\mathcal{R}}$, it follows that $\mathcal{M} = \hat{\mathcal{M}}$.

Next, define the functional $\mathcal{E} : \mathcal{C}_+ \to \mathbb{R}$ by

$$
\mathcal{E}(\psi(\cdot)) = \mathbf{e}^{\mathrm{T}}\psi(0) + \sum_{i=1}^{q_{\mathrm{d}}} \int_{-\tau_i}^{0} \mathbf{e}^{\mathrm{T}}A_{\mathrm{d}i}\psi(\theta)\mathrm{d}\theta, \tag{15.55}
$$

and note that $\dot{\mathcal{E}}(E_t) \equiv 0$ along the trajectories of (15.47). Thus, for all $t \geq 0$,

$$
\mathcal{E}(E_t) = \mathcal{E}(\eta(\cdot)) = \mathbf{e}^{\mathrm{T}}\eta(0) + \sum_{i=1}^{q_{\mathrm{d}}} \int_{-\tau_i}^{0} \mathbf{e}^{\mathrm{T}}A_{\mathrm{d}i}\eta(\theta)\mathrm{d}\theta, \tag{15.56}
$$

which implies that $E_t \to \mathcal{M} \cap \mathcal{Q}$, where $\mathcal{Q} \triangleq \{\psi(\cdot) \in \mathcal{C}_+ : \mathcal{E}(\psi(\cdot)) = \mathcal{E}(\eta(\cdot))\}$. Hence, since $\mathcal{M} \cap \mathcal{Q} = \{\alpha^*\mathbf{e}\}$, it follows that $E(t) \to \alpha^*\mathbf{e}$, where α^* is given by (15.52).

Finally, Lyapunov stability of $\alpha\mathbf{e}$, $\alpha \geq 0$, follows by considering the

Lyapunov-Krasovskii functional

$$V(\psi(\cdot)) = (\psi(0) - \alpha\mathbf{e})^{\mathrm{T}}(\psi(0) - \alpha\mathbf{e}) + \sum_{i=1}^{q_{\mathrm{d}}} \int_{-\tau_i}^{0} (\psi(\theta) - \alpha\mathbf{e})^{\mathrm{T}} A_{\mathrm{d}i}^{\mathrm{T}} Q_i^{\mathrm{D}} A_{\mathrm{d}i}(\psi(\theta) - \alpha\mathbf{e})\mathrm{d}\theta$$

and noting that $V(\psi) \geq \|\psi(0) - \alpha\mathbf{e}\|_2^2$. □

Note that if $q_{\mathrm{d}} = q^2 - q$, $A_{\mathrm{d}} = A_{\mathrm{d}}^{\mathrm{T}}$, and $(A + A_{\mathrm{d}})\mathbf{e} = 0$, then (15.47) can be rewritten as

$$\dot{E}_i(t) = - \sum_{j=1, j\neq i}^{q} \sigma_{ij}[E_i(t) - E_j(t - \tau_{ij})], \; E(\theta) = \eta(\theta), \; -\bar{\tau} \leq \theta \leq 0, \; t \geq 0, \tag{15.57}$$

where $i = 1, \ldots, q$, and $\tau_{ij} \in [0, \bar{\tau}]$, $i \neq j$, $i, j = 1, \ldots, q$, which implies that the rate of energy transfer from the ith compartment to the jth compartment is proportional to the delayed energy difference $E_j(t-\tau_{ij}) - E_i(t)$. Hence, the rate of energy transfer is positive (respectively, negative) if $E_j(t-\tau_{ij}) > E_i(t)$ (respectively, $E_j(t - \tau_{ij}) < E_i(t)$).

Equation (15.57) is a power balance equation that governs the energy exchange among coupled subsystems and is completely analogous to the equations of thermal transfer with subsystem energy playing the role of temperatures. Furthermore, note that since $\sigma_{ij} \geq 0$, $i \neq j$, $i, j = 1, \ldots, q$, energy flows from more energetic subsystems to less energetic subsystems, which is consistent with the second law of thermodynamics requiring that heat (energy) must flow in the direction of lower temperatures.

15.7 Semistability and Equipartition of Energy for Nonlinear Thermodynamic Systems with Time Delay

In this section, we extend the results of Section 15.6 to nonlinear thermodynamic systems with time delay. Specifically, we consider nonlinear nonnegative time delay dynamical systems $\mathcal{G}$ of the form

$$\dot{E}(t) = f(E(t)) + f_{\mathrm{d}}(E(t - \tau_1), \ldots, E(t - \tau_{q_{\mathrm{d}}})), \quad E(\theta) = \eta(\theta), \quad -\bar{\tau} \leq \theta \leq 0, \quad t \geq 0, \tag{15.58}$$

where $E(t) \in \mathbb{R}^q$, $t \geq 0$, $f : \mathbb{R}^q \to \mathbb{R}^q$ is locally Lipschitz continuous and $f(0) = 0$, $f_{\mathrm{d}} : \mathbb{R}^q \times \cdots \times \mathbb{R}^q \to \mathbb{R}^q$ is locally Lipschitz continuous and $f_{\mathrm{d}}(0, \ldots, 0) = 0$, $\bar{\tau} = \max_{i\in\{1,\ldots,q_{\mathrm{d}}\}} \tau_i$, $\tau_i \geq 0$, $i = 1, \ldots, q_{\mathrm{d}}$, and $\eta(\cdot) \in \mathcal{C} = \mathcal{C}([-\bar{\tau}, 0], \mathbb{R}^q)$ is a continuous vector-valued function specifying the initial state of the system. Note that since $\eta(\cdot)$ is continuous it follows from Theorem 2.3 of [203, p. 44] that there exists a unique solution $E(\eta)$ defined on $[-\bar{\tau}, \infty)$ that coincides with η on $[-\bar{\tau}, 0]$ and satisfies (15.58) for all $t \geq 0$. Furthermore, we assume that $f(\cdot)$ is essentially nonnegative and $f_{\rm d}(\cdot)$ is nonnegative so that for every $\eta(\cdot) \in \mathcal{C}_+$, the nonlinear time delay dynamical system $\mathcal{G}$ given by (15.58) is nonnegative.

Next, we consider a subclass of (15.58) corresponding to nonlinear compartmental systems.

Definition 15.7. The nonlinear time delay dynamical system (15.58) is called a *compartmental dynamical system* if $F(\cdot)$ is compartmental, where $F(x) \triangleq f(x) + f_{\rm d}(x, x, \ldots, x)$.

Note that the nonlinear time delay dynamical system is compartmental if $f = [f_1, \ldots, f_q]^{\rm T}$ and $f_{\rm d} = [f_{\rm d1}, \ldots, f_{{\rm d}q}]^{\rm T}$ are given by

$$
\begin{aligned}
f_i(E(t)) &= -\sum_{j=1, j\neq i}^{q} \sigma_{ji}(E(t)), \\
f_{{\rm d}i}(E(t-\tau_1), \ldots, E(t-\tau_{q_{\rm d}})) &= \sum_{j=1, j\neq i}^{q} \sigma_{ij}(E(t-\tau_{ij})),
\end{aligned}
$$

where $\sigma_{ii}(E(\cdot)) \geq 0$, $E(\cdot) \in \mathcal{C}_+$, $\sigma_{ii}(0) = 0$, $i \in \{1, \ldots, q\}$, denotes the instantaneous rate of flow of energy loss of the ith compartment, $\sigma_{ij}(E(\cdot)) \geq 0$, $E(\cdot) \in \mathcal{C}_+$, $i \neq j$, $i, j \in \{1, \ldots, q\}$, denotes the instantaneous rate of energy flow from the jth compartment to the ith compartment, τ_{ij}, $i \neq j$, $i, j \in \{1, \ldots, q\}$, denotes the transfer time of energy flow from the jth compartment to the ith compartment, and $\sigma_{ii}(\cdot)$ and $\sigma_{ij}(\cdot)$ are such that if $E_i = 0$, then $\sigma_{ii}(E) = 0$ and $\sigma_{ji}(E) = 0$ for all $i, j = 1, \ldots, q$, and $E \in \overline{\mathbb{R}}_+^q$. Note that the above constraints imply that $f(\cdot)$ is essentially nonnegative and $f_{\rm d}(\cdot)$ is nonnegative.

The next result generalizes Theorem 15.7 to nonlinear time delay compartmental systems of the form

$$
\dot{E}(t) = f(E(t)) + \sum_{i=1}^{q_{\rm d}} f_{{\rm d}i}(E(t-\tau_i)), \quad E(\theta) = \eta(\theta), \quad -\bar{\tau} \leq \theta \leq 0, \quad t \geq 0, \tag{15.59}
$$

where $f : \overline{\mathbb{R}}_+^q \to \overline{\mathbb{R}}_+^q$ is given by $f(E) = [f_1(E_1), \ldots, f_q(E_q)]^{\rm T}$, $f(0) = 0$, $f_{{\rm d}i} : \overline{\mathbb{R}}_+^q \to \overline{\mathbb{R}}_+^q$, $i = 1, \ldots, q_{\rm d}$, and $f_{\rm d}(0) = 0$. For this result, we assume that $f_i(\cdot)$, $i = 1, \ldots, q$, are strictly decreasing functions.

Theorem 15.8. Consider the nonlinear time delay dynamical system given by (15.59) where $f_i(\cdot)$, $i = 1, \ldots, q$, is strictly decreasing and $f_i(0) = 0$. Assume that $\mathbf{e}^{\rm T}[f(E) + \sum_{i=1}^{q_{\rm d}} f_{{\rm d}i}(E)] = 0$, $E \in \overline{\mathbb{R}}_+^q$, and $f(E) + \sum_{i=1}^{q_{\rm d}} f_{{\rm d}i}(E) = 0$ if and only if $E = \alpha\mathbf{e}$ for some $\alpha \geq 0$. Furthermore, assume there exist nonnegative diagonal matrices $P_i \in \overline{\mathbb{R}}_+^{q\times q}$, $i = 1,\ldots,q_{\rm d}$, such that $P \triangleq \sum_{i=1}^{q_{\rm d}} P_i > 0$,

$$P_i^{\rm D} P_i f_{{\rm d}i}(E) = f_{{\rm d}i}(E), \quad E \in \overline{\mathbb{R}}_+^q, \quad i = 1,\ldots,q_{\rm d}, \tag{15.60}$$

$$\sum_{i=1}^{q_{\rm d}} f_{{\rm d}i}^{\rm T}(E) P_i f_{{\rm d}i}(E) \le f^{\rm T}(E) P f(E), \quad E \in \overline{\mathbb{R}}_+^q. \tag{15.61}$$

Then, for every $\alpha \ge 0$, $\alpha\mathbf{e}$ is a semistable equilibrium point of (15.59). Furthermore, $E(t) \to \alpha^* {\rm e}$ as $t \to \infty$, where α^* satisfies

$$q\alpha^* + \sum_{i=1}^{q_{\rm d}} \tau_i \mathbf{e}^{\rm T} f_{{\rm d}i}(\alpha^* \mathbf{e}) = \mathbf{e}^{\rm T}\eta(0) + \sum_{i=1}^{q_{\rm d}} \int_{-\tau_i}^0 \mathbf{e}^{\rm T} f_{{\rm d}i}(\eta(\theta)){\rm d}\theta. \tag{15.62}$$

Proof. Consider the Lyapunov-Krasovskii functional $V : \mathcal{C}_+ \to \mathbb{R}$ given by

$$V(\psi(\cdot)) = -2\sum_{i=1}^{q} \int_0^{\psi_i(0)} P_{(i,i)} f_i(\zeta){\rm d}\zeta + \sum_{i=1}^{q_{\rm d}} \int_{-\tau_i}^0 f_{{\rm d}i}^{\rm T}(\psi(\theta)) P_i f_{{\rm d}i}(\psi(\theta)){\rm d}\theta. \tag{15.63}$$

Since, $f_i(\cdot)$, $i = 1,\ldots,q$, is a strictly decreasing function it follows that $V(\psi) \ge 2\sum_{i=1}^{q} P_{(i,i)}[-f_i(\delta_i\psi_i(0))]\psi_i(0) > 0$ for all $\psi(0) \ne 0$, where $0 < \delta_i < 1$, and hence, there exists a class $\mathcal{K}$ function $\alpha(\cdot)$ such that $V(\psi) \ge \alpha(\|\psi(0)\|)$.

Now, note that the directional derivative of $V(E_t)$ along the trajectories of (15.59) is given by

$$\begin{aligned}
\dot V(E_t) &= -2f^{\rm T}(E(t))P\dot E(t) + \sum_{i=1}^{q_{\rm d}} f_{{\rm d}i}^{\rm T}(E(t))P_i f_{{\rm d}i}(E(t)) \\
&\quad - \sum_{i=1}^{q_{\rm d}} f_{{\rm d}i}^{\rm T}(E(t-\tau_i))P_i f_{{\rm d}i}(E(t-\tau_i)) \\
&= -2f^{\rm T}(E(t))Pf(E(t)) - 2\sum_{i=1}^{q_{\rm d}} f^{\rm T}(E(t))Pf_{{\rm d}i}(E(t-\tau_i)) \\
&\quad + \sum_{i=1}^{q_{\rm d}} f_{{\rm d}i}^{\rm T}(E(t))P_i f_{{\rm d}i}(E(t)) - \sum_{i=1}^{q_{\rm d}} f_{{\rm d}i}^{\rm T}(E(t-\tau_i))P_i f_{{\rm d}i}(E(t-\tau_i)) \\
&\le -f^{\rm T}(E(t))Pf(E(t)) - 2\sum_{i=1}^{q_{\rm d}} f^{\rm T}(E(t))PP_i^{\rm D}P_i f_{{\rm d}i}(E(t-\tau_i)) \\
&\quad - \sum_{i=1}^{q_{\rm d}} f_{{\rm d}i}^{\rm T}(E(t-\tau_i))P_iP_i^{\rm D}P_i f_{{\rm d}i}(E(t-\tau_i))
\end{aligned}$$

$$
\begin{aligned}
&= -\sum_{i=1}^{q_{\mathrm{d}}}[Pf(E(t)) + P_i f_{\mathrm{d}i}(E(t-\tau_i))]^{\mathrm{T}} P_i^{\mathrm{D}} \\
&\quad \cdot [Pf(E(t)) + P_i f_{\mathrm{d}i}(E(t-\tau_i))] \\
&\leq 0, \quad t \geq 0,
\end{aligned}
\tag{15.64}
$$

where the first inequality in (15.64) follows from (15.60) and (15.61), and the last equality in (15.64) follows from the fact that $f^{\mathrm{T}}(E)Pf(E) = \sum_{i=1}^{q_{\mathrm{d}}} f^{\mathrm{T}}(E)PP_i^{\mathrm{D}}Pf(E)$, $E \in \overline{\mathbb{R}}_+^q$.

Next, let $\mathcal{R} \triangleq \{\psi(\cdot) \in \mathcal{C}_+ : Pf(\psi(0)) + P_i f_{\mathrm{d}i}(\psi(-\tau_i)) = 0, i = 1, \ldots, q_{\mathrm{d}}\}$ and note that, since the positive orbit $\mathcal{O}_\eta^+$ of (15.59) is bounded, $\mathcal{O}_\eta^+$ belongs to a compact subset of $\mathcal{C}_+$, and hence, it follows from Theorem 15.3 that $E_t \to \mathcal{M}$, where $\mathcal{M}$ denotes the largest invariant set (with respect to (15.59)) contained in $\mathcal{R}$. Now, since $\mathbf{e}^{\mathrm{T}}(f(x) + \sum_{i=1}^{q_{\mathrm{d}}} f_{\mathrm{d}i}(x)) = 0, x \in \overline{\mathbb{R}}_+^q$, it follows that

$$
\begin{aligned}
\mathcal{R} \subset \hat{\mathcal{R}} &\triangleq \{\psi(\cdot) \in \mathcal{C}_+ : f(\psi(0)) + \sum_{i=1}^{q_{\mathrm{d}}} f_{\mathrm{d}i}(\psi(-\tau_i)) = 0\} \\
&= \{\psi(\cdot) \in \mathcal{C}_+ : \psi(\theta) = \alpha\mathbf{e},\ \theta \in [-\bar{\tau}, 0],\ \alpha \geq 0\},
\end{aligned}
$$

which implies that $E_t \to \hat{\mathcal{R}}$ as $t \to \infty$.

Next, define the functional $\mathcal{E} : \mathcal{C}_+ \to \mathbb{R}$ by

$$
\mathcal{E}(\psi(\cdot)) = \mathbf{e}^{\mathrm{T}}\psi(0) + \sum_{i=1}^{q_{\mathrm{d}}} \int_{-\tau_i}^{0} \mathbf{e}^{\mathrm{T}} f_{\mathrm{d}i}(\psi(\theta))\mathrm{d}\theta, \tag{15.65}
$$

and note that $\dot{\mathcal{E}}(E_t) \equiv 0$ along the trajectories of (15.59). Thus, for all $t \geq 0$,

$$
\mathcal{E}(E_t) = \mathcal{E}(\eta(\cdot)) = \mathbf{e}^{\mathrm{T}}\eta(0) + \sum_{i=1}^{q_{\mathrm{d}}} \int_{-\tau_i}^{0} \mathbf{e}^{\mathrm{T}} f_{\mathrm{d}i}(\eta(\theta))\mathrm{d}\theta, \tag{15.66}
$$

which implies that $E_t \to \hat{\mathcal{R}} \cap \mathcal{Q}$, where $\mathcal{Q} \triangleq \{\psi(\cdot) \in \mathcal{C}_+ : \mathcal{E}(\psi(\cdot)) = \mathcal{E}(\eta(\cdot))\}$. Hence, $\hat{\mathcal{R}} \cap \mathcal{Q} = \{\alpha^*\mathrm{e}\}$, it follows that $E(t) \to \alpha^*\mathbf{e}$, where α^* satisfies (15.62).

Finally, Lyapunov stability of $\alpha\mathbf{e}$, $\alpha \geq 0$, follows by considering the Lyapunov-Krasovskii functional

$$
\begin{aligned}
V(\psi(\cdot)) &= -2\sum_{i=1}^{q} \int_{\alpha}^{\psi_i(0)} P_{(i,i)}(f_i(\zeta) - f_i(\alpha))\mathrm{d}\zeta \\
&\quad + \sum_{i=1}^{q_{\mathrm{d}}} \int_{-\tau_i}^{0} [f_{\mathrm{d}i}(\psi(\theta)) - f_{\mathrm{d}i}(\alpha\mathbf{e})]^{\mathrm{T}} P_i [f_{\mathrm{d}i}(\psi(\theta)) - f_{\mathrm{d}i}(\alpha\mathbf{e})]\mathrm{d}\theta,
\end{aligned}
$$

and noting that $V(\psi) \geq 2\sum_{i=1}^{q} P_{(i,i)}[f_i(\alpha) - f_i(\alpha + \delta_i(\psi_i(0) - \alpha))](\psi_i(0) -$

$\alpha) > 0$, for all $\psi_i(0) \neq \alpha$, where $0 < \delta_i < 1$. □

Theorem 15.8 establishes semistability and state equipartition for the special case of nonlinear compartmental systems of the form (15.59) where $f(\cdot)$ and $f_{\mathrm{d}i}(\cdot)$, $i = 1, \ldots, q$, satisfy (15.60) and (15.61). For general q-dimensional nonlinear compartmental systems with time delay and vector fields given by (15.59) it is not possible to guarantee semistability and state equipartition. However, semistability without state equipartition may be shown.

For example, consider the nonlinear time delay compartmental dynamical system given by

$$\dot{E}_1(t) = -\sigma_{21}(E_1(t)) + \sigma_{12}(E_2(t-\tau_{12})), \quad E_1(\theta) = \eta_1(\theta), \quad -\bar{\tau} \leq \theta \leq 0, \quad t \geq 0, \tag{15.67}$$

$$\dot{E}_2(t) = -\sigma_{12}(E_2(t)) + \sigma_{21}(E_1(t-\tau_{21})), \quad E_2(\theta) = \eta_2(\theta), \quad -\bar{\tau} \leq \theta \leq 0, \quad t \geq 0, \tag{15.68}$$

where $E_1(t)$ and $E_2(t) \in \mathbb{R}$, $t \geq 0$, $\sigma_{12} : \overline{\mathbb{R}}_+ \to \overline{\mathbb{R}}_+$ and $\sigma_{21} : \overline{\mathbb{R}}_+ \to \overline{\mathbb{R}}_+$ satisfy $\sigma_{12}(0) = \sigma_{21}(0) = 0$ and $\sigma_{12}(\cdot)$ and $\sigma_{21}(\cdot)$ are strictly increasing, $\tau_{12}, \tau_{21} > 0$, $\bar{\tau} = \max\{\tau_{12}, \tau_{21}\}$, and $\eta_1(\cdot), \eta_2(\cdot) \in \mathcal{C}_+ = \mathcal{C}([-\bar{\tau}, 0], \overline{\mathbb{R}}_+)$. Note that (15.67) and (15.68) can have multiple equilibria with all the equilibria lying on the curve $\sigma_{21}(u) = \sigma_{12}(v)$, $u, v \geq 0$. It follows from the conditions on $\sigma_{12}(\cdot)$ and $\sigma_{21}(\cdot)$ that all system equilibria lie on the curve $E_2 = \sigma_{12}^{-1}(\sigma_{21}(E_1))$ in the (E_1, E_2) plane, where $\sigma_{12}^{-1}(\cdot)$ denotes the inverse function of $\sigma_{12}(\cdot)$.

Consider the functional $\mathcal{E} : \mathcal{C}_+ \times \mathcal{C}_+ \to \mathbb{R}$ given by

$$\mathcal{E}(\psi_1, \psi_2) = \psi_1(0) + \psi_2(0) + \int_{-\tau_{12}}^{0} \sigma_{12}(\psi_2(\theta))\mathrm{d}\theta + \int_{-\tau_{21}}^{0} \sigma_{21}(\psi_1(\theta))\mathrm{d}\theta.$$

Now, it can be easily shown that the directional derivative of $\mathcal{E}(\psi_1, \psi_2)$ along the trajectories of (15.67) and (15.68) is identically zero for all $t \geq 0$, which implies that, for all $t \geq 0$,

$$\begin{aligned} \mathcal{E}(E_{1t}, E_{2t}) &= \mathcal{E}(\eta_1, \eta_2) \\ &= \eta_1(0) + \eta_2(0) \\ &\quad + \int_{-\tau_{12}}^{0} \sigma_{12}(\eta_2(\theta))\mathrm{d}\theta + \int_{-\tau_{21}}^{0} \sigma_{21}(\eta_1(\theta))\mathrm{d}\theta. \end{aligned} \tag{15.69}$$

Next, consider the functional $V : \mathcal{C}_+ \times \mathcal{C}_+ \to \mathbb{R}$ given by

$$V(\psi_1, \psi_2) = 2\int_{0}^{\psi_1(0)} \sigma_{21}(\theta)\mathrm{d}\theta + 2\int_{0}^{\psi_2(0)} \sigma_{12}(\theta)\mathrm{d}\theta$$

$$+\int_{-\tau_{12}}^{0}\sigma_{12}^{2}(\psi_2(\theta))\mathrm{d}\theta+\int_{-\tau_{21}}^{0}\sigma_{21}^{2}(\psi_1(\theta))\mathrm{d}\theta, \quad (15.70)$$

and note that the directional derivative of $V(\psi_1,\psi_2)$ along the trajectories of (15.67) and (15.68) is given by

$$\begin{aligned}\dot{V}(E_{1t},E_{2t}) &= -[\sigma_{21}(E_1(t))-\sigma_{12}(E_2(t-\tau_{12}))]^2\\ &\quad -[\sigma_{12}(E_2(t))-\sigma_{21}(E_1(t-\tau_{21}))]^2. \end{aligned} \quad (15.71)$$

Now, using similar arguments as in the proof of Theorem 15.8 it follows that $(E_1(t),E_2(t))\to(\alpha^*,\sigma_{12}^{-1}(\sigma_{21}(\alpha^*)))$ as $t\to\infty$, where α^* is the solution to the equation

$$\begin{aligned}&\alpha^*+\sigma_{12}^{-1}(\sigma_{21}(\alpha^*))+(\tau_{12}+\tau_{21})\sigma_{21}(\alpha^*)\\ &\quad =\eta_1(0)+\eta_2(0)+\int_{-\tau_{12}}^{0}\sigma_{12}(\eta_2(\theta))\mathrm{d}\theta+\int_{-\tau_{21}}^{0}\sigma_{21}(\eta_1(\theta))\mathrm{d}\theta, \end{aligned} \quad (15.72)$$

and $(\alpha^*,\sigma_{12}^{-1}(\sigma_{21}(\alpha^*)))$ is a Lyapunov stable equilibrium state. The above analysis shows that all two-dimensional nonlinear compartmental dynamical systems of the form (15.67) and (15.68) are semistable with system states reaching equilibria lying on the curve $E_2=\sigma_{12}^{-1}(\sigma_{21}(E_1))$ in the (E_1,E_2) plane.

To demonstrate the utility of Theorem 15.8 we consider a nonlinear two-compartment time delay dynamical system given by

$$\dot{E}_1(t) = -\sum_{i=1}^{q_{\mathrm{d}}}[\sigma_i(E_1(t))+\sigma_i(E_2(t-\tau_i))],\quad E_1(\theta)=\eta_1(\theta),\quad -\bar{\tau}\le\theta\le 0,\quad t\ge 0, \quad (15.73)$$

$$\dot{E}_2(t) = \sum_{i=1}^{q_{\mathrm{d}}}[\sigma_i(E_1(t-\tau_i))-\sigma_i(E_2(t))],\quad E_2(\theta)=\eta_2(\theta),\quad -\bar{\tau}\le\theta\le 0, \quad (15.74)$$

where $\sigma_i:\overline{\mathbb{R}}_+\to\overline{\mathbb{R}}_+$, $i=1,\ldots,q_{\mathrm{d}}$, are such that, for every $i=1,\ldots,q_{\mathrm{d}}$,

$$[\sigma_i(E_1)-\sigma_i(E_2)](E_1-E_2)>0,\quad E_1\ne E_2, \quad (15.75)$$

and $\sigma_i(0)=0$, $i=1,\ldots,q_{\mathrm{d}}$. Note that (15.73) and (15.74) capture energy flow balance between the two compartments, and (15.75) is consistent with the second law of thermodynamics; that is, energy flows from the more energetic compartment to the less energetic compartment. Furthermore, since $\sigma_i(0)=0$, (15.75) implies that $\sigma_i(\cdot)$, $i=1,\ldots,q_{\mathrm{d}}$, is strictly increasing.

Now, note that (15.73) and (15.74) can be written in the form of

(15.59) with

$$f(x) = \begin{bmatrix} -\sum_{i=1}^{q_{\rm d}} \sigma_i(E_1) \\ -\sum_{i=1}^{q_{\rm d}} \sigma_i(E_2) \end{bmatrix}, \quad f_{{\rm d}i}(x) = \begin{bmatrix} \sigma_i(E_2) \\ \sigma_i(E_1) \end{bmatrix}, \quad i = 1, \ldots, q_{\rm d},$$

which implies that $f_j(E_j)$, $j = 1, 2$, are strictly decreasing. Next, with $P_i = I_n$, $i = 1, \ldots, q_{\rm d}$, (15.60) and (15.61) are trivially satisfied, and hence, it follows from Theorem 15.8 that $E_1(t) - E_2(t) \to 0$ as $t \to \infty$.

Next, we consider nonlinear compartmental time delay dynamical systems of the form

$$\dot{E}_i(t) = -\sum_{j=1, j\neq i}^{q} \sigma_{ji}(E_i(t)) + \sum_{j=1, j\neq i}^{q} \sigma_{ij}(E_j(t-\tau_i)), \quad E(\theta) = \eta(\theta),$$
$$-\bar{\tau} \leq \theta \leq 0, \quad t \geq 0, \tag{15.76}$$

where $i = 1, \ldots, q$, $\sigma_{ij} : \overline{\mathbb{R}}_+ \to \overline{\mathbb{R}}_+$, $i \neq j$, $i, j \in \{1, \ldots, q\}$, are such that $\sigma_{ij}(0) = 0$ and $\sigma_{ij}(\cdot)$, $i \neq j$, $i, j = 1, \ldots, q$, is strictly increasing. Note that each transfer coefficient $\sigma_{ij}(\cdot)$ is only a function of E_j and not E. In this case, (15.76) can be written in the form given by (15.59) with $q_{\rm d} = q$,

$$f_i(E_i) = -\sum_{j=1, j\neq i}^{q} \sigma_{ji}(E_i), \quad f_{{\rm d}i}(x) = \mathbf{e}_i \sum_{j=1}^{q} \sigma_{ij}(E_j), \quad i = 1, \ldots, q. \tag{15.77}$$

Next, with $P_i = \mathbf{e}_i \mathbf{e}_i^{\rm T}$, $i = 1, \ldots, q$, so that $P = I_q$, it follows that (15.60) is trivially satisfied and (15.61) holds if and only if

$$\sum_{i=1}^{q} \left[\sum_{j=1, i\neq j}^{q} \sigma_{ij}(E_j) \right]^2 \leq \sum_{i=1}^{q} \left[\sum_{j=1, i\neq j}^{q} \sigma_{ji}(E_i) \right]^2, \quad E \in \overline{\mathbb{R}}_+^q. \tag{15.78}$$

In the case where $q = 2$, (15.78) is trivially satisfied, and hence, it follows from Theorem 15.8 that $E_1(t) - E_2(t) \to 0$ as $t \to \infty$. In general, (15.78) does not hold for arbitrary strictly increasing functions $\sigma_{ij}(\cdot)$. However, if $\sigma_{ij}(\cdot) = \sigma(\cdot)$, $i \neq j$, $i, j = 1, \ldots, q$, where $\sigma : \overline{\mathbb{R}}_+ \to \overline{\mathbb{R}}_+$ is such that $\sigma(0) = 0$ and $\sigma(\cdot)$ is strictly increasing, then (15.78) holds if and only if

$$\sum_{i=1}^{q} \left[\sum_{j=1, i\neq j}^{q} \sigma(E_j) \right]^2 \leq \sum_{i=1}^{q} \left[\sum_{j=1, i\neq j}^{q} \sigma(E_i) \right]^2, \quad E \in \mathbb{R}_+^q. \tag{15.79}$$

In this case, since

$$0 \geq (q-1) \sum_{i=1}^{q} \sigma^2(E_i) + (q-2) \sum_{i=1}^{q} \sum_{j=1, j\neq i}^{q} \sigma(E_i)\sigma(E_j)$$

$$-(q-1)^2\sum_{i=1}^{q}\sigma^2(E_i)$$

$$= -(q-2)\sum_{i=1}^{q}\sum_{j=1,j\neq i}^{q}(\sigma(E_i)-\sigma(E_j))^2,$$

(15.79) holds, and hence, it follows from Theorem 15.8 that $E_i(t)-E_j(t)\to 0$ as $t\to\infty$, where $i\neq j$, $i,j=1,\ldots,q$.

Next, we specialize Theorem 15.8 to nonlinear time delay compartmental systems of the form

$$\dot{E}(t) = A\hat{\sigma}(E(t)) + \sum_{i=1}^{q_{\rm d}} A_{{\rm d}i}\hat{\sigma}(E(t-\tau_i)), \quad E(\theta)=\eta(\theta), \quad -\bar{\tau}\leq\theta\leq 0,$$

$$t\geq 0, \qquad (15.80)$$

where $\hat{\sigma}:\overline{\mathbb{R}}_+^q\to\overline{\mathbb{R}}_+^q$ is given by $\hat{\sigma}(x)=[\sigma(E_1),\sigma(E_2),\ldots,\sigma(E_q)]^{\rm T}$, where $\sigma:\overline{\mathbb{R}}_+\to\overline{\mathbb{R}}_+$ is such that $\sigma(u)=0$ if and only if $u=0$, and A and $A_{\rm d}=\sum_{i=1}^{q_{\rm d}}A_{{\rm d}i}$ are as given by (15.48).

Theorem 15.9. Consider the nonlinear time delay system given by (15.80) where $\sigma:\overline{\mathbb{R}}_+\to\overline{\mathbb{R}}_+$ is such that $\sigma(0)=0$ and $\sigma(\cdot)$ is strictly increasing. Assume that $(A+\sum_{i=1}^{q_{\rm d}}A_{{\rm d}i})^{\rm T}\mathbf{e}=(A+\sum_{i=1}^{q_{\rm d}}A_{{\rm d}i})\mathbf{e}=0$ and $\operatorname{rank}(A+\sum_{i=1}^{q_{\rm d}}A_{{\rm d}i})=q-1$. Then, for every $\alpha\geq 0$, $\alpha\mathbf{e}$ is a semistable equilibrium point of (15.80). Furthermore, $E(t)\to\alpha^*\mathbf{e}$ as $t\to\infty$, where α^* satisfies

$$n\alpha^* + \sigma(\alpha^*)\sum_{i=1}^{q_{\rm d}}\tau_i\mathbf{e}^{\rm T}A_{{\rm d}i}\mathbf{e} = \mathbf{e}^{\rm T}\eta(0) + \sum_{i=1}^{q_{\rm d}}\int_{-\tau_i}^{0}\mathbf{e}^{\rm T}A_{{\rm d}i}\hat{\sigma}(\eta(\theta)){\rm d}\theta. \quad (15.81)$$

Proof. It follows from Lemma 15.1 that there exists Q_i, $i=1,\ldots,q_{\rm d}$, such that (15.49) holds with Q_i given by (15.50). Now, since $A=-\sum_{i=1}^{q_{\rm d}}Q_i=-\sum_{i=1}^{q_{\rm d}}P_i^{\rm D}=-P^{-1}$, where $P=\sum_{i=1}^{q_{\rm d}}P_i$, it follows from (15.49) that, for all $E\in\overline{\mathbb{R}}_+^q$,

$$0 \geq 2\hat{\sigma}^{\rm T}(E)A\hat{\sigma}(E) + \hat{\sigma}^{\rm T}(E)\sum_{i=1}^{q_{\rm d}}(Q_i + A_{{\rm d}i}^{\rm T}Q_i^{\rm D}A_{{\rm d}i})\hat{\sigma}(E)$$

$$= -f^{\rm T}(E)Pf(E) + \sum_{i=1}^{q_{\rm d}}f_{{\rm d}i}^{\rm T}(x)P_if_{{\rm d}i}(E),$$

where $f(E)=A\hat{\sigma}(E)$ and $f_{{\rm d}i}(E)=A_{{\rm d}i}\hat{\sigma}(E)$, $i=1,\ldots,q_{\rm d}$, $E\in\overline{\mathbb{R}}_+^q$. Furthermore, since $P_i^{\rm D}P_iA_{{\rm d}i}=A_{{\rm d}i}$, $i=1,\ldots,q_{\rm d}$, it follows that $P_i^{\rm D}P_if_{{\rm d}i}(E)=f_{{\rm d}i}(E)$, $i=1,\ldots,q_{\rm d}$, $E\in\overline{\mathbb{R}}_+^q$.

Now, the result is an immediate consequence of Theorem 15.8 by noting that $\mathbf{e}^{\mathrm{T}}[f(E) + \sum_{i=1}^{q_{\mathrm{d}}} f_{\mathrm{d}i}(E)] = 0$ and $f(E) + \sum_{i=1}^{q_{\mathrm{d}}} f_{\mathrm{d}i}(E) = 0$ if and only if $E = \alpha\mathbf{e}$ for some $\alpha \geq 0$. □

Example 15.1. In this example, we apply Theorems 15.7 and 15.9 to a linear and nonlinear compartmental thermodynamic system with time delay. Specifically, consider

$$\dot{E}_i(t) = -\sum_{j=1, i\neq j}^{q} \sigma_{ji} E_i(t) + \sum_{j=1, i\neq j}^{q} \sigma_{ij} E_j(t - \tau_{ij}), \quad E_i(\theta) = \eta_i(\theta), \quad -\bar{\tau} \leq \theta \leq 0, \quad t \geq 0, \tag{15.82}$$

for all $i = 1, \ldots, q$, or, equivalently, in vector form

$$\dot{E}(t) = AE(t) + \sum_{l=1}^{q_{\mathrm{d}}} A_{\mathrm{d}l} E(t - \tau_l), \qquad E(\theta) = \eta(\theta), \quad -\bar{\tau} \leq \theta \leq 0, \quad t \geq 0, \tag{15.83}$$

where $q_{\mathrm{d}} \triangleq q^2$, $A \in \mathbb{R}^{q\times q}$, and $A_{\mathrm{d}l} \in \mathbb{R}^{q\times q}$, $l = 1, \ldots, q_{\mathrm{d}}$, with

$$A = \operatorname{diag}\left[-\sum_{j=2}^{q} \sigma_{j1}, \ldots, -\sum_{j=1}^{q-1} \sigma_{jq}\right], \tag{15.84}$$

$A_{\mathrm{d}((i-1)q+j)} = \sigma_{ij}\mathbf{e}_i\mathbf{e}_j^{\mathrm{T}}$, and $\tau_{((i-1)q+j)} = \tau_{ij}$, $i, j = 1, \ldots, q$. Note that if $(j, i) \notin \mathcal{E}$, then $A_{\mathrm{d}((i-1)q+j)} = 0$.

Furthermore, it can be easily shown that $(A + A_{\mathrm{d}})^{\mathrm{T}}\mathbf{e} = 0$, where $A_{\mathrm{d}} \triangleq \sum_{l=1}^{q_{\mathrm{d}}} A_{\mathrm{d}l}$, and $\operatorname{rank}(A + A_{\mathrm{d}}) = q - 1$ if and only if for every pair of nodes $(i, j) \in \mathcal{V}$ there exists a *path* from compartment i to compartment j [177]. Here, we assume that the adjacency matrix $\mathcal{A}$ is chosen such that $(A + A_{\mathrm{d}})\mathbf{e} = 0$ so that the linear time delay compartmental dynamical system (15.83) satisfies all the conditions of Theorem 15.7. Hence, it follows from Theorem 15.7 that the compartmental system given by (15.83) achieves state equipartition, that is, $\lim_{t\to\infty} E_i(t) = \lim_{t\to\infty} E_j(t) = \alpha^*$, $i, j = 1, \ldots, q$, $i \neq j$, where α^* is given by (15.52).

Alternatively, it follows from Theorem 15.9 that the nonlinear compartmental dynamical system given by

$$\dot{E}(t) = A\hat{\sigma}(E(t)) + \sum_{i=1}^{q_{\mathrm{d}}} A_{\mathrm{d}i}\hat{\sigma}(E(t-\tau_i)), \quad E(\theta) = \eta(\theta), \quad -\bar{\tau} \leq \theta \leq 0, \quad t \geq 0, \tag{15.85}$$

also achieves state equipartition if $\sigma(\cdot)$ and $\hat{\sigma}(\cdot)$ satisfy the conditions in Theorem 15.9. In this case, $\lim_{t\to\infty} E_i(t) = \lim_{t\to\infty} E_j(t) = \alpha^*$, $i, j = 1, \ldots, q$, $i \neq j$, where α^* is a solution to (15.81). Note that if $\sigma(\theta) = \theta$,

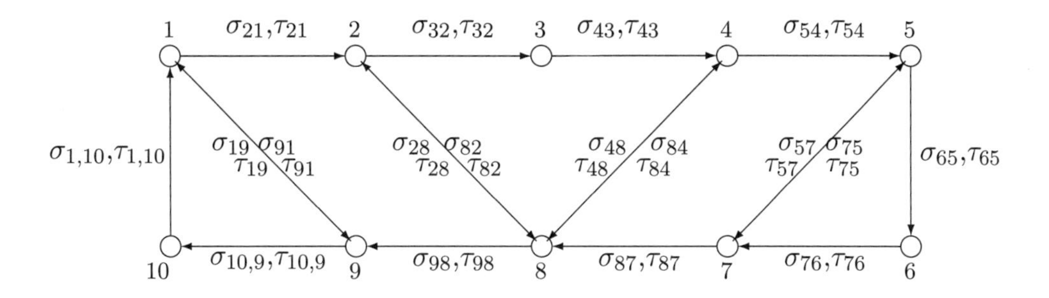

Figure 15.3 Thermodynamic model with undirected and directed heat flow.

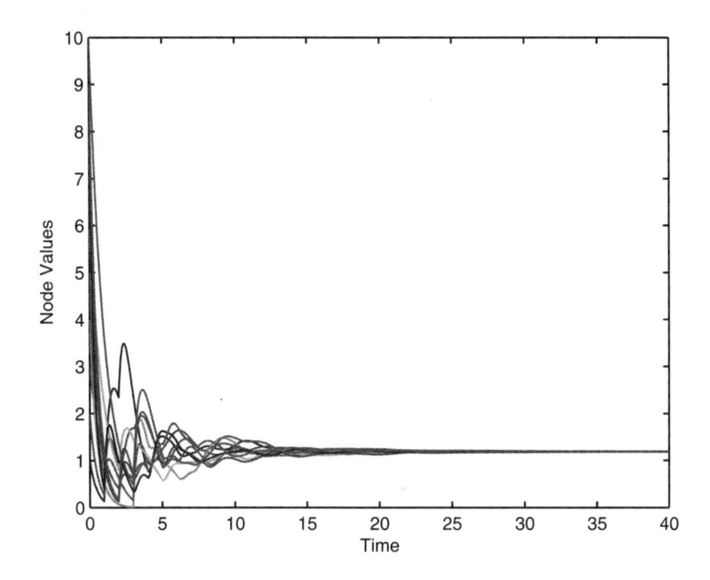

Figure 15.4 Linear thermodynamic model.

(15.85) specializes to (15.83).

To illustrate the two models given by (15.83) and (15.85), consider the compartmental system given by the graph shown in Figure 15.3 where σ_{ij} and τ_{ij} denote the weight and the time delay for each edge shown. Here, we choose $\sigma_{i,j} = 1$ if $(i,j) \in \mathcal{E}$ so that $(A + A_{\rm d})\mathbf{e} = 0$. In addition, it can be easily shown that $\operatorname{rank}(A+A_{\rm d}) = q-1 = 9$. With $E_0 = [1\,2\,3\,4\,5\,6\,7\,8\,9\,10]^{\rm T}$, Figures 15.4 and 15.5 demonstrate state equipartitioning of (15.82) and (15.85) with $\sigma(\theta) = \tanh(\theta)$ in (15.85). △

Next, we establish a sufficient condition for guaranteeing asymptotic stability of the zero solution $E_t \equiv 0$ to (15.80).

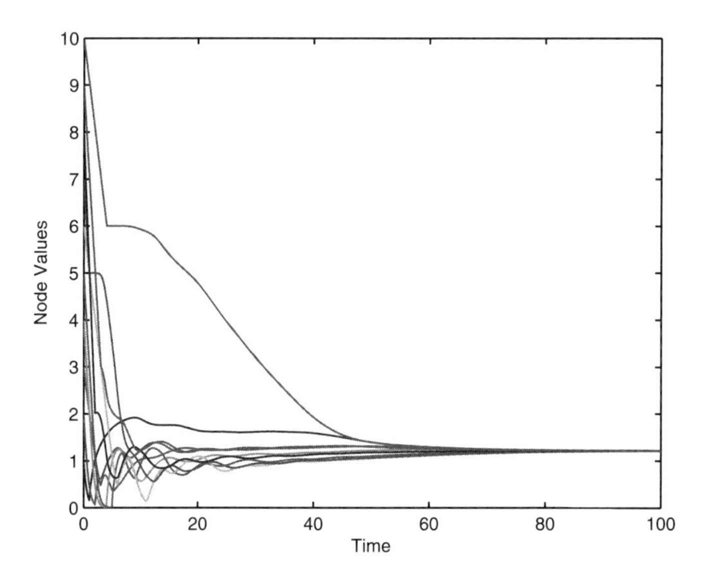

Figure 15.5 Nonlinear thermodynamic model.

Theorem 15.10. Consider the nonlinear nonnegative time delay dynamical system given by (15.80) where $\sigma : \overline{\mathbb{R}}_+ \to \overline{\mathbb{R}}_+$ is such that $\sigma(\cdot)$ is positive, $A \in \mathbb{R}^{q\times q}$ is essentially nonnegative, and $A_{\rm d} \in \mathbb{R}^{q\times q}$ is nonnegative. If there exist $p, r \in \mathbb{R}^q$ such that $p >> 0$ and $r >> 0$ satisfy

$$0 = \left(A + \sum_{i=1}^{q_{\rm d}} A_{{\rm d}i}\right)^{\rm T} p + r, \tag{15.86}$$

then the zero solution $E_t \equiv 0$ to (15.80) is asymptotically stable for all $\bar{\tau} \in [0, \infty)$.

Proof. Consider the Lyapunov-Krasovskii functional given by

$$V(\psi) = p^{\rm T}\psi(0) + \sum_{i=1}^{q_{\rm d}} \int_{-\tau_i}^{0} p^{\rm T} A_{{\rm d}i}\hat{\sigma}(\psi(\theta)){\rm d}\theta, \tag{15.87}$$

and note that $V(\psi) \geq \min_{i\in\{1,\ldots,q\}} p_i\|\psi(0)\|$. Next, using (15.86), it follows that the Lyapunov-Krasovskii directional derivative along the trajectories of (15.80) is given by

$$\begin{aligned}\dot{V}(E_t) &= p^{\rm T}\dot{E}(t) + \sum_{i=1}^{q_{\rm d}} p^{\rm T} A_{{\rm d}i}[\hat{\sigma}(E(t)) - \hat{\sigma}(E(t-\tau_i))] \\ &= p^{\rm T}\left(A + \sum_{i=1}^{q_{\rm d}} A_{{\rm d}i}\right)\hat{\sigma}(E(t))\end{aligned}$$

$$= -r^{\mathrm{T}}\hat{\sigma}(E(t))$$
$$\leq -\beta(\|E(t)\|), \quad t \geq 0, \tag{15.88}$$

where $\beta : \overline{\mathbb{R}}_+ \to \overline{\mathbb{R}}_+$ is a class $\mathcal{K}$ function. Now, it follows from Theorem 15.1 that the zero solution $E_t \equiv 0$ to (15.80) is asymptotically stable. □

Finally, we consider a class of nonlinear compartmental dynamical systems given by

$$\dot{E}_i(t) = -\sum_{j=1}^{q} \sigma_{ji}(E_i(t)) + \sum_{j=1, j\neq i}^{q} \sigma_{ij}(E_j(t-\tau_{ij})), \quad E(\theta) = \eta(\theta),$$
$$-\bar{\tau} \leq \theta \leq 0, \quad t \geq 0, \tag{15.89}$$

where $i = 1, \ldots, q$, and, for each $i, j \in \{1, \ldots, q\}$, $\sigma_{ij}(\cdot)$ satisfies $\sigma_{ij}(0) = 0$ and

$$M_{ij} \leq \frac{\sigma_{ij}(u)}{u} \leq L_{ij}, \quad u > 0, \tag{15.90}$$

where $M_{ij} \geq 0$, $i, j = 1, \ldots, q$. Next, using the inequality [206]

$$\left(\prod_{k=1}^{m+1} \sigma_k^{p_k}\right)^{\frac{1}{P_{m+1}}} \leq \left(\sum_{k=1}^{m+1} p_k \sigma_k^r\right)^{\frac{1}{r}} P_{m+1}^{-\frac{1}{r}}, \tag{15.91}$$

where $\sigma_k \geq 0$, $p_k > 0$, $k = 1, \ldots, m+1$, $r > 0$, and $P_{m+1} = \sum_{k=1}^{m+1} p_k$, we establish sufficient conditions for Lyapunov and asymptotic stability of the nonlinear time delay compartmental system given by (15.89).

Theorem 15.11. Consider the nonlinear time delay dynamical system given by (15.89). Assume that (15.90) holds. If there exist constants $r_k > 0$, $k = 1, \ldots, N$, $b_i > 0$, $i = 1, \ldots, q$, $p_{ij} \in \mathbb{R}$, $q_{ij} \in \mathbb{R}$, $i, j = 1, \ldots, q$, $i \neq j$, such that

$$\sum_{j=1, j\neq i}^{q} \sum_{k=1}^{N} r_k L_{ij}^{\frac{pp_{ij}}{r_k}} + \sum_{j=1, j\neq i}^{q} \frac{b_j}{b_i} L_{ji}^{pq_{ji}} \leq p \sum_{j=1}^{n} M_{ji}, \quad i = 1, \ldots, q, \tag{15.92}$$

where p_{ij} and q_{ij} satisfy $Np_{ij} + q_{ij} = 1$, $i, j = 1 \ldots, q$, $i \neq j$, and $p = 1 + \sum_{k=1}^{N} r_k$, then the zero solution $E_t \equiv 0$ to (15.89) is Lyapunov stable for all $\tau_{ij} \in [0, \infty)$, $i, j = 1, \ldots, q$, $i \neq j$. If, in addition, the inequality in (15.92) is strict, then the zero solution $E_t \equiv 0$ to (15.89) is asymptotically stable for all $\tau_{ij} \in [0, \infty)$, $i, j = 1, \ldots, q$, $i \neq j$. Alternatively, if there exist constants $b_i > 0$, $i = 1, \ldots, q$, such that

$$\sum_{j=1, j\neq i}^{q} \frac{b_j}{b_i} L_{ji} \leq \sum_{j=1}^{n} M_{ji}, \quad i = 1, \ldots, q, \tag{15.93}$$

then the zero solution $E_t \equiv 0$ to (15.89) is Lyapunov stable for all $\tau_{ij} \in$

$[0, \infty)$, $i, j = 1, \ldots, q$, $i \neq j$. Finally, if the inequality (15.93) is strict, then the zero solution $E_t \equiv 0$ to (15.89) is asymptotically stable for all $\tau_{ij} \in [0, \infty)$, $i, j = 1, \ldots, q$, $i \neq j$.

Proof. To show Lyapunov stability of the zero solution $E_t \equiv 0$ to (15.89), consider the Lyapunov-Krasovskii functional given by

$$V(\psi(\cdot)) = \sum_{i=1}^{q} \frac{b_i}{p} \psi_i^p(0) + \sum_{i=1}^{q} \sum_{j=1, j\neq i}^{q} \frac{b_i}{p} L_{ij}^{pq_{ij}} \int_{-\tau_{ij}}^{0} \psi_j^p(\theta) \mathrm{d}\theta, \qquad (15.94)$$

and note that $V(\psi) \geq \sum_{i=1}^{q} \frac{b_i}{p} \psi_i^p(0) \geq \min_{i\in\{1,\ldots,q\}} \frac{b_i}{p} \|\psi(0)\|_p^p$. Then the directional derivative of $V(\psi)$ along the trajectories of (15.89) is given by

$$\begin{aligned}
\dot{V}(E_t) &= \sum_{i=1}^{q} b_i E_i^{p-1}(t) \dot{E}_i(t) + \sum_{i=1}^{q} \sum_{j=1, j\neq i}^{q} \frac{b_i}{p} L_{ij}^{pq_{ij}} (E_j^p(t) - E_j^p(t - \tau_{ij})) \\
&= \sum_{i=1}^{q} b_i E_i^{p-1}(t) \left[-\sum_{j=1}^{q} \sigma_{ji}(E_i(t)) + \sum_{j=1, j\neq i}^{q} \sigma_{ij}(E_j(t - \tau_{ij})) \right] \\
&\quad + \sum_{i=1}^{q} \sum_{j=1, j\neq i}^{q} \frac{b_i}{p} L_{ij}^{pq_{ij}} (E_j^p(t) - E_j^p(t - \tau_{ij})) \\
&\leq \sum_{i=1}^{q} b_i \left[-\sum_{j=1}^{q} M_{ji} E_i^p(t) + \sum_{j=1, j\neq i}^{q} L_{ij} E_j(t - \tau_{ij}) E_i^{p-1}(t) \right] \\
&\quad + \sum_{i=1}^{q} \sum_{j=1, j\neq i}^{q} \frac{b_i}{p} L_{ij}^{pq_{ij}} (E_j^p(t) - E_j^p(t - \tau_{ij})) \\
&= \sum_{i=1}^{q} b_i \left[-\sum_{j=1}^{q} M_{ji} E_i^p(t) + \sum_{j=1, j\neq i}^{q} \prod_{k=1}^{N} \left(L_{ij}^{\frac{p_{ij}}{r_k}} E_i(t) \right)^{r_k} L_{ij}^{q_{ij}} E_j(t - \tau_{ij}) \right] \\
&\quad + \sum_{i=1}^{q} \sum_{j=1, j\neq i}^{q} \frac{b_i}{p} L_{ij}^{pq_{ij}} (E_j^p(t) - E_j^p(t - \tau_{ij})) \\
&\leq -\sum_{i=1}^{q} b_i \left[-\sum_{j=1}^{q} M_{ji} E_i^p(t) + \frac{1}{p} \sum_{j=1, j\neq i}^{q} \sum_{k=1}^{N} r_k L_{ij}^{\frac{pp_{ij}}{r_k}} E_i^p(t) \right] \\
&\quad + \sum_{i=1}^{q} \sum_{j=1, j\neq i}^{q} \frac{b_i}{p} L_{ij}^{pq_{ij}} E_j^p(t) \\
&= -\sum_{i=1}^{q} b_i \left[\sum_{j=1}^{q} M_{ji} - \frac{1}{p} \sum_{j=1, j\neq i}^{q} \sum_{k=1}^{N} r_k L_{ij}^{\frac{pp_{ij}}{r_k}} - \frac{1}{p} \sum_{j=1, j\neq i}^{q} \frac{b_j}{b_i} L_{ji}^{pq_{ji}} \right] E_i^p(t)
\end{aligned}$$

$$\leq -c \sum_{i=1}^{q} E_i^p(t), \tag{15.95}$$

where

$$c \triangleq \min_{1 \leq i \leq q} \sum_{i=1}^{q} b_i \left[\sum_{j=1}^{q} M_{ji} - \frac{1}{p} \sum_{j=1, j \neq i}^{q} \sum_{k=1}^{N} r_k L_{ij}^{\frac{pp_{ij}}{r_k}} - \frac{1}{p} \sum_{j=1, j \neq i}^{q} \frac{b_j}{b_i} L_{ji}^{pq_{ji}} \right] \geq 0 \tag{15.96}$$

and where inequality (15.91) was used in (15.95). This establishes Lyapunov stability of the zero solution $E_t \equiv 0$ to (15.89).

Next, if the inequality in (15.92) is strict, then it follows from (15.96) that $c > 0$. In this case, it follows from (15.95) that

$$c \int_0^t \sum_{i=1}^{q} E_i^p(\theta) \mathrm{d}\theta \leq V(E_0) - V(E_t), \quad t \geq 0, \tag{15.97}$$

which implies that $\int_0^\infty \sum_{i=1}^q E_i^p(\theta)\mathrm{d}\theta < \infty$, and hence, $E_i(t)$ and $\dot{E}_i(t)$ are bounded on $(0, \infty)$. Hence, $E_i^p(t)$ and $\sum_{i=1}^q E_i^p(t)$ are uniformly continuous on $(0, \infty)$. Now, it follows from Barbalat's Lemma [191, p. 221] that $\lim_{t\to\infty} \sum_{i=1}^q E_i^p(t) = 0$, and hence, the zero solution $E_t \equiv 0$ to (15.89) is asymptotically stable.

Finally, to show Lyapunov and asymptotic stability in the case where (15.93) holds, consider the Lyapunov-Krasovskii functional given by

$$V(\psi(\cdot)) = \sum_{i=1}^{q} b_i \psi_i(0) + \sum_{i=1}^{q} \sum_{j=1, j \neq i}^{q} b_i L_{ij} \int_{-\tau_{ij}}^{0} \psi_j(\theta) \mathrm{d}\theta \tag{15.98}$$

and note that $V(\psi) \geq \min_{i \in \{1,\ldots,q\}} b_i \|\psi(0)\|$. Now, the result follows by using similar arguments as above. □

If we set $r_k = p - 1$, $N = 1$, $p_{ij} = \frac{p-1}{p}$, and $q_{ij} = \frac{1}{p}$ in Theorem 15.11, then the following corollary is immediate.

Corollary 15.2. Consider the nonlinear time delay dynamical system given by (15.89). Assume that (15.90) holds. If there exist constants $p \geq 1$ and $b_i > 0$, $i = 1, \ldots, q$, such that

$$\sum_{j=1, j \neq i}^{q} (p-1) L_{ij} + \frac{b_j}{b_i} L_{ji} \leq p \sum_{j=1}^{q} M_{ji}, \quad i = 1, \ldots, q, \tag{15.99}$$

then the zero solution $E_t \equiv 0$ to (15.89) is Lyapunov stable for all $\tau_{ij} \in [0, \infty)$, $i, j = 1, \ldots, q$, $i \neq j$. Alternatively, if the inequality (15.99) is strict, then the zero solution $E_t \equiv 0$ to (15.89) is asymptotically stable for all $\tau_{ij} \in [0, \infty)$, $i, j = 1, \ldots, q$, $i \neq j$.

It follows from Corollary 15.2 that the time delay compartmental dynamical system (15.89) is asymptotically stable as long as M_{ii}, $i = 1, \ldots, q$, are sufficiently large.

15.8 Monotonicity of System Energies in Thermodynamic Processes with Time Delay

Thermodynamic models with time delay can admit nonmonotonic solutions (e.g., underdamped solutions) which introduce oscillations in the system energies that may otherwise be nonoscillatory in the absence of time delay. Since the subsystem energies in certain heat conduction models should monotonically decrease or monotonically increase after the discontinuation of heat addition, it is of interest to determine necessary and sufficient conditions under which these systems possess monotonic solutions.

In this section, we develop necessary and sufficient conditions for identifying nonnegative and compartmental dynamical systems that only admit nonoscillatory and monotonic solutions in the presence of time lags. Specifically, consider the linear time delay dynamical system $\mathcal{G}$ given by

$$\dot{E}(t) = AE(t) + A_{\mathrm{d}}E(t-\tau) + GS(t), \quad E(\theta) = \phi(\theta), \quad -\tau \leq \theta \leq 0, \quad t \geq 0, \tag{15.100}$$

where, for $t \geq 0$, $E(t) \in \mathbb{R}^q$, $S(t) \in \mathbb{R}^m$, $A \in \mathbb{R}^{q\times q}$, $A_{\mathrm{d}} \in \mathbb{R}^{q\times q}$, $G \in \mathbb{R}^{q\times m}$, $\tau \geq 0$, and $\phi(\cdot) \in \mathcal{C} = \mathcal{C}([-\tau, 0], \mathbb{R}^q)$. Here, for simplicity of exposition we have assumed that all energy transfer times between compartments are given by τ. The more general multiple delay case can be addressed as shown in (15.47).

The following definition for nonnegative time delay systems is needed for the main results in this section.

Definition 15.8. The linear time delay dynamical system $\mathcal{G}$ given by (15.100) is *nonnegative* if for every $\phi(\cdot) \in \mathcal{C}_+$, where $\mathcal{C}_+ \triangleq \{\psi(\cdot) \in \mathcal{C} : \psi(\theta) \geq\geq 0,\ \theta \in [-\tau, 0]\}$, and $S(t) \geq\geq 0$, $t \geq 0$, the solution $E(t)$, $t \geq 0$, to (15.100) is nonnegative.

Proposition 15.4. The linear time delay dynamical system $\mathcal{G}$ given by (15.100) is nonnegative if and only if $A \in \mathbb{R}^{q\times q}$ is essentially nonnegative, $A_{\mathrm{d}} \in \mathbb{R}^{q\times q}$ is nonnegative, and $G \geq\geq 0$.

Proof. First, note that the solution to (15.100) is given by

$$\begin{aligned} E(t) &= e^{At}E(0) + \int_0^t e^{A(t-\theta)}[A_{\rm d}E(\theta-\tau) + GS(\theta)]{\rm d}\theta \\ &= e^{At}\phi(0) + \int_{-\tau}^{t-\tau} e^{A(t-\tau-\theta)}A_{\rm d}E(\theta){\rm d}\theta + \int_0^t e^{A(t-\theta)}GS(\theta){\rm d}\theta. \end{aligned} \tag{15.101}$$

Now, if A is essentially nonnegative, then it follows from Proposition 2.4 that $e^{At} \geq\geq 0$, $t \geq 0$; and if $\phi(\cdot) \in \mathcal{C}_+$, $A_{\rm d} \geq\geq 0$, and $G \geq\geq 0$, then it follows that

$$E(t) = e^{At}\phi(0) + \int_{-\tau}^{t-\tau} e^{A(t-\tau-\theta)}A_{\rm d}E(\theta){\rm d}\theta + \int_0^t e^{A(t-\theta)}GS(\theta){\rm d}\theta \geq\geq 0, \quad t \in [0,\tau). \tag{15.102}$$

Alternatively, for all $t > \tau$,

$$E(t) = e^{A\tau}E(t-\tau) + \int_0^\tau e^{A(\tau-\theta)}A_{\rm d}E(t+\theta-2\tau){\rm d}\theta + \int_0^t e^{A(t-\theta)}GS(\theta){\rm d}\theta, \tag{15.103}$$

and hence, since $E(t) \geq\geq 0$, $t \in [-\tau,\tau)$, it follows that $E(t) \geq\geq 0$, $t \in [\tau, 2\tau)$. Repeating this procedure iteratively, it follows that $E(t) \geq\geq 0$, $t \geq 0$, which implies that $\mathcal{G}$ is nonnegative.

Conversely, suppose $\mathcal{G}$ is nonnegative. Now, let $\phi(\theta) = 0$, $-\tau \leq \theta \leq 0$, and let $S(t) = \delta(t-\hat{t})\hat{S}$, $t, \hat{t} \in [0,\tau)$, where $\delta(\cdot)$ denotes the Dirac delta function and $\hat{S} \geq\geq 0$. In this case, since $E(\hat{t}) = G\hat{S} \geq\geq 0$ for all $\hat{S} \in \overline{\mathbb{R}}_+^m$ it follows that $G \geq\geq 0$. Furthermore, with $S(t) = 0$, $\phi(\theta) = 0$, $-\tau \leq \theta \leq 0$, $E(t) = e^{At}\phi(0)$, $t \in [0,\tau)$, and hence, it follows from Proposition 2.4 that if $E(t) \geq\geq 0$, $t \geq 0$, for all $\phi(0) \in \overline{\mathbb{R}}_+^q$, then A is essentially nonnegative.

Finally, suppose, *ad absurdum*, that $A_{\rm d}$ is not nonnegative, that is, there exist $I, J \in \{1,\ldots,q\}$ such that $A_{{\rm d}(I,J)} < 0$. Let $S(t) = 0$, $t \geq 0$, and let $\{v_n\}_{n=1}^\infty \subset \mathcal{C}_+$ denote a sequence of functions such that $\lim_{n\to\infty} v_n(\theta) = e_J\delta(\theta+\eta-\tau)$, where $0 < \eta < \tau$ and $\delta(\cdot)$ denotes the Dirac delta function. In this case, it follows from (15.101) that

$$E_n(t) = e^{A\eta}v_n(0) + \int_0^\eta e^{A(\eta-\theta)}A_{\rm d}E(\theta-\tau){\rm d}\theta, \tag{15.104}$$

where $E_n(\cdot)$ denotes the solution to (15.100) with $\phi(\theta) = v_n(\theta)$, which further implies that $E(\eta) = \lim_{n\to\infty} E_n(\eta) = e^{A\eta}A_{\rm d}e_J$. Now, by choosing η sufficiently small, it follows that $E_I(\eta) < 0$, which is a contradiction. □

Next, we present a definition for the monotonicity of solutions over all time for a time delay nonnegative dynamical system of the form given by (15.100).

Definition 15.9. Consider the linear nonnegative time delay dynamical system (15.100) where $\phi(\cdot) \in \mathcal{C}_+$, A is essentially nonnegative, $A_{\rm d}$ is nonnegative, G is nonnegative, and $S(t)$, $t \geq 0$, is nonnegative. The linear nonnegative time delay dynamical system (15.100) is *monotonic* if there exists a matrix $R \in \mathbb{R}^{q\times q}$ such that $R = \operatorname{diag}[r_1, \ldots, r_q]$, $r_i = \pm 1$, $i = 1, \ldots, q$, and, for every $\phi(\cdot) \in \mathcal{C}_+$, $RE(t_2) \leq\leq RE(t_1)$, $0 \leq t_1 \leq t_2$.

Next, we present necessary and sufficient conditions that guarantee monotonicity for nonnegative dynamical systems with time delay.

Theorem 15.12. Consider the linear time delay nonnegative dynamical system given by (15.100) where $\phi(\cdot) \in \mathcal{C}_+$, $A \in \mathbb{R}^{q\times q}$ is essentially nonnegative, $A_{\rm d} \in \mathbb{R}^{q\times q}$ is nonnegative, $G \in \mathbb{R}^{q\times m}$ is nonnegative, and $S(t)$, $t \geq 0$, is nonnegative. Assume there exists a matrix $R \in \mathbb{R}^{q\times q}$ such that $R = \operatorname{diag}[r_1, \ldots, r_q]$, $r_i = \pm 1$, $i = 1, \ldots, q$, and $RA \leq\leq 0$, $RA_{\rm d} \leq\leq 0$, and $RG \leq\leq 0$. Then the linear nonnegative dynamical system (15.100) is monotonic.

Proof. The proof is a direct consequence of Theorem 4.2 of [194] with B and $u(t)$, $t \geq 0$, replaced by $[A_{\rm d}\ G]$ and $[E^{\rm T}(t-\tau)\ S^{\rm T}(t)]^{\rm T}$, $t \geq 0$, respectively. □

The following result gives necessary and sufficient conditions for monotonicity in the case where $S(t) \equiv 0$.

Theorem 15.13. Consider the linear time delay nonnegative dynamical system given by (15.100) where $\phi(\cdot) \in \mathcal{C}_+$, $A \in \mathbb{R}^{q\times q}$ is essentially nonnegative, $A_{\rm d} \in \mathbb{R}^{q\times q}$ is nonnegative, and $S(t) \equiv 0$. Then the linear time delay dynamical system is monotonic if and only if there exists a matrix $R \in \mathbb{R}^{q\times q}$ such that $R = \operatorname{diag}[r_1, \ldots, r_q]$, $r_i = \pm 1$, $i = 1, \ldots, q$, $RA \leq\leq 0$, and $RA_{\rm d} \leq\leq 0$.

Proof. Sufficiency follows from Theorem 15.12 with $S(t) \equiv 0$. To show necessity, assume that the linear time delay dynamical system given by (15.100) is monotonic. In this case, it follows from (15.100) that

$$R\dot{E}(t) = RAE(t) + RA_{\rm d}E(t-\tau), \quad E(\theta) = \phi(\theta), \quad -\tau \leq \theta \leq 0, \quad t \geq 0,$$

which implies that, for some $t_1 \geq 0$ and $t_2 > t_1$,

$$RE(t_2) = RE(t_1) + \int_{t_1}^{t_2} [RAE(t) + RA_{\rm d}E(t-\tau)]{\rm d}t.$$

Next, suppose, *ad absurdum*, that there exist $I, J \in \{1, \ldots, q\}$ such that $M_{(I,J)} > 0$, where $M \triangleq RA$. Now, let $\phi(\cdot) \in \mathcal{C}_+$ be such that $\phi(t) = 0$, $-\tau < t < \eta - \tau$ and let $\phi(0) = e_J$, where $\tau > \eta > 0$ and $e_J \in \mathbb{R}^q$ is a vector of zeros with one in the Jth component. Next, it follows from continuity of solutions that there exists $0 < \varepsilon \leq \eta$ such that $E_J(t) > 0$, $0 \leq t \leq \varepsilon$. Thus, for every $t \in [0, \eta]$, it follows that

$$\begin{aligned}
[RE(t)]_I &= [RE(0)]_I + \int_0^t [RAE(s) + RA_{\rm d}E(s-\tau)]_I {\rm d}s \\
&= [RE(0)]_I + \int_{-\tau}^{t-\tau} [RAE(s+\tau) + RA_{\rm d}E(s)]_I {\rm d}s \\
&= [RE(0)]_I + \int_{-\tau}^{t-\tau} [RAE(s+\tau)]_I {\rm d}s \\
&> [RE(0)]_I,
\end{aligned} \tag{15.105}$$

which is a contradiction. Hence, $RA \leq\leq 0$.

Finally, suppose, *ad absurdum*, that there exist $I, J \in \{1, \ldots, q\}$ such that $M_{(I,J)} > 0$, where $M \triangleq RA_{\rm d}$. Now, let $\{v_n\}_{n=1}^{\infty} \subset \mathcal{C}_+$ denote a sequence of functions such that $\lim_{n\to\infty} v_n(t) = e_J\delta(t - \eta + \tau)$, where $0 < \eta < \tau$ and $\delta(\cdot)$ denotes the Dirac delta function. Furthermore, let $\phi(\cdot) \in \mathcal{C}_+$ be such that $\phi(t) = \lim_{n\to\infty} v_n(t)$, $-\tau < t < 0$. In this case, it follows from (15.105) that

$$\begin{aligned}
RE_n(t) &= Rv_n(0) + \int_0^t [RAE_n(s) + RA_{\rm d}E_n(s-\tau)]{\rm d}s \\
&= \int_{-\tau}^{t-\tau} [RAv_n(s+\tau) + RA_{\rm d}v_n(s)]{\rm d}s,
\end{aligned}$$

where $E_n(\cdot)$ denotes the solution to (15.100) with $\phi(t) = v_n(t)$. Now, note that $E(t) \triangleq \lim_{n\to\infty} [RE_n(t)]_I = RA_{\rm d}e_J$. Hence, $E_I(t) = [RE(t)]_I > [RE(0)]_I = E_I(0) = 0$. Thus, there exists $n > 0$ such that $\phi(t) = v_n(t)$ implies $[RE(t)]_I > [RE(0)]$ and, hence, $RA_{\rm d} \leq\leq 0$. $\square$

Next, we present a sufficient condition that guarantees monotonicity for nonlinear nonnegative dynamical systems with time delay. In particular, we consider nonlinear time delay dynamical systems $\mathcal{G}$ of the form

$$\dot{E}(t) = AE(t) + f_{\rm d}(E(t-\tau)) + G(E(t))S(t), \quad E(\theta) = \phi(\theta), \quad -\tau \leq \theta \leq 0, \quad t \geq 0, \tag{15.106}$$

where, for $t \geq 0$, $E(t) \in \mathbb{R}^q$, $S(t) \in \mathbb{R}^m$, $A \in \mathbb{R}^{q\times q}$, $f_{\rm d} : \mathbb{R}^q \to \mathbb{R}^q$ is locally Lipschitz and $f_{\rm d}(0) = 0$, $\tau \geq 0$, $G : \mathbb{R}^q \to \mathbb{R}^{q\times m}$, and $\phi(\cdot) \in \mathcal{C}$. For the nonlinear dynamical system $\mathcal{G}$ given by (15.106) the definitions of monotonicity hold with (15.100) replaced by (15.106).

Theorem 15.14. Consider the nonlinear time delay nonnegative dynamical system given by (15.106) where $\phi(\cdot) \in \mathcal{C}_+$, $A \in \mathbb{R}^{q\times q}$ is essentially nonnegative, $f_\mathrm{d} : \mathbb{R}^q \to \mathbb{R}^q$ is nonnegative, $G(E) \geq\geq 0$, and $S(t)$, $t \geq 0$, is nonnegative. Assume there exists a matrix $R \in \mathbb{R}^{q\times q}$ such that $R = \mathrm{diag}[r_1, \ldots, r_q]$, $r_i = \pm 1$, $i = 1, \ldots, q$, $RA \leq\leq 0$, $Rf_\mathrm{d}(E) \leq\leq 0$, and $RG(E) \leq\leq 0$, $E \in \mathbb{R}^q$. Then the nonlinear nonnegative dynamical system (15.106) is monotonic.

Proof. The proof is similar to the proof of Theorem 15.13 and, hence, is omitted. □

Finally, it is important to note that analogous results for discrete-time, hybrid, and stochastic linear and nonlinear thermodynamic systems with time delay can be derived in a similar fashion as presented in this chapter. These extensions are left as an exercise for the reader.

Chapter Sixteen

Conclusion

In this monograph, we have outlined a general dynamical systems theory framework for thermodynamics in an attempt to harmonize it with classical mechanics. The proposed macroscopic mathematical model is based on a nonlinear (finite- and infinite-dimensional) compartmental dynamical system model that is characterized by energy conservation laws capturing the exchange of energy between coupled macroscopic subsystems. Specifically, using a large-scale dynamical systems perspective, we developed some of the fundamental properties of reversible and irreversible thermodynamic systems involving conservation of energy, nonconservation of entropy and ectropy, and energy and temperature equipartition. This model is formulated in the language of dynamical systems and control theory, and it is argued that it offers conceptual advantages for describing nonequilibrium thermodynamic systems.

Using compartmental dynamical systems involving the exchange of energy via intercompartmental flow laws and invoking the two fundamental axioms of the science of heat, namely,

i) *if the energies in the connected subsystems are equal, then energy exchange between these subsystems is not possible*,

and

ii) *energy flows from more energetic subsystems to less energetic subsystems*,

we established the existence of a continuous entropy function for our thermodynamically consistent large-scale dynamical system utilizing the language of modern mathematics within a theorem-proof format. In addition, we proved the global existence and uniqueness of a continuously differentiable entropy and ectropy function for all equilibrium and nonequilibrium states of our dynamical system. Furthermore, the fundamental properties of reversible and irreversible thermodynamics were also established using a system-theoretic dynamical systems approach.

In particular, for our thermodynamically consistent large-scale dynamical system, it was shown that:

i) *The increase in internal energy of a dynamical system equals the heat energy received by the system minus the work expended by the system.*

ii) *The total energy in an isolated dynamical system is constant.*

iii) *For every dynamical transformation in an adiabatically isolated system, the entropy of the final state is greater than or equal to the entropy of the initial state.*

iv) *The entropy of an adiabatically isolated dynamical system tends to a maximum.*

v) *An isolated large-scale dynamical system naturally evolves toward a state of energy equipartition.*

vi) *Although the total energy in an adiabatically isolated dynamical system is conserved, the usable energy is diffused.*

vii) *For an equilibrium of any isolated dynamical system, it is necessary and sufficient that in all possible variations of the state of the system that do not alter its energy, the change in entropy is zero or negative.*

viii) *The entropy of every dynamical system at absolute zero can always be taken to be equal to zero.*

In addition, in our formulation the notion of subsystem thermodynamic temperatures was derived as a direct consequence of the existence of the unique, continuously differentiable subsystem entropies. Hence, thermal equilibrium is an equivalence relation between subsystem energies and does not rely on the subjective notions of hotness and coldness of each subsystem.

As discussed in Chapter 1, the theory of thermodynamics followed two conceptually rather different schools of thought, namely, the macroscopic point of view versus the microscopic point of view. The microscopic point of view of thermodynamics was first established by Maxwell [314] and further developed by Boltzmann [58] by reinterpreting thermodynamic systems in terms of molecules or atoms. However, since the microscopic states of thermodynamic systems involve a large number of similar molecules, the laws of classical mechanics were reformulated so that even though individual atoms are assumed to obey the laws of Newtonian mechanics, the statistical nature of the velocity distribution of the system particles corresponds to the thermodynamic properties of all the atoms together. This resulted in the birth of statistical mechanics. The laws of mechanics, however, as established by Poincaré [96], show that every isolated mechanical system will return arbitrarily close to its initial state infinitely often. Hence, entropy must undergo cyclic changes and thus cannot monotonically increase. This is known as the *recurrence paradox* or *Loschmidt's paradox*.

Loschmidt [290] was among the first to challenge the theory of statistical thermodynamics by pointing out that Boltzmann's theory violated the time-reversal symmetry of the microscopic equations of motion of the system particles. In fact, Poincaré's recurrence theorem prohibits irreversibility of conservative dynamical systems in the classical sense. To the present day, many scientists have attempted to provide an explanation of the recurrence paradox in which a lossless dynamical system that possesses time-reversal symmetry on a microscopic scale breaks this symmetry on a macroscopic scale.

Many scientists have made untenable arguments that despite microscopic reversibility, not all solutions need possess full time-reversal symmetry while others have averted their eyes from Loschmidt's paradox. In light of Poincaré recurrence, the law of (absolute) entropy increase cannot be derived from statistical mechanics and to this day has eluded the deepest thinkers in science. The problem of duplicating the second law of thermodynamics remains one of the hardest and most controversial problems in statistical physics.

In statistical thermodynamics the recurrence paradox is resolved by asserting that, in principle, the entropy of an isolated system can sometimes decrease. However, the probability of this happening, when computed, is incredibly small. Thus, statistical thermodynamics stipulates that the direction in which system transformations occur is determined by the laws of probability, and hence, they result in a more probable state corresponding to a higher system entropy. However, unlike classical thermodynamics, in statistical thermodynamics it is not absolutely certain that entropy increases in every system transformation. Hence, thermodynamics based on statistical mechanics gives the most probable course of system evolution and not the only possible one, and thus heat flows in the direction of lower temperature with only *statistical certainty* and not absolute certainty. Nevertheless, general arguments exploiting system fluctuations in a systematic way [455] seem to show that it is impossible, even in principle, to violate the second law of thermodynamics.

Perhaps the most famous experiment in trying to turn back the course of the universe was Maxwell's thought experiment involving a finite being, which has become known as Maxwell's demon [455]. Namely, this finite being, with "infinitely subtle senses" [205, p. 152], controls a tiny opening in a diaphragm dividing a container of gas at two different temperatures. Of course Maxwell had not meant for his demon to exist, except as a hypothetical being that replaces chance with purpose by using information to reduce entropy, and hence, defying the second law of thermodynamics.

The intelligent demon can be viewed as a computer with a two-state sorting memory that has knowledge of the positions and velocities of the molecules moving before its eyes and chooses whether or not to let them pass through the tiny diaphragm opening in defiance of the statistical certainty of heat flow. Even though Maxwell's demon perplexed many scientists of the nineteenth century, Maxwell had failed to realize that the conversion between information and energy of each molecule is not free. It was Szilárd [430] who realized that each unit of information gives a corresponding increase in entropy, bringing the demise of Maxwell's demon after sixty-two years of imparting a state of turmoil surrounding the second law.

No exception has ever been found to the second law of thermodynamics, making it, along with the first law, one of the most perfect laws of nature. In this regard, Eddington [123, p. 81] writes:

> The law that entropy always increases—the second law of thermodynamics—holds, I think, the supreme position among the laws of Nature. If someone points out to you that your pet theory of the universe is in disagreement with Maxwell's equations—then so much worse for Maxwell's equations. If it is found to be contradicted by observation—well, these experimentalists bungle things sometimes. But if your theory is found to be against the second law of thermodynamics I can give you no hope; there is nothing for it but to collapse in deepest humiliation.

And according to Einstein [134, p. 32]:

> [Thermodynamics] is the only physical theory of a universal content of which I am convinced that ... it will never be overthrown.

The underlying intention of this monograph has been to present one of the most useful and general physical branches of science in the language of dynamical systems theory. In particular, we developed a novel formulation of thermodynamics using a middle-ground systems theory that bridges the gap between classical and statistical thermodynamics. The laws of thermodynamics are among the most firmly established laws of nature, and it is hoped that this monograph will help to stimulate increased interaction between physicists and dynamical systems and control theorists. Besides the fact that irreversible thermodynamics plays a critical role in the understanding of our expanding universe, it forms the underpinning of several fundamental life science and engineering disciplines, including biological systems, physiological systems, chemical reaction systems, queuing systems, ecological systems, demographic systems, telecommunications systems, transportation systems, network systems, and power systems, to cite but a few examples.

The newly developed dynamical systems notion of entropy proposed in this monograph involving an analytical description of an objective property of matter can potentially offer a conceptual advantage over the subjective quantum expressions for entropy proposed in the literature (e.g., Daróczy entropy, Hartley entropy, Rényi entropy, von Neumann entropy, infinite-norm entropy) involving a measure of information. An even more important benefit of the dynamical systems representation of thermodynamics is the potential for developing a unified classical and quantum theory that encompasses both mechanics and thermodynamics without the need for statistical (subjective or informational) probabilities.

There is no doubt that thermodynamics is a theory of universal proportions whose laws reign supreme among the laws of nature and are capable of addressing some of science's most intriguing questions about the origins and fabric of our universe. While from its inception its speculations about the universe have been grandiose, its mathematical foundation has been amazingly obscure and imprecise. A discipline as cardinal as thermodynamics entrusted with some of the most perplexing secrets of our universe demands far more than "physical mathematics" as its underpinning. Even though many great physicists such as Archimedes, Newton, and Lagrange have humbled us with their mathematically seamless eurekas over the centuries, a great many physicists and engineers who have developed the theory of thermodynamics over the last one and a half centuries seem to have forgotten that mathematics, when used rigorously, is the irrefutable pathway to truth.

Our goal with this monograph has been to develop a dynamical systems formalism for classical thermodynamics. As a result, we use system-theoretic ideas to bring coherence, clarity, and precision to an extremely important and poorly understood classical area of science. Our system thermodynamics formalism brings classical thermodynamics within the framework of modern dynamical systems by bringing to bear some of the hallmark analytical tools from dynamical systems and control theory. A dynamical systems formalism of thermodynamics has been long overdue and aligns classical thermodynamics with the development of classical mechanics, which also started as a physical theory concerned mainly with equilibrium systems and with empirical principles initially formulated by the great cosmic theorists of ancient Greece, and later established by physicists such as Copernicus, Brahe, Kepler, and Galileo.

In particular, Archimedean mechanics addressed forces as quasi-static or nearly in equilibrium at all times, and problems involving mechanical systems were solved using the mechanical principles established by Archimedes. Newtonian dynamics subsumes Archimedean mechanics by weakening the underlying hypothesis of the Archimedean theory, and hence, provides a generalization of Archimedean mechanics in much the same way as Einsteinian mechanics is a generalization of Newtonian mechanics. The key leap from natural philosophy—Aristotelian physics—to modern science was sparked by Archimedean mechanics and Hellenistic scientific theories (e.g., solid mechanics, hydrostatics, geometric optics, astronomy, anatomy, physiology, scientific medicine), which cemented the path for modern-day science and with several of these scientific theories being subsumed without change into today's scientific disciplines.

Unlike classical thermodynamics, which remained a physical theory, in the seventeenth through the nineteenth centuries the physical approach of mechanics was replaced by mathematical theories involving abstract geometrical structures (configuration manifolds, Minkowski spacetime, Riemann spacetime), wherein the mechanistic empirical principles were incorporated into topological properties of abstract mathematical spaces. This physical-mathematical bifurcation of mechanics, which was pioneered by giants such as Euler, Hamilton, Lagrange, Huygens, and Newton, along with the fact that classical thermodynamics remained concerned with systems in equilibrium, made it all but impossible to unify classical thermodynamics with classical mechanics, leaving these two classical disciplines of physics to stand in sharp contrast to one another in the one and a half centuries of their coexistence.

While it seems impossible to reduce thermodynamics to a mechanistic world picture due to microscopic reversibility and Poincaré recurrence, our dynamical thermodynamic formulation provides a harmonization of classical thermodynamics with classical mechanics. In particular, our dynamical systems formalism captures all of the key aspects of thermodynamics, including its fundamental laws, while providing a mathematically rigorous formulation for thermodynamical systems out of equilibrium by unifying the theory of heat transfer with that of classical thermodynamics. In addition, the concept of entropy for a nonequilibrium state of a dynamical process is defined, and its global existence and uniqueness is established.

This state space formalism of thermodynamics shows that the behavior of heat, as described by the conservation equations of thermal transport and as described by classical thermodynamics, can be derived from the same basic principles and is part of the same scientific discipline. Finally, classical thermodynamics meets Fourier's theory of heat conduction in the one hundred and fifty years of their coexistence. And for those numerous thermodynamicists who have repeatedly confused statics with dynamics and have been under the illusion that their science of *thermostatics* (i.e., classical thermodynamics) somehow rivals classical mechanics, in consequence, they need not so remain.

Chapter Seventeen

Epilogue

17.1 Introduction

Thermodynamics is universal, and hence, in principle, it applies to everything in nature—from simple engineering systems to complex living organisms to our expanding universe. The laws of thermodynamics form the theoretical underpinning of diverse disciplines such as biology, chemistry, climatology, ecology, economics, engineering, genetics, geology, neuroscience, physics, physiology, sociology, and cosmology, and they play a key role in the understanding of these disciplines.

Modeling the fundamental dynamic phenomena of these disciplines gives rise to large-scale complex dynamical systems[1] that have numerous input, state, and output properties related to conservation, dissipation, and transport of mass, energy, and information. These systems are governed by conservation laws (e.g., mass, energy, fluid, bit) and comprise multiple subsystems or compartments that exchange variable quantities of material via intercompartmental flow laws, and can be characterized as network thermodynamic (i.e., advection-diffusion) systems with compartmental masses, energies, or information playing the role of heat energy in subsystems at different temperatures.

In particular, large-scale compartmental models have been widely used in biology, pharmacology, and physiology to describe the distribution of a substance (e.g., biomass, drug, radioactive tracer) among different tissues of an organism. In this case, a compartment represents the amount of the substance inside a particular tissue, and the intercompartmental flows are due to diffusion processes. In engineering and the physical sciences, compartments typically represent the energy, mass, or information content of the different parts of the system, and different compartments interact by exchanging heat, work energy, and matter.

[1]Complexity here refers to the quality of a system wherein interacting subsystems self-organize to form hierarchical evolving structures exhibiting emergent system properties.

In ecology and economics, compartments can represent soil and debris, or finished goods and raw materials in different regions, and the flows are due to energy and nutrient exchange (e.g., nitrates, phosphates, carbon), or money and securities. Compartmental systems can also be used to model chemical reaction systems. In this case, the compartments would represent quantities of different chemical substances contained within the compartment, and the compartmental flows would characterize transformation rates of reactants to products.

The underlying thread of the aforementioned disciplines is the universal principles of conservation of energy and nonconservation of entropy leading to thermodynamic irreversibility, and thus, imperfect animate and inanimate mechanisms—from the diminutive cells in our bodies to the cyclopean spinning galaxies in the heavens. These universal principles underlying the most perplexing secrets of the cosmos are entrusted to the métier of classical thermodynamics—a phenomenological scientific discipline characterized by a purely empirical foundation.

In this monograph, we combined the two universalisms of thermodynamics and dynamical systems theory under a single umbrella, with the latter providing the ideal language for the former, to provide a system-theoretic foundation of thermodynamics and establish rigorous connections between the arrow of time, irreversibility, and the second law of thermodynamics for *nonequilibrium* systems. Given the proposed dynamical systems formalism of thermodynamics, a question that then arises is whether the proposed dynamical systems framework can be used to shed any new insights into some of the far-reaching consequences of the laws of thermodynamics; consequences involving living systems and large-scale cosmological physics, elementary particle physics and the emergence of damping in conservative time-reversible microscopic dynamics, and the ravaging evolution of the universe and, hence, the ultimate destiny of mankind.

Thermodynamicists have always presented theories predicated on the second law of thermodynamics employing equilibrium or near equilibrium models in trying to explain some of the most perplexing secrets of science from a high-level systems perspective. These theories utilize thermodynamic models that include an attempt in explaining the mysteries of the origins of the universe and life, the subsistence of life, biological growth, ecosystem development and sustenance, and biological evolution, as well as nonliving organized systems such as galaxies, stars, accretion disks, black holes, convection cells, tornados, hurricanes, eddies, vortices, and river deltas, to name but a few examples. However, most of these thermodynamic models are restricted to equilibrium or near equilibrium systems, and hence, are inappropriate in addressing some of these deep scientific mysteries as they involve nonequilibrium nonlinear dynamical processes undergoing temporal and spatial changes.

Given that dynamical systems theory has proved to be one of the most universal formalisms in describing manifestations of nature that involve change and time, it provides the ideal framework for developing a postmodern foundation for nonequilibrium thermodynamics. This can potentially lead to mathematical models with greater precision and self-consistency; and while "self-consistency is not necessarily truth, self-inconsistency is certainly falsehood" [1].

The proposed dynamical systems framework of thermodynamics can potentially provide deeper insights into some of the most perplexing questions concerning the origins and fabric of our universe that require dynamical system models that are far from equilibrium. In addition, dynamical thermodynamics can foster the development of new frameworks in explaining the fundamental thermodynamic processes of nature, explore new hypotheses that challenge the use of classical thermodynamics, and allow for the development of new assertions that can provide deeper insights into the constitutive mechanisms that describe acute microcosms and macrocosms in the ever elusive pursuit of unifying the subatomic and astronomical domains.

Among these processes are the thermodynamics of living systems, the origins of life and the universe, consciousness, and death. The nuances underlying these climacterical areas are the subject matter of this chapter. We stress, however, that unlike the previous chapters in this monograph, which provided a rigorous mathematical presentation of the key concepts of dynamical thermodynamics, in this chapter we give a high-level scientific discussion of these peripheral albeit very important subjects.

17.2 Thermodynamics of Living Systems

The universal irreversibility associated with the second law of thermodynamics responsible for the enfeeblement and eventual demise of the universe has led numerous writers, historians, philosophers, and theologians to ask the momentous question, How is it possible for life to come into being in a universe governed by a supreme law that impedes the very existence of life? This question, followed by numerous erroneous apodoses that life defies the second law and that the creation of life is an unnatural state in the constant ravaging progression of diminishing order in the universe, has been used by creationists, evolution theorists, and intelligent designers to promote their own points of view regarding the existence or nonexistence of a supreme being. However, what many bloviators have failed to understand is that these assertions on the origins and subsistence of life along with its evolution present merely opinions of simpletons and amateur scientists; they have very little to do with a rigorous understanding of the supreme law of nature—the second law of thermodynamics.

Life and evolution do not violate the second law of thermodynamics; we know of nothing in nature that violates this absolute law. In fact, the second law of thermodynamics is not an impediment to understanding life but rather is a necessary condition for providing a complete elucidation of living processes. The nonequilibrium state of the universe is essential in extracting free energy[2] and *destroying* this energy to create entropy and consequently maintaining highly organized structures in living systems. It is important to stress here that even though energy is conserved and cannot be created or destroyed, free energy or, more descriptively, extractable useful work energy, *can* be destroyed. Every dynamical process in nature results in the destruction (i.e., degradation) of free energy.

The entire fabric of life on Earth necessitates an intricate balance of *organization* involving a highly ordered, low entropy state. This in turn requires dynamic interconnected structures governed by biological and chemical principles, including the principle of natural selection, which do not violate the thermodynamic laws as applied to the *whole* system in question. Organization here refers to a *large-scale system* with multiple interdependent parts forming *ordered* (possibly organic) spatiotemporal unity, and wherein the system is greater than the sum of its parts. Order here refers to arrangements involving patterns. Both organization and order are ubiquitous in nature and reinforce energy flow to reduce gradients over time of any work-related potential.

For example, biology has shown us that many species of animals, such as insect swarms, ungulate flocks, fish schools, ant colonies, and bacterial colonies, self-organize in nature. These biological aggregations give rise to remarkably complex global behaviors from simple local interactions between a large number of relatively unintelligent agents without the need for a centralized architecture. The spontaneous development (i.e., self-organization) of these autonomous biological systems and their spatiotemporal evolution to more complex states often appears without any external system interaction. In other words, structural morphing into coherent

---

[2]Here we use the term *free energy* to denote either the Helmholtz free energy or the Gibbs free energy depending on context. Recall that the Helmholtz free energy $F = U - TS$ is the maximal amount of work a system can perform at a constant volume and temperature, whereas the Gibbs free energy $G = U - TS - pV$ is the maximal amount of work a system can perform at constant pressure and temperature. Hence, if pressure gradients can perform useful work and actuate organization (i.e., hurricanes, shock waves, tornados), then the Helmholtz free energy is the most relevant free energy. Alternatively, if pressure is constant and changes in volume need to be accounted for, then the Gibbs free energy is the relevant free energy.

groups is internal to the system and results from local interactions among subsystem components that are either dependent or independent of the physical nature of the individual components. These local interactions often comprise a simple set of rules that lead to remarkably complex and robust behaviors. Robustness here refers to insensitivity of individual subsystem failures and unplanned behavior at the individual subsystem level.

The second law does not preclude subsystem parts of a large-scale system from achieving a more ordered state in time, and hence, evolving into a more organized and complex form. The second law asserts that the *total* system entropy of an *isolated* system always increases. Thus, even though the creation of a certain measure of life and order on Earth involves a dynamic transition to a more structured and organized state, it is unavoidably accompanied by an even greater measure of death and disorder in the (almost) isolated system encompassing the Sun and the Earth. In other words, although some parts of this isolated system will gain entropy ($\mathrm{d}\mathcal{S} > 0$) and other parts will lose entropy ($\mathrm{d}\mathcal{S} < 0$), the total entropy of this isolated system will always increase over time.

Our Sun[3] is the predominant energy source that sustains life on Earth in organisms composed of mutually interconnected parts maintaining various vital processes. These organisms maintain a highly organized state by extracting free energy from their surroundings, subsumed within a larger system, and processing this energy to maintain a low entropy state. In other words, living systems continuously extract free energy from their environment and export waste to the environment thereby producing entropy. Thus, the creation of internal system organization and complexity (negative entropy change) through available energy dispersal (i.e., elimination of energy gradients) necessary for sustaining life is balanced by a greater production of entropy to the surroundings acquiescing to the second law.

The spontaneous dispersal of energy gradients by every complex animate or inanimate mechanism leads to the law of entropy increase; that is, in an isolated system, entropy is generated and released to the environment, resulting in an overall system entropy increase. However, Jaynes' principle of maximum entropy production [232, 429],[4] which states that among all

---

[3]The external source of energy of almost all life on Earth is principally supplied by solar radiation. An interesting exception to this are the deep ocean volcanic vent ecosystems that derive their energy from endogenous heat sources due to radioactive material and chemical gradients emanating from volcanic activities on the ocean floor.

[4]It is important to note here that Jaynes' maximum entropy principle refers to information theory and not thermodynamics. However, Jaynes' formulation addresses information gained from theoretical analysis and empirical data through experimental observation, and hence, applies to thermodynamic experiments. More specifically, given an incomplete set of experimental data, the formulation addresses the problem of how one can construct a well-defined complete probabilistic

possible energy dissipation paths a system may move through spontaneously, the system will select the path that will maximize its entropy production rate, adds yet another level of subtlety to all dissipative structures in the universe, including all forms of life. In particular, since complex organized structures efficiently dissipate energy gradients, nature, whenever possible, optimally transitions (in the fastest possible way) systems into complex organized structures to reinforce energy flow and dissipate energy gradients spontaneously, thus maximizing entropy production. This results in a universe that exists at the *expense* of life and order—the enthroning emblem of the second law of thermodynamics.

The degradation of free energy sources by living organisms for maintaining nonequilibrium configurations as well as their *local* level of organization at the expense of a larger *global* entropy production was originally recognized by Schrödinger [395]. Schrödinger maintained that life was composed from two fundamental processes in nature; namely, *order from order* and *order from disorder*. He postulated that life cannot exist without both processes; and that order from disorder is necessary to generate life, whereas order from order is necessary to ensure the propagation of life.

In 1953 Watson and Crick [460] addressed the first process by observing that genes generate order from order in a species; that is, inherited traits are encoded and passed down from parents to offspring within the DNA double helix. The dynamical systems framework of thermodynamics can address the fundamental process of the order from disorder mystery by connecting systems biology to the second law of thermodynamics. More specifically, a dynamical systems theory of thermodynamics can provide an understanding of how the microscopic molecular properties of living systems lead to the observed macroscopic properties of these systems.

The external source of energy that sustains life on Earth—the Sun—is of course strictured by the second law of thermodynamics. However, it is important to recognize that this energy source is a low entropy energy source and that the energy flow over a twenty-four-hour (day-night) cycle from the Sun to the Earth and thenceforth the Earth to the darkness of space is essentially balanced. This balance is modulo the small amounts of heat generated by the burning of fossil fuels and global warming, as well as the radioactive material from volcanic deep ocean vent ecosystems. This balance can also be affected by Milankovich cycles (i.e., changes in the

---

model that is consistent with the observational data while avoiding the invention of new data that does not exist. Jaynes addresses this problem by defining a measure of ignorance in terms of the informational entropy and then characterizes the probability distribution that *maximizes* the entropy subject to an agreement with the available data. In this case, the probability distribution that maximizes the a priori measure of ignorance will be (on average) the least likely to invent new (and most likely incorrect) data since the corresponding amount of the posterior learned information is maximized.

Earth's climate due to variations in eccentricity, axial tilt, and procession of the Earth's orbit) as well as variations in solar luminosity. If this intricate energy balance did not take place, then the Earth would accelerate toward a thermal equilibrium and all life on Earth would cease to exist.

The subsistence of life on Earth is due to the fact that the Sun's temperature is far greater than that of dark space, and hence, yellow light photons[5] emitted from the Sun's atoms have a much higher frequency than the infrared photons the Earth dissipates into dark space. This energy exchange includes reflection and scattering by the oceans, polar ice caps, barren terrain, and clouds; atmosphere absorption leading to energy dissipation through thermal infrared emissions and chemical pathways; ocean absorption via physicochemical and biological processes; passive absorption and reemission of infrared photons by barren terrain; latent heat generation through hydrological cycles; and the thermal dissipation from the Earth's heat sources.

Since the photons absorbed by the Earth from the Sun have a much higher frequency, and hence, shorter wavelength than those dissipated to space by the Earth, it follows from Planck's quantum formula[6] relating the energy of each photon of light to the frequency of the radiation it emits that the *average* energy flow into the Earth from the Sun by each individual photon is far greater than the dissipated energy by the Earth to dark space. This implies that the number of different photon microstates carrying the energy from the Earth to the darkness of space is far greater than the number of microstates carrying the same energy from the Sun to the Earth. Thus, it follows from Boltzmann's entropy formula that the incident energy from the Sun is at a considerably lower entropy level than the entropy produced by the Earth through energy dissipation.

It is important to stress that the Sun continuously supplies the Earth with high grade free energy (i.e., low entropy) that keeps the Earth's entropy low. The Earth's low entropy is predominantly maintained through phototrophs (i.e., plants, algae, cyanobacteria, and all photosynthesizers) and chemotrophs (e.g., iron and manganese oxidizing bacteria found in igneous lava rock), which get their free energy from solar photons and inorganic compounds. For example, leaves on most green plants have a relatively large surface area per unit volume allowing them to capture high grade free energy from the Sun. Specifically, photon absorption

[5]Photon energy from the Sun is generated as gamma rays that are produced by thermonuclear reactions (i.e., fusion) at the center of the Sun and distributed among billions of photons through the Sun's photosphere.

[6]Planck's work on thermodynamics and blackbody radiation led him to formulate the foundations of quantum theory, wherein electromagnetic energy is viewed as discrete amounts of energy known as *quanta* or *photons*. The Planck quantum formula relates the energy of each photon E to the frequency of radiation ν as $E = h\nu$, where h is the Planck constant.

by photosynthesizers generates pathways for energy conversion coupled to photochemical, chemical, and electrochemical reactions leading to high grade energy. This energy is converted into useful work to maintain the plants in a complex, high-energy organized state via photosynthesis supporting all heterotrophic life on Earth.

Energy dissipation leading to the production of entropy is a direct consequence of the fact that the Sun's temperature ($T_{\rm sun} = 5760\,{\rm K}$) is at a much higher temperature than that of the Earth's temperature ($T_{\rm earth} = 255\,{\rm K}$) and, in turn, the Earth's temperature is higher than the temperature of outer space ($T_{\rm space} = 2.73\,{\rm K}$). Even though the energy coming to the Earth from the Sun is balanced by the energy radiated by the Earth into dark space, the photons reaching the Earth's surface from the Sun have a higher energy content (i.e., shorter wavelength) than the photons radiated by the Earth into the dark sky.

In particular, for every one solar photon infused by the Earth, the Earth diffuses approximately twenty photons into the dark sky with approximately twenty times longer wavelengths [284].[7] And since the entropy of photons is proportional to the number of photons [284], for every solar photon the Earth absorbs at low entropy, the Earth dilutes the energy of the solar photon among twenty photons and radiates them into deep space. In other words, the Earth is continuously extracting high grade free energy from the Sun and exporting degraded energy, thereby producing entropy and maintaining low entropy structures on Earth.

The same conclusion can also be arrived at using a macroscopic formulation of entropy. In particular, note that the energy reaching the Earth $\mathrm{d}Q$ is equal to the energy radiated by the Earth[8] $-\mathrm{d}Q$, $T_{\rm earth} < T_{\rm sun}$, and

$$\mathrm{d}\mathcal{S}_{\rm in} \geq \frac{\mathrm{d}Q}{T_{\rm sun}}, \tag{17.1}$$

$$\mathrm{d}\mathcal{S}_{\rm out} \geq -\frac{\mathrm{d}Q}{T_{\rm earth}}, \tag{17.2}$$

where $\mathrm{d}\mathcal{S}_{\rm in}$ and $\mathrm{d}\mathcal{S}_{\rm out}$ is the change in entropy from the absorption and emission of photons, respectively. Now, under steady Earth conditions, that is, constant average temperature, constant internal energy $U_{\rm earth}$, and

[7] When photons are absorbed by the Earth, they induce electromagnetic transmissions in matched energy absorber bands leading to photochemistry decay and fluorescence, phosphorescence, and infrared emissions.

[8] Here we are assuming that the average temperature of the Earth is constant, and hence, the amount of energy delivered by solar photons to the Earth is equal to the amount of energy radiated by infrared photons from the Earth. If this were not the case, then the internal energy of the Earth $U_{\rm earth}$ would increase, resulting in a rise of the Earth's average temperature.

constant entropy $\mathcal{S}_{\rm earth}$, the change in entropy of the Earth is

$$\mathrm{d}\mathcal{S}_{\rm earth} = \mathrm{d}\mathcal{S}_{\rm photons} + \mathrm{d}\mathcal{S}_{\rm dissp} = 0, \tag{17.3}$$

where $\mathrm{d}\mathcal{S}_{\rm photons} = \mathrm{d}\mathcal{S}_{\rm in} + \mathrm{d}\mathcal{S}_{\rm out}$ and $\mathrm{d}\mathcal{S}_{\rm dissp}$ denotes the entropy produced by all the Earth's dissipative structures. Hence,

$$\mathrm{d}\mathcal{S}_{\rm photons} = -\mathrm{d}\mathcal{S}_{\rm dissp} < 0. \tag{17.4}$$

Thus, $|\mathrm{d}\mathcal{S}_{\rm in}| < |\mathrm{d}\mathcal{S}_{\rm out}|$, and hence, it follows from (17.1) and (17.2) that

$$\left|\frac{\mathrm{d}\mathcal{S}_{\rm out}}{\mathrm{d}\mathcal{S}_{\rm in}}\right| \leq \frac{T_{\rm sun}}{T_{\rm earth}} = \frac{5760}{255} \approx 20, \tag{17.5}$$

which implies that the Earth produces at most twenty times as much entropy as it receives from the Sun. Thus, the net change in the Earth's entropy from solar photon infusion and infrared photon radiation is given by

$$\begin{aligned}
\mathrm{d}\mathcal{S}_{\rm photons} &= \mathrm{d}\mathcal{S}_{\rm in} + \mathrm{d}\mathcal{S}_{\rm out} \\
&\geq \mathrm{d}Q\left(\frac{1}{T_{\rm sun}} - \frac{1}{T_{\rm earth}}\right) \\
&= -\frac{\mathrm{d}Q}{T_{\rm earth}}\left(1 - \frac{T_{\rm earth}}{T_{\rm sun}}\right) \\
&= -0.95\frac{\mathrm{d}Q}{T_{\rm earth}}.
\end{aligned} \tag{17.6}$$

It is important to note here that even though there is a net decrease in the entropy of the Earth from the absorption and emission of photons, this is balanced by the increase in entropy exported from all of the low entropy dissipative structures on Earth. Hence, the exported waste by living systems and dissipative structures into the environment results in an entropy production that does not lower the entropy of the Earth, but rather sustains the Earth's entropy at a constant low level. And to maintain this low level entropy, a constant supply of free energy is needed.

Specifically, since the amount of the Earth's free energy is not increasing under steady Earth conditions, it follows that

$$\mathrm{d}F_{\rm earth} = \mathrm{d}F_{\rm photons} + \mathrm{d}F_{\rm dissp} = 0, \tag{17.7}$$

where $\mathrm{d}F_{\rm photons}$ denotes the amount of free energy received by the Earth through solar photons and $\mathrm{d}F_{\rm dissp}$ is the amount of free energy dissipated by the Earth's dissipative structures (e.g., a myriad of life-forms, cyclones, thermohaline circulation, storms, atmospheric circulation, convection cells), and hence, $\mathrm{d}F_{\rm photons} = -\mathrm{d}F_{\rm dissp}$. Finally, noting that

$$\mathrm{d}F_{\rm earth} = \mathrm{d}U_{\rm earth} - T_{\rm earth}\mathrm{d}\mathcal{S}_{\rm earth} - \mathcal{S}_{\rm earth}\mathrm{d}T_{\rm earth}, \tag{17.8}$$

and assuming steady Earth conditions, it follows from (17.8) that

$$\mathrm{d}F_{\text{photons}} + \mathrm{d}F_{\text{dissp}} = -T_{\text{earth}} \left(\mathrm{d}\mathcal{S}_{\text{photons}} + \mathrm{d}\mathcal{S}_{\text{dissp}}\right). \tag{17.9}$$

Now, equating analogous contributing terms in (17.9) gives $\mathrm{d}F_{\text{photons}} = -T_{\text{earth}}\mathrm{d}\mathcal{S}_{\text{photons}}$, which using (17.6), yields

$$\mathrm{d}F_{\text{photons}} \leq 0.95\mathrm{d}Q. \tag{17.10}$$

Thus, it follows from (17.10) that, under steady conditions, at most 95% of the Sun's energy can be used to perform useful work on Earth.

17.3 Thermodynamics and the Origin of Life

In Section 17.2, we provided a rationalization for the existence of living systems using thermodynamics. The *origins* of life, however, is a far more complex matter. At the heart of this complexity lies the interplay of Schrödinger's two fundamental processes for life; namely, *order from order* and *order from disorder* [395]. These fundamental processes involve complex self-organizing living organisms that exhibit remarkable robustness over their healthy stages of life as well as across generations, through heredity.

The order from disorder robustness in maintaining a living system requires that the free energy flow through the system is such that the negative (local or internal) entropy production rate accumulated by the living system $\mathrm{d}\mathcal{S}_{\text{int}}$ is greater than the *external* entropy rate $\mathrm{d}\mathcal{S}_{\text{ext}}$ generated by the living system from irreversible processes (e.g., diffusion, heat exchange, heat production, chemical reactions) taking place *within* the system, that is, $|\mathrm{d}\mathcal{S}_{\text{int}}| > \mathrm{d}\mathcal{S}_{\text{ext}} > 0$. Alternatively, the order from order robustness is embedded within the genes of the living organism encapsulating information, wherein every cell transmits and receives as well as codes and decodes life's blueprint passed down through eons.

Deoxyribonucleic acid (DNA) is the quintessential information molecule that carries the genetic instructions of all known living organisms. The human DNA is a cellular information processor composed of an alphabet code sequence of six billion bits. DNA, ribonucleic acid (RNA), proteins, and polysaccharides (i.e., complex carbohydrates) are key types of macromolecules that are essential for all forms of life. The macromolecules (DNA, RNA, proteins, and polysaccharides) of organic life form a highly complex, large-scale interconnected communication network.

The genesis and evolution of these macromolecules not only use free energy but also embody the constant storage and transfer of information collected by the living organism through its interaction with its environment. These processes of information storage or, more specifically, preservation of information via DNA replication, and information transfer via message transmission from nucleic acids (DNA and RNA) to proteins, form the basis for the continuity of all life on Earth.

The free energy flow process required to sustain organized and complex life-forms generating negative (local or internal) entropy is intricately coupled with transferring this negative entropy into information. This increase in information is a result of system replication and information transfer by enzyme and nucleic acid information-bearing molecules. The information content in a given sequence of proteins depends on the minimum number of specifications needed to characterize a given DNA structure; the more complex the DNA structure, the larger the number of specifications (i.e., microstates) needed to specify or describe the structure. Using Boltzmann's entropy formula $\mathcal{S} = k \log_e \mathcal{W}$, numerous physicists have argued that this information entropy increase results in an overall thermodynamic entropy increase.

A general relation between informational entropy and thermodynamic entropy is enigmatic. The two notions are consociated only if they are related explicitly through physical states. Storage of genetic information affects work through the synthesis of DNA polymers, wherein data is stored, encoded, transmitted, and restored, and hence, information content is related to system ordering. However, it is important to note that the semantic component of information (i.e., the meaning of the transmitted message) through binary coding elements is not related to thermodynamic entropy.

Even though the Boltzmann entropy formula gives a measure of disorder or randomness in a closed system, its connection to thermodynamic entropy remains tenuous. Planck was the first to introduce the Boltzmann constant k within the Boltzmann entropy formula to attach physical units in calculating the entropy increase in the nondissipative mixing of two different gases, thereby relating the Boltzmann (configurational) entropy to the thermodynamic entropy. In this case, the Boltzmann constant relates the average kinetic energy of a gas molecule to its temperature. For more complex systems, however, it is not at all clear how these two entropies are related despite the fact that numerous physicists use the Boltzmann (configurational) and the Clausius (thermodynamic) entropies interchangeably.

The use of the Boltzmann constant by physicists to unify the thermodynamic entropy with the Boltzmann entropy is untenable. Notable exceptions to this pretermission are Shannon [405] and Feynman [147]. In his seminal paper [276], Shannon clearly states that his information entropy, which has the same form as the Boltzmann (configuration) entropy, is relative to the coordinate system that is used, and hence, a change of coordinates will generally result in a change in the value of the entropy. No physical units are ever given to the information entropy and nowhere is the Boltzmann constant mentioned in his work.[9] Similarly, Feynman [147] also never introduces the Boltzmann constant leaving the Boltzmann entropy formula free of any physical units. In addition, he states that the value of the Boltzmann entropy will depend on the *artificial* number of ways one subdivides the volume of the phase space $\mathcal{W}$.

The second law of thermodynamics institutes a progression from complexity (organization and order) to decomplexification (disorder) in the physical universe. While internal system organization and complexity (negative entropy change) through energy dispersal is a necessary condition to polymerize the macromolecules of life, it is certainly not a sufficient condition for the *origin* of life. The mystery of how thermal, chemical, or solar energy flows from a low-entropy source through simple chemicals leading to the *genesis* of complex polymer structures remains unanswered.

Natural selection applies to living systems which already have the capacity of replication, and hence, the building of DNA and enzyme architectures by processes other than natural selection is a necessary but not sufficient condition for a prevalent explanation of the origin of life. Living organisms are prodigious due to their archetypical hierarchal complex structures. The coupling mechanism between the flow of free energy, the architectural desideratum in the formation of DNA and protein molecules, and information (configurational) entropy necessary for unveiling the codex of the origin of life remains a mystery.

Even though numerous theoretical and experimental models have been proposed in explaining the coupling mechanism between the flow of free energy, the architectural blueprints for the formation of DNA and protein molecules, and the information necessary for codifying these blueprints, they are all deficient in providing a convincing explanation for the origins of life. These models include mechanisms based on chance [276], neo-Darwinian natural selection [101], inherent self-ordering tendencies in matter [151,427], mineral catalysis [462], and nonequilibrium processes [338,371].

Before unlocking the mystery of the origin of life, it is necessary to develop a rigorous unification between thermodynamic entropy, information entropy, and biological organization; these fundamental notions remain disjoint. And this is not surprising when one considers how Shannon chose

[9]In his fifty-five page paper, Shannon introduces a constant K into his information entropy formula in only two places stating that K can be introduced as a matter of convenience and can be used to attach a choice of a unit measure.

the use of the term *entropy* for characterizing the *amount of information*. In particular, he states [444]:

> My greatest concern was what to call it. I thought of calling it information, but the word was overly used, so I decided to call it uncertainty. John von Neumann had a better idea, he told me, You should call it entropy, for two reasons. In the first place, your uncertainty function goes by that name in statistical mechanics. In the second place, and more important, nobody knows what entropy really is, so in a debate you will always have the advantage.

17.4 The Second Law, Entropy, Gravity, and Life

The singular and most important notion in understanding the origins of life in the universe is *gravity*. The inflationary bang that resulted in a gigantic burst of spatial expansion of the early universe followed by gravitational collapse providing the fundamental mechanism for structure formation (e.g., galaxies, stars, planets, accretion disks, black holes) in the universe was the cradle for the early, low entropy universe necessary for the creation of dissipative structures responsible for the genesis and evolution of life.

Inflationary cosmology, wherein initial *repulsive* gravitational forces and *negative* pressures provided an outward blast driving space to surge, resulted in a uniform spatial expansion made up of (almost) uniformly distributed matter.[10] In particular, at the end of the inflationary burst, the size of the universe had unfathomably grown and the nonuniformity in the curvature of space had been distended with all density clusters dispersed leading to the nascency of all dissipative structures, including the origin of life, and the dawn of the entropic arrow of time in the universe.

The stars in the heavens provide the free energy sources for heterotrophs through solar photons, whereas the chemical and thermal nonequilibrium states of the planets provide the free energy sources for chemotrophs through inorganic compounds. The three sources of free energy in the universe are gravitational, nuclear, and chemical. Gravitational free energy arises from dissipation in accretion disks leading to angular momentum exchange between heavenly bodies. Nuclear free energy is generated by binding energy per nucleon from strong nuclear forces and forms the necessary gradients for fusion and fission reactions. Chemical free energy is discharged from electrons sliding deeper into electrostatic potential wells and is the energy that heterotrophic life avulses from organic compounds and chemotrophic life draws from inorganic compounds [336].

---

[10]It follows from general relativity that pressure generates gravity. Thus, repulsive gravity involves an *internal* gravitational field acting within space impregnated by negative pressure. This negative pressure generates a repulsive gravitational field that acts within space [355, 356].

The origin of these sources of free energy is a direct consequence of the gravitational collapse with solar photons emitted from fusion reactions taking place at extremely high temperatures in the dense centers of stars, and thermal and chemical gradients emanating from volcanic activities on planets. Gravitational collapse is responsible for generating nonequilibrium dissipative structures and structure formation giving rise to temperature, density, pressure, and chemical gradients. This gravitational structure of the universe was spawned from a primordial low entropy universe with a nearly homogeneous distribution of matter preceded by inflation and the cosmic horizon problem [284, 355, 356].[11]

Given the extreme uniformity of the cosmic microwave background radiation (i.e., the horizon problem) of the universe, it is important to note that the expansion of the universe does not change the entropy of the photons in the universe. This is due to the fact that the expanding universe is adiabatic since the photons within any arbitrary fixed volume of the universe have the same temperature as the surrounding neighborhood of this volume, and hence, heat does not enter or leave the system. This adiabatic expansion implies that the entropy of the photons does not increase. Alternatively, this fact also follows by noting that the entropy of blackbody photons is proportional to the number of photons [284], and hence, since the number of photons remains constant for a given fixed volume, so does the entropy.

The question that then remains is, How did inflation effect the low entropy state of the early universe? Since the inflationary burst resulted in a smoothing effect of the universe, one would surmise that the total entropy decreased, violating the second law of thermodynamics. This assertion, however, is incorrect; even though gravitational entropy decreased, the increase in entropy from the uniform production of matter—in the form of relativistic particles—resulted in a total increase in entropy.

More specifically, gravitational entropy or, equivalently, *matter entropy* was generated by the gravitational force field. And since matter can convert gravitational potential energy into kinetic energy, matter can coalesce to form clusters that release radiation, and hence, produce entropy. If the matter entropy is low, then the gravitational forces can generate a large amount of entropy via matter clustering. In particular, this matter clustering engenders a nonuniform, nonhomogeneous gravitational field characterized by warps and ripples in spacetime.

---

[11] The cosmic horizon problem pertains to the uniformity of the cosmic microwave background radiation, homogeneity of space and temperature, and a uniform cosmic time. Its basis lies in the hypothesis that inflationary cosmology involved a short period of time wherein superluminal expansion of space took place. Here, it is important to distinguish between superluminal motion *through* space, which would contradict special relativity, and superluminal motion of the dilation of space itself. For details see [174].

The origin of this matter and radiation originated from the release of the enormous potential energy of the *Higgs field*: the energy field suffused through the universe and connected with the fundamental Higgs boson particle. This field gives rise to mass from fundamental particles (e.g., electrons and quarks) combining into composite particles (e.g., protons, neutrons, and atoms) in the Higgs field through the Higgs bosons, which contain relative mass in the form of energy—a process known as the Higgs effect.

Thus, false vacuum gauge boson states decayed into a true vacuum through spontaneous symmetry breaking, wherein the release in potential energy from the Higgs effect was (almost) uniformly distributed into the universe cooling and clustering matter. And even though specific symmetries have vanished throughout different cosmic state transitions, the heat released during these transitions ensures that the overall entropy of the universe steadfastly increases. This has led to the creation of the cosmic evolution of entropy increase and the gravitational origins of free energy necessary for the genesis and subsistence of life in the universe.

Spontaneous symmetry breaking is a phenomenon where a symmetry in the basic laws of physics appears to be broken. Specifically, even though the dynamical system as a whole changes under a symmetry transformation, the underlying physical laws are invariant under this transformation. In other words, symmetrical system states can transition to an asymmetrical lower energy state but the system Lagrangian is invariant under this action. Spontaneous symmetry breaking of the non-Abelian group $\mathrm{SU}(2) \times \mathrm{U}(1)$, where $\mathrm{SU}(2)$ is the special unitary group of degree 2 and $\mathrm{U}(1)$ is the unitary group of degree 1, admitting gauge invariance associated with the *electroweak force* (i.e., the unified description of the electromagnetic and weak nuclear forces) generates the masses of the W and Z bosons (i.e., carrier particles), and separates the electromagnetic and weak forces.

However, it should be noted that, in contrast to the $\mathrm{U}(1)$ Abelian Lie group[12] involving the gauge electromagnetic field, the isospin transformations under the action of the Lie group $\mathrm{SU}(2)$ involving electroweak interactions do not commute, and hence, even though the weak nuclear and electromagnetic interactions are part of a single mathematical framework, they are not fully unified. This is further exacerbated when one includes strong nuclear interactions, wherein the non-Abelian group $\mathrm{SU}(2) \times \mathrm{U}(1)$ is augmented to include the Lie group $\mathrm{SU}(3)$ involving charge in quantum chromodynamics to define local symmetry. The continuing quest in unifying the four universal forces—strong nuclear, electromagnetic, weak nuclear, and gravitational—under a single mathematical structure remains irresolute.

[12] An *Abelian Lie group* is a Lie group on which the defined binary operation is commutative.

The fact that the universe is constantly evolving toward organized clustered matter and gravitational structures is not in violation of the second law. Specifically, even though clustered matter and gravitational structures formed during the gravitational collapse in the universe are more ordered than the early (nearly) uniformly distributed universe immediately following the inflationary burst, the entropy decrease in achieving this ordered structure is offset by an even greater increase in the entropy production generated by the enormous amount of heat and light released in thermonuclear reactions as well as exported angular momenta necessary in the morphing and shaping of these gravitational structures. Organized animate and inanimate hierarchical structures in the universe maintain their highly ordered low entropy states by expelling their internal entropy, resulting in a net entropy increase of the universe. Hence, neither gravitational collapse giving rise to order in parts of the universe nor life violates the second law of thermodynamics.

In this monograph, we established that entropy and time are homeomorphic, and hence, the second law substantiates the arrow of time. And since all living systems constitute nonequilibrium dynamical systems involving dissipative structures, wherein evolution through time is homeomorphic to the direction of entropy increase and the availability of free energy, our existence depends on $\mathrm{d}\mathcal{S} > 0$. In other words, every reachable and observable state in the course of life, structure, and order in the universe necessitates that $\mathrm{d}\mathcal{S} > 0$; the state corresponding to $\mathrm{d}\mathcal{S} = 0$ is an unobservable albeit reachable state.

Since a necessary condition for life and order in the universe requires the ability to extract free energy (at low entropy) and produce entropy, a universe that can sustain life requires a state of low entropy. Thus, a maximum entropy universe is neither controllable nor observable; that is, a state of heat death, though reachable, is uncontrollable and unobservable. In a universe where *energy is conserved*, that is, $\mathrm{d}U_{\mathrm{universe}} = 0$, and the cosmic horizon holds, that is, the uniformity of the microwave background radiation holds, the free energy is given by $\mathrm{d}F_{\mathrm{universe}} = -T\mathrm{d}\mathcal{S}_{\mathrm{universe}}$. Hence, if $\mathrm{d}\mathcal{S}_{\mathrm{universe}} > 0$, then the flow of free energy continues to create and sustain life. However, as $\mathrm{d}\mathcal{S}_{\mathrm{universe}} \to 0$ and $T \to 0$, then $\mathrm{d}F_{\mathrm{universe}} \to 0$, leading to an eventual demise of all life in that universe.

17.5 The Second Law, Health, Illness, Aging, and Death

As noted in Section 17.2, structural variability, complexity, organization, and entropy production are all integral parts of life. Physiological pattern variations, such as heart rate variability, respiratory rate, blood pressure, venous oxygen saturation, and oxygen and glucose metabolism, leading to emergent patterns in our dynamical system physiology are key indicators of youth, health, aging, illness, and eventually death as we transition through life trammeled by the second law of thermodynamics. In physiology, entropy production is closely related to metabolism, which involves the physical and chemical processes necessary for maintaining life in an organism. And since, by the first law of thermodynamics, work energy and chemical energy lead to heat energy, a biological system produces heat, and hence, entropy.

In the central nervous system,[13] this heat production is a consequence of the release of chemical energy through oxygen metabolism (e.g., oxygen consumption, carbon dioxide, and waste production) and glycolysis (glucose converted to pyruvate) leading to the breakdown of macromolecules to form high energy compounds (e.g., adenosine diphosphate (ADP) and adenosine triphosphate (ATP)). Specifically, electrochemical energy gradients are used to drive phosphorylation of ADP and ATP through a molecular turbine coupled to a rotating catalytic mechanism. This stored energy is then used to drive anabolic processes through coupled reactions in which ATP is hydrolyzed to ADP and phosphate, releasing heat. Additionally, heat production occurs during oxygen intake with metabolism breaking down macromolecules (e.g., carbohydrates, lipids, and proteins) to free high-quality chemical energy to produce work. In both cases, this heat dissipation leads to entropy production [5, 399].[14]

Brain imaging studies have found increasing cortical complexity in early life, throughout childhood, and into adulthood [53, 411, 469], followed by decreasing complexity in the later stages of life [246, 431]. Similarly, a rapid rise in entropy production[15] from early life, throughout childhood, and into early adulthood is also seen, followed by a slow decline thereafter [4]. Identical patterns to the rise and fall of entropy production seen through the deteriorating path to aging are also observed in healthy versus ill individuals [399]. Namely, a decrease in cortical complexity (i.e., fractal dimension[16] of the central nervous system) with illness correlating with metabolic decline

[13]The brain has the most abounding energy metabolism in the human body. Specifically, in a resting awake state, it requires approximately 20% of the total oxygen supplied by the respiratory system and 25% of the total body glucose [317, 421]. Cerebral glucose metabolism is critical for neural activity. This can be seen in hypoglycemic individuals with diminished glucose metabolism experiencing impaired cognitive function.

[14]Every living organism that is deprived of oxygen consumption and glycolysis dies within a few minutes as it can no longer produce heat and release entropy to its environment. In other words, its homeostatic state is destroyed.

[15]In addition to the central nervous system, entropy production over the course of a living mammalian organism is directly influenced by dissipation of energy and mass to the environment. However, the change in entropy due to mass exchange with the environment is negligible (approximately 2%), with the predominant part of entropy production attributed to heat loss due to radiation and water evaporation.

[16]A *fractal* is a set of points having a detailed structure that is visible on arbitrarily small scales and exhibits repeating patterns over multiple measurement scales. A *fractal dimension* is an index measure of complexity capturing changes in fractal patterns as a function of scale.

in glucose metabolism is observed in ill individuals [357,407].[17]

This observed reduction in complexity is not limited to the central nervous system and is evidenced in other physiological parameters. For example, the reduction in the degree of irregularity across time scales of the time series of the heart rate in heart disease patients and the elderly [48,170, 171,300,398] as well as the decoupling of biological oscillators during organ failure [168] lead to system decomplexification. Thus, health is correlated with system complexity (i.e., fractal variability) and the ability to effectively and efficiently dissipate energy gradients to maximize entropy production, whereas illness and aging are associated with decomplexification resulting in decreased entropy production and ultimately death.

The decrease in entropy production and the inability to efficiently dissipate energy gradients results from a diminished ability of the living system to extract and consume free energy. This can be due, for example, to diseases such as ischemia, hypoxia, and angina, which severely curtail the body's ability to efficiently dissipate energy and consequently maximize entropy production, as well as diseases such as renal failure, hypotension, edema, cirrhosis, and diabetes, which result in pathologic retention of entropy.

The loss of system complexity or, equivalently, the loss of complex system dynamics, system connectivity, and fractal variability with illness and age results in the inability of the system to optimally dissipate free energy and maximize entropy production. In this case, the system loses the ability to extract and direct free energy from the environment as well as adapt to environmental changes leading to deterioration, atrophy, and eventually death. Conversely, vibrant life extracts free energy at low entropy (i.e., high grade energy) and ejects it at high entropy (i.e., low grade energy). The overall degree of system complexity provides a measure of the ability of a system to actively adapt and is reflected by the ratio of the maximum system energy output (work expenditure) to the resting energy output (resting work output), in analogy to Carnot's theorem. Thus, life is a dynamical thermodynamic system that is part of the cosmic construct that generates structure, complexity, and order to efficiently reduce, degrade, and destroy gradients by exporting entropy into space as it marches towards a natural end equilibrium state with its environment.

[17]There are exceptions to this pattern. For example, most cancers including gliomas, central nervous system lymphomas, and pituitary lesions are hypermetabolic, resulting in a high rate of glycolysis leading to an increased entropy production and fractal dimension [399]. This accession of entropy production, however, is deleterious to the host organism as it results in an overexpenditure of the energy substratum; see [399].

17.6 The Second Law, Consciousness, and the Entropic Arrow of Time

In this section, we present some qualitative insights using thermodynamic notions that can potentially be useful in developing mechanistic models[18] [2] for explaining the underlying mechanisms for consciousness. Specifically, by merging thermodynamics and dynamical systems theory with neuroscience [108, 136], one can potentially provide key insights into the theoretical foundation for understanding the network properties of the brain by rigorously addressing large-scale interconnected biological neuronal network models that govern the neuroelectric behavior of biological excitatory and inhibitory neuronal networks.

The fundamental building block of the central nervous system, the *neuron*, can be divided into three functionally distinct parts, namely, the *dendrites*, *soma* (or cell body), and *axon*. Neurons, like all cells in the human body, maintain an electrochemical potential gradient between the inside of the cell and the surrounding milieu. However, neurons are characterized by excitability—that is, with a stimulus the neuron will discharge its electrochemical gradient, and this discharge creates a change in the potential of neurons to which it is connected (the postsynaptic potential). And this postsynaptic potential, if excitatory, can serve as the stimulus for the discharge of the connected neuron or, if inhibitory, can hyperpolarize the connected neuron and inhibit firing.

An important open question in neuroscience is how information is represented and transmitted in the brain by a network of neurons. Neurons may be thought of as dynamic elements that are excitable, and can generate a pulse or spike whenever the electrochemical potential across the cell membrane of the neuron exceeds a certain threshold. The Hodgkin-Huxley model [217] is the prominent model for characterizing nerve pulse propagation and relies on ion currents through ion channels (resistors) and the lipid membrane (capacitor). This model is a purely electrical model and assumes that proteins alone enable nerve cells to propagate signals due to the ability of various ion channel proteins to transport sodium and potassium ions.

Even though the Hodgkin-Huxley model forms the foundation of modern neurophysiology, it is inconsistent with thermodynamic principles. A revised view of the action potential based on the laws of thermodynamics was recently proposed in [212]. This electromechanical model traces its

[18]Mechanistic models in this context are models that are predicated on dynamical systems theory wherein an internal state model is used to describe dynamic linking between phenotypic states using biological and physiological laws and system interconnections.

origins to the fundamental work of Tasaki [432] and characterizes the action potential as a propagating density pulse (i.e., a soliton). More specifically, the nerve pulse propagation is nearly adiabatic displaying a near reversible heat release as the nerve membrane undergoes a phase transition from fluid lipid to crystalline lipid with changes in the action potential (i.e., voltage pulse) and the nerve cell dimensions.

Since the neocortex contains on the order of 100 billion neurons, which have up to 100 trillion connections, this leads to immense system complexity. Computational neuroscience has addressed this problem by reducing a large population of neurons to a distribution function describing their probabilistic evolution, that is, a function that captures the distribution of neuronal states at a given time [72, 110].

To elucidate how thermodynamics can be used to explain the conscious-unconscious[19] state transition we focus our attention on the *anesthetic cascade* problem [199]. In current clinical practice of general anesthesia, potent drugs are administered which profoundly influence levels of consciousness and vital respiratory (ventilation and oxygenation) and cardiovascular (heart rate, blood pressure, and cardiac output) functions. These variation patterns of the physiologic parameters (i.e., ventilation, oxygenation, heart rate variability, blood pressure, and cardiac output) and their alteration with levels of consciousness can provide scale-invariant fractal temporal structures to characterize the *degree of consciousness* in sedated patients.

The term "degree of consciousness" reflects the intensity of a noxious stimulus. For example, we are often not aware (conscious) of ambient noise but would certainly be aware of an explosion. Thus, the term "degree" reflects awareness over a spectrum of stimuli. For any particular stimulus the transition from consciousness to unconsciousness is a very sharp transition which can be modeled using a very sharp sigmoidal function—practically a step function.

Here, we hypothesize that the degree of consciousness is reflected by the adaptability of the central nervous system and is proportional to the maximum work output under a fully conscious state divided by the work output of a given anesthetized state, once again in analogy to

[19] Here we adopt a scientific (i.e., neuroscience) perspective of consciousness and not a philosophical one. There have been numerous speculative theories of human consciousness, with prominent metaphysical theories going back to ancient Greece and India. For example, Herakleitos proposed that there exists a single, all-powerful, divine consciousness that controls all things in Nature and that ultimate wisdom is reached when one achieves a fundamental understanding of the universal laws that govern all things and all forces in the universe—Εἶναι γὰρ ἓν τὸ σοφόν, ἐπίρτασθαι γνώμην, ὁτέη ἐκυβέρνησε πάντα διὰ πάντων. In Hinduism, Brahman represents the ultimate reality in all existence, wherein each individual's consciousness materializes from a unitary consciousness suffused throughout the universe.

Carnot's theorem. A reduction in maximum work output (and cerebral oxygen consumption) or elevation in the anesthetized work output (or cerebral oxygen consumption) will thus reduce the degree of consciousness. Hence, the fractal nature (i.e., complexity) of conscious variability is a self-organizing emergent property of the large-scale interconnected biological neuronal network since it enables the central nervous system to maximize entropy production and dissipate energy gradients. Within the context of aging and complexity in acute illnesses, variation of physiologic parameters and their relationship to system complexity and system thermodynamics have been explored in [48, 168, 170, 171, 300, 398].

Complex dynamical systems involving self-organizing components forming spatio-temporal evolving structures that exhibit a hierarchy of emergent system properties form the underpinning of the central nervous system. These complex dynamical systems are ubiquitous in nature and are not limited to the central nervous system. As discussed above, such systems include, for example, biological systems, immune systems, ecological systems, quantum particle systems, chemical reaction systems, economic systems, cellular systems, and galaxies, to cite but a few examples. And as noted in Chapter 1, these systems are known as dissipative systems [252,369] and consume free energy and matter while maintaining their stable structures by injecting entropy into the environment.

As in thermodynamics, neuroscience is a theory of large-scale systems wherein graph theory [169] can be used in capturing the (possibly dynamic) connectivity properties of network interconnections, with neurons represented by nodes, synapses represented by edges or arcs, and synaptic efficacy captured by edge weighting, giving rise to a weighted adjacency matrix governing the underlying *directed dynamic* graph network topology [198, 199, 221]. Dynamic graph topologies involving neuron connections and disconnections over the evolution of time are essential in modeling the plasticity of the central nervous system. Dynamic neural network topologies capturing synaptic separation or creation can be modeled by differential inclusions [93, 190] and can provide a mechanism for the objective selective inhibition of feedback connectivity in association with anesthetic-induced unconsciousness.

In particular, recent neuroimaging findings have shown that the loss of top-down (feedback) processing in association with loss of consciousness observed in electroencephalographic (EEG) signals and functional magnetic resonance imaging (fMRI) is associated with functional disconnections between anterior and posterior brain structures [234, 275, 308]. These studies show that topological rather than network connection strength of functional networks correlate with states of consciousness. Specifically, anesthetics reconfigure the topological structure of functional brain networks by suppressing midbrain-pontine areas involved in regulating arousal, leading to a dynamic network topology. Hence, changes in the brain network topology are essential in capturing the mechanism of action for the conscious-unconscious transition rather than network connectivity strength during general anesthesia.

In the central nervous system, billions of neurons interact to form self-organizing nonequilibrium structures. However, unlike thermodynamics, wherein energy spontaneously flows from a state of higher temperature to a state of lower temperature, neuron membrane potential variations occur due to ion species exchanges, which evolve from regions of higher chemical potentials to regions of lower chemical potentials (i.e., Gibbs' chemical potential [189]). And this evolution does not occur spontaneously but rather requires a hierarchical (i.e., hybrid) continuous-discrete architecture for the opening and closing of specific gates within specific ion channels.

The physical connection between neurons occurs in the *synapse*, a small gap between the axon, the extension of the cell body of the transmitting neuron, and the dendrite, the extension of the receiving neuron. The signal is transmitted by the release of a neurotransmitter molecule from the axon into the synapse. The neurotransmitter molecule diffuses across the synapse, binds to a postsynaptic receptor membrane protein on the dendrite, and alters the electrochemical potential of the receiving neuron. The time frame for the synthesis of neurotransmitter molecules in the brain far exceeds the time scale of neural activity, and hence, within this finite time frame, neurotransmitter molecules are conserved.

Embedding thermodynamic state notions (i.e., entropy, energy, free energy, chemical potential) within dynamical neuroscience frameworks [198, 199, 221] can allow us to directly address the otherwise mathematically complex and computationally prohibitive large-scale brain dynamical models [198, 221]. In particular, a thermodynamically consistent neuroscience model would emulate the clinically observed self-organizing, spatiotemporal fractal structures that optimally dissipate free energy and optimize entropy production in thalamocortical circuits of fully conscious healthy individuals. This thermodynamically consistent neuroscience framework can provide the necessary tools involving semistability, *synaptic drive* equipartitioning (i.e., synchronization across time scales), free energy dispersal, and entropy production for connecting biophysical findings to psychophysical phenomena for general anesthesia.

The synaptic drive state accounts for pre- and postsynaptic potentials in a neural network to give a measure of neural activity in the network.

In particular, the synaptic drive quantifies the present activity (via the firing rate) along with all previous activity within a neural population appropriately scaled (via a temporal decay) by when a particular firing event occurred. Hence, the synaptic drive provides a measure of neuronal population activity that captures the influence of a given neuron population on the behavior of the network from the infinite past to the current time instant. For details see [198, 221].

System synchronization refers to the fact that the dynamical system states (i.e., synaptic drives) achieve *temporal coincidence* over a finite or infinite time horizon, whereas state equipartitioning refers to the fact that the dynamical system states *converge* to a common value over a finite or infinite time horizon. Hence, both notions involve state agreement or consensus in some sense. However, equipartitioning involves convergence of the state values to a constant state, whereas synchronization involves agreement over time instants. Thus, state equipartitioning implies synchronization; however, the converse is not necessarily true. It is only true insofar as consensus is interpreted to hold over time instants.

The EEG signal does not reflect individual firing rates but rather the synchronization of a large number of neurons (organized in cortical columns) generating macroscopic currents. Since there exists EEG activity during consciousness, there must be some degree of synchronization when the patient is awake. However, during an anesthetic transition the predominant frequency changes, and the dynamical model should be able to predict this frequency shift.

In particular, we hypothesize that as the model dynamics transition to an anesthetized state, the system will involve a reduction in system complexity—defined as a reduction in the degree of irregularity across time scales—exhibiting synchronization of neural oscillators (i.e., thermodynamic energy equipartitioning). This would result in a decrease in system energy consumption (myocardial depression, respiratory depression, hypoxia, ischemia, hypotension, vasodilation), and hence, a decrease in the rate of entropy production. In other words, unconsciousness is characterized by system decomplexification, which is manifested in the failure to develop efficient mechanisms to dissipate energy thereby pathologically retaining higher internal (or local) entropy levels.

The human brain is *isothermal* and *isobaric*,[20] that is, the tempera-

[20]This statement is not true in general; it is true insofar as the human brain is healthy. The brain is enclosed in a closed vault (i.e., the skull). If the brain becomes edematous (i.e., has an excessive accumulation of serous fluid) due to, for example, a traumatic brain injury, then this will increase the pressure (the intracranial pressure) inside this closed vault. If the intracranial pressure becomes too large, then the brain will be compressed and this can result in serious injury

tures of the subnetworks of the brain are equal and remain constant, and the pressure in each subnetwork also remains constant. The human brain network is also constantly supplied with a source of (Gibbs) free energy provided by chemical nourishment of the blood to ensure adequate cerebral blood flow and oxygenation, which involves a blood concentration of oxygen to ensure proper brain function. Information-gathering channels of the blood also serve as a constant source of free energy for the brain. If these sources of free energy are degraded, then internal (local) entropy is increased.

In the transition to an anesthetic state, complex physiologic work cycles (e.g., cardiac respiratory pressure-volume loops, mitochondrial ATP production) necessary for homeostasis follow regressive diffusion and chemical reaction paths that degrade free energy consumption and decrease the *rate of entropy production*. Hence, in an *isolated* large-scale network (i.e., a network with no energy exchange between the internal and external environment) all the energy, though always conserved, will eventually be degraded to the point where it cannot produce any useful work (e.g., oxygenation, ventilation, heart rate stability, organ function).

In this case, all motion (neural activity) would cease, leading the brain network to a state of unconsciousness (semistability) wherein all or partial[21] subnetworks will possess identical energies (energy equipartition or synchronization) and, hence, internal entropy will approach a local maximum or, more precisely, a saddle surface in the state space of the process state variables. Thus, the transition to a state of anesthetic unconsciousness involves an evolution from an initial state of high (external) entropy production (consciousness) to a temporary saddle state characterized by a series of fluctuations corresponding to a state of significantly reduced (external) entropy production (unconsciousness).

In contrast, in a *healthy* conscious human, entropy production occurs spontaneously and, in accordance with Jaynes' maximum entropy production principle [232, 429], energy dispersal is optimal, leading to a maximum entropy production rate. Hence, low entropy in (healthy) human brain networks is synonymous with consciousness and the creation of order (negative entropy change or the availability of energy to do work) reflects a rich fractal spatiotemporal variability, which, since the brain controls key

if not death. In cases of intracranial pathology (e.g., brain tumors, traumatic injury to the brain, bleeding in the brain), there will be increased edema, and hence, increased intracranial pressure as well. This is exacerbated by increased carbon dioxide as this increases blood flow to the brain and increases the edema fluid load.

[21]When patients lose consciousness, other parts of the brain are still functional (heart rate control, ventilation, oxygenation, etc.), and hence, the development of biological neural network models that exhibit partial synchronization is critical. In particular, models that can handle synchronization of subsets of the brain with the nonsynchronized parts firing at normal levels is essential in capturing biophysical behavior. For further details see [198, 221].

physiological processes such as ventilation and cardiovascular function, is critical for delivering oxygen and anabolic substances as well as clearing the products of catabolism to maintain healthy organ function.

In accordance with the second law of thermodynamics, the creation and maintenance of consciousness (i.e., internal order—negative entropy change) is balanced by the external production of a greater degree of (positive) entropy. This is consistent with the maxim of the second law of thermodynamics as well as the writings of Kelvin [438], Gibbs [161, 162], and Schrödinger [395] in that the creation of a certain degree of life and order in the universe is inevitably coupled with an even greater degree of death and disorder [195].

In a network thermodynamic model of the human brain, consciousness can be equated to the brain dynamic states corresponding to a low internal (i.e., local) system entropy. In this monograph, we have shown that the second law of thermodynamics provides a physical foundation for the arrow of time. In particular, we showed that the existence of a global strictly increasing entropy function on every nontrivial network thermodynamic system trajectory establishes the existence of a completely ordered time set that has a topological structure involving a closed set homeomorphic to the real line, which establishes the emergence of the direction of time flow.

Thus, the physical awareness of the passage of time (i.e., chronognosis–κρόνογνωσις) is a direct consequence of the regressive changes in the continuous rate of entropy production taking place in the brain and eventually leading (in finite time) to an equilibrium state of no further entropy production (i.e., death). Since these universal regressive changes in the rate of entropy production are spontaneous, continuous, and decrease in time with age, human experience perceives time flow as unidirectional and nonuniform. In particular, since the rate of time flow and the rate of system entropy regression (i.e., free energy consumption) are bijective (i.e., one-to-one and onto), the human perception of time flow is subjective.[22]

During the transition to an anesthetized state, the external and internal free energy sources are substantially reduced or completely severed in part of the brain, leading the human brain network to a semistable state corresponding to a state of *local* saddle (stationary) high entropy. Since all motion in the state variables (synaptic drives) ceases in this unconscious (synchronized) state, our index for the passage of time vanishes until the anesthetic wears off, allowing for a gradual increase of the flow of free

[22] All living things are chronognostic as they adopt their behavior to a dynamic (i.e., changing) environment. Thus, behavior has a temporal component, wherein all living organisms perceive their future relative to the present. Behavioral anticipation is vital for survival, and hence, chronognosis is *zoecentric* and not only anthropocentric.

energy back into the brain and other parts of the body. This, in turn, gives rise to a state of consciousness, wherein system entropy production is spontaneously resumed and the patient takes in free energy at low entropy and excretes it at high entropy. Merging system thermodynamics with mathematical neuroscience can provide a mathematical framework for describing conscious-unconscious state transitions as well as developing a deeper understanding on how the arrow of time is built into the very fabric of our conscious brain.

17.7 Conclusion

There is no doubt that thermodynamics plays a fundamental role in cosmology, physics, chemistry, biology, engineering, and the information sciences as its universal laws represent the *condiciones sine quibus non* in understanding nature. Thermodynamics, however, has had an impact far beyond the physical, chemical, biological, and engineering sciences; its outreach has reverberated across disciplines, with philosophers to mathematicians, anthropologists to physicians and physiologists, economists to sociologists and humanists, writers to artists, and apocalypticists to theologians each providing their own formulations, interpretations, predictions, speculations, and fantasies on the ramifications of its supreme laws governing nature. And all have added to the inconsistencies, misunderstandings, controversies, confusions, and paradoxes of this aporetic yet profound subject.

The fact that almost everyone has given their own opinion regarding how thermodynamics relates to their own field of specialization has given thermodynamics an *anthropomorphic* character. In that regard, Bridgman writes [65, p. 3]:

> It must be admitted, I think, that the laws of thermodynamics have a different feel from most other laws of the physicist. There is something more palpably verbal about them—they smell more of their human origin.

The second law has been applied to the understanding of human society as well as highlighting the conflict between evolutionists, who claim that human society is constantly ascending, and degradationists, who believe that the globalization processes in human societies will lead to mediocrity (i.e., equipartitioning), social degradation, and the lack of further progress—a transition toward a cretinoid state of human society in analogy to the thermodynamic heat death [425].

Many have found this to be evident in today's golden age of science and technology as societies find themselves heading towards a path cursed with sterility, waiting on practicality and shackled by the chains of political correctness and bureaucratic incompetence. Puerile entertainers and athletes are recognized as stars and exemplars, whereas epistemologists and accomplished citizens are marginalized. Our culture glorifies triviality, coarseness, and mindlessness, promoting anti-intellectualism and mediocrity. And sadly, this has permeated into education. The signs of decline in education and scholarship are unmistakable, and are nowhere more evident than in many of our secondary educational institutions, wherein fraudulent diplomas serve as nothing more than certificates of attendance resulting in remedial students at best and virtual illiterates at worst.

This self-perpetuating culture of mediocrity and simplism has started to metastasize to higher education. Grade inflation and diminished expectations have resulted in intellectual impotence producing academic cripples, weakening the most valuable resource of our country—the University. Shallow teaching techniques (group testing, classroom flipping, TQM-based teaching, problem-based learning, etc.) sacrificing structure and rigor in mathematics, science, and engineering, have resulted in higher education adapting to the declining motivation and vanishing intellectual commitment of students.[23]

More disturbingly, higher education research is dominated by impact factors and citation analysis, often driven by transient and superficial industry needs, and skewed by funding agencies blindly adopting the most recent fads. Under today's research paradigms, Archimedes' work anticipating Newton and Leibnitz by more than two millennia would have been deemed esoteric and impractical. Yet, two thousand years later, calculus emerged as a mathematical concept that is fundamental to how we understand the world around us and is one of the greatest intellectual achievements of all time.

Thermodynamics and entropy have also found their way into economics, wherein human society is viewed as a *superorganism* with the global economy serving as its substrate [160,419]. In particular, the complex, large-scale material structure of civilization involving global transportation and telecommunication networks, water supplies, power plants, electric grids, computer networks, massive goods and services, etc., is regarded as the substratum of human society, and its economy is driven by ingesting free energy (hydrocarbons, ores, fossil fuels, etc.), material, and resources, and secreting them in a degraded form of heat and diluted minerals resulting in an increase of entropy.

---

[23]These ascertainments can be traced over the chronicle of our civilization and are fundamentally rooted in a nation's socioeconomic structure, fabric, and cultural decline. On this matter, Herakleitos states Εἷς ἐμοὶ μύριοι, ἐὰν ἄριστος ᾖι—To me one is worth ten thousand if he is truly outstanding.

Connections between free energy and wealth, wealth and money, and thermodynamic depreciation reflected in the value of money over time via purchasing power loss and negative interest rates have also been addressed in the literature. Soddy [419] maintained that real wealth is subject to the second law of thermodynamics; namely, as entropy increases, real wealth deteriorates. In particular, there are limits to growth given the finite resources of our planet, and there exists a balance between the economy and our ecosphere. In other words, the rate of production of goods has to be delimited by the sustainable capacity of the global environment.

The economic system involves a combination of components that interact between constituent parts of a larger dynamical system, and these interactions cause the degradation of high grade energy (free energy) into low grade energy and, consequently, an increase in entropy. The vanishing supplies of petroleum and metal ores, deforestation, polluted oceans, cropland degradation by erosion and urbanization, and global warming are all limiting factors in economics. And as we continue to demand exponential economic growth in the face of a geometrically increasing population that exceeds the environment's sustainability, this will inevitably result in a finite-time bifurcation leading to a catastrophic collapse of the ecosphere.

Even though such analogies to thermodynamics abound in the literature for all of the aforementioned disciplines, they are almost all formulated in a pseudoscientific language and a fragile mathematical structure. The metabasis of one discipline into another can lead to ambiguities, dilemmas, and contradictions; this is especially true when concepts from a scientific discipline are applied to a nonscientific discipline by analogy. Aristotle (384–322 B.C.) was the first to caution against the transition or metabasis—$\mu\varepsilon\tau\grave{\alpha}\beta\alpha\sigma\iota\varsigma$—of one science into another, and that such a transition should be viewed incisively.

Perhaps the most famous example of this is Malthus' 1798 essay *On the Principle of Population as It Affects the Future Improvement of Society*. His eschatological missive of a geometrical population growth within an arithmetic environmental growing capacity reverberates among economists from Malthus' time to the present. Heilbroner [211] writes: "In one staggering intellectual blow Malthus undid all the roseate hopes of an age oriented toward self-satisfaction and a comfortable vista of progress." And after reading Malthus, Thomas Carlyle refers to the field of economics as "the dismal science." Of course, Malthus' argument is hypothetical, and his conclusion importunate; such assertions, however, are neither provable nor refutable.

Given the universality of thermodynamics and the absolute reign of its supreme laws over nature, it is not surprising that numerous authors ranging from scientists to divinests have laid claim to this unparalleled subject. And this has, unfortunately, resulted in an extensive amount of confusing, inconsistent, and paradoxical formulations of thermodynamics accompanied by a vague and ambiguous lexicon giving it its anthropomorphic character. However, if thermodynamics is formulated as a part of mathematical dynamical systems theory—the prevailing language of science—then thermodynamics is the irrefutable pathway to uncovering the deepest secrets of the universe.

More specifically, given that energy and entropy, like gravity, are pervasive in our ever changing universe, dynamical thermodynamics is certain to take center stage in unveiling the universe's most inscrutable secrets. In attempting to develop a complete understanding of discreteness and continuity, indeterminism and determinism, and quantum mechanics and general relativity in the pursuit of the ever elusive unification of the subatomic (i.e., Planck lengths and times[24]) and astronomical scales, dynamical systems theory is destined in playing a pivotal role in the advancement toward a unified field theory for explaining all of (the known) physics. Superstring theory,[25] M-theory,[26] and loop quantum gravity[27] are among the leading system theories of twentieth-century physics in attempting to unify quantum mechanics and general relativity. However, a grand unification of the fundamental laws of nature is not possible by omitting the central constituent of the universe—dynamical entropy. And correctly addressing this omnipotent component of the universe requires a seamless theoretical nexus between thermodynamics, dynamical systems theory, general relativity, and quantum mechanics.

---

[24]A Planck length is the length scale (10^{-35} m) wherein the structure of spacetime becomes dominated by quantum effects and the theories of quantum mechanics and general relativity are incompatible. Planck time is the time (10^{-43} sec) it takes for light to travel a distance of one Planck length.

[25]Superstring theory is a theory that uses supersymmetry in which bosons and fermions are treated equally and point particles are replaced by one-dimensional closed strings (loops) or open strings (snippets) of vibrating energy to spread out quantum fluctuations thereby substantially reducing their magnitude. The theory uses higher-dimensional spaces to unify general relativity with quantum mechanics. The theory also suggests that each known subatomic particle has an as yet to be detected counterpart; namely, supersymmetric subatomic particles or *sparticles* (e.g., squarks, selectrons) with string vibrational patterns in the extra dimensional spaces playing a pivotal role in determining the particle properties. For details see [173].

[26]M-theory is an incomplete quantum mechanical theory attempting to unify the strong nuclear, electromagnetic, weak nuclear, and gravitational universal forces. The theory is a refinement of superstring theory postulating multidimensional entities called *supermembranes* or *branes* for spreading out quantum fluctuations and smoothing out the quantum foam (i.e., quantum jitters). M-theory suggests that there are as many as eleven dimensions of spacetime and that the seven extra spatial dimensions are only detectable at the subatomic scale. For details see [173].

[27]Loop quantum gravity is an incomplete theory attempting to unify general relativity and quantum mechanics using elemental loops for quantizing space and time to describe the universe. For details see [416].

Chapter Eighteen

Afterword

The objective of every physical theory is to describe and understand reality; yet the fundamental governing equations through which the physical laws of nature are formulated are invariant under time reversal. That is, from any given moment of time, the fundamental laws of physics treat past and future in exactly the same way. More specifically, these fundamental laws apply to dynamical processes whose reversal (not necessarily in time) is allowed; and it is irrelevant whether or not the reversed dynamical process actually occurs in nature. However, it follows from the second law of thermodynamics that the evolution of every dynamical process in nature and our observable universe is irreversible.

This schism between thermodynamics and the remaining laws of physics has eluded the deepest thinkers in science. The controversy of time, change, motion, and irreversibility in nature can be traced back to the dialogues of Herakleitos and Parmenides[1] and continued by Zeno, Epikouros, Lucretius, Kant, Hegel, Bergson, and Einstein as well as numerous other prominent philosophers and scientists over the centuries. As discussed in Chapter 1, Herakleitos postulated that the universe is in constant change (Τα πάντα ρεί) and that there is an underlying order to this change—the Logos (Λόγος)—affecting stability, or more precisely, semistability in a dynamically changing cosmos.

Parmenides' aphorism on the other hand rejects the notion of *creato ex nihilo* and postulates that reality is unchanging, eternal, motionless, perfect, and monist. These two apposing[2] Ionian and Eleatic schools of thought have

[1]Even though it has been suggested by several authors that Parmenides' philosophy on being is a repudiation of Herakleitos' philosophy on becoming, there does not exist any historical evidence credibly linking these two contemporary giant thinkers of the ancient world.

[2]One can find authors (e.g., [382, 383]) that maintain that the Heraclitean and Parmenidean worldviews are not antagonistic, but rather their contraposition originates from a doxographic tradition that dates back to the writing and interpretation of Plato. This, I maintain, is due to the inability of these authors in understanding the sublimity and subtlety of the ancient Greek text and having to rely on obfuscated translations. Herakleitos' fragments: *through change, equilibrium is reached* (Μεταβάλλον ἀναπαύεται) and *the logos unifies all things as one* (Οὐκ ἐμεῦ ἀλλὰ τοῦ λόγου ἀκούσαντας ὁμολογέειν σοφόν ἐστι, ἓν πάντα εἶναι) are often misinterpreted to support

become known as the Heraclitean and Parmenidean cosmological views of our universe; namely, one involving a dynamic and ever changing universe, and the other involving a timeless and changeless universe.

Parmenides' famous poem *On Nature* (Περὶ Φύσεως) has been repeatedly plagued by numerous inconsistent interpretations by some of the most brilliant minds in philosophy dating back to Aristotle. This can partly be attributed to the fact that after two and a half millennia, over two-thirds of the poem has been lost to time [213]. The thesis of Parmenides' argument can be distilled to *What is, is* (ὅπως ἐστίν), and *What is not, is not* (ὡς οὐκ ἐστίν). Parmenides' central argument focused on how to ensure the reliability of discourse [213]. Yet, many scholars have erroneously projected Parmenides' terse statements on the object of discourse and inquiry to the tangible world, the universe, existence, reality, or monism.[3] This is nowhere more pronouncedly expressed than by Popper [366, pp. 79–80] who interprets Parmenides' doctrine as "The world consists, in reality, of one large, unmoving, homogeneous, solid block of spherical shape in which nothing can ever happen; there is no past or future."

Even though many thinkers and scientists have taken a cosmological interpretation of Parmenides' object of inquiry, reasoning, and judgment, there is no evidence that Parmenides (or Zeno for that matter) denied the existence of time, change, or motion in the *physical universe*. One need only refer to Fragment 4 of his poem to deduce this [213, p. 157].

> Behold things which, although absent, are yet securely present to the mind; for you cannot cut off What IS from holding on to What IS; neither by dispersing it in every way, everywhere throughout the cosmos, nor by gathering it together.

Referring to the two aforementioned statements of "dispersing" and "gathering together," it is clear that Parmenides is discussing change in space, and does not object to change—as long as this change relates to the object of perception (i.e., the physical world) and not the mind (i.e., the metaphysical).

However, given that human inquiry, reasoning, knowledge, and judgment form the underpinnings for understanding existence, physical reality,

---

their views that the cosmos is changeless, albeit formed by changing entities [382, 383]. They, however, completely ignore Herakleitos' other fragments that explicitly support his steadfast view of a changing, dynamic universe. For example, Herakleitos' fragments stating: ... τὸ ἀντίξουν συμφέρον, καὶ ἐκ τῶν διαφερόντων καλλίστην ἁρμονίαν, καὶ πάντα κατ' ἔριν γίνεσθαι and Οὐ ξυνίασι ὅκως διαφερόμενον ἑωυτῷ ὁμολογέει· παλίντροπος ἁρμονίη ὅκωσπερ τόξον καὶ λύρης, as well as the fragments cited in Chapter 1, clearly support his view on a time-bounded universe.

[3] When interpreting Parmenides' doctrine cosmologically, Aristotle links him and his followers to lunatics [24]. Here, however, Parmenides is in good company as Aristotle also refers to Herakleitos as the obscure one or the dark one (ὁ Σκοτεινός).

and the tangible world, it is not surprising that Parmenides' principles have been extrapolated to include our understanding of the universe. Parmenides himself professed early scientific accounts of the cosmos in the last section of his poem—the doxa (δόξα). Parmenidean principles adopted to a timeless, changeless universe have been proposed by many scientists in an attempt to align his principles with Einstein's theory of general relativity and advocating that time, motion, and change in our universe is an illusion (see, for example, [23,382,383]).

As discussed in Section 13.9, general relativity theory of space, time, and gravitation describes matter through gravitational fields via the curvature of spacetime. And in accordance with the principle of covariance the theory is expressed in a form that is independent of the choice of spacetime coordinates using abstract differential geometry. A formulation of general relativity theory, known as the manifold model of spacetime, is proposed in [210, 235] wherein spacetime singularities[4] can be avoided. In this formulation, spacetime is defined as the ontological sum of ordered events. This model can be applied to the whole of the universe wherein all events and all things in nature and the universe that have happened, are happening, and will happen coexist. In other words, all past, present, and future events exist and are equally present at every event instant time but at different locations in space.

The manifold model of spacetime is characterized by a smooth (i.e., infinitely continuously differentiable), $(3+1)$-dimensional, semi-Riemannian world manifold $\mathcal{M}$. That is, $\mathcal{M}$ is covered by subsets whose elements $p=(x,y,z,t)\in\mathcal{M}$ form a bijective map with subsets of $\mathbb{R}^{3+1}$. Furthermore, every element $p\in\mathcal{M}$ represents an event; the converse, however, is not necessarily true. The metric structure of $\mathcal{M}$ capturing distances between events is given by a rank two metric tensor field $g_{\mu\nu} = g_{\mu\nu}(p)$ on the Minkowskian tangent space $T_p\mathcal{M}$, where $p\in\mathcal{M}$. The metric tensor field of the world manifold $\mathcal{M}$ is determined by the rank two energy-momentum tensor $T_{\mu\nu}$ via the Einstein field equations given by (13.138).

Recall that (13.138) gives a set of ten coupled nonlinear partial differential equations for the metric tensor coefficients, which represent the

[4]Spacetime or gravitational singularities appear within general relativity theory wherein the field equations relating spacetime geometry to the matter content of the universe contain singularities where the spacetime curvatures as well as the matter and energy densities become infinite. In this case, classical general relativity theory predicts the formation of super ultradense regions in the universe wherein these singularities are developed during gravitational collapse of massive stars. In addition, these singularities are also considered within cosmogony and cosmology as defining the earliest state of the universe during the big bang. The manifold model of spacetime uses the Penrose-Hawking singularity theorems [210] to avoid singularity occurrence in spacetime. This, however, raises the possibility of violating causality in spacetime, the positivity of energy conditions, or the existence of closed trapped surfaces describing the inner regions of an event horizon in the dynamical evolution of the universe [235].

gravitational potential and its covariant derivatives, that determine the succession of events (i.e., different slices of spacetime) through the Levi-Civita (affine) connection of the world manifold. Hence, once the geometry of the world manifold $\mathcal{M}$ is determined by the metric tensor field $g_{\mu\nu}$ characterizing the energy-momentum distribution $T_{\mu\nu}$, then the structure of the world $\mathcal{M}$ is fixed.

The fact that the $(3+1)$-dimensional manifold model of spacetime predicated on the Einstein field equations allows for the succession of events (i.e., different slices or sections of spacetime), or even the entire history of all events in spacetime, with the world manifold giving the *totality* of such events, many physicists—including Einstein—have claimed that the world (i.e., the history of the universe) is ontologically determined. This is perhaps best formally delineated in [383] where the author argues that the points $p \in \mathcal{M}$ represent events that do not affect the spacetime as a whole, and asserts that the world represented as a $(3+1)$-dimensional manifold is unchanging, eternal, motionless, and monist. In other words, our universe is Parmenidean and the passage of time is an illusion. This belief was also held by Albert Einstein, who maintained that change is a local property and not a global property. More specifically, our perceived three-dimensional worldview gives us the illusion of time and change, but when viewed as a four-dimensional world involving the union of space and time the cosmos is timeless and changeless.

Even though relativity theory dispenses with the notion of absolute space and time leading to the impossibility of defining an absolute simultaneity relation between events, it is completely consistent with the principle of causality in the sense that observed ordered events of physical processes are *not* affected by transformations between inertial frames. Thus, unlike simultaneity, which is observer dependent, the time ordering of events are *not* observer dependent. In this sense, irreversible processes occurring in nature (i.e., the irreversible succession of events) are described by asymmetries in the foliation[5] of the world manifold $\mathcal{M}$ characterizing spacetime [383]. However, the existence of foliations by invariant submanifolds within the world manifold is only possible for integrable dynamical systems[6] (e.g.,

[5]A *foliation* is a pattern of lower-dimensional disjoint, connected, and immersed strips (i.e., submanifolds) $\mathcal{L}$ on $\mathcal{M}$, called the *leaves* of the foliation, which are locally behaved (i.e., admit a continuous bijective mapping $h : \mathcal{L} \to \mathcal{M}$ that satisfies a local flatness condition). The tangent space to the leaves of a foliation forms a vector bundle $T\mathcal{M}$ over $\mathcal{M}$. The leaves of a foliation consist of integrable subbundles of $T\mathcal{M}$ requiring that the existence of a foliation of $\mathcal{M}$ be homotopic to an *integrable distribution* (i.e., a distribution whose Lie bracket of its sections is involutive).

[6]A universal definition of integrability for dynamical systems is elusive. Integrability is an intrinsic property of a given dynamical system through which certain constraints are imposed on the evolution of the system solutions in the phase space. For classical mechanics, these constraints describe system properties implicitly through conserved quantities throughout the motion of the dynamical system (i.e., constants of motion). For general dynamical systems, integrability refers to the existence of invariant foliations whose leaves are embedded submanifolds of a minimal

classical and quantum mechanics).

Given that general relativity theory as developed by Einstein does not account for thermodynamics and entropy, a seamless morphing between metric gravity and quantum gravity, and the fact that, as shown by Poincaré [370], most large-scale dynamical processes in nature are nonintegrable, a Parmenidean worldview of the universe is highly speculative. Nature, through entropy, breaks the symmetry of (duration) time. In this regard, it seems that spacetime itself can change through a multidimensional, entropy-time coupling, that is, a *meta-time*, which governs the rate of time flow with respect to the change in entropy leading to the irreversible succession of events in spacetime.[7] Thus, at every instant of time the present is moving and future events are unknown and objectively nonexistent as duration time advances. In the absence of a quantum field theory for gravitation and the disassociation between gravitation and the other fundamental forces of nature, as well as the absence of a seamless integration of gravitation, global entropy, and dynamical (i.e., nonequilibrium) thermodynamics, the arrow of time as an intrinsic property of reality continues to remain a subject of debate (Ἀγχιβασίην).

We have a very limited understanding of the events after the big bang along with how the universe remains far from a thermodynamic equilibrium state by not only destroying, but also creating complexity and increased organization through nature's abhorrence of gradients. As natural scientists we constantly study idealized entities that serve to approximate the physical world. While science and mathematics provide rigorously reliable tools for studying these idealized entities in our continuing quest in understanding nature and the universe, the central question is whether the universe in *idealized form* is still a universe [213]. No epistemological discipline has a dictum over reality. In this regard Einstein orates[8] [133]: "Insofar as the statements of mathematics refer to reality, they are not certain, and insofar as they are certain, they do not refer to reality." More perspicaciously, Herakleitos decrees that "Nature loves to hide" (Φύσις κρύπτεσθαι φιλεί). When dealing with mathematical thought and structure there cannot be any room for error, whereas when dealing with the universe error is unavoidable.

---

dimension and are invariant under the flow of the dynamical system. In other words, every local one-parameter group associated to a vector field in the Lie algebra generated by the tangent bundle of the dynamical system transforms leaves into leaves.

[7]Here, the time parameter is interpreted as duration time and *not* as an event instant time. A similar notion of duration time within general relativity was also advocated by Prigogine, wherein he showed that nonlinear dissipative structures can define a unique Liouville-type operator that is time asymmetric and can characterize a unique time parameter describing dynamical system evolution.

[8]"Insofern sich die Sätze der Mathematik auf die Wirklichkeit beziehen, sind sie nicht sicher, und insofern sie sicher sind, beziehen sie sich nicht auf die Wirklichkeit."

In the absence of a unified field theory as well as a complete and unfettered unification between the physical, informational, biological, and psychological sciences, the opposing intellections of being (timeless) and becoming (time bounded) as fundamental intrinsic properties of reality remain enigmatic. If, however, dynamical thermodynamics is to continue to play an imperative role in such a grand unification of the aforementioned sciences, then the entropic arrow of time is almost surely certain to emerge as the most fundamental property of the universe.

Bibliography

[1] L. F. Abbott, "Theoretical neuroscience rising," *Neuron*, vol. 60, pp. 489–495, 2008.

[2] J. M. Aerts, W. M. Haddad, G. An, and Y. Vodovtz, "From data patterns to mechanistic models in acute critical illness," *J. Crit. Care*, vol. 29, pp. 604–610, 2014.

[3] G. Amelino-Camelia, "Relativity: Still special," *Nature*, vol. 450, pp. 801–803, 2007.

[4] I. Aoki, "Entropy principle for human development, growth and aging," *J. Theor. Biol.*, vol. 150, pp. 215–223, 1991.

[5] I. Aoki, "Min-max principle of entropy production with time in aquatic communities," *Ecological Complexity*, vol. 3, pp. 56–63, 2006.

[6] T. M. Apostol, *Mathematical Analysis.* Reading, MA: Addison-Wesley, 1957.

[7] T. M. Apostol, *Mathematical Analysis,* 2nd ed. Reading, MA: Addison-Wesley, 1974.

[8] A. Arapostathis, V. S. Borkar, and M. K. Ghosh, *Ergodic Control of Diffusion Processes.* Cambridge, U.K.: Cambridge University Press, 2012.

[9] L. Arnold, *Stochastic Differential Equations: Theory and Applications.* New York, NY: Wiley-Interscience, 1974.

[10] V. I. Arnold, *Mathematical Models of Classical Mechanics.* New York, NY: Springer-Verlag, 1989.

[11] V. I. Arnold, "Contact geometry: The geometrical method of Gibbs' thermodynamics," in *Proceedings of the Gibbs Symposium*, (D. Caldi and G. Mostow, eds.), pp. 163–179, Providence, RI: American Mathematical Society, 1990.

[12] H. Arzelies, "Transformation relativiste de la température et de quelques autres grandeurs thermodynamiques," *Nuovo Cimento*, vol. 35, pp. 792–804, 1965.

[13] J. P. Aubin and A. Cellina, *Differential Inclusions.* Berlin, Germany: Springer-Verlag, 1984.

[14] I. Avramov, "Relativity and temperature," *Russ. J. Phys. Chem.*, vol. 77, pp. 179–182, 2003.

[15] A. Bacciotti and F. Ceragioli, "Stability and stabilization of discontinuous systems and nonsmooth Lyapunov functions," *ESAIM Control Optim. Calculus Variations*, vol. 4, pp. 361–376, 1999.

[16] R. Baierlein, "The elusive chemical potential," *Amer. J. Phys.*, vol. 69, no. 4, pp. 423–434, 2001.

[17] C. Baik and A. S. Lavine, "On the hyperbolic heat conduction equation and the second law of thermodynamics," *Trans. ASME J. Heat Transfer*, vol. 117, pp. 256–263, 1995.

[18] D. D. Bainov and P. S. Simeonov, *Systems with Impulse Effect: Stability, Theory and Applications.* Chichester, U.K.: Ellis-Horwood, 1989.

[19] D. D. Bainov and P. S. Simeonov, *Impulsive Differential Equations: Asymptotic Properties of the Solutions.* Singapore: World Scientific, 1995.

[20] A. W. Balakrishnan, "On the controllability of a nonlinear system," *Proc. Nat. Acad. Sci.*, vol. 55, pp. 465–468, 1966.

[21] R. Balescu, "Relativistic statistical thermodynamics," *Physica*, vol. 40, pp. 309–338, 1968.

[22] J. Bang-Jensen and G. Gutin, *Digraphs: Theory, Algorithms and Applications.* New York, NY: Springer, 2002. Corrected First Edition.

[23] J. Barbour, *The End of Time: The Next Revolution in Physics.* Oxford, U.K.: Oxford University Press, 2000.

[24] J. Barnes, *The Complete Works of Aristotle: The Revised Oxford Translation.* Princeton, NJ: Princeton University Press, 1991.

[25] C. Basaran, "A thermodynamic framework for damage mechanics of solder joints," *ASME J. Electronic Packaging*, vol. 120, pp. 379–384, 1998.

[26] C. Basaran and S. Nie, "An irreversible thermodynamics theory for damage mechanics of solids," *Int. J. Damage Mechanics*, vol. 13, pp. 205–223, 2004.

[27] M. S. Bazaraa, H. D. Sherali, and C. M. Shetty, *Nonlinear Programming: Theory and Algorithms.* New York, NY: Wiley, 1993.

[28] J. D. Bekenstein, "Black holes and entropy," *Phys. Rev. D*, vol. 7, pp. 2333–2346, 1973.

[29] E. T. Bell, *Men of Mathematics.* New York, NY: Simon and Schuster, 1986.

[30] L. Bergström and A. Goobar, *Cosmology and Particle Astrophysics.* Berlin, Germany: Springer-Verlag, 2004.

[31] A. Berman and R. J. Plemmons, *Nonnegative Matrices in the Mathematical Sciences.* New York, NY: Academic, 1979.

[32] A. Berman, R. S. Varga, and R. C. Ward, "ALPS: Matrices with nonpositive off-diagonal entries," *Linear Algebra Appl.*, vol. 21, pp. 233–244, 1978.

[33] D. S. Bernstein, *Matrix Mathematics.* Princeton, NJ: Princeton University Press, 2005.

[34] D. S. Bernstein and S. P. Bhat, "Nonnegativity, reducibility, and semistability of mass action kinetics," in *Proc. IEEE Conf. Decision and Control*, Phoenix, AZ, pp. 2206–2211, 1999.

[35] D. S. Bernstein and S. P. Bhat, "Energy equipartition and the emergence of damping in lossless systems," in *Proc. IEEE Conf. Decision and Control*, Las Vegas, NV, pp. 2913–2918, 2002.

[36] D. S. Bernstein and S. P. Bhat, "Linear output-reversible systems," in *Proc. American Control Conf.*, Denver, CO, pp. 3240–3241, 2003.

[37] D. S. Bernstein and D. C. Hyland, "Compartmental modeling and second-moment analysis of state space systems," *SIAM J. Matrix Anal. Appl.*, vol. 14, pp. 880–901, 1993.

[38] J. Bertrand, *Thermodynamique.* Paris, France: Gauthier-Villars, 1887.

[39] A. Berut, A. Arakelyan, A. Petrosyan, S. Ciliberto, R. Dillenschneider, and E. Lutz, "Experimental verification of Landauer's principle linking information and thermodynamics," *Nature*, vol. 483, pp. 187–190, 2012.

[40] S. P. Bhat and D. S. Bernstein, "Lyapunov analysis of semistability," in *Proc. American Control Conf.*, San Diego, CA, pp. 1608–1612, 1999.

[41] S. P. Bhat and D. S. Bernstein, "Finite-time stability of continuous autonomous systems," *SIAM J. Control Optim.*, vol. 38, pp. 751–766, 2000.

[42] S. P. Bhat and D. S. Bernstein, "Nontangency-based Lyapunov tests for convergence and stability in systems having a continuum of equilibra," *SIAM J. Control Optim.*, vol. 42, pp. 1745–1775, 2003.

[43] S. P. Bhat and D. S. Bernstein, "Geometric homogeneity with applications to finite-time stability," *Math. Control Signals Syst.*, vol. 17, pp. 101–127, 2005.

[44] S. P. Bhat and D. S. Bernstein, "Average-preserving symmetries and energy equipartition in linear Hamiltonian systems," *Math. Control Signals Syst.*, vol. 21, pp. 127–146, 2009.

[45] S. P. Bhat and D. S. Bernstein, "Arc-length-based Lyapunov tests for convergence and stability in systems having a continuum of equilibria," *Math. Control Signals Syst.*, vol. 22, pp. 155–184, 2010.

[46] N. P. Bhatia and G. P. Szegö, *Stability Theory of Dynamical Systems.* Berlin, Germany: Springer-Verlag, 1970.

[47] M. Biggs, *Algebraic Graph Theory,* 2nd ed. Cambridge, U.K.: Cambridge University Press, 1993.

[48] J. Bircher, "Towards a dynamic definition of health and disease," *Med. Health Care Philos.*, vol. 8, pp. 335–341, 2005.

[49] G. D. Birkhoff, "Dynamical systems with two degrees of freedom," *Trans. Amer. Math. Soc.*, vol. 18, pp. 199–300, 1917.

[50] G. D. Birkhoff, "Recent advances in dynamics," *Science*, vol. 51, pp. 51–55, 1920.

[51] G. D. Birkhoff, "Collected mathematical papers," *Am. Math. Soc.*, vol. 1, 2, 3, 1950.

[52] T. S. Biro, *Is There a Temperature? Conceptual Challenges at High Energy, Acceleration and Complexity.* New York, NY: Springer, 2011.

[53] R. E. Blanton, J. G. Levitt, P. M. Thompson, K. L. Narr, L. Capetillo-Cunliffe, A. Nobel, J. D. Singerman, J. T. McCracken, and A. W. Toga, "Mapping cortical asymmetry and complexity patterns in normal children," *Psychiatry Res.*, vol. 107, pp. 29–43, 2007.

[54] D. Blanusa, "Sur les paradoxes de la notion d' énergie," *Glasnik mat.-fiz i astr.*, vol. 2, pp. 249–250, 1947.

[55] G. N. Bochkov and Y. E. Kuzovlev, "General theory of thermal fluctuations in nonlinear systems," *Sov. Phys. JETP*, vol. 45, pp. 125–130, 1977.

[56] G. N. Bochkov and Y. E. Kuzovlev, "Fluctuation-dissipation relations for nonequilibrium processes in open systems," *Sov. Phys. JETP*, vol. 49, pp. 543–551, 1979.

[57] L. Boltzmann, "Über die Beziehung eines Allgemeine Mechanischen Satzes zum zweiten Hauptsatze der Warmetheorie," *Sitzungsberichte Akad. Wiss., Vienna, Part II*, vol. 75, pp. 67–73, 1877.

[58] L. Boltzmann, *Vorlesungen über die Gastheorie,* 2nd ed. Leipzig, Germany: J. A. Barth, 1910.

[59] M. Born, "Kritische Betrachtungen zur traditionellen Darstellung der Thermodynamik," *Physikalische Zeitschrift.*, vol. 22, pp. 218–224, 249–254, 282–286, 1921.

[60] M. Born, *The Born-Einstein Letters.* New York, NY: Walker, 1971.

[61] M. Born, *My Life: Recollections of a Nobel Laureate.* London, England: Taylor and Francis, 1978.

[62] R. Bousso, "A covariant entropy conjecture," *Journal of High Energy Physics*, vol. 7, pp. 1–33, 1999.

[63] J. B. Boyling, "An axiomatic approach to classical thermodynamics," *Proceedings of the Royal Society of London. Series A, Mathematical and Physical Sciences*, vol. 329, pp. 35–70, July 1972.

[64] H. Brezis, "On a characterization of flow-induced invariant sets," *Commun. Pure Appl. Math.*, vol. 23, pp. 261–263, 1970.

[65] P. Bridgman, *The Nature of Thermodynamics.* Cambridge, MA: Harvard University Press, 1941. Reprinted by Peter Smith: Gloucester, MA, 1969.

[66] R. W. Brockett and J. C. Willems, "Stochastic control and the second law of thermodynamics," in *Proc. IEEE Conf. Decision and Control*, San Diego, CA, pp. 1007–1011, 1978.

[67] R. Brown, "A brief account of microscopical observations made in the months of June, July, and August, 1827, on the particles contained in

the pollen of plants; and on the general existence of active molecules in organic and inorganic bodies," *Phil. Mag.*, vol. 4, pp. 161–173, 1827.

[68] J. Brunet, "Information theory and thermodynamics," *Cybernetica*, vol. 32, pp. 45–78, 1989.

[69] S. G. Brush, *The Kind of Motion We Call Heat: A History of the Kinetic Theory in the Nineteenth Century.* Amsterdam, The Netherlands: North Holland, 1976.

[70] Z. Brzezniak, M. Capinski, and F. Flandoli, "Pathwise global attractors for stationary random dynamical systems," *Prob. Th. Rel. Fields*, vol. 95, pp. 87–102, 1993.

[71] H. Buchdahl, *The Concepts of Classical Thermodynamics.* Cambridge, U.K.: Cambridge University Press, 1966.

[72] M. A. Buice and J. D. Cowan, "Field-theoretic approach to fluctuation effects in neural networks," *Phys. Rev. E*, vol. 75, no. 5, pp. 1–14, 2007.

[73] H. Callen, *Thermodynamics.* New York, NY: John Wiley and Sons, 1960.

[74] H. Callen and G. Horwitz, "Relativistic thermodynamics," *Astrophysical Journal*, vol. 39, pp. 938–947, 1971.

[75] S. L. Campbell and N. J. Rose, "Singular perturbation of autonomous linear systems," *SIAM J. Math. Anal.*, vol. 10, pp. 542–551, 1979.

[76] C. Carathéodory, "Untersuchungen über die Grundlagen der Thermodynamik," *Math. Annalen*, vol. 67, pp. 355–386, 1909.

[77] C. Carathéodory, "Über die Bestimmung der Energie und der absoluten Temperatur mit Hilfe von reversiblen Prozessen," *Sitzungsberichte der preußischen Akademie der Wissenschaften, Math. Phys. Klasse*, pp. 39–47, 1925.

[78] A. Carcaterra, "An entropy formulation for the analysis of energy flow between mechanical resonators," *Mechanical Systems and Signal Processing*, vol. 16, pp. 905–920, 2002.

[79] D. S. L. Cardwell, *From Watt to Clausius: The Rise of Thermodynamics in the Early Industrial Age.* Ithaca, NY: Cornell University Press, 1971.

[80] S. Carnot, *Réflexions sur la puissance motrice du feu et sur les machines propres a développer cette puissance.* Paris, France: Chez Bachelier, Libraire, 1824.

[81] J. Casas-Vázquez, D. Jou, and G. Lebon, *Recent Development in Nonequilibrium Thermodynamics.* Berlin, Germany: Springer-Verlag, 1984.

[82] H. B. G. Casimir, "On Onsager's principle of microscopic reversibility," *Rev. Mod. Phys.*, vol. 17, pp. 343–350, 1945.

[83] C. Cattaneo, "A form of heat conduction equation which eliminates the paradox of instantaneous propagation," *Compte Rendus*, vol. 247, pp. 431–433, 1958.

[84] V. Chellaboina, S. P. Bhat, W. M. Haddad, and D. S. Bernstein, "Modeling and analysis of mass-action kinetics: Nonnegativity, realizability, reducibility, and semistability," *IEEE Contr. Syst. Mag.*, vol. 29, pp. 60–78, 2009.

[85] J. H. Christenson, J. W. Cronin, V. L. Fitch, and R. Turlay, "Evidence for the 2π decay of the K_2^0 meson," *Phys. Rev. Lett.*, vol. 13, no. 4, pp. 138–140, 1964.

[86] F. H. Clarke, *Optimization and Nonsmooth Analysis.* New York, NY: Wiley, 1983.

[87] R. Clausius, "Über verschiedene für die Anwendung bequeme Formen der Haubtgleichungen der mechanischen wärmetheorie," *Viertelsjahrschrift der naturforschenden Gesellschaft (Zürich)*, vol. 10, pp. 1–59, 1865. Also in [89, pp. 1–56], and translated in [240, pp. 162–193].

[88] R. Clausius, *Abhandlungungen über die Mechanische Wärmetheorie,* vol. 2. Braunschweig, Germany: Vieweg and Sohn, 1867.

[89] R. Clausius, *Mechanische Wärmetheorie.* Braunschweig, Germany: Vieweg and Sohn, 1876.

[90] R. Clausius, "Über die Concentration von Wärme- und Lichtstrahlen und die Gränze Ihre Wirkung," in *Abhandlungen über die Mechanischen Wärmetheorie*, pp. 322–361, Braunschweig, Germany: Vieweg and Sohn, 1864.

[91] B. D. Coleman, "The thermodynamics of materials with memory," *Arch. Rational Mech. Anal.*, vol. 17, pp. 1–46, 1964.

[92] B. D. Coleman and W. Noll, "The thermodynamics of elastic materials with heat conduction and viscosity," *Arch. Rational Mech. Anal.*, vol. 13, pp. 167–178, 1963.

[93] J. Cortés, "Discontinuous dynamical systems," *IEEE Control Syst. Mag.*, vol. 28, pp. 36–73, 2008.

[94] J. Cortés and F. Bullo, "Coordination and geometric optimization via distributed dynamical systems," *SIAM J. Control Optim.*, vol. 44, pp. 1543–1574, 2005.

[95] S. S. Costa and G. E. A. Matsas, "Temperature and relativity," *Phys. Lett. A.*, vol. 209, pp. 155–159, 1995.

[96] P. Coveney, *The Arrow of Time.* New York, NY: Ballantine Books, 1990.

[97] M. G. Crandall, "A generalization of Peano's theorem and flow invariance," *MRC Technical Summary Report 1228*, 1972.

[98] M. G. Crandall and T. M. Liggett, "Generation of semigroups of nonlinear transformations on general Banach spaces," *Amer. J. Math.*, vol. 93, pp. 265–298, 1971.

[99] H. Crauel, A. Debussche, and F. Flandoli, "Random attractors," *J. Dynamics Differential Equations*, vol. 9, no. 2, pp. 307–341, 1997.

[100] H. Crauel and F. Flandoli, "Attractors of random dynamical systems," *Prob. Th. Rel. Fields*, vol. 100, pp. 365–393, 1994.

[101] F. Crick, *Of Molecules and Men.* Seattle, WA: University of Washington Press, 1966.

[102] G. E. Crooks, "Entropy production fluctuation theorem and the nonequilibrium work relation for free energy differences," *Phys. Rev. E*, vol. 60, no. 3, pp. 2721–2726, 1999.

[103] G. E. Crooks, "Path-ensemble averages in systems driven far from equilibrium," *Phys. Rev. E*, vol. 61, no. 3, pp. 2361–2366, 2000.

[104] R. F. Curtain and H. J. Zwart, *An Introduction to Infinite-Dimensional Linear Systems Theory.* New York, NY: Springer-Verlag, 1995.

[105] C. M. Dafermos, *Hyperbolic Conservation Laws in Continuum Physics.* Berlin, Germany: Springer-Verlag, 2000.

[106] W. A. Day, "Thermodynamics based on a work axiom," *Arch. Rational Mech. Anal.*, vol. 31, pp. 1–34, 1968.

[107] W. A. Day, "A theory of thermodynamics for materials with memory," *Arch. Rational Mech. Anal.*, vol. 34, pp. 86–96, 1969.

[108] P. Dayan and L. F. Abbott, *Theoretical Neuroscience.* Cambridge, MA: MIT Press, 2005.

[109] L. Debnath and P. Mikusiński, *Introduction to Hilbert Spaces with Applications.* San Diego, CA: Academic Press, 1999.

[110] G. Deco, V. K. Jirsa, P. A. Robinson, M. Breakspear, and K. Friston, "The dynamic brain: From spiking neurons to neural masses and cortical fields," *PLoS Comput. Biol.*, vol. 4, no. 8, pp. 1–35, 2008.

[111] T. DeDonder, *L'Affinité.* Paris, France: Gauthiers-Villars, 1927.

[112] T. DeDonder and P. Van Rysselberghe, *Affinity.* Menlo Park, CA: Stanford University Press, 1936.

[113] S. R. de Groot, *Thermodynamics of Irreversible Processes.* Amsterdam, The Netherlands: North-Holland, 1951.

[114] S. R. de Groot and P. Mazur, *Nonequilibrium Thermodynamics.* Amsterdam, The Netherlands: North-Holland, 1962.

[115] R. Diestel, *Graph Theory.* New York, NY: Springer-Verlag, 1997.

[116] G. L. Dirichlet, "Note sur la Stabilité de l'Équilibre," *J. Math. Pures Appl.*, vol. 12, pp. 474–478, 1847.

[117] B. A. Dubrovin, A. T. Fomenko, and S. P. Novikov, *Modern Geometry-Methods and Applications: Part II: The Geometry and Topology of Manifolds.* New York, NY: Springer-Verlag, 1985.

[118] P. Duhem, *Traité dénergétique ou de Thermodynamique générale.* Paris, France: Gauthier-Villars, 1911.

[119] J. Dunkel and P. Hänggi, "Relativistic Brownian motion," *Phys. Rep.*, vol. 471, pp. 1–73, 2009.

[120] J. Dunkel, P. Hänggi, and S. Hilbert, "Non-local observables and lightcone-averaging in relativistic thermodynamics," *Nature Phys.*, vol. 5, pp. 741–747, 2009.

[121] J. Dunning-Davies, "Concavity, superadditivity and the second law," *Found. Phys. Lett.*, vol. 6, pp. 289–295, 1993.

[122] J. Earman, "Irreversibility and temporal asymmetry," *Journal of Philosophy*, vol. 64, pp. 543–549, 1967.

[123] A. Eddington, *The Nature of the Physical World.* London, U.K.: Dent and Sons, 1935.

[124] T. Ehrenfest-Afanassjewa, "Zur Axiomatisierung des zweiten Hauptsatzes der Thermodynamik," *Zeitschrift fur Physik*, vol. 33, pp. 933–945, 1925.

[125] A. Einstein, "Über die von der molekularkinetischen Theorie der Wärme geforderte Bewegung von in ruhenden Flüssigkeiten suspendierten Teilchen," *Annalen der Physik*, vol. 322, no. 8, pp. 549–560, 1905.

[126] A. Einstein, "Über einen die Erzeugung und Verwandlung des Lichtes betreffenden heuristischen Gesichtspunkt," *Annalen der Physik*, vol. 17, no. 6, pp. 132–148, 1905.

[127] A. Einstein, "Zur Elektrodynamik bewegter Körper," *Annalen der Physik*, vol. 17, pp. 891–921, 1905.

[128] A. Einstein, "Kosmologische Betrachtungen zur allgemeinen Relativitaetstheorie," *Sitzungsberichte der Königlich Preussischen Akademie der Wissenschaften*, vol. 8, pp. 142–152, 1907.

[129] A. Einstein, "Über das Relativitätsprinzip und die aus demselben gezogenen Folgerungen," *J. Radioaktivität Elektronik*, vol. 4, pp. 411–462, 1907.

[130] A. Einstein, "Theorie der Opaleszenz von homogenen Flüssigkeiten und Flüssigkeitsgemischen in der Nhe des kritischen Zustandes," *Annalen der Physik*, vol. 33, pp. 1275–1298, 1910.

[131] A. Einstein, "Die grundlage der allgemeinen relativitätstheorie," *Annalen der Physik*, vol. 49, pp. 284–339, 1916.

[132] A. Einstein, *Relativity: The Special and General Theory*. New York, NY: Holt and Company, 1920.

[133] A. Einstein, *Geometrie und Erfahrung*. Berlin, Germany: Julius Springer, 1921.

[134] A. Einstein, "Autobiographical notes," *Albert Einstein: Philosopher-Scientist* (P. A. Schilpp, ed.), Evanston, IL: Library of Living Philosophers, 1949.

[135] P. Erdi and J. Toth, *Mathematical Models of Chemical Reactions: Theory and Applications of Deterministic and Stochastic Models*. Princeton, NJ: Princeton University Press, 1988.

[136] G. B. Ermentrout and D. H. Terman, *Mathematical Foundations of Neuroscience*. New York, NY: Springer-Verlag, 2010.

[137] L. Euler, *Theoria Motuum Lunae.* Saint Petersburg, Russia: Acad. Imp. Sci. Petropolitanae, 1753.

[138] L. C. Evans, *Partial Differential Equations,* 2nd ed. Providence, RI: American Mathematical Society, 2002.

[139] D. J. Evans and D. J. Searles, "Equilibrium microstates which generate second law violating steady states," *Phys. Rev. E*, vol. 50, pp. 1645–1648, 1994.

[140] F. Fagnani and J. C. Willems, "Representations of time-reversible systems," *Journal of Mathematical Systems, Estimation, and Control*, vol. 1, no. 1, pp. 5–28, 1991.

[141] F. Fagnani and J. C. Willems, "Representations of symmetric linear dynamical systems," *SIAM J. Control and Optimization*, vol. 31, no. 5, pp. 1267–1293, 1993.

[142] M. Faraday, *Experimental Researches in Electricity.* London, U.K.: Dent and Sons, 1922.

[143] M. Farkas, *Periodic Motions.* New York, NY: Springer-Verlag, 1994.

[144] M. Feinberg, "Chemical reaction network structure and the stability of complex isothermal reactors I: The deficiency zero and deficiency one theorems," *Chem. Eng. Sci.*, vol. 42, pp. 2229–2268, 1987.

[145] M. Feinberg, "The existence and uniqueness of steady states for a class of chemical reaction networks," *Arch. Rational Mech. Anal.*, vol. 132, pp. 311–370, 1995.

[146] R. P. Feynman and A. R. Hibbs, *Quantum Mechanics and Path Integrals.* Burr Ridge, IL: McGraw-Hill, 1965.

[147] R. P. Feynman, R. B. Leighton, and M. Sands, *The Feynman Lectures on Physics,* vol. I. Reading, MA: Addison-Wesley, 1963.

[148] A. F. Filippov, "Differential equations with discontinuous right-hand side," *Amer. Math. Soc. Transl.*, vol. 42, pp. 199–231, 1964.

[149] A. F. Filippov, *Differential Equations with Discontinuous Right-Hand Sides.* Dordrecht, The Netherlands: Kluwer, 1988.

[150] G. B. Folland, *Introduction to Partial Differential Equations,* 2nd ed. Princeton, NJ: Princeton University Press, 1995.

[151] C. E. Folsome, *The Origin of Life.* San Francisco, CA: W. H. Freeman, 1979.

[152] H. U. Fuchs, *The Dynamics of Heat.* New York, NY: Springer-Verlag, 1996.

[153] G. Gallavotti and E. G. D. Cohen, "Dynamical ensembles in nonequilibrium statistical mechanics," *Phys. Rev. Lett.*, vol. 74, pp. 2694–2697, 1995.

[154] G. Gamov, *My World Line: An Informal Autobiography.* New York, NY: Viking Press, 1970.

[155] T. Gaosheng, Z. Ruzeng, and X. Wenbao, "Temperature transformation in relativistic thermodynamics," *Scientia Sinica*, vol. 25, pp. 615–627, 1982.

[156] T. C. Gard, *Introduction to Stochastic Differential Equations.* New York, NY: Marcel Dekker, 1988.

[157] K. Gatermann and B. Huber, "A family of sparse polynomial systems arising in chemical reaction systems," *J. Symbolic Computation*, vol. 33, pp. 275–305, 2002.

[158] K. Gatermann and M. Wolfrum, "Bernstein's second theorem and Viro's method for sparse polynomial systems in chemistry," *Adv. Appl. Math.*, vol. 34, pp. 252–294, 2005.

[159] C. F. Gauss, *Theoria Attractionis Corporum Sphaeroidicorum Ellipticorum Homogeneorum Methodo Nova Tractata.* Berlin, Germany: Königliche Gesellschaft der Wissenschaften zu Göttingen, 1877.

[160] N. Georgescu-Roegen, *The Entropy Law and the Economic Process.* Cambridge, MA: Harvard University Press, 1971.

[161] J. W. Gibbs, "On the equilibrium of heterogeneous substances," *Trans. Conn. Acad. Sci.*, vol. III, pp. 108–248, 1875.

[162] J. W. Gibbs, "On the equilibrium of heterogeneous substances," *Trans. Conn. Acad. Sci.*, vol. III, pp. 343–524, 1878.

[163] J. W. Gibbs, *The Scientific Papers of J. Willard Gibbs:* Vol. 1, *Thermodynamics.* London, U.K.: Longmans, 1906.

[164] R. Giles, *Mathematical Foundations of Thermodynamics.* Oxford, U.K.: Pergamon, 1964.

[165] G. P. Gladyshev, *Thermodynamic Theory of the Evolution of Living Beings.* New York, NY: Nova Science, 1997.

[166] P. Glansdorff and I. Prigogine, *Thermodynamic Theory of Structure, Stability, and Fluctuations.* London, U.K.: Wiley-Interscience, 1971.

[167] J. Gleick, *The Information.* New York, NY: Pantheon Books, 2011.

[168] P. J. Godin and T. G. Buchman, "Uncoupling of biological oscillators: A complementary hypothesis concerning the pathogenesis of multiple organ dysfunction syndrome," *Crit. Care Med.*, vol. 24, pp. 1107–1116, 1996.

[169] C. Godsil and G. Royle, *Algebraic Graph Theory.* New York, NY: Springer-Verlag, 2001.

[170] A. L. Goldberger, C. K. Peng, and L. A. Lipsitz, "What is physiologic complexity and how does it change with aging and disease?," *Neurobiol. Aging*, vol. 23, pp. 23–27, 2002.

[171] A. L. Goldberger, D. R. Rigney, and B. J. West, "Science in pictures: Chaos and fractals in human physiology," *Sci. Am.*, vol. 262, pp. 42–49, 1990.

[172] M. Goldstein and I. F. Goldstein, *The Refrigerator and the Universe.* Cambridge, MA: Harvard University Press, 1993.

[173] B. Green, *The Elegant Universe: Superstrings, Hidden Dimensions, and the Quest for the Ultimate Theory.* New York, NY: Vintage, 2000.

[174] B. Green, *The Fabric of the Cosmos.* New York, NY: Knoph, 2004.

[175] A. Greven, G. Keller, and G. Warnecke, *Entropy.* Princeton, NJ: Princeton University Press, 2003.

[176] M. Grmela and G. Lebon, "Finite-speed propagation of heat: A nonlocal and nonlinear approach," *Physica A*, vol. 248, no. 3, pp. 428–441, 2000.

[177] J. L. Gross and J. Yellen, *Handbook of Graph Theory.* Boca Raton, FL: CRC, 2004.

[178] A. Grünbaum, "The anisotropy of time," in *The Nature of Time* (T. Gold, ed.). Ithaca, NY: Cornell University Press, 1967.

[179] D. Grünbaum and A. Okubo, "Modeling social animal aggregations," in *Frontiers in Mathematical Biology*, Berlin, Germany: Springer-Verlag, pp. 296–325, 1994.

[180] V. Guillemin and A. Pollack, *Differential Topology.* Englewood Cliffs, NJ: Prentice-Hall, 1974.

[181] C. M. Guldberg and P. Waage, "Etude sur les affinités," *Les Mondes*, vol. 12, pp. 107–113, 1864.

[182] M. Gurtin, "On the thermodynamics of materials with memory," *Arch. Rational Mech. Anal.*, vol. 28, pp. 40–50, 1968.

[183] A. H. Guth, *The Inflationary Universe*. Reading, MA: Perseus Books, 1997.

[184] E. P. Gyftopoulos and G. P. Beretta, *Thermodynamics: Foundations and Applications*. New York, NY: Macmillan, 1991.

[185] E. P. Gyftopoulos and E. Çubukçu, "Entropy: Thermodynamic definition and quantum expression," *Phys. Rev. E*, vol. 55, no. 4, pp. 3851–3858, 1997.

[186] I. Gyori, "Delay differential and integro-differential equations in biological compartment models," *Syst. Sci.*, vol. 8, no. 2–3, pp. 167–187, 1982.

[187] C. C. Habeger, "The second law of thermodynamics and special relativity," *Annals of Physics*, vol. 72, pp. 1–28, 1972.

[188] W. M. Haddad, "Temporal asymmetry, entropic irreversibility, and finite-time thermodynamics: From Parmenides-Einstein time-reversal symmetry to the Heraclitan entropic arrow of time," *Entropy*, vol. 14, pp. 407–455, 2012.

[189] W. M. Haddad, "A unification between dynamical system theory and thermodynamics involving an energy, mass, and entropy state space formalism," *Entropy*, vol. 15, pp. 1821–1846, 2013.

[190] W. M. Haddad, "Nonlinear differential equations with discontinuous right-hand sides: Filippov solutions, nonsmooth stability and dissipativity theory, and optimal discontinuous feedback control," *Comm. App. Analysis*, vol. 18, pp. 455–522, 2014.

[191] W. M. Haddad and V. Chellaboina, *Nonlinear Dynamical Systems and Control: A Lyapunov-Based Approach*. Princeton, NJ: Princeton University Press, 2008.

[192] W. M. Haddad, V. Chellaboina, and E. August, "Stability and dissipativity theory for nonnegative dynamical systems: A thermodynamic framework for biological and physiological systems," in *Proc. IEEE Conf. Decision and Control*, Orlando, FL, pp. 442–458, 2001.

[193] W. M. Haddad, V. Chellaboina, and E. August, "Stability and dissipativity theory for discrete-time nonnegative and compartmental dynamical systems," *Int. J. Control*, vol. 76, pp. 1845–1861, 2003.

[194] W. M. Haddad, V. Chellaboina, and Q. Hui, *Nonnegative and Compartmental Dynamical Systems.* Princeton, NJ: Princeton University Press, 2010.

[195] W. M. Haddad, V. Chellaboina, and S. G. Nersesov, *Thermodynamics: A Dynamical Systems Approach.* Princeton, NJ: Princeton University Press, 2005.

[196] W. M. Haddad, V. Chellaboina, and S. G. Nersesov, *Impulsive and Hybrid Dynamical Systems: Stability, Dissipativity, and Control.* Princeton, NJ: Princeton University Press, 2006.

[197] W. M. Haddad, V. Chellaboina, and T. Rajpurohit, "Dissipativity theory for nonnegative and compartmental dynamical systems with time delay," *IEEE Trans. Autom. Contr.*, vol. 49, pp. 747–751, 2004.

[198] W. M. Haddad, S. P. Hou, J. M. Bailey, and N. Meskin, "A neural field theory for loss of consciousness: Synaptic drive dynamics, system stability, attractors, partial synchronization, and Hopf bifurcations characterizing the anesthetic cascade," in *Control of Complex Systems*, (S. Jagannathan and K. G. Vamvoudakis, eds.), pp. 93–162, Oxford, U.K.: Butterworth-Heinemann, 2016.

[199] W. M. Haddad, Q. Hui, and J. M. Bailey, "Human brain networks: Spiking neuron models, multistability, synchronization, thermodynamics, maximum entropy production, and anesthetic cascade mechanisms," *Entropy*, vol. 16, pp. 3939–4003, 2015.

[200] W. M. Haddad, S. G. Nersesov, and V. Chellaboina, "Energy-based control for hybrid port-controlled Hamiltonian systems," *Automatica*, vol. 39, pp. 1425–1435, 2003.

[201] J. K. Hale, "Dynamical systems and stability," *J. Math. Anal. Appl.*, vol. 26, pp. 39–59, 1969.

[202] J. K. Hale, *Ordinary Differential Equations,* 2nd ed. New York, NY: Wiley, 1980. Reprinted by Krieger: Malabar, FL, 1991.

[203] J. K. Hale and S. M. Verduyn Lunel, *Introduction to Functional Differential Equations.* New York, NY: Springer, 1993.

[204] S. R. Hall, D. G. MacMartin, and D. S. Bernstein, "Covariance averaging in the analysis of uncertain systems," in *Proc. IEEE Conf. Decision and Control*, Tucson, AZ, pp. 1842–1859, 1992.

[205] G. B. Halsted, *The Foundations of Science.* New York, NY: Science Press, 1913.

[206] G. H. Hardy, J. E. Littlewood, and G. Pólya, *Inequalities.* Cambridge, U.K.: Cambridge University Press, 1952.

[207] P. Hartman, "On invariant sets and on a theorem of Wazewski," *Proc. Am. Math. Soc.*, vol. 32, pp. 511–520, 1972.

[208] S. W. Hawking, "Particle creation by black holes," *Commun. Math. Phys.*, vol. 43, pp. 199–220, 1975.

[209] S. W. Hawking, "Black holes and thermodynamics," *Phys. Rev. D*, vol. 13, pp. 191–197, 1976.

[210] S. W. Hawking and G. F. R. Ellis, *The Large Scale Structure of Spacetime.* Cambridge, U.K.: Cambridge University Press, 1973.

[211] R. Heilbroner, *The Worldly Philosophers.* New York, NY: Simon and Schuster, 1953.

[212] T. Heimburg and A. D. Jackson, "On soliton propagation in biomembranes and nerves," *Proceedings of the National Academy of Sciences*, vol. 102, pp. 9790–9795, 2005.

[213] A. Hermann, *To Think Like God: Pythagoras and Parmenides.* Las Vegas, NV: Parmenides Publishing, 2004.

[214] D. J. Hill and P. J. Moylan, "Dissipative dynamical systems: Basic input-output and state properties," *J. Franklin Inst.*, vol. 309, pp. 327–357, 1980.

[215] H. Hinrichsen, "Non-equilibrium thermodynamics and phase transitions into absorbing states," *Advances in Physics*, vol. 49, no. 7, pp. 815–958, 2000.

[216] M. W. Hirsch, "Differential equations and convergence almost everywhere in strongly monotone flows," *Contemporary Math.*, vol. 17, pp. 267–285, 1983.

[217] A. L. Hodgkin and A. F. Huxley, "A quantitative description of membrane current and application to conduction and excitation in nerve," *J. Physiol.*, vol. 117, pp. 500–544, 1952.

[218] R. A. Horn and R. C. Johnson, *Matrix Analysis.* Cambridge, U.K.: Cambridge University Press, 1985.

[219] R. A. Horn and R. C. Johnson, *Topics in Matrix Analysis.* Cambridge, U.K.: Cambridge University Press, 1995.

[220] P. Horwich, *Asymmetries in Time.* Cambridge, MA: MIT Press, 1987.

[221] S. P. Hou, W. M. Haddad, N. Meskin, and J. M. Bailey, "A mechanistic neural mean field theory of how anesthesia suppresses consciousness: Synaptic drive dynamics, bifurcations, attractors, and partial state synchronization," *J. Math. Neuroscience*, vol. 2015, pp. 1–50, 2015.

[222] Q. Hui, W. M. Haddad, and S. P. Bhat, "Semistability, finite-time stability, differential inclusions, and discontinuous dynamical systems having a continuum of equilibria," *IEEE Trans. Autom. Contr.*, vol. 54, pp. 2465–2470, 2009.

[223] G. Hummer and A. Szabo, "Free energy reconstruction from nonequilibrium single-molecule pulling experiments," *Proc. Natl. Acad. Sci.*, vol. 98, no. 7, pp. 3658–3661, 2001.

[224] K. Itô, "Differential equations determining Markov processes," *Zenkoku Shijo Sugaku Danwaki*, vol. 1077, pp. 1352–1400, 1942.

[225] K. Itô, "Stochastic integral," *Proc. Imperial Acad. Tokyo*, vol. 20, pp. 519–524, 1944.

[226] J. A. Jacquez, *Compartmental Analysis in Biology and Medicine,* 2nd ed. Ann Arbor, MI: University of Michigan Press, 1985.

[227] J. A. Jacquez and C. P. Simon, "Qualitative theory of compartmental systems," *SIAM Rev.*, vol. 35, pp. 43–79, 1993.

[228] A. S. Jarrah, R. Laubenbacher, B. Stigler, and M. Stillman, "Reverse-engineering of polynomial dynamical systems," *Adv. Appl. Math.*, vol. 39, pp. 477–489, 2007.

[229] C. Jarzynski, "Equilibrium free-energy differences from nonequilibrium measurements: A master-equation approach," *Phys. Rev. E*, vol. 56, no. 5, pp. 5018–5035, 1997.

[230] C. Jarzynski, "Nonequilibrium equality for free energy differences," *Phys. Rev. Lett.*, vol. 78, no. 14, pp. 2690–2693, 1997.

[231] J. Jauch, "Analytical thermodynamics. Part 1. Thermostatics—General theory," *Foundations of Physics*, vol. 5, pp. 111–132, 1975.

[232] E. T. Jaynes, "Information theory and statistical mechanics," *Phys. Rev.*, vol. 106, pp. 620–630, 1957.

[233] G. Job and F. Herrmann, "Chemical potential—A quantity in search of recognition," *Eur. J. Phys.*, vol. 27, pp. 353–371, 2006.

[234] D. Jordan *et al.*, "Simultaneous electroencephalographic and functional magnetic resonance imaging indicate impaired cortical top-down processing in association with anesthetic-induced unconsciousness," *Anesthesiology*, vol. 119, no. 5, pp. 1031–1042, 2013.

[235] P. S. Joshi, *Global Aspects in Gravitation and Cosmology.* Oxford, U.K.: Oxford Clarendon Press, 1993.

[236] D. Jou, J. Casas-Vázquez, and G. Lebon, *Extended Irreversible Thermodynamics.* Berlin, Germany: Springer-Verlag, 1993.

[237] F. Jüttner, "Das Maxwellsche Gesetz der Geschwindigkeitsverteilung in der Relativtheorie," *Ann. Phys*, vol. 34, pp. 856–882, 1911.

[238] A. J. Keane and W. G. Price, "Statistical energy analysis of strongly coupled systems," *J. Sound Vibration*, vol. 117, pp. 363–386, 1987.

[239] R. P. Kerr, "Gravitational field of a spinning mass as an example of algebraically special metrics," *Physical Review Letters*, vol. 11, no. 5, pp. 237–238, 1963.

[240] J. Kestin, *The Second Law of Thermodynamics.* Stroudsburg, PA: Dowden, Hutchinson and Ross, 1976.

[241] J. Kestin, *A Course in Thermodynamics–Volumes I and II.* New York, NY: McGraw-Hill, 1979.

[242] H. K. Khalil, *Nonlinear Systems.* Upper Saddle River, NJ: Prentice-Hall, 1996.

[243] H. K. Khalil, *Nonlinear Systems,* 3rd ed. Upper Saddle River, NJ: Prentice-Hall, 2002.

[244] R. Z. Khasminskii, *Stochastic Stability of Differential Equations.* Berlin, Germany: Springer-Verlag, 2012.

[245] A. Khinchin, *Mathematical Foundations of Statistical Mechanics.* New York, NY: Dover, 1949.

[246] R. D. King, A. T. George, T. Jeon, L. S. Hynan, T. S. Youn, D. N. Kennedy, and B. Dickerson, "Characterization of atrophic changes in the cerebral cortex using fractal dimensional analysis," *Brain Imaging Behav.*, vol. 3, pp. 154–166, 2009.

[247] Y. Kishimoto and D. S. Bernstein, "Thermodynamic modeling of interconnected systems, I: Conservative coupling," *J. Sound Vibration*, vol. 182, pp. 23–58, 1995.

[248] Y. Kishimoto and D. S. Bernstein, "Thermodynamic modeling of interconnected systems, II: Dissipative coupling," *J. Sound Vibration*, vol. 182, pp. 59–76, 1995.

[249] Y. Kishimoto, D. S. Bernstein, and S. R. Hall, "Energy flow modeling of interconnected structures: A deterministic foundation for statistical energy analysis," *J. Sound Vibration*, vol. 186, pp. 407–445, 1995.

[250] D. E. Koditschek and K. S. Narendra, "Limit cycles of planar quadratic differential equations," *J. Diff. Equat.*, vol. 54, pp. 181–195, 1984.

[251] A. Komar, "Relativistic temperature," *Gen. Rel. Grav.*, vol. 27, pp. 1185–1206, 1995.

[252] D. Kondepudi and I. Prigogine, *Modern Thermodynamics: From Heat Engines to Dissipative Structures.* Chichester, U.K.: John Wiley and Sons, 1998.

[253] S. Kostios, "An algorithm for designing feedback stabilizers of nonlinear polynomial systems," in *Proc. Med. Conf. Contr. Autom.*, Athens, Greece, pp. T23–007, 2007.

[254] P. Kroes, *Time: Its Structure and Role in Physical Theories.* Dordrecht, The Netherlands: Reidel, 1985.

[255] J. Kurchan, "Fluctuation theorem for stochastic dynamics," *J. Phys. A: Math. Gen.*, vol. 31, pp. 3719–3729, 1998.

[256] J. L. Lagrange, *Méchanique Analitique.* Paris, France: Desaint, 1788.

[257] V. Lakshmikantham, D. D. Bainov, and P. S. Simeonov, *Theory of Impulsive Differential Equations.* Singapore: World Scientific, 1989.

[258] J. S. W. Lamb and J. A. G. Roberts, "Time reversal symmetry in dynamical systems: A survey," *Phys. D*, vol. 112, pp. 1–39, 1998.

[259] R. Landauer, "Information is physical," *Physics Today*, vol. 44, no. 5, pp. 23–29, 1991.

[260] L. D. Landau and E. M. Lifshitz, *Statistical Physics.* Oxford, U.K.: Butterworth-Heinemann, 1980.

[261] P. T. Landsberg, "Foundations of thermodynamics," *Review of Modern Physics*, vol. 28, pp. 63–393, 1956.

[262] P. T. Landsberg, "Does a moving body appear cool?," *Nature*, vol. 212, pp. 571–572, 1966.

[263] P. T. Landsberg, "Does a moving body appear cool?," *Nature*, vol. 214, pp. 903–904, 1967.

[264] P. T. Landsberg, "Heat engines and heat pumps at positive and negative absolute temperatures," *J. Phys. A: Math. Gen.*, vol. 10, pp. 1773–1780, 1977.

[265] P. T. Landsberg, "Einstein and statistical thermodynamics I. Relativistic thermodynamics," *European Journal of Physics*, vol. 2, no. 4, pp. 203–207, 1981.

[266] P. T. Landsberg, "Einstein and statistical thermodynamics II. Oscillator quantization," *European Journal of Physics*, vol. 2, no. 4, pp. 208–212, 1981.

[267] P. T. Landsberg, "Einstein and statistical thermodynamics III. The diffusion-mobility relation in semiconductors," *European Journal of Physics*, vol. 2, no. 4, pp. 213–219, 1981.

[268] P. T. Landsberg and K. A. Johns, "The problem of moving thermometers," *Proc. R. Sot. A.*, vol. 306, pp. 477–486, 1968.

[269] P. Langevin, "Sur la théorie du mouvement Brownien," *C. R. Acad. Sci. Paris*, vol. 146, pp. 530–533, 1908.

[270] R. S. Langley, "A general derivation of the statistical energy analysis equations for coupled dynamic systems," *J. Sound Vibration*, vol. 135, pp. 499–508, 1989.

[271] P. S. Laplace, *Oeuvres Complètes de Laplace.* Paris, France: Gauthier-Villars, 1895.

[272] B. Lavenda, *Thermodynamics of Irreversible Processes.* London, U.K.: Macmillan, 1978. Reprinted by Dover: New York, NY, 1993.

[273] A. Le Bot, *Foundations of Statistical Energy Analysis in Vibroacoustics.* New York, NY: Oxford University Press, 2015.

[274] J. L. Lebowitz and H. Spohn, "A Gallavotti-Cohen-type symmetry in the large deviation functional for stochastic dynamics," *J. Stat. Phys.*, vol. 95, no. 1, pp. 333–365, 1999.

[275] H. Lee, G. A. Mashour, S. Kim, and U. Lee, "Simultaneous electroencephalographic and functional magnetic resonance imaging indicate impaired cortical top-down processing in association with

anesthetic-induced unconsciousness," *Anesthesiology*, vol. 119, no. 6, pp. 1347–1359, 2013.

[276] A. L. Lehninger, *Biochemistry.* New York, NY: Worth Publishers, 1970.

[277] P. P. Lévy, *Théorie de L'addition des Variables Aléatoires.* Paris, France: Gauthier-Villars, 1937.

[278] D. C. Lewis, "A qualitative analysis of S-systems: Hopf bifurcations," in *Canonical Nonlinear Modeling: S-Systems Approach to Understanding Complexity* (E. O. Voit, ed.). New York, NY: Van Nostrand Reinhold, 1991.

[279] P. Liberman and C. M. Marle, *Symplectic Geometry and Analytical Mechanics.* Dordrecht, The Netherlands: Reidel, 1987.

[280] E. H. Lieb and J. Yngvason, "The physics and mathematics of the second law of thermodynamics," *Phys. Rep.*, vol. 310, pp. 1–96, erratum, vol. 314, p. 669, 1999.

[281] S.-K. Lin, "Correlation of entropy with similarity and symmetry," *J. Chem. Inf. Comput. Sci.*, vol. 36, pp. 367–376, 1996.

[282] S. K. Lin, "Diversity and entropy," *Entropy*, vol. 1, no. 1, pp. 1–3, 1999.

[283] J. Lindhard, "Temperature in special relativity," *Physica*, vol. 38, pp. 635–640, 1968.

[284] C. H. Lineweaver and C. A. Egana, "Life, gravity and the second law of thermodynamics," *Physics of Life Reviews*, vol. 5, pp. 225–242, 2008.

[285] C. C. Lin and L. A. Segel, *Mathematics Applied to Deterministic Problems in the Natural Sciences.* New York, NY: Academic, 1974.

[286] J. Liouville, "Formules Générales Relatives à la Question de la Stabilité de l'Équilibre d'Une Masse Liquide Homogène Douée d'un Mouvement de Rotation Autour d'un Axe," *J. Math. Pures Appl.*, vol. 20, pp. 164–184, 1855.

[287] R. Sh. Liptser and A. N. Shiryayev, *Theory of Martingales.* Dordrecht, The Netherlands: Kluwer Academic Publishers, 1989.

[288] C. Liu, "Einstein and relativistic thermodynamics in 1952: A historical and critical study of a strange episode in the history of modern physics," *Br. J. Hist. Sci.*, vol. 25, pp. 185–206, 1992.

[289] C. Liu, "Is there a relativistic thermodynamics? A case study in the meaning of special relativity," *Stud. Hist. Phil. Sci.*, vol. 25, pp. 983–1004, 1994.

[290] J. Loschmidt, ""Über den Zustand des Wärmegleichgewichtes eines Systems von Körpern mit Rücksicht auf die Schwere," *Wiener Berichte*, vol. 73, pp. 128–142, 1876.

[291] R. Lozano, B. Brogliato, O. Egeland, and B. Maschke, *Dissipative Systems Analysis and Control.* London, U.K.: Springer-Verlag, 2000.

[292] D. G. Luenberger, *Optimization by Vector Space Methods.* New York, NY: Wiley, 1969.

[293] E. W. Lund, "Guldberg and Waage and the law of mass action," *Journal of Chemical Education*, vol. 42, pp. 548–550, 1965.

[294] A. M. Lyapunov, *The General Problem of the Stability of Motion.* Kharkov, Russia: Kharkov Mathematical Society, 1892.

[295] A. M. Lyapunov, *Probléme Generale de la Stabilité du Mouvement.* Princeton, NJ: Princeton University Press, 1949.

[296] A. M. Lyapunov, *The General Problem of Stability of Motion* (A. T. Fuller, trans. and ed.). Washington, DC: Taylor and Francis, 1992.

[297] R. H. Lyon, *Statistical Energy Analysis of Dynamical Systems: Theory and Applications.* Cambridge, MA: MIT Press, 1975.

[298] S.-K. Ma, *Statistical Mechanics.* Singapore: World Scientific, 1985.

[299] M. C. Mackey, *Time's Arrow: The Origins of Thermodynamic Behavior.* New York, NY: Springer-Verlag, 1992.

[300] P. T. Macklem and A. J. E. Seely, "Towards a definition of life," *Perspect. Biol. Med.*, vol. 53, pp. 330–340, 2010.

[301] H. Maeda, S. Kodama, and T. Konishi, "Stability theory and existence of periodic solutions of time delayed compartmental systems," *Electron. Commun. Jpn.*, vol. 65, no. 1, pp. 1–8, 1982.

[302] X. Mao, *Stochastic Differential Equations and Applications.* New York, NY: Harwood, 1997.

[303] X. Mao, "Stochastic versions of the LaSalle theorem," *Journal of Differential Equations*, vol. 153, pp. 175–195, 1999.

[304] X. Mao and C. Yuan, *Stochastic Differential Equations with Markovian Switching.* London, U.K.: Imperial College Press, 2006.

[305] J. J. Mares, P. Hubik, J. Sestak, V. Spicka, J. Kristofik, and J. Stavek, "Phenomenological approach to the caloric theory of heat," *Thermochimica Acta*, vol. 474, pp. 16–24, 2008.

[306] J. J. Mares, P. Hubk, J. Sestak, V. Spicka, J. Kristofik, and J. Stavek, "Relativistic transformation of temperature and Mosengeil-Ott's antinomy," *Physica E*, vol. 42, pp. 484–487, 2010.

[307] M. Marvan, *Negative Absolute Temperatures.* London, U.K.: Iliffe Books, 1966.

[308] G. A. Mashour, "Consciousness and the 21st century operating room," *Anesthesiology*, vol. 119, no. 5, pp. 1003–1005, 2013.

[309] G. A. Maugin, *The Thermomechanics of Plasticity and Fracture.* Cambridge, U.K.: Cambridge University Press, 1992.

[310] G. A. Maugin, *The Thermomechanics of Nonlinear Irreversible Behaviors.* Singapore: World Scientific, 1999.

[311] J. C. Maxwell, *On the Stability of the Motion of Saturn's Rings.* London, U.K.: Macmillan, 1859.

[312] J. C. Maxwell, "On physical lines of force," *Philosophical Magazine*, vol. 23, pp. 161–175, 281–291, 338–348, 1861.

[313] J. C. Maxwell, "A dynamical theory of the electromagnetic field," *Philosophical Transactions of the Royal Society of London*, vol. 155, pp. 459–512, 1865.

[314] J. C. Maxwell, "On the dynamical theory of gases," *Philos. Trans. Roy. Soc. London Ser. A*, vol. 157, pp. 49–88, 1866.

[315] J. C. Maxwell, "On the equilibrium of heterogeneous substances," *Scientific Papers* (W. D. Niven, ed.), vol. 2, pp. 498–500, Cambridge, U.K.: Cambridge University Press, 1890.

[316] J. Mazur, *Zeno's Paradox.* New York, NY: Penguin Group, 2007.

[317] M. McKenna, R. Gruetter, U. Sonnewald, H. Waagepetersen, and A. Schousboe, "Energy metabolism in the brain," in *Basic Neurochemistry: Molecular, Cellular, and Medical Aspects,* 7th ed., (G. Siegel, R. W. Albers, S. Brady, and D. L. Price, eds.), pp. 531–557. London, U.K.: Elsevier, 2006.

[318] J. Meixner, "On the foundation of thermodynamics of processes," in *A Critical Review of Thermodynamics* (E. B. Stuart, B. Gal-Or, and

A. J. Brainard, eds.), pp. 37–47. Baltimore, MD: Mono Book Corp., 1970.

[319] E. Mendoza, *Reflections on the Motive Power of Fire by Sadi Carnot and Other Papers on the Second Law of Thermodynamics by É. Clapeyron and R. Clausius.* New York, NY: Dover, 1960.

[320] C. D. Meyer and M. W. Stadelmaier, "Singular M-matrices and inverse positivity," *Linear Algebra Appl.*, vol. 22, pp. 139–156, 1978.

[321] S. P. Meyn and R. L. Tweedie, *Markov Chains and Stochastic Stability.* London, U.K.: Springer-Verlag, 1993.

[322] A. A. Michelson and E. W. Morley, "On the relative motion of the Earth and the luminiferous aether," *American Journal of Science*, vol. 34, pp. 333–345, 1887.

[323] A. N. Michel, Y. Sun, and A. P. Molchanov, "Stability analysis of discontinuous dynamical systems determined by semigroups," *IEEE Trans. Autom. Control*, vol. 50, pp. 1277–1290, 2005.

[324] N. Minamide, "An extension of the matrix inversion lemma," *SIAM J. Alg. Disc. Meth.*, vol. 6, pp. 371–377, 1985.

[325] C. W. Misner, K. S. Thorne, and J. A. Wheeler, *Gravitation.* New York, NY: Freeman and Company, 1973.

[326] X. Mora, "Semilinear problems define semiflows in C^k-spaces," *Trans. Amer. Math. Soc.*, vol. 278, pp. 21–55, 1983.

[327] P. M. Morse and H. Feshbach, *Methods of Theoretical Physics.* New York, NY: McGraw-Hill, 1953.

[328] J. Moser and C. L. Siegel, *Lectures on Celestial Mechanics.* Berlin, Germany: Springer-Verlag, 1971.

[329] C. Mugler, *Archimède I. De la sphère et du cylindre. La mesure du cercle. Sur les conoïdes et les sphéroïdes.* Collections des Universités de France, Paris, Les Belles Lettres, 1970.

[330] C. Mugler, *Archimède II. Des spirales. De l'équilibre des figures planes. L'arénaire. La quadrature de la parabole.* Collections des Universités de France, Paris, Les Belles Lettres, 1971.

[331] C. Mugler, *Archimède III. Des corps flottants. Stomachion. La méthode. Le livre des lemmes. Le problème des boeufs.* Collections des Universités de France, Paris, Les Belles Lettres, 1971.

[332] C. Mugler, *Archimède IV. Commentaires d'Eutocius. Fragments.* Collections des Universités de France, Paris, Les Belles Lettres, 1972.

[333] I. Müller, "Die Kältefunktion, eine universelle Funktion in der Thermodynamik viscoser wärmeleitender Flüssigkeiten," *Arch. Rational Mech. Anal.*, vol. 40, pp. 1–36, 1971.

[334] I. Müller and T. Ruggeri, *Rational Extended Thermodynamics.* New York, NY: Springer, 1998.

[335] T. K. Nakamura, "Three views of a secret in relativistic thermodynamics," *Prog. Theor. Phys.*, vol. 128, pp. 463–475, 2012.

[336] K. H. Nealson and P. G. Conrad, "Life, past, present and future," *Phil. Trans. Roy. Soc. B*, vol. 354, pp. 1–17, 1999.

[337] I. Newton, *Philosophiae Naturalis Principia Mathematica.* London, U.K.: Royal Society, 1687.

[338] G. Nicolis and I. Prigogine, *Self Organization in Nonequilibrium Systems.* New York, NY: Wiley, 1977.

[339] S. I. Niculescu, *Delay Effects on Stability: A Robust Control Approach.* New York, NY: Springer, 2001.

[340] H. Nijmeijer and A. J. van der Schaft, *Nonlinear Dynamical Control Systems.* New York, NY: Springer, 1990.

[341] E. Nöether, "Invariante variations probleme," *Kgl. Ger. Wiss. Nachr. Gottingen, Math-Phys. Klasse*, vol. 2, pp. 235–257, 1918.

[342] E. F. Obert, *Concepts of Thermodynamics.* New York, NY: McGraw-Hill, 1960.

[343] E. P. Odum, *Fundamentals of Ecology,* 3rd ed. Philadelphia, PA: Saunders, 1971.

[344] B. Øksendal, *Stochastic Differential Equations: An Introduction with Applications.* Berlin, Germany: Springer-Verlag, 1995.

[345] A. Okubo, *Diffusion and Ecological Problems: Mathematical Models.* Berlin, Germany: Springer-Verlag, 1980.

[346] R. Olfati-Saber and R. M. Murray, "Consensus problems in networks of agents with switching topology and time-delays," *IEEE Trans. Autom. Control*, vol. 49, pp. 1520–1533, 2004.

[347] L. Onsager, "Reciprocal relations in irreversible processes, I," *Phys. Rev.*, vol. 37, pp. 405–426, 1931.

[348] L. Onsager, "Reciprocal relations in irreversible processes, II," *Phys. Rev.*, vol. 38, pp. 2265–2279, 1932.

[349] H. Ott, "Lorentz-Transformation der Wärme und der Temperatur," *Z. Phys.*, vol. 175, pp. 70–104, 1963.

[350] M. N. Özisik and D. Y. Tzou, "On the wave theory of heat conduction," *Trans. ASME J. Heat Transfer*, vol. 116, pp. 526–535, 1994.

[351] B. E. Paden and S. S. Sastry, "A calculus for computing Filippov's differential inclusion with application to the variable structure control of robot manipulators," *IEEE Trans. Circuit Syst.*, vol. 34, pp. 73–82, 1987.

[352] W. Pauli, *Theory of Relativity.* Oxford, U.K.: Pergamon Press, 1958.

[353] M. Pavon, "Stochastic control and nonequilibrium thermodynamical systems," *Appl. Math. Optim.*, vol. 19, pp. 187–202, 1989.

[354] R. K. Pearson and T. L. Johnson, "Energy equipartition and fluctuation-dissipation theorems for damped flexible structures," *Quart. Appl. Math.*, vol. 45, pp. 223–238, 1987.

[355] R. Penrose, *The Emperor's New Mind.* Oxford, U.K.: Oxford University Press, 1989.

[356] R. Penrose, *Road to Reality.* London, U.K.: Vintage Books, 2004.

[357] M. C. Petit-Taboue, B. Landeau, J. F. Desson, B. Desgranges, and J. C. Baron, "Effects of healthy aging on the regional cerebral metabolic rate of glucose assessed with statistical parametric mapping," *Neuroimaging*, vol. 7, pp. 176–184, 1998.

[358] M. Planck, *Vorlesungen über Thermodynamik.* Leipzig, Germany: Veit, 1897.

[359] M. Planck, "Über das Gesetz der Energieverteilung im Normalspectrum," *Annalen der Physik*, vol. 4, pp. 553–563, 1901.

[360] M. Planck, "Zur Dynamik bewegter Systeme," *Ann. Phys. Leipz.*, vol. 26, pp. 1–34, 1908.

[361] M. Planck, *Vorlesungen ber die Theorie der Wärmestrahlung.* Leipzig, Germany: J. A. Barth, 1913.

[362] M. Planck, "Über die Begrundung des zweiten Hauptsatzes der Thermodynamik," *Sitzungsberichte der preußischen Akademie der Wissenschaften, Math. Phys. Klasse*, pp. 453–463, 1926.

[363] H. Poincaré, "Mémoire sur les Courbes Définies par une Equation Différentielle," *J. Mathématiques*, vol. 7, pp. 375–422, 1881. Oeuvre (1880–1890), Gauthier-Villars, Paris, France.

[364] H. Poincaré, "Sur le probléme des trois corps et les équations de la dynamique," *Acta Math.*, vol. 13, pp. 1–270, 1890.

[365] H. Poincaré, "Sur les proprietes des fonctions definies par les equations aux differences partielles," in *Oeuvres*, vol. 1. Paris, France: Gauthier-Villars, 1929.

[366] K. R. Popper, *The World of Parmenides: Essays on the Presocratic Enlightenment.* London, U.K.: Routledge, 1998.

[367] I. Prigogine, *Thermodynamics of Irreversible Processes.* New York, NY: Interscience, 1955.

[368] I. Prigogine, *Introduction to Thermodynamics of Irreversible Processes.* New York, NY: Wiley-Interscience, 1968.

[369] I. Prigogine, *From Being to Becoming.* San Francisco, CA: Freeman, 1980.

[370] I. Prigogine and I. Antoniou, "Laws of nature and time symmetry breaking," *Annals New York Acad. Sci.*, vol. 879, pp. 8–28, 1999.

[371] I. Prigogine, G. Nicolis, and A. Babloyantz, "Thermodynamics of evolution," *Physics Today*, vol. 25, pp. 23–31, 1972.

[372] T. Rajpurohit and W. M. Haddad, "Dissipativity theory for nonlinear stochastic dynamical systems," *IEEE Trans. Autom. Contr.*, vol. 62, pp. 1684–1699, 2017.

[373] N. Ramsey, "Thermodynamics and statistical mechanics at negative absolute temperatures," *Phys. Rev.*, vol. 103, p. 20, 1956.

[374] B. Ratra and P. J. E. Peebles, "Cosmological consequences of a rolling homogeneous scalar field," *Physical Review D.*, vol. 37, no. 12, pp. 3406–3427, 1988.

[375] R. M. Redheffer, "The theorems of Bony and Brezis on flow-invariant sets," *Am. Math. Monthly*, vol. 79, pp. 740–747, 1972.

[376] R. M. Redheffer and W. Walter, "Flow-invariant sets and differential inequalities in normed spaces," *Applicable Anal.*, vol. 5, pp. 149–161, 1975.

[377] H. Reichenbach, *The Direction of Time.* Berkeley, CA: University of California Press, 1956.

[378] R. Resnick, *Introduction to Special Relativity.* New York, NY: Wiley, 1968.

[379] A. Riess *et al.*, "Observational evidence from supernovae for an accelerating universe and a cosmological constant," *Astronomical Journal*, vol. 116, no. 3, pp. 1009–1038, 1998.

[380] R. C. Robinson, *An Introduction to Dynamical Systems: Continuous and Discrete.* Upper Saddle River, NJ: Prentice Hall, 2004.

[381] R. T. Rockafellar, *Convex Analysis.* Princeton, NJ: Princeton University Press, 1970.

[382] G. E. Romero, "Parmenides reloaded," *Foundations of Science*, vol. 17, pp. 291–299, 2012.

[383] G. E. Romero, "From change to spacetime: An Eleatic journey," *Foundations of Science*, vol. 18, pp. 139–148, 2013.

[384] L. Rosier, "Homogeneous Lyapunov function for homogeneous continuous vector field," *Syst. Control Lett.*, vol. 19, pp. 467–473, 1992.

[385] M. A. Rothman, "Things that go faster than light," *Scientific American*, vol. 203, no. 1, pp. 142–152, 1960.

[386] E. J. Routh, *A Treatise on the Stability of a Given State of Motion.* London: Macmillan, 1877.

[387] H. L. Royden, *Real Analysis.* Englewood Cliffs, NJ: Prentice-Hall, 1988.

[388] M. B. Rubin, "Hyperbolic heat conduction and the second law," *Int. J. Eng. Sci.*, vol. 30, pp. 1665–1676, 1992.

[389] L. Russo, *The Forgotten Revolution: How Science Was Born in 300 B.C. and Why It Had to Be Reborn.* Berlin, Germany: Springer-Verlag, 2004.

[390] E. P. Ryan, "An integral invariance principle for differential inclusions with applications in adaptive control," *SIAM J. Control Optim.*, vol. 36, pp. 960–980, 1998.

[391] B. Ryden, *Introduction to Cosmology.* New York, NY: Addison-Wesley, 2003.

[392] R. G. Sachs, *The Physics of Time Reversal.* Chicago, IL: University of Chicago Press, 1987.

[393] I. Samohýl, *Thermodynamics of Irreversible Processes in Fluid Mixtures.* Leipzig, Germany: Teubner, 1987.

[394] A. M. Samoilenko and N. A. Perestyuk, *Impulsive Differential Equations.* Singapore: World Scientific, 1995.

[395] E. Schrödinger, *What Is Life?* Cambridge, U.K.: Cambridge University Press, 1944.

[396] K. Schwarzschild, "Über das Gravitationsfeld eines Massenpunktes nach der Einsteinschen Theorie," *Sitzungsberichte der Königlich Preussischen Akademie der Wissenschaften*, vol. 7, pp. 189–196, 1916.

[397] S. K. Scott, *Chemical Chaos.* Oxford, U.K.: Oxford University Press, 1991.

[398] A. J. E. Seely and P. Macklem, "Fractal variability: An emergent property of complex dissipative systems," *Chaos*, vol. 22, pp. 1–7, 2012.

[399] A. J. E. Seely, K. D. Newman, and C. L. Herry, "Fractal structure and entropy production within the central nervous system," *Entropy*, vol. 16, pp. 4497–4520, 2014.

[400] U. Seifert, "Stochastic thermodynamics: Principles and perspectives," *Eur. Phys. J. B*, vol. 64, no. 3, pp. 423–431, 2008.

[401] U. Seifert, "Stochastic thermodynamics, fluctuation theorems and molecular machines," *Rep. Prog. Phys.*, vol. 75, pp. 1–58, 2012.

[402] K. Sekimoto, "Kinetic characterization of heat bath and the energetics of thermal ratchet models," *J. Phys. Soc. Japan*, vol. 66, pp. 1234–1237, 1997.

[403] K. Sekimoto, "Langevin equation and thermodynamics," *Prog. Theor. Phys. Supp.*, vol. 130, pp. 17–27, 1998.

[404] K. Sekimoto, *Stochastic Energetics.* Berlin, Germany: Springer-Verlag, 2010.

[405] C. E. Shannon, "A mathematical theory of communication," *Bell System Technical Journal*, vol. 27, pp. 379–423, 623–656, 1948.

[406] C. E. Shannon and W. Weave, *The Mathematical Theory of Communication.* Urbana, IL: University of Illinois Press, 1949.

[407] X. Shen, H. Liu, Z. Hu, H. Hu, and P. Shi, "The relationship between cerebral glucose metabolism and age: Report of a large brain pet data set," *PLoS One*, vol. 7, pp. 1–10, 2012.

[408] D. Shevitz and B. Paden, "Lyapunov stability theory of nonsmooth systems," *IEEE Trans. Autom. Control*, vol. 39, pp. 1910–1914, 1994.

[409] I. Shnaid, "Thermodynamical proof of transport phenomena kinetic equations," *J. Mech. Behavior Mater.*, vol. 11, no. 5, pp. 353–364, 2000.

[410] I. Shnaid, "Thermodynamically consistent description of heat conduction with finite speed of heat propagation," *Int. J. Heat and Mass Transfer*, vol. 46, pp. 3853–3863, 2003.

[411] K. K. Shyu, Y. T. Wu, T. R. Chen, H. Y. Chen, H. H. Hu, and W. Y. Guo, "Measuring complexity of fetal cortical surface from MR images using 3-D modified box-counting method," *IEEE Trans. Instrum. Meas.*, vol. 60, pp. 522–531, 2011.

[412] D. D. Siljak, *Large-Scale Dynamic Systems.* New York, NY: North-Holland, 1978.

[413] F. E. Simon, "On the third law of thermodynamics," *Physica*, vol. 10, p. 1089, 1937.

[414] P. W. Smith, "Statistical models of coupled dynamical systems and the transition from weak to strong coupling," *Journal of the Acoustical Society of America*, vol. 65, pp. 695–698, 1979.

[415] H. L. Smith, *Monotone Dynamical Systems.* Providence, RI: American Mathematical Society, 1995.

[416] L. Smolin, *Three Roads to Quantum Gravity.* New York, NY: Basic Books, 2001.

[417] M. Smoluchowski, "Zur kinetischen Theorie der Brownschen Molekularbewegung und der Suspensionen," *Annalen der Physik*, vol. 21, no. 14, pp. 756–780, 1906.

[418] S. L. Sobolev, "Applications of functional analysis in mathematical physics," *Translations of Mathematical Monographs*, vol. 7, Providence, RI: American Mathematical Society, 1963.

[419] F. Soddy, *Wealth, Virtual Wealth and Debt: The Solution of the Economic Paradox.* Sydney, Australia: Allen and Unwin, 1926.

[420] I. S. Sokolnikoff, *Tensor Analysis: Theory and Applications.* New York, NY: Wiley, 1951.

[421] L. Sokoloff, "Energetics of functional activation in neural tissues," *Neurochem. Res.*, vol. 24, pp. 321–329, 1999.

[422] E. D. Sontag, "Structure and stability of certain chemical networks and applications to the kinetic proofreading model of T-cell receptor signal transduction," *IEEE Trans. Autom. Control*, pp. 1028–1047, 2001.

[423] L. A. Sosnovskiy and S. S. Sherbakov, "Mechanothermodynamic entropy and analysis of damage state of complex systems," *Entropy*, vol. 18, pp. 1–34, 2016.

[424] L. A. Sosnovskiy and S. S. Sherbakov, *Mechanothermodynamics*. Cham, Switzerland: Springer, 2016.

[425] O. Spengler, *The Decline of the West*. New York, NY: Oxford University Press, 1991.

[426] J. I. Steinfeld, J. S. Francisco, and W. L. Hase, *Chemical Kinetics and Dynamics*. Upper Saddle River, NJ: Prentice-Hall, 1989.

[427] G. Steinman and M. Cole, "Synthesis of biologically pertinent peptides under possible primordial conditions," *Proc. Nat. Acad. Sci.*, vol. 58, pp. 735–742, 1967.

[428] L. Susskind, "The world as a hologram," *J. Math. Phys.*, vol. 36, no. 11, pp. 6377–6396, 1995.

[429] R. Swenson, "Emergent attractors and the law of maximum entropy production: Foundations to a theory of general evolution," *Syst. Res.*, vol. 6, pp. 187–197, 1989.

[430] L. Szilárd, "Über Die Entropieverminderung in Einem Thermodynamischen System Bei Eingriffen Intelligenter Wesen," *Zeitschrift Für Physik*, vol. 53, pp. 840–856, 1929.

[431] T. Takahashi, T. Murata, M. Omori, H. Kosaka, K. Takahashi, Y. Yonekura, and Y. Wada, "Quantitative evaluation of age-related white matter microstructural changes on MRI by multifractal analysis," *J. Neurol. Sci.*, vol. 225, pp. 33–37, 2004.

[432] I. Tasaki, "Evidence for phase transitions in nerve fibers, cells and synapses," *Ferroelectrics*, vol. 220, pp. 305–316, 1999.

[433] E. F. Taylor and J. A. Wheeler, *Spacetime Physics*. New York, NY: Freeman and Company, 1992.

[434] E. F. Taylor and J. A. Wheeler, *Exploring Black Holes*. San Francisco, CA: Addison Wesley, 2000.

[435] A. Teel, E. Panteley, and A. Loria, "Integral characterization of uniform asymptotic and exponential stability with applications," *Math. Contr. Sign. Syst.*, vol. 15, pp. 177–201, 2002.

[436] D. Ter Haar and H. Wergeland, "Thermodynamics and statistical mechanics on the special theory of relativity," *Phys. Rep.*, vol. 1, pp. 31–54, 1971.

[437] W. Thomson (Lord Kelvin), "Manuscript notes for 'On the dynamical theory of heat'," *Archives of the History of the Exact Sciences*, vol. 16, pp. 281–282, 1851.

[438] W. Thomson (Lord Kelvin), "On a universal tendency in nature to the dissipation of mechanical energy," *Proc. Roy. Soc. Edinburgh*, vol. 20, pp. 139–142, 1852.

[439] F. J. Tipler, "General relativity, thermodynamics, and the Poincaré cycle," *Nature*, vol. 280, pp. 203–205, 1979.

[440] L. Tisza, "Thermodynamics in a state of flux. A search for new foundations," *A Critical Review of Thermodynamics* (E. B. Stuart, B. Gal-Or, and A. J. Brainard, eds.), pp. 107–118, Baltimore, MD: Mono Book Corp., 1970.

[441] R. C. Tolman, "Relativity theory: The equipartition law in a system of particles," *Phil. Mag.*, vol. 28, pp. 583–600, 1914.

[442] R. C. Tolman, *Relativity, Thermodynamics and Cosmology.* Oxford, U.K.: Clarendon, 1934.

[443] E. Torricelli, *Opera Geometrica.* Florence, Italy: Musse, 1644.

[444] M. Tribus and E. C. McIrvine, "Energy and information," *Scientific American*, vol. 224, pp. 178–184, 1971.

[445] C. Truesdell, *Essays in the History of Mechanics.* New York, NY: Springer, 1968.

[446] C. Truesdell, *Rational Thermodynamics.* New York, NY: McGraw-Hill, 1969.

[447] C. Truesdell, *The Tragicomical History of Thermodynamics.* New York, NY: Springer-Verlag, 1980.

[448] J. Uffink, "Bluff your way in the second law of thermodynamics," *Stud. Hist. Phil. Mod. Phys.*, vol. 32, pp. 305–394, 2001.

[449] J. Van der Waals and P. Kohnstamm, *Lehrbuch der Thermostatik.* Leipzig, Germany: J. A. Barth, 1927.

[450] N. G. Van Kampen, "Relativistic thermodynamics of moving systems," *Phys. Rev.*, vol. 173, pp. 295–301, 1968.

[451] N. G. Van Kampen, "Relativistic thermodynamics," *J. Phys. Soc. Jpn.*, vol. 26, pp. 316–321, 1969.

[452] N. G. Van Kampen, *Stochastic Processes in Physics and Chemistry.* Amsterdam, The Netherlands: Elsevier, 1992.

[453] P. Vernotte, "Les paradoxes de la theorie continue de l'équation de la chaleur," *CR Acad. Sci.*, vol. 246, no. 22, pp. 3154–3155, 1958.

[454] L. R. Volevich and B. P. Paneyakh, "Certain spaces of generalized functions and embedding theorems," *Russian Math. Surveys*, vol. 20, pp. 1–73, 1965.

[455] H. C. von Baeyer, *Maxwell's Demon: Why Warmth Disperses and Time Passes.* New York, NY: Random House, 1998.

[456] M. von Laue, *Die Relativitätstheorie.* Braunschweig, Germany: Vieweg, 1961.

[457] K. von Mosengeil, "Theorie der stationären Strahlung in einem gleichförmig bewegten Hohlraum," *Annalen der Physik*, vol. 327, no. 5, pp. 867–904, 1907.

[458] C. Y. Wang, "Thermodynamics since Einstein," *Advances in Natural Science*, vol. 6, no. 2, pp. 13–17, 2013.

[459] J. Warga, *Optimal Control of Differential and Functional Equations.* New York, NY: Academic, 1972.

[460] J. D. Watson and F. H. C. Crick, "Molecular structure of nucleic acids," *Nature*, vol. 171, no. 4356, pp. 737–738, 1953.

[461] N. Wiener, *Nonlinear Problems in Random Theory.* Cambridge, MA: MIT Press, 1958.

[462] A. E. Wilder-Smith, *The Creation of Life.* Wheaton, IL: Harold Shaw, 1970.

[463] J. C. Willems, "Consequences of a dissipation inequality in the theory of dynamical systems," *Physical Structure in Systems Theory* (J. J. van Dixhoorn, ed.), pp. 193–218, New York, NY: Academic Press, 1974.

[464] J. C. Willems, "Dissipative dynamical systems, part I: General theory," *Arch. Rational Mech. Anal.*, vol. 45, pp. 321–351, 1972.

[465] J. C. Willems, "Qualitative behavior of interconnected systems," *Ann. Syst. Research*, vol. 3, pp. 61–80, 1973.

[466] J. C. Willems, "System theoretic models for the analysis of physical systems," *Rieerche Automat*, vol. 10, pp. 71–106, 1979.

[467] F. W. Wilson, "Smoothing derivatives of functions and applications," *Trans. Am. Math. Soc.*, vol. 139, pp. 413–428, 1969.

[468] J. Woodhouse, "An approach to the theoretical background of statistical energy analysis applied to structural vibration," *Journal of the Acoustical Society of America*, vol. 69, pp. 1695–1709, 1981.

[469] Y. T. Wu, K. K. Shyu, T. R. Chen, and W. Y. Guo, "Using three-dimensional fractal dimension to analyze the complexity of fetal cortical surface from magnetic resonance images," *Nonlinear Dyn.*, vol. 58, pp. 745–752, 2009.

[470] Z. J. Wu, X. J. Xie, P. Shi, and Y. Q. Xia, "Backstepping controller design for a class of stochastic nonlinear systems with Markovian switching," *Automatica*, vol. 45, pp. 997–1004, 2009.

[471] C. N. Yang and R. Milles, "Conservation of isotopic spin and isotopic gauge invariance," *Phys. Rev. E*, vol. 96, no. 1, pp. 191–195, 1954.

[472] B. E. Ydstie and A. A. Alonso, "Process systems and passivity via the Clausius-Planck inequality," *Systems Control Lett.*, vol. 30, pp. 253–264, 1997.

[473] T. Yoshizawa, *Stability Theory by Liapunov's Second Method.* Tokyo, Japan: Math. Soc. Japan, 1966.

[474] K. Yosida, *Functional Analysis,* 6th ed. Berlin, Germany: Springer-Verlag, 1980.

[475] C. K. Yuen, "Lorentz transformation of thermodynamic quantities," *Am. J. Phys.*, vol. 38, pp. 246–252, 1970.

[476] H. D. Zeh, *The Physical Basis of the Direction of Time.* New York, NY: Springer-Verlag, 1989.

[477] M. W. Zemansky, *Heat and Thermodynamics.* New York, NY: McGraw–Hill, 1968.

[478] J. M. Ziman, *Models of Disorder.* Cambridge, England: Cambridge University Press, 1979.

Index

Princeton Series in Applied Mathematics

Chaotic Transitions in Deterministic and Stochastic Dynamical Systems: Applications of Melnikov Processes in Engineering, Physics, and Neuroscience, Emil Simiu

Selfsimilar Processes, Paul Embrechts and Makoto Maejima

Self-Regularity: A New Paradigm for Primal-Dual Interior-Point Algorithms, Jiming Peng, Cornelis Roos, and Tamás Terlaky

Analytic Theory of Global Bifurcation: An Introduction, Boris Buffoni and John Toland

Entropy, Andreas Greven, Gerhard Keller, and Gerald Warnecke, editors

Auxiliary Signal Design for Failure Detection, Stephen L. Campbell and Ramine Nikoukhah

Thermodynamics: A Dynamical Systems Approach, Wassim M. Haddad, VijaySekhar Chellaboina, and Sergey G. Nersesov

Optimization: Insights and Applications, Jan Brinkhuis and Vladimir Tikhomirov

Max Plus at Work, Modeling and Analysis of Synchronized Systems: A Course on Max-Plus Algebra and Its Applications, Bernd Heidergott, Geert Jan Olsder, and Jacob van der Woude

Impulsive and Hybrid Dynamical Systems: Stability, Dissipativity, and Control, Wassim M. Haddad, VijaySekhar Chellaboina, and Sergey G. Nersesov

Positive Definite Matrices, Rajendra Bhatia

The Traveling Salesman Problem: A Computational Study, David L. Applegate, Robert E. Bixby, Vašek Chvátal, and William J. Cook

Genomic Signal Processing, Ilya Shmulevich and Edward R. Dougherty

Wave Scattering by Time-Dependent Perturbations: An Introduction, G. F. Roach

Algebraic Curves over a Finite Field, J.W.P. Hirschfeld, G. Korchmáros, and F. Torres

Distributed Control of Robotic Networks: A Mathematical Approach to Motion Coordination Algorithms, Francesco Bullo, Jorge Cortés, and Sonia Martínez

Robust Optimization, Aharon Ben-Tal, Laurent El Ghaoui, and Arkadi Nemirovski

Control Theoretic Splines: Optimal Control, Statistics, and Path Planning, Magnus Egerstedt and Clyde Martin

Matrices, Moments and Quadrature with Applications, Gene H. Golub and Gérard Meurant

Graph Theoretic Methods in Multiagent Networks, Mehran Mesbahi and Magnus Egerstedt

Totally Nonnegative Matrices, Shaun M. Fallat and Charles R. Johnson

Modern Anti-windup Synthesis: Control Augmentation for Actuator Saturation, Luca Zaccarian and Andrew W. Teel

Matrix Completions, Moments, and Sums of Hermitian Squares, Mihály Bakonyi and Hugo J. Woerdeman

Stability and Control of Large-Scale Dynamical Systems: A Vector Dissipative Systems Approach, Wassim M. Haddad and Sergey G. Nersesov

Mathematical Analysis of Deterministic and Stochastic Problems in Complex Media Electromagnetics, G. F. Roach, I. G. Stratis, and A. N. Yannacopoulos

Topics in Quaternion Linear Algebra, Leiba Rodman

Hidden Markov Processes: Theory and Applications to Biology, M. Vidyasagar

Mathematical Methods in Elasticity Imaging, Habib Ammari, Elie Bretin, Josselin Garnier, Hyeonbae Kang, Hyundae Lee, and Abdul Wahab

Rays, Waves, and Scattering: Topics in Classical Mathematical Physics, John A. Adam

Formal Verification of Control Systems Software, Pierre-Loïc Garoche

A Dynamical Systems Theory of Thermodynamics, Wassim M. Haddad